VASCULAR PLANT SYSTEMATICS

VASCULAR PLANT SYSTEMATICS

by

ALBERT E. RADFORD

WILLIAM C. DICKISON, JIMMY R. MASSEY, C. RITCHIE BELL

with

CONTRIBUTIONS

by

Ben W. Smith, Kenneth M. Becker, Theodore J. Crovello, Stanwyn G. Shetler, Laurie S. Radford, Clifford R. Parks, Norton G. Miller, J. Kenneth Moore, Major M. Goodman, Theodore M. Barkley, James W. Hardin, Robert F. Thorne, Leo J. Hickey, John T. Mickel, Lytton J. Musselman, Robert L. Wilbur and John M. Herr, Jr.

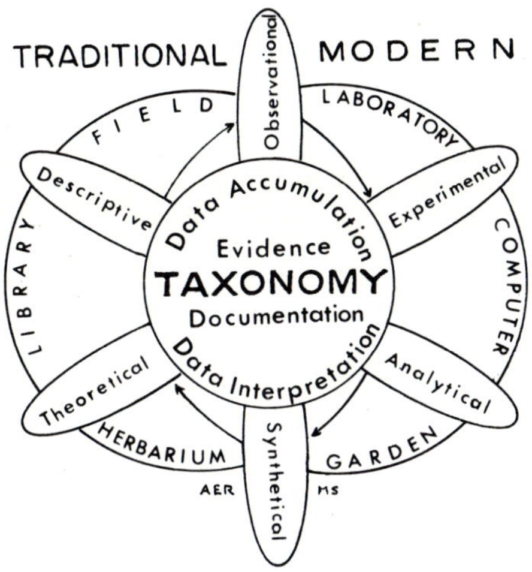

The
"ASPECTUS"

ILLUSTRATIONS
(Except Appendix)

by

MARION S. SEILER

HARPER & ROW, PUBLISHERS
NEW YORK, EVANSTON, SAN FRANCISCO, LONDON

VASCULAR PLANT SYSTEMATICS

For information address Harper & Row, Publishers, Inc., 10 East 53rd Street, New York, N.Y. 10022.

Standard Book Numbers: 06-045308-7 (Clothbound Edition)
06-045309-5 (Paperback Edition)

Library of Congress Catalog Card Number: 74-8068

ACKNOWLEDGMENTS

The authors of Vascular Plant Systematics wish to express their sincere appreciation to the many authors, publishers, editors and societies for permission to use or modify previously published or unpublished materials. Author credits for figures, tables, or other information are given in the accompanying legends. The publishers and journals that have given reprint permission are: American Journal of Botany; American Naturalist; Annals of the Missouri Botanical Garden; Brittonia; Chronica Botanica, Waltham, Mass.; Clarendon Press, Oxford, England; D. C. Heath and Company, Lexington, Mass.; Evolution; Gebrüder Borntraeger, Stuttgart, Germany; Houghton Mifflin Company, Boston, Mass.; John Wiley and Sons, New York, N. Y.; Journal of the Linnean Society; Macmillan Company, New York, N. Y.; McGraw-Hill Book Company, New York, N. Y.; Pergamon Press, New York, N. Y.; Quarterly Review of Biology; Taxon; Wadsworth Publishing Company, Belmont, Calif.

Special acknowledgment is made for permission to use copyrighted material from:

An Introduction to the Embryology of Angiosperms by P. Maheshwari. Copyright 1950 by McGraw-Hill Book Co., Inc. Used with permission of McGraw-Hill Book Company.

Plant Classification by Lyman Benson. Lexington, Mass.: D. C. Heath and Company, 1957.

The Evolution and Classification of Flowering Plants by Arthur Cronquist. Boston, Mass.: Houghton Mifflin Company, 1968.

The authors are indebted to the George R. Cooley Fund and the Department of Botany, University of North Carolina at Chapel Hill, for absorbing all preparation costs for camera-ready copy--typing, drawings, plate preparation, reductions, indexing and xeroxing. All royalties from this publication will be deposited in the George R. Cooley Fund for the support of taxonomic research except those paid to The Arnold Arboretum of Harvard University for the use of the family plates in the Appendix.

This

REFERENCE TEXT *and* SOURCE BOOK

is

DEDICATED

to

ALL STUDENTS OF SYSTEMATICS

PAST, PRESENT, FUTURE

VASCULAR PLANT SYSTEMATICS

TABLE OF CONTENTS

PREFACE

The ultimate taxonomy of species and higher plant taxa will be based on an understanding of the morphology, anatomy, embryology, palynology, ecology, physiology, cytology, genetics, and biochemistry, i.e., the biology, of individuals and populations. To acquire the proper perspective and conspective, the student of plant taxonomy should develop a holistic approach to systematics which involves (1) learning and observing, from all types of evidence, the distinguishing and diagnostic characteristics of plants, (2) discovering the limiting factors in the distribution of plants, (3) discerning the role of individuals in the ecosystem, (4) understanding the evolutionary mechanisms and relationships involved, and (5) determining the phylogenetic trends in diverse taxa. This book, a reference text and source book, is the culmination of an effort to produce in one volume, at a reasonable cost, a comprehensive and balanced approach to plant systematics.

VASCULAR PLANT SYSTEMATICS includes (1) an essentially synoptical treatment of the evidence, principles, and concepts considered fundamental to vascular plant taxonomic studies and research; (2) organized laboratory and field exercises and problems basic to systematics; (3) useable and useful techniques; (4) summaries of terminology pertinent to taxonomy; (5) relevant bibliographies and indices; and (6) information on systematic facilities. This book was written as a reference text for basic courses in taxonomy and as a source book of information, procedures and references for ecosystematics, biosystematics, phylosystematics, and chemosystematics. The intellectual range within the text is from the relatively simple manual "how to" to the comparatively complex philosophical "why do." The material should be adequate for the different emphases at many types of institutions for a series of quarter, semester, or year courses. The different synoptical treatments can be individually amplified with local examples and additional information by the teacher. The phytography, family, community samples, and special studies can be abbreviated, extended, or modified for the different levels of sophistication and depths of study required or anticipated in a local course. The glossaries, techniques, and bibliographies are fundamental tools for beginning and advanced students, as well as useful references for the professional taxonomist, the research worker, the field student, and the teacher of organismal botany or biology.

The authors certainly do not expect any one individual student to master the information in this source book in one course or several courses. Probably very few, if any, professors or teachers of taxonomy feel competent to teach in depth the material in all of these chapters! Furthermore, memorizing the information, applying the procedures, working out the exercises and problems and answering the questions in the text will not necessarily make a student a taxonomist. If a student has interest and motivation, intellectual curiosity, creativity with rational application, and inquisitiveness (what, when, where, how, why) while acquiring this information and methodology, then he will be well on the way to becoming a taxonomic scholar and scientist. Much experience, however, is necessary for a real understanding of most of the principles, concepts, definitions, and information presented in this publication.

Hopefully this presentation will provide the informational basis and organizational flexibility for the stimulation of innovation and improvement in systematics teaching, learning, and research. Desirably, this approach will lead to more extensive and diversified data-gathering which, when coupled with computer technology, should enable taxonomists to realize at a much earlier date a better understanding of the taxonomy of species and higher taxa. The authors, however, will consider their time well spent if this work provides the impetus for sounder taxonomic scholarship and produces added pleasure in "learning about plants" by the amateur as well as the professional.

The major authors have edited and tried to integrate and coordinate all of the presentations in the text. The ultimate responsibility, however, for content and accuracy belongs to each individual contributor. The contributions are as follows by chapter (and section where pertinent):

1. Radford
C II-Bell
C III-Dickison
C IV-Massey
2. Barkley
3. Radford
C-Wilbur
4. L. Radford
5. Radford
6. Radford
C-Hardin
D-Mickel
7. Dickison
D-Hickey
8. Dickison
B-Bell
9. Dickison
10. Smith
11. Smith
12. Bell
13. Parks
14. Radford & Dickison
15. Radford
C VIII-Musselman
16. Radford
D-Thorne
17. Radford
18. Massey
19. Miller
20. Bell
21. Bell
22. Crovello
23. Goodman
24. Radford
25. Massey
26. Bell
27. Dickison
28. Radford
B-Hardin
D-Mickel
29. Becker
30. Massey
31. Massey
32. Moore
33. Shetler
34. Hardin
35. Radford
B III-Bell
36. Radford

We are grateful to James W. Hardin for criticism and review of the many versions of the manuscript and to Robert Ornduff for review of the Draft Copy. The principal authors are indebted to the following individuals for comment and criticism of final drafts of individual chapters: E. L. Core (2, 30); R. L. Wilbur (3); W. O. West III (4); K. M. Becker (6, 27, 28, 30, 36); J. W. Walker (8); Barbara Palser (9); B. G. Brehm (13); J. A. Doyle and R. E. Taggart (19); J. H. Beaman and L. E. Morse (25, 33); R. H. Eyde (27); Ann Farren, G. J. Goodman, Cheryl Lawson, J. F. Matthews (30); J. E. Bryan, Francis Ching, F. C. Galle, R. W. Lighty, G. L. Tarbox, Jr., R. C. Taylor, R. H. Vick and J. A. Witt (32); R. A. Creighton, J. J. Crockett, T. J. Crovello, G. S. Daniels, T. E. Kopfler, Harriet Krauss, R. D. MacDonald, J. F. Mello and F. J. Rohlf (33). Many colleagues and students have provided helpful comments, suggested additional features for the text, made corrections in the manuscript and proposed new ideas for incorporation into the book. All have been a positive stimulus toward completion of the work and we sincerely appreciate their help.

The authors are particularly grateful to Helen Massey for her patience and professional excellence in the typing of the manuscript and camera-ready copy, and to Laurie Radford for assembling the materials and making the initial design for the plates in Chapter 6. We deeply appreciate the editing of the entire book by V. A. Greulach and we are especially indebted to Marion S. Seiler for her superlative illustrations and to Carroll E. Wood, Jr. for the use of the family illustrations and the information on the production of the plates.

TO THE TEACHER AND STUDENT

Taxonomy is a unifying and integrating field of biology (see the "Taxogram"). It is broad, it is involved (see the "Aspectus"). It is not local flora, which is a minor but basic introductory component of the subject (see the "Taxonomic Fundamentals"). The student of taxonomy must be a sound and profound organismal botanist with a basic understanding of the molecular and ultrastructural aspects of plants. The general student should have some botanical background before undertaking a study of systematics. Botanical information necessary for the training of vascular plant taxonomists is presented, essentially, in synoptic form, throughout this book but especially in Chapters 6-16.

Student assignments and responsibilities have to be made clear and concise for effective use of a reference text. Different levels of sophistication have been deliberately included in the text, so the teacher should be selective in the material to be used. The student should not be overwhelmed, but challenged. We are convinced that, ideally, the student should know the objectives and schedule of a systematics course well enough so that with adequate materials and time he can develop his knowledge and experience at his best learning rate. The student should be able to enter the classroom or laboratory at a time of his own choosing while the material is available, obtain an instruction sheet on the topics to be studied during that week or period, review the synoptic information or study a pertinent reference, observe the demonstration material, and then apply the knowledge gained by working out the exercise associated with the subject. Students with an inadequate background or those having difficulty with an assignment could listen to a tape and observe a film loop or strip on the subject or consult the professor or laboratory instructor for aid. The professor or instructor, however, should be available at regularly scheduled times for interpretive discussions or lectures and informational reviews. Real evaluative quizzes should be given frequently enough to insure a responsible amount and rate of learning and an understanding of the significance and relevance of each topic studied. The freedom with responsibility approach is most applicable to a field as broad as plant systematics where no two students have the same background, interest, ability and training for the subject. A reference text and source book can be most effectively used in one or more minicourses or in the modular approach to taxonomic training.

Problem training with problem definition, with data accumulation and analysis, with selection and execution of experimental and analytical procedures, with use of techniques, with documentation, with data interpretation and generalization, and with the drawing of conclusions and writing of a summary should be an integral, even though necessarily limited, part of advanced taxonomic course training. The procedures and exercises in many chapters and the problems in Chapter 35 were designed for this type of work. A species biology problem worked out by the student in determining the character states, by observation and/or from the literature, is an effective teaching device for the study of evidence in the basic systematics course.

PED-AED, i.e., preparation, explanation, demonstration--application, examination, discussion by the teacher is essential. The levels or phases of learning: recall, recognition, application, interpretation, generalization, and evaluation should consciously be applied throughout a course in taxonomy, where applicable, to challenge the mental abilities and intellectual capacities of the student. The holotaxonomic approach should incorporate the observational and descriptive, the experimental and analytical, and the synthetic and theoretical aspects in order to develop the inspective, perspective, and conspective attitude necessary for good taxonomic scholarship. Any course or program in systematics with these three aspects, or a major portion thereof, will be most challenging, meaningful, and effective in the professional and liberal training of any student in taxonomy. AER

Albert E. Radford
Jimmy R. Massey

February, 1974
Chapel Hill, N.C.

William C. Dickison
C. Ritchie Bell

Chapter 1. INTRODUCTION TO TAXONOMY

Systematics is a broad field of science which uses characteristics and data from many disciplines in carrying out its primary objectives of describing, naming, classifying, identifying, and determining relationships of organisms (see the "Taxogram" on the cover). Taxonomy utilizes many intellectual approaches in its study: the descriptive and observational, the analytical and experimental, the synthetical and theoretical. Systematic data are accumulated from research in the field, the laboratory, the garden, the herbarium, and the library. These data may be analyzed with the aid of the computer and may be documented in the form of dried specimens in the herbarium, as living specimens in the botanical garden, or as written information deposited in the library (see the "Aspectus" on the title page). Taxonomy is the pedestal upon which biology is built. Also, in the determination of relationships based upon diverse characteristics derived from many sources, it is the unifying and integrating discipline which represents the peak of biological endeavor.

Plant taxonomy is certainly a relevant field of inquiry for modern man. Plants are our fundamental sources of food and energy, of shelter and clothing, of drugs and beverages, of the oxygen we breathe, and form the aesthetic base of our environment. The characterizing, the naming, the classifying, and the identifying of all organisms, whether actually or potentially useful to man or not, is one duty of the taxonomist or systematist.

The ecologist studying pollutant decomposers has to have the organisms which are involved identified; the geneticist introducing new germ plasm into a crop, for increased production or disease resistance, has to know the characteristics of the source plant; and the chemist analyzing plant material for a cancer-inhibiting drug has to know the name of the organism from which the material was procured. All these scientists will be interested in the classification of related organisms as they search for more effective agents in their work. Related decomposers may prove to be more efficient removers of pollution; closely related species may be a source of better germ plasm for increased production or genes more resistant to disease; and other genotypes may provide cancer-inhibiting drugs that will be more effective, with fewer side-effects. The scientific name is the communication medium to which all information is attached. The taxonomist has to develop a descriptive language for precise and useful characterization of organisms; he has to devise simple schemes for identification; he has to work out a system of classification to show relationships; and he has to have a method for naming all organisms. These systematic activities are basic to the operations of man (see the "Fundamentals" on page twelve).

Systematics is a dynamic science. The duties of the taxonomist are never-ending. The constant stream of data on plant materials and about plants necessitates a reordering of descriptive information, revisions of identification schemes, reevaluation of classification systems, and perception of new relationships for increased effectiveness in communication about plants and for a greater understanding of the plant kingdom. As long as the plant world exists, there will always be more to learn about plants, plant products, and plant taxa, for both practical and theoretical purposes.

The great challenge today is how to preserve the plant life of the world. We need to learn more about these organisms which are basic to our survival, and how to store and retrieve the vast amount of information on plants for man's use. It is the primary duty, and one of the major problems, of the taxonomist to incorporate this body of knowledge into the taxonomy of plants so that this storehouse or data bank of information and classification can be efficiently and effectively used.

The major objectives of plant systematics, phases of development of taxonomy and basic terminology are presented in Section A. An example of a general plant systematics curriculum is presented in Sections C and D to give taxonomic prospective. The organization and objectives of a basic and four advanced courses in systematics are outlined to show the magnitude and composition of a balanced and comprehensive approach to the subject.

Section A. OBJECTIVES, PHASES, BASIC TERMS

I. MAJOR OBJECTIVES OF PLANT SYSTEMATICS

1. To provide a convenient method of identifying, naming, and describing plant taxa.
2. To provide a classification scheme which attempts to express natural or phylogenetic relationships.
3. To provide an inventory of plant taxa; to provide local, regional, and continental floras.
4. To provide an understanding of evolutionary processes and relationships.
5. To provide an integrating and unifying role or focal point in the training of biology students, particularly in regard to the relationships between the many biological fields or types of evidence and the diversity of organisms; to be a synthesizing science as well as a data-gathering science.

II. PHASES OF DEVELOPMENT IN PLANT TAXONOMY

1. Pioneer. The discovery, classification, naming, and identification of plants.
2. Consolidation. The synthesis, mostly based on gross morphology, of field and herbarium knowledge in the preparation of floras, manuals, monographs, and form-based classification systems.
3. Experimental or Biosystematic. The analysis of breeding systems, variation patterns, evolutionary potential, and pertinent work in the chemical, numerical, cytological, anatomical, embryological, and palynological aspects of systematics.
4. Encyclopedic or Holotaxonomic. The analysis and synthesis of all information and types of data in the development of one or more classification systems based on evolutionary or phylogenetic relationships.

III. BASIC TERMS GLOSSARY

Biosystematics. The field of study dealing with variation and evolution, primarily experimental and analytical, and mostly treating the species and infraspecific taxa.

Classification. The assignment of plants to groups within a system or hierarchy of ranks or categories distinguished by structure, origins, and other characteristics.

Artificial System. A strictly utilitarian system based on habit, color, form or other characters of a similar type.

Natural System. A system based on form relationships.

Phylogenetic System. A system supposedly based on racial history relationships; one presumably based on common evolutionary descent.

Description. A characterization; a statement of attributes of a particular taxon, individual plant or its parts.

Ecosystematics. The field of study dealing with floristics and plant communities, primarily observational and descriptive, and mostly treating the species, genus, and family taxa.

Evolution. The continuous genetic adaptation of organisms to their environment; directional change in gene frequency, over time, within a population or series of populations in response to selective factors of the environment; change over time within a population.

Flora. An inventory of plants within an area or region.

Identification or Determination. The act of recognizing or establishing the taxon to which a specimen belongs; the establishment of an identity similar to another previously known element; applying a name to a specimen.

Naming. The designating of a specimen or a group of specimens by one or two words; name is the general term for the designation of a taxonomic group.

Nomenclature. A set or system of names for taxa, with the binomial for the specific name as basic.

Phylogeny. The evolutionary history of a group of related organisms.

Phylosystematics. The field of study dealing with phylogeny and classification, primarily theoretical and synthetic, mostly treating the genus, family, order, and class taxa.

Phytography. Phase of taxonomy dealing with descriptive terminology of the plant and its parts with the objective of providing accuracy and completeness for the characterization of taxa.

Systematics. A science usually considered as dealing with a comprehensive study of diverse organisms with regard to their natural relationships.

Taxon (pl. taxa). Any taxonomic group, as a species, genus, family or higher rank.

Taxonomy. A science usually considered as dealing with principles and procedures of classification. Note: Since there is no unanimity of opinion or etymological basis for applicability of the definitions of taxonomy and systematics, the two terms will be used interchangeably throughout this text.

Section B. BASIC COURSE IN PLANT SYSTEMATICS
(Advanced undergraduate-beginning graduate level)

TAXONOMY OF VASCULAR PLANTS. Prerequisites: General Botany. An extensive introduction to basic principles and concepts of systematics with laboratory and field emphasis on phytography, families, identification, and recognition of community dominants with limited training in advanced techniques.

A. Objectives of Course

1. General
   a. To delineate, define, and understand taxonomic principles and concepts.
   b. To acquire the ability to communicate about, and to identify, plant taxa, and to acquire the knowledge necessary for the preparation of an inventory of local plant taxa.

c. To introduce classification systems and fundamentals of evolution.

2. Specific
   a. To learn basic morphology of vascular plants.
   b. To learn how plants are named and why names are changed.
   c. To learn how to identify plants.
   d. To learn how to write keys to, and descriptions of, local taxa.
   e. To learn how to find maximum diversity within the region.
   f. To learn how to prepare simple floras.
   g. To learn how to collect, prepare, and care for herbarium specimens.
   h. To be able to distinguish life forms and dispersal mechanisms.
   i. To be introduced to basic principles of plant geography.
   j. To be able to recognize different plant communities and habitats.
   k. To be able to recognize families, genera, and dominant species.
   l. To learn how to use the literature of vascular plant systematics.
   m. To understand the concepts of character and evidence.
   n. To be introduced to the more sophisticated embryological, palynological, anatomical, cytological, chemical, and numerical studies and techniques to taxonomy.
   o. To know the dicta fundamental to sound phylogeny.
   p. To analyze one or more systems of classification.
   q. To become acquainted with different methods of data presentation.
   r. To be made aware of processes of evolution and speciation.
   s. To be able to recognize the types of variation in individuals and populations.
   t. To understand the historical development of plant systematics.

B. Organization

LECTURE - DISCUSSION TOPICS

1. Introduction
2. History of Classification
3. Nomenclature & Problems
4. Nomenclature & Epithets
5. Characters & Identification
6. Morphological Evidence & Family Introduction
7. Anatomical Evidence
8. Embryological Evidence
9. Palynological Evidence
10. Cytological Evidence
11. Genetic & Breeding Evidence
12. Chemical & Physiological Evidence
13. Ecological Evidence
14. Computer & Taximetrics
15. Literature
16. Nomenclature and Evidence Review
17. Biogeographic Regions, Flora Provinces, and Communities of eastern North America
18. Distribution Principles & Ranges
19. Flora & Floristics Analysis
20. Speciation & Species Problem
21. Types of Variation & Evolutionary Patterns
22. Basic Processes in Evolution
23. Evidences for Evolution
24. Development of Systems of Classification
25. Phylogenetic Principles
26. Systems of Classification
27. Ecosystematics Problem
28. Biosystematics Problem
29. Phylosystematics Problem
30. Review

LABORATORY - FIELD

1. Plant Habit, Root, Stem
2. Stems & Buds
3. Leaves
4. Leaves
5. Inflorescences & Flowers
6. Flowers & Families

7. Fruits & Families	20. Communities - Habitats Trip
8. Seeds & Families	21. Identification & Herbarium Preparation
9. Families & Identification	22. Communities - Habitats Trip
10. Families & Identification	23. Identification & Herbarium Preparation
11. Families & Identification	24. Communities - Habitats Trip
12. Phytography Quiz 1-11	25. Identification
13. Families & Identification	26. Recognition Quiz 13-25
14. Families & Identification	27. Spectral Analysis
15. Families & Identification	28. Datagrams
16. Families & Identification	29. Relationships & Advancement
17. Families & Identification	30. Herbarium Quiz
18. Communities - Habitats Trip	
19. Identification & Herbarium Preparation	

General requirements for laboratory and field work:

A. 36 families to be studied in course
B. 250 species to be recognized by community and habitat
C. 100 specimen herbarium to be prepared, including at least one representative of each family studied (Do not collect rare or uncommon plants!)

Section C. GENERAL PROGRAM FOR ADVANCED SYSTEMATICS*
(Graduate Level)

	I. Ecosystematics (Floristics and Plant Communities)	II. Biosystematics (Variation and Evolution)	III. Phylosystematics (Phylogeny and Classification)
1.	Observational & Descriptive (Inspective)	Experimental & Analytic (Perspective)	Philosophical & Synthetic (Conspective)
2.	Identification of species in representative genera. Recognition of dominants in major communities.	Variation studies using biosystematic techniques and methods.	Study of systems of classification. Evaluation of taxonomic criteria. Study of phylogenetically significant families.
3.	Species, Genus, Family.	Species, "Microspecies," Infraspecific ranks.	Genus, Family, Order, Class.
4.	Manuals, floras, monographs, revisions.	Current revisions, evolution & biosystematic papers.	Systems, history, recent research papers.
5.	Field, Herbarium, Laboratory.	Garden, Greenhouse, Field, Laboratory.	Library, Herbarium, Laboratory.
6.	Community analysis or Floristic analysis.	Biosystematic analysis of intraspecific variation patterns.	Phylogeny of a higher taxon.
7.	Community Ecology, Geobotany, Geology.	Genetics, Cytology, Statistics, Physiological Ecology	Anatomy, Morphology.

1. Nature of Course 2. Course Content 3. Principal Taxa 4. Literature 5. Place of Work 6. Problem 7. Related disciplines

Training: Basically a problem approach: 1) to an understanding of flora; 2) to the analysis of variation and evolution at the species level; and 3) to

*IV. Principles of Taxonomy (See p. 10)

the theories of evolution of major groups. Students should have research problem experience in all phases of taxonomy regardless of M.A. thesis or Ph.D. dissertation topics. All M.A. students should be required to have the basic course, a pertinent advanced course, and research. All Ph.D. students should be required to have additional advanced taxonomy courses and additional research. Courses or training should be available in special plant groups (e.g., grasses and sedges, trees and shrubs); special topics such as nomenclature; or in special research methods (e.g., chemotaxonomy, population analysis). The training should require a sound background in general botany, 4-5 courses in systematics, 3-5 supporting courses from related disciplines, and seminars.

Facilities: A botanical garden or natural area system should be available which provides many diverse natural habitats as well as experimental beds, tanks, and greenhouses for research problems. Equipment for chemotaxonomic, taximetric, anatomical, and ultrastructural work should be available.

The library should contain all of the major botanical systematics periodicals, systematics indices and floras of the world.

The herbarium should have extensive collections, by county, from the home state; excellent representation, by physiographic province, of the flora from each of the pertinent adjacent states; collections by region, from the remainder of North America; and fairly complete representation of the genera of the world.

I. ECOSYSTEMATICS

FLORISTICS AND PLANT COMMUNITIES. Prerequisites: Basic courses in taxonomy and ecology. Primarily a field study of major types of communities in the local region with emphasis on qualitative analysis and distribution. (5 credit hours).

A. Objectives of Course

a. To know the composition of the major plant communities in the region.
b. To know the distribution of the major plant communities in the region.
c. To know the relationships of the plant communities and their indicator species to the types of habitats in the region.
d. To understand the relationships between the major plant communities of the region and those in North America or on the continent.
e. To be able to recognize the dominant species in the major plant communities in the region.
f. To know the qualitative characteristics of the dominant species and the major plant communities in the region.
g. To understand successional relationships and nature of the climax of the major plant communities in the region or to understand principles of climatogenics, pedogenics, biogenics in an area in time.
h. To be able to analyze the communities qualitatively and quantitatively.
i. To understand basic principles involved in origin, migration, and evolution of major plant communities in the region.
j. To become acquainted with the ecological (biotic, abiotic, spatial, temporal) characteristics of species.

B. Field Organization

	Teams:	A	B	C	D	E	F
I	Instruments	1-2	3-4	5-6	7-8	9-10	11-12
II	Soils	3-4	5-6	7-8	9-10	11-12	1-2
III	Canopy	5-6	7-8	9-10	11-12	1-2	3-4
IV	Subcanopy-Shrubs	7-8	9-10	11-12	1-2	3-4	5-6
V	Herbs-Vines	9-10	11-12	1-2	3-4	5-6	7-8
VI	Cerebration	11-12	1-2	3-4	5-6	7-8	9-10

(I, II,...VI = Duties; A, B,...F = Team Designations; 1, 2,...10 = Field Day Assignments)

Duties: I. Determine elevation, slope degree, slope aspect, canopy height, and age and dbh of largest specimens. II. Determine rock and soil types. III. Determine composition of canopy. IV. Make presence lists of subcanopy and shrub species. V. Make presence lists of herb and vine species. VI. Make presence lists of transgressive and seedling species and determine size classes of transgressives; determine type of sere and stage of succession; determine or postulate pre-, post-, subclimax within area; determine or postulate community type within area over different rock type; collect, collate, summarize data for each community type studied by class.

Schedule: First six weeks - field work in the major plant communities of the region; second six weeks - analysis of field data and working out problem involving spectral analysis of a community or flora and the ecological characteristics of two indicator species; last three weeks - literature studies and reports synthesizing and summarizing the ecosystematics of the plant communities in the region.

II. BIOSYSTEMATICS

VARIATION AND EVOLUTION IN PLANTS. Prerequisites: Basic courses in taxonomy and general genetics. The processes and mechanisms of natural variation and evolution considering the taxonomic aspects of plant variation. (5 credit hours)

A. Objectives of Course

1. General
   a. To provide the student with an understanding of evolutionary mechanisms at the species level.
   b. To develop a working knowledge of field and laboratory methods of measuring, interpreting, and presenting data on natural variation--the "raw material" of organic evolution--at the individual, population and interpopulation levels.
   c. To acquaint the student with methods of experimental design in evolutionary problems.
   d. To provide a sound background in the varied literature of evolution and to develop a professional facility in its use.

2. Specific
   a. To learn the causes of plant variation.
   b. To learn to differentiate between genotypic and phenotypic variation patterns.

c. To understand the nature of adaptation.
d. To learn the evolutionary consequences of different breeding systems.
e. To understand the role and importance of polyploidy in evolution.
f. To understand the evolutionary effects of hybridization.
g. To recognize the evolutionary limitations of apomixis.
h. To become aware of the differences between biotic and abiotic selection.
i. To understand the evolutionary role of polymorphism.
j. To become acquainted with the factors responsible for differences in potential evolutionary rates.
k. To be able to apply evolutionary data taxonomically.

B. Organization

LECTURE - DISCUSSION TOPICS

1. Introduction.
   a. Evolution (definition and history). The evolution of evolutionary thought; biological interaction and study of evolution; non-inheritance of acquired characteristics.
2. Sources of variation.
   a. Mutation: types; rates; effects.
   b. Recombination: linkage; gene "storage" via heterozygosity; effect of type of breeding system; effect of chromosome number (ploidy level); hybridization; migration; cytoplasmic inheritance; mutation vs. recombination as a source of variability.
   c. Isolation, apomixis, and breeding systems.
   d. Hybridization, heterosis, and introgression.
3. Range of variation; patterns of variation; cline.
   a. Phenotypic variation; genotypic variation; polymorphism.
4. Selection and adaptation.
   a. Level, type, and intensity of selective forces. Variation in intensity; catastrophic selection; disruptive selection. Selection of unfavorable gene complex; selection in asexual organisms.
   b. Modulation of selection: homeostasis; adaptive polymorphism; preadaptation.
   c. Abiotic selection: desert; alpine; convergence; migration.
   d. Biotic selection: competition; exclusion and character displacement; parasitism and symbiosis.
5. Evolution as the result of selection.
   a. Evolutionary rates and population structure; "conservative" characters; character evolution vs. organism evolution; population size and structure as factors in evolution; genetic drift and random fixation; progressive evolution and co-evolution (co-adaptation); conservation of organization; retrogressive evolution.
   b. Evolutionary aspects of polyploidy.
   c. Differentiation and phylogeny (species and higher categories).
   d. Differentiation and classification (phylogeny vs. phenetics).

LABORATORY

1. Compile breeding system data on each population.
2. Determine the pollination system, the pollinators, and any floral adaptations to the pollination process.
3. Determine chromosome number and behavior at meiosis.

4. Determine total <u>gamete</u> production per population (to relate to potential mutation rates, and also biotic potential).
5. Determine type and relative effectiveness of any asexual methods of reproduction present.
6. Determine the variation patterns (range, mean, standard deviation, standard error and coefficient of variability) of 3 or more microscopic or macroscopic vegetative characters and 3 or more floral characters at the individual, population, and interpopulation levels.
7. Use the most effective graphic methods to compare the variation patterns (bar graphs, polygonal graphs, scatter diagrams, or others) and then summarize the probable reasons for the differences (or similarities) in the variation patterns and relate this to the evolutionary potential of the species as indicated by your limited sample.

A required report on an analysis of the plants of 3 populations should attempt to account, evolutionarily, for the variation patterns and correlations discovered.

III. PHYLOSYSTEMATICS

<u>PHYLOGENY AND CLASSIFICATION OF FLOWERING PLANTS</u>. Prerequisites: Basic courses in taxonomy and morphology. A comparative study of modern systems of classification based upon morphological and phylogenetic considerations. (3 credit hours)

A. <u>Objectives of Course</u>

a. To learn basic phylogenetic dicta.
b. To evaluate conflicting concepts regarding the morphology and evolution of various angiosperm organs, structures and taxa.
c. To learn basic techniques of gathering and interpreting morphological data.
d. To learn the construction of, and morphological basis for, major systems of classification.
e. To become familiar with the primary literature as regards the general subject of angiosperm phylogeny and classification.

B. <u>Organization</u>

LECTURE - DISCUSSION TOPICS

1. The problem of the origin of the angiosperms.
2. Primitive angiosperms (the "Ranales") and evolutionary trends in flowering plants.
3. Morphological and anatomical evidences.
   a. The systematic and phylogenetic value of wood anatomy.
   b. The nature of the angiosperm flower.
   c. The carpel.
   d. The nature of stamens and the significance of the centrifugal androecium.
   e. Pollen as a taxonomic and phylogenetic character.
   f. Systematic value of other vegetative characters - leaves, nodes, phloem, etc.
   g. The primitive habit of angiosperms.

4. The distribution of primitive angiosperm families and factors affecting their distribution and migration.
    a. Floristic regions and distributional patterns.
    b. The factors of distribution.
    c. Theories affecting routes of migration.
5. Natural versus phylogenetic systems of classification.
6. The systems of Engler, Bessey, and Hutchinson.
7. The systems of Takhtajan, Cronquist, and Thorne.
8. The status of the "Amentiferae."
9. The origin and relationships of the monocotyledons.
10. Discussion of research projects and papers.

LABORATORY

When appropriate, laboratory exercises should supplement the discussion topics to enable the student to have practical experience with the subject matter. Selected phylogenetically significant families will be studied in detail.

COURSE REQUIREMENTS: In keeping with the basic objective of acquainting students with the primary literature, selected references will be distributed and read prior to each discussion period. Each student will be required to select a suitable phylogenetic project which after sufficient library, herbarium and laboratory work will culminate in a report.

IV. PRINCIPLES OF TAXONOMY
(Advanced Systematics)

PRINCIPLES OF TAXONOMY. A traditional, historical and philosophical approach to systematics including discussions and literature reviews of basic taxonomic concepts with problem application in the laboratory, library and herbarium.

TOPICS FOR DISCUSSIONS-SEMINARS-LITERATURE REVIEWS

1. General principles of systematics.
2. Perspective in history.
3. Principles of nomenclature (including problems associated with hybrid and polyploid complexes, ecotypes and apomicts).
4. Concepts of ranks and taxa.
5. Principles of characterization and description.
6. Concepts and weighting of characters and evidence.
7. Significance of types of characters and evidence.
8. Concepts of data, information, systems.
9. Problem definition and design.
10. Choices of data for accumulation.
11. Criteria for selection of analytical procedures.
12. Selection of data presentation methods.
13. Documentation principles and implementation.
14. Data summarization.
15. Principles of identification.
16. Characteristics of identification systems.
17. Principles of classification.
18. Bases of classification systems for organisms, characters and evidence.
19. Biotic, evolutionary and phylogenetic relationships.
20. Variation patterns, their causes and taxonomic implications.

21. The evolutionary causes, and taxonomic consequences of polymorphism.
22. Application of abiotic, spatial, and temporal relationships to systematics.
23. Taxonomic literature and bibliographic aids.
24. Composition of floras, manuals, monographs, revisions.
25. Taxonomic philosophy and standards for different publications.
26. General facilities for taxonomic research.
27. Inspective, perspective, conspective training in systematics and the future of taxonomy.

LABORATORY-LIBRARY-HERBARIUM APPLICATION

A study of problems related to the fundamental aspects of taxonomy with emphasis on classical approaches.

1. Analysis of the ICBN and solution of selected nomenclatural problems (library).
2. Preparation of a monograph or revision of a small genus or species complex (laboratory, herbarium, library).
3. Analysis of selected current and historical taxonomic literature in terms of concepts, principles, methodology, documentation and types of presentation (library).
4. Preparation of biographical sketches of selected taxonomists with emphasis on their itineraries, collections, publications, and concepts of taxa (library and herbarium).
5. Preparation of identification devices and programs (library, herbarium, computer center).

GENERAL LITERATURE
(General taxonomic texts)

Benson, L. 1957. Plant Classification. D. C. Heath and Company. Boston.

Core, E. L. 1955. Plant Taxonomy. Prentice-Hall, Inc. Englewood Cliffs, New Jersey.

Davis, P. H., and V. H. Heywood. 1963. Principles of Angiosperm Taxonomy. D. Van Nostrand, Inc. Princeton.

Lawrence, G. H. M. 1951. Taxonomy of Vascular Plants. The Macmillan Company. New York.

Porter, C. L. 1967. Taxonomy of Flowering Plants, 2nd ed. W. H. Freeman and Company. San Francisco.

SYSTEMATICS
PROSPECTIVE and SUMMARY

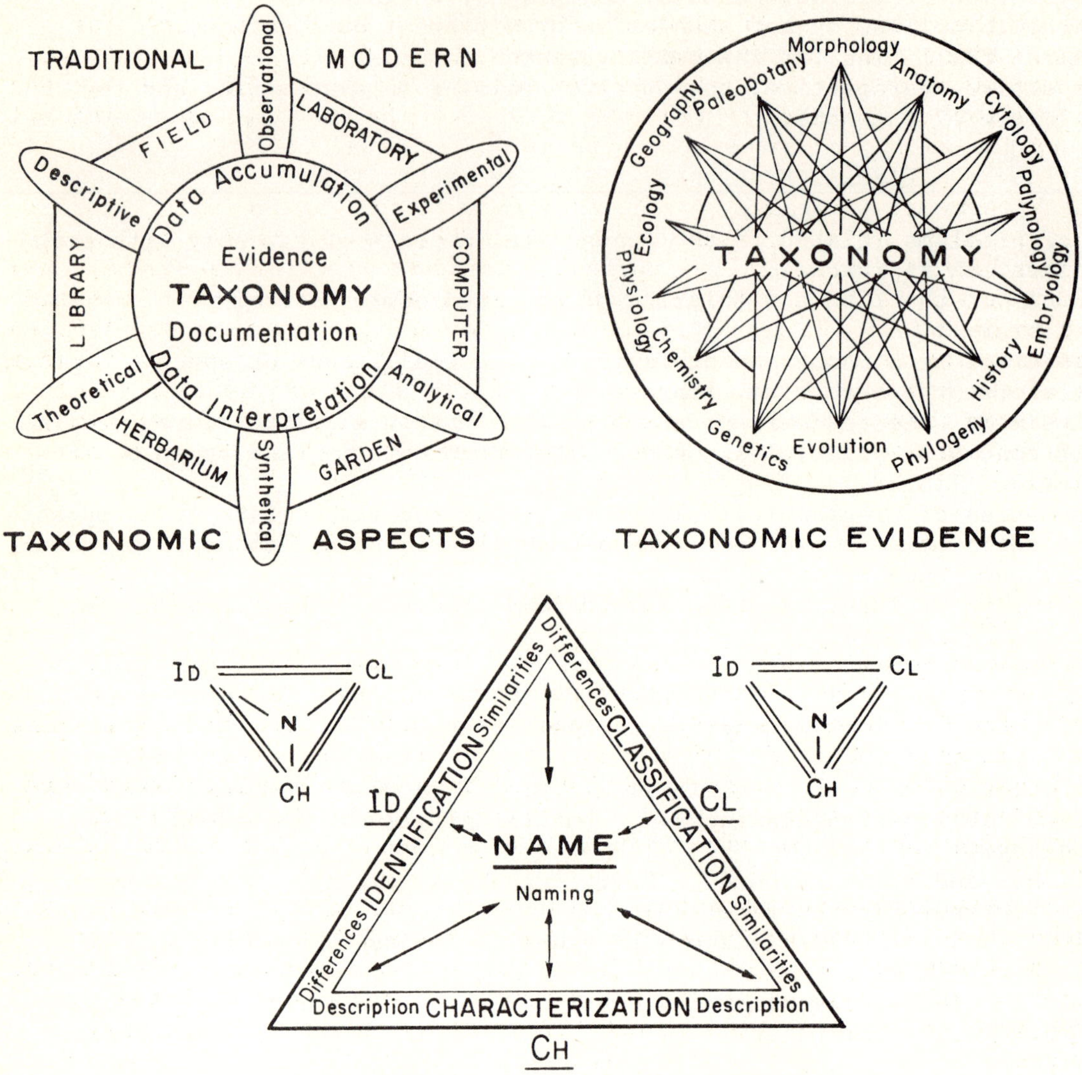

TAXONOMIC FUNDAMENTALS

BASIC PROCEDURES and SYSTEMS
All interrelated and related to the Name

A systematist is one who uses many approaches to fit together appropriately, into an orderly storage and retrieval system with a name reference base, the information about organisms gathered by morphologists, anatomists, cytologists, palynologists, embryologists, geneticists, physiologists, biochemists, ecologists, geographers, paleobotanists and systematists. A systematist is a collector, analyzer and synthesizer of information from all fields of evidence for the characterization, identification and classification of organisms.

Chapter 2. HISTORY OF TAXONOMY*

One of the more noble of mankind's insatiable urges is the quest for knowledge. Since the dawn of history, a fascinating trail has been left as man has sought to learn about his environment and about himself. The purpose of this chapter is to peer briefly down this trail and to offer an occasional interpretation as to how man's knowledge of plants developed.

The history of science always holds a fascination for professionals and the general public alike, and no work of scientific nature is regarded as complete without at least some historical consideration. The result has been an utterly terrifying bibliographic morass for science-historians. Relevant data are to be had in the most diverse sorts of publications, and needless to say, the indexing is often far from adequate. Nevertheless, there are several studies in the history of science that are both detailed and quite readable, and it is from these "secondary sources" that most of the factual material of this chapter has been derived. These sources are itemized in the list of references.

Section A. THE DEVELOPMENT OF CLASSIFICATION

I. THE PRE-LITERATE MAN

Little can be said with any degree of accuracy about the botanical knowledge of pre-literate societies; however, much can be inferred--even the most primitive of ancient men, i.e., those who made their living by gathering food from the landscape, were of necessity practical plant taxonomists. Through experience, these remote ancestors of ours learned which plants were edible and which ones were not. And, (as has been said) those people who failed to learn these elements of botany also failed to become our ancestors.

It is noteworthy that recent studies of existing "primitive" societies have shown them to possess the linguistic mechanisms for accurate distinction between the kinds of plants with which the people are familiar. Furthermore, the distinctions recognized by these people are often as sophisticated, and sometimes more so, than the equivalent distinctions made by professional taxonomists.

II. ANCIENT LITERATE CIVILIZATION

The age of ancient history, when men could read and write but before the advent of printing, saw the rise of the intellectual and social bases of what is now our civilization. It is difficult for us "to put our minds into the first century A.D." to investigate what the mental style of life must have been like, but this is relevant to the history of taxonomy. The advent of writing was most certainly one of the great breakthroughs of ancient man, for it allowed both instant and accurate recall of data to any person skilled in writing. The literate man and his illiterate colleagues regarded the skill chiefly as a means of recording what was being said, and of recalling it

*Contributed by T. M. Barkley, Herbarium, Division of Biology, Kansas State University, Manhattan, Kansas.

accurately at a later date; possibly by another person. The written word simply was fossilized speech, i.e., a preservation of "what people were talking about." To be certain, the more aggressive and quick minds of the times were engaged in polemics and cerebral experimentation, and disputations were committed to script. This was especially the case with the ancient Greeks who had a fully developed sense of intellect. However, it appears that writing was done "for the record" while "publication" was an oral affair. This is a noteworthy point. When we deal with ancient botany, we should read it as material "for the record" and not as new data or new interpretations published with the hope of adding broader vistas to the affairs of men. In other words, the ancient authors are to be regarded as writers of what was held as general knowledge by the sophisticated people of the times. They were offering but little in the way of individual interpretation. This "general knowledge" most assuredly derived from the experiences of pre-literate man.

A. Theophrastos (ca. 370 - 285 B.C.)

Theophrastos of Eresos was the first great botanical writer of classical antiquity and he is the intellectual grandfather of modern botany. As a young man, he went to Athens where he studied with Plato and later with Aristotle. After the deaths of these two great men, Theophrastos became the chief of the Lyceum in Athens and of its gardens and libraries.

Theophrastos is credited with more than 200 written works by the students of Greek scholarship, but most of these works survive as mere fragments or in quotations of other writers. Two of his botanical works have survived and it is upon them that a large measure of his reputation rests; they are: Enquiry into Plants and The Causes of Plants, (both are available in translation). Among the things noted by Theophrastos were: (1) Distinction between external structures (as organs) and internal structures (as tissues). (2) Distinction of different kinds of tissues. (3) Classification of plants into four great groups: trees, shrubs, subshrubs, herbs. (4) Distinction of flowering plants (phanerogams) and nonflowering plants (cryptogams). (5) Recognition of various kinds of sexual and asexual reproduction. (6) A basic understanding of gross anatomy, i.e., that the calyx and corolla are modified leaves, etc. (7) Recognition of the fruit in its modern technical sense (i.e., pericarp).

Furthermore, Theophrastos was acutely aware of some of the biological realities of plants. He was impressed by the fact that many plants (particularly cultivated plants) do not "breed true" from seeds. He was convinced that all species were unstable and were readily changeable. He even gave some definite but speculative suggestions on how to change a few-seeded wheat into more valuable many-seeded wheat. Edward Lee Greene discusses Theophrastos in his Landmarks of Botanical History (1909), and he states that Theophrastos wrote more about more kinds of plants than all of his predecessors, and that he well deserves the recognition he receives. Altogether, Theophrastos wrote about nearly 500 different kinds of plants, and he called them by the names then in common usage. The generic names Crataegus, Daucus, Asparagus, and Narcissus are among those derived directly from the Greek of Theophrastos and they are used today in the same sense that Theophrastos used them.

It is further noteworthy that while associated with Aristotle at the Lyceum, Theophrastos had occasion to become acquainted with Alexander, the son of Philip of Macedon. Alexander, who was to become Alexander the Great, apparently never forgot Theophrastos and during the conquests either he or his associates arranged for materials to be sent back to Athens, thereby allowing Theophrastos to write about such entities as cotton, pepper, cinnamon and bananas.

It is well worth the time and effort for a student of plant taxonomy to read some of the writings of Theophrastos. It gives an insight into what was being done at the Lyceum in Athens (equivalent to the university) during one of the truly golden eras for academic and scholarly matters. It also provides an idea as to what was "common knowledge." The modern reader will be impressed by the relatively high level of botanical sophistication of these ancient people.

Theophrastos is not remembered for literary facility; his sentences are often terse and sometimes rather vague. Some scholars have suggested that the botanical writings of Theophrastos are not meant to be read, but in fact are lecture notes; an entirely plausible explanation.

B. Caius Plinius Secundus (23-79 A.D.)

"Pliny the Elder," as he has come to be known to later generations, deserves a place in the history of taxonomy not because of any elemental contributions on his part, but because he authored one of the most influential compendiums of information ever written. Pliny was, by any standard, "a gentleman and a scholar." He served in the Roman army, was a lawyer, historian, and a friend of the Emperor Vespasian. Pliny died during the eruption of Vesuvius in 79 A.D., his death attributed by some to his being overcome by the sulphurous smoke, while others say he died of heart failure. Pliny wrote voluminously, however his magnum opus is his Natural History, a multi-volume work of which thirty-seven books survive. In his Natural History, Pliny tried to record everything known about the world, and about a quarter of this work dealt with biological subjects. The botany is almost exclusively medical and agricultural in nature. Some of his writing is in error and we may in all charity say that Pliny was gullible and inclined to believe fanciful tales from travelers. However, for more than a thousand years Pliny's Natural History was regarded with almost reverential awe by the peoples of Western Europe. It is no accident that Natural History by Pliny was among the earliest books to be printed by movable type (late 15th century).

C. Pedanios Dioscorides (First Century A.D.)

Presumably Dioscorides was a contemporary of Pliny, but we do not know exactly when he lived. He was of Greek ancestry, a native of the province of Cilicia, and he was a physician in the Roman Army. During the course of his professional labors, he compiled a truly memorable work, the Materia Medica, which presented an account of the various plants and natural materials useful in the treatment of disease. The importance of the work is attested to by ancient writers; however, Dioscorides' fame and influence persists to the present, and the accolades still passed in his direction are justly deserved.

Dioscorides' Materia Medica was derived from first-hand observation by a practicing physician. It discusses with clarity and a degree of precision some 600 species of plants. The book is divided into chapters in a rather helter-skelter way, but many of the plants are grouped together with at least a superficial degree of "natural" relationships, e.g., the mints (Lamiaceae) are grouped together, and the umbels (Apiaceae) are grouped together, etc. Furthermore, the book is written in a straight-forward style, making it possible for one not well-versed in the intricacies of Greek literary rhetoric to still be able to use the book. The plant names are given in Greek, and in at least some versions, they are also in Latin, Hebrew, Turkish (a later addition) and in other languages.

It does us well to call to mind a world in which few people could read. The illiterate masses were not stupid, and oftentimes not ignorant; they simply did not know the art of reading. Into this milieu, the Materia Medica came as a most useful document. It was readily understandable, and it dealt with a subject of perennial concern, viz. human medicine. As a result, the possession of a copy of the Materia Medica was a virtual guarantee of fortune and success. The owner could engage in the practice of pharmacy or medicine because he possessed the most esteemed source book that was available on the subject. The reason this is important to our story here is that shortly after 500 A.D. the Byzantine emperor Flavius Olybrius Anicius had a particularly beautiful manuscript copy of Dioscorides prepared as a gift for his daughter, the Princess Juliana. (This would be regarded as one of the nicest presents that a man could give his daughter). During the ensuing years, the copy prepared for Juliana remained in Constantinople, but the political and military skullduggery between the Austrians and the Ottoman Turks in Constantinople (i.e., Istanbul) led to the transfer of the manuscript from Constantinople to the Imperial Library in Vienna where it remains to this day. The Codex Juliana (as the manuscript is now called), has been studied perhaps in as great detail as any Natural Science book ever written. It is a great link between us and the intellectual heritage of our ancient ancestors.

The Codex Juliana is now available in facsimile and in translation. To the person who has learned the rudiments of Greek, it is a pleasure to look at the illustrations in a facsimile of the Codex Juliana and read the plant names: Anemone, Dipsacon, Aloe, Aristolochia, Phaseolus, etc. Most of the plants in Dioscorides may be recognized with certainty, but a few entities cannot be identified.

To carry the story further, the Codex Juliana attracted the attention of scholars all over Western Europe, and students went to Constantinople and later to Vienna to study it. There is a persistent story that one such student of the Codex Juliana was a Frenchman named Ruel, and a copy of his translation of the Codex came into the hands of Pierandrea Matthioli (Matthiolus), an Italian herbal-writer of the 16th century. Matthioli dutifully recorded the fact that his own herbals (plant-books; see below) were ultimately derived from the Materia Medica of Dioscorides, however the number of medicinal plants (or plants of any sort) known to science had greatly increased in the years between Dioscorides and Matthioli, so it was necessary for Matthioli to extensively revise the organization and the content of the Materia Medica. But it is an undeniable tribute to the fame and influence of Dioscorides that Matthioli in the 16th century would justly proclaim his own work as a derivative of a 1500-year-old text on medicine.

D. Other Ancient Authors

It is an obvious truth that contemporary science has its roots in the Graeco-Roman culture of the Mediterranean Basin, and the botanical works from this region are of prime influence in the development of plant taxonomy.

However, other ancient people dealt with plants in their literatures, and while of secondary importance to us, their works are nonetheless interesting. Virtually every literate society has had works on medicine and agriculture, two subjects close to botany.

The literature of China is particularly rich in botanical works, but is poorly known in the West. There are a series of medical-botanical treatises, always with the words "Pen ts'ao" as part of the title, dating from as far back as the dawn of Chinese civilization. One of these works is attributed to the legendary Emperor Shen-nong who presumably lived ca. 3600 B.C. The oldest extant version of the "Pen ts'ao" dates from ca. 200 B.C., but the information contained in it is much older. The Chinese mastered the art of printing some 400 years before the Europeans, and among the early books printed by the Chinese were versions of the "Pen ts'ao". A particularly famous version is called *Cheng lei pen ts'ao*, and it was printed in 1108 A.D.

Ancient India produced an interesting botanical work presumably sometime prior to the beginning of the Christian era, although it has not been accurately dated. The work is titled the *Vrikshayurveda*, and it is attributed to a writer named Parasara. It is basically a general botany textbook with a discussion of the morphology, classification and in a crude way, the distribution of plants. The morphological materials are presented in comparatively great detail, giving the modern reader the suspicion that the author had some kind of microscope or a good hand-lens.

The Badianus manuscript is a book of interest to Americans. It is an Aztec medical-botany text (in Latin) written by two Aztec Indians in Mexico in 1552. The book itself is relatively simple in a botanical sense, but it is fascinating as a historical document.

III. MEDIEVAL BOTANY

Oceans of ink have been spent detailing the calamity brought to the intellectual life of Europe by the Fall of Rome. The ensuing "Dark Ages" were not devoid of scholarship or human progress, but the general level of culture did sink. It became a matter of inveterate habit for people to assume that the Ancients were infinitely wiser and more cultured than they themselves were. Little was done in the realm of botany (or any other natural science); people preferred to defer to the Ancients. The poverty of botanical work in the period is exemplified by the fact that the most noted contribution of the 9th century is a poem entitled *De cultura Hortorum* by a monk named Wahalfrid Strabo. It is a short work, describing and noting the medicinal uses of the herbs growing in a monastery garden.

The best-remembered natural scientist in Europe of the Middle Ages is Albertus Magnus (ca. 1200-1280 A.D.). He was titled "Doctor Universalis" by his contemporaries, and has been called "the Aristotle of the Middle Ages." St. Albert (he has been elevated to the Sainthood by the Church of Rome) wrote voluminously on many subjects, and his botany is merely a minor contribution. He did not change the course of botanical science, and his influence is hard to assess, however, in an era of general ignorance, he was at least as sophisticated as the ancient Greeks. St. Albert produced a plant classification system that recognized Monocots and Dicots, and generally separated the vascular from the non-vascular plants.

While Europe was somnolent, the near and middle east was approaching a cultural zenith. Muhammad and his followers had successfully converted the population from Persia to Morocco and Spain to the faith of Islam, and groups of brilliant scholars developed in several cultural centers under Muslim suzerainty. The Muslims were practical people, much impressed with Greek learning but not quite prepared to pursue the same path that the Greeks had trod. The Muslim scholars translated and generally curated the speculative and theoretical works of the ancient Greeks, and in large measure, prevented the loss of these works during the Medieval period in Europe. The Muslims made some of their greatest contributions to Natural Science in the realms of medicine, but they did little directly in botany. We should "tip-the-hat-in-passing" to Ibn-Sina, or Avicenna as he is known in the West, who authored the Canon of Medicine, which like the Materia Medica of Dioscorides, was and remains one of the great scientific classics of all time.

One Muslim scholar who dealt chiefly with plants was Ibn-al-'Awwam, a twelfth-century native of Spain. He wrote on the agriculture of his time discussing nearly 600 plants in his works. Among his observations are the sexuality of plants which he seems to have at least superficially understood and the role of certain insects in the pollination of figs.

IV. THE RENAISSANCE AND ITS PROGENY

All of the above named writers, i.e. the ancient and Medieval writers, have one thing in common; each author felt that what he was writing was truly in the realm of "common knowledge" and he was writing it to transmit information. With the coming of the Renaissance, the entire cultural and intellectual fabric of Europe was altered. People sought originality in the arts, in politics, in the sciences, and in all manner of personal expression.

It is a matter of fact (although the causal interrelationships are not clear) that two momentous breakthroughs for Western Man occurred at the time the Renaissance was having its greatest impact. These are the development of printing and the development of the science of navigation. The impact of each was far-reaching in the history of plant taxonomy.

Printing was responsible for making books cheap and literacy common. Anybody who wished could consult the printed works of the ancients without the intervention of a literate scholar. And, anyone who cared to could write a treatise and have a reasonable hope of seeing it printed and distributed. The result was that medically-oriented plant books were quite popular during the early days of printing, and the misconceptions in ancient texts prompted many men to write and publish their own botanical-medical books, or "Herbals" as such books came to be called. The popularity of these herbals offered opportunity for their writers to propose putative relationships between various kinds of plants and to experiment with nomenclatural and reference schemes.

The science of navigation made it possible for a seafarer to set out on a long voyage with a reasonable certainty that he might return from it. This, naturally, led to the successful exploitation of the New World by western Europeans. In turn, this led to a great increase in the number of plants known to Europeans and the practical need to expand the classification scheme employed in the herbals.

A. The Herbalists

In the late 1400's, shortly after the advent of printing, a series of herbals was published under the names of Gart der Gesundheit, or Hortus sanitatis. These books were often mere compilations of the local medical folklore, with crude illustrations, and they were usually published without attribution to an author. Apparently they were profitable for the small print-shops of the day, for many editions published in numerous cities have survived. As a generality, the botanical quality of the Gart der Gesundheit was low.

From about 1500 to 1580, Northwestern Europe was a hotbed of intellectual activity; the general level of education was rising and there was a "Zeitgeist" termed by the philosophers as "The Age of Adventure." This was the time of the great Herbalists who were the first modern men to recognize the inadequacies of the botanical writings of the ancients and to do something about it. Many of the herbal-writers of the era are listed in Chapter 28 along with a concise statement of their contributions. The best known and most influential of the early herbalists were three Germans: Otto Brunfels (1464-1534), Jerome Bock (1489-1554) (Latinized as Hieronymus Tragus), and Leonhart Fuchs (1501-1566). All three were men of letters and scholarship, and Brunfels and Fuchs were practicing physicians. The three have been called the "German Fathers of Botany," which is an apt apellation.

The herbal of Otto Brunfels is called Herbarum vivae Eicones, and it contains many excellent illustrations. Jerome Bock's great herbal is his Neu Kreuterbuch, which contains no illustrations but has a wealth of technically accurate descriptions based on first-hand observations. Fuch's herbal, De historia stirpium included both good descriptions and illustrations.

What separates these three writers from their predecessors is that they were admittedly creating new insights and schemes for viewing the plant kingdom. They admired the erudition of the Ancients, but they were not appealing to the texts of antiquity for authority.

The herbalists of the period 1500-1580 were chiefly motivated by practical considerations, i.e. medicinal and agricultural uses of the plants. Therefore, little emphasis was placed upon the "system" of classification. Furthermore, the number of plants that these herbalists dealt with was not particularly great (less than a thousand) so it was entirely within the limits-of-the-reasonable for a person interested in the subject to learn all of the plants by sight.

B. The 17th Century

Toward the beginning of the 17th century, the great numbers of plants entering the Old World from the New World could not be ignored by the herbalists, and it was becoming obvious that a formalized scheme was required to keep straight the burgeoning numbers of entities in the minds of the botanists. Andrea Caesalpino (1519-1603) was an Italian scientist who deserves our recollection for being one of the early taxonomists to try to base a taxonomic scheme upon reason and logic rather than purely utilitarian concepts. He was philosophically an Aristotelian in his approach, and he based his classification on the assumption that certain features of plant structures were intrinsically more meaningful than others. Such an approach is termed a priori reasoning, and while it may lead to erroneous conclusions, it does provide a way of testing an hypothesis. Furthermore, it is an infinitely better way of erecting a taxonomic scheme than basing it purely upon man's uses of the plants.

Another Aristotelian legacy entered the history of systematics at about this time, and that is the typological concept. In basic (and not quite accurate) terms, it may be defined as the belief that every entity is representable by an ultimate or ideal form, and that there is a "type" embodiment in the mind of man which reflects this ultimate ideal. In effect, this Aristotelian mind-set predisposed the taxonomists of the period to believe in a certain fixity of species, and to regard species or "kinds" of things as discrete and non-varying entities. Subsequent experience has shown this not to be the case at all, however, Aristotelian typology has had an influence in systematic botany that is with us to the present day.

Gaspard Bauhin (1560-1624) a Swiss botanist, put out a monumental opus in his Pinax theatri botanici. This is a register of the different kinds of plants known to exist up to that time, and more importantly, it contains an account of what name the various different botanists and herbalists had used for each plant. In other words, Gaspard Bauhin accounted for the rapidly developing synonymy in systematic botany.

The writings of Gaspard Bauhin are interesting for another reason; he appears to have firmly distinguished between the concept of genus and the concept of species. To a more sophisticated age, the distinction may seem obvious but Bauhin's distinction is of signal importance, for in it is at least tacitly recognized the distinction of different levels (or sets) of hierarchies in taxonomic groupings.

A further point about Gaspard Bauhin is that the Pinax makes use of some binomial nomenclature. It is not used consistently, but with enough regularity so that we know that Bauhin was experimenting with it as a nomenclatural scheme.

Another man who made a noteworthy contribution to the growth of systematic botany throughout the 17th century was the Englishman, John Ray (1627-1705). He published numerous works, but his most significant for us are his Methodus plantarum nova (1682) and the Historia plantarum (3 volumes 1686-1704). Ray produced a system of classification in which plants that looked alike were grouped together, and he recognized the fact that individual characters in themselves may or may not be worthwhile taxonomic features.

In 1700, Joseph Pitton de Tournefort (1656-1708) published his Institutiones rei herbariae which, like the works of John Ray, has left an impress on taxonomy. The Institutiones was less sophisticated than Ray's Historia; it was structured on a mechanical basis, but it was aimed at putting a name on a plant-in-hand. It arranges nearly 9,000 species into about 700 genera, which in turn are grouped into classes. The nomenclatural scheme is imperfect and inconsistent, but it is an important organizational step in the direction of reducing the botanical chaos of the era.

By the end of the 17th century, systematic botany had arrived at the point where a great organizational overhaul was needed. Theoretical studies of the sort engaged in by John Ray had made strides in improving the basis for classification, but the results of the explorations in the New World, the Orient and Africa were literally swamping the European taxonomists with an unmanageable number of plants. Gradually the scheme developed whereby a student of a particular group (e.g., a genus or a family) would provide good descriptions for each of the entities he recognized, along with a description of the group as a whole. This meant that the person wishing to identify a plant could read the descriptions and hopefully find one that would fit his plant; and just as important, it meant that a man with a plant new to science would be aware of the fact because his plant did not "fit the scheme."

C. The Linnaean Period

In the history of plant taxonomy, the 18th century clearly belongs to Carl Linnaeus (1707-1778), the scientist who effectively created the modern referable system of nomenclature. The fame and influence of Linnaeus is so great that his name is commemorated in scientific societies (The Linnaean Society of London), taxonomic journals (Linnaea), formal taxonomy (a linnaeon is a broadly-based concept of species), plus a host of other things. Linnaeus himself was a son of a Swedish parson; he was educated at the University of Uppsala and eventually obtained an M.D. degree at the University of Harderwijk in the Netherlands. After a brief turn at practicing medicine, he returned to Uppsala where he spent the rest of his life as a professor of Natural History. He was a prodigious writer, an inspiring teacher and a great researcher. His life and works have been the subject of many biographical and bibliographical studies.

Linnaeus is to be remembered as the culmination of the botanists who were striving to create a workable system for the classification of plants. Linnaeus divided the plant kingdom on the basis of traditional a priori reasoning; he assumed that reproductive features were ipso facto more important than any other characters for taxonomic purposes. In the case of the flowering plants, the prime divisions are based on the number of stamens that a flower possesses (cf. Chapter 28). This produces incongruities that were apparent even to Linnaeus, (e.g., "Cactus" is classified adjacent to Prunus because of a similar stamen number and placement) but the classification scheme had one crowning achievement; it was usable by anyone basically trained in the field of taxonomy, and it was completely referable. A user of the scheme would repeatedly come up with the same name for the same plant without the necessity of asking an authority. An important facet of the Linnaean system is the consistent use of binomial nomenclature, wherein each entity receives two names: a genus name and a species name, and the correlated requirement that when a person names a plant, he states both of these names together.

In short, Linnaeus provided a solution to the age-old problem of systematics, i.e., how to have a precise, referable and expandable system of nomenclature. The Linnaean nomenclatural scheme was adopted almost immediately by taxonomists in Europe, and it has remained intact to the present. It is not without faults; for example, the treatment of hybrids is cumbersome.

Linnaeus' systematic contributions did not arrive on the scene in one piece. They were proposed in several publications over the years, but the place where he "got it all together" is his Species Plantarum, a two-volume catalog of the plant kingdom, published in May, 1753. This book marks the first consistent use of binomial nomenclature, and it has subsequently been adopted by the botanical community as the starting point for modern botanical nomenclature.

Students interested in the history of science will find Linnaeus and his colleagues an endless source of fascination. Linnaeus served as the major professor for 180 students, and the ultimate contributions and influence of them forms the backbone of modern systematics. My own botanical-genealogy illustrates this: I was a student of Arthur Cronquist, who in turn was a student of Rosendahl, who was a student of Engler, who was a student of Goeppert, who was a student of Beckman, who was a student of Linnaeus. Nearly any contemporary botanist can tell a generally similar story.

Linnaeus, his immediate predecessors and their colleagues were concerned with the mechanics of classification, and were not prime contributors to the understanding of the biological facts of life. They operated under at least the tacit assumption that species were fixed entities with a finite number existing in the world. They did grasp the fact that some plants are more closely related to each other than are other plants, but this fact seems not to have been too weighty in their minds. But one cannot be critical of the Linnaeans, for they brought a large degree of order and stability out of nomenclatural chaos. For this they deserve our frequent recollection and admiration.

D. The Natural System

By the late 1700's a sufficient body of plant information had accumulated to cause the taxonomists to worry about the ultimate purposes of their taxonomy. A readily referable cataloging scheme had been provided by Linnaeus, but it was admittedly little more than a good cataloging scheme. Plants presumed to be related were oftentimes not cataloged together, and there was no "information content" to the scheme. France at this time was undergoing an intellectual ferment with sweeping results in the natural sciences, philosophy, and ultimately in revolutionary politics. It was in the midst of this ferment that a new philosophical basis for taxonomy took root.

Michel Adanson (1727-1806) objected to the arbitrary *a priori* choice of taxonomic characters and the resulting taxonomic scheme; instead he created a taxonomic scheme for several groups of organisms (both plants and animals) based upon the equal use of as many measurable features as possible. His taxonomic schemes have been forgotten, but he is noted for clearly working on the assumption that no single character is intrinsically more important than any other character. The key word here is "intrinsically," for experience in taxonomy shows that indeed some characters are more useful than others, but such conclusions are to be based on empirical data, not on *a priori* reasoning.

The de Jussieu family of France had four members who made noteworthy contributions to botany: Antoine, Bernard, Joseph and Antione-Laurent. In one way or another, all were connected with the Jardin des Plants in Paris, but it is Antoine-Laurent de Jussieu (1748-1836) who attracts our special attention here. Sometime before the French Revolution, it was decided to arrange the plants growing in the Jardin according to the taxonomic scheme of Linnaeus. The plan was never completed, because continual adjustment was made in the planting sequence so that similar plants would be placed together. The ultimate product was a gardenful of plants arranged so that those which seemed to be the most alike were growing next to each other. In 1789 (the year of the French Revolution) Antoine-Laurent de Jussieu published his *Genera plantarum secundum ordines naturales disposita*, which presented a classification scheme derived from experience gained in arranging the garden. This is the first major work to be "natural" in its approach, which in practice meant that plants that looked alike were classed alike.

There was no hard-and-fast scientific principle or easily stated philosophical dictum supporting the idea of a "natural" system. That was to come a century later with the elaboration of the theory of evolution. However, we note that as a matter of historical record, the botanists of the time readily accepted the idea of a natural system with its bases in intuition and empirical observation and they regarded it as a vast improvement over the fundamentally artificial scheme of Linnaeus and his predecessors.

A contemporary of A. L. de Jussieu was J. B. P. de Lamarck (1744-1829), who recognized the philosophical utility of the theory of evolution but who so badly muddled the explanation of it that the theory was regarded as utter nonsense by most members of the scientific community until the middle of the next century.

Another great botanical family, like the de Jussieus, was the Swiss-French family named deCandolle. The father, Augustin Pyramus deCandolle (1778-1841), published in 1813 a book on plant taxonomy called Theorie elementairé, in which the basic principles of the science were given. A. P. deCandolle offered a new arrangement of plants in his classification scheme, but philosophically he was following in the footsteps of A. L. de Jussieu; i.e., it was a natural system aimed at putting related plants together. The truly monumental effort of A. P. deCandolle was the Prodromus systematis naturalis regni vegetabilis, which was an attempt to write a flora for the whole world. A. P. deCandolle wrote at it from 1816 until his death in 1841, whereupon his son Alphonse (1806-1893) worked in the project until 1873. After that a monograph series was published by Alphonse and later his son, Anne Casimir (1836-1918). The Prodromus was never completed, nevertheless for some plant groups it is still the only systematic treatment available. The idea of a world-wide flora remains to be successfully prosecuted.

Throughout the early and middle years of the 19th century, the concept of the "natural" system was firmly entrenched in the minds of taxonomists. The British botanists of the period were active at describing new species and writing colonial floras of the possessions in the British Empire. The last and most complete pre-evolutionary Natural system was conceived at this time by George Bentham (1800-1884), an eminent British "amateur" botanist and Sir Joseph Dalton Hooker (1817-1911) the director of Kew Gardens. Their work, entitled Genera plantarum, was published at intervals between 1862 and 1883. The work is a still-useful compendium of generic descriptions arranged in a natural taxonomic scheme derived from that of A. P. deCandolle (cf. Chapter 28). The generic descriptions are models of completeness and precision, and for the most part were derived from observations of materials in the great herbaria of Britain and with a minimum of data taken from the literature.

V. THE THEORY OF EVOLUTION

In 1859 Charles Darwin published his monumental book, On the Origin of Species, and it made the concept of organic evolution academically respectable. The history of the concept of evolution is interesting in its own as a tale of men groping at a dim but persistent concept, but that is tangential to our story here. What is important is that the Darwin-Wallace theory of evolution as modernized by the incorporation of genetic theory, has basically stood the test of time and has virtually become scientific dogma. Certainly the details are subject to continual revision, but the conceptual schemes of modern biology rest firmly on the foundation of evolutionary theory. Virtually all scientists accept the ideas that gradual change in the nature of life has occurred through the geologic past, and that life on earth consists of populations of organisms which are continually in a state of flux.

A. The Engler-Prantl Approach

A group of German botanists were among the first to struggle with the concept of evolution regarding the development of a taxonomic scheme. S. Endlicher (1805-1849) and later A. W. Eichler (1839-1887) proposed taxonomic schemes, amplified and revised by Adolf Engler (1844-1930) and Karl Prantl (1849-1893). The latter two produced a major work in their *Die natürlichen Pflanzenfamilien*, published in numerous volumes between 1887 and 1915. The original taxonomic scheme was "natural" and it grouped together those plants that the authors believed to be related. By the time the scheme was completely published in the *Pflanzenfamilien* (i.e., 1915) it had acquired a definite progressive character; it started with the simplest plants and worked through the plant kingdom to those which were structurally the most complex. With regard to the flowering plants, it started with the Salicaceae (i.e. *Salix*, the willows, and *Populus*, the cottonwoods and poplars) because they possess flowers of the simplest structure, and it ends with the Compositae, the sunflower family, because of their great floral complexity. Nobody denied that this was a truly magnificent natural system; it was beautifully documented in a multivolumed description of the plant kingdom and it therefore became the standard taxonomic scheme for most botanical institutions and publications. To this day, most "floras" (i.e., books covering the flora of a particular region) have the families arranged in the sequence given by Engler & Prantl in the *Pflanzenfamilien*.

As the impact of the theory of evolution became evident in taxonomy, an effort was made to demonstrate that the progressive sequence of the Engler & Prantl scheme reflected the evolutionary development of the plant kingdom. It was thought by some that simplicity of structure could be equated with primitiveness, however the work of paleobotanists and comparative morphologists has shown this notion to be only partially true.

B. Bessey and His Followers

The work of C. E. Bessey (1845-1915) a professor of Botany at the University of Nebraska, culminates the long road from purely practical taxonomic schemes to the evolutionary schemes of contemporary systematics. Bessey derived a set of "dicta" or propositions based upon his own experience and sources in the literature as to what structural features might be considered primitive, and what might be advanced. Primitive, in this case, refers to being present in the most ancient plants; advanced refers to a feature more recently evolved. Primitive-versus-advanced is not necessarily the same as simple-versus-complex, for a structural feature may become advanced by either reduction toward simplicity or by elaboration toward complexity. Bessey carefully noted his concepts and the sources of them, and then he created a taxonomic system based upon the principle of organic evolution. His goal was to arrange the flowering plants in a scheme reflecting their evolutionary sequence and relationships. The system itself is discussed in Chapter 28, but the point here is that C. E. Bessey (and a German contemporary, Hans Hallier - 1868-1938) added a new dimension to the history of plant taxonomy: viz:, taxonomic schemes should reflect evolution. Such a scheme is said to be "phylogenetic."

Since the time of C. E. Bessey, taxonomic studies have produced numerous alterations and improvements. This is especially true in our understanding of what the primitive land-plants were like, and in the knowledge of the origin of the flowering plants. The most recent taxonomic schemes (e.g., those of the contemporary botanists A. Cronquist, A. L. Takhtajan and R. Thorne, cf. Chapter 28) differ in detail from Bessey, but they remain close to Bessey in the ultimate goal of producing a taxonomy which reflects evolutionary development.

An important point should be noted here concerning the naming of plants. The actual name applied to a plant remains the same, regardless of what kind of a classification scheme is employed. For example, the plant called *Helianthus annuus* (the common weedy sunflower) is called by the same name in any taxonomic scheme, and the fact that this is so is the result of the Linnean system of binomial nomenclature. In modern natural and phylogenetic systems, the identification of a plant implies a knowledge of its natural affinities and its evolutionary relationships. Therefore the modern taxonomic schemes are richer in information content than are the simple mechanical schemes.

Section B. THE KNOWLEDGE OF THE WORLD'S FLORA

I. INTRODUCTION

Books have been written, and more could be written, on how the present understanding of the flora of the world came about. We can only guess at what our pre-literate ancestors knew; but they certainly possessed an intimate knowledge of their environment, including the local flora. Likewise, early commercial enterprise fostered by the development of navigation brought European man into contact with large numbers of previously unfamiliar plants. As a simple exercise in appreciating the impact of these early navigators, one should try to imagine what the dining tables of Europe had on them prior to the exploration of the Americas. Just for starters, there would have been no corn, potatoes or tomatoes. Simply corn alone, and its many by-products, has altered the food habits of European man. The intriguing story, however, is not restricted to the botanical information gained as incidental knowledge to commercial ventures, but it is the story of pure exploration where people traveled far-and-wide simply to find out what was there. The following discussion is of necessity abbreviated and restricted to temperate North America and the reader should realize that the other parts of the world also have their collection of colorful plant collectors.

The roll-call of botanists who made important contributions to the science in North America is long; far longer than space permits simply to list them. Therefore, a few are here discussed in a little greater detail than just listing the names.

A. The Exploratory Period

The first man to do any great amount of botany in what is now the United States was John Clayton (1685-1773), a native of England who lived in Virginia from 1705 until his death. He was a lawyer by profession but a botanist by avocation. He collected plants, wrote descriptions, and corresponded about botany with J. Gronovius (1690-1762), a professor at Leyden in the Netherlands. Clayton's materials were the basis for the publication of the Flora virginica by Gronovius in 1739. It has been said that Gronovius was working on the manuscript for this book at the time when the young Linnaeus visited Leyden after receiving his M.D. degree in the Netherlands. Linnaeus later used the Flora virginica as a major source of data for temperate North American Plants in the preparation of his Species plantarum.

Two other men of essentially the same period as John Clayton were Mark Catesby (1680-1749) and Peter Kalm (1715-1779). The former spent about ten years in Virginia and the Carolinas, and he ultimately published a sumptuous Natural History of Carolina, Florida and the Bahama Islands (1731-1743). Peter Kalm was a student of Linnaeus, and he traveled along the eastern edge of North America between 1748-1751.

John Bartram (1701-1777) and William Bartram (1739-1823), a father and son, deserve discussion in any study of the natural history of North America. They were among the earliest North American natives to become "scientifically respectable" to the naturalists of Europe. The Bartrams were farmers near Philadelphia, and were scholars by taste and avocation. John Bartram was a friend of the notable intellects of his region, including Benjamin Franklin, and he was a charter member of the American Philosophical Society.

He and his son traveled and collected widely in Eastern North America, and William published his Travels through North and South Carolina, Georgia, Florida,, etc., in 1791. In an era in Europe that was intellectually fortified by the philosophy of J. J. Rousseau, the sense of the rustic and the "back to nature" aspect of William Bartram's book made it very popular with the upper classes. The Bartrams carried on an active correspondence with Europeans, plus a large trade in seeds and other plant materials. For years the Bartrams served as the intermediaries for the natural history studies in North America.

Another father-son team was that of the Michaux family; André (1746-1802), the father, and Francois-André (1720-1850), the son. Michaux pere was sent to the New World by the government of France in 1785. Eventually he resided in South Carolina, and his Flora boreali-americana was published in 1803. Michaux fils wrote a 3-volume work of the arborescent flora of North America, publishing it in French in 1810-1813, and in English as Sylva of North America in 1818-1819.

Another botanist of the period was Frederick Pursh (1774-1829) who came to the United States in 1799 from his native Germany. For a number of years he was employed in New York City as the curator-gardener of the Elgin Botanic Garden, the private garden of Dr. David Hosack, who was a prominent physician, botanist and teacher. After some years of plant exploration and the acquisition of specimens, Pursh went back to Europe, where in England in 1814 he published his Flora americae septentrionale, which has remained a botanical classic. This work contained descriptions of plants collected on the Lewis and Clark expedition, as well as other new materials from the western part of the continent.

By the early years of the 1800's, the principles of nomenclature had generally been agreed upon in custom, if not yet in actual codified rules. (Formal rules of nomenclature date from the second half of the 19th century.) Botanists were completely assured by the scientific custom of the time that the discoverer of a new species would "receive credit" for his discovery by having his own name appended to the scientific name if he named it himself, or oftentimes by having the new species named after him if somebody else actually described the new entity. At this distance we can only speculate about what motivated many of the men who are remembered as great in the annals of taxonomy, but it seems evident that a large portion of the taxonomic activity of the middle decades of the 19th century was inspired by the desire to discover and name new species. Two men exemplify this hard-driven work to discover new species: Constantine Rafinesque-Schmaltz (1773-1840) and Thomas Nuttall (1786-1859). Rafinesque (the Schmaltz is dropped in routine reference to the man) was of mixed European ancestry, but he was born in Constantinople. After an early life of wandering and study, he became a professor at Transylvania University, Lexington, Kentucky. His unconventional mannerisms and rather egotistic personaltiy made him something of a local "character" and eventually saddled him with the probably well-deserved reputation as an eccentric. Rafinesque published voluminously and in a wide assortment of periodicals, books, pamphlets, etc. Specimens documenting the new entities that he named have not always been available, making an objective evaluation of his contributions particularly difficult. For a number of years, he was remembered chiefly as a tragi-comic character in the history of North American botany, but the late E. D. Merrill pursued the work of Rafinesque and provided a good basis for a degree of belated respect to this indefatigable but peculiar student of natural history.

Quite a different sort was Thomas Nuttall. He was an Englishman who spent much time in North America, collecting in the West and studying at the Academy of Natural Sciences in Philadelphia. Nuttall had first-hand familiarity with the Mississippi-Missouri river drainage systems (i.e., generally the upper regions of the Louisiana Purchase), and, with admirable precision he described many new species from the region. His collections were properly preserved and are available in several herbaria, in both North America and in Europe. Unlike the work of Rafinesque, there is no question about the utility of the work of Nuttall. The greatest single work of Thomas Nuttall is his Genera of North American Plants (1818) in which the genera are described, species listed and new species treated.

B. The Consolidation Period

In the early 19th century a noticeable change occurred in the botanical procedures in North America. Formerly botanists did the exploring and the explorer often wrote his own contributions on the basis of his own collections, plus the collections of other people insofar as they were available. However, as the frontier moved west and the distances between scientific centers and the unexplored territories became greater, the field work was increasingly done by professional collectors or men at least designated as collectors in government expeditions, while the botanist often remained at home in his herbarium.

One of the noteworthy early collectors was David Douglas, a Scotsman employed by the Horticultural Society of London to collect in the Pacific Northwest of the continent. He was in the areas now called Washington and Oregon in the mid-1820's, where he collected and drew the world's attention to the timber resources of the region. The most valuable of all of the timber trees of the region is now called "Douglas fir," Pseudotsuga menziesii (Mirbel) Franco.

A fortuitous and most productive relationship began with the cooperative efforts of John Torrey (1796-1875) and Asa Gray (1810-1888). Both men were trained as physicians, and both men ultimately were academic botanists; Dr. Torrey at Columbia and Dr. Gray at Harvard. The sequential details of the events in the 1830's and 1840's are beyond the scope of this chapter, but in gross detail it goes thus: Torrey and Gray decided to collaborate on writing a flora of the whole of temperate North America. About the same time, there were numerous government agents and other collectors exploring the vast areas beyond the Appalachian uplift. Many specimens were collected, and ultimately they found their way into the herbaria of John Torrey or Asa Gray. The result was A Flora of North America, by Torrey & Gray, published in several sections between 1838 and 1842. The work established its authors as the chief authorities on the flora of the continent, and this had the effect of attracting more and more specimens to their respective herbaria.

The middle decades of the 19th century saw the establishment of permanent settlements throughout the prairies and plains, plus extensive and detailed exploration of the Cordilleran regions (i.e., from the Rocky Mountains westward). Numerous expeditions were organized with the support of the federal government, and they were charged with surveying the Mexican Boundary, searching for a possible railroad route to the Pacific Ocean, or with some other equally broad-based scheme. These expeditions were well funded and in retrospect, they admirably executed the job of describing a virtually unknown land. The reports of these expeditions still make fascinating reading although they pose bibliographic difficulties because there are numerous authors and titles under which various sections to the reports may be indexed. The plant collections of these expeditions nearly always came to Torrey and Gray, and in later years, just to Asa Gray. Special plant groups were sent to experts for their consideration (e.g., the cactuses went to Dr. G. Engelmann in Saint Louis, the ferns to Prof. D. C. Eaton at Yale, etc.). To Torrey and Gray (and particularly Gray), therefore befell the job of critically analyzing many of the extensive collections from the remote corners of the continent. Consequently, the authorship for many species in the flora of the U. S. is "T. & G.," for Torrey and Gray, or simply "A. Gray," when Gray alone named the plant.

Later in his career, Asa Gray began the preparation of his Synoptical Flora of North America, a project unfinished at the time of his death, but a work of monumental scale, with the sections that were completed still useful to people writing floras. Another project of Asa Gray carries his name to virtually every botanical classroom in the eastern half of the continent. This was the publication (in 1848) of his Manual of Botany. The Manual has been revised by subsequent botanists at The Gray Herbarium of Harvard University; the most recent is the 8th edition, by M. L. Fernald, published in 1950. Throughout the botanical community it is known simply as "Gray's Manual," and in its present form consists of keys, descriptions, and statements of range for the vascular plants native or adventive in the northeastern quarter of the U. S. and adjacent southern Canada.

The preeminence of Asa Gray in the history of American botany is beyond dispute. He is the exemplary embodiment of the "right man in the right place at the right time," for he was both academically prepared and philosophically suited to take complete advantage of the position he occupied.

A lesser known but equally hard-working contemporary of Asa Gray's was Alphonso Wood (1810-1881) who was a man of good scholarship and experience, and who published an immensely popular Classbook of Botany. It went through several editions, and like Gray's Manual, was largely keys and descriptions of the native flora.

C. The "Splitters" of the West

No survey of the history of systematic botany in North America would be complete without mention of Edward Lee Greene (1843-1915) and P. A. Rydberg (1860-1931). These two men are interesting to us because they were able and prolific field workers, they were writers of floras and describers of great numbers of new species, and because they culminate a philosophical approach to the study of the flora of the continent.

E. L. Greene was a clergyman in the mining regions of New Mexico, and he had an unflagging interest in natural history. He sent specimens to Asa Gray at Harvard for identification, but because Professor Gray was otherwise busy and could not spend a great amount of time with these collections, Greene branched out on his own, rapidly becoming versed in the principles of botany, and writing and publishing his own descriptions. Greene eventually left the ministry and became a professional botanist, serving at the University of California at Berkeley, the Federal Government in Washington, and Catholic University of America in Washington. He edited and published two journals, Pittonia and Leaflets of Botanical Observation & Criticism, both of which are almost exclusively restricted to his own writings.

P. A. Rydberg came to the U. S. from Sweden as a mining engineer. Eventually he found his way into botany and a staff position at the New York Botanical Garden. He worked mostly with the flora of the Rocky Mountains and the adjacent prairies and plains regions, producing numerous publications on the flora.

Greene and Rydberg each described many new species which later workers have regarded simply as synonyms of previously published species. The "style" they employed was to study variation, and if a plant seemed to be different, it was given a name. They are the quintessential "splitters." From the vantage of our own hindsight, we should view these two men with a degree of admiration and charity, for they were struggling with the total complexity of a vast and poorly known flora, and at a level of detail impossible to their predecessors. Their systematic philosophy has proved too narrow in the light of modern biological information, but in the time in which they worked Greene and Rydberg provided referable handles for the plants in a huge region.

The Canadian flora was extensively collected and studied by a contemporary of Greene and Rydberg named Joun Macoun (1832-1920). He published his Catalogue of Canadian Plants in three parts between 1883 and 1886. John Macoun's reputation rests today upon his voluminous collections and his critical compilations of floristic lists (i.e., his Catalogue). He did not describe large numbers of new species.

This brief survey of the development of our knowledge of the flora of North America necessarily omits many people and many contributions. In particular, Mexico has received no mention, and neither have the early West Coast explorers nor the Arctic explorers.

The tropical regions had (and still have) groups of gutsy and colorful explorers. Space does not permit a discussion of extra-North American plant exploration, but the reader is referred to the list of references.

Section C. BOTANICAL GARDENS AND HERBARIA

Plant taxonomy has had its origins in the linguistic necessity for precise communication, and this was based as much on the medical uses of plants as on any other single thing. Consequently, among the earliest botanical institutions were the "simples" (i.e., herb and drug) gardens of the ancient and medieval monasteries. The modern academic botanical institution was a product of the Italian Renaissance, wherein some of the "simples" gardens used in training physicians, acquired an identity with a university, complete with a professor as a "director." The oldest of these academic gardens is at Padua, Italy; it has a history as a botanic garden dating from 1533. Nearly as old is the garden at Pisa, dating from 1543. These University gardens eventually justified their existence by disseminating knowledge about plants to the public and therefore they are the intellectual precursors of the modern botanical gardens and herbaria.

As the Renaissance moved across Europe, botanical gardens were created elsewhere, and by the middle 1700's, nearly every national capital and every university had some sort of a botanical facility. Specimens began arriving from the Orient, Africa and the New World, and beginning with the mid-1700's there was a rapid growth in herbaria (i.e., collections of dried plant materials). Naturally, the herbaria in countries with great colonial empires grew the fastest, and came to possess the broadest collections. The result has been the development of several truly great botanical institutions in the world; i.e., places with huge libraries, large herbaria incorporating worldwide collections and extensive collections of living plants in cultivation.

The Royal Botanical Gardens, at Kew, England, is usually regarded as the foremost botanical institution in the world. Its origin goes back to the 1600's, but it became a modern research institution only in 1841, with the appointment of William J. Hooker as the Director. Since then it has actively sought specimens from all over the world, and now has one of the largest herbaria and botanical libraries. The British Museum (Natural History) likewise has extensive facilities for botanical studies, and the fact that the two institutions are in the greater London area makes London something of a Mecca for taxonomists. Other great botanical institutions in Europe are the Paris Natural History Museum, the V. L. Komarov Institute in Leningrad, and the Botanical Garden at Berlin-Dahlem, Germany. The last-named was damaged during World War II, and many specimens were lost.

In North America, the most important early botanical institution was the herbarium and library of the Academy of Natural Sciences in Philadelphia. It was here that specimens were sent during the early years of the Republic, and it was here that visiting European naturalists called, after their arrival in the New World. Thomas Nuttall was an associate of the Academy, and his <u>Genera of North American Plants</u> was published in Philadelphia. The Academy is active to the present day in nearly all phases of Natural History, although it is no longer preeminent in Botany. The herbarium contains more than a million specimens.

There are four other major botanical institutions in the United States: (1) The Gray Herbarium and the Arnold Arboretum of Harvard University are two corporate entities which for practical purposes have combined their herbaria and libraries to form a single research facility. The impetus for founding the botanical tradition at Harvard was provided by Asa Gray, and the tradition of excellence in systematic botany has never waned. The combined herbaria contain over 3 million specimens of vascular plants. (2) The New York Botanical Garden, in New York City, was incorporated in 1891 largely through the efforts of Dr. N. L. Britton, a professor at Columbia University. It contains the materials accumulated by Dr. John Torrey as well as extensive representations from the western half of the continent. The herbarium totals over 3 million specimens of vascular plants and is especially rich in materials from the American tropics. (3) The Botany Department of the Natural History Museum of the Smithsonian Institution, Washington, D. C., is the National Herbarium for the United States. It is supported by the Federal Government and has wide interests in floristic problems in the U. S. The herbarium had its origins in the mid-1800's, but the greatest impetus to its growth and importance in basic research occurred in the 1890's when the herbarium of the U. S. Department of Agriculture was transferred to the Smithsonian Institution. The herbarium now totals about 3 million specimens of vascular plants. (4) The Missouri Botanical Garden, Saint Louis, Missouri, was founded in 1859 by Henry Shaw, a well-to-do merchant of that city. It has been associated with Washington University in Saint Louis, and for many years it has had a high reputation as a graduate school for training plant taxonomists. The herbarium contains some 2 million specimens.

Most major American universities maintain herbaria and the associated library resources for research in plant taxonomy. Some of these university facilities are quite extensive, e.g., both the University of Michigan and the University of California herbaria are in the million-plus specimen category, and several other herbaria are not far behind.

It should be obvious that mere size of a herbarium is not alone an accurate gauge of the utility of the collection. The provenance of the specimens, the quality of the data on each specimen, etc., is of great importance. However, an honest generality would be that doing a major botanical study such as writing a regional flora or monographing a widespread group of plants requires the resources of a major herbarium. Therefore taxonomists often borrow specimens from distant herbaria, and they regularly visit herbaria to consult the collections.

Section D. CONTEMPORARY NORTH AMERICAN TAXONOMY

In the two decades between World Wars I and II, it became evident to the botanical community that most botanical entities in the temperate zone had been formally described, and therefore it was possible to put a referable name on about any specimen submitted for identification. Thus, the period of exploration and description of new entities (sometimes called the period of "alpha taxonomy") had come to an end, at least for the temperate zone. To be certain, "alpha taxonomy" is very much alive in tropical studies, and all evidence suggests that it will be the style of systematics in the tropics for many years to come, so great and complex is the tropical flora. Botanists concerned with the flora of the temperate regions have in general altered the courses of their interests during the recent past, so that under the umbrella of "taxonomy," or "systematic botany," several different facets of the science are found. The following are among them:

1. Floristics. This is the study of the flora of a particular region, which is often a state or a large, natural geographic region. The goal of floristics is to provide accurate keys and descriptions for the plants in the region, and especially in recent years to provide analytical explanations for the historical development of the flora and its distributions. No modern comprehensive flora exists for temperate North America, although there are several recent regional floras of comprehensive scope.

2. Monographic studies. The goal of a monograph is to account for all of the species within a given group, i.e., a genus or section of a genus or a family, etc., regardless of where the plants grow. Detailed monographic studies involving comparative biochemistry, statistical approaches, cytological, etc., have become frequent in recent years. In addition to providing keys, descriptions and an ordering of the nomenclature, a modern monograph also offers an interpretation of the evolution of the group and an explanation of the "biology," e.g., how and why the plants grow where they do with an analysis of the "strategy" for survival.

3. Correlative studies. Both of the above facets have the identification of plant populations as goals. Correlative studies turn the process around and use the expertise of plant taxonomy to further the knowledge of some other field. For example, a plant taxonomist may join forces with an animal behaviorist and a theoretical ecologist to study energy-flow in an ecosystem. Or, a plant taxonomist and a historical geologist may work together to determine the climatic and vegetational history of a region. There still exists a steady opportunity for basic taxonomy in the temperate zone, and there is a continuing place for the skills of the plant taxonomist in contemporary biological science.

SELECTED REFERENCES FOR HISTORY OF TAXONOMY

Further relevant references are cited at the end of Chapter 28, Systems of Classification.

Anderson, E. 1952. Plants, Man and Life. Little, Brown & Co. Boston. [A pleasant book discussing the domestication of crop plants, the origin of agriculture, and the botanical sophistication of primitive peoples.]

Arber, A. 1938. Herbals, Their Origin and Evolution. The University Press. Cambridge, England. [The standard reference on the subject.]

Badianus Manuscript. See Emmart, Emily Walcott.

Brendel, F. 1879, 1880. Historical sketch of the science of botany in North America from 1635-1840. American Naturalist 13: 754-711, 14: 25-38.

Chaytor, H. J. 1950. From Script to Print. W. Heffer. Cambridge, England. [Discusses the impact of printing in Europe.]

Constance, L. 1964. Systematic botany, an unending synthesis. Taxon 13: 257-273.

Core, E. L. 1955. Plant Taxonomy. Prentice-Hall, Inc. Englewood Cliffs, N. J. [Has a well developed and immensely readable section on the history of plant taxonomy.]

Crosswhite, F. S. 1965. Botanical Exploration in North America to 1810. Herbarium of the University of Wisconsin. Madison. [mimeographed]

Davis, P. H., and V. H. Heywood. 1963. Principles of Angiosperm Taxonomy. D. Van Nostrand Co., Inc. Princeton. [Contains a short but good discussion of the philosophical basis of taxonomy.]

Dupree, A. H. 1959. Asa Gray (1810-1888). Harvard University Press. Cambridge. [The standard biography of Gray.]

Eiseley, L. 1958. Darwin's Century. Doubleday & Co. New York. Anchor Books (paperback), 1961. [An account of the origin and impact of Darwinism.]

Emmart, E. W. 1940. The Badianus Manuscript, an Aztec Herbal. Baltimore. [A facsimile of the manuscript, a translation of the text, plus a discussion of its importance.]

Ewan, J. A. (ed.). 1969. A Short History of Botany in the United States. Hafner Publishing Co. New York and London.

________. 1952. Frederick Pursh, 1774-1820, and His Botanical Associates. Proceedings of the American Philosophical Society 96: 599-628.

Fernald, M. L. 1942. Some early botanists of the American Philosophical Society. Proceedings of the American Philosophical Society 86: 63-71.

Gardner, E. J. 1965. History of Biology. Ed. 2. Burgess Publishing Co. Minneapolis.

Garnsey, H. E. F. (translator), and I. B. Balfour (reviser). 1906. Sachs' History of Botany, 1530-1860. Clarendon Press. Oxford. [One of the basic references on the subject.]

Geiser, S. W. 1948. Naturalists of the Frontier. Ed. 2. Southern Methodist University Press. Dallas.

Glass, B., O. Temkin, and W. L. Straus. 1953. Forerunners of Darwin 1745-1859. The Johns Hopkins Press. Baltimore. (Paperback edition 1968). [A collection of essays discussing the intellectual background for the theory of evolution.]

Green, J. R. 1909. A History of Botany, 1860-1900. Clarendon Press. Oxford. [A continuation of the work by J. von Sachs.]

Greene, E. L. 1909. Landmarks of Botanical History. Smithsonian Miscellaneous Collections, Vol. 54. Washington. [A detailed discussion of ancient botany, particularly good regarding Theophrastos.]

Gunther, R. L. 1959. The Greek Herbal of Dioscorides. Hafner Publishing Co. New York.

Irvine, W. 1959. Apes, Angels and Victorians. World Publishing Co. Cleveland. (paperback). [A study of the impact of the theory of evolution.]

Lawrence, G. H. M. 1965. Herbals, Their History and Significance. In: History of Botany, two papers presented at a symposium held at the William Andrew Clark Memorial Library, December 7, 1963. University of California, Los Angeles.

________. 1951. Taxonomy of Vascular Plants. The Macmillan Co. New York.

Leroy, J. F. 1957. Les Botanistes Francais en Amerique du Nord Avant 1850. Centre National de la Recherche Scientifique. Paris. [Contains essays on Michaux and Rafinesque.]

McKelvey, S. D. 1955. Botanical Exploration of the Trans-Mississippi West, 1790-1850. Arnold Arboretum of Harvard University. Jamaica Plain, Massachusetts. [A hefty book with detailed accounts of the field work in the West.]

Nordenskiold, E. 1935. A History of Biology. Tudor Publishing Co. New York.

Pennell, F. W. 1950. Historical botanical collections of the American Philosophical Society and the Academy of Natural Sciences of Philadelphia. Proceedings of the American Philosophical Society 94: 137-151.

Reed, H. S. 1942. A Short History of the Plant Sciences. Chronica Botanica. Waltham, Massachusetts.

Rogers, A. D. III. 1942. John Torrey. Princeton University Press. Princeton. Reprinted 1965 by Hafner, New York. [The standard biography of Torrey.]

Sachs, J. von. See Garnsey & Balfour, and Green, J. R.

Singer, C. 1950. A History of Biology. Abelard-Schuman, Ltd. New York.

Sirks, M. J., and C. Zirkle. 1964. The Evolution of Biology. Ronald Press. New York. [One of the most readable books on the subject; excellent list of references.]

Smith, C. E., Jr. 1956. A century of botany in America. Bartonia 28: 1-30.

Stafleu, F. A. 1971. Lamarck: the birth of biology. Taxon 20: 397-442.

________. 1971. Linnaeus and the Linnaeans. International Association for Plant Taxonomy. Utrecht. (Regnum Vegetabile, Vol. 79).

Steere, W. C. (ed.). 1958. Fifty Years of Botany. Golden Jubilee Volume of the Botanical Society of America. McGraw-Hill Co. New York. [A collection of essays reviewing botanical development.]

Theophrastos. Enquiry into Plants, translated by Sir Arthur Hort. Loeb Classics Library, London, 1916; Harvard University Press, Cambridge, Massachusetts, 1948-1949. [A "pony" edition, with Greek on the left-hand page and an English translation on the right.]

Thomas, J. H. 1969. Botanical explorations in Washington, Oregon, California and adjacent regions. Huntia 3: 5-62.

Chapter 3. NOMENCLATURE

"How strange and chaotic life would become if it were possible to abandon the use of names for the identification of everything we see, make, or handle. The acquisition and dissemination of knowledge would become impossible, and the business of the world could not go on." (Macself in Johnson, 1971)

Man has always been a nomenclaturist; he has used names for plants, animals, and objects; he has categorized plants, animals, and objects with or without special terminology and systems. "Centuries ago each plant was known by a long, descriptive sentence, which was unwieldy, to say the least. Then Caspar Bauhin (1560-1624) devised a plan of adopting two names only for each plant. But it was not until the great Swedish naturalist, Linnaeus (1707-1778), undertook the task of methodically naming and classifying the whole living world 'from buffaloes to buttercups' that the dual name system became permanently established. Linnaeus brought order out of chaos and indexed the vegetable world on a basis so sound and universally acceptable to the peoples of all nations that most of his names are in use to this day." (Johnson, 1971)

Since the publication of "Species Plantarum" by Linnaeus in 1753, the forming of names in Latin for international use has been a fundamental task for botanists or taxonomists. The scientific or specific name of an organism is a binary combination of (1) the genus, or generic name, and (2) the specific epithet. *Quercus alba* L. is the scientific or specific name for white oak; *Picea rubens* Sarg. is the scientific or specific name for red spruce. *Quercus* and *Picea* are the respective generic names and *alba* and *rubens* are the specific epithets. The "L." after *Quercus alba* and "Sarg." after *Picea rubens* are abbreviations for Linnaeus and Sargent, the authors of the two species, respectively. No species name is accurate or complete unless followed by the full or abbreviated name(s) of the author(s). It is necessary to cite the name of the author so that the date of the first valid publication of the name of the taxon can be verified.

Scientific names are italicized in print or underlined when typed or hand-written or -printed. The initial letter of the generic name is always capitalized, the remainder small. All letters in the specific epithet are small except in a few special cases where initial capitals are permitted. However, many botanists today initiate *all* specific epithets with small letters.

The name of a genus is a noun and singular in number, or a word treated as such. The specific epithet is usually an adjective, e.g., in *Quercus alba* L., alba means white; but may be a noun in apposition, e.g., in *Pyrus malus* L., malus is the name of the genus for apple; or a noun in the genitive singular commemorating or honoring a person, e.g., in *Panicum ashei* Pearson, W. W. Ashe is commemorated by having a species named for him by Pearson.

Generic names and specific epithets may be taken from any source whatsoever or composed arbitrarily, but are always treated as Latin. Botanists are enjoined to refrain from forming names that are long and difficult to pronounce in Latin or to adapt to Latin, and to avoid making hybrid names and epithets by combining words of different languages. Latin terminations should be used as far as possible.

Species are grouped into an ascending hierarchy of taxa: the genus, family, order, class, and division with subgroups of each. Taxonomic groups of any rank are known as taxa (singular, taxon). As an example of ascending ranking, Quercus alba L., Quercus laevis Walter, Quercus falcata Michaux, and Quercus bicolor Willd. are four species in the genus, Quercus. Quercus L., Fagus L. and Castanea Miller are three of the genera in the family, Fagaceae; Fagaceae and Betulaceae are the two families in the order, Fagales; Fagales, Urticales, and Piperales are three of the many orders in the class Magnoliopsida, which is in the division Magnoliophyta. (See Section B VI of this chapter for ranks of taxa and their endings and the end of introduction, page 37, for the concepts of ranks.)

Why use this relatively complicated system of names? Supposedly Latin names are so long and unpronounceable. Why not use common names? Benson (1962) states very succinctly why vernacular or common names cannot replace scientific names.

"1. Names in common language are ordinarily applicable in only a single language; they are not universal.
2. In most parts of the world relatively few species have common or vernacular names in any language.
3. Common names are applied indiscriminately to genera, species or varieties.
4. Often two or more unrelated plants are known by the same name, and frequently even in one language a single species may have two to several common names applied either in the same or different localities."

The system of nomenclature has been standardized in a code so that communication about plants, plant parts, and plant products is based on a name acceptable the world over for commerical as well as scientific purposes. This works since (presumably) one kind of plant may not have more than one correct name and no two kinds of plants may have the same name. The scientific name is the communication symbol for plant knowledge and use everywhere.

Modern botanists in all countries use an International Code of Botanical Nomenclature which is a precise and simple system "dealing on the one hand with (1) terms which denote the ranks of taxonomic groups or units, and on the other hand with (2) the scientific names which are applied to the individual taxonomic groups of plants." (See the Preamble to the Code).

For an understanding of the Code the basic provisions of nomenclature, taken directly from the 1972 edition in order to show the organization and fundamental content, are presented in Section A of this chapter. Essential definitions, abbreviations and endings found in the Code and a few additional terms pertinent to an understanding of nomenclature are summarized in Section B. A set of exercises and problems designed to give the student experience with the system and rules of nomenclature are provided in Section C.

The material in this chapter should be adequate for an introduction to how plants are named, and why and how names are changed. The meanings of epithets and names are found in Chapter 4, and additional references on nomenclature are given in the nomenclature section of taxonomic literature in Chapter 30. Even with the information on nomenclature presented in this or any other text, there is no substitute for the actual Code. Every student of taxonomy should have access to the latest edition (1972) and should be introduced to the fundamental provisions of the Code.

Determination of the correct name of a taxon requires careful taxonomic research and a thorough knowledge of the rules and recommendations. Most nomenclatural problems are the result of changes in taxonomic treatments or from shifts of opinion as to the classification of taxa. The rules have to be correctly applied when a taxonomic judgment indicates (1) that a genus should be divided, (2) that a subdivision of a genus should be transferred to another genus, (3) that a species should be transferred to another genus or placed under another generic name, (4) that an infraspecific taxon should be transferred without change of rank to another genus or species, (5) that two or more taxa of the same rank should be united, and (6) that a rank of genus or infrageneric taxon should be changed. Nomenclatural changes result from rejection of illegitimate names such as: (1) a nomenclaturally superfluous name when published, (2) a name which is a later homonym, (3) a name based on a type consisting of two or more entirely discordant elements, (4) a name based on a monstrosity, (5) a name used in different senses which has become a long-persistent source of error, etc. The problem in determining the correct name of a taxon is in the choice of the legitimate name or epithet (one that is in accordance with the rules) for a taxon with a particular circumscription, position, and rank or classification.

CONCEPTS OF RANKS OF TAXA

(See Section B VI for the Hierarchical Order and Endings of Taxa)

The accepted system of nomenclature provides a hierarchical arrangement of the ranks of taxa (species, genus, family, order, class, and division, each with subcategories). The ranks represent levels of relationships required for classification purposes in treating different groups of plants. Conceptually the ranks of taxa cannot be defined precisely, just arranged hierarchically. A group of plants, however, can be circumscribed, characterized, and delimited with diagnostic characteristics as named ranks which is a major part of classification. A genus such as Pinus or Quercus or the family Aceraceae can be sharply defined while other groups at any level may not be easily recognizable, such as many of the genera in the mustard family or many of the species in the genus, Aster. No set number of diagnostic characteristics, no special types of characters, and no special number of entities are involved in the designation of any group of plants as a certain rank or category. Since the concept of evolution is the basic rationale for the taxonomic system, systematists have anticipated a lack of sharp definition of some groups of plants as taxa. Some organisms are rapidly evolving with intermediates of all sorts while others represent dead-ends with the intermediates long since extinct and the remainder are at different stages and conditions of evolutionary development.

Species, the basic rank, has been characterized or defined, by many unidentified authors since the time of Linnaeus, as a name in a book, as a judgment, as what a competent authority thinks it is, as a group of individuals which in sum total of their attributes resemble each other to a degree usually regarded as specific, as the smallest group to which distinctive and invariable characters can be assigned, and as a discrete and immutable entity of divine origin. Some of these characteristics are applicable to a definition of species but all of them combined do not give an adequate introduction to the concept of species. A few definitions by named authors are presented to show the development of the concept.

"A species is the perennial succession of similar individuals perpetuated by generation." De Jussieu (1789), see reference in Chapter 28.

"Species are groups of plants which breed true within their limits of variation." John Ray (1686), see reference in Chapter 28.

"A species is an evolved (and probably evolving) genetically distinctive, reproductively isolated, natural population." Emerson (1938) in Cain (1944), see reference in Chapter 16.

"A species is a recognizable and self-perpetuating population that is more or less isolated genetically as well as by its geographic distribution and its environment." Porter (1959), see reference in Chapter 1.

"A living natural species is a reproducing population or system of populations of genetically closely related individuals." Benson (1962)

A species is "many groups of related populations which live together in the same area, but which remain distinct from each other without exchanging genes." Stebbins (1966), see reference in Chapter 26.

An excellent brief treatment is that of Cronquist (1968):

"Aside from infraspecific taxa, the only taxonomic category with an inherent rank is the species. An exact definition of the species is impossible, and the more precise one attempts to be, the larger the number of species which do not fit the definition. Still, the basic concept is simple enough. A species is the smallest population which is permanently (in terms of human time) distinct and distinguishable from all others. It is the smallest unit which simply cannot be ignored in the scheme of classification. It is the primary taxonomic unit, and it may also be thought of as the basic evolutionary unit. In sexual populations, gene exchange by hybridization within a species is ordinarily rampant, whereas such gene exchange between different species is restricted or even impossible. Interspecific hybrids are not always wholly sterile, but they are not so fertile and so competitively adapted as to swamp out the parents. Although there are some differences in interpretation, a reasonable degree of reproductive isolation from other species, under natural conditions, is an essential specific quality. Without such isolation, the population would lose its identity through interbreeding.

Many species have varieties which are partially segregated from each other, nearly always with a geographic or ecologic correlation, but connected by numerous intermediates. When parts of what had been a single species have diverged enough so that the distinction between them is reasonably sharp and prospectively permanent, the parts have become distinct species. It is only to be expected that at any particular time some species will be in the process of breaking up into separate species. The fact that polyploidy may introduce a barrier to interbreeding without any other significant difference complicates the problem further. The line between strong varieties and weak species is necessarily an arbitrary one, involving subjective taxonomic judgment. The weak species of one taxonomist may be the strong varieties (or subspecies) of another.

Many pages could profitably be devoted to a discussion of how to decide whether a pair (or group) of taxa would better be treated as distinct species or parts of one species, but we are here concerned primarily with the broader aspects of angiosperm taxonomy and evolution. The difficulties in interpretation do not alter the fact that the species is an inherent taxonomic level, the smallest unit which is permanently distinct and distinguishable."

The treatments of natural groups as ranks at the generic, familial, and higher categories is very subjective with many of the recognized higher taxa the result of custom and tradition. Natural groups; e.g., pines (Pinus), oaks (Quercus), clovers (Trifolium), have a certain taxonomic aspect recognizable by the layman and professional alike. Kinds of plants, e.g., the pines, oaks, clovers represent genera or the rank of genus. Historically and traditionally the mints, legumes, composites, umbels, mustards, and others have been recognized at the family level. Orders, classes and divisions have not been recognized by the layman as such but he has known that the fungi, algae, mosses, liverworts, ferns, conifers, and flowering plants were part of a larger group or higher rank of taxa. The professional botanist has known the last named groups as members of divisions or classes or subordinate ranks thereof.

Again from Cronquist (1968):

"It is perfectly clear that natural, recognizable groups of species, and groups of groups, exist. The ranks at which these groups should be received are not inherent in the nature of the group, but depend on subjective individual judgment. The criteria on which that judgment should be based are recondite. They come down to a personal evaluation of the importance of the differences and the size and coherence of the group, in the context of the system as a whole. Since we have already pointed out that individual characters cannot reasonably have a fixed, inherent taxonomic value, any evaluation of the importance of the characters marking a group is likely to be difficult and subject to unresolvable differences of opinion.

In spite of these inherent difficulties, a very considerable agreement has grown up regarding the groups to be recognized as genera and families of angiosperms.--

It should by now be clear that there is good reason to allow custom to enter into the decision on the rank at which a group is received. Since the rank is not inherent in supraspecific groups, it is only by giving some weight to custom that any stability in the taxonomic scheme can ever be achieved. A good operating principle is to maintain the existing classification whenever it can be defended on natural grounds, and to avoid changing the rank of groups if no significant change in the concept of their relationship to each other and to other groups is involved.

On the other hand, if a traditional arrangement runs counter to evolutionary relationships, then it should give way to a more natural arrangement when such can be devised, no matter how heavy the weight of tradition."

Section A. EXCERPTS FROM THE "INTERNATIONAL CODE OF BOTANICAL NOMENCLATURE" 1972

PREAMBLE

Botany requires a precise and simple system of nomenclature used by botanists in all countries, dealing on the one hand with the terms which denote the ranks of taxonomic groups or units, and on the other hand with the scientific names which are applied to the individual taxonomic groups of plants. The purpose of giving a name to a taxonomic group is not to indicate its characters or history, but to supply a means of referring to it and to indicate its taxonomic rank. This Code aims at the provision of a stable method of naming taxonomic groups, avoiding and rejecting the use of names which may cause error or ambiguity or throw science into confusion. Next in importance is the avoidance of the useless creation of names. Other considerations, such as absolute grammatical correctness, regularity or euphony of names, more or less prevailing custom, regard for persons, etc., notwithstanding their undeniable importance, are relatively accessory.

The Principles form the basis of the system of botanical nomenclature.

The detailed provisions are divided into Rules, set out in the Articles, and Recommendations; the notes attached to these are integral parts of them. Examples are added to the rules and recommendations to illustrate them.

The object of the Rules is to put the nomenclature of the past into order and to provide for that of the future; names contrary to a rule cannot be maintained.

The Recommendations deal with subsidiary points, their object being to bring about greater uniformity and clearness, especially in future nomenclature; names contrary to a recommendation cannot, on that account, be rejected, but they are not examples to be followed.

The provisions regulating the modification of this Code form its last division.

The only proper reasons for changing a name are either a more profound knowledge of the facts resulting from adequate taxonomic study or the necessity of giving up a nomenclature that is contrary to the rules.

Division I. Principles

Principle I. Botanical nomenclature is independent of zoological nomenclature. The Code applies equally to names of taxonomic groups treated as plants whether or not these groups were originally assigned to the plant kingdom.

Principle II. The application of names of taxonomic groups is determined by means of nomenclatural types.

Principle III. The nomenclature of a taxonomic group is based upon priority of publication.

Principle IV. Each taxonomic group with a particular circumscription, position and rank can bear only one correct name, the earliest that is in accordance with the Rules, except in specified cases.

Principle V. Scientific names of taxonomic groups are treated as Latin regardless of their derivation.

Principle VI. The Rules of nomenclature are retroactive unless expressly limited.

Division II. Rules and Recommendations

Chapter I. RANKS OF TAXA, AND THE TERMS DENOTING THEM
(See Sect. B VI)*

Article 1. Taxonomic groups of any rank will, in this Code, be referred to as _taxa_ (singular: _taxon_).

Article 2. Every individual plant is treated as belonging to a number of taxa of consecutively subordinate ranks, among which the rank of species (species) is basic.

Article 3. The principal ranks of taxa in ascending sequence are: species (_species_), genus (_genus_), family (_familia_), order (_ordo_), class (_classis_), and division (_divisio_). Thus each species belongs (is to be assigned) to a genus, each genus to a family (certain groups of fossil plants excepted, etc.

Article 4. If a greater number of ranks of taxa is required, the terms for these are made either by adding the prefix sub (_sub_-) to the terms denoting the ranks or by the introduction of supplementary terms. A plant may be assigned to taxa of the following subordinate ranks of the plant kingdom (Regnum Vegetabile): Divisio, Subdivisio, Classis, Subclassis, Ordo, Subordo, Familia, Subfamilia, Tribus, Subtribus, Genus, Subgenus, Sectio, Subsectio, Series, Subseries, Species, Subspecies, Varietas, Subvarietas, Forma, Subforma. --**

Chapter II. NAMES OF TAXA (GENERAL PROVISIONS)
(See Sect. B IV)

Section 1. Definitions

Article 6. _Effective_ publication is a publication in accordance with Arts. 29-31. _Valid_ publication of names is publication in accordance with Arts. 32-45. A _legitimate_ name or epithet is one that is in accordance with the rules. An _illegitimate_ name or epithet is one that is contrary to the rules. The _correct_ name of a taxon with a particular circumscription, position, and rank is the legitimate name which must be adopted for it under the rules (see Art. 11).

*Reference is to Section B of the chapter in this text.
**Indicates that the excerpt of the article is incomplete.

Section 2. Typification
(See Sect. B VII)

Article 7. The application of names of taxa of the rank of family or below is determined by means of nomenclatural types (types of names of taxa). A nomenclatural type (*typus*) is that constituent element of a taxon to which the name of the taxon is permanently attached, whether as a correct name or as a synonym. --

Section 3. Priority

Article 11. Each family or taxon of lower rank with a particular circumscription, position, and rank can bear only one correct name, special exceptions being made for 9 families for which alternative names are permitted (see Art. 18) and for certain fungi and fossil plants (see Art. 59). --

Section 4. Limitation of the Principle of Priority

Article 13. Valid publication of names for plants of the different groups is treated as beginning at the following dates (for each group a work is mentioned which is treated as having been published on the date given for that group):

a. Spermatophyta and Pteridophyta, 1 May 1753 (Linnaeus, *Species Plantarum* ed. 1). --

Article 14. In order to avoid disadvantageous changes in the nomenclature of genera, families, and intermediate taxa entailed by the strict application of the rules, and especially of the principle of priority in starting from the dates given in Art. 13, this Code provides, in Appendices II and III, lists of names that are conserved (*nomina conservanda*) and must be retained as useful exceptions. Conservation aims at retention of those generic names which best serve stability of nomenclature. --

Chapter III. NOMENCLATURE OF TAXA ACCORDING TO THEIR RANK

Section 1. Names of Taxa Above the Rank of Family
(See Sect. B VI)

Article 16. The principles of priority and typification do not affect the form of names of taxa above the rank of family. (See Arts. 10 and 11.) --

Section 2. Names of Families and Subfamilies, Tribes and Subtribes

Article 18. The name of a family is a plural adjective used as a substantive; it is formed by adding the suffix -*aceae* to the stem of a legitimate name of an included genus (see also Art. 10).--

Note 3. The following names, sanctioned by long usage, are treated as validly published: *Palmae* (*Arecaceae*; type, *Areca* L.); *Gramineae* (*Poaceae*; type, *Poa* L.); *Cruciferae* (*Brassicaceae*; type, *Brassica* L.); *Leguminosae* (*Fabaceae*; type, *Faba* Mill. (= *Vicia* L. p.p.)); *Guttiferae* (*Clusiaceae*; type, *Clusia* L.); *Umbelliferae* (*Apiaceae*; type, Apium L.); *Labiatae* (*Lamiaceae*; type, *Lamium* L.); *Compositae* (*Asteraceae*; type, *Aster* L.).

Botanists are authorized, however, to use as alternatives the appropriate names ending in -aceae. --

Section 3. Names of Genera and Subdivisions of Genera

Article 20. The name of a genus is a substantive in the singular number, or a word treated as such. It may be taken from any source whatever, and may even be composed in an absolutely arbitrary manner. --

Section 4. Names of Species

Article 23. The name of a species is a binary combination consisting of the name of the genus followed by a single specific epithet. If an epithet consists of two or more words, these are to be united or hyphened. An epithet not so joined when originally published is not to be rejected but, when used, is to be united or hyphened.

The epithet of a species may be taken from any source whatever, and may even be composed arbitrarily. --

Section 5. Names of Taxa Below the Rank of Species (Infraspecific Taxa)

Article 24. The name of an infraspecific taxon is a combination of the name of a species and an infraspecific epithet connected by a term denoting its rank. Infraspecific epithets are formed as those of species and, when adjectival in form and not used as substantives, they agree grammatically with the generic name. --

Section 6. Names of Plants in Cultivation

Article 28. Plants brought from the wild into cultivation retain the names that are applied to the same taxa growing in nature. --

Chapter IV. EFFECTIVE AND VALID PUBLICATION

Section 1. Conditions and Dates of Effective Publication

Article 29. Publication is effected, under this Code, only by distribution of printed matter (through sale, exchange, or gift) to the general public or at least to botanical institutions with libraries accessible to botanists generally. It is not effected by communication of new names at a public meeting, by the placing of names in collections or gardens open to the public, or by the issue of microfilm made from manuscripts, type-scripts or other unpublished material. Offer for sale of printed matter that does not exist does not constitute effective publication. --

Section 2. Conditions and Dates of Valid Publication of Names

Article 32. In order to be validly published, a name of a taxon must (1) be effectively published (see Art. 29); (2) have a form which complies with the provisions of Arts. 16-17 (but see Art. 18, notes 1, 2, and 3) and Art. H. 7; (3) be accompanied by a description or diagnosis* of the taxon or by a reference (direct or indirect) to a previously and effectively published description or diagnosis of it (except as provided in Art. H. 9); and (4) comply with the special provisions of Arts. 33-45. --

Section 3. Citation of Authors' Names and of Literature For Purposes of Precision (See Sect. B II)

Article 46. For the indication of the name of a taxon to be accurate and complete, and in order that the date may be readily verified, it is necessary to cite the name of the author(s) who first validly published the name concerned unless the provisions of Arts. 19, 22, or 26 apply. --

Chapter V. RETENTION, CHOICE, AND REJECTION OF NAMES AND EPITHETS

Section 1. Retention of Names or Epithets of Taxa Which Are Remodelled or Divided

Article 51. An alteration of the diagnostic characters or of the circumscription of a taxon does not warrant a change in its name, except as may be required (1) by transference of the taxon (Arts. 54-56), or (2) by its union with another taxon of the same rank (Arts. 57, 58, Rec. 57A), or (3) by a change of its rank (Art. 60). --

Article 52. When a genus is divided into two or more genera, the generic name must be retained for one of them or, if it has not been retained, must be reinstated for one of them. When a particular species was originally designated as the type, the generic name must be retained for the genus including that species. --

Article 53. When a species is divided into two or more species, the specific epithet must be retained for one of them or, if it has not been retained, must be reinstated for one of them. When a particular specimen, or figure was originally designated as the type, the specific epithet must be retained for the species including that element. --

Section 2. Retention of Epithets of Taxa Below the Rank of Genus on Transference to Another Genus or Species

Article 54. When a subdivision of a genus* is transferred to another genus or placed under another generic name for the same genus without change of rank, its epithet must be retained or, if it has not been retained, must be reinstated unless one of the following obstacles exists:

(1) The resulting combination has been previously and validly published for a subdivision of a genus based on a different type;

(2) An earlier and legitimate epithet of the same rank is available (but see Arts. 13f, 58, 59); --

Article 55. When a species is transferred to another genus or placed under another generic name for the same genus without change of rank, the specific epithet, if legitimate, must be retained[a] or, if it has not been retained, must be reinstated[b] unless one of the following obstacles exists:

(1) The resulting binary name is a later homonym[c] (Art. 64) or a tautonym[d] (Art. 23);

(2) An earlier legitimate specific epithet is available (but see Arts. 13f, 58, 59, 72).[c] --

Article 56. When an infraspecific taxon is transferred without change of rank to another genus or species, the original epithet must be retained or, if it has not been retained, must be reinstated unless one of the following obstacles exists:

(1) The resulting ternary combination has been previously and validly published for an infraspecific taxon based on a different type, even if that taxon is of different rank;

(2) An earlier legitimate epithet is available (but see Arts. 13f, 58, 59, 72); --

Section 3. Choice of Names When Taxa of the Same Rank Are United

Article 57. When two or more taxa of the same rank are united, the oldest legitimate name or (for taxa below the rank of genus) the oldest legitimate epithet is retained, --

Section 4. Names of Fungi with a Pleomorphic Life Cycle and of Fossils Assigned to Form-Genera

Section 5. Choice of Names When the Rank of a Taxon is Changed

Article 60. When the rank of a genus or infrageneric* taxon is changed, the correct name or epithet is the earliest legitimate one available in the new rank. In no case does a name or an epithet have priority outside its own rank. --

Section 6. Rejection of Names and Epithets

Article 62. A legitimate name or epithet must not be rejected merely because it is inappropriate or disagreeable, or because another is preferable or better known, or because it has lost its original meaning. --

Article 63. A name is illegitimate and is to be rejected if it was nomenclaturally superfluous when published, i.e., if the taxon to which it was applied, as circumscribed by its author, included the type of a name or epithet which ought to have been adopted under the rules. --

Article 64. A name is illegitimate and must be rejected if it is a later homonym, that is, if it is spelled exactly like a name previously and validly published for a taxon of the same rank based on a different type. Even if the earlier homonym is illegitimate, or is generally treated as a synonym on taxonomic grounds, the later homonym must be rejected, unless it has been conserved. --

Article 65. A name is illegitimate and is to be rejected if it is the name of a taxon which on transfer of that taxon from the animal to the plant kingdom becomes, at the time of such transfer, a homonym of a name for a plant taxon.

If a taxon is transferred from the plant kingdom to the animal kingdom, its name or names retain their status in botanical nomenclature for purposes of homonymy. --

Article 66. An epithet of a subdivision of a genus is illegitimate and is to be rejected in the following special cases:

(1) If it was published in contravention of Arts. 51, 54, 57, 58 or 60, i.e. if its author did not adopt the earliest legitimate epithet available for the taxon with its particular circumscription, position and rank.

(2) If it is an epithet of a type subgenus or section which contravenes Art. 22. --

Article 67. A specific or infraspecific epithet is illegitimate and is to be rejected if it was published in contravention of Arts. 51, 53, 55, 56, or 60, i.e. if its author did not adopt the earliest legitimate epithet available for the taxon with its particular circumscription, position, and rank. Such an epithet is also illegitimate if it was published in contravention of Art. 59. --

Article 69. A name is to be rejected if it is used in different senses and so has become a long-persistent source of error. --

Article 70. A name is to be rejected if it is based on a type consisting of two or more entirely discordant elements, unless it is possible to select one of these elements as a satisfactory type. --

Article 71. A name is to be rejected if it is based on a monstrosity. --

Chapter VI. ORTHOGRAPHY OF NAMES AND EPITHETS AND GENDER OF GENERIC NAMES

Section 1. Orthography of Names and Epithets

Article 73. The original spelling of a name or epithet is to be retained, except for the correction of typographic or orthographic errors (but see Art. 14, note 7).

The consonants w and y, foreign to classical Latin, and k, rare in that language, are permissible in Latin plant names. --

Section 2. Gender of Generic Names

Division III. Provisions for Modification of the Code

Appendix I. Names of hybrids and some special categories
Appendix II. Nomina familiarum conservanda
Appendix III. Nomina generica conservanda et rejicienda
Guide for the Determination of Types
Guide to the Citation of Botanical Literature

Section B. SELECTED NOMENCLATURAL TERMS AND ABBREVIATIONS
Based in major part on McVaugh et al. (1968) and the
International Code of Botanical Nomenclature

I. ABBREVIATIONS

Abbreviation	Latin or Greek	English
Ap.	Apomixis (Gr)	Apomict
Auct. non	Auctorum non	Not of author(s)
Cl.	Klon (Gr)	Clone
Comb. nov.	Combinatio nova	Combination new
Corr.	Correctus -a-um	Correct
Emend.	Emendatus -a-um	Emendation (to correct or improve)
Et al.	Et alii (aliorum)	And others
Excl.	Exclusus -a-um	Exclusion (to exclude)
F.	Forma	Form
Mut. char.	Mutatis characteribus	Change of character
Nm.	Nothomorphos (Gr)	Nothomorph
Nom.	Nomen	Name
Nom. cons.	Nomen conservandum	Conserved name
Nom. nov.	Nomen novum	New name
Nom. rejic.	Nomen rejiciendum	Rejected name
Orth. mut.	Orthographia mutata	Orthography change (altered spelling)
P.p.	Pro parte	In part
Pro hybr.	Pro hybrida	As a hybrid
Pro sp.	Pro specie	As a species
Pro syn.	Pro synonyma	As a synonym
S. str.	Sensu stricto	In a strict or narrow sense
S. ampl.	Sensu amplo	In a broad sense
S. lat.	Sensu lato	In a wide sense
Sp.	Species	Species
Spp.	Species	Species (Plural)
Subsp. or ssp.	Subspecies	Subspecies
Stat. nov.	Status novus	Status new
Typ. cons.	Typus conservandus	Type conserved
Var.	Varietas	Variety

II. AUTHOR CITATIONS

Description Alteration. When the alteration of the diagnostic characters, or of the circumscription of a taxon without the exclusion of the type has been considerable, the nature of the change may be indicated by adding such words, abbreviated where suitable, as emendavit (emend.) (followed by the name of the author responsible for the change), mutatis characteribus (mut. char.), pro parte (p.p.), excluso genere or exclusis generibus (excl. gen.), exclusa specie or exclusis speciebus (excl. sp.), exclusa varietate or exclusis varietatibus (excl. var.), sensu amplo (s. ampl.) sensu stricto (s. str.) etc. Examples: Phyllanthus L. emend. Mull. Arg.; Globularia cordifolia L. excl. var. (emend. Lam.). (See I Abbreviations).

Joint Author. When a name has been published jointly by two authors, the names of both should be cited, linked by means of the word et or by an ampersand (&). When a name has been published jointly by more than two authors, the citation should be restricted to that of the first one followed by et al. Examples: Didymopanax gleasonii Britton et Wilson (or Britton & Wilson); Streptomyces albo-niger Hesseltine, J. N. Porter, Deduck, Hauck, Bohonos, & J. H. Williams (Mycologia 46: 19. 1954) should be cited as S. albo-niger Hesseltine et al.

Name Proposal. When a name has been proposed but not validly published by one author and is subsequently validly published and ascribed to him by another author, the name of the former author followed by the connecting word ex may be inserted before the name of the publishing author.
Examples: Havetia flexilis Spruce ex Planch. et Triana or Havetia flexilis Planch. et Triana. - Gossypium tomentosum Nutt. ex Seem. or Gossypium tomentosum Seem. - Lithocarpus polystachya (Wall. ex A. DC.) Rehder or Lithocarpus polystachya (A. DC.) Rehder. (ex = described by)

Original Author. For the indication of the name of a taxon to be accurate and complete, and in order that the date may be readily verified, it is necessary to cite the name of the author(s) who first validly published the name concerned.
Examples: Rosaceae Juss., Rosa L., Rosa gallica L., Rosa gallica var. eriostyla R. Keller, Rosa gallica L. var. gallica.

Pre Priority. The citation of an author who published the name before the starting point of the group concerned is indicated, when such indication is considered useful or desirable, preferably between square brackets.
Examples: Lupinus [Tourn. Inst. 392. pl. 213. 1719] L. Sp. Pl. 721. 1753, Gen. Pl. ed. 5. 322. 1754, or Lupinus [Tourn.] L.

Publication. When a name with a description or diagnosis (or reference to a description or diagnosis) supplied by one author is published in a work by another author, the word in should be used to connect the names of the two authors. In such cases the name of the author who supplied the description or diagnosis is the most important and should be retained when it is desirable to abbreviate such a citation.
Examples: Viburnum ternatum Rehder in Sargent, Trees and Shrubs 2: 37. 1907, Viburnum ternatum Rehder. - Teucrium charidemii Sandwith in Lacaita, Cavanillesia 3: 38. 1930, or Teucrium charidemii Sandwith.

Rank Alteration. When a genus or a taxon of lower rank is altered in rank but retains its name or epithet, the author who first published this as a legitimate name or epithet (the author of the basionym) must be cited in parentheses, followed by the name of the author who effected the alteration (the author of the combination).
Examples: Medicago polymorpha var. orbicularis L. when raised to the rank of species, by Allioni, becomes Medicago orbicularis (L.) All.

Taxon transfer. When a taxon of lower rank than genus is transferred to another taxon, with or without alteration of rank but retains its name or epithet, the author who first published this as legitimate name or epithet (the author of the basionym) must be cited in parentheses, followed by the name of the author who effected the alteration (the author of the combination).
Examples: Cheiranthus tristis L. transferred to the genus Matthiola by Robert Brown becomes Matthiola tristis (L.) R. Br.

III. HYBRIDS

Hybrid. The offspring of two plants or animals of different breeds, races, forms, varieties, subspecies, species, or genera.

Intergeneric Hybrids. Hybrids between species of two or more genera.
Example: X Agropogon (= Agrostis x Polypogon).

Interspecific Hybrids. Hybrids between two species of the same genus. Example: Salix X capreola (= Salix aurita x S. caprea).

IV. NAMES

Accepted. Name accepted by an author when he proposes or adopts it as the correct name for a taxon.

Admissible. Name having such a form that it can enter into botanical nomenclature and be validly published.

Alternative. Names, two or more, published at the same time by the same author for the same taxon.

Ambiguous (nomen ambiguum). Name used in different senses so that it has become a "long-persistent source of error."

Artificial. Name used for a group consisting of specimens that are judged to be too incomplete for taxonomic disposition but are brought together for convenience by the available characteristics, and treated for nomenclatural purposes as a taxon of a particular kind, e.g., an artificial genus. Used principally for fossils, and as an opposite to natural.

Available. Name or epithet that applies to a particular taxon and can be used for it.

Combined. Name of a taxon below rank of genus consisting of the name of a genus combined with one or more epithets.

Confused. Name that is unclear or indistinct.

Common or trivial. See vernacular name.

Conserved (Nomen conservandum). A validly published name that, by decision of an International Botanical Congress, shall be used in spite of the fact that to do so is contrary to the rules.

Correct. Name that must under the rules be adopted for a taxon with a particular rank, position, and circumscription.

Effectively published. Name that was published in printed matter generally available to botanists. See Article 6 in Code.

Generic. Name for a principal category of taxa intermediate in rank between Family and Species.

Illegitimate. Name that is validly published but contrary to one or more articles of the Code.

Invalid. Name which is not valid (this type of name is not synonymous with "illegitimate" - see preceding definition).

Legitimate. Name or epithet that is validly published and in accordance with the rules.

Misapplied. Name applied to a taxon that does not include the type of the name.

Naked (Nomen nudum). Name published without description or diagnosis, or reference to either (not a name in the sense of the Code because not validly published).

New (Nomen novum). Name that is usually a substitute or replacement for an earlier name (not a name being given to a taxon for the first time by an author which is usually indicated by sp. nov.).

Provisional. Name not accepted by an author at time of publication, but proposed in anticipation of the future acceptance of the group concerned, or the future acceptance of the taxon in some particular circumscription, position or rank.

Rejected (Nomen rejiciendum). Name rejected.

Retained. Name or epithet which is continued to be used as a correct name or as the epithet of a correct name.

Scientific. Name composed of a generic name and a specific epithet, a binomial for the species; or a single correct name for a higher taxon.
Superfluous. Name applied to a taxon for which another name or epithet was already available.
Unpublished. Name not published, nomenclaturally equivalent to a name that has not been validly published. Used in reference to herbarium or manuscript names that have not been effectively or validly published.
Validly published. Name published in accordance with Arts. 32-45 in the Code in which the basic provisions are (a) effective publication, (b) publication in the form specified for the name of each category of taxa, (c) publication with a description or diagnosis, or a reference to a previously published description or diagnosis, of the taxon to which the name applies, (d) accompanied by a Latin description or diagnosis or by a reference to a previously and effectively published Latin description or diagnosis of the taxon, and (e) with an indication of the nomenclatural type.
Vernacular. Name used by natives for a plant; a common name.

V. "NYMS"

Autonym. An automatically created tautonym for infrageneric or infraspecific taxa. Examples: *Hypericum* subgenus *Hypericum* section *Hypericum*; *Hypericum perforatum* L. ssp. *perforatum* var. *perforatum*.
Basionym. A specific or infraspecific epithet that has priority and is retained when transferred to a new or different taxon. Examples: *Cypripedium pubescens* Willd. - *Cypripedium calceolus* var. *pubescens* (Willd.) Correll; *Liquidambar peregrina* L. - *Comptonia peregrina* (L.) Coulter.
Homonym. Any of two or more identical names based on different types, only one of which can be legitimate. Example: *Spergula stricta* Sw. (1799) transferred to *Arenaria* cannot become *Arenaria stricta* because of a different species, *Arenaria stricta* Michx. (1803). If transferred, *S.* "stricta" would be a homonym of *A.* "stricta" Michx. (1803).
Hyponym. A name not based on a type species; name rejected for lack of identified type.
Synonym. A rejected name due to a misapplication or difference in taxonomic judgment.
  Nomenclatural or Homotypic Synonym. Different name based on same type, second name synonym of previous name.
  Taxonomic or Heterotypic Synonym. Different name based on different type but taxonomic judgment indicates identity equal to previously described taxon; a metanym.
Tautonym. Illegitimate binomial in which generic name and specific epithet are the same. Example: *Sophia sophia* (L.) Britton. Note: See *autonym* above.
Typonym. A name based on an indicated type specimen, not on a diagnosis or description.

VI. RANKS AND ENDINGS

Rank	Ending	Example
Division*	phyta; mycota (Fungi)	Pterophyta; Eumycota
Subdivision*	phytina; mycotina (Fungi)	Pterophytina; Eumycotina

*Recommended endings.

Rank	Ending	Example
Class*	opsida (Cormophyta); phyceae (Algae); mycetes (Fungi)	Pteropsida; Cyanophyceae; Basidiomycetes
Subclass*	opsidae (Cormophyta); phycidae (Algae); mycetidae (Fungi)	Pteropsidae; Cyanophycidae; Basidiomycetidae
Order	ales	Rosales
Suborder	ineae	Rosineae
Family	aceae	Rosaceae
Subfamily	oideae	Rosoideae
Tribe	eae	Roseae
Subtribe	inae	Rosinae
Genus	us-a-um-es-on, etc.	*Rosa*; *Ipomoea*, *Geum*
Subgenus	Note on subdivision(s) of a Genus:	
Section	The name of a subdivision of a genus is a combination of a	
Subsection	generic name and a subdivisional epithet connected by a	
Series	term (subgenus, section, series, etc.) denoting its rank.	
Subseries	The epithet is either of the same form as a generic name, or a plural adjective agreeing in gender with the generic name and written with a capital initial letter. Examples: *Ricinocarpos* sect. *Anomodiscus*; *Euphorbia* sect. *Tithymalus* subsect. *Tenellae*	
Species	us-a-um-es-on, etc.	
Subspecies	Note on infraspecific taxa:	
Varietas	The name of an infraspecific taxon is a combination of the	
Subvarietas	name of a species and an infraspecific epithet connected	
Forma	by a term denoting its rank. Infraspecific epithets are	
Subforma	formed as those of species and, when adjectival in form and not used as substantives, they agree grammatically with the generic name. Example: *Saxifraga aizoon* var. *aizoon* subvar. *brevifolia* forma *multicaulis* subforma *surculosa* Engler & Irmscher or *Saxifraga aizoon* subforma *surculosa* Engler & Irmscher.	

VII. TYPES

Generitype. The type species of a genus.

Holotype. The one specimen or other element used by the author or designated by him as the nomenclatural type. As long as a holotype is extant, it automatically fixes the application of the name concerned.

Isotype. A duplicate (part of a single gathering made by a collector at one time) of the holotype; it is always a specimen.

Isosyntype. A duplicate of a syntype, not cited in a protologue. All duplicates cited by the author are syntypes.

Lectotype. A specimen or other element selected from the original material to serve as a nomenclatural type when no holotype was designated at the time of publication or as long as it is missing.

Neotype. A specimen or other element selected to serve as nomenclatural type as long as all of the material on which the name of the taxon was based is missing.

Nomenclatural Type. That element with which the name is permanently associated. It is not necessarily the most typical or representative element of a taxon.

*Recommended endings.

Paratype. A specimen cited with the original description other than the holotype or isotype.

Syntype. Any one of two or more specimens cited by the author when no holotype was designated, or any one of two or more specimens simultaneously designated as types.

Topotype. A specimen of a named taxon collected, usually later, from the original type locality; or from the area from which the species was described.

Section C. PROBLEMS IN NOMENCLATURE*

1. Ceuthospora Fries, Syst. Orb. Veg. 1825 emend. Grev., Scot. Crypt. Flora 5: 253. 1827. (A deuteromycete).
   Ceuthospora Crum & Anderson, Jour. Elisha Mitchell Sci. Soc. 74: 31. 1958. (A moss).
   Bryoceuthospora Crum & Anderson, Bryologist 62: 66. 1959, (the same moss).
   What is the valid name of the moss genus?

2. Fissidens Andersoni Grout, Moss Flora N. Am. 1(4): 252. 1939.

3. Carex Nealae R. W. Krauss, Pac. Sci. 4: 267. 1950.

4. Smilax hispida var. montana W. C. Coker, Jour. Elisha Mitchell Sci. Soc. 60: 49. 1944.

5. Exidia Cokeri Olive, Jour. Elisha Mitchell Sci. Soc. 60: 18. 1944.

6. Sabbatia cymosa in the sense of G. Don (Gen. Hist. 4: 207. 1838) as to Don's description but excluding reference to Chironia cymosa Lam., a synonym of Sabatia difformis (L.) Druce.
   Sabbatia paniculata in the sense of authors after 1860 but not that of Michx., the author of the basionym, nor of Pursh who made the combination. The earlier authors (i.e., Michx. and Pursh) were describing the perennial now known as S. difformis (L.) Druce.
   Sabatia brachiata f. candida Fern., Rhodora 39: 443. 1937.
   Sabatia quadrangula Wilbur, Rhodora 57: 22. 1955.

7. Amorpha georgiana Small ex Schneider, Bot. Gaz. 43: 305. 1907. (pro syn.)
   Amorpha georgiana Wilbur, Rhodora 56: 261. 1954 [1955].

8. Achlya diffusa Harvey, Jour. Elisha Mitchell Sci. Soc. 58: 29. 1942.
   Achlya diffusa Harvey ex T. W. Johnson, Achlya 64. 1956.

9. Sabatia Adans., Fam. Pl. 2: 503. 1763.
   Sabbatia (the spelling accepted by most authors between 1800-1900 and many thereafter).
   (named in honor of the Roman botanist, Liberato Sabbati)

10. Sabbatia maculata (Benth.) Benth., Gen. Pl. 2: 812. 1873.
    Sabbatia maculata (Benth.) Benth. ex A. Gray, Proc. Am. Acad. 22: 438. 1887.

11. Tabernaemontana Amsonia L., Sp. Pl. ed. 2. 2: 301. 1762.
    Amsonia Tabernaemontana Walt., Fl. Carol. 98. 1788.

*Contributed by Robert L. Wilbur, Department of Botany, Duke University.

<u>Amsonia</u> <u>latifolia</u> Michx., Fl. Bor. Am. 1: 121. 1803.
<u>Amsonia</u> <u>Amsonia</u> (L.) Britt., Mem. Torr. Club 5: 262. 1894.

12. <u>Lupinus</u> <u>Westiana</u> Small, Torreya 26: 91. 1926.
<u>Lupinus</u> <u>Westianus</u> Small, Man. SE. U.S. 681. 1933.

13. <u>Pleomele</u> <u>lanaiensis</u> Degener, Fl. Haw. fam. 68. Aug. 10, 1932.
<u>Pleomele</u> <u>Fernaldii</u> St. John, Contr. Gray Herb. 65: 39. 1947. (and see Degener, Fl. Haw. fam. 68. 5/28/56).

14. <u>Zemisne</u> Degener & Sherff, Am. Jour. Bot. 22: 708. 1935. (named in honor of Archibald Menzies, botanist of Vancouver's Expedition 1791-1795).

15. <u>Subtribe</u> <u>Fitchiinae</u> Carlquist, U. of Calif. Publ. Bot. 29(1): 5. 1957.

16. <u>Panicum</u> <u>annulum</u> Ashe, Jour. Elisha Mitchell Sci. Soc. 15: 58. 1898.
<u>Panicum</u> <u>Boqueanum</u> Ashe, Jour. Elisha Mitchell Sci. Soc. 16: 85. 1900.

17. <u>Wisteria</u> Nutt., Gen. N. Am. Pl. 2: 115. 1818. (honoring Dr. Caspar Wistar).
<u>Wistaria</u> - the orthography employed by Benth. & Hook., Gen. Pl. 1: 499. 1865, and many other authors.

18. <u>Diplonyx</u> Raf., Fl. Ludov. 101. 1817.
<u>Wisteria</u> Nutt., Gen. N. Am. Fl. Pl. 2: 115. 1818.

19. <u>Wilkesia</u> <u>Gymnoxiphium</u> A. Gray, Proc. Am. Acad. Sci. 2: 160. 1852.

20. <u>Septobasidium</u> <u>Westoni</u> Couch, The Genus Septobasidium, p. 120. 1938.

21. <u>Panicum</u> <u>Shallotte</u> Ashe, Jour. Elisha Mitchell Sci. Soc. 16: 84. 1900.

22. <u>Fagara</u> Sect. <u>Dipetalae</u> Degener & Skottsberg, Brittonia 2: 362. 1937.

23. <u>Tabernaemontana</u> <u>rimulosa</u> Woodson in R. E. Schultes, Bot. Mus. Leafl. Harvard Univ. 18: 180. 1958.
<u>Tabernaemontana</u> <u>rimulosa</u> Woodson ex R. E. Schultes, <u>op</u>. <u>cit</u>. 18: 317. 1959.

24. <u>Helianthus</u> <u>glaucophyllus</u> D. M. Smith, Brittonia 10: 192: 1958; 11: 192. 1959.

25. <u>Phaeosphaera</u> <u>perforata</u> L. A. Whitford, Jour. Elisha Mitchell Sci. Soc. 59: 159. 1943.
<u>Phaeosphaera</u> <u>perforata</u> L. A. Whitford, Trans. Am. Microscop. Soc. 75: 202. 1956.

26. <u>Myoporum</u> <u>sandwicense</u> (A. DC.) A. Gray, Proc. Am. Acad. Sci. 6: 52-53. 1866.
<u>Polycoelium</u> <u>sandwicense</u> A. DC., Prodr. 11: 706. 1847.
see Webster (Pacific Sci. 5: 60. 1951) for one opinion.

27. <u>Cyanea</u> subsection <u>Glabriflorae</u> Rock, Mem. B. P. Bishop Mus. 7(2): 231. 1919.

28. Asarum Lewisii Fern., Rhodora 45: 398. 1943.
Hexastylis pilosiflora Blomquist, Castanea 10: 75. 1945.
Hexastylis lewisii (Fern.) Blomq. & Oosting, Spring Fl. Piedm. N. C. ed. 4. 60. 1948.

29. Sabatia Adans., Fam. Pl. 2: 503. 1763.
Pleienta Raf., New Fl. 4: 92. 1838.
Sect. Pleienta (Raf.) Blake, Rhodora 17: 56. 1915.

30. Rhynchospora V. Glomeratae Small, Man. SE. Fl. 175. 1933.
Rhynchospora Ser. Glomeratae Small, see Gale in Rhodora 46: 104. 1944.

31. Pleienta capitata Raf., Fl. Tell. 3: 30. 1837.
Sabbatia Boykinii A. Gray in Chapm., Fl. S. U.S. 354. 1860.
Sabatia capitata (Raf.) Blake, Rhodora 17: 54. 1915.

32. Chironia angularis var. β angustifolia Michx., Fl. Bor. Am. 1: 147. 1803.
Sabbatia brachiata Ell., Sk. Bot. S.C. & Ga. 1: 284. 1817.
Sabbatia angustifolia (Michx.) Britt., Mem. Torr. Club 5: 259. 1889.

33. Amorpha pumila Michx., Fl. Bor. Am. 2: 64. 1803.

34. Pluchea section Stylimnus (Raf.) DC., Prodr. 5: 450. 1836.
see R. K. Godfrey's account (Jour. Elisha Mitchell Sci. Soc. 68: 238-271. 1952) noting the type of the genus Pluchea and section Stylimnus.

35. Physica sect. Conradia (Malme) M. Choisy, Bull. Soc. Mycol. France 75: 66. 1959.

36. Eugenia anthera Small, Man. SE. Fl. 935, 1506. 1933.

37. Hedyotis centranthoides var. centranthoides Fosberg, B. P. Bishop Mus. Bull. 174: 61. 1943.

38. Hedyotis centranthoides f. centranthoides Fosberg, B. P. Bishop Mus. Bull. 174: 61. 1943.

39. Chilopsis linearis var. originaria Fosberg, Madrono 3: 365. 1936.

40. Kuhnia Schaffneri A. Gray, Proc. Am. Acad. 17: 207. 1882.
Kuhnia microphylla Shinners, Wrightia 1: 127. 1946.

41. Metrosideros polymorpha Gaud. Sect. I. Glabrae J. Rock, Indig. Trees Haw. Is. 327. 1913.

42. Teniola Reese, Bryologist 62: 221. 1959(1960).
Teniolophora Reese, Bryologist 65: 67. 1962.

43. Smyrnium cordatum Walt., Fl. Car. 114. 1788.

44. Polygala corymbosa Michx., Fl. Bor.-Am. 2: 54. 1803.

45. Pycnanthemum aristatum Michx., Fl. Bor.-Am. 2: 8. 1803.

46. Desmodium arenicola (Vail) Hermann, Jour. Wash. Acad. Sci. 38: 237. 1948.
Meibomia arenicola Vail, Bull. Torr. Club 23: 140. 1896.
Hedysarum lineatum Michx., Fl. Bor.-Am. 2: 72. 1803; not L., Syst. ed. X 1070. 1759.
Desmodium lineatum DC., Prodr. 2: 330. 1825.
What is the proper name under Desmodium?

47. Euphorbia nigro-purpurea M. E. Jones, Contr. Western Bot. no. 15: 145. 1929.

48. Asplenosorus Wherry, Am. Fern Jour. 27: 56. 1937.

49. Iris verna var. Smalliana Fern., Rhodora 49: 214. 1947.
Iris verna var. Smalliana Fern. ex M. E. Edwards 71: 213. 1969.

50. Gossypium sect. Longicalyciana Philips & Strickland, Canad. Jour. Genetics and Cytol. 8: 92. 1966.

51. Reedrollinsia J. W. Walker, Contr. Gray Herb. Harvard 202: 24 and 124. 1971.
Reedrollinsia J. W. Walker, Rhodora 73: 461. 1971.

52. Nymphaea pentapetala Walt., Fl. Carol. 155. 1788.
Nelumbium luteum Willd., Sp. Pl. ed. 4. 2: 1259. 1799.
Nelumbo lutea (Willd.) Pers., Syn. Pl. 92. 1807.
Nelumbo pentapetala (Walt.) Fern., Rhodora 36: 23. 1934.
read Gleason's opinion, Phytologia 2: 205-206. 1947.

53. Ericaceae tribe Daboecieae Cox, Amer. Midl. Nat. 39: 238-239. 1948.

NOMENCLATURE LITERATURE

Benson, L. 1962. Plant Taxonomy: Methods and Principles. The Ronald Press Company. New York.

Cronquist, A. 1968. The Evolution and Classification of Flowering Plants. Houghton Mifflin Company. Boston.

Johnson, A. T. 1971. Plant Names Simplified. 2nd ed. W. H. & L. Collingridge Ltd. London.

McVaugh, R., R. Ross, and F. A. Stafleu. 1968. An annotated glossary of botanical nomenclature. Regnum Vegetabile 56: 1-31.

Stafleu, F. A. (chairman of editorial committee). 1972. International Code of Botanical Nomenclature, adopted by the Eleventh International Botanical Congress, Seattle, August 1969. Regnum Vegetabile 82. Utrecht.

St. John, H. 1958. Nomenclature of Plants. The Ronald Press Company. New York.

NOMENCLATURE EXERCISE

(Articles are found in the International Code of Botanical Nomenclature.)

I. For the genus, Rosa, consult a manual for the correct names of the higher taxa to which the genus belongs. Read Articles 16, 17, 18 and 19 to determine the correct endings for the higher taxa found in the manual.

II. In the following indicate why the first name is correct and the second incorrect.

A. Descurainea sophia (L.) Webb
Sophia Sophia (L.) Britton (See Article 23)

B. Rhexia nuttallii James, nom. nov. 1956
Rhexia serrulata Nutt. 1818 not R. serrulata Rich. 1813 (See Article 64)

C. Lipocarpha R. Brown. 1818
Hypaelyptum Vahl. 1806 (See Article 14 and Appendix III)

D. In Hypericum the lectotype and generitype is H. perforatum L.
Hypericum subgenus Hypericum Section Hypericum
Hypericum subgenus Euhypericum Section Perforata (See Article 19)

E. Myrica gale L., Species Plantarum, 1753
Gale belgica Dumortier, Fl. Belgica, 1827 (See Article 13)

III. Give the correct species name for each of the following transfers and changes of rank:

A. Transfer
1. Antirrhinum spurium L. to the genus, Linaria (See Article 55)
2. Melissa calamintha L. to the genus, Calamintha (See Article 55)
3. Spergula stricta to the genus, Arenaria and to Minuartia (See Article 55)

B. Change the rank
1. Ascyrum hypericoides var. multicaule (Michx.) Fernald to a species of Hypericum (See Article 60)
2. Myrica pusilla Raf. to a variety of Morella (See Article 56)
3. Prunus americana var. lanata Sudw. to a species of Prunus (See Article 60)

IV. List the synonyms of Sorbus arbutifolia var. atropurpurea (Britton) Schneider. Consult Gray Herbarium Card Index and Index Kewensis.

EPITHETS AND PRONUNCIATION EXERCISE

(See Chapter 4)

I. After studying prefixes, suffixes, and names of plant parts, select ten epithets from a manual or guide pertinent to your region and give the meaning of each (See Section A).

II. Select ten species from a manual or guide pertinent to your region and indicate the type of epithet; e.g., commemorative, habit, habitat, season (See Section B).

III. Select ten species from a manual or guide pertinent to your region, then syllabalize, accent, and pronounce each specific epithet (See Section C).

Chapter 4. BOTANICAL NAMES*

The scientific names of plants are either Latin words or words that have been latinized from some other language, most often Greek. The student who knows the meanings of scientific terms and their roots will find that he can learn and remember with comparative ease, and with much satisfaction, the names of a great number of plants, as well as some of their characteristics. He need not be a scholar of Latin or Greek, although an elementary course in either could give him a decided advantage. Since the student of taxonomy today often lacks a background in classical languages, Chapter 4 is presented to help him understand how scientific names are formed, and to provide him with lists of words and terms that are commonly used in making plant names. By studying these lists of Latin and latinized words along with their English meanings, he should soon be able to recognize and remember the names of an increasing number of plants, and to understand the meanings of numerous other botanical terms as well.

The discussion of scientific names in the present chapter will be limited to specific epithets only; generic names will not be considered here.[1] The latter part of the chapter will deal with the pronunciation of specific epithets, followed by a syllabilized and accented index. While scientific names are generally underlined in writing, that rule will be set aside here in cases where for clarity or emphasis it seems important to underline only endings or other parts of words.

Section A. FORMATION OF SPECIFIC EPITHETS

I. GENERAL INFORMATION

Each species of plant has only one correct scientific name, peculiar to that species alone. This is called a binomial and consists of a generic name and a specific epithet. For example, *Salix nigra* is the binomial for black willow, *Salix* being the generic name, and *nigra* the specific epithet. Thus, a specific epithet can be defined as the second part of a binomial. It should be understood, however, that it is never correct to use the specific epithet alone to designate a particular species; it must always be combined with a generic name to form the binomial for that species. One cannot say that *nigra* is the scientific name for black willow or for any other species; standing alone, *nigra* is simply the Latin word for "black." When combined with *Salix* it is the name for black willow (*Salix nigra*), or with *Juglans* the name for black walnut (*Juglans nigra*), or with *Fraxinus* the name for black ash (*Fraxinus nigra*), etc.

Specific epithets are formed from nouns, adjectives, participles, gerundives, etc., and by combining such words with a variety of prefixes and suffixes. It should be understood that unlike English nouns, every Latin or latinized noun has gender, either masculine (m.), feminine (f.), or neuter (n.). In most cases, each gender is indicated by a different ending, e.g., most nouns ending in -us are masculine, nouns ending in -a are nearly always

*Contributed by Laurie Stewart Radford, University of North Carolina at Chapel Hill.

[1]See an unabridged dictionary for pronunciation of family and generic names.

feminine, while those ending in -um are neuter. An adjective or other modifier must agree in gender with the noun it modifies. While a noun will have only one nominative ending depending upon its gender, a modifier will often have three different endings, depending upon the gender of the word it is modifying. Thus, the Latin adjective for "hairy" appears with three different endings as it modifies the nouns Lathyrus, Lactuca, and Vaccinium, respectively masculine, feminine, and neuter: Lathyrus hirsutus, Lactuca hirsuta, and Vaccinium hirsutum. It should be noted that occasionally the specific epithet is a noun in apposition, in which case it carries its own gender, regardless of that of the generic name it follows. In the binomial Cypripedium calceolus, a species of lady slipper, the neuter genus name Cypripedium is followed by the masculine noun calceolus (a small shoe), in apposition and retaining its own gender.

Most specific epithets are adjectives, or words used in an adjectival sense and will have masculine, feminine, or neuter endings. The more common Latin endings are:

m.	f.	n.	examples
-us	-a	-um	albus, alba, album (white)
-er	-ra	-rum	niger, nigra, nigrum (black)
-is	-is	-e	brevis, brevis, breve (short)
-r	-ris	-re	acer, acris, acre (pungent)

Modifiers ending in -ans (elegans, elegant), -ens (repens, creeping), -or (bicolor, with two colors), -x (simplex, simple), etc., use the one ending for all three genders. Example: Ranunculus repens, Ludwigia repens, and Trifolium repens.

II. COMMEMORATIVE EPITHETS

The names of persons are sometimes used as specific epithets, generally to honor or commemorate the man or woman who first discovered a particular species. In such cases the specific epithet usually has a genitive rather than a nominative ending. The treatment given here for commemorative names is based on Article 73 of the 1972 Edition of the International Code of Botanical Nomenclature, Recommendation 73 C.

A specific epithet taken from the name of a man should be formed as follows:

1. If the name ends in any vowel except a, (i.e., e, i, o, u, and final y) the letter i is added to the end of the name, e.g., Lilium grayi for Asa Gray or Aster blakei for Joseph Blake.

2. If the name ends in a, the letter e is added: balansae for Mr. Balansa.

3. If the name ends in a consonant, ii is added: Chelone lyonii for John Lyon; Rubus grimesii for Earl Jerome Grimes. If a name ends in er, only one i is used, which is an exception: Setaria faberi for Ernst Faber; Solidago cutleri for Manasseh Cutler.

4. If a name is used as an adjective, it must agree in case and gender with the genus it modifies: Rubus cardianus for Fred Wallace Card; Chenopodium boscianum for Louis Augustin Bosc; Ruellia purshiana for Frederick Traugott Pursh.

5. If a woman's name is used in the substantive form as an epithet, the ending will be the feminine genitive singular for that word: Crataegus coleae for Emma Jane Cole; C. beckwithae for Florence E. Beckwith; Cornus priceae for Sarah Frances Price. If the name is used as an adjective, the same rules apply as for a man's name: Apios priceana for Miss Price.

III. DESCRIPTIVE EPITHETS

Most specific epithets indicate something characteristic about a species, such as color of flowers or fruit (Aronia melanocarpa, black-fruited), habit of growth (Celastrus scandens, climbing), shape of leaf (Desmodium rotundifolium, round-leaved), discoverer of the species (Lilium grayi, for Asa Gray), place of discovery (Solidago roanensis, of Roan Mt., N. C.), etc. Names of plant parts are frequently used in combination with prefixes and suffixes to form a number of specific epithets: biflorus, two-flowered, from the prefix bi (two) and the plant part flos (flower); petiolaris, having a petiole, from plant part petiolus (petiole) and the suffix -aris (indicating possession). It is possible to determine the meanings of a great many epithets simply by referring to the lists of plant parts, prefixes, suffixes, etc., given in the sections that follow.

IV. NAMES OF PLANT PARTS

The most obvious parts of a plant (here arranged alphabetically for convenience), are bud, flower, fruit, leaf, root, seed, seedling, and stem. Each term is followed by its Latin equivalent, and by the Greek where known, indicated by (L.) and (Gr.) respectively. The compound words are usually separated into their basic parts with English meanings, to give the student a better understanding of the word: e.g., carpopodium: carpo- (fruit) + podion (little foot).

A. Bud - gemma (L.); blastos (Gr.)
- primordium - primordium: primus (first) + ordiri (to begin)
- promeristem - promeristema: pro- (before) + meristos (divided)
- scale - squama; lepis

B. Flower - flos (L.); anthos (Gr.)
- androecium - androecium: andro- (male) + oikos (house)
  - stamen - stamen; stemon
  - staminodium - staminodium
- androgynophore - androgynophorum: andro- (male) + gyno- (female) + phoras (bearing)
  - stipe - stipes (stem)
  - column - columna (pillar)
- androphore - androphorum: andro- (male) + phoras (bearing)
  - receptacle - receptaculum (vessel for holding something)
- calyx - calyx
  - sepal - sepalum; sepalon
- carpel - carpellum
  - carpopodium - carpopodium: carpo- (fruit) + podion (little foot)
  - funiculus - funiculus (cord, rope)

locule - loculus (chamber, room)
ovary - ovarium
ovule - ovulum
placenta - placenta (a flat cake)
podogyne - podogynium: podo- (foot) + gyne (female)
stigma - stigma (mark or spot)
stipe - stipes (stem)
style - stylus; stylos
clinanthium - clinanthium
corolla - corolla (a little crown)
petal - petalum; petalon
disc - discus (circular plate)
gynophore - gynophorum: gyno- (female) + phoras (bearing)
gynoecium - gynoecium: gyno- (female) + oikos (house)
hypanthium - hypanthium
base - fundus
limb - limbus
neck - collum
tube - tubus
ovary - ovarium
ovule - ovulum
pedicel - pedicellus
perianth - perianthium
banner - vexillum
base - fundus
beard - barba
bristle - seta
callosity - callositas
carina - carina
claw - unguis
corona - corona
fringe - fimbria
hood - cucullus
horn - cornu
keel - carina
labellum - labellum
ligula - ligula
lip - labium
limb - limbus
lobe - lobus
nectary - nectarium
palate - palatum
petal - petalum
pouch - saccus
sac - saccus
scale - squama; lepis
sepal - sepalum
spur - calcar
standard - vexillum
tepal - tepallum
throat - faux
tube - tubus
vexillum - vexillum
wing - ala
petal - petalum; petalon
pistil - pistillum
polyphore - polyphore: poly- (many) + phoras (bearing)
receptacle - receptaculum (vessel for holding something)
sepal - sepalum
stamen - stamen; stemon
anther - anthera
filament - filamentum (thread)
stigma - stigma (mark or spot)
stylus - stylus; stylos
tepal - tepalum
whorl - verticillus

C. Fruit - fructus (L.); carpos (Gr.)
carpophore - carpophorum: carpo- (fruit) + phoras (bearing)
dissepiment - dissepimentum (a partition)
ectocarp - ectocarpium: ecto- (outside) + carpos (fruit)
endocarp - endocarpium: endo- (within) + carpos (fruit)
funiculus - funiculus (a rope or cord)
mericarp - mericarpium: meri- (part of) + carpos (fruit)
mesocarp - mesocarpium: meso- (middle) + carpos (fruit)
pericarp - pericarpium: peri- (around) + carpos (fruit)
placenta - placenta (a flat cake)

seed - semen; sperma
septum - septum; phragma: (wall or partition)

D. Leaf - folium (L.); phyllon (Gr.)
- blade - lamina
- leaflet - foliolum
- midrib - costa
- petiole - petiolus
- petiolule - petiolulus
- pulvinus - pulvinus (cushion)
- rachilla - rhachilla (little axis)
- rachis - rhachis (axis)
- stipel - stipella (little stalk)
- stipule - stipula (stalk)

E. Root - radix (L.); rhiza (Gr.)

F. Seed - semen (L.); sperma (Gr.)
- aril - arillus (fleshy growth)
- chalaza - chalaza (a lump)
- embryo - embryo
  - coleoptile - coleos: (sheath) + ptilon (feather)
  - cotyledon - cotyledon (cup leaf)
  - epicotyl - epicotylus: epi- (upon) + cotylus (cup)
  - hypocotyl - hypocotylus: hypo- (beneath) + cotylus (cup)
  - plumule - plumula (little feather)
  - radicle - radicula (little root)
- endosperm - endospermium: endo- (within) + sperma (seed)
- hilum - hilum (scar, trifle)
- raphe - rhaphe (seam)
- seed coat - testa

G. Seedling - plantula (L.)
- cataphyll - cataphyllum: catta- (below) + phyllon (leaf)
- collet - from collum (neck)
- eophyll - eophyllum: eos (early, primitive) + phyllon (leaf)
- metaphyll - metaphyllum: meta- (changed) + phyllon (leaf)

H. Stem - caulis (L.)
- bark - cortex
- pith - medulla
- wood - lignum
- branch - ramus

V. PREFIXES

A prefix is an element of one or more letters placed at the beginning of a word (usually a noun or an adjective), to modify its meaning or application. Many prefixes are adverbs or prepositions. In botanical Latin adjectives with modified endings often serve as prefixes. In such cases, the stem of the adjective and usually a connecting vowel (i or o) form the prefix. To illustrate: the adjective altus (tall) is reduced to its stem alt and the vowel "i" added, the result being alti which now can serve as a prefix meaning "having the quality of being tall, or of tallness." When used before the term for stem (caulis), we get the word alticaulis, meaning "tall-stemmed." A hyphen usually follows a prefix, indicating attachment at the beginning of another word.

A. Number

uni- (L.): one. uniflorus, one-flowered
mono- (Gr.): one. monanthos, one-flowered
bi- (L.): two. bifoliatus, two-leaved

di- (Gr.): two. _diphyllus_, two-leaved.
tri- (L.): three. _triangularis_, having three angles
tri- (Gr.): three. _triacanthos_, having three thorns
quadri- (L.): four. _quadrifolius_, four-leaved
tetra- (Gr.): four. _tetraphyllus_, four-leaved
quinque- (L.): five. _quinquefolius_, five-leaved
penta- (Gr.): five. _pentaphyllus_, five-leaved
sex- (L.): six. _sexangularis_, six-angled
hexa- (Gr.): six. _hexagonus_, six-sided
septem- (L.): seven. _septemlobus_, seven-lobed
hepta- (Gr.): seven. _heptapetalus_, seven-petalled
octo- (L.): eight. _octoflorus_, eight-flowered
octo- (Gr.): eight. _octandrus_, with eight stamens
novem- (L.): nine. _novemnervis_, with nine nerves
ennea- (Gr.): nine. _enneaphyllus_, with nine leaves
decem- (L.): ten. _decemlobus_, with ten lobes
deca- (Gr.): ten. _decapetalus_, with ten petals

amphi- (Gr.): double, of two kinds
centri- (L.): one hundred
dicha-, dicho- (Gr.): in two
diplo- (Gr.): double
haplo- (Gr.): single
multi- (L.): many
myrio- (Gr.): countless
oligo- (Gr.): few
pan-, panto- (Gr.): all
pauci- (L.): few
pleio- (Gr.): few
pluri- (L.): several
poly- (Gr.): many, numerous
sesqui- (L.): one and a half
terni- (L.): in threes

B. Position

a-, ab- (L.): away from
ad- (L.): toward, against
ambi- (L.): around, round about
amphi- (Gr.): around, on both sides of
ante- (L.): before, in front of
anti- (L. & Gr.): against
apo- (Gr.): away from, down
co-, com-, con- (L.): together
de- (L.): downwards, away from
di-, dis- (L.): apart from, asunder
dia- (L. & Gr.): through
ecto- (Gr.): out of, from
endo-, ento- (Gr.): inside, within
epi- (Gr.): upon, on top of
ex- (L.): from, out of
extra- (L.): outside
hyper, hypero (Gr.): beyond, above
hypo- (Gr.): beneath
infra- (L.): below
inter- (L.): between, among
intra- (L.): within
intro- (L.): inside
meta- (L. & Gr.): next to, after
ob- (L.): against
para- (Gr.): near, beside
per- (L.): through
peri- (L. & Gr.): around
prae- (L.): before, in front of
sub- (L.): below, under
super-, supra- (L.): above, over
syn-, sys- (Gr.): together, joined
trans- (L.): across, beyond

C. Shape

aniso- (Gr.): uneven, unequal
astro- (Gr.): stellate, star-like
cerato- (Gr.): horn-like
cteno- (Gr.): pertaining to a comb
cyath- (L. & Gr.): cup-like
cyclo- (Gr.): circular
fili- (L.): thread-like
fimbri- (L.): fimbriate
glosso- (Gr.): tongue-like
goni- (Gr.): angled, angular
holo- (Gr.): entire
hetero- (Gr.): various, different
lanci- (L.): lance-shaped
nephro- (Gr.): kidney-shaped

ophio- (Gr.): snake-like
ortho- (Gr.): straight; erect
ovi- (L.): egg-shaped
pachy- (Gr.): thick
rhyncho- (Gr.): having a snout or beak
rhytido- (Gr.): wrinkled; rumpled
scyphi-, scypho- (L. & Gr.): cup-like
spheno- (Gr.): wedge-shaped
tylo- (Gr.): with knots or projections
uro- (Gr.): tailed

D. Size

angusti- (L.): narrow
brachy- (Gr.): short
brevi- (L.) short
crassi- (L.): thick, short
grandi- (L.): large
iso- (Gr.): equal
lati- (L.): wide; broad
lepto- (Gr.): slender
longi- (L.): long
macro- (Gr.): large; giant
mega-, megalo- (Gr.): very large; great
micro- (Gr.): small, little
nano- (Gr.): dwarf
parvi- (L.): small
platy- (Gr.): broad
steno- (Gr.): narrow
tenui- (L.): slender; thin

E. Miscellaneous

a-, an- (Gr.): without, not
actino- (Gr.): rayed; star-like
andro- (Gr.): male
anemo- (Gr.): pertaining to wind
argyro- (Gr.): silvery
archae-, arche- (Gr.): old; primitive
atri-, atro- (L.): black
botry- (Gr.): bunch
callo- (Gr.): beautiful
canio- (Gr.): pertaining to dogs
cardia- (Gr.): pertaining to a heart
carpo- (Gr.): relating to fruit
caryo- (Gr.): nut-like
chlamydo- (Gr.): wearing a cloak
chloro- (Gr.): green
chryso- (Gr.): golden
coelo- (Gr.): pertaining to a hollow
crypto- (Gr.): hidden
cyano- (Gr.): dark blue
dasy- (Gr.): shaggy; hairy
e-, ef-, ex- (L.): without; lacking
erio- (Gr.): woolly
erythro- (Gr.): reddish
eu- (Gr.): good; well
flavi- (L.): yellowish
fusci- (L.): dark or dark brown
galacto- (Gr.): milky
gamo- (Gr.): fused; united
geo- (Gr.): pertaining to earth
gymno- (Gr.): naked; bare
gyno- (Gr.): female
hirti- (L.): hairy with long hairs
homo- (Gr.): like; same
laevi- (L.): smooth
lani- (L.): woolly
lasio- (Gr.): woolly
laxi- (L.): loose; unstrung
leio- (Gr.): smooth
lepido- (Gr.): scaly
leuco- (Gr.): white
lino- (Gr.): made of flax
melano- (Gr.): black; very dark
ne- (Gr.): not; free from
neo- (Gr.): new
non- (L.): not
nudi- (L.): naked
ochro- (Gr.): yellowish
odonto- (Gr.): tooth-shaped
oxy- (Gr.): sharp
paleo- (Gr.): old
phaeo- (Gr.): dark
phanero- (Gr.): easily seen; visible
photo- (Gr.): light
phyllo- (Gr.): pertaining to a leaf
phyto- (Gr.): pertaining to a plant
picro- (Gr.): bitter
podo- (Gr.): of a foot
porphyro- (Gr.): purple
prae- (L.): before; very
pro- (L.): for; instead of
pseudo- (Gr.): false
ptero- (Gr.): winged
ptycho- (Gr.): pertaining to grooves or folds
pyro- (Gr.): firey
pyrrho- (Gr.): fire-red; ruby-red
rami- (L.): pertaining to branches
re- (L.): back
rhizo- (Gr.): pertaining to roots
rhodo- (Gr.): rose-colored

sapro- (Gr.): rotten; decayed
sarco- (Gr.): fleshy
schisto- (Gr.): split; cleft
schizo- (Gr.): deeply divided
sclero- (Gr.): hard
semper- (L.): always
sessili- (L.): sessile
stachyo- (Gr.): spiked
stato- (Gr.): fixed; standing
sticto- (Gr.): spotted
sur- (L.): somewhat; above
tephro- (Gr.): ash gray
trachy- (Gr.): rough
tricho- (Gr.): hairy
viridi- (L.): green
viti- (L.): pertaining to a vine
xero- (Gr.): dry
xylo- (Gr.): woody
zantho-, xantho- (Gr.): yellow
zygo- (Gr.): joined; married

VI. SUFFIXES

A suffix is an element of one or more letters placed at the end of a word to modify its meaning or application. The particular grammatical nature, gender, and meaning of the resulting compound is determined by that particular suffix. To illustrate: by joining the noun herba (herb, plant) and the suffix -arium (place where something is done or kept), we have the compound word herbarium, a noun of neuter gender meaning a collection of dried plants. Compounds thus formed by adding suffixes to a number of word stems can result in nouns or adjectives, but most often they are the latter, as they apply to specific epithets. For this reason, the majority of suffixes listed here are adjectival in nature. The Latin or Greek origin of each suffix is indicated by the symbols (L.) and (Gr.) respectively, followed by the character or quality each denotes, plus an example to illustrate. Masculine, feminine, and neuter endings are given for each suffix, where applicable. A hyphen precedes a suffix, indicating attachment at the end of another word.

-aceus, -a, -um (L.): likeness, resemblance. crustaceus, crust-like.
-aeus, -a, -um (Gr.): belonging to. aetnaeus, pertaining to Mt. Etna.
-alis, -is, -e (L.): possession, or pertaining to. digitalis, pertaining to a finger.
-anus, -a, -um (L.): belonging to; position. virginianus, of Virginia.
-aris, -is, -e (L.): relating to; possession. petiolaris, having a petiole.
-arium (L.): place where something is done or kept. herbarium, collection of dried plants.
-arius, -a, -um (L.): possession, or connection. plumarius, pertaining to plumes.
-ascens (L.): process of becoming; incomplete. violascens, becoming violet.
-aticus, -a, -um (L.): place of growth. aquaticus, growing in water.
-atilis, -is, -e (L.): place of growth. fluviatilis, growing in streams.
-atus, -a, -um (L.): likeness or possession. rostratus, having a beak.
-bilis, -is, -e (L.): ability or capacity. sensibilis, capable of irritability or sensitivity.
-bundus, -a, -um (L.): fullness, abundance. floribundus, full of flowers.
-ellus, -a, -um (L.): diminutive. echinellus, minutely spiny.
-ensis, -is, -e (L.): origin, country or place of growth. alabamensis, from Alabama.
-escens, -is, -s (L.): process of becoming. flavescens, becoming yellow; yellowish.
-estris, -is, -e (L.): place of growth. campestris, growing in fields.
-eus, -a, -um (L.): resemblance in quality or color. roseus, rose-colored.
-eus, -a, -um (Gr.): possessed by or belonging to. giganteus, belonging to giants, thus gigantic.
-icans (L.): almost identical resemblance. candicans, whitish.
-icola (L.): a dweller. saxicola, a dweller among rocks; growing among rocks.

-icus, -a, -um (Gr.): belonging to. virginicus, belonging to Virginia.
-ilis, -is, -e (L.): capacity or ability, property. flexilis, capable of being bent; flexible.
-ineus, -a, -um (L. & Gr.): color or material. stramineus, straw-colored.
-inus, -a, -um (L.): possession or resemblance. velutinus, like velvet.
-oideus, -a, -um (L.); -oides and -odes (Gr.): like, resemblance. helianthoides, resembling the genus Helianthus.
-osus, -a, -um (L.): abundance, fullness. foliosus, full of leaves.
-utus, -a, -um (L.): possession. cornutus, having horns; horned.

Section B. MEANINGS OF SPECIFIC EPITHETS

I. ADJECTIVAL EPITHETS

As previously stated, most specific epithets are adjectival in nature and thus convey something characteristic about a particular species. The listings given here are grouped by characteristics such as color, size, place of growth, etc., and with very few exceptions, are adjectival in nature.

A. Epithets Relating to Color

aeneus. brassy green; bronze
albescens. becoming white; turning white
albicans. whitish
albidus. whitish
albolutescens. whitish-yellow
albus. white
argenteus. silvery
argyreus. silvery
atropurpureus. dark purple
atrorubens. dark red
atroviolaceus. dark violet
atrovirens. dark green
aurantiacus. yellowish-orange
aureus. golden yellow
azureus. sky-blue
brunnescens. becoming deep brown
caerulescens. becoming deep blue
caeruleus. deep blue; sky-blue
caesius. lavender; bluish-gray
calcareus. chalk-white
candicans. becoming pure white
candidus. shining white
canescens. becoming grayish or hoary
cardinalis. cardinal-red
carneus. flesh-colored
castaneus. chestnut-colored
cinereus. ash gray
cinnamomeus. light reddish brown
coccineus. scarlet; deep red
croceus. saffron yellow
cruentus. blood-red; blood-stained
cyaneus. dark, deep blue
dealbatus. covered with white powder; whitened
eburneus. ivory-white
exalbidus. whitish
ferrugineus. rusty; light reddish brown
flavescens. becoming yellow
flavidulus. slightly yellow
flavidus. somewhat yellow
flavovirens. yellowish-green
flavus. pale yellow
fulgidus. shining; bright-colored
fulvus. tawny; dull yellow
glaucescens. becoming sea-green; grayish-blue
glaucus. gray-green, with a bloom
griseus. pearl-gray
helvolus. pale brownish-yellow
hepaticus. liver-colored; dark reddish-brown
hyacinthinus. purplish-blue
incanus. hoary; whitish-gray
incarnatus. flesh-colored
lividus. lead-colored
luridus. dull yellow; brownish-yellow
luteolus. pale yellow
lutescens. becoming pale yellow
luteus. deep yellow
niger. black; glossy black
niveus. snow-white; purest white
ochroleucus. yellowish-white
porphyreus. purple

prasinus. grass-green; bright green
puniceus. crimson; purple-red
purpurascens. becoming purple; purplish
purpureus. purple
roseus. rose-colored; rosy
rubellus. reddish
rubens. reddish
rubiginosus. brown-red
rufidulus. somewhat red
rufus. reddish
stramineus. straw-colored
violaceus. violet
violascens. becoming violet-colored
virens. green
virescens. becoming green
viridescens. becoming green
viridis. green
vitellinus. egg-yolk yellow

B. Epithets Relating to Direction

aquilonius. northern
australis. southern, from the southern hemisphere
austrinus. southern
borealis. northern
meridionalis. southern
occidentalis. western
orientalis. eastern
septentrionalis. northern

C. Epithets Relating to Geography

acadiensis. of Nova Scotia, Canada
aegyptiacus. of Egypt
africanus. of Africa
alabamensis. of Alabama
aleppicus. of Aleppo, N. Syria
allegheniensis. of the Alleghenies
alpinus, alpestris. of the Alps; of high mountains
altamaha. the Altamaha River, Ga.
americanus. of America
amurensis. of the Amur River
anglicus. of England
arabicus. of Arabia
argentinus. of Argentina
asiaticus. Asian
atlanticus. of the Atlantic
australiensis. of Australia
austriacus. of Austria
austro-carolinianus. of S. Carolina
babylonicus. of Babylon
barbadensis. of Barbados
bavaricus. of Bavaria, Western Germany
bermudensis. of Bermuda
bonariensis. of Buenos Aires
brasiliensis. of Brazil
burmanicus. of Burma
californicus. of California
canariensis. of the Canary Islands
canadensis. of Canada
cantabrigiensis. of Cambridge, England
capensis. of the Cape of Good Hope; of a cape
caribaeus. from the Lesser Antilles, West Indies
carolinae-septentrionalis. of North Carolina
carolinianus, carolinensis, carolinus. of the Carolinas, U.S.A.
carthagensis. of Carthage, N. Africa
catawbiensis. from Catawba River, N. C.
cherokeensis. of the Cherokee Country, or Cherokee Co., N. C.
chilensis. of Chile
chinensis, sinensis, cathayanus. of China
cisatlanticus. on this side of the Atlantic
columbianus. of the district of Columbia
corsicus. of Corsica, France
cubensis. of Cuba
curassavicus. of Curacae, Caribbean Sea
domingensis. of Santo Domingo
europaeus. of Europe
florentinus. of Florence, Italy
floridanus. of Florida
gallicus. of France
germanicus. of Germany
gileadensis. of Gilead
graecus. of Greece
groenlandicus. of Greenland
guadalupensis. of Guadeloupe, W. Indies
halepensis. of Aleppo
helveticus. of Switzerland
hispanicus. of Spain

idaeus. of Mt. Ida
illinoensis. of Illinois
indicus. of India or the Indies
islandicus. of Iceland; of an island
italicus. of Italy
jamaicensis. of Jamaica
lancastriensis. of Lancaster, Pa.
ludovicianus. of Louisiana or former Louisiana Territory
marianus, marilandicus. of Maryland
martinicensis. of Martinique, W. Indies
mexicanus. of Mexico
mississippiensis. of Mississippi
monspeliensis. of Montpellier, s. France
neapolitanus. of Naples, Italy
neerlandicus. of the Netherlands
neogaeus. of the New World
norvegicus. of Norway
novae-angliae. of New England
noveboracensis. of New York
oglethorpensis. of Oglethorpe, Ga.
ohiensis. of Ohio
pennsylvanicus. of Pennsylvania
philadelphicus. of Philadelphia
polonicus. of Poland
provincialis. of Provence, Southern France
roanensis. of Roan Mt., N. C.
sibiricus. of Siberia, U.S.S.R.
sinensis. of China
syriacus. of Syria
tennesseensis. of Tennessee
texanus, texensis. of Texas
thapsus. of Thapsus, Sicily
virginianus, virginiensis, virginicus. of Virginia
zetlandicus. of the Shetland Isles, Britain
zeylanicus, ceylanicus, taprobanicus. of Ceylon

D. Epithets Relating to Habit

arborescens. arborescent
ascendens. ascending
caespitosus. cespitose
decumbens. decumbent
dichotomus. dichotomous
erectus. erect
expansus. clambering
fastigiatus. fastigiate
fruticosus. fruticose
furcatus. forked
geniculatus. geniculate
patens. spreading
procumbens. procumbent
prostratus. prostrate
ramosus. branched
repens. creeping; prostrate and rooting
reptans. creeping; prostrate and rooting
scandens. climbing
soboliferus. soboliferous
stoloniferus. stoloniferous
volubilis. twining

E. Epithets Relating to Habitat

agrestis. pertaining to fields or cultivated land
alpinus. of the Alps; growing in an alpine zone
alsodes. of woods
amphibius. living in water and on land
aquaticus. growing in water
arenarius. growing on sand
arenicola. dweller in sandy places
arvensis. pertaining to fields, especially plowed fields
austromontanus. of the southern mountains
campestris. of fields
collicola. dweller in the hills
collinus. dwelling or growing in the hills
cumulicola. dweller on a heap or mound
demersus. growing under water
elodes. of marshes
epihydrus. upon the water
fluviatilis. of rivers
fontinalis. of springs
hypogeus. underground
inundatus. growing in places apt to be flooded
jugosus. mountainous
lacustris. of lakes or ponds

lithophilus. rock-loving; growing on rocks
littoralis. of the seashore
lucorum. sacred thickets
maritimus. of or belonging to the sea
montanus. of the mountains
monticola. dweller in the mountains
nemorosus. of woodlands and groves
orae. coastal; of the coast
paludosus. growing in boggy places
palustris. of marshes or swamps
porophilum. lover of soft rock
pratensis. growing in meadows.
riparius. pertaining to river banks
rivularis. growing by streams
ruderalis. growing among rubbish
rupestris. growing among rocks
sabulosus. growing in sandy places
saltuense. of or pertaining to forests or woodland pastures
sativus. cultivated
saxatilis. growing among rocks
saxicola. dweller among rocks
segetalis. growing in grain fields
silicola. dweller on siliceous or flinty soils
sylvaticus, silvaticus. of the woods
sylvestris, silvestris. growing in the woods
terrestris. growing in dry ground
umbrosus. growing in the shade
vinealis. growing in vineyards

F. Epithets Relating to Seasons

aestivalis. pertaining to summer
aestivus. pertaining to summer
autumnalis. pertaining to fall or autumn
hiemalis. belonging to winter
solstitialis. of summer
vernalis. pertaining to spring
vernus. pertaining to spring

G. Epithets Relating to Size

altissimus. very tall; high
altus. tall; high
angustatus. narrow; slender
depauperatus. reduced
dilatatus. broadened; widened
elatior. taller
elatus. tall
exaltatus. very tall
exiguus. little
giganteus. very large; gigantic
grandis. large
humilis. dwarf
intermedius. half-way; intermediate
magnus. large
major. larger; greater
minor. smaller; less
minus. small; minute
minutus. very small; minute
nanus. dwarf
parvulus. very small
pauxillus. small
perpusillus. very small
praealtus. very tall or high
procerus. very tall
pumilus. dwarf
pusillus. very small; insignificant
reductus. reduced
robustus. stout

II. NOMINATIVE EPITHETS: OLD GENERIC, COMMON, AND ABORIGINAL NAMES

alsine. Greek meaning of a luxuriant plant; chickweed
ammi. Greek for an umbelliferous plant
amomum. an aromatic shrub from which the Romans made fragrant balsam
anagallidea. resembling the genus Anagallis, pimpernel
ananassa. from Ananassa, the pineapple
aparine. old generic name for a plant, cleavers
armoracia. ancient name for horseradish
atamasco. aboriginal name
azedarach. from the Persian meaning free or noble tree
batatos. native name for sweet potato
benzoin. old name for some member of the Lauraceae (Fernald)

bursa-pastoris. shepherd's-pouch, from Bursa, an old generic name (Fernald)
calamintha. old generic name meaning "beautiful mint"
carota. old generic name for carrot
catalpa. aboriginal name
cepa. Latin word for onion
chamaedrys. old generic name meaning "ground-oak"
convolvulus. name for any twining plant; literally, "a caterpillar that wraps itself up in a leaf"
cotula. from the genus Cotula, a composite; Greek for a small cup
cracca. Latin word for pulse or wild vetch
cucubalus. old generic name (Fernald)
cucullaria. old generic name meaning "hood-like"
cyparissias. name used by Pliny for a plant related to Spurge
dracontium. Greek for dragon-wort
elatine. classical name for a low, creeping plant (Fernald)
githago. old generic name
halicacabum. from old generic name Halicacabus, another name for bladder-wort
haspan. native name for Cyperus haspan in Ceylon (Fernald)
helioscopia. an ancient name meaning turning toward the sun (Fernald)
hydropiper. Greek for water-pepper
intybus. Latin word for endive
jalapa. old generic name from Jalapa, Mexico (Fernald)
julibrissin. modification of a Persian name (Fernald)
kali. old generic name from the Persian, a large carpel (Fernald)
labrusca. Latin word for the wild vine, presumedly a grape
lappa. old generic name meaning a "bur"
leucanthemum. old generic name meaning "white flowered"
lupulus. early generic name (Fernald)
lychnitis. Greek for a plant from which wicks were made
mays. aboriginal name
mitreola. old generic name referring to miter-like fruits (Fernald)
mollugo. old generic name for whorled-leaved plants
napus. Latin word for a kind of turnip
negundo. aboriginal name
psyllium. old generic name; Greek for fleabane is psyllion
quamoclit. native Mexican name for a morning-glory
rhoeas. old Greek name meaning "wild poppy"
senega. old generic name given to plant the Seneca Indians used (Fernald)
serpentaria. Latin name for snakeweed
tetrahit. old generic name meaning 4-parted (Fernald)

III. MISCELLANEOUS EPITHETS

acicularis. needle-like
acris. sharp-pointed
affinis. related
alatus. winged
amabilis. lovely
amoenus. lovely; pleasing
anceps. two-headed or edged
annuus. annual
anomalus. abnormal; unlike its kind
apodus. without a foot; sessile
aristatus. awned
aromaticus. aromatic or spicy
axillaris. growing in an axil
baccatus. berry-like
barbatus. bearded
biennis. biennial
bracteosus. with conspicuous bracts
brevibarbis. with short beard
brevicaudatus. short tail-like appendage
brevipilis. short-haired
brevistylus. with short style

bulbiferus. bulb-bearing
caducus. falling off or dropping early
caespitosus. growing in tufts or patches
calcaratus. bearing spurs
campanulatus. bell-shaped
capillaceus. slender; hair-like
capitatus. growing in heads
caudatus. tailed
cernuus. nodding or drooping
ciliatus. fringed with hairs
clandestinus. hidden or concealed
clavellatus. shaped like small club
comosus. bearded; tufted
concolor. uniform in color
cornutus. horned
costatus. ribbed
cuneatus. wedged
dasystachys. woolly-spiked
debilis. weak
deciduus. falling off; shedding
decurrens. running down
densus. thick; dense
dentatus. toothed
desiccatus. dried up
didymus. twins; in pairs
difformis. irregular; unevenly formed
dioicus. dioecious; unisexual
discolor. not uniform in color
dumosus. shrubby; bushy
echinatus. spiny
elegans. elegant
epetiolatus. without a petiole
ericoides. heath-like
farinosus. mealy
fasciculatus. growing in bunches; fascicled
fertilis. capable of producing fruit
filiculmis. with thread-like stem
fistulosus. hollow; cylindrical; tubular
flabellatus. like a small fan
flexuosus. bent alternately in opposite directions
floridus. flowering
fluitans. floating
foetidus. ill-smelling; strong-smelling
frondosus. full of leaves
fruticosus. shrubby; bushy
fungosus. spongy
furcatus. forked
fuscatus. dusky
generalis. pertaining to all
geniculatus. abruptly bent, like a knee joint
glaber. smooth, without hairs
gracilis. slender
gregarius. herded or flocked together
heterophyllus. with leaves of more than one kind
hybridus. designating a hybrid
hystrix. bristly (literally a porcupine
imbricatus. overlapping, like shingles
inflatus. bladdery; swollen
inodorus. without scent or odor
invisus. hateful; detested
laevigatus. smooth, as if polished
lanosus. woolly
laxus. loose; lax
linearis. narrow; linear
lyratus. lyre-shaped
maculatus. spotted or blotched
mollis. soft; pubescent
muricatus. rough with short, hard points
natans. floating on the surface of the water
nitens. shining; polished
nodosus. knotted; knobby
normalis. at right angles; normal
nudatus. exposed; laid bare
odoratus. fragrant
patens. spreading
pendulus. hanging down; nodding
perennis. perennial
petiolatus. having a petiole; petioled
plicatulus. folded into small longitudinal pleats
praecox. developing very early; precocious
prostratus. lying flat; thrown to the ground
pubens. downy
punctatus. marked with dots, depressions or glands
radiatus. spreading from common center
ramosus. branched
recurvus. curved backwards or backwards
regularis. typical; according to rules
reticulatus. netted
rigidus. stiff
rostratus. beaked

rosulatus. in form of a rosette
rugosus. wrinkled
scaber. rough; harsh
scaposus. having a scape
sebiferus. wax- or tallow-bearing
sericeus. silky
serotinus. late to leaf, flower, or appear
setosus. bristly
speciosus. good-looking; beautiful
spectabilis. showy; worth seeing
squarrosus. rough, with outward projecting tips
stans. erect
sterilis. sterile; barren
stipitatus. with a little stalk or stipe
sulcatus. furrowed
tenellus. slender; tender; soft
teres. cylindric; circular in cross section
tinctorius. used for dyeing
tortus. twisted
trivialis. common; ordinary; trivial
truncatus. shortened; cut off at end
tuberosus. tuberous
uncinatus. hooked
velutinus. velvety
ventricosus. swollen, especially on one side
versicolor. variously colored
vulgaris. common

Section C. PRONUNCIATION OF BOTANICAL NAMES

Since botanical names are Latin or Latinized words, it would seem that they should be pronounced as Latin is pronounced. But what sort of Latin? Reformed academic? "Roman" Latin? Church Latin? It has been said that Latin is the universal language of botanists and other scientists; this may be true of written Latin, but wherever spoken, it has been modified somewhat by language usage peculiar to each country in which it has been spoken, so that the same word may sound differently when uttered by a German, a Frenchman, an Italian, an Englishman, an American, etc.

Several methods for pronouncing Latin are in use today. One, the so-called English method, has rules for pronouncing Latin and Latinized words that are generally analogous to those for the pronunciation of English words. The "Roman" method attempts to follow as closely as possible the pronunciation used by the Romans themselves from about 50 B.C. to 50 A.D., as far as can be determined. The Continental method was developed during the Middle Ages from the modern languages of that time, and widely used by the Roman Catholic Church. The vowels had about the same values as those of the Roman method, but the consonants were pronounced as those of the language of the country in which the method was used.

Botanists, at least English-speaking botanists, who have expressed themselves on the matter of pronunciation seem to agree generally that for our purposes it is probably best to use the English sounds for vowels and consonants, while following the rules of classical Latin for accenting. Until someone can devise a simpler method to which most botanists will ascribe, the beginning student can manage very well by checking with a manual like Gray's 7th or 8th Edition for which a grave accent (`) denotes a long vowel and an acute accent (´) a short vowel. If he is curious to know why the vowel is long or short, or why the accent falls where it does, he will need to understand a few conditions and rules. The following may be of help to one so inclined.

I. GENERAL INFORMATION

A. Syllables. Before a word can be accented, it must be divided into syllables. Every Latin word has as many syllables as it has separate vowels or diphthongs.

1. When a single consonant comes between two vowels, it is taken with the vowel that follows it: acer, a-cer.
2. When two consonants come between two vowels, one goes with the first vowel, and the other with the second: albidus, al-bi-dus. Exception: If the first consonant is b, c, d, g, k, p, or t, and the second is either l or r, both consonants go together with the second vowel: glabra, gla-bra. If the two consonants are ch, ph, or th, each pair is counted as one letter and goes with the second vowel: microcephala, mi-cro-ce-pha-la.
3. When there are more than two consonants, all but the first go with the second vowel: absconditus, ab-scon-di-tus.
4. X is always taken with the vowel preceding it: expansus, ex-pan-sus; exalbidus, ex-al-bi-dus.

B. Vowels.

1. Final vowels have the long sound (alsine, al-si-nee), except final a, which has the sound of an unstressed "ah" (verna, vernah).
2. Final es sounds like the English word "ease": alsodes, al-so-deez.
3. Y is always a vowel with the quality of i: diphyllus, di-fil-lus.
4. Two vowels together that do not form a dipthong, are always sounded separately. The first of the two has the short sound: filifolia, fi-li-fo-li-ah. It should be noted that this rule does not hold for words transcribed from the Greek, e.g., the e in Achillēa is long because it is a contraction of the Greek diphthong ei. A diphthong is treated as a long vowel wherever it occurs, even if transcribed by a single letter.

C. Diphthongs.*

1. ae and oe have the sound of long e in "me": laevis, lee-vis; rhoeas, ree-as.
2. au sounds like au in "caudal" or aw in "awful": caudatus, caw-da-tus.
3. ei usually becomes i, and is like English long i, in "kite."
4. eu sounds like u in "neuter," or "neurology": eurycarpus, u-ri-car-pus.
5. Diphthongs are always classed as long vowels.

D. Consonants.

1. c and g have soft sounds of s and j respectively when followed by e, i, y, ae or oe. Examples: Cedrus, see-drus; cyaneus, si-a-ne-us; generalis, je-ne-ra-lis; gynandra, ji-nan-dra. Otherwise, c has the hard sound of k and g the hard sound of g in "go." Examples: candidus, kan-di-dus; glabrus, gla-brus.
2. When a word is begun by one of the following pairs of consonants, the first letter is silent: cn, ct, gn, mn, pn, ps, pt, tm. Examples: Cnicus, ni-kus; Ctenium, te-ni-um; psyllium, sil-i-um.

*Some botanists treat oi as a diphthong, pronouncing it as oi in "oil," not as two separate vowels according to Latin rules. Thus they say Helianthoi-des, rather than Helian-tho-i-des.

3. ch, ph, and th are counted as one letter each. ch has the sound of k, but is silent before th at the beginning of a word; ph is sounded as f; th as th in "thing." Examples: chloranthus, klo-ran-thus; phellos, fel-los.
4. cc followed by i or y sounds like k-si: coccinea, kok-si-ne-a.
5. g before a "soft" g (gg), takes the sound of one soft g.
6. Initial x has sound of z: Xanthium, zan-thi-um; otherwise x sounds like ks: Zanthoxyllum, zan-thok-sil-lum.
7. ci, si and ti, when following an accented syllable and followed by another vowel, often have the sound of sh or zh: Senecio, se-nè-shi-o; Artemisia, ar-te-mí-zhi-a.

II. ACCENTING ACCORDING TO THE RULES OF CLASSICAL LATIN

A. General.

1. The last syllable is never accented. The next to the last syllable is called the penult: de-cum-bens; the third from the last syllable is called the antepenult: de-cum-bens.
2. Words of two syllables are always accented on the first: à-cer.
3. In a word of more than two syllables, the accent falls on the next to the last (penult) if the penult is long. It is long if it ends in a long vowel, a diphthong, or a consonant: alsine, al-sì-ne; amoenus, a-moè-nus; decumbens, de-cúm-bens.
   Note: When the penult ends in a consonant, the vowel is short, although the syllable is long.
4. If the penult is not long (therefore short), the accent falls on the antepenult, the third syllable from the end: dracontium, dra-con-ti-um (penult, ti, is short because it ends in a vowel with short sound).
5. The accent can never be farther from the end than the antepenult. Thus the penult is the crucial factor in accenting. If it ends in a consonant, it has the accent; if it ends in a diphthong it has the accent because a diphthong is always counted as a long vowel; if it ends in a vowel, one must determine whether that vowel is long or short. This requires a Latin dictionary. Most specific epithets are Latin or Greek adjectives, some are nouns in the genitive case, and others are old generic and aboriginal names. There are no rules given for the last two categories, but there are some hints that can be of help in pronouncing the adjectives and genitives:

B. Long Penult. As a rule, the penult is long in the following adjective endings and thus takes the long accent:

-àlis	-ènus	-ìvus	-ùnes
-ànus	-ètus	-oìdes	-ùnus
-àris	-ìnus*	-òsus	-ùrus
-àtus	-ìtus	-òvus	-ùsus
			-ùtus

Examples: cardinàlis, barbàtus, farinòsus

*Exceptions: short in cannábinus, gossýpinus, serótinus, etc; usually long in cyperìnus and salicìnus.

C. Short Penult. In the following adjective endings the vowel in the penult is usually short, thus removing the accent to the antepenult:

-ăcus	-ĕus	-ĭdus	-ĭor
-ĕger	-ĭchus	-ĭlis	-ĭus
-ĕris	-ĭcus	-ĭlus	-ŭus
-ĕrus	-ĭdis	-ĭmus	-y̆us

Examples: sebíferus, hepáticus, cándidus, ambíguus

Compound words ending in the following syllables also have a short vowel in the penult (thus the accent is removed to the antepenult):

-clădus	-gĕra	-lĕpis	-phy̆tum
-cŏla	-gĕrum	-ŏlens	-tŏmus
-cŏlor	-gy̆nus	-pĭlis	-stŏmus
-fĕrum	-fĭdus	-phĭlus	-vĭrens
-fĕra	-fĭlus	-pŏdus	

Examples: cerífera, cumulícola, dichótomus

Note: If the connecting vowels, short i in Latin and short o in Greek, appear in the penult, they do not have the accent because they make the penult short: spinĭfer, spi-nĭ-fer. When the connecting vowels appear in the antepenult in words with short penult, the antepenult receives the accent and it is short: spinĭfera, spi-ní-fe-ra. Of course, in an adjective ending in which the penult has consonant length (ends in a consonant), the penult has the accent and the vowel is short:

-éllus	-éssus	-éstus	-íllus
-énsis	-éster	-fórmis	-úster

Examples: halepénsis, multifórmis

The same rule applies to compound words ending in -cárpus, -róstris, and -phýllus: chlorocárpus, heterophýllus.

D. Commemorative Names. A few comments should be made about the pronunciation of Latinized commemorative names. It is here that the rules are most apt to be disregarded or ignored, chiefly because the person being honored may not be able to recognize his name if the rules are strictly adhered to. An example may suffice to illustrate the dilemma:

> The family name, James, pronounced as one syllable in English, with a long a, Jàmz, becomes in Latinized form a 4-syllable word with a long accent on the antepenult: Ja-mè-si-i. In cases of this sort, many botanists simply pronounce the family name as it sounds and add the proper Latin ending. It now comes out as Jàmes-i-i, which is recognizable, but very un-Latin, because in Latin words the accent cannot be farther back than the antepenult, nor can there be more than one vowel to a syllable. Still many botanists prefer taking the exception to the rules if the name cannot be recognized in its Latinized form.

In conclusion, as was stated by the authors of the 7th edition of Gray's Manual, "Botanical names...are not always capable of easy or consistent pronunciation. From long-established custom they are usually pronounced in English-speaking countries according to the pronunciation of Latin after the English method... The subject is one into which considerations of taste, convenience, and custom enter to such an extent that it is most difficult to lay down definite principles free from pedantry."

SPECIFIC EPITHETS INDEX

*Common usage

SPECIFIC EPITHETS INDEX (cont.)

*Common usage

SPECIFIC EPITHETS INDEX (cont.)

*Common usage

SPECIFIC EPITHETS INDEX (cont.)

EPITHETS LITERATURE

Bennett, C. E. 1963. New Latin Grammar. Allyn and Bacon. New York.

Brown, R. W. 1956. Composition of Scientific Words. Reese Press. Baltimore.

Chandler, C. 1889. Pronunciation of Latin and quasi-Latin scientific terms. Denison University Scientific Laboratory Bulletin 4(2): 161-176.

Fernald, M. L. 1950. Gray's Manual of Botany, 8th ed. American Book Company. New York.

Gleason, H. A. 1932. The pronunciation of botanical names. Torreya 32: 53-58.

International Code of Botanical Nomenclature. 1972. Utrecht, Netherlands.

Jackson, B. D. 1928. A Glossary of Botanic Terms. J. B. Lippincott Company. Philadelphia.

Nybakken, O. E. 1959. Greek and Latin in Scientific Terminology. Iowa State College Press. Ames, Iowa.

Robinson, B. L., and M. L. Fernald. 1908. Gray's New Manual of Botany, 7th ed. American Book Company. New York.

Rydberg, P. A. 1922. Flora of the Rocky Mountains and Adjacent Plains. Privately published by the author. New York.

Stearn, W. T. 1966. Botanical Latin. Thomas Nelson and Sons, Ltd. London.

Woods, R. S. 1966. An English Dictionary for the Use of Taxonomists. Pomona College. Claremont, California.

Chapter 5. CHARACTER, EVIDENCE, CHARACTERIZATION

The concept of character is fundamental to the science of taxonomy. Characters provide the basic information for classification; characters are the features used in identification; characters are used in the determination of relationships; and complexes of characters associated with different organisms give the basis for the naming of taxa. Characterization or description of plants and their component parts is the initial and primary process in systematics. The assignment of attributes to plants and their parts (characterization) precedes, and is a requirement for, the establishment of discontinuities in form and/or function between organisms. Discontinuities have to be present for the delimitation and definition of taxa. Delimitation and definition of taxa are prerequisites to the development of identification and classification schemes. Only well-delimited and -defined taxa should be named.

In general, a character is "any attribute (or descriptive phrase) referring to form, structure or behaviour which the taxonomist separates from the whole organism for a particular purpose such as comparison or interpretation" (Cain and Harrison in Davis and Heywood [1963]). Practically, "a character may be defined as any feature whose expression can be measured, counted or otherwise assessed" (Hedberg in Davis and Heywood [1963]). "A unit or single character is a taxonomic character of two or more states, which within the study at hand cannot be further subdivided logically" (Sokal and Sneath [1963]). Character states are descriptors or characteristics.

The fundamental unit of description is the character state which is a basic component of character. Characters are grouped as character classes which, together, form a type of evidence. For example, "ovate" is a character state of the character, shape; "crenate" is a character state of the character, margin; "acute" is a character state of the character, apex. Shape, margin, apex are characters in the character class, general shape. "Opposite" is a character state for the character, arrangement; "basal" is a character state for the character, position; "ascending" is a character state for the character, orientation. Arrangement, position, orientation are characters in the character class, position-arrangement. These two character classes and others would be components of a type of evidence, morphology. Morphological, anatomical, palynological, etc. evidence (see Chapters 6-19) would form a comprehensive descriptive terminology for plants and their parts.

A precise and complete terminology has to be devised for comprehensive description. Character states have to be well defined for accuracy and ease of application. Characters have to be effectively ordered and ranked for a meaningful comparative classification of descriptive terms. Character classes have to be reasonably delimited but sufficient in number to cover all aspects of the types of evidence used in description. Comprehensive description has to include all types of evidence for diversity of information necessary for determination of true relationships between taxa. Precision in definition of character states, logic in the hierarchical ordering and arranging of characters and character classes, and comprehensiveness of descriptive evidence are fundamental aspects of description or phytography.

Essentially this chapter is an introduction to the character classes, characters and character states described in the succeeding chapters (6-19) on morphological, anatomical, palynological embryological, cytological, genetic, reproductive, chemical, physiological, ecological, geographic, and paleobotanical evidence. The rather complete terminology presented within these chapters has been given for a comprehensive characterization of plant parts, plants, and taxa. Comprehensive description has to include all types of evidence for a real understanding of the taxonomy of plants.

A conscious effort has been made to classify types of evidence into character classes and character states in the chapters on evidence, particularly in Chapter 6. The greater the number of characters perceptively used in description, the "more meaningful," the "more reliable," the "more diagnostic," and the "more critical" will the character states be that are used in determination of relationships between taxa, in classification, in identification; and presumably the more significant the attributes will be that are used in fields related to systematics. Relative values of characters do vary as classifications change with the increasing experience of the classifiers and as new relationships are discovered with additional characters studied.

Definitions of many kinds of general characters are presented below (I A), in contrasting couplets, to introduce the student to the concept of character, to terminology related to character as used in the field of taxonomy, and to the problems involved in defining types of characters. What is "meaningful," "reliable," "good," "diagnostic" in one situation can be "meaningless," "unreliable," "bad," or simply a "descriptive" term under another set of conditions. Types of evidence are presented in I B. For detailed information on character variation analysis see Chapter 22; for data accumulation and use related to characters and character states see Chapter 20; and for the ordering of characters see Chapters 6 and 24.

I. GENERAL TYPES OF CHARACTERS AND EVIDENCE

A. <u>General Types of Characters</u>.

1. a. Adaptive. Characters that respond to the changing environmental situation enabling the organism to survive by an alteration of structure and/or function.
   b. Non-adaptive. Characters that do not change in response to a changing environmental situation.
2. a. Analytic. Characters used in identification, characterization and delimitation of taxa.
   b. Synthetic. Characters of constant nature and wide occurrence used in constitutive and organizational characterization of larger taxa.
3. a. Biological. Characters related to some vital function or behavior.
   b. Fortuitous. Characters not related to some vital function or behavior.
4. a. Continuous. Characters showing uniform intergradation or a cline.
   b. Discontinuous. Characters not showing uniform intergradation or a cline, but separated by breaks in the continuum.
5. a. Cryptic. Characters hidden, microscopic.
   b. Phaneritic. Characters obvious, macroscopic.

6. a. Diagnostic. Characters used to distinguish one taxon from another.
   b. Descriptive. Characters used to give the features or attributes of a taxon.
7. a. Good. Characters easily recognizable, have a narrow range of expression, and are seemingly genetically fixed.
   b. Bad. Characters not easily recognizable, or have a wide range of expression, and are seemingly not genetically fixed.
8. a. Homologous. Characters having a common ancestral origin.
   b. Analogous. Characters similar but not having a common ancestral origin.
9. a. Logically correlated. Character or property which is a logical consequence of another.
   b. Non-correlated. Character or property which is not the logical consequence of another.
10. a. Macro. Characters gross, macroscopic, usually external.
    b. Micro. Characters small, microscopic, usually internal.
11. a. Meaningful. Characters or attributes which are significant or valuable in characterization, identification, or classification.
    b. Meaningless. Characters or attributes which are not significant or valuable in characterization, identification, or classification.
12. a. Ontogenetic. Characters associated with the development of the individual.
    b. Phylogenetic. Characters associated with the development of the taxon over long periods of time.
13. a. Plastic. Characters seemingly variable.
    b. Fixed. Characters seemingly invariable.
14. a. Primitive. Characters possessed by present day taxon and its ancestors.
    b. Advanced. Characters possessed by present day taxon not possessed by its ancestors.
15. a. Qualitative. Characters relating to kinds of forms, structures, behaviors, or functions.
    b. Quantitative. Characters assessed by number and size.
16. a. Reliable. Characters that are consistent, distinct, recognizable and usable by taxonomists.
    b. Unreliable. Characters that are inconsistent, indistinct, unrecognizable, and not usable by taxonomists.
17. a. Specific. Diagnostic characters used in delimiting a species.
    b. Generic. Diagnostic characters used in delimiting a genus.
18. a. Two-state. Characters "all or none," present or absent.
    b. Multistate. Characters that can be expressed by more than "all or none" or present or absent states.
19. a. Variant. Characters that do vary within population samples.
    b. Invariant. Characters that do not vary within population samples.
20. a. Weighted. Characters given a greater value due to presumed importance.
    b. Non-weighted. Characters given same value regardless of importance.

B. <u>General Types of Evidence</u>.
1. a. Vegetative. Characters associated with root, leaf, stem, or bud.
   b. Reproductive. Characters associated with flower, fruit, or seed.
2. a. Chemical. Characters related to compounds or broad classes of compounds found in organisms.
   b. Physical. Characters related to inanimate properties of organisms.

3. a. Exomorphic. Characters external, usually structurally macroscopic.
   b. Endomorphic. Characters internal, usually structurally microscopic.
4. a. Morphological. Characters associated with gross structure.
   b. Anatomical. Characters associated primarily with tissues and cell types.
5. a. Palynological. Characters associated with pollen grains and their development.
   b. Embryological. Characters associated with embryo types and their development.
6. a. Cytological. Characters associated primarily with chromosome structure and number, cellular organelles, and ultrastructure.
   b. Genetical. Characters primarily associated with transmitted traits and breeding behavior.
7. a. Physiological. Characters related to function.
   b. Ecological. Characters related to habitat preference or environmental modifications of structure.
8. a. Paleo. Characters related to the old or fossil.
   b. Neo. Characters related to the new or recent.

GENERAL CHARACTER LITERATURE

Cronquist, A. 1968. Evolution and Classification of Flowering Plants. Houghton Mifflin Company. Boston.
Davis, P. H., and V. H. Heywood. 1963. Principles of Angiosperm Taxonomy. D. Van Nostrand, Inc. Princeton.
Sokal, R. R., and P. H. A. Sneath. 1963. Principles of Numerical Taxonomy. W. H. Freeman and Company. San Francisco.

CHARACTER EXERCISE

Select two consecutive pages, one of which should have a key, from a manual pertinent to the flora of your region and list all of the kinds and classes of characters, with an example of each, found on those two pages. Or select two consecutive pages, one of which should have a key, from a classification system; e.g., Cronquist (1968), and list all of the kinds and classes of characters with an example of each found on those two pages.

Chapter 6. PHYTOGRAPHY - MORPHOLOGICAL EVIDENCE

Phytography deals with the descriptive terminology of plants and their component parts for the purpose of providing an accurate and complete vocabulary for description, identification, and classification. Phytographic studies should (1) provide the student with a vocabulary for intelligent communication about plants and an understanding of the use of relative terms; and (2) help the student observe plants more critically and describe them more precisely. The morphological and developmental features of angiosperms form the core of the terms treated in this chapter.

The classification of terminology herein represents basically a compromise between (1) the traditional approaches to phytography, (2) effective teaching-learning organization, and (3) ordering of characters and character states for increased precision in comparative studies and data banking. The use of types and structural types conforms fairly closely to accepted practice. Major organs and their component parts are basic morphology. The subclassifications, as in fruit types and arrangement, have been made for greater learning effectiveness and for compliance with traditional usage. The treatment of position-arrangement, shapes, and surface have been ordered or ranked to make those classifications comparative. The ratios and fractions of the subsection on shapes, if applied, will make the relative terms or character states more consistent in application.

Section A* is organized basically as the Plant and its major structures (I-XII) which are subclassified into component parts (A), types (B), and structural types (C). Plant types are based on habit; root, stem, leaf, flower, and inflorescence types are based primarily on position and/or arrangement; and the perianth, androecium, and gynoecium types are based on fusion of parts. Structural types includes typical and specialized structures. Section B* is organized as character classes (I-XII) which are subclassified as numerous characters. Sections C and D are glossaries of terms pertinent to gymnosperms and lower vascular plants. The index includes all terms presented in the chapter. Unspecialized and specialized structures are included in the structural type classification for completeness of treatment. All terms within a subject are alphabetized except those in shape where it was deemed best to group related terms. Adjectives or noun forms, usually not both, are used throughout a classification. Some classifications knowingly have incomplete vocabularies but instructions are included for the determinations of the meanings of many additional terms.

The morphological glossary is based primarily upon those found in Lawrence (1951), Stearn (1966), Johnson (1931), Featherley (1954), Correll and Johnston (1971), Swartz (1971), and Porter et al. (1973). Some classifications and definitions of terms are based on special papers which are cited in the literature. A few new terms for perianth position, gynoecial and staminal types are introduced.

Phytography, as an ordered and precise subject, is still in a rather primitive state. "Different terms are often applied to the same character state, and, conversely the same term is used for many character states" (Porter et al., 1973). Seemingly there are too many terms in this chapter, but a real need exists for even more, with precise definition and application. Terms represent concepts, or states of concepts. Generalized concepts provide the basis of classification. Improvement is needed throughout the term-concept-classification system to advance phytography and taxonomy.

*See Contents on next page and Character Classification on the succeeding pages.

SYSTEM OF CHARACTERIZATION

Section A. STRUCTURAL PARTS AND TYPES

Section B. GENERAL CHARACTERS AND CHARACTER STATES

Section D. LOWER VASCULAR PLANT GLOSSARY

CHARACTER CLASSIFICATION

The general groups of characters I-XII are considered character classes which include several types of characters each of which is defined. The states of each character are arranged by order of importance and defined in Section B of this chapter. This arrangement of character states by character and ordering or ranking of types of character within a character class is an attempt to make this classification comparative and precise. See Chapter 5 for basic definitions of characters and kinds of characters as well as types of evidence.

I. POSITION AND ARRANGEMENT (A Character Class with character types 1-8 as listed below. States of each character type are in Section B I of this chapter).
1. Location. Position of organs or parts in relation to their surrounding environment (See Section B I A for the character states of this character type).
2. Position. Location or attachment of organs or parts with respect to other, dissimilar organs or major parts (See Section B I B).
3. Arrangement. Disposition of organs or parts with respect to one another (See Section B I C).
4. Orientation. Divergence of organs or parts with reference to a central axis or point (See Section B I D).
5. Transverse Posture. The relation of the ends of a single organ or part to its center or transverse axis (See Section B I E).
6. Longitudinal Posture. The relation of the margins of a single organ or part to its central longitudinal axis (See Section B I F).
7. General Structural Position. The point, spot, local or general region of or on a single structure; e.g., the apical region of a leaf, the central point of a petal (See Section B I G).
8. Embryonic Position. Position and arrangement of immature organs or parts.
   a. Aestivation or Prefloration. Arrangement, in the strict sense of I 3, of embryonic floral parts. Note: aestivation for the calyx and corolla can be the same or different within a flower bud (See Section B I H).
   b. Ptyxis. The rolling and folding of embryonic leaves in relation to a central point, transverse axis or longitudinal axis; a combination of transverse and longitudinal posture for embryonic parts (See Section B I H).
   c. Vernation. The arrangement, in the strict sense of I 3, of embryonic leaves. Most authors treat ptyxis as this.

II. NUMBER AND SIZE
1. Number. Any value or range of values expressing absolute count, sum or aggregate of individual units; or an expression of relative or comparative units (See Section B II A).
2. Size. Spatial dimensions or proportions of any organ or its parts, usually expressed as number values in reference to length, width, and depth; or an expression of relative or comparative amounts (See Section B II B).
3. Cycly. The absolute number of whorls of parts, such as number of whorls of floral parts or leaves (See Section B II C).

4. Merosity. The absolute number of parts within a whorl, such as the number of petals in the corolla or number of stamens in an androecial whorl (See Section B II D).
5. Fusion. The coalescence or union of like or unlike organs or parts (See Section B II E).
6. Division. The separation, divergence, or forking into segments, parts, lobes, and branches of any organ or its parts (See Section B II F).

III. SHAPE AND OUTLINE
1. Shape. An outline of specific form which is plane or two-dimensional; or a specific figure which is three dimensional (See Section B III A).
2. Apex. The tip of a two-dimensional structure which is usually bounded by the upper 5-10% of the margins; or the top of a three-dimensional structure (See Section B III B).
3. Margin. The border region of the side of a two-dimensional or plane structure (See Section B III C).
4. Base. The lower portion of a two-dimensional structure which is usually bounded by the lower 5-10% of the margins; or the bottom of three-dimensional structure (See Section B III B).

IV. SURFACE, VENATION, TEXTURE
1. Configuration. Surface pattern usually resulting from internal structural form, exclusive of venation (See Section B IV A).
2. Venation. Vein arrangement visible on the surface (See Section B IV B). For internal venation features see Venation in Chapter 7.
3. Surface. Features derived from epidermal outgrowths or excrescences, exclusive of trichome cover types (See Section B IV C).
4. Vestiture or Indument. Trichome cover types (See Section B IV D). For individual trichome types see Trichomes in Chapter 7.
5. Texture. Characteristic physical condition or consistency of an organ or its parts (See Section B IV E).

V. SEX
1. Flower. Sexual condition of the individual flower (See Section B V A).
2. Inflorescence. Arrangement of flowers by sexual condition within an inflorescence; or total diversity of flower sexual conditions within an inflorescence (See Section B V B).
3. Plant. Total diversity of flower sexual conditions within a single plant (See Section B V C).
4. Population or Species. Total diversity of flower sexual conditions within a population (See Section B V C).

VI. POLLINATION AND FERTILIZATION
1. Pollination Type. Agent by which pollen is transferred from the pollen sac or anther to the stigma or ovule (See Section B VI A).
2. Pollination Pathway. Structural route that the pollen tube traverses in reaching the ovule (See Section B VI B).
3. Fertilization Type. Source(s) of male gametes for fertilization (See Section B VI C).

4. Fertilization Pathway. Structural route that the pollen tube traverses to the embryo sac after reaching the ovule (See Section B VI D).

VII. TEMPORAL PHENOMENA

1. Periodicity. Recurrence of growth phenomena at regular time intervals; or seasonality of occurrence of plant growth phenomena (See Section B VII A).
2. Maturation. Relative development time of different plant organs or parts (See Section B VII B).
3. Duration. Length of time that a plant or any of its component parts exists (See Section B VII C).

VIII. GROWTH AND DEVELOPMENT

1. Growth Region. Location of meristematic region on or within an organ or its parts (See Section B VIII A).
2. Growth Periodicity. Recurrence or duration of meristematic activity within an organ or its parts (See Section B VIII B).
3. General Development. Direction of maturation within the plant of its organs or its parts; e.g., from the base upward or from the inside outward (See Section B VIII C).
4. Developmental Shoot Types. Origin and maturation of different shoots (stem with leaves) (See Section B VIII D).

IX. PATTERNS

1. Symmetry. Correspondence of parts in size, form and arrangement on opposite sides of a plane, line, or point (See Section B IX A).
2. Arrangement Systems. Total disposition of organs with respect to each other based on the angle of divergence between succeeding similar parts (See Section B IX B).
3. Branching Patterns. Position of branches on the main axis and the arrangement of the lateral branches (See Section B IX C).

X. SECRETORY AND EXCRETORY STRUCTURES

1. Functional Types. The chemical nature of the substance secreted or excreted (See Section B X A).
2. Positional Types. Location of glands on organs or parts (See Section B X B).

XI. COLORATION

1. Standard. Color determinations based on accepted color charts and codes (See Section B XI A).
2. Distribution. Patterns of color variegation on an organ or parts (See Section B XI B).

XII. GENERAL

1. General. Descriptors applicable to any character.

Section A. STRUCTURE AND SPECIALIZED CHARACTERS

I. THE PLANT

A. Plant Parts

Bud. Immature vegetative or floral shoot or both, often covered by scales.

Flower. Reproductive structure of flowering plants with or without protective envelopes, the calyx and/or corolla; short shoot with sporophylls and with or without sterile protective leaves, the calyx and corolla.

Fruit. Matured ovary of flowering plants, with or without accessory parts.

Leaf. A photosynthetic and transpiring organ, usually developed from leaf primordium in the bud; an expanded, usually green, organ borne on the stem of a plant.

Root. An absorbing and anchoring organ, usually initially developed from the radicle and growing downward.

Seed. Matured ovule of seed plants.

Stem. A supporting and conducting organ usually developed initially from the epicotyl and growing upward.

B. Plant Types

(Classification based on habit)

Herb. A usually low, soft, or coarse plant with annual aboveground stems.

Shrub. A much-branched woody perennial plant usually without a single trunk.

Tree. A tall, woody perennial plant usually with a single trunk.

Vine or Liana. An elongate, weak-stemmed, often climbing annual or perennial plant, with herbaceous or woody texture.

II. ROOTS

A. Root Parts

Root Cap. Parenchymatous, protective apex of root.

Root Hair. Lateral, absorbing outgrowth of the epidermal cell.

Secondary Root. Lateral root with root cap and hairs, derived from the pericycle.

B. Root Types

(Classification based on position and origin)

Adventitious. Arising from organ other than root; usually lateral.

Primary. From radicle of embryo; tip of main axis.

Secondary. From pericycle within the primary or secondary root; lateral.

C. Root Structural Types (Figure 6-1, p. 89)

Aerating or Knee. Vertical or horizontal above-ground roots.

Aerial. Fibrous, adventitious roots, frequently with an adhesive disk; a crampon.

Buttress. Roots with board-like or plank-like growth on upper side, presumably a supporting structure.

Contractile or Pull. Roots capable of shortening, usually drawing the plant or plant part deeper into the soil, usually with a wrinkled surface.

Fascicled. Fleshy or tuberous roots in a cluster.

Fibrous. With fine, threadlike or slender roots.

ROOT STRUCTURAL TYPES and STEM PARTS

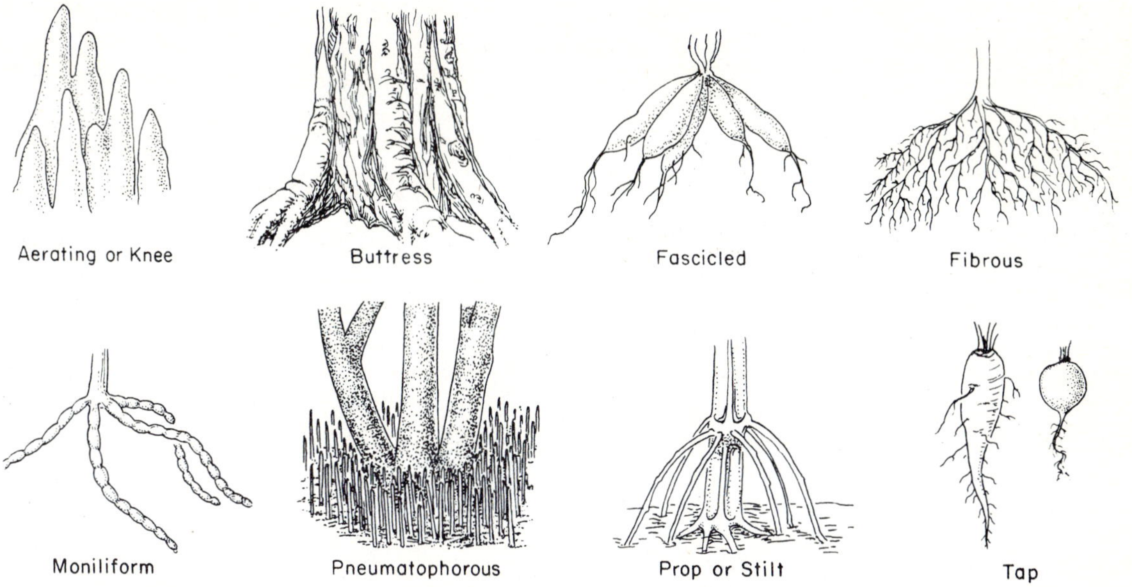

TWIG SURFACE PARTS

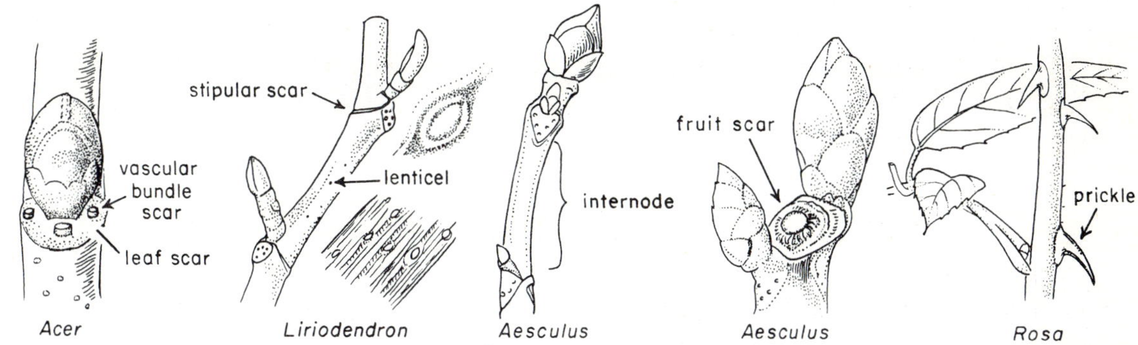

THORNS

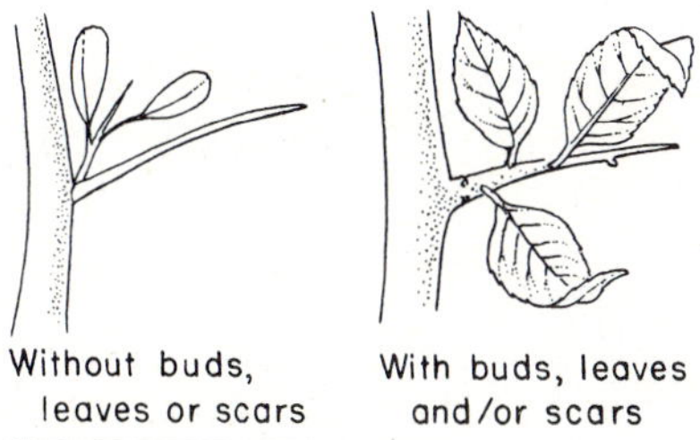

PITH

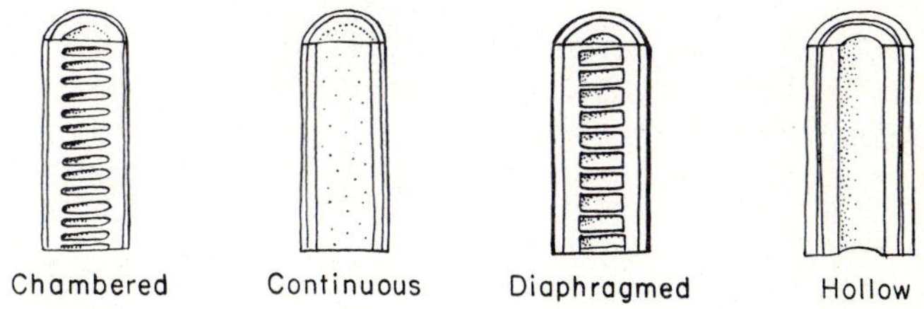

WOOD

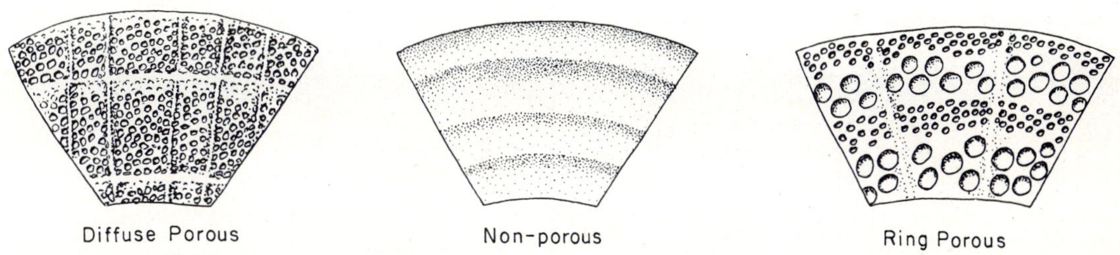

STEM TYPES, STEM STRUCTURAL TYPES and BUD TYPES

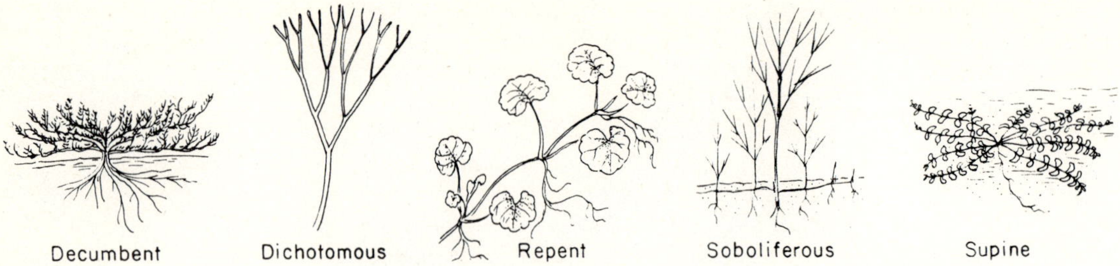

STEM STRUCTURAL TYPES

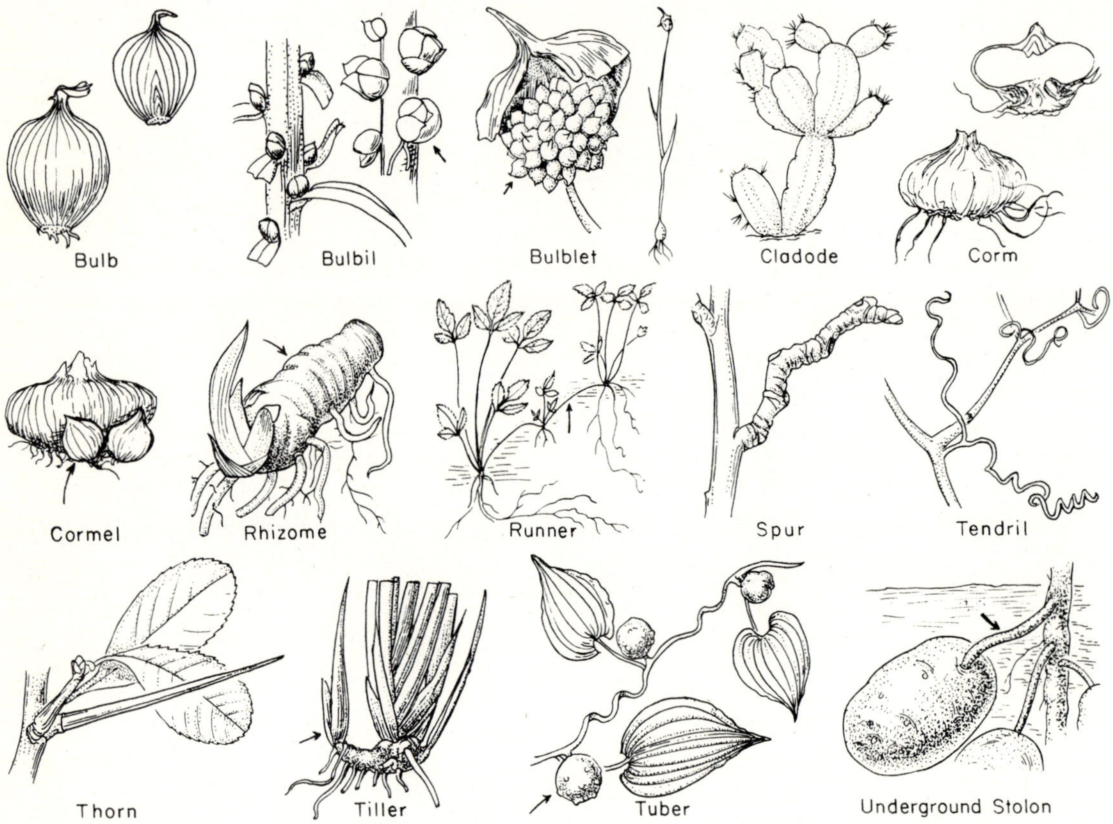

BUD TYPES

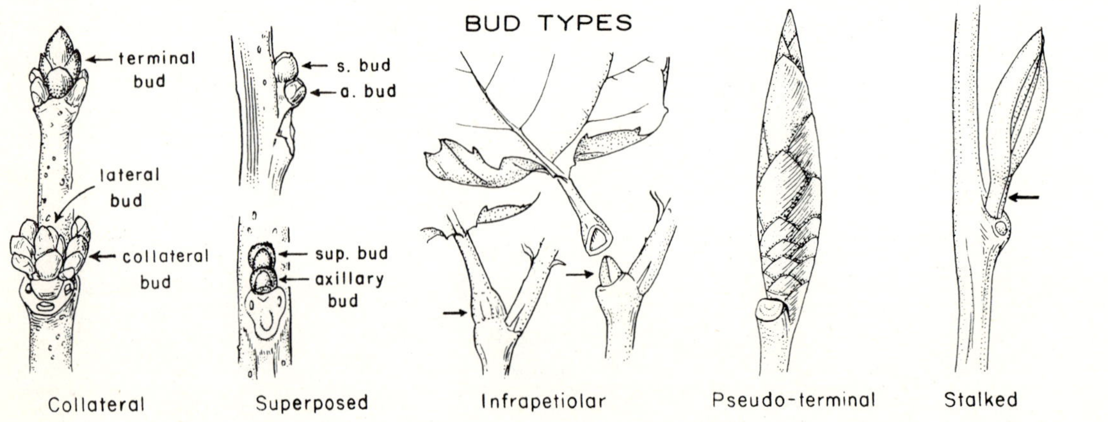

Fleshy. Succulent roots.

Haustorial. Absorbing roots, within host of some parasitic species.

Moniliform. Elongate roots with regularly arranged swollen areas.

Pneumatophorous. With spongy, aerating roots, usually found in marsh plants.

Prop or Stilt. Adventitious, supporting roots usually arising at lower nodes.

Tap. Persistent, well-developed primary root.

Tuberous. Fleshy roots resembling stem tubers.

III. STEMS

A. Stem Parts (Figure 6-1, p. 89)

1. Twig Surface Parts

Bud. An immature shoot.

Internode. A section or region of stem between nodes.

Leaf Scar. A mark indicating former place of attachment of petiole or leaf base.

Lenticel. A pore in the bark.

Node. Region of stem from which a leaf, leaves, or branches arise.

Prickle. A sharp-pointed outgrowth from the epidermis or cortex of any organ.

Stipular Scar. A mark indicating former place of attachment of stipule.

Terminal Bud Scale Scar Rings. Several marks in a ring indicating former places of attachment of bud scales.

Vascular Bundle or Trace Scar. A mark indicating former place of attachment within the leaf scar of the vascular bundle or trace.

2. Major Stem Parts

Bark. Tissues of plant outside wood or xylem.

- Exfoliating Bark. Bark cracking and splitting off in large sheets.
- Fissured Bark. Split or cracked bark.
- Plated Bark. Split or cracked bark with flat plates between fissures.
- Ringed Bark. Split or cracked bark with circular fissures.
- Shreddy Bark. Soft but coarse fibrous bark, usually shallowly fissured.
- Smooth Bark. Bark without fissures.
- Winged Bark. Bark with one or more thin, flat longitudinal expansions or elongate plates.

Pith. Centermost tissue of stem, usually soft.

- Chambered Pith. Solid core of pith cells absent, only distinct partitions present.
- Continuous Pith. With solid core of parenchyma or pith cells.
- Diaphragmed Pith. With solid core of pith cells and distinct partitions.
- Hollow Pith. Disintegrated pith with a large central cavity.
- Spongy Pith. Porous, easily compressible pith.

Wood. Xylem consisting of vessels and/or tracheids, fibers, and parenchyma cells.

- Annual Ring. Usually one year's growth of wood; spring and summer wood.
- Diffuse Porous Wood. Annual rings with vessels or pores more or less evenly distributed.
- Non-porous Wood. Annual rings with tracheids only, no vessels produced in spring or summer wood.
- Ring Porous Wood. Annual rings with vessels or pores usually in the spring wood, in a well-defined circular band.

B. Stem Types (Figure 6-2, p. 90)

(Classification based on habit, direction of growth, or position)

Arborescent. Tree-like in appearance and size.

Ascending. Inclined upward.

Cespitose. Short, much-branched, plant forming a cushion.

Clambering. Sprawling across objects, without climbing structures.

Climbing. Growing upward by means of tendrils, petioles, or adventitious roots.

Columnar. Erect with a stout main stem or trunk.

Decumbent. Reclining or lying on the ground with the tips ascending.

Dichotomous. With equally forked branches or stems.

Eramous. With unbranched stems.

Erect. Upright.

Fastigiate. Strictly erect and parallel.

Fruticose. Shrubby.

Geniculate. Abruptly bent at a node, zigzag.

Procumbent, Prostrate, or Reclining. Trailing or lying flat, not rooting at the nodes; humistrate.

Ramose. Branched.

Repent. Creeping or lying flat and rooting at the nodes.

Soboliferous. With loosely clumped shoots arising some distance apart from rhizomes or underground suckers.

Stoloniferous. With runners or propagative shoots rooting at the tip producing new plants; bearing stolons; sarmentose.

Strict. Stiff and rigid.

Supine. Prostrate, with parts oriented upward.

Trailing. Sprawling on ground, usually with adventitious roots.

Twining. Coiling around an object.

Virgate. Wand-like; long, slender, and straight.

C. Stem Structural Types (Figure 6-2, p. 90)

Bulb. A short, erect, underground stem surrounded by fleshy leaves.

Bulbel. A small bulb produced from the base of a larger bulb.

Bulbil. A small bulb or bulblike body produced on above-ground parts.

Bulblet. A small bulb, irrespective of origin.

Caudex. A short, thick, vertical or branched perennial stem usually subterranean, or at ground level.

Cladode (phylloclad). A flattened main stem resembling a leaf.

Corm. The enlarged, solid, fleshy base of a stem with scales; an upright underground storage stem.

Cormel. Small corm produced at base of parent corm.

Culm. Flowering and fruiting stems of grasses and sedges.

Pachycauly. Short, thick, frequently succulent stems, as in some cacti.

Primocane. The first-year nonflowering stem, as in most blackberries; a turion.

Pterocauly. Winged stems.

Rhizome. A horizontal underground stem.

Rootstock. A term applied to miscellaneous types of underground stems or parts.

Runner or Stolon. An indeterminate, elongate, above-ground propagative stem, with long internodes, rooting at the tip forming new plants.

Sarcocauly. Fleshy stems.

Scape. A naked flowering stem with or without a few scale leaves, arising from an underground stem.

Sclerocauly. Hard, dryish stems.

Spur. A short shoot on which flowers and fruits or leaves are borne.

Sucker. A shoot arising below ground or from an old stem, usually fast-growing and adventitious; surculose.

Tendril. Long, slender, coiling branch, adapted for climbing (most tendrils are leaf structures).

Thorn. A sharp-pointed branch.

Tiller. A grass shoot produced from the base of the stem.

Tuber. A thick storage stem, usually not upright.

Turion. An overwintering bud, as in Lemna.

Underground Stolon. A determinate, elongate, underground propagative stem with long internodes forming a bulb or tuber at the tip.

IV. BUDS

A. Bud Parts

Bud Primordium. Meristematic tissue that gives rise to a lateral bud.

Flower Primordium. Meristematic tissue that gives rise to a flower.

Leaf Primordium. Meristematic tissue that gives rise to a leaf.

Promeristem. Apical growing or meristematic tissue that gives rise to other bud parts.

Scale. Protective leaf on outside of bud.

B. Bud Types (Figure 6-2, p. 90)

(Classification based on position and arrangement)

Accessory. Buds lateral to or above axillary buds.

- Collateral. Bud(s) lateral to axillary bud.
- Superposed. Bud(s) above axillary bud.

Axillary or Lateral. In axils of leaves or leaf scars.

Infrapetiolar or Subpetiolar. Axillary bud surrounded by base of petiole.

Pseudo-terminal. Bud appearing apical but is lateral near apex, developing with death or nondevelopment of terminal bud.

Terminal. At apex or end of stem.

C. Bud Structural Types

(Classification based on composition and cover)

Flower. Contains flower primordia; will give rise to one or more flowers.

Leaf. Contains leaf and stem primordia; will give rise to branch with leaves.

Mixed. Contains flower, leaf, and stem primordia; will give rise to branch with leaves and flower(s).

Naked. Shoot and/or flower primordia not surrounded by scales.

Protected. Shoot and/or flower primordia surrounded by scales.

V. LEAVES

A. Leaf Parts

Blade. Flat, expanded portion of leaf.

Leaflet. A distinct and separate segment of a leaf.

Ligule. An outgrowth or projection from the top of the sheath, as in the Poaceae.

Midrib. The central conducting and supporting structure of the blade of a simple leaf.

Midvein. The central conducting and supporting structure of the blade of a leaflet.

Petiole. Leaf stalk.

Petiolule. Leaflet stalk.

Pulvinus. The swollen base of a petiole or petiolule.

Rachilla. Secondary axis of compound leaf.

Rachis. The main axis of a pinnately compound leaf.

Sheath. Any more or less tubular portion of the leaf surrounding the stem or culm, as in the Poaceae.
Stipels. Paired scales, spines, or glands at the base of petiolule.
Stipules. Paired scales, spines, glands, or blade-like structures at the base of a petiole.

B. Leaf Types

(Figure 6-3, p. 95)

(Classification based primarily on arrangement of leaflets)

Note: This classification is based on discrete segments of leaves or leaflets, but the terms with "compound" are equally applicable to segments, divisions, etc. of any structure with a blade; e.g., palmately divided, pinnately cleft, etc.

Bifoliolate, Geminate, or Jugate. With two leaflets from a common point.
Bigeminate, Bijugate. With two orders of leaflets, each bifoliolate; doubly paired.
Bipalmately Compound. With two orders of leaflets, each palmately compound.
Bipinnately Compound. With two orders of leaflets, each pinnately compound.
Biternate. With two orders of leaflets, each ternately compound.
Compound. With leaf divided into two or more leaflets.
Decompound. A general term for leaflets in two or more orders-bi-, tri-etc.-pinnately, palmately, or ternately compound.
Imparipinnately Compound. <u>Odd-pinnately</u> compound, with a terminal leaflet.
Interruptedly Pinnately Compound. With smaller and larger leaflets alternating along the rachis.
Palmately Compound. With leaflets from one point at end of petiole.
Palmate-pinnate. With first order leaflets palmately arranged, second order pinnately arranged.
Paripinnately Compound. <u>Even-pinnately</u> compound, without a terminal leaflet.
Pinnately Compound. With leaflets arranged oppositely or alternately along a common axis, the rachis.
Simple. With leaf not divided into leaflets.
Tergeminate. With three orders of leaflets, each bifoliolate, or with geminate leaflets ternately compound.
Ternately Compound. With leaflets in three's.
Trifoliolate. With leaflets three, pinnately compound with terminal petiolule longer than lateral; or palmately compound with petiolules equal in length.
Tripalmately Compound. With three orders of leaflets, each palmately compound.
Tripinnately Compound. With three orders of leaflets, each pinnately compound.
Triternate. With three orders of leaflets, each ternately compound.
Unifoliolate. With a single leaflet with a petiolule distinct from the petiole of the whole leaf, as in <u>Cercis</u>.

C. Leaf Structural Types

(Figure 6-3, p. 95)

Bract. Modified leaf found in the inflorescence.
Bracteole or Prophyllum. Small leaf, usually on a pedicel.
Chaff or Pale. Scale or bract at base of tubular flower in composites.
Complete. Leaf with blade, petiole, and stipules.
Cotyledon. Embryonic leaf.
Elaminate. Without blade.
Epetiolate. Without petiole, leaf sessile.

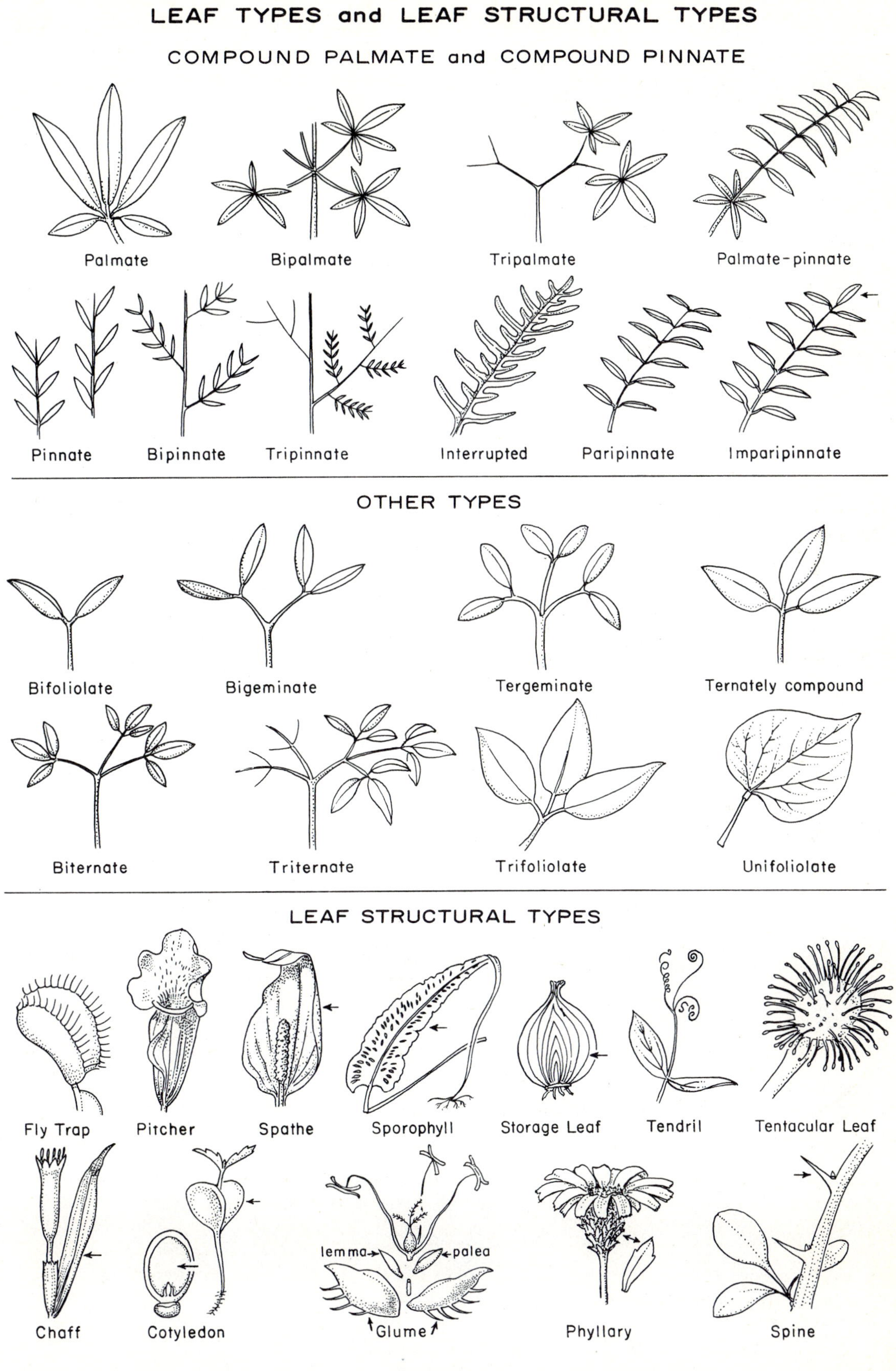
LEAF TYPES and LEAF STRUCTURAL TYPES
COMPOUND PALMATE and COMPOUND PINNATE
Palmate
Bipalmate
Tripalmate
Palmate-pinnate
Pinnate
Bipinnate
Tripinnate
Interrupted
Paripinnate
Imparipinnate
OTHER TYPES
Bifoliolate
Bigeminate
Tergeminate
Ternately compound
Biternate
Triternate
Trifoliolate
Unifoliolate
LEAF STRUCTURAL TYPES
Fly Trap
Pitcher
Spathe
Sporophyll
Storage Leaf
Tendril
Tentacular Leaf
Chaff
Cotyledon
lemma
palea
Glume
Phyllary
Spine

INFLORESCENSE TYPES

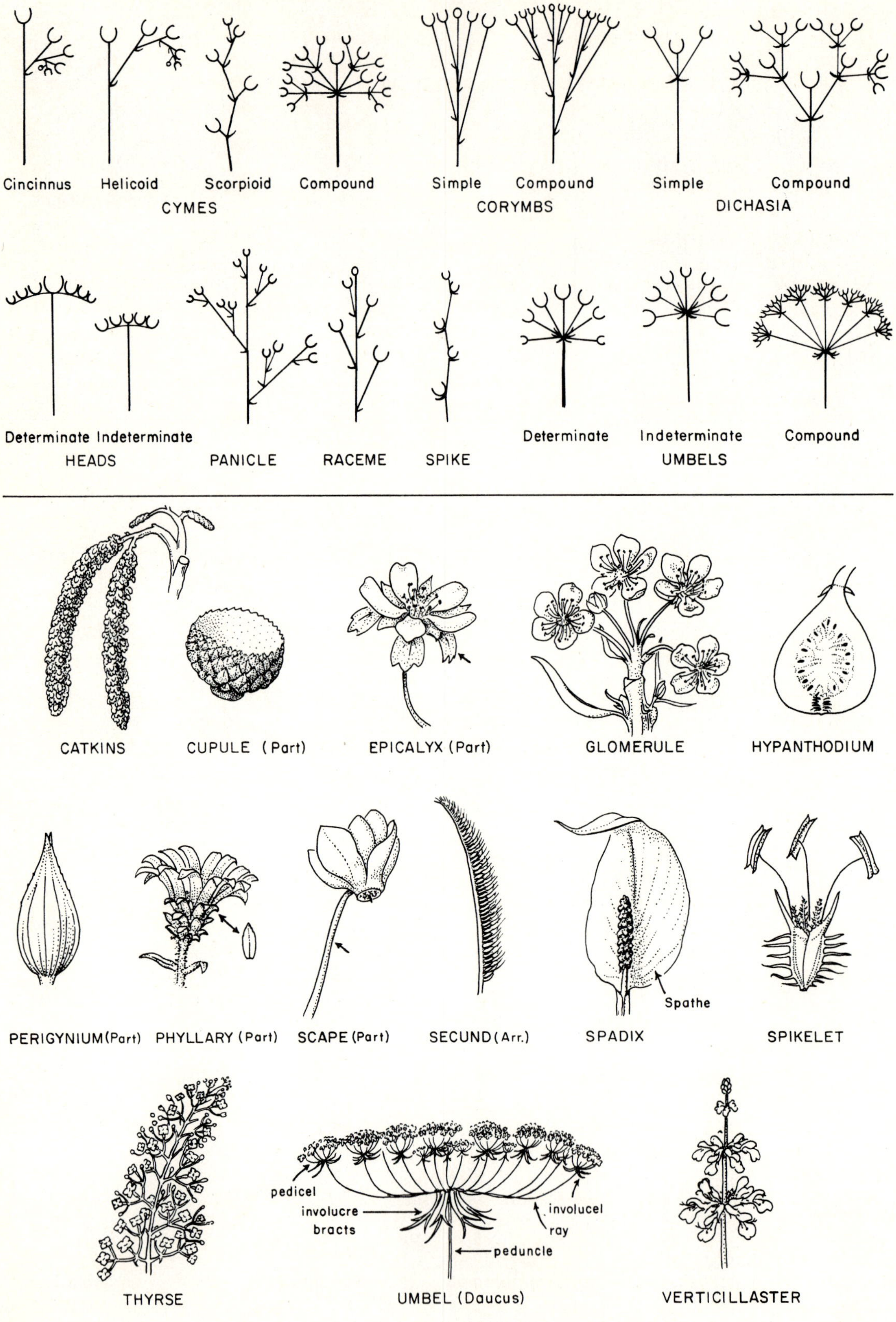

Epetiolulate. Without petiolule, leaflet sessile.
Epicalyx. Group of leaves resembling sepals below the true calyx.
Exstipellate. Without stipels.
Exstipulate. Without stipules.
Fly Trap. Hinged, insectivorous leaf, as in Dionaea.
Glume. Bract, usually occurring in pairs, at the base of the grass spikelet.
Incomplete. Leaf without one or more parts: blade, petiole, stipules.
Lemma. Outer scale subtending grass floret.
Palea. Inner scale subtending grass floret.
Phyllary. One of the involucral leaves subtending a capitulum, as in composites.
Phyllodium. Flattened blade-like petiole or midrib.
Pitcher. Ventricose to tubular insectivorous leaf, as in Sarracenia.
Scale. Small, non-green leaf on bud and modified stem.
Spathe. An enlarged bract enclosing an inflorescence.
Spine. Sharp-pointed petiole, midrib, vein, or stipule.
Sporophyll. A spore bearing leaf.
Storage Leaf. Succulent, fleshy leaf.
Tendril. Usually a coiled rachis or twining leaflet modification.
Tentacular. Glandular-haired or tentacle-bearing insectivorous leaf, as in Drosera.

D. Petiole and Petiolule Structural Types

Channelled or Canaliculate. With a longitudinal groove.
Inflated. Swollen or thickened, as in Eichhornia.
Pericladial. With a sheathing base, as in the Apiaceae.
Petiolate. With a petiole.
Petiolulate. With a petiolule.
Phyllodial. Flattened and blade-like.
Pulvinal. With a swollen base, as in the Fabaceae.
Sessile or Absent. Without petiole or petiolule.
Winged. With flattened blade-like margins.

E. Stipule and Stipel Types

(Classification based primarily on attachment or function)

Adnate. With stipule attached to petiole.
Basal. With stipules attached near base of petiole.
Interpetiolar. With connate stipules from two opposite leaves.
Lateral. With stipules adnate to petiole and free part of stipules located along the petiole.
Median. With stipules adnate to petiole with free part of stipules near middle of petiole.
Photosynthetic. Blade-like and green.
Sheathing or Protective. Enclosing a bud or flower.
Stipellate. With stipels.
Stipulate. With stipules.
Vestigial. Minute; a remnant.

VI. INFLORESCENCES

A. Inflorescence Parts

(Figure 6-4, p. 96)

Bract. Modified, usually reduced, leaf in the inflorescence.
Bracteole or Bractlet. A secondary or smaller bract.
Cupule. Fused involucral bracts subtending flower, as in Quercus.
Epicalyx or Calycle. A whorl of bracts below but resembling a true calyx.
Flower. Modified reproductive shoot of angiosperms.

Involucel. Small involucre; secondary involucre.
Involucre. A group or cluster of bracts subtending an inflorescence.
Pedicel. Individual flower stalk.
Peduncle. Main stalk for entire inflorescence.
Perigynium. Sac-like bract subtending the pistillate flower, as in Carex.
Phyllary. Individual bract within involucre.
Rachilla. Central axis of a grass or sedge spikelet.
Rachis. Major axis within an inflorescence.
Ray. Secondary axis in a compound inflorescence.
Scape. Naked peduncle.
Spathe. A sheathing leaf subtending or enclosing an inflorescence.

B. Inflorescence Types (Figure 6-4, p. 96)

(Classification based primarily on arrangement and development)

Note: Inflorescence types are essentially secondary arrangements. The primary arrangement (alternate, opposite, or whorled) of the individual flowers should be indicated; and the tertiary arrangement of sessile-flowered inflorencences should be noted; e.g., spikelets racemose or heads umbellate. See Chapter 27 for structural evolution of inflorescences. Determinate inflorescences have the central flower maturing first with the arrest of the elongation of the central axis; indeterminate inflorescences have the lateral or lower flowers maturing first without the arrest of the elongation of the central axis.

1. Inflorescences with Sessile Flowers

Ament or Catkin. A unisexual spike or elongate axis with simple dichasia that falls as a unit after flowering or fruiting.
Capitulum or Head. A determinate or indeterminate crowded group of sessile or subsessile flowers on a compound receptacle or torus.
Glomerule. An indeterminate dense cluster of sessile or subsessile flowers.
Hypanthodium. An inflorescence with flowers on wall of a concave capitulum, as in Ficus.
Spadix. Unbranched, indeterminate inflorescence with flowers embedded in the rachis.
Spike. Unbranched, indeterminate, elongate inflorescence with sessile flowers.
Spikelet or Locusta. A small spike; the basic inflorescence unit in grasses and sedges.

2. Unbranched Inflorescences with Pedicellate Flowers

Cincinnus. A tight, modified helicoid cyme in which pedicels are short on the developed side.
Corymb. A flat-topped or convex indeterminate cluster of flowers.
Cymule. A simple, small dichasium.
Helicoid Cyme or Bostryx. A determinate inflorescence in which the branches develop on one side only, appearing simple.
Raceme. Unbranched, indeterminate inflorescence with pedicelled flowers.
Scorpioid Cyme or Rhipidium. A zigzag determinate inflorescence with branches developed on opposite sides of the rachis alternately.
Simple Cyme or Dichasium. A determinate, dichotomous inflorescence with the pedicels of equal length.
Umbel. A determinate or indeterminate flat-topped or convex inflorescence with the pedicels arising at a common point.

3. Branched Inflorescences with Pedicellate Flowers

Compound Corymb. A branched corymb.

Compound Cyme. A branched cyme.

Compound Umbel. An umbel with primary rays or peduncles arising at a common point with a secondary umbel arising from the tip of the primary rays; a branched umbel.

Panicle. Branched inflorescence with pedicelled flowers.

Thyrse. A many-flowered inflorencence with an indeterminate central axis and with many opposite lateral dichasia.

Verticillaster. Whorled dichasia at the nodes of an elongate rachis.

4. General Inflorescence Terms and Types

Cyathium. A pseudanthium subtended by an involucre, frequently with petaloid glands, as in Euphorbia.

Monochasium. A cymose inflorescence with one main axis.

Pleiochasium. Compound dichasium in which each cymule has three lateral branches.

Pseudanthium. Several flowers simulating a simple flower but composed of more than a single axis with subsidiary flowers.

Scapose. With a solitary flower on a leafless peduncle or scape, usually arising from a basal rosette.

Secund. One-sided arrangement.

Solitary. One-flowered, not an inflorescence.

Umbellet. The secondary umbel in a compound umbel.

VII. FLOWERS

A. Flower Parts (Figure 6-5, p. 100)

Accessory Organs. The calyx and corolla.

Androecium. One or more whorls or groups of stamens; all stamens in flower.

Androgynophore. The stipe or column on which stamens and carpels are borne.

Calyx. The lowermost whorl of modified leaves, sepals.

Carpel. The female sporophyll within flower; floral organ that bears ovules in angiosperms; unit of compound pistil.

Clinanthium. The compound receptacle of the composite head.

Corolla. The whorl of petals located above the sepals.

Disc. A discoid structure developed from receptacle at base of ovary or from stamens around the ovary.

Essential Organs. The androecium and gynoecium.

Gynoecium or Pistil. The whorl or group of carpels in the center or at the top of the flower; all carpels in a flower.

Gynophore. The stipe of a pistil or carpel.

Hypanthium. The fused or coalesced basal portion of floral parts (sepals, petals, stamens) around the ovary.

Pedicel. The flower stalk.

Perianth. An aggregation of tepals or combined calyx and corolla.

Petal. A corolla member or segment; a unit of the corolla.

Polyphore. A receptacle or torus bearing many distinct carpels, as in Rosa.

Receptacle or Torus. The region at end of pedicel or on axis to which flower is attached.

Sepal. A calyx member or segment; a unit of the calyx.

Stamen. The male sporophyll within the flower; the floral organ that bears pollen in angiosperms.

FLOWER PARTS

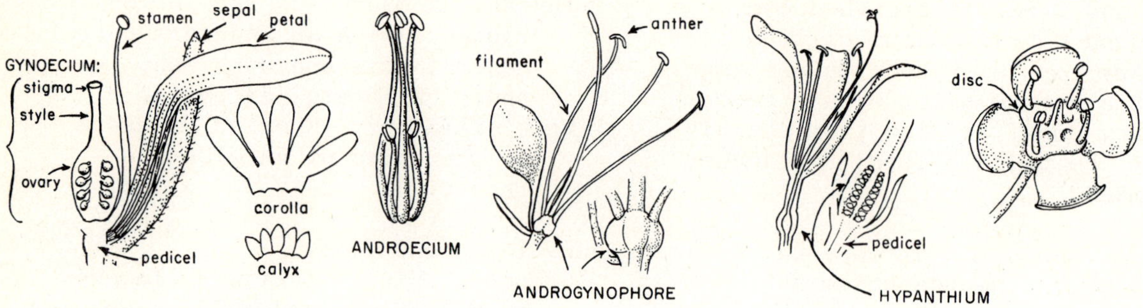

PERIANTH and HYPANTHIUM PARTS

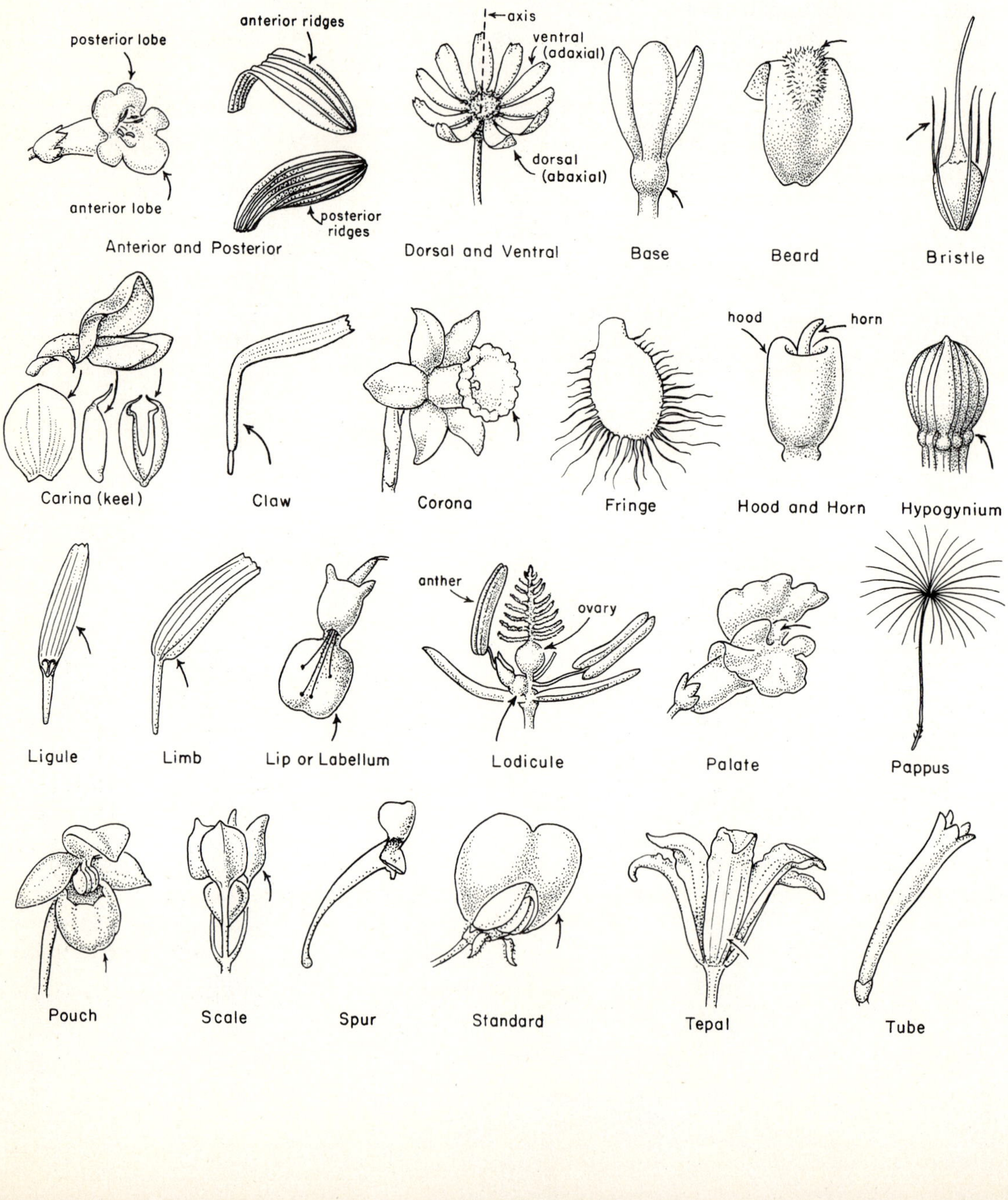

FLOWER TYPES

FLORAL EVOLUTION of ANGIOSPERMS (TYPE CLASSES)

(Adapted from Leppik)

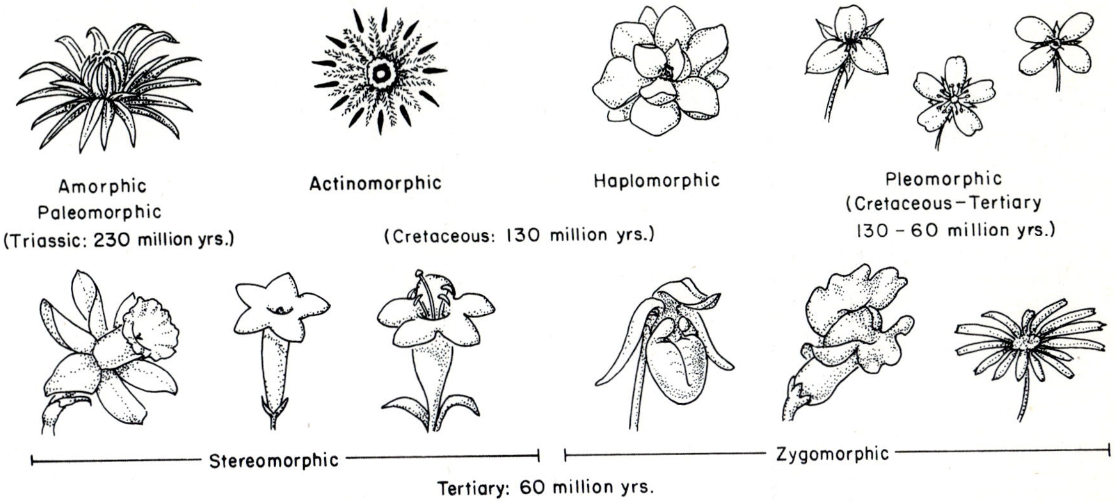

PERIANTH STRUCTURAL TYPES

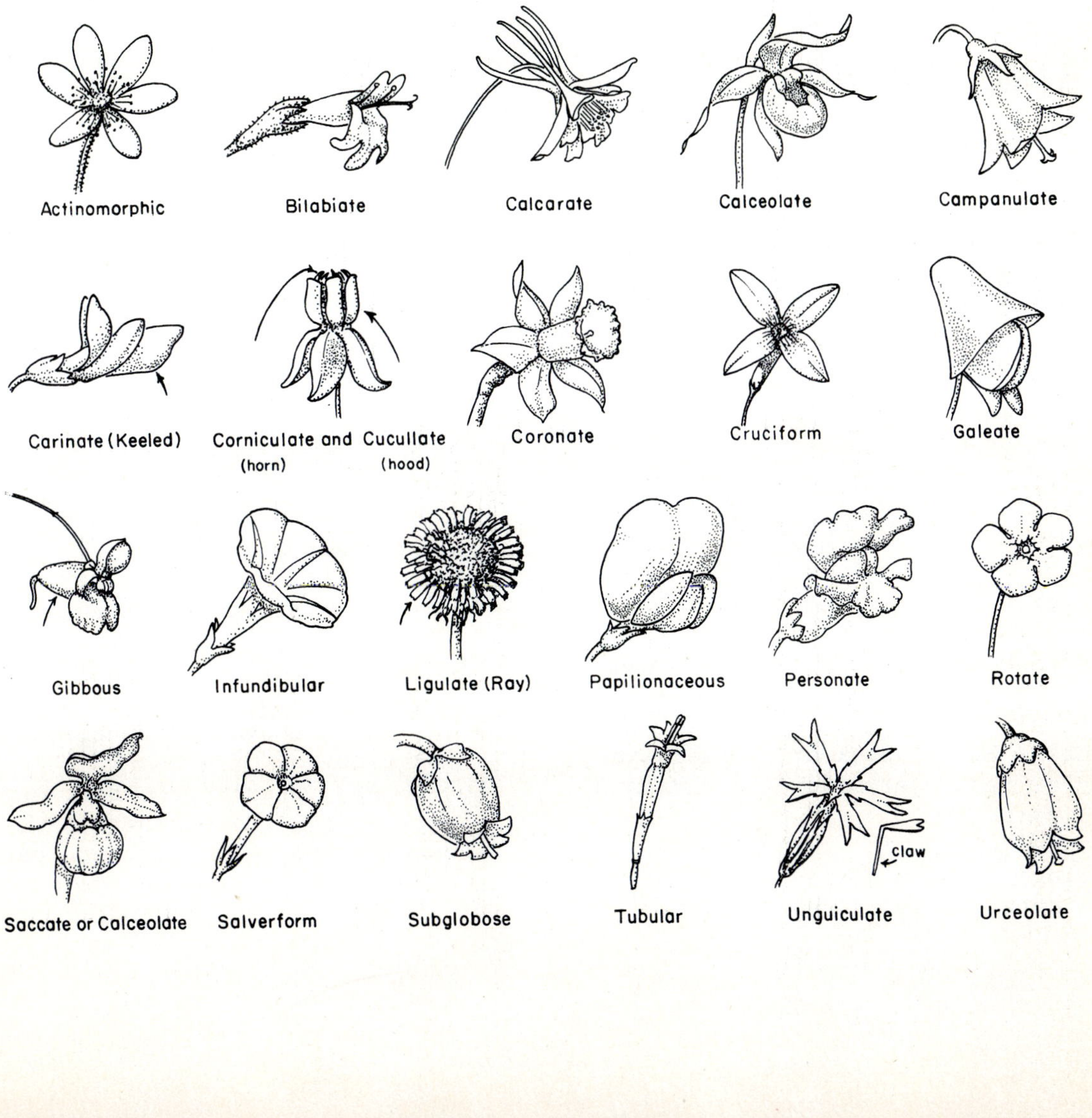

Tepal. A perianth member or segment; term used for perianth parts undifferentiated into distinctive sepals and petals.

Whorl. A cyclic or acyclic group of sepals, or petals, or stamens, or carpels.

B. Flower Types (Figure 6-6, p. 101)

(Classification based on evolutionary flower-pollinator relationships--from Leppik [1957])

Actinomorphic. Flowers with radial symmetry and parts arranged at one level; with definite numbers of parts and size; e.g., Anemone, Caltha.

Amorphic or Paleomorphic. Flowers without symmetry; usually with an indefinite number of stamens and carpels, and usually subtended by bracts or discolored upper leaves; e.g., Salix discolor, Echinops ritro (mostly fossil forms).

Haplomorphic. Flowers with parts spirally arranged at a simple level in a semispheric or hemispheric form; petals or tepals colored; parts numerous; e.g., Nymphaea, Magnolia.

Pleomorphic. Actinomorphic with numbers of parts reduced; e.g., Tripogandra.

Stereomorphic. Flowers 3-dimensional with basically radial symmetry; parts many or reduced, and usually regular; e.g., Narcissus, Aquilegia.

Zygomorphic. Flowers with bilateral symmetry; parts usually reduced in number and irregular; e.g., Cypripedium, Salvia.

VIII. PERIANTH AND HYPANTHIUM

A. Perianth Parts (Figure 6-5, p. 100)

Anterior Lobes. The lobes away from axis, toward the subtending bract; abaxial lobes.

Anterior Ridges, Lines, Grooves. The lines, grooves, ridges in or on the dorsal side, abaxial, within the perianth.

Base. Bottom or lower portion.

Beard. A tuft, line or zone of trichomes.

Bristle. A stiff, strong trichome, as in the perianth of some members of the Cyperaceae.

Callosity. A thickened, raised area, which is usually hard; a callus.

Carina. Keel.

Claw. The long, narrow petiole-like base of a sepal or petal.

Corona. A crown; any outgrowth between the stamens and corolla which may be petaline or staminal in origin.

Dorsal Side. Back or abaxial side, or the lower side of a perianth part.

Faucal Area. The throat area.

Fringe. The modified margin of a petal, sepal, tepal or lip.

Hood. A cover-shaped perianth part, usually with a turned down margin.

Horn. A curved, pointed and hollow protuberance from the perianth.

Hypogynium. Perianth-like structure of bony scales subtending the ovary, as in Scleria and other members of the Cyperaceae.

Keel. The two united petals of a papilionaceous flower; any structure ridged like the bottom of a boat.

Ligule. The strap-shaped portion of a ray or ligulate corolla.

Lip or Labellum. Either of two variously shaped parts into which a corolla or calyx is divided, usually into an upper and lower lip, as in the Lamiaceae and Orchidaceae.

Limb. Expanded portion of corolla or calyx above the tube, throat or claw.

Lobe. Any, usually rounded, segment or part of the perianth.

Lodicule. Abortive perianth part in the Poaceae; hyaline scales at base of ovary in the Poaceae.

Palate. The raised area in the throat of a sympetalous corolla.
Pappus. Bristly or scaly calyx in the Asteraceae.
Petal. A corolla member or segment; a unit of the corolla.
Posterior Lobe. The lobe next to axis, away from the subtending bract; adaxial lobe.
Posterior Ridges, Lines, Grooves. The lines, grooves, ridges in or on the ventral side, adaxial, within the perianth.
Pouch or Sac. A bag-shaped structure.
Scale. Small, scarious to coriaceous flattened bodies within the perianth, as in the Cyperaceae and Asteraceae.
Sepal. A calyx member or segment; a unit of the calyx.
Spur. A tubular or pointed projection from the perianth.
Standard, Banner, or Vexillum. The upper, usually wide petal in a papilionaceous corolla.
Tepal. A member or segment of perianth in which the parts are not differentiated into distinct sepals and petals.
Throat. An open, expanded tube in the perianth.
Tube. The cylindrical part of the perianth.
Ventral Side. Top side or upper side of a perianth part.
Wing. Lateral petals, as in the Fabaceae; a flattened extension, appendage or projection from a perianth part.

B. Hypanthium Parts

Base. Bottom or lower portion of the hypanthium.
Limb. Free, flared portion of the hypanthium.
Neck. Narrowed portion of hypanthium, between the base and a flared limb.
Tube or Casing. Cylindrical part of the hypanthium.

C. Perianth Types

(Classification based primarily on fusion of parts)

Achlamydeous. Without perianth.
Apetalous. No petals or corolla.
Apopetalous or Choripetalous. With separate petals.
Aposepalous or Chorisepalous. With separate sepals.
Asepalous. No sepals or calyx.
Chlamydeous. With perianth.
Dichlamydeous. With perianth composed of distinct calyx and corolla.
Homochlamydeous. With perianth composed of similar parts, each part a tepal.
Sympetalous. With fused petals.
Synsepalous. With fused sepals.

D. Perianth and Hypanthium Structural Types (Figure 6-6, p. 101)

Actinomorphic. With radially arranged perianth parts; ray-like figure.
Bilabiate. Two-lipped, with two unequal divisions.
Calcarate. Spurred.
Calceolate. Slipper-shaped, as in the corolla of *Cypripedium*.
Campanulate. Bell-shaped; with flaring tube about as broad as long and a flaring limb.
Carinate. Keeled.
Corniculate. Horned.
Coronate. Tubular or flaring perianth or staminal outgrowth; petaloid appendage.
Cruciate. Four separate petals in cross form.
Cucullate. Hooded.
Galeate. Helmet-shaped, as one sepal in *Aconitum*.
Gibbous. Inflated on one side near the base.
Globose. Round.
Infundibular. Funnel-shaped.
Ligulate or Ray. Strap-shaped.

Papilionaceous. With large posterior petal (banner or standard) two lateral petals (wings) and usually two connate lower petals (keel); as in the Fabaceae.
Personate. Two-lipped with the upper arched and the lower protruding into corolla throat.
Rotate. Wheel-shaped, with short tube and wide limb at right angles to tube.
Saccate. Pouch-like.
Salverform. Trumpet-shaped. With slender tube and limb nearly at right angles to tube.
Subglobose. Almost round or spherical.
Tubular. Cylindrical.
Unguiculate. Clawed.
Urceolate. Urn-shaped.
Ventricose. Inflated on one side near the middle.

IX. ANDROECIUM

A. Androecial Parts

Stamen. Male sporophyll within the flower; floral organ that bears pollen in angiosperms.
Staminodium. Sterile stamen, may be modified as a nectary or petaloid structure.
Staminal Disc. A fleshy, elevated cushion formed from coalesced staminodia or nectaries.

B. Androecial Types (Figure 6-7, p. 105)

(Classification based primarily on fusion of parts)

Apostemonous. With separate stamens.
Diadelphous. With two groups of stamens connate by their filaments.
Gynandrial or Gynostemial. With fused stamens and carpels (stigma and style) as in the Orchidaceae.
Monadelphous. With one group of stamens connate by their filaments.
Petalostemonous. With filaments fused to corolla, anthers free.
Polydelphous. With several groups of stamens connate by their filaments.
Syngenesious. With fused anthers.

C. Stamen Parts

Anther. Pollen-bearing portion of stamen.
Filament. Stamen stalk.

D. Stamen Structural Types (Figure 6-7, p. 105)

(Classification based primarily on structure of filament and anther)

Note: In this classification intermediate types of stamens do occur; shapes, apices, and bases of anthers should be described separately and independently of stamen type.

Appendicular. Typical stamen with a variously-shaped or modified, protruding connective, as in Viola.
Laminar. Leaf-like stamen without a distinct anther and filament but with embedded or superficial microsporangia, as in Degeneria.
Petalantherous. With a terminal anther and distinctly petaloid filament, as in Saxifraga.
Petaloid. Petal-like stamen without distinct anther and filament but with marginal microsporangia, as in Magnolia nitida.
Filantherous or Typical. Stamen with distinct anther and filament with or without thecal appendages, as in Rhexia or Vaccinium.

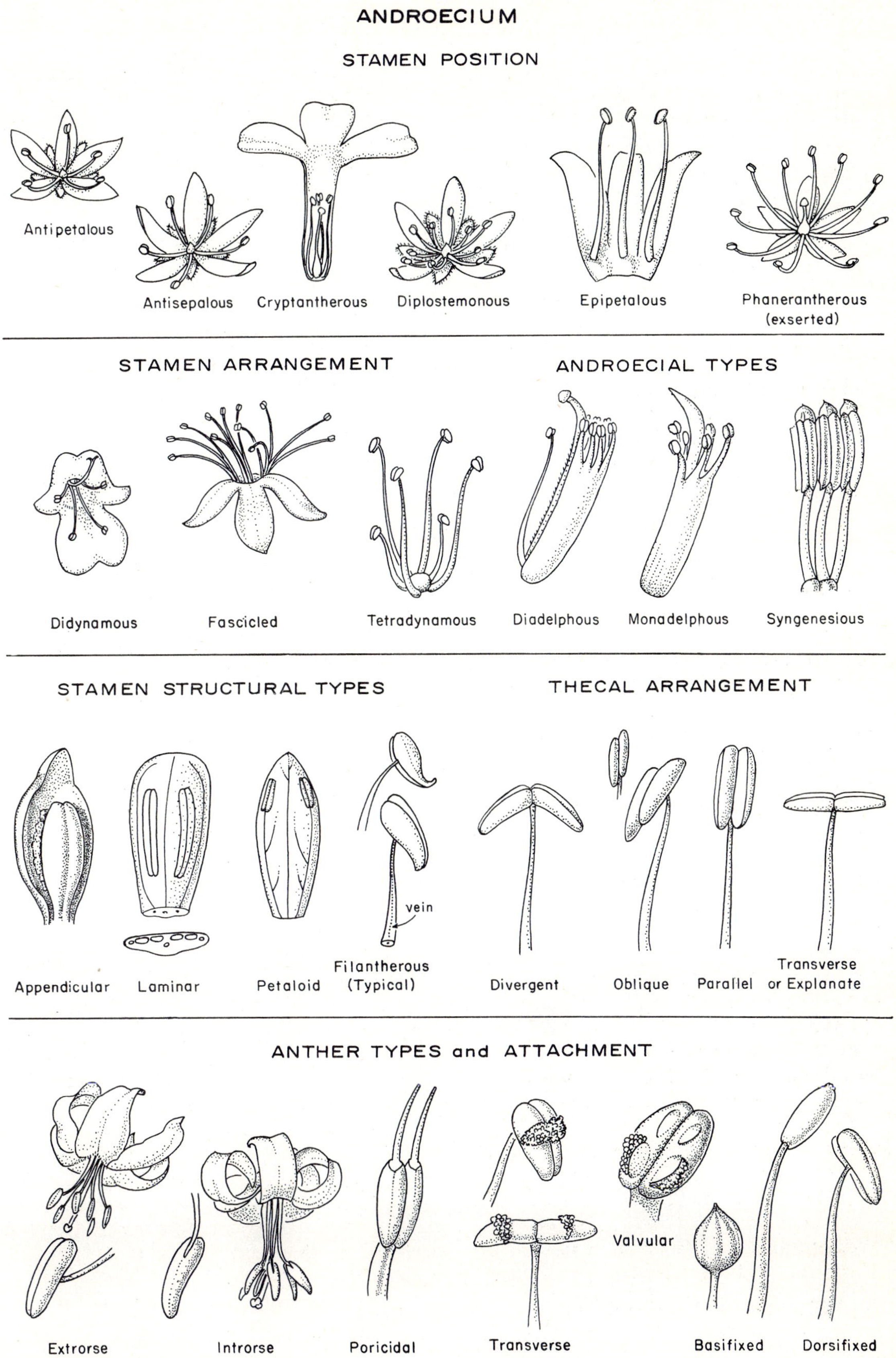
ANDROECIUM
STAMEN POSITION
Antipetalous
Antisepalous
Cryptantherous
Diplostemonous
Epipetalous
Phaneranthérous (exserted)
STAMEN ARRANGEMENT
ANDROECIAL TYPES
Didynamous
Fascicled
Tetradynamous
Diadelphous
Monadelphous
Syngenesious
STAMEN STRUCTURAL TYPES
THECAL ARRANGEMENT
vein
Appendicular
Laminar
Petaloid
Filantherous (Typical)
Divergent
Oblique
Parallel
Transverse or Explanate
ANTHER TYPES and ATTACHMENT
Valvular
Extrorse
Introrse
Poricidal
Transverse
Basifixed
Dorsifixed

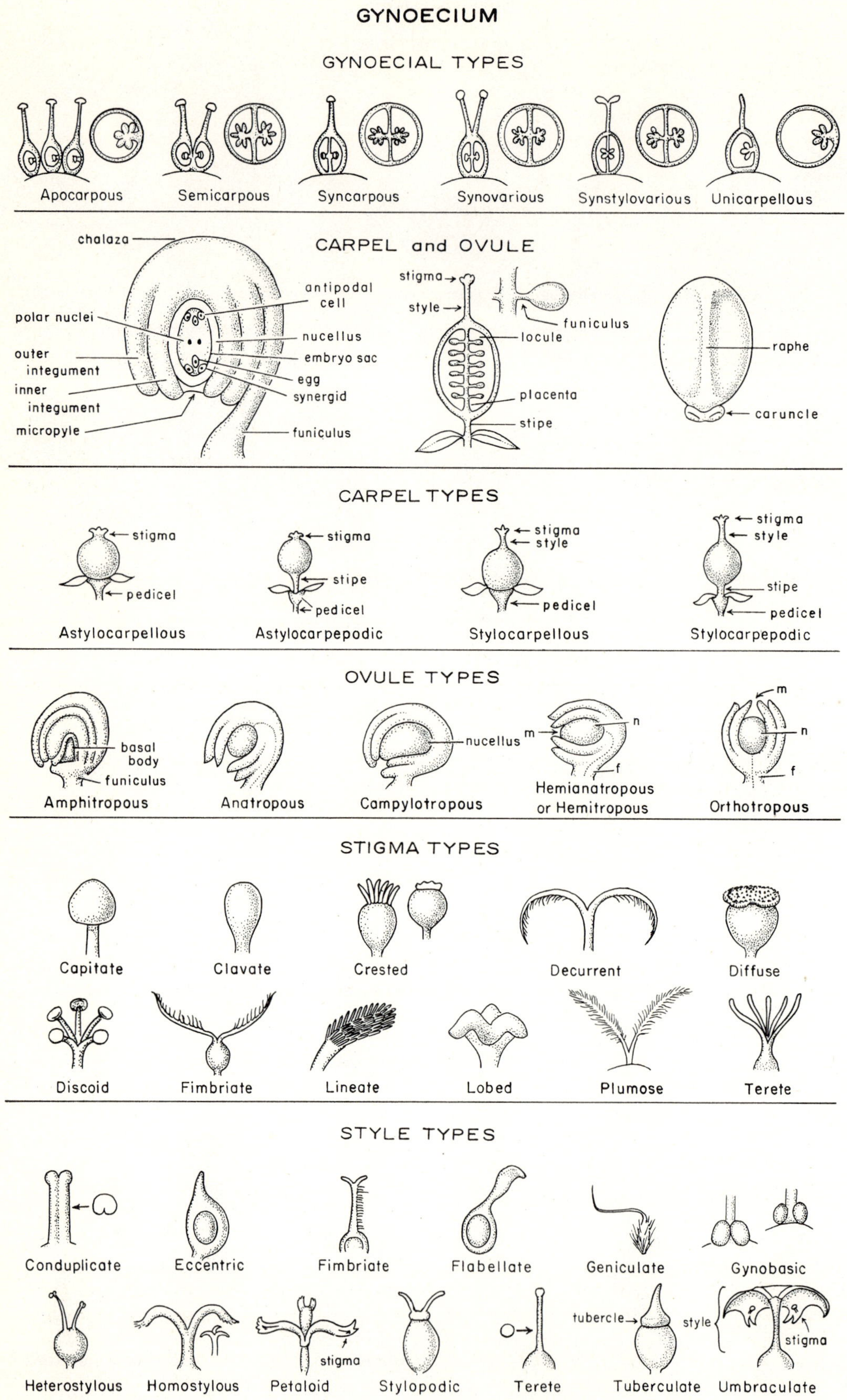
GYNOECIUM
GYNOECIAL TYPES
Apocarpous
Semicarpous
Syncarpous
Synovarious
Synstylovarious
Unicarpellous
CARPEL and OVULE
chalaza
polar nuclei
outer integument
inner integument
micropyle
antipodal cell
nucellus
embryo sac
egg
synergid
funiculus
stigma
style
funiculus
locule
placenta
stipe
raphe
caruncle
CARPEL TYPES
stigma
pedicel
Astylocarpellous
stigma
stipe
pedicel
Astylocarpepodic
stigma
style
pedicel
Stylocarpellous
stigma
style
stipe
pedicel
Stylocarpepodic
OVULE TYPES
basal body
funiculus
Amphitropous
Anatropous
nucellus
Campylotropous
m
n
f
Hemianatropous or Hemitropous
m
n
f
Orthotropous
STIGMA TYPES
Capitate
Clavate
Crested
Decurrent
Diffuse
Discoid
Fimbriate
Lineate
Lobed
Plumose
Terete
STYLE TYPES
Conduplicate
Eccentric
Fimbriate
Flabellate
Geniculate
Gynobasic
Heterostylous
Homostylous
stigma
Petaloid
Stylopodic
Terete
tubercle
style
Tuberculate
stigma
Umbraculate

E. Anther Parts (Figure 6-7, p. 105)

Connective. Filament extension between thecae.
Locule. Compartment of an anther.
Pollen Grain. Young male gametophyte.
Pollen Sac. Male sporangium.
Theca. One half of anther containing two pollen sacs or male sporangia.

F. Anther Types (Figure 6-7, p. 105)
(Classification based on dehiscence)

Longitudinal. Dehiscing along long axis of theca.
- Extrorse. Dehiscing longitudinally outward.
- Introrse. Dehiscing longitudinally inward.
- Latrorse. Dehiscing longitudinally and laterally.

Poricidal. Dehiscing through a pore at apex of theca.
Transverse. Dehiscing at right angles to long axis of theca.
Valvular. Dehiscing through a pore covered by a flap of tissue.

G. Anther Attachment (Figure 6-7, p. 105)

Basifixed. Anther attached at its base to apex of filament.
Dorsifixed. Anther attached dorsally and medially to apex of filament.
Subbasifixed. Anther attached near its base to apex of filament.
Versatile. Dorsifixed but anther seemingly swinging free on the filament.

H. Pollen
(See palynology in Chapter 8)

Dyads. Grains occurring in clusters of two.
Filiform. Thread-like.
Monad. Grains occurring singly.
Pollinia. Grains occurring in uniform coherent masses.
Polyad. Grains occurring in groups of more than four.
Tetrads. Grains occurring in groups of four.

X. GYNOECIUM

A. Gynoecial Parts (Figure 6-8, p. 106)

Carpel. Female sporophyll within flower; floral organ that bears ovules in angiosperms.
Carpopodium. Short, thick, pistillate stalk.
Locule. Ovary cavity.
Ovary. Ovule-bearing part of pistil.
Placenta. Ovule-bearing region of ovary wall.
Stigma. Pollen-receptive portion of pistil.
Stipe. Pistillate stalk.
Style. Attenuated, non-ovule-bearing portion of pistil between stigma and ovary.

B. Gynoecial Types (Figure 6-8, p. 106)
(Classification based on fusion)

Apocarpous. With carpels separate.
Semicarpous. With ovaries of adjacent carpels partly fused, stigmas and styles separate.
Syncarpous. With stigmas, styles, and ovaries completely fused.
Synovarious. With ovaries of adjacent carpels completely fused, styles and stigmas separate.
Synstylovarious. With ovaries and styles of adjacent carpels completely fused, stigmas separate.
Unicarpellous or Stylodious. With solitary, free carpel in gynoecium.

C. Carpel Parts (Figure 6-8, p. 106)

Funiculus. Stalk by which ovule is attached to placenta.
Locule. Ovary cavity.
Ovary. Ovule-bearing part of carpel in simple ovary.
Ovule. Embryonic seed consisting of integument(s) and nucellus.
Placenta. Ovule-bearing region of ovary wall.
Stigma. Pollen receptive portion of carpel.
Stipe, Podogyne, Carpopodium. Basal stalk.
Style. Attenuated portion of carpel between stigma and ovary.

D. Carpel Types (Figure 6-8, p. 106)

(Classification based on presence or absence of style and stipe)

Astylocarpellous. Without a style and a stipe.
Astylocarpepodic. Without a style, with a stipe.
Stylocarpepodic. With a style and stipe.
Stylocarpellous. With a style and without a stipe, the normal carpel.

E. Stigma Types (Figure 6-8, p. 106)

(Classification based on shape)

Capitate. Head-like.
Clavate. Club-shaped.
Crested or Cristate. With a terminal ridge or tuft.
Decurrent. Elongate, extending downward.
Diffuse. Spread over a wide surface.
Discoid. Disc-like.
Fimbriate. Fringed.
Lineate. In lines, stigmatic surface linear.
Lobed. Divided into lobes.
Plumose. Feather-like.
Terete. Cylindrical and elongate.

F. Style Types (Figure 6-8, p. 106)

(Classification based primarily on shape)

Astylous. Style absent.
Conduplicate. Folded with a longitudinal groove.
Cristate. Crested.
Eccentric. Off-center style.
Fimbriate. Fringed.
Flabellate. Fan-shaped.
Geniculate. Bent abruptly.
Gynobasic. Attached at base of ovary in central depression.
Heterostylous. With styles of different sizes or lengths or shapes within a species.
Homostylous. With styles of same sizes or lengths and shapes.
Involute. With margins infolded longitudinally, with groove present.
Petaloid. Petal-like.
Stylopodic. With a stylopodium or discoid base, as in the Apiaceae.
Terete. Cylindrical and elongate.
Tuberculate. With hard, swollen, persistent base or tubercle.
Umbraculate. Umbrella-shaped, as in *Sarracenia*.

G. Ovule Parts (Figure 6-8, p. 106)

Chalaza. End of ovule opposite micropyle.
Embryo Sac. Female gametophyte.
Integuments. Outer covering of ovule; embryonic seed coat.
Micropyle. Hole through integument(s).
Nucellus. Female sporangium within ovule; megasporangium in seed plants.
Raphe. Longitudinal ridge on outer integument.

H. Ovule Types (Figure 6-8, p. 106)

(Classification based on orientation of ovule body in relation to the funiculus and micropyle)

Amphitropous. With body bent or curved on both sides so that the micropyle is near the medially attached funiculus.

Anatropous. With body completely inverted so that funiculus is attached basally near adjoining micropyle area.

Campylotropous. With body bent or curved on one side so that micropyle is near medially attached funiculus.

Hemianatropous or Hemitropous. With body half-inverted so that funiculus is attached near middle with micropyle terminal and at right angles.

Orthotropous or Atropous. With straight body so that funicular attachment is at one end and micropyle at other.

XI. FRUITS

A. Fruit Parts

Carpophore. Floral axis extension between adjacent carpels, as in the Apiaceae.

Ectocarp or Exocarp. Outermost layer of pericarp.

Endocarp. Innermost differentiated layer of pericarp.

Funiculus. Seed stalk.

Mericarp. A portion of fruit that seemingly matured as a separate fruit.

Mesocarp. Middle layer of pericarp.

Pericarp. Fruit wall.

Placenta. Region of attachment of seeds on inner fruit wall.

Replum. Persistent septum after dehiscence of fruits, as in the Brassicaceae.

Retinaculum, Jaculator or Echma. A persistent indurated, hook-like funiculus in the fruits of Acanthaceae.

Rostellum or Beak. Persistent stylar base on fruit.

Seed. A matured ovule.

Septum or Dissepiment. Partition.

B. Fruit Structural Types

(Classification based primarily on origin, texture, and dehiscence; types grouped as simple, aggregate, multiple, accessory.)

1. Simple Fruits

(Fruit derived from the ovary of a solitary pistil in a single flower)

a. Dry Indehiscent Fruit Types (Figure 6-9, p. 112)

(Fruits that do not split open at maturity)

Achene. A one-seeded, dry, indehiscent fruit with seed attached to fruit wall at one point only, derived from a one-loculed superior ovary.

Balausta. Many-seeded, many-loculed indehiscent fruit with a tough, leathery pericarp, as in *Punica*.

Calybium. A hard one-loculed dry fruit derived from an inferior ovary, as in *Quercus*.

Capsule, Indehiscent. Dry fruit derived from a two- or more loculed ovary, as in *Peplis*.

Caryopsis or Grain. A one-seeded dry, indehiscent fruit with the seed coat adnate to the fruit wall, derived from a one-loculed superior ovary.

Cypsela. An achene derived from a one-loculed, inferior ovary.

Nut. A one-seeded, dry, indehiscent fruit with a hard pericarp, usually derived from a one-loculed ovary.

Nutlet. A small nut.

Samara. A winged, dry fruit.

Utricle. A small, bladdery or inflated, one-seeded, dry fruit.

b. Dry Dehiscent Fruit Types (Figure 6-9, p. 112)

(Fruits that split open at maturity)

Capsule. Dry, dehiscent fruit derived from a compound ovary of 2 or more carpels.

Diplotegium. A pyxis derived from an inferior ovary.

Follicle. A dry, dehiscent fruit derived from one carpel that splits along one suture.

Legume. A usually dry, dehiscent fruit derived from one carpel that splits along two sutures.

Loment. A legume that separates transversely between seed sections.

Silicle. A dry, dehiscent fruit derived from two or more carpels that dehisce along two sutures and which has a persistent partition after dehiscence and is as broad as, or broader, than long.

Silique. A silicle type fruit that is longer than broad.

c. Capsule Types (Figure 6-9, p. 112)

(Classification based on type of dehiscence)

Acrocidal Capsule. One that dehisces through terminal slits, or fissures, as in *Staphylea*.

Anomalicidal or Rupturing Capsule. One that dehisces irregularly, as in *Ammannia*.

Basicidal Capsule. One that dehisces through basal slits or fissures, as in some species of *Aristolochia*.

Circumscissle Capsule or Pyxis. One that dehisces circumferentially, as in *Plantago*.

Denticidal Capsule. One that dehisces apically, leaving a ring of teeth, as in *Cerastium*.

Indehiscent Capsule. One that does not dehisce at maturity, as in *Peplis*.

Loculicidal Capsule. One that dehisces longitudinally into the cavity of the locule, as in *Epilobium*.

Operculate Capsule. One that dehisces through pores, each of which is covered by a flap, cap, or lid, as in *Papaver*.

Poricidal Capsule. One that dehisces through pores, as in *Triodanis*.

Septicidal Capsule. One that dehisces longitudinally through the septa, as in *Penstemon*.

Valvular or Septifragal Capsule. One with valves breaking away from the septa, as in *Ipomoea*.

d. Schizocarpic Fruit Types (Figure 6-9, p. 112)

(Fruits derived from a simple, two- or more-locular compound ovary in which the locules separate at fruit maturity simulating fruits derived from the ovaries of simple pistils)

Schizocarpic Achenes. Separating achenes which are one-seeded, dry, indehiscent fruits with seed attached to fruit wall at one point only, derived from a superior ovary, as in *Sidalcea*.

Schizocarpic Berries. Separating berries which have a fleshy pericarp, as in *Phytolacca*.

Schizocarpic Carcerules. Separating carcerules which are dry, few-seeded, indehiscent locules, as in *Althaea*.

Schizocarpic Follicles. Separating follicles which are dry, dehiscent fruits derived from one carpel, splitting along one suture, as in Apocynaceae.

Schizocarpic Mericarps, Cremocarp, or Carpopodium. Separating mericarps which are dry, seed-like fruits derived from an inferior ovary, as in the Apiaceae.

Schizocarpic Nutlets, Cenobium. Separating nutlets which are dry, indehiscent 4-parted fruits with a hard pericarp around a gynobasic style, as in the Boraginaceae and Lamiaceae.

Schizocarpic Samaras. Separating samaras which are winged, dry fruits, as in _Acer_.

e. _Fleshy Fruit Types_ (Figure 6-9, p. 112)

Amphisarca. A berry-like succulent fruit with a crustaceous or woody rind, as in _Lagenaria_.

Berry. Fleshy fruit, with succulent pericarp, as in _Vitis_.

Drupe. A fleshy fruit with a stony endocarp, as in _Prunus_.

Drupelet. A small drupe, as in _Rubus_.

Hesperidium. A thick-skinned septate berry with the bulk of the fruit derived from glandular hairs, as in _Citrus_.

Pepo. A berry with a leathery non-septate rind derived from an inferior ovary, as in _Cucurbita_.

Pyrene. Fleshy fruit with each seed surrounded by a bony endocarp, as in _Ilex_.

2. _Aggregate Fruit Types_ (Conocarpium) (Figure 6-10, p. 113)

(A group of separate fruits developed from carpels of one flower)

Achenecetum. An aggregation of achenes, as in _Ranunculus_.

Baccacetum or Etaerio. An aggregation of berries, as in _Actaea_.

Drupecetum. An aggregation of drupelets, as in _Rubus_.

Follicetum. An aggregation of follicles, as in _Caltha_.

Samaracetum. An aggregation of samaras, as in _Liriodendron_.

3. _Multiple Fruit Types_

(Fruits on a common axis that are usually coalesced and derived from the ovaries of several flowers)

Bibacca. A fused double berry, as in _Lonicera_.

Sorosis. Fruits on a common axis that are usually coalesced and derived from the ovaries of several flowers, as in _Morus_.

Syconium. A syncarp with the achenes borne on the inside of a hollowed-out receptacle or peduncle, as in _Ficus_.

4. _Accessory Fruit Types_ (Figure 6-10, p. 113)

(Fruits derived from simple or compound ovaries and some non-ovarian tissues, as the hypanthium; classification arranged alphabetically; types of accessory structures given in parentheses below.)

Bur (Involucre). Cypsela enclosed in dry involucre, as in _Xanthium_.

Coenocarpium (Various Structures). Multiple fruit derived from ovaries, floral parts, and receptacles of many coalesced flowers, as in _Ananas_.

Diclesium (Calyx). Achene or nut surrounded by a persistent calyx, as in _Mirabalis_.

Glans (Involucre). Nut subtended by a cupulate, dry involucre, as in _Quercus_.

Hip or Cynarrhodion (Receptacle and Hypanthium). An aggregation of achenes surrounded by an urceolate receptacle and hypanthium, as in _Rosa_.

Pome (Receptacle and Hypanthium). A berry-like fruit, adnate to a

FRUIT TYPES

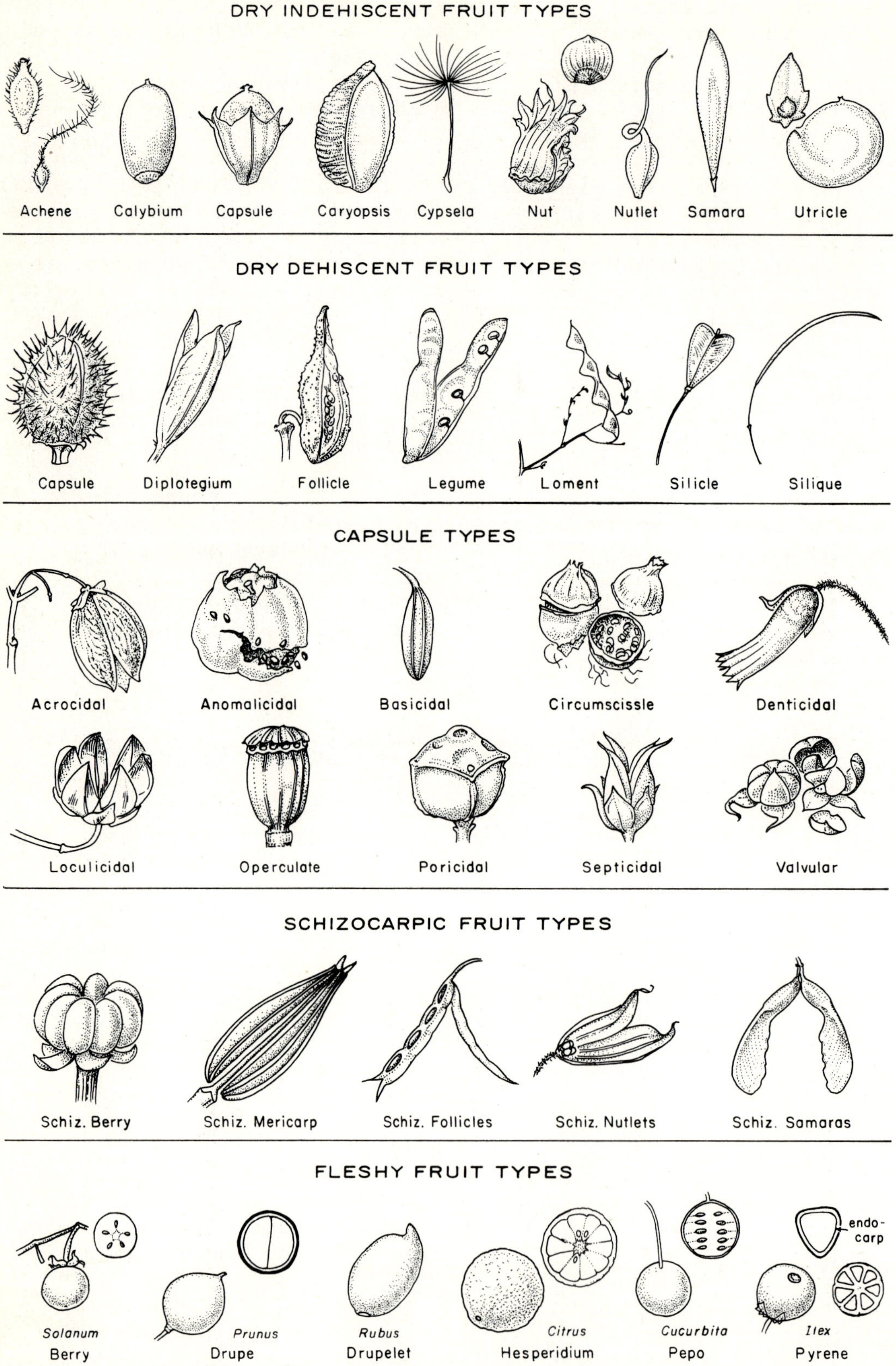
DRY INDEHISCENT FRUIT TYPES
Achene
Calybium
Capsule
Caryopsis
Cypsela
Nut
Nutlet
Samara
Utricle
DRY DEHISCENT FRUIT TYPES
Capsule
Diplotegium
Follicle
Legume
Loment
Silicle
Silique
CAPSULE TYPES
Acrocidal
Anomalicidal
Basicidal
Circumscissle
Denticidal
Loculicidal
Operculate
Poricidal
Septicidal
Valvular
SCHIZOCARPIC FRUIT TYPES
Schiz. Berry
Schiz. Mericarp
Schiz. Follicles
Schiz. Nutlets
Schiz. Samaras
FLESHY FRUIT TYPES
endo-
carp
Solanum
Berry
Prunus
Drupe
Rubus
Drupelet
Citrus
Hesperidium
Cucurbita
Pepo
Ilex
Pyrene

FRUIT TYPES, Cont'd., SEED PARTS and EMBRYO TYPES

AGGREGATE FRUIT TYPES

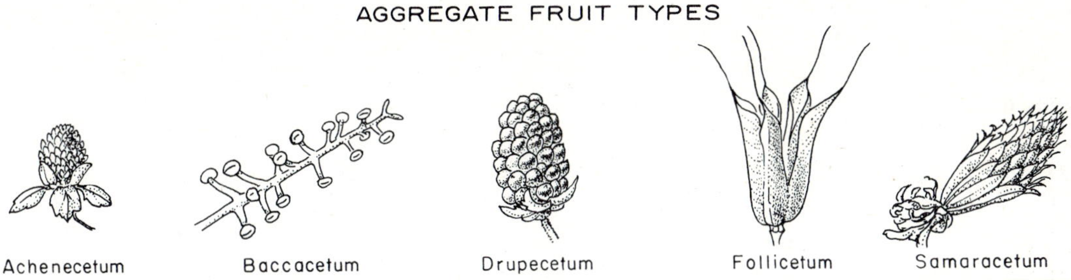

ACCESSORY FRUIT TYPES

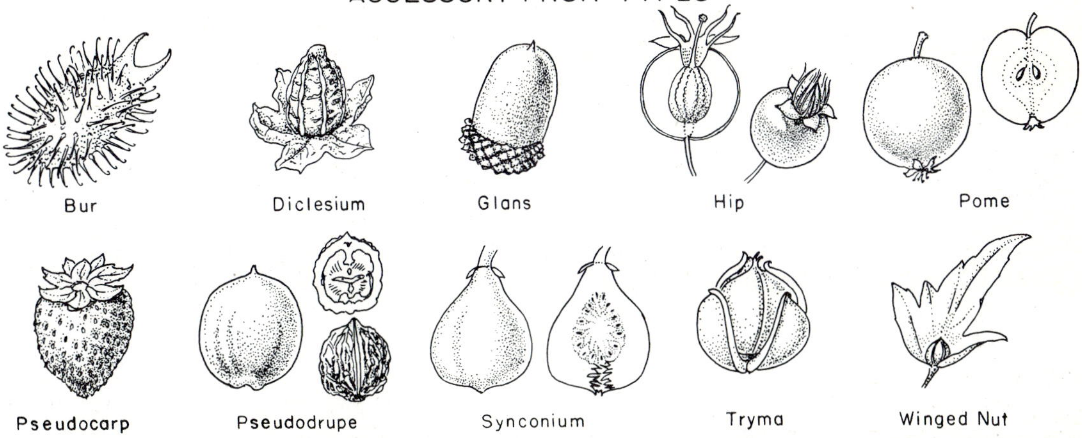

SEED PARTS

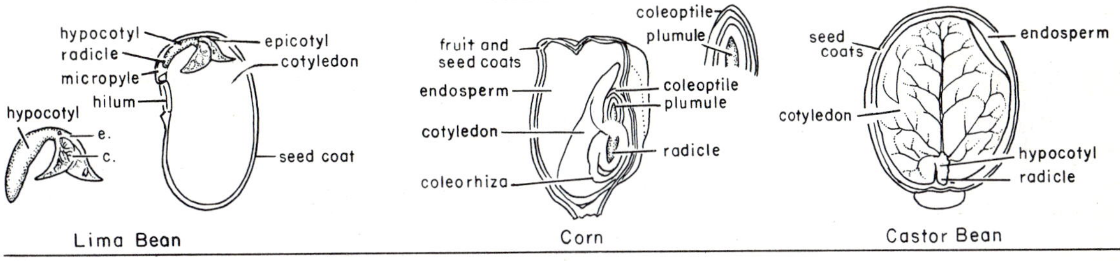

EMBRYO TYPES

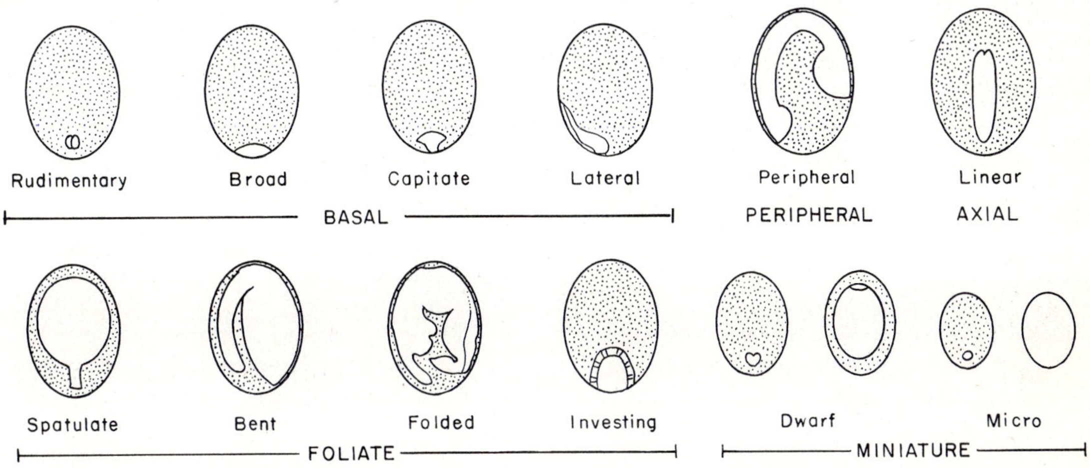

fleshy receptacle, with cartilaginous endocarp, as in Malus.

Pseudocarp (Receptacle). An aggregation of achenes embedded in a fleshy receptacle, as in Fragaria.

Pseudodrupe (Involucre). Two-four loculed nut surrounded by a fleshy involucre, as in Juglans.

Syconium (Receptacle, possibly peduncle). Multiple fruit surrounded by a hollow, compound, fleshy receptacle, as in Ficus.

Tryma (Involucre). Two-four loculed nut surrounded by a dehiscent involucre at maturity, as in most species of Carya.

Winged Nut (Bract). Nut enclosed in a winglike bract, as in Carpinus.

XII. SEEDS AND SEEDLINGS

A. Seed Parts (Figure 6-10, p. 113)

Aril. Outgrowth of funiculus, raphe, or integuments; or fleshy integuments or seed coat, a sarcotesta.

Chalaza. Funicular end of seed body.

Embryo. Young sporophyte consisting of epicotyl, hypocotyl, radicle, and one or more cotyledons.

Endosperm. Food reserve tissue in seed derived from fertilized polar nuclei; or food reserve derived from megametophyte in gymnosperms.

Hilum. Funicular scar on seed coat.

Micropyle. Hole through seed coat.

Perisperm. Food reserve in seed derived from diploid nucellus or integuments.

Raphe. Ridge on seed coat formed from adnate funiculus.

Seed Coat. Outer protective covering of seed.

B. Seed Types

(Classification based on type of nourishing tissue)

Cotylespermous. With food reserve in cotyledon, derived from zygote.

Endospermous or Albuminous. With food reserve in endosperm or albumen, derived from fertilized polar nuclei.

Hypocotylespermous or Macropodial. With food reserve stored in hypocotyl, derived from zygote.

Perispermous. With food reserve in perisperm, derived from diploid nucellus or integuments.

C. Embryo Parts (Figure 6-10, p. 113)

Coleoptile. Protective sheath around epicotyl in grasses.

Coleorhiza. Protective sheath around radicle in grasses.

Cotyledon. Embryonic leaf or leaves in seed.

Epicotyl. Apical end of embryo axis that gives rise to shoot system.

Hypocotyl. Embryonic stem in seed, located below cotyledons.

Plumule. Embryonic leaves in seed derived from epicotyl.

Radicle. Basal end of embryo axis that gives rise to root system.

D. Embryo Types (Figure 6-10, p. 113)

(Classification based primarily on embryo shape, size, and position--adapted from Martin [1946])

Bent. Foliate embryo with expanded and usually thick cotyledons in an axile position bent upon the hypocotyl in a jacknife position.

Broad. Basal, globular or lenticular embryo in copious endosperm.

Capitate. More or less basal headlike or turbinate embryo in copious endosperm.

Dwarf. Axial embryo variable in size relative to seed, small to nearly total size of seed; seeds 0.2-2 mm. long.

Folded. Foliate embryo with cotyledons usually thin and extensively expanded and folded in various ways.

Investing. Axial embryo usually erect with thick cotyledons overlapping and encasing the somewhat dwarfed hypocotyl; endosperm wanting or limited.

Lateral. Basal or baso-lateral embryo, discoid or lenticular, usually surrounded by copious endosperm.

Linear. Axial embryo several times longer than broad, straight, curved or coiled; cotyledons not expanded; endosperm present or absent.

Micro. Axial embryo in minute seeds, less than 0.2 mm. long; minute and undifferentiated to almost total size of seed.

Peripheral. Peripheral embryo large and elongate, arcuate, annular, spirolobal, or straight; cotyledons narrow or expanded; perisperm central or lateral.

Rudimentary. Basal, small non-peripheral embryo in small to large seed; relatively undifferentiated; endosperm copious.

Spatulate. Foliate, erect embryo with variable cotyledons, thin to thick and slightly expanded to broad.

E. Aril Structural Types and Selected Seed Surface Features

1. Aril Structural Types

Arillate. General term for an outgrowth from the funiculus, seed coat or chalaza; or a fleshy seed coat.

Carunculate. With an excrescent outgrowth from integuments near the hilum, as in Euphorbia.

Fibrous. With stringy or cord-like seed coat, as mace in Myristica.

Funicular. With a persistent elongate funiculus attached to seed coat, as in Magnolia.

Sarcous. With the seed coat fleshy.

Strophiolate. With elongate aril or strophiole in the hilum region.

2. Special Seed Surface Features

(See Surface in Section B of this chapter)

Alate. Winged.

Circumalate. Winged circumferentially.

Comose. With a tuft of trichomes.

Coronate. With a crown.

Crested. With elevated ridge or ridges, raphal.

Umbonate. With a distinct projection usually from the side.

Verrucose. Warty.

F. Seedling Parts

(Specialized parts only--adapted from Duke [1969])

Cataphyll. Rudimentary scale leaf produced by seedling, usually in cryptocotylar species.

Collet. External demarcation between hypocotyl and root.

Eophyll. Term applied to first few leaves with green, expanded lamina developed by seedlings; transitional type leaves developed before formation of adult leaves.

Metaphyll. Adult leaf.

G. Seedling Types

(Classification based on position of cotyledons in germination)

Cryptocotylar or Hypogeous. With the cotyledons remaining inside the seed; seed usually remaining below ground.

Phanerocotylar or Epigeous. With the cotyledons emergent from seed, usually appearing above ground.

Section B. GENERAL CHARACTERS AND CHARACTER STATES

I. POSITION AND ARRANGEMENT

A. Location or Environmental Position

(Classification based on position of organs or parts in their surrounding environment)

1. General

Aerial or Epigeous. Above the ground or water; in the air.

Emergent. With part(s) of plant aerial and part(s) submersed; rising out of the water above the surface.

Epipetric. Upon rock.

Epiphytic. Upon another plant.

Floating. Upon the surface of the water.

Submersed. Beneath the surface of the water.

Subterranean or Hypogeous. Below the surface of the ground.

Surficial or Epigeous. Upon or spread over the surface of the ground.

2. Special

(Selected location terms. See habitat prefixes in Chapter 15 and word stems for plant organs and parts in Chapter 4 for meanings of other location terms.)

Aerocaulous. With aerial stems.

Aerophyllous. With aerial leaves.

Amphicarpous. With fruits in two environments; e.g., aerial and subterranean.

Emersifolious. With emergent leaves.

Epirhizous. With roots upon another plant.

Flotophyllous. With floating leaves.

Geoflorous. With subterranean flowers.

Petrorhizous. With roots on rock.

Submersicaulous. With submersed stems.

Surcarpous. With fruits on surface of ground.

B. Position

(Classification based on location of parts or organs with respect to other dissimilar parts, or organs)

1. General

Apical or Terminal. At the top, tip, or end of a structure.

Basal or Radical. At the bottom or base of a structure.

Continuous. Basal, lateral, and terminal.

Discontinuous. Basal and lateral, basal and terminal, or lateral and terminal; not continuous.

Lateral or Axillary. On the side of a structure or at the nodes of the axis.

2. Special

(Classification based on positional terms usually applicable to individual parts)

a. Androecial Position

(See Perianth and Stamen Position, h. and k.)

b. Branch Position

Acrocaulous. With terminal branches.
Basicaulous. With basal branches.
Caulous. With branches more or less evenly spaced along trunk.
Subacrocaulous. With branches at or near tip of main stem.
Subbasicaulous. With branches at or near base of main stem.
Zonocaulous. With branches intermittently spaced along main stem.

c. Cotyledon Position

(Figure 6-12, p. 121)

Accumbent or Pleurorhizal. Reclinate with cotyledon edges against hypocotyl.
Incumbent or Notorhizal. Reclinate with sides of cotyledons against hypocotyl.

d. Flower, Fruit, Inflorescence, Infructescence Position

Acrocaulous. At the tip of the stem.
Amphiflorous, Amphicarpous. Flowers or fruits above and below ground, as in *Amphicarpum*.
Axillary. In axil of leaf.
Basicaulous. Near base of stem.
Cauline or Caulous. On old woody stem.
Epiphyllous. From a phylloclad or peculiar bract, as in *Tilia*.
Geoflorous, Geocarpous. Flowers or fruits below ground, as in *Amphicarpum*.
Infrafoliar. On the stem below the leaves, as in the Arecaceae.
Interfoliar. On the stem between the leaves, as in the Arecaceae.
Leaf-opposed. On stem opposite the base of the leaf, as in *Alchemilla*.
Suprafoliar. On the stem above the leaves, as in the Arecaceae.
Terminal. At or near tip of branch.

e. Leaf Position

Acroramous. Leaves terminal, near apex of branch.
Aphyllopodic. Without blade-bearing leaves at base of plant.
Basiramous. Leaves on lower part of branch.
Cauline or Ramous. Leaves more or less evenly distributed on stem or branch.
Phyllopodic. With blade-bearing leaves at base of plant.
Radical. Leaves basal, near ground, usually from caudex or rootstock.

f. Ovary Position

Inferior. Other floral organs attached above ovary with hypanthium adnate to ovary.
Half-inferior. Other floral organs attached around ovary with hypanthium adnate to lower half of ovary.
Superior. Other floral organs attached below ovary.

g. Ovule Position

(Figure 6-12, p. 121)

(Based on position of ovule in locule and orientation of the micropyle and raphe--adapted from Björnstad [1970])

Epitropous, dorsal. Ovule pendulous or hanging, micropyle above, raphe dorsal (away from ventral bundle).
Epitropous, ventral. Ovule pendulous or hanging, micropyle above, raphe ventral (toward ventral bundle.

Heterotropous. Ovule position not fixed in ovary.

Hypotropous, dorsal. Ovule erect, micropyle below, raphe dorsal (away from ventral bundle).

Hypotropous, ventral. Ovule erect, micropyle below, raphe ventral (toward ventral bundle).

Pleurotropous, dorsal. Ovule horizontal, micropyle toward ventral bundle, raphe above.

Pleurotropous, ventral. Ovule horizontal, micropyle toward ventral bundle, raphe below.

h. Perianth and Androecium Position (Figure 6-11, p. 120)

(Classification based on insertion of Floral Parts--Corolla, Calyx, and Androecium--the androperianth)

Epigyny. The condition in which the sepals, petals, stamens are attached to the floral tube above the ovary with the ovary adnate to the tube or hypanthium.

Epihyperigyny. The condition in which the sepals, petals, stamens are attached to the floral tube or hypanthium surrounding the ovary; a combination perigyny and partly inferior ovary.

Epihypogyny. The condition in which the sepals, petals, stamens are attached about half-way from the base of the ovary to the partly adnate hypanthium tube; half-inferior insertion of parts.

Epiperigyny. The condition in which the sepals, petals, stamens are attached to the floral or hypanthium cup above the ovary with the lower part of the hypanthium completely adnate to the ovary.

Hypanepigyny. The condition in which the sepals, petals, stamens are attached to the elongate floral tube or hypanthium above the inferior ovary, as in Oenothera.

Hypogyny. The condition in which the sepals, petals, stamens are attached below the ovary.

Perigyny. The condition in which the sepals, petals, stamens are attached to the floral tube or hypanthium surrounding the ovary with the tube or hypanthium free from the ovary.

i. Placenta Position (Placentation) (Figure 6-11, p. 120)

Axile. With the placentae along the central axis in a compound ovary with septa.

Basal. With the placenta at the base of the ovary.

Free-central. With the placenta along the central axis in a compound ovary without septa.

Laminate. With the placenta over the inner surface of the ovary wall.

Marginal or Ventral. With the placenta along the margin of the simple ovary.

Parietal. With the placentae on the wall or intruding partitions of a unilocular compound ovary.

Pendulous, Apical, or Suspended. With the placenta at the top of the ovary.

j. Radicle Position

Antitropous. With radicle pointing away from hilum.

Syntropous. With radicle pointing toward hilum.

k. Stamen Position (Figure 6-7, p. 105)

Allagostemonous. Having stamens attached to petal and torus alternately.

Antipetalous. Opposite the petals.

Antisepalous. Opposite the sepals.

Cryptantherous. With stamens included.

Diplostemonous. With stamens in two whorls, outer opposite the sepals, inner opposite petals.

Epipetalous. With stamens attached to or inserted upon petals or corolla.
Episepalous. With stamens attached or inserted upon sepals or calyx.
Obdiplostemonous. With stamens in two whorls, outer opposite petals, inner opposite the sepals.
Phanerantherous. With stamens exserted.

1. Style Position

Gynobasic. At the base of an invaginated ovary.
Lateral. At the side of an ovary.
Subapical. At one side near apex of ovary.
Terminal or Apical. At the apex of the ovary.

C. Arrangement

(Classification based on location of parts in relation to each other)

1. General (Figure 6-16, p. 136)

Alternate. One leaf or other structure per node.
Clustered, Conglomerate, Agglomerate, Crowded, Aggregate. Parts dense, usually irregularly overlapping each other.
Continuous. Symmetry of arrangement even, not broken.
Decussate. Opposite leaves at right angle to preceding pair.
Distichous. Leaves 2-ranked, in one plane.
Equitant. Leaves 2-ranked with overlapping bases, usually sharply folded along midrib.
Fasciculate. Leaves or other structures in a cluster from a common point.
Geminate or Binate. Paired; in pairs.
Imbricate. Leaves or other structures overlapping.
Interrupted or Discontinuous. Symmetry of arrangement broken, with uneven lengths of internodes.
Loose, Distant, or Scattered. Parts widely separated from one another, usually irregularly.
Opposite. Two leaves or other structures per node, on opposite sides of stem or central axis.
Polystichous. Leaves or other structures in many rows.
Rosulate. Leaves in a rosette.
Secund or Unilateral. Flowers or other structures on one side of axis.
Tetrastichous. Leaves or other structures in four rows.
Tristichous. Leaves or other structures in three rows.
Whorled, Radiate, or Verticillate. Three or more leaves or other structures per node.

2. Special

(Classification based on arrangement with special terms applicable to individual plant parts)

a. Stamen Arrangement (Figure 6-7, p. 105)

(See General Arrangement for additional terms)

Didymous. With stamens in two equal pairs.
Didynamous. With stamens in two unequal pairs.
Tetradynamous. With stamens in two groups, usually four long and two short.
Tridynamous. With stamens in two equal groups of three.

POSITION and POSTURE

PERIANTH and ANDROECIAL POSITION

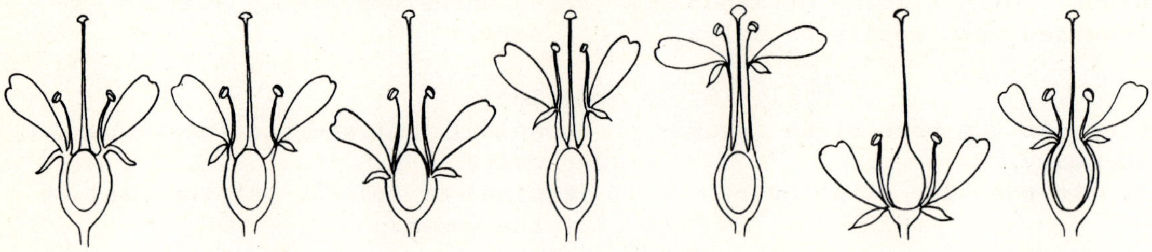

PLACENTAL POSITION

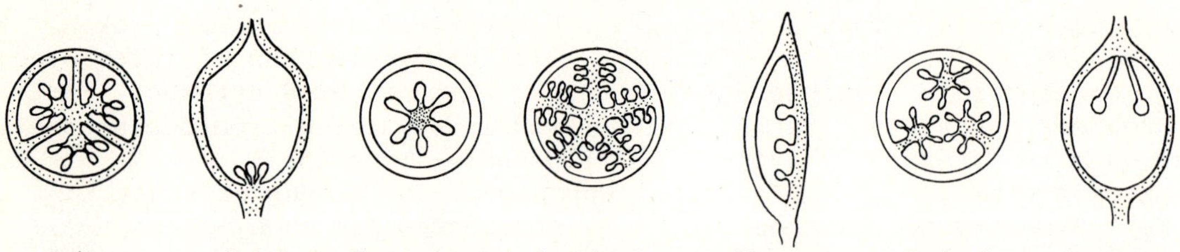

TRANSVERSE POSTURE

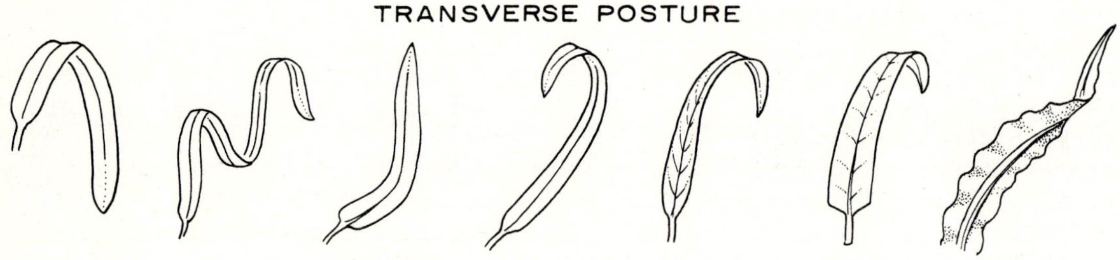

LONGITUDINAL POSTURE

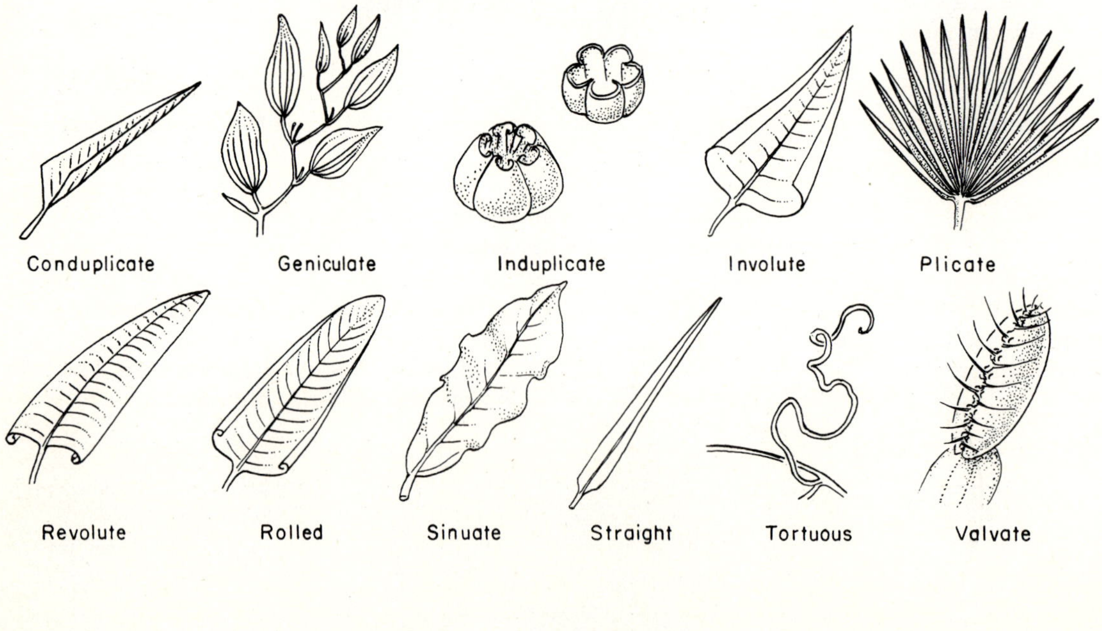

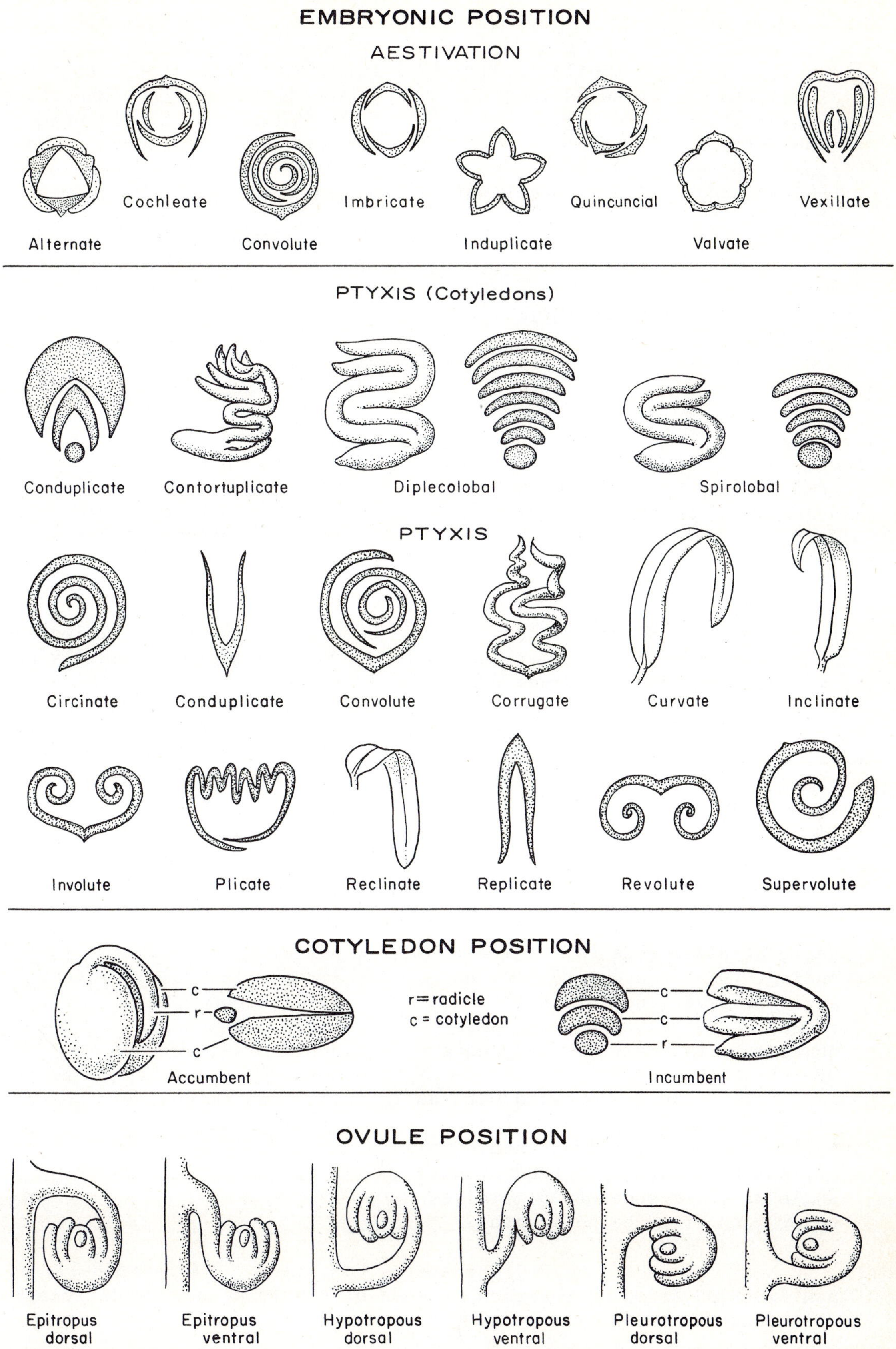
EMBRYONIC POSITION
AESTIVATION
Alternate
Cochleate
Convolute
Imbricate
Induplicate
Quincuncial
Valvate
Vexillate
PTYXIS (Cotyledons)
Conduplicate
Contortuplicate
Diplecolobal
Spirolobal
PTYXIS
Circinate
Conduplicate
Convolute
Corrugate
Curvate
Inclinate
Involute
Plicate
Reclinate
Replicate
Revolute
Supervolute
COTYLEDON POSITION
c
r
c
Accumbent
r = radicle
c = cotyledon
c
c
r
Incumbent
OVULE POSITION
Epitropus dorsal
Epitropus ventral
Hypotropous dorsal
Hypotropous ventral
Pleurotropous dorsal
Pleurotropous ventral

b. Thecal Arrangement (Figure 6-7, p. 105)

(The thecae in this classification can be conjunctive or disjunctive.)

Divergent. Thecae or anther cells divaricate or separated from one another at an acute angle to the connective or filament.

Oblique. Thecae or anther cells lower on one side of connective than the other.

Parallel. Thecae or anther cells along side of the connective or longitudinal to each other.

Transverse or Explanate. Thecae or anther cells with maximum divergence of about 90° from the connective or filament.

D. Orientation

(Classification based on arrangement of parts in relation to vertical angle of divergence from a central axis or point)

1. General

Acroscopic. Facing apically.

Agglomerate, Conglomerate, Crowded, or Aggregate. Dense structures with varied angles of divergence.

Antrorse. Bent or directed upward.

Assurgent. Directed upward or forward.

Basiscopic. Facing basally.

Connivent. Convergent apically without fusion.

Contorted. Twisted around a central axis; twisted.

Declinate. Directed or curved downward.

Deflexed. Bent abruptly downward.

Dextrorse. Rising helically from right to left, a characteristic of twining stems.

Inflexed. Bent abruptly inward or upward.

Patent. Spreading.

Pendulous. Hanging loosely or freely.

Reclinate. Bent down upon the axis, no angle of divergence.

Reflexed. Bent or turned downward.

Retrorse. Bent or directed downward.

Salient, Porrect, or Projected. Pointed outward, usually said of teeth.

Sinistrorse. Rising helically from left to right, a characteristic of twining stems.

Twining. Twisted around a central axis.

2. Special

(Classification based on stated degrees of divergence)

Appressed or adpressed. Pressed closely to axis upward with angle of divergence 15° or less.

Ascending. Directed upward with an angle of divergence of 16-45°.

Depressed. Pressed closely to axis downward with angle of divergence of 166-180°.

Descending. Directed downward with an angle of divergence of 136-165°.

Divergent, Patent, or Divaricate. More or less horizontally spreading with angle of divergence of 15° or less up or down from the horizontal.

Horizontally. Spreading outward at 90° from vertical axis or plane.

Inclined. Ascending at 46-75° angle of divergence.

Reclined. Descending at 106-135° angle of divergence.

Resupinate. Inverted or twisted 180°, as in pedicels in the Orchidaceae.

E. Transverse Posture (Figure 6-11, p. 120)

(Classification based on position of ends of single structure in relation to its center or transverse axis)

Applanate or Plane. Flat, without vertical curves or bends.
Arcuate. Curved like a crescent, can be downward or upward.
Cernuous. Drooping.
Flexuous. With a series of long or open vertical curves at right angles to the central axis.
Geniculate. Abruptly bent vertically, usually near the base.
Incurved. Curved inward or upward.
Lorate. With elongate vertical waves in the margins or sides at right angles to the longitudinal axis.
Recurved. Curved outward or downward.
Squarrose. Usually sharply curved downward or outward in the apical region, as the bracts of some species of Aster.
Undulate. With a series of vertical curves at right angles to the central axis.

F. Longitudinal Posture (Figure 6-11, p. 120)

(Classification based on position of the sides of a single structure in relation to its central axis)

Conduplicate. Longitudinally folded upward or downward along the central axis so that ventral and/or dorsal sides face each other.
Geniculate. Abruptly bent horizontally, usually in series.
Induplicate. Having margins bent inward and touching margin of each adjacent structure.
Involute. Margins or outer portion of sides rolled inward over upper or ventral surface.
Plicate. With a series of longitudinal folds; plaited.
Revolute. Margins or outer portion of sides rolled outward or downward over lower or dorsal surface.
Rolled. Sides enrolled, usually loosely, over upper or lower surfaces.
Sinuate. Long horizontal curves in the body of the structure parallel to the central axis.
Straight. Without a curve, bend, or angle.
Tortuous. Irregularly twisted.
Valvate. Sides enrolled, adaxially or abaxially so that margins touch.

G. General Structural Position

(Pertains to regional locations on a structure)

1. General

Abaxial. Away from the axis; the lower surface of the leaf; dorsal.
Adaxial. Next to the axis; facing the stem; ventral.
Apical. At or near the tip.
Basal. At or near the bottom.
Central. In the middle or middle plane of a structure.
Circumferential. At or near the circumference; surrounding a rounded structure.
Distal. Away from the point of origin or attachment.
Dorsal. Pertaining to the surface most distant from the axis; back of an outer face of organ; lower side of leaf; abaxial.
Marginal. Pertaining to the border or edge.
Medial. Upon or along the longitudinal axis.
Peripheral. On the outer surface or edge.
Proximal. Near the point of origin or attachment.
Subbasal. Near the base.
Subterminal. Near the apex.
Ventral. Pertaining to the surface nearest the axis; inner face of an organ; the upper surface of the leaf; adaxial.

2. Special

(Selected terms for location on a structure. The meanings of many additional terms can be determined from the positional prefixes and word stems of plant organs and parts in Chapter 4.)

Acrocaulous. At tip of stem.
Basipetiolar. At the base of the petiole.
Centroramous. At the center of the branch.
Dorsilaminar. On dorsal side of blade.
Laterospermous. On the side of the seed.
Pericarpous. Around the fruit.
Suprarhizous. On top of the root.
Ventristipular. On ventral side of stipule.

H. Embryonic Position

(Position of immature organs or parts)

1. Aestivation or Prefloration (Figure 6-12, p. 121)

(Classification based on position of embryonic perianth parts. Calyx and corolla may have different aestivation types.)

Alternate. Having structure in two rows or series so that the inner structure has its margins overlapped by a margin from each adjacent outer structure.
Cochleate. Having one hollow or helmet-shaped structure which encloses or covers the others.
Contorted. Having several structures in a whorl or close spiral with one margin covering the margin of an adjacent structure.
Convolute. Having one leaf or perianth part rolled in another, usually twisted apically.
Imbricate. Having margins overlapping.
Induplicate. Having margins bent inward and touching margin of adjacent structure.
Quincuncial. Having five structures, two of which are exterior, two interior, and a fifth with one margin covering interior structure and other margin covered by that of one of the exterior structures.
Valvate. Having margins of adjacent structures touching at edges only.
Vexillate. Having one structure larger than others which is folded over smaller enclosed structures.

2. Cotyledon Ptyxis (Figure 6-12, p. 121)

(Classification based on position of cotyledons in seed)

Conduplicate. Cotyledons folded lengthwise along midrib with one cotyledon covering other and inner cotyledon covering hypocotyl.
Contortuplicate. With weirdly folded corrugate cotyledons.
Diplecolobal. With incumbent cotyledons folded two or more times.
Spirolobal. With incumbent cotyledons folded once.

3. Ptyxis (Figure 6-12, p. 121)

(Classification based on rolling or folding of individual embryonic leaves and arrangement of embryonic leaves within a structure; vernation according to most authors). Adapted from Davis and Heywood (1963).

Circinate. With lamina rolled from apex to base with apex in center of coil.
Conduplicate. With lamina folded once adaxially along midrib or midvein.
Convolute. With one lamina enrolled in another lamina.
Corrugate. With lamina irregularly folded in all directions, wrinkled.
Curvative or arcuate. With lamina folded transversely into an arc.

Implicate. With both lamina margins folded sharply inward.
Inclinate. With lamina folded or curved transversely near the apex.
Involute. With lamina margins enrolled adaxially.
Planate or Plain. With lamina flat, without folds or rolls.
Plicate. With many longitudinal folds in lamina.
Reclinate. With lamina folded or curved backwards from near its base so that embryonic blade is parallel to its petiole, hypocotyl, or stem.
Replicate. With lamina folded once abaxially along midrib or midvein.
Revolute. With lamina margins enrolled abaxially.
Supervolute. With lamina with one edge tightly enrolled and with the other loosely enrolled covering the first, loosely convolute.

II. NUMBER AND SIZE

A. Number

(Pertains to selected terms dealing with numbers. See number prefixes and word stems in Chapter 4 for meanings of additional terms.)

Ancipital. Two-edged.
Bicarpellate. Two-carpelled.
Bidentate. Two-toothed.
Biflorous. Two-flowered.
Bifoliate. Two-leaved.
Bilabiate. Two-lipped.
Bilocular. Two-locular.
Binate. Twinned.
Biseriate. Two-rowed; in two series.
Bisexual. Both sexes in same flower (monoclinous, perfect).
Diadelphous. With two groups of stamens connate by their filaments.
Diandrous. With two stamens per flower.
Dichasium. Cymose inflorescence in which each axis produces a pair of lateral axes.
Dichlamydeous. With two perianth parts, a distinct calyx and corolla.
Diclinous. Having the stamens and carpels in separate flowers, imperfect, either monoecious or dioecious.
Dicotyledonous. With two cotyledons.
Dioecious. With staminate and carpellate flowers on separate plants.
Dipterous. Two-winged.
Dyad. Pollen grains occurring in clusters of two.
Monad. Pollen grains occurring singly.
Monadelphous. With one group of stamens connate by their filaments.
Monocarpellate. One-carpelled.
Monocephalous. One-headed, as in composites.
Monochasium. Cymose inflorescence in which each central axis bears one lateral axis.
Monochlamydeous. With perianth composed of similar parts, each being a tepal; with one whorl of accessory parts.
Monocotyledonous. With one cotyledon.
Monoecious. With staminate and carpellate flowers on same plant.
Monophyllous. One-leaved.
Multicellular. Many-celled.
Multicipital. With many axes or stems from one rootstock or caudex.
Multicostal. Many-ribbed.
Multilocular. Many-locular.
Multiseriate. Many-rowed; in many series.
Multistriate. Many-lined.
Pentagonal. Five-angled.
Pentandrous. With five stamens.
Polyad. Pollen grains in clusters of more than four.
Polyandrous. Many-stamened.
Polycarpellate. Many-carpellate.
Polycephalous. Many-headed, as in composites.
Polydelphous. With several groups of stamens connate by their filaments.

Tetrad. Pollen grains in clusters of four.
Tetragonal. Four-angled.
Tetrahedral. Having the form of a tetrahedron.
Tetralocular. Four-locular.
Tetrandrous. With four stamens.
Tricarpellate. Three-carpellate.
Triflorous. Three-flowered.
Trifoliate. Three-leaved.
Trifoliolate. With three leaflets.
Trigonous. Three-angled.
Triquetrous. Three-angled with the sides usually concave.
Unilocular. One-locular.
Uniseriate. One-rowed; in one series.
Unisexual. With only one sex in each flower.

B. Size

(Selected terms related to size and frequently shape. See size prefixes and word stems of organs and parts in Chapter 4 for meanings of additional terms.)

Ampliate. Enlarged; dilated.
Angustate. Narrow.
Anisocarpous. With unequal carpels.
Anisocotylous. With unequal cotyledons.
Anisolateral. With unequal sides.
Anisopetalous. With unequal petals.
Anisophyllous. With unequal leaves.
Anisostylous. With unequal styles.
Depauperate. Small and usually poorly developed.
Dilated. Widened; expanded.
Dwarf. Very small.
Gigantic. Very large.
Heterandrous. With stamens of different sizes and/or shapes.
Heteroblasty. With juvenile foliage distinctly different from adult foliage in size or shape.
Heterocarpous. With carpels of different sizes and/or shapes.
Heterocladous. With stems of different sizes and/or shapes.
Heteropetalous. With petals of different sizes and/or shapes.
Heterophyllous. With leaves of different sizes and/or shapes.
Heterosepalous. With sepals of different sizes and/or shapes.
Heterostichus. With unequal rows.
Homandrous. With stamens of same size and shape.
Homocarpous. With carpels of same size and shape.
Hypophyllous. With small leaves, as bracts, scales, cataphylls.
Inequilateral. With unequal sides.
Isocotylous. With cotyledons of same size and shape.
Isodynamous. With equally developed structures.
Isopetalous. With petals of same size and shape.
Isophyllous. With leaves of same size and shape.
Isosepalous. With sepals of same size and shape.
Isostichous. With equal rows.
Latiflorous. With broad-flowers.
Leptophyllous. With leaves to 25 sq. mm. in size.
Major. Greater in size.
Minor. Smaller in size.
Minute. Very small.
Nanophyllous. With leaves to 225 sq. mm. in size.
Platycanthous. With flat and usually large spines.
Reduced. Decreased in size.
Robust. Large.

C. Cycly

(Pertains to number of whorls of floral parts, leaves, or stems)

Acarpous. No carpels or carpellate whorl; no pistil.
Achlamydeous. Without perianth.
Apetalous. No petals or corolla.
Aphyllous. Without leaves, no whorls of leaves.
Arhizous. Without roots, no whorls of roots.
Asepalous. No sepals or calyx.
Astemonous or Anandrous. No stamens or androecium.
Chlamydeous. With perianth.

Complete. With four types of floral parts.
Dichlamydeous. With perianth composed of distinct calyx and corolla.
Dicyclic. Two-whorled.
Homochlamydeous. With perianth composed of similar parts, each part a tepal.
Incomplete. One or more types of floral parts absent.
Monocyclic. One-whorled.
Oligotaxy. Reduction in number of whorls.
Pentacyclic. Five-whorled.
Pleiotaxy. Increase in number of whorls.
Polycyclic. Many-whorled.
Tetracyclic. Four-whorled.
Tricyclic. Three-whorled.

D. Merosity

(Pertains to number of parts within whorls of floral parts, leaves, or stems)

Dimerous. Whorl with two members.
Heteromerous or Anisomerous. With different number of members in different whorls.
Isomerous. With same number of members in different whorls.
Monomerous. Whorl with one member.
Oligomerous. With reduction in number of members within whorl.
Pentamerous. Whorl with five members.
Pleiomerous. With increase in number of members within whorl.
Polymerous. Whorl with many members.
Pseudomonomerous. Whorl seemingly with one member which is a fusion product of two or more parts.
Tetramerous. Whorl with four members.
Trimerous. Whorl with three members.

E. Fusion

(Pertains to fusion of members within and between whorls of floral parts)

1. General

(Based on Porter et al., [1973])

Adherent. With unlike parts of organs joined, but only superficially and without actual histological continuity.
Adnate. With unlike parts or organs integrally fused to one another with histological continuity.
Coalesced. With like or unlike parts or organs incompletely separated; partially fused in a more or less irregular fashion.
Coherent. With like parts or organs joined, but only superficially and without actual histological continuity.
Connate. With like parts or organs integrally fused to one another with histological continuity.
Contiguous. Touching but not adnate, connate, adherent, or coherent.
Distinct. With like parts or organs unjoined and separate from one another.
Fasciated. Unnaturally and often monstrously connate or adnate, the coalesced parts often unnaturally proliferated in size and/or number; e.g., inflorescence of Celosia.
Free. Unlike parts or organs unjoined and separate from one another.

2. Special

(Selected terms pertaining to fusion)

Anthocarpous. Having a body of combined floral and fruit parts, as in multiple fruits.
Apocarpous. With separate carpels.
Apopetalous or Choripetalous. With separate petals.
Aposepalous or Chorisepalous. With separate sepals.
Apostemonous. With separate stamens.

Column, Gynostemium or Gynandrium. With fused stamens and carpels (stigma and style) as in Orchis.
Conjugate. Fused pairs, as the fruits of Lonicera.
Diadelphous. With two groups of stamens connate by their filaments.
Hypanthium. Fused floral parts forming an envelope around the ovary.
Monadelphous. With one group of stamens connate by their filaments.
Petalostemonous. With filaments fused to corolla, anthers free.
Polydelphous. With several groups of stamens connate by their filaments.
Sympetalous. With fused petals.
Syncarpous. With fused carpels.
Syncotyly. Cotyledons coalesced, forming a funnel or trumpet.
Synsepalous. With fused sepals.
Syngenesious. With fused anthers.

3. Hypanthium Adnation
(Based on fusion with ovary)

Absent. No hypanthium present.
Adnate. Hypanthium completely fused to ovary.
Free. Hypanthium surrounding but completely free from the ovary.
Free-adnate. Hypanthium fused with ovary and having a free limb around or above ovary.
Partly adnate. Hypanthium adnate to part of the ovary and with no free limb or tube.

F. Division

Bifid. Cut or divided into two lobes or parts.
Bifurcate. Divided into two forks or branches.
Cleft. Cut 1/4-1/2 of distance of middle of structure.
Dichotomous. Divided into two equal parts.
Dimidiate. Divided into unequal halves.
Dissected. Irregularly cut into numerous segments.
Divided. Cut 3/4 to almost entire distance to middle of structure.
Furcate. Forked.
Incised. Margins sharply and deeply cut, usually jaggedly.
Lacerate. Irregularly cut, appearing torn.
Laciniate. Cut into closely parallel ribbonlike or straplike projections.
Lobed. Round-toothed, cut 1/8-1/4 of distance to middle of structure.
Palmatifid. Cut palmately.
Palmatisect. Sectioned or divided palmately into distinct segments.
Parted. Cut 1/2-3/4 of distance to middle of structure.
Pectinate. Having closely parallel toothlike projections; comblike.
Pedate, Bipalmate. Palmately cleft or divided with lateral lobes cleft or divided.
Pinnatifid. Cut pinnately.
Pinnatisect. Sectioned or divided pinnately into distinct segments.
Quadrifid. Cut or divided into four lobes or parts.
Septate. Divided by internal partitions into locules or cells.
Serrated. Cut into sawlike teeth.
Trifid. Cut or divided into three lobes or parts.
Trifurcate. Divided into three forks or branches; three-forked.

III. SHAPES

A. Shapes--Plane and Solid (Figure 6-13, p. 131)

1. Symmetric Figures. Based on terminology in Taxon (Vol. 11) and Stearn (1966).
   A. Elliptic. With widest axis at midpoint of structure and with margins symmetrically curved.

Plane L/W

a. Narrowly elliptic more than 6:1-3:1
b. Elliptic 2:1-3:2
c. Widely elliptic 6:5
d. Circular 1:1
e. Oblate 5:6
f. Transversely elliptic 2:3-1:2
g. Narrowly transversely elliptic 1:3-1:6 or more

Solid L/D

a. Narrowly ellipsoid more than 6:1-3:1
b. Ellipsoid 2:1-3:2
c. Broadly ellipsoid 6:5
d. Spheroid 1:1
e. Obloid 5:6
f. Transversely ellipsoid 2:3-1:2
g. Lenticular 1:3-1:6 or more

B. Oblong. With widest axis at midpoint of structure and with margins essentially parallel.

Plane L/W

a. Linear more than 12:1
b. Narrowly oblong 6:1-3:1 or lorate
c. Oblong 2:1-3:2
d. Widely oblong 6:5
e. Square 1:1
f. Transversely widely oblong 5:6
g. Transversely oblong 2:3-1:2
h. Transversely narrowly oblong 1:3-1:6
i. Transversely linear 1:12 or more

Solid L/D

a. Cylindric or terete more than 12:1
b. Narrowly oblong 6:1-3:1
c. Oblong 2:1-3:2
d. Broadly oblong 6:5
e. Cubical 1:1
f. Transversely broadly oblong 5:6
g. Transversely oblong 2:3-1:2
h. Transversely narrowly oblong 1:3-1:6
i. Transversely cylindrical or terete 1:12 or more

C. Ovate. With widest axis below middle and with margins symmetrically curved; egg-shaped.

a. Lanceolate more than 6:1-3:1
b. Ovate 2:1-3:2
c. Widely ovate 6:5
d. Very widely ovate 1:1
e. Widely depressed ovate 5:6
f. Depressed ovate 2:3-1:2

a. Lanceoloid more than 6:1-3:1
b. Ovoid 2:1-3:2
c. Broadly ovoid 6:5
d. Very broadly ovoid 1:1
e. Broadly depressed ovoid 5:6
f. Depressed ovoid 2:3-1:2

D. Obovate Inversely ovate.

Plane

a. Oblanceolate more than 6:1-3:1
b. Obovate 2:1-3:2
c. Widely obovate 6:5
d. Very widely obovate 1:1
e. Widely depressed obovate 5:6
f. Depressed obovate 2:3-1:2

Solid L/D

a. Oblanceoloid more than 6:1-3:1
b. Obovoid 2:1-3:2
c. Broadly obovoid 6:5
d. Very broadly obovoid 1:1
e. Broadly depressed obovoid 5:6
f. Depressed obovoid 2:3-1:2

E. Rhombic. With widest axis at midpoint of structure, and with straight margins; elliptic but margins straight and middle angled.

Plane L/W

a. Narrowly rhombic more than 6:1-3:1
b. Rhombic 2:1-3:2
c. Widely rhombic 6:5
d. Quadrate rhombic 1:1
e. Transversely widely rhombic 5:6
f. Transversely rhombic 2:3-1:2
g. Narrowly transversely rhombic 1:2-1:6 or more

Solid L/D

a. Narrowly rhomboid more than 3:1
b. Rhomboid 2:1-3:2
c. Broadly rhomboid 6:5
d. Quadrate rhomboid 1:1
e. Transversely broadly rhomboid 5:6
f. Transversely rhomboid 2:3-1:2
g. Narrowly transversely rhomboid 1:3-1:6 or more

F. Trullate. With widest axis below middle and with straight margins; ovate but margins straight and angled below middle; trowel-shaped.

Plane L/W

a. Narrowly trullate more than 6:1-3:1

Solid L/D

a. Narrowly trulloid more than 6:1-3:1

b. Trullate 2:1-3:2
c. Widely trullate 6:5
d. Very widely trullate 1:1
e. Widely depressed trullate 5:6
f. Transversely depressed trullate 2:3-1:2

b. Trulloid 2:1-3:2
c. Broadly trulloid 6:5
d. Very broadly trulloid 1:1
e. Broadly depressed trulloid 5:6
f. Transversely depressed trulloid 2:3-1:2

G. Obtrullate. Inversely trullate.

Plane L/W

a. Narrowly obtrullate more than 6:1-3:1
b. Obtrullate 3:2-2:1
c. Widely obtrullate 6:5
d. Very widely obtrullate 1:1
e. Widely depressed obtrullate 5:6
f. Transversely depressed obtrullate 2:3-1:2

Solid L/D

a. Narrowly obtrulloid more than 6:1-3:1
b. Obtrulloid 3:2-2:1
c. Broadly obtrulloid 6:5
d. Very broadly obtrulloid 1:1
e. Broadly depressed obtrulloid 5:6
f. Transversely depressed obtrulloid 2:3-1:2

H. Triangular. With three sides and three angles.

Plane L/W

a. Linear-triangular more than 12:1
b. Narrowly triangular 6:1-3:1
c. Triangular 2:1-3:2
d. Widely deltate 6:5
e. Deltate 1:1
f. Shallowly deltate 5:6
g. Shallowly triangular 2:3-1:2
h. Very shallowly triangular 1:3-1:6 or more

Solid L/D

a. Subulate more than 12:1
b. Narrowly pyramidal 6:1-3:1
c. Triangular 2:1-3:2
d. Broadly deltoid 6:5
e. Deltoid 1:1
f. Shallowly deltoid 5:6
g. Shallowly pyramidal 2:3-1:2
h. Very shallowly pyramidal 1:3-1:6 or more

I. Obtriangular. Inversely triangular.

Plane L/W

a. Linear-obtriangular or narrowly cuneate more than 12:1
b. Cuneate 6:1-3:1
c. Obtriangular 2:1-3:2 or widely cuneate
d. Widely obdeltate 6:5
e. Obdeltate 1:1
f. Shallowly obdeltate 5:6
g. Shallowly obtriangular 2:3-1:2
h. Very shallowly obtriangular 1:3-1:6 or more

Solid L/D

a. Linear-obpyramidal or narrowly cuneiform more than 12:1
b. Cuneiform 6:1-3:1
c. Obpyramidal or broadly cuneiform 2:1-3:2
d. Broadly obdeltoid 6:5
e. Obdeltoid 1:1
f. Shallowly obdeltoid 5:6
g. Shallowly obpyramidal 2:3-1:2
h. Very shallowly obpyramidal 1:3-1:6 or more

2. Special Plane Figures--Outline

a. Acicular. Needlelike, round or grooved in cross section.
b. Auriculiform. Usually obovate with two small rounded, basal lobes.
c. Cordiform. Heart-shaped.
d. Dimidiate. Inequilateral with one-half wholly or nearly wanting.
e. Falcate. Scimitar-shaped.
f. Filiform. Threadlike, usually flexuous.

SHAPES

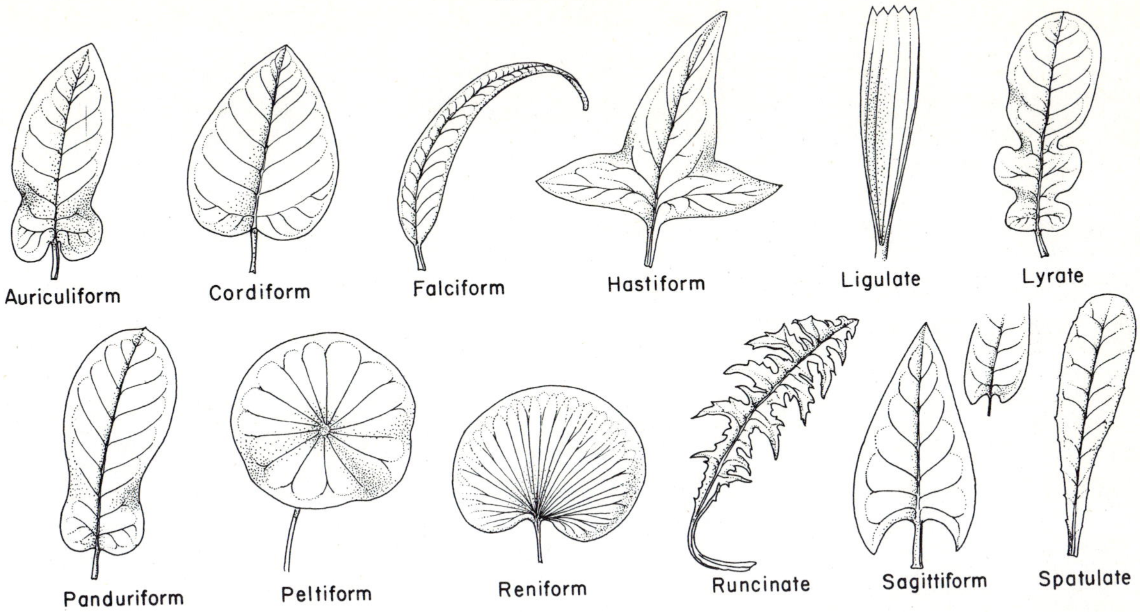

SYMMETRIC PLANE FIGURES

(Adapted from Taxon, 1962)

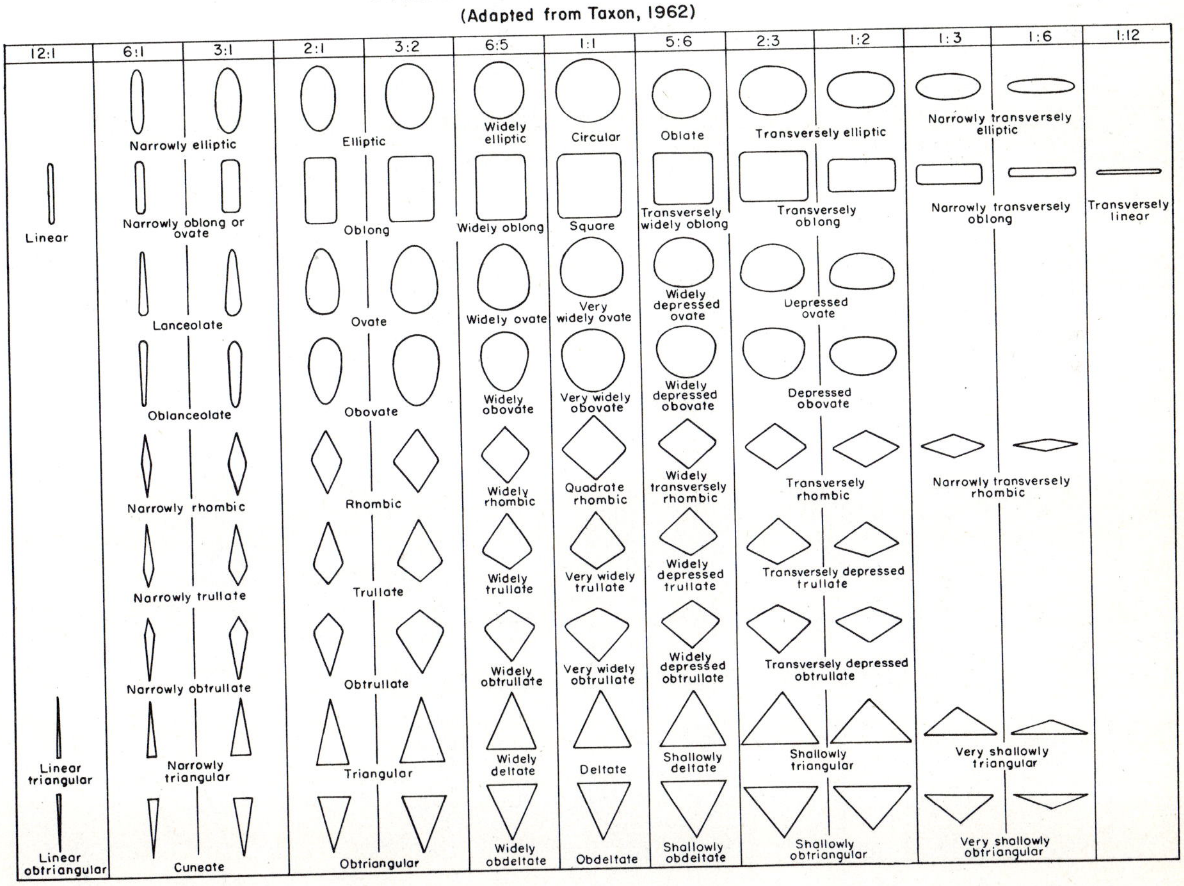

Figure 6-14

SHAPES

g. Hastiform. Triangular with two flaring basal lobes.
h. Lunate. Crescent-shaped, with acute ends.
i. Lyrate. Lyre-shaped; pinnatifid with large terminal lobe and smaller lower lobes.
j. Obcordiform. Inversely cordiform.
k. Panduriform. Fiddle-shaped; obovate with sinus or indentation on each side near base and with two small basal lobes.
l. Peltiform. Rounded with petiole attached to center of blade or apparently to laminar tissue.
m. Rectangular. Box-shaped, longer than wide.
n. Reniform. Kidney-shaped, with shallow sinus and widely rounded margins.
o. Runcinate. Oblanceolate with lacerate to parted margins.
p. Sagittiform. Triangular-ovate with two straight or slightly incurved basal lobes.
q. Spathulate or Spatulate. Oblong or obovate apically with a long attenuate base.

3. Special Solid Figures (See calyx, corolla, and perianth types for special 3-dimensional shapes, Sect. A, VIII) (Figure 6-14, p. 132)

a. Acerose. Needle-shaped; sharp.
b. Annular. Ring-like.
c. Arcuate. Bent like the arc of a circle.
d. Botuliform. Sausage-shaped.
e. Capillate. Hair-shaped.
f. Capitate. Head-like.
g. Clavate. Club-shaped.
h. Cochleate. Snail-shaped.
i. Compressed or Complanate. Flattened.
j. Conical. Having figure of true cone.
k. Coroniform. Crown-shaped.
l. Cotyliform. Cup-shaped.
m. Crateriform. Shallow cup-shaped as the involucre of some species of *Quercus*.
n. Cruciform or Cruciate. Cross-shaped.
o. Cylindric. Long-tubular.
p. Cymbiform. Boat-shaped.
q. Discoid. Orbicular with convex faces.
r. Dolabriform. Axe-shaped.
s. Excentric. One-sided; off-center.
t. Falcate or Seculate. Sickle-shaped.
u. Fistulose. Hollow, as a culm without pith.
v. Flabelliform. Fan-shaped.
w. Fusiform. Spindle-shaped; broadest in middle and tapering to each end.
x. Half-terete. Flat on one side, terete on other; semicircular in cross section.
y. Hippocrepiform. Horseshoe-shaped.
z. Lenticular. Biconvex, usually elongate and flattish.
aa. Lingulate. Tongue-shaped, plano-convex in cross section.
bb. Meniscoidal. Thin and concave-convex.
cc. Napiform. Turnip-shaped.
dd. Navicular. Boat-shaped.
ee. Nodiform or Nodulose. Knotty or knobby, as the roots of most of the Fabaceae.
ff. Obconic. Inversely conical.
gg. Patelliform. Knee-shaped; disk-shaped.
hh. Pisiform. Pea-shaped.
ii. Pyriform. Pear-shaped.
jj. Rectangular. Boxlike, longer than wide.
kk. Spiral. Twisted like a corkscrew.
ll. Stellate. Star-shaped.
mm. Strombiform. Elongate snail-shaped.
nn. Terete. Cylindrical.
oo. Torose. Cylindrical with contractions at intervals.
pp. Turbinate. Top-shaped; obconic.
qq. Turgid. Tumid or swollen.
rr. Umbilicate. Depressed in the center.
ss. Umbonate. Round with a projection in center.

tt. Umbraculiform. Umbrella-shaped.

uu. Vermiform. Worm-shaped.

B. Apices and Bases

(Figure 6-15, p. 135)

(Leaves, petals, sepals, scales, bracts or other flattened structures)

1. Apices and Bases with Sinuses.
   a. Retuse. Lobe rounded; sinus depth to 1/16 distance to midpoint of blade; margins convex.
   b. Emarginate. Lobe rounded; sinus depth 1/16-1/8 distance to midpoint of blade; margins straight or convex.
   c. Cordate (Apex obcordate). Lobe rounded; sinus depth 1/8-1/4 distance to midpoint of blade; margins convex and/or straight.
   d. Cleft. Lobe rounded; sinus depth 1/4-1/2 distance to midpoint of blade; margins convex and/or straight.
   e. Reniform. Lobe rounded; sinus depth variable; outer margin convex or straight, inner concave.
   f. Auriculate. Lobe rounded; sinus depth variable; outer margin concave, inner convex or straight.
   g. Lobate. Lobe rounded; sinus depth variable; outer and inner margins concave.
   h. Sagittate. Lobe pointed and oriented downward or inward in relation to petiole or midrib; sinus depth variable; margins variable.
   i. Hastate. Lobe pointed and oriented outward or divergent in relation to petiole or midrib; sinus depth variable; margins variable.
2. Apices and Bases without Sinuses.
   a. Truncate. Cut straight across; ending abruptly almost at right angles to midrib or midvein.
   b. Rounded. Margins and apex forming a smooth arc.
   c. Obtuse. Margins straight to convex, forming a terminal angle more than 90°.
   d. Acute (Base cuneate). Margins straight to convex forming a terminal angle 45°-90°.
   e. Acuminate (Base narrowly cuneate). Margins straight to convex forming a terminal angle of less than 45°.
   f. Caudate (Base attenuate). Acuminate with concave margins.
   g. Hastate. Margins variable; lobe pointed; oriented outward or divergently in relation to petiole or midrib.
   h. Cuspidate. Acute but coriaceous and stiff.
   i. Spinose or Pungent. Acuminate but coriaceous and stiff.
3. Apices with Midrib, Midvein or Vein Extension.
   a. Apiculate. More than 3:1 l/w, usually slightly curled and flexuous.
   b. Aristate. More than 3:1 l/w, usually prolonged, straight and stiff.
   c. Cirrhous. More than 10:1 l/w, coiled and flexuous.
   d. Mucronate. Less than 3:1 l/w, straight and stiff.
   e. Mucronulate. 1:1 l/w or broader than long; straight.
   f. Muticous. Without a vein extension, awn or hair.
   g. Piliferous. More than 20:1 l/w, hair-like, flexuous.
4. Specialized Bases and Leaf Attachments.
   a. Amplexicaul. Completely clasping the stem.
   b. Clasping. Partly surrounding the stem.
   c. Connate or Connate-perfoliate. Having bases of opposite leaves fused around the stem.
   d. Decurrent. Extending along stem downward from leaf base.

LEAF APICES, ATTACHMENTS AND BASES

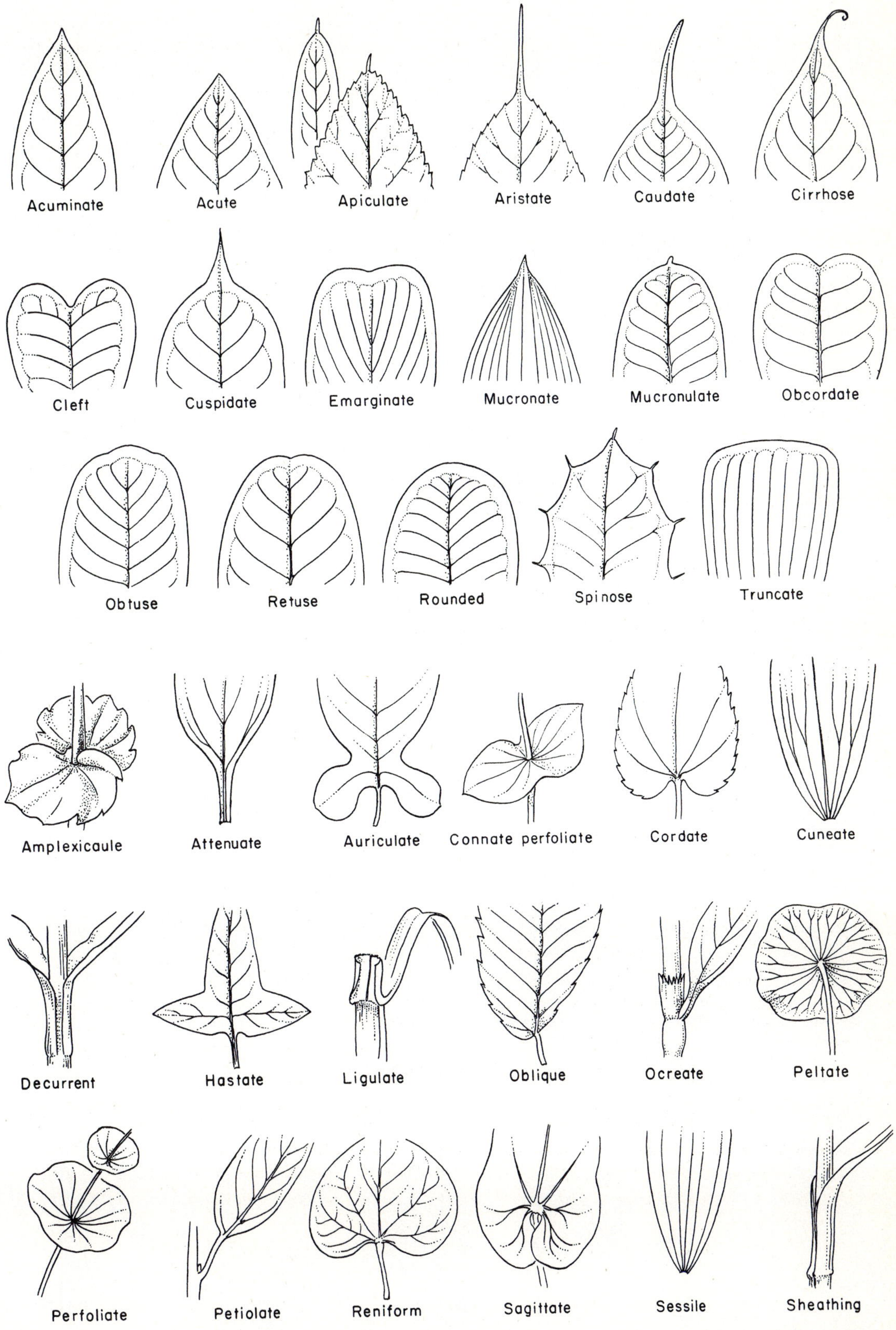

Figure 6-16

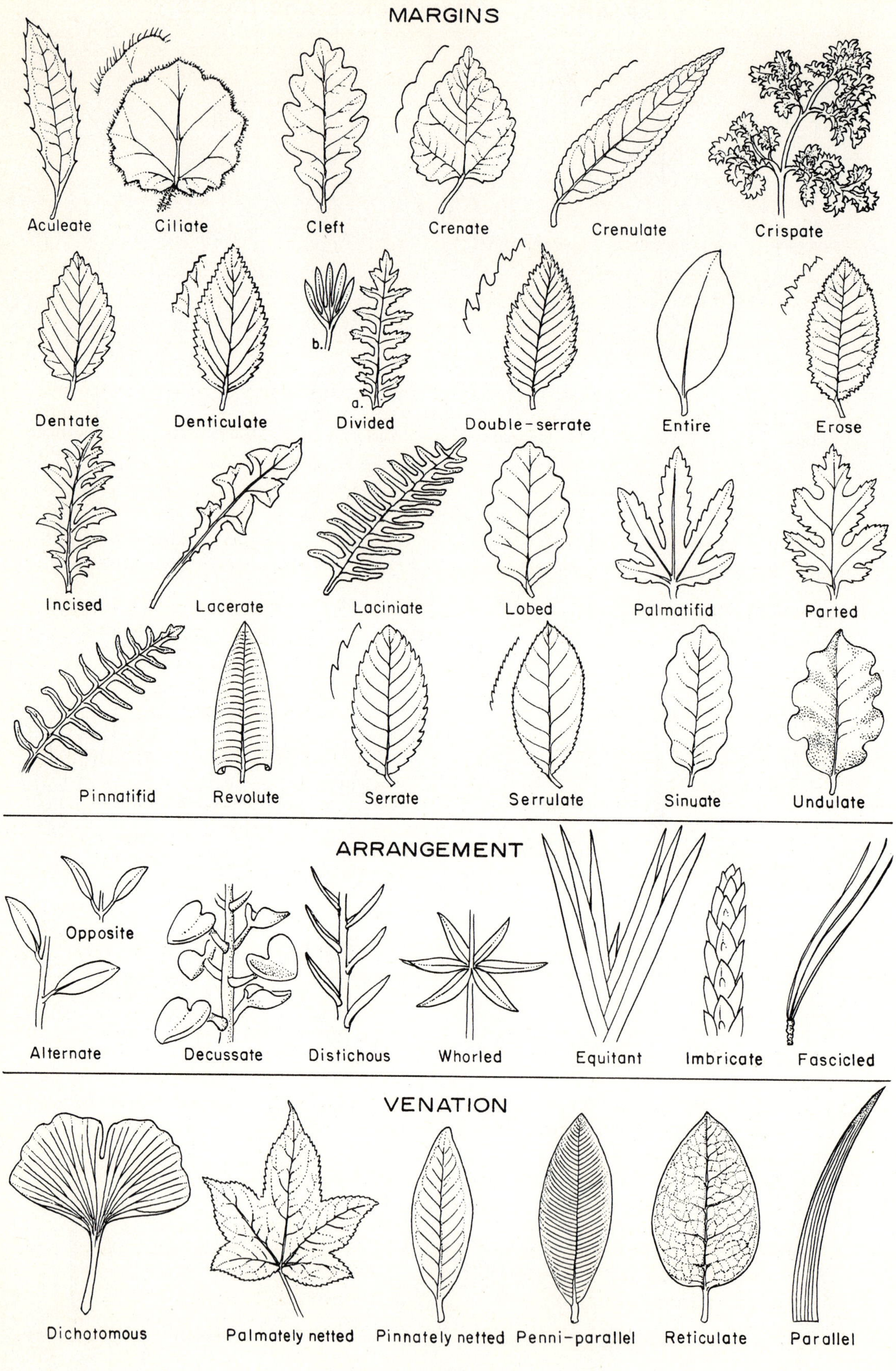
MARGINS
Aculeate
Ciliate
Cleft
Crenate
Crenulate
Crispate
Dentate
Denticulate
b.
a.
Divided
Double-serrate
Entire
Erose
Incised
Lacerate
Laciniate
Lobed
Palmatifid
Parted
Pinnatifid
Revolute
Serrate
Serrulate
Sinuate
Undulate
ARRANGEMENT
Opposite
Alternate
Decussate
Distichous
Whorled
Equitant
Imbricate
Fascicled
VENATION
Dichotomous
Palmately netted
Pinnately netted
Penni-parallel
Reticulate
Parallel

e. Ligulate. Having a tongue-like outgrowth at base of blade or top of sheath.
f. Oblique. Having an asymmetrical base.
g. Ocreate. Having a stipular tube surrounding stem above insertion of petiole or blade.
h. Ocreolate. Diminutive of ocreate; usually applied to bract bases.
i. Peltate. Usually having petiole attached near the center on the underside of blade.
j. Perfoliate. Having base completely surrounding stem.
k. Petiolate. With a petiole.
l. Sessile. Without a petiole.
m. Sheathing. Having tubular structure enclosing stem below apparent insertion of blade or petiole.
n. Surcurrent. Extending along stem upward from leaf base.

C. Margins (Figure 6-16, p. 136)

(Leaves, petals, sepals, bracts, scales or other flattened structure)

Note: For precison in margin description the type (as described below), the symmetry of the individual tooth, the margins of the individual tooth, the apex of the individual tooth, the type of sinus (rounded or angled), the number of teeth per unit of margin measurement, the spacing (regular or irregular) of the teeth, the nature of teeth (simple or compound in two or more size groups) should be indicated or described.

1. Margin Types
   a. Aculeate. Prickly.
   b. Bicrenate or Doubly-crenate. With smaller rounded teeth on larger rounded teeth.
   c. Biserrate or Doubly-serrate. With sharply cut teeth on the margins of larger sharply cut teeth.
   d. Ciliate. With trichomes protruding from margins.
   e. Cleft. Indentations or incisions cut 1/4-1/2 distance to midrib or midvein.
   f. Crenate. Shallowly ascending round-toothed, or teeth obtuse; teeth cut less than 1/8 way to midrib or midvein.
   g. Crenulate. Diminutive of crenate, teeth cut to 1/16 distance to midrib or midvein.
   h. Crispate. Curled; margins divided and twisted in more than one plane.
   i. Dentate. Margins with rounded or sharp, coarse teeth that point outwards at right angles to midrib or midvein, cut 1/16 to 1/8 distance to midrib or midvein.
   j. Denticulate. Diminutive of dentate, cut to 1/16 distance to midrib or midvein.
   k. Divided. Indentations or incisions cut 3/4-almost completely to midrib or midvein.
   l. Entire. Without indentations or incisions on margins; smooth.
   m. Erose. Irregularly, shallowly toothed and/or lobed margins; appearing gnawed.
   n. Filamentose or Filiferous. With coarse marginal fibers or threads.
   o. Fimbriate. Margins fringed.
   p. Fimbriolate. Minutely fimbriate.
   q. Incised. Margins sharply and deeply cut, usually jaggedly.
   r. Involute. Margins rolled inward.
   s. Lacerate. Margins irregularly cut, appearing torn.
   t. Laciniate. Margins cut into ribbon-like segments.
   u. Lobed. Large, round-toothed, cut 1/8-1/4 distance to midvein.
   v. Palmatifid. Cut palmately.
   w. Parted. Indentations or incisions cut 1/2-3/4 distance to midrib.
   x. Pinnatifid. Cut pinnately.
   y. Repand. Sinuate with indentions less than 1/16 distance to midrib or midvein.
   z. Retrorsely Crenate. Rounded teeth directed toward base.

aa. Retrorsely Serrate. Sharp or pointed teeth directed toward base.
bb. Revolute. Margins rolled under.
cc. Serrate. Saw-toothed; teeth sharp and ascending, but cut 1/16-1/8 distance to midrib or midvein.
dd. Serrulate. Diminutive of serrate, but cut to 1/16 distance to midrib or midvein.
ee. Sinuate. Margins shallowly and smoothly indented, wavy in a horizontal plane, without distinctive teeth or lobes, indented 1/16-1/8 distance to midrib or midvein.
ff. Undulate. Margins shallowly and smoothly indented, wavy in a vertical plane.

IV. SURFACE-VENATION-TEXTURE

A. Configuration of Surface

(Classification based on patterns of configuration)

Aciculate. Finely marked as with pin pricks, fine lines usually randomly arranged.
Alate. Winged.
Alveolate. Honey-combed.
Areolate. Divided into many angular or squarish spaces.
Bullate. Puckered or blistered.
Canaliculate. Longitudinally grooved, usually in relation to petioles or midribs.
Cancellate or Clathrate. Latticed.
Corrugate. Ridged.
Costate. Coarsely ribbed.
Fenestrate. With windowlike holes through the leaves or other structures.
Flexuous. Coarsely undulate with folds at right angles to long axis.
Foveolate. Pitted.
Plicate or Plaited. Fluted, longitudinally folded.
Punctate. Covered with minute impressions or depressions.
Pustulate. With scattered blister-like swellings.
Reticulate. Netted.
Ribbed. With longitudinal nerves.
Ringed. With old bud scale scar rings.
Rugose. Covered with coarse reticulate lines.
Ruminate. Coarsely wrinkled, appearing as chewed.
Scarred. With old leaf base, stipular and/or branch scar regions.
Smooth or Plane. Without configuration.
Striate. With longitudinal lines.
Sulcate. With longitudinal grooves.
Tortuous. Having the surface variously twisted.

B. Venation

(Classification based on vein pattern. See Chapter 7, section D for detailed structure and page 194 for illustrations. Based on Hickey [1973])

1. General

Acrodromous. With two or more primary or strongly developed secondary veins diverging at or above the base of the blade and running in convergent arches toward the apex over some or all of the blade length, the arches not basally curved.

Actinodromous. With three or more primary veins diverging radially from a single point at or above the base of the blade and running toward the margin, reaching it or not.

Brochidodromous. With a single primary vein, the secondary veins not terminating at the margin but joined together in a series of prominent upward arches or marginal loops on each side of the primary vein.
Campylodromous. With several primary veins or their branches diverging at or close to a single point and running in strongly developed, basally recurved arches which converge toward the apex, reaching it or not.
Cladodromous. With a single primary vein, the secondary veins not terminating at the margin and freely ramified toward it.
Craspedodromous, Mixed. With a single primary vein, some of the secondary veins terminating at the margin and an approximately equal number otherwise.
Craspedodromous, Simple. With a single primary vein, all of the secondary veins and their branches terminating at the margin.
Eucamptodromous. With a single primary vein, the secondary veins curved upward and gradually diminishing distally within the margin and interconnected by a series of cross-veins without forming conspicuous marginal loops.
Hyphodromous. With a single primary vein and all other venation absent, rudimentary, or concealed within a coriaceous or fleshy blade.
Palinactinodromous. Actinodromous, the primary veins with one or more subsidiary radiations above the primary one.
Parallelodromous. With two or more primary veins originating beside one another at the blade base and running more or less parallel to the apex where they converge.
Reticulodromous. With a single primary vein, the secondary veins not terminating at the margin and losing their identities near the margin by repeated branching, yielding a dense reticulum.
Semicraspedodromous. With a single primary vein, the secondary veins branching just within the margin, one branch from each terminating at the margin and the other forming a marginal loop and joining the superadjacent secondary vein.

2. Special (Figure 6-16, p. 136)

(Traditional classification)

Dichotomous. With veins branching or forking in pairs equally.
Netted or Reticulate. With veins forming a network.
Pinnately Netted. With secondary veins arising from midrib or midvein.
Palmately or Digitately Netted. With three or more primary veins arising from a common point.
Parallel. With veins extending from base to apex, essentially parallel.
Penni-parallel. With veins extending from midrib to margins, essentially parallel.

C. <u>Surface</u> (Figure 6-17, p. 141)

(Classification based usually on epidermal outgrowths or excrescences, but not trichomes)

Asperous. Having a rough surface.
Bristly. Beset with bristles.
Echinate. Covered with spines.
Farinaceous. Mealy.
Glabrate or Glabrescent. Becoming glabrous.
Glabrous. Smooth; devoid of trichomes.
Glandular. Covered with minute, blackish to translucent glands.
Glaucescent. Sparingly or slightly glaucous.
Glaucous. Covered with a bloom or smooth, waxy coating.
Glutinous. Having a shiny, sticky surface.
Granular. Finely mealy, covered with small granules.
Greasy or Unctuous. Slick, oily, slippery to touch.

Mucilaginous. Gummy or gelatinous.
Muricate. Covered with short, hard protuberances.
Muriculate. Minutely muricate.
Opaque. With a dull surface.
Prickly. With prickles.
Pruinose, Frosted, or Sebiferous. With a heavy wax coat.
Pubescent. Covered with dense or scattered trichomes.
Pulverulent. Covered with fine, powdery wax granules.
Ramentaceous. Having many thin scales, as on the epidermis of some ferns.
Resinous. Having a yellowish, sticky, exudate.
Roridulate or Dewy. Covered with waxy platelets, appearing dewy.
Scaberulent. Approaching scabrous.
Scabridulous. Minutely scabrous.
Scabrous. Having a harsh surface.
Shining, Nitid, or Laevigate. Lustrous, polished.
Spiculate. With crystals in or on the surface.
Squamose. Having coarse scales.
Subglabrate. Almost glabrous.
Tuberculate or Verrucose. With a warty surface.
Viscid. Sticky or glutinous.

D. <u>Vestiture or Indument</u> (Figure 6-17, p. 141)
(Classification based on type of hairy cover, see Chapter 7, section E for individual types of trichomes and page 200 for illustrations.)

Aculeate. Prickly.
Arachnoid. Cobwebby.
Barbed. With short, rigid reflexed bristles or processes.
Barbellate. Minutely barbed.
Bearded or Barbate. With long trichomes usually in a tuft, line or zone.
Bristly. Covered with stiff, strong trichomes.
Canescent or Incanous. Covered with dense, fine grayish-white trichomes.
Ciliate. With conspicuous marginal trichomes.
Ciliolate. With tiny or small marginal trichomes.
Comose. With a tuft of trichomes, usually apical.
Downy. Covered with short, weak, soft trichomes.
Floccose. Covered with dense, appressed trichomes in patches or tufts.
Glabrate. Without trichomes.
Glandular. Covered with secretory or excretory trichomes.
Glochidiate. With barbed trichomes, glochids, usually in tufts.
Hirsute. Covered with long, rather stiff trichomes.
Hirsutullous or Hirtellous. Minutely hirsute.
Hispid. Covered with very long, stiff trichomes.
Hispidulous. Approaching hispid, minutely hispid.
Inermous. Unarmed. Without prickles or spines.
Lanate. Covered with long, intertwined trichomes, cottony.
Lanuginose. Cottony, similar to lanate but trichomes shorter.
Lepidote or Squamulose. Covered with minute scales.
Paleaceous. With small membranous scales, chaffy.
Pannose or Felted. With matted, feltlike layer of trichomes.
Papillose. Covered with minute tubercles.
Pilose. With soft, shaggy trichomes.
Puberulent. Minutely pubescent.
Pubescent. Usually with straight, slender trichomes.
Scurfy or Lentiginous. With exfoliating scaly incrustations.
Sericeous. With long, silky trichomes, usually appressed.
Setose, Setaceous. Having setae or bristlelike trichomes.
Strigillose. Diminutive of strigose.
Strigose. Covered with sharp, coarse, bent trichomes usually with a bulbous base.
Tomentose. Covered with dense, interwoven trichomes.
Urent or Stinging. With erect, usually long trichomes that produce irritation when touched.
Velutinous. Covered with dense, straight, long and soft trichomes; pile-like.

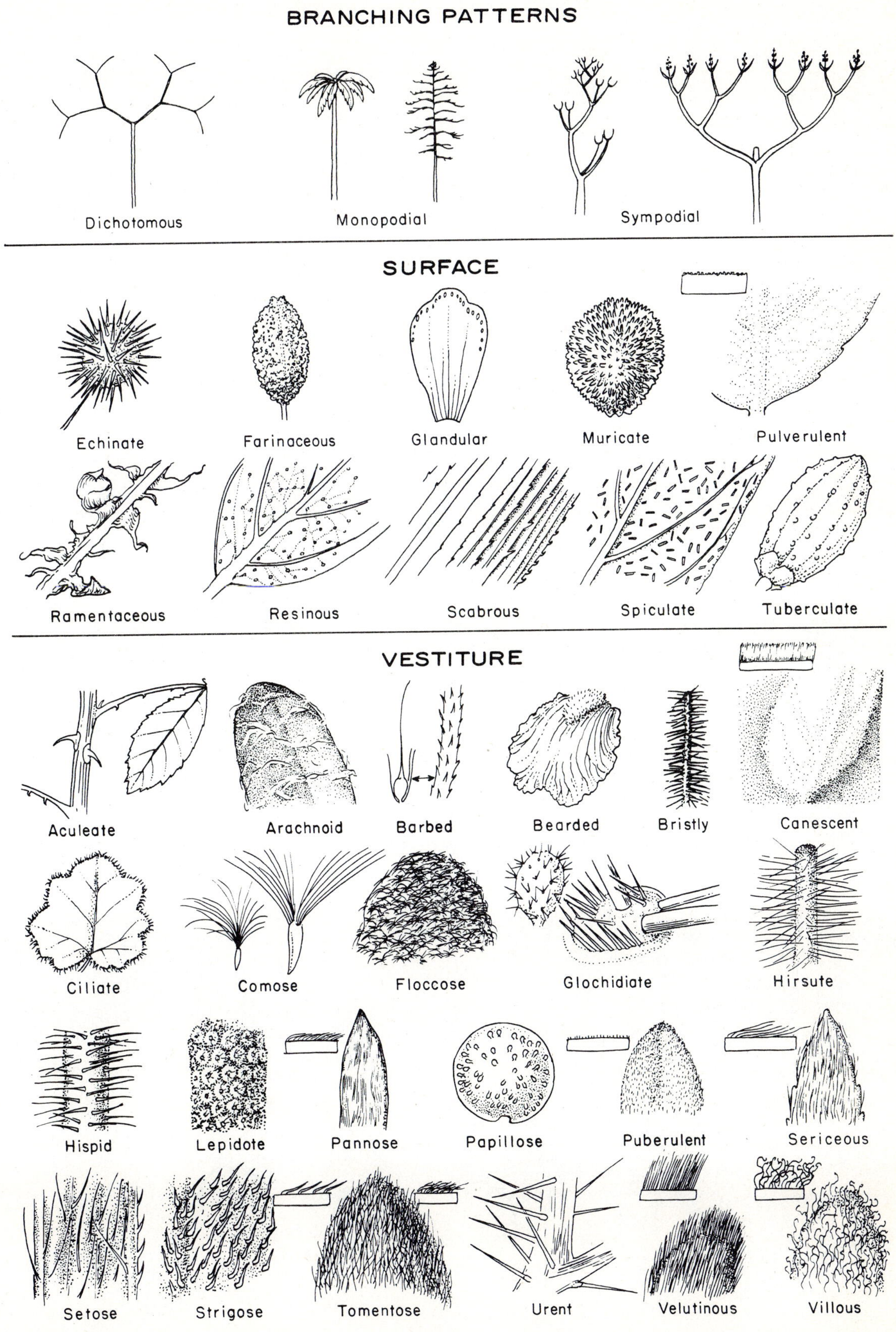
BRANCHING PATTERNS
Dichotomous
Monopodial
Sympodial
SURFACE
Echinate
Farinaceous
Glandular
Muricate
Pulverulent
Ramentaceous
Resinous
Scabrous
Spiculate
Tuberculate
VESTITURE
Aculeate
Arachnoid
Barbed
Bearded
Bristly
Canescent
Ciliate
Comose
Floccose
Glochidiate
Hirsute
Hispid
Lepidote
Pannose
Papillose
Puberulent
Sericeous
Setose
Strigose
Tomentose
Urent
Velutinous
Villous

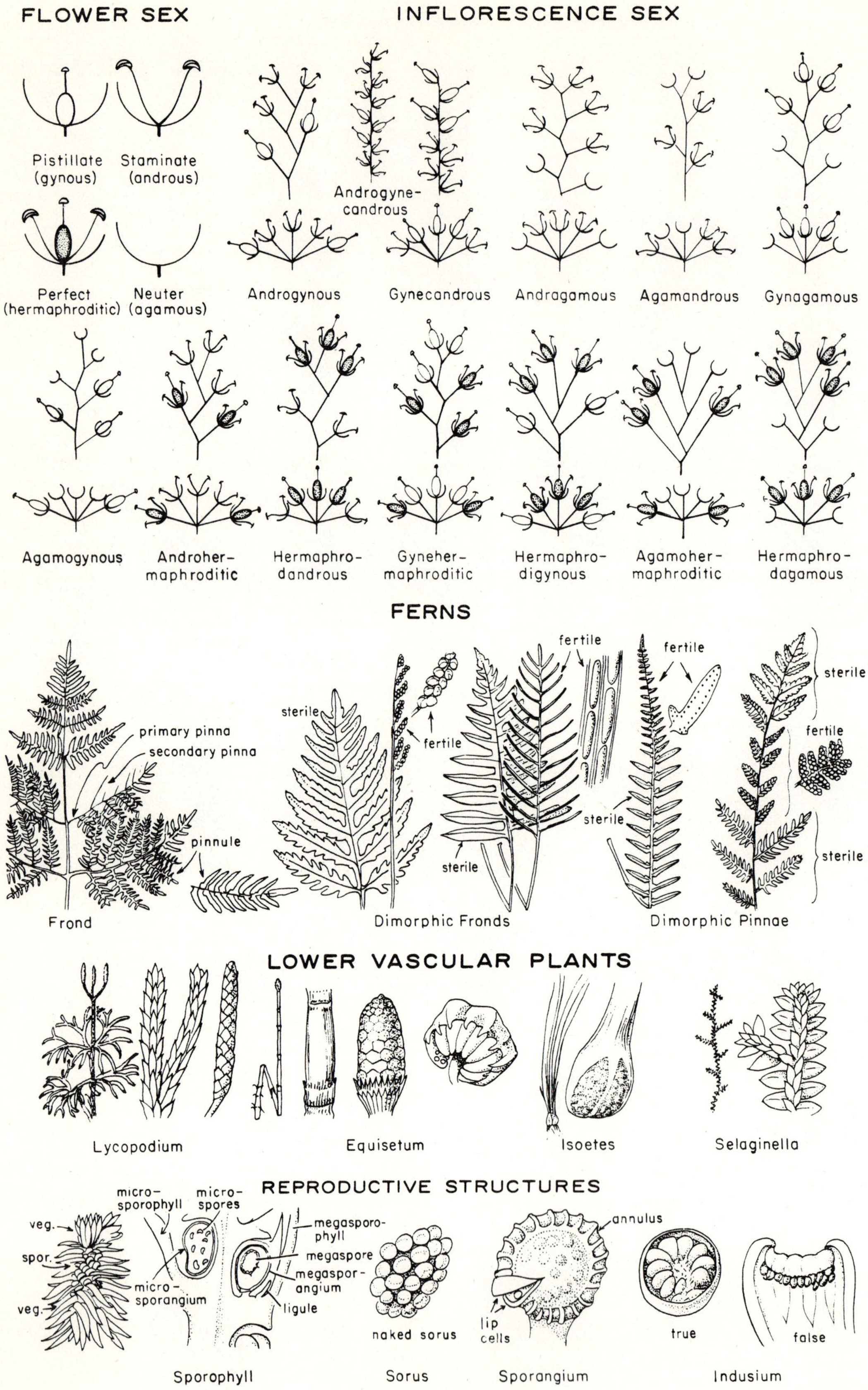
FLOWER SEX
INFLORESCENCE SEX
Pistillate (gynous)
Staminate (androus)
Perfect (hermaphroditic)
Neuter (agamous)
Androgyne-candrous
Androgynous
Gynecandrous
Andragamous
Agamandrous
Gynagamous
Agamogynous
Androhermaphroditic
Hermaphrodandrous
Gynehermaphroditic
Hermaphrodigynous
Agamohermaphroditic
Hermaphrodagamous
FERNS
primary pinna
secondary pinna
pinnule
Frond
sterile
fertile
fertile
sterile
fertile
sterile
Dimorphic Fronds
sterile
fertile
sterile
Dimorphic Pinnae
LOWER VASCULAR PLANTS
Lycopodium
Equisetum
Isoetes
Selaginella
REPRODUCTIVE STRUCTURES
veg.
spor.
veg.
micro-sporophyll
micro-spores
micro-sporangium
megasporophyll
megaspore
megasporangium
ligule
Sporophyll
naked sorus
Sorus
annulus
lip cells
Sporangium
true
false
Indusium

Villosulous. Minutely villous.

Villous. Covered with long, soft, crooked trichomes.

E. Texture

(Classification based on consistency of leaves, scales, petals, fruits and other parts)

Baccate. Juicy and very succulent.
Carnose or Sarcous. Fleshy.
Cartilaginous. Hard and tough but flexible.
Ceraceous. Waxy.
Chartaceous. Papery, opaque and thin.
Coriaceous. Thick and leathery.
Corneous. Horny.
Crustaceous. Hard, thin, and brittle.
Diaphanous. Translucent.
Fibrous. Having loose, woody fibers.
Flaccid. Lax and weak.
Gelatinous. Jellylike; soft and quivery.
Herbaceous. Soft and succulent.
Hyaline. Thin and translucent or transparent.
Incrassate. Thickened.
Indurate. Hardened.
Ligneous. Woody.
Membranous. Thin and semi-translucent; membrane-like.
Osseous. Bony.
Pannose. With a felty texture.
Pellucid. Clear, transparent.
Scarious. Thin and dry, appearing shriveled.
Sclerous. Hard.
Spongy. Cellular; sponge-like.
Suberous. Corky.
Suffrutescent. Woody basally, herbaceous apically.
Woody. Hard and lignified.

V. SEX

A. Flower Sex

(Figure 6-18, p. 142)

(Classification based on sexual condition of the individual flower)

Imperfect or Unisexual. With stamens or carpels absent in the flower.
Neuter or Agamous. Without stamens and carpels in flower; or sex organs abortive.
Perfect or Bisexual. With both stamens and carpels or pistils in the flower.
Pistillate, Carpellate, or Female. With pistils or carpels only in the flower.
Staminate or Male. With stamens only in the flower.

B. Inflorescence Sex

(Figure 6-18, p. 142)

(Classification based on sexual condition of individual flowers in different parts of inflorescence; arrangement of flowers by sexual condition in an inflorescence)

Note: If plant sex is staminate or pistillate or hermaphroditic or neuter, then inflorescence sex would be respectively the same. The term in ( ) below is an example of the sexual condition of one inflorescence type, the head, as in composites.

Androgynecandrous. Inflorescence with staminate flowers above and below pistillate, as in the spikes of some species of *Carex*.
Androgynous. Inflorescence with staminate flowers inside or above and pistillate outside or below. (androgynecephalous)
Agamandrous. Inflorescence with neuter flowers inside or above and staminate outside or below. (agamandrocephalous)
Agamogynous. Inflorescence with neuter flowers inside or above and pistillate outside or below. (agamogynecephalous)

Agamohermaphroditic. Inflorescence with neuter flowers inside or above and hermaphroditic outside or below. (agamohermaphrodicephalous)

Andragamous. Inflorescence with staminate flowers inside or above and neuter flowers outside or below. (andragamocephalous)

Androhermaphroditic. Inflorescence with staminate flowers inside or above and hermaphroditic outside or below. (androhermaphrodicephalous)

Gynagamous. Inflorescence with pistillate flowers inside or above and neuter flowers outside or below. (gynagamocephalous)

Gynecandrous. Inflorescence with pistillate flowers inside or above and staminate outside or below, as in spikes of some species of Carex.

Gynehermaphroditic. Inflorescence with pistillate flowers inside or above and hermaphroditic outside or below. (gynehermaphrodicephalous)

Hermaphrodagamous. Inflorescence with hermaphroditic flowers inside or above and neuter outside or below. (hermaphrodagamocephalous)

Hermaphrodandrous. Inflorescence with hermaphroditic flowers inside or above and staminate outside or below. (hermaphrodandrocephalous)

Hermaphrodigynous. Inflorescence with hermaphroditic flowers inside or above and pistillate outside or below. (hermaphrodigynecephalous)

C. Plant Sex

(Classification based on sexual condition of all flowers on the plant(s) within the species)

Androdioecious. Some plants with staminate flowers and some with perfect flowers.

Andromonoecious. Plant with staminate and perfect flowers.

Diclinous. Plant with imperfect flowers; stamens and carpels in separate flowers.

Dioecious. Plant with all flowers imperfect, but staminate and pistillate on separate plants.

Female, Carpellate, or Pistillate. Plant with pistillate flowers only.

Gynodioecious. Some plants with perfect flowers and some with pistillate.

Gynomonoecious. Plant with pistillate and perfect flowers.

Hermaphroditic or Monoclinous. Plant with all flowers perfect.

Male or Staminate. Plant with staminate flowers only.

Monoecious. Plant with all flowers imperfect, but staminate and pistillate flowers on same plant.

Polygamous. Plant with perfect and imperfect flowers.

Polygamo-dioecious. Plants dioecious, but with some perfect flowers on staminate or pistillate plants or both.

Polygamo-monoecious. Plant monoecious, but with some perfect flowers.

Trioecious. Plants staminate, pistillate or perfect.

D. General Sexual Terms

Agamous or Neuter. Without sex; sexual organs abortive.

Heterocephalous, Heterocymous, Heterospicous. Heads, cymes, spikes with flowers of different sexual conditions. Note: other inflorescence word stems could be used for appropriate inflorescence type.

Heterosexual. Inflorescences or flowers within the plant with different sexual conditions.

Homocephalous, Homocymous, Homospicous. Heads, cymes, spikes with flowers sexually uniform.

Homosexual. Inflorescences or flowers sexually uniform.

VI. POLLINATION AND FERTILIZATION

A. Pollination Types

(Classification based on agent by which pollen is transferred from the pollen sac or anther to the stigma or ovule)

Anemophily. Pollinated by wind.
Anthropophily. Pollinated by man.
Cantharophily. Pollinated by beetles.
Cheiropterophily. Pollinated by bats.
Entomophily. Pollinated by insects.
Hydrophily. Pollinated by water.
Hymenopterophily. Pollinated by bees.
Malacophily. Pollinated by snails or slugs.
Melittophily. Pollinated by bees.
Micromelittophily. Pollinated by small bees.
Myiophily. Pollinated by diptera.
Myrmecophily. Pollinated by ants.
Necrocoleopterophily. Pollinated by carrion beetles.
Ornithophily. Pollinated by birds.
Phalaenophily. Pollinated by moths.
Psychophily. Pollinated by butterflies.
Sapromyiophily. Pollinated by carrion or dung flies.
Sphingophily. Pollinated by hawk moths and nocturnal lepidoptera.

B. Pollination Pathways

Chasmantheric Pollination. Pollen transferred from a normally dehisced anther by a pollinating agent with pollen grain germination on the stigma and subsequent growth of the pollen tube through stigma, style and the ovule into the embryo sac.
Cleistantheric Pollination. Pollen not transferred from a normally dehisced anther by a pollinating agent; pollen grain germinates within the anther with subsequent growth of the pollen tube through the anther wall and ovary wall into the ovule and embryo sac.

C. Fertilization Types

(Refers usually to sources of male gametes in pollen parent)

Allogamy or Xenogamy. Cross-fertilization in plants.
Allautogamy. Cross- and self-fertilization in same plant, as in *Viola*.
Autogamy. Self-fertilization in a single flower.
Geitonogamy. Fertilization of one flower by another on the same plant.

D. Fertilization Pathways

(Classification based on place of entrance of pollen tube into ovule)

Chalazogamy. Pollen tube entrance through chalaza.
Pleurogamy. Pollen tube entrance through side of ovule.
Porogamy. Pollen tube entrance through the micropyle.

VII. TEMPORAL PHENOMENA

A. Periodicity

Aestival. Appearing in summer.
Aianthous or Semperflorous. With flowers appearing throughout the year.
Annotinal or Yearly. Appearing yearly.
Autumnal. Appearing in autumn.
Biferous. Appearing twice yearly.

Biflorous. Flowering in autumn as well as in spring.
Bimestrial. Occurring every two months.
Diurnal. Opening during the day.
Equinoctial. Having flowers which expand and close regularly at particular hours of the day.
Hibernal or Hiemal. Appearing in the winter.
Matutinal. With flowers opening in the morning.
Nocturnal. Opening during night.
Seasonally. Occurring during a seasonal cycle, or each season.
Serotinal. Opening late; appearing in late summer.
Vernal. Appearing in spring.
Vespertine. With flowers opening in the evening or night; appearing or expanding in the evening.

B. Maturation

Anthesis. Time of flowering; opening of flower with parts available for pollination.
Blastocarpous. Germination of seeds while within the pericarp, as in Rhizophora.
Coetaneous. Flowering as the leaves expand; synantherous.
Dichogamous. With maturation of stamens or anther and carpels or stigma at different times.
Homogamous. With maturation of stamens or anther and carpels or stigma at same time.
Hysteranthous. With leaves appearing after flowers.
Precocious. Developing unusually early.
Proanthesis. Flowering before normal period, as spring flowers in the fall.
Protandrous. With stamens or anthers developing before carpels or stigma.
Protantherous. With leaves appearing before flowers.
Protogynous. With carpels or stigma maturing before stamens or anthers.
Synantherous. With leaves and flowers appearing at same time.

C. Duration

Annual. Living one year or less.
Biduous. Lasting two days.
Biennial. Living two years, usually flowering second year.
Bimestrial. Lasting two months.
Caducous. Dropping off very early, usually applied to floral parts.
Caulocarpic. Plants having the stem living for many years, bearing flowers and fruits.
Cladoptosic. Shedding of branches, stems and leaves simultaneously, as in Taxodium.
Deciduous. Persistent for one growing season.
Deliquescent. Softening and wasting away.
Ephemeral. Germinating, growing, flowering and fruiting in a short period, as most desert herbs.
Evanescent. Passing away, disappearing early.
Evergreen. Persistent two or more growing seasons.
Fugacious. Ephemeral, usually applied to plant parts.
Marcescent. Usually ephemeral with persistent remains; withering persistent.
Monocarpic, Hapaxanthic. Perennial or annual, flowering and fruiting once, then dying; fruiting once.
Perennial, Polycarpic. Living more than two years; fruiting more than once.
Persistent. Remaining attached; applied to individual parts.
Pliestesial or Multiperennial. Monocarpic but living several to many years before flowering, as in Agave.
Rhizocarpic. Plants having the roots living for many years with the stems dying annually.
Seasonally Deciduous. Falling after one growing season.
Summer Annual. Germinating in spring or early summer and flowering and fruiting in late summer or early fall, then dying.

Winter Annual. Living less than one year but through the winter; germination usually in late fall, and usually flowering and fruiting in early spring.

IX. GROWTH AND DEVELOPMENT

A. Growth Regions

Adaxial or Abaxial. Growth region localized on top or bottom of a leaf part or any dorsi-ventral structure.

Apical or Terminal. Growth region at the apex of the structure.

Basal. Growth region at the base of a blade, as in grasses.

Intercalary. Growth region near the base of an internode or base of blade.

Interstitial. Growth all-over in an organ, no localized meristems, as in some fruits.

Marginal. Growth region near the edge of a leaf.

B. Growth Periodicity

Accrescent. Growing after flowering or bud development has occurred, as the sepals in Hypericum and bud scales in Carya.

Determinate. Growth of plant parts, the size of which is limited by cessation of meristematic activity during the year.

Indeterminate or Evergrowing. Continual growth of plant parts, not limited by a cessation of meristematic activity.

Intermittent. A renewal and cessation of meristematic activity which produces clusters of stems and/or leaves along an axis.

C. General Development

Acropetal. Developing upward, toward apex.

Basipetal. Developing downward, toward base.

Centripetal. Developing from the outside inward, or from bottom upward.

Centrifugal. Developing from the inside outward, or from top downward.

Determinate. Central flower develops first, arresting development of primary axis.

Indeterminate. Lateral flowers develop first with primary axis continuing to elongate or develop.

D. Developmental Shoot Types

(Classification based on origin and development of shoots)

Dwarf Shoots or Spurs. Shoots that develop from preformed buds which have very short internodal lengths or intervals.

Epicormic Shoots or Water Sprouts. Shoots that develop from dormant lateral buds on the trunk which have very long and frequently variable internodal lengths or intervals.

Flushing Shoots. Shoots that develop from mature terminal buds several times during a season. Terminal bud will develop shoot with new terminal bud which will develop more shoots and a terminal bud which will develop etc.--several times in a season with several flushes of growth.

Heterophyllous Shoots. Shoots that develop from winter buds which do not contain the primordia of all the leaves to develop during the year.

Lammas Shoots. Abnormal late season shoots that develop from the terminal bud, not a recurring phenomenon as in flushing shoots.

Long Bud Shoots. Abnormal buds or shoots which elongate, then have arrested growth without the development of leaves and lateral branches.

Long Shoots. Normal shoots that develop from terminal or axillary buds which have normal internodal lengths or intervals.

Preformed Shoots. Normal shoots that develop from winter buds which contain primordia of all leaves that will expand during the season.

Proleptic Shoots. Abnormal late season shoots that develop from the lateral buds immediately beneath the terminal.

Sucker Shoots. Shoots that develop from adventitious buds on old stumps or roots, usually after cutting or injury, which have elongate internodal lengths and intervals.

Syleptic Shoots. Abnormal shoots that develop from lateral buds before they have reached maturity.

IX. PATTERNS

A. Symmetry

Actinomorphic or Radial. With floral parts radiate from center like spokes on wheel.

Asymmetric. Without regularity in any dimension.

Dorsiventral. Planate and having distinct dorsal and ventral surfaces, the two usually different.

Equilateral. With halves or sides equal in shape and size.

Inequilateral. With halves or sides unequal in shape and size.

Irregular. With floral parts within a whorl dissimilar in shape and/or size.

Regular. With floral parts within a whorl similar in shape and size.

Spherical. With multi-dimensional radial symmetry.

Zygomorphic or Bilateral. With floral parts in two symmetrical halves.

B. Arrangement Systems
(Based on Leppik [1961])

Anthotaxis. Arrangement of sporophylls, primarily reproductive in function.

Carpotaxis. Arrangement of fruits, reproductive in function.

Phyllotaxis. Arrangement of leaves, primarily photosynthetic in function.

Rhizotaxis. Arrangement of roots.

Semataxis. Arrangement of semaphylls (petals, sepals, tepals), primarily advertising (pollinator attracting) in function.

xxxxxxxxxxxxxxxxxxxxxxxxxxxxxxxxxxxxxx

Fibonacci Phyllotaxis. A fundamental type of leaf arrangement expressed as a fraction in which each succeeding fraction is the sum of the two previous numerators and the sum of the two previous denominators, i.e., 1/2, 1/3, 2/5, 3/8, 5/13, 8/21, etc. The numerator represents the number of turns or spirals around a stem before one leaf is directly above another and the denominator represents the number of leaves in the turns or spirals before one is directly above the other. 2/5 phyllotaxy would mean two twists and five leaves before one leaf is directly above the other or an angle of divergence of 144° between succeeding leaves in the stem (2/5 of 360°). According to Leppik anthotaxis and semataxis do not necessarily follow the same pattern, with anthotaxis in *Michelia* cited as being in 2/7, 3/7, 3/8, and 4/10 systems of arrangement.

C. Branching Patterns (Figure 6-17, p. 141)
(Adapted from Hallé [1971])

Dichotomous. Branching into two equal parts.

Flabellate. Fan-shaped branching.

Monopodial. Branching with a main axis and reduced or missing laterals; excurrent.

a. Without lateral branches. Main axis ending in a cluster of large leaves; e.g., *Carica papaya*.
b. With continuous lateral branches. Main axis beset with laterals; e.g., *Trema lamarckiana*.
c. With continuous but self-pruning lateral branches (excurrent). Main axis with upper laterals present and lower having fallen off; e.g., *Pinus*.
d. With short-lived laterals. Main axis beset with phyllomorphic branches, appearing as compound leaves that may be deciduous; e.g., some tropical Rubiaceae, *Pycnanthus*.
e. With spiral or whorled main laterals on main axis and distichous branches; e.g., *Pycnanthus*.
f. With spiral or whorled main laterals and dichotomous branches; e.g., *Phyllanthus*.
g. With tiered or clustered lateral branches. Main axis with periodic lateral branch production; e.g., *Bombax*.

Sympodial. Branching without a main axis but with many, more or less, equal laterals.

a. Leader displacement. Successive laterals take over leader growth producing umbrella-shaped trees; e.g., *Acacia*. *Melia*; deliquescent.
b. Leader displacement by two equal laterals (dichotomous substitution). Production of a forked branching system; e.g., *Manihot*.
c. Leader displacement as a short-shoot (apposition). Usually a single lateral becomes dominant; e.g., *Terminalia*.

Note: Other types of stem growth and branching patterns occur, particularly in tropical woody plants but no adequate classification has been devised.

X. SECRETORY AND EXCRETORY STRUCTURES
(Nectaries, Digestive Glands, Hydathodes, Tanniferous and Ethereal Oil Glands)

Classification based on Structure Note: No satisfactory classification scheme based on structure is available. Some glands are single-celled, most are multicellular; some are sessile, some stipitate; some are capitate, some are variously shaped; some are highly modified structures, some are so simple that they simply secrete the substance through the cuticle.

A. Classification based on Function.

Digestive. Enzyme-secreting glands found mostly on leaves of carnivorous and insectivorous plants.

Ethereal Oil Producing. Mostly aromatic compound producing glands found on various parts of the plant, without definitely known functions.

Nectar Secreting. Sugary compound producing glands found mostly on floral parts that produce attractants for pollinators.

Salt Excreting. Glands that excrete or secrete salt solutions which usually dry during the day and deliquesce at night.

Tannin Bearing. Tannin producing glands found in various parts of the plant, presumably protective in some structures.

Water Excreting. Specialized vein openings (hydathodes) through which water is lost under certain atmospheric conditions.

B. Classification based on Position.

1. Floral Nectaries
   a. Petaliferous. Most frequently at the base of petals.
   b. Sepaliferous. Most frequently at the base of sepals.
   c. Septal. At the junction of the septa in the ovary.
   d. Staminal Disc. Entire disc frequently nectariferous.
   e. Staminal. Most frequently at base of filament.
   f. Staminodial. Frequently entire structure is nectariferous.
2. Extrafloral Nectaries
   a. Bracteal and b. Laminar. May be localized or found over entire structure.
   c. Nodal. At the nodes.
3. Ethereal Oil Glands.
   a. Cauline. May be all over (general) or along the ribs (costal), or in the grooves (canaliculate).
   b. Laminar. May be all over (general) or near apex (acrolaminar) or near base (basilaminar) or marginal.
   c. Petiolar. May be all over (general) or near apex (acropetiolar) or near base (basipetiolar).

XI. COLOR

A. Standardized Colors

Use standardized color charts, such as the Ridgeway for solid color descriptions.

B. Distribution of Colors (Variegation)

(Adapted from Lindley [1848])

Banded. Transverse stripes of one color crossing another.
Blotched. The color disposed in broad, irregular blotches.
Bordered. One color is surrounded by an edging of another.
Clouded. Colors are unequally blended together.
Discoidal. A single large spot of color in the center of another.
Dotted. The color disposed in very small round spots.
Edged. One color is surrounded by a very narrow rim of another.
Marbled. A surface traversed by irregular veins of color; as a block of marble often is.
Ocellated. A broad spot of some color has another spot of a different color within it.
Painted. Colors disposed in streaks of unequal intensity.
Spotted. The color disposed in small spots.
Striped. Longitudinal stripes of one color crossing another.
Tessellated. Color arranged in small squares, so as to have some resemblance to a checkered pavement.
Variegated. The color disposed in various irregular, sinuous, spaces.
Zoned. The same as ocellated, but the concentric bands more numerous.

XII. GENERAL

(Unclassified descriptors applicable to any character, from Porter et al. [1973])

Compound. Composed of two or more anatomically or morphologically equivalent units, whether subdivided into them or an aggregate of them.

Dimorphic. Having two different sizes and/or shapes within the same species.

Evident. Clearly visible macroscopically.

Heteranthous. Having different states in two different sets of flowers, only one state present in each set.

Heterophytous. Having different states in two different sets of plants, only one state present in each set.

Homanthous. Having more than one state within each individual flower, all flowers the same.

Homophytous. Having more than one state within each individual plant, all plants the same.

Monomorphic. All of the same shape and size.

Not Applicable. Not relevant to the taxon at hand; passive null value.

Obscure. Not clearly visible macroscopically, usually owing to incomplete differentiation.

Petaloid. Petal-like in shape, texture and/or color.

Polyanthous. Having different states in several to many (more than three) different sets of flowers, only one state present in each set.

Polyheterophytous. Having different states in several to many (more than three) different individuals or sets of plants, only one state present in each set.

Polymorphic. Having several to many (more than three) different shapes and/or sizes within the same species.

Same. Same as in description of next higher taxon.

Sepaloid. Sepal-like in shape, texture and/or color.

Simple. Not composed of more than one anatomically or morphologically equivalent unit.

Triheteranthous. Having different states in three different sets of flowers, only one state present in each set.

Triheterophytous. Having different states in three different sets of plants, only one state present in each set.

Trimorphic. Having three different shapes and/or sizes within the same species.

Unknown. Relevant to taxon at hand but status not known; information unavailable.

Section C. GYMNOSPERM GLOSSARY*

A. Vegetative Structures

Acicular Leaf. Needle-like, long and slender; e.g., Pinus.

Awl-shaped Leaf. Subulate; narrow, flat, stiff, sharp-pointed, usually less than 1/2 in. long; e.g., Juniperus.

Fascicle. Cluster of needles borne on a minute determinate short shoot in the axil of a primary leaf (bract); e.g., Pinus.

Fascicle Sheath. Closely imbricated bud scales at the base of the fascicle of needles; e.g., Pinus.

Linear Leaf. Narrow, flattened, triangular, or quadrangular leaf usually 1/2-2 in. long; e.g., Taxus, Picea.

Long Shoot. Elongated internodes, rapid annual growth.

*Contributed by James W. Hardin, North Carolina State University, Raleigh.

Multinodal Shoot. Spring shoot developing from the terminal winter bud and producing 2 or more whorls of branches; the cones are partly lateral in the middle of the shoot; e.g., Pinus echinata.

Needle. Acicular; slender, elongated leaf, usually over 2 in long; e.g., Pinus.

Peg (sterigmata). Lateral stem projection to which leaf is attached and persistent after leaf dehiscence; i.e., abscission layer between peg and leaf. Leaf may be sessile; e.g., Picea; or petiolate; e.g., Tsuga, on the peg.

Scale Leaf. Small, usually appressed and imbricate; e.g., Juniperus, Thuja.

Short Shoot. Very short or inconspicuous internodes and growth very slow if at all.

- Determinate. No continued growth; shoot with leaves drops as a unit; e.g., Pinus fascicle.
- Indeterminate. Continued growth by an apical meristem; spur-branch; e.g., Larix, Cedrus, Ginkgo.

Uninodal Shoot. Spring shoot developing from the terminal winter bud and producing only one internode with one whorl of branches at the end; the cones are subterminal at the end of the shoot; e.g., Pinus resinosa.

B. Reproductive Structures

Apophysis. Exposed outer surface of either an ovuliferous scale or megasporophyll as seen when the cone is closed.

Aril. An outgrowth from the stem forming a fleshy covering of the seed; e.g., Taxus, Torreya; or only rudimentary at base of the fleshy seed; e.g., Cephalotaxus.

Bract. Modified leaf subtending the ovuliferous scale; may be distinct or fused to the scale.

Cone (strobilus). Aggregation of sporangia-bearing structures at tip of the stem (either sporophylls or scales in the Gymnosperms).

- Female Cone (megasporangiate strobilus). Bearing ovules or seeds.
- Male Cone (microsporangiate strobilus). Bearing pollen sacs (microsporangia).

Epimatium. Fleshy covering of the seed and more or less fused with the integument; arising from the chalazal end of the ovule like an additional integument; e.g., Podocarpus.

Megasporophyll. Modified leaf bearing ovules; e.g., Zamia.

Microsporophyll. Modified leaf bearing microsporangia or pollen sacs.

Ovuliferous Scale. Highly modified lateral branch in the axil of a leaf (bract), and bearing ovules. May be flat or peltate, woody or fleshy; e.g., Pinaceae.

Receptaculum. A fleshy structure below the seed formed from the bases of bracts and the swollen receptacle or cone axis; e.g., Acmopyle, and some Podocarpus spp.

Umbo. Projection, with or without spine or prickle, on the apophysis of the cone scale.

Section D. LOWER VASCULAR PLANT GLOSSARY*

I. SPOROPHYTE

A. Stems

1. Branching

Axial. With branches arising from buds in leaf axil.

Dichotomous. With branches forking into two more or less equal parts.

Epipetiolar. With branches arising from buds on the petiole.

Monopodial. Having one main axis of growth.

Random. With branches arising from buds without relation to leaves.

Sympodial. With branches more or less equal without a main axis.

2. Phyllotaxis

Distichous. With leaves in two rows.

Polystichous. With leaves in several rows.

3. External Features

Aerial Stem. An erect stem arising from a horizontal rhizome.

Articulate. Generally meaning having a joint as in leaves, leaflets or stems, as in heterophyllous species of *Selaginella*; or having a swollen area, often discolored, at the point of branching of the stem.

Jointed. With stems that can be pulled apart easily at the nodes, as in *Equisetum*.

Node. Point on the stem where leaves are attached; or the point of branching of the stem.

Rameal Sheath. Leaf sheath on the stem joints, as in *Equisetum*.

Tubercules. Silica deposits on the stem ridges, as in *Equisetum*.

4. Internal Features

Canals. As in *Equisetum*.

- Central Canal. The large centrally located air space in the stem.
- Carinal Canal. A canal beneath a stem ridge associated with a vascular bundle.
- Vallecular Canal. A canal beneath a stem groove.

Stele. The central primary vascular system of the stem and associated tissues. Delimited from the cortex by endodermis and pericycle.

- Actinostele. A protostele having a xylem core in the form of radiating ribs, as viewed in transverse section.
- Dictyostele. A dissected solenostele with each individual bundle a meristele.
- Eustele. A dissected siphonostele with phloem only to the outside of the xylem.
- Plectostele. A protostele dissected into anastomosing platelike units.
- Protostele. Stele having a solid column of vascular tissue with xylem centrally located.
- Siphonostele. A stele having vascular tissue in the form of a hollow cylinder, with a central pith.
- Solenostele. A siphonostele having phloem both internal and external to the xylem.

*Contributed by John T. Mickel, New York Botanical Garden.

B. Leaves

1. Duration

Evergreen. Bearing green leaves through the winter.

Marcescent. The leaves of short duration, dying at the end of the growing season.

2. Megaphylls

(Leaves with a branching vein system which are associated with nodal leaf gap)

a. General

Crozier. The coiled developing leaf of a fern.

Frond. The leaf of a fern.

b. Parts of a Leaf

Blade. The expanded portion of a leaf.

Costa. The midvein of a minor division of a fern leaf.

Lamina. The leaf tissue other than the veins or axes.

Pinna. A primary division of a fern leaf.

Pinnule. A secondary division of a fern leaf.

Rachis. The axis of a compound fern blade.

Segment. The ultimate division or unit of a dissected fern leaf.

Stipe. The petiole of a fern leaf.

Stipe Bundles. The vascular bundles of the fern petiole.

c. Venation Patterns

Areoles. The spaces formed by a vein network.

False Veins. Small vein-like areas of thick-walled cells in the leaves of some lower vascular plants.

Free. No veins uniting to form a network.

Included Veinlets. Veins ending inside areoles.

Net. Veins uniting to form a network.

d. Leaf Division

Bipinnate. Twice pinnate.

Pectinate. Pinnatifid with closely set segments; comb-like.

Pinnate. Compound, with the leaflets arranged on both sides of a common axis.

Pinnate-pinnatifid. Pinnate with pinnatifid pinnae.

Pinnatifid. Pinnately cut, more than half way to the midvein.

Simple. Undivided.

e. Blade Architecture

Anadromous. Having the first lobe or segment of a pinna arising basiscopically in compound leaves.

Catadromous. Having the first lobe or segment of a pinna arising acroscopically in compound leaves.

Imparipinnate. Pinnate with a conform terminal leaflet.

Palmate. Radiately lobed or divided.

Pedate. Palmately cut or divided with the lower pair basiscopically exaggerated.

Pinnate. Compound, with the leaflets arranged on both sides of a common axis.

f. Dimorphism (Figure 6-18, p. 142)

(With leaves of two types in ferns, the fertile being of different size, shape or dissection than the vegetative)

Complete Dimorphism. Leaves completely fertile or vegetative.

Partial Dimorphism. Leaves with only a part modified as fertile.

Slight Dimorphism. Fertile and vegetative leaves only slightly different in appearance.

Vegetative Frond. Frond lacking sporangia.

3. Microphylls

(Leaves with only one vein and no leaf gap.)

Axillary leaves. Leaves borne in the axils of branches, as in heterosporous species of _Selaginella_.

Bast bundles (peripheral strands). Bundles of thick-walled cells parallel to the midrib, as in _Isoetes_.

Lacuna. Chamber or internal air space.

Lateral leaf. Leaf on the side of the stem, as in heterophyllous species of _Selaginella_.

Ligule. A small membranous outgrowth or projection at the base of the leaf, appearing above the sporangium in fertile leaves, as in _Selaginella_ and _Isoetes_.

Median leaf. Leaf on top of stem, as in heterophyllous species of _Selaginella_.

Seta. A hair-like extension of the leaf, as in homophyllous species of _Selaginella_.

C. Reproductive Structures (Figure 6-18, p. 142)

Annulus. Thick-walled ring of cells on the sporangium.

Elater. One of four elongate appendages on the spores, as in _Equisetum_.

Eusporangiate. Having the sporangium develop from a great amount of leaf tissue as opposed to only one or a few cells.

Exospore or Exine. Outer spore wall layer.

Fovea. Pit or depression containing the sporangium in the leaf base of _Isoetes_.

Heterosporous. Having two kinds of spores, usually differing in size.

Homosporous. Having spores of only one kind.

Indusium. A flap of tissue covering a sorus.

- False indusium. A folded leaf margin protecting the sorus.
- True indusium. An epidermal outgrowth protecting the sorus.
- Inferior indusium. An indusium attached beneath the sorus with the sporangia appearing above it.

Leptosporangiate. Having the entire sporangium develop from a periclinal division of a superficial cell or small group of cells.

Lip Cells. The line of cells between which the sporangium dehisces.

Massula. A clump of microspores, as in _Azolla_.

Megasporangium. The sporangium in which megaspores are produced.

Megaspore. A spore that gives rise to a female gametophyte.

Megasporophyll. A leaf bearing one or more megasporangia.

Microsporangium. The sporangium in which microspores are produced.

Microspore. A spore that gives rise to a male gametophyte.

Microsporophyll. A leaf bearing one or more microsporangia.

Paraphyses. Hairs or hair-like structures in the sorus.

Perispore or Perine. An outer covering of some fern spores, with different configuration than that of the exospore.
Receptacle. Point on a leaf where sporangia are attached.
Sorus. A cluster of sporangia.
Sporangiophore. The umbrella-shaped sporangium-bearing unit of the strobilus, as in Equisetum.
Sporangium. A spore case.
Spore Shapes.
Monolete. Bean-shaped, with a single scar line.
Trilete. Basically tetrahedral, but often appearing round or triangular, with three scar lines forming a Y.
Sporocarp. A hard, nut-like structure containing the sporangia in heterosporous ferns.
Sporophyll. A leaf bearing sporangia.
Stomium. Lip cell region of a fern sporangium.
Strobilus. Stem with short internodes and spore-bearing appendages; a cone.
Velum. The membranous flap covering the sporangium, as in Isoetes.

II. GAMETOPHYTE

Antheridium. The male sex organ producing the sperm.
Archegonium. The female sex organ producing the egg.
Prothallus. Gametophyte of lower vascular plants.
Rhizoid. A hair-like absorptive organ on gametophytes and rarely on sporophytes.

III. ASEXUAL REPRODUCTION

Apogamy. Producing sporophytes from a gametophyte without fertilization.
Apospory. Producing gametophytes directly from a sporophyte without producing spores.
Bulblet. A small, bud-like vegetative propagule produced on the leaves of some ferns.
Gemma. A vegetative reproductive bud borne on the stem, as in Lycopodium; a multicellular reproductive propagule on gametophytes, as in ferns.

PHYTOGRAPHIC LITERATURE

Bocquet, G. 1959. The campylotropous ovule. Phytomorphology 9: 222-227.
Björnstad, I. N. 1970. Comparative embryology of Asparagoideae-Polygonateae, Liliaceae. Nytt Magasin for Botanikk 17(3-4): 169-207.
Correll, D. S., and M. C. Johnston. 1971. Manual of the Vascular Plants of Texas. Texas Research Foundation. Renner.
Cronquist. A. 1968. The Evolution and Classification of Flowering Plants. Houghton-Mifflin Company. Boston.
Davis, P. H., and V. H. Heywood. 1963. Principles of Angiosperm Taxonomy. D. Van Nostrand, Inc. Princeton.
Duke, J. A. 1969. On tropical tree seedlings 1. Seeds, seedlings, systems, and systematics. Annals Missouri Botanical Garden 56(2): 125-161.
Faegri, K., and van der L. Pijl. 1966. The Principles of Pollination Ecology. Pergamon Press. New York.
Featherly, H. I. 1954. Taxonomic Terminology of Higher Plants. Iowa State College Press. Ames.
Fernald, M. L. 1950. Gray's Manual of Botany. 8th ed. American Book Company. New York.
Gray, P. 1967. The Dictionary of Biological Sciences. Reinhold Publishing Corporation. New York.

Hallé, F. 1971. Architecture and growth of tropical trees exemplified by the Euphorbiaceae. Biotropica 3(1): 56-62.
Hickey, L. J. 1973. Classification of the architecture of dicotyledonous leaves. American Journal of Botany 60(1): 17-33.
Jackson, B. D. 1928. A Glossary of Botanic Terms, 4th ed. J. B. Lippincott Company. Philadelphia.
Johnson, A. M. 1931. Taxonomy of the Flowering Plants. The Century Company. New York.
Lawrence, G. H. M. 1951. Taxonomy of Vascular Plants. The MacMillan Company. New York.
Leppik, E. E. 1957. A new system for classification of flower types. Taxon 6: 64-67.
________. 1961. Phyllotaxis, anthotaxis, and semataxis. Acta Biotheoretica 14: 1-28.
Lindley, J. 1848. An Introduction to Botany, 4th ed. London.
Martin, A. C. 1946. The comparative internal morphology of seeds. American Midland Naturalist 36: 513-660.
Porter, D. M., R. W. Kiger, and J. E. Monahan. 1973. A guide for contributors to Flora North America, Part II: An outline and glossary of terms for morphological and habitat description (Provisional Edition). FNA Report 66. Department of Botany, Smithsonian Institution. Washington.
Rickett, H. W. 1944. The classification of inflorescences. Botanical Review 10: 187-231.
Stearn, W. T. 1966. Botanical Latin. Thomas Nelson and Sons Ltd. London.
Swartz, D. 1971. Collegiate Dictionary of Botany. The Ronald Press Company. New York.
Systematics Association Committee for Descriptive Terminology. 1960. I. Preliminary list of works relevant to Descriptive Biological Terminology. Taxon IX(8): 245-257.
________. 1962. II. Terminology of simple symmetrical plane shapes. Taxon XI(5): 145-156, 245.
Uphof, J. C. Th. 1938. Cleistogamic flowers. Botanical Review 4(1): 21-49.

PHYTOGRAPHIC EXERCISE SUGGESTIONS

I. General

Phytographic studies should be a combination of laboratory and field work. The nature of the subject makes it ideal for individual and group scavenger hunts for different structures and features. The objectives of these studies should be to provide (1) a vocabulary for intelligent communication about vascular plants, (2) an understanding of the use of relative terms and to help the student become (3) a more critical observer, and (4) more precise in his descriptions of vascular plants.

1. Identify the parts of the plants, stems, roots, buds, leaves, inflorescences, flowers, fruits, and seeds in the material on display (or in the field.
2. Identify the types or habits of plants, stems, roots, buds, leaves, inflorescences, flowers, fruits, seeds in the material on display (or in the field).
3. Determine the position and arrangements of stems, roots, buds, leaves, inflorescences, flowers, stamens, carpels, and cotyledons of the material on display (or in the field).

4. Determine the shapes, margins, bases, and apices of the structures on display (or in the field).
5. Determine the surface, texture, vestiture features for the material on display (or in the field).
6. Distinguish the perianthal, androecial, and gynoecial types or structural types of the material on display.
7. Make a comparative study of seeds, embryos, endosperm, and seedlings on display. Identify parts, types, and arrangements of parts.
8. Make a study of numbers of parts, fusion of parts, number of whorls, symmetry of parts, and insertion of parts for the material on display (or in the field).
9. Determine the sex, pollinator, and insect visitor--flower relationships for several species in the field.
10. Determine the periodicity states, maturation times, and growth and development features for several species in the field.

II. Special

Select one species and determine or describe the character state or characteristic for each of the following:

I. Plant
   1. Type
   2. Duration

II. Roots
   1. Type
   2. Structural Type

III. Stems
   1. Bark Type
   2. Wood Ring Type
   3. Pith Type
   4. Texture
   5. Type or Structural Type
   6. Position
   7. Configuration
   8. Surface
   9. Vestiture

IV. Buds
   1. Type
   2. Structural Type
   3. Scale Arrangement
   4. Shape
   5. Surface
   6. Vestiture

V. Leaves
   1. Type or Structural Type
   2. Duration
   3. Position
   4. Arrangement
   5. Texture
   6. Surface
   7. Vestiture
   8. Venation
   9. Ptyxis
   10. Petiole Type
   11. Stipule Type
   12. Blade Shape
   13. Blade Apex
   14. Blade Margin
   15. Blade Base
   16. Laminar Secretory Structure

VI. Inflorescences
   1. Type
   2. Position

VII. Flowers
   1. Sex
   2. Type
   3. Insertion of Parts
   4. # Members in Whorl
   5. # of Whorls
   6. Fusion of Parts
   7. Perianth Type or Structural Type
   8. Aestivation
   9. Nectary Position

VIII. Androecium
   1. Androecial Type
   2. Stamen Structural Type
   3. Anther Attachment
   4. Anther Shape
   5. Anther Apex
   6. Anther Base
   7. Anther Type
   8. Stamen Arrangement
   9. Thecal Arrangement

IX. Gynoecium
1. Gynoecial Type
2. Carpel Arrangement
3. Carpel Type
4. Stigma Type
5. Style Type
6. Ovary Position
7. Ovary Shape
8. Ovule Type
9. Ovule Position

X. Fruits
1. Structural Type
2. Position
3. Placentation

XI. Seeds and Seedlings
1. Seed Surface Feature
2. Cotyledon Position
3. Seed Type
4. Embryo Type
5. Aril Structural Type
6. Eophyll # & Shape
7. Seedling Type

XII. Sex and Pollination
1. Plant Sex
2. Pollination Type
3. Fertilization Type
4. Pollination Pathway
5. Fertilization Pathway

XIII. Periodicity & Maturation
1. Leaf Periodicity
2. Flower Periodicity
3. Fruit Periodicity
4. Leaf Maturation
5. Stamen Maturation
6. Carpel Maturation

XIV. Growth and Development
1. Specialized Growth Regions
2. Growth Periodicity (Stem and Root)
3. Flower Development
4. Stamen Development
5. Floral Symmetry
6. Anthotaxis
7. Carpotaxis
8. Semataxis
9. Phyllotaxis
10. Stem Branching Pattern

PHYTOGRAPHIC INDEX

P

Q

R

Chapter 7. ANATOMICAL EVIDENCE

This chapter deals with anatomy as a systematic tool. As stated by C. R. Metcalfe (1968) anatomy of the vegetative organs of flowering plants can be taxonomically useful in the following ways: (1) the identification of fragmentary material, (2) the preliminary identification of herbarium specimens, and (3) as an aid toward establishing the interrelationships of taxa at and above the species level.

Anatomical characters have been employed for systematic purposes well over a hundred years. The most significant guiding principles of systematic anatomical study resulting from this long history have been eloquently discussed by I. W. Bailey (1951, 1953, 1957), and include: (1) anatomical data tends to be most useful at the level of the genus and at higher taxonomic categories, (2) anatomical characters are no more or less reliable than characters from other parts of the plant, (3) similarities in structural specialization do not necessarily imply close relationship but may be the result of parallel and convergent evolution, (4) anatomical data have proven most reliable in statements of negation of close relationship rather than positive assertions of relationship, (5) the assumption that a single sample of an organ or tissue provides reliable data for the anatomical characterization of a species or genus is invalid, (6) it is essential that more become known regarding the ranges of variability of anatomical characters within the same individual and different individuals of the same species, and (7) only when anatomical information is coupled with evidences from other parts of the plant will a natural classification be attained. It is important, furthermore, to emphasize that anatomical information is frequently taxonomically useful without having obvious evolutionary or phylogenetic interpretation. Information on floral anatomy and evolutionary morphology may be found in Chapter 27. Unless otherwise stated, terms are defined in accordance with K. Esau (1961).

Section A. SECONDARY XYLEM

This section will be concerned with the anatomy of wood. Included are characters of significant systematic value. In no other vegetative tissues of the plant are the trends of evolution as well understood (see Chapter 27). As a result, wood anatomy has assumed considerable importance in interpretations of vascular plant evolution (see Chalk, 1937, 1944; Tippo, 1946). Transverse, radial and tangential sections of wood should be prepared and studied to understand the organization of the tissues. Permanent macerations also should be made to obtain statistical data on cell length and other features where data on individual elements of the secondary xylem are desired.

The following list of diagnostic characteristics for descriptions of dicotyledonous woods has been taken with modifications from Tippo (1941). Included among this list are features of proven taxonomic and phylogenetic value. Terms are mostly defined in accordance with the Multilingual Glossary of Terms Used in Wood Anatomy (Committee on Nomenclature, International Association of Wood Anatomists, 1964). For the most part, categories of measurement follow the suggestions of the Committee on Standardization of Terms of Cell Size (1937, 1939). An abbreviated data work sheet is provided

Figure 7-1. Cell types of the secondary xylem, a-d, wood fibers. a, libriform wood fiber. Note the simple pits and pointed ends of this most advanced of wood fiber types. b-d, fiber-tracheids. Note the progressive loss of the pit border from d to b. e-f, tracheids. The cell figured in e has circular-bordered pits whereas the element pictured in f has scalariform pits. g-k, vessel elements. g, the cell pictured here represents a primitive vessel element type with many-barred scalariform perforation plates that are fully bordered (inset) and scalariform intervascular pitting. h, this cell also has scalariform perforation plates but the number of bars is reduced in number and they are bordered only at the ends (inset). The intervascular pitting is composed of circular bordered pits arranged in an opposite manner. i, this vessel element represents an intermediate stage in evolution with a scalariform perforation plate at one end and a simple perforation plate at the other end. Intervascular pitting is still essentially opposite. j, the cell pictured here is further advanced as evidenced by the exclusively simple perforation plates which are tending to become transverse in orientation. The intervascular pitting is best described as transitional between opposite and alternate. k, this vessel element is the most advanced type illustrated with simple, transverse perforation plates and the alternate intervascular pitting. Also note its short length and increase in width. l-n, crystal types. l, druse crystal. m, prismatic crystal. n, raphides. o, septate libriform wood fiber. p, vasicentric tracheid. Note the irregular shape of this element.

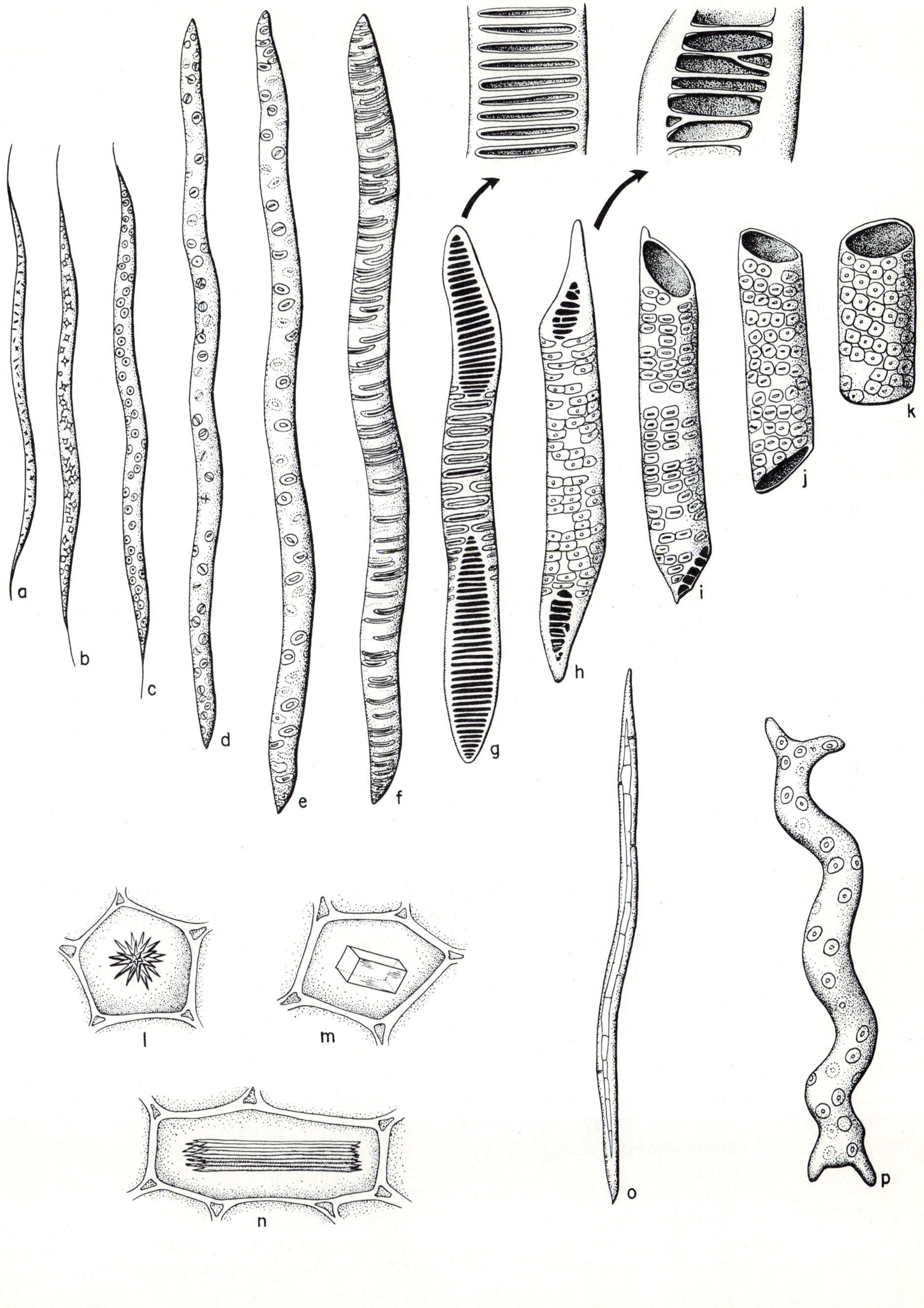
a
b
c
d
e
f
g
h
i
j
k
l
m
n
o
p

to record observations. For more complete wood descriptions a form like the one used by Zamuco (1965) is recommended.

I. GROWTH LAYERS [ANNUAL RINGS] (PRESENT OR ABSENT).

A layer of wood or bark produced apparently during one growing period; frequently, especially in woods of the temperate zones, divisible into early and late wood or bark.

II. TRACHEIDS AND FIBERS (IMPERFORATE TRACHEARY ELEMENTS).

A. Types (see Figure 7-1).
1. Tracheid. An imperforate wood cell with bordered pits to congeneric elements.
2. Vasicentric tracheid. A short, irregularly-formed tracheid in the immediate proximity of a vessel and not forming part of a definite axial row.
3. Vascular tracheid. An imperforate cell resembling in form and position a small vessel member.
4. Fiber. A general term of convenience in wood anatomy for any long, narrow cell of wood or bast other than vessels and parenchyma.
NOTE: Often further qualified as wood fibers or bast fibers; the former including both the tracheids of gymnosperms and libriform wood fibers and fiber-tracheids of woody angiosperms. Also used loosely for wood elements in general.
5. Fiber-tracheid. A fiber-like tracheid; commonly thick-walled with a small lumen, pointed ends, and bordered pit pairs having lenticular to slit-like apertures. This term is applicable to the late wood tracheids of gymnosperms as well as to the fiber-like tracheids of woody angiosperms.
6. Septate fiber-tracheid. A fiber-tracheid with thin transverse walls across the lumen.
7. Libriform wood fiber. An elongated, commonly thick-walled cell with simple pits; usually distinctly longer than the cambial initial as inferred from the length of vessel members and parenchyma strands.
8. Septate wood fiber. A fiber with thin transverse walls across the lumen.
9. Gelatinous fiber. A fiber having a more or less unlignified inner wall with a gelatinous appearance.

B. Wall Thickness, Sculpture and Cell Length
1. Wall thickness.
a. Very thin - lumen much greater than thickness of walls.
b. Thin - lumen greater than thickness of walls.
c. Thick - lumen less than thickness of walls.
d. Very thick - lumen almost completely closed.
2. Size of pits - minute, small, large.
3. Length (measure 100 cells from macerations). Record range, most frequent range, and mean. [μm = micrometer]
a. Very short .... up to 1000μm
b. Short .... 1000 - 1500μm
c. Long .... 1500 - 2000μm
d. Very long .... over 2000μm
4. Spiral Thickenings (Record presence or absence). Helical ridges on the inner face of, and part of, the secondary wall.

III. VESSELS (PERFORATE TRACHEARY ELEMENTS).

A vessel is an axial series of cells (vessel members or elements) that have coalesced to form an articulated tube-like structure of indeterminate length; the pits to congeneric elements are bordered. Among dicotyledons, the primitively vesselless condition is encountered in the Trochodendraceae, Tetracentraceae, Winteraceae, Monimiaceae and Chloranthaceae.

A. Frequency and Distribution (see Figure 7-2).

1. Number per sq. mm. (as seen in transverse section). Count from several fields. Record range, most frequent range, and mean.
   a. Very few .... up to 2
   b. Few 2- 5
   c. Moderately few 5-10
   d. Moderately numerous 10-20
   e. Numerous 20-40
   f. Very numerous .... over 40
2. Pore (vessel) distribution.
   a. Solitary pores. A pore completely surrounded by other elements.
   b. Pore chain. A series or line of adjacent solitary pores.
   c. Pore multiple. A group of two or more pores crowded together and flattened along the lines of contact so as to appear as subdivisions of a single pore. The most common type is a radial pore multiple, in which the pores are in radial files with flattened tangential walls between them. Another type is a pore cluster, in which the grouping is irregular.
3. Diffuse-porous wood. Wood in which the pores are of fairly uniform or only gradually changing size and distribution throughout a growth ring.
4. Semi-ring porous wood. Wood in which the early wood is marked by a zone of (a) occasional large vessels, or (b) numerous small vessels.
5. Ring-porous wood. Wood in which the pores of the early wood are distinctly larger than those of the late wood and form a well-defined zone or ring.

B. Vessel Element Shape, Wall Thickness and Size.

1. Angular or circular (as seen in transverse section).
2. Thin or thick walled.
3. Diameters (as measured in transverse section). Measure 100 cells. Record range, most frequent range, and mean.
   a. Extremely small .... up to 25μm
   b. Very small 25- 50μm
   c. Moderately small 50-100μm
   d. Medium-sized 100-200μm
   e. Moderately large 200-300μm
   f. Very large 300-400μm
   g. Extremely large .... over 400μm
4. Length of vessel elements (total length - tip to tip). Measure cells from macerations. Record range, most frequent range, and mean.
   a. Extremely short .... less than 175μm
   b. Very short 175-250μm

Figure 7-2. Transverse sections of dicotyledonous woods. All X 117. A, *Trochodendron aralioides* (Trochodendraceae), a primitively vesselless angiosperm showing the secondary xylem composed of tracheids (T). Distinction can be noted between tracheids of the late wood (LW) and early wood (EW). B, *Anneslea crassipes* (Theaceae). This wood shows solitary, very angular vessel elements (V) and mostly diffuse axial parenchyma. The imperforate tracheary elements are fiber-tracheids. C, *Dillenia pentagyna* (Dilleniaceae). Although this wood is still of a comparatively primitive nature the vessel elements are tending to become rounded in outline and the axial parenchyma is both paratracheal scanty and apotracheal diffuse to diffuse-in-aggregates. The imperforate tracheary elements are fiber-tracheids (FT). D, *Scytopetalum tieghemii* (Scytopetalaceae). In this wood the vessel elements are distributed as solitary pores and radial pore multiples. Imperforate tracheary elements are thick-walled libriform fibers. E, *Bombax ellipticum* (Bombacaceae). In this section note the rounded vessel elements and the diffuse-in-aggregates parenchyma. F, *Acer saccharum* (Aceraceae). In this comparatively advanced wood the vessel elements occasionally occur in pore clusters. The pores are circular in outline and the imperforate tracheary elements are libriform fibers. Rays are homocellular.

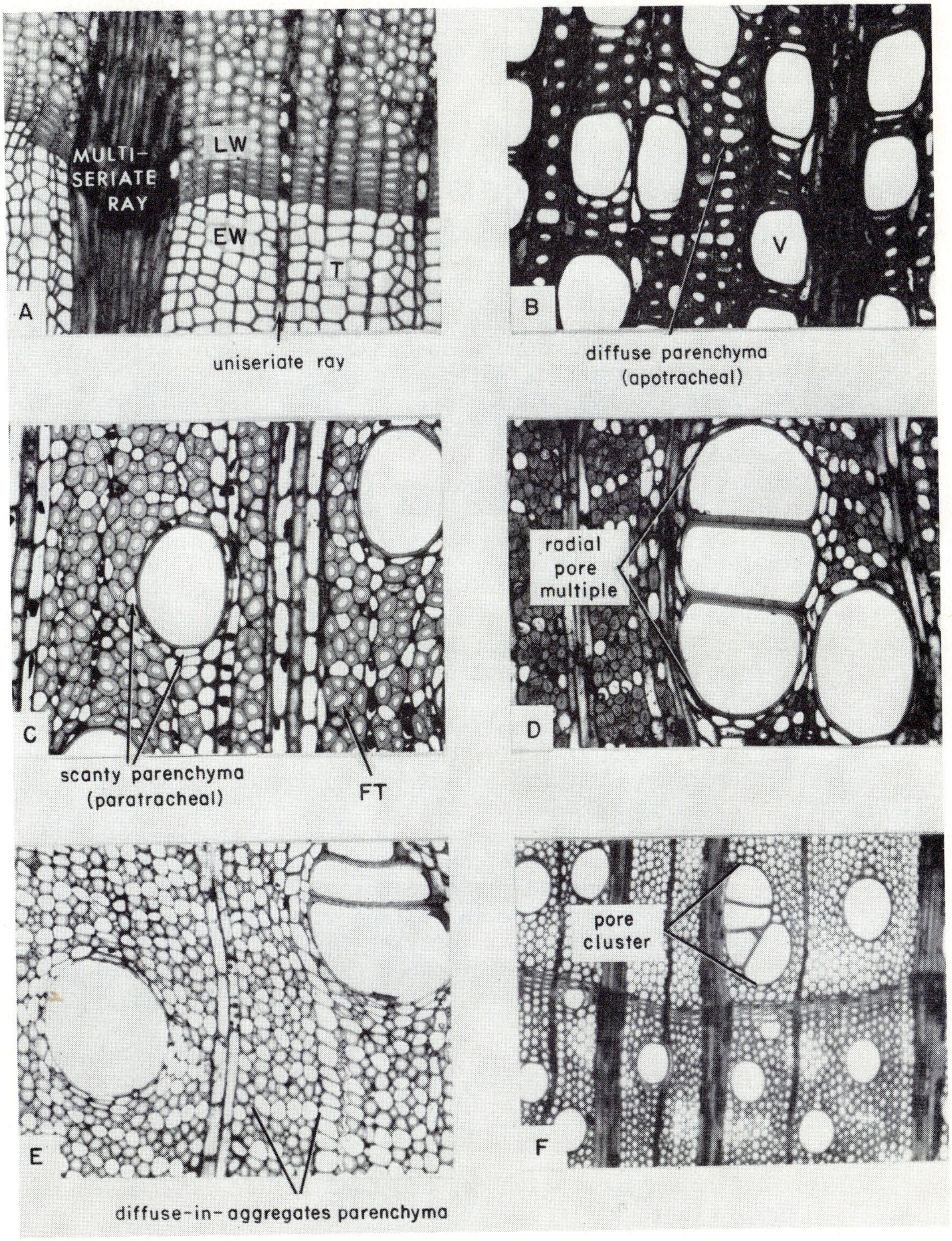
MULTI-
SERIATE
RAY
LW
EW
T
A
uniseriate ray
V
B
diffuse parenchyma
(apotracheal)
C
scanty parenchyma
(paratracheal)
FT
radial
pore
multiple
D
E
diffuse-in-aggregates parenchyma
pore
cluster
F

c. Moderately short 250- 350μm
d. Medium-sized 350- 800μm
e. Moderately long 800-1100μm
f. Very long 1100-1900μm
g. Extremely long .... over 1900μm

C. Perforation plates. A term of convenience for the area of the wall (originally imperforate) involved in the coalescence of two members of a vessel.
  1. Types (see Figure 7-1).
    a. Perforation plate scalariform. A plate with multiple perforations elongated and parallel. The remnants of the plate between the openings are called bars. NOTE: Record number of bars (range, most frequent range, and mean), and whether bars are completely bordered, bordered to middle, bordered at ends, or non-bordered.
    b. Perforation plate simple. A single and usually large and more or less rounded opening in the perforation plate.
    c. Perforation plates simple and scalariform.
    d. Perforation plate reticulate. A plate with multiple perforations having a net-like appearance (as in certain Bignoniaceae).
  2. End walls. Record whether oblique (express as angle) or transverse.

D. Pitting. A recess in the secondary wall of a cell, together with its external closing membrane; open internally to the lumen.
  1. Intervascular pitting. A term used (a) in a wide sense for pitting between tracheary elements, and (b) in a narrower sense in wood anatomy for pitting between vessel members. Record sparse or crowded, shape of pits (circular, square, pentagonal, etc.), and size of pits - Minute (less than 4μm), Small (4-7μm), Medium (7-10μm), Large (10-15μm), and Very Large (over 15μm). (See Figure 7-1).
    a. Scalariform. Pitting in which elongated or linear pits are arranged in a ladder-like series.
    b. Opposite. Multiseriate pitting in which the pits are in horizontal pairs or in short horizontal rows. NOTE: when pits are crowded the outlines of the borders tend to become rectangular in surface view.
    c. Alternate. Multiseriate pitting in which the pits are in diagonal rows. NOTE: when the pits are crowded the outlines of the borders tend to become hexagonal in surface view.
  2. Ray-Vessel pitting. Pitting between a ray cell and a vessel member. Record size, shape, whether scalariform, opposite, alternate, or irregular in distribution, and whether bordered or simple.

E. Other Features.
  1. Tylosis. An outgrowth from an adjacent ray or axial parenchyma cell through a pit cavity in a vessel wall, partially or completely blocking the vessel lumen. Tyloses may be few or many crowded together; thin- or thick-walled; pitted or unpitted; with or without starch, crystals, resins, gums, etc.

IV. RAYS.

A ribbon-like aggregate of cells extending radially in the xylem and phloem.

A. <u>Types</u> (see Figure 7-3).
   1. Heterocellular ray. Ray tissue in which the individual rays are composed of both procumbent and square or upright cells.
   2. Homocellular ray. Ray tissue in which the individual rays are composed wholly of procumbent cells; or all square or upright cells.
   3. Aggregate ray. A group of small, narrow, xylem rays appearing to the unaided eye or a low magnification as a single large ray.

B. <u>Abundance</u>. Number per sq. mm. (as seen in tangential section).
   1. Very few .... up to 2
   2. Few 2- 4
   3. Moderately numerous 4- 7
   4. Numerous 7-10
   5. Very numerous .... over 10

C. <u>Size</u>.
   1. Width (number of cells wide).
      a. Uniseriate ray. A ray one cell wide as seen in tangential section.
      b. Multiseriate ray. A ray two or more cells wide as seen in tangential section. Record range and most frequent range.
   2. Height (number of cells high). Record range and most frequent range.

D. <u>Pitting</u> (between ray cells and other parenchyma cells). Record size and number - few, many, clustered.

E. <u>Specialized Cells and Contents</u>.
   1. Lignified cell walls or nonlignified cell walls.
   2. Intercellular canals. A tubular intercellular space of indeterminate length, generally serving as a repository for resin, gum, etc., secreted by the epithelium.
      a. Lysigenous canal. Originating by dissolution of cells.
      b. Schizogenous canal. Originating by separation of cell walls along the middle lamella.
   3. Latex tubes. A laticifer enclosed by a ray.
   4. Oil cell. A specialized cell of the ray or axial parenchyma containing oil, typically rounded in outline. NOTE: limited to woody dicotyledons and similar to a mucilage cell, except for contents.
   5. Crystalliferous cell. A cell containing one or more crystals.
   6. Sheath cell. One of a series of upright cells on the margins of, and tending to form a sheath around, the procumbent cells of a multiseriate ray as seen in tangential section.
   7. Tile cell. A specialized type of apparently empty upright ray cell of approximately the same height as the procumbent ray cells and occurring in indeterminate horizontal series usually interspersed among the procumbent cells. NOTE: common in certain of the Tiliales and Malvales.
   8. Sclerotic ray cell. Ray cell with thick, often lignified secondary walls.

V. AXIAL PARENCHYMA.

Parenchyma cells derived from fusiform cambial initials.

A. <u>Distribution</u> (as seen in transverse section). (See Figure 7-2).

Figure 7-3. Tangential sections of dicotyledonous woods. All X 117. A, *Eupomatia laurina* (Eupomatiaceae). The wood of *Eupomatia* is comparatively primitive in structure. The multiseriate rays are decidedly heterocellular and composed of nearly all upright cells. Sheath cells (S) are of frequent occurrence. B, *Magnolia grandiflora* (Magnoliaceae). This wood shows both uniseriate and biseriate heterocellular rays. The uniseriate wing on the biseriate rays is frequently reduced to a single cell. C, *Ilex opaca* (Aquifoliaceae). This wood also illustrates the occurrence of both uniseriate and multiseriate rays. The uniseriate rays are composed almost exclusively of upright cells whereas the main body of the multiseriate ray consists of procumbent cells with only the cells of the uniseriate wing having an upright orientation. D, *Quercus alba* (Fagaceae). This species of oak illustrates the very wide nature of some multiseriate rays. The rays are nearly homocellular. E, *Gonystylus warburgianus* (Thymelaeaceae). The rays of this species are predominantly uniseriate and almost homocellular. F, *Populus deltoides* (Salicaceae). The rays of this plant are exclusively uniseriate and homocellular. Two adjoining vessel elements (V) are also shown.

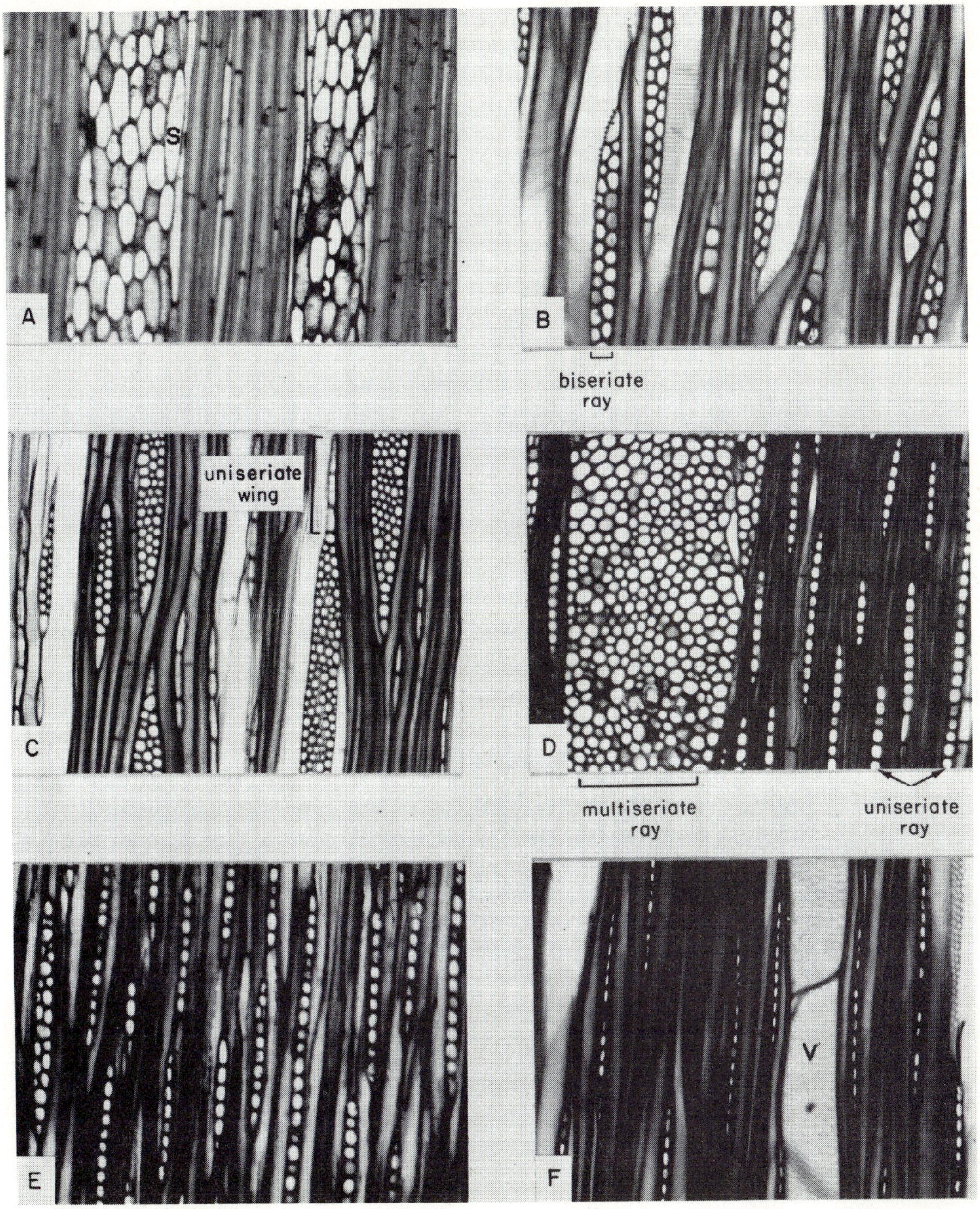
A
S
B
biseriate ray
C
uniseriate wing
D
multiseriate ray
uniseriate ray
E
F
V

1. Apotracheal parenchyma. Axial parenchyma typically independent of the pores or vessels.
   a. Terminal parenchyma. Apotracheal parenchyma cells occurring either singly or forming a more or less continuous layer of variable width at the close of a season's growth.
   b. Diffuse parenchyma. Single apotracheal parenchyma strands or cells distributed irregularly among fibers.
   c. Diffuse-in-aggregates. Apotracheal parenchyma cells that tend to be grouped in short tangential lines from ray to ray.
   d. Banded apotracheal parenchyma. Axial parenchyma forming concentric lines or bands. NOTE: termed Apotracheal Banded, if typically independent of the vessels; Paratracheal Banded, if associated with the vessels.
2. Paratracheal parenchyma. Axial parenchyma associated with the vessels or vascular tracheids.
   a. Scanty parenchyma. Incomplete sheaths or occasional parenchyma cells around the vessels.
   b. Vasicentric parenchyma. Paratracheal parenchyma forming a complete sheath around a vessel, of variable width and circular or slightly oval in cross section.
   c. Aliform parenchyma. Paratracheal parenchyma with wing-like lateral extensions, as seen in cross section.
   d. Confluent parenchyma. Coalesced aliform parenchyma forming irregular tangential or diagonal bands as seen in cross section.

B. <u>Pitting</u> (between axial parenchyma cells). Record size and number--few, many, or clustered.

C. <u>Specialized types</u>.
   1. Fusiform parenchyma cell. An axial parenchyma cell, derived from a fusiform initial without subdivision.
   2. Septate parenchyma cell. An axial or radial parenchyma cell with one or more thin transverse walls across its lumen. NOTE: in these elements the protoplast divides after the formation of the secondary cell wall.
   3. Crystalliferous axial parenchyma cell.
   4. Chambered crystalliferous cell. A crystalliferous cell that is divided into compartments by septa.
   5. Sclerotic parenchyma cell.

VI. OTHER FEATURES.

A. <u>Storied</u>. A term applied to the axial cells and rays in wood when these are arranged in horizontal series on tangential surfaces. NOTE: the term is applied to particular tissues, e.g., "storied parenchyma" or used in a general sense, as in "woods with storied structure."

B. <u>Crystals</u> (present or absent). (See Figure 7-1).
   1. Acicular. A slender, needle-shaped crystal. NOTE: not to be confused with a styloid, which is a columnar crystal.
   2. Crystal Sand. A granular mass of very fine crystals.
   3. Druse. A globular cluster of crystals, sometimes with an organic core, either attached to the wall by a peg or lying free in the cell.

4. Raphide (pl. raphides). A needle-shaped (acicular) crystal occurring typically as one of a closely packed, sheath-like bundle.
5. Styloid. An elongated crystal, typically about four times as long as broad, with pointed or square ends.
6. Prismatic. A solitary crystal either a rhombohedron or octahedron.

C. Intercellular canals. Tubular intercellular space of indeterminate length.

D. Included phloem. Phloem strands or layers included in the secondary xylem of certain dicotyledonous woods.

E. Vestured pits. A bordered pit with the pit cavity wholly or partially lined with projections from the tertiary wall.

F. Fibriform vessel element. A vessel element of relatively small diameter bearing a resemblance to a fiber-tracheid.

G. Disjunctive tracheid. A tracheid partly disjoined laterally from another during differentiation; contact is maintained by means of tubular processes.

H. Disjunctive parenchyma. Axial or radial parenchyma cells partially disjoined during the process of differentiation; contact is maintained by means of tubular processes.

SECONDARY XYLEM LITERATURE

I. Periodicals.

Tropical Woods. New Haven, Conn. Yale University School of Forestry. (This journal began publication in 1925 and terminated with No. 113 in 1960).

II. General References.

Bailey, I. W. 1957. The potentialities and limitations of wood anatomy in the phylogeny and classification of angiosperms. Journal of the Arnold Arboretum (Harvard University) 38: 243-254.

Chalk, L. 1937. The phylogenetic value of certain anatomical features of dicotyledonous woods. Annals of Botany, N.S., 1: 409-428.

________. 1944. On the taxonomic value of the anatomical structure of the vegetative organs of the dicotyledons. 2. The taxonomic value of wood anatomy. Proceedings Linnean Society of London 155: 214-218.

Committee on Standardization of Terms of Cell Size, International Association of Wood Anatomists, 1937. Standard terms of length of vessel members and wood fibers. Tropical Woods 51: 21.

________. 1939. Standard terms of size for vessel diameter and ray width. Tropical Woods 59: 51-52.

Tippo, O. 1941. A list of diagnostic characteristics for descriptions of dicotyledonous woods. Transactions Illinois Academy of Science 34: 105-106.

________. 1946. The role of wood anatomy in phylogeny. American Midland Naturalist 36: 362-372.

Zamuco, I. T. 1965. Bark and wood anatomy of four Philippine bast-fiber producing trees. The Philippine Journal of Forestry 21: 127-146.

TECHNIQUES EMPLOYED IN THE STUDY OF WOOD ANATOMY

I. MACERATIONS.

Maceration is a chemical method of dissociating elements by dissolving out the intercellular cement substance, thus allowing the investigator a clear three-dimensional picture of the cell types comprising the tissues.

1. In macerating dry woods, split the material into slivers no thicker than a toothpick, then boil in water until all pieces are thoroughly saturated.
2. Transfer the slivers to a shell vial and cover them with Jeffrey's macerating fluid (equal volumes of 10% aqueous nitric acid and 10% aqueous chromic acid). Stopper the vial and place in a 60° paraffin oven for 24-48 hours. The time varies with the hardness and thickness of the material, so check regularly at the end of the first day; when the maceration process has progressed sufficiently, the wood elements will separate readily when teased apart with needles.
3. Wash the material thoroughly in running water to remove all traces of the acids, then dehydrate rapidly by decanting and adding increasingly stronger alcohols; e.g., 50, 75, 95, and 100%.
4. Staining is effected by transferring the wood slivers into a 1% solution of light green in 75 parts clove oil and 25 parts absolute alcohol. If a red stain is preferred, stain 4-6 hours in a 1% solution of safranin in 70% alcohol before completing the dehydration series.
5. After about ten minutes the material can be removed from the stain and the excess stain can be removed by blotting with a filter paper. A sliver of wood is then placed on a clean glass slide in a drop of xylol and mounting medium, and the elements can be teased apart and carefully spread around with dissecting needles. Avoid too much material on one slide.
6. Add cover slip to slide and place on a slide warmer to dry.

II. USE OF THE SLIDING MICROTOME.

1. Soft woods can often be sectioned successfully on a sliding microtome without embedding, following their saturation in boiling water. Hard woods may also be cut unembedded either by the use of a jet of super-heated steam directed upon the wood clamped in the microtome, or by preliminary softening of the wood in hydrofluoric acid. Some woods can be sectioned successfully only after being embedded in celloidin.

   Trim the wood carefully into blocks measuring approximately 1 x 1 x 2 cm., and prepare them by boiling in water and pumping in cold water until the wood is thoroughly saturated and sinks.

   Clamp the knife firmly in the knife holder of the sliding microtome and adjust the three clamping screws to give the desired cutting and clearance angles. (The best cutting angle for the knife edge, with reference to the line of travel, usually ranges from 30 to 35°. A greater angle is necessary when cutting harder objects.) Fasten the block of wood in the split-ball clamp, and adjust the universal joint until the desired plane of sectioning is parallel to the plane of travel of the knife. A thickness of 15 to 20μ is satisfactory for most woody subjects. Cut cross, radial, and tangential sections and place them in 95% alcohol in a Syracuse watch glass. In the event that sections tend to curl after removal from the knife, press the sections, with the concave side down, on a slide covered with a film of glycerine-alcohol or use a detergent solution. Then place a dry slide over the sections, weight with lead blocks and cover with water in a Petri dish or finger bowl. The water will render the sections flexible and permit them to flatten, so that they can be properly stained and mounted.

2. Staining.

Run sections down to water rapidly, using a section lifter or camel's hair brush to make the transfers between solutions in watch glasses. Mordant 10-20 minutes in a 4% solution of ferric ammonium sulphate ("iron alum"), followed by at least three washes in tap water. Care should be taken that neither of the first two washes remain very long on the sections or ferric hydroxide will form in the tissues and spoil their staining quality. The last tap water wash should be allowed to stand on the tissues 9-10 minutes, followed by a distilled water rinse to remove the salts dissolved in the tap water. Transfer sections to a watch glass containing 3-4 drops of 0.5% aqueous Heidenhain's hematoxylin for 10-20 minutes, or until cell walls appear grayish-black when observed in water under the microscope. Wash out excess stain in three changes of distilled water. (If sections are overstained, destain in 2% iron alum, washing this out carefully with tap and distilled water as before). Transfer sections to a watch glass containing distilled water plus 4-5 drops of a 1% solution of safranin in 50 or 70% alcohol, and leave in the stain overnight. Use three changes of distilled water, then dehydrate in 50, 70, 95, and absolute alcohol, allowing about 5 minutes in each. Transfer sections to equal proportions of absolute alcohol and xylol, and complete the clearing in two changes of xylol, ca. 5 minutes in each. Mount transverse, radial, and tangential sections all on one slide in balsam. Dry slides on a 50° warming plate for 24 hours, then place lead weights on the cover slips to force out the excess dammar and to further flatten the sections while drying.

Section B. LEAF ANATOMY

Carlquist (1961) has stated that the leaf is "perhaps anatomically the most varied organ of angiosperms, ...". It is not surprising, therefore, that this organ possesses many anatomical features of potential taxonomic significance. Although features of the leaf are often diagnostic at the genus and species level, these characters cannot at our present state of knowledge be readily interpreted evolutionarily or accounted for by environmental conditions. If fresh or preserved specimens are available, carefully prepared freehand sections often reveal considerable anatomical detail; however, paraffin-embedded and sectioned leaves are most desirable. In many instances, epidermal characters can be studied by making simple epidermal peels. Among the characters which have proven to be of systematic value are the following:

I. CUTICULAR CHARACTERS.

Record distribution and orientation of papillae, striae, and rods.

II. EPIDERMIS.

The outer layer of cells primary in origin. If it is multiseriate (multiple epidermis), only the outermost layer differentiates epidermal characteristics.

A. Uniseriate or multiseriate. Record number of cells in latter.
B. Thickness of cell walls. Record thin-walled, thick-walled or unevenly thickened.
C. Sclerification of epidermis. Record presence or absence.
D. Size, shape and contents of cells.
    1. Isodiametric, polygonal or transversely extended in surface views. Anticlinal walls wavy or straight.
    2. Cells in files.
    3. Differentiation into long and short cells (Gramineae).
    4. Distribution of cork cells (Gramineae).
    5. Epidermis papillose. Record occurrence on lower, upper, or both surfaces, and size and shape of papillae.
    6. Epidermis mucilaginous. Record occurrence and nature of mucilage.
    7. Distribution and nature of bulliform and colorless cells (Gramineae). A bulliform cell is an enlarged epidermal cell occurring in longitudinal rows of similar cells in leaves of grasses.

III. STOMATA.

A pore in the epidermis.

A. Subsidiary cells. Epidermal cells associated with a stoma and at least morphologically distinguishable from the epidermal cells composing the ground mass of the tissue. Also called accessory cells.
    1. Classification of angiosperm stomata based on appearance of mature stomata. Definitions after Van Cotthem (1970). (See Figure 7-4).
        a. Anomocytic (irregular-celled, ranunculaceous type). Stoma surrounded by a limited number of cells that are indistinguishable in size, shape, or form from those of the remainder of epidermis.

b. Anisocytic (unequal-celled, cruciferous type). Stoma surrounded by three cells of which one is distinctly smaller than the other two.
c. Paracytic (parallel-celled, rubiaceous type). Stoma accompanied on either side by one or more subsidiary cells parallel to the long axis of the pore and guard cells.
d. Diacytic (cross-celled, caryophyllaceous type). Stoma enclosed by a pair of subsidiary cells whose common wall is at right angles to the guard cells.
e. Tetracytic. Four subsidiary cells are present, two lateral and two terminal. Characteristic of many monocotyledons.
f. Actinocytic. Stoma surrounded by a circle of radially elongate subsidiary cells.
g. Cyclocytic. Stoma surrounded by four or more subsidiary cells which form a ring around each stomata.
h. Hexacytic. Stoma accompanied by six subsidiary cells consisting of two lateral pairs parallel to the long axis of the pore and two polar (terminal) cells; the second lateral pair as long as the stomatal complex. (This type could be described as a modification of the tetracytic type with an additional pair of lateral subsidiary cells.)

2. Classification of angiosperm stomata based on ontogeny.
   a. Mesogenous. Stoma in which the subsidiary cells are produced from the same meristemoid (initial) as the guard cell initials.
   b. Perigenous. Stoma in which the subsidiary cells do not have a common origin with the guard cells, but are formed by cells lying around the meristemoid that divides to form the guard cells.
   c. Mesoperigenous. Stoma in which at least one of the subsidiary cells has a common origin with the guard cells, but the others do not.
   d. Helicocytic. A type of mesogenous stoma in which a helix of four or more subsidiary cells surround the guard cells.
   e. Allelocytic. A type of mesogenous stoma in which an alternating complex of three or more C-shaped subsidiary cells of graded sizes occurs. This pattern may be further divided into:
      1. parallelocytic. Guard cells develop parallel to the subsidiary cells.
      2. diallelocytic. Guard cells develop at right angles to subsidiary cells.
3. Classification of gymnosperm stomata based on ontogeny.
   a. Haplocheilic. Stoma in which the subsidiary cells are not derived from the same meristemoid as the guard cells.
   b. Syndetocheilic. Stoma in which the subsidiary cells are derived from the same meristemoid as the guard cells.

B. Guard cells. In stoma. Two cells which by changes in turgescence open or close the stomatal opening. Variable features include cell shape (symmetrical, slightly asymmetrical, or markedly asymmetrical) and size.

C. Distribution. Stomata may be confined to the upper leaf surface (epistomatic), lower leaf surface (hypostomatic), or occur on both upper and lower surfaces (amphistomatic).

IV. HYPODERMIS.

General term for layer, or layers, of cells beneath the epidermis which are morphologically distinct from underlying cortical layers. Record prescence or absence, location, number of layers, and sclerotic condition.

V. SCLERENCHYMA.

A. <u>Sclereid(s)</u>. A sclerenchyma cell, varied in form, but typically not much elongated, and having thick lignified secondary walls with many pits.
   1. Types based on form (see Figure 7-4).
      a. Astrosclereid. A branched (ramified) type of sclereid.
      b. Brachysclereid. A short, roughly isodiametric sclereid, resembling a parenchyma cell in shape. A "stone cell."
      c. Filiform sclereid. A much elongated, slender sclereid resembling a fiber.
      d. Macrosclereid. A somewhat elongated sclereid with unevenly distributed secondary walls.
      e. Osteosclereid. Sclereid having the shape of a bone, with a columnar middle part and enlargements at both ends.
      f. Trichosclereid. A type of branched sclereid with thin hair-like branches extending into intercellular spaces.
   2. Types based on distribution.
      a. Diffuse sclereid. Sclereid dispersed in the leaf mesophyll.
      b. Terminal sclereid. Sclereid confined to the ends of small veins.

B. <u>Fiber(s)</u>. An elongated tapering sclerenchyma cell with a more or less thick secondary wall, with or without lignin.
   1. Girders. A sclerenchymatous connection between a vascular bundle and the epidermis.
   2. Incomplete (partial) girders. A sclerenchymatous extension from the bundle sheath which does not reach the epidermis.
   3. Strand. Sclerenchyma free from the vascular bundle sheath and which may be subepidermal.

VI. MESOPHYLL.

Photosynthetic parenchyma tissue of a leaf located between epidermal layers.

A. <u>Types</u>.
   1. Bifacial. Palisade parenchyma on one side of the blade and spongy parenchyma on the other side.
   2. Isolateral. Palisade parenchyma on both sides of the blade.

B. <u>Construction</u>.
   1. Relative differentiation into palisade and spongy layers and number of layers in each.
   2. Distribution and shape of chlorenchyma.
   3. Shape of mesophyll cells. Record isodiametric, stellate, thick-walled, etc.
   4. Presence or absence of air-lacunae. Mesophyll "loose" or "compact."

C. <u>Presence, absence and distribution of secretory canals or cavities</u>.

D. <u>Form and distribution of crystals</u> (see Section A. SECONDARY XYLEM).

STOMATAL TYPES

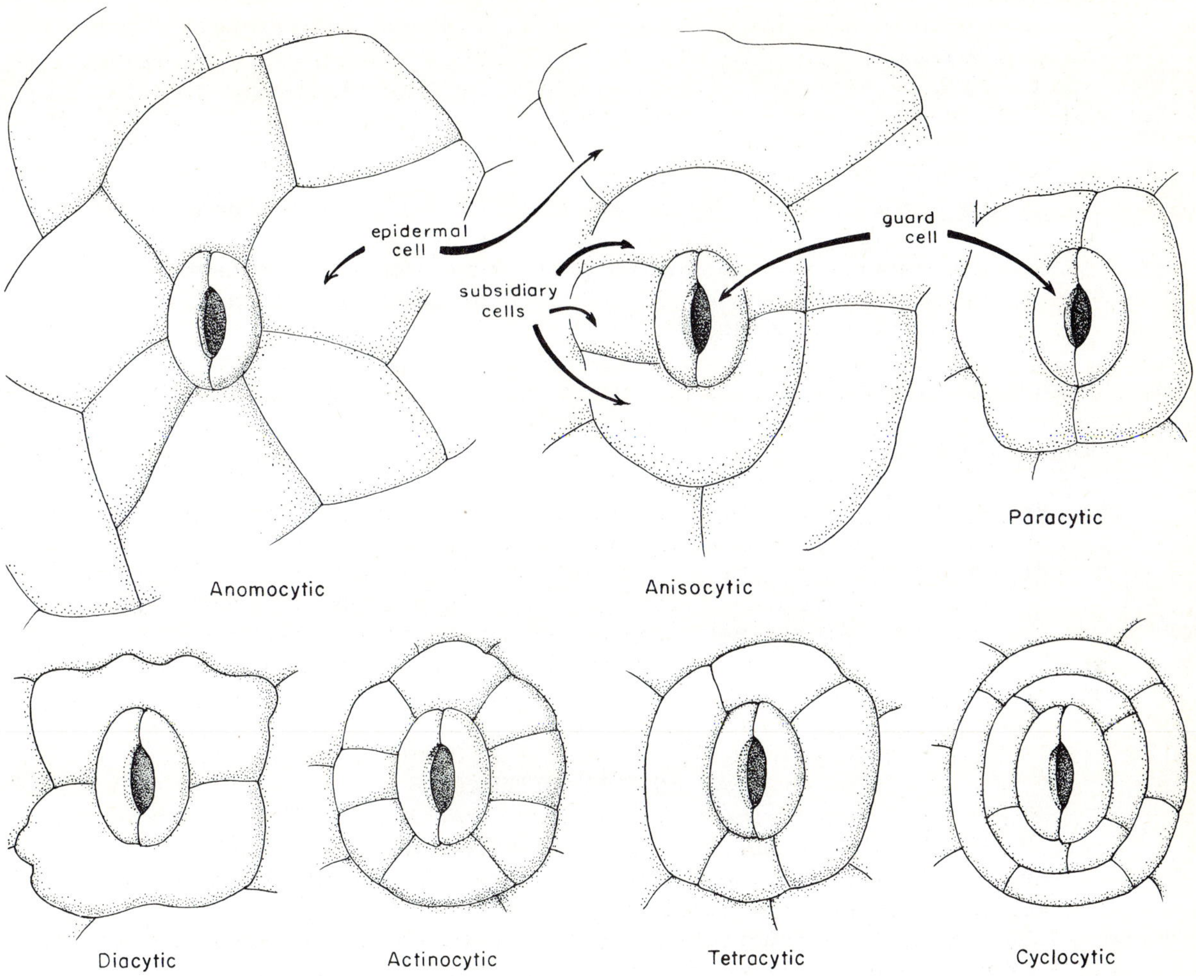

SCLEREID TYPES

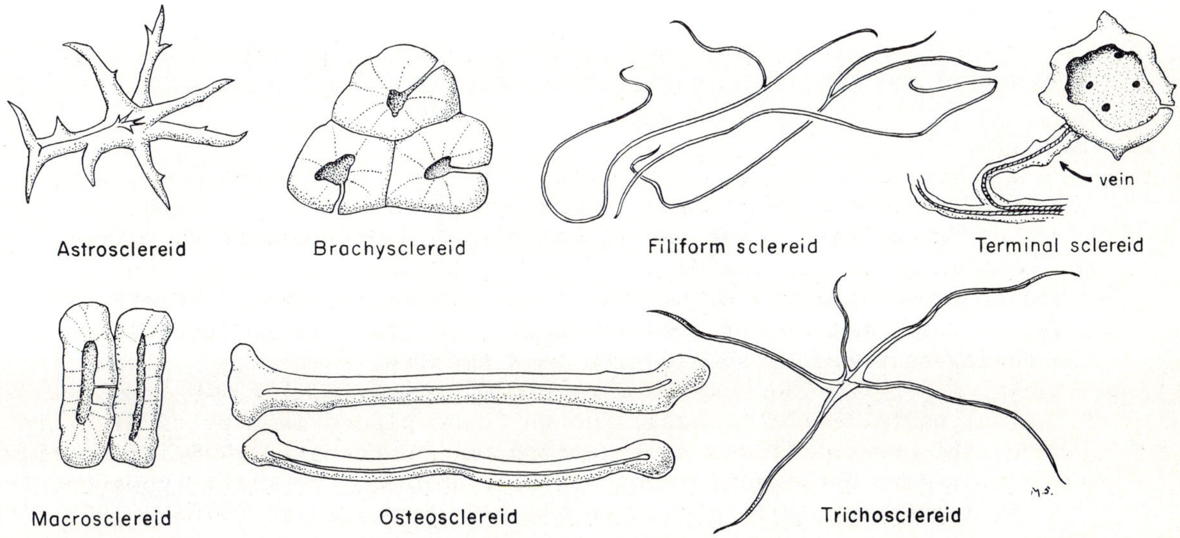

VII. SILICA BODIES.

Record shape and distribution in all tissues. Characteristic of monocotyledons.

VIII. VENATION.

A. Petiole. (see Section C).
B. Lamina. (see Section D). Record major venation pattern, minor venation pattern, including islet size and veinlet termination number, and nature of vein sheathing (parenchymatous, sclerenchymatous, or double sheaths of one or both of these types, and size and shape of sheathing cells).

IX. TRICHOMES. (See Section E).

X. OTHER CHARACTERS.

A. Presence, absence and size of hydathodes. A hydathode is a water pore or water gland, a structure that releases water.
B. Presence or absence of nectaries. A nectary is a multicellular glandular structure secreting a sugary liquid. Occurs on flowers (floral nectary) or on vegetative plant parts (extrafloral nectary).
C. Ecological modification such as presence and structure of stomatal crypts.

LEAF ANATOMY LITERATURE

Bokhari, M. H., and B. L. Burtt. 1970. Studies in the Gesneriaceae of the Old World XXXII: Foliar sclereids in *Cyrtandra*. Notes from the Royal Botanic Garden, Edinburgh 30: 11-21.
Payne, W. W. 1970. Helicocytic and allelocytic stomata: unrecognized patterns in the Dicotyledonae. American Journal of Botany 57: 140-147.
Van Cotthem, W. R. 1970. A classification of stomatal types. Botanical Journal of the Linnean Society 63: 235-246.
________. 1971. Vergleichende morphologische Studien über Stomata and eine neue Klassifikation iherer Typen. Berichte der Deutschen Botanischen Gesellschaft 84: 141-168.

TECHNIQUES EMPLOYED IN THE STUDY OF LEAF AND FLORAL ANATOMY

I. CLEARING.

In addition to the standard microtome sectioning procedures, the technique of "clearing" leaves and floral parts in order to study vasculation patterns is extremely useful and widely employed. This technique is applicable to fresh, preserved, or dried (herbarium) specimens. See Payne, W. W. 1969. A Quick Method for Clearing Leaves. Ward's Bulletin 8(61): 4-5. The following schedule should give good results.

1. Plant parts should be small enough to be placed in a petri dish (in the case of leaves some cutting may be required). Dried or liquid-preserved materials need no special prior treatment; however, fresh leaves should be boiled in alcohol or placed in Carnoys (95%) for a while, to remove chlorophyll before the clearing procedure.

2. Flood tissue with 5% NaOH (or KOH) and place in an oven at about 37° C. for one to many days depending upon the nature of the tissue.
3. After tissue has become transparent, or nearly so, wash briefly in water. NOTE: some plants contain various dark pigments that can be removed at this point by flooding tissue with Stockwell's Bleach. Stockwell's is composed of:

Water	90 cc.
Potassium bichromate	1 g.
Glacial acetic acid	10 cc.
Chromic acid	1 g.

   This solution should remain on the tissue for one to several hours at room temperature until all pigments are removed. Pour off bleach and wash thoroughly.
4. Some find it desirable at this point to use a concentrated chloral hydrate solution which will continue the clearing process if the tissue is not completely transparent. In many cases this is not necessary.
5. Tissue should now begin to be dehydrated by pouring on 30% and 50% ethyl alcohol for 5 minutes each.
6. Staining is accomplished by using 1% Safranin in 50% alcohol. Generally only a few minutes in the stain is sufficient with gentle agitation.
7. Continue dehydration using 70% and 95% alcohols. The alcohols will destain (i.e., remove the safranin) so care should be taken not to leave them on the tissue too long. Remember, however, the tissue <u>must</u> be completely dehydrated.
8. Pour on absolute alcohol for several minutes. Very little destaining will occur in the higher alcohols.
9. Pour on a solution of 1/2 absolute alcohol and 1/2 xylene for several minutes.
10. Clearing is the final step after dehydration is complete using xylene. If tissue is not completely dehydrated it will turn a milky white; if this happens, go back to absolute alcohol.
11. Mount on slide using a synthetic resin or balsam. For some floral parts it is sometimes desirable not to mount tissue on a slide but store in a vial so all aspects and views of the vasculation pattern can be studied.

II. EPIDERMAL PEELS.

Although various techniques have been published for making epidermal impressions (see previously cited paper by Payne, 1970, for a method using cellulose acetate), epidermal cell patterns can frequently be studied by simply taking a fresh or preserved leaf and with the aid of a sharp razor blade peeling off the adaxial and abaxial layer, mounting in water, and observing under the microscope. Since this preparation is a temporary one it is desirable to prepare a sketch of the epidermis using a camera lucida.

TECHNIQUE LITERATURE

Arnott, H. J. 1959. Leaf clearings. Turtox News 37: 192-194.

Lersten, N. R. 1967. An annotated bibliography of botanical clearing methods. Iowa State Journal of Science 41: 481-486.

Payne, W. W. 1969. A quick method for clearing leaves. Ward's Bulletin 8(61): 4-5.

LEAF ANATOMY EXERCISE

		Species I	Species II
Cuticular Characters			
Epidermis	Uniseriate or Multiseriate		
	Thickness of walls		
	Sclerification + -		
	Size, shape, content of cells		
Trichome Type			
Stomatal Type (prepare drawing)			
Hypodermis + - (number of layers if +)			
Sclerenchyma (type and distribution)			
Mesophyll	Bifacial or isolateral		
	Number of palisade layers		
	Shape of cells		
	Crystals (type and distribution)		
Major Venation Type			
Minor Venation Pattern			
Type of Vein Sheathing			
Other Features of Interest			

Section C. PETIOLE ANATOMY

The taxonomic usefulness of petiole vascularization patterns has become increasingly apparent in recent years. The most complete system of classification of major venation patterns in the petiole of woody dicotyledons is that of Howard (1962). This classification is presented below. In studies of this nature, it is important to examine sections throughout the length of the petiole to understand completely the changes which occur in the vascular configuration. It will be noted, accordingly, that the system below considers changes which occur in the vascular tissue from the time it leaves the stem until it becomes relatively stable in the leaf. NOTE: This classification does not account for all of the variation patterns that may exist, but does cover the more common major patterns. Also, terminology with respect to petiole vasculation patterns has not been standardized (see Figure 7-5).

I. NODE 3:3 (SEE SECTION F-I), BUNDLES FREE.

A. Three bundles throughout petiole; e.g., *Pedilanthus*.
B. Lateral traces divide.
   1. Petiole with five free traces; e.g., *Pittosporum*.
   2. Petiole with many traces in "U"-shaped configuration; e.g., *Miconia*.
   3. Free traces in a ring with medullary bundles; e.g., *Aesculus*.
   4. Free traces in a ring without medullary bundles; e.g., *Sambucus*.
C. Median trace divides and the division products assume a dorsal position; e.g., *Hibiscus*.

II. NODE 3:3, BUNDLES FUSE TO FORM AN ARC.

A. Bundles fuse and form flat arc; e.g., *Lonicera*.
B. Bundles fuse and form a flat arc with dorsal free traces; e.g., *Cornus*.
C. Bundles fuse in "U"-shaped arc; e.g., *Betula*.
D. Bundles fuse and invaginate at ends; e.g., *Congea*.

III. NODE 3:3, BUNDLES FUSE TO FORM A SIPHONOSTELE.

A. Siphonostele simple.
   1. Formed by simple fusion of traces; e.g., *Cotinus*.
   2. Formed subsequent to division of the median, the branches of which form dorsal bundles; e.g., *Acer*.
B. Siphonostele with accessory bundles.
   1. One large accessory bundle situated dorsally; e.g., *Hamamelis*.
   2. Small multiple accessory bundles dorsal; e.g., *Carya*.
   3. Accessory bundles in medullary position; e.g., *Bauhinia*.

IV. NODE 3:3, BUNDLES FUSE TO FORM MORE COMPLEX PATTERNS.

A. By invagination forming one or many medullary bundles or plates; e.g., *Quercus* or *Tilia*
B. Siphonostele invaginating forming included or dorsal accessory bundle; e.g., *Fagus*.
C. Siphonostele formed, then lateral invaginations give rise to dorsal, smaller siphonostele or plate over "U"-shaped arc; e.g., *Carpinus*.
D. Polystelic types.
   1. Axillary bud included, petiole not compressed; e.g., *Platanus* type.

COMMON PETIOLE VASCULATION PATTERNS

(Transverse Sections at Base of Lamina)

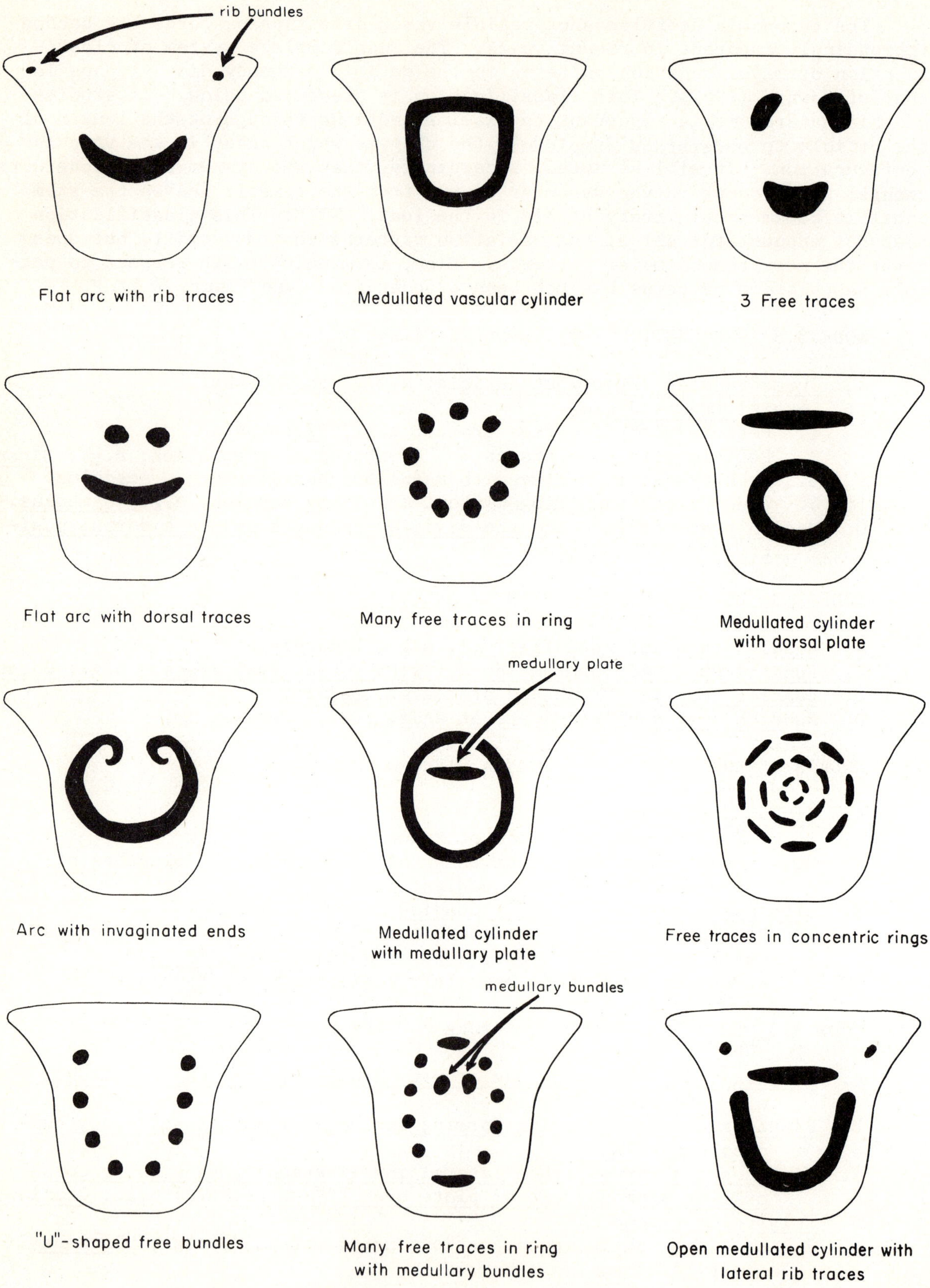

2. Axillary bud not included, petiole compressed; e.g., Populus type.

V. NODE 1:1, SIMPLE TRACE, FLAT, SLIGHTLY CURVED OR "U"-SHAPED.

A. Trace continuous.
   1. Without rib traces; e.g., Allamanda.
   2. With rib traces; e.g., Graptophyllum.
   3. Trace split longitudinally, later fusing; e.g., Eugenia.

B. Trace forming open or closed siphonostele.
   1. Stele open, with terminal rib traces; e.g., Ilex.
   2. Stele open, with lateral rib traces; e.g., Actinidia.
   3. Stele open, without rib traces; e.g., Celastrus.
   4. Stele closed, without rib traces; e.g., Terminalia.

C. Trace invaginates at ends.
   1. Ends inroll; e.g., Lyonia.
   2. Forming siphonostele with medullary plate; e.g., Cordia.
   3. Inverted plate over simple arc; e.g., Capparis.

VI. NODE 2 OR MORE TRACES FROM ONE GAP.

A. Trace bipartite.
   1. In petiole; e.g., Clerodendron.
   2. Below petiole; e.g., Calycanthus.

B. Trace tripartite or more.
   1. Bundles free, forming an arc; e.g., Bougainvillea.
   2. Bundles fused in an arc; e.g., Solandra.
   3. Bundles free, in ring; e.g., Hernandia.

VII. NODE MULTILACUNAR, MANY TRACES FROM EQUAL NUMBER OF GAPS.

A. Traces remain free.
   1. Bundles form ring or "U"; e.g., Ricinus.
   2. Bundle pattern invaginates; e.g., Rhizophora.
   3. By anastomoses forming concentric rings; e.g., Coccoloba.

B. Traces fuse.
   1. Ring simple; e.g., Dendropanax.
   2. Ring with included bundles; e.g., Macaranga.

PETIOLE ANATOMY LITERATURE

Howard, R. A. 1962. The vascular structure of the petiole as a taxonomic character. In: J. C. Garnaud (ed.) Advances in Horticultural Science and Their Application. Vol. III. Pergamon Press. New York.

Schofield, E. K. 1968. Petiole anatomy of the Guttiferae and related families. Memoirs of the New York Botanical Garden 18: 1-55.

Section D. FOLIAR VENATION*

Leaves are a generally neglected organ in taxonomic and comparative morphologic studies due in large part to the lack of a unified, unambiguous and detailed classification of their features. The purpose of this section is to present an inclusive, coherent classification of the leaf venation of dicotyledons. Many of the features to be enumerated are also found in the monocots but no specific attempt was made to extend the classification to them. This system of leaf venation partially incorporates modifications of the earlier classification of von Ettingshausen (1861) (see Figure 7-6).

I. TYPES.

A. Pinnate. With a single primary vein (mid-vein) serving as the origin for the higher order venation.
  1. Craspedodromous. Secondary veins terminating at the margin.
    a. Simple. All of the secondary veins and their branches terminating at the margin.
    b. Semicraspedodromous. Secondaries branching just within the margin, one of the branches terminating at the margin, the other joining the superadjacent secondary.
    c. Mixed. Some of the secondaries terminating at the margin and an approximately equal number of (usually intervening) secondaries otherwise.
  2. Camptodromous. Secondary veins not terminating at the margin.
    a. Brochidodromous. Secondaries joined together in a series of prominent arches.
    b. Eucamptodromous. Secondaries upturned and gradually diminishing apically inside the margin, connected to the superadjacent secondaries by a series of cross veins without forming prominent marginal loops.
    c. Reticulodromous. Secondaries losing their identity toward the leaf margin by repeated branching into a vein reticulum.
    d. Kladodromous. Secondaries freely ramified toward the margin.
  3. Hyphodromous. All but the primary vein absent, rudimentary, or concealed within the coriaceous mesophyll.

B. Parallelodromous. Two or more primary veins originating beside each other at the leaf base and running parallel to the apex where they converge.

C. Campylodromous. Several primaries or their branches, originating at, or close to, a single point and running in strongly developed, recurved arches before converging toward the leaf apex. Vein pattern convergent above and below.

D. Acrodromous. Two or more primary or strongly developed secondary veins running in convergent arches toward the leaf apex. Arches not recurved at base.
  1. Position.
    a. Basal. Acrodromous veins originating at the base of the leaf.
    b. Suprabasal. Acrodromous veins originating some distance above the leaf base.
  2. Development.
    a. Perfect. Lateral acrodromous veins well developed, running at least 2/3 of the distance to the leaf apex.

*Contributed by Leo J. Hickey, Smithsonian Institution.

b. Imperfect. Acrodromous veins thin, not prominent, running less than 2/3 of the distance to the leaf apex.

E. Actinodromous. Three or more primary veins diverging radially from a single point.

F. Palinactinodromous. Primaries having one or more subsidiary points of radiation above the lowest one; e.g., *Platanus*.

(The following categories apply to E and F above.)

1. Position of the first point of primary vein radiation.
    a. Basal
    b. Suprabasal
2. Development.
    a. Perfect. Ramification of the lateral actinodromous veins covering 2/3 of the blade area.
        1. Marginal. Actinodromous veins reaching the margin.
        2. Reticulate. Lateral actinodromous veins not reaching the margin.
    b. Imperfect. Veins originating on the lateral actinodromous primary veins covering less than 2/3 of the blade area.
        1. Marginal.
        2. Reticulate.
    c. Flabellate. Several to many equally fine basal veins diverge radially at low angles and branch apically.

II. ORDERS OF VENATION.

A. Primary Veins. The thickest vein(s) of the leaf, occurring either singly as the midvein, or as a series of veins of relatively equal thickness emerging from the petiole, or as branches of a medial or lateral primary above the base of the blade (suprabasal primaries). In the latter case, both the branch and the continuation of its primary source are of the same relative thickness at the point of branching.

1. Size. Determined midway between the leaf apex and base as the ratio of vein width (vw) to leaf width (LW). $\frac{vw}{LW} \times 100\% = \text{Size}$
    a. Massive. 4%
    b. Stout. 2-4%
    c. Moderate. 1.25-2%
    d. Weak. 1.25%
2. Course.
    a. Straight. (unbranched or branched).
    b. Markedly curved. Bent noticeably into a smooth arc.
    c. Sinuous. Repeated smooth changes in direction of curvature.
    d. Zig-zag. Repeated angular changes in direction.

B. Secondary Veins. The next smallest size class of veins (which arise from the primaries) and their branches of relatively equal thickness at point of branching.

1. Angle of divergence. Measured between the branch and the continuation of the source vein above the point of branching.
    a. Acute. Narrow, less than 45°; Moderate, 45-65°; Wide, 65-80°.
    b. Right angle or approximately so, 80-100°.
    c. Obtuse. Greater than 100°.
2. Variations in angle of divergence.
    a. Divergence angle nearly uniform. The general case, departures from which are often useful distinguishing features.
    b. Upper secondaries more obtuse than lower.

Figure 7-6

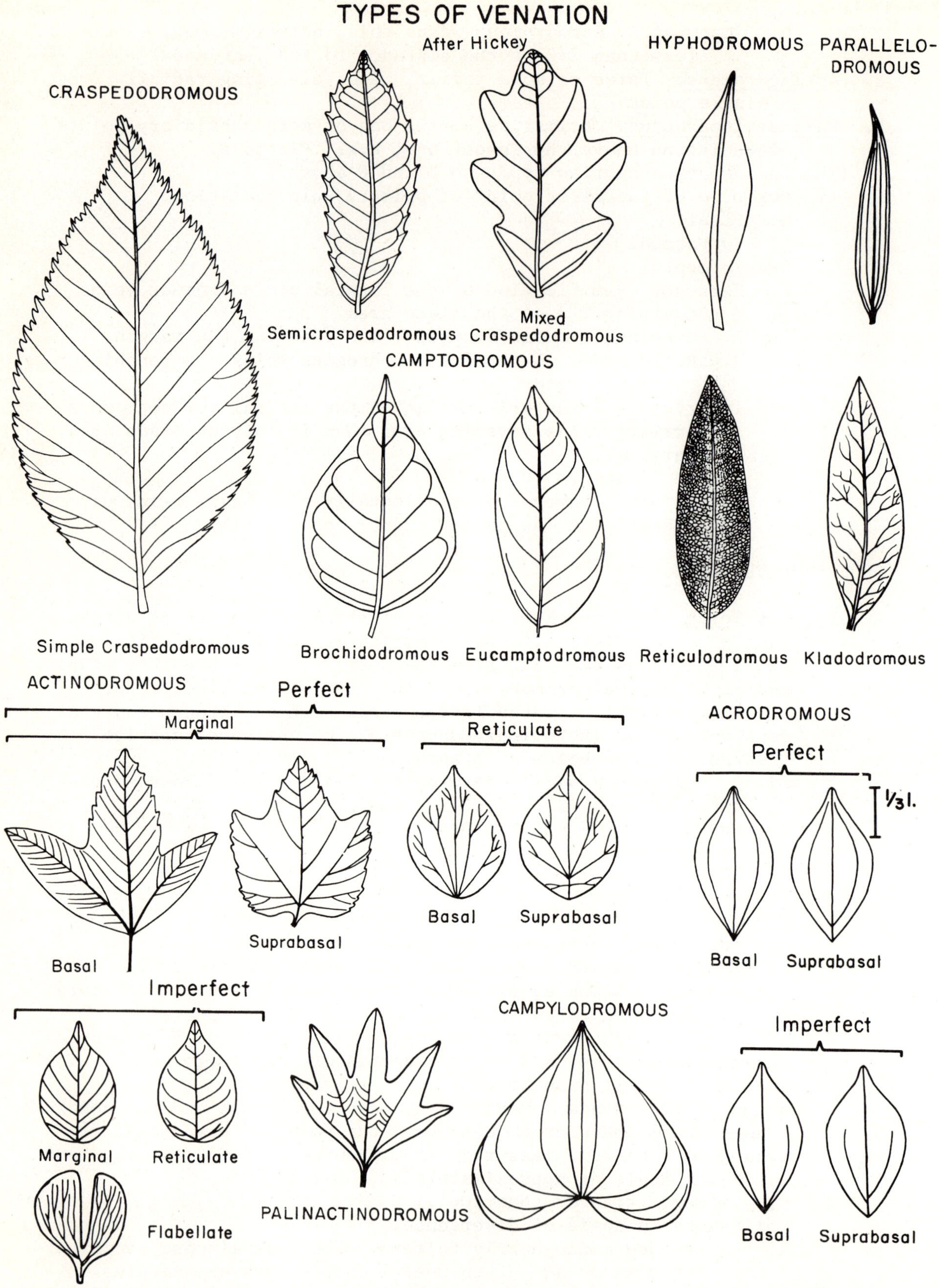
TYPES OF VENATION
After Hickey
CRASPEDODROMOUS
Simple Craspedodromous
Semicraspedodromous
Mixed Craspedodromous
HYPHODROMOUS
PARALLELO-DROMOUS
CAMPTODROMOUS
Brochidodromous
Eucamptodromous
Reticulodromous
Kladodromous
ACTINODROMOUS
Perfect
Marginal
Reticulate
Basal
Suprabasal
Basal
Suprabasal
ACRODROMOUS
Perfect
1/3 l.
Basal
Suprabasal
Imperfect
Marginal
Reticulate
Flabellate
PALINACTINODROMOUS
CAMPYLODROMOUS
Imperfect
Basal
Suprabasal

c. Upper secondaries more acute than lower.
d. Only lowest pair of secondaries more acute than pairs above it.
e. Lower and upper secondaries more obtuse than middle sets.
f. Divergence angle more acute on one side of the leaf than on the other.
g. Divergence angle varies irregularly.

3. Relative thickness of secondary veins.
   a. Thick. Proportionately wide in relation to the primary and tertiary orders or to the secondaries in leaves of similar size.
   b. Moderate. The general case, neither thick nor fine.
   c. Fine to hairlike. Proportionately narrow in relation to the other vein sets or to the secondaries in leaves of similar size.
4. Course. More than one of the following terms may apply.
   a. Straight. Without noticeable deviation in course.
   b. Recurved. Arching basally for part of its course.
   c. Curved (uniformly or abruptly, i.e., with a sharp local increase in degree of curvature).
   d. Sinuous. Repeated smooth changes in direction of curvature.
   e. Zig-zag. Repeated angular changes in direction.
   f. Unbranched. Without second order ramifications.
   g. Branched. With one or more secondary ramifications.
   h. Provided with outer secondary veins.
5. Behavior of loop-forming branches (if any).
   a. Joining superadjacent secondary at acute angle.
   b. Joining superadjacent secondary at right angle.
   c. Joining superadjacent secondary at obtuse angle.
   d. Enclosed by secondary arches. 3° or 4° arches.
   e. Forming an intramarginal vein.
6. Intersecondary veins. Thickness intermediate between that of the second and third order veins; generally originating from the medial primary vein interspersed among the secondaries and having a course parallel, or nearly so, to them. They are of two types:
   a. Simple. Consisting of a single vein segment.
   b. Composite. Made up of coalesced tertiary vein segments for over 50% of its length.
7. Intramarginal vein. A vein closely paralleling the leaf margins and into which the secondary veins are fused. Probably the result of the fusion and straightening of the exmedial brochidodromous secondary arch segments into what appears to be an independent vein.
8. Intercostal areas. Those portions of the leaf blade lying between the secondary veins.

C. <u>Tertiary Veins</u>. The next finest branches of the secondary veins and those branches of equal thickness from the primaries.

1. Angle of origin (defined above). When the predominant angle of tertiary origin on the exmedial (lower) side of the secondaries is compared with that on the admedial (upper) side of the secondaries the following combinations are possible. Recognition of this trait is of diagnostic value.

		Angle on Exmedial (Lower) Side of 2°		
		Acute	Right	Obtuse
Angle on Admedial (Upper) Side of 2°	Acute	AA	RA	OA
	Right	AR	RR	OR
	Obtuse	AO	RO	OO

As a rule, the tertiary angle of origin from the mid-vein equals the tertiary exmedial angle of origin from the secondaries. Departure from this rule is a taxonomically significant feature.

2. Pattern.
   a. Ramified. Tertiary veins branch into higher orders without rejoining the secondaries (although their higher order derivatives may do so).
      1. Exmedial. Branching oriented toward the margin.
      2. Admedial. Branching oriented toward the leaf axis.
      3. Transverse. Branching oriented across intercostal area, the commonest case.
   b. Reticulate. Tertiaries anastamosing with other tertiary veins or with the secondaries.
      1. Random reticulate. Angles of anastamoses vary.
      2. Orthogonal reticulate. Angles of anastamoses predominantly right angles.
   c. Percurrent. Tertiaries from the opposite secondaries joining.
      1. Course.
         a. Simple. Unbranched.
         b. Forked. Giving rise to third order ramification.
         c. Straight. Passing across the intercostal area without a noticeable change in course.
         d. Convex. Curving away from the center of the leaf.
         e. Concave. Curving toward the center of the leaf.
         f. Sinuous. Repeatedly changing direction of curvature.
         g. Retroflexed. Forming a single "S"-shaped curve concave apically and convex basally.
         h. Recurved to the mid-vein.
      2. Relationship to mid-vein.
         a. Approximately at right angles.
         b. Longitudinal. Approximately parallel.
         c. Oblique. Trending in an obtuse or, rarely, an acute angle to the mid-vein.
            (1) Tertiary angle with mid-vein remaining approximately constant.
            (2) Angle decreasing admedially.
            (3) Angle decreasing apically.
      3. Arrangement.
         a. Predominantly alternate. Joining each other with an offset, i.e., an abrupt angular discontinuity.
         b. Predominantly opposite. Joining each other smoothly in a straight or curved path.
         c. Alternate and opposite in about equal proportions.

d. Distant. Interval between veins 0.5 cm. or greater.
e. Close. Interval between veins less than 0.5 cm.

D. Higher order venation. The next finer order of veins originating from the tertiaries and those of equal size from lower order veins is known as the quarternary venation and the veins originating from these and those of equal size from lower orders are the quinternaries.

1. Resolution.
   a. Higher order venation forming a reticulum in which vein orders cannot be distinguished.
   b. Vein orders distinct.
2. Quarternary veins.
   a. Size. Notable only is the departure from the average width in leaves of similar size and shape. Thick or thin.
   b. Course.
      1. Relatively randomly oriented.
      2. Orthogonal. Arising at right angles. Their subsequent courses may not be at right angles.
3. Quinternary veins.
   a. Size (as above). Thick or thin.
   b. Course (as above). Random or orthogonal.
4. Highest vein order of leaf: 3°, 4°, 5°, 6°, 7°.
5. Highest vein order showing monopodial branching: 2°, 3°, 4°, 5°, 6°.
6. Marginal ultimate venation.
   a. Incomplete. Freely ending veinlets directly adjacent to the margin.
   b. Looped. The major portion of the marginal ultimate venation recurved to form loops.
   c. Fimbriate. Higher vein orders fused into a vein running just inside of the margin (fimbrial vein).

E. Veinlets. The freely ending ultimate venation of the leaf or veins of the same order which occasionally cross areoles (see below) to become connected distally.

1. Veinlets none.
2. Simple. Without branches (linear or curved).
3. Branched. Once, twice, three times, etc.

F. Areolation. The smallest areas of the leaf tissue surrounded by veins which form a contiguous field over most of the area of the leaf. Thus, smaller areas occasionally formed when veinlets cross their areoles are excluded. Any order of venation in a leaf from the primaries to the highest order of non-freely ending veinlets can form one or more sides of an areole. However, only the order represented by the veinlets will intrude, or occasionally cross, the islets formed by the non-freely ending veins.

1. Development.
   a. Well-developed. Meshes of relatively consistent size and shape.
   b. Imperfect. Meshes of irregular shape, more or less variable in size.
   c. Incompletely closed meshes. One or more sides of a mesh not bounded by a vein, giving rise to anomalously large meshes of highly irregular shape.
   d. Areolation lacking. As in hyphodromous or succulent leaves. Very rarely venation simply ramifying into the intercostal

spaces with no coherent shape, size or pattern to the areas surrounded by veins.

2. Shape.
   a. Triangular
   b. Quadrangular
   c. Pentagonal
   d. Polygonal. With more than 5 sides.
   e. Rounded
   f. Irregular
3. Size. The intervals chosen appear to represent at least partially natural categories.
   a. Very large. Larger than 2 mm.
   b. Large. 2-1 mm.
   c. Medium. 1-0.3 mm.
   d. Small. Less than 0.3 mm.

FOLIAR VENATION LITERATURE

Ettingshausen, C. von. 1861. Die blattskelete des Dicotyledonen. K. K. Hof. u. Staatsdruckeri, Vienna.

Hickey, L. F. 1973. Classification of the architecture of dicotyledonous leaves. American Journal of Botany 60: 17-33.

Section E. TRICHOMES

Although trichomes have been classically employed for systematic purposes, there has yet to be a classification which satisfactorily accounts for their morphological diversity. Metcalfe and Chalk (1950) in commenting upon plant hairs state, "The various types are not always clearly defined, however, and their value for systematic purposes is lessened by the fact that many kinds occur in families which are generally thought to be unrelated, thus making it reasonably certain that the same type of hair must have been evolved along independent lines." Carlquist (1961) has emphasized the need for describing the entire "trichome complement" of a given taxonomic unit. The following classification is a convenient one although admittedly incomplete. No attempt has been made to include the abundant variations in trichome form within specific families and genera. An example of trichome terminology within a specific taxon (Solanum) is illustrated by the recent treatment of Roe (1971), whose suggestions have been widely adopted in this section. Distinction should always be made between degree of pubescence and the morphology of the individual plant hairs. In view of the absence of complete descriptive terminology, descriptions of trichomes should also be accompanied by a drawing.

I. MORPHOLOGY (see Figure 7-7).

In addition to the general descriptive terms below, individual trichomes should be further characterized by size, shape, orientation of cells, relation to the surrounding cells, thickness of walls, and ornamentation of walls.

A. Simple trichomes; unbranched.
   1. Uniseriate. Consisting of a single row of cells; either unicellular, bicellular, or multicellular.
      a. Filiform. A long, thin-walled trichome which often becomes flexuous.

b. Uncinate. Hooked tip.
c. Barbed. With the tip bent back acutely or with sides reflexed or provided with teeth or minute retrorse barbs.
d. Clavate. Broadest at the apex.
e. Falcate. Sickle-shaped.
f. Moniliform. A multicellular trichome with constrictions between the component cells.
g. Conical. With the figure of a cone; narrowly deltoid in outline.
h. Cylindrical. Typically, a thin-walled trichome with straight sides (parallel) which is many times longer than broad.
i. Barrel-shaped. A thin-walled trichome with a large lumen, typically broader than high.
j. Papillate. Resembling a small nipple-like projection. Often difficult to distinguish from a papillate epidermis.
k. Micro-hair. A term used in reference to grasses. A very small, usually bicellular trichome not readily visible without a microscope. See Metcalfe (1960) for descriptions of various types.
l. Macro-hair. A term used in reference to grasses. Typically a unicellular trichome which can be seen with the naked eye or hand lens. See Metcalfe (1960) for descriptions of various types.

2. Multiseriate. A trichome consisting of several to many adjacent rows of cells.
a. Setose. An erect, straight, stiff trichome which tapers from a stout base; these range in size from small warts to long setae.
b. Hirsute. More slender than setose trichomes, arising more or less perpendicular to the epidermis, but less erect and somewhat flexuous.
c. Paleaceous. Longer than hirsute trichome, flattened toward the broad base, becoming circular in cross section toward the apex.
d. Strigose. Trichome bends abruptly at the swollen base; stiff, mostly appressed to the epidermis, and usually directed toward the apex of the organ on which they occur.
e. Sericeous. Long, silky trichomes mostly appressed to the surface.
f. Loriform. A trichome with a long flexuous tip which often becomes similar to a sericeous hair.

B. <u>Complex trichome; branched</u>.
1. Stellate. A sessile, star-shaped trichome with slender rays projecting from a center like rays of a star.
2. Radiate. A sessile trichome resembling the stellate condition but frequently with ovoid arms which are less than 5 times longer than broad.
3. Peltate. (Scales). With the horizontal rays fused along much of their length (flattened). A central ray may be present or it may be replaced by a knob-like cell or group of cells known as an umbo. Peltate plant hairs may be stalked or sessile.
4. Multangulate. (Tufted). A many-angled hair; a term applied to hairs of two to many rays (usually less than 16) projecting from a center at various angles. When using this rather inclusive term, it is desirable to indicate the approximate range in number of rays per hair if some degree of constancy

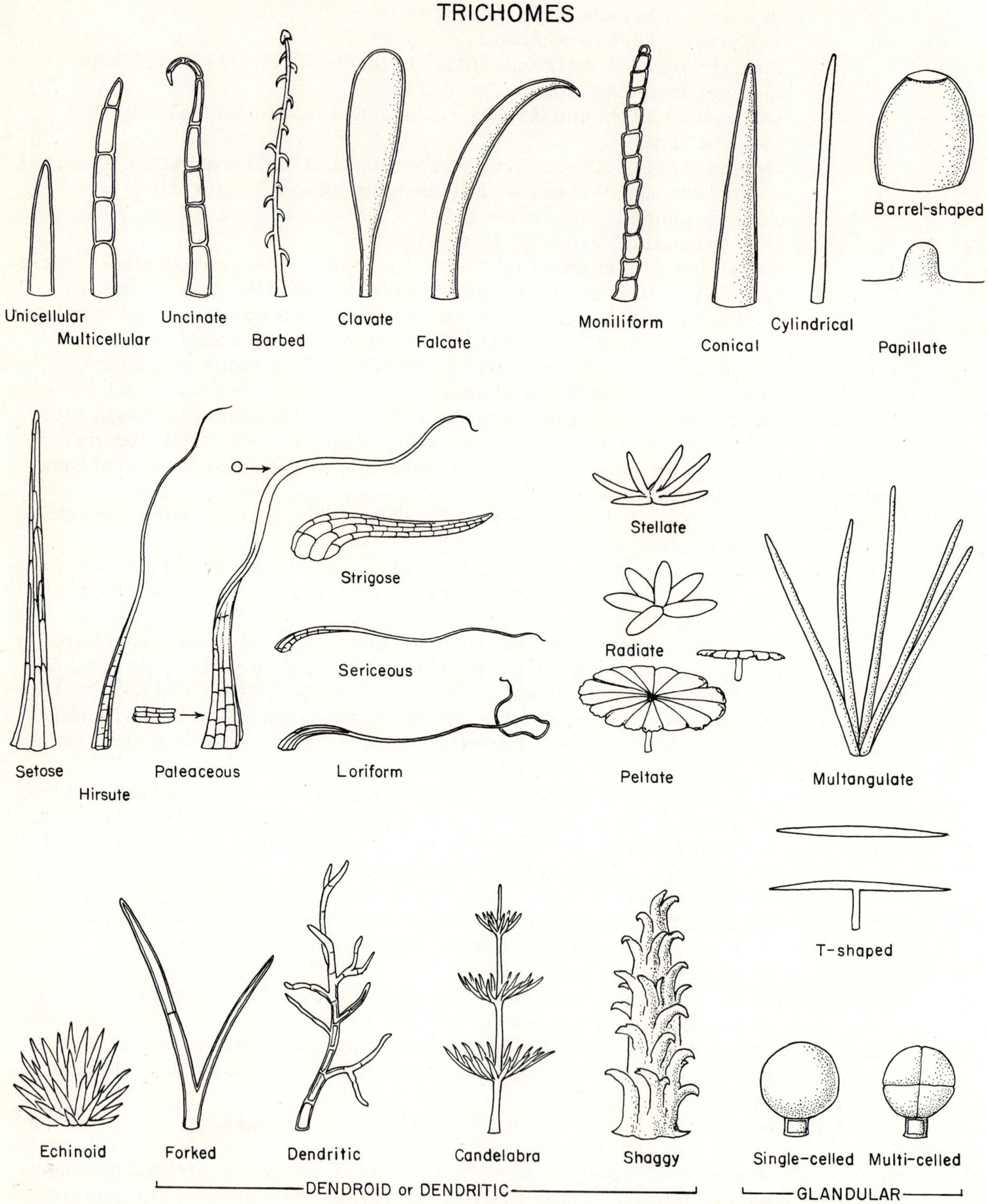
TRICHOMES
Unicellular
Multicellular
Uncinate
Barbed
Clavate
Falcate
Moniliform
Conical
Cylindrical
Barrel-shaped
Papillate
Setose
Hirsute
Paleaceous
Strigose
Sericeous
Loriform
Stellate
Radiate
Peltate
Multangulate
T-shaped
Echinoid
Forked
Dendritic
Candelabra
Shaggy
DENDROID or DENDRITIC
Single-celled
Multi-celled
GLANDULAR

can be observed.

5. Multiradiate. (Tufted). A many-rayed trichome; this term can apply to some hairs described as multangulate but must be restricted to those having many rays. Thus, it might also be applied to or confused with those trichomes described as echinoid.
6. Echinoid or echinate. (Tufted). A trichome consisting of numerous rays (more than 20) projecting at various angles from a center which has the appearance of a solid ball due to the compacted ray bases. The division between echinoid and multangulate is admittedly artificial. Whenever the rays per hair are so numerous as to be conveniently uncountable, the hair is considered echinoid.
7. Dendroid or dendritic. A tree-like trichome with branches or rays along the stalk.
   a. Forked or furcate. Divided into 2 branches.
   b. Dendritic. Usually consisting of a uniseriate, flexuous main stem with few to several divergent branches or rays seemingly distributed at random.
   c. Candelabra type. (Abietiform). The rays originate in whorls at regularly spaced tiers.
   d. Dendritic-echinoid. A branched trichome, modified candelabra type; similar to and a modification of the echinoid hair but with an elongated axis, i.e., a multiseriate stipitate plant hair with many rays projecting from along the stalk without interruption.
   e. Shaggy. A trichome in which the cells making up the stalk are coherent for part of their length but reflexed toward their apices, giving the trichome a branched appearance throughout its entire length.
8. T-shaped hair. A two-armed trichome. A trichome in which the apical cell is orientated transversely with regard to the stalk cells. Either unicellular, bicellular or multicellular.

C. Glandular trichome. A trichome bearing a knob-like secretory swelling (head) at its tip. In some classifications the distinction between glandular and nonglandular trichomes is regarded as basic. NOTE: Unicellular, multicellular and peltate trichomes may be glandular.

1. Single-celled head. (Variations exist in nature of stalk--long or short, uniseriate or multiseriate.)
2. Multi-celled head. Also called colleters. (Variations exist in long or short stalks, anticlinal walls, periclinal walls, or both.)

II. OTHER FEATURES.

Record distribution, mode of attachment, silicification or calcification and sclerification of hairs.

TRICHOME LITERATURE

Carolin, R. C. 1971. The trichomes of the Goodeniaceae. Proceedings of the Linnean Society of New South Wales 96: 8-22.

Heintzelman, C. E., Jr., and R. A. Howard. 1948. The comparative morphology of the Icacinaceae. V. The pubescence and the crystals. American Journal of Botany 35: 42-52.

Netolitsky, F. 1932. Die Pflanzenhaare. In: K. Linsbauer (ed.), Handbuch der Pflanzenanatomie 4: 1-253.

Hunter, G. E. 1966. Revision of Mexican and Central American Saurauia (Dilleniaceae). Annals Missouri Botanical Garden 53: 47-89.

Roe, K. E. 1971. Terminology of hairs in the genus Solanum. Taxon 20: 501-508.

Staudermann, W. Von. 1924. Die haare der Monokotylen. Botanisches Archiv 8: 105-184.

Uphof, J. C. Th. 1962. Plant Hairs. Hummel, K. & K. Staesche. Die Verbreitung der Haartypen in den Naturlichen Verwandtschaftsgruppen. In: Encyclopedia of Plant Anatomy IV(5). W. Zimmerman & P. G. Ozenda (eds.). Gebrüder Borntraeger. Berlin.

Section F. YOUNG STEM ANATOMY

I. NODAL ANATOMY.

The node is that part of the stem to which leaves are attached. Nodal anatomy, accordingly, describes the pattern of vascular continuity between these two organs. Depending upon whether one, three or five or more leaf gaps (in the classical sense) are left in the cauline stele by the departure of vascular tissue (leaf traces) to the leaf, nodes are described as being: (1) Unilacunar, (2) Trilacunar, and (3) Multilacunar. The nodal pattern is often expressed in terms of the number of traces and gaps. For example, a unilacunar node with a single trace would be described as 1:1; a unilacunar node with two departing traces would be 2:1; a trilacunar, three-trace node would be 3:3; and so on. The first figure represents the number of traces, the second figure the number of gaps in the vascular cylinder of the stem (see Figure 7-8).

II. ORIGIN OF PERIDERM.

Periderm is secondary protective tissue derived from the phellogen (cork cambium) and replacing the epidermis, typically in stems and roots. Consists of phellem (cork), phellogen and phelloderm. Record whether originating in a superficial or deep-seated manner.

III. SCLERENCHYMA. (See Section B. LEAF)

A. Sclereids. Record form and location.

B. Perivascular fibers. Fibers located on the outer periphery of the vascular region and originating outside the primary phloem as distinct from primary phloem fibers. Often called pericyclic fibers. Record presence or absence.

IV. WIDTH OF PRIMARY MEDULLARY RAYS.

Tissue located between vascular bundles in stem. Also called interfascicular region. Record broad, narrow or mixed.

V. VASCULAR BUNDLE TYPE.

A. Collateral. A bundle with the phloem on one side of the xylem only.

B. Bicollateral. A bundle with the phloem on two sides of the xylem.

C. Concentric.

1. Amphicribral. A bundle with the phloem surrounding the xylem.
2. Amphivasal. A bundle with the xylem surrounding the phloem.

VI. LOCATION AND DISTRIBUTION OF VASCULAR BUNDLES.

Also record the occurrence of cortical and medullary bundles. A medullary bundle is one located more or less close to the center of the stem in the pith region.

VII. NATURE OF THE GROUND TISSUE.

A. Type and shape of cells.
   1. Parenchyma. A living, nucleate cell, often more or less isodiametric, concerned with one or more of the various physiologic activities of the plant. Record form and size.
   2. Collenchyma. A supporting cell (tissue) composed of more or less elongated living cells with unevenly thickened walls, which are usually interpreted as primary walls.
   3. Sclerenchyma. Cell variable in form and size and having more or less thick, often lignified, secondary walls and lacking a protoplast at maturity.

B. Presence or absence of an endodermoid layer. The layer of ground tissue forming a sheath around the vascular region and having specific wall characteristics - Casparian strips or secondary thickenings.

C. Relative differentiation of cortical zone.

D. Presence and location of crystals and other idioblastic cells.

VIII. RECORD VESSEL ELEMENT TYPE. (See Section A. SECONDARY XYLEM)

IX. PHLOEM.

A. Type of sieve element. The cell in the phloem tissue concerned with mainly longitudinal conduction of food materials. Classified into:
   1. Sieve cell. A type of sieve element that has relatively undifferentiated sieve areas, that is, sieve areas with narrow pore and thin connecting strands; and these sieve areas are rather uniform in structure in all walls, that is, there are no sieve plates.
   2. Sieve-tube member. A sieve element with specialized sieve areas differentiated into sieve plates. One of the cellular components of a sieve tube. Record nature of sieve plate.

B. Phloem sclerenchyma. Record type, abundance and distribution.

C. Idioblastic cells. Record type and distribution.

D. Internal phloem. Primary phloem located internally to the primary xylem. Record presence or absence.

X. PITH DIAPHRAGMS.

A pith in which transverse layers (diaphragms) of firm-walled cells alternate with regions of soft cells. The latter may collapse with age. Then the diaphragms divide the pith chamber into small compartments. Record presence or absence.

XI. ANOMALOUS STRUCTURE.

A. Secondary thickening from successive cambia. Record presence or absence.

NODAL PATTERNS

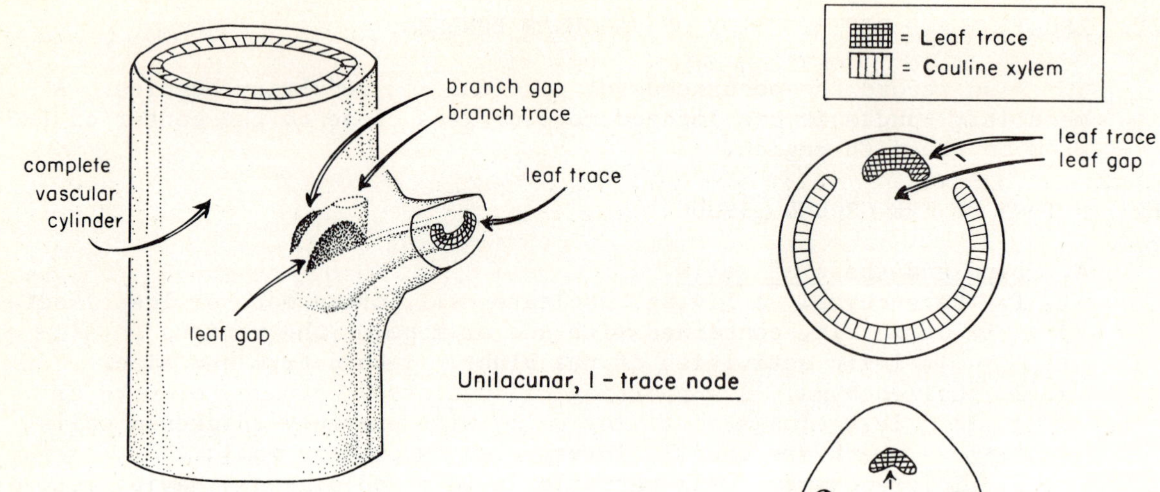

Unilacunar, I - trace node

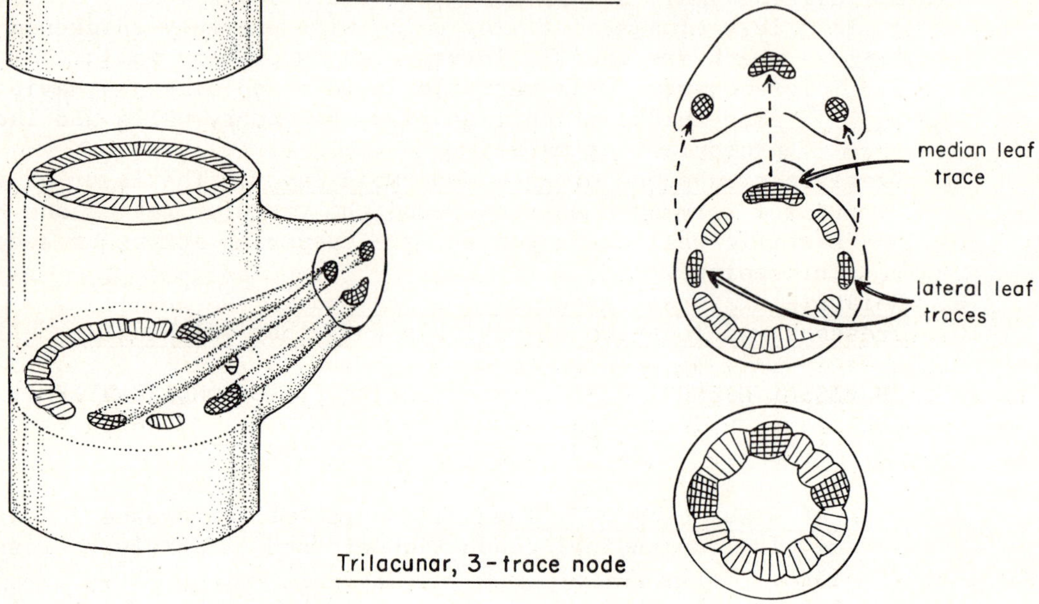

Trilacunar, 3 - trace node

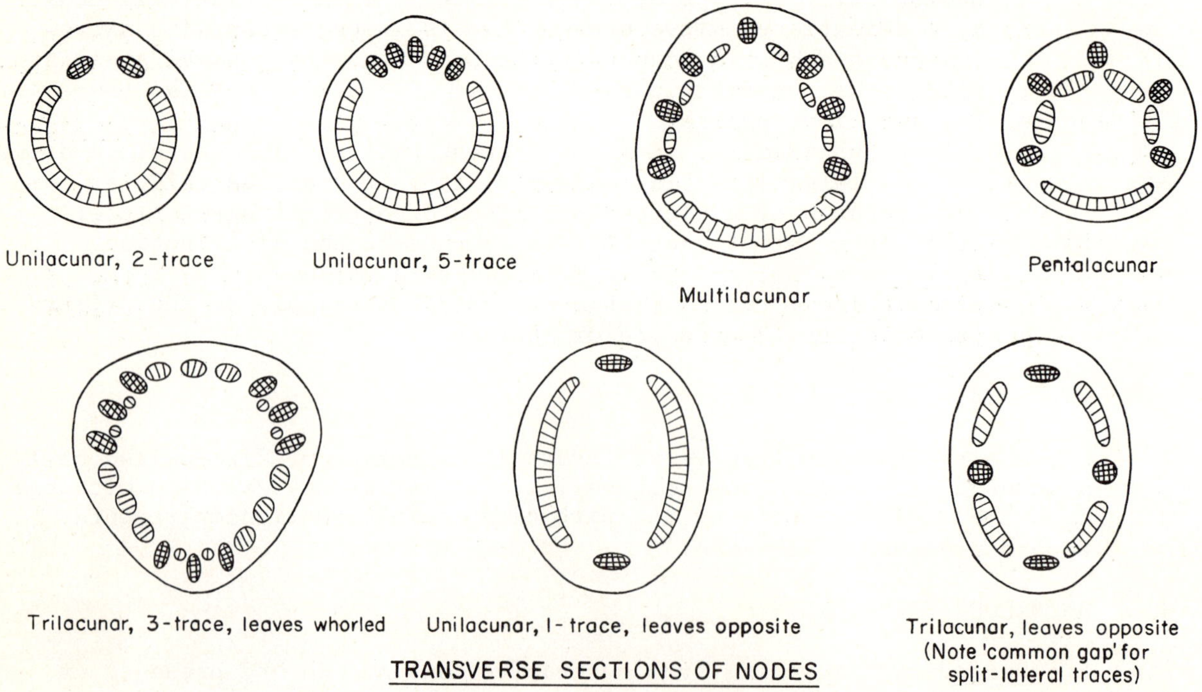

TRANSVERSE SECTIONS OF NODES

B. Included phloem. Phloem strands or layers included in the secondary xylem. Record presence or absence.

FREEHAND METHOD FOR PETIOLAR AND NODAL STUDY

The following procedure for preparing freehand sections of petiole or nodal material is one suggested by the comparative anatomist, William L. Stern.

1. With dried material boil to hydrate. With fluid-preserved material, proceed directly to step 2.
2. Prepare freehand sections of petiole or young stems using a sharp razor blade. Hold material between pieces of elder pith or carrot for support.
3. Sections should be cut directly into a Syracuse watch crystal containing a saturated aqueous solution of phloroglucinol. A few minutes in this solution is usually sufficient. Phloroglucinol is commonly employed to demonstrate lignified tissues, hence it has application to plant anatomy where gross anatomical demonstration will suffice. This is often true when characterizing the vascularization of petioles and in determining the nodal pattern in stems.
4. Place sections directly onto a microscope slide and cover with concentrated hydrochloric acid; cover with a cover slip; blot any excess acid off of the slide.
5. Place slide on microscope for observation; be extremely careful to keep the acid from spilling on the microscope.
6. Lignified tissue will turn bright purple-red immediately, at which time a sketch of the vascular pattern should be made. The hydrochloric acid rapidly decomposes tissues and the staining fades.

YOUNG STEM ANATOMY LITERATURE

Bailey, I. W. 1956. Nodal anatomy in retrospect. Journal of the Arnold Arboretum 37: 269-287.

________, and R. A. Howard. 1941. The comparative morphology of the Icacinaceae. I. Anatomy of the node and internode. Journal of the Arnold Arboretum 22: 125-132.

Philipson, W. R., and M. N. Philipson. 1968. Diverse nodal types in *Rhododendron*. Journal of the Arnold Arboretum 49: 193-224.

Sinnott, E. W. 1914. Investigations on the phylogeny of the angiosperms. I. The anatomy of the node as an aid in the classification of angiosperms. American Journal of Botany 1: 303-322.

Zahur, M. S. 1959. Comparative study of secondary phloem of 423 species of woody dicotyledons belonging to 85 families. Cornell University Agricultural Experiment Station Memoir 358: 1-160.

NODAL AND PETIOLAR ANATOMY EXERCISE

This exercise is an introduction to the diversity encountered in nodal and petiolar vascular anatomy. Specimens used in this exercise should illustrate the basic nodal patterns and a variety of petiolar vasculation types. With the aid of a razor blade, free-hand sections should be made using fresh material at the levels indicated below. It should be remembered, however, that it may be necessary to obtain additional sections in order to completely understand the changes which occur in a complex vascular pattern. In preparing the drawings attention should be given to (1) arrangement of the vascular tissue (separate or fused vascular bundles, scattered bundles, medullary bundles, etc.), (2) outline of the vascular tissue (flattened, crescent-shaped, circular, etc.), and (3) location and abundance of sclerenchyma. Other considerations include relationship of the nodal pattern and the petiole vascular pattern, complexity of the petiole venation and leaf size, and the relationship of the nodal pattern to the presence or absence of stipules.

Drawings of Serial Transverse Sections through the Node and Leaf Illustrating Vasculation Pattern

Species, leaf arrangement, stipules + or -	Level 1 (nodal region)	Level 2 (base of petiole)	Level 3 (mid-petiole)	Level 4 (base of laminia)	Level 5 (mid-rib)

Section G. STELAR ANATOMY

Foster and Gifford (1959) remarked, "Despite the absence of consistent histological boundaries between the cortex and stele, the value of the Stelar Theory as a unified concept has been widely recognized and has led to efforts to classify and interpret phylogenetically the varied types of vascular cylinders that occur in stems and roots." The major types of stelar patterns are reviewed here (see Figure 7-9).

I. THE STELE.

Conceived by Van Tieghem as a morphologic unit of the plant body comprising the vascular system and the associated ground tissue (pericycle, interfascicular regions, and pith); the central cylinder of the axis (stem and root).

II. PATTERNS OF RADIAL PRIMARY XYLEM MATURATION.

A. Exarch. Xylem strand in which the maturation of the cells progresses centripetally, that is, the oldest elements (the protoxylem) are farthest from the center of the axis. Typical of roots in seed plants.

B. Endarch. Xylem strand in which the maturation of the cells progresses centrifugally; that is, the oldest elements (the protoxylem) are closest to the center of the axis. Typical of stems in seed plants, also leaves, in which the oldest xylem is on the adaxial side.

C. Mesarch. Xylem strand in which the maturation of the cells begins in the center and then progresses both centripetally and centrifugally, that is, the oldest elements (the protoxylem) are in the center of the strand. Typical of many ferns.

III. MAJOR STELAR PATTERNS.

A. Protostele. Simplest type of stele, containing a solid column of vascular tissue. Considered by Jeffrey to be the most primitive pattern.

1. Haplostele. A stele in which a cylindrical strand of xylem is surrounded by phloem; e.g., *Rhynia* (extinct).
2. Actinostele. A stele in which a lobed or stellate strand of xylem is surrounded by phloem as viewed in transverse section; e.g., *Lycopodium*.
3. Plectostele. A stele in which bands of xylem alternate with phloem as viewed in transverse section; e.g., *Lycopodium*.

B. Siphonostele. A type of stele in which the vascular system appears in the form of a hollow cylinder; that is, pith is present. Considered by Jeffrey to have arisen from the protostele.

1. Phyllosiphonic. A term used by Jeffrey to describe a siphonostele in which foliar gaps are present; e.g., higher ferns.
2. Cladosiphonic. A term used by Jeffrey to describe a siphonostele in which branch gaps are present but not foliar gaps; e.g., *Equisetum*.
3. Ectophloic siphonostele. A stele having one phloem region external to the xylem and a central pith; e.g., *Botrychium*, *Osmunda*.

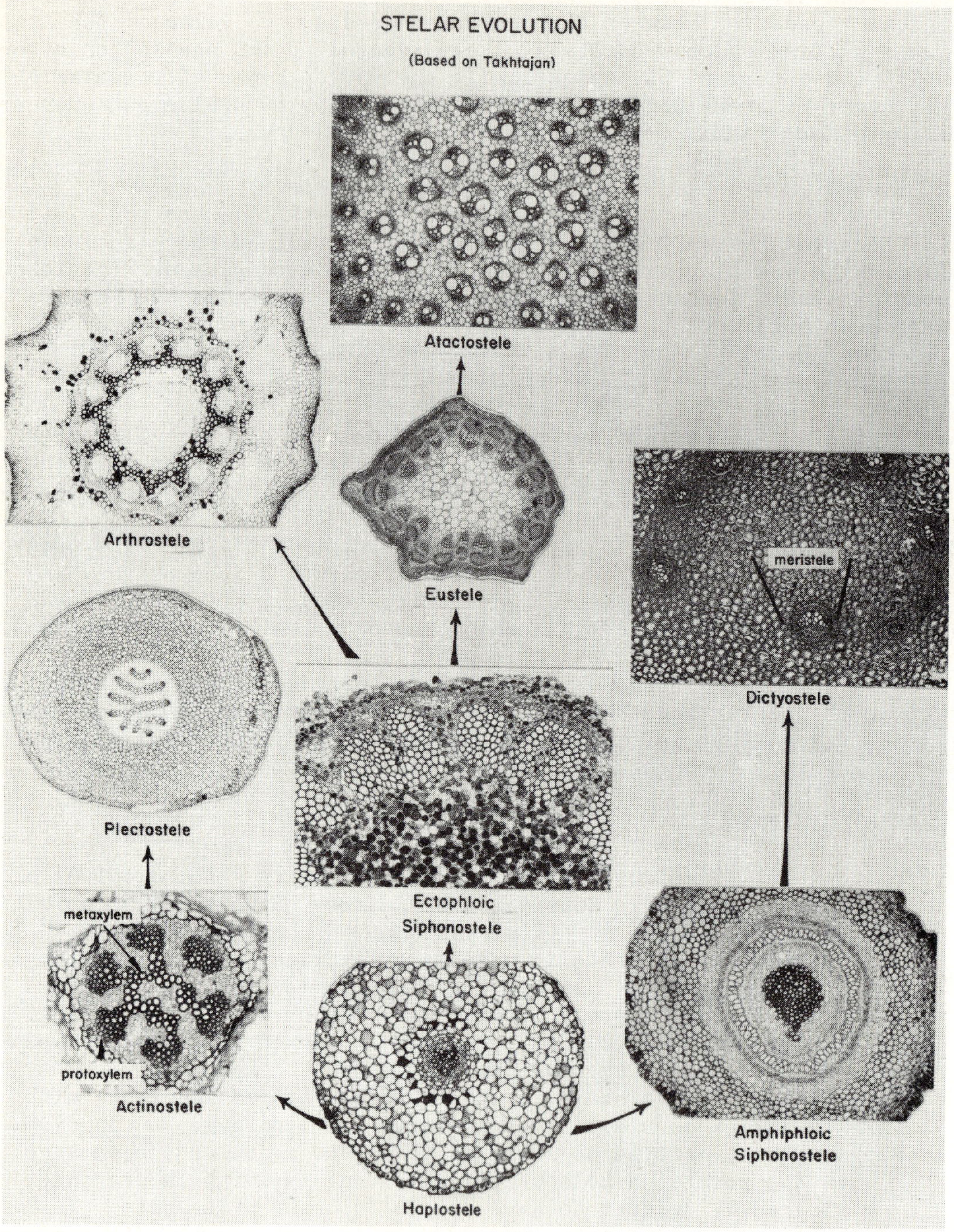
STELAR EVOLUTION
(Based on Takhtajan)
Atactostele
Arthrostele
Eustele
meristele
Dictyostele
Plectostele
Ectophloic
Siphonostele
metaxylem
protoxylem
Actinostele
Amphiphloic
Siphonostele
Haplostele

4. Amphiphloic siphonostele. A stele having a pith and two phloem regions, one external and the other internal to the xylem; e.g., Dennstaedtia.
5. Polycyclic siphonostele. A stele in which the xylem and phloem are arranged in alternate concentric cylinders; e.g., Matonia.

C. Solenostele. As originally defined by Gwynne-Vaughn, an amphiphloic siphonostele interrupted by leaf gaps which close prior to the departure of the succeeding leaf trace.

D. Dictyostele. As originally defined by Brebner, a special type of siphonostele in ferns characterized by dissected concentric vascular bundles. Each individual bundle in this case is termed a meristele.

E. Eustele. As originally defined by Brebner, a dissected siphonostele in which the individual vascular bundles are collateral or bicollateral. This pattern is widespread in dicotyledonous stems.

F. Atactostele. A stele characteristic of many monocots in which the vascular bundles appear "scattered" within the ground tissue.

G. Arthrostele. A term used by Takhtajan to describe the special stelar pattern in Equisetum. Vascular bundles are essentially collateral with primary xylem being endarch or perhaps more commonly mesarch. Patches of lateral xylem may occur on either side of the bundles.

ANATOMICAL LITERATURE

I. Periodicals. (The following periodicals have historically devoted many pages to systematic anatomy and morphology.)

American Journal of Botany. 1914+. Published by the Botanical Society of America.

Botanische Jahrbücher fur Systematik, Pflanzengeschichte und Pflanzengeographie. Stuttgart. 1881+.

Botanical Journal of the Linnean Society. London. 1847+. Published by The Linnean Society of London.

Journal of the Arnold Arboretum. Harvard University. Jamaica Plain, Mass. 1919+.

Phytomorphology. New Delhi. 1951+. Published by The International Society of Plant Morphologists.

II. General References

Ayensu, E. S. 1971. Anatomy of the Monocotyledons. VI. Dioscoreales. C. R. Metcalfe (ed.). The Clarendon Press. Oxford.

Bailey, I. W. 1951. The use and the abuse of anatomical data in the study of phylogeny and classification. Phytomorphology 1: 67-69.

________. 1953. The anatomical approach to the study of genera. Chronica Botanica 14: 121-125.

Carlquist, S. 1961. Handling the anomalous genus. Recent Advances in Botany. Vol. 1. pp. 165-168.

________. 1961. Comparative Plant Anatomy. Holt, Rinehart and Winston. New York.

Chowbury, K. A. 1961. Limitations of anatomical aid to taxonomy of angiosperms. Recent Advances in Botany. Vol. 1 pp. 150-155. University of Toronto. Toronto.

Cutler, D. F. 1969. Anatomy of the Monocotyledons. IV. Juncales. C. R. Metcalfe (ed.). The Clarendon Press. Oxford.

Esau, K. 1961. Anatomy of Seed Plants. John Wiley and Sons, Inc. New York.

________. 1965. Plant Anatomy. 2d ed. John Wiley and Sons, Inc. New York.

Foster, A. S., and E. M. Gifford, Jr. 1959. Comparative Morphology of Vascular Plants. W. H. Freeman and Co. San Francisco.

Fritsch, F. E. 1903. The use of anatomical characters for systematic purposes. The New Phytologist 2: 177-184.

Metcalfe, C. R. 1954. An anatomist's views on angiosperm classification. Kew Bulletin 427-440.

________. 1959. A vista in plant anatomy. In: W. B. Turrill (ed.). Vistas in Botany. Vol. 1. pp. 76-79. Pergamon Press. New York.

________. 1960. Anatomy of the Monocotyledons. I. Gramineae. The Clarendon Press. Oxford.

________. 1961. The anatomical approach to systematics. General introduction with special reference to recent work on monocotyledons. Recent Advances in Botany. Vol. 1. pp. 146-150. University of Toronto Press. Toronto.

________. 1968. Current developments in systematic plant anatomy. In: V. H. Heywood. Modern Methods in Plant Taxonomy. Academic Press. London.

________. 1971. Anatomy of the Monocotyledons. V. Cyperaceae. The Clarendon Press. Oxford.

________, and L. Chalk. 1950. Anatomy of the Dicotyledons. 2 vols. The Clarendon Press. Oxford.

Solereder, H. 1908. Systematic Anatomy of the Dicotyledons. 2 vols. (English translation by Boodle and Fritsch). The Clarendon Press. Oxford.

________, and F. J. Meyer. 1928-1933. Systematische Anatomie der Monokotyledonen. Gebrüder Borntraeger. Berlin. (Only parts 1, 3, 4 and 6 are known to have been published).

Tomlinson, P. B. 1961. Anatomy of the Monocotyledons. II. Palmae. C. R. Metcalfe (ed.). The Clarendon Press. Oxford.

________. 1969. Anatomy of the Monocotyledons. III. Commelinales-Zingiberales. C. R. Metcalfe (ed.). The Clarendon Press. Oxford.

Chapter 8. PALYNOLOGICAL EVIDENCE

The past two decades have seen the study of modern and fossil pollen and spores grow from infancy into a distinct and important botanical discipline, palynology. Application of palynological data has proven to be of valuable assistance in interpreting problems related to stratigraphy, paleoecology, archeology, ethnobotany and higher plant systematics. The ease by which pollen grains can be prepared for study and their morphological diversity makes pollen morphology a taxonomic tool which should be explored by systematists. Pollen data have proven taxonomically useful at all taxonomic levels. Recent advances have been made in palynological research employing the transmission electron microscope and the scanning electron microscope; however, valuable contributions can still be made by the taxonomist using the light microscope. Unfortunately the student will discover that the descriptive terminology used with reference to pollen grains is in a state of confusion. There is still no one glossary which is acceptable to all workers, although a recent attempt toward standardization has been offered by Reitsma (1970), whose suggestions have been widely followed. The following supplement provides the essential terminology necessary for making pollen descriptions of extant angiosperms. A discussion of pollen grain evolution can be found in Chapter 27, section C. Pollen as a tool in fertility studies is discussed in Chapter 12, and aspects of pollen size studies are treated at the end of this chapter.

Section A. MORPHOLOGIC CHARACTERS OF POLLEN GRAINS (See Figure 8-1)

A. Single or Compound Pollen Grains: Pollen Units.

Monad. Solitary pollen grain.

Dyad. Pollen grains united in pairs; e.g., *Scheuchzeria palustris*.

Tetrad. Four coherent pollen grains.

1. Tetrahedral tetrad; e.g., Ericaceae, Droseraceae.
2. Tetragonal tetrad; e.g., Annonaceae (in part).
3. Rhomboidal tetrad; e.g., *Meconopsis*.
4. Linear tetrad; e.g., *Typha* (in part).

Polyad. Compound grains, more than four in each group.

Massula. Walker (1971) suggests restricting this term to those pollen masses less than pollinia in which the number of individual grains is not ascertainable.

Pollinium. An entire pollen mass of a thecal chamber.

Pollinium apparatus. The functional unit of a corpusculum with its two attached arms and pollinia. (See Figure 21-2D for structure and terminology of the specialized pollen apparatus of the Asclepiadaceae.)

B. Polarity.

Apolar. Pollen grain with no distinct polarity (subsequent to tetrad separation).

Distal part. Part of a pollen grain that faces opposite the center of the tetrad during meiosis (i.e., is not in contact with other grains).

Equator. Borderline of the distal and proximal part of a pollen grain.

Equatorial view. A pollen grain viewed when the polar axis is at right angles to the observer.

Heteropolar. Pollen grain in which the two polar faces are different, one face has an aperture, the other has not.

Isopolar. Pollen grain in which the equatorial plane divides the grain into identical halves; i.e., the proximal and distal faces are similar.

Polar. Pollen grain with distinct poles.

Polar axis. An imaginary straight line connecting the distal and the proximal poles.

Polar view. A pollen grain viewed with the polar axis facing the observer.

Pole. The center of both distal and proximal parts.

Proximal part. Part of a pollen grain that faces toward the center of the tetrad during meiosis (i.e., is in contact with other grains).

C. Symmetry.

Bilateral. Pollen grain with two vertical planes of symmetry and the equatorial axes are not equilong.

Radiosymmetry. Pollen grain with more than two vertical planes of symmetry.

D. Shape.

1. Outline of grain in polar view.

a. Non-angular
   1. circular
   2. elliptic
   3. ovate
   4. lanceolate

b. Angular
   1. triangular (3-angled)
   2. quadrangular (4-angled)
   3. quinquangular (5-angled)
   4. sexangular (6-angled)

2. Outline of grain in equatorial view.

a. Non-angular
   1. circular
   2. elliptic

b. Angular
   1. rectangular
   2. rhombic

In angular forms the sides can be described as straight, concave, or convex. Erdtman (1952) described pollen in equatorial view as the ratio between the polar axis (P) and the maximum length of equatorial axis (E) - the P/E ratio. This ratio should be indicated separately. The P/E ratios are sometimes divided into shape classes--spherical, prolate, and oblate (see Faegri and Iversen, 1964).

E. Size. Since the size of pollen grains is greatly affected by the method of preparation, it is very important to indicate the technique(s) followed. The following size classes are taken from Erdtman, G. (1945). Pollen morphology and plant taxonomy. III. Morina, L. with an addition on pollen morphological terminology. Svensk Botanisk Tidskrift 39: 187-191. [μm = micrometer]

Very small grains	< 10μm	Large grains.	50-100μm
Small grains	10-25μm	Very large grains . .	100-200μm
Medium-sized grains . . .	25-50μm	Gigantic grains . . .	> 200μm

F. Number, Location and Size of Germinal Apertures.

Aperture. Any preformed thinning or absence of a part of the exine.

Colpus (pl. colpi). Oblong-elliptic aperture. Length/breadth ratio greater than 2.

Porus (pl. pori). Circular or faintly elliptic aperture.

Terms Relating to Aperture Condition

Note: The number of apertures is indicated by using mono-, di-, tri-, tetra-, penta-, hexa-, and poly-, before terms colpate, colporate, porate, and pororate.

Aperturate. Pollen grain provided with one or more apertures.
Colpate. Pollen grain provided with one or more colpi.
Colporate. Pollen grain with compound apertures, one or more porate colpi.
Costa (pl. costae). Thickening of the nexine near an aperture.
Aperture membrane. That part of the nexine forming the bottom of an aperture. The membrane may be smooth or granulate.
Eurypalynous. Families characterized by a great array of pollen types (different apertures, sculpture, etc.).
Inaperturate. Pollen grain without germinal pores or colpi.
Margo (pl. margines). Zone around a colpus formed by an abrupt thinning or thickening of the sexine or by any other sexine structure different from the remaining sexine.
Operculum (pl. opercula). Closed part of the sexine which is found on the aperture membrane and which is completely or partly isolated from the rest of the sexine by an uncovered part of this membrane.
Pantoaperturate (Periaperturate). Pollen grain with apertures scattered over the whole surface (pantocolpate, pantocolporate, pantoporate, pantopororate).
Porate. Pollen grain provided with one or more pori.
Pororate. Pollen grain with one or more compound pori.
Stenopalynous. Families characterized by slight variation in pollen types.
Syncolpate. Pollen grain with anastomosing colpi, fused to form spirals, rings, etc.
Trichotomocolpate. A three-slit aperture; e.g., Canellaceae.
Vestibulum (pl. vestibula). Cavity inside a porus, caused by a separation of two layers of the exine.
Zonoaperturate (Stephanoaperturate). Pollen grain provided with apertures located in an equatorial zone (zonocolpate, zonocolporate, zonoporate, zonopororate).

G. Sculpturing. External features of the pollen wall.
Baculate. Pollen with pillar-like processes, always longer than broad and higher than 1μm. The process is a baculum (pl. bacula); e.g., *Nymphaea*.
Clavate. Pollen with processes higher than broad, with swollen heads that are slightly tapering towards the base, higher than 1μm. The process is a clava (pl. clavae); e.g., *Ilex*.
Echinate. Pollen with external spine-like processes, always higher than 1μm. The process is an echina (pl. echinae); e.g., *Taraxacum*.
Foveolate. Pollen with pitted surface.
Gemmate. Pollen with processes that are constricted at the base, always higher than 1μm, the diameter is equal to or larger than the height. The process is a gemma (pl. gemmae); e.g., *Juniperus*.
Lumen (pl. lumina). Space between muri.
Murus (pl. muri). Structural elements (ridges) separating lumina in reticulate pollen.
Psilate. Pollen grain with no visible external features; e.g., *Betula*.
Reticulate (pl. reticula). Network (reticular pattern) formed by muri and lumina; e.g., *Illicium*.
Rugulate. Pollen grain provided with an irregular pattern of lumina and muri; e.g., *Nymphoides* (p.p.).
Scabrate. Pollen with elements of different shape, smaller than 1μm. Elements are scabra (pl. scabrae); e.g., *Quercus*.
Striate. Pollen grain provided with a regular pattern of approximately parallel lumina and muri; e.g., *Menyanthes*.

POLLEN MORPHOLOGY

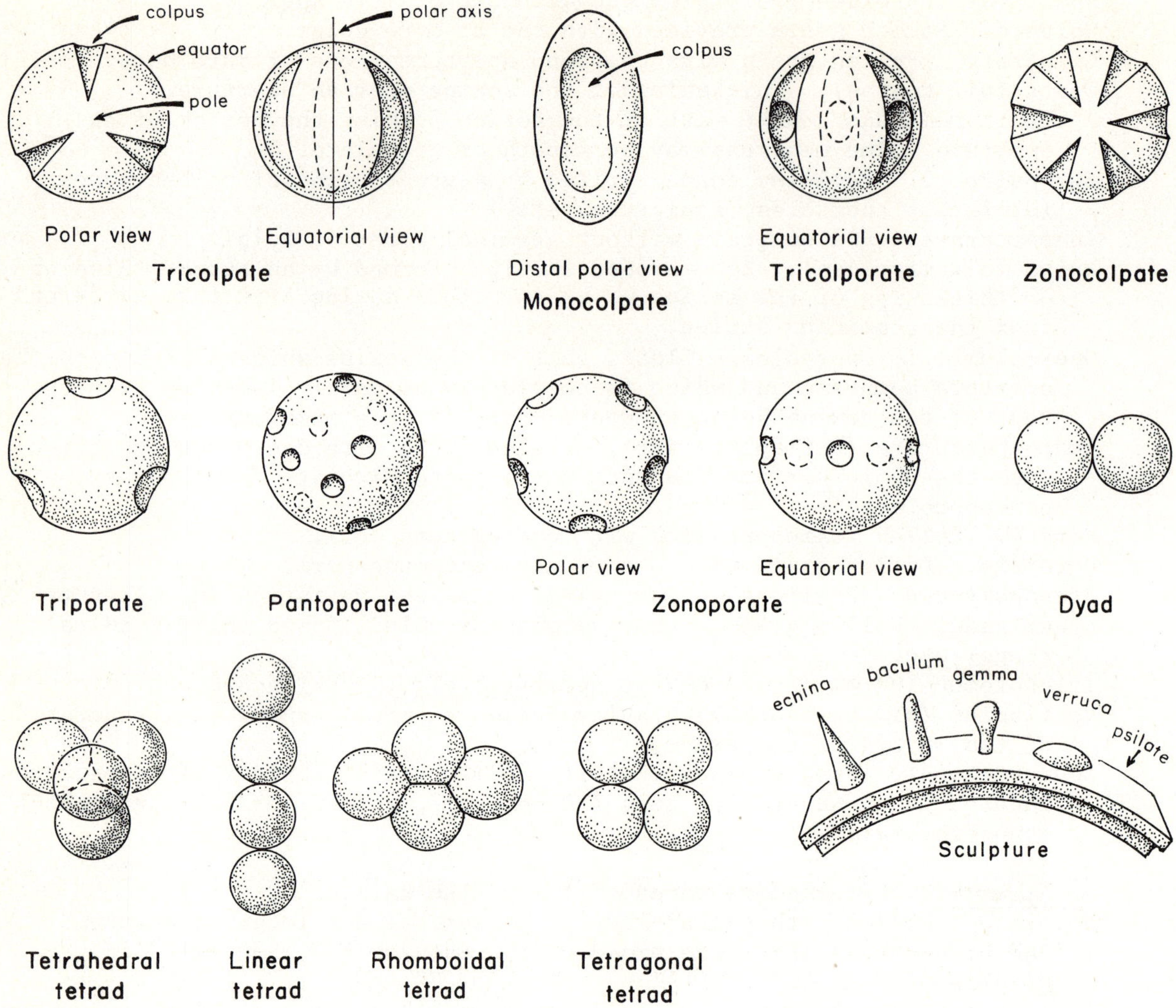

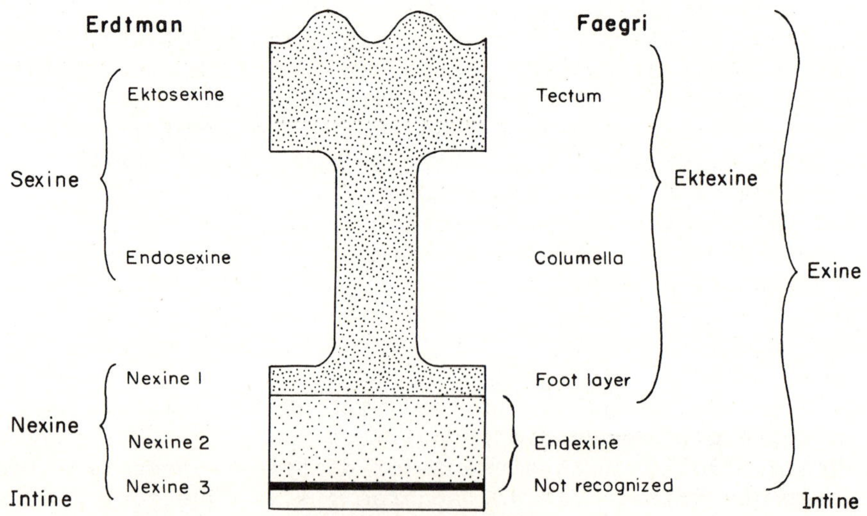

Structure of Pollen Wall
(Modified, after Tschudy and Scott)

M. Seiler

Verrucate. Pollen with wart-like processes always broader than high and always higher than 1μm. The process is a verruca (pl. verrucae); e.g., *Berberis*.

H. Structure and Stratification of Wall. Concepts regarding the stratification of the pollen wall have been and still are in a state of confusion. For a brief review of this subject the reader is referred to the editorial, Review of Paleobotany and Palynology 10(1): 4-37. 1970. The following terminology is only that in common usage.

Columella (pl. columellae). Pillar-like elements, supporting a layer or crowned by a single element. Columellae may be well developed, moderately well developed, or reduced in size.

Exine. The part of a pollen wall outside the intine; is acetolysis resistant.

Intectate. Pollen grain without a tectum or tectal areas.

Intine. The cell wall proper, inner layer of a pollen wall (removed during acetolysis).

Nexine. Inner united homogeneous layer of the exine.

Sexine. Outer layer of the exine.

Tectum. Outermost closed layer of the sexine. The pollen grain is tectate.

NOTE: The exine is divided by some workers on chemical differences into the endexine, an inner homogeneous layer, and an outer ektexine. In order to determine exine structure the student is encouraged to learn the technique of L-O analysis (see Erdtman, 1952, 1969).

I. Nuclear State. Pollen is shed in either the binucleate or trinucleate stage. The phylogenetic significance of this character has been discussed by Brewbaker (1967).

Section B. SCANNING ELECTRON MICROSCOPE POLLEN STUDIES

With the scanning E.M. the intricate surface features of pollen grains can be studied in detail at magnifications up to 20,000X and features that cannot even be seen with the light microscope can be accurately measured and compared. The great depth of field possible (see Fig. 8-2) and the relative ease of preparation (many specimens can be observed in the living condition without either fixation or coating) make this method of pollen study of special value and interest to the taxonomist with access to a scanning electron microscope. NOTE: The S.E.M. has also been used to advantage in studying the minute surface features of such things as seeds, leaves, trichomes and apical meristems. [See Falk, et al., 1971; also Chuang, T., and L. Heckard. 1972. Seed coat morphology in *Cordylanthus* (Scrophulariaceae) and its taxonomic significance. American Journal of Botany 59: 258-265; and Whiffen, T., and A. S. Tomb. 1972. The systematic significance of seed morphology in the neotropical capsular-fruited Melastomataceae. American Journal of Botany 59: 411-422.]

Section C. POLLEN SIZE AND VIABILITY STUDIES

In addition to systematic information that can often be gained from the intricate detail of the exine or outer pollen wall as beautifully revealed by the scanning electron microscope, the role of the pollen grain in the reproductive process makes it a valuable tool in fertility studies and the relatively determinate nature of pollen size (a characteristic shared with guard cells)

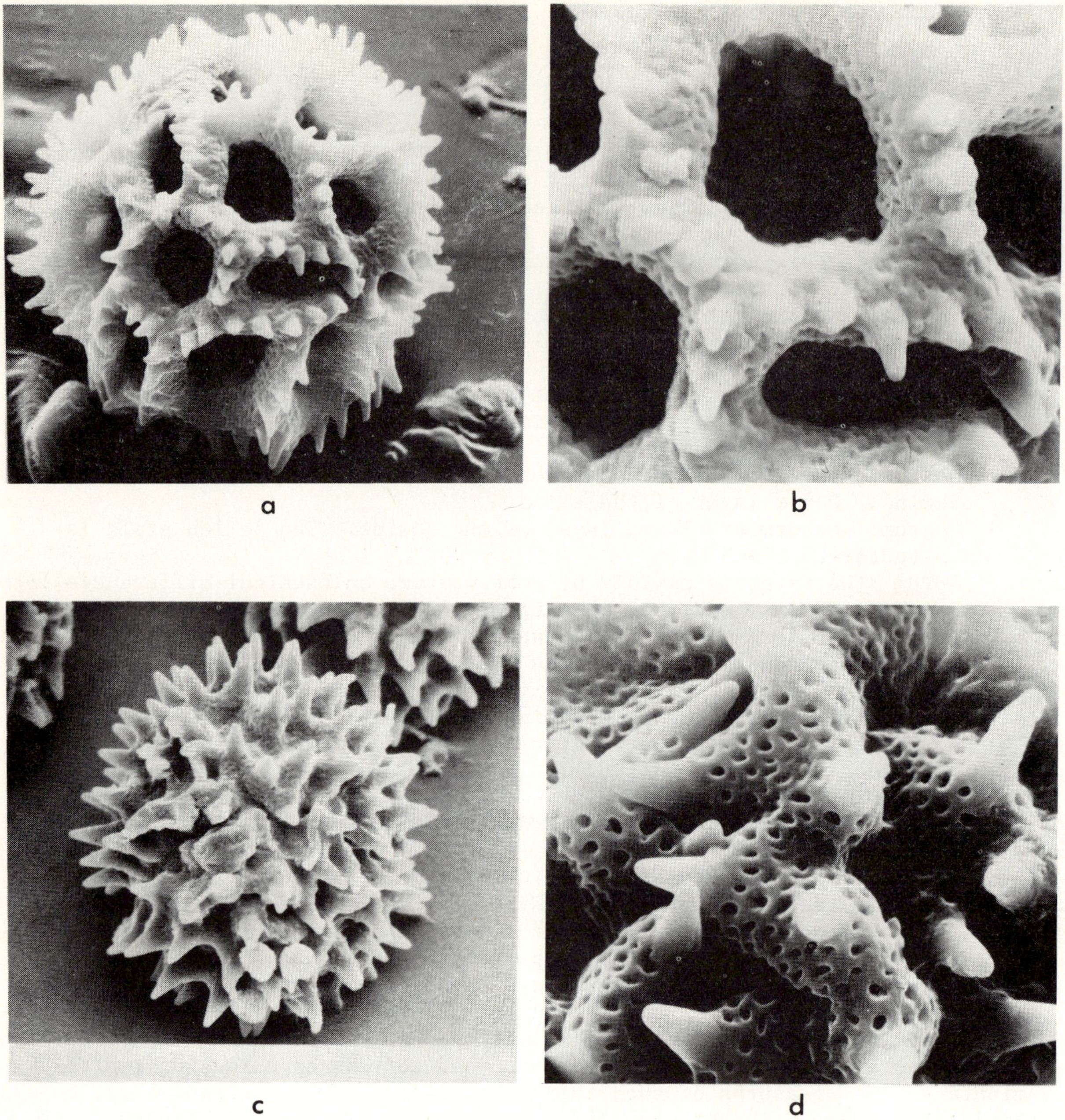

Figure 8-2. Scanning electron photomicrographs of Vernonia (Asteraceae) pollen grains: (a) equatorial view of entire pollen grain of V. cinerea, ca. 30μ in diameter; (b) surface view of same grain at greater magnification to show crests and lacunae, each lacunae is ca. 10μ in diameter; (c) equatorial view of entire pollen grain of V. triflosculosa, ca. 32μ in diameter; (d) surface detail of crests, spines and microapertures of pollen grain of V. lindheimeri, the spines are ca. 5μ long. [From Jones, S. B., Jr. 1970. Used with permission.]

offers some degree of assistance in pollination biology studies and in the possible detection of polyploidy. For these latter two types of observation only standard light microscopes are necessary. Expanded pollen, measured across the largest dimension (usually the diameter) at the highest practical magnification will give the most accurate information in regard to size studies, and the following survey method provides good preliminary information on pollen size variation within one flower, for one plant and for one or more populations. The sets of data can be easily analyzed for comparison by the simple statistical methods shown in Chapter 20, section B; for pollen slide preparation see methods 1-4 in the section below.

Make 5 pollen slides using pollen from five flowers of one plant; this will indicate something of the range of flower to flower pollen size variation in one plant.

Make 5 pollen slides using pollen from one or more flowers from each of five different plants from one population (or geographic area); this will indicate the degree of plant to plant pollen size variation in one population.

Make 5 pollen slides using pollen from one or more flowers from each of five different plants from a second population (or geographic area); this will give a preliminary estimate of the pollen size range of the species in the geographic area sampled.

The same method of slide preparation must be used for all fifteen slides. If the pollen is mounted in one of the glycerine jelly preparations, the cells may continue to swell for several days; thus all measurements should be done after the pollen size has stabilized and at about the same time after the slide was prepared.

Measure 100 pollen grains on each slide and compute the size range, mean pollen size, the standard deviation, and standard error of the mean for the pollen on each slide (see Chapter 20, section B). Also compute the same statistics for each of the three samples (of five slides each) and the total sample of 15 slides. If two or more taxa are involved similar sets of slides and calculations should be made for comparative purposes.

Section D. METHODS OF POLLEN SLIDE PREPARATION

1. Methyl green in glycerine jelly (semi-permanent; grains expanded). Place small amount of pollen in center of slide, add a drop of alcohol and allow to partly evaporate. A second, third, or fourth drop may be added if necessary. The alcohol spreads out and leaves a ring of dissolved oils from the pollen. Wipe oily ring off with cotton moistened in alcohol. Before specimen dries completely place a drop of hot methyl-green glycerine jelly on the slide and stir the pollen into it with a clean needle. Keep slide warm on warming plate, or by passing over flame. Cover with a No. 0 coverslip and heat briefly. The methyl-green stains only the exine. If contrast is desired the cell contents may be stained on the slide before starting the alcohol with a weak aqueous eosine solution; blot off excess stain. Pollen color is most brilliant after about 3 days; good up to a year, then fades, but the pollen can be restained.

Brandt's Glycerol Jelly

Soak 40 gm. sheet of granulated gelatin 2-3 hrs. in cool water.
Drain off excess water.
Melt hydrated gelatin in water bath.
Add 60 ml. of glycerol and 1-2 grams phenol crystals; mix thoroughly.
Filter through glass wool.
Bottle in 50 ml. (or 25 ml.) containers.

To prepare stain (for 50 ml. portion) add 2.5 ml. saturated solution of methyl green in 50% ethanol, and also 2 ml. of a saturated solution of Phloxine (or rose Bengal) in 50% ethanol to 50 ml. of melted glycerine jelly. The jelly-dye mixture keeps only 4-6 months so should be made up fresh; also, reheating causes the jelly to deteriorate, so use of only small quantities of the jelly will prevent excess waste.

To prepare slides:

a. tap pollen onto slide
b. wash off oils and resins with a drop of 70% ethanol
c. remove ring of sludge with a cotton swab moistened in 70% ethanol
d. place a drop of melted jelly on the pollen on the slide
e. stir gently to get even distribution of the pollen (the slide may be kept warm to facilitate mounting, but excess heat will rupture many pollen grains).
f. add #0 or #1 cover slip (after slide is removed from heat, if used)
g. allow slide to cool.

2. Gentian violet in aniline oil (permanent; grains not expanded).
Place pollen on slide and add 2 or 3 drops of aniline oil just tinted purple with gentian violet. Heat <u>gently</u> over small flame until grains are deeply stained. Allow slide to cool to room temperature, draw off excess oil with filter paper; wash by repeatedly adding and blotting off xylol until all oil and unabsorbed dye is removed. Mount in clarite.

3. Aceto-carmine in Glycerine Jelly (for pollen fertility counts).
Aceto-carmine is frequently used to differentiate between good and bad pollen but when quantitative estimates of pollen fertility are required the low viscosity of an aqueous solution is a disadvantage. Empty pollen grains being lighter tend to move towards the edges of the cover slip when it is lowered, so that a random distribution of good and bad pollen is not obtained. Counts, therefore made near the center of the cover slip are biased in favor of good pollen. This error can be avoided by using a viscous mounting medium. The aceto-carmine glycerol jelly described here has the advantage of being liquid in bulk at room temperature, setting to a jelly when used on a slide and giving a permanent preparation which can be scored at any time that is convenient.

The medium is prepared in two stages as follows:

a. Stock jelly. Add 0.25 gm. of finely powdered carmine to 50 ml. of melted Brandt's glycerol jelly. Mix thoroughly so as to obtain an even suspension of carmine particles throughout the jelly when cooled. This concentrated jelly is convenient for storage; small quantities can be used to make the final medium, as required.
b. Aceto-carmine jelly. Add 40 ml. of 45% acetic acid to 20 ml. of melted stock jelly. Boil gently until the solution is clear and all the carmine has dissolved; this can be checked by

examining a drop of the medium under the microscope. Finally, add 4 ml. of a saturated solution of ferric acetate in 45% acetic acid; this causes the medium to acquire a deep port wine color.

Macerate an undehisced anther in a drop of the medium on a slide, remove debris and stir the drop with a needle to distribute the pollen grains evenly. The drop of medium used should be such that no excess will exude round the edges of the cover slip when the latter is lowered into position. The preparation can be used after a few hours by which time staining and differentiation have occurred and the mount is permanent. [Adapted from G. E. Marks, Stain Technology 29: 277, 1954.]

4. Cotton Blue in Lactophenol (for pollen fertility counts).
For temporary or semi-permanent pollen slides Cotton Blue in lactophenol is very convenient. The fertile pollen grains stain a dark blue and sterile grains generally do not stain or stain very faintly. The lactophenol is prepared from: 20 ml. melted phenol, 40 ml. glycerin, 40 ml. water. To the 100 ml. of lactophenol add 1-5 ml. of 1% aqueous solution of Cotton Blue, depending on the intensity of color desired. Prepare slides as in #3.

5. Double Staining of Pollen Grains.
The medium consists of 50 ml. of glycerol jelly to which 2.5 ml. of methyl green and 2 ml. of phloxine, both in 50% alcoholic solutions, have been added. Prior to the application of the jelly-dye mixture the pollen is washed with 70% ethanol to remove adhering oils and resins. The staining reaction is differential and permits the rapid classification of pollen grains. The "functional" pollen expands and stains with both dyes whereas the aborted grains remain shrunken and take only the methyl green wall stain. The same reaction functions with orchid seed. [Adapted from A. Owczarzak, A rapid method for mounting pollen grains, with special regard to sterility studies, Stain Technology 27: 249-251.]

6. Hoyer's Mounting Medium (for pollen and spores).
A general mounting medium that gives expanded, unstained pollen grains for study with the light microscope.

Distilled water	50 gm.	Chloral hydrate	200 gm.
Arabic gum lump	30 gm.	Glycerine	20 gm.

The Arabic gum should be soaked in the distilled water about 24 hours. Add chloral hydrate and let the solution stand until all the material dissolves (this may require several days) before the glycerine is added and the solution is ready for use.

7. Acetolysis Technique.
The acetolysis method for the preparation of pollen grains from dried or preserved plants is the most widely used palynological technique. Nearly all of the world's pollen reference collections have been prepared in this manner. As will be noted, this acid hydrolysis process employs very caustic chemicals which leave only the resistant outer exine wall layer of the pollen grain intact. The pollen of a few angiosperms does not survive this treatment and must be prepared in other ways. Following acetolysis, grains are frequently brownish in

color. Some workers have found it desirable to stain the prepared pollen lightly (e.g., basic fuchsin). Although slight variations exist in the acetolysis technique, the basic procedure as outlined below should give satisfactory results.

CAUTION: THIS PROCEDURE SHOULD BE DONE UNDER A HOOD. DO NOT BREATHE FUMES OR GET ACETOLYSIS MIXTURE ON SKIN OR CLOTHES.

a. Prepare acetylation reagent-9 parts acetic anhydride:1 part concentrated sulfuric acid. Add acid slowly to the anhydride. Prepare just enough for one day's use; it will not keep longer.
b. Remove perianth parts of the flower and powder dried stamens through a fine mesh screen into a funnel placed in a centrifuge tube. Wet or fresh pollen must be dehydrated in glacial acetic acid first. Note: 1 to 6 samples may be run simultaneously in separate tubes.
c. Pour 5-10 cc of the acetylation mixture over the polleniferous contents of each centrifuge tube. Wash pollen off the sides of the funnels with the above mixture also.
d. Place centrifuge tubes in a water bath under a hood. Equip each tube with a glass stirring rod and stir contents at regular intervals. Bring water bath to a boil and cook for 10 minutes. Remove stirring rods and let bath cool.
e. Centrifuge tubes at about 1500+ rpms for 1-2 minutes, then decant. All large debris should be decanted. REMEMBER: IF ACETOLYSIS MIXTURE IS ADDED TO WATER A VIOLENT REACTION WILL OCCUR. DECANT SLOWLY.
f. Wash pollen in tubes with glacial acetic acid, centrifuge and decant.
g. Add distilled water to the tubes, shake thoroughly, centrifuge and decant.
h. Add a few drops of equal parts glycerine and distilled water to the tubes. Allow to stand for ca. 10 minutes, centrifuge and decant.
i. Mount pollen grains. Use a bacteriological loop to remove pollen from centrifuge tubes. The question of the most satisfactory mounting medium has not been resolved. The most widely used medium is glycerine jelly, although some workers use lactic acid or other media. Cover slips may be sealed using paraffin or left unsealed.
j. REMEMBER: AS IN ALL TAXONOMIC RESEARCH, IT IS VERY DESIRABLE TO HAVE A VOUCHER SPECIMEN FROM WHICH POLLEN WAS REMOVED PLACED IN AN HERBARIUM.

PALYNOLOGICAL LITERATURE

I. Periodicals

Grana--An international journal of Palynology. Uppsala, Sweden. 1954+. (until vol. 10(1), 1970, this journal was published as Grana Palynologica, Stockholm, Sweden).

Japanese Journal of Palynology. Shizuoka, Japan. 1967+. [In Japanese]

Journal of Palynology. Lucknow, India. 1965+.

Pollen et Spores. Paris. 1959+. (Bibliographic supplement published regularly).

Review of Palaeobotany and Palynology. Amsterdam, The Netherlands. 1965+.

II. General References

Brewbaker, J. L. 1967. The distribution and phylogenetic significance of binucleate and trinucleate pollen grains. American Journal of Botany 54: 1069-1083.

Cranwell, L. M. 1953. New Zealand Pollen Studies. The Monocotyledons. Harvard University Press. Cambridge.

Erdtman, G. 1943. An Introduction to Pollen Analysis. Chronical Botanica Company. Waltham, Massachusetts.

Erdtman, G. 1952. Pollen Morphology and Plant Taxonomy - Angiosperms. (An Introduction to Palynology, vol. I). Almqvist and Wiksell. Stockholm.

________. 1963. Palynology. In: Vistas in Botany, IV. W. B. Turrill (ed.). MacMillan and Company. New York.

________. 1969. Handbook of Palynology. Morphology-Taxonomy-Ecology. An Introduction to the Study of Pollen Grains and Spores. Munksgaard. Copenhagen.

Faegri, K., and J. Iversen. 1964. Textbook of Pollen Analysis. Ed. 2. Hafner Publishing Company. New York.

Felix, J. C. 1961. An Introduction to Palynology. In: H. N. Andrews, Studies in Paleobotany. John Wiley and Sons. New York.

Germeraad, J. H., and J. Muller. 1970. A computer-based numerical coding system for the description of pollen grains and spores. Review of Paleobotany and Palynology 10: 175-202.

Hyde, H. A., and K. F. Adams. 1958. An Atlas of Airborne Pollen Grains. MacMillan Company. London.

Ikuse, M. 1956. Pollen Grains of Japan. Hirokawa Publishing Company. Tokyo.

Kapp, R. O. 1969. How to Know Pollen and Spores. William C. Brown Company. Dubuque, Iowa.

Kremp, G. O. W. 1965. Morphologic Encyclopedia of Palynology. University of Arizona Press. Tucson.

Reitsma, T. 1970. Suggestions towards unification of descriptive terminology of angiosperm pollen grains. Review of Palaeobotany and Palynology 10: 39-60.

Richard, P. 1970. Atlas pollinique des arbres et de quelques arbustes indigénes du Québec. I-IV. Naturaliste Canadien 97: 1-34; 97-161; 241-306.

Tschudy, R. H. 1969. The Plant Kingdom and Its Palynological Representation. In: R. H. Tschudy and R. A. Scott (eds.). Aspects of Palynology. John Wiley and Sons. New York.

Wodehouse, R. P. 1935 (reprinted in 1960). Pollen Grains: Their Structure, Identification, and Significance in Science and Medicine. McGraw-Hill Company. New York.

World Pollen Flora. 1970-. G. Erdtman (ed.). Hafner Publishing Company. New York. (to be published in parts. Part I appeared in 1970).

III. References Illustrating Taxonomic Application of Pollen Morphology

Dahl, A. O. 1952. The comparative morphology of the Icacinaceae, VI. The pollen. Journal of the Arnold Arboretum 33: 252-295.

Ferguson, I. K., and D. A. Webb. 1970. Pollen morphology in the genus *Saxifraga* and its taxonomic significance. Botanical Journal of the Linnean Society 63: 295-311.

Graham, A., and S. A. Graham. 1967. Pollen morphology and taxonomy of *Cuphea* (Lythraceae). Review Paleobotany and Palynology 3: 155-162.

Jones, S. B. 1970. Scanning electron microscopy of pollen as an aid to the systematics of *Vernonia* (Compositae). Bulletin, Torrey Botanical Club 97: 325-335.

Lewis, W. H. 1965. Pollen morphology and evolution in *Hedyotis* subgenus *Edrisia* (Rubiaceae). American Journal of Botany 52: 257-264.

Skvarla, J. J., and B. L. Turner. 1966. Systematic implications from electron microscopic studies of Compositae pollen - a review. Annals Missouri Botanical Garden 53: 220-256.

Walker, J. W. 1971. Pollen morphology, phytogeography, and phylogeny of the Annonaceae. Contributions from the Gray Herbarium 202. 131 p.

IV. References on Technique

Brown, C. A. 1960. Palynological Techniques. Privately published by the author, Baton Rouge, Louisiana.

Erdtman, G. 1960. The acetolysis method. A revised description. Svensk Botanisk Tidskrift 54(4): 561-564.

Falk, R. H., E. M. Gifford, Jr., and E. G. Cutter. 1971. The effect of various fixation schedules on the scanning electron microscopic image of Tropaeolum majus. American Journal of Botany 58: 676-680.

PALYNOLOGICAL EXERCISE

From slides prepared by the student or provided by the instructor obtain pollen data for one or more species.

	Family ______	Family ______
	Species I ______	Species II ______
1. Pollen units........................	______	______
2. Symmetry............................	______	______
3. Polarity............................	______	______
4. Size (Polar axis and equatorial axis)............................	______	______
5. Aperture Type.......................	______	______
6. Wall Structure......................	______	______
7. Wall Sculpture......................	______	______
8. Nuclear State.......................	______	______

Prepare an accurate drawing showing the above features.

For an exercise dealing with pollen size variation see the pollen exercise used as an example for data accumulation (Chapter 20, Section F).

Chapter 9. EMBRYOLOGICAL EVIDENCE

If broadly defined, the term embryology includes the successive stages of megasporogenesis and microsporogenesis (meiotic development of the functional megaspore and microspore), megagametogenesis and microgametogenesis (development of the female and male gametophyte), and the post-fertilization stages of embryogeny (development stages from the fertilized egg to the mature embryo and its surrounding seed coats). Information derived from embryology has proven to be of systematic value; however, obtaining these data requires considerable effort. Not only do flowers have to be collected at critical periods during their development, but they must be skillfully fixed, sectioned and stained. Björnstad (1970) states that an embryological study can be helpful to the taxonomist in three ways: (1) systematic questions may be decided when two or more positions have been proposed on the basis of other characters, (2) new research may be promoted on a taxon when embryological data make the present systematic position doubtful or improbable, and (3) confirmation of the present systematic position of a taxon by embryological information. The following twelve embryological characters are considered by Maheshwari (1950, 1964) to be taxonomically significant.

A. Anther.
   1. Number and arrangement of loculi.
   2. Structure and thickening of the endothecium ("fibrous" or not).
   3. Nature of tapetum.
      a. Glandular (secretory) tapetum. Tapetum composed of highly vacuolated cells which remain stationary.
      b. Amoeboid (plasmodial) tapetum. Tapetum in which the cells, subsequent to wall breakdown, move within the loculus.
      c. Nuclear condition (uni-, bi-, multinucleate, polyploid).
   4. Nature of periplasmodium.
   5. Type of wall formation (Basic, Dicotyledonous, Monocotyledonous, Reduced).

Basic

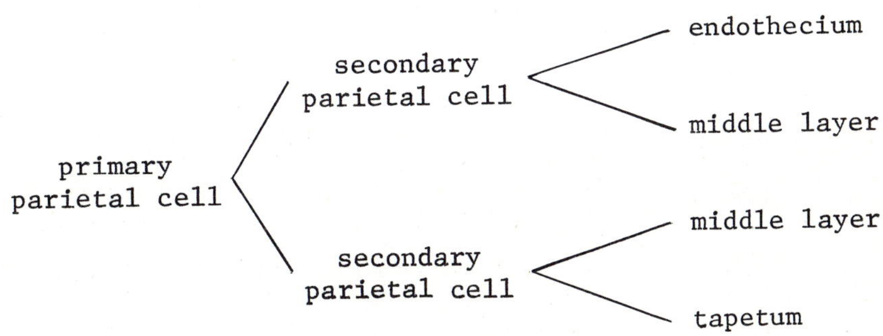

Dicotyledonous

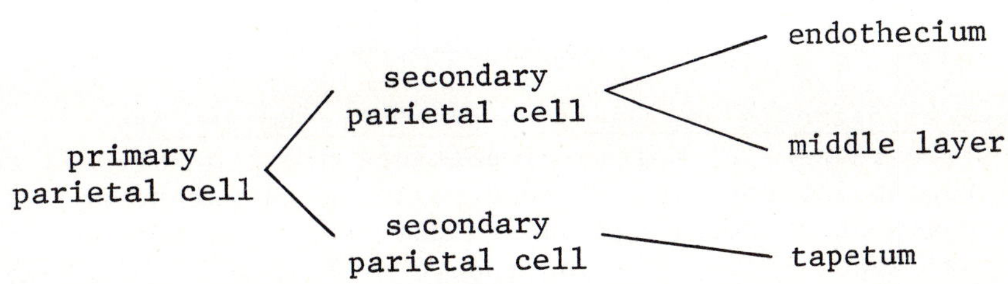

Monocotyledonous

- primary parietal cell
  - secondary parietal cell — endothecium
  - secondary parietal cell
    - middle layer
    - tapetum

Reduced

- primary parietal cell
  - secondary parietal cell — endothecium
  - secondary parietal cell — tapetum

B. Quadripartition of the Microspore Mother Cells.

Variable characters include successive or simultaneous division, tetrad formation by furrowing or cell plate formation, and microspores (pollen) shed as single grains, tetrads, or polyads.

C. Development and Organization of the Pollen Grain. (See Chapter 8, Palynology)

D. Development and Structure of the Ovule.

1. Number and structure of integuments.
2. Method of vasculation (when present, where?)
3. Presence or absence of an aril.
4. Presence or absence of an arillode ("false aril").
5. Presence or absence of a caruncle (strophiole).
6. Presence and place of origin of an obturator.
7. Structure of the micropyle.
8. Ovule type based on nature of the nucellus.
   a. Crassinucellate ovule. An ovule with a massive nucellus in which the megaspore mother cell is deeply embedded.
   b. Tenuinucellate ovule. An ovule in which there is a small amount of nucellus with the megaspore mother cell subepidermal.
9. Presence or absence of an endothelium (integumentary tapetum) and extent of coverage of megagametophyte.

(Items 3–5: Pre- or post-fertilization development)

E. Form and Extent of the Nucellus.

Size and shape, presence or absence of a hypostase, epistase, method of origin of integument or integuments, persistence of nucellus as perisperm, presence or absence of a nucellar cap, and persistence or degeneration during development.

F. Origin and Extent of the Sporogenous Tissue in the Ovule.

Number of archesporial cells (one, few, many) and presence or absence of wall layers, and nature of cell divisions.

G. Megasporogenesis and Development of the Embryo Sac.

Characters of special importance are time of wall formation in megasporogenesis, position of functioning megaspores, and size, shape and arrangement of cells and nuclei composing the embryo sac (female gametophyte) type. Determine and record type (see Figure 9-1).

Figure 9-1

EMBRYO SAC TYPES IN ANGIOSPERMS

(After Maheshwari)

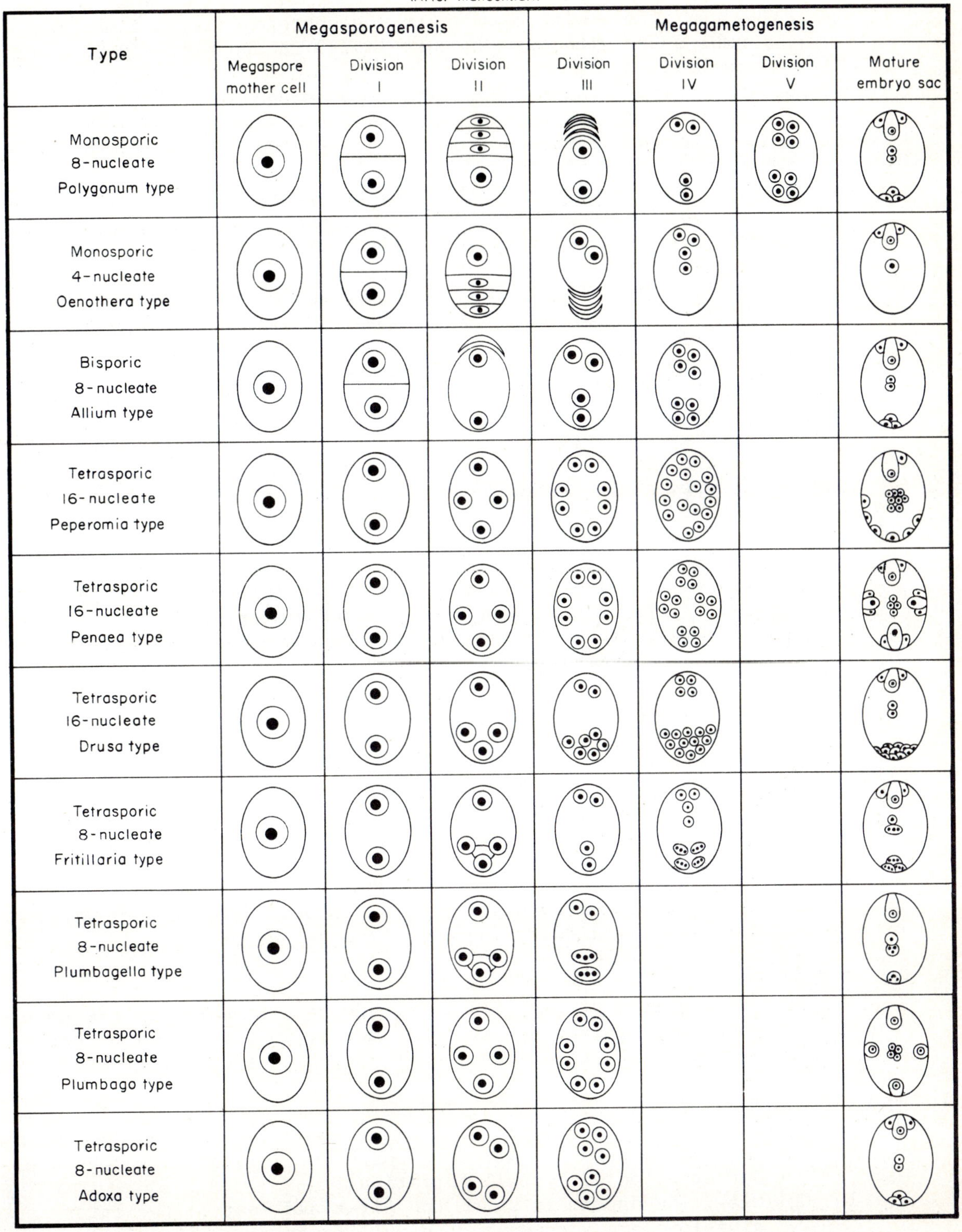

1. Monosporic. Embryo sac formed by the germination of one of the megaspores of the tetrad.
   a. Polygonum type. With 8-nucleate megasporogenesis, the mature embryo sac with egg, 2 synergids, 3 antipodals, and 2 polar nuclei.
   b. Oenothera type. With 4-nucleate megasporogenesis, the mature embryo sac with egg, 2 synergids, and a single polar nucleus.
2. Bisporic. Embryo sac formed from one of the two dyad cells formed after division I of meiosis.
   a. Allium type. With 8-nucleate megasporogenesis, the mature embryo sac with egg, 2 synergids, 3 antipodals, and 2 polar nuclei.
3. Tetrasporic. Embryo sac formed by four spores resulting from division of the megaspore mother cell.
   a. Peperomia type. With 16-nucleate megasporogenesis, the mature embryo sac with egg, 1 synergid, 6 antipodals, and 8 polar nuclei.
   b. Penaea type. With 16-nucleate megasporogenesis, the mature embryo sac with 4 "triads," one at each cardinal compass point around the periphery, and 4 polar nuclei; each triad consisting of an egg cell and 2 synergids, only the egg cell of the micropylar "triad" being functional.
   c. Drusa type. With 16-nucleate megasporogenesis, the mature embryo sac with egg, 2 synergids, 11 antipodals, and 2 polar nuclei.
   d. Fritillaria type. With 8-nucleate megasporogenesis, the mature embryo sac with egg, 2 synergids, 3 triploid antipodals, and 2 polar nuclei, one haploid and one triploid.
   e. Plumbagella type. With 8-nucleate megasporogenesis, the mature embryo sac with egg, 1 triploid antipodal, and 2 polar nuclei, one haploid and one triploid.
   f. Plumbago type. With 8-nucleate megasporogenesis, the mature embryo sac with egg, 4 polar nuclei, and 3 peripheral nuclei, one at the chalazal end and one to each side.
   g. Adoxa type. With 8-nucleate megasporogenesis, the mature embryo sac with egg, 2 synergids, 3 antipodals, and 2 polar nuclei.

H. Form and Organization of the Mature Embryo Sac.

Variable characters include shape, number and arrangement of the nuclei and cells, nature of synergids and antipodals, presence of starch or other food reserve, and formation of embryo sac caeca.

I. Fertilization.

Variable characters of importance include path of entry of the pollen tube (porogamous, chalazogamous, mesogamous), interval between pollination and fertilization, and branching of pollen tube during growth.

J. Endosperm (Post Fertilization Development).

1. Nature of endosperm formation: (see Figure 9-2)
   a. Nuclear. Endosperm formation in which the initial divisions are not accompanied by wall formation.
   b. Cellular. Endosperm in which the first and most of the later divisions are accompanied by wall formation. If cellular, orientation of first walls should be recorded.
   c. Helobial. Endosperm in which the first division is accompanied by wall formation but subsequent divisions are free nuclear, at least in the micropylar chamber.

TYPES OF ENDOSPERM FORMATION

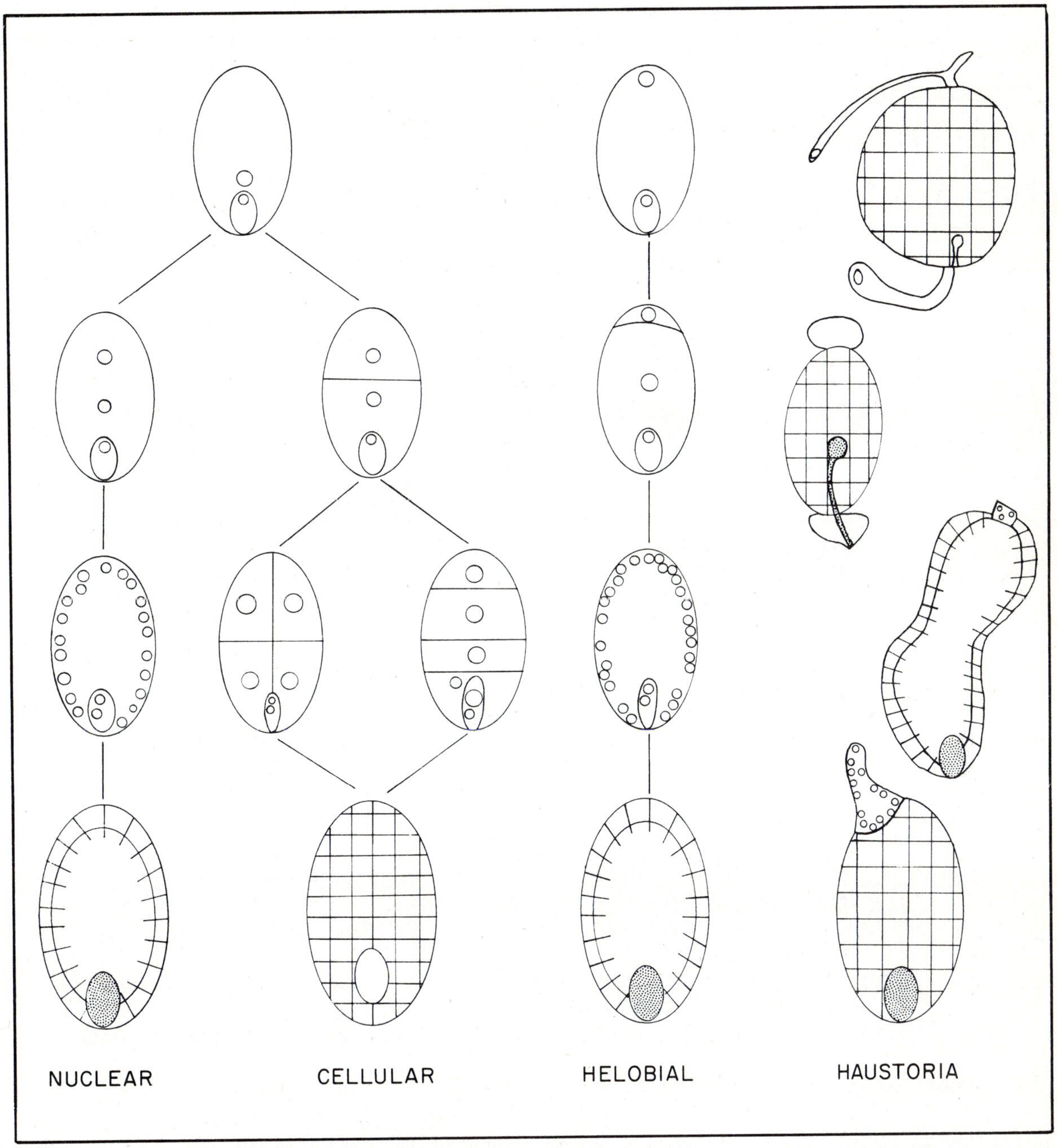

[Used with permission of Dr. Barbara Palser.]

2. Presence of ruminate endosperm.
3. Presence, absence and extent of endosperm haustoria and manner of their formation.
4. Nature of endosperm food reserve, and fate of endosperm in mature seed.

K. Embryogeny Type (Post Fertilization Development).

Variable characters include relation of proembryonal cells to body regions of the embryo, form and organization of the mature embryo (size, straight or curved, degree of differentiation), and presence or absence of suspensor haustoria. Attention should also be given to the source and structure of the seed coat and whether there is persistence of nucellus in the seed or not (perisperm). Johansen (1950) classified the embryos of dicotyledons into six principal types based on their development. A taxonomic distribution of these embryonic types and a pertinent discussion of the relative value of these terms to systematics is given by Wardlaw (1955).

1. Crucifer (Onagrad) type. With the terminal cell of the proembryo dividing by a longitudinal wall, the basal cell playing only a minor part or none in subsequent embryo development.
2. Asterad type. With the terminal cell of the proembryo dividing by a longitudinal wall, the basal and terminal cells both contributing in a major way to subsequent embryo development.
3. Solanad type. With the terminal cell of the proembryo dividing by a transverse wall, the basal cell usually forming a 2- or more-celled suspensor but playing no major part in subsequent embryo development.
4. Caryophyllad type. With the terminal cell of the proembryo dividing by a transverse wall, the basal cell undergoing no division and playing no role in subsequent embryo development, the suspensor, if any, derived from the terminal cell.
5. Chenopodiad type. With the terminal cell of the proembryo dividing by a transverse wall, the basal and terminal cells both contributing in a major way to subsequent embryo development.
6. Piperad type. The zygote divides by a wall which is either vertical or so obliquely orientated that it may be assumed to be essentially longitudinal.

L. Certain Abnormalities of Development (Parthenogenesis, Apogamy, Adventive Embryony, Polyembryony, etc.).

1. Apogamy. Development of a new embryo from a cell of the gametophyte other than the egg.
2. Adventive embryony. Embryo development from diploid cells of the nucellus or the integument.
3. Polyembryony. The occurrence of more than one embryo in a seed.

Embryological information has proved to be of systematic value at all taxonomic levels from the species to class. One of the outstanding examples of a taxon being clearly defined embryologically is the Ericales. In no other order do the following combination of characters occur: no endothecium; glandular, multinucleate tapetum; pollen in tetrads; pollen two-celled; unitegmic; nucellus thin, ephemeral; endothelium present; no parietal cells in ovule; megagametophyte monosporic and 8-nucleate, broad at micropylar end and narrow at chalazal end; fluted hollow style; cellular endosperm, first 4 cells in a linear row; micropylar and chalazal endosperm haustoria; single-layered seed coat; seed with fleshy endosperm, straight embryo; nectary present; anthers introrse by inversion; anther dehiscence by "apical" slit or pore; anthers frequently appendaged; single archesporial cell in ovule; and zygote elongated before division. The failure of a plant to possess a majority of these characteristics is reason enough to cast doubt on its ericalean affinities.

EMBRYOLOGICAL GLOSSARY

Anther. Pollen producing portion of stamen.

Antipodal. One to many cells of the mature embryo sac located at the chalazal end.

Archesporium. One to several hypodermal cells in the ovule primordium distinguished by their larger size, larger nuclei, and dense cytoplasm.

Aril. The secondary outgrowth of an ovule which surrounds the ovule in post-fertilization stages, which is derived from various regions but commonly the funiculus or raphe.

Arillode. A fleshy coat derived from the micropylar rim of the outer integument.

Caruncle. An integumentary protuberance restricted to the micropylar region of a seed and derived from cells of the outer integument (characteristic of euphorbiaceous seed).

Central Cell. The cell of the embryo sac in which the polar nuclei reside.

Embryo Sac. Female gametophyte of flowering plants.

Endosperm. Tissue formed from the fusion of the two polar nuclei and one of the sperm cells; nutritive in function.

Endothecium. The outermost wall layer of an anther, located beneath the epidermis.

Hypostase. A group of thick-walled, lignified cells at the base of the nucellus below the sporogenous tissue.

Integument. An enclosing envelope of an ovule; embryonic seed coat.

Loculus. A pollen-containing chamber of the anther.

Megaspore. A spore formed by meiosis of the megasporocyte and producing the megagametophyte (embryo sac).

Micropyle. Hole through integument(s).

Microspore. A spore formed by meiosis of a microsporocyte and producing the microgametophyte.

Nucellar Cap. A persistent group of thick-walled and lignified cells over the apex of the embryo sac.

Nucellus. Female sporangium within the ovule.

Obturator. A proliferation of tissue in the form of a small protuberance which is generally of placental origin and which grows into space between nucellus and the integument or also between the ovule and the ovary wall; suggested function is to aid in pollen tube conduction to the micropyle.

Ovule. A megasporangium enclosed by an integumentary system; an embryonic seed.

Periplasmodium. The mass resulting from disintegration and coalescence of tapetal cells.

Perisperm. Persistent nucellus in seed; believed to be a storage tissue.

Polar nucleus. One of usually two nuclei of the angiosperm embryo sac that migrates to the center from the poles.

Ruminate endosperm. A type of endosperm which exhibits some degree of irregularity as uneveness in its contour accompanied by irregularity in the inner surface of the seed coat.

Synergid. A sterile cell associated with the egg in the angiosperm embryo sac; a part of the egg apparatus.

Tapetum. The innermost parietal wall layer of an anther, nutritive in function.

EMBRYOLOGICAL LITERATURE

Björnstad, I. N. 1970. Comparative embryology of Asparagoidead - Polygonateae, Liliaceae. Nytt Magasin for Botanikk 17: 169-207.

Davis, G. L. 1966. Systematic Embryology of the Angiosperms. John Wiley and Sons, Inc. New York.

Johansen, D. A. 1950. Plant Embryology. Chronical Botanica Company. Waltham.

Maheswari, P. 1950. An Introduction to the Embryology of Angiosperms. McGraw-Hill. New York.

________. 1964. Embryology in relation to taxonomy. In: W. B. Turrill (ed.), Vistas in Botany IV. MacMillan Company. New York.

Recent Advances in the Embryology of Angiosperms. 1963. P. Maheswari (ed.). International Society of Plant Morphologists. Delhi.

Wardlaw, C. W. 1955. Embryogenesis in Plants. John Wiley and Sons, Inc. New York.

A CLEARING-SQUASH TECHNIQUE FOR THE STUDY OF OVULE AND MEGAGAMETOPHYTE DEVELOPMENT IN ANGIOSPERMS*

The technique outlined here was devised specifically to extend investigation beyond the limits imposed by traditional section methods. Known as the 4 1/2 clearing technique, this procedure incorporates features of both clearing and squash methods and involves phase contrast optics for examination of cell structure in optical sections through whole ovules.

1. Fixation. Ovularies from flowers in all stages of anthesis should be fixed for 24 hrs. in FPA_{50} (formalin, propionic acid, 50% ethanol, 5:5:90), Allen's modified Bouin's fluid, or Randoph's modified Navashin fluid (Johansen, 1940). Other fixatives may prove quite suitable, but some, such as the various Carnoy's fluids, have limited value.
2. Storage. Fixed ovularies should be transferred to 70% ethanol where they may be stored indefinitely.
3. Clearing. The whole ovulary or ovules dissected from the ovulary in 70% ethanol should be placed in standard 4 1/2 clearing fluid composed of lactic acid (85%), chloral hydrate, phenol (crystals), clove oil, and xylene (2:2:2:2:1, by weight) for 2-24 hrs. (Herr, 1971). A concave container such as the Radnoti planchet (Radnotic Glass Technology, Inc., 2614 Seaman Avenue, El Monte, California 91733) is recommended for the clearing process (Figure 9-3). Cleared ovularies appear nearly transparent under the dissecting microscope equipped with transmitted light. The ovules should be dissected from the cleared ovularies and by means of a Pasteur pipette should be transferred with a drop of the fluid to a Raj slide (Herr, 1971, 1973a; Smith, 1973) for microscopic examination.
4. The Raj Slide. Two cover glasses (No. 0-2, 12 or 18 mm. sq.) are affixed with permount or balsam 1-1.5 cm. apart on a standard microscope slide which is then placed on a warming table at 50°C for 3 days to insure

*Contributed by J. M. Herr, Jr., University of South Carolina, Columbia.

adequate hardening. The drop of clearing fluid with several ovules is placed in the center of the 1-1.5 cm. space and a cover glass is then placed over the preparation so to rest on the two mounted cover glasses (Figure 9-4). The support cover glasses must be of sufficient thickness to eliminate pressure of the top cover on the ovules.

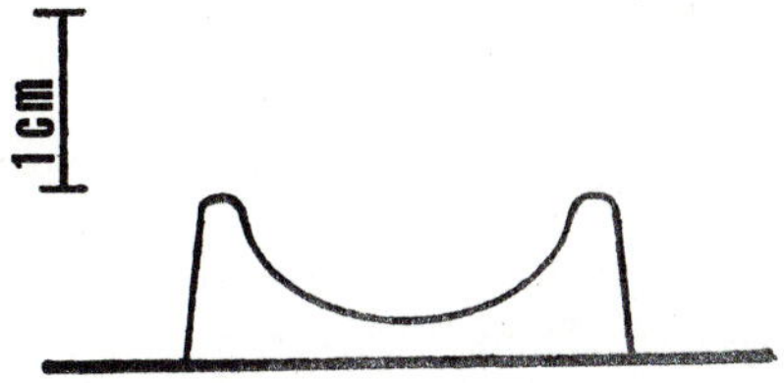

Figure 9-3. Cross-section of Radnoti planchet.

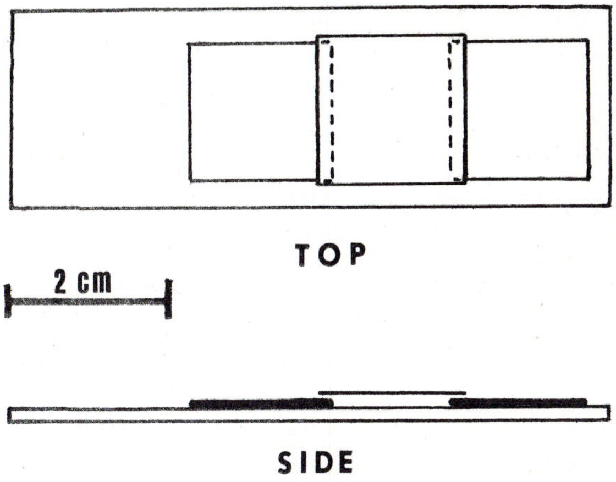

Figure 9-4. The Raj slide.

5. Microscopic Examination. The preparation should be examined with phase contrast optics. Ovules in all stages of anatropous or campylotropous development tend to lie sideways; that is, with the long axis of the nucellus parallel to the slide. Downward focusing, therefore, reveals ovule structure through a series of optical sagittal sections (Figure 9-5). By slight shifting of the cover glass on the support mounts, small ovules can be reoriented for optical sections in the frontal or transverse planes.
6. Squash Procedure. After brief exposure to the clearing fluid, the ovules become quite fragile. If, therefore, the cover glass is lightly and repeatedly pressed with a dissecting needle midway between the support covers, the cells of the ovule gradually become spread apart. This squash procedure, performed under observation with a dissecting microscope, does not disrupt the structural integrity of the individual cells. Furthermore, it provides a method for extending the observations to the detailed features of specific structures such as megasporocytes, megaspores, and developing megagametophytes previously identified within intact ovules (Figure 9-5).
7. Alternative and Supplementary Procedures. The clearing process in some cases may be greatly enhanced if the stored ovularies or ovules are dehydrated or dehydrated and pretreated prior to application of the clearing fluid. In addition, modifications of the standard 4 1/2 fluid may produce an improved clearing image.
   a. Dehydration. In some cases, dehydration of the stored ovularies or ovules to absolute ethanol (70%, 95%, 100% ethanol) is necessary for best results.

Figure 9-5. A, Opitcal sagittal section of an ovule from _Glycine max_ Merr. Megasporocyte during meiosis, telophase I. X 660. (Photograph by Ralph A. George). B, Optical sagittal section of an ovule from _Phaseolus aureus_ Roxb. Dyad, both cells at prophase II. X 600. (Photograph by Glenda P. George). C, Optical sagittal section of an ovule from _Phaseolus aureus_ Roxb. Dyad during meiosis, telophase II. X 600. (Photograph by Glenda P. George). D, Optical sagittal section of an ovule from _Glycine max_ Merr. Preparation partially squashed to separate nucellus from integuments. 4-nucleate megagametophyte with characteristic central vacuole and small chalazal vacuole. X 660. (Photograph by Ralph A. George). E-G, Two optical sagittal sections and an optical cross section of a single ovule from _Phaseolus aureus_ Roxb. All X 380. (Photograph by Glenda P. George). E, Dyad in upper focal plane. F, Second adjacent dyad in lower focal plane. G, Two dyad members, one from each dyad, in optical cross section of the ovule.

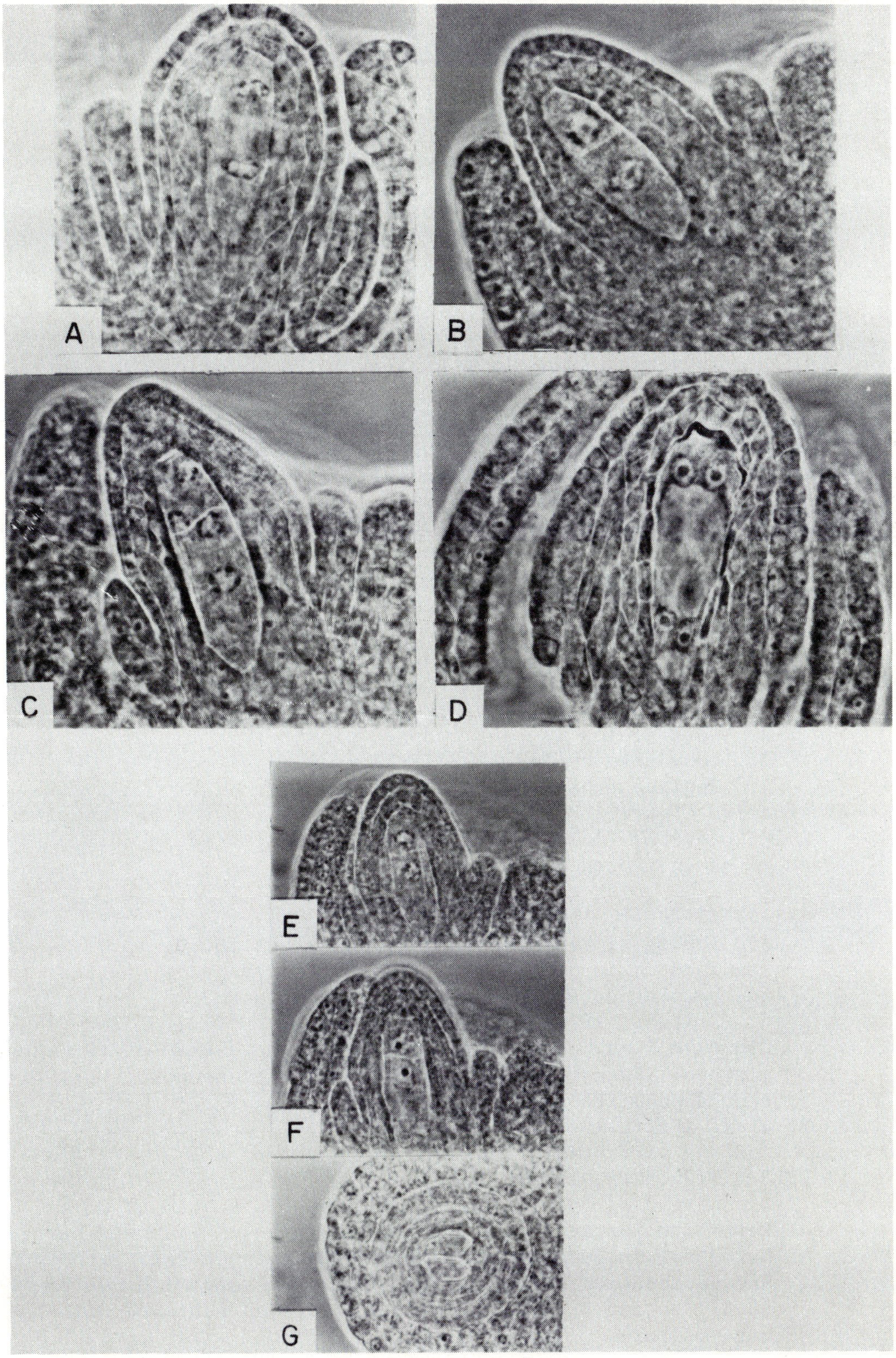
A
B
C
D
E
F
G

b. Clearing Pretreatment. For some species, a pretreatment procedure is required for successful clearing. The following methods have been applied to whole ovularies and ovules from the genera indicated.

Lactic Acid (Oenothera, Ludwigia, Cornus, and Rubus)

1. Dehydrate ovularies or ovules to absolute ethanol.
2. Pass the material through a graded series of absolute ethanol-lactic acid (3:1, 2:2, 1:3) with 10-15 minutes exposure to each fluid.
3. Transfer the material to lactic acid (85%) and incubate in an open container at 50°C for 24 hrs.
4. Replace the lactic acid with the clearing fluid at room temperature.

Lactic Acid-Phenol-benzyl benzoate, 1:1:1 by 1 (Oenothera, Ludwigia, and Cornus)

1. Dehydrate ovularies or ovules to absolute ethanol.
2. Pass the material through a graded series of absolute ethanol-pretreatment mixture (3:1, 2:2, 1:3) with 10-15 minutes exposure to each fluid.
3. Transfer the material to the pretreatment mixture and incubate in a closed container at 60°C for 24 hrs.
4. Replace the pretreatment mixture with the clearing fluid at room temperature.

Potassium Hydroxide (Cornus; Smith, 1973)

1. Transfer ovules from 70% ethanol to a 10% potassium hydroxide solution for two minutes.
2. Wash ovules in four changes of water.
3. Dehydrate ovules to 95% ethanol.
4. Replace ethanol with clearing fluid.

c. Alternative Clearing Mixtures. Several modification of the standard 4 1/2 clearing fluid have applications discussed in detail elsewhere as indicated below.

IKI-4 1/2 (Herr, 1972)
Add 100 mg. iodine and 500 mg. potassium iodide to 9 gm. of the standard mixture. IKI-4 1/2 enhances the phase contrast image and permits detection of accumulated starch.

PP-4 1/2 (Herr, 1973a)
Dissolve 3 mg. potassium permanganate in a gram of the standard mixture. Potassium permanganate-4 1/2 is somewhat unstable and should be mixed for immediate use.

BB-4 1/2 (Herr, 1973b)
Add one part benzyl benzoate by weight to the standard mixture.

PPBB-4 1/2 (Herr, 1973)
Dissolve 3 mg. potassium permanganate in a gram of BB-4 1/2. Potassium permanganate, benzyl benzoate-4 1/2 should be prepared for immediate use. Contrast between nuclei and cytoplasm and delineation of cell layers are increased with use of PP-4 1/2, BB-4 1/2, and PPBB-4 1/2 beyond the level achieved with the standard mixture.

Pretreatment procedures and alternative clearing agents essential for success in some plants produce adverse results in others. For example, all pretreatment procedures listed here adversely affect the clearing image of all legumes investigated (*Phaseolus*, *Glycine*, and *Cassia*). For these genera, however, BB-4 1/2 and PPBB-4 1/2 are superior to the standard clearing mixture.

The original method should be used initially in an investigation with the alternatives and supplements incorporated later as needed. Each modification of the original technique must be assessed for its improved or deleterious effect on the clearing image.

Summary

Ovules or whole ovularies to be cleared in standard 4 1/2 clearing fluid or one of its modifications should be processed by the following steps:

1. Fix material in FPA_{50}, Allen's modified Bouin's fluid, or Randolph's modified Navashin fluid (Johansen, 1940).
2. Store indefinitely in 70% ethanol.
   a. Dehydrate to absolute ethanol.
   b. Pretreat: lactic acid at 60°C for 24 hrs.;
   c. Lactic acid-phenol-benzyl benzoate (1:1:1 by weight) at 60°C for 24 hrs. (closed container); or
   d. 10% KOH for 2 minutes. Water; dehydrate to 95% ethanol.
3. Clear in standard 4 1/2 clearing fluid (lactic acid, chloral hydrate, phenol crystals, clove oil, and xylene; 2:2:2:2:1, by weight).
   a. IKI-4 1/2 (100 mg. iodine, 500 mg. potassium iodide to 9 gm. 4 1/2).
   b. PP-4 1/2 (3 mg. potassium permanganate to 1 gm. 4 1/2).
   c. BB-4 1/2 (one part by weight benzl benzoate to 4 1/2).
4. Mount ovules beneath the supported cover glass of a Raj slide and examine in optical section with phase contrast optics.

OVULE CLEARING TECHNIQUE LITERATURE

Herr, J. M., Jr. 1971. A new clearing-squash technique for the study of ovule development in angiosperms. American Journal of Botany 58(8): 785-790.

________. 1972. Applications of a new clearing technique for the investigation of vascular plant morphology. Journal of the Elisha Mitchell Scientific Society 88(3): 137-143.

________. 1973a. The use of Nomarski interference microscopy for the study of structural features in cleared ovules. Journal of the Society for Advancement of Botany 1: *in press*.

________. 1973b. The cytological effects of several fixatives on the ovules of *Cassia abbreviata* Oliver var. *granitica* Bak. New Botanist 1: *in press*.

Johansen, D. A. 1940. Plant Microtechnique. McGraw-Hill Company. New York and London.

Smith, B. B. 1973. The use of a new clearing technique for the study of early ovule development, megasporogenesis, and megagametogenesis in five species of *Cornus* L. American Journal of Botany 60(4): 322-338.

EMBRYOLOGICAL EXERCISE

From material on display, or preferably from specimens that have been fixed, sectioned or cleared by the student, determine the following selected character states. Ideally, a sufficient number of floral buds should be available so an assessment of different fixing and staining procedures can be made. Accurate illustrations of embryological features are required. It should also be remembered that a number of floral buds in various stages of development may have to be prepared in order to give a complete embryological description of the plant. In view of the general paucity of accurate embryological information for angiosperms as a whole, any information that can be documented is potentially valuable and should be recorded, even though it appears incomplete.

	Species 1	Species 2
Method of Specimen Preparation A. Fixative B. Stain C. Clearing		
Character State 1. Anther a. Endothecium Type b. Tapetum Type c. Type of Wall Formation		
2. Ovule Type		
3. Form of Nucellus		
4. Type of Embryo Sac Development		
5. Type of Endosperm Formation		
6. Embryogeny		
7. Abnormalities if any		

Chapter 10. CYTOLOGICAL EVIDENCE*

Cytology, broadly defined, includes the study of any and all characteristics of cells and thus embraces both morphological and physiological work. Even the morphological aspects of cytology are enormously complex, for they include investigations of the structure of organelles and of cells and their contents both living and nonliving. The degree to which these structural features are faithfully represented in dead or preserved cells depends upon the nature of the structure in the original living system and upon the particular treatments which are employed to preserve them for study. Traditionally, cytological studies have been made with the aid of the compound light microscope; more recently the electron microscope has provided an increasingly useful means of examining the minute details of cells and organelle structure. Clearly, the structural aspects of general cytology are closely related to and inseparable from the cellular aspects of anatomy.

The special branch of cytology which deals with the nucleus of the cell and the hereditary material it contains, the chromosomes, is more accurately designated as karyology. Ever since the essential correctness of the nineteenth century chromosome theory of heredity was experimentally demonstrated in the early decades of this century, karyology has gained increasing importance as an aid in taxonomic work. Karyology deals with the chromosomes of both plants and animals. Except for details which are peculiar to the reproductive processes in each kingdom, the principles of chromosome organization and behavior are the same in all eucaryotic organisms. Certain chromosome phenomena; e.g., polyploidy, are more frequent and better known in plants; others, e.g., duplication and inversion, have been more precisely analyzed in animals. For comprehensive treatments of chromosome cytology, with illustrations and examples from extensive literature, and for clarification of many details and discussions of controversial issues, the reader may be referred to several excellent books by distinguished students of chromosomes all of whom also happen to have been professional botanists (Darlington 1963, 1965; Sharp 1934, 1943; Stebbins 1971; Swanson 1957).

This chapter is designed to accomplish two things:

(1) To provide a brief outline of karyological concepts and principles with emphasis on those aspects which have proved to be most helpful in plant systematics.

(2) To outline the basic rationale and methodology for practical chromosome work in the laboratory for those who wish to add the karyological approach to their multidisciplinary attack upon the problems of vascular plant taxonomy.

The chromosomes are the physical bodies in the cell which contain most but not quite all (as modern investigations of organelles show) of the essential hereditary material, the DNA. This is reason enough to attach special importance to the study of chromosomes, their numbers, form, internal structure, and behavior. It also led the earlier cytotaxonomists to assume that the characters of the chromosomes would be more reliable guides to taxonomic and

*Contributed by Ben W. Smith, North Carolina State University, Raleigh.

evolutionary relationships than many of the traditional morphological characters. This view is still widely held among chromosome cytologists and cytogeneticists, but it has been categorically rejected by a number of highly competent taxonomists. They maintain that the characters of the chromosomes should be given no more (and no less) weight than any other characters which may be available for study.

Each of these interpretations appears to be correct in particular cases, but neither can be accepted as a general evaluation of the usefulness of chromosome studies in resolving taxonomic questions. In various genera and sometimes higher groups, chromosome numbers, size, and form are monotonously similar from species to species even though it is obvious from other characters of the plants that considerable evolutionary divergence has occurred. On the other hand, there are groups such as Asplenium in temperate North America or the Eurasian Triticum, Aegilops, Secale complex, for example, in which chromosome studies, especially when combined with hybridization and genetic analysis, have provided the essential clues for tracing the evolutionary history of the group. Cytogenetics provides a framework in such cases upon which the taxonomist reevaluates the morphological evidence of relationship among the various populations of the group. Davis and Heywood (1963, Chapter 6) present an excellent summary and discussion of this and other issues in cytotaxonomy.

There is no reliable way in which the relative usefulness of chromosome studies in karyologically and genetically unknown material can be predicted with assurance. It is necessary to examine the chromosomes. Since only a small proportion of the total number of species among the angiosperms have been examined, and some of the early work needs to be repeated and documented by more reliable methods, much remains to be done. It is true, however, that irregularities in reproduction such as infertility and sterility, and unusual variation patterns such as extreme uniformity in progeny which suggests some form of apomixis, or in contrast, unusual transgressive variants in segregation which may result from irregularities in chromosome behavior often lead the cytologist to interesting discoveries.

Section A. KARYOLOGY IN THEORY

A few general cytological terms are defined as well as certain cytological-morphological concepts which characterize the cytology of the reproductive process in higher plants. The descriptive terminology of nuclear and chromosome cytology, arranged here in broad categories and when possible in developmental sequences, provides a conceptual outline of the karyological principles which have developed in chromosome work. To obtain a brief characterization of a given topic, each of the following subdivisions should be read in the sequence presented. If a novice in the subject finds that this raises as many questions as it answers, he should refer at once to one of the book-length treatments listed in the General References at the end of this chapter. These books also provide many detailed examples of the use of karyological results in evolutionary and systematic studies. The material presented here also may be utilized as a glossary by referring to the general index.

I. CYTOLOGY

Cell (cyt-, -cyte). Unit of protoplasm, usually containing a single nucleus and cytoplasm bounded by a plasma membrane and in plants usually surrounded by a cell wall.

Cytokinesis (cell division, cell reproduction). Process by which a single cell divides into two daughter cells and thus reproduces itself. Distinct from but often synchronous with nuclear reproduction.
Spindle (achromatic figure). Fibrous structure formed in the cytoplasm and extending through the nuclear region. Active in chromosome movement during nuclear division and in cytokinesis.
Phragmoplast. Extension of fibrous spindle-like structure to the periphery of the cell, active in cell plate formation.
Cell plate. Newly formed cell membrane transverse to the spindle, first observed stage is a thickening of the spindle fibers.
Cytoplasm. The living extra-nuclear portion of the cell including its various organelles and membranes.
Plasma membrane. The membrane which bounds the cytoplasm of the cell and in plants lies just within the cell wall.
Cell wall. The non-living bounding material which is laid down (secreted) outside the plasma membrane and between adjacent cells. The primary wall in plants is usually cellulose; other substances are often added which may be variously thickened, sculptured, or pitted in the cells of different tissues. Some cells characteristically produce a secondary wall.
Organelle. Any of various organized structures within the cell which have specific morphologies and functions such as plastids, mitochondria, chromosomes, nucleoli.
Meiocyte. A cell that is destined to undergo the process of meiosis.
Syngamy (fertilization). The process of cellular and nuclear fusion of the two gametes (egg and sperm) which is one of the two essential processes of sexual reproduction (cf. meiosis).
Zygote. The single cell which results from syngamy--the fertilized egg.
Artifact. An observed structure or apparent condition which is not a true characteristic of the living specimen, but is the result of post mortem changes, cytological processing techniques, or limitations of the equipment and the observer. Sharp distinctions between artifacts and "reality" are often difficult in cytology and require good judgment as well as experience.

CYTOLOGY (Botanical Terms)

Sporocyte. A cell which by the process of meiosis is destined to produce a set of four haploid spores.
Microsporocyte (pollen mother cell). A sporocyte which by meiosis (microsporogenesis) produces the microspores or "male spores." These occur in the anthers of seed plants.
Megasporocyte. A sporocyte which by meiosis (megasporogenesis) produces the usually larger megaspores or "female spores." These occur often singly and more or less embedded in the nucellar tissue of the ovule in seed plants.
Meiospore. General term for any spore produced by the process of meiosis.
Secondary sporocyte dyad. The pair of cells produced by the first division of meiosis in those forms in which cell division occurs synchronously with the first meiotic division.
Spore tetrad. The four meiospores produced from a single primary sporocyte by meiosis. Usually applied to stages at which the four spores remain in contact.
Microspore. Product of microsporogenesis--"male spore."
Megaspore. Product of megasporogenesis--"female spore."

Microspore mitosis (microspore division). Applied to the first mitosis which occurs in the haploid microspore following its formation by meiosis.

Pollen grain. The male gametophyte, applied to the stages including and following the microspore until pollen germination.

Vegetative (tube) cell (tube nucleus). Cell with its haploid nucleus which constitutes the bulk of the developing male gametophyte in seed plants. In angiosperms, one of the products of the microspore division.

Generative cell. Cell enclosed by its membrane within the cytoplasm of the vegetative cell which is destined to produce the two male gametes. In angiosperms, the other product of the microspore division.

Generative division. Division of the generative cell to produce the male gametes. May occur before pollen germination (3-nucleate pollen grains), or in the pollen tube (2-nucleate pollen grains).

Male gametes (sperm). Haploid sperm cells each with its membrane and cytoplasm.

Megagametogenesis. Developmental sequence in the female gametophyte which leads to production of the female gametes (egg cells) in the archegonium or in the embryo sac of angiosperms.

Embryo sac. The female gametophyte of angiosperms.

Gametophyte. The haploid phase of the life cycle beginning with the meiospore, ending with production of the gametes.

Sporophyte. Diploid phase of the life cycle beginning with the zygote, ending with haploid spore production by meiosis.

Alternation of generations. Alternation of diploid ($2\underline{n}$) sporophyte and haploid ($\underline{n}$) gametophyte generations in the life cycles of eucaryotic plants.

II. KARYOLOGY

Nucleus (karyo-, caryo-). Conspicuous spheroidal or ellipsoidal body within the protoplast of a cell, enclosed by a membrane and containing the chromosomes (chromatin), a nucleolus, and nucleoplasm or karyolymph.

Nuclear membrane. The membrane which bounds the nucleus of the cell except during the metaphase and anaphase stages of nuclear reproduction.

Nucleolus. Dense body within the metabolic nucleus largely consisting of RNA which stains differentially. (Nucleoli usually appear in telophase, disappear during prophase.)

Eucaryote. An organism the cells of which contain organized nuclei with nuclear membranes and chromosomes.

Karyokinesis. The process of nuclear division or nuclear reproduction.

Mitosis. General term for the process of nuclear reproduction with particular reference to the reproduction of the chromosomes. Mitosis in somatic cells is used in contrast to meiosis in germ cells or sporocytes.

Prophase. The early stages of mitosis from the first detectable changes in the organization of the chromatin as seen in the metabolic (resting) nucleus to the congression of the contracted chromosomes at the equatorial plate.

Equatorial plate. Stage in mitosis at which the contracted chromosomes come to be arranged across the equator of the spindle with the centromeres of each chromosome lying in a single plane.

Metaphase. Literally "between the phases," originally meaning between prophase and anaphase. By usage metaphase has become the equivalent of the equatorial plate stage, the moment in mitosis when the chromosomes are arranged most nearly in a single plane just before anaphase disjunction begins.

Anaphase. The stages of mitosis during which the daughter chromosomes separate (disjoin) and move apart from the equatorial plate to the two poles of the spindle.

Telophase. Stages of mitosis during which each of the newly formed daughter nuclei becomes enclosed in a new nuclear membrane, the contraction of the chromosomes relaxes, new nucleoli are formed, and eventually the chromatin "reticulum" characteristic of the metabolic nucleus reappears.

Metabolic (resting) phase. The condition of the nucleus in a metabolic cell, a nucleus which is not involved in active mitosis. The contraction of the chromosomes has relaxed so that they form a finely divided mass of very fine strands which often appear as a granular or reticulate "network." The linear integrity of the individual chromosomes is maintained, however, and the apparently anastomosing "network" is an artifact.

Meiosis. Form of nuclear reproduction in which the chromosomes divide only once while the nuclei divide twice producing four nuclei with the reduced (haploid) number of chromosomes; one of the two essential processes of sexual reproduction (cf., syngamy).

Heterotypic division. First of the two successive meiotic divisions.

Leptotene (leptonema). Earliest stage of meiotic prophase. Individual chromosomes are extremely long, fine threads.

Zygotene. Pairing stage of meiotic prophase. Homologous thread-like chromosomes become intimately associated throughout their lengths.

Synizesis. "Knotted thread stage" often associated with early meiotic prophase. The fine threaded nucleus is seen collapsed into a tangled mass. This is an artifact of fixation, but represents a physiological condition in which the fine structural details of the living system are most susceptible to distortion.

Pachytene. Thick thread stage of meiotic prophase. Each of the completely paired homologous chromosomes has also replicated, the chromosomes are four-partite.

Diplotene. The paired chromosomes separate from each other; a looped structure results because the chromosomes are held together at the chiasmata.

Diakinesis. Final stage of meiotic prophase, the chromosome pairs are much contracted, but are still held together by terminalized chiasmata.

Metaphase I (M I). Metaphase of the first meiotic division. Chromosomes become cooriented at the equatorial plate with opposing centromeres on opposite sides of the equatorial plane.

Anaphase I (A I). Anaphase of the first meiotic division. Chromosome pairs disjoin, chiasmata complete terminalization, chromosomes move to the poles. Maximum chromosome contraction.

Telophase I (T I). Telophase of the first meiotic division. Relaxation of chromosome coiling, formation of nuclear membranes.

Interkinesis. The "resting" stage which occurs between the two meiotic divisions. In many forms there is no telophase I, interphase, prophase II sequence; instead metaphase II occurs immediately after the chromosomes reach the poles in anaphase I.

Homeotypic division. The second meiotic division, so named because it is "similar" to a mitotic division.

Prophase II. Prophase of the second meiotic division. Chromosomes are dyads held together at the centromeres with the arms lying free.

Metaphase II (M II). Metaphase of the second meiotic division. The centromeres become oriented at the equatorial plate and disjoin.

Anaphase II (A II). Anaphase of the second meiotic division. Individual chromosomes separate and move to the poles.

Telophase II (T II). Completion of meiosis, relaxation of chromosome coiling, appearance of nuclear membranes around the four haploid nuclei. The spore tetrad results from the completion of cytokinesis.

Chromosome complement. The total haploid, or diploid group of chromosomes contained within a particular cell, or individual.

Karyotype. The mitotic metaphase morphology of any given chromosome complement of an individual, race, or species.

Idiogram. Diagrammatic arrangement of photographs, drawings, or graphical representations of all the individual chromosomes of the karyotype to show the principal morphological features. The chromosomes are usually arranged in a horizontal row of homologous pairs in order according to length and/or centromere position.

Cytogenetics. Investigations which combine and synthesize the methods and results of chromosome cytology with the genetic methods and results of breeding analysis.

Cytotaxonomy. The investigation of chromosomal relationships in populations, races, and species in order to elucidate taxonomic and evolutionary relationships.

Cytogeography. The geographical distribution of chromosomal differences between various populations in relation to taxonomy and evolution.

III. CHROMOSOME - ABERRATIONS, REARRANGEMENTS

Deletion (deficiency). The loss or elimination of a microscopically observable segment of a chromosome.

Duplication. The presence of a chromosome segment which duplicates a portion of a chromosome within the same haploid chromosome complement. Duplications may occur in tandem within the same chromosome.

Inversion. Segment of a chromosome in which the order of the structural elements (and included genetic material) is in reverse to the order found in otherwise homologous chromosomes.

Pericentric inversion. The inverted segment includes the centromere region.

Paracentric inversion. The inverted segment lies entirely within a single chromosome arm.

Inversion bridge and fragment. Characteristic anaphase configuration found in meiosis following crossing-over in a heterozygous paracentric inversion. One of the resulting chromosomes becomes dicentric and forms a bridge as the centromeres migrate toward the poles at anaphase. The other cross-over product is an acentric fragment. The relative position of the point of crossing-over to other cross-overs determines whether the bridge and fragment appear at the first or second anaphase (see literature).

Dicentric chromosome. Aberrant chromosome containing two centromeres.

Translocation. Reciprocal exchange of segments between two non-homologous chromosomes.

Translocation ring. Configuration of 4 chromosomes formed by the meiotic pairing of homologous segments in a heterozygous translocation. Additional heterozygous translocations in the same complement produce additional or larger rings.

Alternate disjunction. Centromeres of adjacent members of a translocation ring proceed to opposite poles at anaphase I; produces zigzag configurations and balanced assortment of chromosomes.

Adjacent disjunction. Adjacent members of the ring proceed to the same pole at anaphase I; produce an open configuration and unbalanced assortment with duplicated and deficient segments in both daughter nuclei.

Structural change. General term for the various rearrangements of chromosome segments which result from chromosome aberrations.

Structural hybridity. Heterozygosity of the chromosome complement for one or more structural changes.

Cryptic structural hybridity. Heterozygosity of the chromosome complement for one or more minute and microscopically undetectable structural changes which presumably account for marked reductions in pairing between chromosomes from different races or species.

Isochromosome. A chromosome resulting from transverse centromere misdivision in which the two arms are identical although arranged in inverted linear order with respect to each other.

Chromosome fragment. A broken portion of a normal chromosome.

- Acentric. No centromere is present, is usually lost after one or a few few mitotic cycles.
- Centric. Centromere is present, may be irregularly or regularly transmitted in mitosis, is much more erratic in behavior and subject to elimination in meiosis.

Ring chromosome. Usually arising as a centric spontaneous or induced chromosome fragment in which the broken ends have rejoined in a ring configuration.

IV. CHROMOSOME - MEIOTIC BEHAVIOR

Pairing. The detailed chromomere by chromomere association of homologous chromosome threads which occurs at zygotene. Commonly also applied to the continued association of homologous chromosomes observed in diplotene, diakinesis, and metaphase I which results from the earlier zygotene pairing.

Synapsis. Process of chromosome pairing.

- Asynapsis. Failure of the process of chromosome pairing (in presumably homologous chromosomes).
- Desynapsis. Precocious separation of previously paired chromosomes before completion of meiotic prophases.

Homologous. Applied to chromosomes and to segments of chromosomes which have the same chromosome ancestry and thus contain essentially the same genetic material.

Non-homologous. Applied to chromosomes and to segments of chromosomes which have different chromosomal ancestry and thus differ in genetic content and arrangement.

Homeologous. Applied to chromosomes in polyploids which are similar in origin (probably once homologous or partially so) but which have become so differentiated that they pair incompletely or not at all.

Chiasma(ta) (crossing over). Chiasma describes the cross shaped configuration which is the observed result of the exchange of segments between homologous chromosomes during meiotic prophase. Crossing-over is the genetic term for the actual exchange of genetic determiners.

Chromatid. A single one of the two daughter chromosome strands following chromosome reduplication in mitosis or meiosis while they remain more or less coiled about each other.

Sister chromatids. The two chromatids produced by the replication of a single original chromosome.

Tetrad chromosome. The four chromatids of a pair of homologous chromosomes, each of which consists of two sister chromatids (late pachytene to M I).

Dyad. The individual chromosome consisting of two chromatids at Anaphase I and metaphase II.

Terminalization (of chiasmata). Process occurring when the chromosomes progressively separate as the centromeres move away from each other. This causes the relational coils of the two chromosomes to unwind at the ends. The nodes of the loops which originally occurred at the points of actual exchange appear to move toward the ends of the chromosomes and eventually disappear.

Congression. Movement of the chromosomes to the equatorial region of the cell just before metaphase.

Orientation. Arrangement of the centromeres of mitotic chromosomes so that each lies in the plane of the equatorial plate.

Coorientation. Arrangement of the two centromeres of each meiotic chromosome pair so that each pair is oriented with the centromeres opposite each other and equidistant from the equatorial plane. More complex coorientations result from associations of three or more chromosomes at M I.

Disjunction. Separation of the chromosomes at anaphase.

Univalent. One unpaired chromosome in meiosis.

Bivalent. A pair of homologous chromosomes in meiosis.

Trivalent. An association of three homologous or partially homologous chromosomes (e.g., in triploids or trisomics).

Quadrivalent. An association of four homologous or partially homologous chromosomes.

Hexavalent. An association of six.

Multivalent. An association of three or more partially paired chromosomes in meiosis. Note: no matter how many chromosomes are involved in a multivalent complex, at any one point intimate association is always that of one to one pairing.

Autosyndesis. In polyploids and their hybrids, the pairing of chromosomes derived from the same parent. In $A_1A_1A_2A_2$ x $A_3A_3A_4A_4$, pairing is A_1A_2 or A_3A_4.

Allosyndesis. In polyploids and their hybrids, the pairing of chromosomes derived from different (opposite) parents. In $A_1A_1B_1B_1$ x $A_3A_3B_3B_3$, pairing is A_1A_3 and B_1B_3.

V. CHROMOSOME - MORPHOLOGY

Chromosome. Thread-like strand of DNA and associated nucleoproteins which replicates itself in each complete mitotic or meiotic cycle and which by coiling appears to contract to a usually rod-like or in the smallest chromosomes to an almost spherical form. Chromosomes and their constituent material, often called chromatin, stain deeply and differentially in basic dyes.

Chromomere. Individual minute, more or less spherical, beadlike body in the chromosome thread.

Centromere (kinetochore, primary constriction). A specialized chromomere to which the chromosomal fibers of the spindle appear attached, which initiates the chromosome movement at diplotene of meiosis and as anaphase begins. In many preparations the centromere is invisible and is represented in a fully contracted chromosome by a (primary) constriction. The relative position of the centromere with respect to the two chromosome arms is one of the most reliable morphological features of the chromosome.

Secondary constriction. A constricted region within one of the chromosome arms.

Satellite. A small terminal portion of a chromosome separated from the remainder of its chromosome arm by a prominent, (often long) secondary constriction. Often associated with a nucleolus organizer.

Chromosome arm. The portion of a chromosome to one side or the other of the centromere (or primary constriction). The proximal region is adjacent to or in the direction of the centromere in contrast to the distal or terminal region of the chromosome arm.

Metacentric (median centromere, V). A chromosome with a median centromere has equal arms and is V-shaped at anaphase.

Isobrachial. Equal-armed chromosome.

Submetacentric (submedian centromere). Chromosome with its centromere near but not exactly at the median point.

Acrocentric (subterminal centromere, J or I). Chromosome with the centromere near one end, thus with one long and one short arm such chromosomes are J- or I-shaped.

Heterobrachial. Any chromosome with arms of unequal length.

Telocentric. The centromere is at the end of the chromosome. Such chromosomes are rare, most I-shaped chromosomes are acrocentric with a minute short arm.

VI. CHROMOSOME - STRUCTURE, CONTENT

Chromatin. The deeply staining material (DNA and nucleoprotein) of the chromosomes.

Achromatic. A region of the contracted chromosome which is not deeply staining such as the centromere region or a secondary constriction.

Euchromatin. Those portions of the chromosomes (usually the majority) which follow the typical mitotic cycle of condensation by increased spiralization and consequently deeper staining through prophase to early anaphase, followed by relaxation of coiling and progressively lighter staining which reaches its lower limit in the resting stage; as opposed to heterochromatin.

Heterochromatin. Chromosome segments or chromosomes which remain contracted and deeply staining or contract early in contrast to the behavior of the euchromatin in the resting and prophase stages. Heterochromatin may also despiralize and appear as lightly stained segments at metaphase when euchromatin is most condensed. The genetic content or at least the function of the heterochromatic regions of chromosomes have been shown to be quite different from that of euchromatin, and there appear to be a number of different kinds of heterochromatin. Heterochromatic chromosomes and segments are useful in characterizing the morphology of particular chromosomes and karyotypes.

Prochromosomes. Small condensed "chromosomes," actually heterochromatic segments of the whole chromosomes which appear in the resting nuclei of many plant species; the number of prochromosomes may correspond to the metaphase chromosome number but accurate determinations are difficult.

Nucleolus organizer. Specific chromomere or chromosome region associated with the formation of a nucleolus, often but not always associated with a long secondary constriction and just proximal to a satellite.

Chromosome knob. Large chromomere, which is more or less heterochromatic, with a characteristic morphology and constant position. They may be interstitial or terminal and are best observed in early prophase.

B-chromosome (accessory chromosome). Supernumerary chromosomes, that are in addition to the normal diploid complement, which are primarily heterochromatic and occur in varying numbers among different individuals of a population. They may be minute (microchromosomes) or moderately large in comparison with the normal members of the complement.

Spiralization. Process of internal coiling in chromosomes which causes changes in apparent chromosome length and staining intensity during mitosis and meiosis.

Allocycly. Differential cycles of coiling and condensation in different chromosomes or in different segments of the same chromosome which lead to observed contrasts in the cytological appearance of heterochromatin and euchromatin.

Autosomes. The typical chromosomes of the complement whose segregation does not usually affect sex determination.

Sex chromosomes (allosomes). Particular chromosomes, which may occur singly, as a more or less heteromorphic pair, or in multiple chromosome complexes, whose segregation is related to sex determination.

VII. CHROMOSOME - NUMBER

Haploid (monoploid, n). One complete set of chromosomes, such as the complement in the nucleus of one of the gametes. The n number of chromosomes in the haploid = monoploid = gametic number of chromosomes.

Diploid (2n). The chromosome complement of a zygote, two complete sets of chromosomes one from each parental gamete - the zygotic number. The 2n diploid chromosome number is the normal somatic chromosome number of any organism which has arisen from a zygote, the organism itself is often referred to as diploid.

Polyploid (-ploid). Any individual (or a cell) which contains three or more complete sets of chromosomes. Various combinations of words or numbers with "-ploid" indicate the number of haploid sets of chromosomes.

triploid = 3n	hexaploid = 6n
tetraploid = 4n	octoploid = 8n
pentaploid = 5n	10-ploid = 10n, etc.

Polyploidy (ploidy). Variations in chromosome number involving more than the diploid number of complete chromosome sets.

Autopolyploid (autoploid). A polyploid in which all of the chromosome sets (genomes) have arisen from the same source, each chromosome is capable of pairing (autosyndesis) with a corresponding chromosome in each of the other sets. Easily applied to experimentally induced polyploids; e.g., a normal diploid is doubled to produce an autotetraploid, this can then be crossed with the original diploid to produce an autotriploid and so on. The same events can and do occur in nature as the result of spontaneous chromosome doubling. In natural populations, however, both within and between related species, every possible degree of similarity from essential identity to marked genic and even structural divergence may have occurred between the parents of the original diploid individual in which the chromosome doubling occurred (see Stebbins, 1971, and others).

Allopolyploid (alloploid). A polyploid in which the original chromosome sets (genomes) have been derived from different sources as the result of hybridization and no effective pairing occurs between the chromosomes of the different genomes. In a strict allotetraploid only bivalents are formed at meiosis and no multivalent associations.

Segmental allopolyploid. A polyploid in which part of the chromosome complement is essentially autopolyploid in origin, part is allopolyploid. One of the frequent and more easily recognized intermediate situations between classical autopolyploidy and allopolyploidy.

Amphidiploid (literally both diploid). The allotetraploid formed by doubling the chromosomes of a sterile F_1 hybrid which contains two haploid genomes which are incapable of effective pairing at meiosis. Fertility is restored after chromosome doubling because each genome has become duplex, bivalents are formed at meiosis, and balanced chromosome assortment ensues.

Genome. A complete haploid set of chromosomes; the haploid chromosome complement of a diploid race or species.

Euploid. A chromosome number which is an exact multiple of a given haploid number.

Basic number = *x*. The number of chromosomes in the original genome (haploid set) from which a polyploid or a group of polyploid forms or species is known or postulated to have arisen.

Polyploid series (2*x*, 3*x*, 4*x*, 6*x*, ...). A group of related races or species with various chromosome numbers all of which are multiples of a single basic number (e.g., the euploid series indicated).

Homoploid series. A group of related species (e.g., the species of *Pinus*, in all of them 2*n* = 24) all of which have the same chromosome number.

Dibasic polyploidy. Allopolyploids with chromosome numbers which result from combinations of two genomes with different basic numbers.

Aneuploid. Chromosome numbers which are not exact multiples of an original haploid number, with one or more chromosomes missing (2*n* - 1) or in addition (2*n* + 1) to the euploid number.

Aneuploid series. A series of aneuploid numbers in a group of individuals, or species, each chromosome number in the series differing from the next by the addition (or subtraction) of one or two chromosomes (e.g., 2*x*, 2*x* + 1, 2*x* + 2, 2*x* + 4, 2*x* + 5, etc.).

Dysploid. Diverse aneuploid chromosome numbers which do not fall into a series or regular pattern.

Monosomic. An individual in which the diploid chromosome complement lacks one member of one of the chromosome pairs, 2*n* - 1. Thus one member of the complement is monoploid or simplex. There are as many different monosomic types possible as the number of chromosome pairs.

Nullisomic. Aneuploid in which both members of a chromosome pair are missing, 2*n* - 2. This is not found at the diploid level, but can occur in polyploids, in which the genetic material of the missing chromosomes may be duplicated elsewhere in the chromosome complement so that no essential functions are lost.

Disomic. The normal diploid (2*n*) condition in which a given chromosome is represented twice in the complement.

Trisomic. Aneuploid in which one of the individual chromosomes is present in triplicate (*n* + 1). As with monosomics there are as many possible trisomic types as there are different chromosomes in the haploid genome. A secondary trisomic occurs when the extra chromosome is an isochromosome.

Tetrasomic. An aneuploid in which a single chromosome pair is duplicated and thus is tetraploid, 2*n* + 2. This is more likely to occur in polyploids.

Section B. KARYOLOGY IN PRACTICE

I. CHROMOSOME METHODS

The objective is to obtain first class microscopic slide preparations of chromosomes which may be observed, counted, measured, photographed, and otherwise studied with the aid of the compound light microscope. Early investigations were based upon the study of serial sections of appropriate meristematic or sporogenous tissues which had been stained by one of the haematoxylin procedures or with an aniline dye, most often crystal violet. These techniques are still available and valid for precise cytological investigations and whenever the faster and more direct squash (smear) methods do not yield satisfactory results. The relatively laborious fixing, embedding, sectioning, staining procedures have a major advantage: they permit the use of numerous chemical reagents in various combinations (fixatives) in order to kill the living system very quickly and to preserve the desired features of its morphological structure with a minimum of distortion (artifact). The chief source of difficulty and error in the method is the necessity of avoiding or possibly of reconstructing chromosome configurations which have been cut and lie in more than one section.

The aceto-carmine squash method was one of several such procedures invented by the cytologist, John Belling, for the study of meiotic chromosomes in the sporogenous tissues of a variety of flowering plants. The basic method has since been adapted to the study of mitotic chromosomes found in various actively growing somatic meristems. A variety of stains have been substituted for carmine by different workers and for different materials. Various combinations of alcohol, acetic acid (or propionic acid) and chloroform are employed in the killing-fixing solution. In the acetic-orcein method, the saturated solution of orcein in 45% acetic acid is applied directly to the living root tips and thus serves both as fixative and stain. The Feulgen nucleal staining reaction when correctly carried out is a specific stain for DNA. Many workers substitute the Feulgen reaction for the orcein or carmine stain, or sometimes combine it with one of these stains in conjunction with the squash technique. (Detailed procedures and variations may be found in Darlington and LaCour, 1962, and in Sharma and Sharma, 1965.)

The squash methods all involve spreading actively dividing cells on a slide so that the cells occur loosely associated in groups in a layer only one cell thick. This is usually easy to accomplish with pollen mother cells (PMC's) and male sporocytes in later stages of meiosis because the cells of the sporogenous tissue usually float out separately when pressed from the anther. To obtain the same sort of "mono-cellular" layer from somatic meristem, a maceration process is necessary. This is usually accomplished by placing the material (root tips, leaf buds, etc.) in 1N HCl solution at 60°C for 5-15 minutes, or by some other combination of heat and weak HCl. The middle lamellae between the cellulose cell walls are destroyed and the cellulose itself is softened. The preparation is completed in each case by adding the cover glass, after which gentle tapping followed by firm pressure is used to flatten and spread the cells.

II. CHROMOSOME MATERIALS

Chromosome number, size, and morphology are best determined in somatic (mitotic) material and may be observed in any vegetative tissue in which the cells are actively dividing. Meristems from fast growing root tips are the most convenient source when the plants can be grown in culture. Root tips are sometimes taken from recently germinated seeds, but this involves a serious difficulty. There is no plant upon which to base identification of the species or strain under study. Everything depends upon the correctness of the label on the seed packet which depends in turn upon care and accuracy of the collector and nature of the original source. Furthermore, if different chromosome results are obtained from the root tips of different seedlings in the same lot, there is no way of associating the chromosome differences with different plants. It is far better to grow plants in separate pots so that a one to one correspondence between observations and individual plants can be maintained. Later the plants may be used in cytogenetic and other experiments, and they may become voucher herbarium specimens which provide taxonomic documentation for the karyological results.

It is impracticable, often impossible, to collect good root tips from a known individual plant which is growing in its natural habitat. Some other tissue must be employed if field collected material is necessary or desirable. Actively growing stem tips, early leaf primordia, and in special cases, young petals, ovules, or other organs can all provide good mitotic chromosome preparations. Herbarium specimens must be collected from the same plants or populations.

Studies of the chromosomes in meiosis are indispensable for determining the presence and nature of chromosome irregularities. Meiosis in hybrids reveals the chromosomal causes of sterility and the extent to which changes in chromosome structure have occurred. In favorable material, e.g., maize, *Lilium*, *Solanum*, detailed chromomere analysis at pachytene provides direct morphological evidence of structural divergence. Analysis of chromosome pairing relations at meiosis permits the classification of the various types of polyploids, and in hybrids, provides the basis for determining the degree of chromosome homology from which the evolutionary relationships of the parents are inferred. Species vary greatly in the usefulness of particular stages of meiosis for detailed study. The excellent pachytene of *Zea mays* is followed by diakinesis and metaphase chromosomes which are so contracted that structural details are hidden. Diakinesis and metaphase are excellent in many species with relatively large chromosomes, but the preceding pachytene is so crowded and visually complex that it defies analysis.

Morphological features of the individual chromosomes are more difficult to observe in the metaphase and anaphase stages of meiosis because the chromosomes are often so much contracted that features easily seen in mitosis are obscure. The chromosomes of many species are small and relatively difficult to analyze even with the aid of the best oil immersion objectives at the limit of useful magnification. Even chromosome number determinations based on meiotic material are subject to error under these conditions unless the material is diploid with completely normal pairing and disjunction. In unknown or cytologically irregular material, these problems can be minimized by correlating the meiotic analysis with mitotic observations.

The sporogenous cells of developing anthers (microsporangia in gymnosperms) are the usual source of meiotic material in seed plants. (Megasporocytes are too inaccessible and too few in number to be generally useful for chromosome studies.) Meiosis occurs one to three weeks before anthesis in many flowering plants, but in various woody species, e.g., conifers, *Cornus*, and in certain herbaceous perennials, e.g., many bulbs, *Trillium*, *Podophyllum*; meiosis may be completed one to several months before the pollen is actually shed. Unless the time of meiosis with respect to other events in the life cycle is already known, a considerable amount of trial and error observation and sampling may be required to determine the proper developmental stage and the optimum time for collection. This can be critical for the success of a given project, particularly if it involves field collections, because the optimum season for many species is restricted to a few weeks of the year.

III. MICROSCOPIC OBSERVATIONS

Chromosome work requires good microscopy; practice and experience are the best routes to proficiency. The equipment does not need to be elaborate or unduly expensive. The basic requirements are a good range of eyepieces and objectives including at least one oil immersion lens (90-100X) for bright field which combined with the appropriate eyepieces (12.5-15X) produces a magnification of 1200 to 1500X. This is near the useful limit for visual work. Phase contrast optics are very useful for difficult subjects, but should not be used as a substitute for care in preparation and precise staining. A graduated mechanical stage adds greatly to the efficiency of systematic slide study and the location and relocation of desired chromosome figures. A good light source and filters including a green one are essential. An attachable or built on photomicrographic camera is indispensable for recording observations, occasionally the camera lucida drawing apparatus is useful.

Chromosome number and other determinations should be based upon replicated independent observations of several figures from the same slide, from different slides, from different individuals, from different populations, whenever this is possible or feasible. The most experienced observer is often misled in his interpretation by points of overlap between chromosomes, by clear constrictions and occasional broken chromosomes which make one chromosome appear as two, by differences in focal plane which obscure arm lengths and size relations, and many other details. The microscopist must often make conscious effort to avoid subjective bias in his interpretations. This is aided by making free hand diagrams including each chromosome in its proper space relation as the figure is studied. Completion of the sketch provides an objective chromosome count and aids objectivity in the evaluation of morphological details.

It is far better to make a number of slides of the same material so that they may be quickly surveyed and only the best selected for careful study, than to spend much time struggling to interpret mediocre preparations. Similarly one passes quickly by the more difficult figures found in a given slide and returns to study them carefully only when nothing better is available. This is a common and reasonable practice in most cases, but it can lead to biased results in the evaluation of meiotic pairing relationships because the more complex pairing configurations are also those which are most difficult to interpret. Frequently, when chromosomes are small and numerous, or somewhat crowded, it is not possible to determine whether a figure is worthy of careful study without the enhanced resolution provided by an oil immersion lens. For such material a 60X oil immersion objective and relatively low power eyepieces often provide the most efficient combination for routine slide scanning.

IV. RECORDS AND DOCUMENTATION

Accurate records are a necessary part of any scientific study; certain complications arise in chromosome work which need to be anticipated. A simple numbering system for identifying field collections and plots in culture with subordinate numbers for individuals will serve to identify the plant material. In addition the collection vials have to be numbered so that they can be correlated with the plants. All slides prepared from the material in a given vial; all notes, drawings, and eventually photographs derived from those particular slides, also have to be identified so that they can be referred with assurance to the particular plant or collection which was the original source of the material.

The plant materials are documented by herbarium specimens: whole plant specimens for taxon identification, and if appropriate, mass collections of particular plant parts to record individual plant variations which are relevant to the cytological results. The chromosome results are documented by photomicrographs (be sure to record magnifications) which illustrate representative observations, supplemented as needed by drawings and/or interpretive diagrams. Representative slides should be made permanent as further documentation of the chromosomal results. The dry ice freezing method is excellent (Conger and Fairchild); various other techniques are available. Sometimes the very cells which are most interesting are destroyed in the permanizing process; the essential records and photography should be completed first.

TECHNIQUES

I. OXYQUINOLINE-ACETIC-ORCEIN SQUASH METHOD FOR ROOT-TIPS

A. Prepare a stock solution of 8-hydroxyquinoline.
This is an almost saturated solution obtained by dissolving 0.3 gm 8-hydroxyquinoline in 1 liter of distilled water. Prepare a day in advance; shake occasionally until completely dissolved. Store in refrigerator. It will remain usable for several weeks.

B. 1. Prepare a stock solution of aceto-orcein.
Dissolve 2.2 grams of orcein in 100 ml. glacial acetic. Bring the glacial acetic acid to a gentle boil and slowly add the orcein. Remove from heat, cool, and filter. (We very much prefer the George T. Gurr, Ltd., London SW,6, England, natural orcein to the other varieties including American synthetic.)

2. For storage: Keep as 100% acetic-acid stock solution.

3. For use: (a) Aceto-orcein staining solution. Dilute to 45% strength acetic acid by adding distilled water. Always refilter when preparing stain for use and also filter this 45% stain every two or three days as precipitate accumulates.
(b) Orcein-HCl solution. Also prepare in a separate dropping bottle a solution containing: 1 pt. N HCl
9 pts. of the aceto-orcein stain (i.e., the 45% acetic acid solution)

C. Root-tips may be obtained:
(a) From germinating seeds--if seeds are plentiful and individual plant identity is not important.
(b) From the pot ball of plants grown in culture. Use seedlings or older plants which have been recently repotted. Allow the plants to grow undisturbed until a network of roots begins to form on the surface of the soil inside the pot. At this stage tap out each plant from its pot, replace immediately, and water thoroughly. Within two to several days return to these same plants to make root-tip collections. The "tapping-out" creates a small air space into which the more active roots grow rapidly during the next few days.

D. Procedure (Adapted from Tjio, J. A., and A. Levan, 1950. The use of oxyquinoline in chromosome analysis. Anales de la Estacion Expt. de Aula Dei 2: 21-64.)

1. Collection:
Collect fresh root-tips directly from the plant into vials containing the aqueous oxyquinoline solution.

2. Pretreatment:
   Place the vials in a cool room, T. & L. say 65° F. or use a 55° room, for 4 hours. Some materials such as Pinus with very large, long chromosomes require a much longer pretreatment--as much as 24-30 hours.
3. Staining:
   a. Place the root-tips in a few drops of the Orcein-HCl (9:1) solution in an ordinary watch glass. Heat gently (until the fumes of HCl begin to rise from the surface) over an alcohol lamp. Repeat the heating at intervals of a minute or more until the sample has been heated 3 or 4 successive times. The heating hastens the macerating effect of the HCl. Leave the material in this solution for 20-30 minutes.
   b. Transfer the root-tips to a large depression slide containing aceto-orcein stain (the 45% solution). Cover to prevent evaporation and leave 20-30 minutes. With very small chromosomes a much longer period in stain, as much as 12-16 hours, e.g., overnight, may be helpful.
4. Slide preparation:
   a. Have clean slides and covers readily available in containers, e.g., keep them immersed in 95% alcohol, and wipe them clean and dry with a fine cheese cloth as needed.
   b. Using a clean slide as a dissecting board, place a stained root-tip under a low power (10X) stereoscopic microscope. With a needle provided with a scalpel-like tip (half-spear preferred) cut away the root-cap, and then excise a small piece of meristem (not more than 1-2 mm in length). Place this bit of tissue in a small drop of stain at the center of a second clean slide. Break the tissue into small pieces by quick strokes with the flat side of the half spear needle. Spread the tissues in a small circular area on the slide and apply cover glass immediately. Observe the result and note any bits of tissue which are still intact; give the larger of these a quick, light tap with a blunt instrument such as the handle of a pair of forceps. If the dissection and teasing are properly done, two or three taps will be sufficient. The cover should now rest in contact with the drop of stain over most of its area.

      Invert the slide on a small blotter. Use a cheesecloth to avoid finger prints and press moderately to hard with the ball of the thumb. The amount of pressure varies with the material--size of cells, softness of the tissue, etc. It must be learned by experience. No amount of pressure will correct for failure to soften the tissue beforehand, or for the presence of foreign matter such as sand grains under the cover!

      CAUTIONS:
      1. DO NOT HEAT ORCEIN PREPARATIONS. In this respect orcein contrasts with carmine.
      2. ALWAYS CLEAN THE HALF-SPEAR NEEDLE AFTER MAKING EACH PREPARATION. NEVER HEAT ONE OF THESE NEEDLES.
      3. Label the slide with the appropriate collection number or other symbol immediately after preparation is complete.
5. Storage and handling of slides:
   a. Aceto-orcein slides may be stored in a moist chamber, in the refrigerator overnight (or sometimes for two or three days)

without sealing. (Additional stain is added as necessary.)

b. They may be sealed with paraffin, with one of the various sealing mixtures, or with finger nail polish and kept (refrigerated) for several weeks. Check at frequent intervals to be sure seal is intact.

c. They may be made permanent by the McClintock method or by the Conger dry ice method. (Conger, A. D., and L. M. Fairchild, 1953. A quick-freeze method for making slides permanent. Stain Technology 28: 281-283.) Alcohol destains orcein (it does not destain carmine) so the slides and covers must be moved through the 95% and 100% alcohol with dispatch.

d. When making permanent--BE SURE TO INSCRIBE PERMANENT SLIDE NUMBER WITH DIAMOND POINT PENCIL--AND MAKE PROPER RECORD--BEFORE RUNNING SLIDE THROUGH PERMANIZING PROCEDURE.

II. FEULGEN TECHNIQUE*

The Feulgen stain, or leuco-basic fuchsin, stains DNA specifically and is used for cytophotometric determinations of DNA in nuclei or chromosomes or simply as a qualitative highly specific stain for chromosomes. Various techniques have been described for staining chromosomes by this procedure, but the basic principle involved in all is: liberation of the aldehyde groups of DNA by hydrolysis, a chemical reaction binding the free aldehyde groups to the leuco-basic fuchsin, followed by the removal of unbound stain from the chromosomes. (See Darlington and LaCour, 1962 for further details of solutions and schedules.)

The following are the essential steps of the Feulgen squash technique as used with somatic tissues in Nicotiana:

1. Fix tissue in Carnoy fixative (6 parts 95% ethanol, 3 parts chloroform, 1 part Glacial acetic acid) or 3:1 (3 parts absolute alcohol, 1 part Glacial acetic acid) for 2-24 hours.
2. Replace fixative with 70% ethanol and squash or store in refrigerator.
3. Hydrolyze tissue in 10% concentrated HCl at 60° for 10 minutes.
4. Rinse in several changes of distilled water.
5. Transfer to leuco-basic fuchsin for 10-15 minutes or until desired staining intensity is reached.
6. Rinse in 3 10-minute changes of SO_2 water followed by 5 minutes in distilled water (the purpose of this is to remove the unbound stain from the chromosomes).
7. Transfer to 45% acetic acid.
8. Squash in 45% acetic acid on slide.
9. Cover with cover glass.

COMMENTS:

1. Rinsing in SO_2 water can be eliminated if quantitative DNA analysis is not important. In that case rinse only 3 5-minute changes of distilled water.
2. The leuco-basic fuchsin and SO_2 water are made using Darlington and LaCour, schedule 4, page 144-145, or prepared solutions can be purchased.
3. The most common causes of failure of this technique are poor fixation, incorrect timing of hydrolysis, improperly prepared staining solutions, or poor samples of the powdered basic fuchsin.

*Contributed by Joyce Burns, North Carolina State University, Raleigh.

III. IRON ACETO-CARMINE METHOD
(Pollen mother cells or somatic meristems)

1. Fixation (three alternatives).
   a. 3 pts 100% ethyl alcohol: 1 pt. glacial acetic acid.
   b. 3 pts 100% ethyl alcohol: 1 pt. chloroform: 1 pt. glacial acetic acid.
   c. Substitute 100% propionic acid for acetic acid both in the fixative and in the staining solution. Better results with some materials.

MIX FIXATIVES IN SMALL QUANTITIES JUST BEFORE USE!

NOTE: The volume of fixative should always exceed gross volume of the plant tissue by 10-50 times. Buds tightly enclosed by protective structures need to be clipped or spread open as they are dropped into the solution.

2. Processing.
   a. Leave in fixative overnight.
   b. Rinse in 95% ethyl alcohol and change to 95% next morning.
   c. Change to 80% late that day.
   d. Change to 70% next morning. Store in refrigerator (2-3 weeks).
   e. Longer storage. Place material in 70%, or material in fixative which does not include chloroform in freezer. It will keep for 2 or more years if tightly closed.
3. Staining.
   a. Prepare a saturated solution of carmine (alum lake) in 45% acetic (or propionic) acid. Bring acid to gentle boil, add excess of powdered carmine and a few crystals of chloral hydrate, boil ca. 1 minute, cool, filter. Do not add iron and solution will keep indefinitely. (See literature for many variations.)
   b. Squash pollen mother cells or meristematic cells in a small drop of stain on a clean slide. For anthers, cut off an end and squeeze pollen mother cells out by pressing out anther. Be sure to remove remains of anther sac. (See aceto-orcein method for further suggestions on squashing procedure.) Iron for the staining reaction comes from the needles used in spreading the material. Be sure to clean needles after each use.

IV. ALCOHOLIC HYDROCHLORIC ACID-CARMINE (SNOW'S STAIN) FOR CHROMOSOMES IN SQUASH PREPARATIONS
(modified after Snow, 1963)

1. Fixation: Use fluids suitable for the aceto-carmine method discussed above.
2. Staining.
   a. Staining solution. Add 4 gm. of carmine and 1 ml. of concentrated HCl to 15 ml. of distilled water. Mix well and boil (GENTLY) for 10 minutes while stirring. Cool the solution. Add 95 ml. of 85% ethanol and filter. Store in stock bottles.
   b. Staining procedure.
      (1) Wash out fixative with several changes of 70% ethanol. Allow an hour for each change. Materials may be stored in the last change and refrigerated if so desired.
      (2) Remove buds or root tips from the 70% alcohol, drain on absorbent paper and place in a vial of stain (2-3 times as much as needed to cover the material). Leave materials in the stain until they are well penetrated. The time varies and must be determined by experience, but for root-tips and anthers, which are exposed directly to the solution, 24 hours is often long enough; compact tissue requires a longer period in the stain. To hasten staining, place capped and labeled vials in a 50-60°C

paraffin oven for 24-48 hours, or longer for some materials.

(3) Pour stain into a stock bottle for reuse (filter to remove any debris before reusing).

(4) Rinse material with water or 70% alcohol and prepare squashes as usual substituting 45% acetic acid in place of aceto-carmine. Excess material for future use may be stored in 70% alcohol.

(5) Dissection of buds and maceration of the tissue, if desirable, should be done in 45% acetic acid.

(6) Differentiate by cautious heating over a small flame (e.g., alcohol burner). (Not necessary for all materials)

(7) Squash in 45% acetic acid or for permanent slides mount in equal parts of Hoyer's mounting medium (see Chapter 8) and 40% acetic acid and squash.

V. QUICK FREEZE METHOD FOR MAKING SLIDES PERMANENT

A. Materials

1. A slice or small block of dry ice (solid CO_2). (Obtainable from dairy produce plants, or manufactured in the laboratory from compressed CO_2).
2. Two Petri dishes containing 95% and 100% ethyl alcohol, respectively, each containing a short glass rod to support the slide.
3. Euparal or other mounting medium with a solvent compatible with residual traces of water (not Canada balsam).

B. Procedure: (Adapted from Conger, A. D., and L. M. Fairchild, 1953. Stain Technology 28: 281-283).

1. Remove temporary sealing material from slide and cover; inscribe permanent slide number and make record.
2. Place slide face up on flat surface of dry ice for 30-60 seconds (hold slide firmly in place by pressing it against the dry ice with the eraser of a pencil); then remove cover glass by sliding a flat dissecting needle (such as a half spear) or edge of a razor blade under a corner. With optimum timing the cover slip will pop off without breaking or damaging the material.
3. Quickly place both slide and cover in the dish of 95% alcohol. Keep slide in original position but invert cover so material side will face up.
4. Agitate slide gently for 15-30 seconds in order to disperse water which had condensed on the glass during the freezing process. Transfer both slide and cover to 100% alcohol; keep "material side" up.
5. After 30-60 seconds, remove slide from dish, drain quickly and wipe excess alcohol from blank ends. Place drop of Euparal on the material; re-invert and mount the cover glass in its original position. Note: The quick changes indicated here are necessary with orcein stained material. Changes can be made more slowly after carmine or Feulgen staining.
6. Allow slides to dry on a flat surface overnight. It is best to store recently processed slides in a horizontal position.

CHROMOSOME TECHNIQUE EXERCISE

A. Preparations:

1. Each student will have read and carefully considered Chapter 10, section B, Karyology in Practice, before attending this laboratory session.

2. Optimum equipment and advance preparation are indicated; procedures must be modified to fit the actual conditions. If the students are to prepare the reagents and collect the materials this must be done at least two or three days in advance of the present exercise.

B. Equipment:
1. Compound microscope with oil immersion objective and mechanical stage (graduated if possible).
2. High intensity light source (filters and filter holders desirable, not essential).
3. Stereoscopic microscope (10X).
4. 60°C oven and refrigerator available.
5. Dissecting instruments, alcohol lamp, vials, slides, cover glasses, slide box, marking pencil, blotters, cheese cloth, Petrie dishes, short glass rods.

C. Reagents:
1. Glacial acetic acid.
2. Ethyl alcohol: 70%, 95%, 100%.
3. Stains (in dropping bottles): aceto-carmine, acetic-orcein, 9:1 orcein-HCl solution.
4. 1N HCl solution.
5. For fresh root tip collections: 8-hydroxyquinoline solution (kept refrigerated).
6. Supply of dry ice slices; finger nail polish, rubber cement for temporary sealing.

D. Materials:
1. Vials containing pre-treated and pre-fixed root tips from several diverse plant species.
2. If feasible, vials of root tips collected 3 hrs. before class begins, undergoing 8-hydroxyquinoline pretreatment in cold room.
3. Vials containing buds at meiotic stages, and representing diverse species and chromosome situations. These may be collected days to months in advance and stored in a freezer.

E. Procedure (Instructor):
1. Demonstrations of slide making procedures.
   a. Root tips, prestained so that they are ready for use as the period begins.
   b. Flower bud material; illustrate dissection and sampling procedures to locate desired meiotic stages.
2. Demonstration of dry ice (or other) method of making squash slides permanent. (This may be postponed until later in the session if desired.)

F. Procedure (Individual Student):
1. Each student will make root tip slides of at least 2 different species. The pre-fixed material is to be stained in aceto-carmine; the fresh pretreated material in acetic orcein. Slides will be examined to verify the presence of well-spread chromosome figures at mitotic metaphase.
2. Each student will find, and make slides of, as many different stages of meiosis as possible from the materials available.

G. Procedure (Group):
3. Each student will demonstrate his best mitotic slide(s) to the instructor and to the class. For greatest interest and variety, different students should have made slides of different species.

4. The instructor will set up demonstration microscopes showing examples of meiotic stages. Students will demonstrate their own meiotic slides to the instructor and to the class and compare with the stages on exhibit.
5. Temporary sealing or practice in making slides permanent as time permits.

NOTE: To succeed this exercise must be based upon pretested materials which are known for excellence as objects of chromosome study and ease of manipulation such as: 1. root tips of onion, hyacinth, rye, barley; 2. buds or inflorescences of *Tradescantia paludosa*, *Rumex hastatulus*, *Podophyllum peltatum*, *Diphylleia cymosa*, rye, maize.

CYTOLOGY LITERATURE

I. General References.

Darlington, C. D. 1963. Chromosome Botany and the Origins of Cultivated Plants. George Allen and Unwin. London.

________. 1965. Cytology. J. and A. Churchill. London.

________, and L. F. LaCour, The Handling of the Chromosomes, 4th ed. George Allen and Unwin. London.

Sharma, A. K., and A. Sharma. 1965. Chromosome Techniques. Theory and Practice. Butterworths. Washington.

Sharp, L. W. 1934. Introduction to Cytology. McGraw-Hill. New York.

________. 1943. Fundamentals of Cytology. McGraw-Hill. New York.

Snow, R. 1963. Alcoholic hydrochloric acid-carmine as a stain for chromosomes in squash preparations. Stain Technology 38: 9-13.

Stebbins, G. L. 1971. Chromosome Evolution in Higher Plants. Addison-Wesley. Reading, Massachusetts.

Swanson, C. P. 1957. Cytology and Cytogenetics. Prentice-Hall. Englewood Cliffs, New Jersey.

II. Periodicals.

(Note: Plant chromosome data appear in numerous biological, botanical, and genetics journals; also see references at end of Chapter 11 and in section II below).

Canadian Journal of Genetics and Cytology. Genetics Society of Canada; Experimental Farm, Ottawa.

Caryologia. International Journal of Cytology, Cytosystematics and Cytogenetics. Published by the University of Florence, Italy.

Chromosoma. Published by Springer Verlag, 175 Fifth Ave., New York.

Cytogenetics. S. Karger, AG, Arnold-Boeklindstr. 25, 4000 Basel, Switzerland.

Cytologia. The International Journal of Cytology. Published in Tokyo.

Stain Technology. A Journal for Microtechnic and Histochemistry; Official publication of the Biological Stain Commission.

III. Chromosome Number Lists and References

Darlington, C. D., and A. P. Wilie. 1956. Chromosome Atlas of Flowering Plants. George Allen and Unwin. London.

Federov, A. A. (ed.). 1969. Chromosome Numbers of Flowering Plants. Academy of Sciences of the USSR, V. L. Komarov Botanical Institute.

Index to Plant Chromosome Numbers. 1956-1964. Vol. 1 & 2 (Nos. 1-9) plus a pre-1956 supplement). University of North Carolina Press. Chapel Hill, North Carolina.

Index to Plant Chromosome Numbers. 1965-. Annually, in Regnum Vegetabile v. 50, 1965; v. 55, 1966; v. 59, 1967; v. 68, 1969... International Bureau Plant Taxonomy and Nomenclature.

IOPB Chromosome number reports. A regular feature in Taxon, published by the International Association for Plant Taxonomy.

Love, A., and D. Love. 1961. Chromosome Numbers of Central and Northwest European Plant Species. Opera Botanica. Vol. 5. Almquist & Wiksell (distributors).

Chapter 11. GENETIC EVIDENCE*

Genetics is concerned with the study of variation, its expression in development, and its transmission in inheritance. A basic undertaking of taxonomy is the recognition, description, and use in classification, of the variation which exists between natural populations of organisms. The analysis of experimental populations provides the geneticist with a powerful method for investigating the underlying causes of variation and the continuing processes of evolutionary change. The analysis of discontinuities in the diversity of natural populations leads the taxonomist to classify; ideally, the classification evaluates the relationships of existing organisms as products of these same evolutionary causes and processes.

Modern genetics extends from viruses to man, from mathematics and biochemistry to psychology and medicine; many of its complexities have no direct relevance to vascular plant systematics. The discoveries of molecular genetics, however, have caused some major and many subtle changes in classical concepts and terminology. Watson (1970) presents a general account of molecular genetics, while Srb, Owen, and Edgar (1965) have integrated the old with the new. King (1968) furnishes definitions, while Rieger, Michaelis, and Green (1968) provide current and authoritative explanations of terms and concepts. Genetic theory is now deeply woven into the fabric of the modern synthetic theory of evolution. Dobzhansky (1970), Grant (1963, 1971), Mayr (1963), Simpson (1953), and Stebbins (1950, 1967, 1970) all discuss at length the genetic and cytogenetic processes in evolution.

Linnaeus, by defining species and arranging them in orderly categories, provided the foundations upon which systematic biology has been built. Darwin, a century later, worked much of his life and entitled his greatest book "On the Origin of Species." The Linnaeans considered species to be immutable, divinely created forms while variations within species were individual imperfections in the realization of the ideal form. Although the recognition of species is crucial to any consideration of their origin from varieties as the result of variation and natural selection, Darwin tended to consider the species as units of convenience in man's description of the genealogical lines of descent.

Twentieth century genetics has resolved many of the earlier contradictions regarding species concepts by demonstrating that species must not be considered as ideal, typological abstractions but as dynamic biological populations in which the individuals vary morphologically and physiologically and within which they interbreed. Controversy still surrounds our modern definitions of species largely because there are many different kinds of biological populations with different breeding systems including some (agamic complexes) which have even lost the capacity for interbreeding. These circumstances often create serious difficulties for the practicing taxonomist because in the absence of detailed genetic and cytogenetic analysis he has to make his decisions regarding the nature of the species populations with which he deals on the basis of indirect circumstantial evidence.

Genetic analysis has demonstrated unequivocally that the species is a real, probably the most significant, unit in the evolutionary differentiation

*Contributed by Ben W. Smith, North Carolina State University, Raleigh.

of biological populations. The species represents the "point of no return" in evolution, the point at which the differentiation of a descendant population becomes irreversible because progressive change prevents its return to the ancestral condition and at which reproductive isolation prevents interbreeding with collateral species populations which have the same ancestry. This is the basis for the biological species concept. It is fundamental for any phylogenetic system of classification.

The impact of genetics upon systematics is two-fold: (1) conceptual and general, or (2) experimental and addressed to specific problems. The experimental populations of genetics usually fall within the bounds of a single taxonomic race or a single species. The general principles discovered in such studies are of great significance to systematics, but the detailed results of genetic analyses in such populations are of little direct value in resolving particular taxonomic questions. On the other hand, systematists have to deal with the relationships of various higher categories between which divergence has proceeded so far that genetic experimentation is impossible. At the lower levels of the taxonomic hierarchy, from race or subspecies to species and genus, genetic experiments of various kinds are possible in natural populations and the methods of genetics are directly applicable to the investigation of taxonomic problems.

Geneticists have developed modes of thinking about variation, its transmission, and its expression in both individuals and populations of organisms which are highly relevant to systematics. The two disciplines are interrelated in many ways, but only a brief summary of the most pertinent genetic concepts can be presented here. This consists of: (1) the basic tenets of Mendelian heredity which arise from the particulate nature of inheritance, (2) the features and processes which account for the occurrence of differing genetic systems which lead in turn to the production of populations with markedly different variabilities and breeding behavior, and (3) concepts from genecology and population genetics (in words rather than mathematical formulations) which are helpful in dealing with populations at the subspecific, specific, and generic levels of the taxonomic hierarchy, the critical stages in evolutionary divergence. Selection and related matters have been deferred to Chapter 26 on evolution. As in Chapter 10, Cytological Evidence, each of the three units which follows is designed to be read in sequence as a synopsis of the subject, but the terms discussed may be found individually by means of the index.

I. BASIC CONCEPTS AND CONSEQUENCES OF PARTICULATE INHERITANCE

Concepts of the ultimate physical units of inheritance have passed through several metamorphoses from the "elements" and unit characters of Mendel, to the genes of Morgan and his associates, to the nucleotide triplets (which are codons for amino acids, which are the building blocks of polypeptide chains), the sequences of which in turn make up the genes of molecular genetics. None of this has altered Mendel's basic conclusion that the ultimate hereditary material is particulate and that it is transmitted essentially unchanged from cell to cell and from generation to generation. There have been many refinements, however, in our understanding of the process.

The genetic units which actually segregate and recombine in each cycle of meiosis and syngamy are cross-over segments, which consist of linearly arranged sequences of linked genes in the chromosome, segments of the DNA macromolecule. The sizes and limits of these mendelizing chromosome segments, and the relative degree of linkage versus disassociation between individual genes, are functions of cross-over position and frequency which are in turn subject to genetic control and can be profoundly modified by structural changes in the chromosomes themselves.

Gene and Locus. The gene is the ultimate unit of Mendelian inheritance, a precise sequence of nucleotide triplets (codons), which occupies a specific chromosomal locus, segregates and recombines as a unit, and which can mutate to various allelic forms which have different effects in inheritance. The locus (plural, loci) is the specific point in the linear order of genes in the chromosome which is occupied by a particular gene.

Allele (Allelomorph). One of the alternative forms of a particular gene which can occur at a locus. When more than two alleles are known for a given locus and gene, they are called multiple alleles.

Heterozygote. A zygote or individual carrying two different alleles of a given gene (e.g., Aa).

Homozogote. A zygote or individual formed from the fusion of gametes each of which carries the same allele of a given gene (e.g., AA).

Dominance and Recessiveness. The dominance in the expression of particular alleles and the recessiveness of others is commonly observed, but precise investigation reveals that dominance is frequently incomplete. Although superficially similar to that of the homozygous dominant allele, character expression in the heterozygote is usually to some degree intermediate between the dominant and the recessive homozygote. Dominance is the general term for inter-allelic interactions in quantitative genetics studies.

Overdominance. Overdominance occurs when the heterozygote (Aa) exceeds both homozygotes (AA or aa) in the genotypic value of the character under study. The expression of the heterozygote falls outside the range of the two homozygotes.

Epistasis. This is the general term for the interaction of non-allelic genes and is commonly employed in the analysis of quantitative inheritance. The epistatic effect of an individual gene refers to the manifestation of its effect over that of non-allelic genes. The converse, hypostatic, refers to the effect upon the expression of the non-allelic genes in such a relationship.

Unit Characters and Pleiotropy. Specific genes often produce spectacular effects upon specific characters. Further study has shown that most genes affect many different characteristics even when one of the effects is more obvious than the others. Also some genes (pleiotropic) have major effects upon a number of seemingly unrelated characters. Since the action of most genes is related to enzyme systems and metabolic pathways, these results are to be expected.

Penetrance and Expressivity. Penetrance is the frequency in percent of dominants or of recessive homozygotes which are actually expressed phenotypically. Expressivity refers to the strength or weakness in phenotypic expression of a given gene. Both qualities depend upon the interaction of genotype and environment.

Qualitative and Quantitative Characters. These terms designate the extremes of a continuous array of character expressions and gene effects. Qualitative characters have been described as differences in kind: major effects upon color, shape, hyper-development or abortion of certain organs, etc., but even such characters have quantitative attributes. Quantitative characters are those which can only be described in metrical terms: numbers, lengths, areas, volumes, etc., and in which the net effect of a single gene among the many affecting the same character is small in relation to the magnitude of non-heritable, environmentally induced modification. Quantitative characters exhibit continuous variation; qualitative characters produce discontinuous variation.

Oligogenes and Polygenes. Terms for the genes assumed to correspond respectively to the qualitative and quantitative character effects. There are many genes which are observed to have only relatively small effects, but other genes (oligogenes) have not only major effects but also minor effects on other characters and act as "polygenes." The polygenic inheritance of recent literature is essentially equivalent to the multiple factor inheritance of earlier authors.

Modifier Genes and Genetic Background. Genes which are known or assumed to modify the primary effects of other genes are often called modifier genes. The effect of any gene or gene complex under study always occurs in the context of all the other genes which are present in the genotype. These other genes constitute the genetic background or residual genotype.

Genotype and Phenotype. The genetic constitution of an individual organism is the genotype. This means the total genetic endowment of the organism present in the original zygote. The phenotype is the expressed characteristics of an individual organism. This means all the characters of the organism which result from the interaction of its genetic endowment (genotype) with the environment during development and throughout the life of the individual. (NOTE: Genotype and phenotype are often used in a much restricted sense for purposes of analysis. Genotype designates the genic or chromosomal constitution of an individual in a particular context; phenotype the character expression in such an individual.)

Segregation and Independent Assortment. Segregation is the separation of allelic pairs, the genes received from the parents, and their distribution to different cells, the gametes. Independent assortment describes the independence of non-allelic genes (on non-homologous chromosomes) in the segregation process; non-alleles on different chromosomes segregate randomly with respect to each other, cf. linkage.

Transgressive Variation. When the range of variability found in a segregating generation such as the F_2 exceeds the range found in either parent or the F_1, the variation is transgressive.

Linkage. The non-random, "linked" inheritance of groups of genes and the characters they control which happen to be spatially associated on the same chromosome. The degree of linkage is relative to the physical distances between the genes on the chromosome. These distances are inferred from the frequencies of crossing-over between the genes as determined from the associations of phenotypic characters in breeding experiments. The cross-over (map) distances between genes can be affected by factors other than physical distance, e.g., structural modification of the chromosomes, or genetic effects on cross-over frequency.

Supergene. A gene complex which is more or less permanently linked and transmitted as a unit is sometimes called a supergene. It is often difficult to distinguish such gene complexes from individual genes by experimental means.

Recombination. The constant reassortment of chromosomes, cross-over segments, and therefore genes which produces new genotypes and thus new phenotypes in each generation of sexually reproducing organisms. Recombination is the immediate source of the variation observed in populations, the raw material upon which the processes of evolution and speciation act.

Mutation. Occurrence of a heritable change in the genotype of an organism which was not inherited from its ancestors; the ultimate source of all genetic variability (cf. recombination). "All genetic changes, except those due to recombination, are mutations" (Dobzhansky, 1970). A mutation may be as minute as the substitution of a single nucleotide pair in the DNA molecule, as great as a major change in chromosome structure or number.

Recurrent Mutation. Precisely the same mutation may occur again and again at the same genetic locus, either in nature or under the influence of a given physical or chemical agent (a mutagen). The frequency of such mutations under a given set of conditions is the mutation rate.

Viability and Lethality. All genes affect in some direct or indirect way the survival, health, and longevity of the organism. Relative viability can be measured by comparison with a standard or "normal" population. Even lethality, in terms of gene action, is not absolute but relative, because death caused by the action of a specific gene or gene complex may be haplontic in spore or gamete, or diplontic in the zygote, or delayed to a late embryonic stage or even to early maturity.

Fitness. Success in producing viable and fertile offspring.

II. THE GENETIC SYSTEM

The Genetic System refers to all the intrinsic (genetic) processes and their concerted operation which affect genetic recombination in a population or species. The major components of the genetic system are the organization of the chromosomes and their behavior in meiosis, and the breeding system. Genetic systems range from that of the outbreeding diploid with complete pairing and unrestricted crossing-over of homologous chromosomes to that of the obligate apomict with no sexual reproduction. Studies of the genetic system of an organism include characterization of the chromosomes and their cytogenetic behavior (See Chapter 10, Cytological Evidence), experimental outbreeding and inbreeding, and an analysis of the reproductive biology of the species, including isolating mechanisms. The results of such a study often show the taxonomist the most logical biological boundaries for species delimitation. Major concepts involved in the analysis of genetic systems may be briefly characterized:

Reproductive Biology. A collective term applied to all of the many mechanisms, agents, and adaptations involved in the reproductive process of a species. In higher plants, for example, reproductive biology includes floral structure, mode and/or agent of pollination, self- or cross-incompatibility, various types of sterility, time and season of flowering, and similar characteristics. Although fundamental to an understanding of the dynamics of any natural population, reproductive biology is inadequately known in most plant species.

Panmixis. The interbreeding of all the actively reproducing members of a population in such a way that mating is at random, i.e., each individual has the same probability of mating with any other individual.

Outbreeding (Cross-breeding). A breeding system in which sexual reproduction involves the mating and union of gametes of different individuals (requires cross-pollination in seed plants). The common mode of sexual reproduction in higher animals. In plants outbreeding may be obligate or facultative. The necessity or tendency for outbreeding may be reinforced by dioecism, self-incompatibility, self-sterility, heterostyly and other mechanisms. It is usually aided by wind, insect, or other pollinating agents. Outbreeding may imply mating between individuals which are less closely related than would occur with random mating.

Inbreeding. A breeding system in which sexual reproduction involves the mating and union of gametes of individuals which are more closely related than would occur with random mating. Selfing, sib-mating, and back-crossing (e.g., parent-offspring matings) are examples of close inbreeding. Self-pollination is rather frequent in plants, especially among annuals. It may be obligate (e.g., cleistogamy) or facultative and is often aided by floral adaptations which enhance the opportunities for selfing.

Heterozygosity and Homozygosity. Applied to the presence of a pair of dissimilar or identical homologous alleles (or chromosomes) respectively in the individual, but also used to characterize the relative numbers of such gene pairs in populations. Strict outbreeding produces and maintains relatively high degrees of heterozygosity; strict inbreeding produces and maintains homozygosity.

Inbreeding Depression and Hybrid Vigor. The inbreeding of normally crossbred organisms usually causes depressed vigor and viability because of the increased homozygosity of deleterious genes; outbreeding of normally inbred organisms (and of experimental inbred lines) often results in hybrid vigor. Wide crosses, i.e., between species or genera, sometimes produce extremely vigorous hybrids, but more often the hybrids are relatively weak and infertile.

Heterosis. Superiority of heterozygous genotypes in one or more characters over the corresponding homozygotes, the phenotypic result of gene interaction in the heterozygotes.

Self-Incompatibility. Genetically controlled physiological interactions which prevent or inhibit (partial self-incompatibility) the completion of self-pollination and/or self-fertilization in certain perfect flowered or monoecious plants (e.g., species of *Nicotiana*, *Hemerocallis*, *Trifolium*) and thus promote cross-breeding. The same genotypic combinations which cause self-incompatibility also cause cross-incompatibility reactions when they occur in separate individuals which may be otherwise unrelated.

Heterostyly. Combined morphological, physiological, and genetic mechanism in certain angiosperms which promotes cross-breeding. Flowers of two, sometimes three kinds which differ in the relative positions of the styles (and anthers) are produced by separate individuals. Successful pollination and fertilization usually occurs only between flowers which differ, e.g., long styled by short styled or the converse, thus cross-pollination is ensured.

Apomixis. The substitution of some form of asexual reproduction for the sexual process amphimixis in an organism derived from sexually reproducing ancestors. Apomixis should not be confused with amixis, the asexual forms of reproduction which are common in procaryotic organisms.

Clone. All the members of a population of cells or organisms derived asexually from a single individual by vegetative (mitotic) reproduction.

Agamospermy. Includes those forms of apomixis in which seeds are produced by one of a wide variety of asexual processes. Agamospermy does not necessarily produce clones because meiotic segregation and recombination can occur in certain kinds of agamospermy.

Agamic Complex. Complex of hybrid and polyploid forms reproducing by agamospermy so that each original form comes to be represented by an apomictic population. These are extremely difficult taxonomically because each "population" is actually the extension of a single individual of hybrid origin.

III. GENECOLOGY AND THE GENETICS OF POPULATIONS

Genecology. Study of the genetic basis of ecological differentiation. "A turning-point in the study of evolutionary taxonomy." (See review and definitions of terms in Davis and Heywood, 1963, Chapter 12.)

Deme Terminology. The suffix "-<u>deme</u>" as originally proposed and correctly employed is defined as "a term, always used in this terminology with a prefix, denoting any group of individuals of a specified taxon." The nature of the association of the group of individuals or the kind of populational unit which they may represent must always be specified by the addition of an appropriate prefix. Second order, compound prefixes may also be used to specify relationships more precisely. The word <u>deme</u> used alone simply means "a group of individuals of a specified taxon" and shows no other relationships unless this is specified in the context (Davis and Heywood, 1963; Heslop-Harrison, 1967). Various basic -deme terms are indicated below. The deme terminology equivalents are also indicated when appropriate in connection with other concepts described in this section.

Topodeme. A deme occurring in a specified geographical area.

Ecodeme. A deme occurring in a specified kind of habitat.

Phenodeme. A deme which differs from others phenotypically.

Genodeme. A deme which differs from others genetically.

Cytodeme. A deme which differs from others cytologically.

Plastodeme. A deme which does not differ from others genotypically, but does differ phenotypically.

Gamodeme. A deme that is a panmictic breeding unit.

Autodeme. A deme composed of predominantly self-fertilizing individuals.

Endodeme. A gamodeme composed of predominantly endogamous (closely inbreeding) dioecious plants or animals.

Clinodeme. One of a series of demes which collectively show a variational trend and thus collectively form a <u>cline</u>.

Ecotype=Ecological Race=Ecogenodeme. A population differing genetically from others which occupies a specific ecological habitat.

Ecospecies=Hologamodeme. All the ecotypes so related that they are able to exchange genes freely without loss of fertility or vigor in the offspring. In many groups this unit is approximately equivalent to the <u>taxonomic species</u>.

Cenospecies=Coenogamodeme. All the ecospecies so related that they may exchange genes among themselves to a limited extent through hybridization. This is equivalent in most cases to the <u>biological species</u>.

Biological Species. Grant (1971): "The sum total of the races that interbreed frequently or occasionally with one another, and that intergrade more or less continuously in their phenotypic characters." Mayr (1963): "Groups of interbreeding natural populations that are reproductively isolated from other such groups." Dobzhansky (1970): "A Mendelian population is a community of individuals of a sexually reproducing species within which matings take place. A biological species is the most inclusive Mendelian population; it is integrated by the bonds of sexual reproduction and parentage." Taxonomic species are sometimes equivalent to biological species but for a variety of reasons, some good, some bad, are often less inclusive.

Comparium=Syngamodeme. All the cenospecies between which hybridization is possible either directly or through intermediaries. Gene exchange between cenospecies is impossible.

Hybrid. In taxonomy, the result of a cross between two taxa, often <u>interspecific</u>. In genetics, the result of a cross between any two individuals or races which differ genetically.

Hybrid Swarm. A population of interspecific or interracial hybrids and their segregates and intercrossed derivatives. Formation of such a population can be the first step in introgressive hybridization, but is not the actual process of introgression which always involves successive backcrossing.

Hybrid Zone. Geographic area or region in which races or species meet and hybridize.

Introgressive Hybridization. A process of successive hybridizations in nature which causes the migration of genetic material from one species into another. An initial interspecific hybridization is followed by a series of successive backcrosses to one of the parents (it can occur independently in both directions). The process of gene migration is often called introgression.

Clinal Variation. A sequence of morphologically interrelated forms which have evolved in response to an environmental gradient and form a more or less continuous series, a cline. The populations in the series are clinodemes. Clinal variation is in contrast to introgression but easily confounded with it.

Isolating Mechanism. Any of a number of intrinsic (genetic or cytogenetic) or extrinsic (geographic, ecological environmental) mechanisms or circumstances which prevent interbreeding and thus isolate two parts of a once continuous population.

Prezygotic Isolation. This results from any geographical, ecological, or temporal circumstance or behavioral, structural, or genetic incompatibility which prevents the formation of hybrid zygotes.

Postzygotic Isolation. This results from mechanisms which reduce the viability or fertility of hybrid zygotes or individuals.

Effective Population Size. The number of individuals of a population which are actually reproducing in any generation, the breeding population.

Mendelian Population. An interbreeding group of individuals sharing a common gene pool.

Panmictic Unit=Gamodeme. A local population whose individual members have the same chance of mating and producing progeny--the smallest Mendelian population.

Gene Pool. The total genetic potentialities in the sum total of the genes in a breeding population from which the genes of the next generation are derived.

Gene Frequency. The proportion of a given allele at a specific locus which occurs with reference to all other alleles at this locus in a given breeding population. Usually expressed in values from 0 to 1.0. A gene frequency of 0.5 indicates that half the loci in the population carry the specified allele.

Gene Flow. Changes in gene frequency which result from the migration of genes or gametes into a breeding population from a different population.

Hardy-Weinberg Law. The principle that both gene and genotype frequencies are constant from generation to generation provided that there is no migration, mutation, or selection with respect to the gene in question, i.e., the population is in Hardy-Weinberg equilibrium. The principle is based upon a large random mating population.

Genetic Drift. This is the loss or the fixation of a gene without or in spite of selection because of the "accidents of sampling." Genetic drift has been too often invoked to account for circumstances for which the investigator had no ready explanation. The random fixation (gene frequency = 1.0) or loss ($f = 0.0$) is most likely to occur when one of two conditions exists: (1) the population size is very small, 10-30 individuals, or (2) the effect of the given gene is very small, in which case the effect of fixation either way would be difficult to detect.

Founder Principle (Ernst Mayr). If a small population invades a new area and becomes the basis of a new evolutionary line, divergence from the original ancestral population will be accelerated because the founder population includes (as the result of sampling) only a limited portion of the ancestral gene pool.

Selection. "Anything tending to produce systematic, heritable change between one generation and the next" - G. G. Simpson.

Fitness versus Flexibility. Fitness is the degree of adaptation of an individual or a population to the environment in which it lives. The ultimate measure of fitness in terms of population genetics is the relative ability to reproduce over generations. Flexibility is the capability of response to change, i.e., the adjustment of adaptation to environmental change. Fitness places a premium on homogeneity with minimal expressed variability; flexibility depends upon a wide range of free or potential genetic variation. The best compromise between these two conflicting demands upon a population appears to be a wide range of variation coupled with a large modal class of individuals which are relatively uniform and near the optimum in adaptive characters.

Natural Selection. "This preservation of favorable individual differences and variations, and the destruction of those which are injurious, I have called Natural Selection or the Survival of the Fittest" - Darwin. Evolutionists now state this in genetic terms: "Natural selection is the differential reproduction of genotypes" - Lerner.

GENETICS LITERATURE

I. Periodicals.

American Naturalist. 1867+. Published for the American Society of Naturalists. University of Chicago Press. Chicago.

Canadian Journal of Genetics and Cytology. 1959+. Published by the Genetics Society of Canada. Ottawa.

Evolution. 1947+. International journal of organic evolution. Published by the Society for the Study of Evolution.

Genetical Research. 1960+. Published by Cambridge University Press. London.

Genetics. 1916+. Official publication of the Genetics Society of America.

Genetics Abstracts. 1968+. Published by Information Retrieval Ltd. London.

Genetika. 1965. Published by Akademiia Nauk SSSR. Moscow.

Hereditas. 1920+. Published for the Mendelian Society of Lund for Scandinavian Association of Geneticists.

Heredity. 1947+. An international journal of genetics. Longman Group Ltd. Edinburgh.

Journal of Heredity. 1910+. Publication of the American Genetic Association. Washington, D. C.

Plant Breeding Abstracts. 1930+. Published by the Commonwealth Agricultural Bureau. Farnham Royal, England.

Silvae Genetica. 1951+. Published by J. D. Sauerlander. Frankfurt an Main.

II. General References.

Davis, P. H., and V. H. Heywood. 1963. Principles of Angiosperm Taxonomy. D. Van Nostrand. Princeton, New Jersey.

Dobzhansky, Th. 1970. Genetics of the Evolutionary Process. Columbia University Press. New York.

Grant, V. 1963. The Origin of Adaptations. Columbia University Press. New York.

________. 1971. Plant Speciation. Columbia University Press. New York.

Heslop-Harrison, J. 1967. New Concepts in Flowering Plant Taxonomy. Harvard University Press. Cambridge, Mass.

King, R. C. 1968. A Dictionary of Genetics. Oxford University Press. New York.

Mayr, E. 1963. Animal Species and Evolution. Belknap Press, Harvard. Cambridge, Mass.

Rieger, R., A. Michaelis, and M. M. Green. 1968. A Glossary of Genetics and Cytogenetics. 3rd ed. Springer Verlag. New York.

Simpson, G. G. 1953. The Major Features of Evolution. Columbia University Press. New York.

Srb, A. M., R. D. Owen, and R. S. Edgar. 1965. General Genetics. 2nd ed. W. H. Freeman & Co. San Francisco.

Stebbins, G. L. 1950. Variation and Evolution in Plants. Columbia University Press. New York.

________. 1967. Adaptive radiation and trends of evolution in higher plants. Evolutionary Biology 1: 101-142.

________. 1970. Variation and evolution in plants: progress during the past twenty years. In: Essays in Evolutionary Biology, Hecht, M. K., and W. C. Steere (eds.). Appleton-Century-Crofts. New York.

Watson, J. D. 1970. Molecular Biology of the Gene. 2nd ed. W. A. Benjamin, Inc. New York.

Chapter 12. REPRODUCTIVE BIOLOGY AND SYSTEMATICS

Despite the vast array of individual variation and the taxonomic problems such variation may cause in some cases, the actual processes of evolution and speciation occur at the population level rather than at the level of the individual organism. This is so because evolution, broadly defined, is a directional change in gene frequency through time, and changes in gene frequency only occur in populations of an organism. (Individuals do not change their gene frequencies, thus they do not adapt in the evolutionary sense, they merely adjust during their life span, within their fixed genetic limits.) Furthermore, all populations, even of a single kind of organism, do not evolve at the same rate because different populations are under different selective pressures that affect, in different ways, the actual or potential evolutionary history of the population. Accordingly, knowledge of population dynamics and biological interrelationships often aids in a better understanding of the evolutionary results of importance to systematics. Two of the most obvious population characteristics of such importance are population size and population breeding structure.

Section A. POPULATION SIZE AND DENSITY

A large interbreeding population, containing many hundreds or thousands of individuals, would normally have a much larger gene pool than a small population of perhaps only a dozen or so individuals. A large gene pool provides a vital storehouse of genetic variation which in turn provides a population with the capability of adapting, over a period of time, to gradual changes in the selective forces of the environment. The gene pool of a small population has proportionally fewer different alleles and, consequently, lacks the adaptive capabilities of the large population under changing environmental conditions. However, just on the basis of chance, a new mutant or a particular new gene combination is much more likely to become established in a small population than in a large population. A large gene pool thus has a strong buffering effect that prevents rapid shifts in gene frequency and possible overspecialization.

The chance (as opposed to selective) fluctuation of a gene's frequency in a population is called drift. If the frequency of a particular gene "drifts" to either 0 or 100%, the gene frequency is fixed and, since selection has presumably played no important role in the process, it is called random fixation. In small populations, random fixation is much more likely, and more frequent, than in very large populations. Indeed, it is very difficult, even with strong selection, to raise the frequency of a particular gene to 100% or to lower it to zero in a large population.

The population structure that provides maximum evolutionary potential is thus neither very large nor very small, but is generally considered to be a series of interbreeding populations of intermediate size. There can be a relatively rapid development of slight genetic differences between some of the populations and then reblending of the slightly divergent gene pools. Such a reticulate breeding structure has the equivalent gene pool of a large population, yet has much more adaptive flexibility without the disadvantages of the small population.

In the establishment of a population at a new site, population size will most likely be small for the first few generations. Thus, regardless of the ultimate size of the population, the gene pool may show some of the effects of drift and random fixation that were involved in the earlier stages of population establishment. Furthermore, the entire fate of a population may depend on the genes brought into the population when it was started. If, for example, a new population is started by a single seed that was somehow transported to the new site, the very limited gene pool will produce a population of rather uniform individuals with, presumably, a relatively low evolutionary potential. The reduction in population variability caused by the small size of the original gene pool provided by the organisms that started the population is called founder effect.

Population size, however, involves more than the gene pool: it involves some form of competition between the individuals of the population. Since populations of organisms tend to occur in relatively specific ecological situations which have finite spatial limits, a second important aspect of population size is population density. In plant populations an increase in density means more direct competition between individuals for moisture, nutrients and light. The larger the population in a given area (i.e., the more dense the population) the more severe the competition between its members and the more harsh the natural selection.

In plants, population size in the spatial sense is established by certain environmental limits which define the ecologically suitable habitats of each species. Some populations, limited to a very specific soil type, type of rock outcrop, pollinator range, or moisture condition may be only a few square meters in area; populations of plants with wider physiological tolerances may occupy many square miles. Many populations of certain plants are more or less linear and follow stream banks, roadsides, coastal dunes and similar linear habitats for considerable distances. However, populations of a given species are never uniformly distributed within the geographic range of the species, nor are the individual plants usually uniformly spaced within the population. Both populations and individuals tend to occur in "clumps" or "patches" within their respective areas. The proximity of one population to another, and the maintenance of a particular plant density within each population, becomes extremely important not only in the matter of pollination and seed production but also in regard to gene flow and attendant individual and population variability which is of evolutionary and systematic importance.

For some plants, such as trees and shrubs that live for many years and get larger each year, population density actually decreases as the population gets older and individual plants continue to die as they get shaded out by the more robust or faster growing individuals in the population. This "self thinning" has also been observed in some annuals that maintain a rather uniform density at maturity despite varied densities of the seed at planting time. This comes about in two ways. In the first case, the first seedlings to germinate produce a water-soluble chemical substance that inhibits the germination of the other seeds in the immediate area. The first seedlings to start growth thus have a strong competitive advantage over the later seedlings that must compete with a larger, established, seedling for light, water, and soil nutrients. The roots of some plants, such as those of the black walnut tree (*Juglans nigra*) are suspected of giving off substances which are toxic to many plants, including their own seedling, thus providing life-long protection against competition. In the second case, germination is not inhibited, dense stands of seedlings are produced, and disease (such as damping off) quickly removes many of the weak seedlings and population density is lowered.

Although an individual plant obviously cannot migrate in response to increasing population density, migration is still an important factor in relation to competition and density in plant populations. There are numerous adaptations of fruits and seeds which promote migration away from the parent plant and often away from the population. For example, the seeds of the magnolia (*Magnolia grandiflora*) have a bright red fleshy covering, called an *aril*. When the seeds are ripe, they hang exposed outside the follicles that make up the cone-like fruit. The fleshy aril contains a water-soluble chemical which inhibits or prevents seed germination; only when the aril is removed will the seed sprout. If a seed falls to the ground under the parent tree, it will not germinate. If a seed is eaten by a bird, however, the fleshy aril and its chemical inhibitor are removed as the seed passes through the digestive tract of the bird. A very hard seed coat prevents the digestion of the seed itself, therefore, when the bird excretes the seed, presumably at some distance away from the parent tree, it will be ready to germinate and the seedling will not be in fatal competition with the parent tree.

A high reproductive rate, if actually translated into high population density rather than into an expanded range, often ceases to be an adaptive advantage and becomes an adaptive liability that lowers the biological fitness of the organism involved. (Also see Chapter 16 on Geography and Plant Distribution.)

Section B. BREEDING SYSTEMS

Many of the evolutionary properties of populations depend upon the heterozygosity maintained by a relatively steady rate of recombination within the population. However, in outbreeding populations chromosomal differences (e.g. inversion, translocation), assortive mating, and, occasionally, varying amounts of intrapopulation sterility provide a certain degree of restriction to complete or random recombination. Many flowers are normally bisexual or hermaphroditic, thus inbreeding at the level of the individual is possible for most higher plants. Conversely, in these plants capable of self-pollination there are found a number of mechanisms that provide for some recombination through either frequent or occasional outcrossing. Since both inbreeding and outcrossing each have survival and evolutionary value under different conditions of natural selection, it is not surprising to find that many plants have evolved slightly ambivalent reproductive systems that allow for the effective operation of both of these divergent breeding mechanisms within a single population or, in some cases, a single individual.

In general three different systems, based on the degree of inbreeding versus the degree of outbreeding, can be recognized in plants: (1) predominant outcrossing, (2) predominant selfing, and (3) mixed selfing and outcrossing. In addition, several highly successful methods of asexual reproduction have evolved among some of the higher plants and the combination of sexual and asexual reproduction adds to the evolutionary flexibility (and often the range of phenotypic variation and taxonomic confusion) of those plant groups in which it occurs. (See Chapter 6-VI, A and B for terms relating to pollination types and pollination.)

In dioecious plants, where one plant bears only male or staminate flowers and another plant bears only female or pistillate flowers, we find maximum obligatory cross-fertilization. However, the enforced immobility of plants makes strict dioecism of dubious survival value, especially in a small population which might consist of plants of only one sex. Selection might therefore favor the production and survival of plants that are functionally dioecious but which may either normally, or under certain environmental conditions, produce a few perfect flowers. Such plants would be polygamodioecious. If some perfect flowers occur on an otherwise female plant the plant would be gynodioecious. The perfect flowers (as well as some of the pistillate flowers) of this gynodioecious plant could be self-pollinated, but most of the seed set would be the result of normal cross pollination from another, staminate, plant. The gynodioecious condition, which occurs in some species of the pink, mint and aster families (Caryophyllaceae, Lamiaceae, Asteraceae) allows for a maximum of recombination and a minimum of homozygosity.

In populations of sexually reproducing organisms maximum genetic recombination, or gene flow, would occur if the population were panmictic, that is, if all individuals in the population could mate at random. This means that any one individual is equally likely to mate with any other individual of the opposite sex in that particular population. Although panmixis is a useful evolutionary concept for theoretical calculations, or as a reference point in experimental work, true panmixis most likely does not occur, or if so, only rarely, in natural populations of higher organisms. Among plant populations panmixis might be most closely approached in a population of wind pollinated plants. However, interplant distances, wind velocity, and the prevailing wind direction would all be factors that would reduce the randomness of wind pollination and therefore the randomness of interbreeding among the individuals comprising the population.

I. OUTBREEDING SYSTEMS

Once an organism becomes evolutionarily adapted to a particular environment, or set of selective forces, it can usually maintain itself indefinitely in a state of ecological equilibrium with the other biotic elements to which it is adapted as long as the total environment itself remains essentially stable. Under such conditions, selection would be for continued stabilization of the organism; increased uniformity rather than increased variability would have the most selective value; homozygosity would tend to increase and heterozygosity would tend to decrease. In plants, which have evolved a much more variable series of reproductive methods than have animals, strong stabilizing selection might also lead to changes in the breeding system itself--there might be a shift to more inbreeding (self-pollination) and possibly to some form of apomixis or asexual reproduction.

Among obligate outbreeding plants it is possible that they might be highly heterozygous genetically even though they are quite uniform phenotypically. Also, heterozygosity itself may increase individual survival values and also population fitness through heterosis and homeostatic effects, while the increased gene pool possible with heterozygosity provides a relatively high degree of evolutionary potential under changing environmental conditions or changes in the environmental selective forces that might be incurred by migration.

With enforced outbreeding, the accumulation of particular alleles, or groups of alleles, can most easily take place only in relatively small groups of interbreeding individuals that are in some way separated from other groups or populations of the same species with which they would normally interbreed. The organisms belonging to separated, or allopatric, populations therefore tend to differentiate more readily from each other than do the sympatric organisms of a single population. If two or more populations of a particular organism are separated or isolated geographically for a long enough time (i.e., for enough generations) it is possible that, in addition to any morphological differences that might develop, certain physiological or genetic changes would evolve that would prevent effective interbreeding between members of the two populations should they now again come into contact at some point. If this occurs, the two forms of the organism are said to be reproductively isolated (see Chapter 26, II) and one of the first evolutionary steps toward the possible formation of a new species has been completed (note Figure 21-15).

For many years, it was believed that because of the strong recombination effect of obligate outbreeding, geographic isolation was a necessary first step in the differentiation of races and the development of other more effective isolating mechanisms. On the other hand, there is now some experimental evidence to indicate that strong disruptive selection, even with maximum crossing or interbreeding of the two divergent forms of a polymorphic population, can be sufficient to maintain, and presumably to isolate, two divergent groups of individuals within the same population. Such sympatric speciation is most likely a rare occurrence in nature, but it should no longer be considered completely impossible. Given a particular combination of generation time, population size and polymorphism and a strong disruptive selective force, sympatric speciation could be realized.

Many of the higher plants are bisexual and capable of self-pollination and self-fertilization. If a single flower bears both male and female reproductive structures, that is, both stamens and pistils, it is said to be a perfect flower. An imperfect flower would be unisexual, bearing either stamens or pistils but not both. If a single plant produces both male and female flowers, as is the case in many wind pollinated species such as corn (Zea mays), oak trees (Quercus sp.), and alder (Alnus), the plant is said to be monoecious. In a modification of the monoecious condition, staminate and perfect flowers may occur on one plant (andromonoecism) or pistillate and perfect flowers may be found on a plant (gynomonoecism). Heterozygosity and evolutionary potential are even more strongly assured if the unisexual flowers are borne on separate plants. Such plants, as indicated previously, are said to be dioecious.

In the case of plants with perfect flowers, self-pollination is not only possible but is, in many cases, the primary type of pollination. Such close and continued inbreeding tends to reduce recombination, reduce heterozygosity, reduce variability, and reduce evolutionary potential. On the other hand, self-pollination makes it possible for a single plant to reproduce and start a population; it can also insure adequate seed set in some plant populations in the absence of sufficient or appropriate specialized pollinators. In monoecious plants, self-pollination is certainly possible but not probable, since the pollen is usually distributed by the wind and the female flowers of one plant are much more likely to receive pollen from the male flowers of a neighboring plant than from the male flowers on their own plant. Here, outbreeding is predominant, and recombination, heterozygosity, and evolutionary potential are accordingly increased.

Outcrossing is obviously obligatory for dioecious plants, and this raises several problems in terms of the optimum sexual composition of a population in terms of the ratio of male and female plants. Very often, the first problem is "solved" by the fact that, as previously mentioned, many dioecious plants do actually produce a few perfect flowers capable of self-pollination. Also, many of the species of dioecious plants are perennial; annual seed production is not essential to survival of the population. In the second case, selective pressures, possibly related to some aspect of conservation of energy within the population, have caused a shift to more female plants and fewer male plants in some species of dioecious plants.

If the male and female parts of a perfect flower, or the male and female flowers of a given monoecious plant, mature a few days apart, the plants cannot self-pollinate and must outcross. If the pistillate flowers (or the pistillate portion of perfect flowers) mature first the plant (or flower) is said to be <u>protogynous</u>. In a <u>protandrous</u> plant (or flower) the reverse is true: the stamens of a given flower mature and shed their pollen before the stigmas of the flower are receptive.

Strong protandry has apparently produced a shift toward the andromonoecious condition in several species of the carrot family (<u>Apiaceae</u>). Here the numerous small flowers are borne in aggregations called umbels. Large compound umbels, as in Queen-Anne's lace or wild carrot (<u>Daucus carota</u>), are made up of many small umbellets. If the individual perfect flowers are protandrous, the first flowers to open in the umbel may be pollinated by flowers nearer the center of the umbel that open later. As this process continues, it is evident that the stigmas of the central flowers of the terminal umbel will be receptive after all of the pollen from the flowers of this umbel is shed. However, stamens from the outer flowers of the lateral umbels (which begin to open as the central flowers of the terminal umbel finish blooming) furnish the pollen necessary to pollinate these central flowers of the primary umbel.

In a number of other plants, self-pollination is still possible, despite strong protandry. In many members of the Asteraceae, for example, the pollen is shed as the stigma, still morphologically and physiologically immature, pushes up through the ring of anthers. The mass of sticky pollen is literally pushed out of the flower, and most of it is usually soon picked up by a pollinator. As the stigma matures, it separates, exposing its pollen-free receptive inner surface. At this stage, pollen from another source, when placed on the stigma, would effect cross-pollination. However, if cross-pollination does not occur within a certain time, the tips of the stigma begin to reflex until the receptive inner surfaces come in contact with the style which may still be covered with pollen that was not picked up by the pollinator. In this way, self-pollination is accomplished and seed production assured even if cross-pollination has not taken place.

In some plants with perfect flowers self-pollination is unlikely because of the arrangement of the anthers and stigmas within the flower. If intraspecific differences in floral structure are responsible for the enforced outcrossing, the incompatibility system is said to be <u>heteromorphic</u>. Heterostyly is a good example of heteromorphic incompatibility and is found in <u>Oxalis</u> (Oxalidaceae), <u>Houstonia</u> (Rubiaceae) and a number of other genera. As the name implies, the flowers of one species are of two forms: some flowers have long styles and some flowers have short styles. In addition, the stamens of the long-styled (or "pin") flowers are well below the stigma while the stamens of the short-styled (or "thrum") flowers are above the stigma. In this way, cross-pollination between "pin" and "thrum" flowers is made more likely than self-pollination. As usual, the system is not absolute, and sometimes "illegitimate cross-pollination" may occur; in this situation, pollen from the low stamens of one long-styled flower reaches the stigma of another long-styled flower rather than the stigma of a short-styled flower, or vice versa. However, even when illegitimate pollinations are made, fertilization may not occur because of secondary morphological or physiological adaptations correlated with the heterostylous condition. "Thrum" pollen is often larger than "pin" pollen, the papillae on the stigmatic surfaces are different in the two flower forms, and there are often differential rates of growth of the pollen tubes, the best growth occurring when the pollen type is the opposite of the style type (i.e. when the pollinations are "legitimate"). To function effectively as a unit, the various genes controlling all of the correlated characteristics associated with the two flower types, "pin" and "thrum," must be closely associated, perhaps as a supergene or block of genes, on a single chromosome. But even such linkage can be variously modified under different selective forces. If the survival value of self-pollination is great enough, heterostyly may be modified to homostyly, in which the styles are of constant length and the stamens are attached to the corolla tube in such a position that self-pollination is both possible and probable. In another modification, tristyly, the anthers of a single flower are at two levels and there are three style lengths correlated with specific stamen positions.

In other plants, which produce perfect flowers of only one form in which self-pollination cannot be prevented, genetically controlled physiological factors may prevent self-fertilization. This condition, called homomorphic self-incompatibility, effectively blocks inbreeding and may be either of two types: gametophytic self-incompatibility in which the inhibition of fertilization is the result of gene action in the pollen grain (the male gametophyte) itself as the pollen tube grows through the style; and sporophytic self-incompatibility in which the inhibition of pollen germination or pollen tube growth is imposed by the gene action of the sporophytic tissue of the stigma. In those cases that have been investigated, there is a correlation between binucleate pollen and gametophytic self-incompatibility and the formation of trinucleate pollen and the sporophytic type of self-incompatibility. In both forms, however, only one or two genes, and their alleles, seem to be involved. Although found in about a dozen unrelated plant families, the genetic basis of the self-sterility mechanism is remarkably similar in all cases: the pollen tube will not grow on a stigma or through a style if the pollen and the stigma or style have a particular allele in common. Thus, a plant of the genotype S_1S_2 would produce two kinds of pollen, S_1 and S_2, neither of which would grow through the S_1S_2 sporophytic tissue of the pistil of the flower of the plant that produced the pollen. Such pollen grains would, however, be effective on a stigma or in a style with another genotype such as S_3S_3 or S_4S_5.

In the ground cherry, Physalis (Solanaceae), self-incompatibility is due to two independently assorting genes, S and Z and their alleles. This imposes an even more stringent selection for outcrossing: in the cross between two plants of the genotypes $S_4S_5Z_1Z_6$ and $S_1S_4Z_2Z_6$ for example, eight genetically different types of pollen would be produced but only 1/4 of the pollen produced by plants of the first genotype would be effective on the stigmas of the plants with the other genotype. The wasted pollen is part of the biological "cost" of maintaining heterozygosity and a greater evolutionary potential in these particular plants.

Although sexual reproduction is the mechanism which insures variability and evolutionary potential through recombination, asexual reproduction has strong survival value under certain conditions in which the production of uniform (but not necessarily homozygous!) preadapted genotypes can maintain, or even increase, a population in a given environment. Or, in cases where various forms of sterility prevent effective sexual reproduction, asexual reproduction is obviously the only method of survival of the organism, and in the higher plants several secondary types of asexual reproduction have evolved that provide for reproduction without the fusion of male and female gametes.

II. ASEXUAL REPRODUCTION

Vegetative propagation or reproduction is quite common among the perennial flowering plants and can take place by the formation of new plants from rhizomes, tubers, stolons and root divisions which become separated from the parent plant. Layering, the rooting of detached branches, propagation from individual bulb scales, and the formation of vegetative propagules of various kinds on the leaves, stems, or in the inflorescences of certain plants can also produce large numbers of genetically identical individuals called a _clone_. If environmental conditions remain stable, the members of a clone may maintain a population for many years. Some clones, such as the long-lived coast redwood (_Sequoia sempervirens_) which reproduces by stump sprouts from old root crowns, and some of the rhizomatous grasses of the American prairies, provide examples of clones that have survived for thousands of years.

Because of the uniform genetic background of the members of a clone, any variability between the members, even within a particular population, will probably be phenotypic variation caused by slight environmental differences. Clone material, therefore, is ideal for studying the morphological and physiological effects of different environments on a particular plant or series of plants from one or more populations from different areas of habitats.

Another form of asexual reproduction, called _agamospermy_, involves the production of viable seeds which have been formed without fertilization, although in some cases pollination is required to trigger the agamospermous process. The embryo in such seeds may develop from an unreduced egg (this is _parthenogenesis_) or some other unreduced cell in the embryo sac, or it may develop from a somatic cell in the ovule. In all cases, the embryo is genetically identical with the seed-producing parent. If there are several different genotypes in the population, these same genotypes will be represented unchanged in the next generation, but possibly in different proportions. Although recombination is not possible, the series of populations maintains a small amount of flexibility through time as selection operates to increase or decrease the percentage of each genotype in each generation as conditions change. Such a situation occurs in the common dandelion (_Taraxacum officinale_) and numerous other, though usually perennial, members of the aster family (Asteraceae), grass family (Poaceae), and rose family (Rosaceae).

However, despite the immediate fitness exhibited by some asexually reproducing organisms, unless some sexual reproduction is possible, the members of a clone, as are most other organisms that result from asexual reproductive processes, are at an evolutionary dead end.

In those organisms that combine some degree of sexual reproduction, which insures recombination and evolutionary potential, with various forms of asexual reproduction, which provides for the rapid multiplication of adapted genotypes, we see excellent examples of evolutionary buffering. Such organisms have generally had long evolutionary histories and will most likely continue to survive under whatever drastic local or world-wide environmental changes that man causes.

III. INBREEDING SYSTEMS

Despite the long-term evolutionary advantages of cross-fertilization, a high degree of homozygosity and a positive pollination mechanism that regularly insures optimal seed production, even in the absence of pollinators, is of considerable selective advantage in some plant populations such as those of many annual plants or colonizing species in new or ephemeral habitats. Here the advantages of immediate fitness (i.e., survival) and the rapid production of numerous offspring preadapted to the environmental conditions of the new or temporary habitat far outweigh the long-term benefits to be derived from outcrossing. It is not surprising that many of our weedy annuals, under the strong selection for immediate fitness, are found to be autogamous, that is, they are self-pollinating, self-fertile, inbreeders.

The value of autogamy in either annual or perennial plants of colonizing species is also obvious. A single seed of a self-pollinating plant is capable of giving rise, in several generations, to an entire colony of plants. In many instances of dispersal, even long distance dispersal, it is not unrealistic to assume multiple introductions of a species into a given area over a period of time. The establishment of pioneer populations from different seed sources provides for some genetic variability in these populations that are disjunct from the main area of distribution of the species.

Since autogamy occurs in several unrelated taxa that are predominantly outcrossing, and since self-pollination is more likely to be present in disjunct populations of a given species, autogamy is considered to be an advanced or derived condition that has arisen independently in different groups of plants. The strong correlation between self-fertilization and colonization has been referred to as "Baker's Law."

A special situation has been described for several plant species which have dimorphic flowers and are regularly cross-pollinated near their center of distribution but which have homomorphic self-pollinated flowers in the peripheral portions of their ranges. This occurrence of populations of self-pollinating plants in an otherwise outcrossing species can sometimes be correlated with environmental factors that greatly reduce the number of pollinators. The weather of the Faeroe Islands, in the north Atlantic off the coast of Norway, is often cloudy and there is much rain. As a consequence the number of insects is quite low and several widespread species of plants that are normally cross-pollinated by insects in continental Europe are adapted to self-pollination (by raindrops!) in the Faeroes. Autogamy is also frequent in plants that inhabit desert or semi-desert areas where flowering may be periodic and pollinators may be scarce. Other normally outcrossing plants may adapt to occasional unfavorable seasons or periodic fluctuations of temperature or moisture by what amounts to facultative autogamy. Under unfavorable environmental conditions only self-pollinated flowers are produced but under more nearly optimum conditions cross-pollination takes place. An interesting extension of this shift from self-pollination to cross-pollination in an individual plant in a single season can be seen in the mouse-tail (Myosurus sp.; Ranunculaceae) in which the flowers are normally self-pollinated. Here the carpels are separate on an elongated receptacle in the center of the flower and the carpels present when the flower opens are pollinated with pollen from the same flower. If the environmental conditions remain static or get worse (e.g., if the habitat gets too dry) no further carpels are produced. However, if the plants remain healthy and have abundant water the receptacle may continue to elongate and produce additional carpels above the already developing achenes formed earlier. The newer carpels will necessarily be cross-pollinated since no more pollen is available from that particular flower.

When we find both inbreeding and outcrossing in two closely related species, in a single species, or even in a single plant, it is quite likely that some floral dimorphism will be correlated with the two pollination types. Under reduced pollinator selection the self-pollinated flowers may be smaller, less showy, and have smaller anthers. The ultimate in this type of dimorphism occurs in those plants, such as violets and henbit, which normally produce both open, showy, cross-pollinated chasmogamous flowers and on the same plant also produce small, often drab cleistogamous flowers that never open but self-pollinate within the closed floral envelope. The cleistogamous flowers of henbit (Lamium amplexicaule; Lamiaceae), a small winter annual weed of lawns and waste places, are usually produced in the late winter or early spring and are followed shortly by, or even intermixed with, the more showy chasmogamous flowers. In the common purple violet (Viola papilionacea; Violaceae), an herbaceous perennial of woodland clearings and open areas, the fragrant and showy chasmogamous flowers appear early in the spring when the days are short and the cleistogamous flowers follow with the longer days of summer as the new crop of leaves mature. Such plants as the violet and henbit thus have very effective mechanisms to insure both outbreeding and inbreeding as a normal part of each reproductive cycle.

However, even in species that are predominantly self-pollinated, there usually remains the capacity for occasional outcrossing and a renewal of heterozygosity. A series of population studies of predominantly self-pollinated species of cultivated plants such as wheat, oats, and lima beans, has shown that normal outcrossing in these closely inbred plants may range from 1-12%. In addition, the heterozygotes appear to be more vigorous and fertile than the homozygotes. This further indicates that self-pollination is a derived type of reproductive process that, under certain current conditions, has a higher selective value than outcrossing despite demonstrated heterosis. (Also see Chapter 11 on Genetics and Chapter 26 on Evolution.)

Section C. REPRODUCTIVE BIOLOGY IN RELATION TO SYSTEMATICS

The direct relationship between the total reproductive biology of a group of related species and their taxonomic treatment can most easily be seen by a brief comparison of the "key" characters used in the separation of species of oaks with the "key" characters used in the separation of species of orchids from one another: in the wind-pollinated oaks, fruits, leaf and bark characters are used in both identification and classification; in the orchids, in which there is often an extremely close morphological relationship between flower structure and a specific pollinator, flower structure is the primary basis for both classification and identification. Even within a group of rather closely related species the selective pressures associated with the vital function of reproduction are such that different methods of reproduction, different pollinators, and different population structures may all play a role in the size of the species gene pool, the total amount of recombination (and variability) present and, ultimately, in the evolutionary production of a number of important biological differences that can be of taxonomic value. By way of a summary, Table 12-1 lists a number of general character differences which may exist between two closely related species, one of which normally outcrosses and the other of which is normally self-pollinated in some fashion.

Table 12-1. General character differences between related xenogamous and autogamous species of plants. (From Ornduff, 1969. Reproductive biology in relation to systematics. Taxon 18: 121-133. Used with permission)

Characters of xenogamous plants	Characters of autogamous plants
Outcrossing	Autogamous
Self-incompatible	Self-compatible
Diploid	Polyploid
Low recombination index	High recombination index
Flowers many	Flowers fewer
Pedicels or peduncles long	Pedicels or peduncles short
Sepals large	Sepals smaller
Corollas rotate	Corollas funnelform, cylindric, or closed
Petals large	Petals smaller
Petals emarginate	Petals less emarginate or entire
Floral color pattern contrasting; bicolored; polycolored	Floral color pattern less contrasting; greenish, whitish, or monocolored
Nectaries present	Nectaries reduced or absent
Flowers scented	Flowers scentless
Nectar guides present	Nectar guides absent
Anthers long	Anthers shorter
Anthers extrorse	Anthers introrse
Anthers distant from stigma	Anthers adjacent to stigma
Pollen grains many	Pollen grains fewer
Anther appendages well developed	Anther tips reduced (Asteraceae)
Pollen presented	Pollen not presented
Pistil long	Pistil shorter
Stamens longer or shorter than pistil	Stamens equal to pistil
Style exserted	Style included
Stigmatic area well defined, pubescent	Stigmatic area poorly defined, less pubescent
Stigma receptivity and anther dehiscence asynchronous	Stigma receptivity and anther dehiscence synchronous
Many ovules per flower (inflorescence)	Fewer ovules per flower (inflorescence)
Many ovules not maturing to seed	All ovules maturing to seed
Some fruits not maturing	All fruits maturing
Ray florets many	Ray florets fewer (Asteraceae)
Ray florets conspicuous	Ray florets inconspicuous or absent (Asteraceae)
Disc corollas pentamerous	Disc corollas tetramerous (Asteraceae)
Narrow distribution	Wide distribution

Table 12-2. Insect Pollination Flower Syndromes (from Baker and Hurd, 1968).

Pollination Class and Insect Type	Anthesis	Predominant Colors	Odor	Flower Shape	Flower Depth	Nectar (honey) Guides	Nectar	Other
1. Cantharophily (Beetles)	Day and night	Variable, usually dull	Strong, fruity or aminoid	Actinomorphic (haplomorphic and actinomorphic acc. to Leppik)	Flat to bowl-shaped (rarely closed)	None	None or open	Pollen or food bodies. Flower parts in large numbers
2. Sapromyophily (Carrion and dung flies)	Day and night	Purple-brown or greenish	Strong, often of decaying protein	Usually actinomorphic	None or deep if of trap type	None	Open or none; sometimes pseudo-nectaries	Often no food provided. Transparent windows or other features contribute to temporary traps. Mobile appendages to flowers or "tails" to petals often present.
3. Myophily (Syrphids and bee-flies)	Day and night	Variable	Variable	Usually actinomorphic	None to moderate	None	None or present	Bee-flies have tongues to 50 mm; syrphids can chew pollen.

4. Melittophily (Bees)	Day and night or diurnal	Variable except pure red	Present; usually sweet	Actinomorphic to zygomorphic; morphology of flowers can be recognized	None to moderate; often a broad tube (gullet)	Present	None or present (usual); open to concealed	Larger bees can open "closed" flowers. The most variable of the syndromes
5. Sphingophily (Hawkmoths)	Usually nocturnal or crepuscular	Usually white or pale (or green)	Strong; usually sweet	Usually actinomorphic; held horizontal or pendent	Deep narrow tube to corolla (or spur)	Usually none	Present; concealed; ample	Rim of corolla may be reflexed and dissected, often no alighting platform. Anthers versatile.
6. Phalaenophily (Small moths)	As in 4	As in 4	May be less strong than in 4	As in 4	Less deep than in 4	None	Present; concealed	Less extreme modifications than in 4
7. Psychophily (Butterflies)	Day and night or diurnal	Variable (pink being very common)	Moderately strong; sweet	Usually actinomorphic; usually held upright	Deep; narrow tube to corolla (or spur)	Often present	Present; concealed; ample	Usually flat wide margin to corolla tube; not dissected

The extent to which pollinators and reproductive methods may modify floral structures within a single family is well illustrated by Verne Grant's studies (1965) on pollination in the Polemoniaceae (where plants in individual species in various genera either self-pollinate or are pollinated by a series of flies, bees, butterflies, birds and bats) and the work of Leppik (1964) on floral evolution in the Ranunculaceae. In each instance floral modifications can be correlated with the type of pollinator, thus indicating something of the role of pollinating mechanisms, and pollinators, in both interspecific isolation and speciation. Some of the more common insect pollinator-flower relationships pointed out by van der Pijl (1960-61) and by Faegri and van der Pijl (1966) have been summarized by Baker and Hurd (1968) and are given in Table 12-2.

The evolutionary, and taxonomic consequences of the various floral adaptations or modifications associated with xenogamy, autogamy, dioecism, heterostyly, apomixis, cleistogamy and other aspects of sexual or asexual reproduction provide an interesting, and almost unlimited, field of biological study for those interested in "getting behind the scenes" in systematic work where data are accumulated and attempts are made to answer the question "why" in regard to certain types and patterns of plant variation upon which we base many of our taxonomic conclusions.

Section D. REPRODUCTIVE BIOLOGY STUDY METHODS

Careful field observations will usually indicate pollinator type and specificity as well as the sequence of anthesis within a single flower, an inflorescence, or for an entire plant. The functional details of flower structure are also often best studied in the field although laboratory observations are obviously required for the details of floral morphology which can be observed fresh (for ultraviolet patterns not otherwise apparent), cleared (for venation patterns and other anatomical detail) or otherwise prepared (for chemical analysis or electron microscope observation). Pollen studies (size, viability, morphology--see Chapter 8) are easily organized into blocks of laboratory activity, as are the cytological studies (see Chapter 10) that are a necessary part of any detailed reproductive biology research. However, compatability studies, tests for apomixis, and experimental hybridization require experimental field plots or suitable greenhouse space.

A series of variously treated caged and uncaged flowers will usually give a preliminary picture of the reproductive patterns within a population of plants. A practical preliminary program involves a total of 25 flowers for each "treatment" in each population. A practical minimum of 5 plants should be involved in each treatment. If each plant has many flowers, then all five treatments (5 flowers each) could be done on one plant and only 5 plants per population would be needed.

25 flowers tagged but otherwise left alone as a test for cross-pollination and as a control to determine normal seed set.

25 flowers caged, tied in bud, or otherwise protected, and then self-pollinated to test for self-compatibility, which would be indicated by full seed set.

25 flowers caged and left alone to test for self-pollination; normal seed set indicates self-pollination and self-fertility (or apomixis).

25 flowers caged and emasculated to test for apomixis; seed set under these conditions would indicate apomixis.

25 flowers caged (and emasculated if they are self-fertile) and crossed with plants from another population; reduced seed set could indicate developing sterility between the two populations being tested.

Nylon netting (or nylon stockings) over wire frames (or frames made by cutting the sides out of 1/2 gallon cardboard milk cartons) easily serves to exclude pollinators from individual flowers or inflorescences. The "cage" should be tied to a stake for support and the nylon netting tied around the stem of the plant below the flower or inflorescence.

In making the various pollinations, pollen from ripe anthers can be tapped onto a glass slide and the stigma rubbed in the pollen or, for smaller flowers, the pollen must be transferred from the slide (or anther) to the stigma with a fine brush. The tools used should be cleaned in 70% alcohol from one pollination operation to the next to prevent the accidental transfer of unwanted pollen.

Actually proving the mechanism of apomictic seed production involves considerable embryological work, but apomixis per se, can be detected readily if self-fertile flowers set seed after emasculation and isolation from pollinators. In the case of self-sterile flowers only caging is necessary in the first stages of the experiment but, if seed forms, it may be the result of a low percent of self-fertility instead of apomixis. Emasculation in subsequent trials would indicate which process was responsible for seed set. Needless to say, emasculation must be done prior to the dehiscence of any of the anthers in a flower and must be done carefully with fine forceps or a small hooked needle to prevent damage to other floral structures. If any pollen has been shed in a particular flower that flower can either be removed from the inflorescence or tagged with a color-coded string around the pedicel to indicate that it may have self-pollinated.

Additional caged and emasculated flowers can be used for attempts at hybridization with related species. Although ease of hybridization may indicate a fairly close relationship the reverse is not necessarily true; however, the production of F_1 hybrids and a study of their morphology and of their meiotic chromosome pairing seldom fails to provide valuable background material to any evolutionary study.

LITERATURE ON REPRODUCTIVE BIOLOGY

Baker, H. G. 1963. Evolutionary mechanisms in pollination biology. Science 139: 877-882.

________, and P. Hurd, Jr. 1968. Intrafloral ecology. Annual Review of Entomology 13: 385-414.

Ehrlich, P. R., and P. H. Raven. 1965. Butterflies and plants: a study in co-evolution. Evolution 18: 586-608.

Faegri, K., and L. van der Pijl. 1966. The Principles of Pollination Ecology. Pergamon Press. Oxford.

Grant, V. 1965. Pollination in the Phlox Family. Columbia University Press. New York.

Grun, P., and A. Radlow. 1961. Evolution of barriers to crossing of self-compatible with self-incompatible species of Solanum. Heredity 16: 137-143.

Leppik, E. E. 1957. Evolutionary relationship between entomophilous plants and anthrophilous insects. Evolution 11: 461-481.

________. 1964. Floral evolution in the Ranunculaceae. Iowa State Journal of Science 39: 1-101.

Levin, D. A., and H. W. Kerster. 1967. Natural selection for reproductive isolation in *Phlox*. Evolution 21: 679-687.

Lewis, H. 1963. The taxonomic problem of inbreeders. Regnum Vegetabile 27: 37-44.

Löve, A. 1963. Biosystematics and classification of apomicts. Feddes repertorium specierum novarum regni vegetabilis 63: 136-149. Berlin.

Ornduff, R. 1969. Reproductive biology in relation to systematics. Taxon 18: 121-133.

Pandey, K. K. 1960. Evolution of gametophytic and sporophytic systems of self-incompatibility in angiosperms. Evolution 14: 98-115.

Ray, P. M., and H. F. Chisaki. 1957. Studies on *Amsinckia*. I. A synopsis of the genus, with a study of heterostyly in it. American Journal of Botany 44: 529-536.

Solbrig, O. 1971. The population biology of dandelions. American Scientist 59: 686-694.

Van der Pijl, L. 1960-1961. Ecological aspects of flower evolution I, II. Evolution 14: 403-416; 15: 44-59.

Vuilleumier, B. S. 1967. The origin and evolutionary development of heterostyly in the angiosperms. Evolution 21: 210-226.

Chapter 13. CHEMICAL EVIDENCE*

The potential importance of chemical evidence in taxonomy was suggested by a number of early taxonomists. DeCandolle, Hoffmann, Hallier and Molisch (Hegnauer, 1969, see: Harborne and Swain) are among those who anticipated this innovation, but it is only in recent decades that rapid methods of physical analysis and structural elucidation of organic compounds have been developed to the degree that reasonably rapid surveys of plant extractives are feasible. As more comprehensive surveys of plant chemistry have become available, taxonomists have shown a growing interest in the application of chemical characters to taxonomic problems.

Virtually every chemical component found in plants, including both the primary metabolites (involved in vital functions in the cell) and the secondary products of metabolism (accumulated in cells but not considered to be involved in vital processes), has been used in chemical comparisons of taxa at various taxonomic levels. This chapter is a brief survey of the applications of chemical evidence to taxonomy. Some specific examples of the taxonomic use of chemical evidence are discussed to illustrate this approach. The scope of chemical evidence ranges from data on the distribution of a single compound to a tabulation of a wide range of chemical constituents; and thus, chemical evidence can range from a single taxonomic character to a system of characters which can rather thoroughly define the group under study.

Section A. HISTORICAL PERSPECTIVE

While there are many pre-twentieth century references to the possible role of chemistry in taxonomy, the work of Abbott (1886) represents one of the earliest attempts to correlate chemistry with the phylogenetic level of development. She found the saponin-containing plants occupied the middle level of Heckel's scheme of plant evolution (multiplicity of floral elements). She concluded that all saponin-containing groups are closely allied and possess, in addition, some other structural similarities. The following quote from Abbott seems much more contemporary than 1886:

> The facts obtained from these studies tend to show a chemical progression in plants, and a mutual dependence between chemical constituents and change in vegetable form... The evolution of chemical constituents in which they follow parallel lines with the evolutionary course of plant forms, the one being intimately connected with the other, and consequently that chemical constituents are indicative of the height of the scale of progression, and are essentially appropriate for a basis of botanical classification. In other words, that the theory of evolution in plant life is best illustrated by the chemical constituents of vegetable form.

In 1909, Greshoff (cited by Gibbs 1958) suggested that chemical characters should be included in a natural classification, and he proposed that

*Contributed by C. R. Parks, University of North Carolina, Chapel Hill. This work was partially supported by National Science Foundation Grant GB30780. The author would like to express his appreciation to Miss E. Dale Johnson for her aid in preparing the manuscript.

every description of a new genus or species should include a short chemical description of that plant taxon.

One of the first successful attempts to combine chemical and morphological evidence in the study of a single genus was the work of Baker and Smith (1920, cited by Gibbs 1958) on the essential oils of Eucalyptus. They suggested that the level of relationship should be reflected in chemical similarities (primitive plant = chemically primitive). According to the morphological and chemical data collected on 176 Eucalyptus species, they divided the genus into three groups differing in both morphological structure and chemical constituents. They concluded that primitive species are those which have feather-veined leaves and high pinene (I) content in their essential oils (terpenes), while more advanced types have intermediate venation and contain pinene and cineole (II). According to their system, the most advanced taxa show butterfly wing venation and contain oils with phellandrene (III), piperitone (IV) and geranyl acetate (V). Despite the fact that more recent research (cited by Alston and Turner 1963) on essential oils in Eucalyptus has provided evidence that more intraspecific variation occurs than was shown by the early workers, the fact remains that the work of Baker and Smith was the first comprehensive chemotaxonomic-morphological study of a complex genus, and in general their conclusions have been a significant contribution to the taxonomy of Eucalyptus.

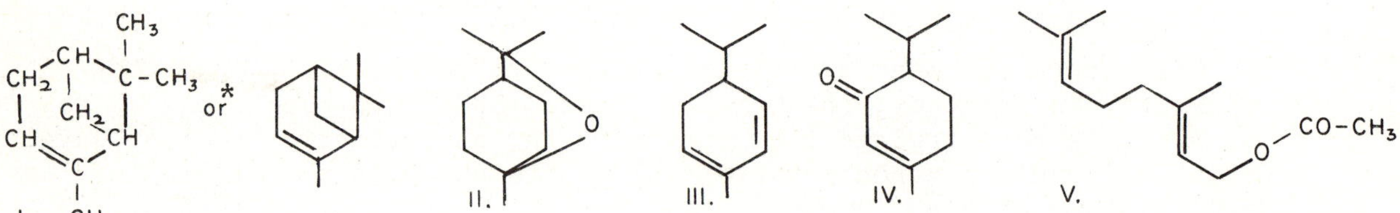

I. α-Pinene II. Cineole III. α-Phellandrene IV. Piperitone V. Geranyl Acetate

As a result of his studies, McNair (1935) set up four quantifiable chemical criteria. His premises were that more closely related taxa produce more similar chemical products, and that more highly evolved members produce larger molecules (alkaloids, volatile oils and glycerides). He further assumed that unsaturation in oils is an advanced character, and that the degree of increase in molecular weight and unsaturation is proportional to phylogenetic advancement. Applying these assumptions he concluded, for example, that the Magnoliaceae is more primitive than the Berberidaceae or the Ranunculaceae; that the monocots are more primitive than the dicots; and that the polypetalous dicots are less advanced than their gamopetalous relatives. His data supported the phylogenetic hierarchy according to Engler and Gilg (1919, cited by McNair

*The abbreviated method of presenting structure will be followed in the remainder of the structures diagrammed in this paper.

1935) in some cases, but in others it was more applicable to the Bessey system. According to McNair's evidence, however, the Ranales were placed too low in both phylogenetic systems. Many reviewers (see Gibbs 1958; and Alston and Turner 1963) doubt McNair's criteria of chemical advancement.

In a later paper McNair (1945) attempted to show that chemical ontogeny can be evidence for chemical phylogeny. One example he used correlated progressive changes in terpene content and aging in Eucalyptus. According to his criteria of chemical advancement of terpenes, more advanced compounds were more typically characteristic of mature trees. He also attempted comparisons involving seed chemical constituents.

The difficulty in obtaining comprehensive chemical information of the type necessary for phylogenetic comparisons between large numbers of plant species proved to be almost impossible until the development of modern physical methods in organic chemistry. Yet despite the great difficulty in chemically characterizing taxa in the early chemotaxonomic surveys, these early contributors must be credited with providing a philosophical foundation for the application of modern chemistry to taxonomy.

Chemical surveys of taxonomically more restricted groupings of plant species have been more successful in providing data for the working phylogenist. The distribution of betalains [formerly known as nitrogen-containing anthocyanins (see XI and XII for basic structures)], along with other evidence, led Lawrence et al. (1939) to suggest that the Cactaceae should be aligned with the Centrospermae; and this disposition is now generally accepted (Mabry and Dreiding 1968, see: Mabry, Alston, and Runeckles). From an extensive survey of anthocyanin content in plant species, Beale et al. (1941) observed that the incidence of pelargonidin pigmentation is more common in tropical species than in those from temperate regions. They also noted that the flowers of the majority of tropical or subtropical species containing cyanidin (VII) or delphinidin (VIII) were appreciably redder than temperate species with the same anthocyanins in their flowers. They concluded that natural selection favors the survival of red-flowered forms in the tropics. It is now well known that pollination vectors which favor red are common in tropical regions.

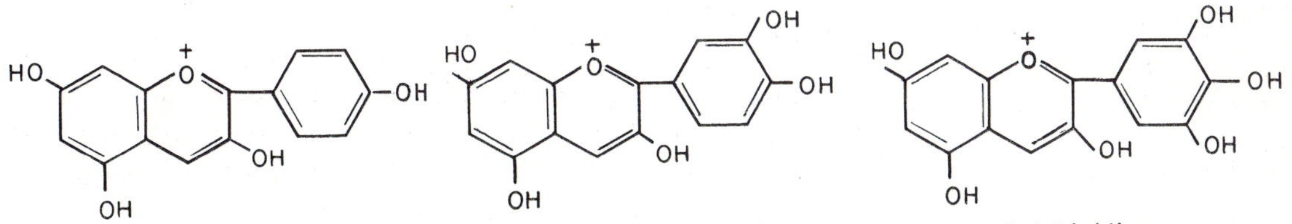

VI. Pelargonidin VII. Cyanidin VIII. Delphinidin

Beale and his collaborators (1941) also surveyed 32 species of Tulipa for anthocyanin content. This represented one of the first efforts to use the distribution of a grouping of compounds in a genus as a taxonomic character. They found that the anthocyanin distribution fully supported the generic structure as determined by morphologists. The chemical constituents selected for analysis, as morphological traits selected, must be chosen in light of the knowledge concerning the broader distribution and selective forces affecting particular classes of compounds.

In the 1940's Mirov (1967) began his detailed analysis of conifer terpenoids. His work, along with that of others, has led to the elucidation of many taxonomic problems in the conifers.

Serological methods have been applied to taxonomy since the turn of the century, but over-all these have contributed relatively little to systematics. Study of the results of the earliest serological investigations proved that serology was a potentially valuable instrument of taxonomic analysis (Chester 1937). The exhaustive serological survey of the plant kingdom led by Mez and his associates (cited by Chester 1937) culminated in the construction of a phylogenetic tree, the "Stammbaum." Though Mez's work stimulated much controversy, relatively little of it has been incorporated into the mainline of systematics. Gell et al. (1960) speculated that the comparatively crude serological tools of that day, coupled with the absence of an additional reliable criterion to aid in evaluating conflicts of interpretation between serological results and morphologically-based taxonomy, led the taxonomists to choose their own time-tested techniques.

Paper chromatography was developed by Consden et al. in 1944. Bate-Smith (1948) showed that the methods of paper chromatography could be applied to plant pigments in his analysis of petal-pigment mutants in _Dahlia variabilis_. These techniques are now widely applied in plant systematic studies.

Section B. THE APPLICATION OF NATURAL PRODUCTS CHEMISTRY TO TAXONOMY AND PHYLOGENY

Different groupings of natural products may have a narrow or wide botanical distribution. The distribution of a compound, or groups of similar compounds with a limited occurrence, may be correlated with a taxa system for which relationships have been proposed. The chemical data may strengthen the hypothesis or provide clues for an alternative interpretation. If correlations are not found, it is necessary to collect additional chemical or botanical data. Extractives which are limited to a few families or orders are proving to be particularly useful in this regard. Certain compounds are limited in this distribution to one or a few species, and these are of little taxonomic value.

Compound classes such as the monoterpenes (see I-V) and the flavonoids (see IX and X) are ubiquitous in their distribution. While the distribution of an individual flavonoid or monoterperoid may have little or no significance above the generic level, the general groupings of terpenoids or flavonoids found in families or even in higher categories may be of considerable significance.

The natural products surveyed in taxonomic studies are usually secondary metabolites. Alston (1966, see: Swain), as well as a number of others, has pointed out that secondary metabolites are subject to low selection pressures, and thus would be conservative characters. While this argument has some obvious validity, studies of plant biology and biochemistry often show that

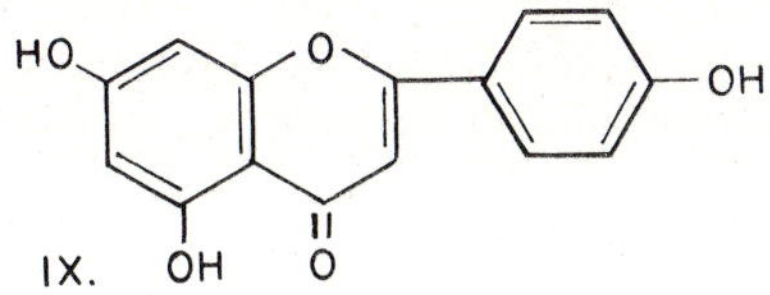

IX. Apigenin (a common flavone)

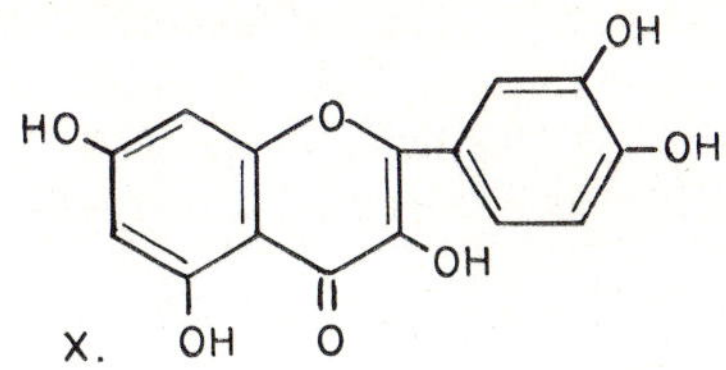

X. Quercetin (a common flavonol)

the secondary constituents are very much an integrated part of the relationship of the plant to its environment and to its metabolism. In fact, secondary constituents may be under high selection pressures. The anthocyanins, for example, which are secondary constituents, are correlated with the type of pollination vector mentioned earlier; therefore, the floral anthocyanins would be expected to be under strong selection pressures.

It is beyond the scope of this brief chapter to completely review the literature. Hegnauer (1962, 1963, 1964, 1966, 1969) has compiled five volumes on chemotaxonomy, and primary papers and reviews continue to be produced at a very rapid rate. Nevertheless, some selected examples are pertinent to our discussion.

Considering the taxonomic importance of instances of compounds with unique distributions in plants, the study of the "nitrogenous anthocyanins" is perhaps the best known example. These red pigments were first recognized as different from the anthocyanins by Schudel in 1918 (see Mabry and Dreiding 1968, see: Mabry et al.), but the structures were not fully resolved until 1962. The betalains, as they are now known, include the red to violet betacyanins (XI) and the yellow betaxanthins (XII). These compounds are distributed only in certain plant families, and this has been recognized for some years as being of importance in establishing the natural relationships of families in the Centrospermae. Ten betalain-containing families are now recognized in this order (Mabry and Dreiding 1968, see: Mabry et al.). One family, the Caryophyllaceae (once thought to be a central member), lacks these compounds. Recent anatomical evidence, coupled with this chemical evidence, supports the removal of this family from the Centrospermae, and provides for its subsequent placement in a separate order, the Caryophyllales (Mabry 1966, see: Swain).

Hegnauer (1969, see: Harborne and Swain) has reviewed phytochemical evidence which can be used in phylogenetic placement of the Cornaceae and its allies. He notes that Takhtajan allies the Cornales with the Araliales in the Umbelliflorae and also considers these to be derived from Rosaceous stocks. He states that Cronquist, on the other hand (unlike most other treatments), allies the Araliales with Rutaceous stocks. Hegnauer found that the rather specific iridoid glycosides (XIII) and the more specific forms of widely distributed compounds found in species of _Cornus_ are very different from the chemical composition of species in the Araliaceae. A wide array of cornalian genera seem to be linked by the occurrence of iridoid glycosides in their tissues. The particular composition of phenolic acids (widely distributed compounds, see XIV) in the Cornales is more like that of the Rosales and less like the Umbelliflorae. Eyde (cited by Hegnauer 1969, see: Harborne and Swain) agrees that floral anatomy does not favor the linking of the Cornaceae and the Nyssaceae in the Umbelliflorae. From a consideration of the chemical evidence, it can be inferred that the Cornales should be excluded from the Umbelliflorae and, instead, should be affiliated with some members of the Rosales. On the other hand, however, there are striking chemical resemblances between the Rutales and the Umbelliflorae. On the basis of the chemical evidence, Hegnauer would link the Cornales with Rosalian stocks and the Umbelliflorae with Rutalian-like ancestors. Hegnauer concludes that comprehensive comparative studies of other types of characters of the taxa concerned should be undertaken.

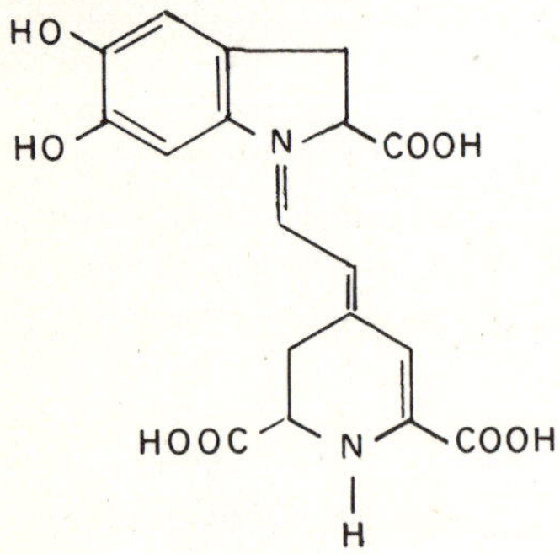

XI. Betanidin
(a betacyanin)

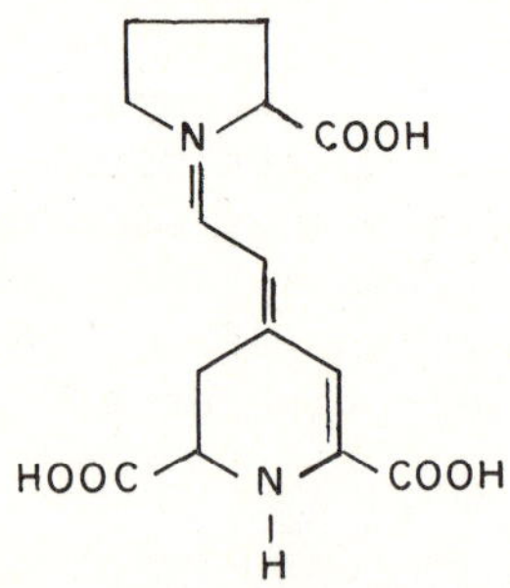

XII. Indicaxanthin
(a betaxanthin)

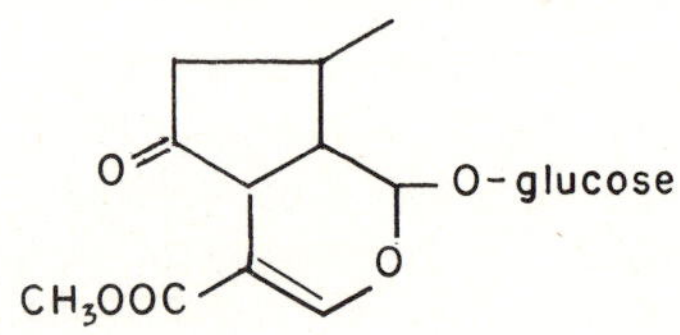

XIII. Cornin
(an iridoid glycoside)

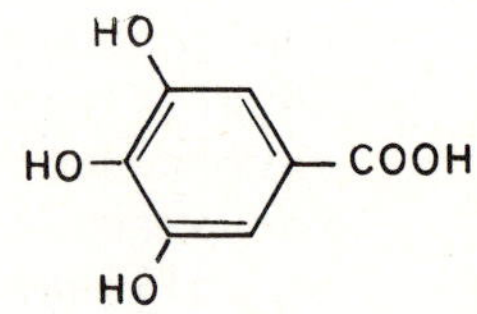

XIV. Gallic acid
(a phenolic acid)

In Table 13-1 some major groupings of plant natural products are listed along with some comment on their botanical distribution. This brief summary is only intended to give some insights into the scope of "natural products chemotaxonomy" and some starting points in the literature. At the end of Table 13-1 is an annotated list of general references which will be of value to the student who wishes to probe further into the area of chemotaxonomy.

Table 13-1. A summary of some groupings of natural products which have been used in chemotaxonomic correlations including some comments on the botanical distribution of these constituents.

Compound grouping	Comments on botanical distribution	Citation
Terpenoids	widely distributed	Weissmann (1966, see: Swain)
Monoterpenes	widely distributed, of taxonomic value primarily below the generic level	numerous reports, see Mirov (1967) or others
Sesquiterpenoids	rather wide distribution, but particularly important and useful in the taxonomy of the Compositae	Herz (1968, see: Mabry et al.) Herout and Sorm (1969, see: Harborne and Swain)

Table 13-1 continued.

Compound grouping	Comments on botanical distribution	Citation
Diterpenes	rather wide distribution in the seed plants, taxonomically useful in specific groups such as the conifers	Erdtman (1968, see: Mabry et al.) Ponsinet et al. (1968, see: Mabry et al.)
Triterpenes	wide distribution in living organisms, taxonomically useful in several angiosperm families such as the Cucurbitaceae, and in the fungi	Ponsinet et al. (1968, see: Mabry et al.)
Carotenoids	the universally distributed photosynthetic carotenoids useful in algal classification; the non-photosynthetic carotenoids in fruits of possible limited value as taxonomic markers in the angiosperms	Goodwin (1966, see: Swain)
Flavonoids	very large number of diverse compounds found throughout the vascular plants and in nearly, if not all, angiosperms; of great taxonomic value below the genus level and of possible use in the classification of higher categories	Bate-Smith (1963, see: Swain); Harborne (1966, see: Swain); Wagner (1966, see: Swain); Williams (1966, see: Swain); Alston (1968, see: Mabry et al.) (extensive literature)
Lignins and lignans	useful in the classification of higher categories	Erdtman (1968, see: Mabry et al.)
Quinones	widely distributed among living organisms, but compounds of this type with limited distribution have potential value in the classification of a number of angiosperm families	Mathis (1966, see: Swain)
Polysaccharides	universally distributed, probably useful in the classification of higher categories - particularly among the algae, comparative data mostly lacking	Percival (1966, see: Swain)
Plant glycosides	great chemical variation in the non-sugar portion of the molecule, varyingly of great usefulness in taxonomy	Paris (1963, see: Swain)

Table 13-1 continued.

Compound grouping	Comments on botanical distribution	Citation
Asperulosides and Aucubins (Iridoid glycosides)	Rubiaceae, Scrophulariaceae, and related families, Plantaginaceae, Cornaceae, and others	Bate-Smith and Swain (1966, see: Swain) Kooiman (1969, 1970)
Ranunculins	found only in the Ranunculaceae	Ruijgrok (1966, see: Swain)
Cyanogenetic compounds	Ranunculaceae and other plant families	Ruijgrok (1966, see: Swain)
Polyalcohols	widely distributed but have taxonomic potential	Plouvier (1963, see: Swain)
Sulfur compounds	a chemically diverse grouping of compounds which in their various forms are widely distributed, the isothiocyanate-producing glycosides are characteristic of families in the Rhoeadales	Kjaer (1966, see: Swain)
Amino acids (non-protein)	Liliaceae and related families, Leguminosae (particularly Papilionatae, and other groups)	Bell (1966, see: Swain)
Alkaloids	a chemically and biosynthetically diverse group of compounds rarely found in the lower vascular plants and irregularly distributed among the angiosperms; highly useful in the taxonomy of some groups	Hegnauer (1966, see: Swain) Schäfer (1964) Price (1963, see: Swain)
Betacyanins and betaxanthins	Centrospermae ("Betanales")	Mabry (1966, see: Swain)
Alkanes (fatty acids and waxes)	widely distributed, possibly of use in classification below the genus level, and of use in organic geochemistry	Shorland (1963, see: Swain) Douglas and Eglinton (1966, see: Swain)
Fatty acid epoxides	found in seven families	Bu'Lock (1966, see: Swain)
Acetylenes	distributed in the basidiomycetes and at least 13 angiosperm families, of particular use in the link between the Compositae and the Umbelliflorae	Bu'Lock (1966, see: Swain) Sørensen (1968, see: Mabry et al.)
Assorted compounds	ferns	Berti and Bottari (1968, see: Reinhold and Liwschitz)

Table 13-1 continued.

Compound grouping	Comments on botanical distribution	Citation
Assorted compounds	lichens	Huneck (1968, see: Reinhold and Liwschitz)
Wide range of chemical approaches	Umbelliferae	Heywood (1971)
Wide range of chemical approaches	Leguminosae	Harborne et al. (1971)

General References:

1. Adams, R. P. 1972. Numerical analysis of some common errors in chemosystematics. Brittonia 24: 9-21. [Errors typical of chemotaxonomic analysis are discussed and analytical procedures are presented which will help to minimize these problems.]

2. Alston, R. E. 1967. Biochemical Systematics, p. 197-305. See: T. Dobzhansky, M. K. Hecht, and W. C. Steere (eds.), Evolutionary Biology, Vol. 1. Appleton-Century Crofts. New York. [This is a critical and detailed review of chemotaxonomy, with particular reference to the use of secondary constituents. The article is very readable and contains an excellent bibliography.]

3. ________. 1965. Comparisons of the importance of basic metabolites, secondary compounds and macromolecules in systematic studies. Lloydia 28: 300-312. [While secondary constituents have been most useful in chemosystematic studies, the analysis of the macromolecule holds great promise for the future.]

4. ________, and B. L. Turner. 1963. Biochemical Systematics. Prentice-Hall. Englewood Cliffs, New Jersey. [The first general textbook to appear in this area of taxonomy includes a discussion of the application and use of chemical data in the field and a review of chemotaxonomic studies arranged by compound groupings.]

5. Fairbrothers, D. E. 1968. Chemosystematics, with emphasis on systematic serology, p. 141-174. See: V. H. Heywood (ed.), Modern Methods in Plant Taxonomy. Academic Press. New York. [This useful review places particular emphasis on serological and electrophoretic methods and includes an excellent bibliography.]

6. Gibbs, R. D. 1965. A classical taxonomist's view of chemistry in taxonomy of higher plants. Lloydia 28: 279-299. [The results of a wide range of simple chemical tests on a spectrum of plant material are provided as examples for the application of chemical data to the placement of taxa in various taxonomic categories.]

7. Harborne, J. B. 1968. Biochemical systematics. The use of chemistry in plant classification, p. 545-588. See: L. Reinhold and Y. Liwschitz (eds.), Progress in Phytochemistry. Vol 1. Interscience Publishers. New York. [This paper is a general introduction to the application of chemical characters to taxonomy with a review of chemotaxonomic studies arranged by compound groupings.]

8. Hawkes, J. G. (ed.). 1968. Chemotaxonomy and Serotaxonomy. Academic Press. New York. [The text contains a compendium of papers in a variety of areas in chemotaxonomy with emphasis on macromolecular approaches.]

9. Hegnauer, R. 1962, 1963, 1964, 1966, 1969. Chemotaxonomie der Pflanzen. Vols. 1-5. Birkhäuser Verlag. Basel. [These volumes offer an exhaustive treatment of plant chemistry arranged by taxonomic groupings.]

10. Levin, D. A. 1971. Plant phenolics: an ecological perspective. The American Naturalist 105: 157-181. [The article discusses phenolic compounds as defensive constituents in plants, with ecological considerations and evaluations.]

Additional books of interest:

Harborne, J. B. (ed.). 1972. Phytochemical Ecology. Academic Press. New York. [A collection of papers is presented in which the ecological roles of a wide array of phytochemicals are discussed.]

________. 1970. Phytochemical Phylogeny. Academic Press. New York. [A collection of papers is presented showing contributions of phytochemistry to the study of phylogeny.]

The extensive survey of a very large number of dicotyledonous and monocotyledonous species for flavonoid and cinnamic acid content by Bate-Smith and his associates (summarized by Bate-Smith, 1969, see: Harborne and Swain) represents a good example of a survey for the distribution of a widely occurring compound grouping for phylogenetic comparison purposes. (The classes of flavonoids include the catechins, XV; leucoanthocyanidins, XVI; flavones, IX; flavonols, X; anthocyanidins, VI-VIII; flavanones, XVII; flavanonols, XVIII; chalcones, XIX; dihydrochalcones, XX; aurones, XXI; and isoflavones, XXII. Two common cinnamic acids are p-coumaric acid, XXIII, and caffeic acid, XXIV.) Some phytochemical correlations may be noted from Bate-Smith's survey of the angiosperm polyphenols. In the dicots, nearly all of the predominately woody families from Casuarinaceae to Ebenaceae contain leucoanthocyanidins, while nearly all of the predominately herbaceous families from the Aristolochiaceae

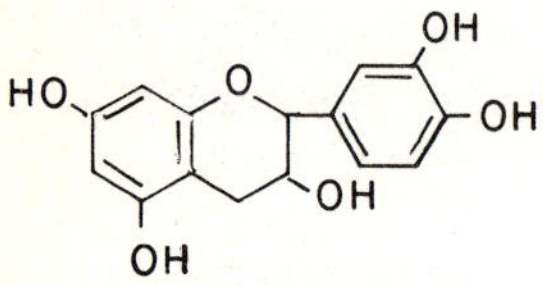

XV. Catechin (a flavan-3-ol)

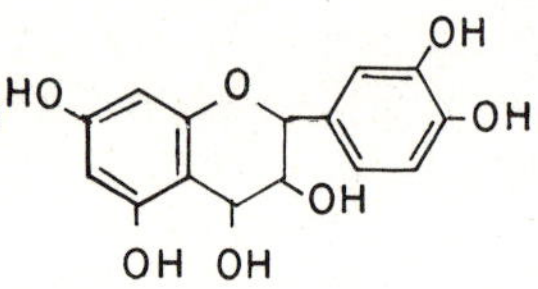

XVI. Leucocyanidin (a flavan-3,4-diol)

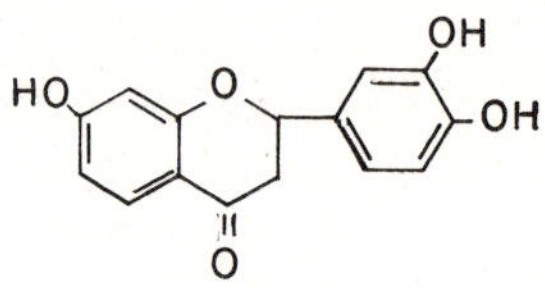

XVII. Butin (a flavanone)

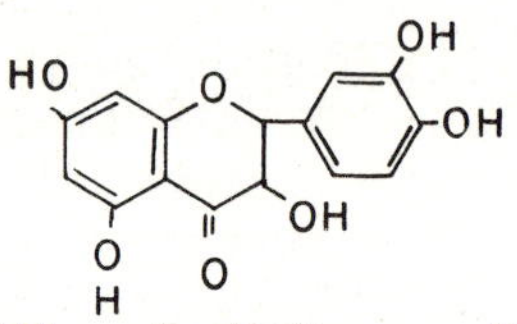

XVIII. Taxifolin (a flavanonol)

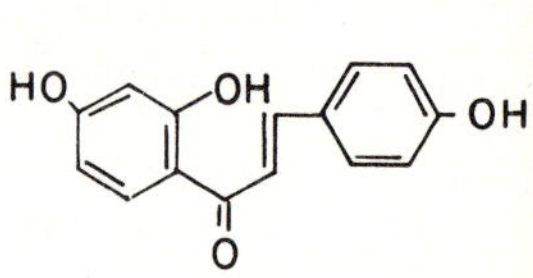

XIX. Dahlia chalcone (a chalcone)

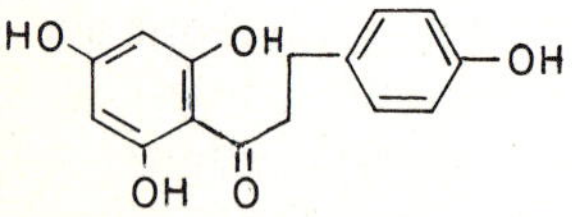

XX. Phloretin (a dihydrochalcone)

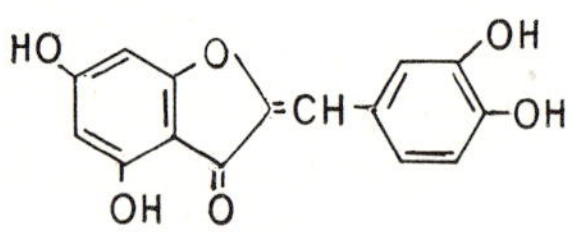

XXI. Aureusidin (an aurone)

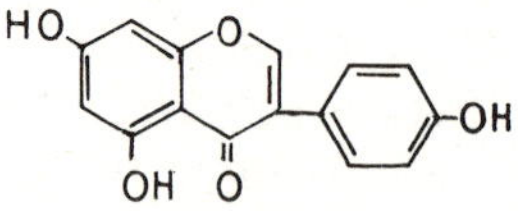

XXII. Genistein (an isoflavone)

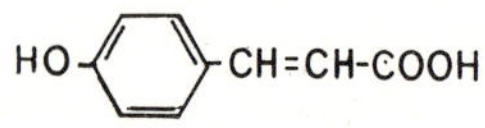

XXIII. p-Coumaric acid

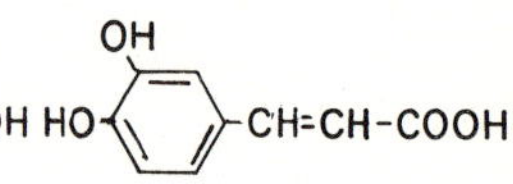

XXIV. Caffeic acid

to Compositae are lacking in these. The trihydroxylated substituted flavonoids (see XXV for myricetin, a flavonol with this type of B ring substitution) follow a similar, but less marked, distribution. Whole families in the Tubiflorae lack, or essentially lack, the flavonols which are replaced by flavones, flavanones or chalcones. Various of the cinnamic acids show similar distributions and correlations.

The polyphenols of monocotyledons are not different from those of the dicotyledons except that ellagic acid (XXVI), widely distributed in the woody dicotyledons, is lacking in the monocotyledons; and the flavonols are generally less frequent in the monocotyledons. Some monocotyledonous families, considered morphologically advanced, have primitive phenolics by analogy to the dicotyledons; and some monocotyledonous families, which show generally

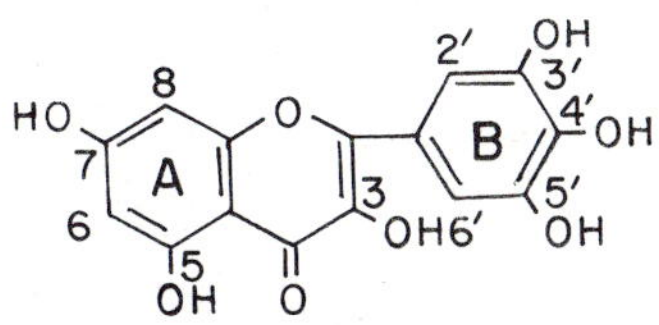

XXV. Myricetin

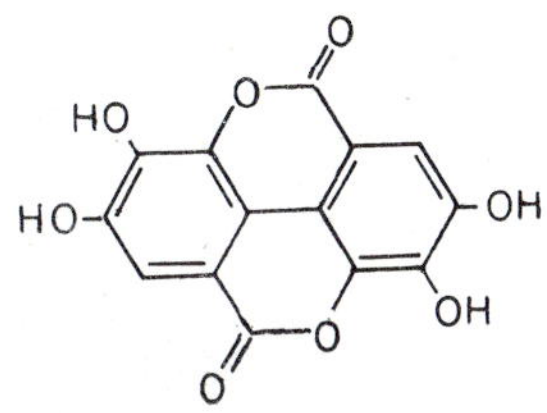

XXVI. Ellagic acid

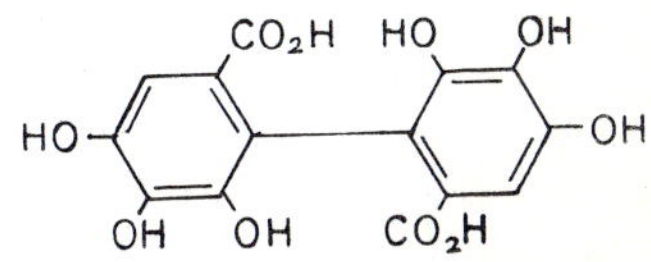

XXVII. Hexahydroxydiphenic acid

advanced phenolic characteristics, have individual genera with phenolics like the primitive dicotyledons. The most advanced monocotyledonous families, such as the orchids and the grasses, are advanced in their polyphenols in a manner similar to the more advanced dicotyledonous families. The ellagitannins, complex derivatives of hexahydroxydiphenic acid (XXVI) (a ring opened form of ellagic acid), are centrally distributed in dicotyledonous orders allied with the Rosales. They are absent in the monocotyledons and in the Ranalian dicotyledons. It can be suggested from this compound distribution that the monocotyledons diverged from ancestral dicotyledons related to the Ranalian complex, rather than from Rosalian ancestors. Some genera, containing primitive phenolic characteristics, belong to families that are otherwise advanced in this respect, thus indicating that the ancestry of these families is traceable through these genera to the primitive angiospermous stock.

Erdtman (1968, see: Mabry et al.) notes that patterns of chemical constituents are more useful to the taxonomist than single chemical traits, and that it is important to investigate in each taxon as many constituents of divergent biosynthetic origin as possible. He and his associates investigated simple phenols, flavonoids, tropolones (XXVIII) and terpenoids in conifers. It has been possible to chemically characterize, to a degree, the conifer families and show many generic relationships. On the basis of the chemical evidence, Erdtman concludes that it is reasonable to merge Ducampopinus with Pinus and to remove Sciadopitys from the Taxodiaceae. Nicotine (XXIX) is found in the Equisetaceae, Lycopodiaceae, Rosaceae, Leguminosae, Asclepiadaceae, Solanaceae and Compositae (Erdtman 1968, see: Mabry et al.). It is not known if this compound is synthesized by the same route in each of these plant families. There is no reason to attach particular importance to the nicotine distribution, however, since other compounds specific to one or a few nicotine-containing families have no parallel distribution to it.

Information concerning the biosynthesis of extractives being used for a phytochemical correlation is important since similar or identical molecules synthesized by different biochemical pathways should not be utilized as evidence of genetic relationships. Erdtman (1968, see: Mabry et al.) points out that anthraquinones occur in both fungi and higher plants. One constituent of lichens and molds, endocrocin (XXX), is known to have a biosynthesis of polyketide origin (acetate pathway); but a similar compound, tectoquinone (XXXI), found in teak wood, <u>Tectona grandis</u>, is derived from an isoprenoid napthalene. Both the polyketide and the mevalonic acid pathway are involved in the natural synthesis of the teak constituent. It is clear that the need to collect data

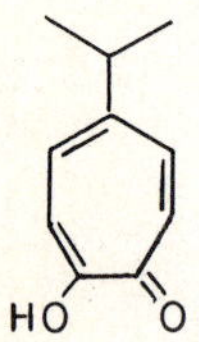

XVIII. γ-Thujaplicin
(a tropolone)

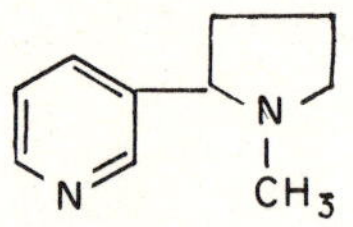

XXIX. Nicotine
(an alkaloid)

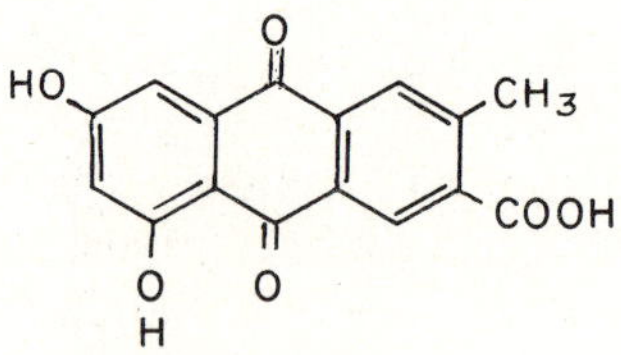

XXX. Endocrocin
(an anthraquinone)

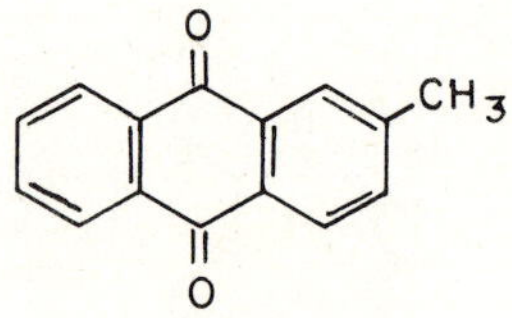

XXXI. Tectoquinone
(an anthraquinone)

on biosynthetic routes does not make the job of the chemotaxonomist any easier, but numerous chemotaxonomists (Alston 1966, see: Swain; Erdtman 1968, see: Mabry et al.; Herout and Šorm 1969, see: Harborne and Swain; Mentzer 1966, see: Swain; Sørensen 1963, see: Swain; and others) have stressed the importance of obtaining this information.

Hegnauer (1969, see: Harborne and Swain) points out that the ability of a taxon to accumulate a particular natural product in measurable quantities is more taxonomically important than the mere presence of that constituent in trace amounts. His opinion is contrary to the working assumption of many chemotaxonomists who labor to show qualitative presence or absence of a constituent. More data and analysis is necessary to determine if this interesting observation can be reinforced.

Herout and Šorm (1969, see: Harborne and Swain) have discussed the problem of negative evidence in chemotaxonomy. Since most phytochemists are not oriented toward taxonomy, they tend not to publish negative evidence which could be useful for taxonomic correlation purposes. Another hindrance, genetically controlled and environmentally induced variation within and between species, can also be confounding in the use of phytochemical data. There are other limitations to the use of natural products chemotaxonomy; but, despite all of them, it promises to become a source for much new information for the taxonomist.

Section C. CHEMICAL EVIDENCE AS A BIOSYSTEMATIC TOOL

While most of the taxonomic conclusions resulting from studies of natural products chemistry have been concerned with the relationships of higher categories, biosystematists using chemical techniques are interested in relationships at or below the generic level. In most cases the chemical data has been treated as a character or grouping of characters and, in the final summation, is used in concert with information from all available sources. This is because the chemical evidence obtained in such studies usually involves the analysis of a single group of related compounds such as the flavonols or the monoterpenes; and while the chemical determinations may be quite detailed and quantitative, a relatively small amount of actual genetic variation may be involved. The addition of chemotaxonomic information to data from other areas, such as cytology, morphology, genetics, etc. can provide a solid foundation for systematic decisions.

I. STEPS WHICH SHOULD BE FOLLOWED IN THE SOLUTION OF A CHEMOTAXONOMIC PROBLEM

A. Selection of chemical procedures.
   1. The facilities available and the investigator's experience in chemistry will determine the amount and kind of chemical analysis which is reasonable.
   2. The pertinent chemical and biosystematic literature should be checked to determine which procedures and compound groupings may be profitably studied. (If adequate work is reported in the literature, the preliminary survey may be abbreviated or omitted entirely.)

B. Preliminary survey.
   1. Run pilot separations on the most distantly related taxa involved in the proposed study to determine if adequate chemical variation exists in the group.
   2. Continue the pilot separations on more closely related taxa in order to make a crude approximation of the level of chemical variation. If a large amount of variation is encountered within a species, it will be clear that extensive sampling of that species will be necessary to define the parameters of chemical variation in that species.
   3. Carry out simple chromatographic and spray reagent tests to determine the general types of compounds which are being isolated.

C. Environmental and physiological variation.
   1. Minor differences in the concentration of constituents are common, but major quantitative or even qualitative differences are sometimes observed. The stability of occurrence of constituents in the same and different organs should be determined from plants which have been grown under the most uniform conditions possible. Organs to be extracted should be collected at a standard physiological stage.

D. Chemical analysis.
   1. Advantages and disadvantages of limiting the chemotaxonomic work to chromatographic separations and chromatographic spray tests.
      a. Benefits: Such a procedure is rapid, and thus allows for the survey of a large number of taxa. The investment in equipment and time is minimal.
      b. Problems: Characterization of structures is minimal; and therefore, matching of constituents isolated from different

taxa may become little more than an educated guess. Lack of information on chemical structure makes it impossible for the investigator to estimate the genetic bases involved in the variation he is measuring.

2. Identification of major constituents by chromatographic and spectral methods increases the reliability of matching chemical constituents in different taxa and provides some information concerning the genetic bases of the variation being studied. Information concerning the structure and distribution of minor constituents may be incomplete or lacking.
3. With full structural determinations of extractives, the significance and distribution of these constituents can best be determined. Of course, the investment in time and equipment by the investigator is much greater.

E. References on techniques of analysis.

1. Geissman, T. A. (ed.). 1962. The Chemistry of Flavonoid Compounds. The Macmillan Company, New York.
2. Harborne, J. B. 1967. Comparative Biochemistry of the Flavonoids. Academic Press, New York.
3. Mabry, T. J., K. R. Markham, and M. B. Thomas. 1970. The Systematic Identification of Flavonoids. Springer-Verlag. New York.
4. Ribèreau-Gayon, P. 1972. Plant Phenolics. Hafner Publishing Company. New York.
5. Stuessy, T. F. (ed.). 1972. References and Lists of Equipment for Systematic Studies with Flavonoids, Mustard Oils, Proteins (Electrophoresis and Serology), and Terpenoids. (Prepared for the A.S.P.T. symposium, "Biochemistry and the Taxonomist," held in August, 1972, in Minneapolis, Minnesota).
6. von Rudloff, E. 1969. Scope and limitations of gas chromatography of terpenes in chemosystematic studies. See: M. K. Seikel and V. C. Runeckles (eds.), Recent Advances in Phytochemistry, Vol. 2. Appleton-Century-Crofts. New York.

II. AN ANNOTATED LIST OF PAPERS RECOMMENDED AS EXAMPLES OF THE APPLICATION OF CHEMICAL DATA IN BIOSYSTEMATIC STUDIES

A. Comparisons at the generic level.

1. Arditti, J. 1969. Floral anthocyanins in species and hybrids of *Broughtonia*, *Brassavola*, and *Cattleyopsis* (Orchidaceae). American Journal of Botany 56: 59-68. [Floral anthocyanin distribution is shown to be correlated with generic groupings, and, based on this and other evidence, suggestions are made which would correct doubtful generic placement of some species.]
2. Bate-Smith, E. C., and T. C. Whitmore. 1959. Chemistry and taxonomy in the Dipterocarpaceae. Nature 184: 795-796. [Genera in this family are shown to have a characteristic chemistry, and on this basis the authors suggest certain species should be moved to different genera with which they have a greater chemical similarity.]
3. Griffiths, L. A. 1960. A comparative study of the seed polyphenols of the genus *Theobroma*. Biochemical Journal 74: 302-305. [*Theobroma* species can be readily distinguished from *Herrania* species by a generic characteristic polyphenol distribution.]

B. Comparisons at the species level.

1. Adesida, G. A., E. K. Adesogan, D. A. Okorie, D. A. H. Taylor, and B. T. Styles. The limonoid chemistry of the genus *Khaya*

(Meliaceae). Phytochemistry 10: 1845-1853. [Limonoid constituents are identified from six species, and are used to characterize the species in this morphologically poorly defined genus.]

2. Albach, R. F., and G. H. Redman. 1969. Composition and inheritance of flavanones in citrus fruit. Phytochemistry 8: 127-143. [Flavanone distribution in *Citrus* taxa is used to aid in the definition of species groups and boundaries, as well as to verify putative hybridity.]

3. Conklin, M. E., and H. H. Smith. 1971. Peroxidase isozymes: a measure of molecular variation in ten herbaceous species of *Datura*. American Journal of Botany 58: 688-696. [The taxonomic affinities within this genus have been reconsidered in the light of peroxidase distribution determined by electrophoretic separations of leaf proteins.]

4. Ellison, W. L., R. E. Alston, and B. L. Turner. 1962. Methods of presentation of crude biochemical data for systematic purposes, with particular reference to the genus *Bahia* (Compositae). American Journal of Botany 49: 599-604. [Chromographic profiles for species of *Bahia* and related genera are compared by polygonal diagrams, affinity values, and synthetic numerical indices.]

5. Emboden, W. A., Jr., and H. Lewis. 1967. Terpenes as taxonomic characters in *Salvia* section Audibertia. Brittonia 19: 152-160. [Nineteen species of *Salvia* can be recognized by the characteristic terpene composition, and this character is also a good indicator of introgression in one case studied.]

6. Gell, P. G. H., J. G. Hawkes, and S. T. C. Wright. 1960. The application of immunological methods to the taxonomy of species within the genus *Solanum*. Proceedings of the Royal Society of London (B) 151: 364-383. [Immunoelectrophoretic techniques were applied to extracts from 37 *Solanum* species. Interpretations of the serological data tend to support conclusions from other biosystematic analyses of *Solanum*.]

7. Giannasi, D. E., and C. M. Rogers. 1970. Taxonomic significance of floral pigments in *Linum* (Linaceae). Brittonia 22: 163-174. [The division of thirty *Linum* species into five subgeneric groupings is supported by the distribution of petal carotenoids and flavonoids.]

8. Johnson, B. L. 1972. Seed protein profiles and the origin of the hexaploid wheats. American Journal of Botany 59: 952-960. [Seed protein profiles were developed electrophoretically for wheat species at three ploid levels, and a monophyletic model for the origin of the hexaploid wheats best fits this data.]

9. ________, and M. M. Thein. 1970. Assessment of evolutionary affinities in *Gossypium* by protein electrophoresis. American Journal of Botany 57: 1081-1092. [With the exception of the placement of a few species, the accepted subgeneric treatment of *Gossypium* is supported by electrophoretic profiles of seed proteins generated for twenty-five species.]

10. McClure, J. W., and R. E. Alston. 1966. A chemotaxonomic study of Lemnaceae. American Journal of Botany 53: 849-860. [Twenty-two taxa could be identified by their specific flavonoid constituents even when morphological intergradation obscured conventional means of determination. Only minor variations in flavonoid glycosides resulted when the plants were grown under various cultural conditions.]

11. Parks, C. R. 1965. Floral pigmentation studies in the genus Gossypium. I. Species specific pigmentation patterns. American Journal of Botany 52: 309-316. [Flower color mutants in three species of Gossypium do not obscure, beyond recognition, the basic pattern of petal pigmentation characteristic of each species.]

12. Payne, W. W., R. W. Scora, and J. Kumamoto. 1972. The volatile oils of Ambrosia (Compositae:Ambrosieae). Brittonia 24: 189-198. [Twenty species from Ambrosia and two other genera contain the same essential oils in different concentrations. This data provided a method for distinguishing species, but little evidence for evolutionary affinities could be gleaned from it.]

13. Roberts, E. A. H., W. Wight, and D. J. Wood. 1958. Paper chromatography as an aid to the taxonomy of Thea Camellias. The New Phytologist 57: 211-225. [The polyphenol constituents of a wide range of tea cultivars are determined, and the taxonomic complexity of this group is underlined; otherwise, the accepted subgeneric treatment of the genus Camellia is generally supported by the results of this study.]

14. Stone, D. E., G. A. Adrouny, and R. H. Flake. 1969. New world Juglandaceae. II. Hickory nut oils, phenetic similarities, and evolutionary implications in the genus Carya. American Journal of Botany 56: 928-935. [All species contain the same unsaturated and saturated fatty acids, but in quite different relative quantities; and the quantitative data supports the morphological division of the genus into two sections.]

15. Stuessy, T. F. 1971. Systematic relationships in the white-rayed species of Melampodium (Compositae). Brittonia 23: 177-190. [An extensive cytological and chromatographic survey conducted within this group provides additional evidence for species delimitations.]

16. Vaughan, J. G., and K. E. Denford. 1968. An acrylamide gel electrophoretic study of the seed proteins of Brassica and Sinapis species, with special reference to their taxonomic value. Journal of Experimental Botany 19: 724-732. [The taxonomic affinities of species in these two genera are reconsidered in light of the results of the comparative study of seed proteins.]

17. Zavarin, E., L. Lawrence, and M. C. Thomas. 1971. Compositional variations of leaf monoterpenes in Cupressus macrocarpa, C. pygmaea, C. goveniana, C. abramsiana and C. sargentii. Phytochemistry 10: 379-393. [The population structure of the Cupressus species was determined by GLC analysis of foliar monoterpenes.]

C. Comparisons below the species level (including hybrids).

1. Alston, R. E., and B. L. Turner. 1963. Natural hybridization among four species of Baptisia (Leguminosae). American Journal of Botany 50: 159-173. [The parentage of putative natural hybrids in Baptisia was determined by chromatographic analysis. (The Baptisia story is presented in detail in a series of related papers including one by T. J. Mabry, J. Kagan, and H. Rosler. 1965. Phytochemistry 4: 487-493, in which the flavonoid constituents referred to in earlier papers are fully identified).]

2. Ashtakala, S. S., and D. F. Forward. 1971. Pigmentation in iris hybrids: the nature and development of flavonoid pigments and their effect on flower color in cultivated hybrids of Iris germanica. Canadian Journal of Botany 49: 1965-1973. [Six strikingly different color variants of Iris germanica were shown to have the same basic flower pigments; however, differences in relative concentrations of pigments were found to be a major factor in determining flower color.]

3. Belzer, N. F., and M. Ownbey. 1971. Chromatographic comparison of Tragopogon species and hybrids. American Journal of Botany 58: 791-802. [Flavonoids were determined for species, species hybrids, and second generation segregates of Tragopogon. From the resulting segregation data, the genetic bases for some of the flavonoid extractives were determined.]

4. Carter, L. C., and B. G. Brehm. 1969. Chemical and morphological analysis of introgressive hybridization between Iris tenax and I. chrysophylla. Brittonia 21: 44-54. [Although not fully obvious from previous studies, it was determined from this analysis that hybridization and backcrossing are occurring in sympatric populations of these Iris species.

5. Cherry, J. P., F. R. H. Katterman, and J. E. Endrizzi. 1970. Comparative studies of seed proteins of species of Gossypium by gel electrophoresis. Evolution 24: 431-447. [The sectional status of the genus Gossypium and the origin of the amphidiploid species are reconsidered in the light of data from an electrophoretic survey of seed proteins.]

6. Crawford, D. J. 1970. Systematic studies on Mexican Coreopsis (Compositae). Coreopsis mutica: flavonoid chemistry, chromosome numbers, morphology, and hybridization. Brittonia 22: 93-111. [On the basis of these four biosystematic approaches, Crawford divided and defined seven varieties of Coreopsis mutica.]

7. ________. 1971. Systematics of the Coreopsis petrophiloides-lucida-teotepecensis complex. American Journal of Botany 58: 361-367. [Using the same methods as in his 1970 paper, he demonstrated that those Coreopsis species are all part of one variable taxon with no discernible subspecific units.]

8. Glennie, C. W., J. B. Harborne, G. D. Rowley, and C. J. Marchant. 1971. Correlations between flavonoid chemistry and plant geography in the Senecio radicans complex. Phytochemistry 10: 2413-2417. [Flavonoids determined for forty-four clones from twenty-two species were not correlated with known morphological or cytological patterns in this group, but rather were correlated with the plant geography of these species.]

9. Levin, D. A., and B. A. Schaal. 1970. Reticulate evolution in Phlox as seen through protein electrophoresis. American Journal of Botany 57: 977-987. [Electrophoretic profiles of seed proteins were successfully used to verify the putative hybridity of some stabilized taxa in Phlox.]

10. ________. 1972. Seed protein polymorphism in Phlox pilosa (Polemoniaceae). Brittonia 24: 46-56. [Evidence is provided to show that seed protein polymorphism in P. pilosa is caused by introgression with P. glaberrima.]

11. Levy, M., and D. A. Levin. 1971. The origin of novel flavonoids in Phlox allotetraploids. Proceedings of the National Academy of Sciences 68: 1627-1630. [Flavonoids not found in diploid progenetors were identified from their tetraploid products, and a mechanism

is presented to explain the origin of these hybrid specific compounds.]

12. Mirov, N. T. 1956. Composition of turpentine of lodgepole X jack pine hybrids. Canadian Journal of Botany 34: 443-457. [The composition of terpenoid constituents in parent trees, putative natural hybrids, and a synthetic hybrid are determined. The putative hybrids are shown to be chemically similar to the synthetic hybrid on the basis of these constituents.]
13. Mooney, H. A., and W. A. Emboden, Jr. 1968. The relationship of terpene composition, morphology and distribution of populations of Bursera microphylla (Burseraceae). Brittonia 20: 44-51. [The variation in morphology and chemistry is explained by the distribution pattern and the presence or absence of adjacent populations of other species of Bursera.]
14. Seeligmann, P., and R. E. Alston. 1967. Complex chemical variation and the taxonomy of Hymenoxys scaposa (Compositae). Brittonia 19: 205-211. [Chemical patterns were found to be geographically consistent, but not related to the morphology or taxonomy of this group of closely related species. Thus, the authors conclude that one variable species is involved.]
15. Wall, J. R., and T. W. Whitaker. 1971. Genetic control of leucine aminopeptidase and esterase isozymes in the interspecific cross Cucurbita ecuadorensis X C. maxima. Biochemical Genetics 5: 223-229. [Similar, but not identical, proteins were found in each species, and each enzyme was found to be under the control of a codominant allele.]

D. Studies concerning the influence of environment on phytochemicals.

1. Abrahamson, W. G., and O. T. Solbrig. 1970. Soil preferences and variation in flavonoid pigments in species of asters. Rhodora 72: 251-263. [Environmentally induced variation in alcohol soluble secondary compounds was considerably greater than that attributed to genetic differences in asters which showed similar soil preferences.]
2. Adams, R. P. 1970. Seasonal variation of terpenoid constituents in natural populations of Juniperus pinchotii Sudw. Phytochemistry 9: 397-402. [Summer to winter variations in the relative composition of volatile terpenoids was observed. Summer collections were found to be more variable than winter collections.]
3. Ball, G. A., Jr., E. O. Beal, and E. A. Flecker. 1967. Variation of chromatographic spot patterns of two species of clonal plants grown under controlled environmental conditions. Brittonia 19: 273-279. [Chromatographic spot variation observed in clones of Spirodela grown under a variety of cultural conditions are ascribed to environmental variation.]
4. Parks, C. R., S. S. Sandhu, and K. R. Montgomery. 1972. Floral pigmentation studies in the genus Gossypium. IV. Effects of different growing environments on flavonoid pigmentation. American Journal of Botany 59: 158-164. [Different growing regimes had little effect on floral flavonols in several lines from each of the four species studied, but evidence is presented that leaf constituents may be less stable.]

Section D. CHROMATOGRAPHY--BASICS AND DEFINITIONS

Chromatography* is a family of separation techniques of great resolving power, and often of considerable complexity, used to separate a wide assortment of molecules.

A. Some physical properties of molecules which make chromatography possible are primarily the following.
   1. Adsorption is the tendency for a molecule and a finely divided solid to become attached.
   2. Solubility is the tendency of a molecule to dissolve in a liquid.
   3. Volatility is the tendency of a molecule to evaporate.
      Two of these properties of molecules (or one repeated twice) are exhibited when the chromatographic mixture is placed in a moving or dynamic experimental situation.

B. Interactions found in the chromatographic separation.
   1. Partition is the tendency of a substance to distribute itself between two non-mixing liquids, a liquid and a solid, a liquid and a gas, etc.
   2. Phases are the combinations of the solids, liquids and gases between which the substance being separated is distributed.
      In actual practice a moving phase passes over a stationary phase and is, in as far as it is possible, in equilibrium with it. The phases are chosen so that the components in the mixture move at varying rates. Chromatographic separations are possible because of these differences in migration rates.

C. Major chromatographic systems.
   1. Principles.
      a. Partition chromatography involves a moving liquid phase over a stationary liquid film on a suitable support.
      b. Adsorption chromatography involves a moving liquid and a solid adsorbing surface (not a support for a liquid film).
      c. Gas-liquid-partition chromatography (gas chromatography or GLC) involves a moving gas and a stationary liquid film.
         Note: In many cases involving the commonly used methods of chromatography (paper, TLC or column), it is not readily possible to determine whether partition or adsorption phenomena are taking place. In fact, partition and adsorption events are often co-occurring.
   2. Practices.
      a. Paper chromatography involves a moving liquid and paper as an adsorbent or support.
      b. Thin-layer chromatography involves a moving liquid and a finely divided solid as the adsorbing or support surface. Structural support is provided by a glass plate to which the finely divided solid is bound.
      c. Column chromatography involves a moving liquid which passes through a glass column packed with a finely divided solid adsorbent.

D. More definitions associated with chromatographic separations.
   1. Adsorbent. Stationary phase in adsorption chromatography.

*Discussion of chromatography is taken from Bobbitt et al. (1968), and it is recommended that the student refer to that readable and concise treatment.

2. Bands. Substances separating and moving through a column in column chromatography.
3. Development. Movement of the moving phase over the adsorbent or support.
4. Elution. Washing the separated substances from the adsorbent or support.
5. Elution analysis. Analysis of a mixture by progressive elution from a column.
6. Effluent. The liquid which flows out of a column.
7. Gel electrophoresis. A type of chromatography in which the solutes migrate in an electrical field.
8. Gradient elution analysis. As in elution analysis, but solvents are changed during development.
9. Irrigant (solvent). The moving phase in paper, TLC or column chromatography.
10. R_f value. The ratio derived in paper and TLC chromatography by dividing the movement of the separated substance by the total movement of the solvent (solvent front).
11. Solutes (collectively-sample). Substances being separated.
12. Spots. Separated substances appearing on paper or TLC chromatograms.
13. Spray reagent. Reagent sprayed on a paper or TLC chromatogram to aid in visualization.
14. Visualization. Treatment of a chromatogram which lends spots visible.

Section E. SOME COMMENTS ON PROTEIN ANALYSIS

While the importance of protein analyses in chemotaxonomy has been pointed out, a discussion of methods of protein analysis is beyond the scope of this general review. The data obtained by electrophoretic separations of crude seed or other protein extracts are used as supportive evidence in systematic interpretations, much as paper chromatograms of unknown phenolic mixtures are used. Various degrees of analytical refinements may be carried out in these studies, ranging from simple purifications of the crude extract to enzyme purifications or even "fingerprinting" of a specific protein from different taxa.

Another major approach to protein analysis is serology. Serological reactions occur when foreign proteinaceous (usually, but not necessarily) substances, antigens, are injected into an animal, and these antigens elicit the formation of other presumably proteinaceous substances, antibodies, in that host animal. The antibodies precipitate (agglutinate) or otherwise affect the antigen. The host is usually a domestic animal, often a rabbit. In the most common serological test, the quantitative precipitin test, aliquots of the antigen extract are mixed in varying dilutions with the antibody preparation (antiserum). The amount of precipitate produced is proportional to the "strength" of the reaction; that is, the greater the similarity of the test substance to the antigen, the stronger (more precipitate produced) is the reaction. Immunoelectrophoresis is a refinement of the precipitin test, and allows for the generation of qualitative data. Extracts are prepared and subjected to agar-gel electrophoresis. Antiserum is added to a trough parallel to the electrical tract of antigen separation, and the antigens and antibodies diffuse into the agar. Where the appropriate antigens and antibodies meet, stabilized precipitates are formed which are detectable by various means. The precipitates become visible as arcs. By this means the reactions of more than one protein in the antigen mixture may be studied. This is not possible by the straight precipitin test. Recently, data from this method have been applied increasingly to the solution of taxonomic problems.

CHEMICAL EVIDENCE LITERATURE

Abbott, H. C. de S. 1886. Certain chemical constituents of plants considered in relation to their morphology and evolution. The Botanical Gazette 11: 270-272.

Alston, R. E., and B. L. Turner. 1963. Biochemical Systematics. Prentice-Hall, Englewood Cliffs, New Jersey.

Bate-Smith, E. C. 1948. Paper chromatography of anthocyanins and related substances in petal extracts. Nature 161: 835-838.

Beale, G. H., J. R. Price, and V. C. Sturgess. 1941. A survey of anthocyanins VII. The natural selection of flower color. Proceedings of the Royal Society of London (B) 130: 113-126.

Bobbit, J. M., A. E. Schwarting, and R. J. Gritter. 1968. Introduction to Chromatography. Reinhold Book Corporation. New York.

Chester, K. S. 1937. A critique of plant serology (continued) Part II. Application of serology to the classification of plants and the identification of plant products. Quarterly Review of Biology 12: 165-190.

Consden, R., A. H. Gordon, and A. J. P. Martin. 1944. Qualitative analysis of proteins: A partition chromatographic method using paper. The Biochemical Journal 38: 224-232.

Gell, P. G. H., J. G. Hawkes, and S. T. C. Wright. 1960. The application of immunological methods to the taxonomy of species within the genus _Solanum_. Proceedings of the Royal Society of London (B) 151: 364-383.

Gibbs, R. D. 1958. Chemical evolution in plants. The Journal of the Linnean Society of London. Botany 56: 49-57.

Harborne, J., B., and T. Swain (eds.). 1969. Perspectives in Phytochemistry. Academic Press. New York.

________, D. Boulter, and B. L. Turner (eds.). 1971. Chemotaxonomy of the Leguminosae. Academic Press. New York.

Hegnauer, R. 1962, 1963, 1964, 1966, 1969. Chemotaxonomie der Pflanzen. Vols. 1-5. Birkhäuser Verlag. Basel.

Heywood, V. H. (ed.). 1971. The Biology and the Chemistry of the Umbelliferae. Academic Press. New York.

Kooiman, P. 1969. The occurrence of asperulosidic glycosides in the _Rubiaceae_. Acta Botanica Neerlandica 18: 124-137.

________. 1970. The occurrence of iridoid glycosides in the _Scrophulariaceae_. Acta Botanica Neerlandica 19: 329-340.

Lawrence, W. J. C., J. R. Price, G. M. Robinson, and R. Robinson. 1939. The distribution of anthocyanins in flowers, fruits and leaves. Philosophical Transactions of the Royal Society of London (B) 230: 149-178.

Mabry, T. J., R. E. Alston, and V. C. Runeckles (eds.). 1968. Recent Advances in Phytochemistry. Vol. 1. Appleton-Century-Crofts. New York.

McNair, J. B. 1935. Angiosperm phylogeny on a chemical basis. Bulletin of the Torrey Botanical Club 62: 515-532.

________. 1945. Some comparisons of chemical ontogeny with chemical phylogeny in vascular plants. Lloydia 8: 145-169.

Mirov, N. T. 1967. The Genus _Pinus_. The Ronald Press Company. New York.

Reinhold, L., and Y. Liwschitz (eds.). 1968. Progress in Phytochemistry. Vol. 1. Interscience Publishers. New York.

Schäfer, C. 1964. Biogenetic and chemotaxonomic considerations about bisbenzylisoquinoline alkaloids, p. 4-7. See: H. R. Arthur (ed.), Symposium on Phytochemistry. Hong Kong University Press.

Swain, T. (ed.). 1963. Chemical Plant Taxonomy. Academic Press. New York.

________. 1966. Comparative Phytochemistry. Academic Press. New York.

Chapter 14. PHYSIOLOGICAL AND ULTRASTRUCTURAL EVIDENCE

Physiological evidence, including the biochemical, developmental and biophysical aspects, provides a basic set of data which are of increasing significance to taxonomy today. This is particularly true for those metabolic characters related to chemical systems, biosynthetic pathways, and primary and secondary products. Chemical characters are used to distinguish hybrids and species as well as to provide evidence in the characterization of larger taxa; e.g., Hatch-Slack pathway of malic acid carboxylation is possessed by many tropical species of Panicoid grasses, and is found in a number of tropical species in the Cyperaceae, Amaranthaceae, Chenopodiaceae, and Portulacaceae. According to Laetsch (1969) these species are also characterized by having a bundle sheath and dimorphic chloroplasts. (See Chapter 13 on chemical evidence.)

As more is learned about plant development, growth features should be of greater value to systematics, particularly in tropical species in which many types of growth patterns and mechanisms occur (see Chapter 6). According to Tomlinson and Gill (1973) "detailed growth analysis is scarcely begun but offers unlimited scope for future work." An adequate and applicable classification of these patterns would immeasurably aid our present understanding of tropical taxonomy. Features of dormancy and germination have been used in identification and characterization of taxa, but a systematic study of these characteristics should result in a more effective and efficient means of identification of seeds and seedlings, as well as provide a sounder basis for classification (see Chapter 6). Development of individual flowers and inflorescences is used in delimitation of larger taxa, e.g., the centrifugal stamens in the Dilleniidae and centripetal stamens in the Rosidae (see Chapter 6). The ovular developmental patterns, gametogenesis, sporogenesis, embryogeny, embryogenesis, and endosperm characters have been used as diagnostic characters in distinguishing larger taxa, e.g., monosporic megasporogenesis in the Onagraceae and helobial endosperm development in the Alismatidae (see Chapters 8 and 9).

The physical processes of diffusion, osmosis, plasmolysis, and imbibition are so wide-spread that they are of no apparent significance to taxonomy. The following biophysical features, however, might be of some taxonomic significance primarily as structural or chemical evidence: uptake against a concentration gradient in many species; selective absorption, as selenium by species of _Astragalus_; accumulation of bases, as in the leaves of most members of the Myrtaceae; or gas exchange features, as through the pneumatophores of some species of _Ludwigia_ or through the epidermis of submerged species of _Potamogeton_. The types of conduction mechanisms or transport flow systems, when worked out, should show taxonomic relationships.

Mechanical functions are so closely related to form ("structure is function") that morphological rather than physiological features have long been used by systematists. The amount and distribution of sclerenchyma and collenchyma are definitely related to supportive systems. Anchorage is related to type of root system, as tap, fibrous, or fascicled. Protection is afforded by stinging trichomes, spines, thorns, prickles, and silicic and oxalic acid crystals. All of these characters are treated as structural (see Chapter 6).

Reproduction, involving reproductive features and pollination-fertilization types, has been used taxonomically from a broad biological standpoint (see Chapters 6 and 12). In general, physiological adaptations are usually expressed as structural characters in taxonomic work.

Even though most physiological evidence is treated by the chemotaxonomist as chemical characters and by the traditional systematist as morphological, anatomical, or cytological, the taxonomist should be aware of new developments in plant physiology. Particularly they should have an awareness of research in the field of physiological mechanisms such as those of, or in, organ initiation and development, mitotic and meiotic control, active absorption, mineral accumulation, and regulatory controls. These mechanisms, which will usually be expressed as chemical or physico-chemical systems, can be of taxonomic value in the definition and delimitation of family and higher taxa as well as providing evidence of fundamental relationships. Developmental response studies will determine the reliability and meaningfulness of many of the diagnostic characters used in classification and identification.

Since physiological evidence has been so disjoined and dismembered in its treatment by taxonomists (and by the authors of this text!) the functional features described in this book are listed here with the chapters in which they occur:

Structural expressions of physiology in Chapter 6 on morphology and Chapter 7 on anatomy.
Developmental aspects of physiology in Chapter 8 on palynology and Chapter 9 on embryology.
Biophysical and reproductive aspects of physiology in Chapters 12 and 15 on reproductive biology and ecology, respectively.
Chemical aspects of physiology in Chapter 13 on chemotaxonomy.

Ultimately, physiological conditions and processes will be directly correlated with structural characteristics at the cellular level. Although cytological evidences have historically been derived mostly from the structure and behavior of chromosomes, i.e., karyology (see Chapter 10), the widespread use of the electron microscope has shown the cell cytoplasm to be a highly organized region composed of many structurally complex and interrelated membranous organelles. Broadly based comparative ultrastructural information on the plastids, endoplasmic reticulum, Golgi apparatus, mitochondria, and vacuoles is available for very few species. In addition, the fine structure of plant cell walls (including pitting) and pollen grain exines (Chapter 8) are areas of increasing importance.

An example of the significance of the electron microscope in providing additional characters to angiosperm classification is illustrated by the investigation of Behnke (1971, 1972, 1973) on cells of the phloem tissue. It has been shown that sieve tube elements can be separated on the basis of the contents of their plastids as observed with the EM. Sieve tube elements contain plastids of either the S-type or the P-type. S-type plastids store starch only whereas plastids of the P-type contain protein alone or protein and starch. It has been demonstrated, furthermore, that there may be uniformity of plastid type within taxonomic groups as high as the rank of class. Monocotyledons are thus far known to possess only P-type plastids. The taxonomic placement of a plant that varies in plastid type from remaining members of the taxon to which it has been assigned is suspect. As more plants are studied the phylogenetic trends in plastid morphology may be clarified.

Although the physiological and ultrastructural characteristics of vascular plants have not proven, and may never prove, to be of the same taxonomic and evolutionary significance as in the Thallophyta (see Klein and Cronquist, 1967), data and information derived from these fields should be critically evaluated in making taxonomic judgments. The synthesis of data from all fields represents one of the real challenges to the systematic botanist.

PHYSIOLOGICAL AND ULTRASTRUCTURAL LITERATURE

Behnke, H. D. 1971. Sieve-tube plastids of Magnoliidae and Ranunculidae in relation to systematics. Taxon 20: 723-730.

________. 1972. Sieve-tube plastids in relation to angiosperm systematics--An attempt towards a classification by ultrastructural analysis. Botanical Review 38: 155-197.

________. 1973. Sieve-tube plastids of Hamamelidae. Electron microscope investigations with special reference to Urticales. Taxon 22: 205-210.

________, and B. L. Turner. 1971. On specific sieve-tube plastids in Caryophyllales. Further investigations with special reference to the Bataceae. Taxon 20: 731-737.

Johnson, Sister Clement, and W. V. Brown. 1973. Grass leaf ultrastructural variations. American Journal of Botany 60: 727-735.

Klein, R. M., and A. Cronquist. 1967. A consideration of the evolutionary and taxonomic significance of some biochemical, micromorphological, and physiological characters in the thallophytes. Part I. Chemical, micromorphological and functional phylogenetic criteria. The Quarterly Review of Biology 42: 108-210.

Laetsch, W. M. 1969. Relationship between chloroplast structure and photosynthetic carbon-fixation pathways. Science Progress Oxford 57: 323-351.

Levitt, J. 1969. Introduction to Plant Physiology. C. V. Mosby Company. St. Louis, Missouri.

Plant Physiology. 1926+. Baltimore: Official publication of American Society of Plant Physiologists.

Steward, F. C. (ed.). 1959. Plant Physiology. (Several volumes). Academic Press, Inc. New York.

Tomlinson, P. B., and A. M. Gill. 1973. Growth habits of tropical trees: some guiding principles. In: Tropical Forest Ecosystems in Africa and South America: A Comparative Review. Meggers, B. J., E. S. Ayensu, and W. D. Duckworth (eds.). Smithsonian Institution Press. Washington, D. C.

Chapter 15. ECOLOGICAL EVIDENCE

Ecology is basic to taxonomy in understanding (1) the distribution of taxa and composition of floras, (2) the genetic and phylogenetic relations of taxa, and (3) the variation in populations and evolutionary adaptations. The three major sections in this chapter are: A. HABITATS which are classified primarily according to edaphic and vegetation relations; B. ENVIRONMENTAL RELATIONSHIPS which are treated as biotic, abiotic, spatial, and temporal; and C. ADAPTIVE FEATURES which presumably are or have been related to environmental composites or mosaics of factors. Distribution principles, ranges, and biomes are described in Chapter 16 on geography; communities are treated in Chapter 17 on field studies.

Section A. HABITATS

Habitat is the "home" of the individual organism or its native environment. Habitat descriptions should be brief but as comprehensive as possible. For pioneer and early successional species the edaphic (A-F and M-P below) and the geomorphic features (Chapter 17) should be indicated. For late successional and climatic species the community or vegetation type (G-L) should be identified since the vegetation represents the mosaic of environmental factors at a given time in a given place. Regional climatic data can usually be obtained from local weather stations, if the locality is given for a collection. Locality data is a basic part of all collection information. The geomorphic feature will give clues to microclimatic conditions. This classification system provides one word adjectives for the major habitats in North America north of Mexico. The terms were selected or adapted from Clements (1902), Gray (1967), and Dansereau (1957).

A. Water - Hydrophytes (hydor: water)
- Benthophilous. (benthos: sea-bottom) Dwelling submerged in and attached to the bottom of estuaries, sounds, seas.
- Crenophilous. (krene: spring) Dwelling in springs and seepages.
- Haloplanktophilous. (hals, halos: salt + planktos: free-floating) Dwelling on the surface of salt water.
- Limnophilous. (limne: lake, pool) Dwelling in lakes.
- Limnoplanktophilous. (limne: lake, pool + planktos: free-floating) Dwelling on the surface of fresh water.
- Oceanophilous. (oceanus*: ocean) Dwelling in the ocean or marine waters.
- Potamophilous. (potamos: river) Dwelling in rivers.
- Rheophilous. (rheos: stream, current) Dwelling in running water.
- Stasophilous. (statos: standing, fixed) Dwelling in stagnant water.
- Tiphophilous. (tiphos: standing water, pool) Dwelling in ponds or pools.

B. Clay or Mud - Argillophytes (argilla*: white clay, potter's clay)
- Limiphilous. (limus*: mud) Dwelling on mud.
- Pelophilous. (pelos: clay) Dwelling on clay.
- Pelochthophilous. (pelos: clay + chthon: soil, earth) Dwelling on mud banks.

*Latin, others are Greek.

C. <u>Sand</u> - Psammophytes (psammos: sand)

Aigialophilous. (aigialos: strand, sea-shore) Dwelling on beach strands or sand flats.

Amathophilous. (amathos: sand, sandy soil) Dwelling on sand hills, plains, or scarps.

Anemodophilous. (anemos: wind) Dwelling in blowouts.

Cheradophilous. (alluvium, silt, detritis) Dwelling on wet sandbars.

Syrtidophilous. (syrtos: washed along by a stream, alluvial) Dwelling on dry sandbars.

Thinophilous. (this, thinos: beach, shore, sand-heap) Dwelling on sand dunes.

D. <u>Rock</u> - Petrophytes (petra: rock)

Actophilous. (acta*: seashore, promontory) Dwelling on rocky seashores.

Calciphilous. (calyx, calcis*: lime) Dwelling on limestone.

Chasmophilous. (chasma: chasm, gulf) Dwelling in rock crevices or joints.

Cremnophilous. (kremnos: precipice, cliff) Dwelling on cliffs.

Granitophilous. (graniticus: pertaining to plants growing on granite. Mod. Latin) Dwelling on granite.

Gypsophilous. (gypsos: white plaster, gypsum) Dwelling on chalk, or alkali or limestone.

Halophilous. (hals, halos: sea, salt) Dwelling on salt flats.

Lithophilous. (lithos: stone) Dwelling on flat rocks.

Petrodophilous. (petra: rock) Dwelling in boulder fields or flats.

Phellophilous. (phelleus: stony ground) Dwelling on stony ground.

Siliciphilous. (silex, silicis*: any hard stone, flint) Dwelling on quartz or sandstone.

E. <u>Humus</u> - Saprophytes (sapros: decayed, rotten)
Oxygeophytes (oxy: sour + geo: earth)

Coprophilous. (kopros: dung) Dwelling in or on dung.

Eurotophilous. (euros, eurotos: mold, decay) Dwelling in duff or humus.

F. <u>Bog</u> - Oxylophytes (oxys: sour)

Sphagnophilous. (sphagnos: a kind of moss) Dwelling in peat or sphagnum.

G. <u>Marsh</u> - Helophytes (helos: marsh)

Helophilous. (helos: marsh) Dwelling in fresh water marshes.

Limnodophilous. (limnodes: marshy) Dwelling in salt marshes.

H. <u>Grassland or Prairie</u> - Psilophytes (psilos: bare, naked)

Coryphophilous. (koryphe: head, top) Dwelling in alpine meadows.

Leimophilous. (leimon: meadow) Dwelling in low grasslands, meadows, or coastal prairies.

Pedionophilous. (pedion: flat, open country) Dwelling in the plains, or short grasslands.

Psilophilous. (spilos: bare, naked) Dwelling in prairies.

I. <u>Savannahs and Open Woodlands</u> - Pyrophytes (pyr, pyros: fire) and Orgadophytes (orgas, orgados: field, meadow)

Orgadophilous. (orgas, orgado: field, meadow) Dwelling in open woodlands.
Pyrophilous. (pyr, pyros: fire) Dwelling in burned-over areas or soils, savannahs.

J. Scrub and Shrub - Fruticiphytes (frutex, fruticis: shrub)
Aithalophilous. (aithales: evergreen) Dwelling in evergreen thickets.
Lochmodophilous. (lochme: thicket) Dwelling in dry thickets.
Ptenothalophilous. (ptenothales: "deciduous") Dwelling in deciduous thickets.
Sclerophilous. (skleros: tough, hard) Dwelling in chaparral.
Xeropoophilous. (xeros: dry + poa: grass) Dwelling in heath balds, thickets, or pocosins.

K. Forests - Dendrophytes (dendron: tree)
Aiphyllophilous. (aiphyllos: evergreen leaf) Dwelling in broad-leaved evergreen forests.
Ancophilous. (ankos: mountain glen, valley) Dwelling in canyon or cove forests.
Conophorophilous. (konos: coen + phoras: bearing) Dwelling in coniferous forests.
Helogadophilous. (helos: marsh, swamp + orgas, orgados: meadow, field) Dwelling in swamp forests.
Hylophilous. (hyle: wood, forest) Dwelling in forests.
Ombrohylophilous. (ombros: rainstorm + hyle: wood, forest) Dwelling in rain forests.
Xerohylophilous. (xeros: dry + hyle: wood, forest) Dwelling in dry, deciduous forests.

L. Tundra - Crymophytes (crymos: icy cold, chill)
Crymophilous. Dwelling in polar barrens.

M. Ice or Snow - Psychrophytes (psychros: cold, frigid)
Psychrophilous. (psychros: cold, frigid) Dwelling in cold places.

N. Desert - Eremophytes (eremis: desert)
Chersophilous. (chersos: dryland) Dwelling in dry waste places.
Eremophilous. (eremis: desert) Dwelling in deserts.

O. Hill or Mountain - Lophophytes (lophos: crest, ridge) or Orophytes (oros: mountain)
Acrophilous. (akron: top, summit) Dwelling on peaks.
Ancophilous. (ankos: mountain glen, valley) Dwelling in canyons.
Lophophilous. (lophos: crest, ridge) Dwelling on hills or slopes.
Orophilous. (oros: mountain) Dwelling on mountains.
Pagophilous. (pagos: hill) Dwelling in foothills.

P. Disturbed - Anthropophytes (anthropos: man)
Agrophilous. (agros: field) Dwelling in cultivated fields.
Aletophilous. (aletes: wanderer) Dwelling on roadsides, waysides, railroads.
Chledophilous. (chledos: mud, dirt, debris) Dwelling in waste places or rubbish heaps.

Section B. ENVIRONMENTAL RELATIONSHIPS

Ecological characteristics as expressions of the environmental relationships of species should be determined or observed. Populations have tolerance limits for each of the factors; e.g., light, heat, moisture, nutrients. Even though it is practically impossible to determine experimentally the regime for any factor, i.e., temperature amplitude or tolerance limits, the environmental conditions for the best growth and reproduction should be noted by every observant field worker. The acuity of observation of plant behavior under different environmental conditions will determine, to a great extent, the degree of success in transplant and field variation studies and in the understanding of clinal relationships and flora or community dynamics; e.g., phenology, periodicity. Species should be characterized ecologically consciously and intelligently, if impossible experimentally, rather than instinctively or intuitively (see Ecology Exercise for ecological characteristics summary). This treatment is basically a classification of factors with definitions of the characteristics.

I. BIOTIC. Pertaining to the living environment.
   A. <u>Genetic</u>. Relations pertaining to hereditary relationships.
      1. Genetic. Pertaining to immediate gene relationships of organisms.
      2. Phenetic. Pertaining to the observable characteristics of organisms, not necessarily closely related genetically.
      3. Phyletic. Pertaining to racial relationships of organisms.
      4. Phylogenetic. Pertaining to racial history relationships of organisms.
   B. <u>Nutritional Relations</u>. Pertaining to nourishment of organisms.
      1. Autoparasite. Parasite living on a parasite; organism dependent upon a parasitic host for its nourishment; e.g., mistletoe on mistletoe.
      2. Autophyte (autotrophe). Plant that makes its own food; a self-nourishing plant.
      3. Hemiparasite. A partial parasite, usually chlorophyll bearing, one that obtains part of its nourishment from its host; e.g., <u>Comandra</u>, <u>Euphrasia</u>.
      4. Parasite. Plant that obtains its nourishment from a living plant, its host.
      5. Saprophyte. Plant that obtains its nourishment from dead organic matter.
   C. <u>Social Relations</u>. Pertaining to the gregariousness of individual organisms within populations; intrapopulational relationships; sociability. This classification of sociability is based on that of Braun-Blanquet (1932) which is difficult to apply without training by an individual who has had experience with the system.
      Soc. 5 Plants occurring in great crowds (pure populations).
      Soc. 4 Plants in small colonies, in extensive patches, or forming carpets.
      Soc. 3 Plants in small patches or cushions.
      Soc. 2 Plants grouped or tufted.
      Soc. 1 Plants growing singly.
   D. <u>Symbiotic Relations</u>. Pertaining to the environmental relationships between two or more species; interpopulational relationships.

1. Mutualism. Situation in which two species are dependent upon one another for survival or perpetuation of each; e.g., alga--fungus in the lichen, yucca-pronuba moth.
2. Commensalism. Situation in which one species helps another without harm or benefit to itself; e.g., heliophyte providing the shade for a sciophyte.
3. Neutralism. Seemingly no relationships between two or more species within a community.
4. Competition. The struggle or rivalry between two or more species for a common requirement or resource; e.g., moisture, light.
5. Amensalism. Situation in which one species definitely inhibits another without harm to itself; allelopathy; e.g., Juglans secretes substance that inhibits development of many species.
6. Parasitism. Situation in which one species definitely harms another by obtaining its nourishment directly from it.

E. Reproductive Relations. Pertaining to the increase in population.
1. Apomixis. Asexual types of reproduction involving either agamospermy or vegetative propagation.
2. Amphimixis. Sexual reproduction involving union of two gametes.

II. ABIOTIC. Pertaining to the physical environment.
(Relative degree of tolerance or amplitude in ecology is expressed by the prefixes "steno-", meaning narrow; "meso-", meaning middle; and "eury-", meaning wide; e.g., stenothermic - mesothermic - eurythermic refers to temperature, stenohydric - mesohydric - euryhydric refers to water, stenophotic - mesophotic - euryphotic refers to light.)

A. Climatic Relations.
1. Temperature Relations.
   Cryotherm or Hekistotherm. Plant growing in minimum heat, as in arctic-alpine zones; tundra plants.
   Microtherm. Plant growing in little heat, as in the cold temperate zones; boreal forest plants.
   Mesotherm. Plant growing in moderate heat, as in the warm temperate zones; deciduous hardwood plants.
   Megatherm. Plant growing in great heat, as in the tropical zone; tropical rainforest plants.
2. Moisture Relations.
   Hydrophyte. Plant growing in water either as free floating, submerged, floating and anchored, or emergent and anchored.
   Hygrophyte. Plant growing under highly humid conditions.
   Mesophyte. Plant growing under medium moisture conditions.
   Xerophyte. Plant growing under dry conditions.
3. Light Relations.
   Heliophyte. Plant growing in full sunlight.
   Sciophyte. Plant growing in the shade.

B. Edaphic Relations.
1. Soil Texture Relations.
   Psammophyte. Plant that grows in the sand.
   Pelosammophyte. Plant that grows in loam, sand and clay.
   Pelophyte. Plant that grows in clay.
   Saprophyte. Plant that grows on dead organic matter.
2. Soil Aeration Relations for Vascular Plants.

Aerobe. Organism whose roots require free oxygen.
Anaerobe. Organism whose roots live in absence of free oxygen.

3. Soil Nutrition Relations.
   Oligotrophe. Organism that lives in soil poor in mineral nutrients.
   Mesotrophe. Organism that lives in soil moderately supplied with mineral nutrients.
   Eutrophe. Organism that lives in soil abundantly supplied with mineral nutrients.
4. Soil pH Relations.
   Acidophyte. Plant living in acid soil, below pH 6.5.
   Basophyte. Plant living in basic soil, above pH 7.5.
   Neutrophyte. Plant living in circumneutral soil, pH 6.5-7.5.
   Calciphyte. Plant living on basic soil or rock abundantly supplied with calcium ions.
   Halophyte. Plant living on basic soil abundantly supplied with chloride ions.

III. SPATIAL RELATIONS.

A. Numerical Relations.

1. Abundance. An estimation of the plentifulness of individuals of a species within a region. One scale of abundance is 1. very rare, 2. Rare, 3. Infrequent, 4. Abundant, 5. Very abundant.
2. Density. An actual count and determination of the number of individuals on a unit area basis; average number of individuals per area sampled.
3. Frequency. The percentage of sample plots in which a species occurs. Frequencies can be grouped into classes based on percentages of sample plots in which species occur such as: A 1-20%, B 21-40%, C 41-60%, D 61-80%, E 81-100%.

B. Distributional Relations.

1. Random Individuals. Dispersed irregularly; without definite pattern.
2. Clumped Random. Aggregates of individuals dispersed irregularly without definite pattern.
3. Uniform. Individuals dispersed in a regular and definite pattern.

C. Areal Relations.

1. Cover. Amount of horizontal space shaded or occupied by an individual. The following cover classes were adapted from Braun-Blanquet (1932) by Helmut Lieth:
   R - one specimen in area, few if any nearby
   + - Cover less than 1%
   Cov. 1 - Cover 1-5%
   Cov. 2 - Cover 5-12.5%
   Cov. 3 - Cover 12.5-25%
   Cov. 4 - Cover 25-50%
   Cov. 5 - Cover 50-100%
2. Basal Area. Number of square feet of cross-sectional surface of individual tree species determined at breast height (4.5') per unit area such as an acre. Foresters use this method of determining the amount of timber per acre of land and usually measure the d.b.h. (diameter breast height) and then convert the

d.b.h.'s to area using standard tables. Basal area is also the amount of ground cover of bunch grasses or rosette plants.

3. Size classes. Refers to the number of individuals of certain diameters within an area or stand. Within a stand of trees, i.e., there might be a group of individuals with a d.b.h. of 4-6", another group with a d.b.h. at 12-15", and still another group with a d.b.h. of 22-24" which would be three size classes within that stand.
4. Stratification. Within most woody communities the species occur in layers with those at the top forming the 1. Canopy, those in a layer just below the canopy forming the 2. Subcanopy, and just below that is a 3. Shrub layer, and just above the ground is a 4. Herb layer, and frequently beneath the herbs is a 5. Moss-lichen layer.

IV. TEMPORAL (TIME) RELATIONS.

A. Life (Life cyclic; sexual)

Adult Sporophyte
Embryo Sporangium
Zygote Sporocyte
Syngamy____Meiosis
Gamete Spore
Sex Organ
Adult Gametophyte

B. Phenology (Annual cyclic). Study of periodicity in plants as related to climatic events; e.g., time of leafing, flowering, fruiting, budding.
   1. Vernal. Phenomena that occur in spring.
   2. Aestival. Phenomena that occur in summer.
   3. Autumnal. Phenomena that occur in fall.
   4. Hibernal. Phenomena that occur in winter.

C. Periodicity (Daily or seasonal cyclic). Diurnal or seasonal rhythms or fluctuations; e.g., temperature, light, moisture regimes or amounts.
   1. Thermoperiodism. Responses of plants to periodic changes in temperature, usually considered on a diurnal basis.
   2. Photoperiodism. Responses of plants to the length of the diurnal light and dark periods, as regards flower initiation, tuberization, leaf abscission, etc. Long-day plants initiate flowers when the day length is greater than a critical minimum (usually 11-13 hours); short-day plants do so when the day length is less than a critical maximum (usually 10-14 hours; day-neutral plants flower independently of day length.
   3. Hydroperiodism. Responses of plants to periodic changes of complete soil saturation by water.

D. Succession (Generation and Community cyclic). Changes in time of communities in a given area from pioneer to climax communities.
   Invaders
   1. Pioneer (Species or Communities). Plant invaders of an area previously uninhabited by plants or into an area disturbed by man.
   2. Transient (Species or Communities). Later invaders into pioneer and subsequent communities prior to climax.

Sere. A group of plant communities that successively occupy the same area from the pioneer to a mesic climax.

1. Hydrosere. A sere in which the pioneers invade water.
2. Halosere. A sere in which the pioneers invade salt water or salt-flats.
3. Psammosere. A sere in which the pioneers invade sand.
4. Pelosere. A sere in which the pioneers invade clay or mud.
5. Lithosere. A sere in which the pioneers invade rock.
6. Subsere. A sere in which the pioneers invade an area disturbed by man; e.g., pasture, old field.

Section C. ADAPTIVE FEATURES

Adaptive features within a flora or community can give clues as to the (1) origin of the flora or community, (2) migration patterns in the area, (3) evolutionary trends within populations, (4) and indications of present or past climatic trends. Life form and leaf size analyses, when compared with norms determined for similar communities elsewhere, can indicate climatic shifts or stability; diaspore and pollination analytical studies can give hints as to origin and evolution of the flora or community; and fidelity and vitality percentages can indicate the youth or maturity and migrational stability within a flora or community. Studies of other adaptive features within a flora or community; e.g., those associated with floral morphology could be of equal evolutionary significance.

I. LIFE FORMS (Lf). Physiognomic characteristics. Based on Raunkiaer (1934).

Life Form	Symbol	Description
Therophytes	Th	Annuals: from seed germination to seeds in one year.
Geophytes	G	Perennials: perennating organs (bulbs, tubers, rhizomes, corms) underground and protected; plants die above ground
Hemicryptophytes	H	Perennials or biennials: perennating buds very close to ground surface (tussock and rosette plants)
Chamaephytes	CH	Perennials: perennating bodies above ground (coarse herbs or shrubs with buds less than 0.5 m above ground).
Phanerophytes	P	Perennials: perennating buds more than 0.5 m above ground; trees (t), shrubs (s), woody vines (v).
Epiphytes	E	Perennials: borne on other plants or structures high above ground (no root contact with soil).
Succulents	S	Perennials, rarely annuals: either stems, leaves or both very fleshy.
Parasites	Pa	Perennials or annuals: non-chlorophyllous plants dependent upon other plants for all nourishment or chlorophyllous plants dependent upon other plants for water and mineral salts.

Life Form	Symbol	Description
Saprophytes	Sa	Annuals or perennials: Living on dead organic matter.

II. DIASPORE TYPES (Dia). Dispersal mechanisms. Based on Van der Pijl (1969) and Ridley (1930).

Diaspore Type or Disseminule	Symbol	Dispersal Agent - Adaptation
Sporochore	Spo	Wind--minute or dust seeds, fruits, or spores.
Pterochore	Pte	Wind--winged seeds or fruits.
Pogonochore	Pog	Wind--plumed seeds or fruits.
Cyclochore	Cyc	Wind--tumbling infructescences or plants.
Epizoochore	Epz	Animal--barbed, hooked or viscid seeds or fruits.
Synzoochore	Syz	Animal--intentionally buried or stored seeds or fruits.
Endozoochore	Enz	Animal--visceral (non-digestible) seeds or fruits.
Ballochore	Bal	Mechanical--expulsion or jactitation seeds or fruits.
Barochore	Bar	Gravity--heavy seeds or fruits.
Hydrochore	Hyd	Water--floating seeds, fruits, or vegetative parts.
Anthropochore	Ant	Man--none.
Atelechore	Ate	No special adaptation--on the spot dispersal.
Polychore	Pol	More than one agent--could be wind and animal adapted fruits on same plants.

Note: Wind means total Wind Dispersed and Animal means total Animal Dispersed Diaspores in Spectrum.

III. FIDELITY. Indicates the degree to which a species is restricted to a particular type of community. In the following classes Fid. 3 and 4 would indicate that a species is a characteristic of that community and Fid. 5 would indicate that the species is an indicator of that community.

Fid. 1. Strangers, appearing accidentally.

Fid. 2. Indifferents, without a decided affinity for any type of community.

Fid. 3. Preferents, present in several types of communities but predominantly in one type.

Fid. 4. Selectives, usually found in one type of community but occasionally found in others.

Fid. 5. Exclusives, found in one type community only, rarely in another.

IV. VITALITY. Indicates the vigor and reproductive capacity of a species within a community. The following classes have been recognized by most ecologists:

Vit. 1. Ephemeral, adventive and not reproducing.
Vit. 2. Vegetative apomicts, reproducing asexually by vegetative propagules but not completing a sexual cycle of any sort.
Vit. 3. Amphimicts, regularly reproducing sexually or occasionally by agamospermy.

V. LEAF SIZE. Leaf size characteristics seem to be related to different environments. The following classification is adapted from Raunkiaer (1934): (Note: these terms have no morphological significance except size.)

L. S. 1. Leptophyll, to 25 sq. mm.
L. S. 2. Nanophyll, to 225 sq. mm. 25 x 9
L. S. 3. Microphyll, to 2,025 sq. mm. 25×9^2
L. S. 4. Mesophyll, to 18,222 sq. mm. 25×9^3
L. S. 5. Macrophyll, to 164,025 sq. mm. 25×9^4
L. S. 6. Megaphyll, larger than 164,025 sq. mm. 25×9^5

VI. POLLINATION.

A. Pollination Types. Classification based on agents by which pollen is transferred from the pollen sac or anther to the ovule or stigma.

1. Anemophily. Pollinated by wind.
2. Anthropophily. Pollinated by man.
3. Cantharophily. Pollinated by beetles.
4. Cheiropterophily. Pollinated by bats.
5. Entomophily. Pollinated by insects.
6. Hydrophily. Pollinated by water.
7. Hymenopterophily. Pollinated by bees.
8. Malacophily. Pollinated by snails or slugs.
9. Micromelittophily. Pollinated by small bees.
10. Myiophily. Pollinated by diptera.
11. Myrmecophily. Pollinated by ants.
12. Necrocoleopterophily. Pollinated by carrion beetles.
13. Ornithophily. Pollinated by birds.
14. Phalaenophily. Pollinated by moths.
15. Psychophily. Pollinated by butterflies.
16. Sapromyiophily. Pollinated by carrion or dung flies.
17. Sphingophily. Pollinated by hawk moths and nocturnal lepidoptera.

B. Flower-Pollinator (Visitor) Relationships. Adapted from Faegri and van der Pijl (1966)

1. Allophily. Flower with no morphological adaptations for guiding pollinators, can be utilized by non-adapted pollinators.
2. Euphily. Flower strongly adapted for utilization by specialized pollinators.
3. Hemiphily. Flower imperfectly adapted for utilization by intermediately specialized pollinators.
4. Monophily. Euphilic flower utilized or visited by a single or closely related species of pollinator.
5. Oligophily. Hemiphilic flower utilized or visited by some related taxa of pollinators.

6. Polyphily. Hemiphilic flower utilized or visited by many different taxa of pollinators.

C. Pollinator (Visitor)-Flowering Species Relationships. Adapted from Faegri and van der Pijl (1966).

1. Allotropy. Pollinator poorly adapted for utilization of flowers; food obtained from flowers form part of a mixed diet.
2. Dystropy. Pollinator (visitor) unadapted or shows no relationship to flower organization, frequently destructive, but may cause pollination.
3. Eutropy. Pollinator fully adapted for utilization of flower, taking main food from flower.
4. Hemitropy. Pollinator has intermediate degree of specialization for utilization of flower.
5. Monotropy. Pollinator strongly adapted for utilization of one flower or flower from closely related species.
6. Oligotropy. Pollinator, with intermediate degree of specialization for utilization of flower, that visits flowers of related taxa (species).
7. Polytropy. Pollinator, with intermediate degree of specialization for utilization of flower, that visits many different taxa of flowers (species).

D. Cleistogamy. Classification based primarily on environmental cleistogamy. Adapted from Uphof (1938).

1. Cleistogamy. Self-pollination within closed flowers, as in *Viola*.
2. Edaphocleistogamy. Self-pollination in subterranean flowers, as in *Commelina*.
3. Hydrocleistogamy. Self-pollination in closed flowers, which are normally expanded, due to submersion, as in *Mayaca*.
4. Hygrocleistogamy. Self-pollination in closed flowers, which are normally expanded, due to great fog density, as in *Liparis*.
5. Oligotropic Cleistogamy. Self-pollination in closed flowers, which are normally expanded, due to low soil nutrients, as in *Specularia*.
6. Psychrocleistogamy. Self-pollination in closed flowers, which are normally expanded, due to low temperatures, as in *Impatiens*.
7. Sciocleistogamy. Self-pollination in closed flowers, which are normally expanded, due to shading, as in *Viola arvensis*.
8. Thermocleistogamy. Self-pollination in closed flowers, which are normally expanded, due to high temperatures, as in *Eranthemum*.
9. Xerocleistogamy. Self-pollination in closed flowers, which are normally expanded, due to drought, as in *Dicliptera*.

VII. ENVIRONMENTAL EFFECTS UPON MORPHOLOGICAL FEATURES (Developmental Responses and Adaptations)

The taxonomist should note the morphological features that may be correlated with different environmental factors such as light, moisture, and soil aeration, primarily in order to better understand structural variation in the field and laboratory. Secondarily, he should incorporate range measurement extremes due to environmental extremes in his description of a particular taxon and also be aware of the fact that in identification, measurement extremes are not always given in manuals due to insufficient observation. Developmental or behavioral responses should be distinguished from genetically fixed adaptations. A few morphological features have been selected to show developmental or adaptive responses to intense light, low moisture, and poor soil aeration as examples of structural-environmental interactions. These observations have been made by many experienced field and laboratory workers, sometimes with little understanding of the real significance or cause of the response. For specific references to experimental and analytical evidence for these and other interactions, consult the ecological texts and periodicals cited in the literature section of this chapter, particularly Daubenmire (1959).

A. Morphological Features Associated with Intense Light (Heliophytic reactions or adaptations)

Intense light results, developmentally, in

1. Thicker stems
2. Reduced internodes
3. Reduced and thicker leaf blades
4. More trichomes per unit area
5. Thicker cuticle
6. Smaller stomata and blade cells
7. Better developed palisade layers
8. More prolific branching

Adaptively, species living in full sunlight exhibit the developmental characteristics cited above.

B. Morphological Features Associated with Low Moisture (Xerophytic reactions or adaptations)

Low moisture results, developmentally, in

1. Reduced size of shoot
2. Occasionally increased size of root system
3. Smaller leaf blade or absence of blade
4. Smaller and fewer stomata that may be confined to grooves or crypts
5. More heavily lignified tissues in blade
6. Thicker cuticle
7. More numerous trichomes per unit area
8. Better developed palisade layer
9. Increase in vascular tissue per unit area
10. Reduction in cell size and increase in thickness of wall
11. Decrease in volume of intercellular space

Adaptively, low moisture has resulted in the development of

a. Ephemeral annuals
b. Non-succulent perennials with storage taproots and sclerotic shoots
c. Non-succulent perennials with extensive fibrous root systems and reduced shoot-transpiring surfaces
d. Succulent perennials

C. Morphological Features Associated with Poor Soil Aeration (Anaerobic reactions or adaptations)

Poor soil aeration results, developmentally, in

1. Roots usually less numerous
2. Root branching less complex
3. Roots usually shorter and occupying less space

4. Roots and underground stems thicker with the development of more lacunar or aerenchyma tissue
5. Root systems shallow with the presence of numerous pneumatophores
6. Shoots smaller or stems are buttressed and with lacunar or aerenchyma tissue

Adaptively, species with roots living under anaerobic conditions exhibit the developmental or behavioral characteristics listed above.

VIII. HETEROTROPHY IN VASCULAR PLANTS*

A. Nutritional Relationships.

1. Parasite. A plant dependent on a host plant for food and/or water. Parasitism has apparently arisen independently at least six times in angiosperms in the following orders of dicotyledons: Laurales (*Cassytha* of the Lauraceae), Santalales (Santalaceae, Loranthaceae, Olacaceae and others), Rafflesiales (Rafflesiaceae, Hydnoraceae, Balanophoraceae), Scrophulariales (Scrophulariaceae, Orobanchaceae), Polemoniales (Lennoaceae, *Cuscuta* of the Convolvulaceae), and Polygalales (Krameriaceae). No parasitic monocotyledons are known. A single report exists of a parasitic gymnosperm, *Podocarpus ustus*, from New Caledonia. The flowers of parasitic angiosperms are among the largest (*Rafflesia* sp.) and smallest (*Pilostyles* sp., *Arceuthobium* sp.) of all known flowers.
   a. Auto-parasite. A plant parasitizing itself as in some Santalaceae; sometimes called self-parasitism or root grafting. Most auto-parasites are hemi-parasites.
   b. Holo-parasite. A plant lacking chlorophyll and thus entirely dependent on its host, as all Orobanchaceae and *Cuscuta*. Holo-parasites are termed stem parasites or root parasites depending on the portion of the host they attack.
   c. Hyper-parasite. A plant that parasitizes other parasites as some Loranthaceae. Hyper-parasitism has sometimes been referred to as epi-parasitism.
   d. Hemi-parasite. A chlorophyll-containing parasite superficially appearing completely autotrophic, as in many members of the Scrophulariaceae. Hemi-parasites may be stem or root parasites. Sometimes hemi-parasites are called semi- or demi-parasites.
2. Saprophyte. A plant lacking chlorophyll and obtaining nourishment from decaying organic matter. All saprophytes which have been carefully studied have been shown to be epi-parasites.
3. Fungal relationships.
   a. Epi-parasite. An achlorophyllous plant containing a fungus continuous with the mycorrhizal associate of surrounding trees as in *Monotropa* of the Pyrolaceae (or Ericaceae) and plants belonging to thirteen or so genera of the Orchidaceae.
   b. Mycorrhiza. The symbiotic relationship between a fungus and the roots of vascular plants. Mycorrhizae are widespread in angiosperms with only a few families (Brassicaceae, Chenopodiaceae, Cyperaceae) and aquatics thought to be

*Contributed by Lytton J. Musselman, University of North Carolina at Chapel Hill.

entirely non-mycorrhizal.

Endomycorrhiza. A fungus with intracellular hyphal infections in the host root cortex; the most common type of mychorrhiza.

Ectomycorrhiza. A fungus with intercellular infections in the host cortex. All members of the Pinaceae are ectomycorrhizal; in angiosperms ectomycorrhizae are usually limited to amentiferous families. Truffles are the fruiting bodies of an ectomycorrhizal fungus.

Ectendomycorrhiza. Fungus with both inter- and intracellular infection in the host cortex. Known only in tree species which are usually ectomycorrhizal.

4. Bacterial relationships. Bacterial root nodules are best known in the Fabaceae but also occur in many other families of angiosperms and Podocarpaceae of the gymnosperms.
5. Algal relationships. Blue green algae are known to inhabit the roots of the cycad genus *Cycas*. Algae are also found in the roots of some angiosperms.
6. Root grafting. The fusion of two or more contiguous roots, usually tree roots, involving the morphological and anatomical union of cambium, phloem, and xylem of the two roots. Only a few reports of root grafting in herbaceous plants exist. This term has also been applied to auto-parasitism in angiosperm root parasites.

B. Morphological Considerations of Heterotrophy.

Heterotrophy, especially parasitism, has resulted in extreme specialization and reduction in the plant body of the parasite. Difficulty arises when trying to define the highly specialized organs of parasites in the usual morphological context.

1. Haustorium. The organ which forms the morphological and physiological bridge between host and parasite, usually considered to be a highly modified root.
   a. Primary haustorium. A haustorium which arises directly from a root apex, considered more advanced than a secondary haustorium.
   b. Secondary haustorium. A haustorium which arises laterally from any root or stem and not from a root apex.
   c. Endophytic system. The greatly reduced vegetative plant body of a holo-parasite which is entirely contained within the host plant. In *Pilostyles* and other members of the Rafflesiaceae the only indication of the presence of the parasite is when the flower buds burst through the host bark.
2. Roots.
   a. Pilot root. A root which gives rise to haustorial roots but does not form haustoria itself, as in *Pholisma* of the Lennoaceae.
   b. Haustorial root. A root producing haustoria, usually applied to parasites which exhibit root dimorphism.
3. Host responses.
   a. Coralloid root. A short, swollen, dichotomously branched mycorrhizal root.
   b. Isophasic growth. Integrated host-parasite growth where the cells of the parasite are carried along by the apical meristem of the host.

c. Woodrose. Placenta-like structure composed of host wood in response to infection by parasite.
d. Nodule. Globose swelling in root which contains bacteria.
e. Witches broom. A malformation of a host by parasite infection resulting in shortened internodes and a proliferation of branches, as *Arceuthobium* on *Pinus*.

ECOLOGICAL LITERATURE

I. Periodicals

Ecology. Official publication of Ecological Society of America. 1920+. Duke University Press. Durham.

Ecological Monographs. Official publication of Ecological Society of America. 1931+. Duke University Press. Durham.

Journal of Ecology. Official publication of British Ecological Society. 1913+. Cambridge University Press. Cambridge.

Vegetatio. Official publication of International Association of Plant Sociologists. 1948+. D. W. Junk, Publishers. Hague.

II. Special References.

Braun-Blanquet, J. 1932. Plant Sociology. (transl. by H. S. Conard and G. D. Fuller). McGraw-Hill Book Company. New York.

Clements, F. E. 1902. A System of Nomenclature for Phytography. Botanische Jahrubucher fur Systematik, Pflanzengeschichte und Pflanzengeographie 31, beiblatt Nr. 70.

Dansereau, P. 1957. Biogeography: An Ecological Perspective. The Ronald Press Company. New York.

Daubenmire, R. F. 1959. Plants and Environment. 2nd ed. John Wiley and Sons, Inc. New York.

Faegri, K., and L. van der Pijl. 1966. The Principles of Pollination Ecology. Pergamon Press. New York.

Gray, P. 1967. The Dictionary of the Biological Sciences. Reinhold Publishing Corporation. New York.

Jackson, B. D. 1928. A Glossary of Botanic Terms. J. P. Lippincott Company. Philadelphia.

Kruckeberg, A. 1969. The implications of ecology for plant systematics. Taxon 18: 92-120.

Odum, E. P. 1971. Fundamentals of Ecology. 3rd ed. W. B. Saunders Company. Philadelphia.

Percival, Mary S. 1965. Floral Biology. Pergamon Press. New York.

Pijl, L. van der. 1969. Principles of Dispersal in Higher Plants. Springer-Verlag. Berlin.

Raunkiaer, C. 1934. The Life Forms of Plants and Statistical Plant Geography. Clarendon Press. Oxford.

Ridley, H. N. 1930. The Dispersal of Plants throughout the World. L. Reeve and Company, Ltd. Ashford.

Stearn, W. T. 1966. Botanical Latin. Thomas Nelson and Sons, Ltd. Edinburgh.

Swartz, D. 1971. Collegiate Dictionary of Botany. Ronald Press Company. New York.

Uphof, J. C. Th. 1938. Cleistogamic flowers. Botanical Review 4(1): 21-49.

ECOLOGICAL CHARACTERISTICS EXERCISE

Select one or two species and work out the character states for each of the following characters:

CHARACTER	Species #1	Species #2
Habitat		
Biotic Relations:		
Nutritional		
Sociability		
Symbiotic		
Climatic Relations:		
Temperature Regime		
Air Moisture Regime		
Light		
Wind		
Edaphic Relations:		
Soil Texture Regime		
Soil Aeration Regime		
Soil Nutrients Regime		
Soil pH Regime		
Spatial Relations:		
Numerical		
Stratificational		
Temporal Relations:		
Phenological		
Successional		
Adaptive Features:		
Life Form		
Leaf Size		
Diaspore Type		
Fidelity		
Vitality		
Pollinating Agent		

Chapter 16. GEOGRAPHIC EVIDENCE AND DISTRIBUTION

An understanding of geographic evidence and principles of distribution is basic to interpreting the origin, migration, and evolution of species and floras. Distribution ranges correlated with differences or similarities in structural features have been and are the bases for the determination of relationships of taxa at various levels. The following quotations show the roles of geographic evidence and distribution in the determination of taxonomic and evolutionary relationships.

> "The maps (on distribution of morphological features in Aconitum) show that certain features seem to vary independently though loosely correlated with each other and with geographical areas. Geographically placed symbols show the correlations of these features. Three groups become identifiable from the maps: (1) 3-lobed leaf and limited pubescence, at low elevations, Maryland to southwestern Georgia and west of the Appalachians in Tennessee and West Virginia; (2) 5-lobed leaf, greater pubescence, in the Appalachians, from western Pennsylvania to eastern Tennessee and northwestern South Carolina and along the outer Piedmont of North Carolina and Virginia; and (3) 5- to 7-lobed leaf, extreme pubescence, and 'muticum' helmet in populations north of the glacial boundary in New York, Ohio, Wisconsin and Iowa. The taxonomic interpretation of these three entities must take into account three characteristics shown by the maps." (Hardin, 1964, Variation in Aconitum of eastern United States, pp. 88-89).

> "The flora of the northern part of North America and that of the northeastern part of Asia are, because of geographical continuity, closely related. There have evidently been exchanges of species and extensions of ranges of species between the two areas until the very recent geological past, such processes probably continuing at the present time. The relationships of these continuous floras have been investigated especially by Hultén in his many excellent papers.
>
> A study of the habit and habitat of the temperate disjunct genera of eastern Asia and eastern North American will also reveal the nature of the vegetation when these genera and their relatives and associates, both extinct and extant, were once the dominant elements of the floras in the geological past. The following points can be especially noted.
>
> In the first place, the greater number of genera is composed of woody plants, only a few consisting exclusively of herbaceous plants. Among the woody plants are trees like Carya, Liriodendron, Sassafras, Gymnocladus, Cladrastis, Stewartia, Halesia, Magnolia, Gordonia, and Catalpa, shrubs like Hamamelis, Pachysandra, Chiogenes, Chionanthus, Mitchella, Illicium, Itea, Pieris, and Lyonia, and woody vines like Decumaria, Schisandra, Wisteria, Parthenocissus, and Campsis." (Li, 1952, Floristic relationships between eastern Asia and eastern North America, p. 404).

The comprehensive field of plant geography includes studies of origin, distribution, adaptation, and associations or groupings of plants. The treatment in Section A of this chapter is a synoptic descriptive summary of the types of associations such as biogeographic regions, biomes, and flora provinces. The fundamental principles of distribution and definitions of classified geographic terms pertinent to taxonomy are presented in Section B. In Sections C and D major and minor disjunct ranges are discussed. The ranges presented in these sections are used only as examples to show some of the major disjunct or discontinuity ranges in the world and some of the distribution patterns in North America. This, in no sense, is intended to be an exhaustive summary of the subject (consult the references cited for more detailed information). The dynamic aspects of biogeography and the subject of adaptations are not described in this text except in a limited way in Chapter 15. The plant communities in eastern North America are summarized in Chapter 17.

Section A. BIOGEOGRAPHIC REGIONS, BIOMES, FLORA PROVINCES

No hard and fast definitions have been devised for the distinctions between region, biome, and province but for the sense in which the terms are used here the following definitions are applicable: (1) Biogeographic Region - major, more or less continuous land mass with major geologic or ecologic barriers to migration between adjacent regions, such as the Atlantic and Pacific Oceans between the Nearctic and the Palearctic Regions and the Sahara Desert between the eastern Palearctic and Ethiopian Regions; (2) Biome - a group of associations or organisms within a region with a definite physiognomy, such as the needle-leaved trees (conifers) of the Taiga; and (3) Flora Province - region which includes the total range of the dominants of several related types of vegetation or communities, such as the southeastern swamp forests with cypresses, gums, and cottonwoods.

I. BIOGEOGRAPHIC REGIONS
Based on Polunin (1960)

1. Nearctic. North America north of Mexico.
2. Palearctic. Eurasia - North Africa north of the Sahara and e-w mountains such as the Himalayas.
3. Neotropical. South America and Caribbean Regions.
4. Ethiopian. Africa south of the Sahara Desert.
5. Oriental. Asia south of the Himalayas and a major part of the East Indies.
6. Australian. Australia, New Zealand, and South Pacific Islands.
7. Antarctic. Antarctica, southern South America.

II. MAJOR BIOMES OF THE WORLD

1. Tundra and Ice
2. Taiga (Boreal Coniferous Forest)
3. Temperate Deciduous Forest
4. Scrub and Sclerophyll Forest
5. Grassland
6. Desert
7. Tropical Rain Forest
8. Temperate Rain Forest

Biome Distribution in Relation to Altitude and Latitude
in
Eastern North America
(Temperature Relations)

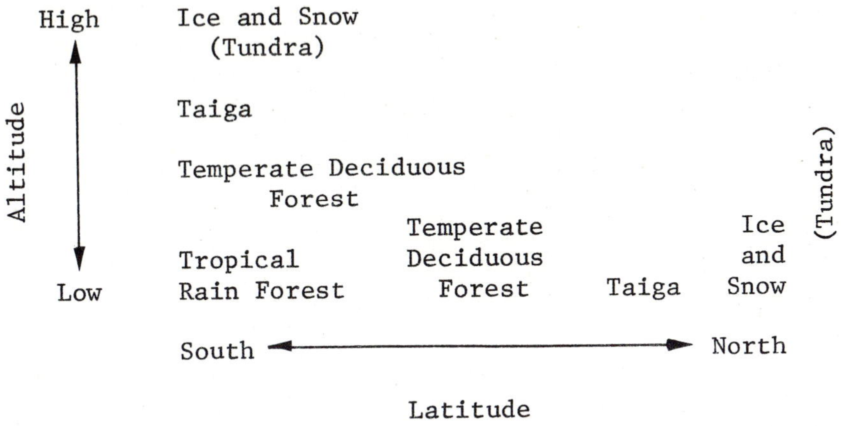

Biome Distribution in Relation to Moisture
in
Temperate North America
(SW = summer wet, winter wet; Sw = summer wet, winter dry;
sW = summer dry, winter wet; sw = summer dry, winter dry)

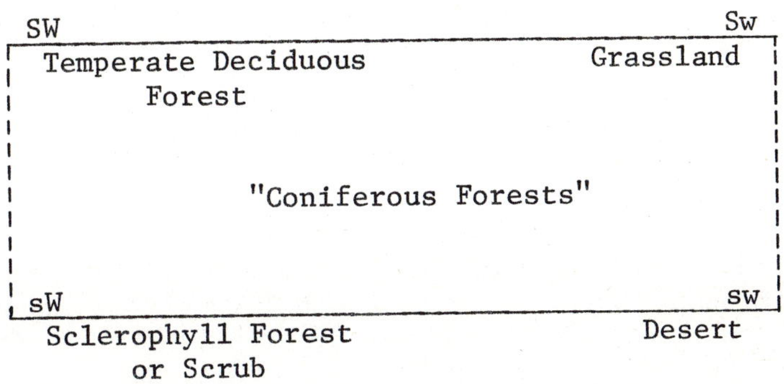

III. MAJOR FLORA PROVINCES OF NORTH AMERICA NORTH OF MEXICO
Based on Shantz and Zon (1935), Shreve and Wiggins (1964), Weaver and Albertson (1956)

<u>Province</u> <u>Dominant or Characteristic Plants</u>

Tundra - Arctic and Alpine (Sedges, Grasses, Saxifrages, Lichens, Dwarf Willows, Gentians, etc.)

Coniferous Forests

Boreal Forests (Spruce, Fir, Tamarack)

Northeastern Pine Forests (Jack, Red, White Pine)

Western
Subalpine (Engelmann Spruce, Alpine Fir, Limber & Lodgepole Pines)
Montane (Ponderosa Pine, Douglas Fir)
Northwestern Mixed Conifer Forests (Larch, Redwood, Western Cedar, Western White Pine, Western Hemlock, Sitka Spruce)
Southwestern Pinyon-Juniper Forests (Pinyon Pine, Juniper)
Southeastern Pine Forests (Longleaf, Loblolly, Slash Pines)

Hardwood Forests
Northern Aspen (Aspen, Birch)
Northeastern Hardwood and Southern Appalachian Mixed Mesophytic (Maple, Beech, Birch, Hemlock)
Eastern Oak-Hickory Forests (Many species of Oaks and Hickories)
Southeastern Broadleaved Evergreen-Hammock and Bay (Live Oak, Red Bay, American Olive, Holly, Magnolia)
Southeastern Swamp Forests (Bald Cypress, Gums, Cottonwoods, Ashes)
Subtropical Mangrove Forests (Buttonwood, Red, Black and White Mangroves)
Western Broad Sclerophyll Forests (Oaks, Incense-cedar, California Bay)
Western Woodlands (Oaks, Pines, Redbud, Gooseberry)

Chapparal (Chamiso, Manzanita, Mountain Mahogany, Wild-lilac, etc.)

Southwestern Savanna (Mesquite, Desert Grasses)

Grasslands
Prairie (Tall Grasses: Bluestem, Indian, Needle)
Mixed Prairie (Mid and Short Grasses: Bunch, Needle, Slender Wheat)
Plains (Short Grasses: Grama, Galleta, Wire, Western Wheat)
Pacific Prairie (Bunch Grasses: Wire, Wheat, Poa)

Desert
Northern Desert Scrub (Salt Sage, Sagebrush, Rabbit Brush, Winter Fat)
Salt or Alkali Desert Scrub (Greasewood, Pickleweed, Samphire, Salt Sage)
Southern Desert Scrub (Creosote, Bur Sage, *Ocotillo*)
Arboreal Desert (*Cercidium*, *Olneya*, *Jatropha*, *Bursera*)
Succulent Desert (*Pachycereus*, *Carnegia*, *Lophocereus*, *Opuntia*-Cacti)

Section B. DISTRIBUTIONAL PRINCIPLES AND TERMINOLOGY

General agreement on the principles of plant distribution was reached many years ago by geographers, ecologists, and taxonomists. The concepts dealing with origin, establishment, dispersal, barriers to migration, and continuous and discontinuous ranges are accepted by individuals working with distribution problems. The criteria for place of origin and center of distribution, however, are debatable; continuous range is an impossibility but is accepted and understood as a concept in the sense of general distribution throughout a climatic regime under the proper edaphic conditions which are always disjunct; the agents of dispersal: wind, water, and animals, are agreed upon but determining the actual effective agent for a given species in a local area or region occasionally presents a dilemma; barriers to migration: climatic, edaphic, and biotic, are the broad general classes of obstacles to dispersal but determining the actual barrier to migration of a species in a limited area can be a problem. Determining the limiting factor(s) for establishment of a species within an area is a traditional problem for experimental ecologists.

I. DISTRIBUTION PRINCIPLES
Based on Good (1964)

1. Plant distribution is primarily controlled by distribution of climatic conditions.

2. Plant distribution is secondarily controlled by distribution of edaphic factors.
3. Plant migration is brought about by transport of dispersal phases of individual plants.
4. Species ranges are limited by genetically controlled tolerances.
5. Flora movements have occurred in the past and apparently are still occurring.
6. Flora movements have occurred at higher latitudes and altitudes due to variation and oscillation in climate.
7. Flora composition has varied with change in the relative distribution of land and sea.
8. Barriers to migration of plants are physiographic, climatic, edaphic, or biotic.

II. AREA AND DISTRIBUTION DEFINITIONS
Based on Cain (1944)

1. Allopatric Distribution. Distribution in which closely related species or subspecies occupy different range or area.
2. Center of Area or Place of Origin. Area with greatest differentiation within a taxon;
   area with greatest abundance of individual organisms;
   area with maximum size of individuals;
   area with greatest number of convergent lines of dispersal;
   area with least dependence upon a restricted habitat;
   area with the greatest number of dominant genes.
3. Continuous Distribution. Areas with wide-spread distribution of one or more species.
4. Discontinuous Distribution. Areas with disjunctions in the ranges of one or more species.
5. Dispersal. The spreading abroad of an organism from one part of an area to another; in plants by means of wind, water, animals, or mechanical devices.
6. Disseminule or Diaspore. The agents of dispersal characteristic of different species, such as plumed seeds, winged fruits.
7. Endemic Species. Species limited to a local area or restricted geographically.
8. Establishment. The successful germination, development, and reproduction of an individual plant or population within an area.
9. Origin. The source from which anything arises, such as species, population, community.
10. Relict Area. Area occupied by remnant of an earlier flora.
11. Sympatric Distribution. Distribution in which closely related species or subspecies do occupy the same range or area.

III. RANGES ACCORDING TO SIZE OF AREA

1. Local. Very limited range, usually applied to endemics within a small area.
2. Regional. Range covers part of a vegetation climax region or physiographic province.
3. Continental. Range usually covers more than one climax region or two or more major physiographic provinces.

IV. GENERAL TYPES OF DISCONTINUITY DISTRIBUTION

1. Altitudinal. Distribution pattern broken into two or more non-adjoining elevation parts.
2. Bipartite. Distribution pattern broken into two parts, a large and a small within the same general region or area.
3. Bipolar. Distribution pattern broken into two parts, one at each pole.
4. Diffuse. Distribution pattern broken into many small more or less equal parts, usually corresponding to local discontinuous climatic patterns or edaphic conditions.
5. Disjunct. Distribution pattern broken into two or more parts; a large and one or more small parts not in the same region or area.

V. WORLD-WIDE TYPES OF CONTINUOUS DISTRIBUTION (GENERIC LEVEL)
Based on Good (1964)

1. Cosmopolitan. World-wide.
2. Circumpolar. Frigid Zone.
3. Circumboreal. North Temperate Zone.
4. Circumaustral. South Temperate Zone.
5. Pantropical. Tropical Zone.

VI. ORIGIN AND ESTABLISHMENT OF PLANTS

1. Adventive. Species that is imperfectly naturalized; a species that reproduces spontaneously for a while and then disappears.
2. Allochthonous. Native species that has not originated within the area; a past immigrant that is now reproducing spontaneously within the area.
3. Autochthonous. Native species that has originated within the area and is reproducing spontaneously.
4. Domesticated. Species planted by man but not reproducing spontaneously.
5. Indigenous or Native. Species occurring within an area and not propagated or introduced accidentally by man.
6. Introduced. Species accidentally or purposely propagated in an area by man.
7. Naturalized. Introduced species that is now permanently established and reproducing spontaneously.
8. Sporadic. Repeatedly introduced species that never becomes permanently established.

VII. DISSEMINULES OR DIASPORES AND DISPERSAL
(See Chapter 15 for definitions and classification)

A. DIASPORE TYPES

1. Atelechore
2. Anthropochore
3. Ballochore
4. Barochore
5. Cyclochore
6. Endozoochore
7. Epizoochore
8. Hydrochore
9. Pogonochore
10. Polychore
11. Pterochore
12. Sporochore
13. Synzoochore

B. CLASSIFICATION OF PLANTS ACCORDING TO PARTS NORMALLY DISPERSED

1. Carpophyllophyte. Fruit and bract or accessory flower part of plant. (Ironwood, clover, linden)
2. Carpophyte. Fruit part of plant. (Oats)
3. Embryophyte. Embryo or plantlet part of plant. (Mangrove, duckweed)
4. Holophyte. Entire, above-ground part of plant. (Russian thistle)

5. Inflophyte. Infructescence (old inflorescence) part of plant. (Tumble grass)
6. Spermophyte. Seed part of plant. (Jewelweed, larkspur)
7. Sporophyte. Spore part of plant. (Club-moss, Christmas fern)

Section C. RANGES AND DISTRIBUTIONAL PATTERNS

The two basic problems in all of plant distribution are (1) the explanation for the distribution of closely related but disjunct taxa in the world and (2) the explanation for local distribution patterns. What is the evidence for a Gondwana Land and Laurasia? Did closely related species that are now disjunct have a continuous distribution at one time? Have the present major barriers to migration such as the Polar Ice Cap, the Sahara Desert, and the Atlantic and Pacific Oceans been in existence since the advent of the angiosperms? Are refugia such as the Southern Appalachians and Southeastern Asia real or figments of the imagination? Has there been parallel evolution in the same taxa in various parts of the world? From a local distributional pattern standpoint, for example, in the flora of the Carolinas, how can a prairie element, a boreal element, a subtropical element, a montane element in the outer piedmont, a sand hill refugium element, a coastal plain element in the lower mountains, and southwestern Georgia-northwestern Florida lime sink element be explained? The answer to these and similar questions are basic to an understanding of the origin, migration, and evolution of species, floras and communities.

Local, regional, intracontinental, and intercontinental distributional patterns and ranges of species and genera provide basic clues to the complex geographical and historical relationships of floras. Floristic components are from many genetic sources from many directions over a long period of time. An analysis of the distributional patterns within a region can lead to an understanding of the floristic relationships of the vegetation of a given area with that of contiguous as well as in disjunct regions of the earth. A knowledge of floristic relationships can lead to an understanding of the evolutionary and taxonomic relationships of similar taxa. Geographic evidence is fundamental to the determination of systematic affinities of taxa.

As an example, the analysis of the distributional patterns of genera of seed plants in eastern North America is presented in Table 16-1 to show the floristic relationships between the Southern Appalachians and western North America (from Wood, 1971). Some intracontinental and endemic ranges on the generic level involved in the study by Wood (1971) are given in Table 16-2.

Detailed mapping of ranges of species is necessary in geographic work. Mapping by physiographic province, rock type, soil type, or climatic regimes will provide explanations for local distributional patterns which will give insight into taxonomic, floristic and evolutionary relationships of various taxa in contiguous and disjunct regions.

In Table 16-3 an analysis of distributional patterns within some Blue Ridge escarpment gorges in the Southern Appalachians is given to show the variety of relationships that can be derived from a local study within a restricted area (from Cooper and Hardin, 1971).

The major intercontinental disjunctions in the geographic ranges are presented in Section D.

Table 16-1. Partial analysis of distributions of genera of seed plants occurring in eastern North America (from Wood, 1971).

Genera occurring in southern Appalachians			557
Genera represented in both southern Appalachians and western North America		365*	
Genera ± continuously represented between southern Appalachians and western North America	237		
Genera with taxa with disjunctions between southern Appalachians and western North America	158*		
Genera with disjunction(s) in range of one or more species: 48 Genera of various distributions with one or more taxa in southern Appalachians and disjunct taxa of same rank in western North America: 46 Genera restricted to North America (or nearly so) with related taxa in southern Appalachians and western North America: 20 Genera with relict Tertiary distribution, with related taxa in southern Appalachians and western North America: 46			
Genera _not_ represented in both southern Appalachians and western North America		192	
Endemic genera, restricted to eastern North America or nearly so	52		
Tertiary relict genera, most restricted to eastern North America and eastern Asia	57		
Genera with other distributions	83		
Genera not occurring in southern Appalachians			475
Genera with one or more representatives in both eastern and western North America		144	
Genera with other distributions		331	
Total indigenous genera			1032
Nonindigenous naturalized genera			384
Total genera of seed plants			1416

*Because some genera appear in more than one category, the subtotals will add up to more than these figures.

Table 16-2. Some intracontinental and endemic ranges in North America (from Wood, 1971).

1. Genera represented in both Southern Appalachian and Western North America. (Partial List)

Abies	Orchis	Salvia	Ambrosia
Juniperus	Oxalis	Stachys	Antennaria
Typha	Geranium	Vaccinium	Aster
Potamogeton	Verbena	Asclepias	Bidens

2. Endemic Genera represented in Southern Appalachians but not in Western North America. (Partial List)

Orontium*	Sanguinaria	Galax	Peltandra
Medeola*	Leiophyllum*	Comptonia*	Amianthium*
Nestronia*	Dalibarda*	Xanthorrhiza*	Epifagus*
Hydrastis*	Neviusia*	Monotropsis*	Diervilla

*Monotypic

3. Relict Tertiary Genera, most restricted to eastern North America and eastern Asia but not in Western North America. (Partial List)

Symplocarpus	Saururus	Astilbe	Shortia
Aletris	Liriodendron	Decumaria	Pieris
Convallaria	Hepatica	Apios	Phryma
Tipularia	Pyrularia	Panax	Triosteum

4. Genera restricted to North America (or nearly so) with related taxa in Southern Appalachians and Western North America. (Partial List)

Thuja	Robinia	Dodecatheon	Pycnanthemum
Camassia	Ceanothus	Frasera	Trichostema
Xerophyllum	Dirca	Phlox	Collinsia
Eriogonum	Oxypolis	Hydrophyllum	Sericocarpus

5. Genera either known to be Relicts of Arcto-tertiary Floras or showing Relict Patterns of Distribution with survivors in Eastern and Western North America and in one or more areas notable as Refugia for Arcto-tertiary Relicts. (Partial List)

Taxus	Disporum	Tofieldia	Asarum
Tsuga	Erythronium	Trillium	Berberis
Hystrix	Narthecium	Ostrya	Calycanthus
Clintonia	Stenanthium	Aristolochia	Dicentra

Table 16-3. Some important distributional patterns of vascular plants in escarpment gorges (from Cooper and Hardin, 1971).

I. Species reaching their upper elevational limits

A. Species most common on coastal plain

Itea virginica
Decumaria barbara
Callicarpa americana
Lorinseria areolata
Viburnum nudum
Calopogon pulchellus

B. Species chiefly on the piedmont and coastal plain

Platanus occidentalis
Fraxinus pennsylvanica
Morus rubra
Carya cordiformis
Liquidambar styraciflua
Asimina triloba
Carya pallida
Quercus falcata
Quercus stellata
Quercus marilandica
Diospyros virginiana

II. Species reaching their lower elevational limits

Carya ovata
Pinus pungens
Comptonia peregrina
Leucothöe recurva

III. Appalachian endemics

A. Present

Shortia galacifolia
Clethra acuminata
Stuartia ovata
Hydrangea arborescens ssp. *radiata*
Diervilla sessilifolia
Krigia montana
Leucothoe recurva

B. Absent

Cladrastis lutea

IV. Characteristic species of mixed mesophytic forest

A. Absent, but to be expected

Cercis canadensis
Acer pensylvanicum
Ostrya virginiana
Magnolia tripetala

B. Present, but of very restricted distribution

Betula lutea (above gorges)
Acer saccharum
Aesculus octandra
Magnolia acuminata
Halesia carolina

V. Problematic distributional patterns

Fagus grandifolia
Tilia heterophylla
Quercus rubra
Carya glabra and *ovalis*

Section D. MAJOR DISJUNCTIONS IN THE GEOGRAPHIC RANGES OF SEED PLANTS*

Among the most fascinating aspects of plant geography are the very wide, intercontinental disjunctions in the natural geographic ranges of many seed-plant groups. Most intriguing to the geographer are the possible explanations for these wide gaps in range, e.g., the floristic links between Africa and America, and their possible bearing upon the theory of continental displacement. Most enlightening to the systematists are the evidences of taxonomic relationships shown in the analyses and explanations for these distributional patterns.

This treatment is an attempt to review and classify the major intercontinental disjunctions in the geographic ranges of seed plants; six guiding principles are followed where possible. Sixteen categories and 34 subcategories of disjunct ranges of intercontinental magnitude (Table 16-1) are classified according to continents and latitudinal bands involved. Widely distributed taxa are, by their very nature, disjunct; they are therefore classified along with the more traditionally discontinuous types. Thus, for the most part, only endemics and taxa with smaller intracontinental disjunctions are omitted from this review.

Generalized distribution maps are presented for selected taxa characteristic of some of the discontinuous types. Other taxa are listed for each of the categories and subcategories, and compendia of good distribution maps are cited for ready reference to additional examples. Possible explanations for the major discontinuities are discussed.

Finally, a statistical summary is given of the numbers of disjunct taxa and the largest disjunct categories. Among the Angiospermae (the flowering plants), approximately 78 per cent of the 324 families, 24 per cent of the genera, and 1 per cent of the species are widely disjunct. The largest number of disjunct genera are found among the African-Eurasian-(Pacific) (600), Asian-Pacific (460), Pacific (370), North American-South American (ca. 360), Pantropical (334), North Temperate (316), and Subcosmopolitan (125) categories; the largest number of disjunct families among the Subcosmopolitan (90), Pantropical (59), North Temperate (20), African-Eurasian-(Pacific) (17), and North American-South American (13) categories.

To achieve the goals in this section of (1) reviewing what is known about the major distributional disjunctions, (2) classifying their patterns, (3) mapping or listing examples of each major type, (4) citing useful compendia of range maps of seed plant taxa, and (5) discussing the possible explanations for the major discontinuities, I have been guided by several principles, or general considerations, which are herewith briefly stated. All should be self-evident but are all too often ignored. Some have recently been stated by Wood (1971) in his excellent paper on floristic relationships between the southern Appalachians and western North America, an intracontinental range disjunction not considered here.

*Contributed by Robert F. Thorne, Rancho Santa Ana Botanic Garden, Claremont, California. This is a condensed version of "Major Disjunctions in the Geographic Ranges of Seed Plants" in The Quarterly Review of Biology 47(4): 365-411. Used with permission.

GUIDING PRINCIPLES

1. Only taxa that have been reliably revised should be seriously considered.

2. Only accurate distributional data should be used in constructing distribution maps.

3. Almost all taxa have discontinuous ranges.

4. The current distribution of a taxon may not indicate its past distribution.

5. Coincidence in range disjunction, although certainly suggestive, does not mean that two taxa have similar dispersal histories.

6. The smaller the rank of a taxon, the more instructive is its disjunct range.

Classification of Intercontinental Disjunctions in Range

In classifying the major types of intercontinental discontinuities in the ranges of seed plants (Table 16-4), I have had to make rather arbitrary decisions to limit the number of categories. Eurasia, Africa, and Australia are considered to be the Old World continents, although Australia is occasionally treated as a very large island. North and South America, the New World, are mostly treated as separate continents, but sometimes it is useful to treat the two as a unit. The Pacific-Indian Ocean basins and the Atlantic-Caribbean basins are both considered of continental rank with respect to marine seed plants. Australasia, as a phytogeographical area, here includes Australia and Tasmania, New Zealand, New Caledonia, and usually New Guinea, all with their satellite islands.

I. EURASIAN-NORTH AMERICAN

Because Eurasia and North America lie across several major latitudinal belts, it seems best to subdivide this areal type into Arctic, Boreal, and Temperate.

1. Arctic.

Wide-ranging plants that display this far-northern distribution pattern, by geography, are necessarily disjunct between Eurasia and North America across the Arctic Sea, and often also across the North Atlantic between Europe-Iceland and Greenland-North America and across the Bering Sea between Alaska and Asia. These are plants which are believed by Hultén (1937) and Löve (1962) to have survived maximum Pleistocene glaciation in refugia mostly north of the ice sheets. With the melting back of the maximum ice, they have quickly spread into their present far northern ranges.

TABLE 16-4

MAJOR DISJUNCTIONS IN THE GEOGRAPHIC RANGES OF SEED PLANTS

I. Eurasian-North American
- 1. Arctic
  - 1a. Circum-Arctic
  - 1b. Beringian-Arctic
  - 1c. Amphi-Atlantic-Arctic
- 2. Boreal
  - 2a. Circum-Boreal
  - 2b. Beringian-Boreal
  - 2c. Amphi-Atlantic-Boreal
- 3. Temperate
  - 3a. Circum-North Temperate
  - 3b. North and South Temperate
  - 3c. Fragmentary North Temperate
    - 3c-1. Amphi-Atlantic Temperate
    - 3c-2. Mediterranean-American
    - 3c-3. Eurasian-Eastern and Western American
    - 3c-4. Eurasian-Eastern American-Mexican
    - 3c-5. Asian-Eastern and Western American
    - 3c-6. Eurasian-Eastern American
    - 3c-7. Asian-Eastern American
    - 3c-8. Eurasian-Western American
    - 3c-9. Asian-Mexican Highland
  - 3d. Wide Intracontinental Disjuncts

II. Amphi-Pacific Tropical

III. Pantropical

IV. African-Eurasian (-Pacific)
- 1. African-Mediterranean
- 2. African-Eurasian
- 3. African-Eurasian-Malesian
- 4. African-Eurasian-Pacific
- 5. African-Eurasian-Australasian
- 6. Indian Ocean-Eurasian (-Pacific)

V. Amphi-Indian Ocean

VI. Asian-Pacific
- 1. Asian-Papuan
- 2. Asian-Papuan-Melanesian
- 3. Asian-Papuan-Pacific Basin
- 4. Asian-Papuan-Australasian

VII. Pacific Ocean

VIII. Pacific-Indian-Atlantic Oceans

IX. American-African

X. North American-South American
- 1. Tropical
- 2. Temperate
- 3. Bipolar

XI. South American-Australasian
- 1. South American-Australasian
- 2. South American-Australasian-Asian
- 3. South American-Australasian-Madagascan

XII. Temperate South American-Asian

XIII. Circum-South Temperate

XIV. Circum-Antarctic

XV. Subcosmopolitan

XVI. Anomalous

1a. Circum-Arctic. The circum-Arctic type of disjunction comprises the Arctic circumpolar and the circumpolar, Arctic-montane plants of Hultén (1937). The circum-Arctic species are found in northern Alaska and Canada, the Canadian archipelago, Greenland, Iceland, northern Scandinavia, Spitzbergen, Franz Josef Land, Novaya Zemlya, the Siberian Arctic, and often farther south at alpine levels in the Eurasian and North American mountains, although each is usually lacking from one or more of these areas. Few species are continuously circumpolar. Hultén (1963) counts 352 species as circumpolar or more or less circumpolar. He counts another 107 taxa as having a more or less circumpolar range but represented by different races or slightly different species on both sides of the Atlantic. He considers this flora an old one, and thinks it probable that it "may have been spread by wind blowing over the frozen Polar Sea or by floating on ice" (p. 72).

Only a relatively few small genera are largely restricted to circum-Arctic (and alpine) ranges, namely Arctagrostis, Arctous, Braya, Diapensia, Dryas, Dupontia, Loiseleuria, and Oxyria. Honkenya, Koenigia, and Phippsia are also essentially circum-Arctic, but they reappear in southern South America and will be discussed below. The species Diapensia lapponica L. illustrates this circum-Arctic type of distribution.

1b. Beringian-Arctic. This subgroup consists of those plants which are mainly distributed on the Arctic shores of North America and Eurasia on either side of the Bering Sea but which do not reach both sides of the North Atlantic. Included are the northern Beringia radiants and Arctic-Pacific plants of Hultén (1937). Hultén suggests that these plants probably survived the Pleistocene glaciations in refugia in Alaska-Yukon, the northern part of the present Bering Sea, and in northeastern Siberia. Some formerly may have had more complete circum-Arctic ranges that were much reduced by glaciation in eastern North America, Greenland, and the European Arctic.

One plant that shows the Beringian-Arctic type of disjunction is Diapensia lapponica L. subsp. obovata (Fr. Schm.) Hult., separated by considerable gaps in northern Canada and Siberia from the amphi-Atlantic D. lapponica L. subsp. lapponica.

1c. Amphi-Atlantic-Arctic. This group is the reverse of the Beringian group of Arctic disjuncts, for the species are mainly distributed on both sides of the North Atlantic. They have received much critical attention from Hultén (1958, 1963). Hultén (1958) maps 278 species, but these include Arctic, boreal, and temperate species, 129 of the Arctic and 148 of boreal or temperate distribution. Hultén (1963) explains many of these amphi-Atlantic plants as former circum-Arctic species whose ranges have been much reduced by changing climatic conditions. Some may have spread, however, from centers on one side or the other of the Atlantic. Certainly some of the maritime and aquatic species can be accounted for by long-distance dispersal by sea currents or by birds. There seems little need, if any at all, for a Late Tertiary or Quaternary land-bridge across the North Atlantic to account for plants having this type of disjunct range.

Diapensia lapponica L. subsp. lapponica and Salix herbacea L. illustrate this type of range discontinuity.

2. Boreal.

Plants considered to have a disrupted boreal range in North America and Eurasia are those whose northern limits generally lie south of the Arctic shores and whose upper altitudinal limits do not reach the alpine levels of the Arctic-alpine species. They are absent usually from Greenland and often from Iceland. Many have been mapped by Hultén (1958, 1968) and by Meusel, Jager, and Winert (1965). Unlike the Arctic disjuncts, these plants apparently survived maximum glaciation in refugia south of the ice (Hultén, 1937).

2a. Circumboreal. Plants having this wide distribution in the boreal belt of Eurasia and North America have been called boreal circumpolar plants by Hultén (1937), and they are very well known to temperate botanists. They are especially abundant in the coniferous forests, bogs, marshes, and oligotrophic lakes of this zone.

Some of the more familiar species of this circumboreal disjunct group are Adoxa moschatellina L., Calla palustris L., Caltha palustris L. (s.l.), Calypso bulbosa (L.) Rchb. f., Carex chordorrhiza Ehrh., C. disperma Dew., C. rostrata Stokes, Cypripedium calceolus L., Drosera rotundifolia L., Juniperus communis L., Menyanthes trifoliata L., Moneses uniflora (L.) Fray, Monotropa hypopitys L., Myrica gale L., Ranunculus reptans L., Rhynchospora alba (L.) Vahl, Rubus idaeus L., Sparganium angustifolium Michx., Subularia aquatica L., Veronica scutellata L., and Viburnum opulus L. (s.l.).

2b. Beringian-Boreal. Taxa of this subgroup have much less extensive ranges than the circumboreal species. They are primarily distributed on either side of the Bering Sea in eastern Eurasia and western North America, but they do not generally reach the Arctic shores of Beringia and they do extend farther south on either side of the Pacific Ocean. Oplopanax, with two or three closely related species in North America, Japan, and Korea, is representative of this discontinuous type. In North America O. horridus (Sm.) Miq., the infamous Devil's Club of the wet coniferous forests of the Pacific Northwest, illustrates a well-known intracontinental disjunction by its jump from Montana to Isle Royale and adjacent islands on the north side of Lake Superior. At least eight additional genera, Achlys, Chamaerhodos, Glehnia, Leptarrhena, Lysichitum, Fauria, Smelowskia, and Thellungiella, have a Beringian-boreal type of discontinuous distribution.

Some of the boreal species or species-pairs with ranges in eastern Asia and western America, a few reaching across boreal America to the Atlantic Coast, are Achlys triphylla (Smith) DC.-A. japonica Maxim., Achillea sibirica Ledeb., Aruncus sylvester Kostel., Carex macrocephala Willd., Cassiope lycopodioides (Pall.) D. Don, and Cornus canadensis L.

2c. Amphi-Atlantic-Boreal. This subgroup, along with the more northerly amphi-Atlantic-Arctic group, has been thoroughly discussed by Hultén (1958, 1963). Because of their occurrence on both sides of the Atlantic Ocean, species having this type of disjunction are well known to European and American botanists. Rhynchospora fusca (L.) Ait. is just one of some 148 species, most of which belong to this group, that have been mapped by Hultén (1958).

3. Temperate.

This very large group of disjunct plants is certainly the best known and most frequently studied of all, as might be expected from the heavy concentration of the world's botanists in the North Temperate zone. To this group belong the many genera that dominate the temperate deciduous forests of Eurasia and North America. It is an ancient group, mostly originating no later than the widespread early Tertiary forests of the northern hemisphere. Many genera have retained their wide ranges of Tertiary times; others have become reduced to fragments of their former vast ranges. These latter have received much more attention because of their rather romantic, highly discontinuous ranges, but surely they are no more informative to the geographer than the genera that retain most of their former ranges. Both kinds of genera, however, together with the fairly extensive fossil record for the woody relicts, tell us much about the vicissitudes the Tertiary forests must have experienced to the present time.

The more fragmented North Temperate disjuncts are today preserved in the scattered deciduous forests of the North Temperate zone, primarily in eastern Asia (Japan, Korea, Taiwan, and China), the lower Himalayan slopes, the Near Eastern Caspian-Caucasus-Black Sea areas, Europe (including the Mediterranean region), eastern North America, the Mexican and Central American highlands, and western North America.

3a. Circum-North Temperate. By analyzing various regional temperate floras, Willis' *Dictionary* (1966), and Engler's *Syllabus* (1964), I have been able to list 118 wide-ranging, primarily temperate genera, 62 that are treated here as circum-North Temperate, and 56 that are equally widely distributed in the North Temperate zone but are also represented in the temperate areas of one or more of the southern continents. Most of the 118 possess relatively uninterrupted ranges across North America and across Eurasia. The 62 assigned to the circum-North Temperate group mostly do not cross the equator. The few that do, like *Acer*, *Alnus*, *Arabis*, *Campanula*, *Corydalis*, *Draba*, *Juniperus*, *Pinus*, *Quercus*, *Rosa*, and *Sedum*, are confined in the southern hemisphere to the tropical mountains. Many of these genera are by no means restricted to the temperate zone but have one or more species represented in the Arctic or boreal zones (as *Arenaria*, *Armeria*, *Betula*, *Chimaphila*, *Draba*, *Juniperus*, *Mertensia*, *Orchis*, *Parnassia*, *Pedicularis*, and *Silene*) or in the tropics (as *Acer*, *Lonicera*, *Pinus*, and *Quercus*).

3b. North and South Temperate. Presumably most of the preceding circum-North-Temperate genera have achieved their far-reaching ranges through the normal dispersal capacities of their member species. Their species, at least in total, must have a wide amplitude of tolerance to often severe environmental conditions in the continental heartlands. Many of the 56 widely distributed North Temperate genera that have successfully migrated across the tropics into temperate areas of one or more of the three southern continents have likewise probably moved by normal short-range dispersal along relatively continuous mountain chains, as *Descurainia*, *Epilobium*, *Gentiana*, *Lotus*, *Ribes*, *Salix*, *Saxifraga*, and *Trifolium*. On the other hand, other genera show very large discontinuities across the tropics, as in *Anemone*, *Euphrasia*, *Geum*, *Phalaris*, *Primula*, and *Sparganium*. These amphitropical genera may have achieved such large disjunctions by the extinction of montane species formerly spanning the gaps, or perhaps more likely, by having disseminules transported from one temperate zone to the other by migrating birds.

3c. Fragmentary North Temperate. Organized under this subheading is a variety of widely discontinuous groupings that have considerably less extensive ranges in Eurasia and North America than the groups discussed above. Their disjunctions in one or both continents probably represent severe restriction of ranges that were relatively continuous during more favorable Tertiary climates. Most of the taxa thus are Tertiary relicts. Long-distance dispersal, however, is more than likely for some of the maritime, marsh, and aquatic plants.

3c-1. Amphi-Atlantic Temperate. Since the warmer parts of Europe and North America lie widely separated by the Atlantic Ocean, with few islands of any consequence between the continents to serve as stepping-stones, it is not surprising that this type of discontinuity is very rare in comparison with the relatively frequent Arctic and boreal amphi-Atlantic disjuncts. It is best represented by Corema, with C. alba (L.) D. Don along the western coast of the Iberian Peninsula and on the Azores and C. conradii Torr., the only other species, along the Atlantic coast of North America from Newfoundland to New Jersey.

Of all the amphi-Atlantic species mapped by Hultén (1958), only 6 angiosperms are primarily temperate amphi-Atlantic disjuncts. They are Drosera intermedia Hayne, Eriocaulon septangulare With., Juncus tenuis Willd., Limosella subulata Ives, Ranunculus hederaceus L., and Spartina patens (Ait.) Muhl. It is surely no coincidence that these are all aquatic, maritime, or ruderal species.

3c-2. Mediterranean-American. Approximately 35 genera have their maximum development in the xerothermic, Mediterranean climates of the Old World Mediterranean region and southwestern North America of the New World, some of them also ranging disjunctly to other areas of Mediterranean climate in South Africa, Australia, or Chile. This kind of discontinuity has been of special interest to Californians because the Mediterranean climate in North America is centered about southern California. Four families, the Cistaceae, Cneoraceae, Ephedraceae, and Resedaceae; 18 groups of vicarious species or species-pairs, listed in Stebbins and Day (1967); and one species, Styrax officinalis L., also possess this kind of major disjunction.

3c-3. Eurasian-Eastern and Western American. Of the North Temperate genera with fragmentary ranges this category of wide disjuncts has ranges most similar to the nearly continuous circum-North Temperate genera. At least 24 genera, although with rather large gaps on one or the other continent, and often on both, are found in Europe, one or more areas of Asia, and in both eastern and western North America. Among these relicts are such well-known temperate genera as Aesculus, Amelanchier, Cercis, Corylus, Juglans, Ostrya, Platanus, Staphylea, Styrax, and Taxus.

3c-4. Eurasian-Eastern American-Mexican. Hardly distinct from the above group, the genera in this subcategory are not found in the western United States and Canada but are found in the highlands of Mexico (and often in Guatemala or farther south). The 6 genera that have this special type of disjunct range are Carpinus, Clethra (this genus is slightly aberrant since it is absent from Europe, but is found in Madeira), Fagus, Halenia, Tilia, and Ulmus. Some of the Eurasian-Eastern and Western American disjunct genera discussed above, like Amelanchier, Juglans, Ostrya, Platanus, Styrax, and Taxus, are also represented in the Mexican highlands and often farther south.

3c-5. Asian-Eastern and Western American. Another large category of Tertiary relicts are those genera, at least 38, that are represented in both eastern and western North America, but are now absent from Europe and mostly also from the Near East. They usually have their centers of development in eastern and southeastern Asia. Among these genera are Aralia, Calycanthus, Chamaecyparis, Clintonia, Dicentra, Disporum, Leucothoë, Physocarpus, Smilacina, Thuja, Torreya, Toxicodendron, Trillium, and Tsuga. Excellent maps showing the disjunct American ranges of some of these and some other genera have recently been published by Wood (1971).

3c-6. Eurasian-Eastern North American. There seem to be only four genera, Castanea, Convallaria, Epigaea, and Polygonatum, that occur both in Europe and Asia and only in the eastern part of North America. Presumably the east-west orientation of mountain chains and seas in Europe, which prevented the escape of temperate species southward before the Pleistocene glaciers, was almost as disastrous to the survival of Tertiary genera in Europe as was the severe deterioration of climate in western North America.

3c-7. Asian-Eastern American. This is the largest category of Tertiary relict temperate genera, the genera having found haven only in eastern North America and in Asia, most of them only in eastern and southeastern Asia. I have listed 74 genera in this much-studied group, early brought to our attention by Asa Gray (1846, 1859), and more recently discussed by Li (1952) and by Wood (1971). Among the more conspicuous or better known of these numerous genera are: Arisaema, Carya, Catalpa, Caulophyllum, Cladrastis, Croomia, Gordonia, Hamamelis, Illicium, Liquidambar, Liriodendron, Magnolia, Nelumbo, Nyssa, Panax, Podophyllum, Sassafras, Saururus, Schisandra, Shortia, Stewartia, Symplocarpus, Wisteria, and Zizania. Some of these are as correctly treated in the amphi-Pacific tropical disjunct category, for their ranges in either America or Asia or in both are as tropical as they are warm temperate. Among such genera are Berchemia, Clethra, Gordonia, Illicium, and Schisandra. Two genera, Arundinaria and Symplocos, were placed in the amphi-Pacific tropical disjunct category because all but one of their American species are in the tropics, although often at montane elevations.

Because so many of the Asian-Eastern American disjuncts have preservable parts, they have left a rather rich fossil record in areas where they are now extinct. The distribution of fossils in such genera as Liriodendron, Magnolia (see Meusel, 1969), Nelumbo, and Nyssa (Eyde and Barghoorn, 1963), indicates that these highly discontinuous temperate-to-tropical genera once had a nearly continuous distribution across Eurasia and North America in the Tertiary. Presumably the same is true for the herbaceous disjuncts, whose chances of being preserved as fossils are much less.

3c-8. Eurasian-Western North American. A small group of primarily temperate genera are the nine restricted to western North America and Eurasia, mostly eastern Asia. Because they are few and the disjunction is relatively less spectacular, this group has undeservedly received little attention. Heterocodon, Paeonia, and several generic-pairs have been discussed by Stebbins (1940). According to Stebbins (1938) the two western American species of Paeonia are closer to P. delavayi Franch. of southwestern China than to the herbaceous species of Europe. The other genera with this type of disjunct range are Adenocaulon, Boschniakia, Calocedrus, Mahonia, Photinia, and Pseudotsuga. They are little different from the nine genera earlier listed as having a Beringian-boreal distribution. In fact, this Eurasian-western North American group could just as well be called Beringian-temperate or even amphi-Pacific North Temperate.

3c-9. Asian-Mexican Highland. Hardly distinct from the preceding group of genera are the six genera found only in eastern Asia and the Mexican highlands or farther south. Several of the preceding group, such as Adenocaulon, Mahonia, Photinia, and Pseudotsuga, occur also in the cool Mexican or Guatemalan highlands. The six that do not occur in North America north of Mexico are Abelia, Deutzia, Langsdorffia, Leibnitzia, Mitrastemon, and Sarcococca. These could also be absorbed, perhaps as well, in the amphi-Pacific tropical group of discontinuous genera discussed next below.

In addition to the numerous genera, many listed above, that have a fragmentary North Temperate distribution, there are at least 18 families and three subfamilies so restricted: Calycanthaceae, Cistaceae, Cneoraceae, Datiscaceae, Ephedraceae, Hippocastanaceae, Hydrastoideae, Illiciaceae, Juglandaceae, Mitrastemonoideae, Nyssaceae, Paeoniaceae, Platanaceae, Podophylloideae, Resedaceae, Saururaceae, Schisandraceae, Stemonaceae, Styracaceae, Taxaceae, and Taxodiaceae.

3d. Wide Intracontinental Disjuncts. Within the temperate zone of North America on the one hand and of Eurasia on the other there are taxa with spectacular disjunctions that are not considered here because they are intracontinental. In distances involved, however, they are often more widely discontinuous than those that we are attempting to classify here. Such North American endemic disjuncts as Camassia, Dirca, Dulichium (with fossil record in Europe), Iliamna, Sullivantia, and Xerophyllum, are mapped and many others are discussed by Wood (1971).

II. AMPHI-PACIFIC TROPICAL

This discontinuous group includes all those taxa that are found both in tropical America and the tropical lands on the western borders of the Pacific Basin. A total of 89 genera, 4 tribes or subtribes, 3 or 4 subfamilies, and 8 to 11 families of flowering plants are amphi-Pacific tropical in their distribution. The number of taxa is inexact because of the overlap with categories discussed above. Clethra, Magnolia, Mitrastemon, and Nelumbo are particularly troublesome, and probably rightly should be included here. Van Steenis (1962) has presented a useful list of "amphi-transpacific genera and other affinities" organized in four latitudinal groups. Some of the most noteworthy amphi-Pacific tropical genera are: Anaxagorea, Cleyera, Endiandra, Hedyosmum, Meliosma, Passiflora, Pentapanax, Perrottetia, Persea, Phoebe, Picrasma, Rhynchoglossum, Saurauia, Sloanea, Spathiphyllum, Symplocos, Talauma, Turpinia, and Xylosma. All of these and 27 more reach southeastern Asia, including the Malesian islands west of New Guinea. Sixteen additional genera range much farther to the Mascarene Islands, Madagascar, or even eastern tropical Africa. Among them are Aphananthe, Calliandra, Callicarpa, Calophyllum, Clusia, Elaeocarpus, Glochidion, Protium, and Suriana. Sixteen genera reach only to Australasia; these include Batis, Bredemeyera, Epistephium, Iresine, Licania, Lindenia, Muehlenbeckia, Nicotiana, and Sicyos. Four genera, Allagoptera, Brachistus, Leucaena, and Pritchardia, reach neither Australasia nor Asia but do reach Polynesia or as far west as the Fiji and Solomon Islands.

Of the larger categories, the families Actinidiaceae (including Saurauia), Clethraceae, Magnoliaceae, Nelumbonaceae, Sabiaceae (including Meliosoma), Staphyleaceae, and Symplocaceae, although each has some warm temperate representatives, are essentially amphi-Pacific tropical in distribution. The

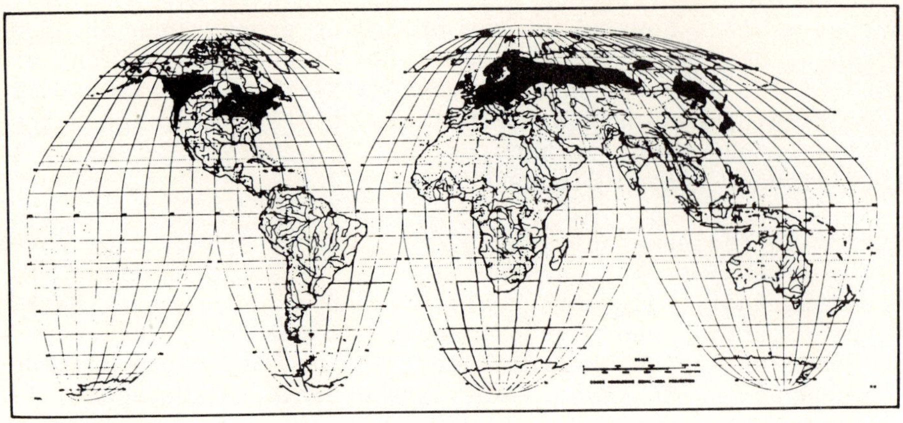

GENERALIZED DISTRIBUTION OF THE CIRCUM-BOREAL BOG SPECIES, *Scheuchzeria palustris*

The typical subsp. *palustris* occurs in Eurasia and the subsp. *americana in* North America. Modified from Meusel, Jager, and Winert (1965) and Hultén (1968); base map as in Fig. 2.

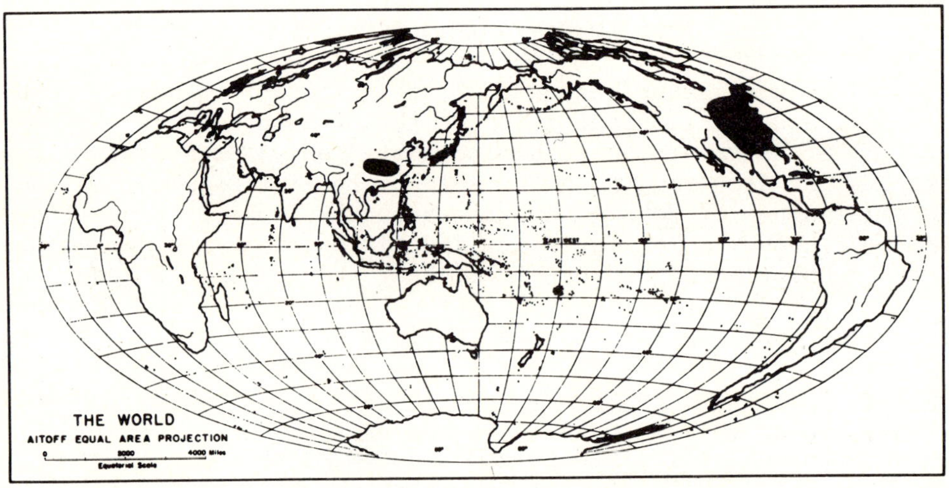

GENERALIZED DISTRIBUTION OF THE ASIAN-EASTERN AMERICAN GENUS *Hamamelis*, THE WITCH-HAZELS

Based in part on Li (1952); base map as in Fig. 1.

Figure 16-1. Generalized distribution of <u>Scheuchzeria</u> and <u>Hamamelis</u>. Used with permission.

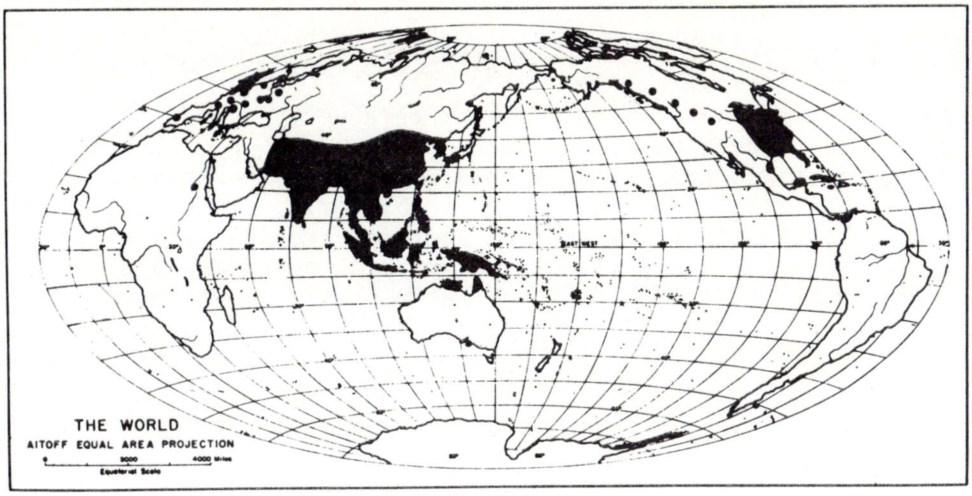

PAST AND PRESENT KNOWN DISTRIBUTION OF THE ASIAN-EASTERN AMERICAN, OR PERHAPS BETTER AMPHI-PACIFIC TROPICAL, GENUS *Nelumbo*, WATER-LOTUS

Modified from Good (1964); base map as in Fig. 1. The black circles represent fossil records.

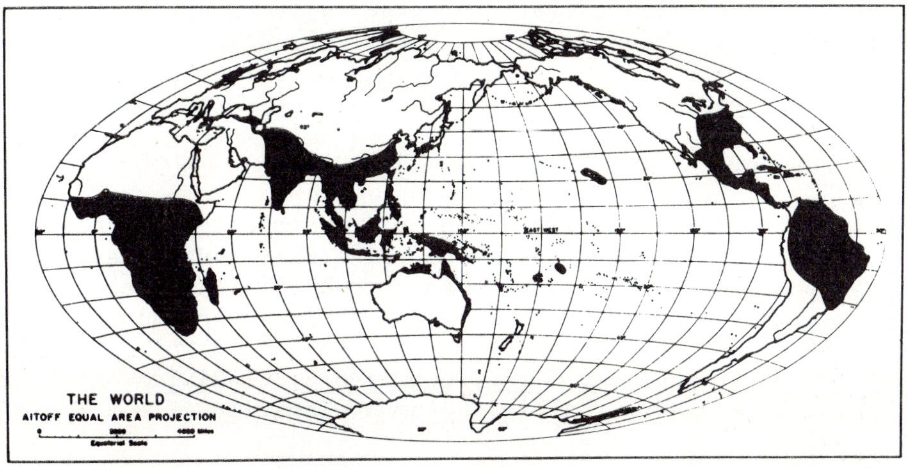

GENERALIZED DISTRIBUTION OF THE PANTROPICAL GENUS *Diospyros* (S. L.) OF THE EBENACEAE

Much modified from Fernald (1931); base map as in Fig. 1.

Figure 16-2. Distribution of Nelumbo and Diospyros. Used with permission.

Bataceae reach only Australasia, but the Trigoniaceae, Elaeocarpaceae, and Chloranthaceae are represented not only in Asia but in Madagascar as well.

III. PANTROPICAL

The same rather conservative approach to genera, as was used elsewhere in this survey with other taxa, has been used to draw up a list of 231 genera of seed plants that are truly pantropical, i.e., are represented by indigenous species in all the major tropical areas of the world. These major tropical areas are defined as including tropical America, tropical Africa, tropical Asia including Malesia, and tropical Australia. An additional 103 genera are not believed to have indigenous species in Australia (Burbidge, 1963) but occur elsewhere in all the major tropical areas. Thus, we can accept a total of 334 widely distributed tropical genera (many with some temperate species), a rather tiny total out of possibly 12,500 currently accepted genera of seed plants. Fifty-nine angiospermous families are also essentially pantropical.

IV. AFRICAN-EURASIAN (-PACIFIC)

An analysis of Willis' Dictionary (1966) produced 555 currently accepted genera that range from mainland Africa to Eurasia or beyond. They are readily divisible into six subcategories.

1. African-Mediterranean.

Only about 9 genera are restricted essentially to Africa south of the Sahara and to the Mediterranean region, a few extending north into Europe or east into the Orient. These are Althenia, Capnophyllum, Coris, Cytinus, Desmazeria, Erica, Hyacinthus, Romulea, and Tolpis.

2. African-Eurasian.

Thirty-eight genera, including such well-known plants as Adenium, Aeonium, Aloë, Catha, Cotyledon, Huernia, Hyphaene, Fissenia, and Sanseverinia, and the family Barbeyaceae, extend from Africa only to Arabia. At least 200 genera and 3 families, Trapaceae, Tamaricaceae, and Moringaceae, reach farther into Asia, usually at least to India, but do not extend into Malesia. These include Balanites, Borassus, Boswellia, Cicer, Commiphora, Gladiolus, Moringa, Ochna, Pedalium, Phoenix, Salvadora, Scilla, Sesamum, Tamarindus, and Trapa.

3. African-Eurasian-Malesian.

Another 106 genera have representatives in Malesia, which here is defined to include all the island groups from Malaya to the New Hebrides. A few of these are Angraecum, Asparagus, Azima, Cnestis, Combretodendron, Ellepanthus, Ensete, Irvingia, Phrynium, Pothos, Quisqualis, Sauromatium, Stromboisia, and Wrightia. Six families have a similar disrupted range: Ancistrocladaceae, Ctenolophonaceae, Dipsacaceae, Dipterocarpaceae, Pandaceae, and Salvadoraceae.

4. African-Eurasian-Pacific.

Another 100 discontinuous genera have representatives also in Fiji, Tonga, Samoa, or more distant islands in the Pacific Basin, as Alangium, Barringtonia, Bruguiera, Canarium, Cerbera, Claoxylon, Elatostema, Flagellaria, Gardenia,

Grewia, *Jasminum*, *Korthalsella*, *Leea*, *Lumnitzera*, *Myristica*, *Olea*, *Pandanus*, *Pittosporum*, *Premna*, *Syzygium*, *Taeniophyllum*, *Xylocarpus*, and *Zeuxine*. Six families extend disjunctly from Africa to Fiji or the Pacific Basin: Alangiaceae, Casuarinaceae, Cycadaceae, Flagellariaceae, Pandanaceae, and Pittosporaceae.

5. African-Eurasian-Australasian.

Ninety-six additional genera extend south into Australasia, some with representatives on New Caledonia (*Acanthus*, *Acridocarpus*, *Aponogeton*, *Ceriops*, *Olax*, *Sonneratia*, and *Tristellateia*) or New Zealand (*Gastrodia*, *Glossostigma*, *Iphigenia*, *Pelargonium*, and *Sebaea*). Actually 205 of the 555 African-Eurasian-Pacific disjunct genera have representatives on the continent of Australia, although many were included in other totals above because they extend farther east into the Pacific.

6. Indian Ocean-Eurasian (-Pacific).

Forty-seven genera that do not reach mainland Africa do have discontinuous ranges from Madagascar, the Mascarene, Comoro, or Seychelle Islands to Eurasia, mostly to southeastern and eastern Asia and to Malesia, and often to Australia or the oceanic islands of the Pacific Basin. Some of the better known of these genera are *Allantospermum*, *Alyxia*, *Balanophora*, *Carallia*, *Deeringia*, *Dillenia*, *Elaeocarpus*, *Erythrospermum*, *Foetidia*, *Geniostoma*, *Haloragis*, *Hedychium*, *Hiptage*, *Myoporum*, *Nepenthes*, *Ochrosia*, *Pipturus*, *Sandoricum*, *Timonius*, and *Vateria*. When added to the mainland African-Eurasian (-Pacific) disjunct genera, slightly more than 600 genera share this common pattern of disjuction.

In view of the relatively large number of discontinuous families and genera reaching Eurasia and beyond from Africa and adjoining islands, one must seek an explanation other than long-range dispersal to account for this very common type of disjunction. Some of the maritime, aquatic, and fleshy-fruited taxa on islands most likely are dispersed over large distances by sea currents and birds. However, many of the wide-ranging taxa have probably migrated over land or over small water gaps by normal methods of short-distance dispersal at a time when much of northern Africa, Arabia, and southern Asia were much moister than they are at present. In addition, there may have been much more land between the continental Seychelle Islands and Ceylon and India in the past, over which some of the taxa might have migrated. The usual Gondwanaland, continental-drift, tectonic-plate movement, and similar explanations have little if any bearing here because such geological phenomena must have taken place, if ever, long before even these relict distributions were initiated.

V. AMPHI-INDIAN OCEAN

Very few taxa are restricted just to Africa and Australasia, thus disjunct across the Indian Ocean. Only 18 genera apparently are so discontinuous between mainland Africa and Australasia, mostly Australia, with another 11 having representation in Australasia and only Madagascar, the Mascarene, or the Seychelle Islands. This total is likely to be reduced as species are found to be naturalized in one area or the other or as careful revisions show the representation on either side of the Indian Ocean to be generically distinct or to belong to larger, wider-ranging genera. This rather rare type of major discontinuity directly contradicts the rather loose speculation of some authors that there is a close relationship between the floras of Australia and Africa. Inasmuch as 15 of the 29 genera have Madagascan, Mascarene, or Seychelle representatives and 13 have insular populations east of the Indian Ocean other than those in Australia-Tasmania, long-distance dispersal would seem to be the logical explanation for these disjuncts. _Adansonia_, _Anacampseros_, _Chrysithrix_, _Dietes_, and _Triraphis_ are primarily African or Madagascan; whereas _Arthropodium_, _Bubbia_, _Caesia_, _Cassinia_, _Cunonia_, _Helipterum_, _Hibbertia_, _Humea_, _Keraudrenia_, _Macadamia_, _Rulingia_, _Soulamea_, and _Villarsia_ are primarily Australian, Papuan, or New Caledonian. Other genera are more evenly divided (_Australina_, _Brachyachne_, _Bulbinella_, _Costularia_, _Entolasia_, _Grammatotheca_, _Potamophila_, and _Wurmbea_), or are primarily oceanic (_Astelia_, _Cohnia_, and _Cossinia_).

VI. ASIAN-PACIFIC

Another large discontinuous category, but one somewhat less wide-ranging usually than the African-Asian (-Pacific) type of distribution, is represented by about 460 genera and five families. These taxa have species in mainland Asia and others in eastern Malesia to the Bismarcks, in the Solomons, in the New Hebrides, in Fiji-Tonga-Samoa, or on more distant islands in the Pacific Basin. These disjuncts are best treated in four subcategories.

1. Asian-Papuan.

Nearly 200 genera range from Asia to New Guinea or the Bismarck Archipelago: _Anisoptera_, _Hopea_, _Ixonanthes_, _Kadsura_, _Lithocarpus_, _Pentaphragma_, _Ploiarum_, _Pometia_, _Sarcandra_, _Shorea_, _Sycopsis_, _Walsura_, and _Zanonia_ among other genera, and the two families Crypteroniaceae and Pentaphragmataceae.

2. Asian-Papuan-Melanesian.

Another 130 genera reach beyond the Bismarcks to the Solomon Islands (46), the Santa Cruz and New Hebrides Islands (15), Fiji (29), Tonga (19), or Samoa (21), including _Aegiceras_, _Ailanthus_, _Alpinia_, _Amoora_, _Artocarpus_, _Caryota_, _Dischidia_, _Dysoxylum_, _Hanguana_, _Hoya_, _Hydnophytum_, _Koelreuteria_, _Livistona_, _Madhuca_, _Mangifera_, _Melicope_, _Metroxylon_, _Nypa_, _Palaquium_, _Semecarpus_, _Tristania_, and _Xanthophyllum_ and the family Daphniphyllaceae.

3. Asian-Papuan-Pacific Basin.

Well beyond the Andesite Line in the Pacific Basin, about 40 genera have representation on the islands of Polynesia (many on the distant islands of the Hawaiian chain): e.g., _Alectryon_, _Alstonia_, _Berrya_, _Cyrtandra_, _Dendrobium_, _Fagraea_, _Freycinetia_, _Gahnia_, _Melastoma_, _Planchonella_, and _Wikstroemia_.

4. Asian-Papuan-Australasian.

To the south 92 additional genera have ranges terminating in Australia (62), New Caledonia (23), or New Zealand (6), including _Aegialitis_, _Apostasia_, _Bassia_, _Cardiopteris_, _Curcuma_, _Helicia_, _Melaleuca_, _Neolitsea_, _Philydrum_, _Polyosma_, _Siphonodon_, _Stemona_, _Stylidium_, _Vanda_, _Tetrameles_ and _Zingiber_, and the two families Cardiopteridaceae and Philydraceae. Actually, if one counts those genera credited to the Solomons and islands farther east in the Pacific, 195 of the Asian-Pacific disjunct category are believed to have indigenous species in Australia.

For many of these taxa, reaching only to eastern Malesia, Australia, or the other continental islands, normal methods of short-distance dispersal are probably adequate to explain the currently disjunct ranges. It is assumed that in the past, particularly during periods of great tectonic activity and during the glacial lowering of sea-levels, much more land at times connected Malesia to mainland Asia, and Australia to Malesia. For various good biogeographical considerations, short over-water hops may have been necessary for disseminules to reach such presently isolated continental islands as New Caledonia, New Zealand, the Solomons, New Hebrides, Fiji, and Tonga (Thorne, 1965), 1969). Beyond the Andesite Line lie only islands of oceanic, volcanic origin. These must have been reached by disseminules carried over long distances by birds, sea or air currents, or by man, although numerous atolls and guyots, flat-topped sea-mounts, in the western Pacific Basin do indicate that once there were more high islands available there for stepping-stones (Thorne, 1963).

VII. PACIFIC OCEAN

Approximately 370 genera and 7 families that reach neither the Asiatic nor the American mainland have discontinuous distributions between at least two Pacific island groups (Australia here being treated as a very large Pacific island). These are truly Pacific taxa. Some have relatively restricted ranges that hardly merit comparison with taxa showing intercontinental disjunctions. Such are the 44 genera disjunct between western Malesia (east to the Moluccas) and eastern Malesia (New Guinea and the Bismarcks); 19 genera between Australia and New Caledonia; 48 between Australia and New Zealand; 5 between Australia and the New Hebrides or Fijis; 16 between New Guinea and New Caledonia (mostly also on intermediate Melanesian islands or in Queensland); one, *Carpodetus*, between New Guinea and New Zealand; 43 between western Malesia or New Guinea and the Fiji, Tonga, or Samoan Islands. Similar western Pacific disjuncts are the Stackhousiaceae and the 5 genera that each represent monogeneric families: *Balanops*, *Byblis*, *Corynocarpus*, *Eupomatia*, and *Galbulimima*. Somewhat wider gaps in range are shown by the 39 genera distributed from western Malesia to Australia (18), to New Caledonia (11), and to New Zealand (10).

More wide-ranging in the Pacific are at least 37 genera that have representation from western Malesia or New Guinea to the Hawaiian or other Polynesian islands (22), including *Ascarina*, *Astronia*, *Coprosma*, *Cyathodes*, *Faradaya*, *Lepinia*, *Neubergia*, *Olearia*, *Pseudomorus*, *Santalum*, and *Trimenia* (Trimeniaceae); from Australia to the Hawaiian or other Polynesian islands (*Nestegis* and *Corokia*); or are disjunct within the Pacific Basin (14): *Apetahia*, *Charpentiera*, *Cheirodendron*, *Crossostylis*, *Earina*, *Fitchia*, *Meryta*, *Nesoluma*, *Oparanthus*, *Pelea*, *Pelagodoxa*, *Phyllostegia*, *Reynoldsia*, and *Sclerotheca*.

Some of the widely distributed genera of the Pacific-Indian Ocean areas discussed under categories IV and V could as well be treated here as a subcategory, Pacific-Indian Ocean disjuncts. At least 14 genera range from the Indian Ocean islands, including Madagascar, to Australia or islands beyond: *Arthropodium*, *Astelia*, *Bleekeria*, *Bubbia*, *Cohnia*, *Cossinia*, *Geniostoma*, *Hibbertia*, *Humea*, *Keraudrenia*, *Macadamia*, *Pipturus*, *Rulingia*, and *Soulamea*.

Because its species are entirely restricted to the cool, shallow coastal waters of the Pacific Ocean and its tributary seas and gulfs along both the North American and Asiatic continents, Phyllospadix of the Zosteraceae must be considered a Pacific genus. It could, however, with some justification have been included under Category I as a Eurasian-Western North American (Beringian-Temperate or Amphi-Pacific North Temperate) disjunct. Similarly, the more tropical sea-grasses, Enhalus and Thalassodendron, of shallow waters of the Indian and western Pacific Oceans belong here in the Pacific-Indian Oceans subcategory of wide disjuncts.

VIII. PACIFIC-INDIAN-ATLANTIC OCEANS

Instead of being intercontinental disjuncts separated by the oceans, the taxa of this category are interoceanic disjuncts separated by the continents. By definition, relatively few taxa belong to this group. Most notable are the widely distributed genera of marine phanerogams. Zostera (s.l.) forms marine meadows in shallow, cool waters off the coasts of all the continents on either side of the equator to the subarctic or subantarctic. The more tropical Halodule, Halophila, Syringodium, and Thalassia have extraordinarily large and discontinuous ranges in the shallow waters of the Indian Ocean-Red Sea-western Pacific Ocean and the Caribbean Sea and Gulf of Mexico, all extending via the Gulf Stream to Bermuda in the Atlantic. Halodule and Halophila extend south to Brazilian coasts (den Hartog, 1972) and one species of Halophila entered the eastern Mediterranean Sea from the Red Sea via the Suez Canal. The eastern Atlantic Ocean seems to be strangely lacking in these genera except for records of Halodule from Mauritania, Senegal, and Angola (den Hartog, 1970). Cymodocea (s.s.) and Posidonia, the sole genus of the Posidoniaceae, have rather different disjunct patterns. Neither is found in tropical American waters, but each has a species that ranges from the Atlantic coasts of North Africa or of the Iberian Peninsula through the Mediterranean Sea and one or more species in Australian waters. Posidonia is extratropical, but the more tropical Cymodocea ranges widely in the Indian Ocean-Red Sea areas and into the western Pacific as far as the Ryukyu Islands and New Caledonia.

Perhaps the only terrestrial angiosperm possessing a somewhat similar pattern of disjunction is the palm genus Pritchardia, assuming that the Cuban genus Colpothrinax is indeed congeneric with it. The combined range is Fiji, Tonga, Tuamotus, Hawaii, and Cuba (van Balgooy, 1971), a truly remarkable, if not anomalous, range. Most of the circumtropical strand plants reach continental as well as island shores, hence are technically and just barely excluded from this category.

IX. AMERICAN-AFRICAN

Since I have published a lengthy discussion of this disjunct category (Thorne, 1973), I shall give just a brief summary here. Twelve angiospermous families, most of them largely American, are limited essentially to America and Africa (including Madagascar): Bromeliaceae, Cactaceae, Canellaceae, Caricaceae, Humiriaceae, Hydnoraceae, Loasaceae, Mayacaceae, Rapateaceae, Turneraceae, (Figure 16-6), Velloziaceae, and Vochysiaceae. Seven subfamilies, including Napoleonoideae of the Lecythidaceae, Siparunoideae of the Monimiaceae, and Strelitzioideae of the Musaceae; 10 tribes and subtribes, including the rafflesiaceous Cytineae and the leguminous Swartzieae; and 111 genera, as Annona, Barbacenia, Chrysobalanus, Elaeis, Guarea, Hyptis, Menodora, Ocotea, Pitcairnia,

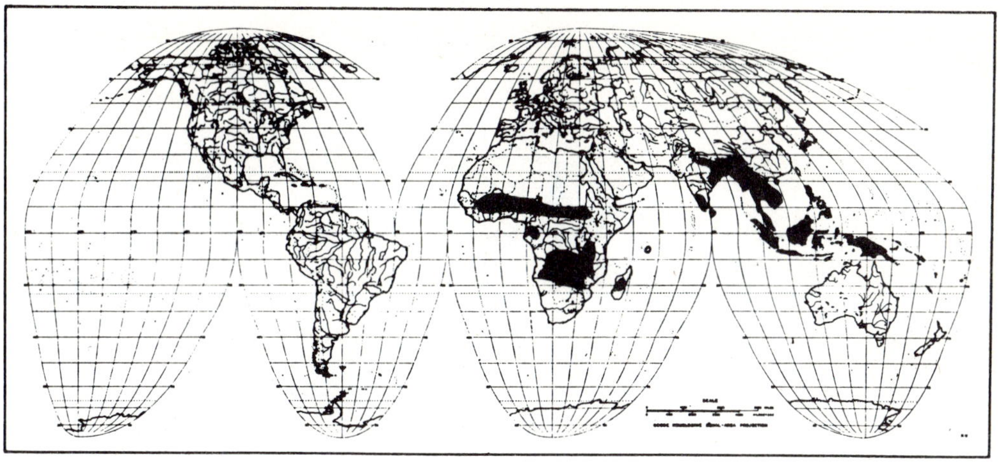

ORIGINAL GENERALIZED MAP OF THE TROPICAL AFRICAN-ASIAN-MALESIAN FAMILY DIPTEROCARPACEAE
Drawn by Chris Davidson; base map as in Fig. 2.

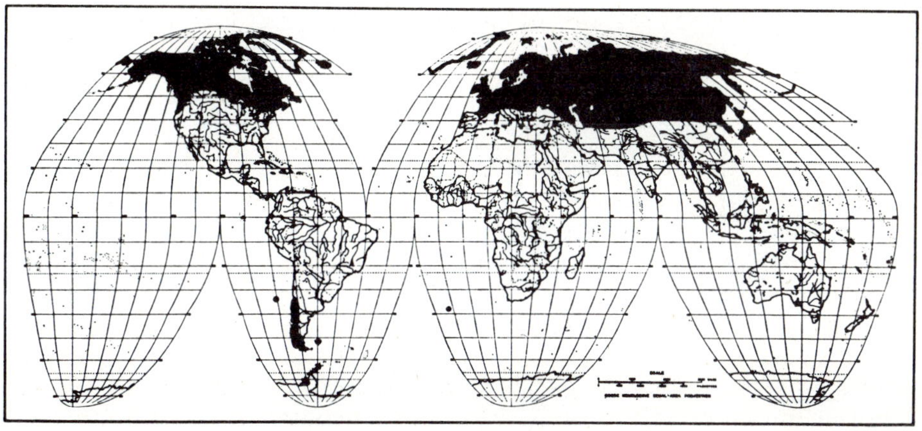

MUCH GENERALIZED DISTRIBUTION OF THE CIRCUM-ARCTIC, CIRCUM-BOREAL, AND BIPOLAR GENUS *Empetrum*, THE CROWBERRIES
Based largely upon Good (1964); base map as in Fig. 2.

Figure 16-3. Generalized distribution of the Dipterocarpaceae and Empetrum. Used with permission.

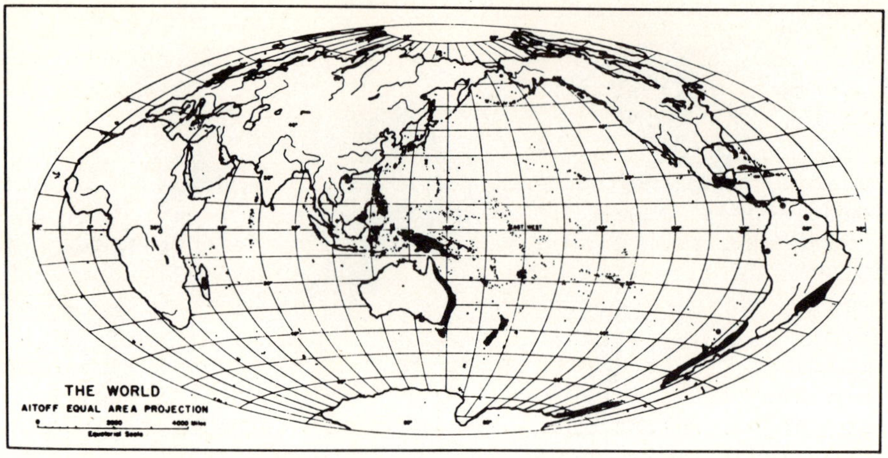

DISTRIBUTION OF THE AMERICAN-AUSTRALASIAN-MADAGASCAN FAMILY WINTERACEAE
Modified from Smith (1943a, b); base map as in Fig. 1.

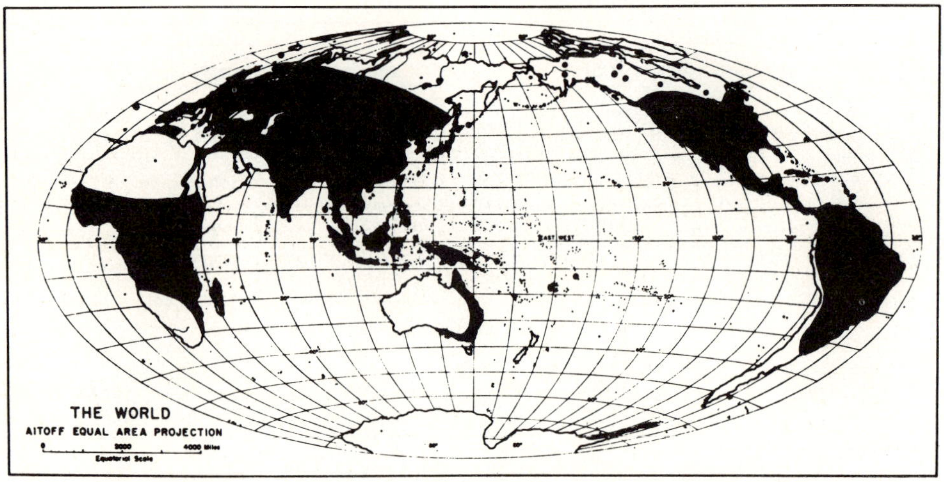

MUCH GENERALIZED DISTRIBUTION OF THE SUBCOSMOPOLITAN AQUATIC SPECIES *Ceratophyllum demersum*
Modified in part from Hultén (1968); base map as in Fig. 1.

Figure 16-4. Distribution of the Winteraceae and Ceratophyllum. Used with permission.

Raphia, Sacoglottis, Symphonia, and Vismia, are restricted to America and Africa. In analyzing the "Flora of West Tropical Africa" (Hutchinson and Dalziel, 1927-1936, 1954-1968), I listed 108 species believed to be indigenous just in America and Africa. These, probably recent immigrants from one continent to the other, are almost all aquatic, semi-aquatic, maritime, riparian, or ruderal plants.

The floristic relationships of Africa to Eurasia-Australasia and of America to Eurasia-Australasia are much stronger than those linking Africa and America. The vast number of taxa on the two continents that have been unable to span the Atlantic far outweighs the relatively few amphi-Atlantic taxa. The greatly varying degree of divergence characterizing the American-African disjuncts reflects roughly the varying lengths of time the various taxa have been separated on the two continents. Consequently, only long-distance dispersal is adequate to explain the widely spaced events of immigration across the Atlantic that must have continued throughout the history of the dispersal of seed plants in the southern hemisphere. In view of all these facts, neither continental displacement via the currently popular tectonic-plate movements nor trans-Atlantic land-bridges are realistic explanations for most members of this category of major discontinuities.

X. NORTH AMERICAN-SOUTH AMERICAN

The taxa discontinuous between North and South America are best considered under several subcategories, depending upon whether their areas are primarily tropical, temperate (amphitropical), or bipolar.

1. Tropical.

Possibly 3,000 seed-plant genera are endemic in tropical America (Good, 1964). Surely, a very large percentage of these, not analyzed for this study, are common to North and South America. This is not surprising considering the massively effective land-bridge furnished since Pliocene time by Central America and the diagramatically placed West Indian islands linking Florida to northern South America. In subtropical Florida (Long and Lakela, 1971) alone there are 150 neotropical genera that are disjunct between the southern tip of the peninsula and the Antilles, Mexico, or South America. Of 228 seed-plant families indigenous in South America, all but 27 are represented also in Mexico, and 7 more have reached at least Panama in Central America. Twelve largely tropical families are restricted to the two continents: Bixaceae, Brunelliaceae, Cannaceae, Cyclanthaceae, Cyrillaceae, Krameriaceae, Lennoaceae, Marcgraviaceae, Martyniaceae, Theophrastaceae, Tropaeolaceae (primarily Andean), and Zamiaceae. The North American Garryaceae and South American Quiinaceae both reach the Greater Antilles and Middle America; and 8 of the 12 American-African tropical families reach Mexico, at least, and most of them reach Florida or farther north in North America. The Sarraceniaceae, partly tropical but largely temperate, are classical in their disjunction between the Guyana Highlands (Heliamphora), California-Oregon (Darlingtonia), and eastern North America (Sarracenia).

2. Temperate.

The amphitropical American disjuncts have recently received considerable attention (Constance, Heckard, Chambers, Ornduff, and Raven, 1963; Thorne and Lathrop, 1970). At least 65 primarily temperate American genera are widely discontinuous across the American tropics, including Agoseris, Amsinckia, Bahia, Blennosperma, Camissonia, Chorizanthe, Clarkia, Cryptantha, Downingia, Gaillardia, Galvezia, Gaura, Gilia, Haplopappus, Lasthenia, Lepuropetalon, Lesquerella, Madia, Monanthochloe, Orthocarpus, Oxytheca, Phacelia, Plectritis, and Schedonnardus. At least 90 more widely ranging temperate genera show similar disjunctions within the New World, as Adenocaulon, Androsace, Antennaria, Calandrinia, Deschampsia, Ephedra, Elatine, Fagonia, Fragaria, Frankenia, Gleditsia, Lepechinia, Lilaeopsis, Lotus, Lupinus, Menodora, Mimulus, Myosotis, Myosurus, Orobanche, Plagiobothrys, Sanicula, Taraxacum, and Thlaspi. Raven (1963) lists 128 species, species-pairs, or species-groups of seed plants that have disjunct ranges in temperate North and South America. He treats in another group the temperate disjuncts, many of them woody, that occur in the deserts of the two continents. Some of these taxa are especially prominent in the American Southwest: Atamisquea emarginata Miers, Errazurizia, Fagonia, Gutierrezia, Helietta, Hoffmanseggia, Koeberlinia spinosa Zucc., Larrea, Menodora, Prosopidastrum, and Prosopis. Raven lists also 22 species or species-pairs of grasses and forbs that are likewise desert disjuncts.

Raven (1963) concluded that the amphitropical disjuncts probably migrated by long-distance dispersal in late Pliocene or Pleistocene time or even more recently. The largely herbaceous temperate and bipolar disjuncts mostly have dispersed from North to South America; whereas, the more woody desert disjuncts mostly evolved in South America or perhaps in part diverged from a common tropical ancestor. The considerations upon which he bases his logical conclusions are that the unbalanced assemblage of disjuncts, mostly self-compatible and autogamous, are quite unrepresentative of the distinctive floras of the two extratropical areas, which have been distinct since the middle Cretaceous. Further, the disjuncts on either side of the tropics are closely related, often conspecific, and they occupy mostly open plant communities, as grasslands and vernal marshes (Thorne and Lathrop, 1970), easily penetrated by migrants.

An alternative explanation with some merit is that the discontinuous taxa may have migrated by normal short-distance dispersal or by stepping-stone movements along the more or less continuous western American mountain system. Certainly this explanation is valid for the dozens of genera like Ribes that are nearly continuous in the American highlands from Alaska to Fuegia. Some of the other genera with relatively continuous montane distribution across the American tropics are Alchemilla, Alnus, Berberis, Caltha, Carex, Castilleja, Epilobium, Gentianella, Geum, Hydrangea, Montia, Thalictrum, Valeriana, and Viburnum. Unlike these, however, most of the temperate taxa discussed above are unrepresented on the Andean cordillera across the American tropics. Most are not montane at all, and many are adapted to the Mediterranean climate of the Pacific coastal areas. Thus, long-distance dispersal rather than continuous migration is strongly indicated for them.

3. Bipolar.

Also discussed by Raven (1963) were the bipolar disjuncts. He listed 26 species or species-pairs of seed plants with bipolar distribution, all but two of them circum-Arctic or circumboreal as well. Nearly a dozen genera conform to this pattern: Armeria, Catabrosa, Empetrum, Euphrasia, Hippuris, Honkenya, Koenigia, Littorella, Monolepis, Phippsia, and Primula. Long-distance dispersal from northern North America in Pleistocene or later times seems to be indicated here, as with the temperate disjuncts. However, at least Euphrasia is an exception to this probable route. The South American species of this genus, according to Du Rietz (1960), are more closely related to the southern and western Pacific species than they are to the North American species.

XI. SOUTH AMERICAN-AUSTRALASIAN

At least 48 genera and 7 families of seed plants are essentially restricted to temperate South America (some passing into the tropics along the Andes) and Australasia, with a few following tropical highlands to southeastern Asia or rarely reaching Madagascar. An additional 28 more widely distributed genera and 2 families are linked to Australasia by common species or closely related species represented in the two regions. The largely Pacific genera, Coprosma, Haloragis, and Santalum, reach the Juan Fernandez Islands but not the South American mainland. These taxa are perhaps better analyzed by treating them in three minor subcategories.

1. South American-Australasian.

This is the largest subcategory, containing those taxa with representatives only in temperate South America (a few passing along the western American cordillera to Mexico) and the Australasian area, including New Zealand and its subantarctic islands, Tasmania, Australia, New Caledonia, and New Guinea (a few reaching the mountains of Borneo or Taiwan or even, in the other direction, the Polynesian and Hawaiian mountains). Groups that possess this spectacular type of discontinuity are the Araucariaceae and Eucryphiaceae and at least 36 genera: Gevuina, Lomatia, Oreocallis, and Orites, Amphibromus, Araucaria, Aristotelia, Azorella, Celmisia, Colobanthus, Discaria, Donatia, Eucryphia, Fuchsia, Gaimardia, Griselinia, Hebe, Hypsela, Jovellana, Laurelia, Libertia, Marsippospermum, Muehlenbeckia, Nothofagus, Oreomyrrhis, Ourisia, Pernettya, Phyllachne, Pseudopanax, Rostkovia, Schizellema, Selliera, Tetrachondra, Trichocline, Uncinia, and Vittadinia. Three American amphitropical disjuncts, Calandrinia, Lilaeopsis, and Plagiobothrys, are also disjunct from temperate South America to Australasia. Most of the other 25 widely ranging genera mentioned above reach only Australasia, and hence also belong in this subcategory.

2. South American-Australasian-Asian.

This subcategory is quite artificial, since it includes wide disjuncts almost identical to the preceding except that they have species successfully established on the mainland of southeastern Asia. Four primarily Australasian families have thus reached at least mainland Asia to the west and temperate South America to the east. They are the Centrolepidaceae, Coriariaceae, Epacridaceae, and Stylidiaceae. Seven genera that likewise fit in this subcategory are Coriaria, Dacrydium, Euphrasia (with also extensive holarctic distribution), Gaultheria, Lagenifera, Leptocarpus, and Oreobolus. Coriaria is somewhat anomalous here, for in addition to one species reaching Tahiti and north along the Andes and Middle American mountains to Mexico, it has another in the Mediterranean region.

3. South American-Australasian-Madagascan.

Likewise little different from the two preceding subcategories are those that extend their wide discontinuities from South America to Australasia and on to the Mascarene Islands or Madagascar, also in some cases involving Malesia or southeastern Asia. The one family, Winteraceae, with this type of disjunct range has not been found on the Asiatic mainland but members have reached north to Mexico in America, north to Borneo and the Philippines in Malesia, and south to the highlands of Madagascar. It is worthy of note here that this vesselless family in most of its features is the most primitive living family of the Angiospermae. Five genera also with these enormous gaps across the Pacific and Indian Oceans are Abrotanella and Astelia (to the Mascarenes), Nertera and Weinmannia (to Madagascar), and Dianella (on Madagascar and also in mainland East Africa).

For some of the plants of this temperate South American-Australasian disjunct category, especially those with fleshy, bird-dispersed fruit that have far-flung distributions on oceanic islands of the Pacific and Indian Oceans, long-distance dispersal by birds is the rather obvious explanation for the discontinuities. Genera with fruit that must be especially attractive to birds are Aristotelia, Astelia, Coprosma, Coriaria, Dacrydium, Dianella, Fuchsia, Gaultheria, Griselinia, Nertera, Pernettya, and Pseudopanax. Other genera that appear to require and be capable of long-distance dispersal are Abrotanella, Azorella, Colobanthus, Gaimardia, Haloragis, Lagenifera, Leptocarpus, Oreobolus, Oreomyrrhis, Ourisia, Rostkovia, Schizellema, Selliera, Tetrachondra, Vittadinia, Uncinia, and Weinmannia. There are several taxa, however, which because of large, dry or heavy cones, fruits, or seeds are not logically explained by bird-carriage or other forms of long-distance dispersal. Among these are Araucaria, Eucryphia, Laurelia, Nothofagus, and the four genera of Proteaceae.

Many authors (listed in Thorne, 1963) have suggested that Antarctica, much warmer, forested, and probably a more extensive archipelago in Cretaceous and Tertiary times, must have served formerly as a "sweepstakes" migration route for many seed plants and animals. Distances between Antarctica and Tasmania-Australia, Antarctica and New Zealand, and Antarctica and South America may have been much less in the past. Biological evidence that Antarctica ever furnished a continuous land-bridge like Beringia, however, is unconvincing. Plate tectonics, now currently accepted by many geologists, has most recently been suggested (Raven and Axelrod, 1972) as the ultimate answer to many of the problems of Australasian-temperate South American paleobiogeography. According to this hypothesis, about 100 million years before the present (B. P.), Antarctica linked South America with New Zealand and with Australia, furnishing an effective pathway for migration of temperate forest plants and associated animals. Later the various tectonic plates moved apart to their present positions, causing considerable tectonic disturbance in the southwestern Pacific. This is a rather attractive hypothesis which one hopes the geologists and paleontologists can in time substantiate.

XII. TEMPERATE SOUTH AMERICAN-ASIAN

This type of disjunction in which the representatives of a taxon are separated by the entire width of the Pacific Ocean is expectedly quite rare. The outstanding example of this huge discontinuity is the family Lardizabalaceae, with 2 genera, Boquila and Lardizabala, in central Chile and 5 in eastern Asia from Japan and Korea to the Himalayas. A few other examples of this South American-Asian disjunction were discussed by Stebbins (1940). He suggested that the largely Andean mutisioid genera Hyaloseris and Proustia are possibly closely related to Catamixis of the northwestern Himalaya and Nouelia of Yunnan, to Hesperomannia of Hawaii, and perhaps to Stifftia of Brazil. He considers to be even more closely related to the mutisioid Leucomeris of southeastern Asia the genus Gochnatia of the Andes and Brazil north to Mexico and Texas. Hypochoeris of the Lactuceae has speciated heavily in southern South America and has a secondary center in the Mediterranean Sea area and a few species across Europe to northeastern Asia. There are no native species of either Hypochoeris or of the Lardizabalaceae in North America. Stebbins also suggested that the bignoniaceous perennial herb Argylia of Andean Peru to temperate South America has its closest relatives in the equally herbaceous Incarvillea of the Sino-Himalayan region.

Of all the disjunct categories discussed, this is the most difficult to explain. The species of Lardizabalaceae have baccate fruit, but to explain this discontinuity by direct bird-carriage between Chile and eastern Asia rather strains credulity. The distribution of Coriaria may be instructive here. Coriaria, with fruit a pseudo-drupe of cocci surrounded by persistent fleshy petals at maturity, is greatly favored by many birds and is almost certainly dispersed by them. If the northern South and Central American, Pacific, and Mediterranean species were deleted from the map, it would show a distribution pattern similar to that of the Lardizabalaceae. However, if the latter family dispersed like Coriaria across the southern Pacific, it should have left some traces along the way. One can as well adduce the bipolar distribution of Empetrum. If the circumpolar, European, and North American parts of its range are eliminated, a similar pattern again emerges. Surely, however, as in Coriaria, Empetrum's disjunct distribution has recently been achieved by long-distance dispersal with birds as vectors. The temperate South American-Asian disjunction appears to be, at least in part, a relict pattern. I would expect that fossils of the Lardizabalaceae will ultimately be identified from North American Cretaceous or Tertiary deposits.

XIII. CIRCUM-SOUTH TEMPERATE

Considering the paucity of land in relation to water in the south temperate zone, it is understandable that relatively few taxa are present in, and largely restricted to, the temperate portions of the three southern continents. Five families, the Cunoniaceae, Gunneraceae, Podocarpaceae, Proteaceae, and Restionaceae, and two subfamilies, Escallonioideae of the Saxifragaceae and Luzuriagoideae of the Liliaceae, are largely circum-south temperate in their distribution. Yet all of these have representatives in the tropics, at least in the tropical highlands, and all reach north of the equator. The 9 genera which likewise have a primarily circum-south temperate type of discontinuous distribution are Acaena, Carpha, Cotula, Carpobrotus, Gunnera, Podocarpus, Pratia, Wahlenbergia, and Tetragonia.

Some of these taxa are well equipped for bird or sea-current dispersal over large distances, as Acaena, Carpobrotus, Gunnera, Podocarpus, Tetragonia, and the luzuriagoids. Many members, however, of the Cunoniaceae, Proteaceae, and Escallonioideae are much less well adapted for long-distance dispersal and would seem to require more nearly complete land connections for normal short-range or stepping-stone dispersal. Their present wide ranges may have resulted from migration over a once warmer and more extensive Antarctic archipelago between South America and Australia. Their South African presence is more likely due to migration from Australia through the Indonesian-Asian and East African highlands. The often called-upon breakup of a possible

Gondwanaland, if it ever existed, must have taken place far too long ago to have had much impact upon the present distribution of seed plants.

XIV. CIRCUM-ANTARCTIC

The land areas defined here as circum-Antarctic are the Grahamland peninsula of Antarctica, the southern tip of South America (especially Tierra del Fuego), and the high latitude islands surrounding the Antarctic, such as the Falklands, South Georgia, Tristan da Cunha, Gough, Marion, Crozet, Heard, Kerguelen, St. Paul, New Amsterdam, and Macquarie Islands, mostly within or near the northern limit of drift ice. The subantarctic lands have a very small flora, as might be expected from the strong cold winds, small and remote land areas, and great amounts of permanent ice. Yet their meager flora of perhaps 100 species (Good, 1964) is a rather characteristic, constant one, with perhaps 30 species with wide distributions in the circum-Antarctic region (Dawson, 1958). Among these species are *Acaena adscendens* Vahl, *Azorella selago* Hook.f., *Callitriche antarctica* Engelm. ex Hegel, *Montia fontana* L., *Myriophyllum elatinoides* Gaud., and *Ranunculus biternatus* Sm. in Rees. During the Pleistocene ice age most of these circum-Antarctic lands were covered with ice, as were Macquarie, Kerguelen, Heard and South Georgia Islands and Grahamland. Consequently, the circumpolar flora must be post-Pleistocene and must have reached these remote islands by long-distance dispersal, presumably via oceanic birds (Falla, 1960; Holdgate, 1960) mostly from southern South America and New Zealand.

XV. SUBCOSMOPOLITAN

Plant taxa that are not primarily temperate nor tropical and that have indigenous representatives on all the continents (including Australia) can be described as subcosmopolitan. Truly cosmopolitan taxa that have representatives on all parts of the earth's surface do not, of course, exist. The most nearly cosmopolitan family of seed plants is the Poaceae, which has species in all habitable parts of all the continents and even in Grahamland of Antarctica. Ninety families (among those listed by Thorne, 1968) and about 125 genera of seed plants are subcosmopolitan in their distribution and in their disjunction. It is certainly no coincidence that at least 72 of the subcosmopolitan genera are largely aquatic or at least have species that inhabit shallow water, wet, open places, maritime situations, or riparian habitats. A few of these are *Callitriche*, *Carex*, *Crassula*, *Drosera*, *Elatine*, *Eleocharis*, *Juncus*, *Lemna*, *Montia*, *Myriophyllum*, *Najas*, *Nymphaea*, *Potamogeton*, *Ranunculus*, *Ruppia*, *Triglochin*, *Typha*, *Utricularia*, *Vallisneria*, and *Wolffia*. An additional 26 genera have species with weedy tendencies, as *Amaranthus*, *Cenchrus*, *Centaurea*, *Cuscuta*, *Euphorbia*, *Gnaphalium*, *Lepidium*, *Oxalis*, *Senecio*, *Setaria*, and *Urtica*. Thus, about four-fifths of the genera have members that by habitat are readily dispersible. The remaining genera are small-seeded herbs or woody plants with mostly succulent, bird-dispersed fruit.

Very few species can be described as subcosmopolitan, and these are almost all of necessity aquatics, like *Ceratophyllum demersum* L. Some other aquatics that also have vast if somewhat sporadic distributions on all or nearly all the continents are *Anagallis minimus* (L.) Krause, *Brasenia schreberi* J. F. Gmel., *Callitriche verna* L., *Chenopodium glaucum* L., *Cladium mariscus* L. (s.l.), *Cyperus flavescens* L., *Juncus bufonius* L., *J. effusus* L., *Lemna gibba* L., *L. minor* L., *L. trisculca* L., *Ludwigia palustris* (L.) Ell, *Montia fontana* L.,

Myriophyllum spicatum L., Najas marina L., Phragmites australis (Cav.) Trin., Potamogeton pectinatus L., Ranunculus trichophyllus Chaix., Ruppia cirrhosa (Petag.) Grande, Scirpus maritimus L., Spirodela polyrhiza (L.) Schleid., Triglochin palustris L., and Zanichellia palustris L. Several of these are mapped by Meusel, Jager, and Winert (1965). There can be little doubt of the efficacy of water and shore birds in the dispersal of these species about the world.

XVI. ANOMALOUS

The final category of major types of discontinuous taxa is a wastebasket for those few genera that I am unable to fit comfortably into the preceding categories. In some instances, further exploration and discovery of the taxon, as is surely to be expected with the rafflesiaceous parasite Pilostyles, may clarify its position, which, in this case, is already approaching circumtropical. Myrica also falls a little short of being subcosmopolitan because of its absence from Australasia. If the circumboreal disjunct Gale belgica Dum. (Myrica gale L.) is removed from Myrica, the genus is also unrepresented in Europe and boreal North America. The present disjunct range of Myrica is certainly a relic of a once worldwide distribution. The santalaceous Thesium falls somewhat short of being circumtropical because of its extremely weak presence in southeastern Asia-Australasia and in southern Brazil (map in Meusel, Jager, and Winert, 1965). It might better be considered an African-Eurasian-Australasian disjunct with a small outlier in southern Brazil, as in the family Pedaliaceae, in which only Rogeria reaches the New World with a species in Brazil (so also with the crassulaceous Kalanchoë). Likewise, the American-African-Madagascan Urera of the Urticaceae transgresses its pattern by having a species in the Hawaiian Islands. Perhaps the most anomalous disjunction of all is found in the chiefly Indomalesian-American genus Thismia of the Burmanniaceae. Among these fleshy saprophytes the small section Rodwaya has one species, T. rodwayi F. v. Muell, in Victoria, Tasmania, and New Zealand; and another T. americana N. E. Pfeiff., found a couple of times many years ago in a moist prairie in Chicago, Illinois.

STATISTICAL SUMMARY

Some authors, vide Good (1964), divide taxa geographically into three major categories, depending upon whether they are widely distributed about the world, are discontinuous, or are endemic in a limited area. This division is artificial, since widely distributed taxa, because of the juxtaposition of continents and seas, are almost always discontinuous in their areas. Thus, we are left practically with only disjunct vs. endemic taxa. The latter present no great phytogeographic problems other than the formerly much greater and often discontinuous ranges of many of the relict endemics, or epibiotics. Therefore, our interest here is centered on the discontinuous taxa.

The larger the category of the taxon, the greater the likelihood that its distribution will be more or less wide-ranging and hence disjunct. Of 324 families that I currently recognize [a total somewhat larger than in my Synopsis (Thorne, 1968)] 254, or 78.4 per cent, have intercontinental or equivalent discontinuities. If the 273 recognized additional subfamilies are added to the family total, 462 of 597 families and subfamilies, about the same percentage, 77.4 per cent, have major discontinuities in their ranges. Of the perhaps 12,500 angiosperm genera usually recognized (Good, 1964), a total of nearly 3,000 are counted in this study as having major range disjunctions. Thus only 24 per cent of flowering plant genera are intercontinentally disjunct. Species have not been thoroughly analyzed for major disjunctions, but a rough estimation from this study would be about 2,000 with wide natural range disjunctions. If 225,000 is accepted as a reasonable estimate of the named valid species of flowering plants (Good, 1964), fewer than 1 per cent of angiospermous species are widely disjunct.

Of the many categories of major disjunctions, those encompassing the largest number of genera are: African-Eurasian-(Pacific), 600; Asian-Pacific, 460; Pacific, 370; North American-South American, ca. 360 (surely a serious underestimate since Middle America is disregarded here); Pantropical, 334; North Temperate, 316; Subcosmopolitan, 125; African-American, 111; and Amphi-Pacific Tropical, 85. Those with the largest number of seed-plant families are: Subcosmopolitan, 90; Pantropical, 59; North Temperate, 20; African-Eurasian-(Pacific), 17; North American-South American, 12; African-American, 12; Tropical Amphi-Pacific, 11; South American-Australasian, 7; and Pacific, 7.

GEOGRAPHIC EXERCISE

Select two species for geographic study and determine the following distribution characteristic of each (consult local and regional manuals, Good [1964], and Blake and Atwood [1942, pt. 1 and 1962, pt. 2]).

DISTRIBUTION CHARACTER	Ref.	Species #1	Species #2
1. Biogeographic Region	A-I		
2. Biome	A-II		
3. Flora Province	A-III		
4. Community Type	Chap. 15		
5. Habitat Type	Chap. 15		
6. Genus Range Pattern	C or D		
7. Genus Range Size	B-III		
8. Genus Discontinuity Distribution Pattern	C		
9. Center of Distribution	B-II		
10. Origin of Establishment in Local Area (Genus)	B-VI		
11. Diaspore Type	Chap. 15		
12. Plant Dispersal Type	B-VII		

GEOGRAPHIC LITERATURE

Balgooy, M. M. J. van. 1966. Distribution maps of Pacific plants. In: C. G. G. J. van Steenis and M. M. J. van Balgooy (eds.). Pacific Plant Areas. Vol. 2. Blumea, Suppl., 5: 1-312.

________. 1971. Plant-geography of the Pacific. Blumea Suppl., 6: 1-222.

________, and D. Hou. 1966. *Perrottetia* H. B. K. In: C. G. G. J. van Steenis and M. M. J. van Balgooy (eds.). Pacific Plant Areas, Vol. 2. Blumea, Suppl., 5: 1-312.

________, and P. W. Leenhouts. 1966. *Fagraea* Thunb. In: C. G. G. J. van Steenis and M. M. J. van Balgooy (eds.). Pacific Plant Areas. Vol. 2. Blumea, Suppl., 5: 1-312.

Blake, S. F., and A. C. Atwood. 1942-1961. Geographical guide to floras of the world. Parts 1-2. U. S. Department of Agriculture Miscellaneous Publication 401. Washington, D. C. (Part 1 reprinted by Hafner Publishing Company, 1963).

Burbidge, N. T. 1960. The phytogeography of the Australian region. Australian Journal of Botany 8: 75-212.

________. 1963. Dictionary of Australian Plant Genera. Angus and Robertson. Sydney.

Cain, S. A. 1944. Foundations of Plant Geography. Harper and Brothers. New York.

Constance, L. 1963. Introduction and historical review. In: Amphitropical Relationships in the Herbaceous Flora of the Pacific Coast of North and South America: A Symposium. Quarterly Review of Biology 38: 109-177.

________, L. R. Heckard, K. L. Chambers, R. Ornduff, and P. H. Raven. 1963. Amphitropical relationships in the herbaceous flora of the Pacific Coast of North and South America: A Symposium. Quarterly Review of Biology 38: 109-177.

Cooper, A. W., and J. W. Hardin. 1971. Floristics and vegetation of the gorges on the southern Blue Ridge escarpment. In: The Distributional History of the Biota of the Southern Appalachians, Part II: Flora, P. C. Holt (ed.). Research Division Monograph 2, Virginia Polytechnic Institute and State University. Blacksburg, Virginia.

Croizat, L. 1952. Manual of Phytogeography. Uitgeveri; Dr. W. Junk, the Hague.

Dansereau, P. M. 1957. Biogeography; an ecological perspective. Ronald Press Company. New York.

Dawson, J. W. 1958. Interrelationships of the Australasian and South American floras. Tuatara 7: 1-6.

Dressler, R. L. 1954. Some floristic relationships between Mexico and the United States. Rhodora 56: 81-96.

Du Rietz, G. E. 1960. Remarks on the botany of the southern cold temperate zone. In: C. F. A. Pantin (ed.), A Discussion on the Biology of the Southern Cold Temperate Zone. Proceedings Royal Society, London, Series B. Biological Science 152: 429-682.

Endress, P. K. 1969. *Molinadendron*, eine neue Hamamelidaceen-Gattung aus Zentralamerika. Botanische Jahrbucher 89: 353-358.

Engler, A. 1964. Syllabus der Pflanzenfamilien. 12th ed. H. Melchior (ed.). Gebruder Borntraeger. Berlin.

Eyde, R. H., and E. S. Barghoorn. 1963. Morphological and paleobotanical studies of the Nyssaceae, II. The fossil record. Journal of the Arnold Arboretum 44: 328-376.

Eyre, S. R. 1963. Vegetation and Soils; a World Picture. Aldine Publishing Company. Chicago.

Falla, R. A. 1960. Oceanic birds as dispersal agents. In: C. F. A. Pantin (ed.). A Discussion on the Biology of the Southern Cold Temperate Zone. Proceedings Royal Society (London), Series B. Biological Science 152: 429-682.

Fernald, M. L. 1925. Persistence of plants in unglaciated areas of boreal America. Memoirs of the American Academy of Arts and Science 15: 239-342.

________. 1931. Specific segregations and identities in some floras of eastern North America and the Old World. Rhodora 33: 25-63.

Florin, R. 1963. The distribution of conifer and taxad genera in time and space. Acta Horti Bergiani 20: 121-312.

Fosberg, F. R. 1948. Derivation of the flora of the Hawaiian Islands. In: E. C. Zimmerman (ed.). Insects of Hawaii. Vol. 1. University of Hawaii Press. Honolulu.

Fryxell, P. A. 1967. The interpretation of disjunct distributions. Taxon 16: 316-324.

Gams, H. 1927. Die Gattung *Trapa* L. In: E. Hannig and H. Winkler (eds.). Die Pflanzenareale. Vol. 1(3). Gustav Fischer. Jena.

Gleason, H. A., and A. Cronquist. 1964. The Natural Geography of Plants. Columbia University Press. New York.

Good, R. 1964. The Geography of Flowering Plants. 3rd ed. Wiley and Sons. New York.

Gray, A. 1846. Analogy between the flora of Japan and that of the United States. American Journal Science and Arts II, 2: 135-136.

________. 1859. Diagnostic characters of new species of phaenogamous plants, collected in Japan by Charles Wright, Botanist of the U. S. North Pacific Exploring Expedition. (Published by request of Captain John Rodgers, Commander of the Expedition.) With observations upon the relations of the Japanese flora to that of North America, and of other parts of the Northern Temperate Zone. Memoirs American Academy Arts and Science II, 6: 377-452.

Hardin, J. W. 1964. Variation in *Aconitum* of eastern United States. Brittonia 16(1): 80-94.

Hartog, C. den. 1966. *Thalassia hemprichii* (Ehrenb.) Aschers. In: C. G. G. J. van Steenis and M. M. J. van Balgooy (eds.). Pacific Plant Areas. Vol. 2. Blumea, Suppl., 5: 1-312.

________. 1970. The sea-grasses of the world. Verhandelingen Koninklijke Neerlandica Academiens Wetenschappen 59(1): 1-275.

________. 1972. The sea-grasses of Brazil. Acta Botanica Neerlandica 21: 512-516.

Hernandez-X., E., E. H. Crum, W. B. Fox, and A. J. Sharp. 1951. A unique vegetational area in Tamaulipas. Bulletin Torrey Botanical Club 78: 458-463.

Holdgate, M. W. 1960. Final discussion. In: C. F. A. Pantin (ed.). A Discussion on the Biology of the Southern Cold Temperate Zone. Proceedings Royal Society (London), Series B. Biological Science 152: 429-682.

Hou, D., and C. G. G. J. van Steenis. 1963. *Bruguiera gymnorrhiza* (L.) Lamk. In: C. G. G. J. van Steenis (ed.). Pacific Plant Areas. Vol. 1. National Institute of Science and Technology. Manila.

Hultén, E. 1937. Outline of the History of Arctic and Boreal Biota During the Quaternary Period. Bokforlags Akteibolaget Thyle. Stockholm.

________. 1958. The amphi-Atlantic plants and their phytogeographical connections. Kung. Svenska Vetenskapsakad. Handl., Fjarde Series 7(I): 1-340. 279 maps.

________. 1963. Phytogeographical connections of the North Atlantic. In: A. Löve and D. Löve (eds.). North Atlantic Biota and Their History. Pergamon Press. Oxford.

Hultén, E. 1968. Flora of Alaska and Neighboring Territories. Stanford University Press. Stanford.

Hutchinson, J., and J. M. Dalziel. 1927-1936. Flora of West Tropical Africa. 2 vols. Crown Agents. London.

________. 1954-1968. Flora of West Tropical Africa. 2nd ed. R. W. J. Keay and F. N. Hepper (eds.). 3 vols. Crown Agents. London.

Li, H. L. 1952. Floristic relationships between eastern Asia and eastern North America. Transactions American Philosophical Society, New Series 42: 371-429.

Long, R. W., and O. Lakela. 1971. A Flora of Tropical Florida. University of Miami Press. Coral Gables.

Löve, D. 1962. Plants and Pleistocene--problems of the Pleistocene and Arctic. McGill University Museums Publication 2: 17-39.

McVaugh, R. 1952. Suggested phylogeny of _Prunus serotina_ and other wide-ranging phylads in North America. Brittonia 7: 317-346.

Meusel, H. 1969. Beziehungen in der Florendifferenzierung von Eurasien und Nordamerika. Flora, Abt. B 158: 537-564.

________, E. Jager, and E. Winert. 1965. Vergleichende Chorologie der Zentraleuropaischen Flora. Karten. Gustav Fischer. Jena.

________, and R. Schubert. 1971. Beitrage zur Pflanzengeographie des Westhimaljas. I. Teil: Die Arealtypen. Flora 160: 137-194.

Miranda, F., and A. J. Sharp. 1950. Characteristics of the vegetation in certain temperate regions of eastern Mexico. Ecology 31: 313-333.

Ostenfeld, C. H. 1915. On the distribution of the sea-grasses. A preliminary communication. Proceedings Royal Society Victoria 27: 179-191.

________. 1927a. Meeregräser 1. Marine Hydrocharitaceae. _In_: E. Hannig and H. Winkler (eds.). Die Pflanzenareale. Vol. 1(3).

________. 1927b. Meeresgräser 2. Marine Potamogetonaceae. _In_: E. Hannig and H. Winkler (eds.). Die Pflanzenareale. Vol. 1(4).

Polunin, P. 1960. Introduction to Plant Geography. Longmans, Green and Company, Ltd. London.

Raven, P. H. 1963. Amphitropical relationships in the floras of North and South America. Quarterly Review of Biology 38: 151-177.

________, and D. I. Axelrod. 1972. Plate tectonics and Australasian paleobiogeography. The complex biogeographic relations of the region reflect its geologic history. Science 176: 1379-1386.

Schofield, W. B. 1969. Phytogeography of northwestern North America: bryophytes and vascular plants. Madroño 20: 155-207.

Setchell, W. A. 1920. Geographical distribution of the marine spermatophytes. Bulletin of the Torrey Botanical Club 47: 563-579.

________. 1935. Geographic elements of the marine flora of the North Pacific Ocean. American Naturalist 69: 560-577.

Shantz, H. L, and R. Zon. 1935. Natural Vegetation in Atlas of American Agriculture. U. S. Government Printing Office. Washington, D. C.

Sharp, A. J. 1946a. Informe preliminar sobre algunos estudios fitogeograficos efectuados in Mexico y Guatemala. Revista de la Sociedad Mexicana de Historia Natural 7: 35-39.

________. 1946b. _Pinus strobus_ south of the United States. Journal of the Elisha Mitchell Scientific Society 62: 229-230.

Shelford, V. E. 1963. The Ecology of North America. University of Illinois Press. Urbana, Illinois.

Shreve, F., and I. W. Wiggins. 1964. Vegetation and Flora of the Sonoran Desert. Stanford University Press. Stanford, California.

Sleumer, H. 1955. Proteaceae. Flora Malesiana 5: 147-206.

Smith, A. C. 1943a. The American species of Drimys. Journal of the Arnold Arboretum 24: 1-33.

________. 1943b. Taxonomic notes on the Old World species of Winteraceae. Journal of the Arnold Arboretum 24: 119-164.

Stebbins, G. L., Jr. 1938. The western American species of Paeonia. Madroño 4: 252-260.

________. 1940. Additional evidence for a holarctic dispersal of flowering plants in the Mesozoic era. Proceedings Sixth Pacific Scientific Congress 4: 649-660.

________, and A. Day. 1967. Cytogenetic evidence for long continued stability in the genus Plantago. Evolution 21: 409-428.

Steenis, C. G. G. J. van. 1962. The land-bridge theory in botany. Blumea 11: 235-372.

________ (ed.). Pacific Plant Areas. Vol. 1. National Institute of Science and Technology. Manila.

________, and M. M. J. van Balgooy (eds.). 1966. Pacific Plant Areas. Vol. 2. Blumea, Suppl., 5: 1-312. 147 maps.

Steenis-Kruseman, M. J. van. 1966. Bibliography of Pacific and Malesian plant maps of phanerogams, first supplement. In: C. G. G. J. van Steenis and M. M. J. van Balgooy (eds.). Pacific Plant Areas. Vol. 2. Blumea, Suppl., 5: 1-312.

Stern, F. C. 1946. A Study of the Genus Paeonia. Royal Horticulture Society, London.

Steyermark, J. A. 1950. Flora of Guatemala. Ecology 31: 368-372.

Stone, D. E. 1972. New World Juglandaceae, III. A new perspective of the tropical members with winged fruits. Annals of the Missouri Botanical Garden 59: 297-321.

Thorne, R. F. 1963. Biotic distribution patterns in the tropical Pacific. In: J. L. Gressitt (ed.). Pacific Basin Biogeography. Bishop Museum Honolulu.

________. 1965. Floristic relationships of New Caledonia. University of Iowa Studies Natural History 20(7): 1-14.

________. 1968. Synopsis of a putatively phylogenetic classification of the flowering plants. Aliso 6(4): 57-66.

________. 1969. Floristic relationships between New Caledonia and the Solomon Islands. Philosophical Transactions Royal Society (London), B, 255: 595-602.

________. 1973a. Floristic relationships between tropical Africa and tropical America. In: B. J. Meggers, E. S. Ayensu, and W. D. Duckworth (eds.). Tropical Forest Ecosystems in Africa and South America: A Comparative Review. Smithsonian Institution Press. Washington, D. C.

________. 1973b. Major disjunctions in the geographic ranges of seed plants. The Quarterly Review of Biology 47(4): 365-411.

________, and E. W. Lathrop. 1970. Pilularia americana on the Santa Rosa Plateau, Riverside County, California. Aliso 7(2): 149-155.

Vester, H. 1940. Die Areale und Arealtypen der Angiospermen-Familien. Botanisches Archiv 41: 203-275, 295-356, 520-577.

Weaver, J. E., and F. W. Albertson. 1956. Grasslands of the Great Plains. Johnson Publishing Company. Lincoln, Nebraska.

Willis, J. C. 1966. A Dictionary of the Flowering Plants and Ferns. 7th ed. Revised by H. K. Airy Shaw. Cambridge University Press. Cambridge.

Wood, C. E., Jr. 1971. Some floristic relationships between the southern Appalachians and western North America. In: P. C. Holdt (ed.). The Distributional History of the Biota of the Southern Appalachians, Part II, Flora, Research Division Monograph 2. Virginia Polytechnic Institute and State University. Blacksburg, Virginia.

Chapter 17. FLORISTIC STUDIES

The plant taxonomist must have a holistic or ecosystematic approach to have an understanding of individuals and populations in the field. This approach involves (1) learning and observing the distinguishing and diagnostic characteristics of taxa, (2) discovering the limiting factors in the distribution of species, (3) discerning the role of individuals in the ecosystem, (4) detecting evolutionary mechanisms at work, and (5) determining the phenetic and phylogenetic relationships of diverse organisms. A field student should study the ecological characteristics--biotic, abiotic, spatial, temporal--of species and should become familiar with the environmental relationships of the individuals within the population. The dynamic role of individual organisms in the environmental processes--Climatogenics, Pedogenics, Biogenics, Phylogenetics--should be observed and determined whenever feasible. And a knowledge of "Habitats" should be routinely applied in identifying geomorphic, geologic, and vegetational features in the field.

To help the student acquire the holistic and ecosystematic perspective are inclusions in this chapter on habitats in North America north of Mexico, a summary outline of ecological characteristics of species and populational relationships, and a community sample study by means of community lists, mosaics or patterns, and profiles. The flora or community spectral analysis will show some of the diversity within the plant assemblage and give some indication of origin and evolution of the flora or community.

Enough field trips should be taken during a study for the student to understand fundamental biotic, abiotic, spatial, and temporal relationships; to learn how to find the maximum diversity in the region; how to collect and care for specimens; how to distinguish life forms and dispersal mechanisms; how to classify or identify different communities and habitats; how to recognize families, genera, and dominant species in the field; and to be introduced to fundamental population and geographic principles. If four trips are taken during a course, habitats, patterns, and profiles could be emphasized on one; life forms, dispersal, and sociability on a second; duration, phenology, and taxon diversity on a third; and distribution, origin, and population concepts on a fourth (See Chaps. 15, 16). Recognition of species should be a basic part of each trip and on each successive trip the emphasis of the preceding trip should be reviewed.

Sound field work combines cerebration with vision. Good floristics practice involves the making of presence lists by community or habitat; classifying the community or identifying the habitat; indicating the locality and date of study; noting slope, exposure, and altitude; and determining the rock or soil type. In addition to collecting equipment, standard accessories for field work should be an altimeter, a Brunton compass or clinometer and standard compass, topographic or political maps, geologic or soil maps, and waterproof writing equipment.

Successful field work involves adequate preparation by the student on well-presented assignments and objectives by the teacher prior to the actual field trip. Herbarium specimens arranged by community or family should be available for study so that the student does not have to spend a major portion of his field time learning which name tag goes with each plant.

Profiles of the communities within the area to be visited should also be available early enough for study so that maximum effectiveness can be made of the all too-limited field time. A meaningful, organized field experience should add to our basic knowledge of species biology; should be a conceptual and observational challenge for all systematists; and should be fundamental training for ecologists and geographers as well as taxonomists.

Section A. HABITATS AND ECOLOGICAL CHARACTERISTICS

In describing habitat the geomorphic feature should be indicated; e.g., flat, dome, lake and the mineral or rock composition should be given where possible; e.g., gneiss, alkali, syenite. For the later successional stages the vegetation feature should be noted. The ecological characteristic and populational relationships are included as an organized topical summary for ecological review.

I. GEOMORPHIC OR GEOLOGIC FEATURES

Consult the annotated glossary at the end of this chapter, or any general geology text glossary for definitions of geomorphic and geologic features and Chapters 14 and 15 and any general ecology text for vegetation information. An excellent topographic form summary is in Lahee (1961).

Bluffs and Cliffs
- Basalt
- Dolomite
- Gneiss
- Quartzite
- Sandstone
- Glacial
- Fault Scarp
- Sink
- Stream

Depressions and Sinks
- Gabbro Depressions
- Granite Depressions
- Lava Sinks
- Lime Sinks
- Playas

Dikes and Sills
- Diabase
- Pegmatite
- Syenite
- Quartz

Domes
- Gneiss
- Granite

Flats
- Alkali
- Marl
- Muck
- Mud
- Peat
- Sand

Glacial Depositional Features
- Drumlins
- Eskers
- Kames
- Outwash Plains
- Valley Trains
- Moraines
  - Terminal
  - Lateral
  - Ground

Glacial Erosional Features
- Aretes
- Cirques
- Horns
- Valleys

Gravitational Features
- Boulder Fields
- Talus Slopes

Igneous Instrusions
- Batholiths
- Dikes
- Laccoliths
- Sills
- Stocks

Islands
- Barrier and Estuarine
- Coral and Limestone
- Volcanic

Lakes, Pools, Ponds
- Fluvial: oxbows, plunge basins potholes
- Glacial: cirque, kettle, moraine lakes
- Longshore: bay-mouth bar ponds
- Meteoritic: bays, pools
- Pluvial: dune ponds, playa lakes
- Solution: lime sink lakes
- Tectonic: fault or earthquake lakes
- Volcanic: caldera or flow lakes

Marine Depositional Features
- Barrier Islands
- Bars
- Spits
- Continental Shelves

Marine Erosional Features
- Wave-cut cliffs
- Wave-cut benches

Mineral Deposits - Examples
- Kaolin
- Marl
- Olivine
- Soapstone

Monadnocks
- Granite
- Sandstone

Mountains
- Fault
- Volcanic
- Fold

Plains
- Continental Shelf (old)
- Deltaic
- Lagoon, sound, estuary (old)
- Flood
- Loess
- Outwash

Plateaus
- Mesa
- Piedmont
- Peneplane

Ridges and Peaks
- Dolomite
- Quartzite
- Gneiss
- Slate
- Sandstone

Rock Barrens
- Granite
- Serpentine
- Limestone
- Shale
- Sandstone

Salt and Brackish Water
- Bays
- River Embayments
- Estuaries
- Sounds
- Lagoons

Sands
- Bars
- Ridges
- Dunes
- Scarps
- Hills
- Strands

Slopes
- Gentle
- Sheltered
- Open
- Steep

Streams
- Branches
- Fresh, brackish, salt
- Creeks
- Black-water, mud-water
- Rapids
- Tidal
- Rivers

Stream Depositional Features
- Flood plains
- Mud Flats
- Deltaic plains
- Playas
- Outwash plains
- Sand Bars

Stream Erosional Features
- Canyons
- Ravines
- Cliffs
- Valleys
- Draws

Valleys and Gorges
- Fluvial
- Solution
- Glacial
- Volcanic

Volcanic Flows
- Felsic or Rhyolitic Flow
- Mafic or Basaltic Flow

II. VEGETATIONAL FEATURES

Aquatic herb: fresh, brackish, salt
Balds: grass, heath
Bogs
Desert Succulents
Forests: needle
Forests: deciduous, broad-leaved
Forests: evergreen, broad-leaved
Glades
Marshes: fresh, brackish, salt
Prairie
Savannah
Shrub or Scrub: deciduous
Shrub or Scrub: evergreen
Tundra

III. ECOLOGICAL CHARACTERISTICS AND POPULATIONAL RELATIONSHIPS (See Ecological Evidence, Chapter 15, for Definitions)

A. Biotic B. Abiotic C. Spatial D. Temporal

A-1 Genetic	A-2 Nutritional	A-3 Symbiotic	A-4 Reproductive
Genetic	Autotrophic	Commensalistic	Amphimictic
Phenetic	Parasitic	Mutualistic	Apomictic
Phyletic	Saprophytic	Antagonistic	

B-1 Climatic		B-2 Edaphic	
Cryothermic	Euryphotic	Acidophilous	Psammophilous
Microthermic	Mesophotic	Basophilous	Argillophilous
Mesothermic	Stenophotic	Neutrophilous	Argillopsammophilous
Megathermic	Hydrophytic	Calciphilous	Saprophilous
Heliophilous	Hygrophytic	Halophilous	Eutrophic
Sciophilous	Mesophytic	Aerobic	Mesotrophic
	Xerophytic	Anaerobic	

C-1 <u>Distributional</u>
- Random
- Clumped Random
- Uniform

C-2 <u>Numerical</u>
- Abundance
- Density
- Frequency

C-3 <u>Areal</u>
- Basal Area
- Cover
- Size-Classes

C-4 <u>Stratificational</u>
- Canopy
- Herb
- Subcanopy
- Shrub

D-1 <u>Age Diversity</u> (Life Cycle)
- Birth-Germination
- Childhood-Seedling
- Youth-Sapling
- Adult-Adult
- Old Age-Death

D-2 <u>Phenological</u> (Annual Cyclic)
- Vernal
- Aestival
- Autumnal
- Hibernal

D-3 <u>Periodic</u> (Daily Seasonal Cyclic)
- Hydroperiodic
- Photoperiodic
- Thermoperiodic

D-4 <u>Successional</u> (Generation Cyclic)
- Pioneer
- Transitional
- Climactic

Section B. SAMPLE COMMUNITY LISTS, PATTERNS, PROFILES

In this section three approaches to a study of communities or vegetation are presented: (1) community lists by province, (2) vegetation patterns based on temperature and moisture (altitude and slope), and (3) profiles showing the distribution of communities in a local region with the major dominants listed. A sample species recognition list based on the flora of several local communities is included as an illustration of species that could be studied in four afternoon field trips in the spring. A community report format is included as a guide to the writing of community analyses.

I. SUMMARY OF MAJOR PLANT COMMUNITIES IN EASTERN UNITED STATES BY MAJOR PHYSIOGRAPHIC PROVINCE

Symbols: A, aquatic; B, bog; C, old field and disturbed communities with 1, 2 etc. = age; D, annual/pioneer communities; DF, deciduous forest; HF, evergreen forest; PF, coniferous forest; G, grassland; M, marsh; R, rock barrens; S, shrub; DS, deciduous shrub or scrub; HS, evergreen shrub or scrub; T, tundra; etc.

A. <u>Communities of Coastal Plain: Atlantic and Gulf Coastal Plain, Keys, Mississippi Embayment</u>

1. Dune - Grass (DG)
2. Sand Flat - Grass (SG)
3. Dune Scrub (DS)
4. Maritime Hammock or Southeastern Broad-leaved Evergreen Forest (MHF)
5. Tropical Hammock (THF)
6. Salt Water-Aquatics (SA)
7. Salt Flat - Annuals (SD)
8. Salt Marsh (SM)
9. Salt Shrub (SS)
10. Mangrove Swamp (MF)
11. Brackish Water (BA)
12. Brackish Marsh (BM)
13. Freshwater - Aquatics (FA)
14. Freshwater Marsh (FM)
15. Freshwater Shrub (FS)
16. Swamp Forest (SDF)
17. Bog (B)
18. Pocosin or Evergreen Shrub Bog (SB)
19. Bay Forest or Bayhead (BHF)
20. Savannah & Prairie (TG)
21. Pine Flatwoods (LPF)
22. Upland Coastal Pine Forests (UPF)
23. Upland Scrub (US)
24. Turkey Oak-Wire Grass (TODF-G)
25. Blackjack Oak-Grass (BODF-G)
26. Intermediate Oak-Hickory Forests (OHDF)
27. Mesic Oak-Hickory Forests (OHDF)
28. Mesic Hardwood Forests (MDF)
29. Inland Hammock Forests (IHF)
30. Lime Sink Aquatics (LA)
31. Cypress Domes, Strands, Forests (CPF)
32. Old Field Communities (Succession (OFC)

VEGETATION PATTERNS IN THE SOUTHERN APPALACHIANS*

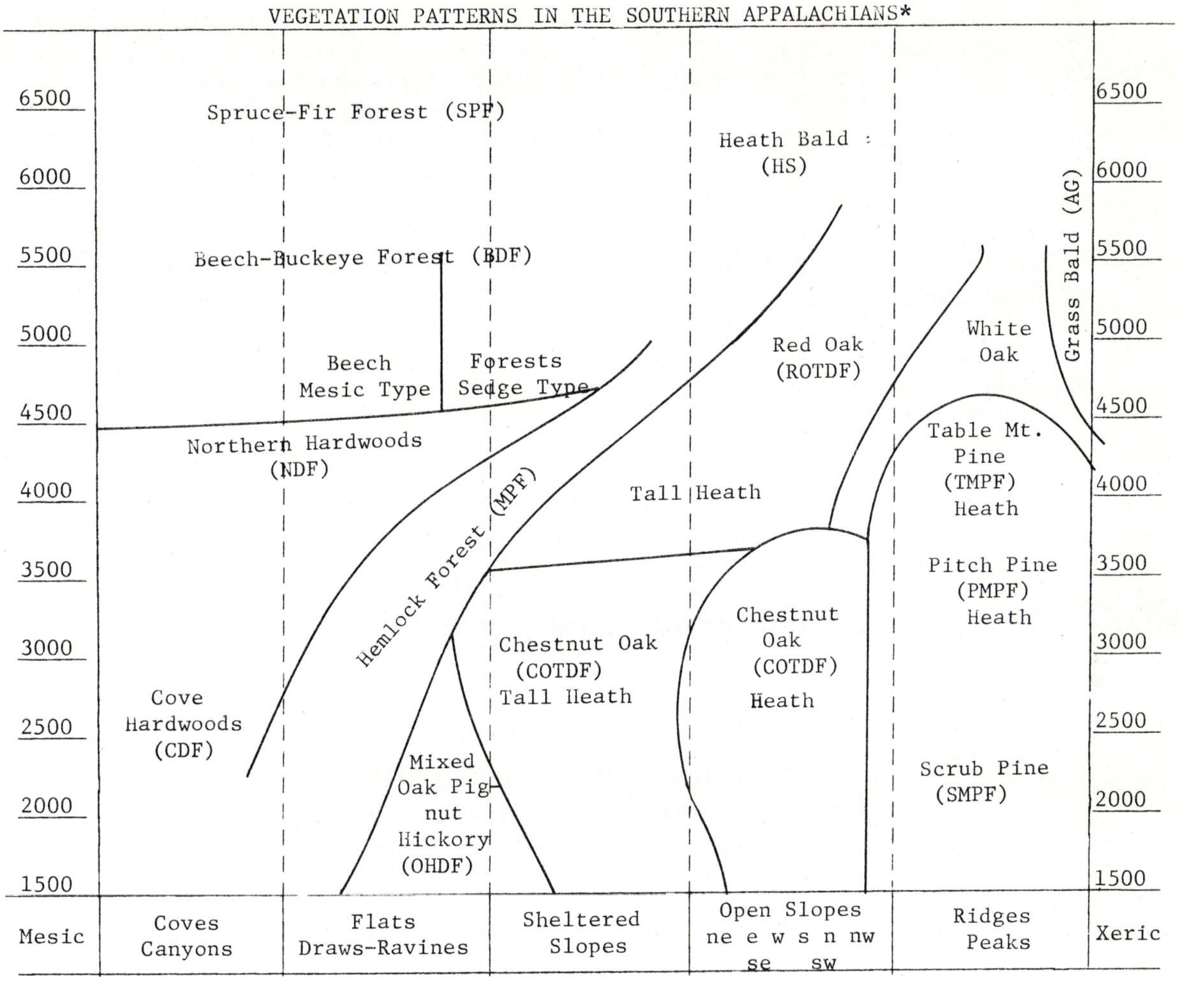

Adapted from Whittaker (1956)

I. Eastern Forest System
- A. Mesic non-quercine forests
  1. Cove and northern hardwood forest
  2. Eastern hemlock forest
  3. Gray beech-buckeye forest
- B. Intermediate oak forests
  4. Mixed oak-pignut hickory forest
  5. Chestnut oak-tall heath forest
  6. Chestnut oak-heath
  7. Red oak-chestnut forest
  8. White oak-heath forest
- C. Xeric Pine forests
  9. Virginia pine forest
  10. Pitch pine-heath
  11. Table Mt. pine-heath
- D. Extreme exposure type
  12. Grassy bald

II. Boreal Forest System
- A. Subalpine forests
  13. Red spruce forest
  14. Fraser fir forest
- B. Extreme exposure type
  15. Heath bald

*This vegetation pattern is based on the distribution of the communities in the Great Smoky Mountains but is basically applicable to the Blue Ridge, Cumberland and Alleghany Plateaus, Cumberland and Alleghany Mountains, and the Ridge and Valley Province. In the Piedmont and Interior Low Plateau with Altitudes 500'-1200' the lower elevation (1500-2000) communities are still present with the mixed oak-hickory complex spreading over most of the meso-xeric center and the cove hardwoods and hemlock limited to the most mesic sites and the chestnut oak-heath and pine-heath being limited to the most xeric sites. A hydromesic alluvial woods community is present from 500' to approximately 2000' in the major river systems.

B. Communities of Piedmont: Piedmont and Interior Low Plateaus

1. Freshwater Aquatics (FA)
2. Freshwater Marsh (FM)
3. Shrub Marsh (FS)
4. Alluvial Woods or Swamp Forest (SF)
5. Prairie (TG)
6. Upland Hardwood Forests (See Vegetation Patterns)
7. Upland Pine Forests (PP)
8. Flatrock Barrens (R)
9. Relict Montane Forests
10. Old Field Communities (Succession) (OFC)

C. Communities of Mountains: Blue Ridge, Alleghany, Cumberland, Catskills, Adirondacks, Green and White, Cumberland Plateau, Alleghany Plateaus, Ridge and Valley

1. Freshwater Aquatics (FA)
2. Freshwater Marsh (FM)
3. Shrub Marsh (FS)
4. Alluvial Woods (SF)
5. Northern Hardwoods (NDF)
6. Cove Hardwoods (CDF) (See Vegetation Patterns)
7. Oak-Hickory Forests (See Vegetation Patterns)
8. Montane Pine Forests (MPF)
9. Hemlock Forests (HPF)
10. Bogs (B)
11. Shrub Bogs (SB)
12. Shrub Balds (HS)
13. Grass Balds (AG)
14. Alpine Tundra (AT)
15. Old Field Communities (Succession) (OFC)

II. COMMUNITY REPORT FORMAT

COMMUNITY
elevation

PHYSIOGRAPHIC PROVINCE
Geomorphic Feature(s); County, State
Locality (distance and direction from a permanent landmark)

Slope: aspect___, degree___. Canopy ht.___. Species______. DBH___, Age___
Soils: series______________, texture______________, pH ts___, pH ss___
A_1__________________, A_2______________, B ______________, C ______________
Rock Type _____________________. Geologic Formation_______________

****

CANOPY (Using percentage composition, list species in order of descending percentages. Using quarter point, list species by importance value. Using the releve', list species by descending cover values)
SUBCANOPY (For presence only, list species alphabetically)
SHRUBS (For presence only, list species alphabetically)
VINES (For presence only, list species alphabetically)
HERBS (List as Forbs, Graminoids, and Ferns. For presence only, list species alphabetically in each group)

****

NOTES:

Indicate the following in this sequence: Sere type; succession stage; seedlings (list by species); transgressive size classes (list by species). Disturbance (type and nature). Adjacent moisture class communities. Adjacent altitudinal communities.

Where and when relevant comment on factors that seem to be controlling community type, e.g.: prevailing wind, salt spray, tidal amplitude, water current, type of substratum, rock or soil nonconformity. Make other comments that will add to an understanding of the community.

III. PROFILES

Coastal Plain: Sandhill Flora
Spout Springs, Harnett Co., N. C.

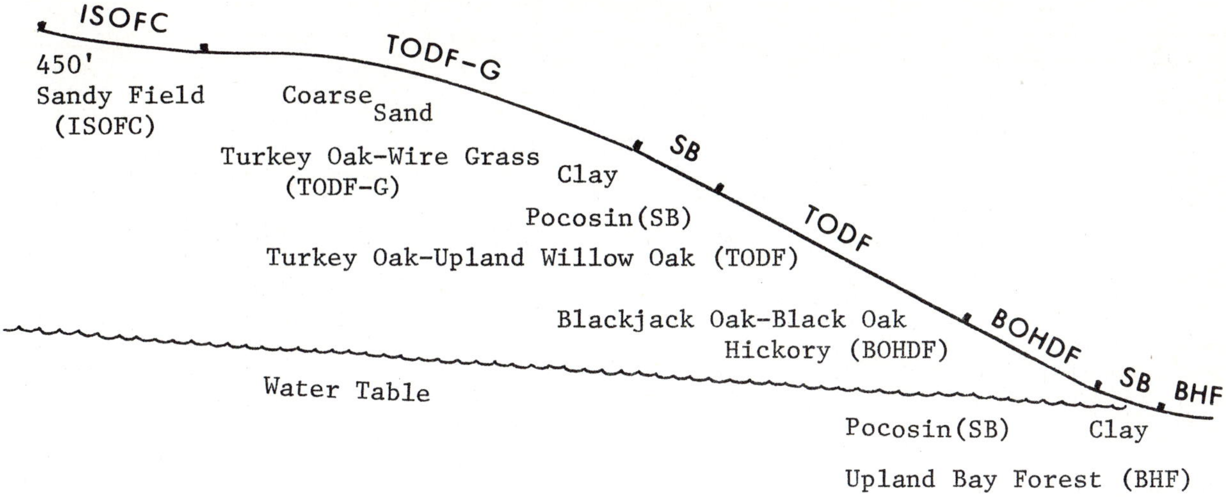

<u>Sandy Field</u> (1SOFC)

h-Striga lutea, Witchweed
h-Linaria canadensis, Toadflax
h-Rumex hastatulus, Sorrel
h-Oenothera laciniata, Evening P.
h-Krigia virginica, Dwarf Dandelion
h-Specularia perfoliata, Evening Looking-glass
g-Festuca octoflora, Fescue

<u>Turkey Oak-Wire Grass</u> (TODF-G)

t-Quercus laevis, Turkey O.
t-Quercus incana, Upland Willow O.
t-Pinus palustris, Long-leaf P.
t-Diospyros virginiana, Persimmon
s-Gaylussacia dumosa, Dwarf H.
s-Vaccinium tenellum, Lowbush B.
g-Aristida stricta, Wire Grass
h-Lupinus diffusus, Lupine
h-Baptisia cinerea, Wild Indigo
h-Cnidoscolus stimulosus, Spurge Nettle
h-Tragia urens, Nettle
h-Arenaria caroliniana, Sandwort
h-Stipulicida setacea, Wiry Sandwort
h-Pyxidanthera barbulata var. brevifolia, Pixie Moss
h-Aureolaria pectinata, Yellow Foxglove

<u>Blackjack Oak-Black Oak-Hickory</u> (BOHDF)

t-Quercus marilandica, Blackjack O.
t-Quercus velutina, Black O.
t-Carya tomentosa, Mockernut H.
t-Pinus palustris, Long-leaf P.
t-Pinus taeda, Loblolly P.
t-Diospyros virginiana, Persimmon
s-Ilex glabra, Gallberry
s-Vaccinium tenellum, Lowbush B.
s-Lyonia mariana, Staggerbush
s-Rhus toxicodendron, Poison Oak
h-Tephrosia virginiana, Goat's Rue
g-Aristida stricta, Wire Grass

<u>Pocosin</u> (Evergreen Shrub Bog) (SB)

t-Pinus serotina, Pond P.
s-Clethra alnifolia, White Alder
s-Gaylussacia frondosa, Huckleberry
s-Lyonia ligustrina, Male-berry
s-Lyonia mariana, Staggerbush
s-Rhododendron atlanticum, Azalea
s-Vaccinium atrococcum, Highbush B.
s-Vaccinium corymbosum, Highbush B.
s-Ilex coriacea, Gallberry
s-Ilex glabra, Gallberry
s-Rhus vernix, Poison Sumac
s-Cyrilla racemiflora, Titi
s-Amelanchier obovalis, Shadbush
s-Sorbus arbutifolia, Chokeberry
s-Styrax americana, Storax
v-Smilax laurifolia, Catbrier
h-Polygala lutea, Bachelor's-Button
h-Sarracenia purpurea, Pitcher Plant
h-Xyris flexuosa, Hard-head
h-Eriocaulon decangulare, Hat-pin

Piedmont: River Valley Flora
Deep River, 2 miles west of Moncure, Lee Co., N. C.

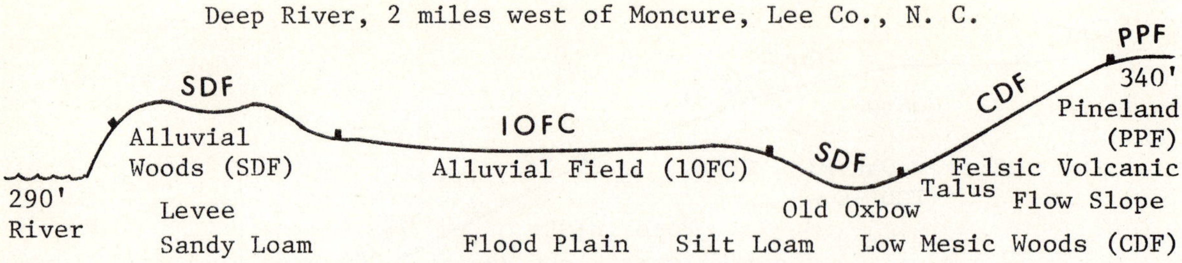

Piedmont Pine Forest (PPF) Successional

t-Pinus taeda, Loblolly Pine
t-Pinus strobus, White P.
t-Pinus echinata, Short leaf P.
t-Ilex opaca, American Holly
t-Cornus florida, Flowering Dogwood
s-Kalmia latifolia, Mountain Laurel
s-Vaccinium corymbosum, High-bush B.
s-Vaccinium vacillans, Low-bush B.
s-Viburnum prunifolium, Blue Haw
s-Hypericum lloydii, St. John's Wort
v-Gelsemium sempervirens, Yellow Jessamine
h-Potentilla canadensis, Cinquefoil
h-Chimaphila maculata, Spotted Wintergreen
h-Houstonia caerulea, Bluet

Low Mesic Woods (CDF)

t-Fagus grandifolia, Beech
t-Quercus rubra, Red Oak
t-Acer saccharum spp. floridanum, Southern Sugar Maple
t-Magnolia tripetala, Umbrella tree
t-Hamamelis virginiana, Witch Hazel
t-Aesculus sylvatica, Dwarf Buckeye
s-Euonymus americanus, Strawberry B.
s-Hydrangea arborescens, Hydrangea
s-Dirca palustris, Leatherwood
s-Prunus americana, American Plum
s-Staphylea trifolia, Bladdernut
h-Dicentra cucullaria, Dutchmans B.
h-Anemone lancifolia, Windflower
h-Thalictrum thalictroides, Anemonella
h-Hepatica americana, Hepatica
h-Isopyrum biternatum, Isopyrum
h-Claytonia virginiana, Spring-beauty
h-Stellaria pubera, Giant Chickweed
h-Sanguinaria canadensis, Bloodroot
h-Dentaria laciniata, Toothwort
h-Hexastylis arifolia, Wild Ginger
h-Saxifraga virginiensis, Saxifrage
h-Uvularia perfoliata, Bellwort
h-Erythronium americanum, Trout Lily
g-Poa cuspidata, Early Bluegrass
g-Carex nigromarginata
g-Carex artitecta
g-Luzula acuminata, Wood Rush
f-Asplenium rhizophyllum, Walking F.
f-Cystopteris fragilis, Fragile F.
f-Adiantum pedatum, Maiden-hair F.
f-Polystichum acrostichoides, Christmas F.

Alluvial Field Weeds (1OFC)

h-Stellaria media, Chickweed
h-Cerastium glomeratum, Mouse-ear Ch.
h-Sagina decumbens, Pearlwort
h-Barbarea verna, Winter Cress
h-Cardamine hirsuta, Winter Cr.
h-Rorippa islandica, Yellow Cress
h-Arabidopsis thaliana, Mouse-ear Cr.
h-Sibara virginica, Sibara
h-Ranunculus abortivus, Buttercup
h-Ranunculus pusillus, Buttercup
h-Ranunculus parviflorus, Buttercup
h-Geranium carolinianum, Geranium
h-Valerianella locusta, Corn Salad
h-Lamium amplexicaule, Hen-bit
h-Veronica arvensis, Veronica
h-Veronica peregrina, Veronica
h-Callitriche deflexa, Starwort
h-Muscari racemosum, Grape Hyacinth
g-Poa annua, Speargrass
g-Poa chapmaniana, Bluegrass

Alluvial Woods (SDF)

t-Betula nigra, River Birch
t-Salix nigra, Black Willow
t-Fraxinus pensylvanica, Green Ash
t-Carpinus caroliniana, Ironwood
t-Platanus occidentalis, Sycamore
s-Lindera benzoin, Benzoin
v-Smilax rotundifolia, Cat-brier
h-Asarum canadense, Wild Ginger

Piedmont: Granite Flatrock Flora
Mitchell's Mill, 2 miles n.e. of Rolesville,
Wake Co., N. C.

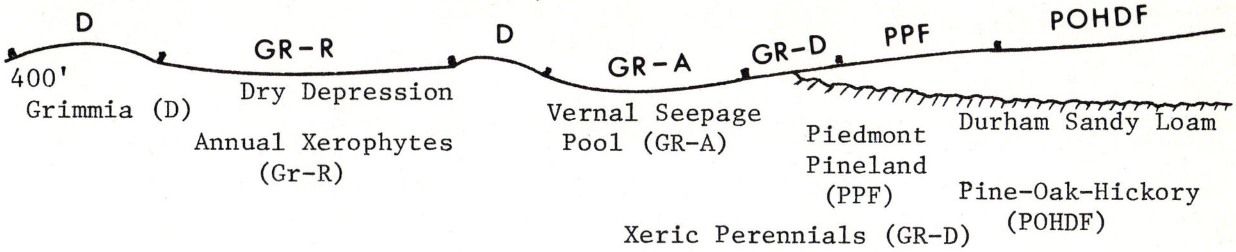

Annual Xerophytes (GR-R)

h-Sedum smallii, Diamorpha
h-Portulaca smallii, Rock Portulaca
h-Arenaria groenlandica var. glabra, Sandwort
h-Crotonopsis elliptica, Rushfoil
h-Hypericum gentianoides, Pineweed
h-Diodia teres, Buttonweed
g-Cyperus granitophilus

Vernal Hydrophytes (GR-A)

h-Lindernia monticola, False Pimpernel
g-Agrostis elliottiana, Bent Grass
g-Bulbostylis capillaris
f-Isoetes melanopoda, Quillwort

Xeric Perennial Herbs (GR-D)

h-Talinum teretifolium
h-Commelina erecta, Dayflower
h-Polygonatum biflorum, Solomon's Seal
h-Heuchera americana, Alum Root
h-Senecio smallii, Ragwort
h-Yucca filamentosa, Beargrass
h-Opuntia compressa, Prickly Pear
g-Andropogon scoparius, Little Bluestem
g-Panicum laxiflorum
f-Selaginella rupestris, Rock Clubmoss
f-Woodsia obtusa, Wood Fern
f-Cheilanthes lanosa, Hair-lip Fern
f-Asplenium platyneuron, Ebony Spleenwort

Piedmont Pineland (PPF)

t-Juniperus virginiana, Red Cedar
t-Pinus virginiana, Scrub P.
t-Ulmus alata, Winged Elm
s-Vaccinium arboreum, Sparkleberry
s-Rhus aromatica, Fragrant Sumac
s-Callicarpa americana, French Mulberry
v-Smilax rotundifolia, Catbrier
v-Lonicera sempervirens, Coral Honeysuckle
h-Chimaphila maculata, Spotted Wintergreen
h-Euphorbia corollata, Spurge

Pine-Oak-Hickory Forest (POHDF)

t-Pinus taeda, Loblolly P.
t-Pinus echinata, Shortlead P.
t-Quercus stellata, Post O.
t-Quercus velutina, Black O.
t-Quercus marilandica, Blackjack O.
t-Quercus falcata, Southern Red O.
t-Quercus alba, White O.
t-Carya tomentosa, Mockernut Hickory
t-Ilex opaca, American Holly
t-Diospyros virginiana, Persimmon
t-Cornus florida, Flowering Dogwood
t-Oxydendrum arboreum, Sourwood
s-Vaccinium stamineum, Deerberry
s-Chionanthus virginicus, Fringe Tree
s-Euonymus americanus, Strawberry Bush
s-Viburnum rufidulum, Black Haw
v-Vitis rotundifolia, Muscadine
v-Parthenocissus quinquefolia, Virginia Creeper
v-Smilax glauca, Catbrier
h-Cypripedium acaule, Pink Ladyslipper
h-Silene caroliniana, Wild Pink
h-Chimaphila maculata, Spotted Wintergreen
h-Chrysogonum virginianum, Green and Gold

Piedmont: Ridge and Slope Flora
Occoneechee Mountain near Eno River, south of Hillsborough,
Orange County, N. C.

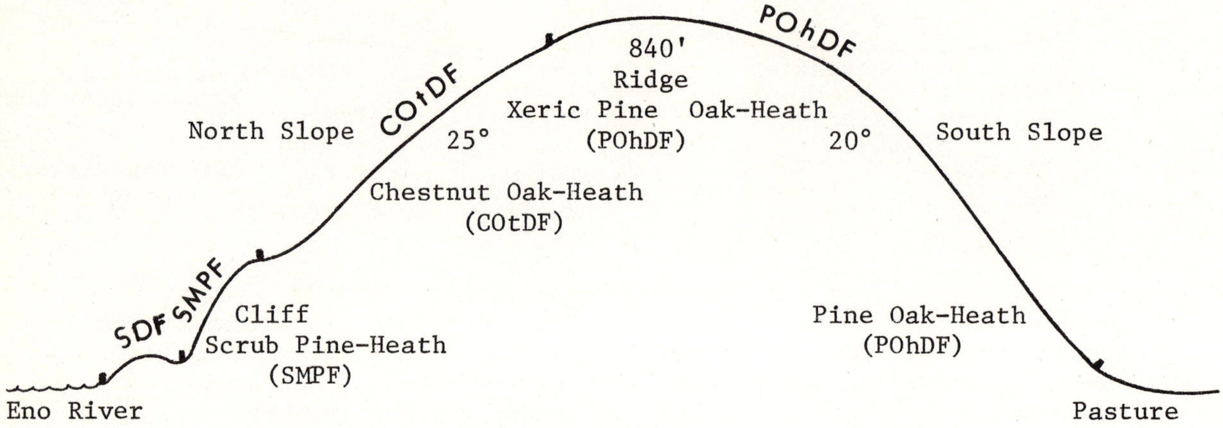

Alluvial Woods (SDF)

t-Betula nigra, River Birch
t-Carpinus caroliniana, Ironwood
t-Quercus shumardii, Swamp Red O.
t-Fagus grandifolia, Beech
t-Liriodendron tulipifera, Poplar
t-Acer saccharum spp. floridanum Southern Sugar Maple
s-Lindera benzoin, Spicebush
s-Alnus serrulata, Tag Alder
h-Erythronium americanum, Trout Lily
h-Podophyllum peltatum, Mandrake
h-Dentaria heterophylla, Toothwort
f-Polystichum acrostichoides, Christmas F.

Scrub Pine-Heath (SMPF)

t-Pinus virginiana, Scrub Pine
t-Oxydendrum arboreum, Sourwood
t-Acer rubrum, Red Maple
s-Kalmia latifolia, Mountain Laurel
s-Rhododendron catawbiense, Rosebay
s-Gaylussacia baccata, Huckleberry
s-Lyonia mariana, Staggerbush
v-Smilax glauca, Catbrier
h-Galax aphylla, Galax
h-Epigaea repens, Trailing Arbutus
h-Gaultheria procumbens, Wintergreen
h-Hexastylis virginica, Wild Ginger

Chestnut Oak-Heath (COtDF)

t-Quercus prinus, Rock Chestnut O.
t-Nyssa sylvatica, Black Gum
t-Acer rubrum, Red Maple
t-Oxydendrum arboreum, Sourwood
s-Styrax grandifolia, Storax
s-Vaccinium vacillans, Blueberry
s-Vaciniuum stamineum, Deerberry
s-Lyonia ligustrina, Male-berry
s-Rhododendron nudiflorum, Wild Azalea
s-Fothergilla major, Witch Alder
s-Robinia hispida, Bristly locust
s-Viburnum acerifolium, Maple-leaved V.
h-Chimaphila maculata, Spotted Wintergreen
h-Galax aphylla, Galax
h-Hypoxis hirsuta, Yellow Star-grass
h-Hieracium venosum, Rattlesnake Weed
h-Hexastylis arifolia, Wild Ginger
g-Carex nigromarginata
f-Osmunda cinnamomea, Cinnamon F.

Xeric Pine, Oak-Heath (POhDF)

t-Pinus virginiana, Scrub Pine
t-Quercus marilandica, Blackjack O.
t-Quercus stellata, Post O.
t-Quercus velutina, Black O.
t-Quercus prinus, Rock Chestnut O.
t-Prunus serotina, Choke Cherry
t-Acer rubrum, Red Maple
t-Oxydendrum arboreum, Sourwood
t-Cornus florida, Flowering Dogwood
s-Gaylussacia baccata, Huckleberry
s-Vaccinium vacillans, Blueberry
s-Vaccinium stamineum, Deerberry
s-Kalmia latifolia, Mountain Laurel
h-Chimaphila maculata, Spotted Wintergreen
h-Epigaea repens, Trailing Arbutus

Coastal Plain: Barrier Island Flora
Dare Co., N. C.

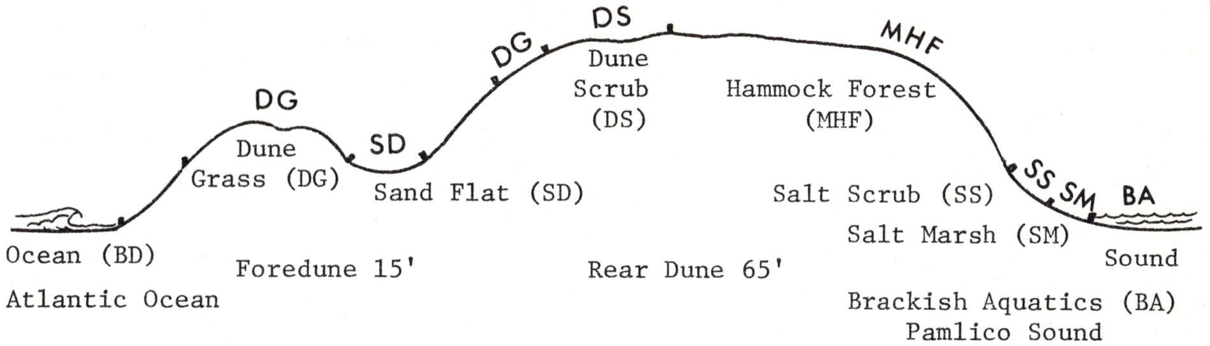

<u>Beach Strand Succulents</u> (BD)

h-Cakile maritima, Sea Kale
h-Euphorbia polygonoides, Spurge
h-Hydrocotyle bonariensis, Pennywort
h-Polygonum glaucum, Knotweed

<u>Dune Grassland</u> (DG)

g-Uniola paniculata, Sea Oats
g-Ammophila breviligulata, Beach Grass
h-Hydrocotyle bonariensis, Pennywort
h-Erigeron pusillus, Horseweed
h-Oenothera humifusa, Evening Primrose
s-Iva imbricata, Iva

<u>Sand Flat Herbs</u> (SD)

g-Muhlenbergia filipes, Muhly
g-Chloris petraea, Finger Grass
g-Cenchrus tribuloides, Sandspur
h-Gaillardia pulchella, Gaillardia
h-Gnaphalium obtusifolium, Rabbit tobacco

<u>Dune Scrub</u> (DS)

v-Parthenocissus quinquefolia, Virginia Creeper
v-Smilax auriculata, Catbrier
v-Vitis aestivalis, Fox Grape
v-Rhus radicans, Poison Ivy
s-Morus rubra, Red Mulberry
s-Callicarpa americana, French Mulberry
s-Zanthoxylum clava-herculis, Tooth-ache Tree
g-Uniola paniculata, Sea Oats

<u>Maritime Hammock</u> (MHF)

t-Quercus virginiana, Live Oak
t-Quercus laurifolia, Laurel Oak
t-Ilex opaca, American Holly
t-Osmanthus americana, American Olive
t-Cornus florida, Flowering Dogwood
s-Ilex vomitoria, Yaupon
v-Gelsemium sempervirens, Yellow Jessamine
h-Mitchella repens, Partridge Berry
h-Galium hispidulum, Bedstraw

<u>Salt Shrub</u> (SS)

s-Baccharis halimifolia, Goundsel
s-Myrica cerifera, Wax Myrtle
s-Iva frutescens, Marshelder
s-Borrichia frutescens, Sea Ox-eye

<u>Salt Marsh</u> (SM)

h-Salicornia europaea, Glasswort
h-Suaeda linearis, Suaeda
h-Limonium carolinianum, Sea Lavender
h-Sesuvium portulacastrum, Sea Purslane
g-Spartina patens
g-Spartina alterniflora
g-Distichlis spicata
g-Juncus roemerianus, Rush

<u>Brackish Aquatics</u> (BA)

h-Ruppia maritima, Widgeon Grass
h-Zostera marina, Eel Grass

Blue Ridge: Montane Flora
Black Mountains, Yancey Co., N. C.

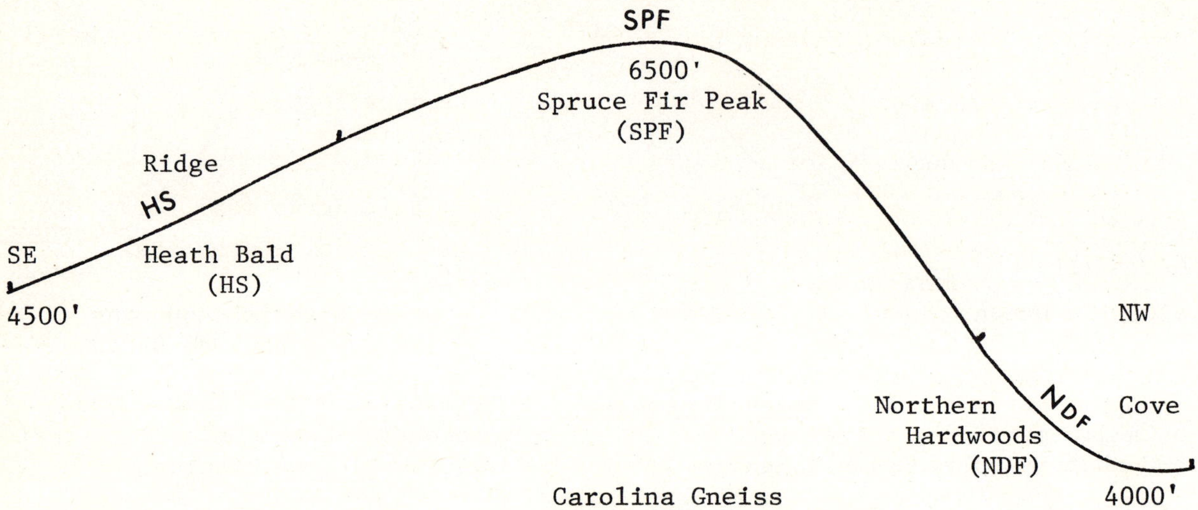

Carolina Gneiss
Black Mountains

<u>Heath Bald</u> (HS)

s-Rhododendron catawbiense, Purple L.
s-Rhododendron minus, Rosebay
s-Kalmia latifolia, Mountain Laurel
s-Leucothoe recurva, Fetterbush
s-Vaccinium constablaei, Blueberry
s-Gaylussacia baccata, Huckleberry
s-Menziesia pilosa, Minnie-bush
s-Viburnum cassinoides, Witherod
s-Cornus alternifolia, Alternate-leaved Dogwood
s-Robinia hispida, Rose-acacia
s-Leiophyllum buxifolium, Sand Myrtle
v-Smilax hispida, Catbrier
h-Galax aphylla, Galax
h-Krigia montana, Dandelion
h-Hypericum graveolens, St. John's-wort
h-Houstonia serpyllifolia, Bluet
h-Angelica triquinata, Angelica
g-Scirpus cespitosus
g-Carex debilis
g-Carex brunnescens
g-Danthonia spicata

<u>Spruce-Fir Forest</u> (SPF)

t-Abies fraseri, Fraser's Fir
t-Picea rubens, Red Spruce
t-Betula lutea, Yellow Birch
t-Betula papyrifera, Paper Birch
t-Sorbus americana, Mountain Ash
t-Prunus pensylvanica, Fire Cherry
t-Acer spicatum, Mountain Maple
s-Ribes rotundifolium, Gooseberry
s-Viburnum alnifolium, Moosewood
s-Menziesia pilosa, Minnie-bush
s-Sambucus pubens, Elderberry
s-Rhododendron catawbiense, Purple L.
s-Vaccinium erythrocarpon, Mt. Cranberry
h-Oxalis acetosella, Wood-shamrock
h-Aster acuminatus, Aster
h-Clintonia borealis, Clintonia
h-Chelone lyoni, Turtlehead
g-Carex intumescens
g-Deschampsia flexuosa

<u>Northern Hardwood Forest</u> (NDF)

t-Acer saccharum, Sugar Maple
t-Aesculus octandra, Buckeye
t-Tilia americana, Basswood
t-Betula lutea, Yellow Birch
t-Acer pensylvanicum, Striped M.
t-Halesia carolina, Silverbell
s-Ribes cynosbati, Gooseberry
s-Leucothoe axillaris, Dog-hobble
v-Aristolochia macrophylla, Dutchman's Pipe
h-Eupatorium rugosum, White Snakeroot
h-Claytonia caroliniana, Springbeauty
h-Laportea canadensis, Stinging Nettle
h-Viola canadensis, Violet
g-Carex plantaginea
g-Carex pensylvanica

IV. SPECIES FOR RECOGNITION QUIZ
(Sample)

FERNS AND FERN ALLIES
Lycopodium lucidulum, Shining Clubmoss (CDF)
Selaginella arenicola, Sand Spikemoss (TODFG)
Equisetum praealtum, Scouring Rush (OFC)
****
Adiantum pedatum, Maidenhair Fern (CDF)
Athyrium asplenioides, Lady Fern (CDF)
Camptosorus rhizophyllus, Walking Fern (CDF)
Cystopteris fragilis, Fragile Fern (CDF)
Osmunda cinnamomea, Cinnamon Fern (CDF)
Polypodium virginianum, Northern Polypody (CDF)
Polystichum acrostichoides, Christmas Fern (CDF)
Pteridium latiusculum, Bracken Fern (UP) (TODF-G)

GYMNOSPERMS
Chamaecyparis thyoides, White Cedar (BHF)
Juniperus virginiana, Red Cedar (PPF)
Pinus echinata, Short-leaf Pine (PPF)
P. palustris, Long-leaf Pine (UPF)
P. pungens, Table Mountain Pine (TMPF)
P. rigida, Pitch Pine (TMPF)
P. serotina, Pond Pine (BHF)
P. strobus, White Pine (PMPF)
P. taeda, Loblolly Pine (PPF)
P. virginiana, Scrub Pine (PPF)
Tsuga canadensis, Canadian Hemlock (SPF)
T. caroliniana, Carolina Hemlock (SPF)

MONOCOTS
Poaceae
Anthoxanthum odoratum, Sweet Vernal Grass (OFC)
Alopecurus carolinianus, Foxtail Grass (OFC)
Aristida stricta, Wire Grass (TODF-G)
Arundinaria gigantea, Cane (BHF)
Bromus catharticus, Brome Grass (OHF)
Festuca sciurea, Fescue Grass (OFC)
Poa annua, Annual Blue Grass (OFC)

Cyperaceae
Carex nigromarginata (POHDF)

Araceae
Arisaema triphyllum, Jack-in-the-Pulpit (CDF)

Juncaceae
Luzula acuminata, Wood Rush (CDF)

Liliaceae
Erythronium americanum, Trout Lily (CDF)
Muscari racemosum, Grape Hyacinth (OFC)
Polygonatum biflorum, Solomon's Seal (CDF)
Smilacina racemosa, False Solomon's Seal (CDF)
Smilax glauca, Cat-brier
S. laurifolia, Bamboo (SB)
S. rotundifolia, Cat-brier (SB)
Trillium grandiflorum, Trillium (CDF)
Uvularia perfoliata, Bellwort (CDF)

Amaryllidaceae
Zephyranthes atamasco, Atamasco Lily (FM)

Iridaceae
Iris cristata, Dwarf Iris (POHDF)

Orchidaceae
Goodyera pubescens, Rattlesnake Plantain (MODF)
Orchis spectabilis, Showy Orchis (CDF)

Dicots
Salicaceae
Salix sericea, Silky Willow (FS)

Myricaceae
Myrica heterophylla, Wax Myrtle (BHF)

Juglandaceae
Carya tomentosa, Mockernut Hickory (POHDF)

Betulaceae
Alnus serrulata, Tag Alder (FS)
Betula nigra, River Birch (SDF)
B. lenta, Cherry Birch (CDF)
Carpinus caroliniana, Ironwood (SDF)
Corylus americana, Hazelnut (SDF)

Etc.

Section C. FLORA AND VEGETATION--OBJECTIVES, AIDS, AND ANALYSIS

The objectives, the inclusions and aids, the spectra, and diversity types are suggestions for comprehensive and basic organization, observation, analysis, and synthesis in flora work. Flora and vegetation objectives and inclusions are given to show the similarities, differences, and complimentary nature of these two aspects of field study. In addition to learning what's where when or making a list of plant species within a given area at a given time, i.e., a simple flora, a field botany student should become acquainted with plant habit (Life Form), growth patterns (Sociability), periodicity (Phenology), duration, reproduction, spatial relations or patterns of occurrence (Population and Community), and distributional relations (Habitat and Geography). Flora study should add basic information about species biology to systematics.

I. OBJECTIVES OF FLORA AND VEGETATION STUDY

A. Flora (By Physiographic Region; Community; Soil, Rock, Habitat Type)
   1. Determine and list species within study area by pattern, i.e., community or habitat. (See Flora Summary)
   2. Analyze flora for life form, diaspore type, sociability, duration, and phenology. (See Spectra 1,2,3,4, and 5)
   3. Analyze flora for indigenous and introduced species, disjuncts, relicts, range extensions, and new variations. (See Spectrum 8)
   4. Determine indicator species for soil and rock types and species that are exclusive to restricted communities or habitats. (See Spectrum 7)
   5. Indicate unique features within a flora, e.g., unusually large specimens of a species, largest fruit, smallest flower, peculiar pollinators, largest sympatric group, rarest taxa, endemics, most variable population, largest population, oldest population, largest tree, oldest plant, etc. (See Flora Summary and Spectrum 6)
   6. Compare flora of study area with that of larger immediate area and disjunct climatically similar areas as to species and larger taxa similarities and/or differences. (See Flora Summary and Spectrum 6)
   7. Interpret flora as to origin, migration, and evolution. (See Flora Summary and Spectra)

B. Vegetation (By Physiographic Region; Soil, Rock, Habitat Type)
   1. Become familiar with flora of study area and prepare a voucher specimen of each species.
   2. Determine patterns (communities) in which flora occurs.
   3. Determine distribution of patterns (communities).
   4. Analyze and map distribution of patterns (communities) in relation to slope, exposure, elevation, etc.
   5. Indicate unique pattern compositions and distributions.
   6. Compare community composition and distribution of communities with those in similar climatic regions.
   7. Interpret communities as to origin, evolution, and migration.

II. STANDARD INCLUSIONS IN FLORA AND VEGETATION STUDY (In addition to flora lists and field, herbarium, and laboratory data)

1. Location and geography of Study Area.
2. History of botanical exploration.

3. Survey of applicable physiography and topography.
4. Resumé of pertinent pedologic and geologic data.
5. Summary of meaningful climatological data.
6. Review of previous pertinent work.
7. Description of present and past land use or abuse.
8. List of cited references.

III. RECONNAISSANCE AND PLANNING AIDS FOR FLORA AND VEGETATION STUDY

1. Topographic Maps (1)
2. Geologic Maps (1)
3. Soil Surveys (2).
4. Aerial Photographs (1,2,6)
5. Forestry & Vegetation Maps (2)
6. Hydrologic Data & Hydographic Maps (1,4,5)
7. Climatological Data Summaries (3)
8. Special Geological Reports (1,4)
9. County Road Maps (6)

Sources: (1) U. S. Geological Survey; (2) U. S. Department of Agriculture; (3) U. S. Weather Bureau; (4) State Geological Surveys or Departments of Conservation & Development; (5) Corps of Engineers, U. S. Army; (6) State Highway Department.

IV. FLORA SUMMARY

Species	Lf	Dia	Soc	Dur	Phe	Di	Communities or Habitat

Note: Species should be listed according to a Manual or System, or systematically by family. See flora spectra classification and explanation for meaning of symbols, and Section B for plant community symbols. This flora summary arrangement can be easily programmed for computer analysis in a variety of ways.

V. DOCUMENTATION

In flora and vegetation work a set of specimens representing each species studied or identified should be deposited in an herbarium as vouchers for documentation. Standard label data should be included with each specimen, and ecological and/or systematic data pertaining to each species acquired during the study should be added to the voucher specimen. A strip label containing the ecological and/or systematic information pertinent to the specimen should be added to the bottom of the herbarium sheet. The ecological and/or systematic information should not be indicated by symbols or code but written out so that future students will have no trouble understanding the information on the label.

Quantitative methods used in analysis should be documented with any modifications of standard procedures carefully described. The types of instruments used in acquisition or analysis of data should be indicated. Literature used as a basis for any part of the work; e.g., physiographic province nomenclature based on Fenneman, 1938, should be cited.

Documentation should be complete and comprehensive enough for the reader of the report to be able to distinguish original observations, analyses, and syntheses from those previously published.

VI. FLORISTIC OR COMMUNITY SPECTRAL ANALYSIS
(A sample form)

1. Life Form (Lf)

	P-t	P-s	P-v	Ch	G	Th	E	S	Pa	Sa	H
# of Species											
% Total Flora											

2. Diaspore (Dia)

	Spo	Pte	Pog	Cyc	Epz	Syz	Enz	Bal	Bar	Hyd	Ant	Ate	Pol
# of Species													
% Total Flora													

3. Sociability (Soc)

	Soc 1	Soc 2	Soc 3	Soc 4	Soc 5
# of Species					
% Total Flora					

4. Duration (Dur)

					Evergreen	
	Ann	Bien	Per	Decid	Broad	Needle
# of Species						
% Total Flora						

5. Phenology (Phe)

Leaf (L) - Flower (Fl) - Fruit (Fr)

	V	Ae	Au	H	All	Ds	Ws
# of Species							
% Total Flora							

6. Taxon (Tax)

Div./Class

	Psi	Lyco	Sphe	Pter	Pino	Magn	Magt	Lilt
# of Species								
% Total Flora								

Sub-Class

	Mag	Ham	Car	Dill	Ros	Ast	Alis	Arec	Com	Lil
# of Species										
% Total Flora										

7. Community (Examples)

	DG	SF	DS	SA	SDF	BHF	TG	LPF	etc.
# of Species									
% Total Flora									

8. Distribution (Di)

	Bor	Cp	App	Tro	Pra	All	Aut	Dis	Rel	Spo	Nat	Dom
# of Species												
% Total Flora												

Note: Many other spectral analyses could be made, e.g.: those based on fidelity, vitality, leaf size, leaf shape, corolla types, fruit types, pollinating vectors, etc.

Spectral Analysis Symbol Explanation

1. Life Form (Lf)

P-t, Phanerophyte, tree; P-s, Phanerophyte, shrub; P-v, woody vine; Ch, Chamaephyte; G, Geophyte; Th, Therophyte; E, Epiphyte; S, Succulent; Pa, Parasite; Sa, Saprophyte; H, Hemicryptophyte.

2. Diaspores (Dia)

Spo, Sporochore; Pte, Pterochore; Pog, Pogonochore; Cyc, Cyclochore; Epz, Epizoochore; Syz, Synzoochore; Enz, Endozoochore; Bal, Ballochore; Bar, Barochore; Hyd, Hydrochore; Ant, Anthropochore; Ate, Atelechore; Pol, Polychore.

3. Sociability (Soc)

Soc 1--Plants growing in one place, singly; Soc 2--Plants grouped or tufted; Soc 3--Plants in small patches, or cushions; Soc 4--Plants in small colonies, in extensive patches, or forming carpets; Soc 5--Plants in pure populations.

4. Duration (Dur)

Ann (1)--Annual; Bien (2)--Biennial; Per (3)--Perennial; Decid (4)--Deciduous-leaved; Evergreen Broad-leaved (5); Evergreen Needle-leaved (6). Note: Annuals could be classified as winter; spring-summer; summer-fall.

5. Phenology (Phe)

Leaf (L)--Time of Leafing; Flower (Fl)--Time of Flowering; Fruit (Fr)--Time of Fruiting. V(1)--Vernal or Spring; Ae(2)--Aestival or Summer; Au(3)--Autumnal or Fall; H(4)--Hibernal or Winter; All(5)--All Year; Ds(6)--Dry Season; Ws(7)--Wet Season.

Note: Phenology could be indicated by three numbers, the first, the leafing period; the second, the time of flowering; and the third, the time of fruiting, e.g., 1-2-3 means leafing in spring, flowering in summer, fruiting in the fall. Also instead of spring, summer, etc., phenology could be indicated by month, e.g., (1) Jan., (2) Feb., (12) Dec., etc.

6. Taxon (Tax)

See Classification System in Cronquist (1968)

Psi--Psilotophyta; Lyco--Lycophyta; Sphe--Sphenophyta, Pter--Pterophyta; Pino--Pinophyta; Magn--Magnoliophyta; Magt--Magnoliatae; Lilt--Liliatae.

Mag--Magnoliidae; Ham--Hamamelidae; Car--Caryophyllidae; Dill--Dilleniidae; Ros--Rosidae; Ast--Asteridae; Alis--Alismatidae; Arec--Arecidae; Com--Commelinidae; Lil--Liliidae.

7. Communities

See types of communities and symbols in Section B.

8. Distribution (Di)

For definitions, see Geography in Chapter 16.

Bor--Boreal; Cp--Coastal Plain; App--Appalachians (which includes the Piedmont, Blue Ridge, Ridge & Valley, Cumberland-Alleghany Mountains, Cumberland-Alleghany Plateaus, and Interior Low Plateaus); Tro--Tropical; Pra--Prairie; All-Allochthonous; Aut--Autochthonous; Dis--Disjunct; Rel--Relict; Spo--Sporadic; Nat--Naturalized, and Dom--Domesticated. Indigenous-native and Introduced-brought in by man.

VI. TYPES OF DIVERSITY IN FLORA OR COMMUNITY (See Flora Spectra Classification and Origin and Establishment in Chapter 16)

1. Community (Habitat)-Area. Number of types of communities per unit area.
2. Community-Species. Number of species per community type.
3. Diaspore-Flora or Community. Number of types of diaspores within flora or community.
4. Diaspore-Species. Number of species per diaspore type within flora or community.
5. Distribution-Flora or Community. Number of prevalent ranges represented within flora or community--e.g., species primarily prairie or boreal in distribution, etc.
6. Distribution-Species. Number of species per range type within flora or community--e.g., number of species primarily with prairie distribution or boreal distribution, etc.
7. Duration-Flora or Community. Number of annuals, biennials, perennials, or evergreen, deciduous species within flora or community.
8. Duration-Species. Number of species per duration type within flora or community.
9. Life Form-Flora or Community. Number of types of life forms within flora or community.
10. Life Form-Species. Number of species per life form type within flora or community.
11. Phenology-Flora or Community. Number of phenological types within flora or community.
12. Phenology-Species. Number of species of each phenological type within flora or community.
13. Sociability-Flora or Community. Number of sociability types within flora or community.
14. Sociability-Species. Number of species per sociability type within flora or community.
15. Species-Area. Number of species per unit area.
16. Species-Equitability. Number of individuals per species within a community or area.
17. Species-Number. Number of species per individual unit--e.g., number of species per number of stems or per number of separate plants.
18. Species-Stratum. Number of species per stratum within a community or flora.
19. Taxon-Flora or Community. Number of divisions, classes, subclasses etc. (ranks) represented in the flora or community.
20. Taxon-Higher Taxon. Number of taxa per next larger taxon within a flora or community--e.g., number of genera within a family, number families within an order, etc.; or number of higher taxa within flora or community.
21. Taxon-Species. Number of species within an order, subclass, etc. within a flora or community--e.g., number of monocot species within flora or community.

COMBINATION DIVERSITY. Any joining of two or more of the diversities listed above--e.g., number of life form types per taxon or range type within the flora or community; number of species with a diaspore type and duration type within the flora or community.

Note: Many other types of diversity are present in a flora or community, e.g., different fruits, pollinating agents, sexual conditions, etc. (See Phytographic Glossary).

SELECTED TOPOGRAPHIC FORMS GLOSSARY

(Adapted from Lahee [1961]; see Lahee [1961] or geomorphology texts [e.g., Easterbrook, 1969] for additional forms and information about all land or topographic forms and geomorphic features.)

A. Basins. Topographic depressions rimmed on all sides.
- Barrier Basin. A depression formed by damming due to lava flow, morainal debris, mudflow, landslide debris, or fluvial deposition.
- Blow-out. Usually an interdunal wind-blown depression.
- Solution Basin, Sinkhole, Gabbro Depression. Depression formed due to chemical dissolution of underlying rock.

B. Beaches. Shoreline deposits of sand, pebbles or rock made by action of waves and longshore currents.
- Barrier Beach. Sand reef built up offshore by deposition of churned bottom sands.
- Bayhead Beach. Accumulated debris from neighboring headlands deposited at the head of a bay.
- Baymouth Bar. Barrier beach built across the entrance to a bay.

C. Benches and Terraces. Relatively flat gently or horizontally inclined surfaces bounded by a steeper ascending slope on one side and a steeper descending slope on the opposite side. Benches are usually considered forms in solid rock and terraces forms in unconsolidated materials.
- River Terrace. A gently sloping or horizontal depositional surface on a flood plain.
- Wave-cut Bench. An erosional flat rock floor which terminates landward at the foot of a seacliff and seaward in a shore terrace.
- Wave-built Terrace. Gently ascending, beachlike form above high water level.

D. Bluffs, Cliffs, Scarps, Precipices, Escarpments. Vertical or very steep face of rock or partially consolidated sediments.
- Fault Cliff or Scarp. Cliffs of varying height formed by faulting with dislocation of bedrock structures.
- Glacial Cliff. Steep bluffs left after glacial scouring or ice quarrying and sapping.
- Landslide Cliff. Steep scars left by the fall of masses of rock or unconsolidated materials.
- Sea Cliff. Erosion bluff formed by undercutting by sea waves.
- Solution Cliff. Steep slopes left by the sudden collapse of the floor of a limesink.
- Stream Cliff. Erosion bluff formed by lateral swinging and subsequent undercutting by a stream.
- Volcanic Cliff. Abrupt edges of lava flows, usually formed at the lava front; or steep slopes resulting from a volcanic explosion.
- Weathering Cliff. Bluff resulting from normal weathering in horizontal or inclined layers of rock of differential resistance.

E. Hills, Ridges, Mountains, Peaks. Relief forms that project upward and are surrounded on all sides by lower land.
- Block Mountains. Ridges due to faulting, usually with one slope gentle and the other steep along the fault scarp.
- Volcanic Cones. Peaks built from lava or volcanic fragments or ejectaments.
- Dunes. Ridges or hills built by wind-blown sand.

Drumlins. Oval-shaped hills formed from glacial drift under a moving ice sheet.
Eskers. Narrow, elongate, winding ridges of mixed sand and gravel formed in stream bed tunnels under a glacier.
Moraines. Lateral, terminal, or recessional ridges or irregular deposits laid down by valley glaciers or ice sheets.
Talus Ridges. Deposits of rock fragments at the foot of a rocky slope.
Monadnock. A residual hill or mountain on a peneplain.
Mesa. A flat-topped mountain consisting of horizontal beds bounded by steep erosional scarps.
Butte. An isolated hill or mountain formed on a reduced or eroded mesa.
Domes. Rounded knobs formed by erosion or exfoliation of granitic rocks.
Hogback. Sharp ridge usually formed from erosion of steeply inclined strata or beds.

F. Lakes, Pools, Ponds, Bays. Water-filled depressions.
Fluvial Lakes or Ponds, Pools. Water-filled depressions due to stream activity.
Oxbow. Water-filled, cut-off meander on a mature flood plain.
Plunge Basin. Water-filled depression formed at the base of a waterfall.
Pothole. Sand-scoured, smooth-sided depressions formed in hard rock by a rapidly down-cutting stream.
Glacial Lakes. Water-filled depressions formed by glacial activity.
Cirque Lake, Pond, Pool. A blocked, water-filled amphitheater-like valley or part thereof at the foot of a mountain glacier.
Kettle. Water-filled depression formed by melting ice left in glacial drift.
Morainal Lake. Water-filled depression behind a morainal dam that blocks drainage.
Longshore Ponds or Bays. Bays or ponds formed by baymouth bars truncating or blocking a stream or current.
Meteoritic Lakes, Pools, Ponds. Water-filled depressions formed by meteoritic impact or shockwaves.
Pluvial Ponds, Pools. Rainfilled depressions.
Rock Barren Pools. Ephemeral, erosional water-filled depressions in granite and other hard rocks.
Dune Pools. Rainfilled depressions in blowouts or interdunal areas.
Playa Lakes. Ephemeral lakes in old lake bed areas.
Solution Lakes, Ponds, Pools. Water-filled depressions formed in rock that has been chemically dissolved.
Tectonic Lakes, Ponds, Pools. Water-filled depressions due to blocking by fault or earthquake movement of materials.
Volcanic Lakes. Water-filled depressions caused by stream blockage by lava flows or volcanic cone blowouts or eruptions producing a deep depression.

G. Plains and Flats. Comparatively flat forms with low inclinations usually covering many acres or square miles.
Alluvial Fan or Cone. Gently sloping deposit formed where a stream leaves mountains with an abruptly reduced load-carrying capacity.
Coastal Plain. Sea floor exposed due to relative uplift of position in respect to sea level.
Deltaic Plain. River deposit made by a stream in a large body of water when the stream loses its load-carrying power.

Estuarine Flat, Mud Flat. Small exposed plain left at low tide.

Estuarine or Lagoon Plain. Large flat left after evaporation or drainage of the estuary or lagoon.

Flood Plain. Alluvial flats and terraces formed from sediments deposited along a stream.

Lake-floor Plain. Flat left after evaporation or drainage of a lake.

Outwash Plains. Glacio-fluvial deposits in front of a continental ice sheet. Glacial delta plains would be built by deposition in a glacial lake; frontal aprons are meltwater outwash deposits that would correspond to alluvial plains; sand plains are glacial outwash composed mostly of sand.

Peneplain. An erosional plain; a landscape of low relief formed by long-continued erosion.

Playa. Plain underlain by clays alternating with salt which were deposited in broad, ephemeral intermittently water-filled basins.

Till Plain. Uniform deposits of glacial materials in lands of low relief.

Volcanic Plain. Surfaces of broad sheets of lava and/or volcanic ejectamenta.

H. <u>Valleys</u>. Open topographic depressions with or without streams.

Glacial Valley. A rock-walled valley which is U-shaped in cross-section and bears evidence of glacial abrasion.

Fault Valleys. A valley formed from a fault in which blocks are tilted so that a depression lies at the base of the fault; a valley formed from rifts along multiple fractures; a valley formed by the slipping downward of a block between two fault scarps; a valley formed over a block as two lateral blocks are uplifted.

Solution Valley. A valley formed over coalesced sink holes; a corrosion valley over soluble rock.

Stream Valley. Valley formed by the down-cutting of a stream usually with a valley train or flood plain in the depression.

Gully, Gorge, Ravine, Canyon. V-shaped destruction valleys with comparatively steep sides.

Hanging Valley. A tributary glacial valley with a floor higher than that of the trunk valley.

SELECTED ROCKY SUBSTRATES
(Adapted from Porter et al. [1973])

Basalt. Dark gray to black, fine-grained, extruded igneous rock.

Caliche. Calcareous hardpan.

Coralline Rock. Limestone derived from the calcareous deposits of corals.

Dolomite. Marble or compact limestone rich in magnesium and calcium carbonates, often with a crystalline-granular structure.

Gneiss. Laminated or foliated metamorphic rock similar to granite in composition.

Granite. Fine- to coarse-textured intruded igneous crystalline rock.

Hardpans. Relatively impermeable hardened strata beneath soils, cemented by a variety of leached or deposited materials.

Igneous Rock. Magmatic; originally molten, issuing from the interior of the earth and solidified following intrusion or extrusion.

Lava. Igneous rocks extruded from volcanos.

Limestone. Sedimentary rock composed primarily of calcium carbonate.

Marl. A loose, crumbling calcareous sedimentary deposit laid down under still water.

Metamorphic Rock. Sedimentary or igneous in origin, the basic structure changed by heat and/or pressure following deposition or intrusion.
Quartz. Crystalline silicon dioxide.
Quartzite. Metamorphosed sandstone.
Rhyolite. Acidic volcanic rock with a crystalline goundmass.
Sandstone. Sedimentary rock composed mainly of sand more or less firmly cemented by other minerals.
Sedimentary Rock. Composed of particulate matter deposited, accumulated, and compacted under water.
Serpentine. Consisting mainly of hydrated magnesium silicate and low in calcium.
Shale. Finely laminated sedimentary rock composed of clay, mud, and/or silt, the constituent minerals essentially unchanged since deposition, the lamina parallel to the bedding plane.
Slate. Dense, fine-grained metamorphic rock formed by the compression of shale, developing more or less parallel cleavage lamina in planes at any angle to the bedding plane.
Soapstone. Soft, soapy-feeling rock composed mainly of talc, chlorite, and often magnetite.
Talus (Scree). Rock fragments derived from the weathering of a cliff or slope and accumulating at its base in a sloping pile.
Tuff (Tufa). Porous and usually stratified rock composed of consolidated volcanic ash.
Volcanic Rock. Igneous rock solidified after extrusion from a volcano or similar vent.

GENERAL FLORISTIC LITERATURE

Braun, E. L. 1950. Deciduous Forests of Eastern North America. The Blakiston Company. Philadelphia.
Daubenmire, R. F. 1968. Plant Communities. Harper & Row. New York.
Lahee, F. H. 1961. Field Geology. 6th ed. McGraw-Hill Book Company. New York.
Porter, D. M., R. W. Kiger, and J. E. Monahan. 1973. A Guide for Contributors to Flora North America, Part II. An Outline and Glossary of Terms for Morphological and Habitat Description (Provisional Edition). Department of Botany, Smithsonian Institution. Washington, D. C.
Smith, R. L. 1966. Ecology and Field Biology. Harper & Row. New York.
Whittaker, R. H. 1956. Vegetation of the Great Smoky Mountains. Ecological Monographs 26: 1-80.
________. 1972. Evolution and measurement of species diversity. Taxon 21: 213-251.

FIELD EXERCISES
(A limited number of suggestions)

I. Spectral Analyses--by class or individual student.
   1. List the species of a community, habitat, or area and make a spectral analysis.
   2. Make comparative spectral analyses of the floras; e.g., of a south slope and a north slope; montane and piedmont selected communities; ridge and cove communities.

II. Diversity Studies--by class or individual student; e.g., determine leaf, root, stem types in a habitat; leaf or floral variation within a population; flowering or fruiting times for canopy and herb layer species in a climax community (phenology).

III. Ecosystematics Problem (See Chapter 35).

Chapter 18. COLLECTION AND FIELD PREPARATION OF SPECIMENS

Herbarium specimens are permanent records of a species (or population) as it occurred at a given time and place. The future value and use of any specimen is largely dependent on the care with which the collector selects, collects and prepares his specimens. The following directions and suggestions on specimen preparation, field equipment and field records are given to assist collectors in preparing high quality herbarium specimens accompanied by adequate field notes.

I. COLLECTING OBJECTIVES AND PLANNING

The specimens to be collected depend on the objectives of the collector, and the types of material to be collected will determine the techniques used, amount of material collected, and type of field data recorded. The following suggestions can greatly aid the collector in getting maximum benefit from valuable field time.

1. Outline objectives for a specific expedition.
2. Prepare a list of all equipment to be utilized on a field expedition.
3. Obtain localities from herbarium specimens if specific materials are needed. It is best, when possible, to have several localities in the event some areas have been disturbed. Check flowering and fruiting dates on the specimens in the herbarium.
4. Check local weather conditions and seasonal progress in the areas to be visited.
5. Collect duplicate specimens (replicates) except in the case of very rare or protected plants (check state laws and conservation lists). Other plants to be avoided include those noxious weeds or parasites and their host plants under state or federal quarantine. Duplicates provide sufficient material for dissection or, if needed, verification by a specialist.
6. Prepare voucher specimens (typical herbarium specimens) when collecting materials for cytological, anatomical, and other studies.
7. Obtain collecting permits and other necessary permission in advance, if possible, if a trip is planned to a specific area such as a national park, foreign country, or other area. It is also advisable to obtain permission for collecting on private lands whenever possible.
8. Collect and document as many different species as possible in floristic studies. Do not overlook the inconspicuous (so-called belly plants) or those plants which are difficult to collect or identify. Collect thoroughly in each habitat.

II. SUPPLIES AND EQUIPMENT

1. Field press. A press typically consists of two hardwood frames, with each frame made as follows:

   4 wood strips 3/4 x 1/4 x 18"
   5 wood strips 3/4 x 1/4 x 12"

   The five short strips should be equally spaced on the four longer strips which are also equally spaced and nailed, or riveted securely at the intersection of the strips. The completed frame should be 12 x 18 inches. Repeat for the second frame.

Some collectors prefer the cheaper and often stronger 12" x 18" plywood press boards cut from 3/8" fir plywood. A number of types of presses may be purchased from biological supply houses.

2. Driers (blotters). Excellent driers may be made by cutting sheets 11 x 16 inches from light weight builders deadening felt (unsaturated) or from heavy blotting paper. Driers are also available from biological supply houses.
3. Newsprint. Cut paper 22 x 16 inches and fold to 11 x 16 inches. Many use newspapers as found on the newsstand but unused newsprint may be purchased in rolls from local printers. Biological supply houses usually offer precut papers.
4. Press straps (webbing straps). A pair of strong web straps (parachute or cinch type) with claw buckles are excellent for field purposes. Sash cord or rope is often also used. The minimum length for press straps is four feet.
5. Field notebook. A pocket-size book which will not disintegrate when wet and pencil or pen with water-proof ink are necessary items. Some prefer to keep two books--one taken to the field and another which remains in a safe place and into which field notes are copied. If field labels are used, one is placed in the paper with the specimen and a copy kept by the collector. Field notebooks should be permanent.
6. String tags. Waterproof string tags are useful for labeling plants which are not pressed immediately following collection.
7. Diggers and clippers. Both pruning shears or garden clippers and digging tools are necessary. A trowel (preferably with a steel shank), geologist pick, dandelion digger or heavy sheath knife are excellent for field use. A small trench shovel and pocket knife are also useful.
8. Hand lens. For field observations and identifications a small 5x or 10x lens is desirable. These are generally available from bookstores and biological supply houses.
9. Collecting bottles. Glass or plastic bottles with leak-proof screw caps are often desirable for collecting some materials. The size used depends upon the materials to be collected. Small vial-type bottles are ideal for collecting floral buds, flowers for clearing and other materials to be preserved in liquid preservatives.
10. Liquid preservative. The type of solution used will, of course, depend on future use and type of material. For general anatomical purposes and materials such as wood, leaves, flowers and the like a mixture of Formalin -Acetic Acid-Alcohol (FAA) is widely used and may be prepared in the following manner: Ethyl alcohol (70%)...........90 cc.
Formalin (commercial strength). 5 cc.
Glacial Acetic acid........... 5 cc.

Cytological materials are often fixed in a 6:3:1 mixture of Chloroform, 95% ethyl alcohol and glacial acetic acid or in Carnoy's Fluid (3:1 absolute ethyl alcohol and glacial acetic acid). Carnoy's containing 95% ethyl alcohol is useful for clearing leaves (see Chapter 7). When using the 6:3:1 mixture the glacial acetic acid should not be added until materials are ready to be fixed.

After materials have been in fixative for 24-48 hours the fixative should be discarded and materials stored in 70 percent alcohol. At Kew a solution of 50% alcohol, 5% Formalin, 5% glycerol and 40% water is used for storing plant materials. Each vial should contain a field label. Use pencil and slips of bond paper; the numbers should correspond to those in the field notebook.

11. Vascula and collecting bags. Plant materials not pressed in the field immediately may be stored in a metal container (vasculum) in folds of wet paper. Due to the cost and bulkiness of this container many now prefer to use plastic bags (turkey or fertilizer) or rubber-lined canvas bags (military laundry bags). Local conditions will determine in part the type of bag to be used, but avoid exposure to sun, particularly if clear plastic bags are used. All materials should be carefully labeled to avoid confusion when materials are pressed.
12. Waxed paper. A roll of waxed paper available in supermarkets is useful in pressing many plants, particularly those that are viscid or which deliquesce. A single sheet placed over the material can greatly facilitate the removal of materials from the pressing paper.
13. Manila coin envelopes. Small coin envelopes are excellent for collecting seeds, individual flowers, leaves, pollen, pollinators and other materials which cannot or should not be pressed. These should always be carefully labelled with the collection number.
14. Trays, cans, jars for living materials. It is often desirable to collect living materials for experimental work in the garden or greenhouse. A variety of containers may be used and will depend on the type of material to be collected and the length of time materials must be stored.
15. Cardboard storage boxes. Merrill cases or other cardboard containers which can be purchased flat make excellent storage for dry materials removed from the press in the field. It is advisable to use an insecticide or repellent if materials are to be stored for an extended period of time.
16. Insecticides and repellents. Moth crystals or flakes (naphthalene or paradichlorobenzene) may be used as repellents. Considerable quantities of paradichlorobenzene in an air-tight container may be used as an insecticide. An excellent treatment is to place PDB in the folds of the specimen paper, tie specimens into bundles, and seal in a plastic bag.
17. Maps. Highway, topographic, and geologic maps are often very useful in locating localities for particular species. Detailed county road maps can be obtained, usually at modest cost, from state highway departments.
18. Camera. Photographs are of considerable value when collecting woody or other materials where entire plants are not pressed.
19. Color Charts. Several color charts are available for determination of flower colors in the field, e.g., Horticulture Society Color Chart, Nickerson Color Fan, Horticultural Color Chart, etc.

III. SELECTION OF MATERIAL

Vigorous, typical specimens are to be selected. Avoid insect damaged plants. Specimens should be representative of the population but should include the range of variation of the plants, not those that best fit the press. Roots, bulbs, and other underground parts should be carefully excavated and the dirt removed with care. In most cases flowering and/or fruiting materials are necessary for identification purposes. Many collectors prefer to add extra flowers and fruits to their collections when possible to avoid dissection of the specimen proper. Plants too large for a single sheet may be divided and pressed as a series of sheets (see discussion below). In collecting large herbs, shrubs and trees, different types of foliage, flowers and fruits should be collected from the same plant. Collect sufficient material to fill an herbarium sheet and still leave enough room for the label. Bark and wood samples are often desirable additions when collecting woody plants. Proper identification of many plants depends on several different characteristics--some roots, others seeds or mature fruits, some flower color (which should be noted in the fieldbook).

The following selected suggestions which are largely adapted from DeWolf (1968), Fogg (1940), Fosberg and Sachet (1965), and Smith (1971) are given to assist in the selection and collection of particular kinds of plants.

Table 18-1. Materials for identification purposes.

TAXON	ESSENTIAL OR DESIRABLE MATERIALS FOR IDENTIFICATION PURPOSES	SUGGESTIONS FOR NOTE-TAKING AND PRESSING
Acanthaceae	Flowers and fruits important.	Flowers often detach easily after collecting. Press immediately.
Agrimonia	Underground parts useful.	
Alismataceae	Flowers and fruits (essential).	Note presence of perfect and/or staminate and pistillate flowers and position of fruiting pedicels.
Allium	Flowers, seeds, bulbs (with coat).	
Amaranthaceae	Ripe fruits (essential in some).	Note whether plants are monoecious or dioecious. Collect both staminate and pistillate plants when possible.
Amelanchier	Flowers, fruits and leaves from same plant necessary.	Note habit (erect or stoloniferous).
Annonaceae	Flowers necessary (these may open precociously and then grow considerably before and during anthesis).	
Apiaceae (Umbelliferae)	Mature fruits and basal leaves.	
Araceae	Fruiting material alone of little value; flowers, inflorescences and underground parts of great value; also leaves (often dimorphic).	Peel epidermis away on one side to speed killing and drying or use alcohol or formaldehyde to kill and preserve in the field. Note glaucescence, sap color, vesture. For large aroids see suggestions in Fosberg and Sachet (1965).
Araliaceae	Flowers and fruits.	

TAXON	ESSENTIAL OR DESIRABLE MATERIALS FOR IDENTIFICATION PURPOSES	SUGGESTIONS FOR NOTE-TAKING AND PRESSING
Arecaceae	Flowers and fruits, inflorescence axis and bracts, leaves, petiole, leaf base, notes on stem.	See Tomlinson (in Fosberg and Sachet, 1965) for details on collecting specimens.
Asclepiadaceae	Flowers and/or fruits.	Note flower color, number of leaves and arrangement, latex color, fruiting pedicels (erect or deflexed.
Asteraceae (Compositae)	Ripe fruits and flowers, basal and median cauline leaves and underground parts.	Note color of ray and disk flowers; press some heads exposing upper surface and others with lower surface exposed. Split heads and press.
Balanophoraceae	Tuber surface important.	Plants sometimes dioecious.
Balsaminaceae	Flowers and fruits desirable.	Flowers fragile and tend to agglutinate on drying under pressure. Liquid preserved flowers desirable. Note flower color and markings.
Bamboos	Flowers and fruits rare but desirable. Culm sheaths from mid-culm nodes, portion of culm and branch, rhizome.	Record node number (starting at base of plant) from which material is taken. Collect from 4th to 5th node. (See suggestions of McClure in Fosberg and Sachet for additional handling).
Begoniaceae	Both staminate and pistillate flowers and ripe fruits.	Careful notes on colors, plants fleshy and often darken on drying.
Betula	Fruiting catkins essential.	Notes on bark desirable.
Boraginaceae	Flowers and ripe fruits.	Split and press some flowers flat.
Brassicaceae	Flowers and fruits desirable; underground parts and rosettes.	
Burseraceae	Fruit.	
Cactaceae	Flowers, stem with grooves, tubercles and spines.	Photographs desirable. Stems may be cut, hollowed and pressed including both transverse and longitudinal sections. Stems may be dipped in boiling water or formalin to kill tissue and allow more rapid drying. Handle carefully (preferably with gloves and forceps) but do not remove or cut all spines.

TAXON	ESSENTIAL OR DESIRABLE MATERIALS FOR IDENTIFICATION PURPOSES	SUGGESTIONS FOR NOTE-TAKING AND PRESSING
Campanulaceae	Flowers.	Corolla shape important, make careful notes and/or drawings.
Capparidaceae	Flowers desirable (these open precociously and then grow considerably before and during anthesis.)	
Caprifoliaceae	Ripe fruits.	
Carex	Mature perigynia, underground parts.	
Caryophyllaceae	Flowers and fruits.	Note number of styles and corolla color.
Casuarinaceae	Staminate and pistillate flowers and ripe fruit.	Dioecious or monoecious.
Celtis	Mature fruits.	Shape and color of fruit important.
Coccoloba	Leaves from adventitious shoots and from mature twigs; staminate and pistillate inflorescences.	
Commelinaceae	Flowers, spathe, underground parts.	Flowers should be pressed between tissue and waxed paper and/or preserved in liquid. Note corolla and anther color. Spread spathe and press flat.
Convolvulaceae	Flowers and ripe fruits desirable.	Split flowers and dry flat. Note corolla color and markings.
Cornaceae	Mature fruits.	Note color of branchlets and pith.
Crataegus	Flowers and fruits from same tree.	Note anther color.
Cucurbitaceae	Flowers and ripe fruits.	Dioecious or monoecious. Note corolla shape and color and color of mature fruits.
Cuscuta	Flowers and fruits.	Note nature of petals and their appendages, shape of fruit, and host plant.
Cyperaceae	Ripe fruits and underground parts.	
Dilleniaceae	Ripe fruits.	
Dioscoreaceae	Staminate and pistillate inflorescences, axillary bulbils, underground parts and base of aerial stem; mature fruits if possible.	Note direction of twining of stems.
Dipterocarpaceae	Ripe fruits.	
Epacridaceae	Ripe fruits.	

TAXON	ESSENTIAL OR DESIRABLE MATERIALS FOR IDENTIFICATION PURPOSES	SUGGESTIONS FOR NOTE-TAKING AND PRESSING
Ericaceae	Flowers and fruits.	Note flower color and waxy "bloom" on fruit.
Euphorbiaceae	Staminate and pistillate flowers, ripe fruits.	Note color of glands. (Sap may cause dermatitis.)
Fabaceae (Leguminosae)	Flowers and fruits essential; underground parts desirable.	Note flower color.
Fagaceae	Ripe fruits, mature leaves.	Collect staminate and pistillate flowers when possible. Note shape and size of plant.
Fraxinus	Ripe fruits nearly essential, staminate and pistillate; flowers desirable.	May be dioecious.
Galium	Fruits often essential; underground parts.	Note flower color and habit of plant (erect or reclining).
Gesneriaceae	Flowers and fruits.	Note colors, plants often fleshy and darken upon drying.
Hydrocharitaceae	Flowers and fruits.	Note position of flowers (floating or submerged).
Iridaceae	Flowers and mature fruits, underground parts.	Press flowers immediately and use tissue and waxed paper.
Isoetes	Fertile specimens.	Note habitat and whether plant was emersed or submerged. Split some plants vertically.
Juglandaceae	Fruits, leaves.	Note bark characters, size and shape of tree.
Juncaceae	Fruits.	Note number of stamens, shape of leaf (flat or terete); expose ligule.
Lamiaceae (Labiatae)	Flowers and mature nutlets, base of stem and underground parts.	Split open some flowers and press flat; note flower color and markings.
Lemnaceae	Flowers and fruits desirable.	Note number of roots per "frond." Float specimens out on paper; place in packets.
Lepidium	Flowers and fruits.	Note presence or absence of petals.
Liliaceae	Fruits alone useless, flowers and underground parts most desirable.	Note if leaves are flat or round. Make longitudinal section of bulbs; leave bulb coats intact.
Loranthaceae	Flowers and fruit.	Note fruit color and host plant.
Malvaceae	Flowers and mature fruits, underground parts helpful.	Note flower color; split open some flowers and press flat.

TAXON	ESSENTIAL OR DESIRABLE MATERIALS FOR IDENTIFICATION PURPOSES	SUGGESTIONS FOR NOTE-TAKING AND PRESSING
Melastomataceae	Flowers and fruits.	Petals fugacious.
Menispermaceae	Staminate and pistillate flowers; fruits desirable.	Dioecious.
Moraceae	Fruits desirable.	Frequently dioecious; note fruit color.
Musaceae	Axis of inflorescence and ripe fruit.	Photograph entire plant when possible.
Myristicaceae	Ripe fruits desirable.	Dioecious.
Myrtaceae	Fruits.	
Najadaceae	Fruits.	Float specimens out on white paper.
Nepenthaceae	Pitchers of full grown cauline leaves essential.	
Orchidaceae	Flowers and fruits; fruits alone almost useless.	Note flower color, markings, and fragrance. Preserve some flowers in liquid.
Orobanchaceae	Ripe fruits desirable.	Note host plant.
Pandanaceae	Ripe fruit, leaf tips.	Note stem diameter; photograph or habit sketch very desirable.
Picea	Mature cones.	Note color of foliage.
Piperaceae	Fruits.	Note colors (plants often darken on drying).
Poaceae (Gramineae)	Mature fruits, underground parts and stolons.	Note anther color. Press culms to show sheath and ligule.
Polemoniaceae	Flowers, fruits and underground parts.	Note flower color and split corollas open and press flat.
Polygalaceae	Flowers and fruits and seeds; underground parts.	Note color, habit, leaf dimorphism, if not pressing entire plant.
Polygonaceae	Fruits and underground parts.	
Potamogeton	Fruits and stipules.	Press so stipules are clearly displayed.
Pteridophyta	Fertile fronds, rhizome or rootstock, sterile fronds if different from the fertile. Tree ferns - petiole, leaf bases and leaf scars.	
Ranunculaceae	Fruits and underground parts.	
Ribes	Flowers and fruits.	Note color of fruits.
Rosaceae	Fruits; flowering and sterile shoots.	Note habit of plant.
Rubiaceae	Fruits very desirable.	
Rubus	Habit important, twigs of sterile and fruiting branches.	Note if canes are arching or not, rooting at tip or not.

TAXON	ESSENTIAL OR DESIRABLE MATERIALS FOR IDENTIFICATION PURPOSES	SUGGESTIONS FOR NOTE-TAKING AND PRESSING
Salix	Mature staminate and pistillate catkins and leaves from the same plant.	
Sapindaceae	Fruits desirable.	
Sapotaceae	Fruits desirable.	
Scrophulariaceae	Flowers.	Split flowers and press flat; make careful notes on color and markings; many must be pressed immediately; commonly darken on drying; note host where applicable.
Smilax	Fruits and vegetative material.	Note fruit color and whether main axis is smooth or spiny.
Solanaceae	Flowers and fruits.	Split open some flowers and press flat.
Styracaceae	Fruits desirable.	
Symplocaceae	Ripe fruits.	
Utricularia	Flowers, fruits, leaves.	Float plants out on white paper, spread leaves. Excavate terrestrial species carefully.
Vaccinium	Flowers, fruit.	Note fruit color and plant habit.
Viburnum	Ripe fruit.	
Zingiberaceae	Inflorescence and underground parts.	Inflorescence best preserved in liquid.

IV. PRESSING PLANT SPECIMENS

A. Arranging and Preparing Specimens. After the specimens have been dug or cut they should be pressed as soon as possible (see V for information on field storage of fresh specimens). The care given a specimen in pressing will largely determine its future value. Specimens should be placed in a single fold of newsprint or other suitable absorbent lightweight paper. Plants too large to fit the 11" x 16" fold of paper may be bent into a "V", "N" or "M" figure. Bruise the stem before bending and it will be less apt to break. Specimens should not protrude from the fold of paper. Protruding parts will likely have to be removed when specimens are mounted. A specimen may be trimmed to reduce bulk and expose certain characters advantageously if sufficient material (e.g., leaf petioles, branch bases, etc.) is left so that the pattern of branching, leaf arrangement, and other features are readily discernible. When pressing large plants, make several sheets rather than a single sheet with a crumpled mass of material. Arrange specimens in such a way that some upper and some lower surfaces of the leaves are exposed. Spread flowers or inflorescences to show as many surfaces or views as possible. Section some flowers longitudinally and press flat to exhibit the inner parts and thereby reduce the need for dissection of the finished specimens.

Excessively bulky and fleshy parts such as stems and fruits may be split and both parts included. Fruits should be sectioned when possible in such a way that both transverse and longitudinal sections are included. Succulent plants (e.g., cacti) may be split and the fleshy inner parts removed. Some collectors prefer dipping these plants in boiling water, xylol or benzene before pressing. Salt may be applied to cut surfaces to hasten drying. Each sheet should bear a collector's number which refers to notes in the collector's field notebook. When preparing a series of sheets for parts of one specimen, number each sheet with the collection number and the number of parts (e.g., #429, sheet 1 of 3); duplicates should simply bear the collection number. Do not include more than one species in a single paper. Collect sufficient material to make a full sheet but avoid crowding or overlapping of specimens in the specimen paper. See Smith (1971) for an excellent illustrated account of specimen preparation.

B. Arranging the Press. Specimens (in the specimen paper) are placed between two driers in the press. A ventilator is often inserted before the next specimen paper and driers are added. For each specimen pressed a drier unit (2 driers with ventilator between them) is added to the press. The usual sequence is ventilator, drier, specimen in specimen paper, drier, ventilator, drier, etc. As the successive specimens are added and the press built, every effort should be made to keep the press level for even distribution of pressure when the press straps or weights are applied. This will mean the use of alternate corners of the sheet for bulky roots or other parts. Sheets or pads of plastic sponges are very useful when placed over or around bulky specimens.

V. FIELD STORAGE OF SPECIMENS

Although it is desirable to press collections immediately, it may not always be practical. Delicate materials should be pressed as soon as possible and other specimens properly stored. Vascula, plastic bags, or rubberized bags can be used for storage if specimens are first wrapped in moistened paper. Specimens may be kept in good shape without spoilage in the containers if they are kept moist and are not packed too tightly. Supersaturation with water or drying out will spoil specimens. The specimen bags should be kept as cool as possible and a conscious effort should be made to park a specimen-loaded vehicle in the shade on field trips.

Upon return from the field, specimens in bags should be put into a coldroom until the plants are pressed. Although immediate pressing is again desirable, plants can be kept in this manner for several days.

Plastic gallon or half-gallon milk cartons or jugs make excellent water containers for field use. It is a good practice to wrap specimens collected at one locality in an uncut, numbered (with a wax pencil before paper is soaked), double sheet of newspaper so that one end is open and the other closed, bind with a rubber band, tag with a locality number which is entered in the field notebook, moisten and place upright in a bag, and then place the bag upright in the field vehicle. Moisten by sprinkling water over the open top of specimen bags as needed. If it is necessary to keep wet materials for long periods of time, as is necessary for collectors in tropical regions, several preservatives may be used:

1. 2 parts commerical formaldehyde (40%):3 parts water.
2. 1 part formaldehyde:2 parts 70% alcohol.

3. 40-50% alcohol.
4. 1-2% aqueous solution of Oxyquinoline Sulfate.

Specimens are dipped, sprayed or brushed with these solutions and enclosed in airtight packages. For additional details consult Fosberg and Sachet (1965), Lawrence (1951), and Smith (1971).

VI. DRYING FIELD COLLECTIONS

Plants should be placed in the press and the press closed and tightened. The faster the drying process the less difficulty with mold, mildew and loss of color. Plants should be sweated in the field press for 12-24 hours and the press opened. Any last arrangement of the specimens must be made at this time and the wet driers changed. For exceptional results driers should be changed at least three times during the first 48 hours. In many areas blotters may be dried in the sun (usually one hour is sufficient). If specimens are to be dried without artificial heat, blotters should be changed daily until specimens are dry. Automobile luggage carriers are excellent means of drying specimens provided ventilators with open ducts are used between the blotters.

If artificial heat is used, there should be maximum airflow through the press. Use doublefaced corrugated cardboard or aluminum ventilators. Never attempt to dry specimens in an oven. A metal or wooden box with an open top which will accommodate a press sideways (corrugations pointing up) and equipped with an electric heater with a fan makes an excellent drying chamber. A collapsible drying frame (either a wooden box or metal frame with canvas skirt) may be used in the field and a camp stove or lantern used as a heat source. Electric heating coils or light bulbs may also be used as heat sources but a fan should be installed either in or above the chamber. Special drying cabinets are sold but most lack sufficient ventilation for proper drying of specimens.

VII. DATA TO ACCOMPANY SPECIMENS

A. Field Notes. As mentioned earlier every collector should keep a field book. This is not simply a road log. Each species collected at a given place and time should be given a collection number. The best system to use is a chronological one beginning with number one and continuing from there. Avoid elaborate numbering systems with prefixes and cryptic notations or abbreviations. Do not use the same number for any other collection. All duplicates or replicates should bear the same collection or collector's number. Although some abbreviations may be useful and efficient in the field, these should be fully written out when permanent labels are made from the field notes. A specimen without field data (at least locality and date) is of little scientific value.

Data to be recorded in the field notebook should include collector's number (for reference), exact locality, approximate altitude, nature of the habitat (type of soil, moisture conditions, slope exposure, light conditions), associated species, and other pertinent information. With reference to the plant proper, record those features which will not be evident from the pressed specimen (height, branching, depth of root system, odor, etc.) and those features which may be lost in drying; e.g., flower color. Flower color may best be determined by using a color chart. The more complete the field notes, the more complete the permanent label can be and the greater the information content of the specimen.

B. Permanent Label. The permanent label is the label affixed to the mounting sheet with a specimen. Information included in addition to the name of the plant and authority (e.g., *Claytonia caroliniana* Michaux) must come from the collector's field notebook. Do not abbreviate or use symbols. Specimens are now used throughout the world and symbols and abbreviations are often difficult to translate. Be specific in giving localities. If local names are used, give some reference to a city, major highway, or easily located reference point. Minimum data for labels should include date, locality, county, state, collector and collection number. Labels, unless done by offset printing, should be typewritten using a carbon ribbon or written in longhand using india or other permanent ink. Paper should be of high rag content--preferably 100%. See sample labels in Chapter 31, page 774.

VIII. IDENTIFICATION OF COLLECTIONS

Materials should be identified using the appropriate manuals, floras, and monographs (see Chapter 30). If it is necessary to have materials identified or verified by a specialist, one of the duplicates is sent with a label to the specialist. Generally the specialist will keep the specimen unless he has agreed to do otherwise. In the event that a duplicate specimen is identified by someone else, the collector should enter the plant name on the label followed by the name of the specialist and the date that the duplicate was determined: e.g.,

Ipomopsis rubra (L.) Wherry
duplicate determined by Verne Grant, 1974

Once materials have been labeled and identified, they are ready to be mounted (see Chapter 31).

COLLECTING LITERATURE

Davis, P. H., and V. H. Heywood. 1963. Principles of Angiosperm Taxonomy. Van Nostrand Company, Inc. New York.

DeWolf, G. P., Jr. 1968. Notes on making an herbarium. Arnoldia 28: 69-111.

Fogg, J. M., Jr. 1940. Suggestions for collectors. Rhodora 42: 145-157.

Fosberg, F. R., and M. Sachet. 1965. Manual for tropical herbaria. Regnum Vegetabile 39. Utrecht.

Franks, J. W. 1965. A Guide to Herbarium Practice. Handbook for Museum Curators, Part E, Section 3. The Museums Association. London.

Knudsen, J. W. 1972. Collecting and Preserving Plants and Animals. Harper and Row. New York.

Lawrence, G. H. M. 1951. Taxonomy of Vascular Plants. The Macmillan Company. New York.

Pike, R. B. 1964. Plant pressing with plastic sponges. Rhodora 66: 172-176.

Smith, C. E., Jr. 1971. Preparing herbarium specimens of vascular plants. Agriculture Information Bulletin No. 348. U. S. Government Printing Office. Washington, D. C.

Chapter 19. PALEOBOTANY AND SYSTEMATICS*

Of four basic questions a systematic botanist can ask about a species or other taxon: what is it? when did it arise? where did it originate? and how did it acquire its present distribution? (see Smith, 1969), the last three can be answered most convincingly using paleobotanical data. Material pertinent to these questions and to the important role of paleobotany in plant classification is presented below, but as these matters also relate to topics covered elsewhere in this book, the reader is referred to Chapter 16 (Geography) and Chapter 27 (Structural Evolution and Phylogeny) for information not included here.

A comprehensive review of paleobotany will not be found in this chapter. Rather, a set of concepts and associated terminology, a brief outline of how the plant fossil record is used, and a series of bibliographic references potentially useful to botanists unfamiliar with the literature of paleobotany and geology are presented. Tertiary and Quaternary paleobotany are stressed because these subjects deal largely with the angiosperms, the plant group of primary focus in this book.

The oldest generally accepted angiosperm fossils are from deposits middle to late Early Cretaceous in age, although some believe angiosperms originated long before this.

I. SOME BASIC TERMINOLOGY AND CONCEPTS IN PALEOBOTANY

A. Plant fossils occur in sedimentary rocks mainly as:

1. Petrifications--three dimensional structural detail usually well preserved in a matrix of silica, calcite, dolomite, or other substances.
2. Compressions--flattened, often coalified plant structures with variable amount of cellular organization preserved; may be removable intact from matrix using mineral acids.
3. Impressions--essentially no organic matter present, external features evident as an imprint in the rock; both compressions and impressions leave two halves, an obverse and a reverse.
4. Casts and molds--infilled cavities, associated organic matter may be partly or completely decayed before infilling, no anatomical details preserved; provide data on external (and sometimes internal), three-dimensional form.

Only slightly compressed wood and plant structures of other kinds are frequent in lignites (brown coals). Other grades of coal may contain well-preserved but compressed plant structures.

Nonpetrified, more or less carbonized plant remains from Quaternary deposits are often called subfossils.

Compounds characteristic of various metabolic pathways in plants have been found in organic residues extracted from sedimentary rocks of diverse ages, and as such they provide biochemical evidence of the

*Contributed by Norton G. Miller, The University of North Carolina at Chapel Hill.

nature and origin of life processes through time (see Eglinton & Calvin, 1967).

B. Mega- and microfossils. A distinction is made between megafossils and microfossils, the former being seeds, fruits, leaves, wood, etc., the latter pollen, spores, and other small structures that require a compound microscope for routine study. In North America, leaf compressions and impressions are among the most common Cretaceous and Tertiary plant megafossils. Numerous leaf floras have been described from the western United States and from sites on the Gulf and Atlantic coastal plains. Compressions of flowers and fruits are infrequent, although fruits (minus outer layers) may be abundant in lignites (e.g., the early Tertiary Brandon Lignite, Vermont). Determination of the spore and pollen content of rocks may provide data to support or refute questionable identifications based on leaves.

C. Nomenclature of fossil plants follows the principles, rules, and recommendations of the International Code of Botanical Nomenclature (Stafleu et al., 1972). A species based on fossil material may be placed in a genus with extant species. An organ-genus is used when a genus of fossil plants is assignable to a family (either extant or extinct; but, if extant, not a modern genus); a form-genus when a genus is unassignable to a family. When proposing new names, particularly species in extant genera with known fossil records, both paleobotanists and neobotanists should check the standard catalogs (e.g., Gray Herbarium Index, Index Kewensis, the catalogs of Andrews, Knowlton, LaMotte, see part B3 in General References) to avoid publishing illegitimate names.

II. GEOLOGIC CONSIDERATIONS

A. Geologic time scale. That portion of earth history richest in fossils comprises three Eras, the Paleozoic, Mesozoic, and Cenozoic (see Table 19-1). Each Era is divided into successively smaller units of time called Periods, Epochs, and Ages. Rocks belonging to these time units comprise time-rock units and are called Systems, Series, and Stages, respectively. One speaks of Early, Middle, and Late Jurassic time, and of Lower, Middle, and Upper Jurassic rocks. The relative age of rocks is determined by their position in a stratigraphic sequence. Achievement of an absolute time scale for geologic events depends largely on radiometric dating.

B. Stratigraphy. This is a geological discipline dealing with the distribution and age of stratified rocks and with episodes in earth history that can be read from such rocks. A basic unit of stratigraphy is the formation. Formations have an easily recognized lithology and as such find wide usage in descriptive studies and mapping. Formational names are binomials and often include the name of the locality at which they were first recognized. If a formation consists of only one kind of rock, its lithology is part of the name, e.g., Dakota Sandstone (Cretaceous, plant localities in Kansas, Nebraska, Minnesota). The term Formation is part of the name if the lithology varies, e.g., Fort Union Formation (Paleocene, widespread plant localities, upper Great Plains).

C. Correlation. Determination of mutual time equivalents is called correlation, and two units occurring in different outcrops are correlative if they are considered time equivalents of each other. Methods used to correlate strata include tracing individual beds, lithologic similarity, radiometric dating (e.g., lead-uranium-thorium, potassium-argon, carbon-14 methods), and

Table 19-1. The Phanerozoic Time Scale

Era	Period	Epoch	Age of Base in Millions of Years
Cenozoic	Quaternary	Holocene	
		Pleistocene	1.5-2
	Tertiary	Pliocene	c. 7
		Miocene	26
		Oligocene	37-38
		Eocene	53-54
		Paleocene	65
Mesozoic	Cretaceous		136
	Jurassic		190-195
	Triassic		225
Paleozoic	Permian		280
	Carboniferous systems: Pennsylvanian		
	Carboniferous systems: Mississippian		345
	Devonian		395
	Silurian		430-440
	Ordovician		c. 500
	Cambrian		570

Ages from: The Quarterly Journal of the Geological Society of London 120s: 260-262 (1964).

comparing assemblages of contained fossils. Because of the universality of progressive organic evolution, paleontologic data are commonly used to determine time equivalents on a world-wide basis.

D. <u>Continental vs. marine deposits</u>. Plant fossils, particularly Tertiary leaf floras, are found predominantly in sediments deposited on continents or in intermediate, near-shore marine environments (e.g., estuaries). Only infrequently do plant megafossils occur in marine sediments, which may contain spores, pollen, or other kinds of plant microfossils and often abundant animal remains. Many of the well known Tertiary floras of the western United States are from extinct fresh water lakes. Those whose sediments include diatomite or volcanic ash often have remarkably well-preserved plant remains, although cellular details and cuticles may be absent. Glacial and nonglacial continental deposits of Pleistocene age are often rich in plant fossils and have been extensively studied.

E. <u>Depositional environments</u>. This subject, of importance to those relating extinct floras and vegetation to past environments, draws heavily on the methodology and principles of sedimentary petrology. Sediment origin (including potential fossil origin), transport, deposition (both physical and chemical aspects of this process), and post-depositional history may relate to the nature of a fossil flora. For example, on both geologic and biologic grounds, the Eocene Wilcox flora of the southeastern United States is considered a strand flora. A consequence of this kind of depositional environment is that plants of the uplands are poorly represented and a biased sample of the total flora is present.

Further information on topics covered in A-E is found in Dunbar and Rodgers (1957), Kummel (1970), and Pettijohn (1957).

F. <u>Paleogeography</u>. The geographical bearings of continents through time, position and elevation of now eroded mountain chains, and distribution of epicontinental (epeiric) seas are kinds of paleogeographical data useful to plant systematists concerned with possible geographic isolation, migration routes, origin of floristic provinces, and related topics. Explanations for phytogeographic patterns often incorporate paleogeographical data.

1. Land bridges. A general review of the land bridge concept and its use in floristic botany is provided by Van Steenis (1962), with special reference to his extensive studies of tropical plants. Unfortunately this report is marred by the author's rejection of continental drift. Land bridges fall into two general categories.
   a. Isthmian links, e.g., Central America [development of the Isthmus of Panama near the end of the Pliocene ended the isolation of North and South America that had prevailed earlier in the Cenozoic (see Hallam, 1972)], Beringia [see evidence summarized in Hopkins (1967) for late Tertiary and Quaternary land connections between Siberia and Alaska].
   b. Insular stepping stones, e.g., the various islands of the South Pacific Ocean.
2. Continental drift. Postulation of now sunken land connections between continents to explain distribution patterns in flowering plants has to some extent been rendered obsolete by the general acceptance of the Theory of Plate Tectonics which provides a mechanism for continental drift (see Dewey, 1972). However certain authors believe that angiosperms evolved too recently for their present distribution to have been seriously affected by continental drift. Geologic data on the time and direction of continental movements and how present continental masses originally fitted together are still accumulating, but some workers accept the existence of a single pre-drift super-continent, Pangaea (Dietz & Holden, 1970), while others two super-continents, Gondwana (South America, Africa, India, Antarctica, Australia) and Laurasia (North America, Eurasia). According to the Dietz and Holden model some of the major events affecting New World paleogeography include:
   a. Breakup of Pangaea (beginning in Middle Triassic time).
   b. Laurasia split away from Gondwana (by late Triassic time).
   c. North Atlantic basin well developed but North America and Eurasia still attached via Greenland, initiation of South Atlantic by rifting (by Late Jurassic time).
   d. Widening of South and North Atlantic, rifting initiated between Greenland and Eurasia (by late Cretaceous time).
   e. Separation of North America, Greenland, and Eurasia (during Cenozoic time).

The breakup of Gondwana has been used by Jardine and McKenzie (1972), Raven and Axelrod (1972), and Schuster (1972) to explain distribution patterns of disjunct biota among land masses of the Southern Hemisphere.

The following references contain useful paleogeographic information for various parts of the world: Harrington (1962), Khudoley and Meyerhoff (1972), Schuchert (1955), Termier and Termier (1960).

III. SOME WAYS FOSSIL PLANTS ARE USED IN BOTANICAL STUDIES

A. <u>Morphological research</u>. Investigations on both external and internal structure of plant fossils are commonly undertaken. Apart from providing a wealth of descriptive information, such studies:

1. Establish and document evolutionary trends.
2. Permit recognition and circumscription of extinct plants or plant

groups, e.g., the Pteridospermae (seed ferns) by F. H. Oliver and D. H. Scott, Rhynia by R. Kidston and W. H. Land, the Progymnospermopsida (probable ancestral stock of many gymnospermous groups) by C. B. Beck.

3. Enable interpretation of problematic structures found in extant plants, e.g., clarification of conifer cone structure (see Florin, 1955).

Many current concepts in plant morphology have been strongly influenced by paleobotanical data.

B. Plant classification. Data obtainable through study of fossil plants should be used extensively in the formulation of a phylogenetic system of plant classification, a major goal of systematic botany. Proposed phylogenies, however, generally are based on other, usually less satisfactory criteria because antecedents of most extant taxa remain unknown or appear to be absent from the fossil record. This may be true even in groups that have a reasonably good record. For example, though the main concern of this book, the flowering plants, "have the most extensive fossil history of any group of comparable rank, fossil angiosperms have probably had the least influence of any fossil plants (except some of the lower plants) on classification of their living representatives. The greater part of this record consists of vegetative organs (leaves, wood) which are yet of little use and have in face been rejected as a basis for either fossil or living angiosperm classification" (Chesters, 1964, p. 283). Nevertheless, the fossil record provides otherwise unobtainable information that will continue to influence classification, particularly the definition and refinement of higher taxonomic categories. Some examples illustrating the influence of paleobotanical research on classification follow (see Delevoryas, 1964, 1969).

1. Recognition of the Lycopsida and Pteropsida (incl. ferns and seed plants) as opposed to the Pteridophyta and Spermatophyta of earlier systems (by E. C. Jeffrey).
2. Separation of the Sphenopsida from the Lycopsida (by D. H. Scott).
3. Re-evaluation of the relationships of gymnospermous plants indicating that as a formal concept the "Gymnospermae" is unnatural (see Florin, 1955).
4. Partitioning the heterogeneous Psilophytales into three morphologically distinct groups of Devonian land plants, the Rhyniophytina, Zosterophyllophytina, and Trimerophytina (by H. P. Banks).

C. Paleoecology. Plant fossils are commonly used in paleoecological research, the goal of which is "to reconstruct an accurate picture of the life and life conditions of the past by inferences derived through a study of modern biota that are comparable with fossil ones" (Cain, 1944, p. 30). Acceptance of uniformitarianism, which includes the concept that "the present is a key to the past," is fundamental to the paleoecological method. As one progresses back in time, the more tentative paleoecological conclusions become. Reasons for this include an increase with time in the number of extinct taxa represented in the fossil record and the problem of dealing with races which may have had different ecological tolerances than those currently living.

D. Floras and vegetation. Studies of Tertiary and Quaternary floras (a taxonomic listing of species) and vegetation (combinations and abundance of species in a flora, many include physiognomy also) have gone hand in hand, but they are distinct concepts that involve a somewhat different research approach and consequently yield different kinds of data.

1. Tertiary and Quaternary floras (paleofloristics). The identification of species (and other taxa) in widespread, isochronous deposits allows maps to be prepared which summarize distribution patterns through time. Maps of this kind may be useful in explaining present ranges of taxa, postulating centers of origin and/or dispersal, paths of migration, and related topics. In the Northern Hemisphere a Tertiary fossil record has helped establish the relict distributional status of numerous genera including: Acer, Carpinus, Castanea, Cercidiphyllum, Fagus, Nyssa (Figure 19-1), Platanus, Juglans, and many other woody angiosperms; and the conifer genera (see Ferguson, 1967) Metasequoia, Sequoia, Taxodium, and Tsuga to name a few. Identifications of both pollen and megafossils are used. Leopold and MacGinitie (1972) have presented detailed synopses of the past and present distribution of Platycarya, Pterocarya, and Engelhardtia (Juglandaceae) and Pachysandra (Buxaceae).

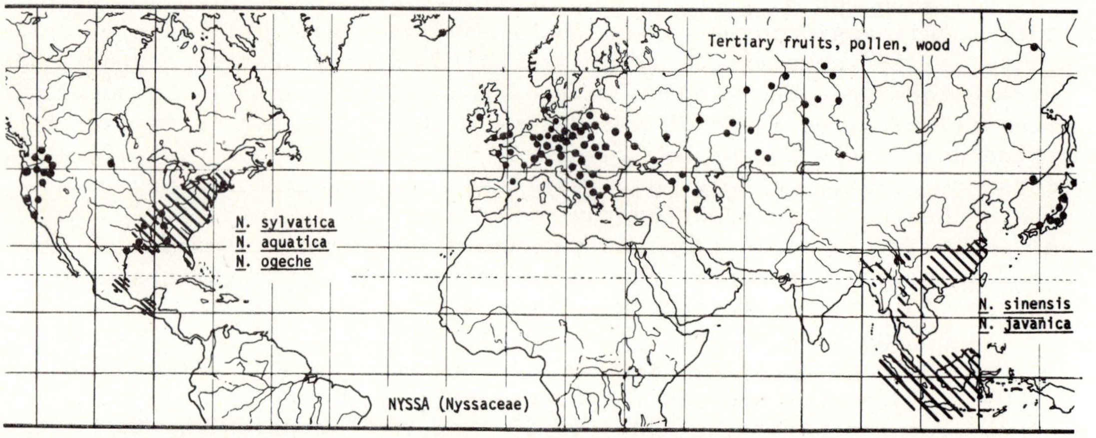

Figure 19-1. Present and Tertiary distribution of Nyssa. Ranges of the five extant species are mapped. (From Wood, 1972.)

Species accurately identified also provide a basis for the recognition of floristic boundaries and evidence that floristic provinces (and consequently vegetation types) have not been stationary through time. For example, southward latitudinal displacement of tundra and boreal floras during late phases of the Pleistocene glaciation has been documented by pollen and megafossil identifications in both Europe and North America. Determination of lineages (phylads) that may have bearing on floristic history as well as providing data on the history of speciation within genera depend on accurately identified species. The occurrence of fossil material of an extant species beyond its present range has climatic implications, a topic that will be explored further below.

2. Vegetation. Recognition of a vegetation type or types from a fossil flora involves not only identifying species but also a quantitative assessment of the abundance of these species in a fossil assemblage. This allows dominance to be established and provides a basis for recognizing communities more or less equivalent to those in existence today. Because plant megafossils generally can be transported for only short distances and remain identifiable, they may be less useful in deciphering aspects of

vegetational history than pollen and spores which often become widely dispersed and buried in perfect condition and thus serve to document both local and more distant communities.

Paleovegetation maps based on data from numerous individual studies have been prepared for various times during the Tertiary and Quaternary and for various regions. Two types of maps are generally attempted: those showing the entire composition of the vegetation (using pie-diagrams, histograms, etc.) and those in which the frequency and distribution of species are presented as a series of maps, one map per species.

a. Tertiary vegetation. The Geoflora concept. Three major long-lived units of vegetation (Geofloras) have traditionally been recognized (Chaney, 1947, 1959) in studies of plant fossils in North American Tertiary deposits, which while abundant in the western part of this continent are rare north of the glacial boundary and to a lesser extent elsewhere in the East. The supposed migration of these units through time and space has found wide application in explaining phytogeographic patterns in diverse plant groups.

1) Arcto-Tertiary Geoflora. A temperate vegetation that occupied high latitude northern land masses around the world early in Tertiary time. Its migration southward in response to cooling and drying, and in North America its extinction in midcontinent regions (a probable result of decreased precipitation) and restriction to the moister coastal areas has been postulated. During these times certain genera were eliminated from all or various parts of North America but survived at places in Eurasia where they occur at present.

2) Neotropical-Tertiary Geoflora. A warm temperate to subtropical vegetation occurring south of the Arcto-Tertiary area in mid- and low latitudes early in the Tertiary. In western North America its migration southward occurred during Oligocene and Miocene times and by the Pliocene only a few members of this geoflora persisted north of Mexico and in southern Florida. Its present region of survival includes Mexico, Central America, northern South America, and the Caribbean Islands.

3) Madro-Tertiary Geoflora. Vegetation types included in this concept are live oak-conifer woodland, chaparral, arid subtropic scrub, plains (desert) grassland, and subdesert to desert (Axelrod, 1958). This geoflora appears to have originated from subtropical to warm temperate groups in southwestern North America (where it now occurs) in response to an increase in area exposed to dry climate.

The concept of an early Tertiary (Eocene), high latitude Arcto-Tertiary Geoflora has been rejected by Wolfe (1969, 1972, and references cited therein) who points out that temperate plant assemblages from this region, originally thought to be from Eocene deposits, are actually Miocene and Pliocene in age. Eocene and Paleocene (and some Oligocene) floras in Alaska (and elsewhere in the far north) are mainly tropical and subtropical and are not temperate in character. The postulated stability and duration of geofloras has also been questioned from a more theoretical standpoint, viz., "It is extremely improbable from a genetic and physiological viewpoint that many lineages could have remained in association throughout the Tertiary; that is, that a given vegetational type remained floristically unchanged." (Wolfe, 1969)

It is of interest that many early Tertiary plant assemblages from western North America have an overall paleotropical affinity, while those from southeastern North America are predominantly neotropical (see Leopold & MacGinitie, 1972).

b. Quaternary vegetation. In North America paleobotanical studies of Pleistocene and early Holocene vegetation have mainly involved the use of pollen and spores for paleoecological purposes, although seeds, seed-like fruits, and wood at times have provided important supplemental information. The common use of radiocarbon dating (age determinations of organic material up to 30-40,000 years old, sometimes older) has permitted rather precise conclusions to be made about the development and composition of vegetation types recognized in this period of time. A brief synopsis of some pertinent concepts as they relate to North American studies follows:

1) The Pleistocene is divided into four glacials (Nebraskan, Kansan, Illinoian, Wisconsin) and four interglacials (Aftonian, Yarmouth, Sangamon, the present). Glacials are divided into stades (ice advancing) and interstades (ice retreating). Both interglacial and interstadial deposits have yielded significant assemblages of fossil plants. The late phases of the Wisconsin are often informally called full-glacial (active glaciation) and late-glacial (general ice retreat). The Holocene (post-glacial) began approximately 10,000 years ago according to many workers.

2) An assemblage of pollen grains from one stratigraphic level is called a pollen spectrum and series of spectra comprise pollen diagrams. Numerical values for various taxa in a spectrum are generally relative frequencies (percentages) calculated using as the percentage base the total number of arboreal pollen (AP) and nonarboreal pollen (NAP) at each level with that of aquatics and spores of pteridophytes excluded. Prior to the early 1960's it was customary to exclude all NAP from the percentage base even if it was tabulated.

3) A few diagrams based on absolute pollen influx (in grains/$cm.^2$/yr.) have appeared in recent years, and because such data are freed of constraints imposed by percentage calculations, influx diagrams have allowed important new interpretations to be made.

4) Comparison of fossil pollen spectra with surface pollen assemblages in an attempt to identify extant analogs of past vegetation types is at present a frequently used methodology. Various statistical techniques are employed in making correlations.

5) In North America more is known of extinct vegetation late in the Quaternary than earlier, and Wright's review (1971) is a recent synopsis of this topic. Much pertinent information will also be found in various INQUA volumes (see Wright & Frey, 1965). The following points pertain to the sequence in glaciated eastern and mid-continent North America (see Figure 19-2).

a) Based on megafossils and pollen, late-glacial tundra (zone T of New England zonation; zone designations vary from region to region, pollen assemblage zones named and treated in accordance with the Code of Stratigraphic

Nomenclature [see American Commission on Stratigraphic Nomenclature, 1961] have come into current use) has been recognized in Massachusetts, Michigan, Minnesota, and elsewhere.

b) Zone A forests rich in spruce (and with jack and/or red pine in the East) replaced the tundra at sites where it is recognized or alternately is the oldest vegetation type at others.

c) The abrupt end of the spruce zone according to pollen analysts marks the beginning of the postglacial and zone B. The species of trees replacing spruce differed between the upper Midwest and the East with white pine and birch being important in New England, white pine, birch, and oak in parts of New York State, Oak in Ohio, and birch and alder in Minnesota.

d) Later in postglacial time deciduous or deciduous-coniferous forests of variable composition come into dominance. The first of these, as recognized at New England sites, with oak and hemlock predominating (zone C-1) is followed by oak and hickory (zone C-2) and by oak and chestnut (zone C-3). Zone C-2 has been traditionally interpreted as a warm/dry interval between 4000 and 1500 years ago. The composition of zone C forests shows considerable regional variation.

Vegetation types recognized south of the glacial boundary in the East and elsewhere in North America are summarized in the references cited above.

E. Climate. Because present vegetational formations and climate have been shown to be intimately related, climates of the past can be inferred once former vegetation types and their analogs are determined. Paleoclimatological research is multidisciplinary with important data contributed by botanists, geochemists, geologists, meteorologists, oceanographers, and others.

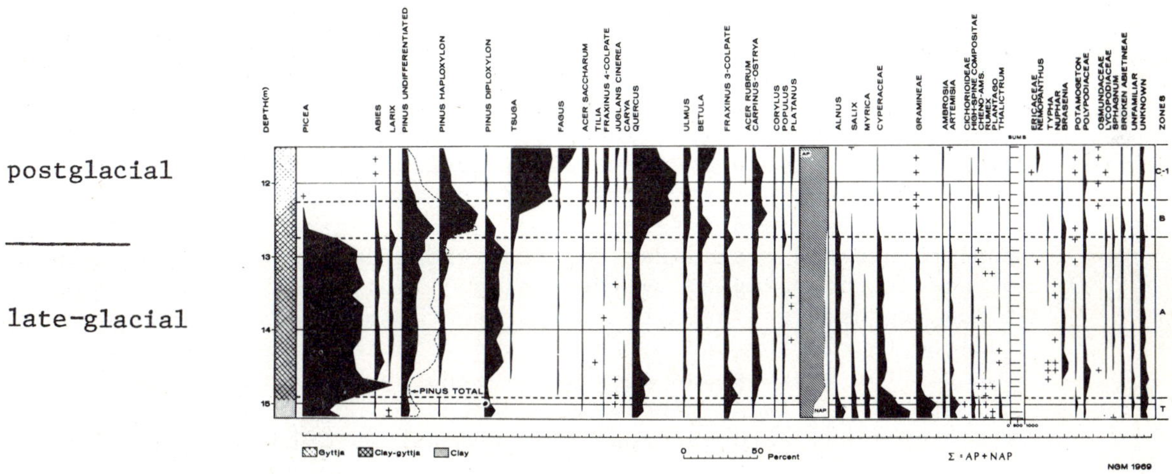

Figure 19-2. Lowest segment of relative frequency diagram from Allenberg Bog (Cattaraugus County, New York) showing late- and early postglacial pollen sequence and zonation. Items to the right of the sums column were omitted from the percentage base. Pollen from trees is graphed at left; percentages from nontrees are between the sums column and the cumulative AP-NAP curve. (From N. G. Miller, New York State Museum and Science Service Bulletin 420, 1973.)

Morphological and anatomical studies of fossil plants can sometimes lead to conclusions about past climates. For example, growth rings and cycles of leaf scars are often taken to indicate periods favorable for plant growth alternating with ones not favorable.

1. Tertiary climates. Two methodologies involving fossil plants (primarily leaf floras) are commonly used to determine Tertiary climates and climatic change (see Wolfe, 1971).
   a. Floristic analysis. Estimating various temperature parameters (incl. mean annual temperature and range of temperature) based on the assumption that thermal requirements of extant genera recognized in a fossil flora have not changed. This involves linking the occurrence of a genus to some part of an extant vegetational formation and this part to climate. Accurate identifications are essential.
   b. Analysis of foliar physiognomy. Relating the percent of species with entire leaf margins in a fossil flora (identifications need not be made) to comparable data assembled for extant vegetation types and using the identified vegetation type to infer climate. It is not possible to determine precise temperature regimes by this method. Numbers of species with drip tips and percent lianous taxa may provide supplemental data. The following percentages of species with entire leaves have been determined for the indicated extant mesic vegetation types (Wolfe, 1969):

      76+%: Tropical rain forest
      57-75%: Paratropical rain forest
      40-56%: Subtropical rain forest
      10-35%: Temperate forest

      Tertiary temperature fluctuations inferred by this method include: "cooling during the Paleocene, maximum warmth in the later Eocene, a strong decline in the Oligocene, and warming in the Middle Miocene" (Wolfe, 1971).
2. Quaternary climates. Pollen analysts reconstruct Quaternary climates by a several step process which generally involves determination of vegetation type from assemblages of fossil pollen, finding extant analogs, and inferring past climatic parameters (usually temperature and precipitation) from those deemed characteristic of the region occupied by the analog. Individual species sometimes are used as indicators if present ranges can be causually related to climate and if former distributions are thoroughly documented for short spans of time. Although knowledge of Quaternary climates in at least glaciated portions of North America is in a state of flux, with evidence that changes in climate in one region were not in phase with changes in another, there is general agreement that an important change took place at about 10,000 years ago (synchronous disappearance of spruce forest over a wide area) and that the succeeding postglacial was marked by climatic amelioration (at least during what has been termed the hypsithermal). General climatic deterioration (cooler, wetter) during the past 1500 years is recognized by some.

IV. SOME LABORATORY TECHNIQUES

A few methods used to study fossil plants, especially those of Tertiary and Quaternary age, are summarized here. Paleobotanical techniques used primarily in morphological and anatomical studies of Carboniferous plants are not presented. Many of these are reviewed in Andrews (1961) and Darrah (1960). Various chapters in Kummel and Raup (1965) also contain helpful information.

A. Reference collections. A complete, thoroughly authenticated and vouchered collection of extant plants or parts of plants obviously should be used as a basis for identifying fossils suspected of belonging or being closely related to extant species. Such a collection can take many forms. Pollen analysts take pains to assemble extensive series of prepared pollen slides, usually several for nearly every species in a flora. Paleobotanists working with seeds or seed-like fruits have collections of these structures for comparative purposes. It is essential that a voucher from which pollen, seeds, etc. were taken be placed in an herbarium so that identifications can be documented at a later date if necessary.

Both the number of proposed new taxa and the accuracy of identification of Tertiary leaves have been criticized, and to remedy the situation, a new look at leaf architecture in extant plants has recently been undertaken (see Hickey, 1973). One outcome of this is the collection of cleared leaves, primarily of woody dicots, being assembled at the Menlo Park, California, office of the U. S. Geological Survey through the efforts of J. A. Wolfe. When complete it is anticipated that the collection will contain leaves from about 4500 genera and over 10,000 species. The collection is open to study by botanists not associated with the Survey and is vouchered by specimens in herbaria.

B. Megafossil techniques.

1. Transfers. These are used with compressions and are particularly useful if the embedded side of a leaf or other structure needs to be revealed. The specimen is glued to a microscope slide and the matrix surrounding the fossil is dissolved in acid (usually hydrofluoric) after the slide has been coated with wax. The specimen is then cleared (if necessary) and mounted for microscopic study.
2. Cuticle isolations. Several techniques are in use. That of Dilcher (1960) follows:

   -scrape cuticle off rock into container of 5% sodium hypochlorite solution
   -bleach (time variable), wash several times with water after bleaching
   -stain (e.g., 2% aqueous safranin)
   -alcohol dehydration, xylene
   -mount on slides.

Comparative cuticles from herbarium specimens may be prepared in the same way.

A procedure for clearing leaves from extant species is found in Chapter 7 (p. 186).

C. Microfossil techniques (pollen and spores). The object is to obtain a more or less pure residue of microfossils from samples under investigation and the technique employed varies depending on sediment lithology. If calcium carbonate is present, exposure to hydrochloric acid is one of the first steps; if the sample is siliceous, hydrofluoric acid is used. Heavy minerals are removed by gravity separation using one of several liquids whose specific gravity can be adjusted (e.g., $ZnCl_2$, bromoform). Extraneous organic matter is removed by sieving following treatment in a solution of weak alkali (peat). Cellulose is dissolved during acetolysis (see chapter 8). Coals and well indurated sedimentary rocks generally require special treatment. Palynological techniques are outlined in Brown (1960), Faegri and Iversen (1964), and Kummel and Raup (1965).

GENERAL GEOLOGICAL REFERENCES

The following sources of paleobotanical and geological data have been selected to guide botanists unfamiliar with these subjects to information of potential usefulness in research. The emphasis is North American. Those tracing the geological history of a genus or family will find the catalogs listed of great value.

A. Geology.

1. Glossary.

Gary, M., R. McAfee, Jr., and C. L. Wolf (eds.). 1972. Glossary of geology. xiv + 805 pp. + A-1 to A-52. American Geological Institute. Washington, D. C.

2. General Bibliographies.

Long, H. K. 1971. A Bibliography of Earth Science Bibliographies. 19 pp. American Geological Institute. Washington, D. C. [Listings by state.]

Ward, D. C. 1967. Geologic Reference Sources. University of Colorado Studies Earth Science 5. xii + 114 pp. [A bibliographical guide to the literature of geology, a general section, listings for subdisciplines of geology and for political divisions of the world.]

3. Bibliographies--North American.

Nickles, J. M. 1923. Geologic Literature on North America 1785-1918. Pt. 1 (Bibliography). United States Geological Survey Bulletin 746; Pt. 2 (Index). Ibid. 747 (1924). [Index organized by subject under political region (state, etc.).]

__________. 1931. Bibliography of North American Geology 1919-1928. Ibid. 823.

Thom, E. M. 1944. Bibliography of North American geology 1929-1939, Pt. 1 (Bibliography), Pt. 2 (Index). Ibid. 937.

King, R. R., et al. 1957. Bibliography of North American Geology 1940-1949, Pt. 1 (Bibliography), Pt. 2 (Index). Ibid. 1049 (in 2 vols.). [Further issues of the bibliography can be found in the following U. S. Geological Survey Bulletins: 985 (for 1950), 1025 (1951), 1035 (1952-53), 1054 (1954), 1065 (1955), 1075 (1956), 1095 (1957), 1115 (1958), 1145 (1959), [1195 (1950-59)], 1196 (1960),

1197 (1961), 1232 (1962), 1233 (1963), 1234 (1964), 1235 (1965), 1266 (1966), 1267 (1967), 1268 (1968), and 1269 (1969).] See also "Bibliography and Index of Geology" in section 4 below.

4. Bibliographies--primarily other than North American.

Geological Society of America. 1934-68. Bibliography and Index of Geology Exclusive of North America. Vol. 1 (for 1933)--Vol. 32. [Geological literature by author (with an index organized by general subjects and by topics under countries in all but last 2 volumes which follow the format continued in the "Bibliography and Index of Geology" (q.v.); short abstracts included in all but the earliest volumes.]

________. 1969-. Bibliography and Index of Geology. Vol. 33-. [Continues the "Bibliography and Index of Geology Exclusive of North America"; world-wide coverage, issued monthly; literature citations with descriptor abstracts arranged in 21 interest fields (incl. paleobotany), also subject and author indices.]

5. Indices to Geologic Names in North America.

Keroher, G. C., and others. 1966. Lexicon of Geologic Names of the United States for 1936-1960. 3 vols. United States Geological Survey Bulletin 1200. ["A compilation of the geologic names (i.e., geologic formations and other units of stratigraphic classification and nomenclature) of the United States, its possessions, the Trust Territory of the Pacific Islands, and the Panama Canal Zone."]

Wilmarth, M. G. 1938. Lexicon of Geologic Names of the United States (including Alaska). 2 vols. <u>Ibid</u>. 896.

B. <u>Paleobotany</u>.

1. Treatise

Boureau, É. (ed.). 1964-. Traité de paléobotanique. Masson et Cie. Paris. [Vols. II--Bryophyta, Psilophyta, Lycophyta (1964), III--Sphenophyta, Noeggerathiophyta (1964), IV (Fasc. 1)--Filicophyta (1970) have appeared; to be completed in nine volumes.]

2. Bibliography.

Boureau, É. (ed.). 1956-. World report on palaeobotany. I. Regnum Vegetabile 7; II. <u>Ibid</u>. 11 (1958); III. <u>Ibid</u>. 19 (1960); IV. <u>Ibid</u>. 24 (1962); V. <u>Ibid</u>. 35 (1964); VI. <u>Ibid</u>. 42 (1966); VII. <u>Ibid</u>. 57 (1968); VIII. <u>Ibid</u>. 78 (1971); IX. <u>Ibid</u>. 89 (1973). [Alphabetical listing of publications by author, the following headings are used in all except Vols. IV and V: General, Pre-Carboniferous, Carboniferous and Permian, Mesozoic, and Tertiary and Quaternary.]

3. Catalogs.

Andrews, H. N., Jr. 1970. Index of Generic Names of Fossil Plants, 1820-1965. United States Geological Survey Bulletin 1300. 354 pp. [Alphabetical catalog of generic names of fossil plants based on the U. S. Geological Survey's "Compendium Index" (q.v.) and having the same coverage; references to original sources included.]

Jongmans, W. J., and S. J. Dijkstra. 1913-. Fossilium catalogus II. Plantae. Pt. 1 (Lycopodiales)-. Publisher varies. G. Feller, Neubrandenburg; W. Junk, Berlin or 's Gravenhage. [Pt. 78 (Psilophyta) issued in 1970; series of greatest usefulness to students of lycopods, sphenophytes, ferns, seed-ferns, and cycads; treatments of the following angiosperm families have appeared: Juglandaceae (1915), Betulaceae (1916), Ulmaceae (1922), Sapindaceae (1928), Anacardiaceae (1935), Cornaceae (1938), Vitaceae 1939), Rosaceae (1942), Symplocaceae (1950), Celastraceae (1955); one part treating dicot wood issued (1931).]

Knowlton, F. H. 1898. A Catalogue of the Cretaceous and Tertiary Plants of North America. United States Geological Survey Bulletin 152. 247 pp.

________. 1919. Catalogue of Mesozoic and Cenozoic Plants of North America. *Ibid*. 696. 815 pp. [Supplemented by: LaMotte, R. S., Supplement to "Catalogue of Mesozoic and Cenozoic Plants of North America," 1919-1937. *Ibid*. 924. 330 pp. (1944).]

LaMotte, R. S. 1952. Catalogue of the Cenozoic Plants of North America through 1950. Geological Society America Memoir 51. 381 pp. [Alphabetical index of fossil plants (pollen, spores, and diatoms excluded) by genus and species from Paleocene through end of Pleistocene, ages and complete bibliographic citations given.]

U. S. Geological Survey. Compendium Index of Paleobotany. [An extensive (ca. 150,000 entries), up-to-date file to original sources by genus and species (diatoms, pollen and spores, and many other types of plant microfossils excluded), world-wide coverage, all ages; for a discussion of the usefulness of this index, see R. H. Eyde, Brittonia 24: 111-116 (1972); not published but available for consultation at the Paleobotany Library of the U. S. Geological Survey, U. S. National Museum, Washington, D. C. 20240; mail requests for information from the index are *not* honored.]

C. *Palynology*.

1. Bibliography.

Van Campo, M., and C. Millerand (eds.). Bibliographie supplément à Pollen et Spores. [Presently an annual supplement to the journal Pollen et Spores, papers listed and indexed, the first entry in Vol. 1: 92 (1959), entry 12,986 in Vol. 13 (1971) and continuing.]

2. Catalogs. Before using pollen and spore reports workers are urged to read J. M. Schopf's analysis (see below) of the state of systematics and nomenclature in pre-Quaternary palynology.

Potonié, R. 1956. Synopsis der Gattungen der Sporae dispersae. I. Teil: Sporites. Beihefte Geologisches Jahrbuch 23. 103 pp. *11 pls*.; II. Teil: Sporites (Nachträge), Saccites, Aletes, Praecolpates,

Polyplicates, Monocolpates. Ibid. 31. 114 pp. 11 pls. (1958); III. Teil: Nachträge Sporites, Fortsetzung Pollenites mit Generalregister zu Teil I-III. Ibid. 39. 189 pp. 9 pls. (1960); IV. Teil: Nachträge zu allen Gruppen (Turmae). Ibid. 72. 244 pp. 15 pls. (1966). [A valuable catalogue of spores found isolated in sedimentary rocks, extensive nomenclatural information and original literature citations are included; genera arranged in an artificial supra-generic classification that employs categories not sanctioned by the I.C.B.N.]

Potonié, R. 1962. Synopsis der Sporae in situ. Die Sporen der fossilen Fruktifikationen (Thallophyta bis Gymnospermophyta) im natürlichen System und im Vergleich mit den Sporae dispersae. Beihefte Geologisches Jahrbuch 52. 204 pp. 19 pls.

Potonié, R. 1967. Versuch der Einordnung der fossilen Sporae dispersae in das phylogenetische System der Pflanzenfamilien. Forschungsberichte Landes Nordrhein-Westfalen 1761. 310 pp. [Part I, Thallophyta to Gnetales, Part II, Angiospermae; a synopsis of records of spore and pollen types in extinct and extant plant groups.]

Schopf, J. M. 1969. Systematics and nomenclature in palynology, pp. 49-77. In: R. H. Tschudy and R. A. Scott (eds.). Aspects of Palynology. vii + 510 pp. Wiley-Interscience. New York. [See especially pp. 49-51.]

Traverse, A., and W. Spackman (Current eds.). 1957-. Catalog of Fossil Spores and Pollen. Pennsylvania State University. University Park, Pennsylvania. Vol. 1. 182 sheets + index, looseleaf. [Vol. 35 (1972) has appeared; republication of original descriptions and, if available, illustrations of new pollen and spore types following a standardized format; three indices have appeared (every 10th volume), authorship, stratigraphic and geologic terms, and taxonomic names are indexed.]

PALEOBOTANICAL LITERATURE CITED

American Commission on Stratigraphic Nomenclature. 1961. Code of stratigraphic nomenclature. Bulletin of the Association of Petroleum Geologists 45: 645-665.

Andrews, H. N., Jr. 1961. Studies in Paleobotany. xii + 487 pp. John Wiley & Sons. New York. [Chapter 18.]

Axelrod, D. A. 1958. The evolution of the Madro-Tertiary Geoflora. The Botanical Review 24: 433-509.

Brown, C. A. 1960. Palynological Techniques. vi + 188 pp. Publ. by author. Baton Rouge, Louisiana.

Cain, S. A. 1944. Foundations of Plant Geography. xiv + 556 pp. Harper & Brothers. New York.

Chaney, R. W. 1947. Tertiary centers and migration routes. Ecological Monographs 17: 139-148.

________. 1959. Composition and interpretation, pp. 1-134. In: Miocene Floras of the Columbia Plateau. Carnegie Institution of Washington Publication 617. viii + 237 pp. 44 pls.

Chesters, K. I. M. 1964. Fossil plant taxonomy, pp. 239-297. In: W. B. Turrill (ed.). Vistas in Botany. Vol. 4. viii + 314 pp. Macmillan. New York.

Darrah, W. C. 1960. Principles of Paleobotany. vii + 295 pp. Ronald Press. New York. [Chapter 3.]

Delevoryas, T. 1964. The role of palaeobotany in vascular plant classification, pp. 29-36. In: V. H. Heywood and J. McNeill (eds.). Phenetic and Phylogenetic Classification. xi + 164 pp. The Systematics Association. London.

________. 1969. Paleobotany, phylogeny, and a natural system of classification. Taxon 18: 204-212.

Dewey, J. F. 1972. Plate tectonics. Scientific American 226(5): 56-68.

Dietz, R. S., and J. C. Holden. 1970. The breakup of Pangaea. Scientific American 223(4): 30-41.

Dilcher, D. L. 1960. Cuticular analysis of Eocene leaves of *Ocotea obtusifolia*. American Journal of Botany 50: 1-8.

Dunbar, C. O., and J. Rodgers. 1957. Principles of Stratigraphy. xii + 356 pp. John Wiley & Sons. New York.

Eglinton, G., and M. Calvin. 1967. Chemical fossils. Scientific American 216(1): 32-43.

Faegri, K., and J. Iversen. 1964. Textbook of Pollen Analysis (2nd Rev. Ed.). 237 pp. Munksgaard. Copenhagen.

Ferguson, D. K. 1967. On the phytogeography of Coniferales in European Cenozoic. Palaeogeography, Palaeoclimatology, Palaeoecology 3: 73-110.

Florin, R. 1955. The systematics of the gymnosperms, pp. 323-403. In: A Century of Progress in the Natural Sciences, 1853-1953. x + 807 pp. California Academy of Sciences. San Francisco.

Hallam, A. 1972. Continental drift and the fossil record. Scientific American 227(5): 56-66.

Harrington, H. J. 1962. Paleogeographic development of South America. Bulletin of the American Association of Petroleum Geologists 46: 1773-1814.

Hickey, L. 1973. Classification of the architecture of dicotyledonous leaves. American Journal of Botany 60: 17-33.

Hopkins, D. M. (ed.). 1967. The Bering Land Bridge. xiii + 495 pp. Stanford University Press. Stanford, California.

Jardine, N., and D. McKenzie. 1972. Continental drift and the dispersal and evolution of organisms. Nature 235: 20-24. [Paleogeography of Southern Hemisphere continents including reconstructions of their positions during the Cenozoic.]

Khudoley, K. M., and A. A. Meyerhoff. 1972. Paleogeography and geological history of Greater Antilles. Geological Society of America Memoir 129. xv + 199 pp.

Kummel, B. 1970. History of the Earth, An Introduction to Historical Geology (Ed. 2). xix + 707 pp. W. H. Freeman. San Francisco.

Kummel, B., and D. Raup (eds.). 1965. Handbook of Paleontological Techniques. xiii + 852 pp. W. H. Freeman. San Francisco.

Leopold, E. B., and H. D. MacGinitie. 1972. Development and affinities of Tertiary floras in the Rocky Mountains, pp. 147-200. In: A. Graham (ed.). Floristics and Paleofloristics of Asia and Eastern North America. xii + 278 pp. Elsevier Publishing Company. Amsterdam.

Pettijohn, F. J. 1957. Sedimentary Rocks (ed. 2). xvi + 718 pp. Harper & Brothers. New York.

Raven, P. H., and D. I. Axelrod. 1972. Plate tectonics and Australasian paleobiogeography. Science 176: 1379-1386.

Schuchert, C. 1955. Atlas of paleogeographic maps of North America. xi + 177 pp. John Wiley & Sons. New York.

Schuster, R. M. 1972. Continental movements, "Wallace's Line" and Indo-malayan-Australasian dispersal of land plants: Some eclectic concepts. The Botanical Review 38: 3-86.

Smith, A. C. 1969. Systematics and appreciation of reality. Taxon 18: 5-13.

Stafleu, F. A. et al. (eds.). 1972. International code of botanical nomenclature. 426 pp. International Association for Plant Taxonomy. Utrecht.

Termier, H., and G. Termier. 1960. Atlas de paléogéographie. 99 pp. Masson & Cie. Paris. [World-wide coverage but pre-drift continental configurations not given.]

Wolfe, J. A. 1969. Neogene floristic and vegetational history of the Pacific Northwest. Madroño 20: 83-110.

Wolfe, J. A. 1971. Tertiary climatic fluctuations and methods of analysis of Tertiary floras. Palaeogeography, Palaeoclimatology, Palaeoecology 9: 27-57.

________. 1972. An interpretation of Alaskan Tertiary floras, pp. 201-233. In: A. Graham (ed.). Floristics and Paleofloristics of Asia and Eastern North America. xii + 278 pp. Elsevier Publishing Company. Amsterdam.

Wood, C. E., Jr. 1972 [1973]. Morphology and phytogeography: The classical approach to the study of disjunctions. Annals of the Missouri Botanical Garden 59: 107-124. [Other papers in this issue of the Annals also treat disjunctions.]

Wright, H. E., Jr. 1971. Late Quaternary vegetational history of North America, pp. 425-464. In: K. K. Turekian (eds.). The Late Cenozoic Glacial Ages. xii + 606 pp. Yale University Press. New Haven, Connecticut.

Wright, H. E., Jr., and D. G. Frey (eds.). 1965. The Quaternary of the United States. x + 922 pp. Princeton University Press. Princeton, New Jersey. [The Proceedings of the VII Congress of the International Association for Quaternary Research, for which the Wright and Frey volume was prepared, have been published in sixteen volumes. Nearly all of these contain information pertinent to botanical studies, but of special value is Vol. 7, Quaternary Paleoecology, E. J. Cushing and H. E. Wright, Jr. (eds.), Yale University Press, 1967.]

Van Steenis, C. G. G. J. 1962. The land-bridge theory in botany with particular reference to tropical plants. Blumea 11: 235-372.

Figure 19-3

CONTINENTAL DRIFT

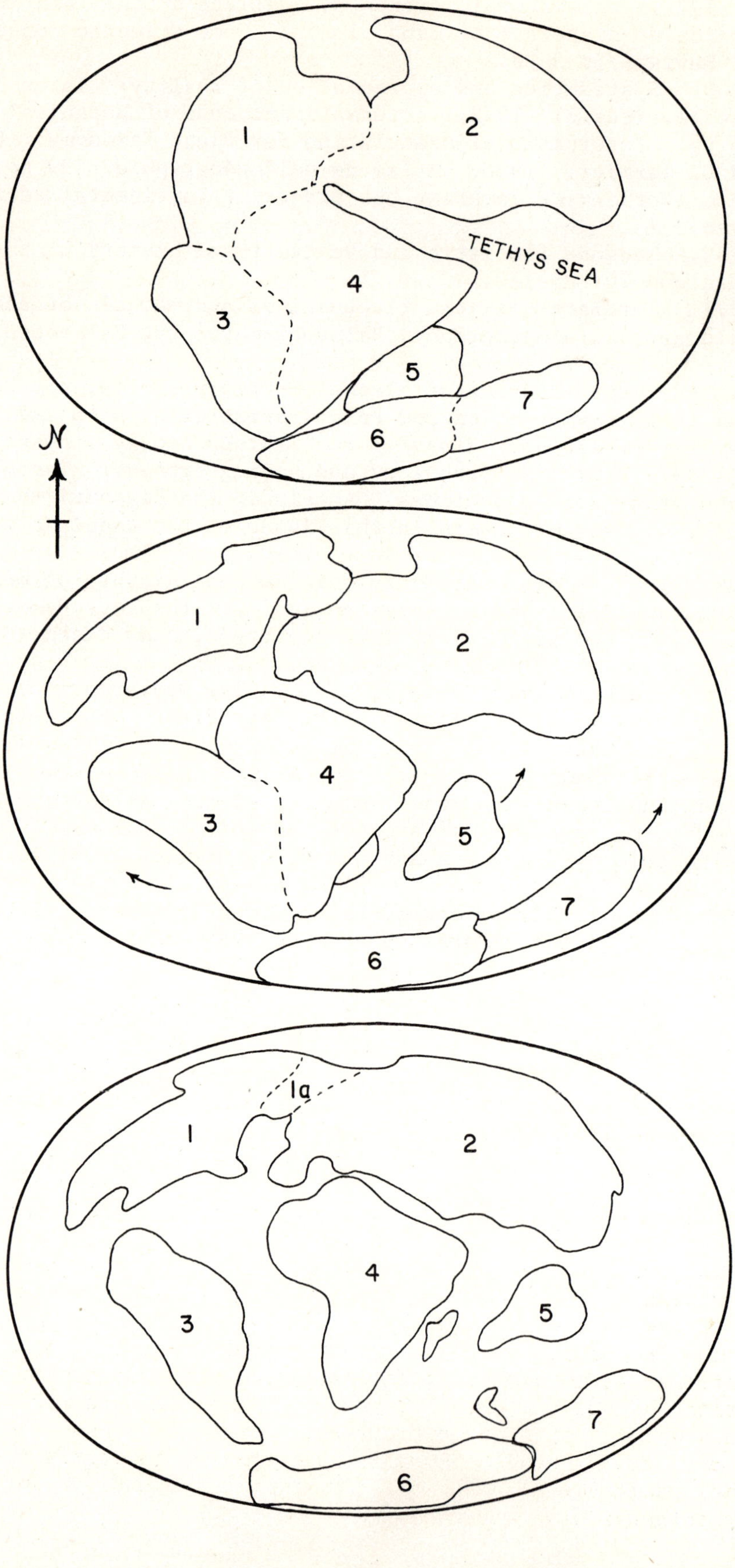

The 200,000,000 year migration and formation of the continental land masses from the presumed breaking of one or two large land masses, Laurasia (1 & 2) and Gondwana (3-7), into:

1. North America
2. Eurasia
3. South America
4. Africa
5. India
6. Antarctica
7. Australia

Chapter 20. DATA ACCUMULATION AND ANALYSIS

Knowledge is acquaintance with facts; information is knowledge derived from observation, reading or instruction and consists of either facts or data, or both. Data (datum, singular) can be either actual facts ("hard data") or assumptions (based on knowledge) that are used, often together, to make inferences which may result in an hypothesis or, with more evidence, a theory. In systematic botany, original observation leads to the accumulation of new facts and thus increases our supply of data and our knowledge about the plants actually observed. These observation, or facts, are often extrapolated in various ways (e.g., statistically, philosophically) to make assumptions about evolutionary relationships, and probable evolutionary pathways, that have a direct bearing upon the taxonomic rank (i.e., classification) of a particular plant or group of plants. Data accumulation thus involves bringing together and interrelating all of the new and old facts and ideas about a given topic; data analysis (comparative, statistical, philosophical, or other) then indicates what logical conclusions, and extrapolations, might be made from the total information available that will hopefully provide a better biological, and thus systematic, understanding of the plants under study. Both data accumulation and data analysis may involve many different approaches and can be carried on at different levels of sophistication and degree of exactness. However, the mere fact that a large amount of data has been accumulated and run through a highly complex computer analysis program does not necessarily mean that the taxonomic value of the data has been increased significantly or that previously unsuspected taxonomic changes are in order. Indeed, the keen observations and "taxonomic intuition" of many of the earlier taxonomists (see Chapter 2) resulted in conclusions that are more often than not upheld by more intense study and the accumulation of additional data made possible by modern technology.

For much of the world's flora we have very scant data of any sort! Less than 5% of the higher plants have been studied biosystematically, and thousands of species are known only from relatively meager herbarium materials which, at best, can usually provide only incomplete morphological data, perhaps a bit of ecological data (if the label is complete!) and, with proper preparation and observation might yield a small amount of anatomical or biochemical data. Additional data, or "evidence," of all kinds (see Chapters 5-16) are thus badly needed to help fill in the many interesting and biologically important evolutionary details necessary to a more complete and more accurate understanding of plant relationships at all taxonomic levels.*

*Note: As data become available from one set of studies after another, and as certain relationships are indicated by one group of data while other possible relationships may be indicated by another set of data, the value of the taxonomically neutral "deme" terminology (see Chapter 11, III; also Evolution Glossary) becomes apparent. If chromosome studies show features of similarity, or of relationship, between the populations or cytological "races" investigated these plants can be referred to, and considered evolutionarily, as a "cytodeme." The plants of similar ecological requirements form, irrespective of some of their other attributes, an "ecodeme"; a group of interbreeding individuals is a "gamodeme," and so on. The deme concept and terminology thus frees the taxonomist from the more rigid "species concept" during (and sometimes after!) his research and thus makes for a more realistic and precise concept and treatment of evolutionary pathways and relationships.

No matter how carefully a research project is designed to collect specific data, there will be times when unexpected and interesting questions arise or when previously unexpected correlations seem likely and more, or additional types of, data are needed. The often routine and mechanical accumulation of raw data may represent up to 95 percent of the research time on a particular project and its importance cannot be overrated because without "good" (or "valid") data there can be no valid conclusions, no interesting synthesis of information and no sound basis for any further research on the given subject. The old saying that "good research is 95 percent perspiration and 5 percent inspiration" still holds true. Therefore it is of utmost importance that every effort be made to insure that the data-gathering phase of any research project be done as carefully and meticuously as possible to insure the validity of the ultimate conclusions.

Section A. DATA ACCUMULATION

I. ORGANIZATION

The effective and efficient collection and use of data depend in large part upon good organization of the research project, facility with the many field and laboratory techniques and procedures necessary for the collection, preparation, and analysis of the plant material, and a knowledge of methods of data presentation that will enable your results to be easily reviewed by others. To this end a preliminary survey of the literature and herbarium material, coupled (if possible) with field study, should precede any major taxonomic research effort. Such a preliminary survey (which may well take a year!) should lead to:

1. A rather precise definition of the problem and some idea as to its practicality in terms of the time and resources available.
2. A list of specific information needed and the reasons for its need. Also, the production of a suitable data sheet for ease of record keeping on an individual plant basis should have evolved from the preparation of this list (see sample data sheets at end of chapter).
3. A concept of particular methods and approaches, either direct or indirect, field or experimental, necessary to get the information needed.
4. Definite figures as to the number of populations to be studied, the number of samples needed, and the realistically optimum sample size needed to get valid data.

Many new items that may appear on the list mentioned in #2 above (and the appropriate methods concerning them) have been covered in previous chapters pertinent to the various topics. Some of the additional descriptive as well as experimental approaches to data accumulation through current biosystematic research are given in the following sections of this chapter along with brief notes on the methods involved.

In the collection of biological data of any sort it is very important that the material used for any comparisons or correlations be comparable and that adequate documentation, in the form of herbarium specimens, as well as photographs, drawings, data sheets and other types of evidence, be provided. In the matter of comparable material the state of the material (fresh, dried, or preserved) is an important factor: measurements of fresh flowers or leaves, for example, may not be comparable with measurements from dried material because of the shrinkage; and color, scent and even texture are easily changed by different treatments. Comparable material also means comparable in a developmental sense--mature structures should be used whenever possible. A mass sample of leaves, for example, might consist of one leaf from a particular node on 100 plants in the population--not just 100 leaves. Here again, it is important to know how much within-plant variation exists before establishing a proposed sample size and sampling method.

II. DATA ACCUMULATION

1. Where does one get data? For the systematist there are five possible sources of information: the field, the herbarium, the library, the garden, and the laboratory. It is important to realize the value--and shortcomings--of information from each source. The library (see Chapter 30) may be the source of an idea for a research problem as one peruses a long list of synonyms in a monograph or notes in an atlas a peculiar distribution pattern for a particular plant or notes in a journal article some unexplained and striking similarity in form, chromosome number, chemical content, ecology or some other aspect of a particular species. Further checking through various biological abstracts and indices may turn up more bits and pieces of related information that can either answer the questions raised or can add to the interest of an embryonic research project. If questions still remain, or if certain kinds of new questions arise, the next place to check is the herbarium. Here one can quickly get information on many aspects of general morphology of a given kind of plant and, if the label data are complete, some information on ecology, phenology, and distribution. Obviously the more specimens, their degree of completeness, and the wider the geographic area they represent, the more information one can get from a preliminary look at the herbarium material.

Most of the early taxonomists were excellent naturalists and keen observers. Their detailed descriptions often leave little room for improvement and their analytical minds often put any anomalous situation into reasonable biological perspective. Today, with more basic knowledge, better communication, and better technology, it is possible to add a new dimension, but rarely a new concept or explanation, to the many interesting systematic problems that confront us. Change does not necessarily mean improvement. To paraphrase Robert Woodson, whose work on the Asclepiadaceae spanned a professional lifetime, no systematist should either add to, or subtract from, the work of another without first putting in an equal amount of study on the organism or organisms involved! This, then, usually means one must leave the library and the herbarium and go into the field to observe first hand and to gather specific materials for further laboratory, garden or herbarium study. Or, if the problem first occurred in the course of field, herbarium or garden work, the library, and then the laboratory, would be the next phases of any preliminary investigations.

Unless properly organized well ahead of time field work (see Chapter 18) can be enormously wasteful of time, energy and research funds. All too often several hours or even days of travel to reach a particular collecting site are essentially wasted because all of the necessary material and data were not taken and a second trip must be made--perhaps a year later, when the plants are again in bloom or fruit. Be prepared! Have a check-list of what material you need, or think you may need and, if the population size will permit, get plenty of material. Within reasonable limits, it is much better to have material that you don't use on a research project than to need material and not have it available!

Before the field trip begins go over the check-list and be sure that you have all of the collecting equipment and materials necessary to do the job. A few tips:

a. Always label every plant, vial, or collection of any sort in the field as soon as collected and also enter the data in a field notebook. Decide ahead of time the codes and numbers you want to use and if individual plants or specimens are to be tagged have the numbered tags prepared ahead of time and strung on a wire in such a way (in reverse order) that they can be taken off in order from 1--onward. This will prevent errors in numbering plants as well as save valuable field time. Labels can also be punched or coded with colored plastic tape if necessary to keep material from different species separate. Do not trust any "permanent" ink; use wax, or soft lead, pencils for making field tags.

b. Living material carefully potted in the field and watered with liquid fertilizer (20-20-20, premixed and in gallon plastic jugs) survives much better than material carelessly dug and planted later, perhaps after drying out a bit in a hot car; many plants, dug with a small ball of earth around their roots, can be wrapped in newspaper, labeled, and put in a waxed cardboard 1/2 gallon milk container where they will "ride" for several days (if given a cup of liquid fertilizer) and can be potted up later with ease.

c. Have collecting vials and bottles already filled to the proper limit (1/2-3/4 full) with the appropriate preservative or fixative. Use only good glass vials and jars with screw caps. Heat will pop out corks and plastic stoppers and the material will be lost. Some plastic vials melt when certain fixatives or other chemicals are put in them. Use a soft lead (#1 or 2) pencil to write collection data on small (1/2" x 2") strips of bond paper to slip into each jar or vial of material. Wax pencil marks on glass, or the plastic tops, is not permanent! A vial of precut label slips will save time in the field.

d. Material to be used for any sort of comparison should be comparable in developmental stage and/or position on the plant.

e. Although material for herbarium specimens can often be transported for a day or two and then held another day or so in a cold room when properly wrapped and wet down in the field, cytological material, some chromatography samples, pollen samples and other items for specialized study should be fixed, preserved, or otherwise prepared in the field at the time of collecting if at all possible in order to keep from introducing another variable--deterioration--into the study.

2. What are the types of data? Data can be either quantitative (objective or measurable by a uniform standard) or qualitative (subjective or the result, in part, of a judgment). Quantitative data is more precise, easier to work with statistically, easier to duplicate or compare, and often easier to understand or explain than qualitative data. However, for some characters it is sometimes much easier, and just as effective, to "score" a variable even if it is somewhat subjective rather than get an absolute or objective measurement

or count. Such a situation occurs when there is a great range of variability (as in some leaf sizes or shapes or in certain degrees of trichome density) and actual measurement would be either very difficult or meaningless, or both. Here"classes" of the character state are established, perhaps by making a few actual measurements or counts, and subsequent samples are merely compared visually with the representative material, photograph or drawing typical of each class and then assigned a score or class value. However, the more objective the data the more objective the research!

3. Sample size. The type of investigation will determine to some degree the amount and kind of material needed. For a preliminary survey of variation in a character, or a group of characters, a single plant (which can often be a herbarium specimen) from each of a dozen or so localities may be sufficient. If the research is to be a more detailed study of variation in a particular character over the geographic range of a species, sufficient samples should be taken to give relatively uniform representation of populations from all parts of the range of the species. Depending on the frequency of occurrence of the plants within their range, the total size of the range, and the variability found in the preliminary studies, anywhere from five to one hundred or more populations might be sampled, and each sample might consist of material from one to five separate plants or might consist of a single mass sample from each population representing five, ten, or even 100 plants if the population size and variation pattern so indicate. In general, the larger the sample the more representative the sample and less extrapolation is necessary in final interpretations of the results. On the other hand a realistic time budget must also be established for a given research project. Here the objectives of the research, the degree of reliability desired, and the technology available for use in the data gathering as well as data analysis stages must be considered. For example, if a study involving chromosome numbers of plants from 100 populations is being undertaken, each collection will require, perhaps, up to four hours of laboratory time to get a chromosome count. If five such counts are desired from each population the laboratory time required for the study increases accordingly--to 2000 hours, or essentially a year's work at 40 hours per week. And there is no machine to do data gathering. By contrast, a pollen size study involving measurement of 100 pollen grains from each of five plants from each of 100 populations might involve only about 15 minutes time per slide for a total of 125 hours of laboratory time to get and tabulate measurements of 50,000 pollen grains. By use of mass samples, with the pollen from the five plants of each population mixed on one slide, the laboratory time can be reduced to about 25 hours. If a "Coulter Counter" is available (and if it is properly calibrated and its operation understood), a single mass sample of 50,000 pollen grains can be measured in less than two minutes; or the 100 mass samples could be run in an hour or so. Here technology makes the data gathering relatively fast and easy.

Often it is possible to determine the realistic sample size needed from preliminary observations or from the early data on population analysis. When observations or tests show the pattern of variation in a character to have more or less stabilized additional measurements or observations are not essential even though they may be highly desirable. The same can be said for the number of characters being investigated: if preliminary studies and measurements indicate that data from five characters or ten characters will support the same conclusions as data from 30 or 40 characters it is not necessary to run a full analysis on the additional characters. However, character correlations, or lack of such correlations, cannot under any circumstances, be assumed past the hypothesis stage of the research but must be shown to exist, through adequate preliminary observation and documentation, before they have any value in a systematic study.

Sample size will also be governed by whether or not the data being gathered are to be used for predictive purposes (where the final answer is not known for certain) or descriptive purposes (in which previous observation has set the general descriptive limits). In the first case large random samples are called for whereas in the second case smaller "representative" samples are quite adequate and much more realistic to work with in every respect. In the selection of the representative sample consideration should be given to proportionate representation of the different naturally occurring states of the character under investigation; however, rare extreme forms might be noted separately rather than included in the general data if such inclusion would distort the variation pattern in an abnormal way. Because of such occasional distortions it is well to keep in mind that the statistical significance of a given set of data is not always directly related to the biological significance of the data.

4. Changes in variation ranges and patterns. Ranges of variation and variation patterns usually change as the source of the sample material changes from the individual plant to the population and from a single population to a series of populations or the species as a whole. However, it is often valuable to know something of the amount and kind of variation at each of the three levels in establishing sample size and in evaluating the data on variation within and between species and populations. (See the exercise on pollen size variation, Section B, III A-3, further on. The same type of analysis of variation--individual, population, species--can, and should, be done for any character being studied.) If, for example, a single plant, or a single population shows most of the range of variability in a particular character, this high degree of uniformity, perhaps the result of founder effect or a rather strong selective force, would make continued or extensive work with this character seem uncalled for in any study of variation.

III. DATA FROM GENERAL VARIATION STUDIES

All organisms vary to some extent and all variation occurs within individual genetic limits. If these genetic limits are very broad for a particular character, the primary phenotypic control of this character is most likely environmental; if the genetic limits are very narrow, primary phenotypic control will likely be genetic. Since both evolutionary and taxonomic studies are based on individual and population variation, it is often important to know the source, and range, of the variability of a character in a population or a group of populations. A number of methods exist to get this information; all involve careful planning, careful observation and adequate record keeping to amass sufficient data on comparable material to warrant later valid conclusions. In addition, appropriate experimental design is necessary to produce results which can help explain, or further substantiate, the conclusions. Among such experimental approaches to morphological and developmental variation are the following:

A. Reciprocal transplants. Move plant A from habitat A to the same hole in habitat B from which plant B was taken; transplant plant B to habitat A in the hole from which plant A was taken. If the transplants retain their characteristics over several seasons their differences are genetic; however, if transplant A now comes to resemble transplant B, and vice versa, their phenotypic variation is primarily the result of their environments.

B. Uniform environment. This is often a more practical method to study variation since plants from a number of populations are all brought together in an experimental garden at one location in an essentially "uniform" environment which may or may not be optimum for one or more of the plants. Phenotypic changes, over several seasons, indicate the range of each genotype (plant) and thus whether or not the variation is primarily environmental or genetic.

C. Growth chamber studies. Where the equipment is available these provide, often in a relatively short time, rather precise data on the total range of phenotypic expression of a series of plant samples under different environments. Growth chamber studies, when appropriately designed, can show clearly specific cause and effect correlations between different combinations of environmental factors (such as temperature and light intensity) and morphological and physiological variation, the latter easily evidenced by changed developmental responses. With variously programmed growth chambers it is also possible to show the potential range of variability of a plant under environmental conditions that do not currently exist in its natural habitat (e.g., higher or lower temperatures, longer or shorter photoperiods) and which may be indicative of prior adaptations to prior environments.

D. Analysis of mass samples. It is often possible to determine the actual and potential range of variation of a character within a species, or a series of populations, found in more or less similar habitats, by an analysis of mass samples. For statistical relevance, such samples should involve measurements or observations on 15 to 100 or more plants (depending on population size) and from each of three or more populations. If the plants or samples are collected completely at random a larger sample will be required to give a picture of total variation than would be the case if "representative" samples (of median and both extreme forms) were selected for measurement. Or, to put it another way: a much larger sample is needed for "predictive statistics" than "descriptive statistics." It might also be well to remember that because of certain genetic and evolutionary facts (e.g., linkage; polymorphism) and the ever present possibility of sample error, a character may show "statistical significance" out of proportion to its "biological significance!"

A combination of both reciprocal transplants and uniform environment studies (with the experimental plots at different altitudes) were used in the classic studies of *Achillea* and *Potentilla* by Clausen, Keck and Hiesey (1948). In general, cloned material, which insures genetic uniformity, is the best for any studies of environmental effect on phenotypic expression. For reciprocal transplants perennials are obviously better suited than annuals; either perennials or annuals can be used effectively in uniform garden or growth chamber experiments, or in mass sample analysis.

E. Data from breeding studies. Since variation patterns, and evolutionary rates, are often related to the breeding system of a species of plants (see Chapter 12), it is important to know whether the plants of a population normally reproduce by outcrossing, self-pollination, or apomixis.

This information can often be easily obtained by a series of observations and experiments as outlined in Chapter 12, Section D.

F. Pollen data. Pollen size studies and pollen viability studies (see Chapter 8, Section C) can often make a valuable contribution to any systematic study involving actual or suspected hybridization or polyploidy. Pollen data are generally easy to gather with a minimum of supplies, equipment, and training and provide good material for simple statistical analysis. The gathering and analysis of sets of pollen data also provides an interesting "term project" type of investigation which, if continued over a number of years, would be cumulative and could provide a valuable block of biological information about certain plants, or populations of plants, which might then be correlated with environmental fluctuations or changes.

G. Chromosome data. Chromosome number, chromosome morphology and chromosome behavior at meiosis are all important bits of biosystematic data often necessary to an understanding of intra- and inter-specific evolutionary pathways. See Chapter 10 on Cytological Evidence and Chapter 11 on Genetic Evidence for concepts and methods associated with the various aspects of chromosome data.

Section B. TREATMENT OF VARIATION DATA

I. THE USE OF SIMPLE STATISTICAL METHODS

Since variation is universal it is necessary for anyone working with biological material to be able to deal effectively with the patterns of variation found within and between individuals, populations, and species of organisms. This leads, in floristics for example, to the necessary inclusion of size ranges in species descriptions (e.g., leaves 7-18 cm long) and, in biosystematics, to the further inclusion of the mean, or average, and some simple notation (the standard deviation) as to the extent of sample variation around the mean (e.g., range 7-18 cm, $\bar{x} = 13.2 \pm 2.3$). It should be noted that the larger the standard deviation (σ) relative to the mean the greater the variability in a sample. Thus, if the mean leaf length for plants of one population is 13.2 cm and the standard deviation is ± 2.3 cm, the plants of the population are more uniform, as to leaf length, than the plants of a second population in which the leaf length range is again 7-18 cm and the mean 13.2 cm but the standard deviation is 3.4. Likewise the figures $\bar{x} = 86.2 \pm 9.3$ indicates less variability (in whatever is being measured) than do the figures $\bar{x} = 12.8 \pm 3.4$.

In addition to the mean there are two other "central tendencies" that may be used on occasion. These are the mode and the median. Although each of these three terms designate a central tendency it is important to realize that they do not fall at the same point of a distribution graph unless the distribution is perfectly symmetrical--a rare situation when dealing with biological material where distribution curves are usually skewed, or asymmetical, as shown in the example below.

mode = 8 cm
mean = 6.9 cm
median = 6 cm

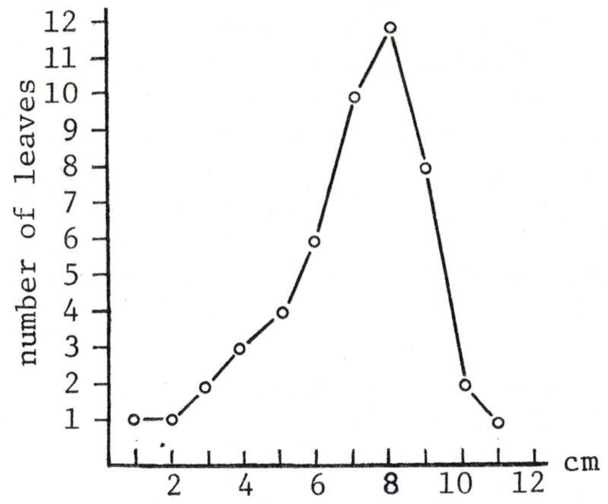

Graph of leaf length frequencies (above) and the same data used to determine the mean leaf length (right)

Class Value v (Leaf length in cm)	Frequency f (Tabulation of number of leaves of each size)	fv (Frequency times value)
1	1	1
2	1	2
3	2	6
4	3	12
5	4	20
6	6	36
7	10	70
8	12	96
9	8	72
10	2	20
11	1	11
	50	346
	Total number of leaves = n	sum of fv

$$\bar{x} = \frac{\Sigma fv}{n} \quad \text{or} \quad \bar{x} = \frac{346}{50} \quad \text{or} \quad \bar{x} = 6.9$$

If the data on leaf length are carried through a few additional computations, as shown on the following computation sheet, we will have additional valuable information about the pattern of leaf length variation in the plant (or population) under study: the standard error of the mean, the standard deviation, and the coefficient of variability. These first two figures are normally given along with the range and the mean and the four are often graphed for ease of comparison with other, similar, sets of figures representing another plant or another population. When so graphed the various items of data are indicated as follows:

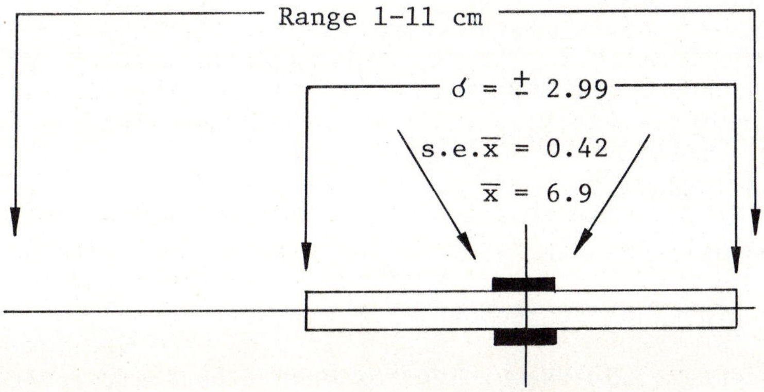

Figure 20-1. Graphic representation of standard statistical figures.

In some instances, as in the examples given below for plants of several populations of three species and a hybrid of Physalis (see Hinton, Fred, 1970, Brittonia 22: 14-19), the standard error of the mean is not graphed. It would, however, be given in the tables of figures that support the graphic presentation.

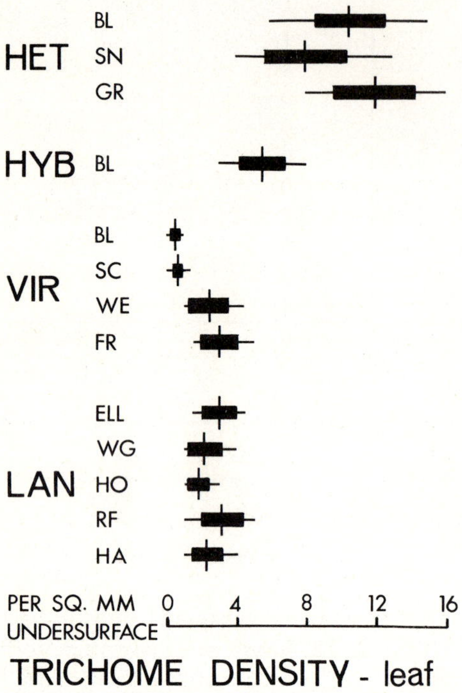

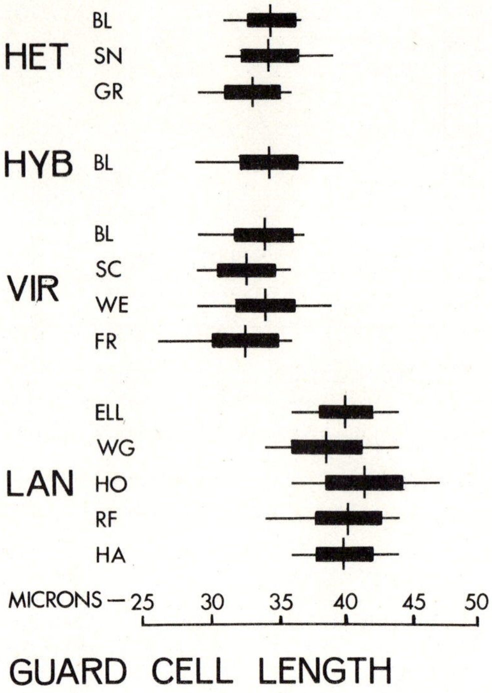

II. SYMBOLS AND DEFINITIONS

$\bar{x}$ = mean

Σ = sum of

v = class value

f = frequency

n = number (of individuals)

σ = standard deviation

d = deviation of class from mean

V = coefficient of variability

S.E. = standard error

P.E. = probable error

Abscissa: the horizontal line of graph.

Average deviation: $= \frac{\Sigma fd}{n}$ (not as good as σ).

Class value or class center: average value for each class; if, for example, a variable series of leaf-length figures are to be treated it might be desirable, for convenience sake, to lump the leaves into size classes (e.g., 20-24 cm, 25-29 cm, 30-34 cm, etc.) and then compute the mean by use of the median class values (v) which would be 22, 27, 32 etc. for the above class ranges.

Coefficient of variability (V): used to compare amount of variability in material measured in different units (e.g., length vs. weight). It is a ratio of the standard deviation (σ) to the mean ($\bar{x}$) of the same population reduced to a percentage. $V = \frac{100\ \sigma}{\bar{x}}$

Deviation of Class from Mean: (d) difference between class value and mean.

Mean: the ordinary average, the sum of the scores in a series divided by the number of scores; $\bar{x}$ (bar x); each class value (v) is multiplied by class frequency (f), the sum (Σ) of these products is divided by the number of individuals (n). Thus: $\bar{x} = \frac{\Sigma fv}{n}$

Median: the middle score in a series of scores after they have been arranged in order from lowest to highest. There are as many scores above it as below it. It is not the middle of the range (except in rare cases where the two coincide).

Mode: is the scale value at which the frequency curve is highest.

Ordinate: the vertical line of graph.

Probable error (P.E.): same as standard error except σ is multiplied by the constant .6745 and the mean of another random sample would fall within the limit set by the P.E. in one out of two cases (50-50 chance).

Range: a single indicator of variability; the extreme scores or values in a set of observations or, by definition, the spread between the largest and smallest scores.

Standard Deviation: (σ or sigma) indicates the amount of variation in a sample; e.g., two samples of ten leaves each, might have identical means, but one sample would range in length from 40 to 50 cm while the second, more variable, ranged from 20 to 70 cm and would have a larger standard deviation. The smaller the relative value of σ the more closely the sample is grouped around the mean. With normal frequency distribution approximately 2/3 of a given sample will fall within the range covered by one σ on each side of the mean; occasionally it is desirable to include a greater percentage of a sample in a given set of figures and then two standard deviations are added on each side of the mean. In this latter case, if the curve is skewed, the doubled standard deviation may exceed the actual range found in the sample. Reference back to Figure 20-1 will clearly show that an increase in the standard deviation figure from 2.99 to 5.98 would extend the standard deviation bar beyond the actual upper range limit of 11 cm. $\sigma = \frac{\Sigma fd}{n}$

Standard error (s.e.): the range within which the mean of another random sample from the same population would fall in two cases out of three.

$s.e.\bar{x} = \frac{\sigma}{n}$ (The greater the s.e. or p.e. the less reliable the data.)

Variance or sum of squares: as was the standard deviation, the variance is an indication of dispersal around the mean. It is, as one name implies, the sum of the squared deviations, or differences, from the mean divided by n-1, or $S = \frac{\Sigma fd}{n-1}$, or for example, $S = \frac{245}{49} = 5.0$. Note that the standard deviation is expressed in the same units as the mean and range but that the variance is not; the variance, however, is the more useful figure in many other mathematical treatments of the data.

III. THE DETERMINATION OF VARIATION IN A SINGLE PLANT AND A SINGLE POPULATION

Any character of a plant is the expression of the physiological processes of the plant acting within the limits set by its genetic constitution. Thus, while the various parts of a single plant have identical genetic constitutions, the physiological processes from one part of the plant to the next are not usually identical, and differences in size and development result. The following method is designed to show some of the variability within one plant and between plants of one population and to show how this variability may be shown by a simple statistical treatment involving the mean, standard deviation, standard error of the mean and the coefficient of variability.

1. The material for the following work should all be taken from the same plant to illustrate the range of variability within an individual.

   (a) Using a minimum of 50 leaves construct a graph showing the number of leaves (ordinate) and the number of sinuses, or serrations, or lateral veins, etc. per leaf (abscissa).

   (b) Using the same leaves construct two bar graphs showing the variation in: (a) the length of the lamina
   (b) the width of the lamina
   using at least ten class intervals.

   (c) Now make a slide (see Chapter 8, Section C) of the pollen from a single flower (or inflorescence if more practical) of the plant you are working with and use an ocular micrometer to measure as accurately as possible the largest dimension of 100 pollen grains. Record the frequency distribution of the measurements. This can be done in "eye-piece units" for convenience, but be sure to calibrate the "eye-piece units" of the ocular micrometer with a stage micrometer marked off in .01 mm units (so that your measurements will be in standard metric units and can be compared with data gathered by others) and convert your final statistical calculations to these standard units.

   (d) Use the data you have to determine the coefficient of variability for each of the three leaf characters and the pollen size. Which of the four characters would be most satisfactory for use in a key? Why?

2. Now repeat the exercise on a population basis by using equal amounts of material from a minimum of 5 plants from a single population; if material from more than 5 plants is available it will, of course, give a more accurate "statistical description" of the population variation.

COMPUTATION SHEET

Material: Terminal Leaf Length
Mass sample, Population 1
4 July 1970

Magnification: IX
(centimeters)

Class Range	Class Value V	Frequency (f) tabulation	totals	fV	Deviation of class from mean d	d^2	fd^2
	1	I	1	1	-6	36	36
	2	I	1	2	-5	25	50
	3	II	2	6	-4	16	32
	4	III	3	12	-3	9	27
	5	IIII	4	20	-2	4	16
	6	𝍸 I	6	36	-1	1	6
	7	𝍸 𝍸	10	70	0*	0	0
	8	𝍸 𝍸 II	12	96	+1	1	12
	9	𝍸 III	8	72	+2	4	32
	10	II	2	20	+3	9	18
	11	I	1	11	+4	16	16
			$n = \Sigma f = 50$	$\Sigma fV = 346$		$fd^2 = 245$	

Mean:

$$\bar{x} = \frac{\Sigma fV}{n} = \frac{346}{50} = 6.9$$

Standard deviation:

$$\sigma = \sqrt{\frac{\Sigma fd^2}{n}} = \sqrt{\frac{245}{50}} = \sqrt{8.8} = 2.99$$

Standard error of mean:

$$\text{s.e. } \bar{x} = \frac{\sigma}{\sqrt{n}} = \frac{2.99}{7} = 0.42$$

Coefficient of variability:

$$V = \frac{100\ \sigma}{\bar{x}} = \frac{299}{6.9} = 43.3$$

*NOTE: For ease of calculation of d^2 and fd (above) an <u>assumed mean</u> of 7 was used, rather than the actual mean of 6.9, in order to work with whole numbers. The s.e.$\bar{x}$ so obtained is adequate for preliminary use. When dealing with decimals, calculations are generally carried to one more place than occurs in the individual base measurements.

Impatiens____________________ Population No.____________________

Date____________________ Locality____________________

Plant number															
Spur length mm															
Sac length mm															
Angle of spur/sac															
Corolla spotting															
Opening width mm															
No. Seeds/capsule															
No. Fruits/plant															
Leaf length															
Leaf width															
Serrations/cm															
Internode diam. mm															
Node diam. mm															
Longest internode cm															
$\bar{x}$ pollen diam. μ															

Data Sheet, Solanum Project

Population #	________	Species	________
Plant No.	________	Date Collected	________
Herbarium #	________	County and State	________

Notes: ____

____

____

Habit	________	Flower color	________
Height	________	Flowers/Infl.	________
Leaf color	________	Corolla diam.	________
Infl. type	________	Lobe length	________
Peduncle length	________	Sepal orient.	________
Leaf shape	________	Fruit color	________
W/L ratio	________	Fruit diam.	________
Basal lf. angle	________	Seeds/fruit	________
Margins	________	Seed size	________
Pubescence	________	Seed shape	________
Trichome density	________	Concretions	________
Sepal length	________	Chromosome #	________
Style type	________	Pollen diam.	________
Style length	________	Guard cell size	________
Anther length	________	Stem XS	________
________		Stem pubesc.	________
________		________	
________		________	
________		________	
________		________	

EXPERIMENTAL AND ANALYTICAL LITERATURE

I. General References

Bell, C. R. 1967. Plant Variation and Classification. Wadsworth Publishing Company. Belmont, California.

Briggs, D., and S. M. Walters. 1969. Plant Variation and Evolution. McGraw-Hill. New York.

Clausen, J., D. D. Keck, and W. Hiesey. 1948. Experimental studies on the nature of species: III. Environmental responses of climatic races of *Achillea*. Carnegie Institution of Washington Publication #581.

Cochran, W. G. 1953. Sampling Techniques. John Wiley & Sons, Inc. New York.

Mode, E. B. 1951. Elements of Statistics. Prentice Hall, Inc. Englewood Cliffs, New Jersey.

Postlethwait, S. N., and N. J. Enochs. 1967. Tachyplants suited to instruction and research. Plant Science Bulletin 13(2): 1-5.

Sokal, R. R., and F. J. Rohlf. 1969. Biometry. W. H. Freeman and Company. San Francisco.

Solbrig, O. T. 1970. Principles and Methods of Plant Biosystematics. The Macmillan Company. New York.

Also see the references at the end of Chapter 12 on Reproductive Biology and Systematics, Chapter 21 on Data Presentation and Documentation, Chapter 22 on Analysis of Character Variation in Systematics, Chapter 23 on Numerical Aids to Taxonomy, and Chapter 26 on Variation and Evolutionary Relationships.

Chapter 21. DATA PRESENTATION AND DOCUMENTATION

Although detailed description is the usual method of presentation of much taxonomic information, nothing can take the place of accurate drawings and clear photographs in the presentation of morphological data, or precise charts and graphs in illustrating clearly the trends shown by large amounts of tabular or statistical data. A page of line drawings costs no more to print than a page of printed material. Photographs, and of course color photographs, cost considerably more to reproduce than line drawings. Photographs should generally not be on the same plate as line drawings. Regardless of type--photograph or drawing--graphic material that is not adequately labeled loses much of its value and impact. Where specific detail is involved the actual voucher or herbarium specimen associated with the figure should be given in the legend. All data should be referable to a pertinent standard herbarium voucher, or other acceptable form of documentation (also see Chapters 18 & 22).

I. DRAWINGS

1. Line Drawings. The most common form of botanical illustration is by means of line drawings. Although such drawings may appear deceptively simple, good drawings require considerable skill. There is more art than mechanical ability in just drawing an ink line, or stippling a drawing. Top notch drawings by a professional artist are an essential investment in scientific accuracy.

   Detailed drawings of unfamiliar objects require more labels than less detailed drawings of standard objects (compare Figures 21-1 and 21-2). Appropriate scales or other indications of magnification or reduction are essential. Also, each figure, plate, chart, graph, or table should carry a legend unless all pertinent information needed to understand the material is included as part of the figure, plate, chart, graph or table. The legends for Figures 21-1 and 21-2 (from unpublished thesis of Drapalik, D. J. 1969, used with permission) read as follows:

   Figure 21-1. Typical seeds, seedlings, leaves, and fruits of the four groups of *Matelea* known from the southeastern USA. Each group is represented by a vertical column. The "Odontostephana group" is represented by *M. baldwyniana*, and each other group is represented by its one and only species.

   Figure 21-2. Floral Structure of *Matelea baldwyniana*.
   A. Top view showing the relative positions of the gynostegium, corona, corolla lobes, and calyx lobes.
   B. Side view of the gynostegium and corona with two-fifths of the upper corona removed.
   C. Radial section of the corona, gynostegium, and carpels.
   D. Abaxial view of a pollinium apparatus. The cellular patterns of the pollinium walls are not portrayed. The outline was drawn with a camera lucida.

   Drawings may be effectively combined with maps to show character variation over a geographic range. Such a pictorialized distribution

Figure 21-1

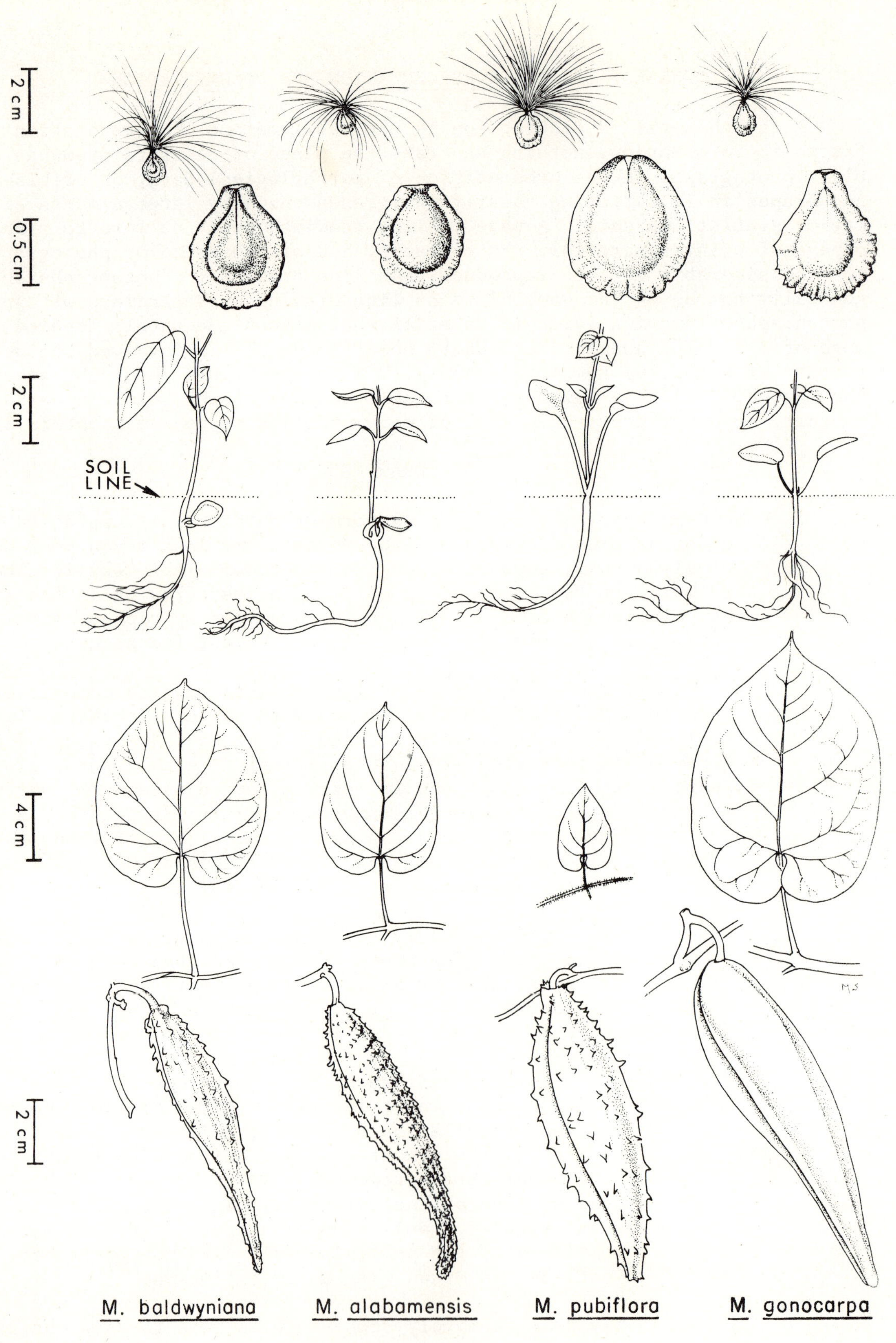
2 cm
0.5 cm
2 cm
SOIL LINE
4 cm
2 cm
MS
M. baldwyniana
M. alabamensis
M. pubiflora
M. gonocarpa

Figure 21-2

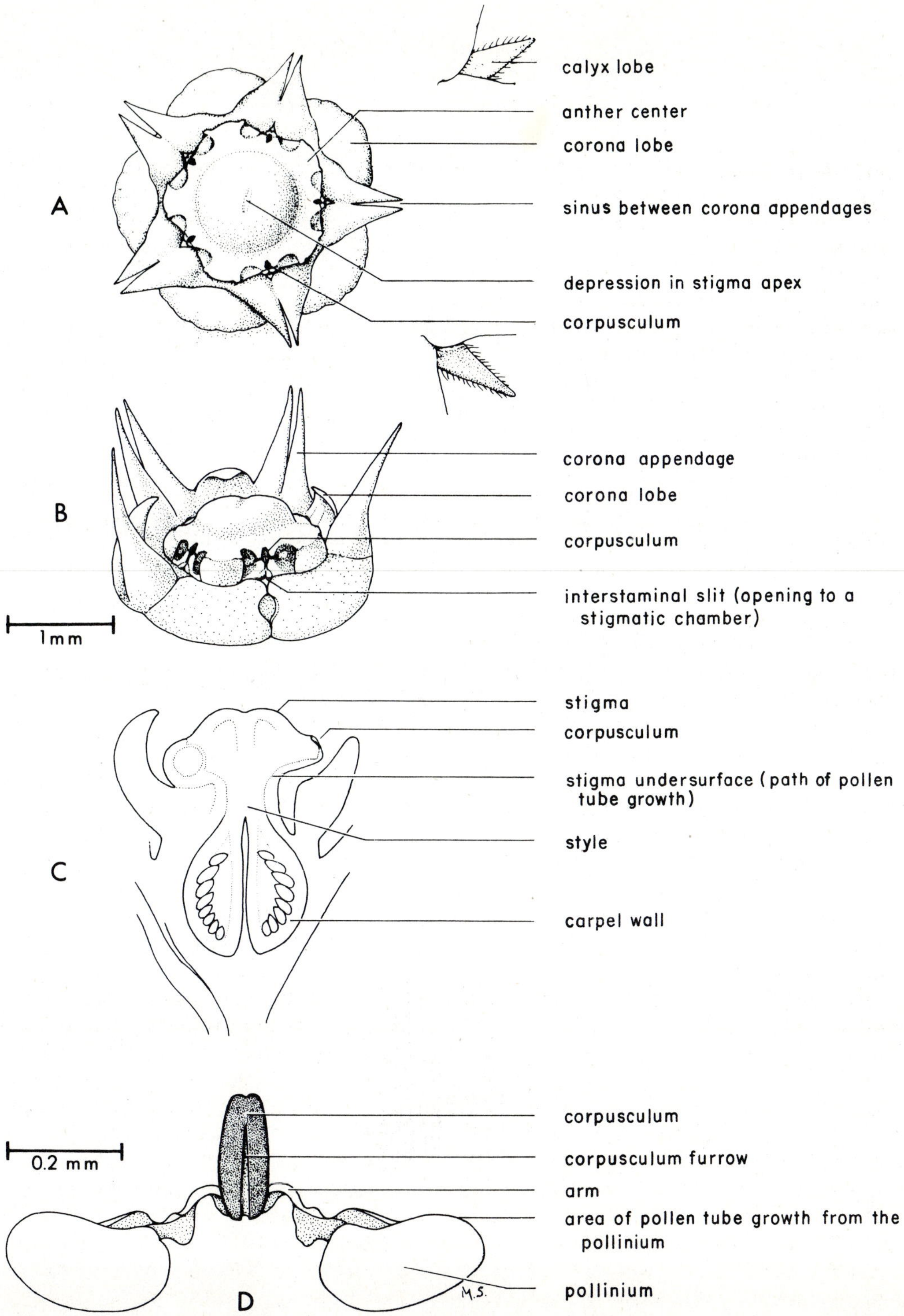
calyx lobe
anther center
corona lobe
A
sinus between corona appendages
depression in stigma apex
corpusculum
B
corona appendage
corona lobe
corpusculum
interstaminal slit (opening to a stigmatic chamber)
1mm
C
stigma
corpusculum
stigma undersurface (path of pollen tube growth)
style
carpel wall
0.2 mm
corpusculum
corpusculum furrow
arm
area of pollen tube growth from the pollinium
pollinium
D
M.S.

Figure 21-3

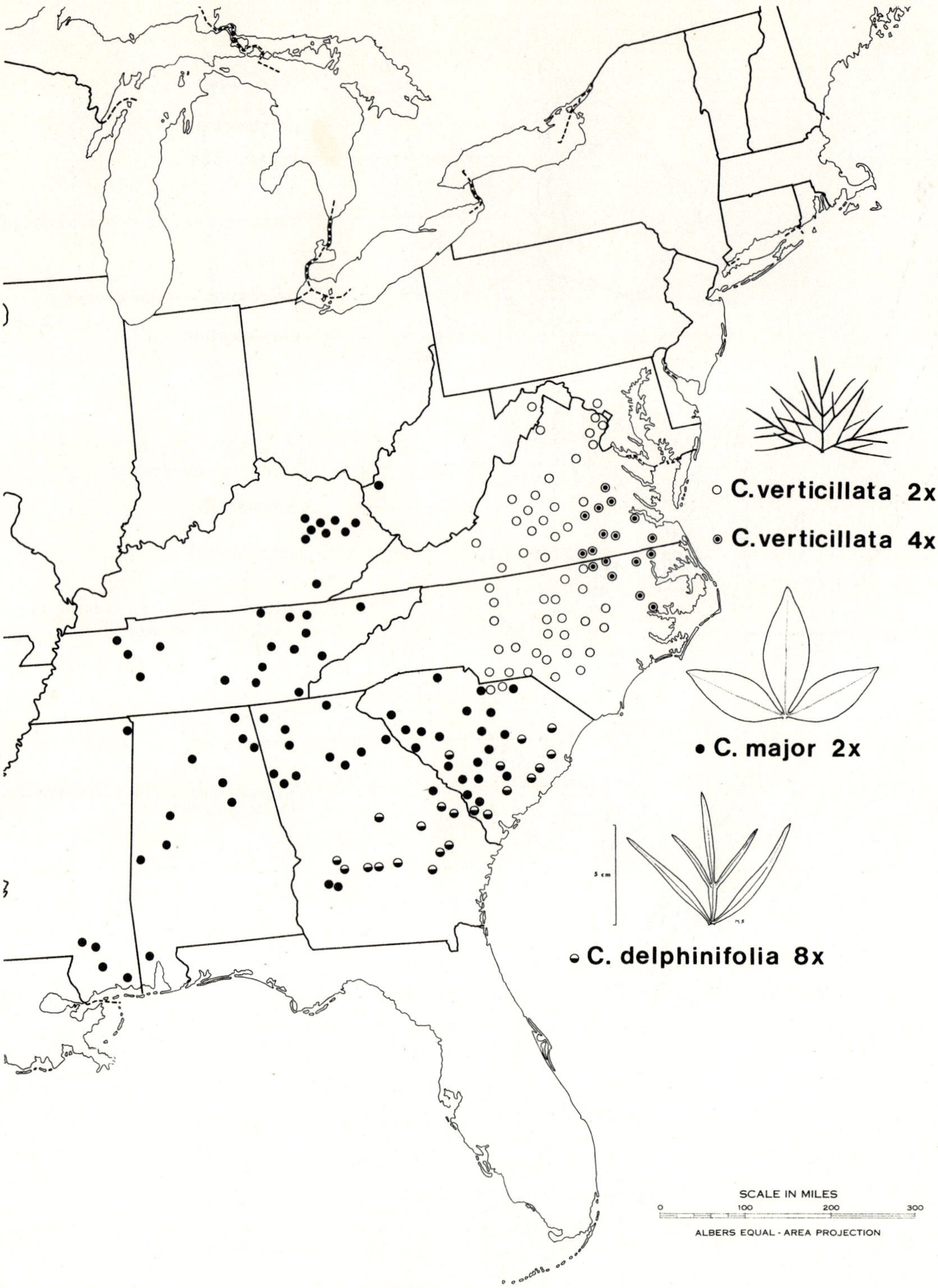

COREOPSIS DISTRIBUTION

Figure 21-4. Xeroxed "drawing" of a pressed plant of _Eleocharis_ _obtusa_.

map for different species and chromosome races of Coreopsis are shown in Figure 21-3 (from unpublished work of Anna M. Mueller, used with permission). Documentation, always essential to any taxonomic study, would be in the form of an extant (and available) herbarium collection to correspond with each dot on the map. The collection data for this group of specimens could be presented in the form of a table (in an appendix, if necessary) and should include for each species the following minimal information in columns, as shown:

Collector	Collection Number	Date	County	State	Herbarium

2. Drawings from xerox copies of plant material. For those plant structures that are more or less laminar, or that can be flattened without undue loss of character, an extremely accurate outline can be obtained merely by having the material xeroxed (Figure 21-4). The xerox copy can then be traced to get an outline drawing, or it can be inked in to get a silhouette. This method is especially valuable for getting comparative drawings of seedlings and of irregular or highly divided leaves, petals, inflorescences or other plant parts. The natural size drawings can be photographically reduced as necessary for publication.

3. Camera Lucida drawings. Camera Lucida drawings, or drawings made with any other projection device attached to a standard light microscope, are both accurate and easy to produce with little or no artistic talent. Such drawings are commonly used to illustrate either meiotic or mitotic chromosomes (Figure 21-5 A and B). Where chromosome morphology is important the somatic chromosomes may be redrawn to scale and with the centromeres on a line, in a stylized form known as an idiogram (Figure 21-5 C).

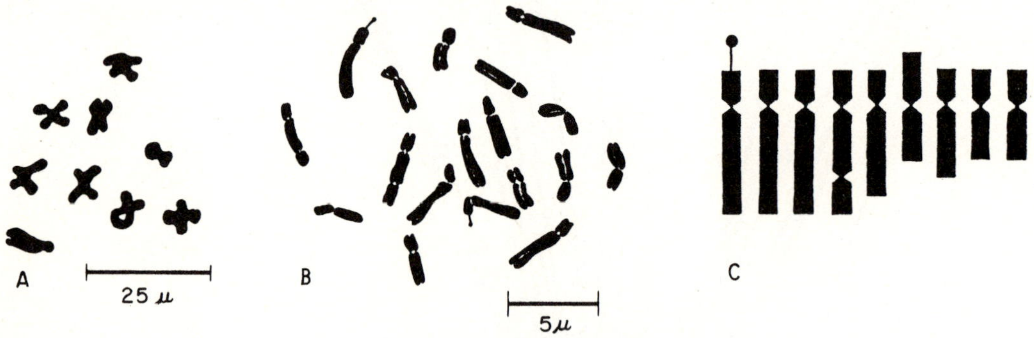

Figure 21-5. Chromosomes of Erigeron simplex: A, meiotic; B, mitotic; C, ideogram. [From unpublished data of Stephen Spongberg. Used with permission]

II. PRESENTATION OF CHROMATOGRAPHIC DATA

Chromatographic information (see Chapter 13) is often presented as shown in Figures 21-6 and 7; when the number of different interspecific chromatographic spots is the main point of the illustration the spots common to all species are not shown (Figure 21-7). If the chromatographic data involve

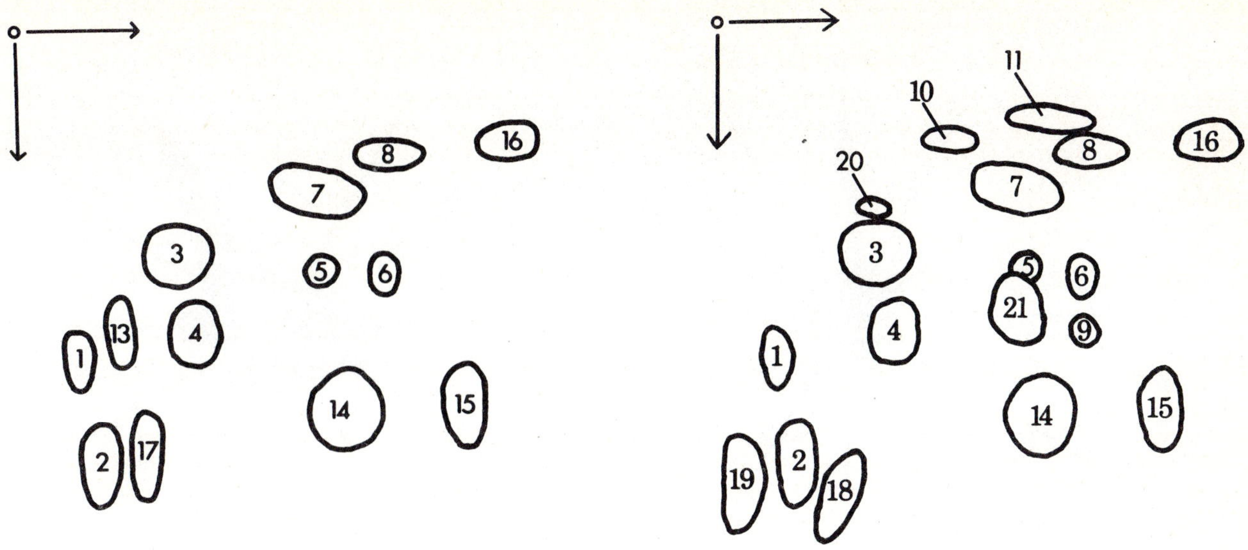

Physalis heterophylla **Physalis lanceolata**

Figure 21-6. Drawing of chromatographs of 2 species of Physalis showing all of the spots obtained on each chromatograph. Equivalent numbers indicate equivalent spots.

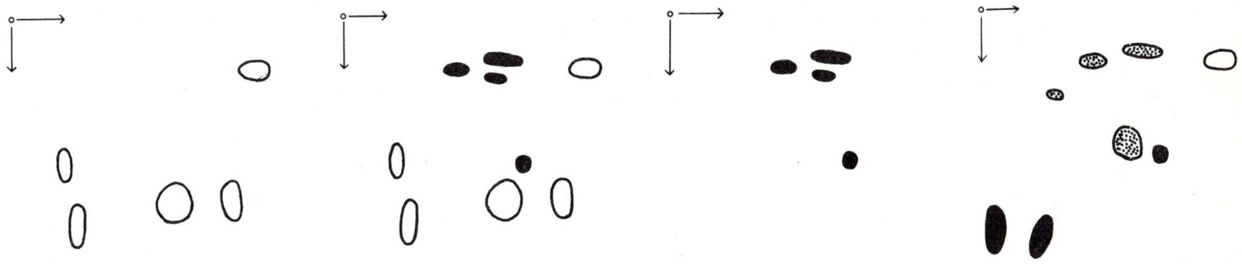

P. heterophylla hybrid P. virginiana P. lanceolata

Figure 21-7. Chromatographs, showing variation in leaf extracts of three species and one hybrid of Physalis, clearly indicate that Physalis lanceolata is not, as once postulated, a hybrid between P. virginiana and P. heterophylla. Note that in this presentation of chromatographic data only those spots that are unique to the plants of a species are shown - the many pigment spots the related plants may have in common are omitted from the chromatograph for the sake of clarity. Compare with the "total spot" drawings shown in Figure 21-6. [From Hinton, W. F. 1970. The taxonomic significance of Physalis lanceolata (Solanaceae) in the Carolina Sandhills. Brittonia 22: 14-19. Used with permission.]

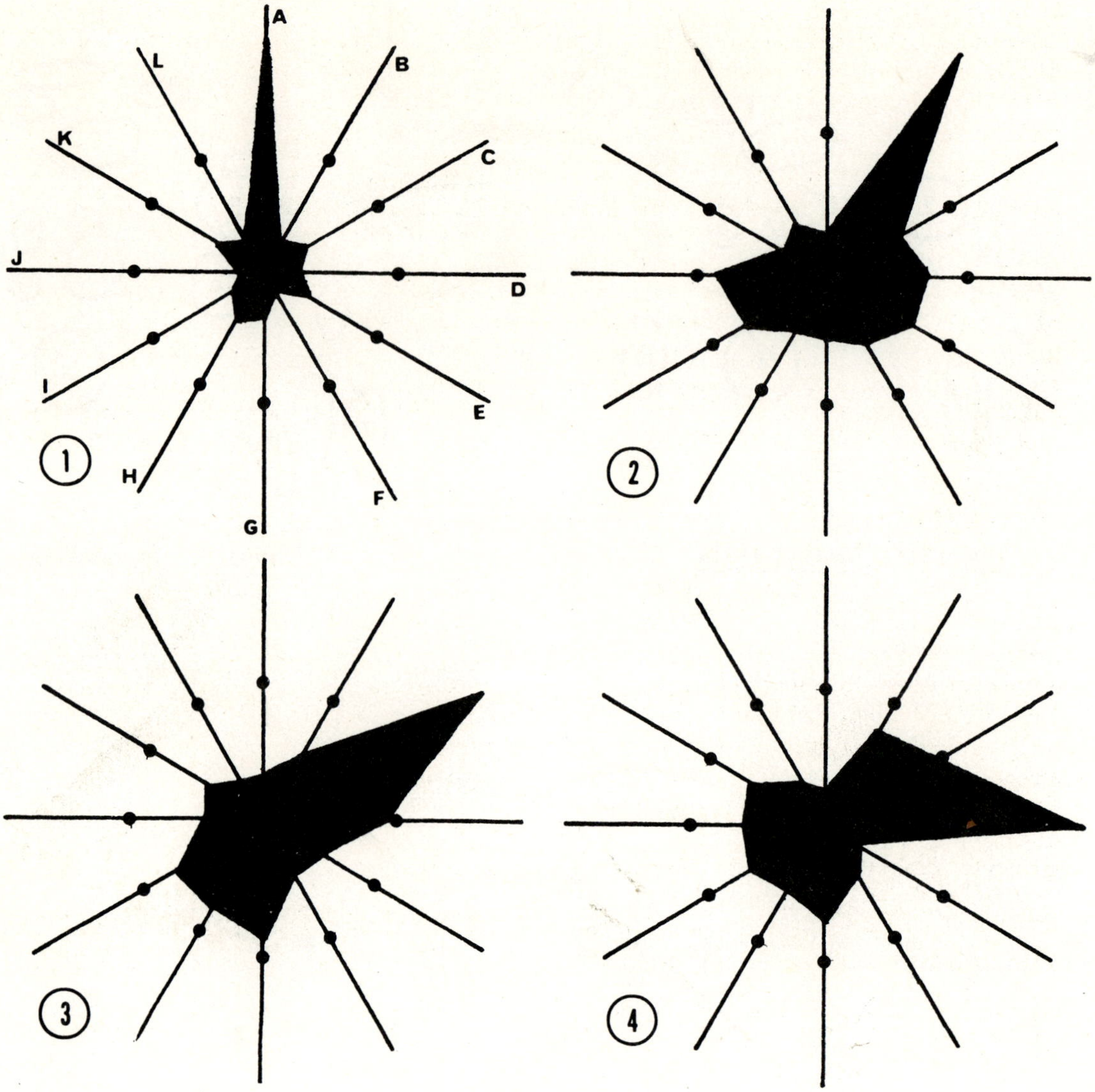

Figure 21-8. Paired Affinity Indices of Arctic and Alpine Erigeron. Dots on each arm = 50% mark, outer end of each arm = 100% affinity. Each arm, as lettered in graph 1, represents one taxon, as follows:

A. Erigeron uniflorus spp. eriocephalus
B. E. simplex spp. simplex
C. E. simplex spp. petiolaris
D. E. simplex spp. flettii
E. E. grandiflorus, northern alpine race
F. E. grandiflorus, southern alpine race
G. E. melanocephalus
H. E. humilis
I. E. aureus
J. E. hyperboreus
K. E. lanatus
L. E. vagus

Graph 1. Erigeron uniflorus ssp. eriocephalus
2. E. simplex ssp. simplex
3. E. simplex ssp. petiolaris
4. E. simplex ssp. flettii

[From unpublished data of Stephen Spongberg. Used with permission.]

a large number of spots and if the use of the information is to illustrate the degree of chromatographic similarity or difference between species, a "paired affinity index" is computed for each pair of species involved in the study (see formula, below) and the resulting percentages are then plotted on the appropriate arms of a polygonal graph (see Figure 21-8).

$$\text{Paired Affinity Index (PA)} = \frac{\text{spots in common for species A and B}}{\text{total spots for A + B}}$$

Since, obviously, a species will show a 100% affinity with itself, the "peak" of the polygon will shift from species to species (arm to arm) around the graph; it is the general relationships shown on the other arms of the graph that are of interest. Through a summation of the PA indices of one taxon with each of the others to which it is being compared a numerical expression of chromatographic similarity (called a Group Affinity Value) can be obtained, as shown by Ellison, et al. (1962).

III. POLYGONAL GRAPHS AND HISTOGRAMS

Polygonal graphs, with up to 16 arms, each representing a character, are also frequently used to compare graphically the variation patterns within or among a series of populations or species. The graphs (Figure 21-10) can be composed of many lines with each line representing an individual plant, or a single line which represents the mean values for a population or a species. When the polygons so formed by the data for a population are placed in the proper position on an outline map the pattern of geographic variation within a species is quickly evident. A "master graph," with each arm clearly labeled as to what character it represents and the units of measurement clearly marked should be given with each set of polygonal graphs presented.

Histograms and regular line graphs (Figure 21-9) are commonly used to illustrate frequency distribution.

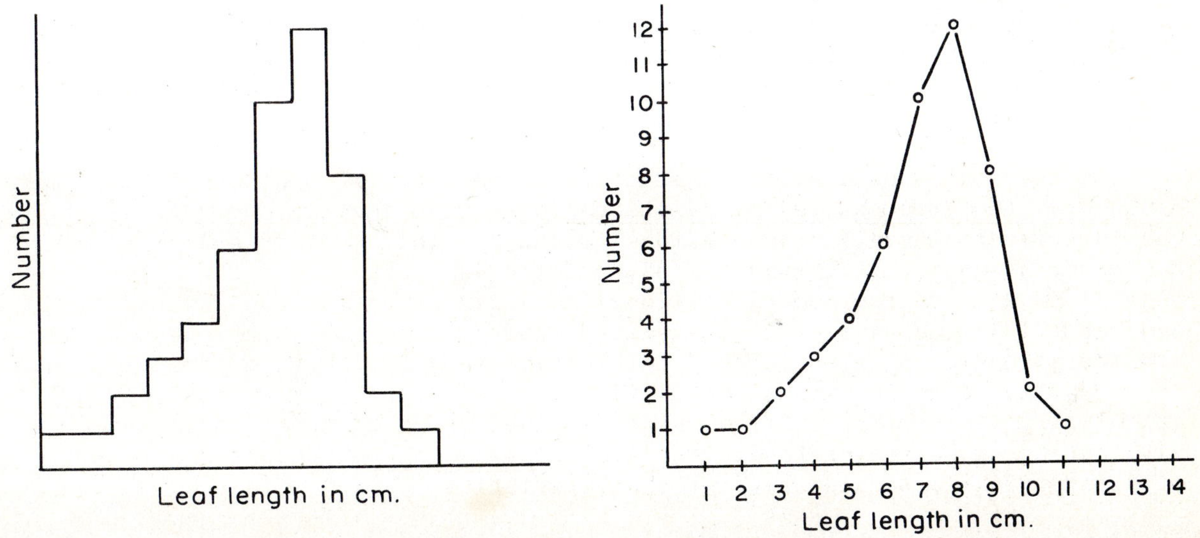

Figure 21-9. Frequency distribution of the data given in Table 20-1 by histogram (left) and line graph (right).

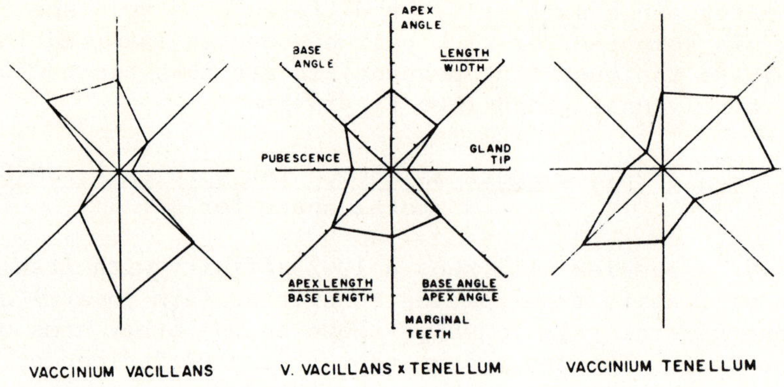

Figure 21-10. Polygonal graphs of eight leaf characters (see center graph) from plants of two species of Vaccinium (Ericaceae) and plants of a presumed hybrid between them. [From unpublished data of Sabina Mueller. Used with permission.]

IV. HYBRID INDEX

A series of measurements, scores, or size classes can be used to construct a numerical Hybrid Index (see Table 21-1) in which the basic characteristics of a group of individual plants are scored on a 1-2-3 basis (or a 1-3-5 basis if a more accurate scoring of intermediates will result), and then summed for each plant. If the non-linked characteristics are involved in the index the plants of one taxon will have a total score of 10 (if they are typical for the taxon), plants of a second taxon would have a score of 50 (if typical) and intermediate plants would have intermediate scores. If a histogram is made of the Hybrid Index scores for a population the distribution pattern will readily illustrate the composition of the population in terms of the different variants (Figure 21-11).

V. SCATTER DIAGRAMS

If the states of a second character are used on the ordinate instead of frequency, the graph shows the relationship between the two characters for the individual plants (or populations) plotted. This type of graph is called a scatter diagram. If, through a pictorial code, each dot on the scatter diagram is made to represent additional characters, as was done by Edgar Anderson (1949) we have a "Pictorialized Scatter Diagram" which presents a considerable amount of correlated information (about either individual plants in a population or individual population of a species) very effectively and quickly (Figure 21-12). Similar pictorialized or coded spots can be used with outline maps (refer to Figure 21-3) to show the geographic distribution of plants with particular sets of characteristics.

Character	Index Value 1	Index Value 3	Index Value 5
Leaf width: length ratio	0.8-1.0	0.4-0.7	0.1-0.3
Leaf pubescence	glabrous	puberulent	tomentose
Stomate density mean per 0.1 mm^2	153	256	311
Plant height	1-3 dm	4-7 dm	8-11 dm
Flower color	red	pink	white
Sepal margin	entire	serrate	fimbriate
Pollen size (mean)	32-36 μ	37-41 μ	42-48 μ
Style length	2-4 mm	2-8 mm	9-12 mm
Stigma shape	globose	ellipsoid	cylindrical
Seed per capsule	1-9	10-19	20-29
Total Index Value	10	30	50
	Taxon R	Taxon S	Taxon T

Table 21-1. Hybrid Index [From Bell, C. R. 1967. Plant Variation and Classification. Wadsworth Publishing Company. Used with permission.]

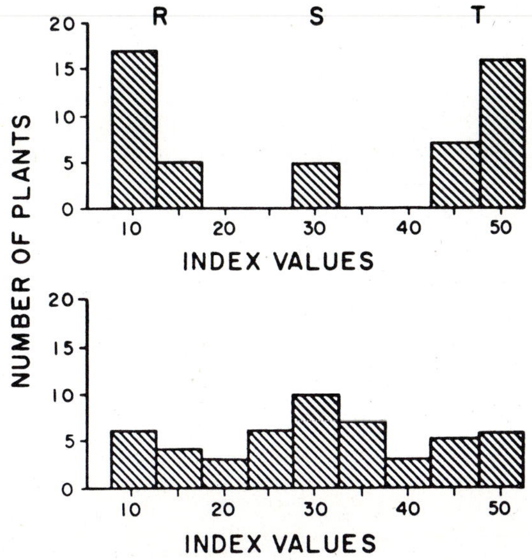

Figure 21-11. Hybrid Index Histogram. Two types of distribution patterns of Hybrid Index values for 50 plants from one population that would show two entities and their presumed hybrid (top), and variation in one entity (bottom). A similar graph could also represent the Hybrid Index averages of samples from 50 different populations; the top figure would then probably best be interpreted as representing three distinct taxa, but the bottom figure would again represent the variation in a single taxon over its geographic range with R and T only representing the extremes in a normal variation curve. [From Bell, C. R. 1967. Plant Variation and Classification. Wadsworth Publishing Company. Used with permission.]

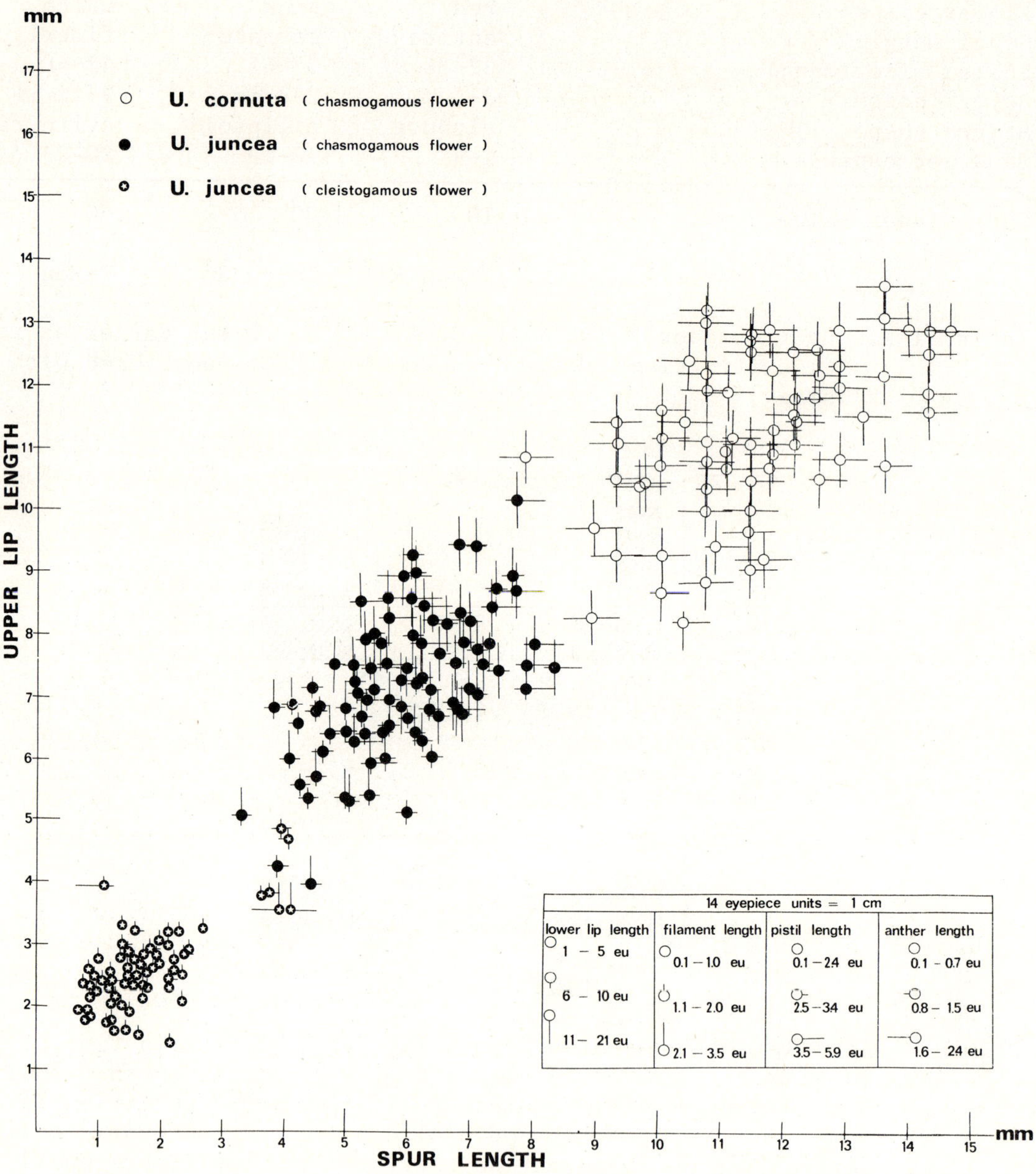

Figure 21-12. A scatter diagram of Utricularia characters. [From Kondo, K. 1972. A comparison of variability in Utricularia cornuta and Utricularia juncea. American Journal of Botany 59:23-37. Used with permission.]

Utricularia Spur Length, (in mm). (CH = chasmogamous flower, CL = cleistogamous flower).

Species	Population	Number Measured	$\overline{X}$	S.D.	S.E.$\overline{X}$	C.V.	Range
U. cornuta	U-11	50	8.4	0.91	0.13	10.9	5.8-10.7
	U-12	50	8.9	0.78	0.11	8.9	6.9-11.4
	U-15	50	11.3	1.05	0.15	9.3	8.6-13.9
	U-17	50	12.9	0.89	0.13	6.9	10.1-14.6
	U-18	50	11.0	0.85	0.12	7.7	8.6-12.5
	U-19	50	10.2	0.79	0.11	7.8	7.6-12.1
	U-20	50	11.6	0.63	0.09	5.4	10.4-12.9
	U-21	50	10.9	0.88	0.12	8.0	9.0-12.5
	U-22	50	11.4	1.14	0.16	10.1	6.5-13.2
U. juncea	U--2(CH)	50	5.6	0.93	0.10	16.6	4.0-8.3
	(CL)	50	1.6	0.64	0.09	40.0	0.5-4.1
	U--6(CH)	50	5.3	0.89	0.12	16.8	2.9-7.4
	(CL)	50	1.8	0.64	0.05	20.0	0.7-4.3
	U-13(CH)	50	4.6	1.00	0.14	21.9	3.1-7.8
	(CL)	50	1.6	0.57	0.07	36.4	0.5-4.1
	U-14(CH)	50	6.6	0.78	0.11	12.0	0.5-8.4
	(CL)	50	1.4	0.71	0.07	50.0	0.7-4.1
	U-23(CH)	50	4.8	1.00	0.14	20.8	3.0-7.8
	(CL)	50	1.4	0.64	0.09	45.7	0.5-4.3

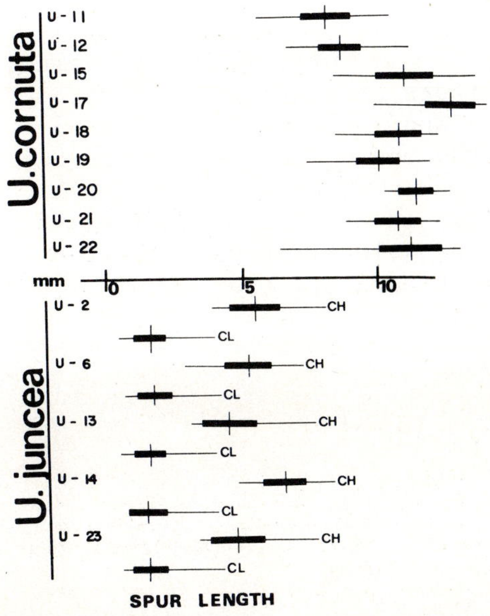

Table 21-2. Comparison of single versus double spacing of tabular material (above) and a graphic summary of the data (left). [From Kondo, K. 1972. A comparison of variability in Utricularia cornuta and Utricularia juncea. American Journal of Botany 59: 23-37. Used with permission.]

VI. TABULAR MATERIAL.

VI. TABULAR MATERIAL

Although tabular material is often summarized graphically in some form, the actual tables of specific data should accompany the charts and graphs derived from them whenever possible, perhaps as an appendix to the body of the research paper. In the tables themselves more details of the analysis of a character can be given (e.g., the coefficient of variability, number of measurement, standard error of the mean) than are generally shown on graphs. Also, the actual figures are easier to compare in detail than are the graphs (compare Table 21-2 and the accompanying graph). Tables, as figures, graphs and all other inclusions, should be fully labeled and captioned to be self-explanatory. Also, the entries (species, populations, etc.) in a series of tables and graphs should always be in the same order to facilitate comparison.

VII. CROSSING DIAGRAMS AND PHENOGRAMS

The graphic illustration of concepts of relationship can take many forms from the phylogenetic "cactus" chart of Bessey (see page 602) dealing with the orders of angiosperms to the simple lines and branches, determined by computer, of a "correlation phenogram" for plants within a single genus as shown in Figure 21-13. In between these seemingly objective diagrams of monophyletic relationships are the many admittedly speculative (to greater or lesser degrees) evolutionary diagrams that seek to express the often rather involved reticulate evolution found in a "species complex" or a "polyploid complex" as shown in Figure 21-14.

Crossing diagrams, which show the relative degree of gene exchange or hybridization possible between the members of a homogamodeme (a series of more or less interfertile plants belonging to several species, varieties, races or populations) are often used to summarize the results of breeding experiments. In these diagrams the thickness of the line connecting two parent sources indicates the relative degree of interfertility between them (Figure 21-15). The direction of the cross or crosses is indicated by an arrow on the line from the pollen parent to the female parent.

There is an almost infinite variety of graphic form used by individual taxonomists to express plant relationships as indicated by the data collected during a specific research program. Since it is not possible here to treat in detail the many combinations of forms and types of drawings, charts, graphs, and photographic techniques, reference should be made to several current volumes of the periodicals listed at the end of Chapter 22 for further ideas on data presentation. As to the actual preparation of the material, there are numerous references on the methodology of both line drawing and photography which should be consulted for the details of preparation. A few such references are given in the following Literature List.

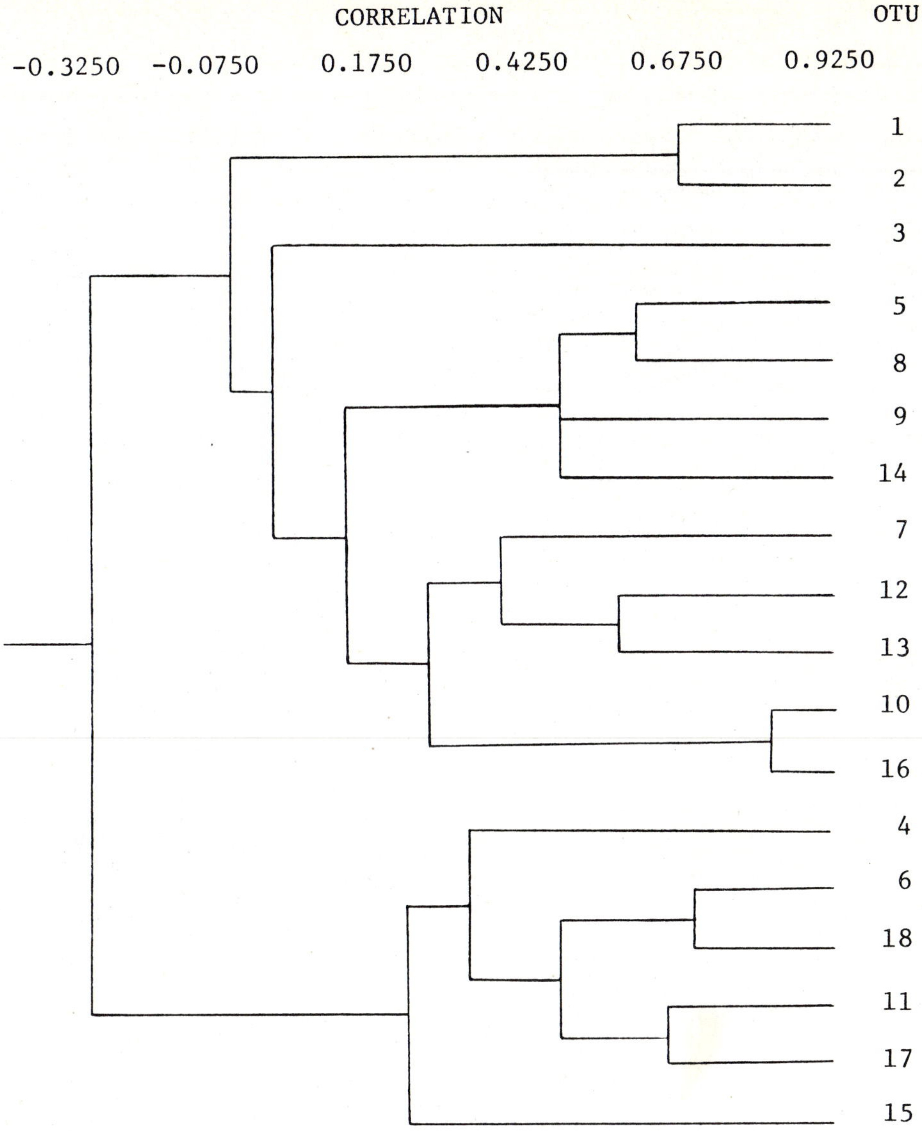

Figure 21-13. A correlation phenogram of 18 sets (OTU's) of plants of Haplopappus based on 31 characters. [From Jackson, R. C. and T. J. Crovello. 1971. Brittonia 23:54-70. Used with permission.]

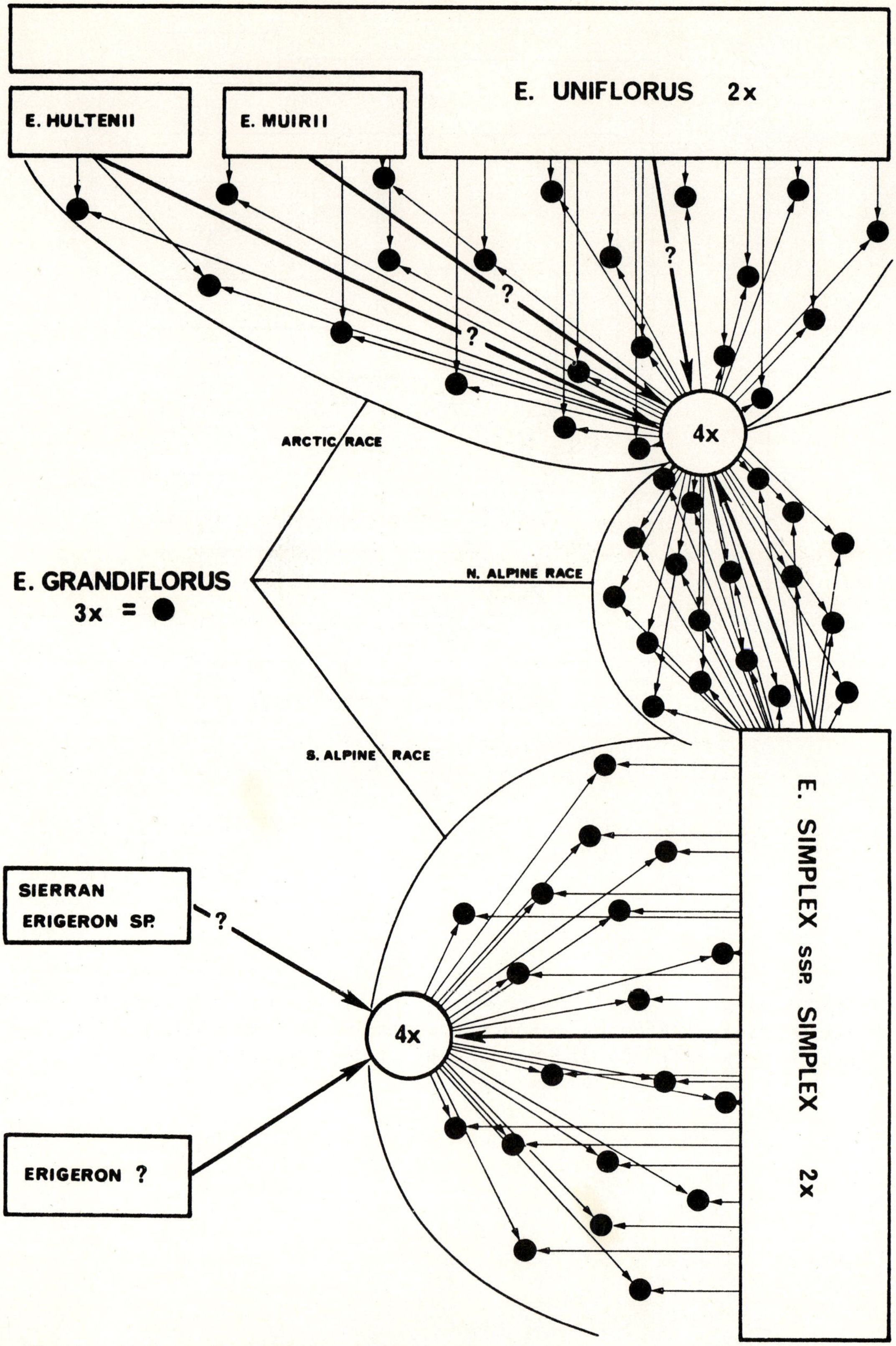

Figure 21-14. Interrelationships within the Erigeron uniflorus-simplex complex. [From unpublished data of Stephen Spongberg, used with permission.]

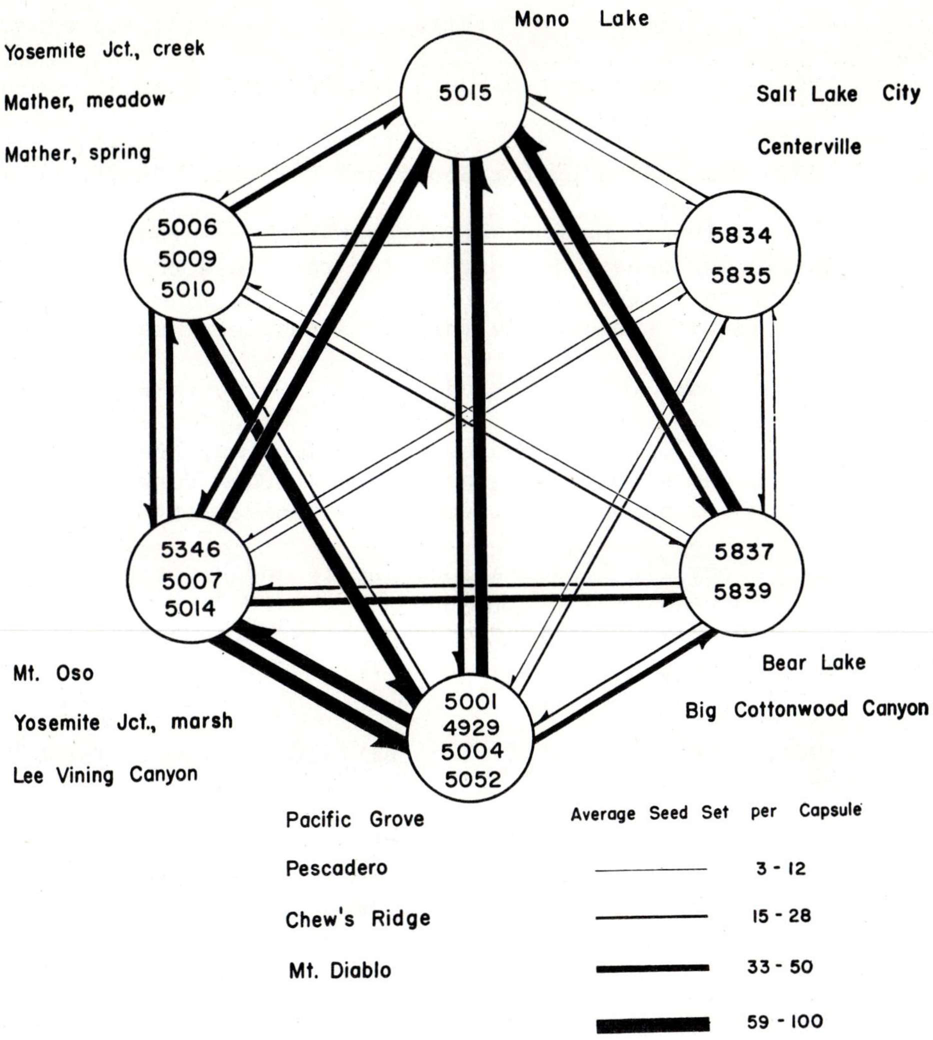

Figure 21-15. Summary of seed set in various crosses between plants of different populations of Mimulus guttatus. [From Vickery, Robert J., Jr. 1959. Evolution 13: 300-310. Used with permission.]

DATA PRESENTATION AND DOCUMENTATION LITERATURE

Anderson, E. 1949. Introgressive Hybridization. John Wiley and Sons, Inc. New York.

Blaker, A. A. 1965. Photography for Scientific Publication. Freeman Books. San Francisco.

Blunt, W. J. W. 1950. The Art of Botanical Illustration. Collins. London.

Bracegirdle, B. 1970. Photography for Books and Reports. David and Charles, Pubs. Newton Abbot, England.

Ellison, W. L., et al. 1962. Methods of presentation of crude biochemical data for systematic purposes, with particular reference to the genus Bahia (Compositae). American Journal of Botany 49: 599-604.

Engel, C. E. (ed.). 1968. Photography for the Scientist. Academic Press. New York.

Hodgkiss, A. G. 1970. Maps for Books and Thesis. Universe Books, Inc. New York.

Knudsen, J. W. 1966. Biological Techniques: Collecting, Preserving, and Illustrating Plants and Animals. Harper and Row. New York.

Lloyd, B. 1935. Handbook of Botanical Diagrams. University of London Press. London.

Papp, C. S. 1968. Scientific Illustration, Theory and Practice. William C. Brown Company. Dubuque, Iowa.

Poole, L., and G. Poole. 1965. Scientists Who Work With Cameras. Dodd, Mead and Company, Inc. New York.

Rose, G. G. (ed.). 1963. Cinemicrography in Cell Biology. Academic Press. New York.

Zweifel, F. W. 1961. A Handbook of Biological Illustrations. The University of Chicago Press. Chicago.

The Consumer Markets Division of Eastman Kodak Company, Rochester, New York 14650, publishes excellent inexpensive handbooks which cover every aspect of photography from basic information on films, lighting, developing and enlarging to highly technical information on specialized photographic methods.

Chapter 22. ANALYSIS OF CHARACTER VARIATION IN SYSTEMATICS*

Knowingly or not, all systematists are engaged in the analysis of character variation. The purpose of this paper is to outline approaches to the study of character variation in systematics. Perhaps this chapter will be of greatest value if it motivates us to get out of our usual "thought-ruts." Some people have not re-evaluated their ideas, purposes, or methods of analysis since their graduate days. For example, many people still use Anderson's pictorialized scatter diagram as the only approach to the study of character variation due to introgression. As creatures of habit, we tend to think along the path of least resistance. So, if one explanation is given for a set of phenomena, it is easiest from that point on not to keep searching for novel explanations. The same concept is true, to some degree, in methodology: while I have tried to include all major areas of systematics in this review, I am aware that my own background and interests influence the emphasis given to each area and method. Hopefully, a glimpse at the many methods of studying character variation used in systematics today will have a stimulating effect and permit the development of new paths of thought and research. One cannot help being impressed with the potential benefits to both the descriptive and explanatory phases of systematics when its workers consider the use of the contents and methods of each other's fields as well as those developed in ecology, other areas of biology and areas outside of biology, e.g., psychology, business, and engineering.

I find it interesting that with the proposed Flora North America Program and a growing number of projects to "computerize herbaria" (see Crovello, 1972 for references), botanists have forged ahead of zoologists in the use of computers for information retrieval. Unfortunately, plant systematists are far behind in the use of computers to analyze character variation. Perhaps this book will help to alleviate the situation.

NECESSARY PRIMITIVE NOTIONS

To minimize misunderstanding, certain essential concepts, or primitive notions, must be agreed upon before any comparison of approaches can be made. An <u>object</u> is an individual organism or a collection of organisms grouped by some common property. Sokal & Sneath (1963) designated the object an operational taxonomic unit, i.e., the unit to be classified. A <u>system</u> is a collection of objects, their properties, the interactions among them, and all "environmental" forces acting upon them. A <u>character</u> is a property of an object or of a system that cannot logically be subdivided further. This compares with Sokal & Sneath's (1963) operational definition of character. A <u>character state</u> is the particular expression of a character of a given object or of a system. For example, in a taxonomic study of willows, each taxospecies may arbitrarily be designated as an object. Each has the character stamen number. The character state for object (species) A may be one stamen; for object (species) B it may be five stamens, etc., depending on the individuals observed in each taxospecies.

The choice of objects, of systems, of characters, and of character states is arbitrary, and the choices vary with the investigator and with the purpose

*Contributed by Theodore J. Crovello, University of Notre Dame.

of the study. Perhaps the areas of systematics that would most strongly deny arbitrariness are those using characters of organisms whose genetic inheritance has been worked out. But even here, one does not use all genetically controlled characters, nor does the choice of character states that coincide with genetic fact mean that this is the most useful choice. For example, in a systematic study of North American Cardamine, it may be of more value to recognize only two character states of the character leaf pubescence: pubescent and not pubescent, without regard for the several genetically different types of pubescence that are possible.

We must also consider the two primitive notions of character space and object space. Character space is an abstraction that describes a system, each dimension of which is a character of the system or a function of one or more characters of the system. Object space describes a system, each dimension of which is an object of the system or a function of one or more objects of the system. In both spaces if more than three dimensions are involved, the space is considered a hyperspace.

I shall adopt the view that any systematist works with a system. Taxonomists work with systems. To some, the system consists only of objects, the specimens or populations being classified. For others, the specimens are linked due to evolutionary processes. The biosystematist interested in pollination systems or in a comparison of life tables of populations of Trillium certainly works with a system. Thus all are interested either in the description of the system or in the explanation of how the system, with its particular pattern, arose and maintains itself. Sometimes a system is so well known or considered in such a simple way that no formal systems description is undertaken. I shall consider the analysis of character variation as first generating a description of a system, e.g., a species complex, and then contributing to an explanation of its origin and maintenance.

I. THE CHARACTER VARIATION ANALYSIS PROCESS

Being a function of the investigator's purpose, of his resources, and of the specific biological system under study, the procedure used to analyze character variation will vary with the investigation. Nevertheless, each character variation analysis can be considered as one instance derived from a general, multistage process. This section presents the general procedure in detail. It emphasizes that the procedure is a multistage decision process and that each stage (and its set of substages) requires quite often arbitrary decisions as to how to carry out the given stage. Table 22-1 lists the steps of the character variation analysis process.

A. Purpose of the Analysis

1. Clear statement of the purpose. A concise, clear statement of the purpose of the planned character variation analysis is essential. Frequently, investigators assume that their reasons for studying character variation are both obvious and clearcut. Decisions as to the "best" method for the study of character variation and the "best" characters to use are partly a function of one's purpose and may change with different purposes. Having stated the purpose of the study, the analyst of character variation then must consider the following points.

TABLE 22-1. The Character Variation Analysis Process

Stage	Substages or Methods
A. Purpose of the analysis	1. Clear statement of the purpose 2. Value of the study to science 3. Relevance of the study to society 4. Feasibility of the character variation analysis 5. Potential of the study to fulfill its stated purpose
B. Data accumulation	1. Relevant information search 2. Choice of the system 3. Choice of objects 4. Choice of characters 5. Choice of parameters of the characters 6. Choice of character states 7. Choice of character state codes 8. Filling in the basic data matrix: Sampling designs Design of experiments 9. Verification of the basic data matrix
C. Mathematical analysis and summarization (See Table 22-2 for details)	1. Statistics 2. Similarity estimation 3. Description of structure in a reduced number of dimensions 4. Decisions and optimization 5. Geometry
D. Summarization graphics (See Table 22-3 for details)	1. Frequency distributions 2. Scatter diagrams 3. Directed graphs 4. Undirected graphs 5. Geographic maps 6. Miscellaneous
E. Evaluation of results and reanalysis	Extensional expansion of data Intensional expansion of data
F. The dynamic data bank	An individual's data bank A network of data banks

2. Value of the study to science. Many argue that any knowledge is of value to the advancement of science. But two facts must be considered: some kinds of knowledge are more valuable than others at any given point in the development of science, and only a finite amount of human and other resources are available for systematics. As a result, decisions must be made as to what studies should be carried out. But, as is true for most of the stages and substages of the character variation analysis process, an objective, nonarbitrary decision function does not exist to determine the relative values to science of different character variation studies in a given field or subfield.
3. Relevance of the study to society. As evolutionary problems in systematics and ecology become more important to the future of society, there is pressure to make every study socially relevant. This demand has two distinct sources. The first is economic. If granting agencies give money only for proposed character variation studies that are relevant to society's problems, then most researchers who want to remain active will move to such research. Secondly, some potential students of systematics will refuse to enter this career unless its relevance to society is made clear.
4. Feasibility of the character variation analysis. Most problems in systematics are complex and cannot be separated into isolated components for analytical laboratory study as easily as some problems concerning systems in molecular and cell biology. It is often difficult to know in advance the amount of time, equipment, and manpower that will be necessary. Most projects require years to complete--e.g., a study of strategies of evolution within and among genera of the North American Brassicaceae or a monograph of the genus <u>Ranunculus</u>. Granting agencies and administrators usually demand frequent reports and do not fund projects for long periods of time all at once. This is unfortunate if it splinters the research effort, but on the other hand it compels the principal investigator to stop periodically to rethink his project and to justify its value, feasibility, and relevance.
5. Potential of the study to fulfill its stated purpose. At least two aspects of the substage must be considered, the quality and the quantity of the proposed character variation analysis. By the quality of the research I mean the ability of the results to provide a definite choice between the alternative hypotheses formulated for the purpose of the study. That is, will the research be germane to the purpose, or will it just give us more information of uncertain value about the system under study? By the quantity of research I mean whether, although it is germane research, the character variation analysis program will have sufficient power to discriminate between alternative hypotheses or whether it will result only in inconclusive evidence about the hypotheses under test.

B. <u>Data Accumulation</u>

1. Relevant information search. Significant research in systematics requires accurate knowledge about the biological system or phenomena chosen for study (e.g., the family Brassicaceae, or

strategies of weed populations), and at least a working knowledge of the current methods used to study such systems.

It is probably not possible to list all information sources of value to systematists. Also, our information sources are undergoing rapid change--the amount and kinds of information are increasing, and the available methods of retrieval and analysis of such sources are improving constantly, thanks to advances in computer science. The following comments are made in the light of what could be available to workers on a widespread scale, because the technology exists today.

Let me refer to the information sources collectively as the systematic data bank. I am concerned with the structure and content of the data bank. The actual computer hardware (remote terminals, central processors) and software (special information retrieval programs, simulation languages) exist today to handle the data bank.

The data bank will obtain input mostly from four sources: 1. the literature, 2. museum and herbarium collections, 3. botanical gardens and arboreta, and 4. raw data. Every scientist has a set of file drawers overflowing with index cards that contain references that he thinks may be of value to him. They may have additional descriptors or key words beyond the usual information (author, date, title, citation). A new card file quickly enters into an exponential rate of growth, but this is short-lived since the worker realizes that to keep up with the relevant literature in detail, he would have no time left for anything else. Often at this stage, the original set of important references is adulterated by the addition of references that are not too valuable to his research, but that were sent to him as unrequested reprints that he feels compelled to catalog. The result is an unwieldy file. What could be the best summary of the past literature in a field to incoming graduate students becomes quite useless.

The above problem can be attacked from two extremes by using computerized data processing. Attempts can be made to capture the entire biological literature in a form suitable for electronic data processing. This data bank then can be searched, usually with the aid of a researcher's interest-profile, to produce a current or a cumulative list of germane articles. Such a service must be performed by a large commercial, government, or non-profit scientific organization. (Note scope and use of Biological Abstracts in Chapter 30.)

At the other extreme, a worker can simultaneously type the information about a reference on his file card and produce a punched paper tape or, more recently, a magnetic tape that can be read into a computerized data bank. The results can include a periodically updated alphabetical listing, a key-word-in-context permutation of all or selected title words of all or selected articles, etc. He will have immediate and useful recall of his extensive bibliography, with the possibility of it being rearranged in any of several different, previously expensive ways. For example, my current personal bibliography file is on magnetic tape and contains 7,000 references of interest to me. Each reference has attached to it one or several deep-index subject numbers reflecting my interest (e.g., 60 is numerical taxonomy, 70 is evolution, 93 is flowering plants, 99 indicates that I have the reprint). Simply by key punching on a search request card the numbers 60 and 93 in the proper way I get a printout of just those references dealing with numerical taxonomy of flowering plants. It takes less than one minute of computer time to search these 7,000 references and to print out the "hits" in whatever order I request. If someone else were doing this we ought to be able to trade bibliography data banks very easily, thus expanding our retrospective literature searches.

The second major source of input to a data bank of value to systematists is specimen data from herbaria and museums. All collections are arranged according to one primary organizing character, usually the taxon name. This is best for most systematic work. But biosystematists, evolutionists and ecologists want information arranged in a different way. For example, an evolutionist might be interested in convergent patterns of evolution of late fall flowering time, which would require him to look through all specimens in the collections. Or, a plant geographer might ask what specimens in a 1,000,000 specimen herbarium are from the Channel Islands off California. An ecologist working with a regional planner might wish to know what plants used to grow in the Wabash Valley before the beginning of heavy industry and exactly where they might be collected now. A biosystematist may desire a list of all previous collections of a taxon. A computer printout from different herbaria would save considerable time. We can provide such a service now for the 65,000 specimens in the Greene Herbarium at Notre Dame (Crovello, 1972). Similar projects are underway at Boulder, Capetown (Hall, 1972), Mexico City (Gómez-Pompa and Nevling, 1973), Ottawa and Saskatoon (Argus and Sheard, 1972) and Washington.

Deciding what collections and how much information to capture, and how to do it, is itself a complex multistage decision process that can only be mentioned here. Suffice it to say that systematists are active in planning and implementing the computerization of specimen collections and systematic literature. In addition to those mentioned above, Crovello & MacDonald (1970) list over 40 such projects.

The third data source are botanical gardens and arboreta. Natural history preserves in urbanized areas perhaps should also be included here rather than in the next category. A good example is the work of Lindsey et al. (1969). Botanical gardens and arboreta have been a source of seed and other living material for centuries. The American Horticultural Society presently supports a computerized Plant Records Center. Soon this should allow a systematist to make just one inquiry and receive information on the availability of plant material of a given taxon in all of the major botanical gardens and arboreta in the United States (see discussion in Chapter 32).

The fourth major input source to the data bank involves raw experimental data and data from any natural system. They may be data from a relatively undisturbed woods or from sidewalk samples from a metropolitan area. The raw data that form the basis for the summary tables and conclusions of journal articles are lost, either immediately because the author does not keep data records very long or, more frequently, because on his retirement the data records are so old and the details so vaguely remembered that they are discarded as useless.

The preservation of raw data in easily retrievable form has important uses. These include the reanalysis of the same data for the same or a different purpose with better or more recent methods, and as a base to allow building more firmly on the earlier work of others. For example, Crovello (1968a) measured over 200 characters on over 500 specimens of willow from California and analyzed this bank of data to make formal taxonomic decisions. These data are deposited on punched cards in the Herbarium of the University of California at Berkeley. A future worker may wish to reconsider the taxonomy of the group or to analyze the genus to detect evolutionary phenomena. The original 200 by 500 data bank may serve his purpose in itself or, more probably, it can supplement the data that he will collect. In other words, his conclusions will be based on the detailed data accumulated by an earlier worker. Hopefully the new conclusions will be sounder, since they are derived from an expanded data base.

2. Choice of the system. Choice of a system is sometimes dictated by a supervisor, by interest, by its relative availability for manipulation and testing, by social and political relevance, or by a combination of these. In this substage, as in the next four, the analyst of character variation must ask continuously what effect the choice of different systems will have on the final outcome. At one extreme, a chosen system may be unable to provide any answers to satisfy the purpose. And only one system can serve the purpose. On the other hand, the purpose may be such that any system will provide answers. Then the problem is to decide how general the results are for other systems.
3. Choice of objects. The problem that confronts the investigator here is to decide at what level of organization to analyze objects. In taxonomy should one consider each plant, each population, or each species as an object?

   Such decisons are examples of perhaps the most common compromise function in systematics. The compromise is between accuracy (or resolution) and practicality (or summarization). For example, presentation of almost any hierarchy, not just the taxonomic one, demands a compromise between the accuracy with which relations among the objects are presented and the degree of summarization that is thought necessary to permit a biologist to understand the salient relationships among groups of objects. Any time different objects are grouped together and treated as similar, inaccuracy is present.
4. Choice of characters. In some disciplines this has produced heated debate. Smith (1965) called numerical taxonomists existentialists because they will use any character for which data are available. Other taxonomists would rather choose some subset of the total available characters according to some criterion such as alleged phylogenetic importance, relative closeness to the DNA level (protein characters versus morphological character), or discriminating value. Actually, numerical taxonomists also choose characters but they do not add the additional criteria above, so they have a greater number available for use.

   What criteria are used to select characters besides those in the definition given earlier? The characters must be measurable with the available resources, be present in at least a majority of the objects, and be germane to the purpose of the study.

5. Choice of parameters of the characters. Two parameters, the mean and the variance (or the standard deviation), appear most frequently in character variation analyses. Means are estimates of central tendency, or of location of the sample, while the variance is an estimate of the degree of variation about the point of central tendency. Use and comparison of means should be only a first step in the analysis. Comparison of the degree of variation among samples, using the variance or any other such estimate, should allow the analyst to formulate and to try to answer a completely different set of meaningful character variation questions.

   For example, geographic variation studies using the mean demonstrate clinal patterns. This is explained by the different forces present from one end of the transect to the other. But what if the variance also shows a cline? What if the peak of the variance cline is geographically different from that of the mean? What would this mean to such biologically important characters as selective pressure? Population and ecological geneticists have used the variance as an essential part and tool of their research analysis. Workers in systematics should at least consider using the variance as they have the mean. The next step is to generalize and ask what other parameters, statistical or not, could be used with value in a character variation analysis.
6. Choice of character states. If a character is to be measured, what set of values should be used to record the state of the character in a given individual? For continuous characters such as leaf length, it might be the nearest millimeter or the nearest centimeter. More difficult decisions involve qualitative attributes, such as leaf shape. Here, an investigator's operational procedure would be simply to recognize those patterns as distinct that best serve the purpose of his analysis. This rule sounds terribly subjective and unscientific, but, to determine character states for qualitative characters, no formal decision function exists that considers one's material and purpose any better.
7. Choice of character state codes. This is a practical procedure that is becoming increasingly important. As an example, suppose samples are collected on different days over several years, as are the millions of plant specimens in our herbaria. Date of collection is a valuable character for analyses of phenology and should be considered as a possible source of confounding variation. Assuming that everyone records the day, month, and year, should it be recorded as 20 November, 1972, or November 20, 1972, or 11/20/72, or 20/11/72? It is relatively unimportant as long as one is working in isolation. But if you wish to swap data banks with a colleague, other programming may be necessary. Still the problem is not serious. However, consider the automated aspects of the proposed Flora North America Program. Its data bank would easily number in the millions of pieces of information. Character state code compatibility then takes on an expensive and time-consuming aspect. By recording the date as year, month, day, with two digits for each, the computer could arrange them in chronological order by simply sorting all six columns at once, rather than having to make several passes.
8. Filling in the basic data matrix. Once all of the necessary decisions have been made in the above substages, the character by object basic data matrix can be filled in. As understood by numerical taxonomists, this is a two dimensional table, with one reading per character per object. I think it valuable, however, to consider expansion into higher dimensions by considering more than one parameter. Although

mentioned above in a different context, the immediate value of using a three-dimensional basic data matrix would be in a better description of the objects and a more reliable estimate of their phenetic relationship. Similarly, the more parameters available, the better the basic data matrix should be for systems models such as pollination simulations.

Expansion of the basic data matrix into other dimensions also could provide new data on the same objects but at different times and could accommodate data for the same objects but from different places. The latter is a function of the scope of the objects. If they were specimens or site samples, they would be irrelevant. But if for some reason one chose species as objects, then they would be germane. This is an important concept in the formation of basic data matrices. A basic data matrix based on a given number of characters and objects can always be reduced in dimensions and in the magnitude of any dimension (e.g., the number of columns can be reduced). Either type of reduction can occur in at least two ways: either by simple deletion, or by combining according to some function or qualitative rule. For example, consider a numerical taxonomic study where the objects are individual site collections of one species and information is present on the mean, variance, and range for many sites of the species over several years. For a certain analysis, all of this information for each species may be expressed as a simple number, resulting in a simple two-dimensional character by species basic data matrix. The point to understand is that such condensation of the matrix is possible and should be useful for given purposes. To paraphrase, we can change the structure of the basic data matrix, but this change is always in one direction. We cannot begin with a character by species matrix and obtain a character by site matrix with real data. (We could generate random sites if the mean and variance vectors were available for each species, but this presupposes earlier information.) So, the analyst of character variation must be sure that the original objects and characters are sufficiently homogeneous to answer all of the questions and purposes for which they are intended.

After the objects, such as local populations of a species, are chosen, the design of experiments must be considered before sampling and actual data accumulation begins. Books on designs involving analysis of variance are available; e.g., Cochran and Cox (1957). Unfortunately, designs for other statistical procedures are rarely discussed. For example, one can read how to carry out a general systems analysis, but details often are learned only by trial and error.

After a design is chosen, sampling designs must be carried out. Most character variation analyses require sampling at several levels of organization, and different sampling procedures may be required at different levels of the sampling hierarchy. For example, to measure gene flow within a flowering-plant species by pollen-tagging experiments, one first chooses the local populations to be investigated, perhaps by a stratified sampling design. Within each site, choice of individual plants to monitor may be by random sampling.

Probably no extensive basic data matrix will ever be free from the problems of missing data. Even when each cell is represented by at least one value, some will be more reliable than others. Few studies in any area of systematics have evaluated the effect of missing data. In numerical taxonomy, Crovello (1968b) compared estimates of similarity based on a 43 character by 30 species matrix of California willows. In one analysis, a comparison was made between a full data matrix and the same matrix from which data had been removed in the same pattern as shown by Munz's flora. The OTU by OTU relevance of the latter was only 48 percent. Yet the resultant species by species similarity matrices were highly correlated (0.887) and did not differ significantly in their means and variances. This suggests that taxonomic data matrices with missing data still are valuable.

9. Verification of the basic data matrix. Whether the data are obtained from experiments by the analyst himself, from some data bank, or from the literature, some degree of effort is necessary to verify the validity of the values of the resulting matrix. This may range from simple verification of experimental results, which have been punched on tab cards or recorded in code on magnetic tape, to extensive checking of the reliability of data obtained from other sources.

C. Mathematical Analysis and Summarization

Methods used in mathematical analysis are not the same in all phases of systematics. Methods vary most between descriptive stages (taxonomy) and explanatory stages (studying the dynamics of populations or the processes of evolution).

Table 22-2 contains an expanded listing of methods of mathematical analysis and summarization that have been used or have potential in character variation analyses. They are grouped into somewhat artificial categories. Other groupings are possible, but they are probably less satisfactory than the one chosen. I am unable to produce an absolutely satisfactory grouping because the subfields of mathematics are related in a complex pattern. Even the simple two-level hierarchy of Table 22-2 causes distortions in relationship among the entries.

Even a brief sketch of each of the basic entries of Table 22-2 would require the space of a book, not an article. Thus, in the following paragraphs I shall not consider items whose use is obvious and agreed upon. References in Tables 22-2 and 22-3 were selected if they were recent, comprehensive, unique, botanical, or readily available. They are not meant to be complete. Rather, they should serve as examples of recent uses of the technique and as basic guides to the literature. Extensive bibliographies of a broad field like character variation analysis in systematics must await computerized information retrieval. References to major reviews and introductions appear next to the titles of the major categories. Sometimes the reader may have to cross into other fields to find a suitable treatment. Within each category the number of references will vary. Some readers may find it interesting to compare these tables with their analogs in Crovello (1970). Those in the 1970 paper have more items and more uses, indicating perhaps that zoological systematists are further into quantitative character analysis than are botanists.

TABLE 22-2. Methods of Mathematical Analysis and Summarization

A. Statistics (Blackith and Reyment, 1971; Sokal and Rohlf, 1969)

Transformations	Multiple regression
Basic statistics	General polynomial regression and trend surface analysis
Nonparametric methods	Analysis of variance
Analysis of frequencies	Analysis of covariance
Linear correlation	Multivariate analysis of variance
Curvilinear correlation	Evolutionary and population allometry
Multiple correlation	
Linear regression	

B. Similarity estimation (Blackith and Reyment, 1971; Mayr, 1965; Sneath and Sokal, 1973)

Association and correlation coefficients: Anderson, 1949; Rhodes, et al., 1968)
Distance coefficients (Crovello, 1968a, 1969)
Information theory (Hawksworth, et al., 1968)
Cladistics (Dayhoff, 1969; Fitch and Margoliash, 1967; Kluge and Farris, 1969)

C. Description of structure in a reduced number of dimensions (Cole, 1969; Morrison, 1967; Sneath and Sokal, 1973; Williams and Dale, 1965)

Visual rearrangement of a matrix:
- Character by object basic data matrix
- Similarity or distance matrices (Figure 22-1)

Principal component analysis (Prance et al., 1969)
Factor analysis
Canonical analysis
Multidimensional scaling (Blackith and Reyment, 1971)
Cluster analysis (Jackson and Crovello, 1971)
Hyperfootballs (Reyment, 1969)
Character analysis (Crovello, 1968d)

D. Decisions and optimization (Blackith and Reyment, 1971)

Decision theory, pattern recognition:
- Discriminant functions (Burnaby, 1966; Goodman, 1967)
- Decision trees
- Taxonomic keys (Bossert, 1969; Goodall, 1968; Hall, 1970; Morse et al., 1968)

Optimization theory (Rosen, 1967; Ruben, 1967)

E. Geometry (Busacker and Saaty, 1965)

Euclidean geometry
Non-Euclidean geometry
Graph theory and networks (Gabriel and Sokal, 1969; Moss, 1967; Prance et al., 1969; Wirth et al., 1966)
Topology (Sneath, 1967; Thompson, 1942)

1. Statistics. Today statistics enjoys wide acceptance among systematists. The concurrent development of digital computers and of improved methods of automated data accumulation are causing an even wider use of statistics and of more involved statistical methods. Interestingly, parametric statistics as we know it may soon be succeeded by the methods of data analysis (Sokal and Rohlf, 1969: Chap. 17). To give one example, several years ago no character variation analyst with extensive data on 10 characters would construct the 45 scatter diagrams of each pair, let alone repeat the procedure using several transformations. Today, even a second-generation computer can complete the task in minutes.

   To use statistics intelligently in plant systematics requires serious study. Fortunately several clear texts are available which contain more than "cookbook" procedures. Except for basic descriptive and inferential statistics, plant systematists have not used more advanced techniques of Category A. This is unfortunate because such methods can provide fresh insights into complex systems, and plant population series are complex.

   The final item in Category A of Table 22-2 is Allometry. Perhaps this is out of place here, but it is difficult to fit it anywhere, since it is not a method but an area of systematics. Yet it combines mathematical methods in such a way and has such great promise for the future that it deserves mention.

2. Similarity Estimation. This has been used mostly by taxonomic workers, but it has great potential for evolutionary systems analysts whenever they wish to summarize relationships among many characters, objects, or components of a system.

   The normal use of similarity estimation begins with a character by object basic data matrix. Some coefficient, or a distance measure in a chosen metric, is calculated between all pairs of objects and results in an object similarity (or distance) matrix. In taxonomy each object (or OTU) is a taxon, or even a local population or an organism. While such similarity estimates are of great value in summarizing relationships among objects, they are averages or summations of some sort. This means that the particular characters in which two objects differ are not revealed, nor is there any indication as to which pairs of objects differ only slightly in many characters and which pairs differ considerably, but in only a few characters.

3. Description of Structure in a Reduced Number of Dimensions. This approach attempts to summarize relationships among objects in fewer than the original number of the smaller of the two dimensions of the basic data matrix. For example, in a study of 50 characters and 20 objects, the various methods listed in Category C of Table 22-2 would seek to minimize distortion of the relationships in a new coordinate system with less than 20 independent dimensions (Gower, 1966).

   Choice of a particular method from Category C is important since different methods will give somewhat different estimates of relationships among the objects. Usually, an object similarity matrix obtained from some method of Category B, a character by character correlation, or a variance-covariance matrix is required for analysis by a method of Category C. Exceptions include those studies that simply rearrange the character by object basic data matrix.

   We can group the methods of Category C of Table 22-2 by several criteria. The methods easily separate into those that reduce the

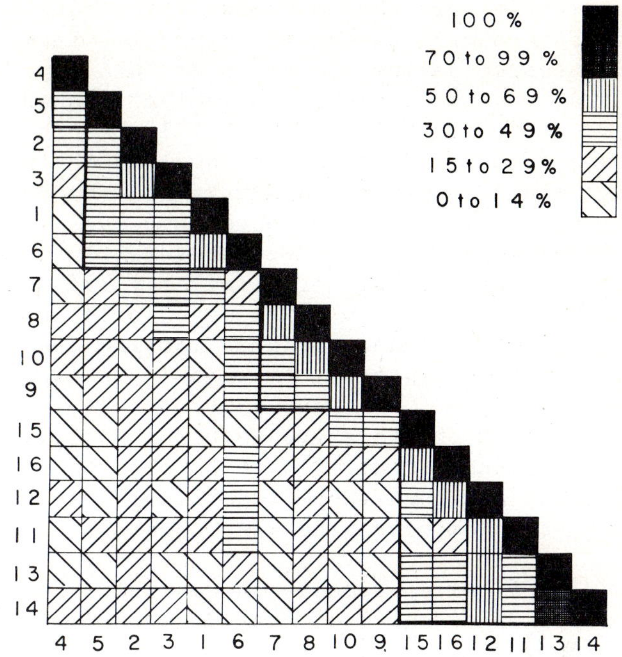

Figure 22-1. Similarity matrix shaded to show the percent of similarity between pairs of 16 OTU's. [Walraven, W. C. 1970. A statistical analysis of sixteen taxa of Rhynchosia (Leguminosae) in the United States. Brittonia 22: 85-92. Used with permission.]

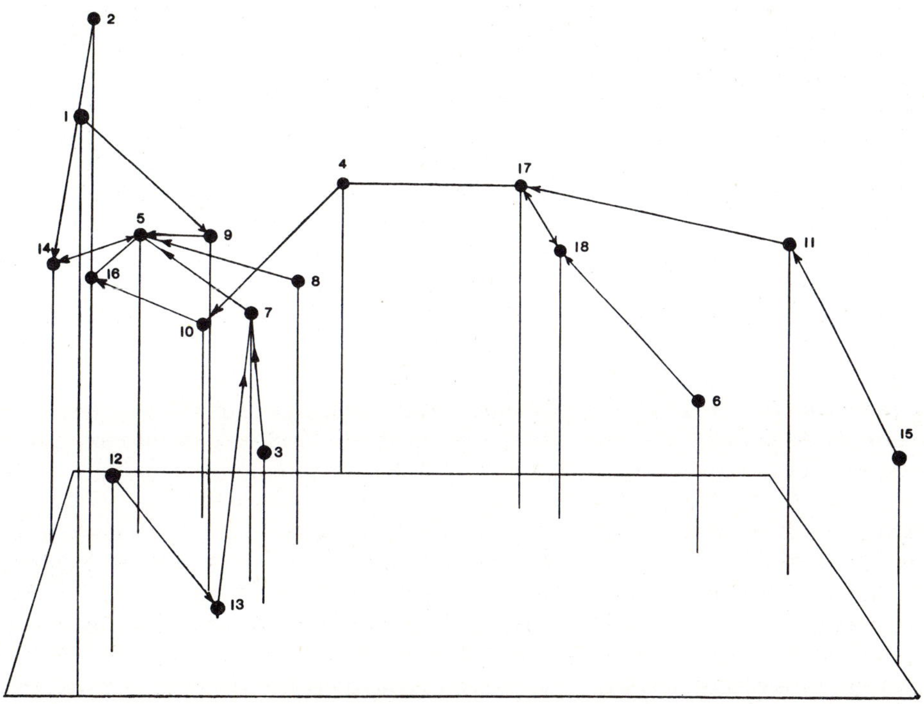

Figure 22-2. 3-dimensional graph and PRIM network based on 29 characters. [From Jackson, R. C., and T. J. Crovello. 1971. A comparison of numerical and biosystematic studies in Haplopappus. Brittonia 23: 54-70. Used with permission.]

original high number of dimensions to one meaningful dimension, and those that retain more than one meaningful dimension. Many phenograms, cladograms, and other simple graphs fall into the former category. Cluster analysis is a term used to define one group of such approaches (Sokal and Sneath, 1963). In contrast, methods such as principal components analysis usually produce more than one meaningful dimension. If we begin with an original character space with a number of dimensions equal to the number of characters, the result of a principal components analysis will be a new set of axes in the same character space, but with a reduced number of independent axes along which significant variation occurs. We can refer to the space of this new set of axes as reduced character space. Similarly, if we carry out a principal components, or other, analysis on an object similarity table (through the derived correlation or variance-covariance matrix) the reduced space can be referred to as reduced object space.

Results of the above types of analysis may yield a phenogram, cladogram, or set of reduced axes in which objects can be positioned. Usually, such summaries are then inspected visually and, from the observed pattern of variation, taxonomic or other decisions are made.

Two final approaches in Category C are of interest because their purpose is often different from that of numerical taxonomy. Reyment (1969) has championed methods that analyze variation and covariation around some multivariate point or points of central tendency. Consider a bivariate scatter diagram containing a set of observed points. Using a multivariate test for the location of the multivariate parametric mean, we can construct an equal frequency confidence region at a given probability level (Sokal and Rohlf, 1969). It will be elliptical in shape. With three or more characters, this ellipse becomes a sphere or hypersphere. Staying with two-dimensional ellipses for simplicity, imagine the 95% confidence regions (our ellipses) of two populations. The chances are great that the origin of each ellipse will be located at a different point in this two-dimensional character space, that the size of each ellipse will differ, and that the orientation of the major axis will differ. The important point is to discover what this means at the biological level. For example, if the ellipses from two populations differ in size but not in orientation, does this mean there are similar selective forces at different intensities? Or, if the size is the same but the ellipses are oriented in different directions, does this mean there are different selective forces working at the same intensity? Other combinations are possible, even in two dimensions, but the possibilities increase as the number of characters (or perhaps principal axes, etc.) increase.

Inverse or character analysis is a term that refers to studies similar to the above but inverted; i.e., the main concern is not to study relationships among objects but among characters. Many uses of factor analysis in systematics obtain this as an intermediate by-product of normal analysis. But whereas the taxonomist is interested in object relationships, when he changes hats and becomes an evolutionist, this previous by-product can be of prime value to him in gaining insights into the causes of the pattern of variation previously described. For example, why do characters cluster? One reason is that they have been selected over time to meet an environmental selective force. As a by-product, then, one should be able to see the salient adaptive complexes of characters at a fairly efficient rate. A character correlation complex actually may consist of a minimum of one adaptive complex, but if two environmental selective forces are correlated in their intensities over several sites, then the evolutionist's task is to assign the characters in a cluster to their proper selective force (Blackith, 1960; Crovello, 1968d). The potential of this method for inter-taxon comparisons is immense.

This is a good time to consider the misuse of some of the methods of Category C of Table 22-2. Seal (1964) states several times in his text that numerical taxonomists particularly have misused such methods as principal components analysis. His argument is that the assumptions of this technique require that the characters' values, the basic data matrix, be sampled from one multivariate normally distributed population. Clearly, in a taxonomic study involving many different species and some discontinuous and non-normally distributed characters, this does not hold. Similarly, Seal and others criticize the use of a distance coefficient that does not weight each character according to its correlation with other characters. Most coefficients in use in numerical taxonomy do not consider relative correlation. Mahalonobis distance, D^2, is suggested as being more valuable, since it takes correlation into account, as well as the covariation within groups.

Given such criticisms on mathematical statistical grounds, how can anyone decide to use principal components analysis in a study of many species or to use a distance coefficient that does not consider correlations among characters? I think this decision in the multistage decision process in systematics is largely a function of the purpose of the character variation analysis. Seal's criticism of the use of principal components analysis in systematics is valid if the worker wants to indulge in statistical inferences about the results (e.g., if he desires to establish probabilistic frequency ellipses about each object or each character). But much will remain outside the realm of parametric statistics since it seems easier to use randomization tests and other nonparametric methods with the computer to determine probability levels. If we see principal components study as another way of looking at our systematic system and if it gives biological insight at a high rate, then I maintain that its use is justified. Of course, one does not use principal components analysis alone. It is complemented by phenograms, which are most accurate at the tips (versus principal components results, which are better in expressing relationships among clusters of objects), and by networks superimposed on the first three principal components axes.

The use of distance coefficients that do not consider character correlations should depend on one's biological purpose, not on whether there exists a readily available repertoire of significance tests. Perhaps because of training, most systematists think about relationships, including distances, in a Euclidean space, where the Phythagorean theorem holds true. They also think about one character per dimension. Now, if one's purpose is to estimate the distance between objects in the context of a given set of characters, then a distance measure that does not take correlations into account may be more useful. Of course then there is a large amount of redundancy of information in terms of information theory, but for some purposes, such redundancies are important.

I suggest the following philosophy for many stages of our character variation decision process: for any stage with two or more alternative methods, always use as many methods as possible. My reason is that every technique provides some new, non-redundant information. If method X gives five units of biological insight into the study and Y gives one unit, but a different one, I would be an inefficient scientist if I ignored Y. So, in the case discussed above, my decision would be to calculate D^2, as well as a distance coefficient that does not take correlations into account. In general, there has been too little comparison of different methods, especially statistical comparisons (as opposed to visual examination of phenograms, etc.). Work that has been done (e.g., Crovello, 1968c, 1969; Gower, 1967; Moss, 1967; Prance et al., 1969) indicates that changes in the index or the clustering method used do produce different, but not unrelated, results. These results display roughly the same relationships. For example, several available studies analyzed data via a character factor analysis and by the use of an unweighted distance coefficient (Crovello, 1968c; Moss, 1967). Then if one also calculates the unweighted distance between objects even over just the first three factors, the resulting distance matrix is highly correlated with that using distances based on all of the characters. Using 131 morphological characters, Crovello (1968c) obtained correlation values of .842 to .870 in a study of 30 taxospecies of willow.

4. Decisions and Optimization. These are a concern of systematists interested in description or in modeling an evolutionary system. The particular decision or optimization methods chosen themselves constitute a decision process. Decisions are necessary in taxonomy in the formal delimitation of taxa (classification) and, later, as a separate procedure in identification of unknown specimens. There exist few rules on the classification of organisms--which may be a good thing! Most taxonomists will recognize two sets of organisms as distinct if the members of each are relatively close to each other with respect to their position in a character space and are relatively removed from the members of the other set. In other words, a set of organisms is put into one class if the within-group variance is relatively low and the among-groups variance is relatively high. Numerical taxonomy, especially cluster analysis and methods to discover relationships in reduced character or object space, is a valuable aid in gaining insights into the differential density of the set of objects in the original character space and in deciding on class boundaries.

Every human being is a taxonomist. Most do not realize it and others do not care to admit it. Every day people group objects and even other people into clusters and make decisions based on it. During the last few years we have come to realize that workers in diverse fields who use computers to build sounder classifications have created methods that either are very similar or are different but very applicable in other fields. Thus it is refreshing to attend the annual meeting of the interdisciplinary Classification Society and hear about the methods used to cluster United Airlines passengers, languages, countries and even breakfast cereals! Such meetings certainly stimulate us to get out of our normal thought-ruts.

Once classes have been decided upon, subsequent organisms must be assigned to one of the classes. This is the interest of the interdisciplinary Pattern Recognition Society.

Discriminant functions (Blackith and Reyment, 1971; Seal, 1964) were developed by Fisher in the 1930's when Edgar Anderson presented him with the problem of how best to discriminate between species of

Iris. The procedure assigns characters differential weights according to their relative ability to discriminate among classes. That is, given a large basic data matrix with objects partitioned into classes, the procedure locates the optimal characters to use in identification. Anderson's (1949) hybrid index is a crude method in terms of statistics, but as Goodman (1967) found recently, for given situations it may be just as useful as discriminant functions. Just why the hybrid index and its complementary technique of the pictorialized bivariate scatter diagram are sufficient to separate closely related classes is an interesting biological question in terms of the minimum number of characters necessary to separate sets of organisms that occupy different niches. Of course, if the number of characters or classes increases above 10, or 3, respectively, Anderson's methods are difficult to use.

5. Geometry. Many relationships can better be described and analyzed by geometric methods, although algebraic representation of the same system is usually possible. Graph theory in general and networks in particular are receiving increased attention as activities in numerical taxonomy increase. A graph is a geometric structure consisting of points (vertices) in space connected by a system of curves (edges). A *directed* graph is one in which every edge has been oriented, or given a direction. An *undirected* graph lacks such orientation. Finally, a graph is *open* if one cannot return to a vertex without using at least one edge twice. In contrast, a *closed* graph is one wherein one can return to at least one vertex without traversing any edge more than once. Readers may expand their knowledge of graph theory beyond these simplified definitions by reference to Busacker and Saaty (1965). In the above terms most evolutionary crossing polygons are closed, undirected graphs; most phenograms and cladograms (without reticulate evolution) are open, directed graphs, and some nearest-neighbor graphs are closed, undirected graphs. A network is a directed graph without any loops. A shortest spanning tree is any shortest possible network connecting a set of points. Table 22-3 includes references to each of the four types of graph. Why try to express relationships in different formal systems, such as graph or set theory? One reason is that the simple representation in the new format is efficient. It also provides a valuable insight into otherwise abstruse relationships.

Another area of geometry that should receive more attention is topology, or "rubber sheet geometry." D'Arcy Thompson (1942) demonstrated its value to systematics when, by regular transformations of points on a grid placed over a drawing of one species of fish, for example, he could produce exactly the shape of another species of fish. Recently Sneath (1967) has expanded this work in two directions by considering three dimensions simultaneously instead of two, and by avoiding the use of an arbitrary standard "species" like Thompson's. The availability of digital computers and incremental plotters eliminates the need for selecting a standard. All combinations can be tried quickly. Coupled with studies of allometry, this technique is just waiting to be used by plant systematists to help to understand what it takes developmentally to get from one plant habit to another, or a flower or leaf shape of one species to that of another.

Computers are responsible for the recent rise of activity in multivariate character variation. Many of the studies referenced in this article would be impossible without them. Sterling and Pollack (1968) provide a good introduction to the digital computer and describe some of its uses. In systematics the digital computer is being used in two major ways: to carry out calculations and then, often graphically, to summarize the results; they are also used to retrieve information from large data banks such as described earlier under data accumulation. Analog computers are the other main type of computer. These are specially designed to solve systems of differential equations. Their major advantage over digital computers is that the user can observe the effect of change of parameters on the system simply by changing the value of any of several potentiometers. Choice of the type of computer to use is a function of the model itself, of the purpose, of the amount of data, and other considerations. Hybrid computers, combinations of digital and analog computers, are available. Often, qualitative scanning of results of a systems analysis is made on an analog computer and then, when more accuracy is desired for a specific combination of parameter values, the digital computer is employed.

D. Summarization Graphics

Table 22-2 was difficult enough to present, but Table 22-3, containing the types of summarization graphics used in character variation analysis, is even more so. Ideally, each type listed in Table 22-3 should also be represented by a figure in the text, but the cost and space needed would be prohibitive. Several examples are given, however, both here and in chapter 21. The references provided next to each item in Table 22-3 were chosen primarily from publications that should be on the shelves of any systematist, or in his institution's library. Every entry is given a reference because all entries could be used by plant systematists. Also, unless one looks at an actual figure, it is difficult to understand its classification as to type and also how it differs from others.

1. Frequency Distributions. Frequency distributions (category A of Table 22-3) are among the basic tools of any systematist. They are graphs whose response variable is either the relative or absolute frequency of the different states of the characters on one or more other axes. For convenience they can be separated primarily into univariate, bivariate, and multivariate frequency distributions. Univariate distributions are further recognized in Table 22-3 as discontinuous or continuous, depending on the nature of the character being analyzed. The most common type of discontinuous, univariate frequency distribution is the bar diagram.

When we turn to continuous univariate frequency distributions, histograms and frequency polygons are most familiar. A frequency polygon is a frequency distribution with the frequency of each recognized state plotted as a point on the graph and the set of such points connected to its nearest neighbor on each side by a straight line or by a higher degree curve. Types of frequency polygons important in systematics include relative frequency distributions (probability density functions), frequency distributions of the states of an index or other compound character, such as an environmental gradient or discriminant function.

A Dice-gram is a type of frequency distribution that has been compressed into one dimension or otherwise topologically transformed. Sokal (1965) has commented on the visual bias involved in such graphs.

TABLE 22-3. Summarization Graphics*

A. Frequency Distributions

Univariate frequency distributions:

Discontinuous:
- Bar diagrams (Davis and Heywood, 1963: 326)
- Sets of horizontal lines of uniform thickness, one per species (MacArthur and Connell, 1966: 167)

Continuous:
- Histograms (Mayr, 1963: 188; Sokal and Sneath, 1963: 214)
- Frequency polygons (Sokal and Rohlf, 1969: 30)
- Sets of shaded frequency distributions (Odum, 1956: 152)

Bivariate frequency distributions:

- With isolines, but flat (Seal, 1964: 102)
- With different-sized circles, but flat (Sokal and Sneath, 1963: 214)
- With isolines, but in perspective (Sokal and Rohlf, 1969: 501)

Multivariate frequency diagrams

B. Scatter Diagrams

Unreduced axes:

Bivariate scatter diagrams:
- Plain (Mayr, 1969: 172)
- Differential shading (MacArthur and Connell, 1966: 154)
- Equal frequency ellipses (Sokal and Rohlf, 1969: 527)
- Regression lines and other curves (Sokal and Rohlf, 1969)
- Isolines (Grieg-Smith, 1965: 193)
- Pictorialized scatter diagrams (Davis and Heywood, 1963: 477)
- Illustrated scatter diagrams (Smith, 1966: 84)
- Directed networks (Allee et al., 1949: 208)

Trivariate scatter diagrams:
- Bar diagrams (McKell et al., 1969)
- Regression and other response surfaces (MacArthur and Connell, 1966: 161)
- Triangular graphs (MacArthur and Connell, 1966: figs. 1-16)
- Perspective drawings (Figure 22-2)

Reduced axes:

Reduced character space axes:
- Principal components axes (Figure 22-2)
- Discriminant functions (DuPraw, 1965)

TABLE 22-3 (Continued)

Canonical axes (Oxnard, 1969)
Stereograms (Rohlf, 1968)

Reduced character space axes and networks (Figure 22-2)

Reduced object space axes (Crovello, 1968c)

Character space and reduced character space combined (Davis and Heywood, 1963: 66)

C. Directed graphs

Open

Branched (dendograms, trees):

Taxonomic phenograms:
Two-dimensional (Figure 22-3)
Three-dimensional (Waterman and Morowitz, 1965: 270)

Cladograms:
Plain (Dayhoff, 1969; Morishima, 1969)
Illustrated (Figure 22-3)
Wagner trees (Figure 22-3)
Cladochronograms
Typological diagram (Mayr, 1963: 599)

Phylograms:
Two-dimensional (Eaton, 1970)
Three-dimensional (Sokal and Sneath: 234)
Illustrated (Davis and Heywood, 1963: 70)
Character cladograms (Davis and Heywood: 189)

Closed:

Flow charts (Sokal and Crovello, 1970)

Some nearest-neighbor taxonomic graphs (Figure 2; Wirth et al., 1966)

Directed crossing experiments (Figure 22-4)

D. Undirected graphs

Open: Some nearest-neighbor taxonomic graphs

Closed (Gabriel and Sokal, 1969)

TABLE 22-3 (Continued)

Some nearest-neighbor taxonomic graphs (Prance et al., 1969)
Crossing experiment or pollen fertility polygons (Figure 22-4)
Character correlation polygons (Clausen, 1967)
Polygram (Davis and Heywood, 1963: 331)
Polygonal graphs, Polygraphs (Figures 21-8 and 21-10)
E. Geographic maps
Profile diagrams:
Plain (MacArthur and Connell, 1966: 76)
With frequency distribution (Lewis and Taylor, 1967: 223)
With glyphs (Kendeigh, 1961: 28)
Uniform, discrete symbols by computer (Perring and Walters, 1962)
Uniform symbols of different size (Mayr, 1963: 263)
Different symbols (Mayr, 1963: 344)
Glyphs (Figure 22-6)
Actual character values (MacArthur and Connell, 1966: 53)
Isophenes (Adams, 1970)
Dice-grams (Sokal, 1965: 355)
Multiple comparison homogeneous lines (Sokal, 1965: 358)
Undirected networks (Figure 22-7)
Directed networks; cladograms (Figure 22-8)
Bar diagrams (Figure 22-9)
Pie diagrams (Mayr, 1963: 431)
Differential shading (Mayr, 1969: 191)
Character polygons (Figure 22-10)
F. Miscellaneous
Photographs and drawings
Differential shading of cells of a matrix (Figure 22-1)
Pie diagrams (Odum, 1956: 380)
Skyline plots and sequence diagrams (Prance et al., 1969; Wirth et al., 1966)
Contour diagrams (Moss, 1967)

*Parentheses next to each category or item contain references to publications that use the item or explain it. Each reference is separated from the next by a semicolon. If a colon appears, the information immediately to its right refers to specific pages or chapters.

Bivariate frequency distributions are natural extensions of the above and are most frequently presented in two dimensions with changes in frequencies indicated by isolines or by different-sized circles. With the growing availability of computer graphics software packages (e.g., Rohlf, 1969), we may expect perspective drawings to become more common in systematics. Multivariate frequency distributions are put in as a separate category only to serve as a stimulus to readers to consider the shape that such graphs might take--besides those in two dimensions that have several different scales for the abscissa.

2. Scatter Diagrams. These are graphs in one or more dimensions on which is located a set of points, each indicating the position of an object (Category B of Table 22-3; also see Figure 21-12). The axes may be regular characters or they may be some combination of characters, such as a principal components axis. Within this concept of scatter diagrams, the frequency distributions discussed above are just a particular type of scatter diagram. I separate them only because they are so often used and are such an involved entity in themselves.

   For convenience only, I separate the scatter diagrams listed in Table 22-3 into those involving unreduced (regular characters) axes only and those involving at least one reduced axis. The meaning of the different types and subtypes should be clear from their names and from the examples cited. The only very recent type is a figure that contains reduced character (or object) space axes and networks. For example, consider a styrofoam model or a perspective drawing that displays the positions of a set of objects in the first three axes of a principal components space. Assume the position of each object is represented by a small circle or a sphere. As mentioned earlier, this gives the best (according to the least-squares criterion) three-dimensional view of the relationships among the objects. But to be completely accurate, we really need more than three dimensions. One way to see what would happen to the object relationships if we could see further into the hyperspace is to draw a directed line on the drawing from each object to its nearest neighbor in the hyperspace. This nearest-neighbor information is available from the object by object similarity or distance matrix calculated earlier. Rohlf (1970) projected the shortest spanning tree down to three dimensions of a principal components analysis. I feel strongly that for many purposes in systematics, this is the best possible summary of descriptive relationships in a multidimensional character space especially if the graph ends are directed (Figure 22-2). Naturally one could add a second network of directed edges (using dashes) to indicate the second closest neighbor, or perhaps the farthest removed. The only problem is that as successive networks are added, the figure becomes harder to comprehend.

3. Directed Graphs. Categories C and D of Table 22-3 more fully consider graphs and networks. As defined in the discussion of Table 22-2, graphs can be separated into open or closed and directed or undirected graphs. Busacker and Saaty (1965) reserve the term network for open, directed graphs, but the use of the term is not uniform. Unlike Mayr (1965) I retain the term dendrogram as a general term to mean any treelike diagram. Busacker and Saaty (1965) define a tree as a graph that is connected and has no circuits. A tree would include our open, directed and open, undirected graphs. To discover methods to construct

graphs that require involved calculations, such as phenograms and cladograms, consult the references of Table 22-2, Sections C and D.

Category C presents uses of directed graphs in character variation analysis. The first type of open, directed graphs are dendrograms. Most dendrograms listed in Table 22-3 do not have explicit directed edges. But they are implied either by direction of descent, as in cladograms, or by the property of inclusion, as in phenograms. Undirected cladograms would belong in Category D of Table 22-3, but phylogenetic systematists usually are willing to give directions. Whether they are always warranted is a different matter.

Phenograms are directed graphs that reflect phenotypic relationships. The result of cluster analysis, they may be of two or more dimensions. They may be directly derived from an object-similarity or distance matrix which is based on a character by object basic data matrix. Or they may be derived from a similarity or distance matrix from reduced character or object space.

Cladograms, graphs depicting the branching sequences of evolution, have been a major systematic tool for years. Cladograms can be drawn in two dimensions, since the only property they reflect is the branching sequence of evolution. Over the last five years many methods of constructing them by computer have been introduced. At regular intervals Dayhoff (1969) recomputes cladograms on the basis of the expanding data bank of known macromolecular sequences. Cladochronograms are included for completeness to emphasize that if one also has accurate time estimates of the branching events, he no longer has just a cladogram. A typological diagram is a peculiar cladogram which shows series of changes over time but without branching.

The term phylogram is reserved only for those graphs that summarize information on the cladistics, chronistics, and phenetics of the group under study. If information on any of these essential parts is missing, then the graph is not a phylogram, but a cladogram or phenogram. Some phylograms are naked, in the sense that only the extant objects are known. Others are full, in that taxa at branch points in the past are known in detail. Paradoxically, naked phylograms are much easier to fit into the formal taxonomic system than full phylograms. With naked phylograms we are not required to make arbitrary decisions to determine with which branch the ancestral types should be placed when we are delimiting formal taxa. The taxonomic hierarchy is characterized by the property of inclusion of subsets within sets. Thus, species are in a genus, genera in a family, etc. But phylograms are characterized by taxa of the same category at each branch point as, for example, in the well-known horse sequence, where *Parahippus* supposedly gave rise to *Merychippus* and it gave rise to *Hipparion*, *Nannippus*, and so forth (Eaton, 1970). I cannot understand why systematists wish to distort supposed phylogenetic relations by making them fit into a taxonomic hierarchy that is defined by the principle of inclusion. It seems more sensible to present two separate graphs, one a taxonomic tree (phenetics) and the other a phylogram or a cladogram. Then one can see where they disagree and can ask the important evolutionary question, why do they disagree?

What about reticulate evolution? The definition of cladogram given above is based on open branching processes and does not consider fusing processes which are thought to be common events at the microevolutionary level. Algorithms to construct cladograms and other closed, directed graphs at the species level and below should be a fruitful field of future research.

A character cladogram is a graph that depicts successive changes in the states of a character, or of a suite of characters, over evolutionary time. Examples include the change of angiosperm ovary and placentation types (Lawrence, 1951: 74,76) or of amino acid sequences in a protein. Much can be learned about the processes of evolution by the construction of character cladograms. Since only one or a few characters are considered in a character cladogram, the problem of possible reversals in the evolution of the character states is more acute. Systematists have neglected allometry and the relation of form and function of a set of adaptively correlated characters.

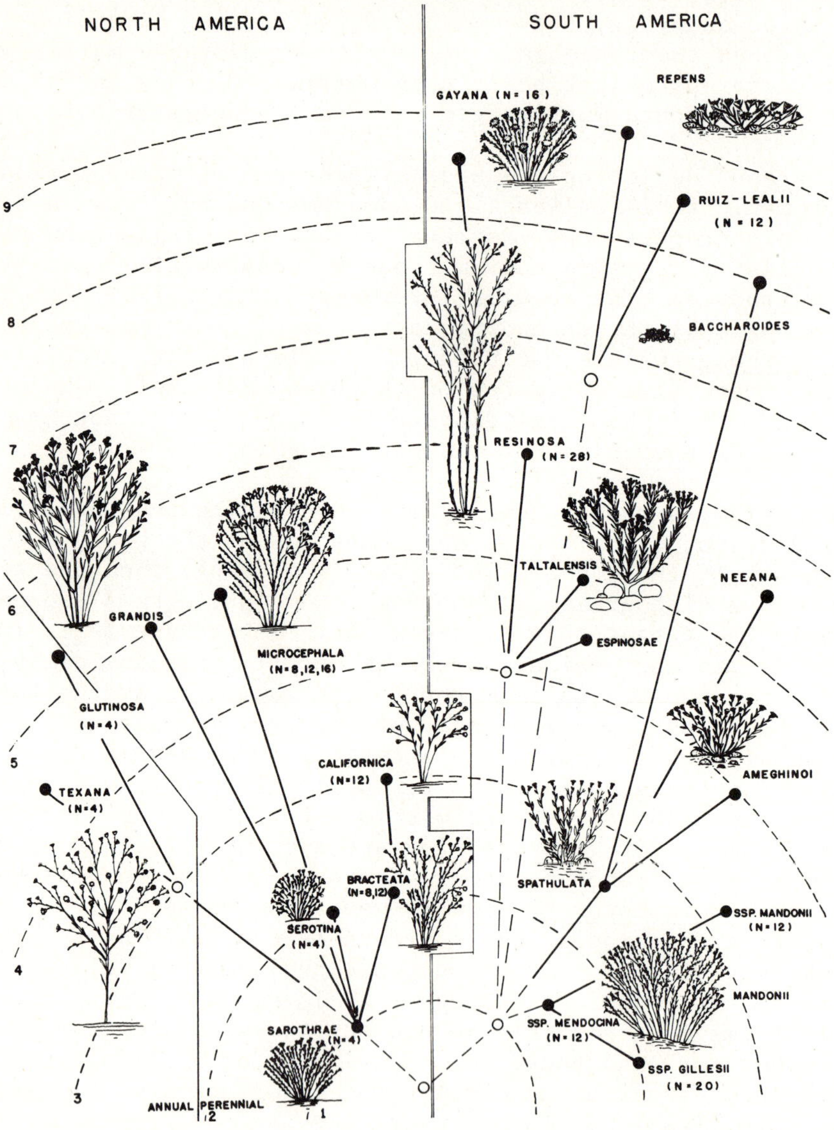

Figure 22-3. Wagner "target" diagram indicating a probable phylogeny for Gutierrezia according to a non-computerized method. Habit of plants drawn to scale; broken lines indicate less well established relationships. [From Solbrig, O. T. 1970. The phylogeny of Gutierrezia. Brittonia 22(3): 217-229. Used with permission.]

Many taxonomic keys used to identify unknown specimens are open, directed graphs.

Flow charts that serve as a preliminary step in complex computer programming are common in systematic studies. The flow chart devised by Sokal and Crovello (1970) to delimit biological species is also a closed, directed graph.

A closed, directed graph sometimes represents the nearest neighbor relationship among a set of objects that a systematist is studying. Undirected graphs are used by some, but this is less informative since a nearest-neighbor line connecting two objects is not necessarily true for both directions. One may be closest to a third object. The results of any directed, crossing experiment can be presented as a closed, directed graph (Figure 22-4). Naturally, a logical refinement of such a graph would be to use different symbols (solid lines, dashes, dotted lines) on one graph to indicate the strength of the crossing relationships, or simply the order of the relationship. Readers should realize that directed graphs have more information than undirected graphs. So where possible, use a directed graph.

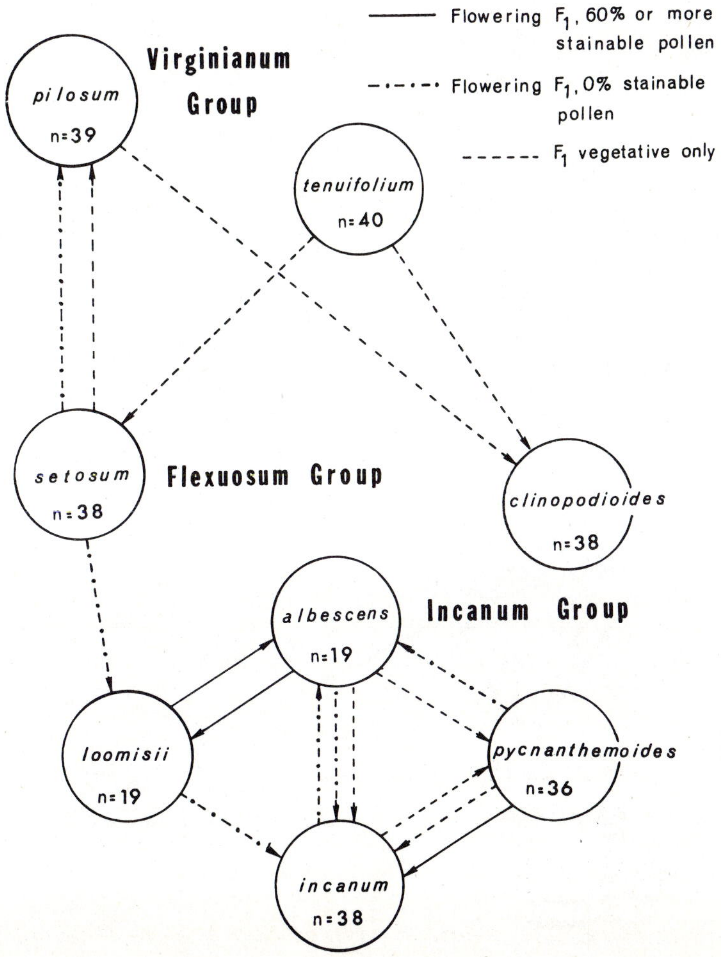

Figure 22-4. Crossing diagram of artificial hybridizations. Arrows point from pollen parent to pistillate parent. [From Chambers, H. L., and K. L. Chambers. 1971. Artificial and natural hybrids in Pycnanthemum (Labiatae). Brittonia 23: 71-88. Used with permission.]

4. Undirected Graphs. Category D of Table 22-3, undirected graphs, also can be separated into open and closed graphs. Both open and closed, undirected networks are exemplified by some nearest-neighbor diagrams. In these and all previous references, the reader should be aware that successive nearest-neighbor graphs are possible and often are quite useful. Character correlation polygons are another example in which sets of lines are added in successive operations. Wirth, Estabrook and Rogers (1966) alter this procedure by showing successive clustering stages.

Polygonal graphs are polygons embedded in a coordinate system from whose origin extend several radii, each calibrated with a different character's character states. The polygon is formed by joining the one state of each character possessed by the object (See Figure 21-10).

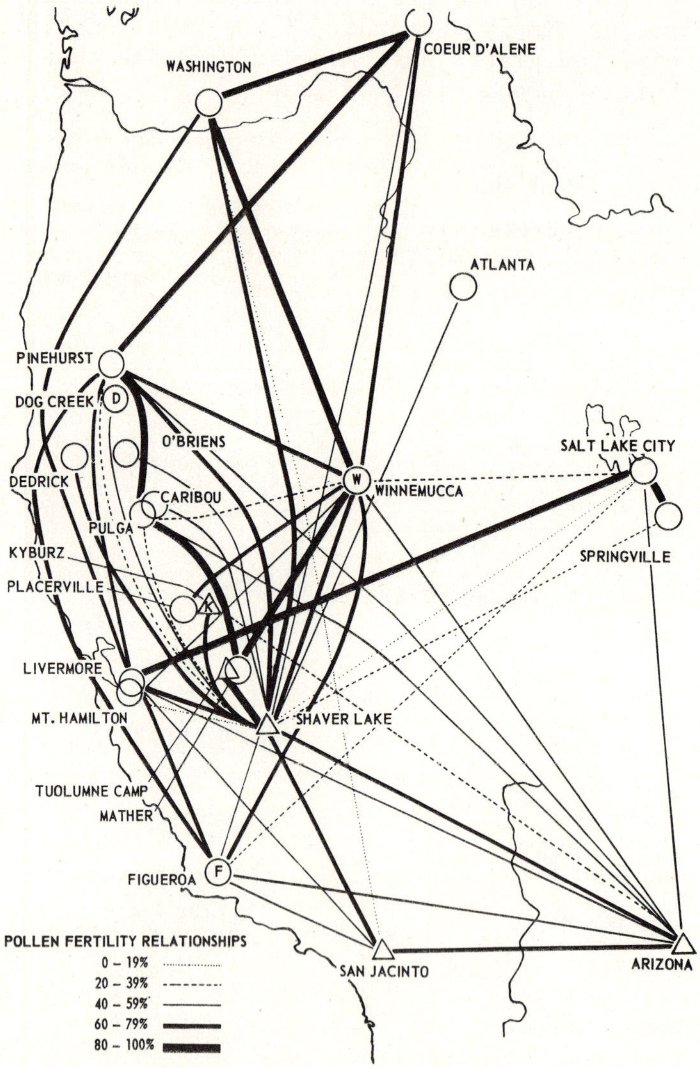

Figure 22-5. Pollen fertility relationships among various populations of Clarkia rhomboidea. [From Mosquin, T. 1966. Toward a more useful taxonomy for chromosomal races. Brittonia 18: 203-214. Used with permission.]

5. Geographic Maps. Category E, geographic maps, is extensive. Fortunately, the meaning of each item is fairly evident and I shall forego reviewing them here. The references given in Table 22-3 should be adequate for understanding.
6. Miscellaneous. The first three entries of Category F of Table 22-3 also are self-explanatory. Skyline plots and sequence diagrams are interesting, different methods clearly presented in the references cited. Finally, Moss's (1967) contour diagrams are quite different from all other methods of depicting relationships among objects. Interested readers should consult his paper for details.

E. Evaluation of Results and Reanalysis

Evaluation of results is a serious step in character variation analysis. First, the researcher ascertains what the results mean to his original purposes and hypotheses. Have they fulfilled and supported, or have they negated, them? Second, at this stage the analyst must integrate his findings with those of others. This is becoming increasingly difficult because of the expanding number of productive workers. Perhaps a computerized data bank will help to alleviate this problem.

Reanalysis, either by the same or by a subsequent investigator, is a necessary endeavor that serves as a constant check on our theories and on the accuracy of our facts. Reanalysis is justified by an intensional or an extensional expansion of the data. Intensional expansion involves collection of more of the same, whether it be more temperature readings from the same place at the same time of year or more individuals from sites already studied. Extensional expansion involves the collection of data from new systems, or from the field or from a data bank. The distinction between intensional and extensional data is somewhat arbitrary since it can be negated with certain examples, but I think it has some value.

F. The Dynamic Data Bank

We now turn to the final stage in our multistage decision process (Table 22-1). Until recently, at the close of a character variation analysis, information not published was usually stored in one's office and then discarded when the space it occupied was needed for a later project. The outstanding exception has been the maintenance of collections of animals and plants in museums and herbaria throughout the world. Earlier in this article I outlined the different sources of information of actual or potential value to systematists.

I do not wish to belabor the obvious. Given the potential of just the computing equipment available today, what I call the dynamic data bank could become a reality. Consider the experimental data from every character variation analysis that has been irretrievably filed in experiment notebooks that are being discarded regularly. Today, such data often can be put onto punched cards or tape to speed up the initial analysis. As a by-product of this process, the data can be deposited into a regional data bank. Also, optical scanners are now commercially available which "read" any of 2000 different type fonts and put them directly onto

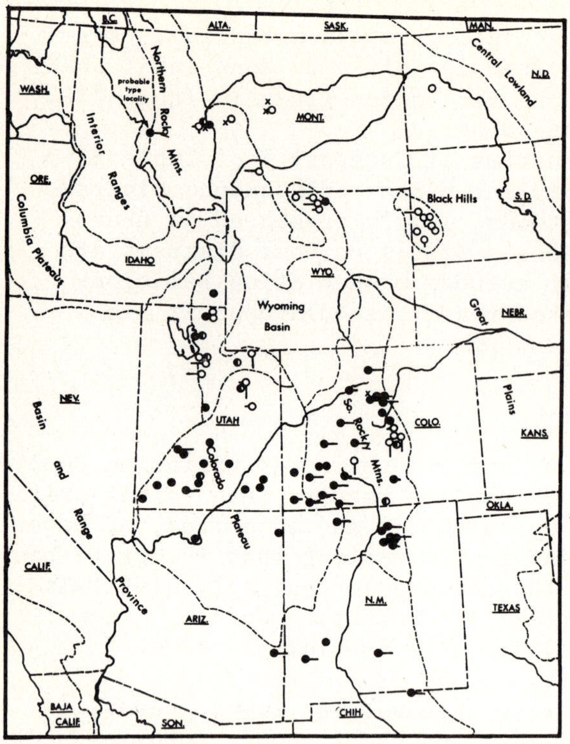

Figure 22-6. Distribution of *Clematis columbiana*. Closed circles with arms to right, extreme form of var. *columbiana*; closed circles without arms, "*repens*" form of var. *columbiana*; half-closed circles, intermediates between var. *columbiana* and var. *tenuiloba*; open circles without arms, extreme form of var. *tenuiloba*; open circles with arms to left, var. *tenuiloba* with coarsely divided leaves; open circles with arms downward, var. *tenuiloba* with elongated aerial stems; x's, probable hybrids between *C. columbiana* and *C. occidentalis*. Glyphs with short arms represent collections which were variable with respect to the trait indicated. [From Pringle, J. S. 1971. Taxonomy and distribution of *Clematis*, Sect. Atragene (Ranunculaceae), in North America. Brittonia 23: 361-393. Used with permission.]

Figure 22-7. Correlation coefficients of the terpene composition spectrum of seven populations of *Bursera microphylla*. Absence of line between localities indicates no correlation. [From Mooney, H. A., and W. A. Emboden, Jr. 1968. The relationship of terpene composition, morphology and distribution of populations of *Bursera microphylla* (Burseraceae). Brittonia 20: 44-51. Used with permission.]

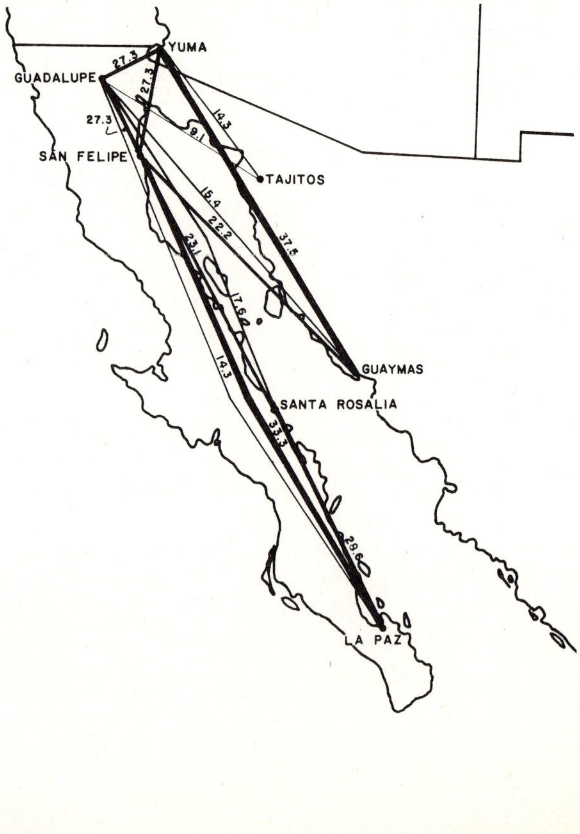

Figure 22-8. Hypothetical evolutionary relationships among the white-rayed taxa of Melampodium. [From Stuessy, T. D. 1971. Systematic relationships in the white-rayed species of Melampodium (Compositae). Brittonia 23: 177-190. Used with permission.]

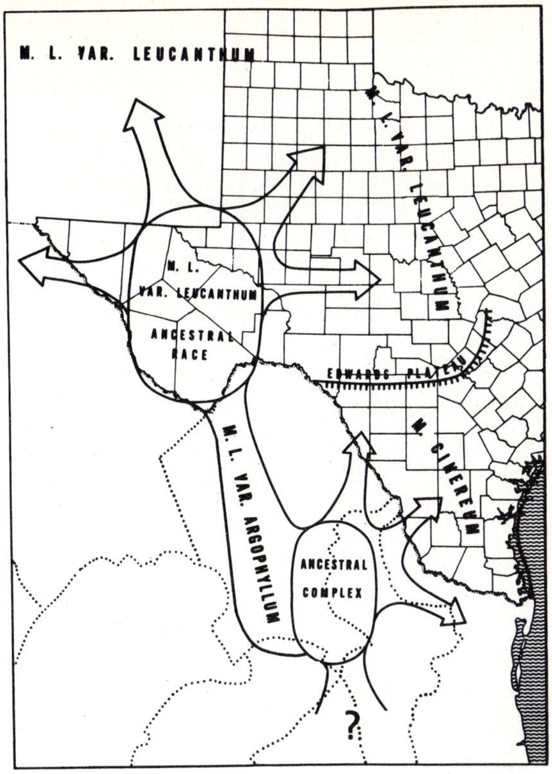

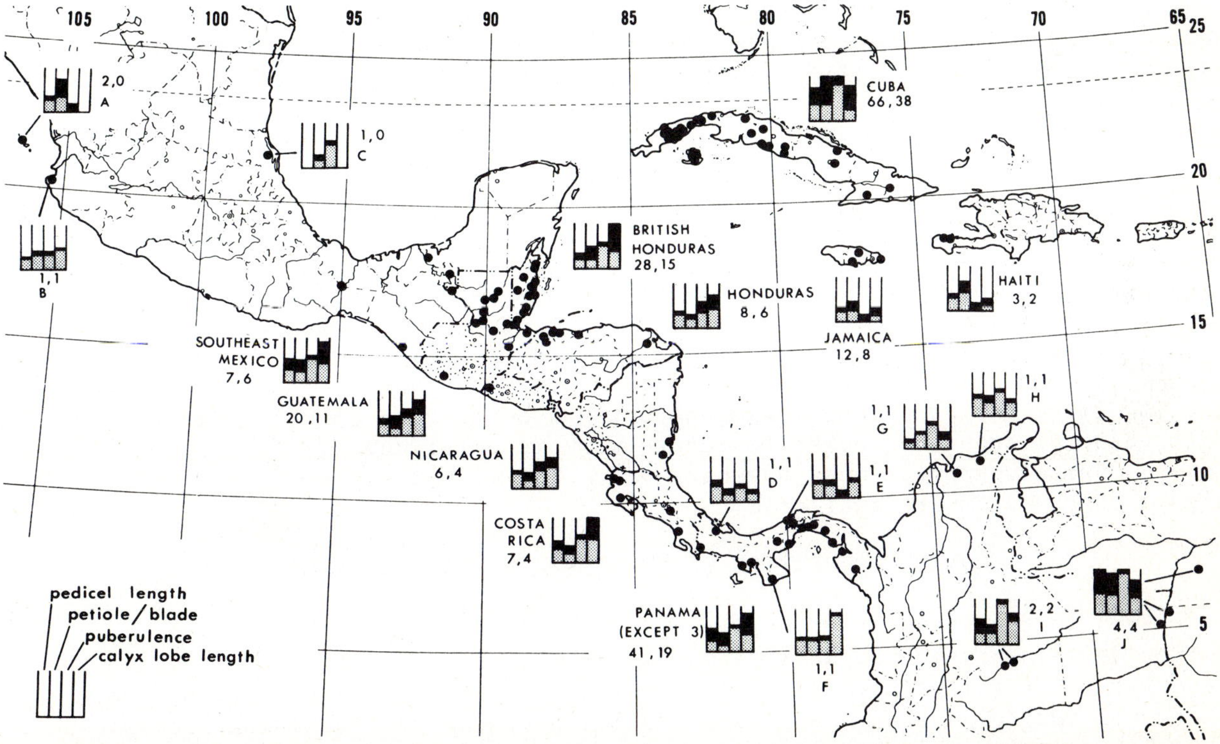

Figure 22-9. [From Morley, T. 1971. Geographic variation in a Widespread neotropical species, Mouriri myrtilloides (Melastomataceae). Brittonia 23: 413-424. Used with permission.]

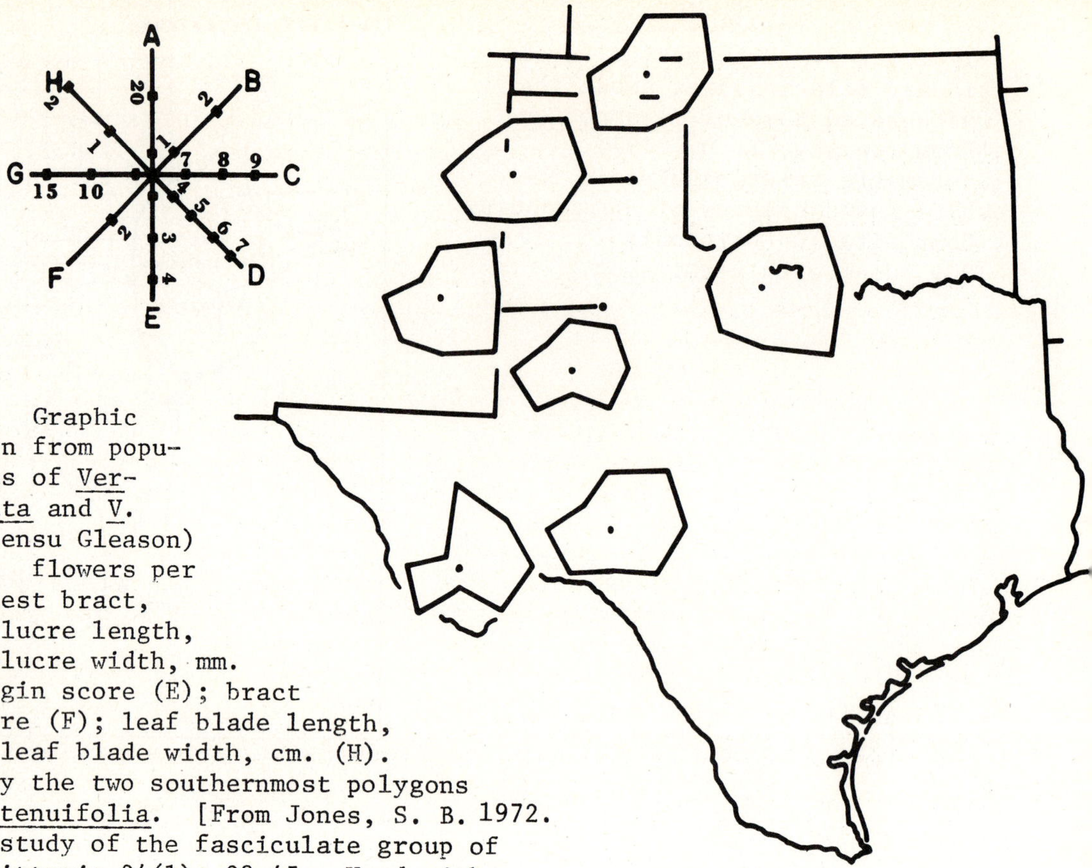

Figure 22-10. Graphic representation from population samples of *Vernonia marginata* and *V. tenuifolia* (sensu Gleason) of means for: flowers per head (A); widest bract, mm. (B); involucre length, mm. (C); involucre width, mm. (D); leaf margin score (E); bract tip shape score (F); leaf blade length, cm. (G); and leaf blade width, cm. (H). Geographically the two southernmost polygons represent *V. tenuifolia*. [From Jones, S. B. 1972. A systematic study of the fasciculate group of *Vernonia*. Brittonia 24(1): 28-45. Used with permission.]

a magnetic tape. Not only would this make published material contained in floras and faunas available to researchers for new and valuable analyses, but also, with increasing publication costs, such a data bank system might gradually replace formal journals as we know them. Such procedures are not without problems, however (Crovello, 1968b; Keller and Crovello, 1973). The big problem in systematics, and in biology in general, has been the lack of a large, driving force actively campaigning to bring about a data bank system. The full value of such data banks to future work is still unexplored. Clearly, the taxonomic community must organize now. The methods of implementing such a data bank system are available.

COMMON PROPERTIES OF CHARACTER VARIATION ANALYSES

Even in the above brief survey of character variation analysis in systematics certain common properties emerge. The purposes of the analysis are the description of the pattern of variation in nature and its explanation. Most character variation analyses in systematics have the same building blocks: i.e., characters, objects, and system.

Another common property that emerges is that any character variation analysis is a multistage decision process and the decisions made at each stage are arbitrary to some degree. All use the same general process, and systematists should benefit from the study of methods used in analyses in areas other than their own specialty. This also helps us to get out of our thought-ruts.

Optimization is a property of all systematic studies. But optimization can occur at all stages of a character variation analysis as well as at different levels within one stage. Taxonomists wish to maximize the validity of a phylogram or of a phenogram to reflect true relationships. Evolutionists usually consider models of evolutionary processes as tending towards optimal levels, either in isolation or, more realistically, with all processes considered at once. Of course, everyone attempts to optimize the use of his own time.

Many have observed that most knowledge is organized into hierarchies. Any taxonomic system is a hierarchy. Aside from its general hierarchial arrangement there are more specialized hierarchies in the system. These include a hierarchy of purposes, or objectives. In any complex character variation analysis, decisions must be made as to the direction research efforts should take, and this is determined by the relative importance of the purposes and of the hierarchial relation among them. The problem with character variation analysis in systematics is that it continually suggests so many interesting side projects that it is difficult to keep the main purpose in mind. Data hierarchies were discussed earlier. These are formed by partitioning the original character by object data matrix into more compact matrices. Thus, character information on several species in many sites may be compressed into a character by species table. Character hierarchies are possible. Davis and Heywood (1963) refer to multiple character which can be divided into several smaller characters. For example, type of leaf pubescence (a character) can be further separated into characters like density of pubescence, length of hairs, number of cells per hair, and hair color. Sampling procedure hierarchies occur when different sampling techniques must be combined in a successive procedure. To obtain a fair estimate of the characteristics of a species, a stratified sampling design may be constructed throughout equal areas of the species' range, but then sampling within each area will be random. Decision hierarchies are found in a complex systems analysis where decisions are made in several substages and the resulting total outcome is used to make a decision at the major stage. Also, in any study of an evolutionary system one can think of a control strategy hierarchy. For example, if a species such as marijuana is to survive in cultivated areas, what are the control strategies of the species as a whole, of each population in different areas of the world, and of each individual in a given population? I realize that this last is anthropomorphic to say the least. But if it provides a worker with insights into the systems of interest to him, it is a valuable concept. Last, and perhaps most important, is the concept of a hierarchy of character variation systems. That is, how can we group separate character variation analyses to provide maximum value to future work in systematics?

We also should not be so concerned with the analysis of variation that we fail to be aware of the opposite situation, invariance. In any system certain characters are invariant over all or part of the set of objects. What does this mean in terms of the action of environmental selective forces, of control strategies, and other biological events and evolutionary responses?

COMMON PROBLEMS OF CHARACTER VARIATION ANALYSIS

Most readers are aware of the problems of character variation analysis. Among those already mentioned is the effect of arbitrary decision-making (concerning the weighting of characters, the clustering of objects, and the choice of the characters and parameters to be measured in a system). The problem of data compatibility at all levels (choice of characters, character states, etc.) also was discussed. An additional problem is determining whether an inference about some specific part of the character variation study, or the total study, is warranted or unwarranted. Also, how does one overcome redundancy in data accumulation beyond that which is necessary for verification? The time element is a problem. How can one insure that data collected separately, at different times and places by different people with different instruments, are at all comparable?

ANALYSIS OF CHARACTER VARIATION LITERATURE

Adams, R. P. 1970. Contour mapping and differential systematics of geographical variation. Systematic Zoology 19: 385-390.

Allee, W. C., et al. 1949. Principles of Animal Ecology. W. B. Saunders. Philadelphia.

Anderson, E. 1949. Introgressive Hybridization. Wiley. New York.

Argus, G. A., and J. W. Sheard. 1972. Two simple labeling and retrieval systems for herbaria. Canadian Journal of Botany 50: 2197-2209.

Blackith, R. E. 1960. A synthesis of multivariate techniques to distinguish patterns of growth in grasshoppers. Biometrics 16: 28-40.

Blackith, R. E., and R. A. Reyment. 1971. Multivariate Morphometrics. Academic Press. New York.

Bossert, W. 1969. Computer techniques in systematics. In: Systematic Biology. Publ. No. 1692. National Academy of Science, 595-605. Washington, D. C.

Burnaby, T. P. 1966. Growth invariant descriminant functions and generalized distances. Biometrics 22(1): 96-110.

Busacker, R. G., and T. L. Saaty. 1965. Finite Graphs and Networks: An Introduction and Applications. McGraw-Hill Book Company. New York.

Clausen, J. 1967. Biosystematic consequences of ecotypic and chromosomal differentiation. Taxon 16: 271-79.

Cochran, W. G., and G. M. Cox. 1957. Experimental Designs. Wiley. New York.

Cole, A. J. (Ed.). 1969. Numerical Taxonomy. Academic Press. New York.

Crovello, T. J. 1968a. A numerical taxonomic study of the genus *Salix*, section Sitchenses. University of California Publication Bot. 44: 1-61.

________. 1968b. The effect of missing data and of two sources of character values on a phenetic study of the willows of California. Madrono 19: 301-15.

________. 1968c. The effect of alteration of technique at two stages in a numerical taxonomic study. University of Kansas Science Bulletin 47(12): 761-85.

________. 1968d. Key communality cluster analysis as a taxonomic tool. Taxon 17: 241-58.

________. 1969. Effects of change of characters and of number of characters in numerical taxonomy. American Midland Naturalist 81: 68-86.

________. 1970. Analysis of character variation in ecology and systematics. Annual Review of Ecology and Systematics 1: 55-98.

________. 1972. Computerization of specimen data from the Edward Lee Greene Herbarium (ND-G) at Notre Dame. Brittonia 24: 131-141.

Crovello, T. J., and R. D. MacDonald. 1970. Index of EDP-IR projects in systematics. Taxon 19: 63-79.

Davis, P. H., and V. H. Heywood. 1963. Principles of Angiosperm Taxonomy. Oliver and Boyd. London.

Dayhoff, M. O. 1969. Atlas of Protein Sequence and Structure, Vol. 4. National Biomedical Research Foundation. Washington, D. C.

Dupraw, E. J. 1965. Non-Linnaean taxonomy and the systematics of honeybees. Systematic Zoology 14: 1-24.

Eaton, T. H. 1970. Evolution. W. W. Norton. New York.

Fitch, W. M., and E. Margoliash. 1967. Construction of phylogenetic trees. Science 155: 279-84.

Gabriel, K. R., and R. R. Sokal. 1969. A new statistical approach to geographic variation analysis. Systematic Zoology 18: 259-78.

Gómez-Pompa, A., and L. I. Nevling. 1973. The use of electronic data processing methods in the Flora Vera Cruz Program. Contrib. Gray Herb. No. 203.

Goodall, D. W. 1968. Identification by computer. Bioscience 18(6): 485-88.

Goodman, M. M. 1967. The identification of hybrid plants in segregating populations. Evolution 21: 334-40.

Gower, J. C. 1966. Some distance properties of latent root and vector methods used in multivariate analysis. Biometrika 53: 325-38.

________. 1967. A comparison of some methods of cluster analysis. Biometrics 23(4): 623-38.

Grieg-Smith, P. 1964. Quantitative Plant Ecology. Butterworths. London.

Hall, A. V. 1970. A computer based system for forming identification keys. Taxon 19: 12-18.

________. 1972. Computer based data banking for taxonomic collections. Taxon 21: 13-26.

Hawksworth, F. G., G. F. Estabrook, and D. J. Rogers. 1968. Application of an information theory model for character analysis in the genus Arceuthobium (Viscaceae). Taxon 17: 605-19.

Jackson, R. C., and T. J. Crovello. 1971. A comparison of numerical and biosystematic studies in Haplopappus. Brittonia 23: 54-70.

Keller, C., and T. J. Crovello. 1973. Problems and procedures in the use of floristic data in a computerized data bank. Proceedings Indiana Academy of Science 82: (in press).

Kendeigh, S. C. 1961. Animal Ecology. Prentice Hall. Englewood Cliffs, New Jersey.

Kluge, A., and J. S. Farris. 1969. Quantitative phyletics and the evolution of anurans. Systematic Zoology 18(1): 1-32.

Lawrence, G. H. M. 1951. Taxonomy of Vascular Plants. Macmillan and Company. New York.

Lewis, T., and L. R. Taylor. 1967. Introduction to Experimental Ecology. Academic Press. New York.

Lindsey, A. A. et al. 1969. Natural Areas in Indiana and Their Preservation. American Midland Naturalist Press. Notre Dame, Indiana. 594 pp.

MacArthur, R. H., and J. Connell. 1966. The Biology of Populations. Wiley. New York.

Mayr, E. 1963. Animal Species and Their Evolution. Harvard University Press. Cambridge.

________. 1965. Numerical phenetics and taxonomic theory. Systematic Zoology 14(2): 73-97.

________. 1969. Principles of Systematic Zoology. McGraw-Hill. New York.

McKell, C. M., C. Duncan, and C. H. Muller. 1969. Competitive relationships of annual ryegrass (Lolium multiflorum). Ecology 50: 653-57.

Morishima, H. 1969. Phenetic similarity and phylogenetic relationshipa among strains of Oryza perennis estimated by methods of numerical taxonomy. Evolution 23: 429-43.

Morrison, D. F. 1967. Multivariate Statistical Methods. McGraw-Hill. New York.

Morse, L. E., J. H. Beaman, and S. G. Shetler. 1968. A computer system for editing diagnostic keys for Flora North America. Taxon 17: 479-483.

Moss, W. W. 1967. Some new analytic and graphic approaches to numerical taxonomy with an example from the Dermanyssidae (Acari). Systematic Zoology 16(3): 177-207.

Odum, E. P. 1956. Fundamentals of Ecology. W. B. Saunders. Philadelphia.

Oxnard, C. E. 1969. Mathematics, shape and function: a study in primate anatomy. American Science 56: 75-96.

Perring, F. H., and S. M. Walters. 1962. Atlas of the British Flora. Thomas Nelson. London.

Prance, G. T., D. J. Rogers, and F. White. 1969. A taximetric study of an Angiosperm family: generic delimitation in the Chrysobalanaceae. New Phytology 68: 1203-1234.

Reyment, R. A. 1969. Biometrical techniques in systematics. In: Systematic Biology. Publication No. 1692. National Academy of Science, 542-587. Washington, D. C.

Rhodes, A. M., W. P. Bemis, T. W. Whitaker, and S. G. Carmer. 1968. A numerical taxonomic study of *Cucurbita*. Brittonia 20: 251-266.

Rohlf, F. J. 1968. Stereograms in numerical taxonomy. Systematic Zoology 17: 246-255.

________. 1969. GRAFPAC, Graphic Output Subroutines for the GE635 Computer. Kansas State Geological Survey Computer Contributions No. 36. Lawrence.

________. 1970. Adaptive hierarchical clustering schemes. Systematic Zoology 19: 58-82.

Rosen, R. 1967. Optimality Principles in Biology. Plenum. New York.

Rubin, J. 1967. Optimal classification into groups: an approach for solving the taxonomy problem. Journal Theoretical Biology 15(1): 103-144.

Seal, H. L. 1964. Multivariate Statistical Analysis for Biologists. Wiley. New York.

Smith, H. 1965. Phenetic taxonomy: an example of existentialism in science. Systematic Zoology 14: 148-149.

Smith, R. L. 1966. Ecology and Field Biology. Harper and Row. New York.

Sneath, P. H. A. 1967. Trent-surface analysis of transformation grids. Journal of Zoology London 151: 65-122.

________, and R. R. Sokal. 1973. Numerical Taxonomy: The Principles and Practice of Classification. W. H. Freeman Company. San Francisco.

Sokal, R. R., and T. J. Crovello. 1970. The biological species concept: a critical evaluation. American Naturalist 104: 127-153.

________, R. R., and F. J. Rohlf. 1969. Biometry. W. H. Freeman Company. San Francisco.

________, and P. H. A. Sneath. 1963. Principles of Numerical Taxonomy. W. H. Freeman Company. San Francisco.

Sterling, T. D., and S. V. Pollack. 1968. Introduction to Statistical Data Processing. Prentice Hall. Englewood Cliffs, New Jersey.

Thompson, D. A. W. 1942. On Growth and Form. Cambridge University Press. New York.

Waterman, T. H., and H. J. Morowitz (eds.). 1965. Theoretical and Mathematical Biology. Blaisdell. New York.

Williams, W. T., and M. B. Dale. 1965. Fundamental problems in numerical taxonomy. Advanced Botanical Research 2: 35-68.

Wirth, M., G. F. Estabrook, and D. J. Rogers. 1966. A graph theory model for systematic biology, with an example for the Oncidunae (Orchidaceae). Systematic Zoology 15(1): 59-69.

Chapter 23. NUMERICAL AIDS IN TAXONOMY[1]*

The use of quantitative methods in plant taxonomy was pioneered by Edgar Anderson (see especially his 1949 review) in the 1930's. The techniques used at that time were simple but effective. A few carefully chosen characters were usually studied on relatively small numbers of plants from at most several populations. Today computers make similar sorts of comparisons possible for many characters using large population sizes and studying many populations simultaneously. Since the late 1950's, Sokal, Sneath, and their co-workers have actively publicized the use of such techniques in a wide variety of organisms.

The objective of numerical aids in taxonomy is to clarify and illustrate degrees of relationship or similarity in an objective and repeatable manner. While some have seemingly argued in the past that numerical techniques would replace traditional taxonomy, the well-worn expression "garbage in → garbage out" perhaps best summarizes many problems not only those of numerical taxonomy.

The problem most often encountered in the use of numerical techniques in taxonomy is the fact that many techniques exist, the techniques may result in different conclusions, and little basis is available for choosing the "best" technique(s). Certain methods are fairly basic, however, and, if clearly understood, can be used for most of the situations that the taxonomist will encounter. A comprehensive survey of almost all available techniques has recently been published by Crovello (1970).

Section A. SAMPLING

While taxonomists wish to make conclusions about populations, they actually work with samples from such populations. These samples may be herbarium specimens, live plants growing in a garden or greenhouse, or wild plants growing in their "normal" environment. In many cases it is possible only to sample certain subpopulations rather than the entire population about which it is wished to make conclusions. Clearly if one's observations are to be pertinent to a population, a sample representative of the population must be studied. Random sampling is generally inappropriate since a small random sample from a large population will not include the extremes of variability which are necessary to properly delimit taxa. An extreme example of the inappropriateness of random sampling is in the collection (and preservation) of varieties of cultivated plants. Random samples of such plants would consist almost completely of the most widely used commercial types. The botanically and historically most interesting types would probably not be collected at all, and in a particular region the sample would most likely consist of essentially the same genotypes collected repeatedly. Common sense and a reasonable knowledge of the range of variation possible in the taxon under

[1]This title was suggested by Hall (1969) as an alternative to either numerical taxonomy or taximetrics.

*Contributed by Major M. Goodman, North Carolina State University, Raleigh.

study are probably the major prerequisites for assembling samples of plants representative of the populations about which inferences are to be made.

Section B. DESCRIPTIVE STATISTICS

Each character, X, in each sample can be described as having an average value or mean, which may be calculated from class values and frequencies as in Chapter 20 (Section B I) or directly as simply the sum of the values of character X for the N different members of the sample divided by the number of individuals in the sample.

$$\bar{x} = \frac{\sum_{i=1}^{N} X_i}{N}$$

The measure of variation used is usually not the range, which depends solely upon the two extreme individuals, but the variance, which may be calculated using frequencies as in Chapter 20 (Section B I) or directly from the raw data as

$$s_X^2 = \sum_{i=1}^{N} (X_i-\bar{x})^2 \ / \ (N-1),$$

or alternatively the standard deviation, the square root of the variance. A simple bar graph of the frequencies of occurrence of the various possible values of X is often informative. The shape of such a graph is often bell-shaped. In such a case the population is said to be approximately normal:

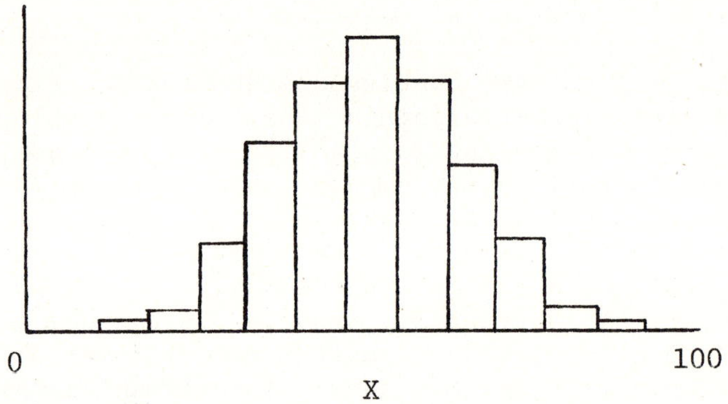

Under such circumstances the interval $(\bar{x} - s_X)$ to $(\bar{x} + s_X)$ includes about 2/3 of the sample. If the population frequencies are not symmetrical about the mean (i.e., skewed) then the range and mode, the most commonly occurring value, should be presented.

When two characters X and Y are being studied, then the closeness of the relationship between them can be measured by the correlation co-efficient

$$p = \frac{\sum_{i=1}^{N} [(X_i-\bar{x})\ (Y_i-\bar{y})]}{[\sum_{i=1}^{N} (X_i-\bar{x})^2]\ [\sum_{i=1}^{N} (Y_i-\bar{y})^2]} .$$

The correlation coefficient can range in value from -1 to +1. When p is close to -1, then high values of X are associated with low values of Y and vice versa. When the correlation is close to +1 then high values of X are associated with high values of Y; conversely low values of X and Y tend to occur together. When the correlation is close to zero, there is very little linear relationship between X and Y. (Under most circumstances non-linear relationships may be safely ignored except in comparisons involving functional relationships such as that between the diameter and surface area of a tree trunk.)

When several populations are being studied at the same time, several types of means, variances, and correlations can be considered. First there are the overall mean and variance for each character and the overall correlation for each pair of characters. Second, each population has its own means, variances, and correlation coefficients. These latter variances and correlations can be averaged over populations to give "pooled" or "common" within population variances and correlations. Third, there are the variances and correlations which can be obtained from the population means. These are called the Among Population variances and correlations. (See any introductory statistical methods textbook, especially the section on nested analysis of variance, for the computational details.)

EXAMPLE 23-1

The samples below were obtained from 6 imaginary populations representative of real data, but simplified for arithmetic purposes.

Sample I

X	Y
10	24
1	16
2	17
8	23
4	20

Sample II

X	Y
10	35
8	37
6	39
12	33
14	31

Sample III

X	Y
1	13
7	17
3	17
9	13
5	15

Sample IV

X	Y
1	49
29	31
15	40
3	35
27	45

Sample V

X	Y
62	94
50	86
48	100
60	90
55	80

Sample VI

X	Y
97	100
93	100
80	100
91	95
89	105

The mean of X for sample I is: $(10 + 1 + 2 + 8 + 4)/5 = 5 = \bar{X}_1$

For Y for sample I: $(24 + 16 + 17 + 23 + 20)/5 = 20 = \bar{Y}_1$

The variance of X for sample I is estimated as:

$$[(10-5)^2 + (1-5)^2 + (2-5)^2 + (8-5)^2 + (4-5)^2]/(5-1) =$$

$$(25 + 16 + 9 + 9 + 1)/4 = 60/4 = 15 = s^2_{X_1}$$

For Y in sample I:

$$[(24-20)^2 + (16-20)^2 + (17-20)^2 + (23-20)^2 + (20-20)^2]/(5-1)$$

$$= (16 + 16 + 9 + 9 + 0)/4 = 50/4 = 12.5 = s^2_{Y_1}$$

The numerator of the correlation coefficient between X and Y for sample I is:

$$\sum_{i=1}^{5} (X_i-\bar{x})(Y_i-\bar{y}) = (10-5)(24-20) + (1-5)(16-20)$$

$$+ (2-5)(17-20) + (8-5)(23-20) + (4-5)(20-20) =$$

$$(5)(4) + (-4)(-4) + (-3)(-3) + (3)(3) + (-1)(0) = 20 + 16 + 9 + 9 = 54$$

The denominator we have mostly calculated above. We need $\sum_{i=1}^{5} (X_i-\bar{x})^2 = 60$ from our calculations of $s^2_{X_1}$ and $\sum_{i=1}^{5} (Y_i-\bar{y})^2 = 50$ from our calculations of $s^2_{Y_1}$. Thus:

$$p_1 = \frac{54}{\sqrt{(60)(50)}} = \frac{54}{\sqrt{3000}} \approx \frac{54}{54.8} \approx 0.99$$

Section C. CHOICE OF CHARACTERS

Characters which vary greatly among populations in relation to their variability within populations are clearly more useful for systematic purposes than are those which show lesser variability among populations in proportion to their variance within populations. Similarly characters which are less affected by the environments in which the plants are found are usually more useful than those which are highly subject to environmental effects. Analyses of variance can be used to select characters which have such desirable properties (see Goodman and Paterniani, 1969, for an example). Standard computer programs are available which perform such analyses and provide estimates of the proportion of total variation attributable to differences among populations, differences among collection sites, and differences among years or seasons of the collections. Such programs are found in the UCLA Biomed (see Dixon, 1968), the TSAR (see Anonymous, 1970), the SAS (see Service, 1972), and the NTS (see Rohlf et al., 1971) software packages, at least one of which should be available at most major U. S. computing centers. In simple cases it is often unnecessary to do more than carefully examine your data. For example for only a few taxa with data collected in only a few years or locations, the data (either raw data or means) might look like:

		Character X			
		Taxon A	Taxon B	Taxon C	Average
Year (or Season or Location)	1	5	10	15	10
" " " " "	2	10	15	20	15
" " " " "	3	7	13	16	12
" " " " "	4	8	13	18	13
Average		7.5	12.75	17.25	12.5

		Character Y			
		Taxon A	Taxon B	Taxon D	Average
Year (or Season or Location)	1	5	10	15	10
" " " " "	2	12	10	9	10.3
" " " " "	3	4	10	15	9.7
Average		7	10	13	10

Clearly despite the fact that year to year (and/or season to season and/or location to location) sampling does affect character X, it is still a useful character effectively separating taxa A, B, and C. However, character Y is quite unreliable (it does not consistently separate the taxa; a comparison of taxon A collected in year 2 compared to taxon D in year 1 is more similar than taxon A is to itself in years 1 and 2).

Since the amounts of time, effort, and money available for any study are limited, the number of characters which can be studied is also limited. There is little reason to study both members of a pair of highly correlated characters when a third character with lower correlation is available and has an equivalent proportion of its variation attributable to differences among populations. In most studies, the number of available mutually non-(or slightly) correlated characters with high proportions of their variability attributable to differences among populations is large in relation to the amount of time available for measurement. Under such circumstances a choice of which characters to use arises. Common sense suggests that the characters should as much as possible depict the entire plant, not some particular feature.

Section D. MATHEMATICS NEEDED BY TAXONOMISTS

In addition to making possible comparisons among large numbers of individuals and populations, computers make possible the use of correlated characters, whereas some Andersonian techniques were really appropriate only for non-correlated characters. The understanding of the procedures used for correlated characters does require some knowledge of basic undergraduate mathematics.

The use of mathematical techniques in plant systematics has been limited by the general lack of mathematical training received by taxonomists. It seems reasonable to expect that in the future almost all students expecting to pursue graduate work in any field of science will complete at least a year of calculus and a semester of matrix algebra. At present, however, most biologists possess little more mathematics than somewhat rusty high school algebra. Even that is quite sufficient for most currently used procedures in numerical taxonomy (taximetrics). This rather fortuitous circumstance results from the fact that the more complex mathematical procedures usually arise due to correlations among the characters being utilized. By choosing non-correlated characters, the mathematics becomes much less difficult. Thus a small amount of preliminary field work designed to determine which characters are essentially uncorrelated with other characters can often make up for a gap in one's knowledge of higher mathematics. Such preliminary studies can often also be used to select characters which not only differ from population to population but which also yield consistent results from year to year and from region to region.

Section E. GROUPING POPULATIONS

Once the characters have been chosen, the populations sampled, and the specimens measured, then data are available to use in grouping the populations. This is usually a two-step process. First some measure of similarity or dissimilarity between members of each possible pair of populations studied is chosen. The number of possible similarity measures is very large indeed (most are found in Sokal and Sneath, 1963), but two general kinds have been widely used.

I. MATCHING AND CORRELATION COEFFICIENTS

Instead of the correlation coefficient described earlier, other measures of agreement between characters are sometimes used. These include matching coefficients and various types of non-parametric procedures usually resulting in some form of non-parametric correlation coefficient (see Kendall, 1962, for details). These usually result in values ranging from 0 for no agreement to +1 for complete agreement (and sometimes -1 for perfect disagreement). As an example, consider two characters, say length of a leaf and width of the same leaf, each divided into three classes: large, medium, and small:

Numbers of Leaves with Small, Medium, and Large Lengths and Widths

	Small	Medium	Large
Length	15	25	13
Width	17	23	13

A total of 53 leaves were examined. Suppose, for each leaf, we score it as a one if length and width fall in the same class, say small, and 0 if they fall in different classes such as small length but medium width. Then the average score for all 53 leaves would be a matching coefficient whose possible range would be from zero (for no matches) to one (for all 53 matches).

It is possible to use matching and correlation coefficients to measure the similarity between pairs of taxa as well as between pairs of characters. The data for the first taxon is treated as the X variable and the data from the second taxon is treated as the Y variable. Then the matches between X and Y are tallied or the correlation coefficient between X and Y is calculated. See Sokal and Sneath (1963), but it should be noted that the characters are usually standardized by subtracting the character mean and subsequently dividing each observation by the standard deviation among population means for that character before calculating the correlation or matching coefficient.

II. DISTANCE MEASURES

The simplest measure of the square of the distance between two taxa i and j is:

$$D_{ij}^2 = [(X_{1i}-X_{1j})^2 + (X_{2i}-X_{2j})^2 + \ldots + (X_{ni}-X_{nj})^2] = \sum_{k=1}^{n} (X_{ki}-X_{kj})^2 ,$$

where X_{kj} is character k of taxon j. Usually the characters are standardized to avoid compounding various scales of measurement such as millimeters, meters, leaf numbers, etc. The standardization most commonly used is to divide each character by its among taxa standard error. Other standardizations can also be employed (see Goodman, 1969).

Such a simple measure of distance works quite well for characters which are reasonably non-correlated. However, when the value of one character can be reasonably predicted on the basis of the knowledge of the values of some of the other characters, the equation above results in distances which differ more amongst themselves than is justified on the basis of the data. To avoid such a situation, it is possible to adjust or transform the characters so that the adjusted characters are not correlated before calculating the distances. The resulting distances are called generalized distances (Mahalanobis, 1936); the procedures involved require a knowledge of matrix algebra and will not be discussed at this point.

EXAMPLE 23-2.

For the six samples examined previously (Example 23-1), we had

Sample	I	II	III	IV	V	VI	Mean	Variance
$\bar{X}$	5	10	5	15	55	90	30	1220
$\bar{Y}$	20	35	15	40	90	100	50	1310

To standardize these scores, we can subtract 30 from each value of X and then divide each value by $\sqrt{1220} \approx 34.9$. For Y we subtract 50 and divide by $\sqrt{1310} \approx 36.2$. Thus our standardized scores are:

Sample	I	II	III	IV	V	VI	Mean	Variance
$\bar{x}_s$	-0.72	-0.57	-0.72	-0.43	0.72	1.72	0	1
$\bar{y}_s$	-0.83	-0.41	-0.97	-0.28	1.10	1.38	0	1

(Actually due to small sounding errors the means and variances of the standardized scores do differ trivially from 0 and 1, respectively.) Since the correlation between the sample means of X and Y is large (about 0.96), a decision should be made as to whether to use just X, just Y, both X and Y, or even neither X nor Y. In practice we should probably choose either X or Y or a new "character" Z which combines X and Y in an "optimum" manner to be discussed later. Any one of the three methods should give about the same results. For now, we will ignore this problem and calculate the distances from both characters:

$$D_{I,II}^2 = [-.72 - (-0.57)]^2 + [-0.83 - (-0.41)]^2 = (0.15)^2 + (0.42)^2 =$$

$$0.0225 + 0.1764 \approx 0.20 .$$

Thus $D_{I,II} \approx \sqrt{0.20} \approx 0.45$.

The complete distance table is:

	I	II	III	IV	V	VI
I	--	0.45	0.14	0.62	2.41	3.29
II	0.45	--	0.58	0.19	1.99	2.91
III	0.14	0.58	--	0.75	2.52	3.39
IV	0.62	0.19	0.75	--	1.38	2.72
V	2.41	1.99	2.52	1.38	--	1.34
VI	3.29	2.91	3.39	2.72	1.34	--

Clearly samples I, II, III, and IV differ little, while V and VI differ from each other as well as differ greatly from the other samples. The same conclusions are evident from either X or Y alone, but the technique can be used for many characters at once, where simple inspection of the data may not be sufficient for reaching conclusions.

III. CLUSTERING PROCEDURES

Once measures of between taxa similarity or distance have been obtained, several techniques can be used to group the taxa. The most widely used one first groups those taxa which are mutually most similar or closest together to form groups, then proceeds to group the groups together on the same basis, treating remaining individual taxa as groups. This results in a tree-like diagram called a dendrogram, where the taxa are located at the branch tips and where a pair of branches join together at the average distance (or average similarity) between all possible pairs of taxa located on the two branches, the first member of a pair being from one branch, the second member from the other. In Example 23-2, the first step would be to group samples I and III, which have the smallest distance, 0.14. Second, samples II and IV would be grouped with their distance of 0.19. Then samples V and VI are grouped at a distance of 1.34, since they are closer to each other than to any other sample. The average distance from {I,III} to {II, IV} is (0.45 + 0.62 + 0.58 + 0.75)/4 = 0.60, which is much less than the distance of either to {V, VI}, so {I, III} and {II, IV} are grouped at 0.60. Finally the average distance from {I, II, III, IV} to {V, VI} is (2.41 + 1.99 + 2.52 + 1.38 + 3.29 + 2.91 + 3.39 + 2.72)/8 = 2.58 which completes the dendrogram.

EXAMPLE 23-3.

Example of a Dendrogram

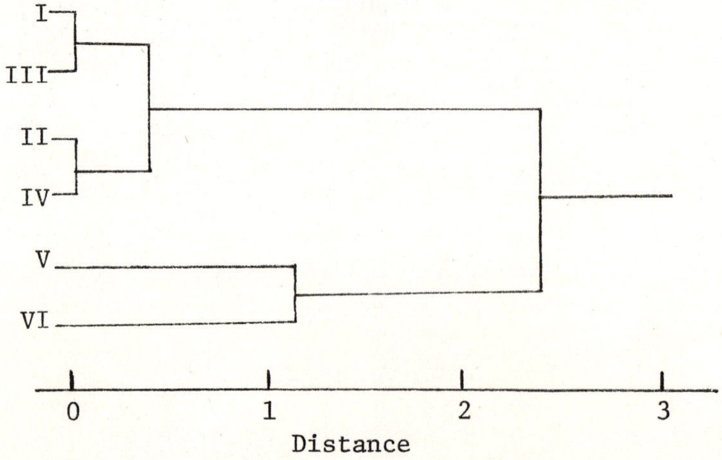

A quick inspection of the data indicates that either the use of X alone or Y alone would have resulted in a dendrogram of essentially the same form.

Instead of average distances, smallest or largest distances between the members of one group and those of another can be used for clustering. However, the agreement between the original table of distances and the distances read from the dendrogram obtained is much less than is obtained using average distances. In Example 23-3, using smallest distances between members of different groups would change the grouping of {I, III} with {II, IV} to 0.45 rather than 0.60, and of {I, II, III, IV} with {V, VI} to 1.38 rather than 2.58, but would leave the form of the dendrogram unchanged. Using largest distances would result in a dendrogram of the same form also, but with {I, III} grouping with {II, IV} at 0.75 and with {I, II, III, IV} grouping with {V, VI} at 3.39.

Rather than averaging distances, the means themselves can be averaged after each clustering and distances recomputed. The latter technique is called centroid clustering. Results generally differ trivially from distance averaging.

Several theoretically appealing alternatives to these procedures have been proposed. One is to form k groups in such a manner that the variation among the within group distances is as small as possible in relation to the among group variation. A second is to find the dendrogram for which the dendrogram distances fit closest to the actual distances among the taxa. Several criteria of closeness of fit could be used, but the usual one is least squares

$$\text{Min} \sum_{i,j} (D_{ij}-d_{ij})^2 ,$$

which is the minimum sum of the squares of differences between the distances, D_{ij}, and the dendrogram distances, d_{ij}. Neither procedure has proved to be practical in the case of more than a few taxa. Gower (1967), Wishart (1969), and Cormack (1971) have recently reviewed a number of alternative clustering procedures.

Section F. PRINCIPAL COMPONENTS

When it is necessary to work with sets of highly correlated characters, the simple application of distance equations is often unsatisfactory. Calculating distances from several highly correlated characters is essentially equivalent to calculating them from a single one of the correlated characters and then multiplying the distances obtained by an unknown constant. When several sets of highly correlated characters plus a few relatively uncorrelated ones must be dealt with, then the distance equations are clearly inappropriate.

To remedy this situation, the characters can be transformed to a new set of uncorrelated characters. Principal components are simply the new characters obtained by transforming the original characters to an uncorrelated set. If X and Y are two correlated characters,

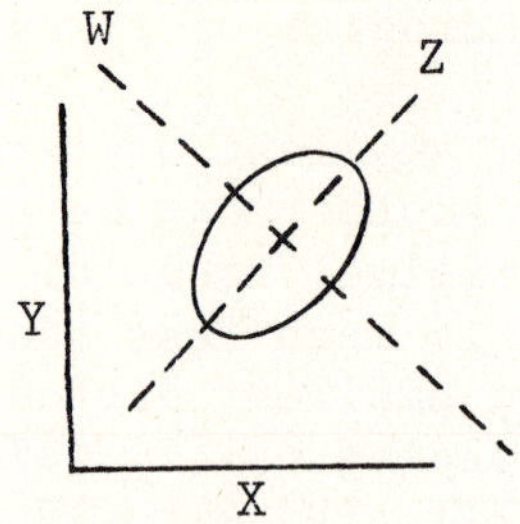

then the W-Z axes define the principal components. An individual plant or a taxon mean has values of W and Z that correspond to its values of X and Y. Theoretically these values could be read from a graph or calculated arithmetically, but in practice computer programs provide them directly. W and Z are uncorrelated (knowing the value of W for an individual tells us nothing about that individual's value of Z and vice versa). Usually W and Z are also standardized, by dividing each by its standard error, so that each has about the same range:

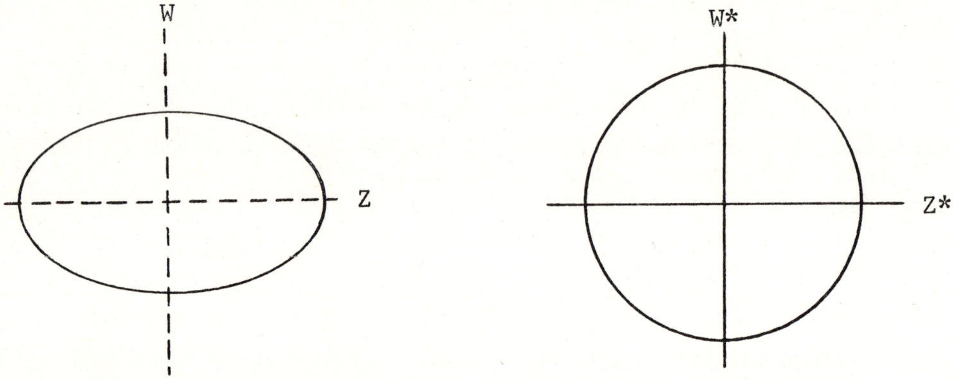

where W* and Z* are the standardized forms of W and Z.

The computations involved require a knowledge of matrix algebra, but an understanding and use of the technique does not. Computer programs such as the SAS, UCLA Biomed, and NTS software packages can first calculate the correlations among the characters used and then make the transformation of the data.

The first principal component (here Z) has the greatest variance and the last one (here W) has the least. Thus, sometimes the first several principal components can be used to summarize most of the variation among the taxa, rather than the full set of characters initially studied. Since the sum of the variances of the principal components is the same as the number of characters studied, if the original characters are "standardized" to mean zero and variance one (by subtracting the mean from each and then dividing by the standard deviation), the proportion of this sum attributable to each of the principal components is easily calculated. If the first two components account for most of the among taxa variability, then a plot of the taxon means, possibly with their within taxon standard errors, for the first two principal components should illustrate quite well the inter-taxa relationships. See Rao (1964) and Goodman (1972) for further discussion. In any case, distances between taxa can be computed using the principal components which account for most of the variation among taxon means. Often the last few principal components are little more than "noise," if the characters are highly correlated.

Section G. CANONICAL VARIATE ANALYSIS

When the problem at hand is to assign individuals to their proper group or when the groups in a dendrogram cluster together so closely that little reliance can be placed upon the particular arrangement obtained (see Rohlf, 1968), canonical variate analysis can often be profitably used. It is closely related to principal components analysis. The first canonical variate, Y_1, is a linear transformation of the original characters

$$Y_1 = b_1X_1 + b_2X_2 + \dots + b_nX_n ,$$

where X_i is character i. The b_i are determined in such a manner that Y_1 has the greatest possible variation among taxa in proportion to its variation within taxa. The second canonical variate has the greatest possible variation among taxa in proportion to the within taxa variation of all possible variates constructed by

$$Y_2 = c_1X_1 + c_2X_2 + \dots + c_nX_n ,$$

where the correlation among and within taxa of Y_1 and Y_2 is zero. In a similar manner, Y_3 is formed so as to have zero among and within taxa correlation with both Y_1 and Y_2 and to maximize the among/within taxon variation on that basis. By using the first two canonical variates, the taxon means and within taxon standard deviations can be plotted and unidentified individuals assigned to their nearest neighbor. Similarly by using the first few canonical variates as input data for the construction of a dendrogram, the relationships among groups will be emphasized over the relationships among individual pairs of taxa.

In the case of only two groups, the first canonical variate is identical to the discriminant function of Fisher (1936), first used for a case of hybridization in *Iris* discovered by Edgar Anderson. These techniques (discriminant functions, canonical analysis) are still widely used in studies of hybridization among plant, especially forest tree, species. (A recent example is Smouse, 1972).

Section H. MISCELLANEOUS

There are a number of specialized topics which will only be mentioned here. There has been much discussion about the use of transformations of the characters (see Wright, 1952) and the use of ratios, rather than the original data (see Sokal, 1965). Tests of significance are discussed in all statistical methods books, but most taxonomic comparisons are among populations known to be different. In that case, the level of significance of a difference will probably depend mostly on the number of specimens measured. Significance is not synonymous with importance. The test for differences among within population covariance matrices is given in Morrison (1967). Such matrices do usually differ significantly, but this may not be important for multivariate studies which are based on samples of approximately equal size. Regression and covariance analyses can be used to adjust the data to eliminate (or at least ameliorate) the effects of variation in characteristics for which little or no variation would be preferable. These might be sex or age differences, location or altitudinal differences, size or weight differences, etc. Standard statistical methods texts give details and examples.

Gower (1966, 1972a,b) has shown how various analytical techniques for measuring similarity may be related and has developed a simple technique for measuring the degree of agreement between sets of analyses based upon either different characters or different measures of relationship or both. Recent examples of taximetric applications of general interest outside the area of plant taxonomy are those of Horne (1967), Fitch and Margoliash (1967, 1970), and the symposium edited by Weiner and Huizinga (1972), while Farris (1970, 1972) has formalized a technique suggested by Wagner and widely used by botanists for deriving tentative phylogenies.

Examples of possible problems encountered in the application of numerical techniques to botanical data can be found in Heiser et al.'s (1965) analysis of Solanum species and hybrids. An informed critique of the philosophy behind much of the early (and indeed of some of the current!) work in numerical taxonomy has been written by Johnson (1970). He also has contrasted the mathematical simplicity being applied to the biological complexity under study, and he has done so in very specific terms. It is a paper well worth reading. A short review of the whole field of numerical taxonomy has recently appeared (Gower, 1969), and edition two of Sokal and Sneath's (1963) basic text has just appeared (Sneath and Sokal, 1973). Another comprehensive review (Blackith and Reyment, 1971) is a book of applied examples (along with extensive computer programs), but the amount of explanation is often minimal. It has an excellent current reference list and can serve as an introduction to the vast literature which has accumulated in recent years.

Section I. ARITHMETIC

To speed up calculations you may wish to code your data by subtraction, addition, multiplication, or division. (For example you might prefer to work with 1, 5, 3, 8, ..., rather than 1001, 1005, 1003, 1008, ..., etc.). After coding your data in that manner and calculating your means and variances, the following rules apply to convert them to their original units of measurement.

Means:
- If you added a constant K, subtract it.
- If you subtracted a constant K, add it.
- If you multiplied by a constant K, divide by it.
- If you divided by a constant K, multiply by it.

Variances:
- If you added or subtracted a constant K, forget it.
- If you multiplied by a constant K, divide by K^2.
- If you divided by a constant K, multiply by K^2.

Finally the following equation is exceedingly helpful in the calculation of variances:

$$\sum_{i=1}^{N} (X_i-\bar{x})^2 = \sum_{i=1}^{N} X_i^2 - (\sum_{i=1}^{N} X_i)^2/N ,$$

and a corresponding equation is very helpful in the calculation of correlation coefficients:

$$\sum_{i=1}^{N} [(X_i-\bar{x})(Y_i-\bar{y})] = \sum_{i=1}^{N} X_iY_i - (\sum_{i=1}^{N} X_i)(\sum_{i=1}^{N} Y_i)/N .$$

NUMERICAL AIDS LITERATURE

I. General Texts or Reference Books

Anderson, E. 1949. Introgressive Hybridization. John Wiley and Sons, Inc. New York.

Anonymous. 1970. TeleStorage and Retrieval System User's Manual. Duke University Computation Center. Durham, North Carolina.

Blackith, R. E., and R. A. Reyment. 1971. Multivariate Morphometrics. Academic Press. London and New York.

Dixon, W. J. (ed.). 1968. BMD Biomedical Computer Programs, 2nd ed. University of California Press. Berkeley and Los Angeles.

Kendall, M. G. 1962. Rank Correlation Methods. 3rd ed. Charles Griffin and Company, Ltd. London.

Morrison, D. F. 1967. Multivariate Statistical Methods. McGraw-Hill Book Company. New York.

Ostle, B. 1963. Statistics in Research. 2nd ed. Iowa State University Press. Ames, Iowa.

Rohlf, F. J., J. Kishpaugh, and D. Kirk. 1971. NT-SYS. Numerical Taxonomy System of Multivariate Statistical Programs. State University of New York. Stony Brook, New York.

Service, J. 1972. A User's Guide to the Statistical Analysis System. Student Supply Stores. North Carolina State University. Raleigh, North Carolina.

Sneath, P. H. A., and R. R. Sokol. 1973. Numerical Taxonomy: The Principles and Practice of Numerical Classification. W. H. Freeman. San Francisco.

Snedecor, G. W., and W. G. Cochran. 1967. Statistical Methods. 6th ed. Iowa State University Press. Ames, Iowa.

Sokol, R. R., and P. H. A. Sneath. 1963. Principles of Numerical Taxonomy. W. H. Freeman. San Francisco.

Weiner, J. S., and J. Huizinga (eds.). 1972. The Assessment of Population Affinities in Man. Oxford University Press. London.

II. Periodicals

Applications of numerical techniques can be found in most botanical, as well as agricultural and forestry, journals. Periodicals such as The American Naturalist, Genetics, and Evolution also frequently carry articles dealing with numerical taxonomy. The following are more specifically aimed at the application of numerical techniques to biology:

Biometrics. 1945+. F. A. Graybill, Editor. Department of Statistics, Colorado State University, Fort Collins, Colorado. Publication of the Biometric Society, Washington, D. C.

The Classification Society Bulletin. 1970+. P. H. A. Sneath, Editor. M. R. C. Microbial Systematics Unit, University of Leicester, Leicester LEI 7RH, England. Publication of the Classification Society.

Systematic Zoology. 1952+. A. J. Rowell, Editor. Museum of Invertibrate Paleontology, Department of Geology, University of Kansas, Lawrence, Kansas. Publication of the Society of Systematic Zoology, New Haven, Connecticut.

III. References

Cormack, R. M. 1971. A review of classification. J. Royal Statistical Society A 134: 321-367.

Crovello, T. J. 1970. Analysis of character variation in ecology and systematics. Annual Review of Ecology and Systematics 1: 55-98.

Farris, J. S. 1970. Methods for computing Wagner trees. Systematic Zoology 19: 83-92.

________. 1972. Estimating phylogenetic trees from distance matrices. The American Naturalist 106: 645-668.

Fisher, R. A. 1936. The use of multiple measurements in taxonomic problems. Annals of Eugenics 7: 179-188.

Fitch, W. M., and E. Margoliash. 1967. Construction of phylogenetic trees. Science 155: 279-284.

________. 1970. The usefulness of amino acid and nucleotide sequences in evolutionary studies. Evolutionary Biology 4: 67-109.

Goodman, M. M. 1969. Measuring evolutionary divergence. Japanese Journal of Genetics 44 (Suppl. 1): 310-316.

________. 1972. Distance analysis in biology. Systematic Zoology 21: 174-186.

________, and E. Paterniani. 1969. The races of maize: III. Choices of appropriate characters for racial classification. Economic Botany 23: 265-273.

Gower, J. C. 1966. Some distance properties of latent root and vector methods used in multivariate analysis. Biometrika 53: 325-338.

________. 1967. A comparison of some methods of cluster analysis. Biometrics 23: 623-637.

________. 1969. A survey of numerical methods useful in taxonomy. Acarologia 11: 357-375.

________. 1971. Statistical methods of comparing different multivariate analyses of the same data. In: Hodson, F. R., D. G. Kendall, and P. Tautu (eds.). Mathematics in the Archaeological and Historical Sciences. University Press. Edinburgh.

________. 1972. Measures of taxonomic distance and their analysis. In: The Assessment of Population Affinities in Man. Oxford University Press. Oxford.

Hall, A. V. 1969. Group forming and discrimination with homogeneity functions. In: Cole, A. J. (ed.). Numerical Taxonomy. Academic Press, Inc. Ltd. London.

Heiser, C. B., J. Soria, and D. L. Burton. 1965. A numerical taxonomic study of Solanum species and hybrids. American Naturalist 99: 471-487.

Horne, S. L. 1967. Comparisons of primate catalase tryptic peptides and implications for the study of molecular evolution. Evolution 21: 771-786.

Johnson, L. A. S. 1970. Rainbow's end: The quest for an optimal taxonomy. Systematic Zoology 19: 203-239.

Mahalanobis, P. C. 1936. On the generalized distance in statistics. Proceedings National Institute of Science, India 2: 49-55.

Rao, C. R. 1964. The use and interpretation of principal component analysis in applied research. Sankyā Series A 26: 329-358.

Rohlf, F. J. 1968. Stereograms in numerical taxonomy. Systematic Zoology 17: 246-255.

Smouse, P. E. 1972. The canonical analysis of multiple species hybridization. Biometrics 28: 361-371.

Sokal, R. R. 1965. Statistical methods in systematics. Biological Review 40: 337-391.

Wishart, D. 1969. Mode analysis: A generalization of nearest neighbor which reduces chaining effects. In: Cole, A. J. (ed.). Numerical Taxonomy. Academic Press, Inc. Ltd. London.

Wright, S. 1952. The genetics of quantitative variability. In: E. C. R. Reeve and C. H. Waddington (eds.). Quantitative Inheritance. Pp. 5-41. Her Majesty's Stationery Office. London.

EXERCISES

23.B.1. The table below derived from the data presented in Example 23-1 contains four incorrect means. Find them and correct them.

	$\bar{X}$	$\bar{Y}$	s_X^2	s_Y^2	p
Sample I	5	20	15	12.5	0.99
Sample II	15	40	10		-1.0
Sample III	5	15	10	4	0
Sample IV	10	35	170	53	
Sample V	55	90			
Sample VI	90	100		12.5	-0.11

23.B.2. Complete the table above.

23.B.3. The overall mean in this experiment for character X is $(10 + 1 + 2 + 8 + 4 + 10 + 8 + 6 + \ldots + 80 + 91 + 89)/30 = 30$. What is the overall mean for Y?

23.B.4. The overall variance for X is

$$[(10-30)^2 + (1-30)^2 + \ldots + (89-30)^2]/29 = 1090.6 \quad .$$

What is the overall variance for Y? What is the corresponding overall correlation between X and Y?

23.B.5. The average within sample variance for X is

$$(15 + 10 + 10 + 170 + \ldots)/6 = 47.$$

What is the average within sample variance for Y?

23.B.6. The variance of the sample means for character X is $[(5-30)^2 + (15-30)^2 + (5-30)^2 + (10-30)^2 + (55-30)^2 + (90-30)^2]/(6-1)$ $= (625 + 225 + 625 + 400 + 625 + 3600)/5 = 6100/5 = 1220.$ How does this compare to the average within sample variance for X? Make the same comparison for Y. Compare these values with the values of the overall variance of X and Y.

23.B.7. Compute the correlation among the sample means of X and Y. The numerator is calculated as $[(5-30)(20-50) + (15-30)(40-50) + (5-30)(15-50) + (10-30)(35-50) + (55-30)(90-50) + (90-30)(100-50)]$. Compare this to the overall correlation between X and Y.

23.B.8. Note that you were not asked to calculate the average correlation within samples. For technical reasons beyond the scope of this text the best estimate of the average within sample correlation is not $(0.99 + (-1.0) + 0 + \ldots + (-0.11))/6$. Instead we average the six values of $\sum_{i=1}^{5} (X_i-\bar{x})(Y_i-\bar{y})$ and call that $\overline{\Sigma xy}$, average the values of $\sum_{i=1}^{5} (X_i-\bar{x})^2$ and call that $\overline{\Sigma x^2}$, and average the values of

$\sum_{i=1}^{5} (Y_i-\bar{y})^2$ and call that $\overline{\Sigma y^2}$. Then the average within sample

correlation coefficient between X and Y is $\dfrac{\overline{\Sigma xy}}{\sqrt{\overline{\Sigma x^2}\ \overline{\Sigma y^2}}}$.

Calculate the latter value.

23.E.1. From the dendrogram in Example 23-3 read off the distances between all possible pairs of samples (for example, the dendrogram distance between I and VI is 2.58). Make a table of these dendrogram distances.

23.E.2. Make tables for the dendrogram distances based on smallest distances and largest distances as well.

23.E.3. Compare these three tables with the original distance table on which the dendrograms are based. Which table compares most closely?

(These should be required of all those who have not previously used computers.)

23.F.1. Using the table of sample means of characters X and Y which we have calculated for samples I, II, ..., VI of Exercise 23.B.1., punch the means of sample I on an IBM card. Place the decimal point for character X in column 5 and the decimal point for character Y in column 10. (If you have not used a keypunch before, you will find it works very much like a typewriter. First experiment with the "Feed," "Register," "Release," and "Duplicate" keys. You will quickly get the knack of simple keypunching). Place the means for samples II through VI on separate cards, being sure to place the decimal points in the proper columns.

23.F.2. Locate a program for computing principal components (SAS is suggested), punch the necessary control cards, and run the program.

23.F.3. Call the principal components W and Z, respectively. Calculate the means of W and Z and their variances. Which is more variable? If W has not been standardized, standardize it.

23.F.4. From the standardized values of W, calculate the distances between the sample means I, II, III, IV, V, and VI. Compare these distances to those obtained earlier in Example 23-2.

23.G.1. Punch, as in Exercise 23.F.1., values of X and Y for each member of samples I, II, ..., VI of Example 23-1. The values of X and Y for the first individual go on the first card, etc., as before. Be certain the decimal points are in the columns specified in the previous exercise.

23.G.2. Using the UCLA Biomed program titled "Stepwise Multiple Discriminant Function Analysis" or some other program which does canonical variate analysis (not the same as canonical correlation analysis), punch the necessary control cards and run the program. Be sure to get your output plotted if possible. Experiment with the various options your program permits. How do the plotted distances compare with those previously calculated? (You may have to plot the means yourself if you use a program other than the one suggested).

23.G.3. Study the entire output of your program. Note the points which you cannot understand. Consult a statistician or numerical taxonomist to clarify those points.

Chapter 24. DESCRIPTION AND DESCRIPTIVE FORMAT

Plant description should convey an image or impression of plant characters, attributes and nature. Description of plants and their parts and taxa provide the basic information for all of taxonomy, i.e., characterization, identification and classification, which form the core of floras, manuals, revisions, monographs, and systems of classification. Descriptive terminology constitutes the framework for determination of relationships between taxa.

No absolutely comprehensive description has ever been written for any species or higher taxon. In data banking, systems are being devised for input and retrieval of all descriptive information as it is produced by various specialists. Monographic treatments are fairly comprehensive for selected types of evidence such as morphological, cytological, ecological but definitely not for all fields of evidence (see Chapters 6-19). Manuals usually contain descriptive terminology relevant to diagnosis and accurate identification, with little else. Simple floras have a more limited descriptive content, frequently not much more than geographic and ecological data for each species listed.

The major problem in descriptive writing is making the description relevant to the type of publication contemplated or in the selection of pertinent characters with the appropriate character states for each taxon described. Another problem is the following of a descriptive sequence, established by the author or editor, for consistent and comparative treatment. All taxa of coordinate rank should have the same characters described in the same sequence. Still another problem will be the bridging of the terminology gap due to an inadequate vocabulary or inadequacy of definition of the terms in the glossary. Many other problems will have to be solved or resolved as one describes parts, plants, and taxa accurately and relevantly.

A general descriptive format is included to encourage consistent key and description writing and for the orderly use of terminology as well as to help the student understand the descriptive sequences in previously published floras and manuals. This example of format is based on that now being used in the writing of the "Flora of the Southeastern United States" (Radford et al., 1967) but it is adaptable, with necessary changes in geography and ecology, for use in writing floras or descriptions of species for any region (Sections A-F).

This format introduces the student to some of the problems in format design and to some of the difficulties involved in the use of any consistent sequence of descriptors. For the understanding of the descriptive language used in any flora, manual, revision, or monograph, the basic phytographic glossary or vocabulary should be part of the descriptive format or, at a minimum, the basic source of terms should be indicated.

For a flora or manual covering a large area in which two or more authors are participating, a Guidebook with an explanation of taxonomic philosophy, concepts, and policy as well as plans, procedures, form and style of treatment (format) is an absolute requirement. Many of the major and minor problems related to descriptions of taxa can be eliminated with appropriate guidelines. The most comprehensive set of instructions ever published for a taxonomic project are those in "A Guide for Contributors to Flora North America

(FNA)" by Shetler et al. (1973). The taxonomic philosophy, concepts and policies for Flora North America are presented in Section G to show the philosophical framework and organizational detail necessary for the making of descriptive treatments, according to a Format, of the taxa within a large flora or manual.

DESCRIPTIVE FORMAT

Keys and Descriptions

1. Family number and name (a, b, f). No authority. Common name (c).
2. Family description (d).
3. Key to Genera (e).
4. Generic number and name (a, f, i). Authority. Common name (c).
5. Generic description (d, f).
6. Key to Species (e).
7. Species number and name (i). Authority. Common name (c).
8. Species description (d).
9. Infraspecific taxon number and name (j). Authority. Common name (c).
10. Infraspecific taxon description (g, h, j).

(a) If a family or genus has a single species within our area, the family and/or generic descriptions are omitted. The species description is modified to include the family and/or generic characteristics for our material.

(b) Family numbers and names are according to Section F (included in part only).

(c) The common name, if available, should be the most commonly used one (rarely more) for the taxon within our area.

(d) All taxon descriptions should follow the detailed arrangement in Section A, examples in Section B, abbreviations in Section C, geography-ecology in Section D, and synonymy in Section E.

(e) Keys to all taxa are to be strictly dichotomous, indented, and each couplet successively numbered (see Section B for examples and Chapter 25 for key writing). Each line or heading of a couplet should start with the same noun or morphological character. All characters used should have the contrasting condition in the second couplet entry. Nouns usually come before adjectives. No more than three characters should be used within a couplet to distinguish any two taxa or groupings. The character easiest to determine should be used first in the couplet. Vegetative, flower and fruit characters should be used whenever possible. Quantitative measurements or ratios should be used, not qualitative statements. Positive statements should be used whenever possible, negative only when the positive contrasting characters occupy too much space. If combined flowering and fruiting or seasonal and sexual keys become too cumbersome, write keys for flowering material and fruiting material, for winter condition and summer condition, for male and female plants and so forth.

(f) All marking, punctuation and manuscript spacing should be according to the samples in Section B.

(g) Briefly note only the distinctive features of the infraspecific taxon, then follow with habitat, distribution, synonymy, etc.

(h) Keys should not be written for infraspecific taxa unless the infraspecific taxon is the only representative in our area.

(i) Genera should be successively numbered in each family beginning with one; species should be successively numbered in each genus beginning with one; each infraspecific taxon should have the species number and each successive one indicated, a, b, c, (4a, 4b, 4c, etc.).

(j) Infraspecific taxa should not be included unless well-defined or easily delimited. If the author desires, reference can be made to previously described infraspecific taxa in the remarks section of each species.

Section A. FORMAT

I. GENERAL COMMENTS

In writing family, generic, species and infraspecific descriptions, the order as given in "B" of Format Details should be followed for the characters considered. In groups where it is not practical to follow this order essentially the author should submit his format to the chairman of the editorial board for approval before final writing of the manuscript. Descriptive terminology should conform to that of Lawrence, G. H. M., 1951, "Taxonomy of Vascular Plants," The Macmillan Company, New York, or preferably that in Chapter 6.

Unless otherwise stated in descriptions it is assumed that the plant is hermaphroditic and autotrophic; the woody plant is perennial; stems are erect; roots are fibrous; woody plant leaves are deciduous; leaves are simple, green, petiolate; flowers are perfect, regular, pedicellate; floral parts are separate for sepals, petals, stamens and fused for carpels; the ovary is superior and the other floral parts are inserted hypogynously. Authors should be conscious of these descriptive assumptions so that exceptions are always indicated in the descriptions.

Key characters should be included in descriptions and amplified where necessary. Family characters common to all genera within the family should be included in the family description and not in each generic description; generic characters common to all species should be included in the generic description and not in each species description. In families with two or more genera and genera with two or more species, structures described should be compared for the closely related taxa within the tribe, section or natural group to eliminate uncertainty, e.g., if genus "a" or species "a" have opposite leaves and cordate stipules, then genera "b," "c," "d" or species "b," "c," "d" should have leaf arrangement and stipule shape described.

Each page of manuscript must be numbered consecutively within the family and genus treatment. Pages for the family description and the keys to the genera should have two parts, the family number and the page number separated by a hyphen, e.g., 140-1, 140-2, 140-3. Page numbers for the treatment of each genus and its species will have three parts; the family number, the genus number, and the page number again all separated by hyphens. The page number starts with 1 again for each genus, e.g., 140-1-1, 140-2-1, 140-2-2, 140-2-3, 140-3-1, etc. Family numbers can be obtained from Section F and the generic numbers within a family are assigned by each author for his own contribution. Each family and each genus should start at the top of a new page. All page numbers should be in the upper, right-hand corner.

Note the following: Fractions should be written "1/2," "1/4," etc., not $\frac{1}{2}$, $\frac{1}{4}$, etc., except after whole numbers where the decimal system should be used: "1.5X" or 1.25X," not 1 1/2X or 1 1/4X. "Times" is to be indicated by capital X; "no. 13" should be used rather than "#13." When using "more" or "less" in measurements follow measurement with appropriate phrase, e.g., 7-9 cm or more long, 3-6 mm or less wide. For width of flat objects use the term "wide," e.g., leaves 2-3 mm wide, not 2-3 mm broad; on "3-dimensional" width measurements use "broad" or "in diam.," e.g., nutlet 2 mm broad; pome 2 cm in diam.; for height use the term "tall": 2-3 dm tall, not high; for prostrate or spreading structures use "long" for length. In hyphenation use "3-nerved" and "2-3 nerved," not 3 nerved or 2-3-nerved. Where the number of parts described per plant is more than one use the plural, if only one, use singular, e.g., for the lily, "leaves" rather than "leaf" and "bulb" rather than "bulbs" would be described.

II. FORMAT DETAILS

A. Manuscript Order

Note Punctuation: period (.) always at end of entry descriptions marked with *.

1. General modifiers , (large, small, coarse, evergreen, parasitic, epiphytic, insectivorous, etc.)

2. Sex of plant , (monoecious, etc.)

3. Plant duration , (annual, perennial)

4. Habit ; or . (tree, shrub, vine, herb)

5. Stem ; or . (nodes, internodes)

*6. Roots .

*7. Leaves .
 a. duration ,
 b. structure , (simple, compound)
 c. arrangement , position ;
 d. blade ; sheath ; ligule ;
 e. petiole ;
 f. stipules .

*8. Inflorescence .
 a. type ; or no punctuation
 b. position ; (terminal, axillary)
 c. peduncle, and/or rachis ;
 d. bracts ; or . bractlets ; or .
 e. pedicel .

*9. Flowers .
 a. sex ; (perfect, imperfect)
 b. symmetry ;
 c. calyx , sepals ;
 d. corolla , petals ;
 e. stamens , anthers , filaments ;
 f. carpels, stigma , style , ovary (position first) , ovule ;
 g. receptacle ; (hypanthium)
 h. pedicel .

10. Fruits ; or ,
 a. type ,
 b. receptacle , (hypanthium)
 c. pedicel ; or .

*11. Seeds .
 a. type ,
 b. seed coat , embryo , endosperm ;
 c. placentation ,

12. (n=) No punctuation [or (3n=) or (5n=) in the case of triploids and pentaploids]

*13. Flowering data .

14. Habitat data ; (see Section D)

*15. Distribution . (see Section D)
 a. province : or .
 b. state .

16. [Distribution in adjacent states] No period unless bracket last item in description.

*17. Synonymy . (see Section E)

*18. Remarks .

*19. Major reference(s). (Literature citation as in American Journal of Botany)

B. Descriptive Sequence for Each Structure Treated

(Applies most often to leaf blade, calyx and corolla)

1. number of parts ,
2. fusion of parts ,
3. habit , (erect, spreading, etc.)
4. color , odor ,
5. shape or form , (unequal also)
6. size, (length, width)
7. texture and/or surface ,
8. venation ,
9. vestiture ,
10. apex ,
11. margin ,
12. base ; or .

General: male before female; chasmogamy before cleistogamy; combined structure before single; tube before lobes; upper surface (use "above"), before lower surface (use "beneath" or "below").

Section B. SPACING, MARKING, PUNCTUATION, FORMAT EXAMPLES

97-1-1

Period at end of name entry only if authority is abbreviated

No punctuation if authority is written in full

2 spaces

97. LARDIZABALACEAE

3 lines between headings in order to allow room to mark for type face

6 spaces

1. AKEBIA Dcne.

1 space

1. A. quinata (Houttuyn) Dcne.--Monoecious, woody, twining vine. Leaves deciduous or semi-evergreen, palmately compound, alternate, exstipulate; leaflets 5, elliptic or weakly obovate, 2-8 cm long, 1-4 cm wide, retuse, entire, terminal leaflet usually largest; petiolules 4-15 mm long. Inflorescence axillary, racemose. Flowers fragrant, functionally unisexual, 3-merous, petaloid sepals and petals light purple; staminate flowers proximal, smaller, stamens 6; the pistillate distal, 1 cm or more broad, carpels 3-10, separate. Follicles purple or violet, fleshy, to 8 cm long. (n=16) Late spring. Established in low woodlands and on roadbanks; mts. and pied. of NC. [Tex, Pa] A rare escape from cult.

No period after ( ) or [ ] unless last entry; 2 spaces to start of next entry

2 spaces

78. FUMARIACEAE[1]

5 spaces

Annual, biennial or perennial, usually glabrous and glaucous erect or scandent herbs. Leaves basal or cauline, pinnately or ternately parted or divided, alternate, exstipulate. Flowers perfect; sepals 2, bractlike, often caducous; corolla bilaterally symmetrical or irregular, petals 4, 1 or 2 of the outer petals basally spurred or saccate, the 2 inner petals usually connate apically over the stigma; stamens 6, diadelphous; stigma lobed, style persistant, ovary bicarpellate, 1-locular.

3 lines

period

period after number and name

1. Corolla 2-spurred.
    2. Vines; petals persistent.................... 1. Adlumia.
    2. Acaulescent herbs; petals deciduous......... 2. Dicentra.

2 spaces

1. Corolla 1-spurred.
    3. Ovary and fruit elongate; fruit dehiscent;
        seed crested.......................... 3. Corydalis.
    3. Ovary and fruit globose; fruit indehiscent;
        seed crestless........................ 4. Fumaria.

3 spaces

6 spaces

do not run keys
past right hand
column of numbers

[1] By C. Ritchie Bell, University of North Carolina

2 spaces 1 space

2. DICENTRA Bernh.

period at end of entry only when genus entry ends in abbrev. authority.

Perennial, scapose, glabrous, often glaucous herbs from scaly rhizomes or bulblet-bearing rootstocks. Leaves basal, long-petiolate, pinnately or ternately decompound, ultimate segments minutely apiculate. Inflorescence a raceme or panicle. Sepals 2, reduced, often caducous; corolla zygomorphic, petals 4, the outer 2 spurred or saccate at base, the inner 2 spatulate, crested dorsally, distally connate; stamens 6, in 2 groups of 3, each group fused and opposite an outer petal. Capsules valvate. Seeds lustrous black, reniform to suborbicular, ca. 2 mm broad, crested. Stern, K. R. 1961. Revision of *Dicentra*. Brittonia 13:1-57.

1. Inflorescence a panicle; flowers more than 18 mm
   long, dark pink.................................1. *D. eximia*.
1. Inflorescence a raceme; flowers less than 18 mm
   long, white or light pink.
   2. Spurs 1/3 or less of total corolla length;
      corolla base cordate........................2. *D. canadensis*.
   2. Spurs 1/2 or more of total corolla length;
      corolla base sagittate......................3. *D. cucullaria*.

Section C. ABBREVIATIONS

A. Nomenclature

sp., spp.	species (sing.), species (pl.)
ssp., sspp.	subspecies (sing.), subspecies (pl.)
var., vars.	variety (sing.), varieties (pl.)
f.	form (sing.), forms (pl.)

Authority abbreviations--follow those in the "Manual of the Vascular Flora of the Carolinas." In case of doubt, write out in full.

Common name for species which duplicates that of genus, use capital first letter for generic common name; e.g., OAK for *Quercus*, WHITE O. for *Q. alba*.

B. Description

Without period		With period	
mm	millimeter(s)	ca.	circa, about
cm	centimeter(s)	cult.	cultivated
dm	decimeter(s)	diam.	diameter
m	meter(s)	freq.	frequent

Flowering times--No abbreviations; do not use months. Spell out spring, summer, fall, all year; late or early can be used as modifiers where appropriate.

C. Distributions

Compass directions (small letters, no periods): n, ne, e, se, s, w, nw, sw

SE (caps, no period) = general distribution throughout all states of the southeastern United States

States (standard abbreviations without periods or internal spacing): Ala, Ark, Del, Fla, Ga, Ky, La, Md, Miss, NC, SC, Tenn, Va, WVa, Tex, Okla, Mo, Ill, Ind, Ohio, Pa, NJ

Physiographic provinces (caps, no period):

SA - Southern Appalachian Highlands
BR - Blue Ridge
RV - Ridge and Valley
CP - Coastal Plain (both Atlantic and Gulf and Mississippi Embayment)
AC - Atlantic Coastal Plain
GC - Gulf Coastal Plain
ME - Mississippi Embayment
PP - Piedmont Plateau
IP - Interior Low Plateau
CU - Cumberland Plateau
AP - Alleghany Plateau
CA - Cumberland-Alleghany Plateau
OH - Ozark - Ouachita Highlands
OZ - Ozarks
OU - Ouachitas

D. Synonyms from Manuals

S - Small, 1933
F - Fernald, 1950
G - Gleason & Cronquist, 1963
R - Radford et al., 1968

incl. If name used includes synonym; e.g., "incl. *A. discolor* Pursh--F, S" under *Aesculus pavia*.

Journal Abbreviations: Follow those in Schwarten, L. & H. W. Rickett. 1958. Abbreviations of titles of Serials cited by Botanists. Bulletin of the Torrey Club 85: 277-300.

Section D. GEOGRAPHY--ECOLOGY

I. GEOGRAPHY

Distribution sheets, one per species, for plotting data can be obtained by writing Albert E. Radford, Department of Botany, University of North Carolina, Chapel Hill, N. C. 27514. See Sample. Distribution is to be by major physiographic provinces, unless restricted (see below), within the southeastern states, and by state within the adjacent states. Each county in the list on the sample is keyed, by type face, to one of the three major provinces: Mountains (Bold Face), Piedmont or Interior Low Plateaus (Italics), and Coastal Plain (Regular). The symbols for the major provinces are: mts. for mountains which includes all subprovinces under SA and OH (see back of sample sheet); pied. for the piedmont and interior low plateaus; and cp. for the coastal plain which includes the subprovinces AC, GC, and ME (see Section C for abbreviations).

Province data will follow semicolon after habitat; e.g., Rich woods; mts. If a species is found in two provinces the provinces will be connected by "and," e.g., Rich woods; mts. and pied. If a species is found in all major provinces in the southeastern states, "all prov." is used, e.g. Rich woods; all prov. If a species is typically found in one or two provinces and occasionally in one or two others, province data should be preceded by "chiefly"; e.g., Rich woods; chiefly mts. and pied. This means that the species is occasionally found in the cp., but primarily in mts. and pied. If "chiefly mts." had been indicated, then the species also might be found occasionally in the pied. and/or cp.

Distribution by states within the southeast will be a separate entry with the states alphabetized; e.g., Rich woods; mts. Ala, Ark, NC, SC. If distribution is in each state of the southeastern United States, the symbol "SE" is used. If distribution is throughout the southeastern states except one, two, or three states, then SE is followed by "except" and the one, two, or three states listed alphabetically.

Fields; all prov. SE except Fla and WVa.
Fields; all prov. SE except Del, Fla, WVa.

Distribution in adjacent states will be a separate entry to be shown by abbreviations, in brackets [], from southwest to northeast, e.g., Tex, Okla, Mo, Ill, Ind, Ohio, Pa, NJ. If the species is present in all eight adjacent states, use the symbol [ALL].

Rich woods; mts. Ky, Va, WVa. [Ind, Ohio, Pa, NJ]
Fields; all prov. SE [ALL]

If distribution is restricted, then the author should give precise data using the appropriate symbols from Section C.

Juncus trifidus--Rocky ledges; BR. NC.
Heuchera rudgei--Rocky woods; IP. Ky and Tenn. [Ohio]
Avicennia nitida--Mangrove swamps; Keys. Fla.
Sedum pusillum--Granite outcrops; PP. Ala, Ga, SC.

If a species is found in two or three major provinces within the southeastern United States but not in the two or three major provinces for each state listed, then the province(s) in which it is found within a state is given after the state symbol.

Planta herba--Rich woods; mts. and pied. Ala mts., Ga, NC, SC. This means that the species is found in the mts. and pied. provinces of the three state area, but in Ala only in the mts.

Planta herba--Fields; all prov. Ala, Ark cp., Del, Fla, Ga, Ky. This means that the species is found in all three major prov. of the southeastern states listed but in Ark only in the cp.

Distribution information is to be based on specimens examined by the

Table 24-1. Counties and Physiographic Provinces of the Southeastern United States.

Species____________________

ALA.

☐ Auta
☐ Bald
☐ Barb
☐ Bibb
☐ **Blou**
☐ Bull
☐ Butl
☐ **Calh**
☐ *Cham*
☐ **Cher**
☐ Chil
☐ Choc
☐ Clar
☐ *Clay*
☐ *Cleb*
☐ Coff
☐ *Colb*
☐ Cone
☐ *Coos*
☐ Covi
☐ Cren
☐ **Cull**
☐ Dale
☐ Dall
☐ **DKal**
☐ Elmo
☐ Esca
☐ **Etow**
☐ Faye
☐ Fran
☐ Gene
☐ Gree
☐ Hale
☐ Henr
☐ Hous
☐ **Jack**
☐ **Jeff**
☐ Lama
☐ *Laud*
☐ *Lawr*
☐ *Lee*
☐ *Lime*
☐ Lown
☐ Maco
☐ *Madi*
☐ Mare
☐ Mari
☐ **Mars**
☐ Mobi
☐ Monr
☐ Mont
☐ **Morg**
☐ Perr
☐ Pick
☐ Pike
☐ *Rand*
☐ Russ
☐ **SCla**
☐ **Shel**
☐ Sumt
☐ **Tald**
☐ *Talp*
☐ Tusc
☐ **Walk**
☐ Wash
☐ Wilc
☐ **Wins**

ARK.

☐ Arka
☐ Ashl
☐ **Baxt**
☐ **Bent**
☐ **Boon**
☐ Brad
☐ Calh
☐ **Carr**
☐ Chic
☐ Clar
☐ Clay
☐ **Cleb**
☐ Clev
☐ Colu
☐ **Conw**
☐ Crai
☐ **Craw**
☐ **Crit**
☐ Cros
☐ Dall
☐ Desh
☐ Drew
☐ **Faul**
☐ **Fran**
☐ **Fult**
☐ **Garl**
☐ Gran
☐ Gree
☐ Hemp
☐ **HSpr**
☐ Howa
☐ **Inde**
☐ **Izar**
☐ Jack
☐ Jeff
☐ **John**
☐ Lafa
☐ Lawr
☐ Lee
☐ Linc
☐ LRiv
☐ **Loga**
☐ Lono
☐ **Madi**
☐ **Mari**
☐ Mill
☐ Miss
☐ Monr
☐ **Mont**
☐ Neva
☐ **Newt**
☐ Ouac
☐ **Perr**
☐ Phil
☐ **Pike**
☐ Poin
☐ **Polk**
☐ **Pope**
☐ Prai
☐ **Pula**
☐ **Rand**
☐ SFra
☐ **Sali**
☐ **Scot**
☐ **Sear**
☐ **Seba**
☐ Sevi
☐ **Shar**
☐ **Ston**
☐ Unio
☐ **VBur**
☐ **Wash**
☐ **Whit**
☐ Wood
☐ **Yell**

DEL.

☐ **Kent**
☐ NCas
☐ Suss

FLA.

☐ Alac
☐ Bake
☐ Bay
☐ Brad
☐ Brev
☐ Brow
☐ Calh
☐ Char
☐ Citr
☐ Clay
☐ Coll
☐ Colu
☐ Dade
☐ DSot
☐ Dixi
☐ Duva
☐ Esca
☐ Flag
☐ Fran
☐ Gads
☐ Gilc
☐ Glad
☐ Gulf
☐ Hami
☐ Hard
☐ Hend
☐ Hern
☐ High
☐ Hill
☐ Holm
☐ IRiv
☐ Jack
☐ Jeff
☐ Lafa
☐ Lake
☐ Lee
☐ Leon
☐ Levy
☐ Libe
☐ Madi
☐ Mana
☐ Mari
☐ Mart
☐ Monr
☐ Nass
☐ Okal
☐ Okke
☐ Oran
☐ Osce
☐ PBea
☐ Pasc
☐ Pine
☐ Polk
☐ Putn
☐ SJoh
☐ SLuc
☐ SRos
☐ Sara
☐ Semi
☐ Sumt
☐ Suwa
☐ Tayl
☐ Unio
☐ Volu
☐ Waku
☐ Walt
☐ Wash

GA.

☐ Appl
☐ Atki
☐ Baco
☐ Bake
☐ *Bald*
☐ *Bank*
☐ *Barr*
☐ **Bart**
☐ BHil
☐ Berr
☐ Bibb
☐ Blec
☐ Bran
☐ Broo
☐ Brya
☐ Bull
☐ Burk
☐ *Butt*
☐ Calh
☐ Camd
☐ Cand
☐ *Carr*
☐ **Cato**
☐ Char
☐ Chat
☐ Chae
☐ **Chaa**
☐ **Cher***
☐ *Clar*
☐ Clay
☐ *Clat*
☐ Clin
☐ *Cobb*
☐ Coff
☐ Colq
☐ *Colu*
☐ Cook
☐ *Cowe*
☐ Craw
☐ Cris
☐ **Dade**
☐ **Daws***
☐ Deca
☐ *DKal*
☐ Dodg
☐ Dool
☐ Douy
☐ *Doug*
☐ Earl
☐ Echo
☐ Effi
☐ *Elbe*
☐ Eman
☐ Evan
☐ **Fann***
☐ *Faye*
☐ **Floy**
☐ **Fors***
☐ *Fran*
☐ *Fult*
☐ **Gilm**
☐ Glas
☐ Glyn
☐ **Gord**
☐ Grad
☐ *Gree*
☐ *Gwin*
☐ **Habe***
☐ **Hall***
☐ *Hanc*
☐ *Hara*
☐ *Harr*
☐ *Hart*
☐ *Hear*
☐ *Henr*
☐ Hous
☐ Irwi
☐ *Jack*
☐ *Jasp*
☐ JDav
☐ Jeff
☐ Jenk
☐ John
☐ *Jone*
☐ *Lama*
☐ Lani
☐ Laur
☐ Lee
☐ Libe
☐ *Linc*
☐ Long
☐ Lown
☐ **Lump***
☐ McDu
☐ McIn
☐ Maco
☐ *Madi*
☐ Mari
☐ *Meri*
☐ Mill
☐ Mitc
☐ *Monr*
☐ Mont
☐ *Morg*
☐ **Murr**
☐ Musc
☐ *Newt*
☐ *Ocon*
☐ *Ogle*
☐ *Paul*
☐ Peac
☐ **Pick**
☐ Pier
☐ *Pike*
☐ **Polk**
☐ Pula
☐ *Putn*
☐ Quit
☐ **Rabu***
☐ Rand
☐ Rich
☐ *Rock*
☐ Schl
☐ Scre
☐ Semi
☐ *Spal*
☐ *Step*
☐ Stew
☐ Sumt
☐ *Talb*
☐ *Tali*
☐ Tatt
☐ Tayl
☐ Telf
☐ Terr
☐ Thom
☐ Tift
☐ Toom
☐ **Town***
☐ Treu
☐ *Trou*
☐ Turn
☐ *Twig*
☐ **Unio***
☐ *Upso*
☐ **Walk**
☐ *Walt*
☐ Ware
☐ *Warr*
☐ Wash
☐ Wayn
☐ Webs
☐ Whee
☐ **Whit***
☐ **Whid**
☐ Wilc
☐ *Wilk*
☐ Wiln
☐ Wort

KY.

☐ *Adai*
☐ *Alle*
☐ *Ande*
☐ Ball
☐ *Barr*
☐ *Bath*
☐ **Bell**
☐ *Boon*
☐ *Bour*
☐ **Boyd**
☐ *Boye*
☐ *Brac*
☐ **Brea**
☐ *Brec*
☐ *Bull*
☐ *Butl*
☐ *Cald*
☐ Call
☐ *Camp*
☐ Carl
☐ *Carr*
☐ **Cart**
☐ *Case*
☐ *Chri*
☐ *Clar*
☐ **Clay**
☐ *Clin*
☐ *Crit*
☐ *Cumb*
☐ *Davi*
☐ *Edmo*
☐ **Elli**
☐ *Esti*
☐ *Faye*
☐ *Flem*
☐ **Floy**
☐ *Fran*
☐ Fult
☐ *Gall*
☐ *Garr*
☐ *Gran*
☐ Grav
☐ *Gray*
☐ *Gren*
☐ **Grep**
☐ *Hanc*
☐ *Hard*
☐ **Harl**
☐ *Harn*
☐ *Hart*
☐ *Hend*
☐ *Henr*
☐ Hick
☐ *Hopk*
☐ **Jack**
☐ *Jeff*
☐ *Jess*
☐ **John**
☐ *Kent*
☐ **Knot**
☐ **Knox**
☐ *Laru*
☐ **Laur**
☐ **Lawr**
☐ **Lee**
☐ **Lesl**
☐ **Letc**
☐ *Lewi*
☐ *Linc*
☐ *Livi*
☐ *Loga*
☐ *Lyon*
☐ McCn
☐ **McCy**
☐ *McLe*
☐ *Madi*
☐ **Mago**
☐ *Mari*
☐ Mars
☐ **Mart**
☐ *Maso*
☐ *Mead*
☐ **Meni**
☐ *Merc*
☐ *Metc*
☐ *Monr*
☐ *Mont*
☐ **Morg**
☐ *Muhl*
☐ *Nels*
☐ *Nich*
☐ *Ohio*
☐ *Oldh*
☐ *Owen*
☐ **Owsl**
☐ *Pend*
☐ **Perr**
☐ **Pike**
☐ *Powe*
☐ *Pula*
☐ *Robe*
☐ *Rock*
☐ *Rowa*
☐ *Russ*
☐ *Scot*
☐ *Shel*
☐ *Simp*
☐ *Spen*
☐ *Tayl*
☐ *Todd*
☐ *Trig*
☐ *Trim*
☐ *Unio*
☐ *Warr*
☐ *Wash*
☐ *Wayn*
☐ *Webs*
☐ **Whit**
☐ **Wolf**
☐ *Wood*

LA.

☐ Acad
☐ Alle
☐ Asce
☐ Assu
☐ Avoy
☐ Beau
☐ Bien
☐ Boss
☐ Cadd
☐ Calc
☐ Cald
☐ Came
☐ Cata
☐ Clai
☐ Conc
☐ DSot
☐ EBRo
☐ ECar
☐ EFel
☐ Evan
☐ Fran
☐ Gran
☐ Ibea
☐ Ibee
☐ Jack
☐ Jeff
☐ JDav
☐ Lafa
☐ LFou
☐ LSal
☐ Linc
☐ Livi
☐ Madi
☐ More
☐ Natc
☐ Orle
☐ Ouac
☐ Plaq
☐ PCou
☐ Rapi
☐ RRiv
☐ Rich
☐ Sabi
☐ SBer
☐ SCha
☐ SHel
☐ SJam
☐ SJBa
☐ SLan
☐ SMar
☐ SMay
☐ STam
☐ Tang
☐ Tens
☐ Terr
☐ Unio
☐ Verm
☐ Vern
☐ Wash
☐ Webs
☐ WBRo
☐ WCar
☐ WFel
☐ Winn

MD.

☐ **Alle**
☐ AAru
☐ *Balt*
☐ BCit
☐ Calv
☐ Caro
☐ *Carr*
☐ Ceci
☐ Char
☐ Dorc
☐ *Fred*
☐ **Garr**
☐ *Harf*
☐ *Howa*
☐ Kent
☐ *Mont*
☐ PGeo
☐ QAnn
☐ SMar
☐ Some
☐ Talb
☐ **Wash**
☐ Wico
☐ Worc

MISS.

☐ Adam
☐ Alco
☐ Amit
☐ Atta
☐ Bent
☐ Boli
☐ Calh
☐ Carr
☐ Chic
☐ Choc
☐ Clai
☐ Clar
☐ Clay
☐ Coah
☐ Copi
☐ Covi
☐ DSot
☐ Forr
☐ Fran
☐ Geor
☐ Gree
☐ Gren
☐ Hanc
☐ Harr
☐ Hind
☐ Holm
☐ Hump
☐ Issa
☐ Itaw
☐ Jack
☐ Jasp
☐ Jeff
☐ JDav
☐ Jone
☐ Kemp
☐ Lafa
☐ Lama
☐ Laud
☐ Lawr
☐ Leak
☐ Lee
☐ Lefl
☐ Linc
☐ Lown
☐ Madi
☐ Mari
☐ Mars
☐ Monr
☐ Mont
☐ Nesh
☐ Newt
☐ Noxu
☐ Okti
☐ Pano
☐ Pear
☐ Perr
☐ Pike
☐ Pont
☐ Pren
☐ Quit
☐ Rank
☐ Scot
☐ Shar
☐ Simp
☐ Smit
☐ Ston
☐ Sunf
☐ Tall
☐ Tate
☐ Tipp
☐ Tish
☐ Tuni
☐ Unio
☐ Walt
☐ Warr

☐ Wash
☐ Wayn
☐ Webs
☐ Wilk
☐ Wins
☐ Yalo
☐ Yazo

N.C.

☐ *Alam*
☐ *Alex*
☐ **Alle***
☐ *Anso*
☐ **Ashe***
☐ **Aver***
☐ Beau
☐ Bert
☐ Blad
☐ Brun
☐ **Bunc***
☐ *Burk*
☐ *Caba*
☐ *Cald*
☐ Camd
☐ Cart
☐ *Casw*
☐ *Cata*
☐ *Chat*
☐ **Cher***
☐ Chow
☐ **Clay***
☐ *Clev*
☐ Colu
☐ Crav
☐ Cumb
☐ Curr
☐ Dare
☐ *Davn*
☐ *Dave*
☐ Dupl
☐ *Durh*
☐ Edge
☐ *Fors*
☐ *Fran*
☐ *Gast*
☐ Gate
☐ **Grah***
☐ *Gran*
☐ Gree
☐ *Guil*
☐ Hali
☐ Harn
☐ **Hayw***
☐ **Hend***
☐ Hert
☐ Hoke
☐ Hyde
☐ *Ired*
☐ **Jack***
☐ John
☐ Jone
☐ *Lee*
☐ Leno
☐ *Linc*
☐ *McDo*
☐ **Maco***
☐ **Madi***
☐ Mart
☐ *Meck*
☐ **Mitc***
☐ *Mont*
☐ Moor
☐ *Nash*
☐ NHan
☐ Nort
☐ Onsl
☐ *Oran*
☐ Paml
☐ Pasq
☐ Pend
☐ Perq
☐ *Pers*
☐ Pitt
☐ **Polk***
☐ *Rand*
☐ Rich
☐ *Robe*
☐ *Rock*
☐ *Rowa*
☐ *Ruth*
☐ Samp
☐ Scot
☐ *Stan*
☐ *Stok*
☐ *Surr*
☐ **Swai***
☐ **Tran***
☐ Tyrr
☐ *Unio*
☐ *Vanc*
☐ *Wake*
☐ *Warr*
☐ Wash
☐ **Wata***
☐ Wayn
☐ *Wilk*
☐ Wils
☐ *Yadk*
☐ **Yanc***

S.C.

☐ *Abbe*
☐ Aike
☐ Alle
☐ *Ande*
☐ Bamb
☐ Barn
☐ Beau
☐ Berk
☐ Calh
☐ Char
☐ *Cher*
☐ *Ches*
☐ Ched
☐ Clar
☐ Coll
☐ Darl
☐ Dill
☐ Dorc
☐ *Edge*
☐ *Fair*
☐ Flor
☐ Geor
☐ *Grev*
☐ *Grew*
☐ Hamp
☐ Horr
☐ Jasp
☐ Kers
☐ *Lanc*
☐ *Laur*
☐ Lee
☐ Lexi
☐ *McCo*
☐ Mari
☐ Marl
☐ *Newb*
☐ *Ocon*
☐ Oran
☐ *Pick*
☐ Rich
☐ *Salu*
☐ *Spar*
☐ Sumt
☐ *Unio*
☐ Will
☐ *York*

TENN.

☐ **Ande**
☐ *Bedf*
☐ *Bent*
☐ **Bled**
☐ **Blou**
☐ **Brad**
☐ **Camp**
☐ *Cann*
☐ Carr
☐ **Cart***
☐ *Chea*
☐ Ches
☐ **Clai**
☐ *Clay*
☐ **Cock**
☐ *Coff*
☐ Croc
☐ **Cumb**
☐ *Davi*
☐ Deca
☐ *DKal*
☐ *Dick*
☐ Dyer
☐ Faye
☐ **Fent**
☐ *Fran*
☐ Gibs
☐ *Gile*
☐ **Grai**
☐ **Gree**
☐ **Grun**
☐ **Hamb**
☐ **Hami**
☐ **Hanc**
☐ Harm
☐ Harn
☐ **Hawk**
☐ Hayw
☐ Hend
☐ Henr
☐ *Hick*
☐ *Hous*
☐ *Hump*
☐ *Jack*
☐ **Jeff**
☐ **John**
☐ **Knox**
☐ Lake
☐ Laud
☐ *Lawr*
☐ *Lewi*
☐ *Linc*
☐ **Loud**
☐ **McMi**
☐ McNa
☐ *Maco*
☐ Madi
☐ **Mari**
☐ *Mars*
☐ *Maur*
☐ **Meig**
☐ **Monr**
☐ *Mont*
☐ *Moor*
☐ **Morg**
☐ Obio
☐ *Over*
☐ *Perr*
☐ *Pick*
☐ **Polk***
☐ *Putn*
☐ **Rhea**
☐ **Roan**
☐ *Robe*
☐ *Ruth*
☐ **Scot**
☐ **Sequ**
☐ **Sevi***
☐ Shel
☐ *Smit*
☐ *Stew*
☐ **Sull**
☐ *Sumn*
☐ Tipt
☐ *Trou*
☐ **Unic***
☐ **Unio**
☐ **VBur**
☐ *Warr*
☐ **Wash**
☐ *Wayn*
☐ Weak
☐ *Whit*
☐ *Will*
☐ *Wils*

VA.

☐ Acco
☐ *Albe*
☐ **Alle**
☐ *Amel*
☐ *Amhe*
☐ *Appo*
☐ *Arli*
☐ **Augu**
☐ **Bath**
☐ *Bedf*
☐ **Blan**
☐ **Bote**
☐ *Brun*
☐ **Buch**
☐ *Buck*
☐ *Camp*
☐ Caro
☐ *Carr*
☐ CCit
☐ *Char*
☐ Ches
☐ **Clar**
☐ **Crai**
☐ *Culp*
☐ *Cumb*
☐ **Dick**
☐ Dinw
☐ Esse
☐ *Fair*
☐ *Fauq*
☐ *Floy*
☐ *Fluv*
☐ *Fran*
☐ **Fred**
☐ **Gile**
☐ Glou
☐ *Gooc*
☐ **Gray***
☐ *Gree*
☐ Grev
☐ *Hali*
☐ Hano
☐ Henr
☐ *Heny*
☐ **High**
☐ IoWi
☐ JCit
☐ KaQu
☐ KGeo
☐ KWil
☐ Lanc
☐ **Lee**
☐ *Loud*
☐ *Loui*
☐ *Lune*
☐ *Madi*
☐ Math
☐ *Meck*
☐ Midd
☐ **Mont**
☐ Nans
☐ *Nels*
☐ NKen
☐ Norn
☐ Nord
☐ *Nott*
☐ *Oran*
☐ **Page**
☐ *Patr*
☐ *Pitt*
☐ *Powh*
☐ *PEdw*
☐ PGeo
☐ *PWil*
☐ **Pula**
☐ *Rapp*
☐ Rich
☐ **Roan**
☐ **Rock**
☐ **Rocm**
☐ **Russ**
☐ **Scot**
☐ **Shen**
☐ **Smyt**
☐ Sout
☐ *Spot*
☐ *Staf*
☐ Surr
☐ Suss
☐ **Taze**
☐ **Warr**
☐ **Wash**
☐ West
☐ **Wise**
☐ **Wyth**
☐ York

W.VA.

☐ **Barb**
☐ **Berk**
☐ **Boon**
☐ **Brax**
☐ **Broo**
☐ **Cabe**
☐ **Calh**
☐ **Clay**
☐ **Dodd**
☐ **Faye**
☐ **Gilm**
☐ **Gran**
☐ **Gree**
☐ **Hamp**
☐ **Hanc**
☐ **Hard**
☐ **Harr**
☐ **Jack**
☐ **Jeff**
☐ **Kana**
☐ **Lewi**
☐ **Linc**
☐ **Loga**
☐ **McDo**
☐ **Mari**
☐ **Mars**
☐ **Maso**
☐ **Merc**
☐ **Mine**
☐ **Ming**
☐ **Mono**
☐ **Monr**
☐ **Morg**
☐ **Nich**
☐ **Ohio**
☐ **Pend**
☐ **Plea**
☐ **Poca**
☐ **Pres**
☐ **Putn**
☐ **Rale**
☐ **Rand**
☐ **Ritc**
☐ **Roan**
☐ **Summ**
☐ **Tayl**
☐ **Tuck**
☐ **Tyle**
☐ **Upsh**
☐ **Wayn**
☐ **Webs**
☐ **Wetz**
☐ **Wirt**
☐ **Wood**
☐ **Wyom**

☐ ILL.
☐ IND.
☐ MO.
☐ N. J.
☐ OHIO
☐ OKLA.
☐ PENN.
☐ TEX.
☐ L. I.

	Coastal Plain			Piedmont	Int. Low	Mountains					
	CP			Plateau	Plateau	SA				OH	
						CA					
	AC	GC	ME	PP	IP	CU	AP	BR	RV	OZ	OU
Ala.	—	☐	—	☐	☐	☐	—	—	☐	—	—
Ark.	—	—	☐	—	—	—	—	—	—	☐	☐
Del.	☐	—	—	☐	—	—	—	—	—	—	—
Fla.	☐	☐	—	—	—	—	—	—	—	—	—
Ga.	☐	—	—	☐	—	—	—	☐	☐	—	—
Ky.	—	—	☐	—	☐	☐	☐	—	—	—	—
La.	—	☐	☐	—	—	—	—	—	—	—	—
Md.	☐	—	—	☐	—	—	☐	☐	☐	—	—
Miss.	—	☐	☐	—	*	—	—	—	—	—	—
N. C.	☐	—	—	☐	—	—	—	☐	—	—	—
S. C.	☐	—	—	☐	—	—	—	*	—	—	—
Tenn.	—	—	☐	—	☐	☐	—	☐	☐	—	—
Va.	☐	—	—	☐	—	—	—	☐	☐	—	—
W. Va.	—	—	—	—	—	—	☐	☐	☐	—	—

*In a small portion of a few counties but not a major portion of any—essentially absent.
Reference: Fenneman, N. M. 1938. "Physiography of Eastern United States." McGraw-Hill Book Co., N. Y.

Standard =Coastal Plain: CP (including AC, GC, ME).

Italic =Piedmont: PP. and Interior Low Plateau: IP

Boldface =Mountains: RV; CA (including CU, AP); OH (including OZ, OU).

Boldface* =Blue Ridge: BR.

author in several major herbaria and from publications by competent authorities. Do not base distribution on general ranges given in manuals.

These geographic instructions do not cover all distribution possibilities so each author is enjoined to be consistent in describing his range exceptions.

II. ECOLOGY--HABITATS

A. Notes

Designation of frequency or density should not be used unless author is entirely confident of statements such as rare, infrequent, common, abundant.

Endemism should be stated where certain, e.g., *Shortia galacifolia*, endemic to BR of Ga, SC, NC.

Specific habitats should be used for highly localized species when definitely known: i.e., Sand dunes (*Uniola paniculata*), Shale barrens (*Oenothera argillicola*), Limestone or dolomite outcrops (*Asplenium resiliens*), Hammocks (*Clusia flava*), etc.

B. Habitat Designations
(See Chapter 17)

Aquatic

river
stream
springhead
ditches
lake
pond or pool
marsh
bog

Use modifiers where certain:

brackish
marine shores
acid (pH 5 & below)
rapid
brackish marsh
peat bog
shrub bog
cypress swamp

Fields (use modifiers when needed)

wet
dry
low
upland
sandy
clay
loam
rocky
old
cult.

Balds

heath balds
grassy balds

Disturbed habitats

waste places
roadsides
railroads
fence rows, hedge rows
gardens
pastures
meadows

Savannahs (use modifiers when needed)

wet
dry
pine

Outcrops (use modifiers when needed)

granite
sandstone
limestone

Woods (use modifiers to describe condition and also characteristic species where needed).

evergreen
deciduous
low (alluvial or swamp)
rich
dry
dry sandy
dry clayey
open
dense

pine, spruce-fir, longleaf pine
oak woods, mixed hardwoods, oak-pine
Taxodium, *Nyssa*
oak woods, mixed mesophytic
oak-hickory, oak-pine
turkey oak
blackjack-post oak
pine
cove hardwoods

Section E. SYNONYMY

Synonyms are to be included only when the names employed in the Flora of the SE differs from those used in the recent manuals of the Eastern United States. These recent manuals and their abbreviations are in chronological order by author:

S J. K. Small. Manual of the Southeastern Flora. 1933.

F M. L. Fernald. Gray's Manual of Botany, 8th edition. 1950.

G H. A. Gleason and A. Cronquist. Manual of the Vascular Plants of the Northeastern United States and Adjacent Canada. 1963.

R A. E. Radford, H. E. Ahles and C. R. Bell. Manual of the Vascular Flora of the Carolinas. 1968.

A) Previous segregates accepted in the above recent manuals but not by the contributor should be listed as follows:

(Under *Cassia nictitans* L.)

"Incl. *Chamaecrista multipinnata* (Pollard) Greene--S; *Cassia n.* var. *hebecarpa* Fernald, *C. n.* var. *leiocarpa* Fernald--F, G."

B) Taxa previously included as synonyms in the above recent manuals but regarded as distinct by the contributor should be listed in synonymy as follows:

(Under *Desmodium nuttallii* (Schindler) Schubert)

". . .*Meibomia viridiflora* (Nuttall) Kuntze--S, in part; *D. viridiflorum* (L.) DC.--G, in part."

(Under *Cassia obtusifolia* L.)

". . .*Emelista tora* (L.) Britton & Rose--S; *C. tora* L.--F, G."

C) Taxa previously passing under the name of other taxa should be indicated as follows:

(Under *Amorpha georgiana* Wilbur)

"*A. cyanostachya* Curtis--sensu S, not Curtis"

(Under *Aeschynomene indica* L.)

"*A. virginica* (L.) BSP--sensu S, in large part."

(Under *Sabatia stellaris* Pursh)

"*S. campanulata* (L.) Torrey--sensu S, in part."

D) Obvious synonyms from the above recent manuals should be listed:

(Under Diamorpha smallii Britton)

"incl. D. cymosa (Nuttall) Britton ex Small--S, Sedum smallii (Britton) Ahles--R."

(Under Psoralea lupinellus Michaux)

"Rhytidomene lupinellus (Michaux) Rydberg--S."

(Under Lotus helleri (Britton)

"Acmispon helleri (Britton) Small--S."

Note: Synonymy is not as mechanically categorized as the above might indicate. The above is meant merely as a tentative guide and the contributor is encouraged to convey the facts accurately and concisely. Constributors are strongly urged to publish their nomenclatural research in full elsewhere. Scholarship demands more than the legal minimum of the basionym.

Section F. VASCULAR FAMILIES OF SOUTHEASTERN UNITED STATES
(Selected)

1. Acanthaceae - Long
2. Aceraceae - James
3. Aizoaceae - Ward
4. Alismataceae - DePoe
5. Amaranthaceae - Massey
6. Amaryllidaceae - Browne
7. Anacardiaceae - Long
8. Annonaceae - Kral
9. Apiaceae - Bell
10. Apocynaceae - Bell
11. Aquifoliaceae - Clark
12. Araceae - Ward
13. Araliaceae - Bell
14. Arecaceae - Read
15. Aristolochiaceae - Bell
16. Asclepiadaceae - Bell
17. Aspidiaceae - Evans
18. Aspleniaceae - Evans
19. Asteraceae - Shinners
20. Azollaceae - Evans
21. Balsaminaceae - Ward
22. Bataceae - Bell
23. Begoniaceae - Smith
24. Berberidaceae - Bell
25. Betulaceae - Hardin
26. Bignoniaceae - Bell
27. Blechnaceae - Evans
28. Boraginaceae - Hommersand

Section G. TAXONOMIC PHILOSOPHY, CONCEPTS, POLICIES
(From Shetler et al. [1973] A Guide for Contributors to Flora North America [FNA] [Provisional Edition])

I. TAXONOMIC PHILOSOPHY

The Editorial Committee will encourage in principle a broad species concept and taxonomic practices that promote stability in the taxonomy and nomenclature of North American plants. To this end, it is expected that the authors will keep changes in rank to a minimum and make them only after adequate taxonomic investigation and publication of the results in an appropriate journal or monograph. Where current authorities offer a choice of rank for a particular taxon, the author's decision should be made in the light of the taxon's variation throughout its whole geographic range. Radical treatments involving excessive species-splitting or species-lumping relative to past practice will be given the closest scrutiny by the Editorial Committee, and the authors of such treatments must be prepared to defend them on objective grounds. In no case will the pages of the Flora be open for publication of new taxa or new combinations. Such publication should be done in the recognized journals, where appropriate evidence and documentation can be presented.

II. TAXONOMIC CONCEPTS AND POLICIES

The contributor will define the limits of all taxa within the group he is treating, while the Editorial Committee, in consultation with the specialists, will be responsible for delimiting taxa of coordinate or higher rank. In practice, this means that in most cases the contributor will be responsible for the limits of infrageneric taxa, mainly the species, and the Editorial Committee will be responsible for setting the limits of genera and higher taxa. Contributors dealing with whole families naturally will be responsible for delimiting all included genera. All decisions by contributors will be subject to the usual editorial review.

The contributors and Editors should take into account all available evidence--morphological, geographical, ecological, cytological, chemical--in delimiting genera and species, but the taxa so recognized must be definable ultimately on a practical morphological basis.

A. Genera. The Editorial Committee will strive to achieve a reasonably uniform generic concept from group to group, but especially within families. Although there can be no fixed criteria for delimiting genera, due consideration should be given to historical precedent and to the importance of maintaining nomenclatural stability. Familiar genera should not be merged or split on the basis of personal preference or other trivial grounds, but only on the basis of carefully weighed new evidence that is documented fully in the literature. Furthermore, generic alignments should be made in the light of treatments in Flora SSSR and *Flora Europaea*. For the sake of advancing common understanding, every effort should be made to eliminate artificial generic distinctions between the FNA and comparable European or Russian treatments of the same or closely related species.

B. Species. The FNA Manual above all is to be a practical work, and the primary units of the Flora, the species, must be practical entities with consistent, if small, morphological distinctions. At the same time it is understood that no two species, notwithstanding their equivalent rank, ever are equivalent in their origin, nature, and extent of evolutionary differentiation. Even within genera the species, though definable by morphological criteria, will illustrate different theoretical or evolutionary situations. All available biosystematic evidence should be used in delineating species in the FNA treatments. Some examples of different kinds of species, viewed theoretically, that may be recognized in the FNA Manual, as in *Flora Europaea*, are discussed in the *Flora Europaea* guidebook (Heywood 1958, Appendix E). At least the following kinds of species may be recognized whenever the morphological requirements can be met:

- Highly distinct, widely distributed plants with little variation.
- Species with different ploidy levels in which the levels are not recognizable morphologically or in which the variation is so nearly continuous that infraspecific taxa cannot be recognized.
- Species in which aneuploidy occurs, but in which the different chromosome numbers are not correlated with morphological features of the plants.

- Species distinguished by minor morphological characters, but which are distinct ecologically from related species.
- Polymorphic species with morphologically definable subspecies or varieties.
- Polymorphic species in which clinal variation occurs and in which infraspecific taxa cannot be distinguished satisfactorily.
- Species that are widely separated and morphologically different from related species with which they are highly interfertile.
- Asexually reproducing plants that are morphologically distinctive and which occupy a significant geographical range.

All recognized species are to receive full and comparable treatment in accordance with the FNA guidelines, and the Editors will not include any species in the Data Bank that are not recognized clearly by the contributor himself on a coordinate basis with all other species recognized by him. The recognition and treatment of microspecies is discouraged, although microspecies may be grouped as complexes into broader species. In such cases, orthodox binomials must be used for the broad species (e.g., *Campanula rotundifolia* L. *s.l.*, not *Campanula rotundifolia* complex), and the microspecies must be treated as taxonomic synonyms, not as a list of included microspecies. To summarize, there can be only one level of species. Species may be aggregated, however, to simplify keying or to show close similarities, provided that the "aggregate," "group," or "complex" can be keyed out and diagnosed as a unit on a practical morphological basis and can serve, therefore, as a useful though informal supraspecific category.

C. *Infraspecific Taxa*. Only major infraspecific taxa that are morphologically definable entities will be included in the FNA Manual. Either the category *subspecies* or the category *variety* may be used, depending on the history and knowledge of the group, but both may not be used in the same species. Furthermore, each contributor should use one category or the other consistently throughout the genus or other group that he is treating, except that individual authors treating medium-sized or large families may be permitted to vary usage from genus to genus if, in the Family Editor's judgment, such inconsistency is justified. The use of subspecies is preferred, because in the Editorial Committee's view this is the more meaningful evolutionary category, but historically the category variety has often been used in about the same sense. To exclude varieties from the Flora would encourage wholesale publication of new combinations, a result that in itself would not contribute to our knowledge of the given taxa and would be contrary to the Committee's goal of promoting nomenclatural and taxonomic stability. The Committee is attempting, therefore, to strike a balance in the use of subspecies and varieties between total inconsistency and total but artificial uniformity, realizing that the guidelines set forth here will cause a certain amount of recombination to bring about the required consistency within groups. In general, the contributor should be guided insofar as possible by existing knowledge and nomenclatural precedent. In groups where subspecies are recognized, traditional varieties should be dropped altogether if they cannot be defended on good evolutionary grounds as subspecies.

The following kinds of taxonomic groups may be treated as subspecies or varieties in the Flora:

- geographically isolated populations that differ from one another by fewer characters than do species of the same genus and can be

demonstrated to have some degree of interfertility.
- ecologically specialized populations, particularly with respect to certain edaphic or climatic conditions, when there also is a morphological means of distinction.
- cytological variants, if they have characteristic distribution patterns and can be recognized by features other than chromosome number alone.
- chemical variants, with concomitant morphological and geographical or ecological attributes.

A key to infraspecific taxa must be provided when either subspecies or varieties are recognized. All formally recognized infraspecific taxa must be treated in the standard way, including full description. Taxa that are not accepted by the author himself may not be included.

Any variation not meriting treatment at the subspecific or varietal level (e.g., clinal or local ecotypic variation, minor variation in color or pubescence), is not to receive formal recognition. Specifically, the formal use of the category form will not be permitted. Informal comment on infraspecific variation must be brief, germane, and without the haphazard use of such terms as "cultivar," "form," "type," and "race"; otherwise, it will be pruned rigorously or eliminated.

D. Other Categories. Infrafamilial and infrageneric categories sanctioned by the International Code of Botanical Nomenclature (ICBN; Stafleu and Voss 1972) may be intercalated between family and genus or between genus and species, respectively, provided they are formally described and serve a useful purpose in the treatment. The judicious use of such categories is encouraged as a means of subdividing large families and genera into more natural units, thereby making the treatment of such a family or genus more useful as a systematic conspectus. Their use also allows for more concise treatment, because characters that are common to a whole group of species or genera can be described under the appropriate infrageneric or infrafamilial category and can be omitted from the descriptions of the individual species or genera. Excessive subdividing is discouraged, however, because the treatments must remain practical. Generally, a single rank between family and genus and between genus and species should be sufficient. Ordinarily, this will be subfamily or tribe between family and genus, and subgenus or section between genus and species, depending on traditional usage within the group. The ranks subtribe, subsection, series, and, particularly, subseries should be used sparingly if at all. Usually, the infrafamilial and infrageneric taxa will not be included in the keys.

Experimental categories, e.g., coenospecies, ecospecies, ecotype, deme, may not be used in the formal nomenclature. Likewise, the term "cultivar" and other special terms in use for cultivated plants will not be permitted as designations of formal categories. If essential to an understanding of the contributor's treatment, the current biosystematic concept of the group may be summarized tersely in pertinent experimental terminology under the section "Observations." Expansive hypothetical discussions in relationships will be condensed to their essence or eliminated by the Editors.

E. Hybrids. Intergeneric and interspecific hybrids may be included in the Flora only when they are common, self-propagating taxa worthy of treatment as ordinary genera or species. Such hybrids must be given regular generic or binomial names prefixed by the multiplication sign (x), in accordance with

Appendix I of the ICBN. Formulas to express hybrid parentage, as provided by the ICBN, may not be used in the formal nomenclature, but may be used sparingly in an explanation of the parentage under "Observations" when such explanation is absolutely necessary.

The indiscriminate naming of hybrids is to be avoided, and in general the Editorial Committee will permit the inclusion of designated hybrid taxa only when their hybrid origin has been thoroughly established and widely accepted and when such taxa clearly represent entities in the North American flora that must be recognized. Long lists of putative hybrids will be eliminated. If it is essential to an understanding of the group, then the significance of hybridization may be explained by simple comment under "Observations."

F. Apomicts. The taxonomy of groups in which apomixis occurs must be approached with special caution. To attempt to describe or even mention all apomicts in such critical genera as *Rubus* and *Taraxacum* is out of the question, and the Editors will not accept treatments in which apomictic plants have been named and described indiscriminately. Insofar as is possible, apomicts should be referred to their closest sexually-reproducing species. However, when an apomictic plant is widely distributed and morphologically distinct on a constant basis, it may be treated as a species. If it is essential to an understanding of the group, then the significance of apomixis may be discussed briefly under "Observations," but without giving a long list of named apomicts.

G. Alien Species. The Editorial Committee will be guided by a liberal policy on the inclusion of alien (introduced) species. Contributors will be given wide latitude to decide which species to include, on the condition that all included aliens be incorporated into the key and described in full. Aliens that are thoroughly naturalized, as judged by the contributor, are to be included automatically. These include not only weeds and accidental introductions but also garden plants that have escaped and are propagating themselves well beyond the immediate vicinity of where they are cultivated. In addition, imperfectly naturalized or casual aliens (adventives, waifs, escapes; i.e., aliens that apparently can propagate themselves in the wild, but which are not known to be spreading nor to be long persisting without constant reintroduction) should be included when in the contributor's opinion they are of special interest or are likely to be encountered frequently enough in the flora to pose problems in identification. Even cultivated perennials (especially trees and shrubs) may be included if they are species that often are found persisting untended in the wild in stands or as individuals. Finally, there are taxonomic groups in which it may be desirable to treat certain of the purely cultivated species because they are trees or crop plants that are planted or cultivated in continuous stands or on an extensive scale. Cultivated plants, i.e., species that will not reproduce in the wild or persist without tending or management, should be included sparingly, however, and only when their inclusion serves a real purpose. Any cultivated species that are included must be keyed and described in the usual manner.

H. Rare and Endangered Species. A special effort will be made to identify those taxa native to the FNA territory which, on the basis of their entire worldwide range, are known or believed to be rare or endangered or to have become extinct in the wild recently. In their treatments, contributors should include recently extinct species (e.g., *Franklinia alatamaha*), describing and keying them, and also should attempt to designate the rare and endangered species. Criteria for recognizing and classifying rare and endangered species are given in a later section of the Guidebook. All introduced, naturalized, and cultivated plants are to be excluded from consideration. A species that is rare or endangered in one or two states, but is common elsewhere in its geographic range, does not qualify for recognition in the FNA rare and endangered list. Primary attention is to be given to species that are endemic to the FNA territory, especially if they are locally endemic or relict species.

A list of rare and endangered species is urgently needed for North America, but none has ever been compiled on a continental basis. The need for at least a national list for the United States has been emphasized repeatedly by those who have prepared environmental impact statements and studies on the location and preservation of natural areas and ecological preserves, and by concerned citizens and conservation organizations. Preserving rare and endangered species has become a matter of policy at national and international levels. The first step in doing so is to provide scientifically accurate information. The designation of rare and endangered species is the subject of recent bills before the U. S. Congress. The International Union for Conservation of Nature and Natural Resources is concerned about endangered plants on an international level. Certain rare and endangered plants are being considered for international protection to prevent their export and import. FNA is the appropriate body to compile a scientifically authoritative list of endangered plants for the United States (less Hawaii) and Canada. Persons concerned about endangered species on the state or local level are encouraged to use the same criteria and classification scheme set forth for the FNA Data Bank. There is a definite need for a hierarchy of lists of endangered species -- local, state, national, and international.

I. Nature and Scope of Descriptions. All descriptions must be comparative as well as diagnostic. The primary object of a flora is to provide a rapid means of correctly identifying the plants of some specified region. For this reason descriptions in Floras usually include only the most diagnostic characters of the plants treated. Because these characters differ greatly from group to group, the descriptions in a Flora rarely are comparable from one family or genus to another or even among species of the same genus.

The FNA Manual is intended to serve the usual purpose of a Flora, and, accordingly, it will include only diagnostic descriptions, which must be as concise as is consistent with clarity and utility. The FNA Data Bank is intended to serve the additional, critically important, new function of correlation (comparison), made possible by the computer-based Information System in which the Data Bank is maintained. Through the query language of the FNA System, it is possible to retrieve the data in any of a virtually limitless array of permutations. Thus, simple to highly complex correlations of morphological, ecological, geographical, and other parameters are possible within and among taxonomic groups at the same or at different levels.

Reliable, meaningful correlation is possible, however, only when the descriptions are all comparable in content, format, and terminology. Otherwise, the problems of missing data and unstandardized formats and terms largely subvert retrieval strategy and make generalization from query responses hazardous, if not impossible. Therefore, the Editorial Committee has defined a minimum set of characters that are common to most vascular plants for which data must be supplied consistently and comparatively in every description by every contributor over the full range of taxa treated, except where the characters are not applicable. This common requirement is intended to guarantee some measure of comparability and completeness in the FNA Data Bank from the outset.

Each contributor will be expected to provide data not only for all required comparative characters, but also for those characters that he considers to be the most diagnostic for the taxon in question, which often may not be among the required set. The guidelines for description are to be followed for the diagnostic characters as well as for the comparative characters. The full descriptions will be entered into the Data Bank and included with the treatments when they are published individually, but only the diagnostic characters will be printed in the concise Manual.

All descriptions are to be based on the diversity and variation of the taxa as they are represented in the FNA territory. An exception is to be made only when this procedure would lead to a distorted description of the taxon, because its FNA representatives are highly unrepresentative of the taxon as a whole. If the diversity or variability found outside the FNA area is taken into consideration, this should be indicated clearly, character by character. It is assumed that the need for taking exception to the above rule will be minimal at the species level and will increase upward through the hierarchy of taxonomic categories.

Chromosome Number is an optional character, to be given at the end of the morphological description when reliable information can be supplied. Contributors are encouraged to include chromosome number in their descriptions whenever possible, especially if number is taxonomically important or critical. However, all counts must be verified in the primary sources and must be based on material of known wild origin within the FNA territory. It is intended that in due course the FNA Data Bank will become the central repository for authentic, critically evaluated information on the chromosome numbers of North American plants. In the interest of producing the FNA Manual rapidly, however, available information on chromosome number is not being requested on a mandatory basis at this stage, because of the extra effort that would be required of the contributors and Editors to search out and verify the data. An unverified FNA compilation would not add significantly to the secondary compilations of chromosome numbers already available to botanists. It is the hope of the Editorial Committee that a separate effort to compile a complete file of documented chromosome numbers can be undertaken once the FNA Manual is published, and sooner if possible.

The FNA Data Bank also is designed so that in the long run it can include comprehensive ecological characterizations of the taxa. For the immediate purposes of the FNA Manual, however, the ecological data will be restricted for practical reasons to characterizations of habitat preference and flowering time. The contributor should provide a brief description of habitat and flowering time for each taxon.

As a final note on description, it should be emphasized that these requirements apply to all taxa at the level of family or below. Orders and higher taxa will not be described. The treatments are to be comparable between taxa of different ranks as well as between taxa of the same rank. Furthermore, each taxon should be described so as to encompass all of its subordinate taxa. To the maximum degree possible, the computer will be used to check whether descriptions take into account the characters and character states of all subordinate taxa.

MANUAL FORMAT EXERCISE

Write a dichotomous key to ten species of angiosperms. The species can be from one to ten genera in one to ten families. Write complete descriptions for two species and a diagnostic description for one of the two species. Consult the format instructions.

DESCRIPTIVE FORMAT LITERATURE

Heywood, V. H. 1958. The Presentation of Taxonomic Information--A Short Guide for Contributors to Flora Europaea. Leicester University Press. Leicester.

Radford, A. E., C. R. Bell, J. W. Hardin, and R. L. Wilbur. 1967. Contributors' Guide for the Vascular Flora of the Southeastern United States. Privately published. Chapel Hill.

Shetler, S. et al. 1973. Guide for Contributors to Flora North America (FNA) (Provisional Edition). Smithsonian Institution. Washington, D. C.

Chapter 25. PLANT IDENTIFICATION

Identification is a basic activity, and one of the primary objectives of systematics. Although identification is a separate activity or process, in practice it involves both classification and nomenclature. Identification is simply the determination of the similarities or differences between two elements, i.e., two elements are the same or they are different. The comparison of an unknown plant with a named specimen and the determination that the two elements are the same also involves classification, i.e., when one correctly decides that an unknown belongs to the same group (species, genus, family, etc.) as a known specimen, the information stored in classification systems becomes available and applicable to the material at hand. Both processes--identification and classification--involve comparison and judgment and require a definition of criteria of similarities. Identification is, therefore, a basic process in classification with nomenclature playing an essential role in the retrieval of information and as a means of communication. According to Blackwelder (1967) "identification enables us to retrieve the appropriate facts from the system (classification) to be associated with some specimen at hand" and is "...better described as the recovery side of taxonomy." In practice one commonly identifies a plant by direct comparison or the use of keys and arrives at a name. The practical aspects and methods of plant identification and identification systems are discussed in this chapter. For further information see Harrington and Durrell's book How to Identify Plants.

Section A. TRADITIONAL IDENTIFICATION METHODS

The traditional methods of identification include (1) expert determination, (2) recognition, (3) comparison, and (4) the use of keys and similar devices. For a thorough and technical discussion of specimen identification see Sneath and Sokal (1973).

In terms of reliability or accuracy the best method of identification is expert determination. In general the expert will have prepared treatments (monographs, revisions, synopses) of the group in question, and it is probable that the more recent floras or manuals include the expert's concepts of taxa. Although of great reliability, this method presents problems by requiring the valuable time of experts and creating delays for identification. Recognition, according to Morse (1971) approaches expert determination in reliability. This is based on extensive, past experience of the identifier with the plant group in question. In some groups this is virtually impossible. A third method is by comparison of an unknown with named specimens, photographs, illustrations or descriptions. Even though this is a reliable method, it may be very time consuming or virtually impossible due to the lack of suitable materials for comparison. The reliability is, of course, dependent on the accuracy and authenticity of the specimens, illustrations, or descriptions used in the comparison. The use of keys or similar devices (synopses, outlines, etc.) is by far the most widely used method and does not require the time, materials, or experience involved in comparison and recognition.

Keys in the traditional sense are a type of taxonomic literature (see Figures 25-1 & 25-2). Keys are devices consisting of a series of contrasting or contradictory statements or propositions requiring the identifier to make comparisons and decisions based on statements in the key as related to the material to be identified. The first modern type keys (dichotomous) clearly designed for identification were those of Lamarck in his Flore Française in 1778 (see Voss, 1952, for an interesting account of the history of keys and phylogenetic trees in systematic biology). The comments of A. P. de Candolle in his dedication to Lamarck in the third edition of this flora concerning keys are equally appropriate today:

> "As to the artificial method, I have without hesitation, given preference to the one which you have contrived, and which consists of leading the student to the name of the plant by always forcing him to choose between two contradictory characters: in this analytic method I have permitted myself only the slight changes necessitated by the increase in the number of plants described. There, after your example, I have sought to distinguish the plants by the easiest and most apparent characters; and when these characters were not constant, I have tried to foresee their aberrations and to arrive at the same name by different routes; but this ease in the distinguishing of plants is very different in different families: in some, such as the crucifers, it is impossible to distinguish the genera without examination of the fruit. . . . When beginners undergo these difficulties in the use of the analytic method, I beg them, before blaming it, to reflect that the most accomplished botanists meet with the same embarrassment, and that no method can make the work easier to students than it is to the masters. . . . But when the pupil knows the name, let him take care not to think he knows the thing! Referred by a number in the analytic method to the description, he will find in this second part the details which put together constitute the whole science." (Quoted from translation by Voss, 1952)

Section B III of this chapter includes a discussion of the construction of identification keys and the application of computers to this process.

The following lists of suggestions are for the use and construction of traditional dichotomous keys.

Suggestions for the Use of Keys

1. Select appropriate keys for the materials to be identified. The keys may be in a flora, manual, guide, handbook, monograph, or revision (see Chapter 30). If the locality of an unknown plant is known, select a flora, guide, or manual treating the plants of that geographic area (see Guides to Floras in Chapter 30). If the family or genus is recognized, one may choose to use a monograph or revision. If locality is unknown, select a general work. If materials to be identified were cultivated, select one of the manuals treating such plants since most floras do not include cultivated plants unless naturalized.

2. Read the introductory comments on format details, abbreviations, etc. before using the key.

3. Read both leads of a couplet before making a choice. Even though the first lead may seem to describe the unknown material, the second lead may be even more appropriate.

4. Use a glossary to check the meaning of terms you do not understand.

5. Measure several similar structures when measurements are used in the key, e.g., measure several leaves not a single leaf. Do not base your decisions on a single observation. It is often desirable to examine several specimens.

6. Try both choices when dichotomies are not clear or when information is insufficient, and make a decision as to which of the two answers best fits the descriptions.

7. Verify your results by reading a description, comparing the specimen with an illustration or an authentically named herbarium specimen.

Suggestions for Construction of Keys

1. Identify all groups to be included in a key.

2. Prepare a description of each taxon (see Chapter 24 for details for description and descriptive format).

3. Select "key characters" with contrasting character states. Use macroscopic, morphological characters and constant character states when possible. Avoid characteristics that can only be seen in the field or on specially prepared specimens, i.e., use those characteristics that are generally available to the user.

4. Prepare a Comparison Chart (see Figure 25-3).

5. Construct strictly dichotomous keys.

6. Use parallel construction and comparative terminology in each lead of a couplet.

7. Use at least two characters per lead when possible.

8. Follow key format (indented or bracketed; see Figures 25-1 and 25-2).

9. Start both leads of a couplet with the same word if at all possible and successive leads with different words.

10. Mention the name of the plant part before descriptive phrases, e.g., leaves alternate not alternate leaves or flowers blue not blue flowers.

11. Place those groups with numerous variable character states in a key several times when necessary.

12. Construct separate keys for dioecious plants, for flowering or fruiting materials and for vegetative materials when pertinent.

A DICHOTOMOUS KEY TO SELECTED GENERA OF SAXIFRAGACEAE

Shrub or woody vine.
  Woody vine; petals 7 or more .... 3. Decumaria
  Shrub; petals 4 or 5.
    Leaves alternate or on short spur branches.
      Leaves pinnately veined; ovary superior; fruit a capsule .... 1. Itea
      Leaves palmately veined; ovary inferior; fruit a berry .... 2. Ribes
    Leaves opposite.
      Petals usually 4; stamens 20-40; fruit longitudinally dehiscent, not ribbed .... 4. Philadelphus
      Petals usually 5; stamens 8-10; fruit poricidally dehiscent, 10- to 15-ribbed .... 5. Hydrangea
Herbs.
  Staminodia present; petals more than 10 mm long .... 6. Parnassia
  Staminodia absent; petals less than 10 mm long.
    Leaves ternately decompound .... 7. Astilbe
    Leaves simple.
      Flowers solitary in leaf axils, or in short, leafy cymes.
        Sepals 4; carpels 2 .... 8. Chrysosplenium
        Sepals 5; carpels 3 .... 9. Lepuropetalon
      Flowers in racemes or panicles.
        Petals pinnatifid or fringed; stem leaves opposite .... 10. Mitella
        Petals not pinnatifid or fringed; stem leaves alternate or absent.
          Ovary 1-celled.
            Inflorescence paniculate; stamens 5 .... 11. Heuchera
            Inflorescence racemose; stamens 10 .... 12. Tiarella
          Ovary 2-celled.
            Stamens 5; leaves palmately lobed .... 13. Boykinia
            Stamens 10; leaves not palmately lobed .... 14. Saxifraga

---

Figure 25-1. Example of an indented (yoked) key. [From Radford, A. E., H. E. Ahles, and C. R. Bell. 1968. Manual of the Vascular Flora of the Carolinas. University of North Carolina Press. Chapel Hill, North Carolina. Used with permission.]

A DICHOTOMOUS KEY TO SELECTED GENERA OF SAXIFRAGACEAE

1. Shrub or woody vine .... 2.
1. Herbs .... 6.
  2. Woody vine; petals 7 or more .... _Decumaria_.
  2. Shrub; petals 4 or 5 .... 3.
3. Leaves alternate or on short spur branches .... 4.
3. Leaves opposite .... 5.
  4. Leaves pinnately veined; ovary superior; fruit a capsule ...... _Itea_.
  4. Leaves palmately veined; ovary inferior; fruit a berry .... _Ribes_.
5. Petals usually 4; stamens 20-40; fruit longitudinally dehiscent, not ribbed .... _Philadelphus_.
5. Petals usually 5; stamens 8-10; fruit poricidally dehiscent, 10-15 ribbed .... _Hydrangea_.
  6. Staminodia present; petals more than 10 mm long .... _Parnassia_.
  6. Staminodia absent; petals less than 10 mm long .... 7.
7. Leaves ternately decompound .... _Astilbe_.
7. Leaves simple .... 8.
  8. Flowers solitary in leaf axils, or in short, leafy cymes .... 9.
  8. Flowers in racemes or panicles .... 10.
9. Sepals 4; carpels 2 .... _Chrysosplenium_.
9. Sepals 5; carpels 3 .... _Lepuropetalon_.
  10. Petals pinnatifid or fringed; stem leaves opposite .... _Mitella_.
  10. Petals not pinnatifid or fringed; stem leaves alternate or absent .... 11.
11. Ovary 1-celled .... 12.
11. Ovary 2-celled .... 13.
  12. Inflorescence paniculate; stamens 5 .... _Heuchera_.
  12. Inflorescence racemose; stamens 10 .... _Tiarella_.
13. Stamens 5; leaves palmately lobed .... _Boykinia_.
13. Stamens 10; leaves not palmately lobed .... _Saxifraga_.

Figure 25-2. Example of a bracketed key. [Modified from Radford, A. E., H. E. Ahles, and C. R. Bell. 1968. Manual of the Vascular Flora of the Carolinas. University of North Carolina Press. Chapel Hill, North Carolina. Used with permission.]

PLANT IDENTIFICATION EXERCISE

1. Identification of an unknown. Select an unknown specimen and identify it by keying in an appropriate manual, flora, or monograph. Verify your results by reading a description, by comparing with an illustration or by checking with your instructor.

2. Preparation of a comparison chart. Select 5 or more specimens from the group provided by your instructor. Identify each by keying. Verify your results. Prepare a description of each similar to those in a flora or manual. Be sure characters and character states are in the same order. Select contrasting character states and prepare a comparison chart (see Figure 25-3).

3. Construction of keys. Construct a dichotomous key to these specimens using the information in the comparison chart.

COMPARISON CHART

	Decumaria	Itea	Ribes	Parnassia	Heuchera	Saxifraga
Habit	Woody vine	Shrub	Shrub	Herb	Herb	Herb
Leaf arrangement	Opposite	Alternate	Alternate or on spur shoots	Basal (Rosulate)	Basal (Rosulate)	Basal (Rosulate)
Petal Number	7-10	5	5	5	5	5
Locule Number	7-10	2	1	1	1	2
Stamen Number	7+	5	5	5 (staminodia 5)	5	10
Fruit Type	Capsule	Capsule	Berry	Capsule	Capsule	Capsule

Figure 25-3. A comparison chart used in the construction of keys (for six of the genera in Figures 25-1 and 25-2).

Section B. RECENT AND NEW IDENTIFICATION METHODS*

I. POLYCLAVES

Polyclaves of various kinds allow one to select the characteristics for use in identifying each specimen, taking his choices from some character set and repeating an elimination process until a tentative identification is made. A printed data table, chart, or matrix giving the status of various taxa for useful characteristics is readily used as a polyclave by listing the possible taxa on scratch paper and crossing out those which do not agree with the specimen's characters. Such data tables appear irregularly in the taxonomic literature, often for only the more difficult groups involved but occasionally for all the treated taxa, as done for medical bacteria by Cowan and Steel (1965). For large groups, the diagnostic tables are not only more powerful than the equivalent key, but also take less space to print. Lists of taxa having various characters were among the first nontabular polyclaves. These resemble the inverted files common in computerized information systems, where entries are listed according to their characteristics, rather than characteristics by entries. Lists of taxa lacking specified features have also been produced; this modification expedites use as one may then jot down the possible taxa and rapidly cross off those differing from the specimen. Polyclaves are readily mechanized, as shown by the familiar edge-punched cards and the less familiar window keys, as well as various mechanical devices. The possibility of a computerized polyclave was noted by Sokal and Sneath (1966) and by Williams (1967); implementation is straightforward once suitable data formats have been devised. The computerized polyclave system we developed at Michigan State University (Morse, in press) uses a General Electric Mark II timesharing system, but has also been tested on several other computers. Other computerized polyclaves include those of Boughey, Bridges, and Ikeda (1968), Dybowski and Franklin (1968), Goodall (1968), and Walker et al. (1968). As yet no comparison of these many parallel approaches has been made.

If an identification method requires, in general, use of all the characters in some list, it is no longer a polyclave, for no user options remain in selecting characters. Several such character-set methods have been devised, mostly statistical. Some computerized "keys" are also based on this model (e.g., Gyllenberg, 1965; Bogdanescu and Racotta, 1967; and Rypka, 1971). Although well intended, character-set algorithms seem to offer no advantages over polyclaves. However, <u>pattern recognition</u> methods, using no formal characters, might be employed effectively for identification in some cases. Such techniques linked with optical scanners, spectroscopy, or chromatography could even offer fully automated identification.

II. TAXA, CHARACTERS, AND DATA MATRICES

An objective identification system requires <u>a priori</u> information of three kinds: the pertinent taxa, the useful differentiating characters of these taxa, and the taxon/character data themselves.

*Adapted from "Specimen Identification and Key Construction with Time-Sharing Computers" by Larry E. Morse (Harvard University, Cambridge, Massachusetts), in Taxon 29: 269-282 (1971), with extensive revisions by Mr. Morse. Used with permission.

The hierarchy of natural taxa is the foundation of biological information retrieval as well as systematic synthesis, since data can be stored, retrieved, and studied at any level of generalization (Morse, 1974). In data processing, we may usually regard a taxon as a group of one or more individuals or lower taxa judged sufficiently similar to each other to be treated together formally as a single evolutionary or informational unit at a particular level in the taxonomic hierarchy, and sufficiently different from other such groups of the same rank to be treated separately from them. Traditionally each taxon (except the highest) belongs to one and only one taxon of the next higher rank, implying each individual belongs to exactly one species (and has one name) in any particular taxonomic treatment of its group. Taxa are either monothetic or polythetic. For monothetic taxa, possession of a certain set of diagnostic characters is both necessary and sufficient for membership. Polythetic taxa are more loosely circumscribed, since presence of only "a large number" (rather than all) of a list of characters is required for membership. In other words, the members of a polythetic taxon exhibit overall similarity. Modern systematists employ the polythetic taxon concept in most of their work, as has been widely recognized in the past decade. However, dichotomous keys and other monothetic methods of identification are still used. Theoretically, identification schemes for modern biology ought to assure that no single-character difference, either in population variability or observer error, can result in a misidentification. Polythetic polyclaves offer a solution; here, no possibility is eliminated until several differences have accumulated between the taxon description and the unknown specimen. The appropriate threshold depends on the variability of the taxa, but toleration of just one or two differences often gives a marked improvement in identification success.

Our character model was originally developed for a list-structure representation of dichotomous keys (Morse, Beaman, and Shetler, 1968). It centers on the concept of character couplets (or two-state characters), and also allows coding of "dependent characters" (Williams, 1969). In our model, each character couplet represents two contrasting alternatives about possible features of an individual population. (For convenience we often label the possibilities A and B, or true and false.) In other words, the character couplets are equivalent to the pairs (couplets) of contrasting leads in dichotomous keys, specifying a part of the organism and asking which of two alternative modifications of it is present. Such a question implies the specified part itself is present--the couplet "petals red" versus "petals yellow" is meaningless for apetalous plants as well as those having petals of different colors. A couplet can thus be inapplicable or noncomparable, often coded NC in numerical-taxonomy data. An additional code is needed when a specimen is variable in its expression of the character states of a couplet; more precisely, variable means sometimes A and sometimes B. At other times the state of a couplet is not known for a specimen, perhaps because the character is difficult to determine or not recorded, or the specimen is incomplete, poorly preserved, or at the wrong stage of development; these cases are coded unknown. All binary-couplet characters can be readily encoded with these five character states, listed below with our code numbers for them:

0 = unknown	3 = false, or B
1 = true, or A	4 = inapplicable
2 = variable	

Each character couplet in our model is a hierarchy of two binary (two-state) characters, one implicit and one explicit. Expression of either state of the explicit character depends on the presence of one particular state of the implied more general character--petal color depends on petal presence. As well illustrated in a diagram by Dale (1968), larger character hierarchies can be built by combination of a number of such couplets.

Extensions to multi-state and quantitative characters are, of course, necessary before our methods can be seriously considered for handling taxonomic information in general. For this we implemented a file system involving three kinds of taxon/character data: two-state characters, as described above; multi-state characters, and quantitative (numeric) characters (Morse, Peters, and Hamel, 1971). Multi-state and quantitative characters may also be coded as couplets for use with the current programs. The former can always be expressed as a list of yes/no binary characters, while the ranges of quantitative characters may be segmented as appropriate. Details of the MSU programs are available elsewhere (Morse, in press).

For identification and allied procedures, the preparation of data matrices as taxon/character rather than specimen/character entries is preferable. In compiling effective taxon/character descriptions it is of course important to study an adequate sampling of specimens and other information, including the relevant literature, as otherwise the data cannot indicate the nature of the taxon as a whole. Yet if enough examples of a taxon were examined, virtually all the characters studied would eventually be marked variable, rendering the whole description useless. However, with polythetic identification methods the occasional oddity can be safely ignored in preparing a taxon description. For polythetic taxa, the code variable means noticeably variable, while A (or B) means usually A (or B), perhaps 80% (Cowan and Steel, 1965) or 85% (Lapage et al., 1973), or a number calculated from the data (Möller, 1962). In our present system, taxon/character data are presented in the format illustrated elsewhere (Morse, in press; Shetler et al., 1971). Briefly, such files consist of a taxon list, a character list, and a coded taxon/character data matrix, with additional fields for various parameters used by the programs. Allowance is also made for extensive documentation.

Hierarchical matrices aid in processing of taxonomic data since different characters may be used for different subgroups of taxa. In a hierarchical taxonomic matrix system, each taxon-description line may refer to another complete matrix differentiating subordinate taxa, permitting programs to work down through the hierarchy much as one uses a key to families, then a key to genera, and then a key to species. This method also allows the user to query the data bank beginning at any level he desires--if he knows his specimen is a *Rhododendron*, he need not first verify that it is an angiosperm!

In preparing identification data, one should include only characters of potential value in making identifications. A number of computer algorithms are available for character-set selection, but subjective screening appears ordinarily adequate for choosing a manageable set of characters for further study. The traditional key characters of the group offer an initial list, but new diagnostic characters should be sought, especially in difficult groups. Guidelines for character selection, offered in a multitude of works, include the following: presence in the usual material to be identified, ease of observation and interpretation, distinctness between the various states, independence from other characters used, and tolerance to environmental influences. Identification characters must not only be well expressed in nature but should also be clearly expressible verbally or pictorially. Occasionally one should describe what the user will think he is seeing, whether or not this be the actual case morphologically. The previous experience of the intended users should also be considered, and especially in educational works the characters might be easily learned and remembered in association with their taxa.

III. CONSTRUCTION OF IDENTIFICATION KEYS

The presentation of dichotomous keys is usually considered a mandatory part of scholarly taxonomic publication, yet until recent years few advances were made in key-constructing theory over the methods of Grew or Lamarck, or even Aristotle and Theophrastus (Voss, 1952). In his study of the theory of keys, Osborne (1963) concludes that the most efficient key is the one in which each dichotomy divides the remaining taxa symmetrically, or as nearly so as possible, assuming the characters have equal probabilities of misinterpretation and the taxa are equally abundant. However, in reality some characters are far easier to use than others, and some taxa are keyed out more often.

The relative conveniences or ease of use of the various characters should be a major consideration in key construction, as one wants the easiest and fastest key which is sufficiently accurate. Ledley and Lusted (1960) suggest the characters (disease symptoms in their paper) be grouped into numbered sets or blocks of similar difficulty. In making a diagnosis they employ all available characters from one block before using any from the next. We allow nine such blocks, with the characters in block one hardest and those in block nine easiest; any characters labelled zero are completely ignored by some routines. In our programs we treat the block codes, or character convenience values, as natural logarithms of a convenience-of-use number. Thus the nine symbols represent a range of about three to eight thousand. When rating characters, one should assign codes such that the intended users would prefer to study any three characters at some level than any single character at the next lower level, seven to one for a two-level difference, twenty to one for three levels, and so forth. The conveniences may, of course, be revised with experience or a change in user groups. Hall (1970) considers conveniences which vary from taxon to taxon, but finds it easier to rate each character only once, instead of once per taxon.

If the relative frequencies of the various taxa can be estimated for a situation, then the taxa can be considered accordingly in determining the evenness of division at a dichotomy, in effect taking each expected occurrence as a separate object to be identified. Determining such frequencies is not at all easy, since we are interested in how often a taxon will be collected and keyed out, a quite different value than how often it occurs in the field.

The intra-taxon variability of the characters is also an important consideration in keys. In our methods, characters are coded as either variable or as (nearly) constant, leaving selection of a variability threshold to the researcher preparing the data. An alternative approach often used involves stating the variability of each taxon for each character as a quantitative entry in the data matrix, usually as a percentage. This percentage matrix would be valuable in a general taxonomic information system, especially while data were being collected and variability patterns were developing. However, the true-false-variable matrices present their data more clearly, and take far less effort to prepare. Also, we know no programs which utilize the percentages directly in constructing keys.

Many texts recommend the use of data tables in writing keys, but other mechanical aids are rarely described. Metcalf (1954) presents an index-card technique, and Peters (1969) used a computer to help incorporate additional taxa in keys. While developing our key-constructing techniques we prepared a key-editing program (Morse, Beaman, and Shetler, 1968), but work on the "advanced system" outlined there was suspended in favor of our current research on key construction and identification procedures. Key editing is important in large projects such as Flora Europaea and the planned Flora North America, where numerous minor improvements are made in the keys during editing. However, if taxa are added or deleted, it is often better to develop an entirely new key.

The possibility of computerized key construction is often mentioned, yet we know of only four programs for construction of biological keys, namely those of Möller (1962), Hall (1970), and Pankhurst (1970, 1971), as well as our own (Morse, 1971 and in press). Pankhurst (1974) provides a comparison of some of these programs. Möller's method requires complete binary data, and has attracted little attention. Hall's program utilizes quantitative data, printing a numeric version of the key which must then be rewritten before use. Pankhurst's algorithm differs from ours primarily in his use of rigid character-convenience blocks and his employment of the attribute-value rather than hierarchical-couplet character concept. His program, like ours, prints the key directly. The production version of our MSU programs is described below; these allow mixed-data key construction with our new data matrices. Several matrix-reduction and monothetic-devisive algorithms in the literature resemble key construction: the KEYCALC program by Niemalä, Hopkins, and Quadling (1968) is typical. Also, some aspects of decision-tree and game-tree research in computer science could contribute to the theory of keys.

In preparing a key, one usually divides the initial group of taxa by a character couplet into two subgroups, each of which is independently divided into further subgroups, and so forth, until every taxon is distinguished from all others. Indeed, the further subdivision of a subgroup can be considered as construction of a full key to that local group, suggesting a concise recursive algorithm for computerization: construct the key by dividing the original taxa into two subgroups by the best possible character couplet, then consider each of these subgroups separately, dividing them similarly. It is readily established that this procedure will produce a key if one can be made at all, but the optimality of such a key apparently remains uninvestigated. The four key-construction programs cited above all operate on this principle.

For use of this algorithm one needs a means for selecting the "best" character couplet for a given dichotomy, given the appropriate taxon/character data for a set of potentially useful characters. In our program, a preliminary screening of the characters is followed by a detailed evaluation of the potentially useful ones. In the preliminary scan, a character is eliminated from further consideration if it is coded as unknown or as inapplicable for any taxon in the local group, or if it is coded in only one way within that group. In other words, the remaining characters are all coded partly true and partly false; true and variable; false and variable; or true, false, and variable.

For keys these are obviously the only characters of immediate interest. Next, a combined measure of ease of use and dichotomizing power is determined for each character. A tally is made for each state (true, false, and variable), giving the sums of the taxon frequencies of the taxa so coded for a character. In effect, these three tallies give the total number of encounters expected under each state of a character. The tally for variable is then multiplied by the sum of the other two tallies, and half this produce is added to twice the product of the true and false tallies, giving a "dichotomizing value."

$$DV = 2q_T q_F + (1/2)q_V(q_T + q_F),$$

where q_T, q_F, and q_V are the respective tallies. High values of DV indicate well-divided dichotomies. In considering the character conveniences, these ratings are treated in exponential (values 3 through 8103) rather than condensed (1-9) form. The "best" couplet is found by maximizing the product of the character-convenience exponential and the dichotomizing value. The character couplet having this product greatest is then used for that dichotomy of the key, and the taxa in the local group are sorted according to their recorded status for that character, placing variable taxa in both subgroups. Note the character conveniences have a powerful but not exclusive effect, as a 5-5 split at convenience 6 (7254) as well as a 6-4 (or worse) split at 5 (7004).

If the indented style is desired, the program prints the key as it selects the dichotomies. However, for the bracketed style the key must be stored (in our list-structure representation) until completed, since division of variable taxa affects the numbering. It is then printed in one operation. The list-structure condensation can also be printed if desired, perhaps for later use in editing.

Ordinarily a batch-processing computer is adequate for key construction since no user interaction during processing is necessary. However, on-line key construction allows the effects of changes in the data to be seen quickly. An on-line key-editing system would also be possible, but most changes can be made with the text-editing software usually supplied with time-sharing systems. For small groups, these methods may also be used manually.

IV. DYNAMIC POLYCLAVES

Polyclaves need not be automated, but might be computerized when a large, rapidly developing information base is involved; otherwise, imaginative publication methods suffice (e.g., Ogden, 1953; Leenhouts, 1966, Archbald, 1967; Duke, 1969; Hansen and Rahn, 1969; Shultz, 1973). Here on-line (usually time-sharing) computers offer a clear advantage over batch-processing systems, since the on-line system allows the user to submit additional data during the execution of a program. This permits a dialogue or conversation between the user and the computer, with the machine printing questions and awaiting responses before continuing the processing. Such conversations can be about as terse or verbose as desired; we have taken a middle course, writing out most questions and taxon names in full, but sometimes using numerical codes for character states and couplets, as well as for several program options. Numeric coding of lengthy answers saves printing or typing time and reduces the chances for errors, assuming the user can copy a short number more accurately than a long descriptive phrase. For regular users of the system a faster abbreviated terminology is planned. As yet, the ideal of free-form language input is not practical for this or almost any information-retrieval system (Simmons, 1970).

Following selection of a group of possible taxa by the choice of the initial data file, our polyclave algorithm consists of three steps, repeated as necessary: (1) request the user to give one or more characteristics of his specimen; (2) eliminate all possibilities inconsistent with this partial description; and (3) print the results of the elimination, either an identification or some other action, and recycle to the first step if necessary. Several alternatives are available, including production of a useful-characters list and deletion of the effects of the last character set submitted. Since a user may start with a data matrix at any level in the taxonomic hierarchy, continuation to a subsequent matrix may be possible after a family, genus, or other higher-level taxon is identified. After each identification, several diagnostic or peculiar characters of the taxon are listed as an immediate check of the suggested identification. Particularly with higher-level matrices, "false positives" may occasionally occur when no single species has a character combination suggested by the generalized description of a taxon.

Polythetic identification, where one difference no longer implies elimination, is available as an option on our computer system. Here the program tallies the number of differences and eliminates a taxon only when its tally exceeds a user-determined value, commonly one, two, or three. Although it allows for greater taxon variability or user error, the polythetic polyclave is slower than the monothetic one since more characters must be submitted to assure complete elimination of all the other taxa.

When a list of useful characteristics is desired part way through a polyclave procedure, a subroutine selects the best characters to continue dividing the set of remaining possibilities. Actually, a portion of the key-constructing algorithm is used to determine the first such character, and repetition of the procedure (ignoring this character) gives the next-best character, and so forth. With the polyclave, of course, such recommended characters are merely suggestions which need not be employed. No matter which state the specimen shows, the recommended character will eliminate about half the possibilities. Any other character, on the average, will delete fewer taxa because it eliminates a larger number only when in its rarer state. However, if the user happens to notice his specimen displays a rare character, he can eliminate a large number of possibilities at once and identify his specimen much more rapidly. Since the ability to recognize rare characters and realize their power comes only through training and experience, the expert delights in the efficiency and power of a polyclave, while the neophyte is lost in its multitude of choices and prefers the supervision and security of the traditional dichotomous key.

PLANT IDENTIFICATION LITERATURE

Archbald, D. 1967. Quick-Key Guide to Trees: Trees of Northeastern and Central North America. Doubleday. Garden City, New York.

Blackwelder, R. E. 1967. Taxonomy, A Text and Reference Book. John Wiley & Sons, Inc. New York.

Bogdanescu, V., & R. Racotta. 1967. Identification of mycobacteria by overall similarity analysis. Journal of General Microbiology 48: 111-126.

Bossert, W. 1969. Computer techniques in systematics. In: Systematic Biology. National Academy of Science Publication 1692. Washington, D. C.

Boughey, A. S., K. W. Bridges, and A. G. Ikeda. 1968. An automated biological identification key. Museum of Systematic Biology. University of California (Irvine).

Cowan, S. T., and K. J. Steel. 1965. Manual for the Identification of Medical Bacteria. Cambridge University Press. Cambridge, England.

Dale, M. B. 1968. On property structure, numerical taxonomy, and data handling. In: Heywood, V. H. Modern Methods in Plant Taxonomy. Academic Press. London and New York.

Duke, J. A. 1969. On tropical tree seedlings. I. Seeds, seedlings, systems, and systematics. Annals of the Missouri Botanical Garden 56: 125-161.

Dybowski, W., and D. A. Franklin. 1968. Conditional probability and the identification of bacteria: a pilot study. Journal of General Microbiology 54: 215-229.

Goodall, D. W. 1968. Identification by computer. BioScience 18: 485-488.

Gyllenberg, H. 1965. A model for computer identification of micro-organisms. Journal of General Microbiology 39: 401-405.

Hall, A. V. 1970. A computer-based system for forming identification keys. Taxon 19: 12-18.

Hansen, B., and K. Rahn. 1969. Determination of angiosperm families by means of a punched-card system. Dansk Botanisk Arkiv 26: 1-46 + 172 punched cards.

Harrington, H. D., and L. W. Durrell. 1957. How to Identify Plants. Swallow Press. Chicago.

Lapage, S. P., S. Bascomb, W. R. Willcox, and M. A. Curtis. 1973. Identification of bacteria by computer: general aspects and perspectives. Journal of General Microbiology 77: 273-290.

Ledley, R. S., and L. B. Lusted. 1960. The use of electronic computers in medical data processing: aids in diagnosis, current information retrieval, and medical record keeping. Institute of Radio Engineers Transactions on Medical Electronics 7: 31-47.

Leenhouts, P. W. 1966. Keys in biology: a survey and a proposal of a new kind. Proceedings Koninklijke Nederlandse Akademie Van Wetenschappen 69 (ser. C): 571-596.

Metcalf, Z. P. 1954. The construction of keys. Systematic Zoology 3: 38-45.

Möller, F. 1962. Quantitative methods in the systematics of Actinomycetales. IV. The theory and application of a probabilistic identification key. Giornale di Microbiologia 10: 29-47.

Morse, L. E. 1974. Computer-assisted storage and retrieval of the data of taxonomy and systematics. Taxon 23, in press.

________. (in press). Computer programs for specimen identification, key construction, and description printing using taxonomic data matrices. Publications of the Museum, Michigan State University (Biological Series). East Lansing, Michigan.

________, J. H. Beaman, and S. G. Shetler. 1968. A computer system for editing diagnostic keys for Flora North America. Taxon 17: 479-483.

________, J. A. Peters, and P. B. Hamel. 1971. A general data format for summarizing taxonomic information. BioScience 21: 174-180, 186.

Niemelä, S. I., J. W. Hopkins, and C. Quadling. 1968. Selecting an economical binary test battery for a set of microbial cultures. Canadian Journal of Microbiology 14: 271-279.

Ogden, E. C. 1953. Key to the North American species of Potamogeton. New York State Museum Circular 31.

Osborne, D. V. 1963. Some aspects of the theory of dichotomous keys. New Phytologist 62: 144-160.

Pankhurst, R. J. 1970. A computer program for generating diagnostic keys. Computer Journal 13: 145-151.

________. 1971. Botanical keys generated by computer. Watsonia 8: 357-368.

Pankhurst, R. J. 1974. Automated identification in systematics. Taxon 23, in press.
Peters, J. A. 1969. Discussion. In: Systematic Biology. National Academy of Science Publication 1692. Washington, D. C.
________, and B. B. Collette. 1968. The role of time-sharing computers in museum research. Curator 11: 65-75.
Rypka, E. W. 1971. Truth table classification and identification. Space Life Sciences 3: 135-156.
Shetler, S. G., J. H. Beaman, M. E. Hale, L. E. Morse, J. J. Crockett, and R. A. Creighton. 1971. Pilot data processing systems for floristic information. In: J. L. Cutbill. Advances in Data Processing for Biology and Geology. Academic Press. London and New York.
Shultz, L. M. 1973. Random-access key to the genera of Colorado wildflowers. University of Colorado Museum. Boulder, Colorado.
Simmons, R. F. 1970. Natural language question-answering systems: 1969. Communications of the Association for Computing Machinery 13: 15-30.
Sneath, P. H. A., and R. R. Sokal. 1973. Numerical Taxonomy: The Principles and Practice of Numerical Classification. W. H. Freeman and Company. San Francisco.
Sokal, R. R., and P. H. A. Sneath. 1966. Efficiency in taxonomy. Taxon 15: 1-21.
Voss, E. G. 1952. The history of keys and phylogenetic trees in systematic biology. Journal of the Scientific Laboratories of Denison University 43: 1-25.
Walker, D., P. Milne, J. Guppy, and J. Williams. 1968. The computer-assisted storage and retrieval of pollen morphological data. Pollen et Spores 10: 251-262.
Williams, W. T. 1967. The computer botanist. Australian Journal of Science 29: 266-271.
________. 1969. The problem of attribute-weighting in numerical classification. Taxon 18: 369-374.

POLYCLAVE EXERCISE

Preparation of a simple polyclave. Select a family of ten or more genera. Construct a master sheet listing an abbreviation or number for each genus and print in red on a 5 x 8 card. This list is the taxa master list underlay. By examining herbarium specimens or using an appropriate manual or monograph select and compile a list of 50 or more characters and character states for each genus. Duplicate (in black) the taxa master list underlay to equal the number of characters selected. These will become character state overlay cards. Label each card with a specific character state from the character state list. Search each specimen or description for the character state in question and punch out the generic name or number listed on the card for each genus having the character state in question. Punch a card for each character state, e.g., if one character state selected is leaves opposite, punch out the names of all genera known to have opposite leaves. To use the polyclave select a specimen identified as belonging to the family in question. Notice character states and begin sorting through overlay cards until you find a card with a characteristic that matches one of the cards you have prepared. Place the character overlay card over the taxa master list. Those names or numbers appearing in red through the holes will represent all genera possessing the character in question. Continue sorting and overlaying until all cards have been used or until a single genus name or number appears.

Chapter 26. VARIATION AND EVOLUTIONARY RELATIONSHIPS

Evolutionary studies provide the essential concepts and core of information necessary for the realistic evaluation of experimental and descriptive data concerned with species origins and relationships and, by extrapolation, the origins and relationships of higher ranks. More specifically, organic or biological evolution is responsible for the variation and change in organisms through time. The operation of the basic evolutionary mechanisms--mutation, recombination, and selection--in different environments today accounts for the current patterns of variation of interest and importance to taxonomists.

Individual organisms do not exist apart from their environment and their genetic history but are the result of present and past interactions between an array of genetic backgrounds and a long sequence of environments. Thus different kinds of organisms are not, as once thought, the result of special creation but are known to be the product of different evolutionary lines which themselves result from the basic interactions between heredity and environment. These interactions have produced, through time, all of the variations in biological form and function found in the fossil record and all of the continuous and discontinuous variation patterns which make possible the current concepts of taxonomic categories as well as the ecological concept of plant communities.

An awareness of biological variation and also an awareness of different kinds of organisms dates back into man's early history. Attempts to account for the different kinds of organisms, biologically or scientifically, covered a century or more before the answer--natural selection--was independently discovered but jointly proposed by Darwin and Wallace in 1859. Even though the fact of universal variation was a key point in the Darwin-Wallace hypothesis, the genetic mechanisms of variation--mutation and recombination--were not discovered and understood for several more decades. It was more than half a century after Darwin that cytological and genetic data began to be generally applied to taxonomic problems. The result was "Cyto-taxonomy." As the mechanisms of selection became better understood through experimental work, as more was learned about the evolutionary role of the population breeding structure, and as more was learned as to the causes of natural variation, taxonomy in much of the world became concerned with "speciation" and thus became primarily a study of evolution---or "Biosystematics." The results of similar evolutionary studies by the plant ecologists interested in variation, adaptations, and community interactions, often found ready application in taxonomy. Indeed, the final use of application of information from biosystematic studies (in either taxonomy or ecology) may often be secondary in importance to the further demonstration or elucidation of evolutionary processes either in general or in the specific organism studied.

The broad taxonomic consequences of more detailed evolutionary studies, and especially those which have used computer-assisted comparisons of the data, have been a clearer understanding of the kinds of relationships that exist between and among organisms and a more precise, and useful, set of definitions associated with such terms as "relationship," "affinity," and "similarity." As pointed out by Sneath and Sokal (1973) the unqualified word "relationship" may mean, in our language, either relationship by descent (a truly phylogenetic relationship) or any of one or more types of relationships based on a common character or similarity (a phenetic relationship). A phenetic relationship may exist quite apart from any genetic relationship or, on the other hand, it may be the result of genetic relationship, and thus also be cladistic--the term applied to genetic relationships that express or reflect recent common ancestry. Cladistic relationships, if plotted against a relative time scale, are also chronistic relationships and the resulting dendrogram that would illustrate such a situation is called a cladogram (see Table 22-3-C, p. 470). The relationships usually sought by evolutionary studies are phyletic or phylogenetic, but the relationships usually shown are, at best, phenetic. However, even when no genetic relationship is implied, phenetic similarities (or "affinities" or "relationships") may be the result of similar evolutionary processes. For example, the evolutionary process of convergence might produce strong phyletic similarities between genetically unrelated organisms.

The experimental and analytical approaches necessary to sound evolutionary research involve variation studies in morphology, anatomy, phytochemistry, embryology and physiology; studies of evolutionary mechamisms involve genetics, cytology, and ecology. Because of this holistic nature of biosystematics, facility with many of the methods, and an understanding of much of the vocabulary, treated in previous chapters will be essential to the demonstration and understanding of evolutionary (or other) relationships between and within plant species. Special note should be made of the material in Chapters 10, 11 and 12 (cytology, genetics, and reproductive biology), the section on adaptive features in Chapter 15 (ecology), and the material on data accumulation and analysis in Chapters 20, 21 and 22.

A very high degree of specialization, and subsequent interrelationship, also exists within the broad field of evolutionary studies. Each of Darwin's five generalized features of organic evolution are today major specialized, but highly reticulated, lines of evolutionary research. The following outline attempts to relate some of the present evolutionary concepts, mechanisms, and terminology to the five primary aspects of classical evolution. For definition of the terms see the Evolution Glossary at the end of this chapter and also the glossaries for Chapters 11, 12, and 15, all of which are indexed under the general subject and terms index beginning on page 881. For examples, see pertinent items from the lists of general references for this, and related, chapters. For the more frequently used methods of data accumulation, analysis of variation, and methods of presenting the data, see Chapters 20, 21, 22, and 23.

I. VARIATION

Variation studies, and a careful analysis of the information gathered (see Chapters 20-22), are the first steps in any evolutionary or biosystematic study. Once the type and magnitude of variation are known it is far more likely that further specific work will enable one to go beyond the descriptive phase and account for the variation observed and thus learn something of the actual biological relationships involved.

A. Types of variation and variation patterns.
   1. By primary control.
      a. genetic (an essential aspect of the life cycle, e.g., reproductive characters; usually has a relatively low amount of variability, e.g., a low coefficient of variability).
      b. environmental (a less essential character in the life cycle, e.g., leaf lengths; usually has a relatively high amount of variability, e.g., a high coefficient of variability).

2. By uniformity.
   a. continuous or clinal (the cline might be altitudinal due to temperature, or other environmental factors but would not permit a clear separation of the series of populations into discrete units or subcategories).
   b. discontinuous (isolation and adaptations to different environments can produce discrete variation patterns with few or no intermediate forms and thus easily separable, on one or more characters, into categories).

3. By form.
   a. monomorphic
   b. dimorphic
   c. polymorphic
   d. heteromorphic.

B. Sources of variation in natural populations (also see Chapter 11, Genetics).
   1. Mutation
      a. gene mutation
      b. chromosomal mutation
         deletion
         duplication
         inversion
         aneuploidy
         polyploidy
   2. Recombination
      a. crossing over at meiosis
      b. random assortment of homologs at meiosis
      c. out-crossing (limited or panmictic)
      d. hybridization
      e. introgression
   3. Genetic interaction
      a. epistasis
      b. transgressive variation
      c. euheterosis

C. Limitations on variation in natural populations.
   1. Self-fertilization.
   2. Asexual reproduction (apomixis).
   3. Stabilizing selection.
   4. Fixation.
   5. Genetic homeostasis.

II. REPRODUCTION

Evolutionary survival depends more on the reproductive methods and reproductive potential of the population than on individual "fitness." Furthermore, since different reproductive methods have a direct influence on the amount of genetic recombination (and thus often on the total amount of population or species variation) a thorough understanding of the reproductive biology of a species, or series of congeners, is essential to an understanding of both the variation patterns and the relationships involved in any biosystematic study (see Chapter 12).

A. Sexual reproduction (high evolutionary potential).
   1. Mechanisms that favor or insure outcrossing or allogamy and thus maximum heterozygosity and variability (either actual or potential).
      a. plants monoecious
      b. plants dioecious
      c. protandry
      d. protogyny
      e. heterostyly
      f. self sterility
      g. chasmogamy

   2. Mechanisms that favor or insure selfing or autogamy and thus maximum homozygosity and genetic uniformity.

a. self fertility
b. cleistogamy
c. flowers perfect
d. autogamy

B. Apomixis or asexual reproduction (low evolutionary potential); here there is no recombination and maximum phenotypic uniformity, variability is phenotypic and under environmental control.

1. Vegetative.
   a. rhizomes, stolons, runners
   b. bulbils
   c. fragmentation ("cuttings")
   d. layering
2. Parthenogenetic.
   a. apospory
   b. apogamy
   c. adventitious embryony
   d. agamospermy

(Note that very few plants are completely restricted to a single reproductive method: many violets, for example, have chasmogamous flowers that may be outcrossed, cleistogamous flowers that self pollinate, and rhizomes which provide effective asexual reproduction. Also, even monoecious and dioecious plants often produce a few perfect flowers that may be self-pollinated and many "inbreeding species" show some degree of outcrossing. The more reproductive pathways an organism has the greater its long range evolutionary potential and also its short range probability of survival.)

C. Isolating mechanisms (interspecific).

1. Prezygotic (isolation which prevents zygote formation).
   a. geographical (the result of geographic separation of the entities)
   b. ecological (the result of ecological separation of the entities, even if sympatric)
   c. seasonal (phenological separation of the entities)
   d. indirect ethological (isolation caused by the animal pollinator preference)
   e. mechanical (structural differences, such as corolla tube size or anther placement, that prevent crossing)

(Note: Except for geographical isolation the above forms of isolation may not be complete and occasional hybrids may be formed; it is not uncommon for two species of plants to be isolated by a combination of two (or more) mechanisms.)

2. Postzygotic (zygote formed, but effective gene exchange between entities still prevented).
   a. failure of embryo to develop
   b. hybrid inviability or sterility
   c. reduced fitness and reproductive potential in F_2 and later generations.

III. COMPETITION AND SURVIVAL

Competition is a test of preadaptation and illustrates the often indirect effect of natural selection. Seed size, viability, and dispersal mechanisms, for example, can in no way be of any selective advantage to the plant that produced the seed! But seed viability per se is only one factor in survival as is shown by the intricate physiological adaptations of the relatively long-lived seeds of desert annuals that enable them to "measure" the rainfall and germinate only after a given amount of rain has fallen in a short enough time period to insure adequate soil moisture for a reproductive life span.

The seed dormancy in cocklebur (*Xanthium* sp.; Asteraceae) shows another germination pattern presumably of adaptive value. Here the seeds of a single annual plant contain different amounts of dormancy producing auxin with the result that a few seeds from this one plant germinate each year over a period of years rather than all at the first opportunity. Every organism is to some degree in competition with every other organism in the area. It is in competition with other organisms of its own kind as well as with organisms belonging to other species. These organisms are in competition for energy (light or food), water, raw materials, space, and all of the other elements of the environment necessary to survival and reproduction.

Some Evolutionary Results of Competition

Within a population	Between species
Genetic coadaptation	Organism coadaptation
Self-thinning	Competitive exclusion
Decrease in variability (stabilizing selection)	Increase in variability (niche subdivision)
	Mimicry
	Succession
Survival of the fittest	Survival of the fittest

For an outline of the aspects of adaptation related to the biotic selective forces (competition) see V-D, below).

IV. NATURAL SELECTION

Natural selection is the "steering mechanism" of evolution. It may vary in type, kind, and intensity and may have varied results--from adaptation and survival to extinction. Artificial selection, or the selection of specific characters or sets of characters by man, is the basic tool of plant and animal breeding programs essential to modern agriculture, and is an excellent example of "applied evolution" from which we can draw many valid conclusions concerning evolution in nature.

Natural selection, the evolutionary process that results in variation and speciation, may be generally classified either by results, by selective agent or by intensity, as follows:

1. By result.
   a. stabilizing
   b. directional
   c. disruptive
2. By agent.
   a. abiotic
   b. biotic
3. By intensity.
   a. Darwinian selection
   b. catastrophic

A. Types of Selection Compared

1. Centripetal, stabilizing or normalizing selection--the selection that occurs in every normal population eliminating extreme variations, maintaining the adaptation of the group somewhere near its optimum for the environment in which it lives.
2. Balancing selection--a type of stabilizing selection in which two or more discontinuous forms are maintained in the same interbreeding population (balanced polymorphism) because each form is of advantage to the population part of the time, or in part of the environmental circumstances under which the population lives, such as seasonal fluctuations in temperature.

3. Centrifugal, disruptive or diversifying selection--selection away from the original mode of a population in two or more directions either as a response to a more or less sudden change in the original environment (short duration) or as a form of balancing selection which produces a complex polymorphism.
4. Linear or directional selection--continued selection in a given direction producing an evolutionary trend, a sequence of adaptive changes in response to an environmental gradient.

B. Comparative Features of Abiotic and Biotic Selection

Abiotic	Biotic
1. Evolution direction fixed (dependent on unalterable selective force, such as climate).	1. Evolution direction not fixed (dependent upon interaction with another evolving organism).
2. May often result in parallel evolution.	2. Rarely results in parallel evolution.
3. Population size not important.	3. Population size important.
4. Biological interactions general.	4. Biological interactions very specific.
5. Does not produce niche subdivision.	5. Produces niche subdivision.
6. Produces low species density.	6. Produces high species density.

C. Comparative Features of Darwinian and Catastrophic Selection

Darwinian	Catastrophic
1. Operates on all individuals in all populations.	1. Operates on marginal individuals in marginal populations.
2. Involves gradual change in the selective force rarely exceeding range of tolerance of most individuals.	2. Involves relatively sudden changes in the extremes of the selective force that may exceed the tolerances of most individuals.
3. Seldom produces drastic decrease in population density.	3. Often produces drastic decrease in population density.
4. Large gene pool maintained; fixation rare.	4. Gene pool often drastically decreased; fixation likely.
5. Selection is homeostatic (multidirectional).	5. Selection is often unidirectional.
6. Rate of evolution slow, long cycle.	6. Rate of evolution faster, shorter cycle.

V. ADAPTATION

Adaptation, which produces diversity and speciation, is the result of natural selection. However, organisms can only adapt within the genetic limits of their existing gene pool. Therefore, different organisms, sometimes even when closely related, may adapt in different ways to a particular selective force. However, a high degree of adaptation (e.g., overspecialization) can lead to a decrease in evolutionary potential and ultimate extinction. The adaptation, when possible, of different populations of an organism to different selective factors of their environments produces polymorphism which may, in turn, lead to speciation.

A. Results of Adaptation at the level of the individual.
  1. Developmental homeostasis (canalization).
  2. Correlation pleiades.
  3. Euheterosis.

B. Results of adaptation at the population level.
  1. Heterozygosity.
  2. Genetic coadaptation.
  3. Genetic homeostasis.

C. Some possible results of adaptation to (usually) abiotic selective forces.
  1. Ecotypes.
  2. Ecoclines.
  3. Ecological equivalents.
  4. Parallel evolution (convergence).
  5. Periodicity.

D. Some possible results of adaptation to (usually) biotic selective forces.
  1. Allopatry (competitive exclusion).
  2. Succession (competitive exclusion).
  3. Character displacement.
  4. Coadaptation or coevolution (species)
  5. Symbiosis.
     a. mutualism
     b. commensalism
  6. Parasitism.
     a. facultative
     b. obligate
  7. Mimicry.
     a. Batesian
     b. Muellerian
     c. Mertensian
     d. automimicry

For a summary of evolutionary mechanisms we can paraphrase the neoDarwinians. It is clear from the above outline that there is indeed "universal variation," both genetic and phenotypic; through various sexual and asexual means of reproduction there is an "overproduction of offspring" and the potential of a geometric increase in the numbers of organisms; however, there is "competition" for space and raw materials, thus the numbers of organisms of each kind generally tend to remain constant; those organisms which have the proper combinations of characters (i.e., are best "adapted") tend to survive in higher proportions than the "weaker" or unadapted organisms; "adaptation" can take many forms and may lead to greater uniformity or greater diversity, depending upon the organism and the circumstances; if adaptation results in directional, hereditary variation we have evolution. Since biological variation, at different levels, is at once the alpha and omega of organic evolution it is important that its causes, and results, be understood.

EVOLUTION GLOSSARY

Abiotic Selection. Selection primarily by the physical environment, usually unidirectional. (See parallel evolution.)

Adaptation. The result of natural selection; the evolution of a character, or group of characters that favor survival under certain selective forces.

Adaptive Radiation. Evolutionary diversification within a group of related species or other taxa brought about by adaptations to different selective forces through time or migration.

Adaptive Zone. The "place" or "niche" of an organism in the evolutionary reticulum of its community; the set of conditions to which the offspring of an organism will be preadapted.

Advanced. An organism, character or condition presumed to be more specialized or evolved than another to which it is compared.

Adventitious Embryony. The production of an embryonic sporophyte by mitotic divisions from the tissues of the parental sporophyte without an intervening gametophytic generation.

Agamospermy. The formation of seeds without fertilization; pollen or male gametes may be necessary to stimulate division of the zygote.

Allogamy. Cross-fertilization.

Allometry. Different growth rates in different parts of the same organism.

Allopatric. Populations or species with separate and nonoverlapping geographic ranges.

Allopatric Speciation. The differentiation of geographically isolated populations into separate species.

Allopatry. The occurrence or origin of species or populations in different geographical regions.

Apogamy. The production in plants of a sporophyte directly from a gametophyte cell without the production of a zygote derived by fusion of gametes.

Apomict. An organism that reproduces by asexual means; it may be facultative or an obligate apomict.

Apomixis. Asexual reproduction, including vegetative propagation.

Apospory. The condition found in some higher plants in which a diploid embryo-sac is formed directly from a somatic cell of the nucellus or chalaza and an embryo is then formed without fertilization.

Autodeme. A group of individuals of a taxon composed of predominantly self-fertilizing individuals.

Autogamy. Self-fertilization; persistent autogamy results in an increase in homozygosity and the division of the population into a number of "pure lines."

Automimicry. Mimicry within a species.

Allaesthetic Selection. Disruptive selection via differential pollination.

Batesian Mimicry. The superficial resemblance of a harmless organism (the mimic) to a sympatric and more plentiful organism (the model).

Biosystematics. The systematic study, usually including reproductive biology and chromosome studies, of the variation and evolution of a species or species complex.

Biotic Potential. The total reproductive potential of an individual or a population; it is directly related to the number of functional gametes, and seeds, produced.

Biotic Selection. Selection primarily by the biotic environment. See competitive exclusion and character displacement.

Bradytelic Evolution. Evolution at a relatively slow rate as compared with most other known or presumed evolutionary rates.

Canalisation. The property, possessed by developmental pathways, of achieving a standard phenotype despite genetic or environmental disturbance.

Catastrophic Selection. A harsh selective force (e.g., fire, severe drought) acting on an ecologically marginal population so there are only a few survivors and a small gene pool remaining; results in operation of the Founder Principle.

Character Displacement. The increase in the differential of a character of two competing species in an area of sympatry.

Cleistogamy. The flower self pollinates, without opening in the bud.

Cline. A character gradient, often related to geographic distribution; a more or less uniform series of variants from one extreme to the other in a population or series of populations; a continuous variation pattern.

Clone. A group of individual organisms produced by vegetative asexual reproduction (rhizomes, stolons, budding, cuttings, etc.) and thus all of the same genotype.

Coadaptation. The adaptive response of two organisms to each other; e.g., flower structure and pollinator.

Coenogamodeme. A unit composed of all the hologamodemes which are considered to be capable of exchanging genes to some extent, but not with freedom, under a specified set of conditions.

Coenospecies. An ecological category approximating a Linnaean genus.

Commensalism. A relationship between individuals of two species in which one species benefits but does not harm the second, or host, species.

Comparium. A genetic term to designate a group of coenospecies which can hybridize either directly or through intermediates.

Competitive Exclusion. The effect of strong biotic selection which eliminates one of two competing sympatric organisms; environmental modification by one organism that results in the exclusion of another; e.g., plant succession.

Convergence. A type of parallel evolution resulting in similar form or structure in different organisms; e.g., ecological equivalents; the morphological similarity of desert-dwelling plants of the Euphorbiaceae and Cactaceae.

Correlation Pleiades. The tendency of two or more characters of an individual that are under the same selective force to have correlated patterns of variation.

Cross Pollination. Pollination between two different plants of the same species; outcross.

Cytodeme. A group of individuals of a taxon differing from others cytologically, e.g., in chromosome number.

-deme. A biologically and taxonomically neutral suffix, which should not be used alone, that denotes a group of individuals of a specified taxon; the prefix indicates the context of the relationship or similarity between the individuals. (See topodeme, gamodeme, cytodeme, etc.; also Chapter 11,III).

Dimorphism. The occurrence of two morphological, or other, forms of a given organism; e.g., sexual dimorphism.

Diplospory. The phenomenon in some higher plants in which a diploid embryo-sac is formed directly from a megaspore mother-cell; an embryo is then formed without fertilization.

Directional Selection. Selection occurring when the environment is changing in a systematic fashion, leading to a regular directional change of the adaptive characteristics of a gamodeme.

Disjunct. A geographically isolated population or species outside of the range of other similar populations or species.

Disruptive Selection. Selection (usually biotic) which breaks up a weakly polymorphic population into a number of differently-adapted variants.

Ecocline. A variational trend correlated with an ecological gradient.

Ecodeme. A group of individuals of a taxon occurring in a specified kind of habitat.

Ecospecies. An ecological category approximating a Linnaean species.

Ecotype. An ecological category indicating an infra-specific group of plants adapted to a particular habitat.

Endemic. A population or species with narrow physiological or other restrictions which limit it to a special habitat or a very restricted geographic range, or both.

Environmental Variance. That portion of the phenotypic variance caused by differences in the environments to which the individuals in a population have been exposed.

Epigenotype. The series of interrelated developmental pathways through which the adult form is realized.

Epistasis. The interaction of nonallelic genes.

Euheterosis. The condition in which the heterozygotes in a population are at an adaptive or selective advantage over the homozygotes.

Evolution. A directional change in gene frequency, over time, in response to selective factors of the environment; the continuous genetic adaptation of organisms to their environment.

Expressitivity. A measure of the uniformity of the phenotypic expression of a gene in a particular environment.

Feedback. The influence of the result of a process upon the functioning of the process; success in producing viable and fertile offspring.

Fitness. The survival value and reproductive capability of a given genotype relative to other genotypes in a population.

Fixation. When the frequency of a particular allele reaches zero or 100% in a (usually small) population.

Founder Principle. The establishment of a new population in isolation so that its gene pool is not identical with that of the parent population; these differences are enhanced as different evolutionary pressures in the areas occupied by the two populations must therefore operate in different genetic environments; the result is increased divergence.

Gamodeme. A group of individuals of a specified taxon which, within the limits of the breeding system, can interbreed.

Gap. A discontinuity in variation.

Genecology. The study of intraspecific variation in plants in relation to environment.

Gene Flow. The distributional "movement" of genes within and among populations.

Gene Frequency. The percentage of an allele in a population; the number of loci at which a given allele is found within a population, divided by the total number of loci at which it could occur.

Gene Pool. The total genetic information possessed by a population.

Genetic Assimilation. The incorporation into the genotype, by a selective process, of characteristics appearing in ontogeny as a response to the environment.

Genetic Coadaptation. Adaptation of blocks or combinations of genes to each other in an interbreeding population.

Genetic Drift. Random fluctuation of gene frequency (usually due to sampling error inherent in the genetic mechanism); present in all populations, its effects are most evident in very small populations.

Genetic Homeostasis. The tendency of populations under selection to maintain, or regress toward the adaptive mean rather than an adaptive peak.

Genetic Load. The average number of potential deaths per individual due to genetic causes, such as lethals and semilethals, in a population.

Genetic System. All the factors that affect genetic recombination in an organism.

Genotype. Refers either to the total genetic component of an organism or, in a different sense, to a specific allele, in a particular individual.

Genotypic Variation. Genetic variation; differences in genotypes within a population (or species) are a result of mutation and recombination.

Gynogenesis. Reproduction by parthenogenesis in which stimulation by a sperm is necessary for the development of the egg.

Haplophase. That part of the life cycle in which the gametic chromosome number is found in reproductive cells.

Hardy-Weinberg Law. The law stating that, in a panmictic population, in the absence of selection, mutation, migration and genetic drift, the frequency of autosomal genes at a given locus remains constant and that, after one generation, the frequency of genotypes at the locus reaches equilibrium.

Heritability. The genetic variance divided by the phenotypic variance; an estimation of the degree of resemblance between offspring and parent.

Hermaphroditic. Possessing the phenotype of both sexes in organisms with male-female differentiation.

Heteroblastic Change. The transition from a juvenile to an adult form accompanied by a more or less abrupt change in morphology.

Heterogametic. Producing gametes with differing chromosome complements.

Heteromorphic. With many forms; polymorphic.

Heterotrophic. Requiring organic carbon for nutrition.

Higher Categories or Ranks. Taxonomic categories or ranks above the level of genus in the established hierarchy of our taxonomic system.

Hologamodeme. A group of individuals of a taxon which are believed to interbreed with a high level of freedom under a specified set of conditions. This term is preferred to 'biological species,' which relates to a taxonomic category.

Homomorphic. With a single form; monomorphic.

Horotelic Evolution. Rates of evolution falling within the distribution (asymmetrical with a mode nearer the upper than the lower end) most commonly found when evolutionary rates are plotted in a frequency distribution.

Hybrid. An individual resulting from cross fertilization between organisms belonging to different taxa, usually different species.

Hybrid Zone. A zone of sympatry and hybridization between two species which may, despite hybridization within the zone, be a barrier to extensive gene flow or introgression between the species involved.

Hybridization. The formation of a zygote through the fusion of gametes from organisms belonging to two different taxa.

Imprinting. The imposition of a stable behavior pattern by exposure, during a particular period in development, to one of a restricted set of stimuli.

Inbreeder. A plant that is normally self-fertilized.

Induction. Determination of the developmental fate of one cell mass by another.

Introgression. Incorporation of genetic material from one population into that of another through hybridization followed by repeated backcrossings; usually restricted to populations regarded as distinct by taxonomists.

Introgressive Hybridization. Genetic modification of one species by another by gene flow through fertile hybrids and backcrosses; introgression.

Isolating Mechanism. Any mechanism or condition which prevents gene exchange between the individuals of two or more populations; isolation may be either prezygotic or postzygotic and may be of several types.

Macroevolution. Evolutionary events viewed through the perspective of geologic time, such as the evolution of different and distinctive organisms.

Matroclinous (Maternal) Inheritance. The condition when a hybrid is closely similar to its seed parent.

Meiotic Drive. A higher probability of one allele at a locus being included in the gametes than that of others.

Mendelian Population. A reproductive group sharing a common gene pool.

Meristic Variation. Variation in numbers of parts or of organs.

Mertensian Mimic. When one deadly and one harmless organism both mimic a moderately poisonous or distasteful model.

Mesopolyploid. A polyploid of intermediate biological or geological age. (See paleopolyploid, neopolyploid).

Microevolution. Evolutionary events usually viewed over a short period of time, such as changes in gene frequency within a population over a few generations.

Migration. The dispersal and establishment of organisms beyond their place of origin; a periodic movement of individuals; both result in the transfer of genetic information among populations.

Mimicry. The superficial resemblance of one organism by another, presumably affecting the actions of predators.

Monomorphic. One form.

Morphogenesis. The process leading to the development of the characteristic mature form of an organism.

Mullerian Mimicry. The superficial resemblance, and thus reinforcement of the pattern on predators, of two or more harmful organisms.

Mutualism. A type of symbiotic relationship between two organisms of different species which benefits both of the organisms involved.

Natural Selection. Nonrandom (differential) reproduction and survival of genotypes without the conscious intervention of man.

Neopolyploid. A relatively recent polyploid, the parent(s) of which are still extant.

Niche. The status (way of life) of an organism in a community; e.g., its position in the food chain or relationships with the physical substrate.

Ontogeny. The development of an individual.

Organizer. A portion of an embryo that determines the developmental fate of the cell masses with which it comes into contact.

Outbreeder. A plant that is normally cross fertilized; outcross.

Outcross. Pollination between two different plants of the same species; cross pollination.

Overdominance. The result of the heterozygote being more extreme than either homozygote.

Paleopolyploid. An ancient polyploid the parent(s) of which are extinct.

Panmixis. Completely random mating within a population.

Parallel Evolution. The result of strong directional selection producing similar adaptations in related organisms.

Parallelism. Evolutionary convergence along similar pathways among closely related organisms.

Parasitism. When one organism obtains food or raw materials, or both from another living organism (the host) with no apparent benefit and often harm to the host; some organisms are obligate parasites and cannot live without their host, other organisms, facultative parasites, may be autotropic under some conditions.

Parthenogenesis. The development of an individual from an unfertilized gamete.

Penetrance. A measure of the proportion of individuals homozygous for a gene that shows its phenotypic effect.

Periodicity. The periodic occurrence of a given biological phenomenon, presumably as a result of natural selection.

Phenetic. Presumed natural (but not phyletic) relationship based on current total similarity; relationships indicated by the use of unweighted characters, as in numerical taxonomy.

Phenocopy. An environmentally produced specialized phenotype that resembles the phenotype of a different, more specialized, genotype; facultative dimorphism.

Phenology. Biological processes that are correlated with the seasons, e.g., flowering, mating, dormancy.

Phenotype. The actual expressed characters or observable form or physiology of an organism.

Phenotypic Variance. The total variance observed in a character.

Phenotypic Variation. Morphological or physiological variation; the result of the action of different environments on one or more genotypes.

Phyletic Evolution. Any change occurring sequentially in a single line of descent.

Phylogenetic. Pertaining to phylogeny, the evolutionary history of an organism; relationship through lineal descent from a common ancestral form.

Plasticity. The degree to which a given genotype may vary phenotypically under different environments; also, the total variation (both genotypic and phenotypic) of a species or population.

Polymorphism. The occurrence at a given stage of development, of two or more distinct variants of a species or population.

Polyploid. An individual with 3 or more (haploid) sets of chromosomes. (See additional ploidy and chromosome number terms in Chapter 10, Section A VII, p. 246.)

Polyploid Complex. A series of interrelated polyploids, often with morphological similarities, involving one or more levels of polyploidy.

Polytopic. The occurrence of a species, or other taxonomic group, in two or more separate areas.

Population. A group of individuals of one kind in a given area at a given time; often used as a synonym for mendelian population.

Population Structure. The sum of all the factors that govern the pattern in which gametes from various individuals unite with each other.

Postzygotic Isolation. The prevention of gene flow between hybridizing populations (or species) caused by the failure of the hybrids, or their derivatives, to be viable or reproductively competitive.

Preadaptation. The historical (evolutionary) or fortuitous ability of an organism to survive in a particular environment in which it finds itself.

Prezygotic Isolation. The prevention of gene flow between populations (or species) caused by geographical, ecological, seasonal, mechanical, ethological or other differences which variously act to prevent cross fertilization.

Primitive. An organism or character judged to be less changed from a presumed common ancestral state than another with which it is compared.

Pseudocopulation. A mode of pollination in some orchids in which structures of the flower resemble a female insect and male insects attempting copulation transfer pollen from one flower to another.

Pseudogamy. The phenomenon found in some apomictic plants, whereby pollination is necessary for seed development.

Pure Line. A lineage of individuals originating from a single homozygous ancestor.

Quantum Evolution. Rapid evolutionary change resulting in what a taxonomist would regard as a new higher taxon.

Ramet. An individual belonging to a clone.

Random Sample. A subset of a population selected or obtained in such a way that all variants in the population are equally likely to be included in the sample.

Ratio-Cline. Clinal variation occurring in polymorphic species, in which successive populations show progressive change in the proportion of the variants.

Recapitulation. The idea that in the course of development an individual passes through stages similar in form to adults of its presumptive ancestors; often stated as "ontogeny recapitulates phylogeny."

Recombination. The recombining of genetic material through crossing over and random assortment at meiosis and the fusion of genetically different gametes.

Recombination Index. A measure of the number of new gene combinations that can be produced in a given time; of Darlington: the average number of chiasmata plus the genetic chromosome number.

Recombination System. The interaction of a series of genetically controlled processes which more or less determines the amount of recombination possible in a species; the system may be "open," "restricted," or "closed."

Replicate. To duplicate repeatedly.

Reproductive Cells. The gametes and their immediate predecessors from which they are produced by division.

Restitution Nucleus. A nucleus formed by the fusion of two daughter nuclei, usually in a cell which did not divide after its nucleus divided.

Reticulate Evolution. Evolution in which phyletic lines merge as well as diverge.

Sample. A subset of a population; a group of items picked by some procedure and from which one hopes to learn certain things about the population.

Sampling Error. Variation due to random elements in the sampling process.

Selection. The action of environmental factors on each member of a population that results in the differential reproduction of genotypes.

Selection Coefficient. The measure of the disadvantage of a given genotype relative to other genotypes in the population.

Self-fertile. Capable of self fertilization.

Sexual Reproduction. Reproduction involving the fusion of gametes produced by a prior meiotic process.

Specialized. Applying to an organism or character judged to be more changed from a presumed common ancestral state than another with which it is compared.

Speciation. The splitting process of evolution that results in the existence of different kinds of organisms that are classified as species by taxonomists.

Species. A category in the taxonomic system; a group of organisms judged by taxonomists (by diverse criteria) to be worthy of formal recognition as a distinct kind, category, or rank.

Spontaneous Generation. Abiogenesis.

Sporophyte. The diploid spore-producing phase of the life-cycle of a plant.

Stabilizing Selection. Selection in which genotypes closer to the mean for a character have an advantage over those at the extremes.

Structural Hybridity. Heterozygosity for a chromosomal rearrangement.

Subspecies. A geographic subdivision of a species deemed worthy of formal recognition by a taxonomist.

Succession. The sequence of transient communities occurring in an area before the climax community is established.

Symbiosis. Living together; see mutualism, commensalism and parasitism.

Sympatric. Occurring in the same geographic area.

Sympatric Speciation. Speciation without geographic isolation; see divergent selection.

Sympatry. Populations (or species) occupying the same geographical area.

Syngamodeme. A unit composed of all coenogamodemes which are connected by the ability of some of their members to form viable but sterile hybrids under a specified set of conditions.

Systematic Pressure. One of the non-random evolutionary pressures: selection, mutation, or migration.

Tachytelic Evolution. Evolution at a relatively rapid rate, e.g., at a much faster rate than horotelic evolution.

Taxon. A unit of taxonomic classification such as family, genus, or species.

Territoriality. The defense of an area against organisms of the same or similar kind.

Topocline. A geographical variational trend which is not necessarily correlated with an ecological gradient.

Topodeme. A group of individuals of a taxon occurring in a specified geographical area.

Transgressive Variation. When the heterozygote (or individual segregating) is more phenotypically extreme than either homozygote (or parental form); overdominance.

Variant. A neutral term applied to any definable individual or group of individuals.

Variation Pattern. The type, form and extent of variation in one population or taxon as compared with that in another population or taxon.

Viability. The capability for living or continuing to develop; often, but incorrectly, used as synonymous with fitness.

Vicariads. Closely allied but allopatric taxa.

Vivipary. In plants, vegetative reproduction in which propagules replace flowers in the inflorescence; in animals, the production of living young rather than eggs.

Xenogamy. Outcrossing; outbreeding.

EVOLUTION LITERATURE

I. Periodicals.

American Journal of Botany. 1914+. Published by the Botanical Society of America.

Brittonia. 1949+. Published for the American Society of Plant Taxonomists by the New York Botanical Garden.

Ecology. 1920+. Published by the Ecological Society of America.

Evolution. 1946+. Published by the Society for the Study of Evolution.

Index to Plant Chromosome Numbers. 1956. Published annually by the International Association of Plant Taxonomists as a volume of Regnum Vegetabile.

Journal of Ecology. 1913+. Published by the British Ecological Society.

The American Naturalist. 1867+. Published for the American Society of Naturalists by the University of Chicago Press.

II. General References.

Bell, C. R. 1967. Plant Variation and Classification. Wadsworth Publishing Company. Belmont, California.

Briggs, D., and S. M. Walters. 1969. Plant Variation and Evolution. McGraw-Hill. New York.

Brosseau, G. E. (ed.). 1967. Evolution: A Book of Readings. William C. Brown Company. Dubuque, Iowa. [A collection of basic reprints.]

Dressler, R. L. 1968. Pollination by Euglossine bees. Evolution 22: 202-210.

Ehrlich, P. R., and R. W. Holm. 1963. The Process of Evolution. McGraw-Hill. New York.

________, and P. H. Raven. 1969. Differentiation of populations. Science 165: 1228-1232.

Evolutionary Biology. 1967. A book of review articles published annually by Appleton-Century Crofts. New York. Edited by T. Dobzhansky, M. K. Hecht, and W. C. Steere.

Faegri, K., and L. van der Pijl. 1971. The Principles of Pollination Ecology. Pergamon Press. New York.

Glass, B., et al. (eds.). 1959. Forerunners of Darwin. The Johns Hopkins Press. Baltimore.

Grant, V. 1968. The Origins of Adaptations. Columbia University Press. New York.

________. 1971. Plant Speciation. Columbia University Press. New York.

Hadley, E. B., and D. A. Levin. 1967. Habitat differences of three Liatris species and their hybrid derivatives in an interbreeding population. American Journal of Botany 54: 550-559.

Harper, J. L. 1961. Approaches to the Study of Plant Competition. Symposia of the Society for Experimental Biology XV, Mechanisms in Biological Competition.

Levin, D. A., and H. W. Kerster. 1967. An analysis of interspecific pollen exchange in Phlox. American Naturalist 101: 387-400.

Levins, R. 1968. Evolution in Changing Environments. Princeton University Press. Princeton, New Jersey.

Lewis, H. 1962. Catastrophic selection as a factor in speciation. Evolution 16: 257-271.

Nelson, A. P. 1967. Racial diversity in Californian Prunella vulgaris. The New Phytologist 66: 707-746.

Ornduff, R. (ed.). 1969. Papers on Plant Systematics. Little, Brown and Company. Boston.

Solbrig, O. T. 1970. Principles and Methods of Plant Biosystematics. The MacMillan Company. New York.

Sneath, P., and R. R. Sokal. 1973. Numerical Taxonomy. W. H. Freeman and Company. San Francisco.

Stebbins, G. L. 1950. Variation and Evolution in Plants. Columbia University Press. New York.

________. 1966. Processes of Organic Evolution. Prentice Hall, Inc. Englewood Cliffs, New Jersey.

________. 1969. The Basis of Progressive Evolution. University of North Carolina Press. Chapel Hill.

Tax, S. (ed.). 1960. Evolution After Darwin. Vol. I: The Evolution of Life. The University of Chicago Press. Chicago.

Wicker, W. 1968. (translated by R. D. Martin). Mimicry in Plants and Animals. McGraw-Hill Book Company. New York.

Williams, G. C. 1966. Adaptation and Natural Selection. Princeton University Press. Princeton, New Jersey.

Chapter 27. STRUCTURAL EVOLUTION AND PHYLOGENY

The word phylogeny refers to the origin and evolutionary history of organisms in time. Phylogenetic characters, accordingly, are characters associated with taxa over long periods of time and which are interpretable evolutionarily. Some botanists have argued that attempts to construct phylogenetic relationships among angiosperms are pure speculation in view of the incomplete fossil record of flowering plants, the conflict between phylogenies based on different kinds of evidence, the varying rates of evolution in both structures and groups of plants, and the apparent inability of botanists to decide what characters are primitive. As noted by Solbrig (1970), however, the term phylogeny can be used to represent both the actual evolutionary history of a taxon and the inferred history as deduced from available data. Davis and Heywood (1963) suggest that most phylogenetic studies involve interpretations of evolutionary trends in organs or tissues and, although this may result in an indication of the relative advancement of a taxon relative to other plants, it tells nothing about the derivations of the organisms concerned. It is important, therefore, to make the distinction between a natural (phenetic) classification and a phylogenetic classification (see also Chapter 28). A phenetic classification is one in which plants are grouped together on the basis of their overall similarity whereas a phylogenetic system attempts to interpret plant structure and function in terms of time and evolutionary change. Phylogeneticists are concerned with how particular morphologies have been attained. We believe it is essential that the student of systematics have thorough training in vascular plant morphology, anatomy and evolutionary principles not only to evaluate effectively conflicting concepts regarding the nature and evolution of angiosperm organs, structures, and taxa, but also to help answer the four basic questions of systematic biology as outlined by A. C. Smith (1969)--(1) What is it? (i.e., a circumscription of the taxon and an indication of how it differs from other taxa); (2) When did it arise? (3) Where did it originate? and (4) How did the taxon acquire its present distribution?

Section A. ORIGIN OF THE ANGIOSPERMS (Darwin's Abominable Mystery)

I. PROBLEMS OF INTERPRETATION

Prior to a consideration of possible ancestors of the angiosperms and their time and place of origin, the following questions confronting the botanist in his search for the origins of angiosperms must be answered: (1) What is an angiosperm? (2) Would all features considered unique to flowering plants be fossilized? (3) Since structures evolve at different rates when does a plant become an "angiosperm"? and (4) Were the angiosperms monophyletic or polyphyletic? The majority of botanists now favor a monophyletic origin and use the following arguments to support their position--the common possession by angiosperms of a similar type of embryo sac; possession of endosperm formed by union of sperm and two polar (female) nuclei; similar pollen types in primitive dicotyledons and monocotyledons; and the common possession of sieve tube elements and companion cells as well as stamens and carpels. The statistical probability of all of these characters arising together more than one time is extremely small.

II. MAJOR THEORIES ON THE ANCESTORS OF ANGIOSPERMS

There is no general agreement as to which group of plants represents the immediate ancestors of the angiosperms. A wide range of putative ancestors, or close relatives, has been suggested at one time or another and include: algae, Filicales, Coniferales, Gnetales, Bennettitales, Cycadales, Pteridosperms, Pentoxylales, and Glossopteridaceae. As noted above, studies on the origin of the angiosperms are further complicated by the suggestion that flowering plants may be polyphyletic. The more prominent theories on the ancestors of angiosperms are briefly discussed below.

A. Coniferales-Amentiferae Theory. Comparisons were made in the early literature between the strobili of the Coniferales and the naked, wind pollinated flowers of the Amentiferae which were regarded as the most primitive angiosperms. According to this interpretation, the stamen with its paired microsporangia was regarded as homologous to the microsporophyll of many conifers with its two microsporangia. Anatomical data have shown the Amentiferae to be advanced, however, and no support is found for this idea today.

B. Gnetalean theory. The combination of vessel-bearing wood, broad leaves with reticulate venation in *Gnetum*, and angiosperm-like reproductive morphology, particularly in *Ephedra*, has prompted suggestions that this group was the ancestor of flowering plants. This theory would derive the catkin-bearing angiosperms from an *Ephedra*-like ancestor with the magnolialian type flower resulting from further condensation of the catkin. The vessel elements of the Gnetales with perforations derived from circular-bordered pits rather than scalariform pits negates this idea according to most botanists.

C. Bennettitalean theory. This theory is based upon the supposed resemblance of the Mesozoic cycadeoid cone to a primitive flower of the *Magnolia* type. The bisexual fructification of *Cycadeoidea* had a basal whorl of perianth-like bracts and a whorl of pinnately compound microsporophylls below an ovuliferous region. Thus, there existed a superficial similarity to a flower. Along with other difficulties, it has been shown that the cone of *Cycadeoidea* never possessed expanded microsporophylls but rather had a compound microsporangiate synangium which seems, to some, to present insurmountable objections to this theory.

D. Pteridosperm theories. The discovery of the Caytoniales, a group of Mesozoic seed ferns, caused considerable interest when comparisons were made between their female reproductive structures and the angiosperm carpel. Hamshaw Thomas (1934) initially concluded that the basal lip of the *Caytonia* "fruit" functioned like a stigma, and as a result the primitive carpel might also have had a gynobasic style. A follicle in this interpretation would represent a more advanced fruit type. Subsequent study of *Caytonia* demonstrated that the pollen grains came into direct contact with the micropyle, which refuted any connection between this extinct group and modern angiosperms. Although there are points of comparison, the late Paleozoic seed ferns are often regarded as possible angiosperm ancestors primarily on the basis that all other groups can be eliminated. There are, however, objections to this idea such as the nature of the microsporangiate fructification and tracheid pitting that have not been satisfactorily resolved.

E. _Cycadales theory_. Recent evidence from fossil cycads indicates that a few primitive cycads bore 2 rows of specialized ovules that were partly covered by a lamina, thus representing an early stage in carpel evolution. These data would lend support to the opinion that the carpel of _Tasmannia_ and _Drimys_ represents the primitive condition in angiosperms, although the idea that the Cycadales were angiosperm ancestors has not been widely accepted.

III. TIME AND PLACE OF ORIGIN

With regard to the time of origin of flowering plants there are basically two opposing schools of thought. One group advocates a Cretaceous origin (or late Jurassic) whereas others strongly support a pre-Cretaceous origin of flowering plants. Scott, Barghoorn and Leopold (1960) reviewed the fossil record and concluded there were no valid angiospermous fossil remains prior to the Cretaceous. The pollen record is generally in agreement with this conclusion. Many have argued, however, that since the angiosperm fossils of the Cretaceous seem, to some, rather advanced in morphology and since these fossils appear rather diversified, the flowering plants must have had a long pre-Cretaceous history, perhaps evolving in situations which were not conducive to fossilization. Various Mesozoic angiosperm fossils have been reported, however, these materials have always come under serious interpretational doubt or otherwise have been shown to be incorrectly dated.

Major theories on the place of origin and subsequent migration of angiosperms include: (1) Angiosperms had a holarctic origin with subsequent southward migration; (2) The flowering plants attained much of their present distribution by long-distance dispersal; (3) Angiosperms arose in the general area of Southeastern Asia-Indomalaysian tropics and reached their present distribution by Antarctic land connections; i.e., migration was over land surfaces not much different from those existing today; and (4) Angiosperms evolved in pre-Cretaceous times and the process of continental drift was the major factor in establishing present distributional patterns.

The geological and biological evidence supporting sea-floor spreading (continental drift) is now so impressive that few dispute it, however, as related to the early evolution of angiosperms the important consideration is what were the relative positions of the continents in geological time. The term Gondwanaland refers to the southern land mass formed by Antarctica, Australia, southern Africa, India, and South America prior to their separation. The northern group of continents prior to their separation is known as Laurasia. Axelrod (1970) is of the opinion that the flowering plants evolved in mild uplands at low latitudes with the breakup of Gondwanaland resulting in a trend toward widespread, more equitable climate which favored the invasion of the lowlands by the early angiosperms. Smith (1970) argues for an origin of angiosperms in the general area of southeastern Asia, adjacent Malaysia, at a time the Gondwanaland and Laurasia land masses were undergoing initial fragmentation by seafloor spreading. Migration was subsequently over three principal routes: (1) Northward overland to northern Asia and Europe, some elements crossing the Bering Strait and then moving southward, (2) Westward along the coast of the Indian Ocean. This took place during moist periods or along insular chains skirting the coast of Madagascar and Africa, and (3) Southward to western and Central Australia. Subsequent migration into the West Antarctic Archipelago resulted in angiosperms reaching South America by means of insular stepping stones. In a recent discussion of the evolution and relationship of the floras of Africa and South America, Smith (1973) suggested that continental drift cannot be used in explaining the plant distributions of these land areas.

IV. ORIGIN OF THE MONOCOTYLEDONS

The origin of the monocotyledons is a subject which has long been debated. At one time or another the monocots have been regarded as ancestral to dicotyledons and polyphyletic in origin. The current widely held view is that the monocots were derived from dicotyledons in a monophyletic fashion, although there has not been uniform agreement on which group of monocots represents the ancestral stock.

As in dicots, the primary xylem vessels of monocotyledons probably originated more than once from scalariformly pitted tracheids (Cheadle, 1953). However, the phylogenetic sequence is different; vessels first arose in the late metaxylem of roots, as evidenced from the fact that the most specialized vessel elements always occur in the root metaxylem, and only subsequently extended into the stem and leaf. Within any one organ, vessel element origin and specialization occur first in the late metaxylem and then progressively in the early metaxylem and protoxylem. The distribution of vessel elements in monocotyledonous families shows they may be restricted within the plant body to the root, to the root and stem, or occur in all parts. In the opinion of Bailey (1944) and Cheadle (1953) the independent origins and specializations of vessel elements in monocots and dicots indicate that if the angiosperms are considered to be monophyletic, the monocotyledons must have arisen from a vesselless woody dicotyledon. Cronquist (1969), however, does not accept the position that vessels evolved independently in monocots and dicots and, therefore, that monocots arose from a primitively vesselless ancestor. In his view, monocots had an aquatic origin (which profoundly influenced their evolutionary history) from dicotyledonous ancestors resembling the modern Nymphaeales. That is, they had apocarpous flowers without a specialized perianth, monocolpate pollen, herbaceous habit; i.e., did not have a functional cambium, had laminar placentation, lacked a functional vessel system either through being lost or never having had one, and were aquatic. The aquatic habit resulted in the loss of a cambium that subsequently resulted in the reduction and loss of vessels. In keeping with this viewpoint, the Alismatidae are considered a near-basal side-branch of the Liliatae. Kosakai, Moseley, and Cheadle (1970) reported the occurrence of primitive vessel elements in the primary xylem of roots of Nelumbo and seriously questioned the idea that monocots were derived from the Nymphaeaceae. These authors emphasized that the Butomaceae and Alismataceae have advanced vessel elements only in the root metaxylem. Accordingly, it is difficult for them to believe, in view of what is known of vessel evolution, that the putatively primitive Alismatidae evolved advanced vessels in an aquatic environment yet gave rise to terrestrial monocots that had more primitive vessel elements in the metaxylem of their roots. The vessel story, according to them, favors the origin of the Alismatidae from terrestrial forms and does not support an aquatic "ranalian" ancestry of the monocotyledons. Suggestions have been made from time to time, e.g., Tomlinson (1970), that the woody, arborescent and perhaps branched habit represents the primitive form in the monocotyledons and not the herbaceous condition.

V. PRIMITIVE ANGIOSPERM FAMILIES

The families most widely believed by evolutionary botanists to possess the greatest number of primitive characters are listed below. It must be remembered that evolution proceeds at different rates in different tissues and thus these families exhibit a mosaic evolutionary pattern in which both primitive and advanced characters occur together. At one time these families were nearly all relegated to the large order "Ranales," however, contemporary systems of classification have created additional orders to accommodate them. The approximate size and distribution of each of the 39 families is presented (after Smith, 1967).

	Number of genera	Number of species	Distribution
1. Winteraceae	7	95	Indomalaysia, eastern Australia, Melanesia, New Zealand; Madagascar, South and Central America
2. Magnoliaceae	12	250	Tropical and temperate S. E. Asia and Indomalaysia; S. E. North America through West Indies and Central America to eastern Brazil
3. Degeneriaceae	1	1	Fiji
4. Eupomatiaceae	1	2	Southeastern Australia to New Guinea
5. Himantandraceae	1	2	New Guinea; Moluccas, eastern Australia
6. Illiciaceae	1	42	Japan, China, and Himalayas, Philippines, Borneo, Malaya, Sumatra; S. E. United States, eastern Mexico, Cuba, Haiti
7. Schisandraceae	2	47	Sakhalin and S. E. Siberia to India and into Malaysia to Amboina and Java; S. E. United States
8. Annonaceae	122	2,100	Predominantly pantropical, extending into subtropical and temperate regions
9. Myristicaceae	16	380	Pantropical
10. Canellaceae	6	20	Tropics and subtropics but lacking in Asia and Indomalaysia
11. Aristolochiaceae	7	600	Tropics and subtropics, extending into temperate areas
12. Austrobaileyaceae	1	2	Northeastern Australia
13. Chloranthaceae	5	70	Tropics and subtropics but not in African region
14. Amborellaceae	1	1	New Caledonia
15. Trimeniaceae	2	7	New Guinea and eastern Australia to New Caledonia and Fiji
16. Monimiaceae	33	450	Tropics and subtropics, predominantly of Southern Hemisphere, with centers in South America, Malaysia, and Madagascar-Mascarene regions
17. Calycanthaceae	2	9	North America, eastern Asia and eastern Australia
18. Lactoridaceae	1	1	Juan Fernandez Island
19. Gomortegaceae	1	1	Chile
20. Lauraceae	47	2,250	Pantropical, abundant in Central and South America and in Malaysia; also subtropical with extensions into temperate regions

	Number of genera	Number of species	Distribution
21. Hernandiaceae	2	52	Pantropical
22. Gyrocarpaceae	2	17	Pantropical
23. Piperaceae	12	2,000	Tropical, abundant in tropical America and Malaysia
24. Saururaceae	4	5	Himalayas to East Asia and Philippines; Atlantic North America and Southwestern United States and Mexico
25. Lardizabalaceae	8	30	Indian to eastern and S. E. Asia; also in Chile
26. Sargentodoxaceae	1	1	China and southeastern Asia
27. Menispermaceae	70	425	Tropical, with slight northern and southern extensions into extra-tropical areas
28. Ranunculaceae	47	2,000	Cosmopolitan; predominantly Northern Hemisphere, frequently in mountains, ascending to 5,700 m. in Himalayas
29. Circaeasteraceae	1	1	Himalayas to northwestern China
30. Berberidaceae	14	650	Northern Hemisphere, with only <u>Berberis</u> and <u>Mahonia</u> extending into Southern Hemisphere
31. Papaveraceae	28	250	Mostly subtropical and temperate regions of Northern Hemisphere, with centers in western North America and eastern Asia; rare in Southern Hemisphere
32. Fumariaceae	19	425	Mostly subtropical and temperate regions of Northern Hemisphere, centering in Eurasia; also in southern Africa
33. Nymphaeaceae	8	80	Cosmopolitan
34. Ceratophyllaceae	1	6	Cosmopolitan
35. Trochodendraceae	1	1	Japan and Formosa
36. Tetracentraceae	1	1	China and Burma
37. Eupteleaceae	1	2	China to Assam
38. Cericidiphyllaceae	1	2	Japan and China
39. Eucommiaceae	1	1	China

LITERATURE ON ORIGIN OF ANGIOSPERMS

Andrews, H. N. 1963. Early seed plants. Science 142: 925-931.

Arber, E. A. N., and J. Parkin. 1908. Studies in the evolution of the angiosperms. The relationship of the angiosperms to the Gnetales. Annals of Botany 22: 489-516.

________ and ________. 1907. The origin of the angiosperms. Journal of Linnean Society (Botany) 38: 29-80.

Axelrod, D. I. 1952. A theory of angiosperm evolution. Evolution 6: 29-60.

________. 1959. Poleward migration of early angiosperm flora. Science 130: 203-207.

________. 1970. Mesozoic paleogeography and early angiosperm history. Botanical Review 36: 277-319.

Bailey, I. W. 1944. The development of vessels in angiosperms and its significance in morphological research. American Journal of Botany 31: 421-428.

________. 1949. Origin of the angiosperms: Need for a broadened outlook. Journal of the Arnold Arboretum 30: 64-70.

Bancroft, H. 1930. The arborescent habit in angiosperms. The New Phylologist 29: 153-169; 227-275.

Benson, M. 1904. The origin of flowering plants. The New Phytologist 3: 49-51.

Bews, J. W. 1927. Studies in the ecological evolution of angiosperms. The New Phytologist 26: 1-21; 65-84; 129-148; 209-248; 273-294.

Camp, W. H., and M. M. Hubbard. 1963. On the origins of the ovule and the cupule in the lyginopterid pteridosperms. American Journal of Botany 50: 235-243.

Cheadle, V. I. 1953. Independent origin of vessels in the monocotyledons and dicotyledons. Phytomorphology 3: 23-44.

Cronquist, A. J. 1969. Broad features of the system of angiosperms. Taxon 18: 188-193.

Doyle, J. A. 1969. Cretaceous angiosperm pollen of the Atlantic coastal plain and its evolutionary significance. Journal of the Arnold Arboretum 50: 1-35.

Erdtman, G. 1948. Did dicotyledonous plants exist in early Jurassic times? Geologiska Föreningens I Stockholm Förhandlingar 70: 265-271.

Greguss, P. 1966. The polyphyletic origin of angiospermae. Advancing Frontiers Plant Science 15: 37-49.

Harris, T. 1960. The origin of the angiosperms. Advancement of Science 67: 207-213.

Hughes, N. F. 1961. Fossil evidence and angiosperm ancestry. Science Progress 49: 84-102.

Just, T. 1948. Gymnosperms and the origin of angiosperms. Botanical Gazette 110: 91-103.

Karsten, G. 1918. Zur phylogenie der angiospermen. Zeitschrift fur Botanik 10: 369-388.

Kosakai, H., M. F. Moseley, Jr., and V. L. Cheadle. 1970. Morphological studies of the Nymphaeaceae. V. Does Nelumbo have vessels? American Journal of Botany 57: 487-494.

Li, H. L. 1960. A theory of the ancestry of angiosperms. Acta Biotheoretica 13: 185-202.

Maguire, B. 1970. On the flora of the Guayana Highland. Biotropica 2: 85-100.

Mamay, S. H. 1969. Cycads: fossil evidence of Paleozoic origin. Science 164: 295-296.

Melville, R. 1966. Continental drift, Mesozoic continents and the migrations of the angiosperms. Nature 211: 116-120.

________. 1969. Leaf venation patterns and the origin of the Angiosperms. Nature 224: 121-125.

________. 1970. Links between the Glossopteridae and the angiosperms. In: Second Gondwana Symposium. Council for Scientific and Industrial Research, Scientia. Pretoria, South Africa.

________. 1971. Some general principles of leaf evolution. South African Journal of Science 67: 310-316.

Meeuse, A. D. J. 1963. The multiple origin of the angiosperms. Advancing Frontiers of Plant Science 1: 105-127.

Puri, V. 1967. The origin and evolution of angiosperms. Journal Indian Botanical Society 46: 1-14.

Schopf, J. M. 1970. Relation of floras of the southern hemisphere to continental drift. Taxon 19: 657-824.

Schuster, R. M. 1972. Continental movements, "Wallace's line" and Indomalayan Australasian dispersal of land plants: some eclectic concepts. The Botanical Review 38: 3-86.

Scott, R. A., E. S. Barghoorn, and E. B. Leopold. 1960. How old are the angiosperms? American Journal of Science 258: 284-299.

________, P. L. Williams, L. C. Craig, E. S. Barghoorn, L. J. Hickey, and H. D. MacGinite. 1972. "Pre-cretaceous" angiosperms from Utah: Evidence for tertiary age of the palm woods and roots. American Journal of Botany 59: 886-896.

Smith, A. C. 1967. The presence of primitive angiosperms in the Amazon Basin and its significance in indicating migrational routes. Atas do Simpósio sôbre a Biota Amazônica 4 (Botânica): 37-59.

________. 1969. Systematics and appreciation of reality. Taxon 18: 5-13.

________. 1970. The Pacific as a key to flowering plant history. Harold L. Lyon Arboretum Lecture Number 1.

________. 1973. Angiosperm evolution and the relationship of the floras of Africa and America. In: Tropical Forest Ecosystems in Africa and South America: A Comparative Review. B. F. Meggers, E. S. Ayensu, W. D. Duckworth (eds.). Smithsonian Institution Press. Washington, D. C.

Sporne, K. R. 1971. The mysterious origin of flowering plants. Oxford Biology Readers 3, F. F. Head and O. E. Lowenstein (eds.). Oxford University Press. Oxford.

Thomas, H. H. 1934. The nature and origin of the stigma. A contribution towards a new morphological interpretation of the angiosperm flower. The New Phytologist 33: 173-198.

________. 1936. Paleobotany and the origin of the angiosperms. Botanical Review 2: 397-418.

________. 1958. Palaeobotany and the evolution of the flowering plants. Proceedings of the Linnean Society of London 169: 134-143.

Tidwell, W. D., S. R. Rushforth, J. L. Reveal, and H. Behunin. 1970. *Palmoxylon simperi* and *Palmoxylon pristina*: Two pre-Cretaceous angiosperms from Utah. Science 168: 835-840.

________, ________, and A. D. Simper. 1970. Pre-Cretaceous flowering plants: further evidence from Utah. Science 170: 547-548.

________, D. A. Medlyn, and G. F. Thayn. 1972. Fossil palm materials from the Tertiary Dipping Vat Formation of central Utah. Great Basin Naturalist 32: 1-15.

Tomlinson, P. B. 1970. Monocotyledons--towards an understanding of their morphology and anatomy. In: R. D. Preston (ed.). Advances in Botanical Research. Academic Press. New York.

Wolfe, J. A., and E. S. Barghoorn. 1960. Generic change in tertiary floras in relation to age. American Journal of Science 258: 388-399.

Section B. PHYLOGENETIC DICTA, PRINCIPLES AND GENERAL EVOLUTIONARY TRENDS IN FLOWERING PLANTS

For the most part, the basic dicta and principles presented in this section form the basis of modern phylogenetic thought and classification. In a discussion of phylogeny the terms primitive and advanced with respect to different character states cannot be avoided. Sporne (1948) used the term "primitive character" to mean one which is possessed by a present day taxon and was also possessed by its ancestors. An "advanced character," on the other hand, is one possessed by a present day taxon and not possessed by its ancestors, that is, it replaced an ancestral character during evolution. A fundamental problem that confronts an evolutionary botanist involved in phylogenetic pursuits is to determine the direction and rate in which structural modifications have occurred or are occurring, i.e., to identify primitive and advanced characters within the taxonomic units under study, and to determine whether this direction has changed. Because the erratic fossil record is of comparatively little assistance in this regard, other approaches must be employed (see Sporne, 1956). The most successful methods in determining advancement or primitiveness, as related to comparative morphology, are termed by Sporne (1956) the doctrines of association, correlation, and common ground plan.

That vessel elements evolved from tracheids appears certain. Advantageously, the sequence involves characters that are measurable and, therefore, can be dealt with statistically. Bailey and Tupper (1918) provided the earliest systematic evidence for this trend in a comparison of length of tracheary elements of vascular cryptogams, gymnosperms, and angiosperms. Frost (1930a, b; 1931) extended this work and initiated a detailed statistical analysis of specialization of the secondary xylem in dicotyledons. Frost made two basic assumptions that were proved by subsequent tests of association and correlation. The first was that the primitive state of the derived structure will resemble the ancestral structure if it can be determined from which structure the other has evolved. Since it was assumed that vessel elements evolved from tracheids, Frost deduced that the most primitive vessel elements were those which most resembled tracheids (doctrine of association). If this were not true, then the assumption of a direct relationship is incorrect, or the structures are so widely separated in the evolutionary scale that the tracheid-like condition of the vessel element has been lost. The doctrine of association has been criticized because of assumptions that are made prior to its application. Frost's second assumption, the doctrine of correlation, takes advantage of the fact that primitive characters are frequently associated with other primitive characters. Thus, if one character can be demonstrated to be primitive (e.g., scalariform perforation plates on the vessel end wall) then characters correlated with this feature are probably also primitive. The doctrine of the common ground plan assumes that any character present in all or most individuals of a taxon, or the most common character state among a group of related plants, has probably been inherited unmodified from a common ancestor and must be the ancestral or primitive state.

Phylogenetic speculation invoking the concepts of tissue conservatism, recapitulation of phylogeny during ontogeny, and teratology generally have met with criticism and limited success. To date, only the secondary xylem of fossil angiosperms has revealed significant phylogenetic information.

Wagner (1962) has pointed out that in the determination of the most probable pathways of phylogeny within a group of plants there is one overriding evolutionary phenomenon which must be taken into account. This is that common ancestry gives rise to divergent derivatives. Some derivatives will closely resemble their ancestor whereas other derivatives will become highly modified and not resemble their ancestor. If this pattern is valid three basic assumptions are required: (1) plants which have a large number of similar characteristics in common have the same common ancestor (similarity corresponds to recency of common ancestry); (2) evolution proceeds in various directions. Different lines will change in different characters and different character-complexes (evolutionary divergence); and (3) evolution takes place at different rates at various times and in different lines. Some forms will remain fairly stereotyped, resembling the common ancestor; others will change radically in the same period from the common ancestor (inequality of evolutionary rates).

It must be emphasized that evolutionary interpretations of this type must be made and evaluated in light of certain basic principles and pitfalls in phylogenetic study. For example, it is often necessary to determine whether simplicity of structure reflects a primitive condition or is the result of secondary specialization. Similarity of structure may not imply or demonstrate close genetic relationship because similar structures may have arisen independently and more than one time in widely unrelated taxa. Morphological similarity between individual organs and entire organisms has frequently resulted from the convergence between two distinct phyletic lines in response to adaptations to similar habitats, or morphological similarity has sometimes resulted from parallelism. Similarity resulting from common ancestry is termed patristic similarity whereas the actual pathway or relative position of divergences is spoken of as cladistic relationship. Both patristics and cladistics are measurements of phylogeny involving time as opposed to strictly phenetic considerations. Thus, morphological resemblances have been attained patristically or by convergent evolution. From a taxonomic standpoint it is the degree of morphological divergence that is significant and not the time at which the divergence occurred. Another fundamental basis of phylogenetic interpretation is homologous and analogous structures. Homologous structures refer to structures similar in evolutionary and developmental origin regardless of function whereas superficially similar structures not the result of common ancestry are analogous. An organ or structure is homologous with another because of what it is, not because of what it does (de Beer, 1971). Understanding which structures and organs are homologous is a basic goal of comparative morphological study and provides evidence of affinity between organisms that have evolved with structural modification from a common ancestry. Deciding whether a particular taxonomic group is monophyletic, having had a single origin, or polyphyletic, having had multiple origins, is a complex problem not easily resolved in the absence of a complete fossil record. If true phylogenetic relationships are to be expressed, it is widely believed that higher taxonomic categories should be monophyletic.

The ultimate goal of phylogenetic study is to determine, with the highest degree of probability, the pathways of relationship. Wagner (1961, 1962) has offered one method (The Groundplan/Divergence Method) of reconstructing the best estimate of phylogeny utilizing all characters. This approach involves three basic steps (quoted from Wagner, 1961): (a) systematic or comparative analysis of the plants in question to find and understand their contrasting characters; (b) determination of ground plans to find the character states common to all or most of the plants in order to deduce the most probable ancestral or primitive states; and (c) phylogenetic synthesis to assemble the taxa according to their respective deviations from the basic ground plan and from each other. Specific steps are as follows: (1) to compare and study all the variable characters among the taxa; (2) to determine the generalized or primitive conditions on the principle that characters found in most or all of a number of related taxa are inherited essentially unchanged from the common ancestor, using data also from related taxonomic groups of the same level. (If no obvious trend can be determined in a given character that character may be used only for grouping purposes.); (3) to assign for each character the value 0 for the generalized or primitive condition, and 1 for the specialized or secondary condition (the intermediate states being assigned the value 0.5); (4) to list in tabular form the taxa and for each give the divergence values from the ground plan, both for individual characters and in total; and (5) to determine the mutual character groupings between taxa and then arrange them in sequence according to these groupings on a concentric chart or graph, the radii and branchings to be determined by the mutual character complexes, and the distances by the divergence indices. It should be noted that this procedure attempts to understand evolutionary trends in individual structures and to correlate these in cladograms that illustrate the degree of divergence from ancestors, those characters which have undergone modification, and the direction in which the evolution has progressed.

Cronquist (1968) has stated, "It thus appears that a workable taxonomic system cannot provide a perfect reflection of evolution, no matter how abundant the evidence on which it is based. Indeed, the more abundant the evidence, the better understanding of phylogeny, the clearer it becomes that the correlation between phylogeny and taxonomy must be general rather than exact. However, the phylogenetic concept still provides the underlying rationale for the natural system. Taxonomy can provide only a somewhat muddy reflection of evolution, but a reflection all the same."

I. DICTA (After Bessey, 1915)

A. General Dicta

1. Evolution is not always upward, but often it involves degradation and degeneration.
2. In general, homogeneous structures (with many and similar parts) are lower, and heterogeneous structures (with fewer and dissimilar parts) are higher.
3. Evolution does not necessarily involve all organs of the plant equally in any particular period, and one organ may be advancing while another is retrograding.
4. Upward development is sometimes through an increase in complexity, and sometimes by a simplification of an organ or a set of organs.
5. Evolution has generally been consistent and when a particular progression of retrogression has set in, it is persisted in to the end of the phylum.
6. In a phylum the holophytic (chlorophyll-green) plants precede the colorless (hysterophytic) plants, and the latter are derived from the former.
7. Plant relationships are up and down the genetic lines, and these must constitute the framework of phylogenetic taxonomy.

B. Dicta having special reference to the general structure of the flowering plants

8. The stem structure with collateral vascular bundles arranged in a cylinder is more primitive than that with scattered bundles, and the latter are to be regarded as derived from the former.
9. Woody stems (as of trees) are more primitive than herbaceous stems, and herbs are held to have been derived from trees.
10. The simple, unbranched stem is an earlier type, from which branching stems have been derived.
11. Historically the arrangement of leaves in pairs on the stem is held to have preceded the spiral arrangement in which the leaves are solitary at the nodes.
12. Historically simple leaves preceded branched ("compound") leaves.
13. Historically leaves were first persistent ("evergreen") and later deciduous.
14. The reticulate venation of leaves is the normal structure, and the parallel venation of some leaves is a special modification derived from it.

C. <u>Dicta having reference to the flowers of flowering plants</u>

15. The polymerous flower structure precedes, and the oligomerous structure follows from it, and this is accompanied by a progressive sterilization of sporophylls.
16. Petaly is the normal perianth structure, and apetaly is the result of perianth reduction (aphanisis).
17. The apochlamydeous perianth is earlier and the gamochlamydeous perianth is derived from it by a symphysis of the members of perianth whorls.
18. Actinomorphy is an earlier structure than zygomorphy, and the latter resulted from a change from a similar to a dissimilar growth of the members of the perianth whorls.
19. Hypogyny is the more primitive structure, and from it epigyny was derived later.
20. Apocarpy is the primitive structure, and from it syncarpy was derived later.
21. Polycarpy is the earlier condition, and oligocarpy was derived from it later.
22. The endospermous seed is primitive and lower, while the seed without endosperm is derived and higher.
23. Consequently, the seed with a small embryo (in endosperm) is more primitive than the seed with a large embryo (in scanty or no endosperm).
24. In earlier (primitive) flowers there are many stamens (polystemonous) while in later flowers there are fewer stamens (oligostemonous).
25. The stamens of primitive flowers are separate (apostemonous), while those of derived flowers are often united (synstemonous).
26. The condition of powdery pollen is more primitive than that with coherent or massed pollen.
27. Flowers with both stamens and carpels (monoclinous) precede those in which these occur on separate flowers (diclinous).
28. In diclinous plants the monoecious condition is the earlier, and the dioecious later.

II. PRINCIPLES (After Thorne, 1958)

1. Existing species have descended with change from pre-existing species, and are therefore the products of evolutionary forces.
2. Ancestral conditions and trends of specialization are often recognizable in the organs, tissues, and cells of living and fossil angiosperms.
3. The primitive, ancestral condition of any given characteristic can be no more specialized than its condition in a derived, existing species most primitive for that characteristic.
4. The presence of vestigial rudiments of organs, or sometimes the presence of vestigial vascular supply to greatly modified or missing organs, often furnishes evidence of evolutionary reduction, loss, fusion, or other major modification of structures.
5. The prevalence of parallel and convergent evolution in habit, function, and structure is predictable consequence of the relatively limited means angiosperms have for effective reproduction and for adaptation to available environmental niches.
6. All parts of plants at all stages of their development may produce evidence that is valuable in establishing relationships.
7. Evolution may tend toward elaboration and diversity or toward reduction and simplicity.

8. The rate and direction of evolution may vary in the different organs and tissues of plants.
9. Most existing angiosperms are highly specialized and greatly modified from their primitive, generalized ancestors.
10. Evolutionary trends are sometimes reversible under the influence of changes in environmental factors.
11. Once lost, organs usually are not regained.
12. New angiospermous structures have arisen as modifications of or as outgrowths from pre-existing structures.
13. The sporadic or restricted occurrence of unusual or uncommon characteristics lacking apparent evolutionary significance is often an indication of relationship when correlated with other characteristics.
14. The occasional attainment of certain characteristics of certain levels of evolutionary development is frequently valuable in determining the affinities of families and orders.
15. Embryos and seedlings of related though dissimilar plants often resemble each other more than do the adult plants because of their apparent retention of primitive characteristics.

III. EVOLUTIONARY TRENDS IN ANGIOSPERMS

A. General trends (Expanded from Smith, 1967)

	Primitive	Advanced
1.	Tropical plants	Temperate plants
2.	Woody plants	Climbing or herbaceous plants
3.	Wood homoxylous (vesselless)	Wood heteroxylous (vesselled)
4.	Perennial habit	Biennial or annual habit
5.	Terrestrial habit	Aquatic, epiphytic, saprophytic or parasitic habit
6.	Vascular bundles in cylinder (Dicotyledons)	Vascular bundles scattered (Monocotyledons)
7.	Chlorophyll present	Chlorophyll absent
8.	Evergreen plants	Deciduous-leaved plants
9.	Stipules present	Stipules absent
10.	Node unilacunar with two traces	Node variously altered
11.	Leaves spiralled	Leaves opposite or whorled
12.	Leaves simple	Leaves compound
13.	Flowers hermaphrodite	Flowers unisexual
14.	Flowers solitary	Flowers in inflorescences
15.	Flowers entomophilous	Flowers anemophilous
16.	Floral parts spirally imbricate	Floral parts whorled or valvate
17.	Flowers polymerous	Flowers oligomerous
18.	Perianth undifferentiated	Perianth differentiated into calyx and corolla, or further reduced
19.	Flowers with petals	Flowers without petals
20.	Petals free	Petals connate
21.	Actinomorphy	Zygomorphy
22.	Hypogyny	Perigyny, epigyny
23.	Stamens many	Stamens few
24.	Stamens separate	Stamens connate
25.	Stamens 3-veined microsporophylls with linear embedded sporangia	Stamens variously modified

26.	Pollen monocolpate (or derived)	Pollen tricolpate (or derived)
27.	Carpels many	Carpels few
28.	Carpels free	Carpels connate
29.	Carpels conduplicate, unsealed, styleless	Carpels variously modified
30.	Placentation laminar	Placentation submarginal to axile to parietal, etc.
31.	Fruits single	Fruits aggregate
32.	Fruit a follicle	Fruit a capsule, berry, drupe, etc.
33.	Seeds large; embryo small; endosperm abundant	Seeds small; embryo well developed; endosperm little or none
34.	Cotyledons 2	Cotyledons 1 or 3 (or more)
35.	Anatropous ovules	Other types
36.	Two integuments	One integument
37.	Nuclear endosperm	Cellular or helobial endosperm
38.	Crassinucellar ovule	Tenuinucellar ovule
39.	Long, slender sieve cells with scattered sieve areas	Shorter sieve tube elements with specialized sieve plates
40.	Chromosome number low; n = 7	Higher numbers

B. Sporne (1954, 1971) has discussed the statistical correlation between floral and vegetative characters in the dicotyledons. According to him the most reliable indicators of primitiveness are the following:

1. Woody habit
2. Presence of secretory cells
3. Leaves alternate
4. Stipules present
5. Flowers actinomorphic
6. Petals free
7. Stamens pleiomerous
8. Carpels pleiomerous
9. Seeds arillate
10. Two integuments
11. Integumentary vascular bundles
12. Nuclear endosperm
13. Carpels free
14. Axile placentation
15. Unisexual flower
16. Leuco-anthocyanins present
17. Small number of seeds
18. Scalariform perforation plates
19. Scalariform intervascular pitting
20. Apotracheal wood parenchyma
21. Unstoried wood
22. Pollen binucleate

LITERATURE ON PHYLOGENETIC DICTA

Bailey, I. W., and W. W. Tupper. 1918. Size variations in tracheary cells. I. A comparison between the secondary xylems of vascular cryptogams, gymnosperms, and angiosperms. Proceedings American Academy of Arts and Sciences 54: 149-204.

Bessey, C. E. 1915. The phylogenetic taxonomy of flowering plants. Annals of Missouri Botanical Garden 2: 109-164.

Frost, F. H. 1930a. Specialization in secondary xylem in dicotyledons. I. Origin of vessels. Botanical Gazette 89: 67-94.

________. 1930b. Specialization in secondary xylem in dicotyledons. II. Evolution of end wall of vessel segment. Botanical Gazette 90: 198-212.

________. 1931. Specialization in secondary xylem in dicotyledons. III. Specialization of lateral wall of vessel segment. Botanical Gazette 91: 88-96.

Smith, A. C. 1967. The presence of primitive angiosperms in the Amazon Basin and its significance in indicating migrational routes. Atas do Simpósio sôbre a Biota amazônica 4 (Botânica): 37-59.

Sporne, K. R. 1948. Correlation and classification in dicotyledons. Proceedings of Linnaean Society of London 160: 40-47.

________. 1954. Statistics and the evolution of dicotyledons. Evolution 8: 55-64.

________. 1956. The phylogenetic classification of the angiosperms. Biological Reviews Cambridge Philosophical Society 31: 1-29.

Thorne, R. F. 1958. Some guiding principles of angiosperm phylogeny. Brittonia 10: 72-77.

________. 1963. Some problems and guiding principles of angiosperm phylogeny. The American Naturalist 97: 287-305.

Wagner, W. H. 1961. Problems in the classification of ferns. In: Recent Advances in Botany 1, sect. 9: 841-844.

________. 1962. The synthesis and expression of phylogenetic data. In: Benson, L., 1962, Plant Taxonomy, Methods and Principles. The Ronald Press Company. New York.

Section C. STRUCTURAL EVOLUTION

Since the fossil record is incomplete, the evolutionary history of angiosperms, for the most part, can only be inferred from data existing in extant plants. The objectives of this section are to reconstruct a hypothetical primitive flowering plant from available information and determine in which direction or directions structural modifications have occurred.

I. VEGETATIVE MORPHOLOGY AND ANATOMY

A. Habit. The ancestral dicotyledonous angiosperm is widely believed to have been perennial, woody, and arborescent. The evidences for the primitiveness of the woody habit has been summarized by Constance (1955): (1) the prevalence of the woody habit in gymnosperms; (2) its correlation with anatomical and floral characters regarded as primitive and, conversely, the association of advanced anatomical and floral characters with the herbaceous habit; (3) the predominance of woody plants in moist tropical and subtropical areas, habitats which are widely believed to be modern equivalents of the climatic conditions widespread at the period of angiosperm inception; (4) the revelation of the persistence of cambium in seedling and herbaceous stems; (5) the supposed correlation between fleshy fruits and woody habits; and (6) the lack of undoubted fossil herbaceous angiosperms in older geologic strata. The viny habit along with epiphytism, saprophytism, and parasitism are derived conditions.

B. Secondary Xylem. During the past fifty years the value of wood anatomy, and vessel element evolution in particular, in the study of the phylogeny and classification of angiosperms has been clearly established. In no other vegetative tissues of the plant are the trends of evolution as clearly defined. These trends were recognized entirely without reference to existing taxonomic systems and thus the relative primitiveness or advancement of the plants in which they occur. The trends of evolution of the vessel element have been firmly established through extensive and intensive comparative anatomical studies on both fossil and living material, statistical analyses (i.e., measurements and correlations), and ontogenetic studies. Therefore, the evolutionary trend from tracheids to vessel elements is the most reliable

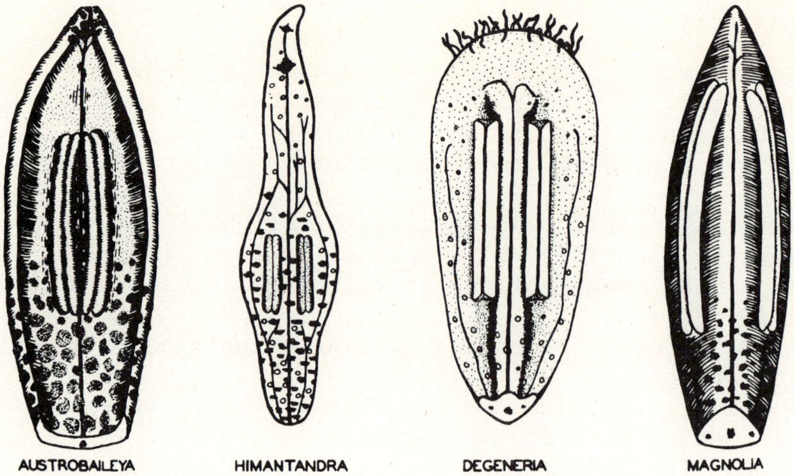

Figure 27-1. Stamens of Representative Magnolialian Genera. (Canright, J.E. 1952)

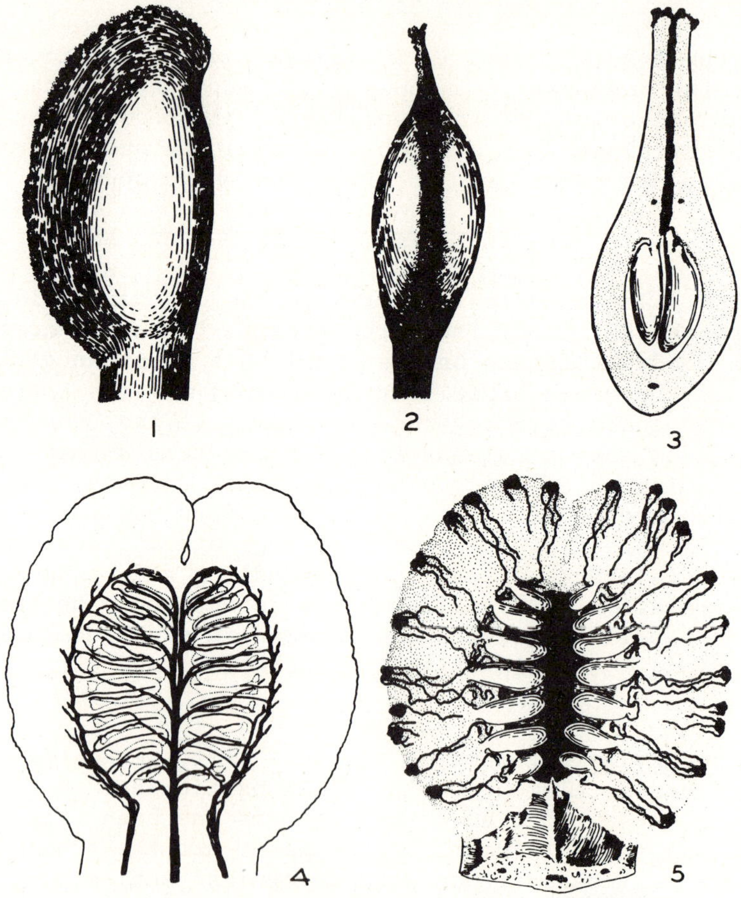

Figure 27-2. The Conduplicate Carpel of Tasmannia (Winteraceae). 1. Side view. 2. Ventral view, showing the paired stigmatic crests. 3. Transverse section, showing penetration of pollen tube as three major carpellary bundles. 4. Cleared, unfolded lamina, showing vasculature. 5. Unfolded lamina, showing placentation, distribution of glandular hairs (stippled), at the course of pollen tubes. (Bailey, I. W., & B. G. L. Swamy. 1951)

tool known in the study of phylogeny. This is true because this trend is both largely unidirectional and irreversible.

It has been shown that the angiosperm vessel element (in secondary wood) has been phylogenetically derived from a scalariform-pitted tracheid by the loss of pit membranes in the regions of tracheid overlap. Therefore, vessel elements which are most tracheid-like, viz., elongate, narrow, many scalariform perforations in the end walls, etc., are the most primitive type. Other known major trends of vessel element specialization are: (1) change in perforation plates from scalariform, many barred, and fully bordered to few barred and borders absent to more advanced porous oblique and transverse porous plates; (2) angle of end wall highly oblique and tapering (much overlap) to transverse; (3) reduction in vessel element length; (4) change in appearance on the transverse section from angular to circular; (5) increase in pore diameter; (6) change in intervascular pitting from scalariform and bordered to opposite and alternate multiseriate; and (7) change in vessel distribution from solitary to extensive aggregate groupings of vessels. Concomitant with these changes in the vessel element are modifications in the fusiform cells of the vascular cambium from long initials with overlapping ends that divide in a pseudo-transverse manner to advanced initials which undergo radio-longitudinal division and are reduced in length.

Although the above trends of vessel evolution are believed to be firmly established, Beck (1970) has pointed out that circular bordered pits had evolved by Middle Devonian time and occurred in either the primary or secondary xylem tracheids of coniferophytic progymnosperms and primitive pteridosperms. In his opinion the acceptance of the scalariform bordered pit as primitive in seed plants should be reconsidered. To quote Beck, "In view of the early appearance of the circular bordered pit in the progymnosperms and the preponderant occurrence of this type in the secondary xylem of cycadophytic and coniferophytic gymnosperms, it seems not unreasonable to suggest that the circular bordered pit might be the basic, primitive type in the gymnosperm-angiosperm line of evolution. This is not to say, necessarily, that the circular bordered pits of some angiosperms might not have evolved from the scalariform pits of their ancestors, whether gymnosperm or angiosperm. In these groups this might be interpreted to have been accomplished through neoteny. The circular bordered pits of other groups of angiosperms might simply reflect their ancestry from gymnosperms with circular bordered pits."

Other well known major phylogenetic trends in the secondary xylem, as revealed by statistically significant correlations using features of vessel elements, include the evolution from tracheids to fiber-tracheids to advanced libriform fibers through a progressive increase in cell length (as compared to vessel elements) with a concomitant reduction in pit size and number and eventual elimination of the tracheid pit border. Primitive woods have both uniseriate and multiseriate rays that are heterocellular and high-celled; the multiseriate rays have long uniseriate wings. Specialization results in the reduction and loss of either the multiseriate or uniseriate rays (or both), the rays becoming homocellular and reduced in height with the uniseriate wings of multiseriate rays reduced to a single cell. The condition of stratified rays formed by a storied cambium is an extreme advancement. Apotracheal diffuse axial parenchyma is regarded as the primitive state from which more advanced paratracheal types (vasicentric, aliform, confluent) were derived, however, Bailey (1957) points out that the available data do "...not provide a single linear series of increasing structural specialization." The phylogenetic significance of the absence of wood parenchyma is unclear.

Additional features of wood structure are believed to represent sporadically occurring divergent trends of specialization whose presence in woods does not imply a close genetic relationship. These include helical thickenings on the walls of tracheary elements and fibers, imperforate tracheary elements which become septate or retain a living protoplast, formation of vascular tracheids and vasicentric tracheids, excessive thickening of walls and widening of vessel element diameter (as in lianas), various reticulate perforation plates or other modifications of perforation plates, and extensive aggregations of vessels in diversified patterns (Bailey, 1957). Furthermore, Carlquist (1962) has shown that primitive features of the primary xylem may occasionally be extended into the secondary xylem (paedomorphosis or juvenilism) in plants that are otherwise specialized.

The great value of wood anatomy as a guide to understanding angiosperm evolution lies not only in the fact that trends toward specialization are well defined, but that all stages in evolutionary specialization are observable in extant angiosperms. Therefore, there is not the slightest doubt that the absence of vessels in eleven genera of woody dicotyledons represents the retention of a primitive feature. The largest number of vesselless genera (7) occur in the Winteraceae, whereas the Trochodendraceae, Tetracentraceae, Amborellaceae, and Chloranthaceae contain a single vesselless genus each. However, based on the sole evidence of lack of vessels, these genera are not necessarily the most primitive angiosperms, since evolution proceeds independently in all organs and tissues of plants. The absence of vessels in plants belonging to families, as Cactaceae, clearly the result of extreme specialization (Bailey, 1966). The importance of how plant habit and the factors of the environment affect wood structure and the trends of xylem evolution have been emphasized recently by Carlquist (1966, 1969a&b). The single most important guiding principle of wood anatomy and phylogeny (Bailey, 1944) emphasizes the impossibility (unlikelihood) that a taxon with specialized xylem gave rise directly to a taxon with less specialized xylem. Accordingly, wood anatomy is more useful in negations than assertions of positive relationship.

Other features of the secondary xylem are diagnostic in so far as they are restricted to certain families or genera; however, their phylogenetic significance remains dubious. Distinctive tile ray cells are confined to certain genera of the Bombacaceae, Sterculiaceae, and Tiliaceae whereas upright sheath cells located on the margins of multiseriate rays are of comparatively widespread occurrence. Vestured pits are reported for twenty-four families of dicotyledons (Bailey, 1933) with a concentration in the Myrtales. Solitary prismatic crystals are relatively widespread, although such forms as raphides, druses, and crystal sand have a more restricted distribution, and, therefore, are of greater taxonomic interest (Metcalfe and Chalk, 1950).

C. Secondary Phloem. Unlike the xylem, information on the comparative anatomy and evolutionary trends in phloem tissue is meager (Esau et al., 1953). For example, witness the absence of phloem data in Metcalfe and Chalk (1950). The subject of phloem is thoroughly reviewed in Esau (1969). Suggested trends of specialization relate mostly to sieve elements in secondary phloem of dicotyledons and metaphloem elements of monocotyledons, and have by no means been demonstrated to be unidirectional or irreversible. Composing the secondary phloem are the principal conducting cells, the sieve elements, in combination with various amounts of phloem parenchyma, sclerenchyma, and ray parenchyma. With regard to sieve elements, Cheadle and Whitford (1941) recognized two basic types, the sieve cell and sieve tube element. The sieve cell is interpreted as a less specialized element with relatively undifferentiated sieve areas (i.e., narrow pores and thin connecting strands), lacking highly developed sieve plates and companion cells. More advanced sieve tube elements have specialized sieve plates on the end walls, have accompanying companion cells, and are superimposed on one another in a longitudinal series forming a sieve tube. Sieve plates may be compound, that is, composed of several highly differentiated sieve areas, or simple and composed of a single sieve area. Also, end walls range in orientation from highly oblique to transverse. Although the distinction between these cell types is not always sharp, gymnosperms and lower vascular plants possess sieve cells whereas angiosperms are characterized by sieve tube elements and associated companion cells.

The sieve cell is generally regarded as the primitive type of conducting element from which the sieve tube element arose. Major trends of sieve tube element specialization (Esau et al., 1953) are believed to involve (1) localization of specialized sieve areas on the end walls, (2) change in orientation of end walls from very oblique to transverse, (3) progressive change from compound to simple sieve plates, and (4) progressive decrease in conspicuousness of the sieve areas on the side walls, although the relation between specialization of lateral sieve areas and the sieve tube element is not consistent (Zahur, 1959). Esau (1969) has stressed that comparative data of lateral sieve areas should come only from walls between contiguous sieve elements. The same general trends have also been agreed upon for monocotyledons (Cheadle, 1948; Cheadle and Whitford, 1941). Measurements of cell length, width, and size of sieve pore area has yet to demonstrate the same phylogenetic significance proposed for tracheary tissue. However, Zahur (1959) is of the opinion that there has been a phylogenetic decrease in sieve tube element length (realizing secondary septation can occur) accompanied by decrease in the length of the end walls and increase in size of sieve plate pores. Following earlier workers, Zahur classified sieve tube elements into three categories of increasing specialization: I, elements long (400 μm or more), with very oblique sieve plates containing 10 or more sieve areas; II, elements of medium length, with oblique sieve plates containing 2 to 10 sieve areas; and III, elements short with slightly oblique to transverse, simple sieve plates. It is clear that there is rather uniform agreement that the most primitive sieve tube elements are ones that most resemble sieve cells. As far as is known, unspecialized sieve cells occur in only one angiosperm, _Austrobaileya scandens_ (Austrobaileyaceae) (Bailey and Swamy, 1949; Srivastava, 1970).

Zahur (1959) also classified companion cells into types reflecting ascending evolutionary advancement. Type A are cells that are much shorter than the sieve tube elements and occur singly, usually at the corners of sieve type elements. Type B are cells as long as the sieve tube element, and Type C are cells that are septate in addition to being as long as the sieve tube element. Zahur found companion cell type to be constant within families and, occasionally, orders with the greatest variability occurring in families whose "naturalness" has been questioned.

Although indecisive, Zahur regarded abundant, extremely variable and irregular parenchyma to be advanced whereas scanty parenchyma showing little variation in size probably represented a primitive condition because of its association with polypetalous families. Zahur could find no bases for deciding whether the presence or absence of phloem sclerenchyma was structurally primitive or advanced. However, Eames (1961) regarded the presence of phloem fibers to be an advanced state.

D. Nodal Anatomy. Since the first comprehensive study of nodal structure and evolution by Sinnott (1914), nodal anatomy has assumed much importance in discussions relating to angiosperm phylogeny. Whether the primitive nodal pattern was trilacunar (Sinnott, 1914), multilacunar (Ozenda, 1949), unilacunar, two-trace (Marsden and Bailey, 1955; Canright, 1955), or trilacunar or multilacunar with two traces diverging from the median gap (Takhtajan, 1969) has yet to be answered. The primary evidence used to support the primitive nature of the 2:1 node are its occurrence (or modifications thereof) in members of the Magnoliidae (Calycanthaceae, Lactoridaceae, Illiciaceae, Amborellaceae, Lauraceae, Chloranthaceae, Monimiaceae, Austrobaileyaceae, Schisandraceae, and Trimeniaceae), the distribution of this type in vascular plants other than angiosperms, and its presence in the cotyledonary node of many dicots even though the mature foliar nodes of the same plant may be tri- or multilacunar (i.e., cotyledonary nodes are conservative and thus reflect ancestral conditions).

Recent papers by Conde and Stone (1970) and Stone (1970) have described an amazing diversity of cotyledonary nodal patterns in the Juglandaceae in which nodal evolution appears correlated with functional demands of the seedling. These authors were able to confirm the 2:1 cotyledonary node as primitive in Juglandaceae and suggested the diversification to more complex patterns as functionally related to independent shifts within the family from epigeal to hypogeal seedlings. Furthermore, a suggestion was made that the apparent conservatism of cotyledonary nodes in dicotyledons as a whole may be directly related to the prevalence in many groups of primitive, epigeal germination, and thus, cotyledonary nodal anatomy should be considered independently of the mature foliar nodes.

Although most families tend to have a uniform nodal anatomy, some families and even genera have variable nodal structure. Trilacunar nodes occur in the majority of dicotyledons (Sinnott, 1914) whereas multilacunar nodes are relatively uncommon in dicotyledonous families (Howard, 1970), but are found in members of primitive orders such as Magnoliales, Piperales, and Trochodendrales, and advanced orders such as Umbellales and Asterales. The unilacunar node has an interesting distribution, namely the Laurales (sensu Takhtajan), Caryophyllales, Ericales, Diapensiales, Ebenales, Primulales, Myrtales, and a majority of the families in the Asteridae. Some orders show transitions in nodal structure (e.g., the Theales) where Schofield (1968) described two possible lines of evolution originating from the ancestral trilacunar condition in the Dilleniaceae. One line maintained the trilacunar node, and eventual amplification culminated in the multilacunar node of the Caryocaraceae. The second line involved reduction to the unilacunar nodes of the Marcgraviaceae and Theaceae.

Namboodiri and Beck (1968) have recently proposed that the eustele of gymnosperms evolved directly from a protostele by gradual medullation and longitudinal dissection followed by formation of a cylinder of longitudinal sympodial bundles from which leaf traces diverge radially. In this interpretation the primary vascular system of seed plants is not directly comparable to that of ferns so that the classic filicinean-type leaf gaps may not occur in eusteles. This concept has also been extended to angiosperms (Slade, 1971; Devadas and Beck, 1972). Benzing (1967) concluded that the unilacunar, one-trace or trilacunar, three-trace nodal pattern was probably primitive.

E. Leaves. Although anatomical leaf characters are generally useful in the delimitation of taxa, the majority of these characters cannot be readily interpreted ecologically or evolutionarily. Suggested trends of foliar evolution have generally not been supported by convincing evidence. The ancestral angiosperm is envisioned as having alternate, simple, entire, stipulate, and pinnately net-veined leaves. This is one modern dictum which is markedly opposed to Bessey's original opinions.

Trends of evolution of stomata have long been unclear. Baranova (1972) has recently advocated paracytic types as primitive within the angiosperms on the basis of the nearly uniform occurrence of this pattern in the Magnoliales (*sensu* Takhtajan). The only taxa of the order that do not have this stomatal pattern are *Liriodendron tulipifera* (Magnoliaceae) and *Bubbia perrieri* (Winteraceae), both advanced in their respective families. Correlated with paracytic stomata in the Magnoliales are thickened lamellae on the outer walls of the guard cells and thick-walled epidermal cells with pores in the outer wall, both characters are regarded by Baranova (1972) as primitive in angiosperms as a whole. Since it is well known that similar mature stomatal patterns can arise by different developmental pathways, an extension of this work utilizing the ontogenetic classification of stomatal types by Fryns-Claessens and Van Cotthem (1973) would be desirable. It should also be noted that paracytic stomata are found in a large number of very unrelated and often advanced families.

Preliminary data by Hickey (1971) show that families otherwise regarded as primitive tend to have venation patterns in which the secondary veins have an irregular course in the lamina, and intercostal areas lack uniform size and shape, whereas more advanced families tend to have more regular venation patterns which follow consistent courses. The phylogenetic significance of the open dichotomous venation in the only two angiosperms which possess this feature--*Kingdonia* and *Circaeaster*--remains unclear (Foster, 1961). Cronquist (1968) regarded this as a modification that simulates a primitive pattern. Although there is considerable anatomical diversity in petiolar vasculation, trends of evolution in this region of the plant are not readily discernible in angiosperms as a whole. Also, phylogenetic trends in trichomes for angiosperms as a whole have been suggested (see Carlquist, 1961) but they have not had significant influence on phylogenetic thought.

II. FLORAL MORPHOLOGY AND ANATOMY

Although the flower has historically been a principal source of taxonomic data, and the literature dealing with floral morphology and anatomy is voluminous, there has not been agreement on a world-wide basis regarding interpretations of floral morphology, anatomy (which is often used as synonymous with floral vascular anatomy) and evolution of floral structures. The following treatment presents some of the basic tenets and concepts of floral morphological study.

A. Taxonomic and evolutionary value and interpretations of floral morphology. Among the characters of floral morphology of systematic importance are the vascular supply, i.e., number of traces and nodes, elaborateness of vascularization, position of microsporangia and ovules in relation to the venation, and stages of morphological alterations that result in syncarpy, cohesion, etc. (Carlquist, 1961). The classic interpretation of the basic nature of the vascular supply to the floral parts is (a) sepals are 3-trace structures;

(b) petals are 1-trace structures; (c) stamens are 1-trace structures; and (d) carpels are 3-trace structures. Two fundamental premises regarding the nature of the floral vascular system that have long been a cornerstone of anatomical investigation have been outlined by Moseley (1967). First, the vascular system is nearly always more conservative than the organs it supplies. Changes in the vascular patterns tend to lag behind morphological alterations; hence, the nature of the changes, often not discernible by gross morphological studies, may be revealed by an examination of the internal characters of the vascular anatomy, such as: nodal lacunae, bundle orientation, and vestigial traces. Secondly, the pattern of the vascular system may reveal the former boundaries, relative positions, numbers, and categories of organs, or their parts, which may now be obscured by reduction, cohesion, and adnation. In addition, there is wide spread belief that a condition of fused vascular bundles, in general, is more advanced than a condition of unfused bundles. Eyde (1971) has stated that the value of floral morphological study to taxonomy and phylogeny lies in the fact a consideration of floral vasculature may engender an extra measure of confidence when external features are difficult to interpret, and in the fact that there is a general evolutionary tendency among the principal floral vascular bundles to unite in various ways. Also, this author emphasizes that similarities and differences in floral vascular patterns can be taxonomically significant characters even though they offer no plausible evolutionary scheme. Carlquist (1970) and Schmid (1972) have recently reviewed various concepts of floral morphological study.

B. Morphological nature of the flower. The classical theory of the flower can be traced back to the ideas of Goethe in the latter part of the 18th century. This theory states the flower is a determinate branch or shoot whose parts (bracts, sepals, petals, stamens, and carpels) are modified leaves. That is, floral parts and leaves are homologous. According to this idea the general trends in floral organography include:

	Primitive	Advanced
1.	Simple bisexual flower	Compound unisexual modified flower

(Numerous arguments have occurred in the literature as to whether the primitive flower was unisexual or bisexual. The generally accepted view today is that the primitive flower was bisexual.)

2.	Radial symmetry (actinomorphy)	Bilateral symmetry (zygomorphy)
3.	Numerous parts spirally arranged	Fewer parts with whorled arrangement

(Stebbins [1967] has pointed out that evolution of floral parts has proceeded both upward and downward in number. Thus, the general trend toward fewer floral parts is not always valid.)

4.	Floral parts free	Floral parts fused (connation and adnation)
5.	Undifferentiated perianth	Differentiated perianth or complete absence of perianth
6.	Ovary superior (hypogynous)	Ovary inferior (epigynous)

(The general trend from hypogyny to epigyny is well established in flowering plants. The first documented reverse change--from epigyny to hypogyny--has been reported in Tetraplasandra gymnocarpa (Araliaceae) by Eyde and Tseng [1969].)

Two major theories have been suggested to explain the inferior ovaries of angiosperms. It is apparent that the inferior ovary has evolved several times among different groups and available evidence suggests that this has occurred by two opposing methods. The appendicular theory essentially states that there has occurred an extensive fusion (both connation and adnation) of the outer floral whorls to one another and to the ovary wall. Evidence is derived from the course of the vascular bundles supplying the ovules. The ovular vascular supply originates in acropetal fashion moving up the ovary and does not descend to supply the ovules. Indications are this type of inferior ovary is the prevalent type in angiosperms. The receptacular theory, in contrast, assumes the carpels have "sunk" into the tissue of the floral receptacle. Evidence is seen in the descending course of the ovular vascular supply.

Nectaries almost certainly have arisen and evolved independently in different groups of angiosperms. Fahn (1953) has suggested that there has occurred a general trend among angiosperms for the floral nectaries to migrate in an acrocentripetal manner: in the most primitive cases the nectaries are borne on the perianth parts. As evolutionary advancement occurs the nectaries migrate to the inner floral parts and in the pistil from the base of the ovary to the base of the style.

C. Inflorescence. The evolution of inflorescences is poorly understood. A few authors have traced lines of specialization within individual taxa but the general trends for angiosperms as a whole are only speculative. At one time or another solitary flowers, dichasial cymes, and panicles have been regarded as primitive types. The widely accepted current idea is that the solitary flower represents the primitive situation in angiosperms as evidenced by the presence of this character in various "ranalean" genera.

D. Perianth. The currently widely accepted interpretation is that the primitive dicotyledonous flower possessed an undifferentiated perianth whose component parts (tepals) were thick and foliaceous in nature. Differentiation of the perianth into calyx and corolla probably took place at many different times.

Anatomical evidence (predominance of three vascular strands) indicates that sepals are most likely the equivalent of sterile bracts. Subtending floral bracts are of foliar origin whereas petals have apparently arisen in two ways. In some flowers petals have arisen from modification from bracts or tepals, in which case sepals and petals are morphologically equivalent. In other plants petals have been derived from sterile stamens (staminodes) and thus are not morphologically equivalent to sepals and bracts.

E. Carpels. Major opposing theories on the morphological nature of the carpel fall into three major categories: (a) appendicular (foliar) theories based on the sporophyll concept; (b) axial theories, which are based on the concept that ovules were originally, and in some instances still are, axis-borne (cauline); and (c) sui generis theories, which state the floral parts do not find homologies in either leaves or stems but are entirely new entities.

1. Classical theory. The carpel is considered fundamentally foliar in nature. The primitive carpel is envisioned as a 3-veined (dorsal and two ventrals) megasporophyll with a lobed stigmatic crest and numerous ovules borne along its inturned margins. The ovules are vascularized by traces derived from the ventral bundles.

2. Conduplicate theory. This concept, which resulted from studies on the primitive dicotyledons (genus Tasmannia), is the most widely accepted among American botanists. The primitive carpel is described as cladified, stipitate, conduplicate, open, 3-veined and with many ovules borne on its adaxial surface. Advancement produces a carpel that is uncladified, closed, with a definite raised stigma, and with few (1-2) ovules adaxially or basally attached.

3. The "phyllospory-stachyospory" theory. This interpretation states that the appendages commonly called carpels are of two basic natures: foliar and cauline. Of primary consideration is the relationship between ovules, placentas and other carpel parts. A fundamental distinction is made between the classic idea in which marginal ovules and placentas are parts of leaves (phyllosporous condition) and the recognition of situations where the ovule and placenta are axillary cauline structures subtended by a sterile appendage (stachyosporous condition). The term stegophyll is used for sterile structures that may surround and thus protect stachyosporous ovules.

4. The Peltate Carpel. This concept has been widely accepted among non-English speaking botanists but has received little attention in America. According to Eames (1961), "The peltate theory presents a morphological origin for the carpel more complex than the simple folding or inrolling of the margins of the lamina. Peltate form in the carpel is assumed to have arisen by the turning upward (ventrally) of the basal lobes of the lamina and their fusion, margin to margin, as in the formation of peltate leaves. Where the two marginal meristems meet, they unite, forming a transverse meristem, the cross zone." The peltate carpel is thus considered homologous with leaves, and ovules are borne on the wall formed by the cross zone. Vasculation is by a dorsal-median bundle and a ventral-median one. The latter is the result of fusion of two ventral bundles of the classic carpel.

5. The Gonophyll theory. According to this theory the basic component of the ovary is a sterile leaf with an epiphyllous (adnate fertile sporangium-bearing) branch. This entire complex is termed a gonophyll. The sterile leaf is a tegophyll and its fertile branch system an androphyll (if male), gynophyll (if female) or androgynophyll (if bisexual). Repetition of the basic gonophyll unit gives an "inflorescence." Diverse floral structure is explained by "modifications" of the basic gonophyll plan. Melville (1962), for example, states, "All superior types of ovary examined could be formed by a simple infolding, or down-folding, of the gonophyll blade to effectively enclose the ovuliferous branches." This idea has failed to gain wide support.

6. Meeuse's theory. This theory has as its basic premise the idea that angiosperm pistils are the homologues of ovuliferous cupules. The cupule is regarded as an organ of independent origin. The sporophyll or foliar nature of the carpel is adamantly rejected.

7. Theory of Carpel Polymorphism. The unspecialized carpel has three functions--it is receptive, reproductive and protective. Evolution has led to the specialization of different forms of carpels for different functions. Three forms of carpels may be therefore recognized: (1) valve carpel, (2) solid carpel, and (3) semi-solid or pseudovalve carpel. This division of labor means that there are usually at least two types of carpels present in any one ovary. The theory of carpel polymorphism, as proposed by Saunders (1925), has been widely criticized and is not generally accepted as valid.

F. Placentation. Although the taxonomic value of placentation type is clear, interpretations of the phylogenetic trends within the various patterns has been considerably less so. In part, this confusion has been the result of inconsistent use of terms. Correlated with the belief that the conduplicate carpel is primitive, there is wide-shared belief that the laminar condition of the ovules is the most primitive pattern in apocarpous gynoecia. In more advanced syncarpous gynoecia, axile placentation apparently gave rise to all other types.

G. Stamens. The classical (ranalean) theory regards the stamen as foliar in nature. The primitive stamen of the classical school is exemplified by the stamens of *Degeneria* (Degneriaceae), *Magnolia* (Magnoliaceae), *Himantandra* (Himantandraceae) and others. The primitive microsporophyll is characterized as cladified, 3-veined, with linear sporangia embedded between the veins on the abaxial or adaxial surface. The advanced stamen is uncladified, 1-veined, with clear differentiation into anther and slender filament.

The telome theory envisions the origin of the stamen from a much reduced branch system. Evidence is found in the androecial vasculation of some flowers in which a strong vascular bundle (stamen trunk) constitutes the central axis of a telome system which supplies several stamens by branches. This theory maintains the flattened stamens of some "ranalean" genera can be explained through a reduction and webbing of a branch system.

The sequence of stamen differentiation has in recent years been interpreted as having phylogenetic significance. Although the observation of "centrifugal stamens" or the "centrifugal androecium" has been reported for over one hundred years. E. J. H. Corner (1946) was the first to fully develop its potential taxonomic significance. It was his idea that those plants with centrifugal stamen initiation represented an evolutionary line quite distinct from those with the more common centripetal pattern. In Cronquist's system, the Dilleniidae are characterized by centrifugal stamens whereas the Rosidae have centripetal stamens.

H. Pollen. It is widely believed that the monocolpate pollen grain with a long distal aperture is the primitive condition in angiosperms. Evidences are seen in the occurrence of this type of grain among gymnosperms, its distribution in the fossil record, and its persistence in such otherwise comparatively primitive genera as *Magnolia* and *Degeneria*. Outside the "ranalean" complex this type of grain is not found in dicotyledons, although it is the characteristic pollen form in the monocotyledons. The general trend in pollen evolution is from large, relatively unornamented, monocolpate pollen with a long, distal aperture to smaller, ornamented, tricolpate and derived types. A persistent problem with this trend was how to explain the origin of the tricolpate condition from a monocolpate ancestor. Evidence now favors at least two origins of the tricolpate condition--(1) the furrows of the tricolpate grain arose *de novo* from inaperturate precursors; and (2) the tricolpate grain arose from a monocolpate grain through a trichotomosulcate intermediary.

I. Embryology. Interpretations of the angiosperm ovule have been numerous and often widely divergent. As of yet no one interpretation seems to satisfactorily account for the morphology of this structure. The primitive angiosperm ovule is described as bitegmic, anatropous, and crassinucellate primarily as a result of the prevalence of this type among plants otherwise regarded as primitive. The uniformity of the mature embryo sac is certainly one of the most distinguishing features of flowering plants. The monosporic, eight-nucleate (Polygonum type) is regarded as the primitive embryo sac condition because: (1) it is the most common pattern among angiosperms; and (2) the largest number of divisions are involved in its formation, i.e., there is a general trend toward reduction in gametophyte size and development. There is no general agreement as to whether the primitive endosperm type was Cellular, Nuclear, or Helobial.

J. Fruit and Seed. Ideas regarding the evolutionary trends in fruits and seeds are not well supported by morphological data. Various types of seeds and fruits have evolved many times in many different evolutionary lines. The current general assumption is that the primitive seed was large and had an undifferentiated small embryo embedded in abundant endosperm. Takhtajan (1969) also envisioned the primitive seed as possessing a seed coat of the "Magnolian-type," i.e., with a strongly developed peripheral layer of parenchymatous cells. Some botanists, including Corner (1949) in his famous Durian Theory, considered the ancestral seed to possess a brightly colored red or yellow aril (this was associated with a fleshy fruit and short seed viability). Martin (1946) regarded the time of inception of dormancy as significant with primitive seeds characterized by early dormancy or at least prior to extensive embryo development. Eames (1961) believed the absence of dormancy was primitive. The primitive fruit is usually described as a many-seeded follicle that opens at maturity.

K. Relationship of plants and insects. Many botanists believe that the ancestral angiosperms were pollinated by beetles (cantharophily). Evidence for this is seen in the existence of this relationship among primitive families (e.g., Magnoliaceae, Winteraceae, Annonaceae, Eupomatiaceae, Calycanthaceae, etc.) and in the long geologic history of beetles. From the cantharophilous condition more advanced insect pollination relationships and wind pollination arose. E. E. Leppik (1957) has stressed the importance of considering the parallel evolutionary development of flowers and insects, i.e., pollinators. This author recognizes six evolutionary levels of floral development. At the primitive level flowers do not possess any special symmetry or particular colors. Numerous stamens and carpels are often surrounded by a large number of bracts. Insects do not require any special sensory abilities to distinguish these flowers. At the simple level flowers are slightly more advanced and possess simple colors such as yellow, white, or greenish. There is no definite symmetry or number of parts. These flowers are visited primarily by beetles and flies. The third level is distinguished by flowers that are actinomorphic and of definite size. Floral parts are in definite numbers and levels. Colors are frequently pure, like white, blue, red, and yellow. The pollinators of these flowers presumably have the ability to distinguish colors other than the simple ones. On the numerate level flowers show distinctive figure numerals. Former radiate flowers are characterized by a reduction in number of petals. "Protected" level flowers have features of the simple and third levels but contain protected nectaries. Pollinators of these flowers must have the ability to distinguish three-dimensional patterns in order to locate the nectar. Flowers on the highest level are typically zygomorphic with various combinations of odor and color.

LITERATURE ON STRUCTURAL EVOLUTION

Bailey, I. W. 1933. Structure, distribution and diagnostic significance of vestured pits in dicotyledons. Journal Arnold Arboretum 14: 259-273.

________. 1944. The development of vessels in angiosperms and its significance in morphological research. American Journal of Botany 31: 421-428.

________. 1953. Evolution of tracheary tissue of land plants. American Journal of Botany 40: 4-8.

________. 1957. The potentialities and limitations of wood anatomy in the phylogeny and classification of angiosperms. Journal Arnold Arboretum 38: 243-254.

________. 1966. The significance of the reduction of vessels in the Cactaceae. Journal Arnold Arboretum 47: 288-292.

________, and B. G. L. Swamy. 1949. The morphology and relationships of Austrobaileya. Journal Arnold Arboretum 30: 211-226.

________, and ________. 1951. The conduplicate carpel of dicotyledons and its initial trends of specialization. American Journal of Botany 38: 373-379.

Baranova, M. 1972. Systematic anatomy of the leaf epidermis in the Magnoliaceae and some related families. Taxon 21: 447-469.

Barghoorn, E. S. 1940. The ontogenetic development and phylogenetic specialization of rays in the xylem of dicotyledons. I. The primitive ray structure. American Journal of Botany 27: 918-928.

Beck, C. 1970. The appearance of gymnospermous structure. Biological Reviews, Cambridge Philosophical Society 45: 379-400.

Benzing, D. 1967. Developmental patterns in stem primary xylem of woody Ranales. American Journal of Botany 54: 805-820.

Brown, W. H. 1938. The bearing of nectaries on the phylogeny of flowering plants. Proceedings of the American Philosophical Society 79: 549-595.

Canright, J. E. 1952. The comparative morphology and relationships of the Magnoliaceae. I. Trends of specialization in the stamens. American Journal of Botany 39: 484-497.

________. 1953. The comparative morphology and relationships of the Magnoliaceae - II. Significance of the pollen. Phytomorphology 3: 355-365.

________. 1955. The comparative morphology and relationships of the Magnoliaceae: IV. Wood and nodal anatomy. Journal of the Arnold Arboretum 36: 119-140.

Carlquist, S. 1962. A theory of paedomorphosis in dicotyledonous woods. Phytomorphology 12: 30-45.

________. 1966. Wood anatomy of Compositae: a summary, with comments on factors controlling wood evolution. Aliso 6: 25-44.

________. 1969a. Wood anatomy of Lobelioideae (Campanulaceae). Biotropica 1: 47-72.

________. 1969b. Wood anatomy of Goodeniaceae and the problem of insular woodiness. Annals Missouri Botanical Garden 56: 358-390.

________. 1970. Toward acceptable evolutionary interpretations of floral anatomy. Phytomorphology 19: 332-362.

Cheadle, V. I. 1948. Observations on the phloem in the Monocotyledoneae II. Additional data on the occurrence and phylogenetic specialization in structure of the sieve tubes in the metaphloem. American Journal of Botany 35: 129-131.

________, and N. B. Whitford. 1941. Observations on the phloem in the Monocotyledoneae. I. The occurrence and phylogenetic specialization in structure of the sieve tubes in the metaphloem. American Journal of Botany 28: 623-627.

Conde, L. F., and D. E. Stone. 1970. Seedling morphology in the Juglandaceae, the cotyledonary node. Journal Arnold Arboretum 51: 463-477.

Corner, E. J. H. 1946. Centrifugal stamens. Journal Arnold Arboretum 27: 423-437.

________. 1949. The durian theory of the origin of the modern tree. Annals of Botany 13: 367-414.

Devadas, C., and C. B. Beck. 1972. Comparative morphology of the primary vascular system in some species of Rosaceae and Leguminosae. American Journal of Botany 59: 557-567.

Doyle, J. A. 1969. Cretaceous angiosperm pollen of the Atlantic coastal plain and its evolutionary significance. Journal of the Arnold Arboretum 50: 1-35.

Eames, A. J. 1931. The vascular anatomy of the flower with refutation of the theory of carpel polymorphism. American Journal of Botany 18: 147-188.

Ehrendorfer, F., F. Krendl, F. Habeler, and W. Sauer. 1968. Chromosome numbers and evolution in primitive angiosperms. Taxon 17: 337-353.

Esau, K. 1969. The Phloem. Encyclopedia of Plant Anatomy. V. (2). Gebrüder Borntraeger. Berlin.

________, V. I. Cheadle and E. M. Gifford, Jr. 1953. Comparative structure and possible trends of specialization of the phloem. American Journal of Botany 40: 9-19

Eyde, R. H. 1971. Evolutionary morphology: Distinguishing ancestral structure from derived structure in flowering plants. Taxon 20: 63-73.

________, and C. C. Tseng. 1969. Flower of _Tetraplasandra gymnocarpa_: Hypogyny with epigynous ancestry. Science 166: 506-508.

Fahn, A. 1953. The topography of the nectary in the flower and its phylogenetic trend. Phytomorphology 3: 424-426.

Foster, A. S. 1961. The phylogenetic significance of dichotomous venation in angiosperms. Recent Advances in Botany 2: 971-975.

Frost, F. H. 1930. Specialization in secondary xylem in dicotyledons. I. Origin of vessels. Botanical Gazette 89: 67-94.

________. 1930. Specialization in secondary xylem in dicotyledons. II. Evolution of end wall of vessel segment. Botanical Gazette 90: 198-212.

________. 1931. Specialization in secondary xylem in dicotyledons. III. Specialization of lateral wall of vessel segment. Botanical Gazette 91: 88-96.

Fryns-Claessens, E., and W. Van Cotthem. 1973. A new classification of the ontogenetic types of stomata. The Botanical Review 39: 71-138.

Guédès, M. 1971. Carpel peltation and syncarpy in _Coriaria rusicifolia_ L. The New Phytologist 70: 213-227.

Hickey, L. J. 1971. Evolutionary significance of leaf architectural features in the woody dicots. American Journal of Botany 58: 469. [Abstract]

Hiepko, P. 1965. Das zentrifugale androecium der Paeoniaceae. Berrichte der Deutschen Botanischen Gesellschaft 77: 427-435.

Howard, R. A. 1970. Some observations on the nodes of woody plants with special reference to the problem of the "split-lateral" versus the "common gap." _In_: New Research in Plant Anatomy. N. K. B. Robson, D. F. Cutler, and M. Gregory (eds.). Supplement 1 to the Botanical Journal of the Linnaean Society 63. Academic Press. London.

Kaplan, D. R. 1967. Floral morphology, organogenesis and interpretation of the inferior ovary in _Downingia bacigalupii_. American Journal of Botany 54: 1274-1290.

Kribs, D. A. 1935. Salient lines of structural specialization in the wood rays of dicotyledons. Botanical Gazette 96: 547-557.

Kribs, D. A. 1937. Salient lines of structural specialization in the wood parenchyma of dicotyledons. Bulletin of the Torrey Botanical Club 64: 177-186.

Lam, H. J. 1948. A new system of the Cormophyta. Blumea 6: 282-289.

Leppik, E. E. 1957. Evolutionary relationship between entomophilous plants and anthophilous insects. Evolution 11: 466-481.

Macior, L. W. 1971. Co-evolution of plants and animals--systematic insights from plant-insect interactions. Taxon 20: 17-28.

Marsden, M. P., and I. W. Bailey. 1955. A fourth type of nodal anatomy in dicotyledons, illustrated by Clerodendron trichotomum Thunb. Journal of the Arnold Arboretum 36: 1-49.

Martin, A. C. 1946. The comparative internal morphology of seeds. The American Midland Naturalist 36: 513-660.

Meeuse, A. D. J. 1971. Interpretative gynoecial morphology of the Lactoridaceae and the Winteraceae--a re-assessment. Acta Botanica Neerlandica 20: 221-238.

Melville, R. 1962. A new theory of the angiosperm flower: I. The gynoecium. Kew Bulletin 16: 1-50.

________. 1963. A new theory of the angiosperm flower. II. The androecium. Kew Bulletin 17: 1-63.

Moseley, M. F., Jr. 1967. The value of the vascular system in the study of the flower. Phytomorphology 17: 159-164.

Muller, J. 1970. Palynological evidence on early differentiation of angiosperms. Biological Reviews 45: 417-450.

Namboodiri, K. K., and C. B. Beck. 1968. A comparative study of the primary vascular system of conifers. III. Stelar evolution in gymnosperms. American Journal of Botany 55: 464-472.

Ozenda, P. 1949. Recherches sur les dicotylédones apocarpiques. Publications des Laboratoires l'Ecole Normal Superieure, Série Biologie Fasc. II. Paris.

Puri, V. 1952. Placentation in angiosperms. Botanical Review 18: 603-651.

Raven, P. H., and D. W. Kyhos. 1965. New evidence concerning the original basic chromosome number of angiosperms. Evolution 19: 244-248.

________, ________, and M. S. Cave. 1971. Chromosome numbers and relationships in Annoniflorae. Taxon 20: 479-483.

Rickett, H. W. 1944. The classification of inflorescences. Botanical Review 10: 187-231.

Saunders, E. R. 1925. On carpel polymorphism. I. Annals of Botany 39: 123-167.

Schmid, R. 1972. Floral bundle fusion and vascular conservatism. Taxon 21: 429-446.

Schofield, E. K. 1968. Petiolar anatomy of the Guttiferae and related families. Memoirs New York Botanical Garden 18: 1-55.

Sinnott, E. W. 1914. Investigations on the phylogeny of the angiosperms. 1. The anatomy of the node as an aid in the classification of angiosperms. American Journal of Botany 1: 303-322.

Slade, B. F. 1971. Stelar evolution in vascular plants. The New Phytologist 70: 879-884.

Srivastava, L. M. 1970. The secondary phloem of Austrobaileya scandens. Canadian Journal of Botany 48: 341-359.

Stebbins, G. L. 1966. Chromosomal variation and evolution. Science 152: 1463-1469.

________. 1967. Adaptive radiation and trends of evolution in higher plants. In: T. Dobzhansky, M. Hecht, and W. C. Steere (eds). Evolutionary Biology 1: 101-141. Appleton-Century Crofts. New York.

Stone, D. E. 1970. Evolution of cotyledonary and nodal vasculature in the Juglandaceae. American Journal of Botany 57: 1219-1225.

Wilson, C. L. 1942. The telome theory and the origin of the stamen. American Journal of Botany 29: 759-764.

Wilson, T. K. 1964. Comparative morphology of the Canellaceae. III. Pollen. Botanical Gazette 125: 192-197.

Wodehouse, R. P. 1936. Evolution of pollen grains. Botanical Review 2: 67-84.

Zahur, M. S. 1959. Comparative study of secondary phloem of 423 species of woody dicotyledons belonging to 85 families. Cornell University Experiment Station Memoir 358.

GENERAL STRUCTURAL REFERENCES

Bierhorst, D. W. 1971. Morphology of Vascular Plants. The Macmillan Company. New York.

Carlquist, S. 1961. Comparative Plant Anatomy. Holt, Rinehart and Winston. New York.

Constance L. 1955. The systematics of the angiosperms. In: A Century of Progress in the Natural Sciences 1853-1953. California Academy of Science. San Francisco.

Cronquist, A. 1968. The Evolution and Classification of Flowering Plants. Houghton-Mifflin Company. Boston.

Davis, P. H., and V. H. Heywood. 1963. Principles of Angiosperm Taxonomy. D. Van Nostrand, Inc. Princeton.

de Beer, G. 1971. Homology, An Unsolved Problem. Oxford Biology Readers II. (F. F. Head and O. E. Lowenstein, eds.). Oxford University Press. Oxford.

Eames, A. 1961. Morphology of the Angiosperms. McGraw-Hill, Inc. New York.

Meeuse, A. D. J. 1966. Fundamentals of Phytomorphology. Ronald Press Company. New York.

Metcalfe, C. R., and L. Chalk. 1950. Anatomy of the Dicotyledons. Clarendon Press. Oxford.

Smith, A. C. 1969. Systematics and appreciation of reality. Taxon 18: 5-13.

Solbrig, O. T. 1970. The phylogeny of Gutierrezia: An eclectic approach. Brittonia 22: 217-229.

Takhtajan, A. 1969. Flowering Plants--Origin and Dispersal. (Engl. transl. C. Jeffrey). Smithsonian Institution Press. Washington, D. C.

Chapter 28. SYSTEMS OF CLASSIFICATION*

The philosophical bases of plant classification are reviewed and criteria used by botanists in devising their systems are discussed in this chapter. We consider here only the classification systems developed in Western Civilization, realizing, of course, that other civilizations (and "uncivilized societies") have produced classification systems independently.

DEFINITION, GENERAL PRINCIPLES, TRENDS OF CLASSIFICATION

The term classification is used for both: (1) the process of classification and (2) the product of this process, i.e., the system of classification. Biological classification is the process of grouping together like organisms, and the subsequent placing together of these groups into larger groups. Due to discontinuities (of greater or lesser magnitude and greater or lesser degrees of completeness) in the pattern of variation in nature, discrete units in the biological world can be recognized. In order to refer to these units, they must be named, and an accepted system of nomenclature must be established. The system of classification contains categories (species, genus, family, order, class, etc.) into which the units, groups of units, and groups of groups are placed. The system of classification thus needs a hierarchy of categories into which the named units and groups are placed.

The recognition of similarities and differences between organisms is a basic human activity, and, in combination with a system of nomenclature, is essential for accurate communication (see discussion of this topic in Chapter 2). Modern systems of classification attempt to serve two needs: (1) to present our current state of knowledge regarding the relationships (totality of similarities and differences, sometimes with phylogenetic interpretation) of organisms, and (2) to provide, in combination with a system of nomenclature, a retrieval system for information about organisms. In this context, the advantages and disadvantages of a linear vs. a two- or three-dimensional presentation of the classification should be recognized. Also, a general system of classification should be devised with practicality in mind. The classification is made for use and must serve not only taxonomists and other scientists but laymen as well.

Two principles of classification which taxonomists have learned by hard experience should be noted here: (1) in the process of classification, the taxonomist should assess the totality of similarities and differences of the organisms being studied, and (2) after groups have been recognized, characters may be weighted with respect to their value in defining the group and differentiating the group from other groups (*a posteriori* weighting).

Considerable controversy exists today between the "pheneticists," i.e., those who would generally reject our ability to make correct "phylogenetic interpretations" when devising classification schemes (especially at the higher levels), and the "phylogenists," i.e., those whose system-making combines use of purely phenetic information with phylogenetic interpretation.

*Contributed by Kenneth M. Becker, The New York Botanical Garden.

Davis and Heywood's Principles of Angiosperm Taxonomy (1963) is an example of a work with a strong phenetic point of view, while Cronquist's Evolution and Classification of Flowering Plants (1968) is an example of a work with a strong phylogenetic point of view.

The following trends in the development of systems of plant classification through the centuries (which can be followed in Section A of this chapter) should be noted: (1) the gradual development of the hierarchy of categories as we know it today; (2) the gradual recognition of the major groups of plants--especially with regard to vascular vs. non-vascular plants, cryptogams vs. phanerogams, gymnosperms vs. angiosperms, monocots vs. dicots, and the gradual recognition of more and more angiosperm families as natural groups; (3) the gradual use of an increasing number of characters in classification and (4) changes in philosophy in the transitions between the various periods in the history of plant classification systems.

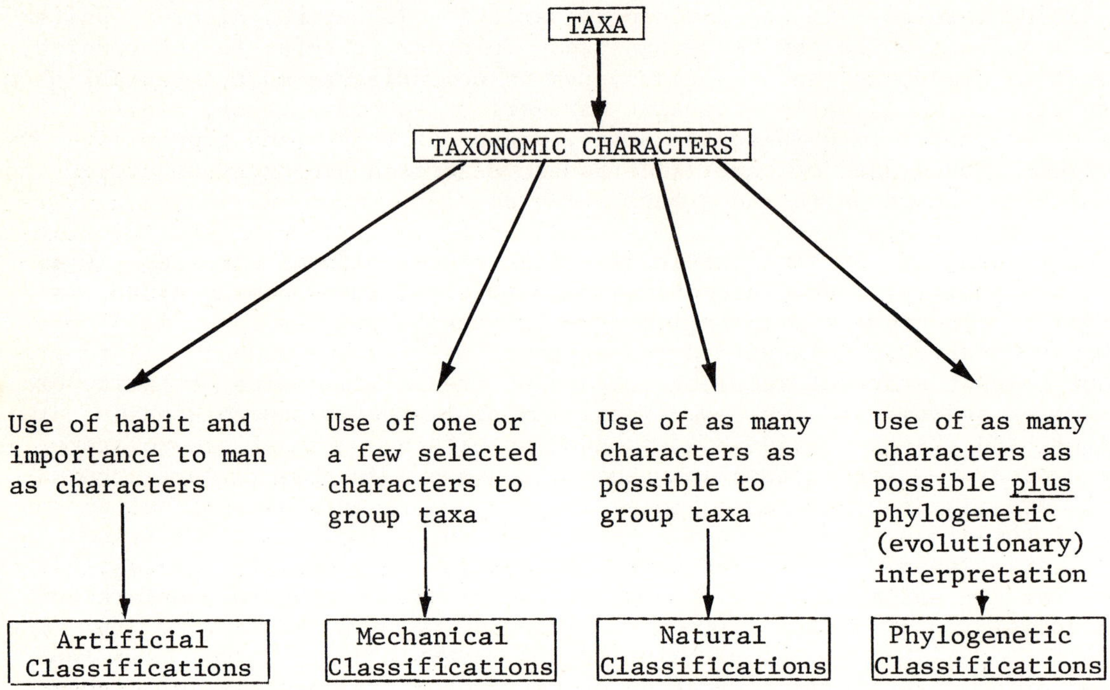

Figure 28-1. From taxa to system in the development of various kinds of general biological classifications.

Section A. BASES OF CLASSIFICATION

Contemporary plant classification systems are the result of the labors and insights of many workers throughout the centuries. Although there are, of course, no sharp lines dividing the various "Periods," the history of plant classification may be broken down as follows:

Period	Workers	Philosophical Base, Criteria Used, Etc.
Ancients to ca. 1500	Theophrastus Dioscorides Pliny Albertus Magnus	Wrote to transmit what was generally known about plants. Classification based on habit. A few natural groups recognized.
Herbalists to ca. 1580	Brunfels Bock Fuchs Cordus Mattioli Dodoens l'Ecluse l'Obel Bankes Turner Gerard	Motivated by medical and commercial considerations--the systems *per se* were not the primary concern. Classification based primarily on habit (see discussion of l'Obel's system). Advent of printing stirred new interest in plants, and 'mystical respect' for writings of the ancients gave way to a period of original researches into plant structure.
Mechanical Systems	Caesalpino Bauhin, J. Bauhin, G. Jung Morison Rivinus Ray Magnol Tournefort Camerarius	Primary concern was information retrieval--the system itself was considered important. Classification based mainly on form, but includes also a number of other kinds of criteria. *A priori* weighting of values of individual features for classification purposes was the rule (see discussion of Tournefort's system). Hierarchy of categories improved; more and more natural groups recognized.
to ca. 1760	Linnaeus	Artificial sexual system, based on numerical relations of floral parts, was a great improvement over earlier systems for purely practical reasons.
Natural Systems to ca. 1880	Adanson Lamarck deJussieu, B. deJussieu, A. deCandolle, A. Brown Lindley Endlicher Brongniart Braun Bentham & Hooker	Attempted to put seemingly related plants (plants that "look alike") together, using as many characters as possible (note change in attitude from mechanical systems). System of A. deJussieu a turning point in the history of plant classification.
1859	Darwin	Produced no system of his own, but (with A. R. Wallace) adopted, recast, expanded, and fostered the Theory of Evolution previously developed, to greater or lesser degrees, by Lamarck, Lyell, et al. Whereas pre-Darwinian natural system-makers wished their systems to reflect the Divine Plan which was thought to underlie all of nature, most post-Darwinian

"phylogenetic" system-makers thought that, since plants and plant-groups are the products of evolution, it was their task to determine the course of this evolution, and have their systems reflect it.

"Phylogenetic" Systems

Eichler
Baillon
Engler-Prantl
Warming
Wettstein
van Tieghem
Wernham
Bessey
Hallier
Rendle
Mez
Hayata
Burtt-Davy
Pulle
Skottsberg
Tippo
Gundersen
Hutchinson
Benson
Soó
Novák
Emberger
Deyl
Kimura
Melchior
Takhtajan
Cronquist
Thorne
Banks
Bierhorst

Attempt to integrate what is known about the evolutionary history of plant groups into the system. Most are similar to natural systems, but attempt to arrange plant groups according to their routes of descent. It became obvious to several workers in the early days of the phylogenetic period that the natural system-makers were recognizing evolutionarily-related groups of plants. Two main schools of thought developed as to origin of angiosperms and as to which character-states were considered primitive:

<u>Englerian</u>	<u>Besseyan/Hallierian</u>
primitive flower apetalous, unisexual, anemophilous. Ancestors of angiosperms were coniferoid or gnetoid gymnosperms with unisexual strobili.	primitive flower with perianth of many, free, equal parts, bisexual, insect-pollinated. Ancestors of angiosperms were cycadophytes.

See Chapter 27, Structural Evolution and Phylogeny.

Some contemporary systems take into account, in addition to morphological characters, data from the fields of anatomy, embryology, chemotaxonomy, palynology, cytology, etc.

Comprehensive revisions of the classification of vascular plants as a whole have appeared in recent years.

I. SYNOPSIS OF DEVELOPMENT OF PLANT CLASSIFICATION

A. <u>Period of the Ancients--ca. 300 BC to 1500 AD</u>

1. Theophrastus
300 BC

Classification based on habit. Plants divided into trees, shrubs, undershrubs, herbs, then into cultivated and wild kinds. Recognized: annuals/biennials/perennials, determinate/indeterminate inflorescences, hypogyny/perigyny/epigyny, apetaly/polypetaly/gamopetaly, monocots/dicots (seed, stem, and leaf differences), roots vs. rhizomes. Described and classified ca. 480 kinds of plants. Considered trees to be the most highly developed plants. Groups strictly artificial. Used no categories as

			we know them today. Used speculative philosophy and methods of Plato and Aristotle (especially the rule of the Excluded Middle: any given object is either A or not A).
2.	Dioscorides	60 AD	Dealt with medicinal plants, and recognized ca. 600 species. Some groups equal more or less natural families (recognized a series of labiates and a series of umbels).
3.	Pliny	70	<u>Natural History</u> was an encyclopedic compilation.
4.	Albertus Magnus	1250	Accepted Theophrastus' classification for the most part. Recognized difference in stem structure between monocots and dicots. Criteria used for major categories: leafy vs. non-leafy; leafy plants divided into monocots and dicots; dicots divided into herbaceous and woody kinds.
5.	"Gart der Gesundheit" <u>Hortus Sanitatis</u>	1485 1491	Compilations with crude descriptions and woodcuts. No definite natural arrangement evident (more or less alphabetical).

B. <u>Period of the Herbalists--1500 to 1580</u>

1.	Brunfels, Otto	1530	Recognized Perfecti (flowers visible when held at arm's length) and Imperfecti (flowers not visible).
2.	Bock, Jerome (Hieronymus Tragus)	1539	Divided plants into trees, shrubs, and herbs. Recognized 567 species, and gave concise descriptions.
3.	Fuchs, Leonhart	1542	Fine engravings. Described New World plants.
4.	Mattioli, Pierandrea	1544	Excellent illustrations.
5.	Cordus, Valerius	1561	502 species, excellent descriptions.
6.	Dodoens, Rembert	1583	Excellent descriptions and illustrations.
7.	l'Ecluse, Charles de	1601	Treated plants of America, Spain, Portugal, Austria, Hungary.
8.	l'Obel, Mathias de	1581	Classification based upon general form, growth habit, economic uses, and width of leaf (separated monocots and dicots).

grasslike plants with narrow leaves → broader leaved bulbous and rhizomatous monocots → dicot herbs → shrubs → trees

(supposed "simpler" forms) ("most perfect")

9\. Bankes, Rycharde — 1525 — Dealt with medicinal plants.

10\. Turner, William — 1551, 1562, 1568 — Alphabetical arrangement by Latin name.

11\. Gerard, John — 1597 — Adopted l'Obel's classification (based upon superficial resemblances and uses to man).

The quality of description and illustration improved greatly during the period of the Herbalists. There was a large increase in the number of plants known; the discovery of the New World had an important impact during this period. There was some attempt to group closely related plants together. Some larger natural families were recognized, such as composites, umbels, grasses, rushes, etc. See Arber (1938), and also Chapter 2 (History).

C. Period of Mechanical Systems--1580 to ca. 1760

1\. Caesalpino, Andrea — 1583 — Sought a classification based upon reasoning, not upon utilitarian considerations. Used basically an Aristotelean approach (*a priori* weighting of characters). Organs of fructification considered more important than habit. Classification based upon woodiness vs. herbaceousness, then on characters of fruits, seeds, and embryos. Fruits classified on basis of position, locule number, and seed number. Classification stresses number, position, and form of parts. Recognized: inferior/superior ovary, bulbs present/absent, sap milky/watery, activities of plants. Gave some technical descriptions. Had a good concept of genera, and described ca. 1520 plants. Recognized Fagales, legumes, umbels, crucifers, composites, borages as natural groups.

2\. Bauhin, Jean — 1650 — Descriptions were good diagnoses; described ca. 5000 plants. Recognized crucifers, labiates.

3\. Bauhin, Gaspard — 1623 — General arrangement as in l'Obel and Gerard (based on texture and form). Neglect of floral and fruit characters in arrangement of larger groups. Recognized distinction between genera and species, and described ca. 6000 species in his *Pinax*. Often used binomial nomenclature. Placed together Umbelliferae, some composites, crucifers, solanads, poppies, labiates, cucurbits, but put duckweeds among cryptogams and water ferns with mosses.

4\. Jung, Joachin — ca. 1650 — Argued against fundamental division into trees/shrubs/herbs.

5\. Morison, Robert — 1680 — System similar to that of Caesalpino; woody/herbaceous; divided into groups based on fruit and seed characters.

6\. Rivinus, Augustus Q. — 1690 — System with emphasis on corolla.

7. Ray, John — 1703

Herbae
- Imperfectae (basically cryptogams)
- Perfectae
  - Dicots
  - Monocots

Arborae
- Monocots
- Dicots

Monocots and dicots classified on fruit type, then leaf characters, flower type. Believed all parts of the plant should be used for classification. Emphasized importance of cotyledon number. Said: "Nature refuses to be forced into the fetters of a precise system." System direct antecedent of that of deJussieu; classification was aimed at practicality. Gave clear keys to genera, and recognized ca. 18,000 species. Distinguished between inherited and environmentally caused differences, and noted the significance of the former for taxonomy. Used anatomical information gathered by Grew, Malpighi, et al. Recognized composites, umbels, mints, crucifers, legumes, etc. Compositae divided into four classes based on relatively unimportant characters, and his Bacciferae includes many unrelated genera having fruits with fleshy pericarp.

8. Magnol, Pierre — 1689 — Used conspicuous characters of roots, stems, flowers, seeds. First concept of modern _families_ (listed 76 in 1689).

9. Tournefort, Joseph Pitton — 1694, 1716 — Classification based on tree/herb, then apetalous/petalous/gamopetalous, then flowers regular/irregular. System widely used until ca. 1760, when it was superseded by that of Linnaeus (used until 1780 in France, whence superseded by that of deJussieu). Author of modern _genus_ concept, and is considered to have had a good concept of genera. Gave generic descriptions, and recognized 698 genera, 10,146 species. Perhaps gave too much emphasis to vegetative characters in delimiting genera. His important system is largely artificial, and is generally considered inferior to that of Ray.

10. Camerarius, Rudolf — 1694 — Demonstrated sexuality in plants.

11. Linnaeus, Carl — 1735, 1737, 1753 — Produced a "Sexual System" based upon the numerical relations of floral parts. Recognized 24 Classes, based on number, union, and length of stamens. Classes subdivided into Orders on basis of number of styles; unisexuality also considered. Linnaeus' category _Classis_ more or less corresponds to what we would now call an order, his category _Ordo_ corresponds to families in the modern sense. Linnaeus' system was later modified by other workers (Willdenow, Gmelin, Schreber, et al.) in many other, later editions of his works, up to about 1805. The Sexual System was of great utility in identification, but was deliberately and admittedly artificial. It was widely adopted until superseded by system of deJussieu. Influenced in producing his Sexual System by Aristotelean reasoning, but not by Aristotelean reliance upon habit to delimit major groups. Used (1) the principles of Logical Division (as did Theophrastus) and (2) _a priori_ reasoning as to the most important characters (i.e., those of the flower).

In his Philosophia Botanica (1751), Linnaeus enumerated 67 "natural orders." Some represent natural groups, such as palms, orchids, grasses, conifers, composites, borages, etc., but some "natural orders" are quite mixed, with monocots and dicots appearing together. Linnaeus was influenced here by the empirical method of natural science, i.e., he developed his natural system based upon years of observational experience, rather than on an *a priori* basis.

LINNAEUS' SEXUAL SYSTEM (*Systema Naturae*, 1735; *Genera Plantarum*, 1737; *Species Plantarum*, 1753)

Class	Subclass (Ordo)	Examples of Species
I. Monandria	Monogynia	*Canna indica*
(Stamen one)	Digynia	*Cinna arundinacea*
II. Diandria	Monogynia	*Ligustrum vulgare*, *Chionanthus virginica*
(Stamens two)	Digynia	*Anthoxanthum odoratum*
	Trigynia	*Piper nigrum*
III. Triandria	Monogynia	*Valeriana officinalis*, *Melothria pendula*
(Stamens three)	Digynia	*Phalaris canariensis*, *Triticum aestivum*
	Trigynia	*Mollugo verticillata*
IV. Tetrandria	Monogynia	*Dipsacus fullonum*, *Plantago major*
(Stamens four)	Digynia	*Cuscuta americana*
	Tetragynia	*Ilex cassine*
V. Pentandria	Monogynia	*Heliotropium indicum*, *Lysimachia quadrifolia*
(Stamens five)	Digynia	*Asclepias syriaca*, *Carum carvi*
	Trigynia	*Rhus vernix*, *Alsine media*
	Tetragynia	*Parnassia palustris*
	Pentagynia	*Aralia spinosa*, *Suriana maritima*
VI. Hexandria	Monogynia	*Tillandsia utriculata*, *Pontederia cordata*
(Stamens six)	Digynia	*Oryza sativa*
	Trigynia	*Rumex verticillatus*, *Medeola virginiana*
	Tetragynia	*Petiveria alliacea*
	Polygynia	*Alisma cordifolia*
VII. Heptandria (7)	Monogynia	*Aesculus pavia*
VIII. Octandria	Monogynia	*Tropaeolum major*, *Gaura biennis*
(Stamens 8)	Digynia	*Galenia africana*
	Trigynia	*Polygonum persicaria*
	Tetragynia	*Paris quadrifolia*
IX. Enneandria	Monogynia	*Laurus benzoin*
(Stamens 9)	Trigynia	*Rheum rhaponticum*
	Hexagynia	*Butomus umbellatus*
X. Decandria	Monogynia	*Cercis canadensis*, *Melia azedarach*
(Stamens 10)	Digynia	*Tiarella cordifolia*
	Trigynia	*Silene virginica*
	Pentagynia	*Oxalis stricta*, *Agrostemma githago*
	Decagynia	*Phytolacca americana*
XI. Dodecandria	Monogynia	*Asarum canadense*, *Rhizophora mangle*
(Stamens	Digynia	*Agrimonia eupatoria*
12-19)	Trigynia	*Euphorbia chamaesyce*
	Pentagynia	*Glinus lotoides*
	Dodecagynia	*Sempervivum arboreum*
XII. Icosandria	Monogynia	*Cactus opuntia*, *Prunus virginiana*
(>19, attached	Digynia	*Crataegus crusgalli*
to calyx)	Trigynia	*Sorbus domestica*

	Pentagynia	*Pyrus communis*
	Polygynia	*Rosa canina*
XIII. Polyandria (Stamens >19, attached to receptacle)	Monogynia	*Capparis spinosa*, *Podophyllum peltatum*
	Digynia	*Paeonia officinalis*
	Trigynia	*Delphinium ajacis*
	Tetragynia	*Tetracera volubulis*
	Pentagynia	*Aquilegia canadensis*
	Hexagynia	*Magnolia virginiana*
	Polygynia	*Annona glabra*, *Anemone virginiana*
XIV. Didynamia (Stamens didynamous)	Gymnospermia	*Ajuga reptans*
	Angiospermia	*Scrophularia marilandica*
XV. Tetradynamia (Stamens tetradynamous)	Siliculosa	*Draba verna* } Cruciferae
	Siliquosa	*Erisymum officinale* } Cruciferae
XVI. Monadelphia (Stamens in one bundle)	Pentandria	*Waltheria indica*
	Decandria	*Geranium maculatum*
	Polyandria	*Sida rhombifolia*
XVII. Diadelphia (Stamens in two bundles)	Hexandria	*Fumaria officinale*
	Octandria	*Polygala incarnata*
	Decandria	*Erythrina herbacea*
XVIII. Polyadelphia (Stamens in several bundles)	Pentandria	*Theobroma cacao*
	Icosandria	*Citrus aurantium*
	Polyandria	*Hypericum perforatum*
XIX. Syngenesia (Stamens with united anthers)	Polygamia Aequalis	*Tragopogon virginicum*
	Polygamia Superflua	*Artemisia annua*
	Polygamia Frustranea	*Helianthus annuus*
	Polygamia Necessaria	*Chrysogonum virginianum*
	Monogamia	*Lobelia cardinalis*, *Viola canadensis*
XX. Gynandria (Stamens adnate to pistil)	Diandria	*Orchis spectabilis*
	Triandria	*Sisyrinchium bermudiana*
	Tetrandria	*Nepenthes distillatoria*
	Pentandria	*Passiflora incarnata*
	Hexandria	*Aristolochia serpentaria*
	Decandria	*Helicteres angustifolia*
	Polyandria	*Arum triphyllum*
XXI. Monoecia (Plants monoecious)	Monandria	*Zannichellia palustris*
	Diandria	*Lemna minor*
	Triandria	*Typha latifolia*, *Zea mays*
	Tetrandria	*Urtica dioica*
	Pentandria	*Ambrosia trifida*
	Hexandria	*Zizania aquatica*
	Heptandria	*Guettarda speciosa*
	Polyandria	*Myriophyllum spicatum*, *Quercus phellos*
	Monadelphia	*Pinus taeda*, *Ricinus communis*
	Syngenesia	*Momordica cylindrica*
	Gynandria	*Andrachne fruticosa*
XXII. Dioecia (Plants dioecious)	Monandria	*Najas marina*
	Diandria	*Salix babylonica*
	Triandria	*Empetrum nigrum*
	Tetrandria	*Viscum album*, *Myrica cerifera*
	Hexandria	*Smilax lanceolata*
	Octandria	*Populus alba*

	Ennaeandria	_Mercurialis procumbens_
	Decandria	_Carica papaya_
	Polyandria	_Cliffortia trifoliata_
	Monadelphia	_Juniperus virginiana_
	Syngenesia	_Ruscus aculeatus_
	Gynandria	_Clutia retusa_
XXIII. Polygamia	Monoecia	_Celtis occidentalis_, _Veratrum nigrum_
(Plants polygamous)	Dioecia	_Nyssa aquatica_
	Polyoecia	_Ficus carica_
XXIV. Cryptogamia	Filices	_Equisetum arvense_, _Adiantum pedatum_
(Flowers concealed)	Musci	_Lycopodium inundatum_, _Hypnum cuspidatum_
	Algae	_Marchantia polymorpha_, _Ulva confervoides_
	Fungi	_Hydnum imbricatum_

D. Period of Natural Systems--1760 to ca. 1880

1. Adanson, Michel — 1763 — Rejected artificial classifications. Emphasized equal weighting of characters, and believed that all organs should be taken into account. Rejected _a priori_ reasoning. Described groups more or less equivalent to modern families and orders. Produced 65 single-character systems from which he hoped to produce a natural system (this was largely a failure--criticized by deJussieu and deCandolle). "Grandfather of Numerical Taxonomy."

2. deJussieu, Bernard — 1759 — Rearranged the plants in a garden at LaTrianon, Versailles, according to his own system, which was not based on habit and was not purely artificial.

3. deJussieu, Antoine-Laurent — 1789 — Adopted and expanded the system of B. deJussieu. "First complete system which can claim to be a natural one"--Rendle. Weighted characters in the following order: embryo > sexual parts > fruit = perianth > vegetative characters. Reintroduced concepts of hypogyny/perigyny/epigyny. Differentiated, described, and named 100 "natural orders"--corresponding to most major families. Some related families (such as palms, lilies, amaryllids, irises) grouped together.

SYSTEM OF A. deJUSSIEU (_Genera Plantarum_, 1789)

		Class	Order (Family) Examples
I. Acotyledones		I	Fungi, Algae, Hepaticae
II. Monocotyledones	Stamina hypogyna	II	Aroideae, Typhae
	perigyna	III	Palmae, Junci
	epigyna	IV	Musae, Orchides

			Class	Order (Family) Examples
III. Dicotyledones	Apetalae	Stamina epigyna	V	Aristolochiae
		perigyna	VI	Elaeagni, Lauri
		hypogyna	VII	Amaranthi
	Monopetalae	Corolla hypogyna	VIII	Acanthi, Vitices
		perigyna	IX	Guaiacanae, Ericae
		epigyna {antheris connatus	X	Cichoraceae
		epigyna {antheris distinctis	XI	Dipsaceae
	Polypetalae	Stamina epigyna	XII	Araliae, Umbelliferae
		hypogyna	XIII	Cruciferae, Acera
		perigyna	XIV	Cacti, Myrti
	Diclines irregulares		XV	Euphorbiae, Urticae

4. deCandolle, Augustin P., Alphonse, and Casimir — 1813, 1873 — An expansion and elaboration of deJussieu's system. Arrangement of dicots improved (united Apetalae and Diclines). Downgraded use of physiological characters, emphasized instead the importance of morphological and anatomical characters. Used _a priori_ weighting. Ranunculaceae at base, arrangement of system based on reduction and fusion of floral parts. Linear sequence recognized as necessarily artificial. Alteration of "original symmetry" by abortion, degeneration, and cohesion. System looks evolutionary but was pre-Darwinian, and it is generally believed that deCandolle did not accept the Lamarckian concept of modification over time. By 1840 had replaced Linnaeus' Sexual System.

SYSTEM OF AUGUSTIN PYRAMUS AND ALPHONSE deCANDOLLE. 1819 and 1873. _Prodromus Systematis Naturalis Regni Vegetabilis_. (1819 is modified below with the treatment of the dicots based on actual publication in succeeding volumes and 1873 summary).

I. Vasculares (Vascular plants with cotyledons)
- Class I Dicotyledoneae (includes Conifers)
  - Subclass I Thalamiflorae (Polypetalous and hypogynous)
    - Orders (Families) 1-54 Ranunculaceae, Dilleniaceae, Magnoliaceae, Rutaceae, Simarubeae, Ochnaceae, etc.
  - Subclass II Calyciflorae (Perigynous or epigynous and polypetalous or sympetalous)
    - Orders (Families) 55-118 Celastrineae, Rhamnaceae, Bruniaceae, Samudeae, Homalineae, Pyrolaceae, Monotropeae, etc.
  - Subclass III Corolliflorae (Gamopetalous and hypogynous)
    - Orders (Families) 119-155 Lentibulariaceae, Primulaceae, Aegiceraceae, Solanaceae, Plantaginaceae, etc.
  - Subclass IV Monochlamydeae (Calyx only present)
    - Orders (Families) 156-207 Phytolaccaceae, Salsolaceae, Basellaceae, Amarantaceae, Euphorbiaceae, etc.
- Class II Monocotyledoneae (includes Cycads)
  - Subclass I Phanerogamae (Liliaceae, Iridaceae, Orchidaceae, etc.)
  - Subclass II Cryptogamae (Ferns, etc.)

II. Cellulares (plants without cotyledons)
  A. Mosses, Liverworts
  B. Lichens, Fungi, Algae
  (Ferns placed in Cellulares in last arrangement of deCandolle system).

5. Brown, Robert — 1827 — Proposed no system of his own. Pointed out nature of gymnosperms (naked seeds) and their distinctness from angiosperms. Opposed Linnaean system. Introduced many new "natural orders," and proposed the grouping of "natural orders" (families) into what we would now call orders.

6. Lindley, John — 1830 — Introduction to the Natural System of Botany is first comprehensive natural system published in English. Recognized distinction between gymnosperms and angiosperms. Dicots divided into Polypetalae, Apetalae, Achlamydeae, and Monopetalae--these groups were considered artificial. Treatment of monocots is precursor of Hutchinson's system on basis of perianth. Keys to, and discussions of, families are considered very good. In his Vegetable Kingdom (1846), Lindley overemphasized the importance of the parasitic habit of some dicots and the importance of net vs. parallel venation among the monocots.

7. Endlicher, Stephan — 1840 — Cryptogams clearly separated from seed plants. Gymnosperms separated from, but placed close to, dicotyledons. Thallophyta: no stem and root vs. Cormophyta: stem and root present. Lower dicots were apetalous groups, a practice followed in later ("Englerian") systems (as opposed to beginning with Ranalian groups as in the deCandolle-Bessey/Hallierian systems). Dicots divided into Apetalae, Gamopetalae, Dialypetalae. Some parasitic angiosperms placed with the cryptogams. Treatment of monocots differed from that of Lindley, similar to that of Bentham and Hooker (see below).

8. Brongniart, Adolphe Théodore — 1843 — Basic division into cryptogams and phanerogams (seed plants). In the monocots, the presence or absence of endosperm in the seed considered important. His "dicots" included both gymnosperms and angiosperms. Angiospermous dicots divided into Gamopetalae and Dialypetalae. His attempt to distribute apetalous forms among the petaliferous forms to which they appeared related is considered a major advance.

9. Braun, Alexander — 1853 — Influenced by Endlicher and Brongniart. Placed monocots and dicots together and separated from gymnosperms. The monocots were arranged to show a progression from the simple to the complex--began with Lemnaceae, ended with Orchidaceae. Dicots: apetalae, sympetalae, polypetalae. Influenced Eichler, Engler, Wettstein, et al.

10. Bentham, George & Hooker, Joseph D. — 1862-1883 — System largely deCandollean, and pre-evolutionary in conception. Groupings above family level differ considerably from Prodromus of deCandolle. Both monocots and dicots start with polypetalous families. Fusion and reduction of parts were basis of organization. Note: (1) maintenance of all apetalous groups as a unit, and separated from the petaliferous groups to which many are probably more related, (2) the Series in the Monochlamydeae and Monocotyledones often contain quite unrelated elements,

and (3) placement of gymnosperms between dicots and monocots. Accepted 200 families; all genera studied anew for Genera Plantarum. System became very popular in English-speaking countries, and used for herbarium arrangement in Great Britain.

SYSTEM OF BENTHAM AND HOOKER (Genera Plantarum, 1862-1883, with some representative examples; some spellings modernized).

Dicotyledones Polypetalae
Series I. Thalamiflorae
Cohort 1. Ranales - Ranunculaceae, Magnoliaceae, Annonaceae, etc.
(Order) 2. Parietales - Sarraceniaceae, Papaveraceae, Violaceae, etc.
3. Polygalinae - Pittosporaceae, Polygalaceae, etc.
4. Caryophyllinae - Caryophyllaceae, Portulacaceae, etc.
5. Guttiferales - Hypericaceae, Dipterocarpaceae, etc.
6. Malvales - Malvaceae, Sterculiaceae, Tiliaceae.
Series II. Disciflorae (hypogynous disc)
7. Geraniales - Linaceae, Geraniaceae, Rutaceae, etc.
8. Olacales - Olacineae, Ilicineae, etc.
9. Celastrales - Celastraceae, Rhamnaceae, etc.
10. Sapindales - Sapindaceae, Anacardiaceae, etc.
Series III. Calyciflorae (floral cup)
11. Rosales - Rosaceae, Leguminosae, Saxifrageae, Droseraceae, etc.
12. Myrtales - Rhizophoraceae, Lythraceae, Onagraceae, etc.
13. Passiflorales - Loasaceae, Cucurbitaceae, Passifloraceae, etc.
14. Ficoidales - Cactaceae, Ficoideae.
15. Umbellales - Umbelliferae, Araliaceae, Cornaceae.
Dicotyledones Gamopetalae
Series I. Inferae (inferior ovary)
Cohort 1. Rubiales - Caprifoliaceae, Rubiaceae.
(Order) 2. Asterales - Valerianae, Dipsaceae, Calyceraceae, Compositae
3. Campanales - Stylideae, Campanulaceae, etc.
Series II. Heteromerae (carpels >2)
4. Ericales - Vaccinieae, Ericaceae, Epacrideae, etc.
5. Primulales - Primulaceae, Plumbaginaceae, etc.
6. Ebenales - Sapotaceae, Ebenaceae, Styracaceae
Series III. Bicarpellatae (carpels = 2, rarely 1 or 3).
7. Gentianales - Oleaceae, Apocynaceae, Gentianaceae, etc.
8. Polemoniales - Polemoniaceae, Hydrophyllaceae, Boraginaceae, etc.
9. Personales - Scrophularineae, Orobanchaceae, Acanthaceae, etc.
10. Lamiales - Verbenaceae, Labiatae, Selagineae, etc.
Dicotyledones Monochlamydeae (Apetalous)
Series I. Curvembryeae - Orders (Families) Nyctagineae, Batideae, etc.
Series II. Multiovulatae Aquaticae - Order (Family) Podostemaceae
Series III. Multiovulatae Terrestres - Orders (Families) Nepenthaceae, etc.
Series IV. Micrembryeae - Orders (Families) Piperaceae, Chloranthaceae, Monimiaceae, etc.
Series V. Daphnales - Orders (Families) Laurineae, Proteaceae, Elaeagnaceae, etc.
Series VI. Achlamydosporeae - Orders (Families) Loranthaceae, Santalaceae, etc.
Series VII. Unisexuales - Orders (Families) Euphorbiaceae, Urticaceae, Cupuliferae, etc.
Series VIII. Ordines anomali - Orders (Families) Empetraceae, Ceratophyllaceae, etc.

Gymnospermae

Monocotyledones

Series I. Microspermae - Orders (Families) Orchideae, Burmanniaceae, etc.
Series II. Epigynae - Orders (Families) Scitamineae, Bromeliaceae, etc.
Series III. Coronarieae - Orders (Families) Liliaceae, Pontederiaceae, etc.
Series IV. Calycinae - Orders (Families) Juncaceae, Palmae, etc.
Series V. Nudiflorae - Orders (Families) Typhaceae, Cyclanthaceae, etc.
Series VI. Apocarpae - Orders (Families) Alismaceae, Naiadaceae, etc.
Series VII. Glumaceae - Orders (Families) Eriocauleae, Gramineae, etc.

11. Hofmeister, Wilhelm — 1851 — Embryological researches helped supply the basis for the division of the plant kingdom into the Thallophyta, Bryophyta, Pteridophyta, Gymnospermae, and Angiospermae.

Asa Gray, during the formative period of American botany, used a combination of the systems of deCandolle and Bentham & Hooker for use in floras of North America.

E. Period of Phylogenetic Systems - 1880 to present

1. Eichler, August W. — 1890 — Cryptogams: Thallophyta, Bryophyta, Pteridophyta
Phanerogams: Gymnosperms, Angiosperms
- Monocotyleae
- Dicotyleae
  - Choripetalae
  - Sympetalae

Began dicots with Amentiferae (as did Endlicher). United apetalous and polypetalous dicots. Accepted theory of evolution, and attempted to arrange plants in a series from the supposedly primitive to the more advanced; equated simplicity with primitiveness. Classification in his Blüthendiagramme (1878) largely artificial.

2. Baillon, Henri — 1867-1895 — Took a broader view of some families than most authors (as Ericaceae, etc.).

3. Engler, Adolf & Karl Prantl — 1892
4. Engler & E. Gilg — 1924
5. Engler & L. Diels — 1936
6. Melchior, H. — 1964

Classification based on that of Eichler.
Bacteria; Algae; Fungi; Lichenes; Bryophyta; Pteridophyta; Embryophyta Siphonogama (seed plants)
- Gymnospermae
- Angiospermae
  - Class Monocotyledoneae
  - Class Dicotyledoneae
    - Subclass Archichlamydeae
      - Apetalae (no petals)
      - Choripetalae (separate petals)
    - Subclass Metachlamydeae (Sympetalae) (united petals)

Phylogenetic principles: the most primitive flowering plants, in both the Monocotyledoneae and Dicotyledoneae, have apetalous, simple flowers in strobilus-like inflorescences

```
monocots                      dicots
   ↑                             ↑
Pandanales                  Amentiferae
primitive                    primitive
        ↖                    ↗
          polyphyletic
          from unknown
      gymnospermous stocks
```

Engler was well aware of the widespread occurrence of parallelism and convergence in the angiosperms, and did not produce a "phylogenetic tree" showing lines of descent (it was only later that Engler claimed his system to be phylogenetic).

Progressive morphological (not necessarily phylogenetic) series:

```
   no         sepaloid         sepals             sepals
perianth ──→  bracts  ──→       and      ──→       and
               only         free petals       united petals
```

This series was thought to have taken place in several phyletic lines in both the monocots and dicots. The system emphasizes apetaly, choripetaly, and sympetaly, with hypogyny, perigyny, and epigyny considered as secondary criteria. Associated some apetalous groups with related petaliferous groups. Gymnosperms clearly set off from angiosperms. Englerian evolutionary principles challenged by Bessey (1897), Hallier (1905), Arber and Parkin (1907), and other authors. System has been very widely used in Europe and America, and many herbaria and floras are arranged according to one or another of the various editions of the *Syllabus*. System used in 20 volume *Die natürlichen Pflanzenfamilien*. The latest (12th) edition of the *Syllabus der Pflanzenfamilien* proposes, but does not execute, many changes in sequence and position of groups based on modern knowledge gathered from the fields of anatomy, chemistry, embryology, etc. The monocots are considerably revised. Some of the evolutionary principles upon which the Englerian systems rest are, in the opinion of most contemporary workers, obsolete.

7. Warming, Eugenius — 1895 — Similar to Engler's system, but with some changes in positioning of groups.

8. Wettstein, Richard von — 1911, 1935 — Similar to Engler's system, but: (1) monocots considered to be derived from Ranales, (2) dicots thought to consist of several independent lines, (3) admitted more cases of perianth reduction than did Engler. Supposed primitive characters of the Amentiferae: woody, flowers unisexual, wind pollinated, apetalous, vascular bundles occasionally present in ovule, gametophytes may have a larger number of cells than is typical of angiosperms (*Casuarina*)--all gymnospermous characteristics. Believed fossil evidence supported amentiferous ancestry. Casuarinales considered primitive, Ephedroid origin for *Casuarina* flower theorized.

9. van Tieghem, Philippe 1898
   i. Liorhizeae Monocotyleae (monocots max. pars.)
   ii. Liorhizeae Dicotyleae
   iii. Climacorhizeae (dicots max. pars.)

These classes are based on the ontogeny of the root apex. Class ii contains the Nymphaeaceae and Gramineae. Class iii is subdivided on the basis of the presence of a "perfect seed" at maturity (Seminées--most dicots) or its absence (Inseminées--mostly parasitic groups). Overreliance on characters of root, ovules, and integuments. System basically an historical curiosity, but provided useful data.

10. Wernham, T. J. 1911 At least seven lines of polypetalous dicots thought to give rise to Sympetalae.

11. Bessey, Charles Edwin 1897 1915 Definitely conceived his system to be phylogenetic. Considered angiosperms to be monophyletic from a Bennettitalean (cycadeoid) type ancestor with bisexual strobili (see "Bessey's Dicta"--Chapter 27, p. 563). Produced a diagram purportedly showing lines of descent in the angiosperms (see "Bessey's Cactus"--p. 602). Ranales sensu lato (flower with many, free, equal parts) considered primitive. Emphasized ovary position (hypogyny vs. perigyny or epigyny) over petal fusion. Recognized 32 orders and 300 families. Accepted tribes of Compositae at family level. Bessey's contribution probably lies more in his set of "phylogenetic dicta" than in his system per se. His system was original in many respects, while also drawing from both the Bentham & Hooker and Engler & Prantl systems. The elimination of the Polypetalae, Gamopetalae, and Monochlamydeae as major groups of dicotyledons should be noted. Clements, Schaffner, and Pool supported and elaborated upon the Bessey system in the early part of the century. The phylogenetic dicta which Bessey propounded have, with modifications, provided the theoretical base and guiding principles for most contemporary systems of angiosperm classification (in this regard credit must also be given to Hallier, see below).

Bessey may be criticized for assuming that perigynous and epigynous forms constitute a single evolutionary line, and in the implication of direct ancestry of contemporary orders from one another in his "Cactus." Also, many supposed family relationships in his system are contrary to current evidence. The descent of most hypogynous sympetalous forms from a caryophyllalean-type ancestor, among other "problems," is particularly difficult to accept today.

SYSTEM OF BESSEY (Phylogenetic Taxonomy of Flowering Plants. Annals of the Missouri Botanical Garden 2: 109. 1915).

Class: ALTERNIFOLIAE (MONOCOTYLEDONEAE)
- Subclass: Alternifoliae-Strobiloideae
  - Order: Alismatales: Alismataceae, Butomaceae, Triuridaceae, Scheuchzeriaceae, Typhaceae, Sparganiaceae, Pandanaceae, Aponogetonaceae, Potamogetonaceae
  - Order: Liliales: Liliaceae, Stemonaceae, Pontederiaceae, Cyanastraceae, Philydraceae, Commelinaceae, Xyridaceae, Mayacaceae, Juncaceae, Eriocaulonaceae, Thurniaceae, Rapateaceae, Naiadaceae
  - Order: Arales: Cyclanthaceae, Araceae, Lemnaceae
  - Order: Palmales: Palmaceae
  - Order: Graminales: Restionaceae, Centrolepidaceae, Flagellariaceae, Cyperaceae, Poaceae

Subclass: Alternifoliae-Cotyloideae
Order: Hydrales: Vallisneriaceae
Order: Iridales: Amaryllidaceae, Haemodoraceae, Iridaceae, Velloziaceae, Taccaceae, Dioscoreaceae, Bromeliaceae, Musaceae, Zingiberaceae, Cannaceae, Marantaceae
Order: Orchidales: Burmanniaceae, Orchidaceae

Class: OPPOSITIFOLIAE (DICOTYLEDONEAE)
Subclass: Oppositifoliae-Strobiloideae
Superorder: Strobiloideae-Apopetalae-Polycarpellatae
Order: Ranales: Magnoliaceae, Calycanthaceae, Monimiaceae, Cercidiphyllaceae, Trochodendraceae, Leitneriaceae, Anonaceae, Lactoridaceae, Gomortegaceae, Myristicaceae, Saururaceae, Piperaceae, Lacistemaceae, Chloranthaceae, Ranunculaceae, Lardizabalaceae, Berberidaceae, Menispermaceae, Lauraceae, Nelumbaceae, Cabombaceae, Ceratophyllaceae, Dilleniaceae, Winteranaceae
Order: Malvales: Sterculiaceae, Malvaceae, Bombacaceae, Scytopetalaceae, Chlaenaceae, Gonystylaceae, Tiliaceae, Elaeocarpaceae, Balanopsidaceae, Ulmaceae, Moraceae, Urticaceae
Order: Sarraceniales: Sarraceniaceae, Nepenthaceae
Order: Geraniales: Geraniaceae, Oxalidaceae, Tropaeolaceae, Balsaminaceae, Limnanthaceae, Linaceae, Humiriaceae, Erythroxylaceae, Zygophyllaceae, Cneoraceae, Rutaceae, Simarubaceae, Burseraceae, Meliaceae, Malpighiaceae, Trigoniaceae, Vochysiaceae, Polygalaceae, Tremandraceae, Dichapetalaceae, Euphorbiaceae, Callitrichaceae
Order: Guttiferales: Theaceae, Cistaceae, Guttiferaceae, Eucryphiaceae, Ochnaceae, Dipterocarpaceae, Caryocaraceae, Quiinaceae, Marcgraviaceae, Flacourtiaceae, Bixaceae, Cochlospermaceae, Violaceae, Malesherbiaceae, Turneraceae, Passifloraceae, Achariaceae, Caricaceae, Stachyuraceae, Koeberliniaceae
Order: Rhoeadales: Papaveraceae, Tovariaceae, Nymphaeaceae, Moringaceae, Resedaceae, Capparidaceae, Brassicaceae
Order: Caryophyllales: Caryophyllaceae, Elatinaceae, Portulacaceae, Aizoaceae, Frankeniaceae, Tamaricaceae, Salicaceae, Podostemonaceae, Hydrostachydaceae, Phytolaccaceae, Basellaceae, Amaranthaceae, Chenopodiaceae, Polygonaceae, Nyctaginaceae, Cynocrambaceae, Batidaceae
Superorder: Strobiloideae-Sympetalae-Polycarpellatae
Order: Ebenales: Sapotaceae, Ebenaceae, Symplocaceae, Styracaceae, Fouquieriaceae
Order: Ericales: Clethraceae, Ericaceae, Epacridaceae, Diapensiaceae, Pirolaceae, Lennoaceae
Order: Primulales: Primulaceae, Plantaginaceae, Plumbaginaceae, Myrsinaceae, Theophrastaceae
Superorder: Strobiloideae-Sympetalae-Dicarpellatae
Order: Gentianales: Oleaceae, Salvadoraceae, Loganiaceae, Gentianaceae, Apocynaceae, Asclepiadaceae
Order: Polemoniales: Polemoniaceae, Convolvulaceae, Hydrophyllaceae, Borraginaceae, Nolanaceae, Solanaceae
Order: Scrophulariales: Scrophulariaceae, Bignoniaceae, Pedaliaceae, Martyniaceae, Orobanchaceae, Gesneraceae, Columelliaceae, Lentibulariaceae, Globulariaceae, Acanthaceae
Order: Lamiales: Myoporaceae, Phrymaceae, Verbenaceae, Lamiaceae

Subclass: Oppositifoliae-Cotyloideae
Superorder: Cotyloideae-Apopetalae
Order: Rosales: Rosaceae, Malaceae, Prunaceae, Crossosomataceae, Connaraceae, Mimosaceae, Cassiaceae, Fabaceae, Saxifragaceae, Hydrangeaceae, Grossulariaceae, Crassulaceae, Droseraceae, Cephalotaceae, Pittosporaceae, Brunelliaceae, Cunoniaceae, Myrothamnaceae, Bruniaceae, Hamamelidaceae, Casuarinaceae, Eucommiaceae, Platanaceae
Order: Myrtales: Lythraceae, Sonneratiaceae, Punicaceae, Lecythidaceae, Melastomataceae, Myrtaceae, Combretaceae, Rhizophoraceae, Oenotheraceae, Halorrhagidaceae, Hippuridaceae, Cynomoriaceae, Aristolochiaceae, Rafflesiaceae, Hydnoraceae
Order: Loasales: Loasaceae, Cucurbitaceae, Begoniaceae, Datiscaceae, Ancistrocladaceae
Order: Cactales: Cactaceae
Order: Celastrales: Rhamnaceae, Vitaceae, Celastraceae, Buxaceae, Aquifoliaceae, Cyrillaceae, Pentaphylacaceae, Corynocarpaceae, Hippocrateaceae, Stackhousiaceae, Staphyleaceae, Geissolomataceae, Penaeaceae, Oliniaceae, Thymelaeaceae, Hernandiaceae, Elaeagnaceae, Myzodendraceae, Santalaceae, Opiliaceae, Grubbiaceae, Olacaceae, Loranthaceae, Balanophoraceae
Order: Sapindales: Sapindaceae, Hippocastanaceae, Aceraceae, Sabiaceae, Icacinaceae, Melianthaceae, Empetraceae, Coriariaceae, Anacardiaceae, Juglandaceae, Betulaceae, Fagaceae, Myricaceae, Julianaceae, Proteaceae
Order: Umbellales: Araliaceae, Apiaceae, Cornaceae
Superorder: Cotyloideae-Sympetalae
Order: Rubiales: Rubiaceae, Caprifoliaceae, Adoxaceae, Valerianaceae, Dipsacaceae
Order: Campanulales: Campanulaceae, Goodeniaceae, Stylidiaceae, Calyceraceae
Order: Asterales: Helianthaceae, Ambrosiaceae, Heleniaceae, Arctotidaceae, Calendulaceae, Inulaceae, Asteraceae, Vernoniaceae, Eupatoriaceae, Anthemidaceae, Senecionidaceae, Carduaceae, Mutisiaceae, Lactucaceae

12. Hallier, Hans — 1905, 1912 — Independently developed ideas similar to those of Bessey. System was designed to be phylogenetic. Saw angiosperms as monophyletic from a Bennettitalean-type ancestor with some Marattialean affinities. Regarded dicots as older and more primitive than monocots. Monocots thought to have been derived from stocks ancestral to the Lardizabalaceae. Polycarpy and spirality of parts considered primitive, syncarpy and cyclicity as advanced. Placed large significance on ovule morphology and position. Defended use of many kinds of characters for classification. His realignment of families displayed many insights, such as the alliance of Salicaceae with Flacourtiaceae, Cactaceae with Aizoaceae, Cucurbitaceae with Passifloraceae, etc. Four primary divisions of dicots: Protérogènes (5 orders), Anonophylales (5 orders), Rhodophyles (7 orders), Ochnigènes (12 orders). His 1905 and 1912 treatments differ quite a bit; recognized 213 families in 1912.

13. Rendle, Alfred Barton — 1930 — Basically an Englerian-type system (held to amentiferous primitiveness), but with some borrowings from Bentham and Hooker. Did not claim his arrangement to be strictly phylogenetic. Treated monocots as more primitive

than dicots. Grouped dicots into three grades corresponding to levels of differentiation of the perianth (compare with use of grade concept by Benson):

Monochlamydeae (simple type of flower, no petals, occasionally with calyx; some thought to be primitively, others secondarily, simple; considered by Rendle to be polyphyletic)

Dialypetalae (Ranales, with free petals (tepals), many spirally arranged parts → Umbelliflorae, with cyclic, reduced parts; considered by Rendle to be polyphyletic)

Sympetalae (higher grades of floral development; pentacyclic → tetracyclic → tetracyclic and epigynous).

Phyletic criteria: wind pollination and woodiness considered primitive. Many taxa were tentatively assigned positions awaiting further data.

14. Mez, Karl C. 1926 Produced a "phylogenetic tree" based (in part) on serological data (see Chapter 13, p. 288).

15. Hayata, Bunzō 1921 Produced a "dynamic system" of angiosperm classification, which was basically a rearrangement of Engler's system. Hayata's arrangement points out the reticulate nature of character variation in the angiosperms.

16. Burtt-Davy, Joseph 1937 Cooperated with the wood anatomist Chalk to produce the following classification of dicots: Amentiferae/Polystemonae (Divs. Apocarpicae, Syncarpicae)/Oligostemonae (Divs. Pentacyclicae, Tetracyclicae).

17. Pulle, August A. 1938 1950 A modified Englerian classification. Dicots patterned after Wettstein; divided into 8 series. Considered the Sympetalae to be polyphyletic and divided them up among the various series. Did not consider the Monochlamydeae to be a natural group. Revised his classification in 1950, incorporating such changes as merging Cactales with Centrospermae.

18. Skottsberg, Carl 1940 A modified Englerian classification, with Wettsteinian influence. Monocots considered derived from an unknown primitive dicot. Considered apocarpy and polycarpy as primitive, syncarpy and monocarpy as advanced in both monocots and dicots. Apetalous forms considered polyphyletic. Amentiferae placed after Rosales, but not redistributed among petaliferous families to which they appear to be more closely related (i.e., case of Salicaceae). Began dicots with Casuarinaceae, but did not consider them to be ancestral to other dicots.

19. Gundersen, Alfred 1950 System only for dicots. Divided into 10 informal, more or less tentative groups: Magnoliflorae, Cistiflorae, Thea Group, Rosaeflorae, Ulmus Group, Malva Group, Geranium Group, Dianthiflorae, Jasminiflorae, Rubiflorae. Considered "Magnoliflorae" basal to other groups. Took much modern data into account (work of Sprague, et al.).

20. Hutchinson, John 1926 1934 1959 1969 Basically in the Bentham-Bessey tradition. Angiosperms considered monophyletic, possibly from an ancestral group of gymnosperms related to the cycadeoids. Divides dicots into those

RELATIONSHIPS of the ORDERS

(after Bessey, 1915)

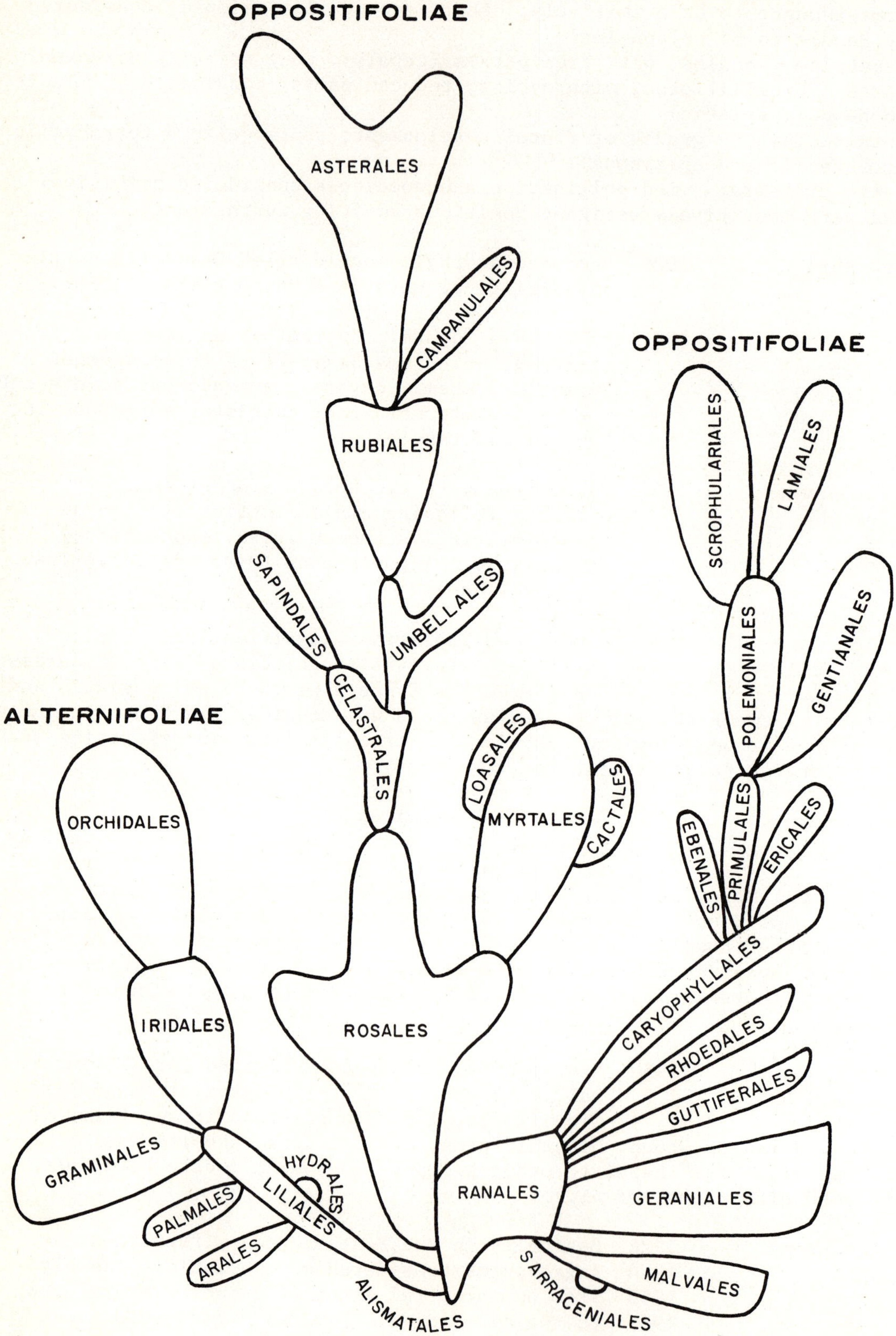

RELATIONSHIPS of the ORDERS of DICOTYLEDONS
(after Hutchinson)

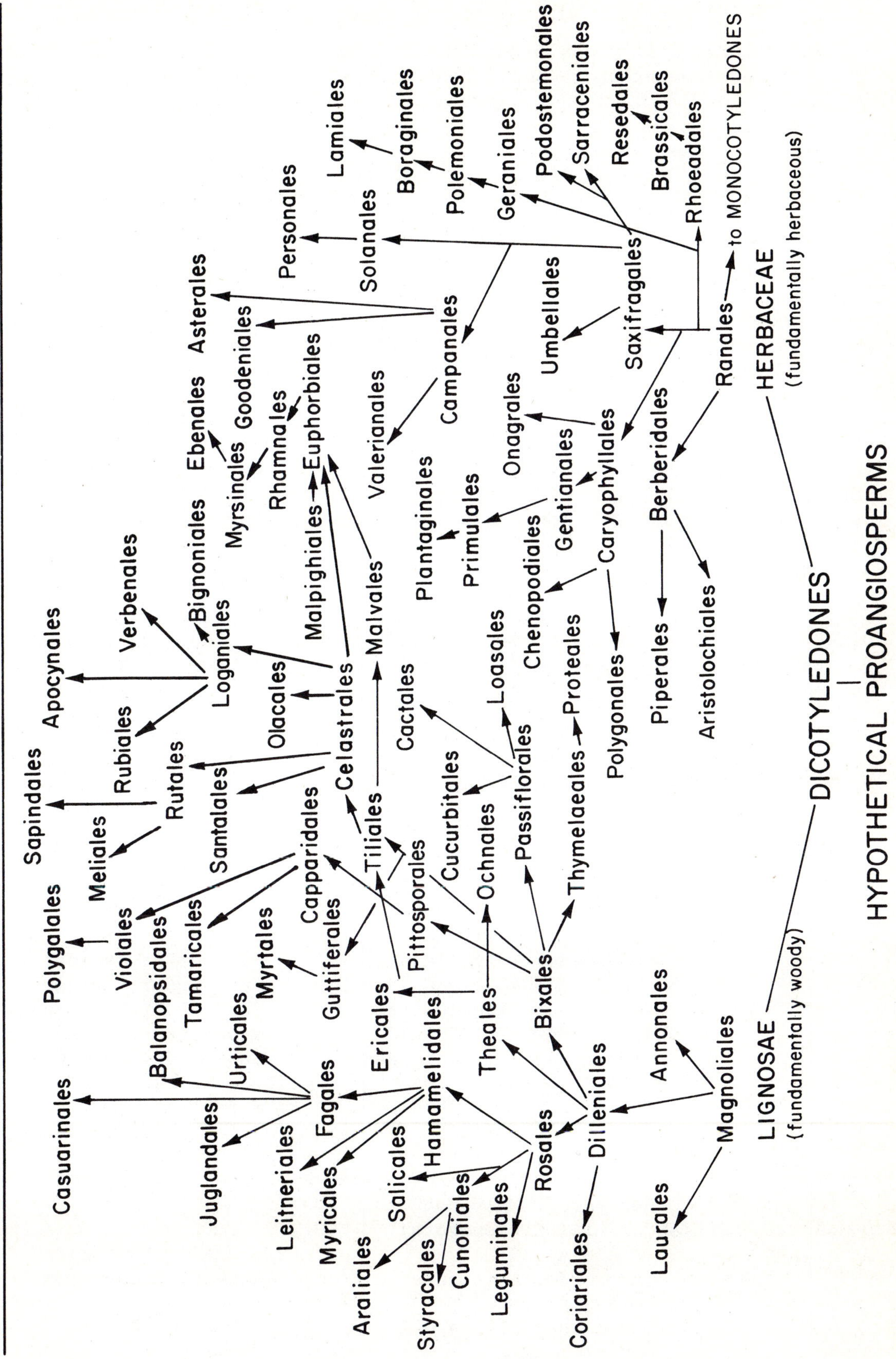

considered "fundamentally and predominantly" woody (Lignosae) and those considered "fundamentally and predominantly" herbaceous (Herbaceae). Parallel lines of development were considered to have occurred within these two main groups.

Herbaceae ← Ranales (Ranunculales)   Magnoliales → Lignosae

This is considered an unnatural division by most workers (note wide separation of Umbelliferae from Araliaceae, Labiatae from Verbenaceae, Bignoniaceae from Scrophulariaceae, etc.). Gamopetalae considered polyphyletic, rather than diphyletic as in Bessey's system. Revision of the monocots more widely accepted, similar to that of Lindley, based more on nature of perianth than on ovary position. Often did not give reasons for changes in the disposition of orders and families (however, here we do not downgrade the value of a good taxonomic intuition). Relied for the most part on morphological characters, to the exclusion of much recent information from the fields of anatomy, embryology, chemistry, cytology, palynology, etc. Excellent knowledge of the world flora. Accepted a large number of orders and families--a "splitter." See chart of Hutchinson's system, p. 603.

21. Benson, Lyman — 1957 — Basically a Besseyan-type system. Dicots divided into five informal polyphyletic grades "representing stages in floral evolution." (See chart, p. 605).

OTHER recent Ranalian-based systems include those of Soó (1953), Novák (1954), and Emberger (1960). Deyl (1955) and Kimura (1956) have presented classifications for the monocotyledons.

22. Takhtajan, Armen — 1943, 1953, 1959, 1966, 1969 — A system in the Bessey-Hallier tradition; strongly influenced by Hallier. The division into subclasses is considered a major conceptual advance by many (for characters of Subclasses see discussion of Cronquist's system below).

<u>MAJOR GROUPS IN TAKHTAJAN'S SYSTEM (1969)</u>

CLASS MAGNOLIATAE (dicots)
- SUBCLASS MAGNOLIIDAE
  - Superorder Magnolianae
- SUBCLASS RANUNCULIDAE
  - Superorder Ranunculanae
- SUBCLASS HAMAMELIDIDAE
  - Superorder Hamamelidanae
- SUBCLASS CARYOPHYLLIDAE
  - Superorder Caryophyllanae
- SUBCLASS DILLENIIDAE
  - Superorder Dillenianae
  - Superorder Ericanae
  - Superorder Malvanae
- SUBCLASS ROSIDAE
  - Superorder Rosanae
  - Superorder Rutanae
  - Superorder Aralianae
  - Superorder Celastranae
  - Superorder Proteanae
- SUBCLASS ASTERIDAE
  - Superorder Lamianae
  - Superorder Asteranae

CLASS LILIATAE (monocots)
- SUBCLASS ALISMIDAE
  - Superorder Alismanae
- SUBCLASS LILIIDAE
  - Superorder Lilianae
- SUBCLASS COMMELINIDAE
  - Superorder Juncanae
  - Superorder Commelinanae
- SUBCLASS ARECIDAE
  - Superorder Arecanae

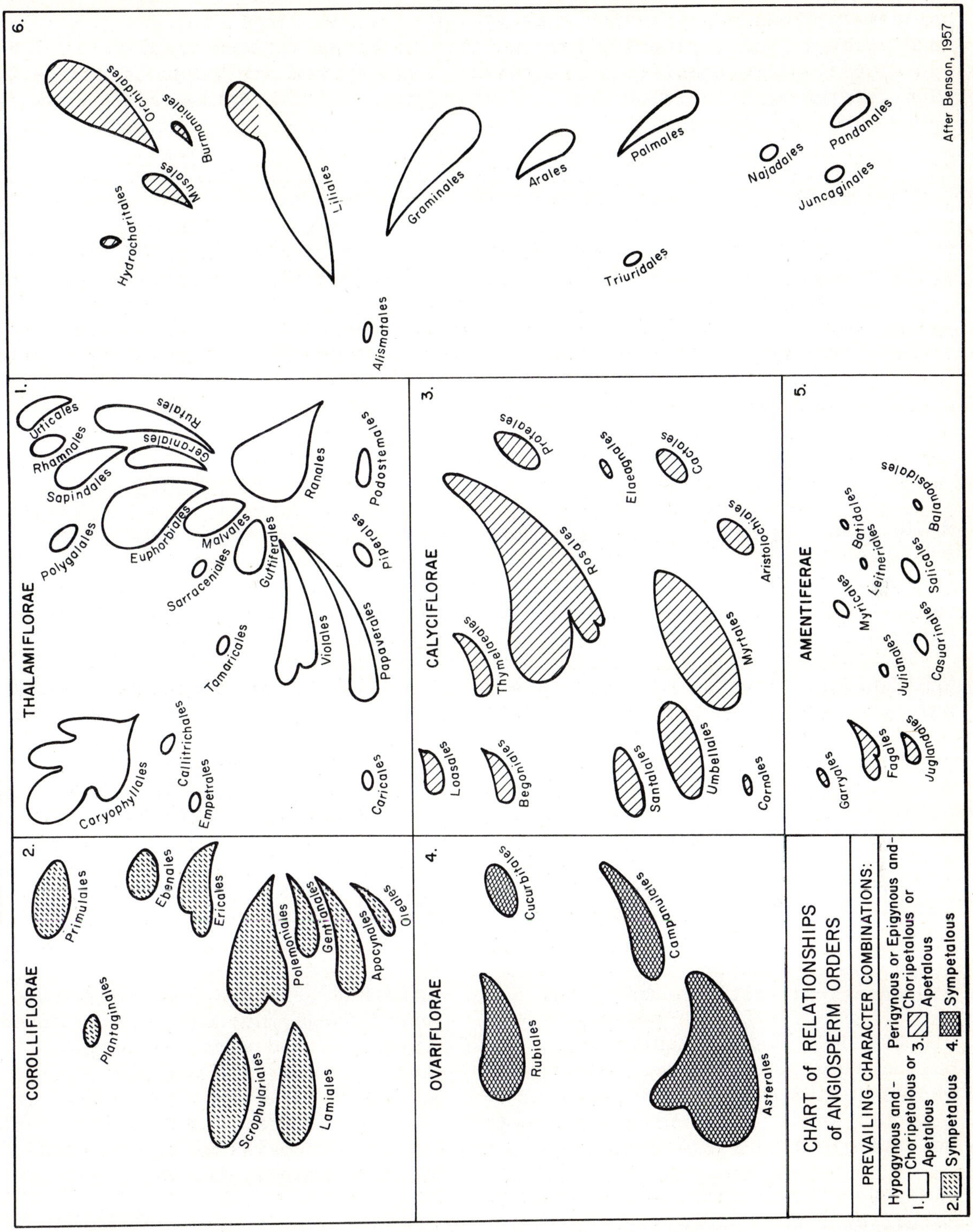
DICOTYLEDONS
MONOCOTYLEDONS
1. THALAMIFLORAE
Urticales
Rhamnales
Sapindales
Rutales
Geraniales
Polygalales
Euphorbiales
Malvales
Sarraceniales
Guttiferales
Ranales
Podostemales
Piperales
Tamaricales
Violales
Papaverales
Caryophyllales
Callitrichales
Empetrales
Caricales
2. COROLLIFLORAE
Primulales
Ebenales
Ericales
Plantaginales
Polemoniales
Gentianales
Apocynales
Oleales
Scrophulariales
Lamiales
3. CALYCIFLORAE
Proteales
Elaeagnales
Cactales
Rosales
Aristolochiales
Thymelaeales
Myrtales
Loasales
Begoniales
Santalales
Umbellales
Cornales
4. OVARIFLORAE
Cucurbitales
Campanulales
Rubiales
Asterales
5. AMENTIFERAE
Batidales
Balanopsidales
Leitneriales
Myricales
Salicales
Julianales
Casuarinales
Garryales
Fagales
Juglandales
6.
Orchidales
Burmanniales
Musales
Hydrocharitales
Liliales
Alismatales
Graminales
Arales
Palmales
Triuridales
Najadales
Pandanales
Juncaginales
After Benson, 1957
CHART of RELATIONSHIPS of ANGIOSPERM ORDERS
PREVAILING CHARACTER COMBINATIONS:
Hypogynous and -
1. Choripetalous or Apetalous
2. Sympetalous
Perigynous or Epigynous and -
3. Choripetalous or Apetalous
4. Sympetalous

A "splitter" at the ordinal and family levels who takes much recent palynological, anatomical, embryological, paleobotanical, chemical, and cytological data into account. An English summary of this important system is given in Takhtajan (1969). More complete expositions of his system are given in the 1959 and 1966 references (in German and Russian respectively). Active system-making (often in cooperation with A. Cronquist) is ongoing.

23. Cronquist, Arthur 1957 1965 1968 Conceptually similar to Takhtajan's system, differing in detail. Subclass Ranunculidae merged with Magnoliidae. Neither much of a lumper nor much of a splitter. Groups considered polyphyletic are split up and placed with taxa to which they appear to be phyletically related (note treatment of Amaryllidaceae), and many small, somewhat anomalous or unusual groups are retained in the larger families to which they appear most closely related. Much modern data taken into account (see Table 28-2, below).

CHARACTERS OF THE SUBCLASSES (Cronquist, 1968. Evolution and Classification of Flowering Plants)

Dicotyledons

MAGNOLIIDAE--Apocarpous, polypetalous (or apetalous); perianth evident, stamens numerous, centripetal; pollen binucleate, often uniaperturate; ovules bitegmic, crassinucellate.

HAMAMELIDAE--Flowers reduced, often unisexual, often in catkins; perianth poorly developed or wanting; never with numerous seeds on parietal placentae.

CARYOPHYLLIDAE--Stamens centrifugal, pollen trinucleate; ovules bitegmic, crassinucellate, often campylotropous or amphitropous; seeds often with perisperm; plants usually with betalains or free-central to basal placentation; mostly herbs.

DILLENIIDAE--Stamens centrifugal, pollen binucleate (except Cruciferae); ovules seldom campylotropous or amphitropous; seeds seldom with perisperm; no betalains; many species trees.

ROSIDAE--Stamens numerous, centripetal, or fewer and opposite or alternate petals; ovules bitegmic or crassinucellate; rarely sympetalous.

ASTERIDAE--Sympetalous; stamens few, never opposite petals, ovules unitegmic and tenuinucellate; carpels usually 2.

Monocotyledons

ALISMATIDAE--Mostly apocarpous; aquatic, always herbs; pollen trinucleate; endosperm wanting or not starchy; stomate subsidiary cells mostly 2.

ARECIDAE--Flowers numerous, small, with spathe, sometimes on spadix; leaves broad, petiolate, usually net-veined; ovary superior or sometimes sunken in axis of spadix; subsidiary cells 4 (3-2).

COMMELINIDAE--Flowers with well-defined sepals and petals, or perianth chaffy or bristly or lacking, not on a spadix; subsidiary cells 2-4 or more.

LILIIDAE--Flowers few to numerous, often large, not usually subtended by spathe and never on spadix; leaves linear and parallel to broader and net-veined; ovary superior or (more often) inferior; subsidiary cells usually 0, less often 2, rarely more.

RELATIONSHIPS of DICOTYLEDONS

(Adapted from Cronquist, 1968)

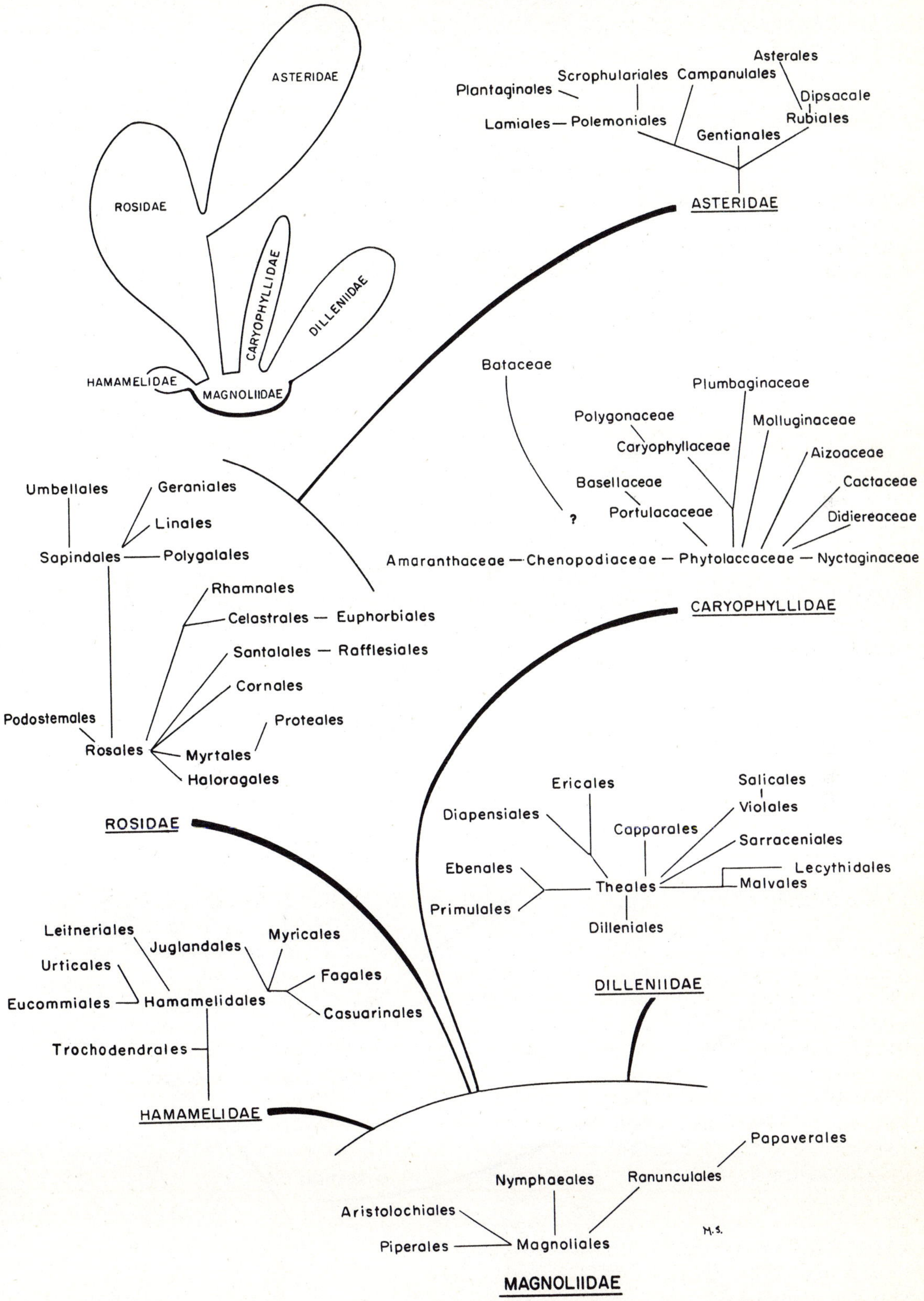

Figure 28-6

RELATIONSHIPS of MONOCOTYLEDONS

(Adapted from Cronquist, 1968)

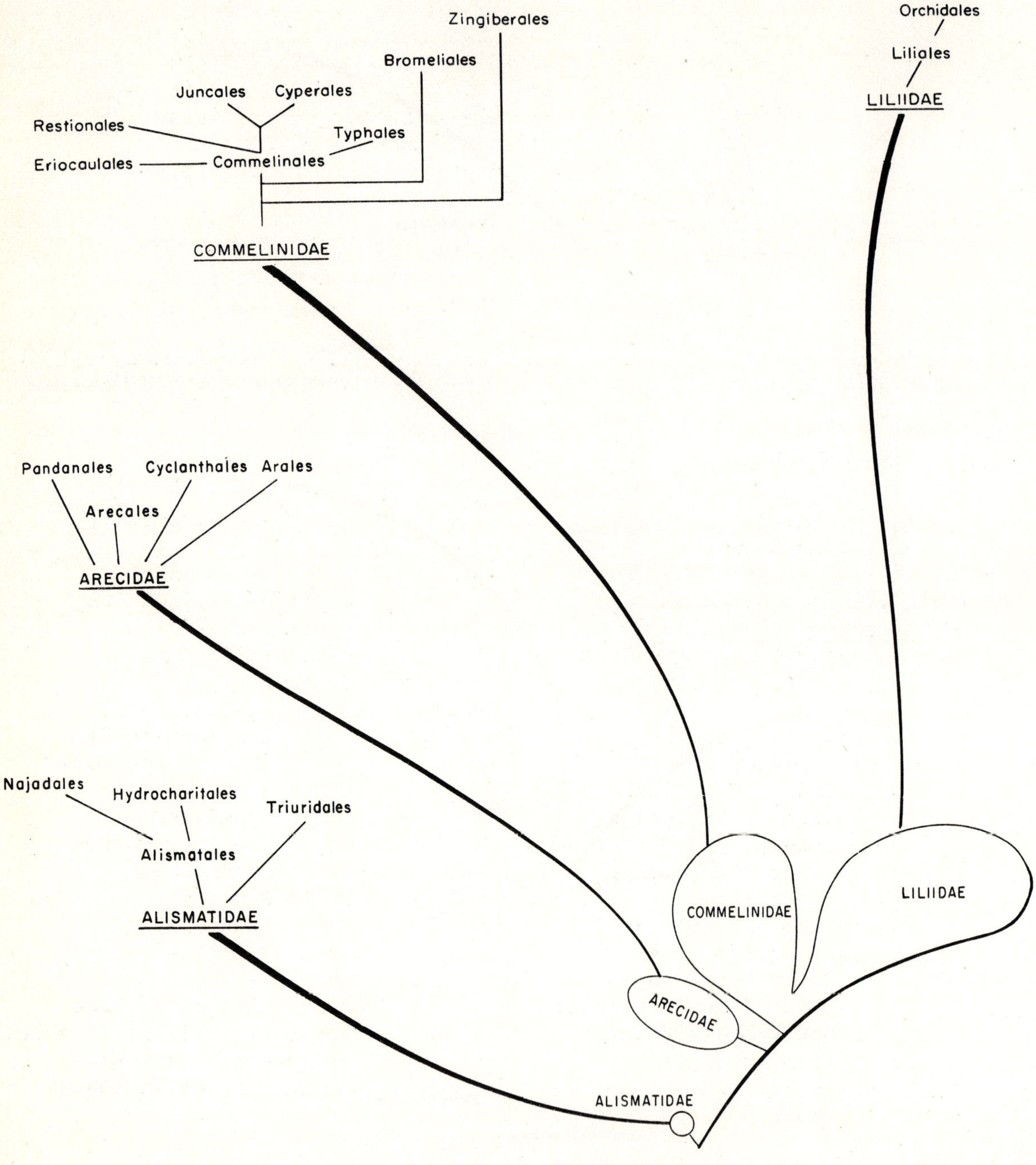

24. Thorne, Robert — 1968 — A system in the Bessey-Hallier tradition presented with the following major taxa (from Thorne, 1968):

SUBCLASS DICOTYLEDONEAE

Superorders: Annoniflorae, Nymphaeiflorae, Sarraceniiflorae, Theiflorae, Rafflesiiflorae, Cistiflorae, Malviiflorae, Santaliflorae, Geraniiflorae, Rutiflorae, Chenopodiiflorae, Hamamelidiflorae, Rosiflorae, Proteiflorae, Myrtiflorae, Gentianiflorae, Corniflorae, Lamiiflorae, Asteriflorae

SUBCLASS MONOCOTYLEDONEAE

Superorders: Alismiflorae, Triuridiflorae, Liliiflorae, Ariflorae, Commeliniflorae

Takes a broader view of families than most. Aims at having a uniform size of morphological gap separating groups of equal rank (thus places Amaryllidaceae in Liliaceae, etc.).

II. EXAMPLES OF RECENT OVERALL CLASSIFICATIONS OF VASCULAR PLANTS

A. Tippo, O. 1942. A modern classification of the plant kingdom. Chronica Botanica 7: 203-206.

- Phylum Tracheophyta (Tracheata)
  - Sub-phylum Psilopsida
    - Class Psilophytinae
      - Order Psilophytales
      - Order Psilotales
  - Sub-phylum Lycopsida
    - Class Lycopodineae
      - Order Lycopodiales
      - Order Selaginellales
      - Order Lepidodendrales
      - Order Pleuromeiales
      - Order Isoetales
  - Sub-phylum Sphenopsida
    - Class Equisetineae
      - Order Hyeniales
      - Order Sphenophyllales
      - Order Equisetales
  - Sub-phylum Pteropsida
    - Class Filicinae
      - Order Coenopteridales
      - Order Ophioglossales
      - Order Marattiales
      - Order Filicales
    - Class Gymnospermae
      - Sub-class Cycadophytae
        - Order Cycadofilicales (Pteridospermae)
        - Order Bennettitales
        - Order Cycadales
      - Sub-class Coniferophytae
        - Order Cordaitales
        - Order Ginkgoales
        - Order Coniferales
        - Order Gnetales
    - Class Angiospermae
      - Sub-class Dicotyledoneae
        - Order Magnoliales (Ranales)
        - Order Ericales, etc.
      - Sub-class Monocotyledoneae

B. Banks, H. P. 1968. The early history of land plants. Symposium on Evolution and Environment. E. T. Drake (ed.). Yale University Press. New Haven.

Division - Tracheophyta
  Subdivision - Rhyniophytina
    Order - Rhyniales
      Family - Rhyniaceae
        Cooksoniaceae
  Subdivision - Zosterophyllophytina
    Order - Zosterophyllales
      Family - Zosterophyllaceae
        Gosslingiaceae
  Subdivision - Psilophytina - Psilotales
  Subdivision - Lycophytina
  Subdivision - Spenophytina
  Subdivision - Trimerophytina
  Subdivision - Pterophytina
      Class - Cladoxylopsida
      Class - Coenopteridopsida
      Class - Filicopsida
      Class - Progymnospermopsida
      Class - Cycadopsida
      Class - Coniferopsida
      Class - Gnetopsida
      Class - Angiospermopsida

C. Bierhorst, D. W. 1971. Morphology of Vascular Plants. The Macmillan Company. New York.

Division - Tracheophyta
  Class - Rhyniopsida
    Order - Rhyniales
  Class - Zosterophyllopsida
    Order - Zosterophyllales
  Class - Lycopodiopsida (Lycopsida)
    Order - Asteroxylales
    Order - Lycopodiales
    Order - Protolepidodendrales
    Order - Selaginellales
    Order - Lepidodendrales
    Order - Isoetales
  Class - Cladoxylopsida
    Order - Cladoxylales
  Class - Equisetopsida (Sphenopsida)
    Order - Hyeniales
    Order - Pseudoborniales
    Order - Spenophyllales
    Order - Equisetales
  Class - Coenopteridopsida
    Order - Coenopteridales
  Class - Filicopsida
    Order - Noeggerathiales
    Order - Filicales
    Order - Marsileales
    Order - Salviniales
  Class - Ophioglossopsida
    Order - Ophioglossales
  Class - Marattiopsida
    Order - Marattiales
  Class - Aneurophytopsida
    Order - Aneurophytales
    Order - Archeopteridales
    Order - Protopityales
  Class - Cycadopsida
    Order - Pteridospermales
    Order - Cycadales
    Order - Cycadeoidales
    Order - Caytoniales
  Class - Coniferopsida
    Order - Cordaitales
    Order - Coniferales
    Order - Taxales
    Order - Ginkgoales
  Class - Gnetopsida
    Order - Ephedrales
    Order - Gnetales
    Order - Welwitschiales
  Class - Angiospermopsida
    Subclass - Dicotyledonidae
    Subclass - Monocotyledonidae

Table 28-1. Comparison of treatment of Cronquist's Subclasses with some other recent systems.

I. DICOTYLEDONS:

Subclass	More or less comparable to group traditionally known as:
MAGNOLIIDAE	Ranales
HAMAMELIDAE	Amentiferae s.s.
CARYOPHYLLIDAE	Centrospermae
DILLENIIDAE ROSIDAE	Seen as natural groups separately derived from Magnoliidae, the same sorts of evolutionary advances occurring in both groups, but with different frequencies - Cronquist
ASTERIDAE	Sympetalae

A. Takhtajan (1969):	Magnoliidae divided into two subclasses - Magnoliidae, Ranunculidae (includes Sarraceniaceae) Hamamelidae - Takhtajan accepts more orders, some differences in sequence. Caryophyllidae - Takhtajan includes Theligonales Dilleniidae - Takhtajan places Euphorbiales and Thymeleales here. Rosidae - Takhtajan places Oleales here, includes Umbellales in Cornales, includes Nepenthaceae and Droseraceae here. Asteridae - Takhtajan gives Dipsacales a basal position, includes Rubiaceae in Gentianales, shuffles some families between Polemoniales and Scrophulariales.
B. Thorne (1968):	The identities of the Magnoliidae, Caryophyllidae, and Hamamelidae are essentially maintained; Rosidae, Dilleniidae, Asteridae mixed.
C. Hutchinson (1959):	All the dicot subclasses, except for the Hamamelidae, are dispersed throughout the system.
D. Melchior (1964):	Informally recognizes, in broad outlines, the Magnoliidae, Hamamelidae, Caryophyllidae, and Asteridae as natural groups; Rosidae and Dilleniidae mixed together.

II. MONOCOTYLEDONS: Subclasses: ALISMATIDAE, ARECIDAE, COMMELINIDAE, LILIIDAE

A. Takhtajan:	Takhtajan's treatment of the monocots is similar to that of Cronquist. Some differences: In the Alismatidae, Takhtajan removes the Triuridales to the Liliidae; Takhtajan places the Typhales in the Arecidae rather than in the Commelinidae; includes the Zingiberales in the Liliidae, rather than in the Commelinidae.

Table 28-1, continued

B.	Thorne:	Thorne's 5 superorders of monocots compare closely with Cronquist's subclasses, thus: Alismatidae ± equivalent to Alismiflorae + Triuridiflorae Arecidae equivalent to Ariflorae Commelinidae equivalent to Commeliniflorae Liliidae equivalent to Liliiflorae
C.	Hutchinson:	Alismatidae kept together, but Cronquist's other three monocot subclasses are mixed together.
D.	Melchior:	Helobiae comparable to Alismatidae of Cronquist, but Cronquist's other three monocot subclasses are mixed together.

Table 28-2. Examples of "Modern" Criteria Considered by Cronquist (from Cronquist, 1968).

Criterion	Level or groups where used
Anatomical--	
Stratification of phloem into fibrous & non-fibrous tissue	Sarcolaenaceae/Malvales
Internal phloem	Caryophyllales; Malvales vs. Lecythidales; in Rosidae (Myrtales); Gentianales
Xylem vesselless	within Magnoliidae; Nymphaeales
Ray width	Ranunculales
Wood anatomy	Columelliaceae
Wood parenchyma diffuse vs. vasicentric	Umbellales vs. Cornales
Unusual vascular bundles	Thurniaceae
Nodal anatomy	subclass, ordinal levels; Umbellales vs. Cornales
Stem with anomalous secondary thickening	Caryophyllidae
Central axis of gynoecium with or without vascular bundles	Umbellales vs. Cornales
Schizogenous secretory system	Myrsinaceae; Umbellales vs. Cornales
Laticifers	Musaceae
Secretory canals	Guttiferae, Sapindales
Latex system	Papaverales/Capparales; Sapotaceae; Celastrales; within Gentianales
Elongate idioblasts	Trochodendrales
Ethereal oil cells	Magnoliidae
Foliar glands	Leitneriales, Myricales, Juglandales
Myrosin cells	Capparales; Tropaeolaceae

Table 28-2, continued

Mucilage cells	Malvales
Raphide sacs	in Zingiberales
Raphides, crystal sand	Dilleniaceae
Silica bodies in leaf epidermis	Thurniaceae
Stellate hairs	Malvales; Styracaceae
Colleters (specialized glandular trichomes)	Rubiaceae-Gentianales
Guard cell symmetry/asymmetry	within Zingiberales (Lowiaceae)
Subsidiary cell number	subclasses of monocots
Nectary type	Malvales; monocots vs. dicots; within Liliidae
Embryological--	
General embryological studies	Agavaceae
Ovules unitegmic vs. bitegmic	subclass level and below
Ovules tenuinucellate vs. crassinucellate	subclass level and below
Endosperm vs. perisperm	Magnoliidae; Caryophyllales
Endosperm cellular vs. nuclear	in Polemoniales; Rubiales-Gentianales-Dipsacales
Endosperm helobial	in Alismatidae, Arecidae
Endosperm starchy vs. nonstarchy	subclasses of monocots
Endosperm haustoria	Ericales/Diapensiales
Integumentary tapetum	within Asteridae
Chalazosperm	Taccaceae vs. Cyanastraceae
Embryo sac bisporic vs. monosporic	in Alismatidae
Possible free-nuclear stage in embryo development	*Paeonia*
Cytological--	
Chromosome number	Agavaceae/Xanthorrhoeaceae
Karyotype	Agavaceae
Chemical--	
Cyanogenetic compounds	within Violales
Alkaloids, glucosides	Gentianales
Gentiopicrin	Gentianaceae
Meliatin	Menyanthaceae
Tryptophane alkaloids	Rubiaceae-Gentianales
Isoquinoline alkaloids, especially berberin	Ranunculales, Papaverales
Betalains	Caryophyllidae
Unusual fatty acids	within Geraniales
Serology	Papaverales/Capparales; Cornales; Caprifoliaceae-Cornaceae
Palynological--	
Pollen binucleate vs. trinucleate	subclass level; Sapindales; in Polemoniales

Table 28-2, continued

Pollen triaperturate vs. uniaperturate	subclass (Magnoliidae)
Pollen non-aperturate	Scheuchzeriaceae
Other pollen characters	Eucommiales, Hamamelidales, Casuarinales; Malvales; Staphyleaceae; in Restionales; Ruppiaceae, Zosteraceae
Physiological--	
Adaptation to NO_3 deficiency	Sarraceniales
Adaptation to water stress	Caryophyllidae; some Violales
Mycorrhizal habit	Ericales; Triuridales; Orchidales
Other--	
Position of vessels in plant	Commelinidae; Liliidae
Petals sepal vs. staminodial-derived	Paeonia
Androecial ontogeny	delimitation of dicot subclasses
Gynoecial ontogeny	Lauraceae/Myristicaceae
Pollination method	subclass (Magnoliidae, Hamamelidae); subclasses of monocots
Fossil pollen record	Hamamelidae
Carpel sealing/unsealing	Magnoliidae

Section B. COMPARISON OF CLASSIFICATIONS

Presented in this section is a comparison, in tabular form, of the flowering plant classification systems of Cronquist, Takhtajan, Thorne, Engler, Hutchinson, and Bentham & Hooker. The linear sequence of orders and families is that presented in Cronquist (1968). All families accepted in any of the six above-mentioned systems are accounted for, as well as all families for which a description is given in Airy Shaw (1966).

The system of Cronquist is that presented in The Evolution and Classification of Flowering Plants (1968). Several changes in the system which are anticipated by Cronquist are indicated in footnotes.

The system of Takhtajan is that presented in Flowering Plants: Origin and Dispersal (1969). Takhtajan's Systema et Phylogenia Magnoliophytorum (1966) was consulted for the disposition of certain segregate families not treated in the 1969 book.

The system of Thorne is that presented in Aliso 6(4) (1968). Inasmuch as that system is as yet available only in outline form, the treatment of taxa not specifically mentioned by Thorne follows that in Melchior (1964), as was indicated by Thorne himself.

The system of Engler is that presented in Syllabus der Pflanzenfamilien (Melchior, ed., 1964). Dalla Torre and Harms' Genera Siphonogamarum (1900-1907) was consulted for the disposition of certain segregate families not treated in Melchior (1964).

The system of Hutchinson is that presented in The Families of Flowering Plants (1959). A number of changes in the ordinal placement of certain dicotyledonous families, as well as the addition of several new families, which were incorporated in Hutchinson's Evolution and Phylogeny of Flowering Plants (1969), are accounted for in the table and in footnotes.

The system of Bentham and Hooker is that presented in Genera Plantarum (1862-1883). In the Polypetalae and Gamopetalae of these authors, cohorts are taken to correspond to orders in the modern sense, and their "orders" to families in the modern sense. However, in the case of the Monochlamydeae (Incompletae) and Monocotyledonae, their series are treated as orders in the modern sense, since the series in the latter two major groups were not divided into cohorts. The correspondence of the categories of Bentham and Hooker to those of contemporary authors is not exact. Bentham and Hooker recognized at the subfamilial (subordinal, in their terminology) level of "tribe" many groups which are today considered families (cf. the "tribes" of Geraniaceae, Najadaceae, Scitamineae, etc.).

Airy Shaw, in Willis' A Dictionary of the Flowering Plants and Ferns, ed. 7, (1966), gives descriptions for a number of families not accepted in any of the six systems. These segregate families are included in the table and are indicated by the symbol (AS+). In most cases, description of a family by Airy Shaw indicates his recognition of family status for the taxon. A number of families not accepted in any of the six systems nor by Airy Shaw are also included.

Families not accepted in any of the six systems are indicated by [brackets]. The generally accepted placement of these groups is indicated.

A question mark (?) indicates uncertainty regarding the position of a taxon, or to the level at which it should be received. A dash (--) indicates that the taxon was not treated.

The alphabetical index to 648 family names which follows the table includes well-known synonyms and a number of variant spellings not included in the body of the table.

FOOTNOTES

a - subsequent to the nomenclatural sessions at the 11th International Botanical Congress in Seattle in 1969, both Cronquist and Takhtajan have adopted -opsida endings for classes. Thus: Magnoliopsida and Liliopsida, rather than Magnoliatae and Liliatae.
b - Cronquist would now refer this family to the Euphorbiales.
c - excluding Rhabdodendron? (Takh.).
d - perhaps belongs in Portulacaceae (Takh.).
e - Hectorella in Portulacaceae, Lyallia in Caryophyllaceae? (B. & H.).

f - Sauvagesieae in Violaceae (B. & H.).
g - as for _Rhaptopetalum_ (B. & H.).
h - affinities of _Hua_ obscure (Cronq.).
i - including _Afrostyrax_? (Takh.).
j - in Bixineae, which also includes Bixaceae (B. & H.).
k - as Samydaceae (Flacourtiaceae p.p.) (B. & H.).
m - including Paropsieae? (Takh.).
n - near Flacourtiaceae (Bixales)? (Hutchinson, 1969).
o - in Bixineae, which includes part of Flacourtiaceae (B. & H.).
p - including _Oceanopapaver_ (Takh.).
q - Oxystylidaceae (Capparales) is a segregate from Capparaceae (Hutchinson, 1969).
r - Prionotaceae is a segregate from Ericaceae and Epacridaceae (Hutchinson, 1969).
s - abberrant in Dioscoreaceae (Cronq.).
t - as treated in Takhtajan (1966), not mentioned in Takhtajan (1969).
u - Cronquist would now refer this family to the Trochodendrales.
v - excluding _Polygonanthus_, including _Corokia_; _Carpodetus_ very isolated (Takh.).
w - excluding _Stylobasium_? (Takh.)
x - Cronquist would now place this family near the Solanaceae.
y - placed in Saxifragales (position uncertain) in Hutchinson (1969).
z - placed in Myrtales in Hutchinson (1969).
aa - placed in Onagrales in Hutchinson (1969).
bb - position of _Corokia_ uncertain (Cronq.).
cc - excluding _Corokia_; _Griselinia_ very isolated (Takh.).
dd - Campylostemonoideae, Cassinoideae, and Tripterygioideae very isolated within the family (Takh.).
ee - as treated in Hutchinson (1969) (but places _Nymania_ in Meliaceae).
ff - _Dapania_ in Geraniaceae (Oxalideae), _Sarcotheca_ in Linaceae (Ixonantheae).
gg - excluding _Peltanthera_ (Takh.).
hh - Salpiglossidaceae is a segregate from Solanaceae (Hutchinson, 1969).
ii - Thorne (personal communication) would now treat as distinct family.
jj - including _Peltanthera_ (Takh.).
kk - including _Nyctanthes_ (Takh.).
ll - _Oftia_ is very isolated within the family (Takh.).
mm - in Theales? (Takh.).
nn - _Acorus_ is very isolated within the family (Takh.).
oo - the proper name for this family is Surianaceae (Cronq.).

A - as treated in Hutchinson (1969).
B - among _Genera anomala_ (B. & H.).
C - as treated in Dalla Torre & Harms.
D - affinities uncertain (Cronq.).
E - segregation as a separate family probably justified (Melchior, 1964).
F - among _Genera dubia_ (B. & H.).

A COMPARISON OF ANGIOSPERM CLASSIFICATION SYSTEMS

CRONQUIST	TAKHTAJAN	THORNE	ENGLER	HUTCHINSON	BENTHAM & HOOKER
CLASS A: MAGNOLIATAE[a] (Dicotyledonae)					
Subclass I: MAGNOLIIDAE					
Order 1: MAGNOLIALES					
1 Austrobaileyaceae	9 Laurales	12 Annonales	50 Magnoliales	13 Laurales	--
2 Lactoridaceae	16 Laurales	18 Annonales	72 Piperales	7 Magnoliales	in Piperaceae
3 Magnoliaceae	1 Magnoliales	4 Annonales	40 Magnoliales	1 Magnoliales	4 Ranales
4 Winteraceae	8 Magnoliales	1 Annonales	43 Magnoliales	3 Magnoliales	in Magnoliaceae
5 Degeneriaceae	2 Magnoliales	5 Annonales	41 Magnoliales	in Winteraceae	--
6 Himantandraceae	3 Magnoliales	7 Annonales	42 Magnoliales	6 Magnoliales	--
7 Annonaceae	5 Magnoliales	8 Annonales	44 Magnoliales	10 Annonales	5 Ranales
8 Myristicaceae	7 Magnoliales	9 Annonales	46 Magnoliales	18 Laurales	143 Micrembryae
9 Canellaceae (synonym: Winteranaceae)	6 Magnoliales	10 Annonales	47 Magnoliales	4 Magnoliales	16 Parietales
10 Illiciaceae	29 Illiciales	2 Annonales	49 Magnoliales	2 Magnoliales	in Magnoliaceae
11 Schisandraceae	30 Illiciales	3 Annonales	48 Magnoliales	5 Magnoliales	in Magnoliaceae
12 Eupomatiaceae	4 Magnoliales	6 Annonales	45 Magnoliales	11 Annonales	in Annonaceae
13 Amborellaceae	10 Laurales	14 Annonales	52 Magnoliales	in Monimiaceae	in Monimiaceae
14 Trimeniaceae	11 Laurales	15 Annonales	51 Magnoliales	14 Laurales	in Monimiaceae?
15 Monimiaceae	12 Laurales	16 Annonales	53 Magnoliales	12 Laurales	144 Micrembryae
[a Atherospermataceae (AS+), b Siparunaceae, c Hortoniaceae: in Monimiaceae]					
16 Gomortegaceae	13 Laurales	19 Annonales	55 Magnoliales	16 Laurales	in Euphorbiaceae?
17 Calycanthaceae	17 Laurales	17 Annonales	54 Magnoliales	26 Rosales	3 Ranales
18 Lauraceae	18 Laurales	20 Annonales	56 Magnoliales	15 Laurales	145 Daphnales
[a Cassythaceae: in Lauraceae]					
19 Hernandiaceae	14 Laurales	21 Annonales	57 Magnoliales	17 Laurales	in Lauraceae
a in Hernandiaceae	19 Gyrocarpaceae (Laurales)	22 Annonales	in Hernandiaceae	in Hernandiaceae	in Combretaceae
Order 2: PIPERALES					
20 Chloranthaceae	15 Laurales	13 Annonales	71 Piperales	266 Piperales	142 Micrembryae
21 Saururaceae	20 Piperales	24 Annonales	69 Piperales	265 Piperales	in Piperaceae
22 Piperaceae	21 Piperales	23 Annonales	70 Piperales	264 Piperales	141 Micrembryae
[a Peperomiaceae (AS+): in Piperaceae]					
Order 3: ARISTOLOCHIALES					
23 Aristolochiaceae	22 Aristolochiales	11 Annonales	73 Aristolochiales	260 Aristolochiales	140 Multiovul. terr.
Order 4: NYMPHAEALES					
24 Nymphaeaceae	26 Nymphaeales	31 Nymphaeales	67 Ranunculales	250 Ranales	8 Ranales

CRONQUIST	TAKHTAJAN	THORNE	ENGLER	HUTCHINSON	BENTHAM & HOOKER
a in Nymphaeaceae	25 Cabombaceae (Nymphaeales)	in Nymphaeaceae[ii]	in Nymphaeaceae	253 Ranales	in Nymphaeaceae
b in Nymphaeaceae	27 Barclayaceae (Nymphaeales)	in Nymphaeaceae	in Nymphaeaceae	in Nymphaeaceae	in Nymphaeaceae
[c Euryalaceae (AS+): in Nymphaeaceae]					
[d Nupharaceae: in Nymphaeaceae]					
25 Nelumbonaceae	31 Nelumbonales	in Nymphaeaceae[ii]	in Nymphaeaceae	in Nymphaeaceae	in Nymphaeaceae
26 Ceratophyllaceae	28 Nymphaeales	32 Nymphaeales	68 Ranunculales	252 Ranales	165 "Anomalous fam."
Order 5: RANUNCULALES					
27 Ranunculaceae	35 Ranunculales	28 Berberidales	62 Ranunculales	249 Ranales	1 Ranales
a in Ranunculaceae	36 Glaucidiaceae (Ranunculales)	in Ranunculaceae	in Ranunculaceae	in Helleboraceae	in Ranunculaceae
b in Ranunculaceae	37 Hydrastidaceae (Ranunculales)	in Ranunculaceae	in Ranunculaceae	in Helleboraceae	in Ranunculaceae
c in Ranunculaceae	39 Podophyllaceae (Ranunculales)	in Berberidaceae	in Berberidaceae	251 Ranales	in Berberidaceae
d in Ranunculaceae	in Ranunculaceae	in Ranunculaceae	in Ranunculaceae	248 Helleboraceae (Ranales)	in Ranunculaceae
28 Circaeasteraceae	38 Ranunculales	in Ranunculaceae[ii]	in Ranunculaceae	258 Berberidales	in Chloranthaceae
[a Kingdoniaceae (AS+): in Circaeasteraceae (Cronq.); in Ranunculaceae? (Takh.)]					
29 Berberidaceae	41 Ranunculales	29 Berberidales	63 Ranunculales	259 Berberidales	7 Ranales
a in Berberidaceae	40 Nandinaceae (Ranunculales)	in Berberidaceae	in Berberidaceae	257 Berberidales	in Berberidaceae
[b Leonticaceae (AS+): in Berberidaceae]					
30 Lardizabalaceae	32 Ranunculales	25 Berberidales	65 Ranunculales	255 Berberidales	in Berberidaceae
a in Lardizabalac.	33 Sargentodoxaceae (Ranunculales)	26 Berberidales	64 Ranunculales	254 Berberidales	--
31 Menispermaceae	34 Ranunculales	27 Berberidales	66 Ranunculales	256 Berberidales	6 Ranales
32 Coriariaceae	241 Rutales?	209 Rosales	144 Sapindales	23 Coriariales	56 "Anomalous fam."
33 Corynocarpaceae	296 Celastrales	208 Rosales	158 Celastrales	181 Celastrales	in Anacardiaceae
34 Sabiaceae	250 Sapindales?	155 Rutales	150 Sapindales	217 Sapindales	54 Sapindales
[a Meliosmaceae (AS+): in Sabiaceae]					
Order 6: PAPAVERALES					
35 Papaveraceae	42 Papaverales	30 Berberidales	95 Papaverales	267 Rhoeadales	10 Parietales
[a Pteridophyllaceae (AS+): in Papaveraceae]					
36 Fumariaceae	44 Papaverales	in Papaveraceae	in Papaveraceae	268 Rhoeadales	in Papaveraceae
a in Fumariaceae	43 Hypecoaceae (Papaverales)	in Papaveraceae	in Papaveraceae	in Fumariaceae	in Papaveraceae

[Takhtajan's orders Illiciales, Nelumbonales, Ranunculales, Papaverales and Sarraceniales constitute his subclass Ranunculidae]

CRONQUIST	TAKHTAJAN	THORNE	ENGLER	HUTCHINSON	BENTHAM & HOOKER
Subclass II: HAMAMELIDAE					
Order 7: TROCHODENDRALES					
37 Tetracentraceae	47 Trochodendrales	180 Hamamelidales	58 Magnoliales	49 Hamamelidales	--
38 Trochodendraceae	46 Trochodendrales	179 Hamamelidales	59 Magnoliales	8 Magnoliales	in Magnoliaceae
Order 8: HAMAMELIDALES					
39 Cercidiphyllaceae	48 Cercidiphyllales	182 Hamamelidales	61 Magnoliales	9 Magnoliales	--
40 Eupteleaceae	49 Eupteleales	181 Hamamelidales	60 Magnoliales	in Trochodendraceae[A]	in Magnoliaceae
41 Platanaceae	52 Hamamelidales	185 Hamamelidales	102 Rosales	52 Hamamelidales	156 Unisexuales
42 Didymelaceae	50 Didymelales	107 Euphorbiales	6 Leitneriales	--	near Leitneria?
43 Hamamelidaceae	51 Hamamelidales	184 Hamamelidales	103 Rosales	50 Hamamelidales	64 Rosales
[a Altingiaceae (AS+): in Hamamelidaceae]					
[b Disanthaceae: in Hamamelidaceae]					
[c Rhodoleiaceae (AS+): in Hamamelidaceae]					
44 Myrothamnaceae	53 Hamamelidales	218 Pittosporales	104 Rosales	51 Hamamelidales	in Hamamelidaceae
Order 9: EUCOMMIALES					
45 Eucommiaceae	54 Eucommiales	183 Hamamelidales	12 Urticales	73 Urticales	--
Order 10: URTICALES					
46 Ulmaceae	55 Urticales	99 Urticales	11 Urticales	68 Urticales	in Urticaceae
[a Celtidaceae: in Ulmaceae]					
47 Barbeyaceae	59 Barbeyales	Urticales[ii]	in Ulmaceae	72 Urticales	--
48 Moraceae	56 Urticales	100 Urticales	13 Urticales	70 Urticales	155 Unisexuales (48–50)
49 Cannabaceae	57 Urticales	in Moraceae	in Moraceae	69 Urticales	
(or Cannabiaceae, Cannabinaceae, Cannabidaceae)					
50 Urticaceae	58 Urticales	101 Urticales	14 Urticales	71 Urticales	
Order 11: LEITNERIALES					
51 Leitneriaceae	67 Leitneriales	164 Leitneriales	5 Leitneriales	58 Leitneriales	157 Unisexuales
Order 12: JUGLANDALES					
52 Rhoipteleaceae	65 Juglandales	161 Rutales	10 Urticales	64 Juglandales	--
53 Picrodendraceae[b]	174 Euphorbiales	in Euphorbiaceae	135 Rutales	66 Juglandales	in Simaroubaceae?
54 Juglandaceae	66 Juglandales	162 Rutales	3 Juglandales	65 Juglandales	158 Unisexuales
Order 13: MYRICALES					
55 Myricaceae	64 Myricales	163 Myricales	2 Juglandales	59 Myricales	159 Unisexuales
Order 14: FAGALES					
56 Balanopaceae	63 Balanopales	189 Balanopales	4 Balanopales	60 Balanopsidales	154 Unisexuales
57 Fagaceae	61 Fagales	187 Fagales	9 Fagales	62 Fagales	161 Unisexuales (as Cupuliferae) (57–58a)
58 Betulaceae	62 Betulales	188 Fagales	8 Fagales	61 Fagales	
a in Betulaceae	in Betulaceae	(in Betulaceae)	in Betulaceae	63 Corylaceae (Fagales)	

CRONQUIST	TAKHTAJAN	THORNE	ENGLER	HUTCHINSON	BENTHAM & HOOKER
[b Carpinaceae (AS+): in Betulaceae]					
Order 15: CASUARINALES					
59 Casuarinaceae	60 Casuarinales	186 Casuarinales	1 Casuarinales	67 Casuarinales	160 Unisexuales
Subclass III: CARYOPHYLLIDAE					
Order 16: CARYOPHYLLALES					
60 Phytolaccaceae	68 Caryophyllales[c]	165 Chenopodiales	26 Centrospermae	279 Chenopodiales	134 Curvembryae
a in Phytolaccaceae	in Phytolaccaceae	in Phytolaccaceae	in Phytolaccaceae	278 Barbeuiaceae (Chenopodiales)	in Phytolaccaceae?
b in Phytolaccaceae	in Phytolaccaceae	in Phytolaccaceae	in Phytolaccaceae	281 Agdestidaceae (Chenopodiales)	in Phytolaccaceae[B]
c in Phytolaccaceae	in Phytolaccaceae	in Phytolaccaceae	in Phytolaccaceae	282 Petiveriaceae (Chenopodiales)	in Phytolaccaceae
d in Phytolaccaceae	in Phytolaccaceae	in Phytolaccaceae	in Phytolaccaceae	90 Stegnospermaceae (Pittosporales)	in Phytolaccaceae[B]
e in Phytolaccaceae	in Phytolaccaceae	in Phytolaccaceae	28 Achatocarpaceae (Centrospermae)	79 Bixales	in Amaranthaceae
f in Phytolaccaceae	59 Gyrostemonaceae (Caryophyllales)	166 Chenopodiales	27 Centrospermae	280 Chenopodiales	in Phytolaccaceae
61 Nyctaginaceae	71 Caryophyllales	167 Chenopodiales	29 Centrospermae	86 Thymelaeales	130 Curvembryae
62 Didiereaceae	78 Caryophyllales	170 Chenopodiales	38 Centrospermae	224 Sapindales	--
63 Cactaceae	75 Caryophyllales	169 Chenopodiales	39 Cactales	113 Cactales	80 Ficoidales
64 Aizoaceae (or Ficoidaceae, Mesembryanthemaceae)	73 Caryophyllales	168 Chenopodiales	31 Centrospermae	274 Caryophyllales	81 Ficoidales
a in Aizoaceae	74 Tetragoniaceae (Caryophyllales)	in Aizoaceae	in Aizoaceae	in Aizoaceae	in Aizoaceae
65 Molluginaceae	72 Caryophyllales	in Aizoaceae	30 Centrospermae	272 Caryophyllales	in Aizoaceae
66 Caryophyllaceae	81 Caryophyllales	176 Chenopodiales	34 Centrospermae	273 Caryophyllales	23 Caryophyllinae
a --	80 Hectorellaceae[d] (Caryophyllales)	in Caryophyllaceae	in Caryophyllaceae?	in Cruciferae[A]	in Portulacaceae & Caryophyllaceae[e]
b in Caryophyllac.	in Caryophyllaceae	in Caryophyllaceae	in Caryophyllaceae	277 Illecebraceae (Polygonales)	131 Curvembryae
[c Alsinaceae: in Caryophyllaceae]					
67 Portulacaceae	76 Caryophyllales	171 Chenopodiales	32 Centrospermae	275 Caryophyllales	24 Caryophyllinae
68 Basellaceae	77 Caryophyllales	172 Chenopodiales	33 Centrospermae	287 Chenopodiales	in Chenopodiaceae
69 Chenopodiaceae	83 Caryophyllales	173 Chenopodiales	36 Centrospermae	283 Chenopodiales	133 Curvembryae
a in Chenopodiaceae	79 Halophytaceae (Caryophyllales)	174 Chenopodiales	in Chenopodiaceae	in Chenopodiaceae	--
b in Chenopodiaceae	in Chenopodiaceae?	in Chenopodiaceae	35 Dysphaniaceae (Centrospermae)	in Chenopodiaceae	in Illecebraceae
70 Amaranthaceae	82 Caryophyllales	175 Chenopodiales	37 Centrospermae	284 Chenopodiales	132 Curvembryae

CRONQUIST	TAKHTAJAN	THORNE	ENGLER	HUTCHINSON	BENTHAM & HOOKER
Order 17: BATALES					
71 Bataceae	70 Caryophyllales	178 Batidales	101 Batales	286 Chenopodiales	135 Curvembryae
Order 18: POLYGONALES					
72 Polygonaceae	84 Polygonales	177 Chenopodiales	25 Polygonales	276 Polygonales	136 Curvembryae
Order 19: PLUMBAGINALES					
73 Plumbaginaceae	85 Plumbaginales	67 Plumbaginales	237 Plumbaginales	296 Primulales	100 Primulales
Subclass IV: DILLENIIDAE					
Order 20: DILLENIALES					
74 Dilleniaceae	87 Dilleniales	34 Theales	76 Guttiferales	19 Dilleniales	2 Ranales
75 Paeoniaceae	89 Paeoniales	35 Theales	77 Guttiferales	247 Ranales	in Ranunculaceae
76 Crossosomataceae	88 Dilleniales	191 Rosales	78 Guttiferales	21 Dilleniales	in Dilleniaceae?
Order 21: THEALES					
77 Ochnaceae	90 Theales	45 Theales	82 Guttiferales	145 Ochnales	42 Geraniales[f]
a in Ochnaceae	91 Lophiraceae (Theales)	in Ochnaceae	in Ochnaceae	in Ochnaceae	in Dipterocarpaceae
b in Ochnaceae	93 Strasburgeriaceae (Theales)	49 Theales	84 Guttiferales	144 Ochnales	--
[c Wallaceaceae: in Ochnaceae]					
78 Rhopalocarpaceae (or: Sphaerosepalaceae)	163 Malvales	94 Malvales	194 Violales	147 Ochnales	--
79 Sarcolaenaceae (or Chlaenaceae, Rhodolaenaceae)	162 Malvales	48 Theales	172 Malvales	146 Ochnales	31 Guttiferales
80 Dipterocarpaceae	92 Theales	50 Theales	85 Guttiferales	148 Ochnales	30 Guttiferales
81 Caryocaraceae	101 Theales	41 Theales	87 Guttiferales	142 Theales	in Theaceae
82 Theaceae	97 Theales	38 Theales	86 Guttiferales	135 Theales	29 Guttiferales
[a Sladeniaceae (AS+): in Theaceae]					
[b Ternstroemiaceae (AS+): in Theaceae]					
[c Camelliaceae (AS+): in Theaceae]					
d in Theaceae	99 Pentaphylaceae (Theales)	44 Theales	156 Celastrales	139 Theales	in Theaceae
e in Theaceae	100 Tetrameristaceae (Theales)	in Theaceae	in Theaceae	140 Theales	in Ochnaceae
f in Theaceae	102 Asteropeiaceae (Theales)	in Theaceae	in Theaceae	in Theaceae[A]	in Flacourtiaceae[B]
g in Theaceae	103 Pelliceraceae (Theales)	in Theaceae	in Theaceae	138 Theales	in Theaceae
h in Theaceae	107 Bonnetiaceae (Theales)	in Theaceae	in Theaceae	134 Theales	in Theaceae
83 Stachyuraceae	113 Violales	37 Theales	186 Violales	53 Hamamelidales	in Theaceae

CRONQUIST	TAKHTAJAN	THORNE	ENGLER	HUTCHINSON	BENTHAM & HOOKER
84 Actinidiaceae	141 Ericales	36 Theales	81 Guttiferales	137 Theales	in Theaceae
a in Actinidiaceae	140 Saurauiaceae (Ericales)	in Actinidiaceae	in Actinidiaceae	136 Theales	in Theaceae
85 Marcgraviaceae	98 Theales	40 Theales	88 Guttiferales	141 Theales	in Theaceae
86 Quiinaceae	104 Theales	46 Theales	89 Guttiferales	161 Guttiferales	in Guttiferae
87 Elatinaceae	110 Theales?	53 Theales	198 Violales	271 Caryophyllales	26 Guttiferales
88 Medusagynaceae	105 Theales?	206 Rosales	80 Guttiferales	143 Theales	--
a --	106 Oncothecaceae?[mm]	--	?[c]	in Ebenaceae?	--
89 Guttiferae (or: Clusiaceae)	108 Theales	52 Theales	90 Guttiferales	159 Guttiferales	28 Guttiferales
a in Guttiferae	109 Hypericaceae (Theales)	in Guttiferae	in Guttiferae	158 Guttiferales	27 Guttiferales
Order 22: MALVALES					
90 Elaeocarpaceae	159 Malvales	95 Malvales	171 Malvales	in Tiliaceae	in Tiliaceae
91 Scytopetalaceae	161 Malvales	47 Theales	177 Malvales	115 Tiliales	in Olacaceae[g]
92 Tiliaceae	160 Malvales	96 Malvales	173 Malvales	116 Tiliales	34 Malvales
93 Sterculiaceae (or: Byttneriaceae)	164 Malvales	93 Malvales	176 Malvales	117 Tiliales	33 Malvales
94 Bombacaceae	165 Malvales	97 Malvales	175 Malvales	119 Tiliales	in Malvaceae
95 Malvaceae	166 Malvales	98 Malvales	174 Malvales	120 Malvales	32 Malvales
a [h]	167 Huaceae[i] (?Malvales)	--	in Styracaceae?	126 Malpighiales	--
Order 23: LECYTHIDALES					
96 Lecythidaceae	220 Myrtales	54 Theales	211 Myrtiflorae	163 Myrtales	in Myrtaceae
[a Asteranthaceae (AS+): in Lecythidaceae; 163a Myrtales? (Hutch.)[A]]					
[b Barringtoniaceae (AS+): in Lecythidaceae; 162a Myrtales? (Hutch.)[A]]					
[c Foetidiaceae (AS+): in Lecythidaceae]					
[d Napoleonaceae (AS+): in Lecythidaceae]					
Order 24: SARRACENIALES					
97 Sarraceniaceae	45 Sarraceniales	33 Sarraceniales	92 Sarraceniales	308 Sarraceniales	9 Parietales
98 Nepenthaceae	212 Nepenthales	55 Nepenthales	93 Sarraceniales	263 Aristolochiales	138 Multiovul. terr.
99 Droseraceae	211 Nepenthales	198 Rosales	94 Sarraceniales	307 Sarraceniales	63 Rosales
Order 25: VIOLALES					
100 Flacourtiaceae	111 Violales	70 Cistales	183 Violales	76 Bixales	{17 Parietales[j] / 73 Passiflorales[k]}
[a Homaliaceae, b Samydaceae, c Neumanniaceae (AS+): in Flacourtiaceae]					
d in Flacourtiaceae	112 Lacistemataceae (Violales)	in Flacourtiaceae	in Flacourtiaceae	80 Bixales[l]	163 "Anomalous fam."
101 Peridiscaceae	114 Violales	72 Cistales	184 Violales	118 Tiliales[n]	in Bixineae
102 Scyphostegiaceae	119 Violales	73 Cistales	187 Violales	187 Celastrales	--
103 Violaceae	115 Violales	74 Cistales	185 Violales	99 Violales	15 Parietales
104 Turneraceae	121 Passiflorales	77 Cistales	188 Violales	104 Loasales	75 Passiflorales

CRONQUIST	TAKHTAJAN	THORNE	ENGLER	HUTCHINSON	BENTHAM & HOOKER
105 Passifloraceae	120 Passiflorales[m]	79 Cistales	190 Violales	107 Passiflorales	76 Passiflorales
106 Malesherbiaceae	122 Passiflorales	78 Cistales	189 Violales	106 Passiflorales	in Passifloraceae
107 Bixaceae	116 Violales	75 Cistales	193 Violales	74 Bixales	17 Parietales[o]
a in Bixaceae	117 Cochlospermaceae (Violales)	in Bixaceae[ii]	195 Violales	77 Bixales	in Bixineae
108 Cistaceae	118 Violales	76 Cistales	192 Violales	75 Bixales	14 Parietales
109 Tamaricaceae	136 Tamaricales	87 Tamaricales	196 Violales	97 Tamaricales	25 Caryophyllinae
110 Frankeniaceae	138 Tamaricales	88 Tamaricales	197 Violales	96 Tamaricales	22 Caryophyllinae
111 Dioncophyllaceae	95 Theales	(in Flacourtiaceae)	83 Guttiferales	in Flacourtiaceae	--
a --	96 Diegodendraceae (?Theales)	--	--	145a Ochnales[A]	--
112 Ancistrocladaceae	94 Theales	134 Geraniales	91 Guttiferales	149 Ochnales	in Dipterocarpaceae
113 Fouquieriaceae	137 Tamaricales	114 Solanales	254 Tubiflorae	98 Tamaricales	in Tamaricaceae
114 Hoplestigmatac.[D]	338 Polemoniales	263 Lamiales	244 Ebenales	78 Bixales	--
115 Achariaceae	123 Passiflorales	80 Cistales	191 Violales	108 Passiflorales	in Passifloraceae
116 Caricaceae	124 Passiflorales	81 Cistales	199 Violales	112 Cucurbitales	in Passifloraceae
117 Loasaceae	339 Polemoniales	85 Cistales	200 Violales	105 Loasales	74 Passiflorales
118 Begoniaceae	127 Begoniales	83 Cistales	202 Violales	110 Cucurbitales	78 Passiflorales
119 Datiscaceae	126 Begoniales	84 Cistales	201 Violales	111 Cucurbitales	79 Passiflorales
[a Tetramelaceae (AS+): in Datiscaceae]					
120 Cucurbitaceae	125 Cucurbitales	82 Cistales	203 Cucurbitales	109 Cucurbitales	77 Passiflorales

Order 26: SALICALES

121 Salicaceae	139 Salicales	86 Salicales	7 Salicales	57 Salicales	162 "Anomalous fam."
[a Populaceae: in Salicaceae]					

Order 27: CAPPARALES

122 Tovariaceae	131 Capparales	in Capparaceae	98 Papaverales	95 Capparidales	in Capparaceae
123 Capparaceae (or: Capparidaceae)	128 Capparales[p]	89 Capparidales	96 Papaverales	93 Capparidales	12 Parietales
a in Capparaceae	in Capparaceae	in Capparaceae	in Capparaceae	Oxystylidaceae[q]	in Capparaceae
b in Capparaceae	in Capparaceae	in Capparaceae	in Capparaceae	Cleomaceae[A] (Cruciales)	in Capparaceae
c in Capparaceae	129 Koeberliniaceae (Capparales)	in Capparaceae	in Capparaceae	172 Celastrales	in Simaroubaceae
[d Canotiaceae (AS+): in Koeberliniaceae (Capparaceae)]					
e in Capparaceae	130 Pentadiplandrac. (Capparales)	in Capparaceae	in Capparaceae	179 Celastrales	--
f (in Capparaceae)	135 Emblingiaceae (Capparales)?	in Capparaceae	in Capparaceae	in Flacourtiaceae[A]	in Capparaceae
124 Cruciferae (or: Brassicaceae)	133 Capparales	92 Capparidales	97 Papaverales	269 Cruciales	11 Parietales
125 Resedaceae	134 Capparales	91 Capparidales	99 Papaverales	270 Resedales	13 Parietales
126 Moringaceae	132 Capparales	90 Capparidales	100 Papaverales	94 Capparidales	57 "Anomalous fam."

CRONQUIST	TAKHTAJAN	THORNE	ENGLER	HUTCHINSON	BENTHAM & HOOKER
Order 28: ERICALES					
127 Cyrillaceae	147 Ericales	43 Theales	155 Celastrales	175 Celastrales	48 Olacales
128 Clethraceae	142 Ericales	42 Theales	229 Ericales	150 Ericales	in Ericaceae[B]
129 Ericaceae	143 Ericales	56 Ericales	231 Ericales	152 Ericales	94 Ericales
a in Ericaceae	in Ericaceae	in Ericaceae	in Ericaceae	157 Vacciniaceae (Ericales)	95 Ericales
b in Ericaceae & Epacridaceae	in Ericaceae & Epacridaceae	in Ericaceae & Epacridaceae	in Ericaceae & Epacridaceae	Prionotaceae[r] (Ericales)	in Ericaceae & Epacridaceae
130 Epacridaceae	144 Ericales	57 Ericales	233 Ericales	153 Ericales	97 Ericales
131 Empetraceae	148 Ericales	58 Ericales	232 Ericales	177 Celastrales	164 "Anomalous fam."
132 Pyrolaceae	145 Ericales	in Ericaceae	230 Ericales	151 Ericales	in Ericaceae
133 Monotropaceae	146 Ericales	in Ericaceae	in Pyrolaceae	155 Ericales	96 Ericales
Order 29: DIAPENSIALES					
134 Diapensiaceae	150 Diapensiales	201 Rosales	228 Diapensiales	154 Ericales	98 Ericales
Order 30: EBENALES					
135 Sapotaceae	155 Ebenales	60 Ebenales	238 Ebenales	207 Ebenales	103 Ebenales
a in Sapotaceae	in Sapotaceae?	in Sapotaceae	239 Ebenales	208 Sarcospermaceae (Ebenales)	in Sapotaceae
[b Boerlagellaceae (AS+): in or near Sapotaceae?]					
136 Ebenaceae	154 Ebenales	59 Ebenales	240 Ebenales	206 Ebenales	104 Ebenales
137 Styracaceae	151 Ebenales	63 Ebenales	241 Ebenales	41 Styracales	105 Ebenales
138 Lissocarpaceae	153 Ebenales	62 Ebenales	242 Ebenales	40 Styracales	in Styracaceae
139 Symplocaceae	152 Ebenales	61 Ebenales	243 Ebenales	42 Styracales	in Styracaceae
Order 31: PRIMULALES					
140 Theophrastaceae	157 Primulales	65 Primulales	234 Primulales	204 Myrsinales	in Myrsinaceae
141 Myrsinaceae	156 Primulales	64 Primulales	235 Primulales	203 Myrsinales	102 Primulales
a in Myrsinaceae	in Myrsinaceae[t]	in Myrsinaceae	in Myrsinaceae	205 Aegicerataceae (Myrsinales)	in Myrsinaceae
142 Primulaceae	158 Primulales	66 Primulales	236 Primulales	295 Primulales	101 Primulales
[a Coridaceae (AS+): in Primulaceae]					
Subclass V: ROSIDAE					
Order 32: ROSALES					
143 Eucryphiaceae	178 Saxifragales	205 Rosales	79 Guttiferales	160 Guttiferales	in Rosaceae
144 Cunoniaceae	176 Saxifragales	202 Rosales	109 Rosales	31 Cunoniales	in Saxifragaceae
a in Cunoniaceae	181 Brunelliaceae (Saxifragales)	203 Rosales	108 Rosales	22 Dilleniales	in Simaroubaceae
b in Cunoniaceae	in Cunoniaceae	in Cunoniaceae	in Saxifragaceae[E]	38 Baueraceae (Cunoniales)	in Saxifragaceae[B]
145 Davidsoniaceae	177 Saxifragales	204 Rosales	110 Rosales	in Cunoniaceae	--
146 Pittosporaceae	193 Saxifragales	211 Pittosporales	111 Rosales	88 Pittosporales	18 Polygalinae

CRONQUIST	TAKHTAJAN	THORNE	ENGLER	HUTCHINSON	BENTHAM & HOOKER
147 Byblidaceae	194 Saxifragales	212 Pittosporales	112 Rosales	89 Pittosporales	in Droseraceae
a in Byblidaceae	192 Roridulaceae (Saxifragales)	214 Pittosporales	113 Rosales	in Byblidaceae	in Droseraceae
148 Columelliaceae	350 Scrophulariales	in Saxifragaceae	274 Tubiflorae	329 Personales	120 Personales
149 Hydrangeaceae	190 Saxifragales	in Saxifragaceae	in Saxifragaceae[E]	33 Cunoniales	in Saxifragaceae
a in Hydrangeaceae	in Hydrangeaceae	in Saxifragaceae	in Saxifragaceae[E]	32 Philadelphaceae (Cunoniales)	in Saxifragaceae
150 Grossulariaceae	189 Saxifragales	in Saxifragaceae	in Saxifragaceae[E]	34 Cunoniales	in Saxifragaceae
a in Grossulariac.	182 Escalloniaceae[V] (Saxifragales)	in Saxifragaceae	in Saxifragaceae[E]	37 Cunoniales	in Saxifragaceae
b in Grossulariac.	184 Tetracarpaeaceae (Saxifragales)	in Saxifragaceae	in Saxifragaceae[E]	in Escalloniaceae	in Saxifragaceae
c in Grossulariac.	185 Iteaceae (Saxifragales)	in Saxifragaceae	in Saxifragaceae[E]	in Escalloniaceae	in Saxifragaceae
d in Grossulariac.	186 Brexiaceae (Saxifragales)	in Saxifragaceae	in Saxifragaceae[E]	in Escalloniaceae	in Saxifragaceae
e in Grossulariac.	187 Phyllonomaceae (or: Dulongiac.) (?Saxifragales)	in Saxifragaceae	in Saxifragaceae[E]	in Escalloniaceae	in Saxifragaceae
f in Grossulariac.	188 Pterostemonaceae (?Saxifragales)	in Saxifragaceae	in Saxifragaceae[E]	30 Cunoniales	in Rosaceae
g in Grossulariac.	191 Montiniaceae (Saxifragales)	in Saxifragaceae	in Saxifragaceae[E]	in Escalloniaceae	in Onagraceae[B]
151 Bruniaceae	195 Saxifragales	215 Pittosporales	114 Rosales	56 Hamamelidales	65 Rosales
152 Alseuosmiaceae	in Caprifoliaceae?	(in Caprifoliaceae)	in Caprifoliaceae?	in Caprifoliaceae[A]	in Caprifoliaceae
153 Crassulaceae	197 Saxifragales	194 Rosales	105 Rosales	298 Saxifragales	62 Rosales
154 Cephalotaceae	198 Saxifragales	195 Rosales	106 Rosales	299 Saxifragales	in Saxifragaceae[B]
155 Saxifragaceae	199 Saxifragales	196 Rosales	107 Rosales	300 Saxifragales	61 Rosales
a in Saxifragaceae	196 Penthoraceae (Saxifragales)	in Saxifragaceae	in Saxifragaceae[E]	in Crassulaceae[A]	in Crassulaceae
b in Saxifragaceae	200 Vahliaceae (Saxifragales)	in Saxifragaceae	in Saxifragaceae[E]	302 Saxifragales	in Saxifragaceae
c in Saxifragaceae	201 Francoaceae (Saxifragales)	in Saxifragaceae	in Saxifragaceae[E]	303 Saxifragales	in Saxifragaceae
d in Saxifragaceae	202 Eremosynaceae (Saxifragales)	in Saxifragaceae	in Saxifragaceae[E]	301 Saxifragales	in Saxifragaceae
e in Saxifragaceae	203 Parnassiaceae (Saxifragales)	in Saxifragaceae	in Saxifragaceae[E]	305 Saxifragales	in Saxifragaceae
f (in Saxifragac.)?[u]	179 Paracryphiaceae (? Saxifragales)	--	--	--	--
g (in Saxifragaceae)	183 Tribelaceae (Saxifragales)	(in Saxifragaceae)	in Saxifragaceae	in Escalloniaceae[A]	in Pittosporaceae[F]
h (in Saxifragaceae)	Lepuropetalaceae ? in Parnassiac.	in Saxifragaceae	in Saxifragaceae	in Saxifragaceae	in Saxifragaceae

CRONQUIST	TAKHTAJAN	THORNE	ENGLER	HUTCHINSON	BENTHAM & HOOKER
156 Rosaceae	204 Rosales	190 Rosales	115 Rosales	24 Rosales	60 Rosales
[a Malaceae, Pomaceae: in Rosaceae]					
[b Prunaceae, Amygdalaceae, Drupaceae: in Rosaceae]					
157 Neuradaceae	206 Rosales	in Rosaceae	116 Rosales	in Rosaceae	in Rosaceae
158 Chrysobalanaceae	205 Rosales[w]	in Rosaceae	117 Rosales	in Rosaceae	in Rosaceae
159 Leguminosae	207 Mimosaceae (Fabales) 208 Caesalpiniaceae (Fabales) 209 Fabaceae (Fabales)	193 Rosales	119 Rosales	28 Mimosaceae (Leguminales) 27 Caesalpiniaceae (Leguminales) 29 Papilionaceae (Leguminales)	59 Rosales
Order 33: PODOSTEMALES					
160 Podostemaceae	213 Podostemales	200 Rosales	122 Podostemales	309 Podostemales	137 Multiovulatae aquaticae
[a Tristichaceae (AS+): in Podostemaceae]					
Order 34: HALORAGALES					
161 Haloragaceae (or: Halorrhagidaceae)	227 Hippuridales	252 Cornales	217 Myrtiflorae	291 Lythrales	66 Rosales
[a Myriophyllaceae: in Haloragaceae]					
162 Gunneraceae	223 Hippuridales	in Haloragaceae[ii]	in Haloragaceae	in Haloragaceae	in Haloragaceae
163 Hippuridaceae[x]	229 Hippuridales	253 Cornales	219 Myrtiflorae	in Haloragaceae	in Haloragaceae
164 Theligonaceae[D] (or: Cynocrambaceae)	85 Theligonales	Chenopodiales[ii]	218 Myrtiflorae	285 Chenopodiales	in Urticaceae
Order 35: MYRTALES					
165 Sonneratiaceae	215 Myrtales	in Lythraceae	209 Myrtiflorae	165 Myrtales	in Lythraceae
166 Lythraceae	214 Myrtales	221 Myrtales	204 Myrtiflorae	288 Lythrales	71 Myrtales
167 Penaeaceae	224 Myrtales	227 Myrtales	179 Thymelaeales	84 Thymelaeales	148 Daphnales
168 Crypteroniaceae	180 Saxifragales	223 Myrtales	206 Myrtiflorae	39 Cunoniales[y]	in Lythraceae
169 Thymelaeaceae	175 Thymelaeales	109 Euphorbiales	181 Thymelaeales	85 Thymelaeales	147 Daphnales
a in Thymelaeaceae	in Thymelaeaceae	in Thymelaeaceae	in Thymelaeaceae	82 Aquilariaceae (Thymelaeales)	in Thymelaeaceae
b in Thymelaeaceae	in Thymelaeaceae	in Thymelaeaceae	in Thymelaeaceae	81 Gonystylaceae (Thymelaeales)	in Thymelaeaceae[B]
170 Trapaceae (or: Hydrocaryaceae)	226 Myrtales	224 Myrtales	205 Myrtiflorae	290 Lythrales[aa]	in Onagraceae
171 Dialypetalanthac.	330 Gentianales	--	208 Myrtiflorae	236 Rubiales	--
172 Myrtaceae	221 Myrtales	228 Myrtales	207 Myrtiflorae	162 Myrtales	69 Myrtales
a in Myrtaceae	in Myrtaceae	(in Myrtaceae)	in Myrtaceae	199 Heteropyxidaceae (Rhamnales)	in Lythraceae[B]
[b Psiloxylaceae (AS+): in or near Myrtaceae]					

CRONQUIST	TAKHTAJAN	THORNE	ENGLER	HUTCHINSON	BENTHAM & HOOKER
173 Punicaceae	216 Myrtales	222 Myrtales	210 Myrtiflorae	166 Lythrales[z]	in Lythraceae[B]
174 Onagraceae	225 Myrtales	230 Myrtales	215 Myrtiflorae	289 Lythrales[aa]	72 Myrtales
175 Oliniaceae	223 Myrtales	226 Myrtales	216 Myrtiflorae	35 Cunoniales	in Lythraceae[B]
176 Melastomataceae	222 Myrtales	229 Myrtales	212 Myrtiflorae	168 Myrtales	70 Myrtales
[a Memecylaceae (AS+): in Melastomataceae]					
177 Combretaceae	219 Myrtales	225 Myrtales	214 Myrtiflorae	167 Myrtales	68 Myrtales
Order 36: PROTEALES					
178 Elaeagnaceae	315 Elaeagnales	103 Rhamnales	182 Thymelaeales	200 Rhamnales	149 Daphnales
179 Proteaceae	316 Proteales	220 Proteales	15 Proteales	87 Proteales	146 Daphnales
Order 37: CORNALES					
180 Rhizophoraceae	217 Myrtales	246 Cornales	213 Myrtiflorae	164 Myrtales	67 Myrtales
a (in Rhizophorac.)	218 Anisophylleaceae (Myrtales)	(in Rhizophoraceae)	in Rhizophoraceae	in Rhizophoraceae	in Rhizophoraceae
[b Polygonanthaceae: in Anisophylleaceae (Rhizophoraceae) (Takh.)]					
181 Davidiaceae	278 Cornales	in Nyssaceae	223 Umbelliflorae	in Nyssaceae	--
182 Nyssaceae	279 Cornales	248 Cornales	222 Umbelliflorae	46 Araliales	in Cornaceae
183 Alangiaceae	280 Cornales	250 Cornales	221 Umbelliflorae	44 Araliales	in Cornaceae
184 Cornaceae[bb]	276 Cornales[cc]	249 Cornales	224 Umbelliflorae	43 Araliales	84 Umbellales
a in Cornaceae	281 Mastixiaceae (Cornales)	in Cornaceae	in Cornaceae	in Cornaceae	in Cornaceae
b (in Cornaceae)	282 Helwingiaceae (Cornales)	in Cornaceae	in Cornaceae	in Araliaceae	in Araliaceae
c in Cornaceae	283 Toricelliaceae (Cornales)	in Cornaceae	in Cornaceae	in Cornaceae	in Cornaceae
185 Garryaceae	277 Cornales	251 Cornales	225 Umbelliflorae	45 Araliales	in Cornaceae
Order 38: SANTALALES					
186 Medusandraceae[D]	310 Santalales	122 Santalales	24 Medusandrales	193 Olacales	--
187 Dipentodontaceae[D]	309 Santalales	71 Cistales	17 Santalales	192 Olacales	--
188 Olacaceae	302 Santalales	123 Santalales	16 Santalales	188 Olacales	46 Olacales
a in Olacaceae	303 Schoepfiaceae (Santalales)	in Olacaceae	in Olacaceae	in Olacaceae	in Olacaceae
b in Olacaceae	305 Octoknemaceae (Santalales)	(in Olacaceae)	in Olacaceae	190 Olacales	--
c in Olacaceae	306 Erythropalaceae (Santalales)	(in Olacaceae)	in Olacaceae	185 Celastrales	in Olacaceae
d in Olacaceae	in Olacaceae	(in Olacaceae)	in Olacaceae	191 Aptandraceae (Olacales)	in Olacaceae
189 Opiliaceae	304 Santalales	124 Santalales	18 Santalales	189 Olacales	in Olacaceae
190 Grubbiaceae	149 Ericales	217 Pittosporales	19 Santalales	195 Santalales	in Santalaceae
191 Santalaceae	308 Santalales	125 Santalales	20 Santalales	196 Santalales	151 Achlamydosporeae
192 Loranthaceae	312 Santalales	127 Santalales	22 Santalales	194 Santalales	150 Achlamydosporeae
a in Loranthaceae	in Loranthaceae	128 Viscaceae (Santalales)	in Loranthaceae	in Loranthaceae	in Loranthaceae

CRONQUIST	TAKHTAJAN	THORNE	ENGLER	HUTCHINSON	BENTHAM & HOOKER
193 Misodendraceae	311 Santalales	126 Santalales	21 Santalales	197 Santalales	in Santalaceae
194 Balanophoraceae	314 Santalales	129 Santalales	23 Balanophorales	198 Santalales	152 Achlamydosporeae
195 Cynomoriaceae	313 Santalales	130 Santalales	220 Myrtiflorae	in Balanophoraceae	in Balanophoraceae
Order 39: RAFFLESIALES					
196 Mitrastemonaceae	in Rafflesiaceae	in Rafflesiaceae	in Rafflesiaceae	in Rafflesiaceae[A]	--
197 Hydnoraceae	24 Rafflesiales	69 Rafflesiales	75 Aristolochiales	261 Aristolochiales	in Rafflesiaceae
198 Rafflesiaceae (or: Cytinaceae)	23 Rafflesiales	68 Rafflesiales	74 Aristolochiales	262 Aristolochiales	139 Multiovul. terr.
Order 40: CELASTRALES					
199 Geissolomataceae	294 Celastrales	216 Pittosporales	178 Thymelaeales	83 Thymelaeales	in Penaeaceae[B]
200 Hippocrateaceae	291 Celastrales	in Celastraceae	162 Celastrales	184 Celastrales	in Celastraceae
201 Celastraceae	290 Celastrales[dd]	118 Santalales	160 Celastrales	180 Celastrales	49 Celastrales
[a Chingithamnaceae: in Celastraceae (Cronq.)]					
[b Lophopyxidaceae (AS+): in Celastraceae? (Takh.)]					
c in Celastraceae	295 Goupiaceae (Celastrales)	in Celastraceae	in Celastraceae	183 Celastrales	in Celastraceae
202 Siphonodontaceae (or: Capusiaceae)	292 Celastrales	in Celastraceae	in Celastraceae	186 Celastrales	in Celastraceae
203 Stackhousiaceae	293 Celastrales	119 Santalales	163 Celastrales	182 Celastrales	50 Celastrales
204 Salvadoraceae	289 Celastrales	131 Oleales	164 Celastrales	171 Celastrales	107 Gentianales
205 Aquifoliaceae	286 Celastrales	39 Theales	157 Celastrales* *incl. Sphenostemonaceae?	170 Celastrales	47 Olacales
a in Aquifoliaceae	287 Phellineaceae (Celastrales)	(in Aquifoliaceae)	in Aquifoliaceae	in Aquifoliaceae	in Rutaceae?
206 Icacinaceae	288 Celastrales	120 Santalales	166 Celastrales	176 Celastrales	in Olacaceae
[a Irvingbaileyaceae: in Icacinaceae (Takh.)]					
207 Cardiopteridaceae	307 Santalales	121 Santalales	167 Celastrales	174 Celastrales	in Olacaceae
208 Dichapetalaceae (or: Chailletiaceae)	172 Euphorbiales	108 Euphorbiales	180 Thymelaeales	25 Rosales	45 Geraniales
Order 41: EUPHORBIALES					
209 Buxaceae	168 Euphorbiales	110 Euphorbiales	165 Celastrales	54 Hamamelidales	in Euphorbiaceae
a in Buxaceae[D]	169 Simmondsiaceae (Euphorbiales)	(in Buxaceae)?	in Buxaceae	in Buxaceae	in Euphorbiaceae
[b Stylocerataceae (AS+): in Buxaceae]					
210 Euphorbiaceae	171 Euphorbiales	104 Euphorbiales	130 Geraniales	133 Euphorbiales	153 Unisexuales
[a Androstachydaceae (AS−): in Euphorbiaceae]					
[b Bischofiaceae (AS+): in Euphorbiaceae]					
[c Hymenocardiaceae (AS+): in Euphorbiaceae]					
[d Peraceae (AS+): in Euphorbiaceae]					
[e Stilaginaceae (AS+): in Euphorbiaceae (near Icacinaceae? (Airy-Shaw)]					
[f Uapacaceae (AS+): in Euphorbiaceae (near Anacardiaceae?)]					
[g Scepaceae (AS+): in Euphorbiaceae]					

CRONQUIST	TAKHTAJAN	THORNE	ENGLER	HUTCHINSON	BENTHAM & HOOKER
211 Daphniphyllaceae	170 Euphorbiales	210 Pittosporales	131 Geraniales	55 Hamamelidales	in Euphorbiaceae
212 Aextoxicaceae	297 Celastrales?	106 Euphorbiales	152 Sapindales	178 Celastrales	in Euphorbiaceae
213 Pandaceae	173 Euphorbiales	105 Euphorbiales	159 Celastrales	169 Celastrales	--
Order 42: RHAMNALES					
214 Rhamnaceae	298 Rhamnales	102 Rhamnales	168 Rhamnales	201 Rhamnales	51 Celastrales
215 Leeaceae	300 Rhamnales	in Vitaceae	170 Rhamnales	in Vitaceae	in Vitaceae
216 Vitaceae (or Vitidaceae, Ampelidaceae)	299 Rhamnales	247 Cornales	169 Rhamnales	202 Rhamnales	52 Celastrales
Order 43: SAPINDALES					
217 Staphyleaceae	242 Sapindales	207 Rosales	161 Celastrales	221 Sapindales	in Sapindaceae
218 Melianthaceae	248 Sapindales	156 Rutales	151 Sapindales	214 Sapindales	in Sapindaceae
219 Greyiaceae	249 Sapindales	199 Rosales	in Melianthaceae	36 Cunoniales	in Sapindaceae
220 Connaraceae	210 Connarales	192 Rosales	118 Rosales	20 Dilleniales	58 Rosales
221 Stylobasiaceae[oo]	235 Rutales?	(in Rosaceae)	in Chrysobalanaceae[E]	in Rosaceae[A]	in Rosaceae
a in Stylobasiaceae	in Simaroubaceae?	149 Surianaceae (Rutales)	in Simaroubaceae	in Simaroubaceae	in Simaroubaceae
222 Akaniaceae	245 Sapindales	157 Rutales	138 Rutales	222 Sapindales	in Sapindaceae
223 Sapindaceae	244 Sapindales	154 Rutales	148 Sapindales	215 Sapindales	53 Sapindales
[a Dodonaeaceae: in Sapindaceae]					
b in Sapindaceae	247 Bretschneiderac. (Sapindales)	160 Rutales	147 Sapindales	in Sapindaceae	--
224 Hippocastanaceae	246 Sapindales	159 Rutales	149 Sapindales	220 Sapindales	in Sapindaceae
225 Aceraceae	243 Sapindales	158 Rutales	146 Sapindales	219 Sapindales	in Sapindaceae
226 Burseraceae	233 Rutales	152 Rutales	136 Rutales	211 Rutales	43 Geraniales
227 Anacardiaceae	230 Rutales	153 Rutales	145 Sapindales	218 Sapindales	55 Sapindales
a in Anacardiaceae	232 Podoaceae (Rutales)	(in Anacardiaceae)	in Anacardiaceae?	216 Sapindales	in Sapindaceae
[b Blepharocaryaceae (AS+): in Anacardiaceae(?)]					
[c Pistaciaceae (AS+): in Anacardiaceae]					
228 Julianaceae	231 Rutales	in Anacardiaceae	154 Julianiales	223 Sapindales	in Anacardiaceae?
229 Simaroubaceae	234 Rutales	148 Rutales	134 Rutales	210 Rutales	41 Geraniales
a in Simaroubaceae	240 Kirkiaceae (?Rutales)	in Simaroubaceae	in Simaroubaceae[E]	in Simaroubaceae[A]	--
b in Simaroubaceae	260 Balanitaceae (Geraniales)	in Simaroubaceae	in Zygophyllaceae	131 Malpighiales	in Simaroubaceae
c in Simaroubaceae	in Ixonanthaceae?	in Simaroubaceae	in Simaroubaceae[E]	125 Irvingiaceae (Malpighiales)	in Simaroubaceae?
230 Cneoraceae	237 Rutales	150 Rutales	133 Rutales	173 Celastrales	in Simaroubaceae
231 Rutaceae	236 Rutales	147 Rutales	132 Rutales	209 Rutales	40 Geraniales
[a Rhabdodendraceae (AS+): in Rutaceae (?) (Takh.); near Centrospermae? (Airy-Shaw)]					
232 Meliaceae	238 Rutales	151 Rutales	137 Rutales	213 Meliales	44 Geraniales
a --	239 Aitoniaceae (?Rutales)	(in Meliaceae)	in Meliaceae[C]	in Sapindaceae[ee]	in Sapindaceae
[b Flindersiaceae (AS+): in/near Meliaceae, Rutaceae]					
[c Ptaeroxylaceae (AS+): in/near Meliaceae, Sapindaceae; in Simaroubaceae (Hutch.)[A]]					

CRONQUIST	TAKHTAJAN	THORNE	ENGLER	HUTCHINSON	BENTHAM & HOOKER
233 Zygophyllaceae	259 Geraniales	136 Geraniales	127 Geraniales	132 Malpighiales	38 Geraniales
a in Zygophyllaceae	258 Nitrariaceae (Geraniales)	in Zygophyllaceae	in Zygophyllaceae	in Zygophyllaceae	in Zygophyllaceae
b in Zygophyllaceae	261 Peganaceae (Geraniales)	in Zygophyllaceae	in Zygophyllaceae	in Zygophyllaceae	in Rutaceae
Order 44: GERANIALES					
234 Oxalidaceae	262 Geraniales	137 Geraniales	124 Geraniales	332 Geraniales	in Geraniaceae --ff
a in Oxalidaceae	256 Lepidobotryaceae (Geraniales)	(in Oxalidaceae)	in Oxalidaceae	130 Malpighiales	
b in Oxalidaceae	in Oxalidaceae	(in Oxalidaceae)	in Oxalidaceae	212 Averrhoaceae (Rutales)	in Geraniaceae
c (in Oxalidaceae)	263 Hypseocharitaceae (Geraniales)	(in Oxalidaceae)	in Oxalidaceae	in Oxalidaceae	in Geraniaceae
235 Geraniaceae	264 Geraniales	138 Geraniales	125 Geraniales	330 Geraniales	39 Geraniales
a in Geraniaceae	265 Dirachmaceae (Geraniales)	(in Geraniaceae)	in Geraniaceae	114 Tiliales	--
b in Geraniaceae	266 Vivianiaceae (Geraniales)	(in Geraniaceae)	in Geraniaceae	91 Pittosporales	in Geraniaceae
c in Geraniaceae	267 Biebersteiniaceae (Geraniales)	(in Geraniaceae)	in Geraniaceae	in Geraniaceae[A]	in Geraniaceae
d in Geraniaceae	in Geraniaceae[t]	(in Geraniaceae)	in Geraniaceae	127 Ledocarpaceae (Malpighiales)	in Geraniaceae
236 Limnanthaceae	270 Geraniales	141 Geraniales	123 Geraniales	331 Geraniales	in Geraniaceae
237 Tropaeolaceae	268 Geraniales	140 Geraniales	126 Geraniales	333 Geraniales	in Geraniaceae
238 Balsaminaceae	269 Geraniales	139 Geraniales	153 Sapindales	334 Geraniales	in Geraniaceae
Order 45: LINALES					
239 Humiriaceae	254 Geraniales	in Linaceae	in Linaceae[E]	123 Malpighiales	36 Geraniales
240 Erythroxylaceae	255 Geraniales	135 Geraniales	129 Geraniales	128 Malpighiales	in Linaceae
241 Linaceae	252 Geraniales	133 Geraniales	128 Geraniales	124 Malpighiales	35 Geraniales
a in Linaceae	251 Hugoniaceae (Geraniales)	in Linaceae	in Linaceae	in Linaceae	in Linaceae
b in Linaceae	253 Ixonanthaceae (Geraniales)	in Linaceae	in Linaceae	121 Malpighiales	in Linaceae
c in Linaceae	in Malpighiaceae	in Linaceae	in Linaceae	129 Ctenolophonaceae (Malpighiales)	--
Order 46: POLYGALALES					
242 Malpighiaceae	257 Geraniales	142 Geraniales	139 Rutales	122 Malpighiales	37 Geraniales
243 Trigoniaceae	271 Polygalales	145 Geraniales	140 Rutales	102 Polygalales	in Vochysiaceae
244 Vochysiaceae	272 Polygalales	146 Geraniales	141 Rutales	103 Polygalales	21 Polygalinae
245 Tremandraceae	275 Polygalales	213 Pittosporales	142 Rutales	92 Pittosporales	19 Polygalinae
246 Xanthophyllaceae	in Polygalaceae	(in Polygalaceae)	in Polygalaceae	in Polygalaceae[A]	in Polygalaceae
247 Polygalaceae	273 Polygalales	143 Geraniales	143 Rutales	100 Polygalales	20 Polygalinae
[a Diclidantheraceae:	in Polygalaceae]				
248 Krameriaceae	274 Polygalales	144 Geraniales	120 Rosales	101 Polygalales	in Polygalaceae

CRONQUIST	TAKHTAJAN	THORNE	ENGLER	HUTCHINSON	BENTHAM & HOOKER
Order 47: UMBELLALES (ARALIALES)					
249 Araliaceae	284 Cornales	254 Cornales	226 Umbelliflorae	47 Araliales	83 Umbellales
250 Umbelliferae (or: Apiaceae)	285 Cornales	in Araliaceae	227 Umbelliflorae	311 Umbellales	82 Umbellales
[a Hydrocotylaceae (AS+): in Umbelliferae]					
Subclass VI: ASTERIDAE					
Order 48: GENTIANALES					
251 Loganiaceae	322 Gentianales	231 Gentianales	246 Gentianales	226 Loganiales	110 Gentianales
a in Loganiaceae	323 Antoniaceae[gg] (Gentianales)	(in Loganiaceae)	in Loganiaceae	228 Loganiales	in Loganiaceae
b in Loganiaceae	324 Potaliaceae (Gentianales)	(in Loganiaceae)	in Loganiaceae	225 Loganiales	in Loganiaceae
c in Loganiaceae	321 Desfontainiaceae (Gentianales)	in Loganiaceae	247 Gentianales	in Potaliaceae	in Loganiaceae
d in Loganiaceae	in Loganiaceae	(in Loganiaceae)	in Loganiaceae	229 Spigeliaceae (Loganiales)	in Loganiaceae
e in Loganiaceae	in Loganiaceae	(in Loganiaceae)	in Loganiaceae	230 Strychnaceae (Loganiales)	in Loganiaceae
252 Gentianaceae	328 Gentianales	234 Gentianales	248 Gentianales	293 Gentianales	111 Gentianales
253 Apocynaceae	326 Gentianales	233 Gentianales	250 Gentianales	233 Apocynales	108 Gentianales
a in Apocynaceae	325 Plocospermataceae (Gentianales)	(in Loganiaceae)	in Loganiaceae[c]	232 Apocynales	in Loganiaceae
254 Asclepiadaceae	327 Gentianales	in Apocynaceae	251 Gentianales	235 Apocynales	109 Gentianales
a in Asclepiadaceae	in Asclepiadaceae	in Apocynaceae	in Asclepiadaceae	234 Periplocaceae (Apocynales)	in Asclepiadaceae
Order 49: POLEMONIALES					
255 Nolanaceae	341 Scrophulariales	in Solanaceae	262 Tubiflorae	323 Solanales	in Convolvulaceae
256 Solanaceae	340 Scrophulariales	111 Solanales	263 Tubiflorae	321 Solanales	116 Polemoniales
a in Solanaceae	in Solanaceae	in Solanaceae	in Solanaceae	Salpiglossidaceae (Solanales)[hh]	in Solanaceae
b --	in Nolanaceae?[t]	in Solanaceae	264 Duckeodendraceae (Tubiflorae)	in Ehretiaceae	--
[c Goetzeaceae (AS+): in Solanaceae (Takh.)]					
257 Convolvulaceae	333 Polemoniales	112 Solanales	255 Tubiflorae	322 Solanales	115 Polemoniales
[a Dichondraceae, b Humbertiaceae (AS+): in Convolvulaceae]					
[c Cardiopterygaceae (AS+): near Convolvulaceae (Airy-Shaw)]					
258 Cuscutaceae	334 Polemoniales	in Convolvulaceae	in Convolvulaceae	337 Polemoniales	in Convolvulaceae
259 Menyanthaceae	329 Gentianales	235 Gentianales	249 Gentianales	294 Gentianales	in Gentianaceae
260 Polemoniaceae	332 Polemoniales	113 Solanales	253 Tubiflorae	335 Polemoniales	112 Polemoniales
a in Polemoniaceae	in Polemoniaceae	(in Polemoniaceae)	in Polemoniaceae	238 Cobaeaceae (Bignoniales)	in Polemoniaceae
261 Hydrophyllaceae	335 Polemoniales	260 Lamiales	256 Tubiflorae	336 Polemoniales	113 Polemoniales
262 Lennoaceae	[337 Polemoniales	262 Lamiales	258 Tubiflorae	156 Ericales	99 Ericales

CRONQUIST	TAKHTAJAN	THORNE	ENGLER	HUTCHINSON	BENTHAM & HOOKER
Order 50: LAMIALES					
263 Boraginaceae	336 Polemoniales	261 Lamiales	257 Tubiflorae	338 Boraginales	114 Polemoniales
a in Boraginaceae	in Boraginaceae	in Boraginaceae	in Boraginaceae	242 Ehretiaceae (Verbenales)	in Boraginaceae
[b Cordiaceae, c Wellstediaceae (AS+), d Heliotropiaceae: in Boraginaceae]					
264 Callitrichaceae	361 Lamiales	265 Lamiales	260 Tubiflorae	292 Lythrales[aa]	in Haloragaceae
265 Verbenaceae	358 Lamiales	264 Lamiales	259 Tubiflorae	243 Verbenales	127 Lamiales
a in Verbenaceae	in Verbenaceae[t]	in Verbenaceae	in Verbenaceae	244 Stilbeaceae (Verbenales)	in Verbenaceae
b in Verbenaceae	in Verbenaceae	in Verbenaceae	in Verbenaceae	245 Chloanthaceae (Verbenales)	in Verbenaceae
[c Avicenniaceae (AS+): in Verbenaceae]					
[d Dicrastylidaceae (AS+): in Verbenaceae (synonym of Chloanthaceae)]					
[e Symphoremaceae (Symphoremataceae) (AS+): in Verbenaceae]					
[f Plagiopteraceae (AS+): in Verbenaceae?]					
266 Phrymaceae	360 Lamiales	in Verbenaceae	278 Tubiflorae	246 Verbenales	in Verbenaceae
267 Labiatae (or Lamiaceae, Menthaceae)	359 Lamiales	266 Lamiales	261 Tubiflorae	342 Lamiales	128 Lamiales
[a Tetrachondraceae (AS+): in Labiatae]					
Order 51: PLANTAGINALES					
268 Plantaginaceae	356 Scrophulariales	241 Bignoniales	279 Plantaginales	297 Plantaginales	129 "Anomalous fam."
Order 52: SCROPHULARIALES					
269 Buddlejaceae	342 Scrophulariales[jj]	in Loganiaceae	265 Tubiflorae	227 Loganiales	in Loganiaceae
a --	343 Retziaceae (Scrophulariales)	(in Loganiaceae)	in Loganiaceae[C]	321a Solanales[A]	in Solanaceae?
270 Oleaceae	301 Oleales[kk]	132 Oleales	245 Oleales	231 Loganiales	106 Gentianales
271 Scrophulariaceae	344 Scrophulariales	240 Bignoniales	266 Tubiflorae	324 Personales	117 Personales
272 Myoporaceae	353 Scrophulariales[ll]	239 Bignoniales	277 Tubiflorae	339 Lamiales	125 Lamiales
273 Globulariaceae	354 Scrophulariales	in Scrophulariac.	267 Tubiflorae	341 Lamiales	in Selaginaceae
a in Globulariaceae	in Scrophulariaceae[t]	in Scrophulariac.	in Scrophulariac.	340 Selaginaceae (Lamiales)	126 Lamiales
274 Gesneriaceae	349 Scrophulariales	245 Bignoniales	273 Tubiflorae	326 Personales	121 Personales
275 Orobanchaceae	351 Scrophulariales	242 Bignoniales	275 Tubiflorae	327 Personales	118 Personales
276 Bignoniaceae	345 Scrophulariales	236 Bignoniales	268 Tubiflorae	239 Bignoniales	122 Personales
277 Acanthaceae	355 Scrophulariales	244 Bignoniales	270 Tubiflorae	325 Personales	124 Personales
[a Mendonciaceae (AS+): in Acanthaceae]					
[b Thunbergiaceae (AS+): in Acanthaceae]					
278 Pedaliaceae	346 Scrophulariales	237 Bignoniales	271 Tubiflorae	240 Bignoniales	123 Personales
a in Pedaliaceae	347 Trapellaceae (Scrophulariales)	(in Pedaliaceae)	in Pedaliaceae	in Pedaliaceae	--
b in Pedaliaceae	348 Martyniaceae (Scrophulariales)	238 Bignoniales	272 Tubiflorae	241 Bignoniales	in Pedaliaceae
279 Lentibulariaceae	352 Scrophulariales	243 Bignoniales	276 Tubiflorae	328 Personales	119 Personales
280 Hydrostachyaceae	357 Scrophulariales	219 Pittosporales	121 Hydrostachyales	310 Podostemales	in Podostemaceae

CRONQUIST	TAKHTAJAN	THORNE	ENGLER	HUTCHINSON	BENTHAM & HOOKER
Order 53: CAMPANULALES					
281 Pentaphragmatac.[D]	in Campanulaceae	115 Campanulales	286 Campanulales	in Campanulaceae	in Campanulaceae
282 Sphenocleaceae	364 Campanulales	in Campanulaceae	285 Campanulales	in Campanulaceae	in Campanulaceae
283 Campanulaceae	362 Campanulales	116 Campanulales	284 Campanulales	315 Campanales	93 Campanales
a in Campanulaceae	363 Lobeliaceae (Campanulales)	in Campanulaceae	in Campanulaceae	316 Campanales	in Campanulaceae
284 Stylidiaceae	366 Campanulales	197 Rosales	289 Campanulales	319 Goodeniales	91 Campanales
a in Stylidiaceae	365 Donatiaceae (Campanulales)	in Stylidiaceae	in Stylidiaceae	304 Saxifragales	in Saxifragaceae
285 Brunoniaceae	368 Campanulales	in Goodeniaceae	288 Campanulales	318 Goodeniales	in Goodeniaceae
286 Goodeniaceae	367 Campanulales	117 Campanulales	287 Campanulales	317 Goodeniales	92 Campanales
Order 54: RUBIALES					
287 Rubiaceae	331 Gentianales	232 Gentianales	252 Gentianales	237 Rubiales	86 Rubiales
a in Rubiaceae	in Rubiaceae?	in Rubiaceae	269 Henriqueziaceae (Tubiflorae)	in Rubiaceae[A]	in Rubiaceae
[b Naucleaceae (AS+): in Rubiaceae]					
Order 55: DIPSACALES					
288 Caprifoliaceae	317 Dipsacales	255 Dipsacales	280 Dipsacales	48 Araliales	85 Rubiales
[a Sambucaceae (AS+): in Caprifoliaceae; (?) Takh.]					
[b Carlemanniaceae (AS+): in Caprifoliaceae; (?) Takh.]					
289 Adoxaceae	318 Dipsacales	256 Dipsacales	281 Dipsacales	306 Saxifragales	in Caprifoliaceae
290 Valerianaceae	319 Dipsacales	257 Dipsacales	282 Dipsacales	312 Valerianales	87 Asterales
291 Dipsacaceae	320 Dipsacales	258 Dipsacales	283 Dipsacales	313 Valerianales	88 Asterales
[a Morinaceae (AS+): in Dipsacaceae]					
[b Triplostegiaceae (AS+): in Dipsacaceae]					
292 Calyceraceae	369 Calycerales	259 Dipsacales	290 Campanulales	314 Valerianales	89 Asterales
Order 56: ASTERALES					
293 Compositae (or: Asteraceae, Carduaceae)	370 Asterales	267 Asterales	291 Campanulales	320 Asterales	90 Asterales
[a Ambrosiaceae (AS+): in Compositae]					
[b Cichoriaceae: in Compositae]					
CLASS B: LILIATAE[a] (Monocotyledonae)					
Subclass VII: ALISMATIDAE					
Order 57: ALISMATALES					
294 Butomaceae	371 Alismales	268 Alismales	293 Helobiae	343 Butomales	in Alismaceae
295 Limnocharitaceae	372 Alismales	in Alismataceae	in Butomaceae	in Butomaceae	in Alismaceae
296 Alismataceae	373 Alismales	269 Alismales	292 Helobiae	345 Alismatales	196 Apocarpae

CRONQUIST	TAKHTAJAN	THORNE	ENGLER	HUTCHINSON	BENTHAM & HOOKER
Order 58: HYDROCHARITALES					
297 Hydrocharitaceae	374 Hydrocharitales	270 Alismales	294 Helobiae	344 Butomales	169 Microspermae
[a Elodeaceae: in Hydrocharitaceae]					
Order 59: NAJADALES					
298 Aponogetonaceae	377 Najadales	271 Zosterales	296 Helobiae	352 Aponogetonales	in Najadaceae
299 Scheuchzeriaceae	375 Najadales	272 Zosterales	295 Helobiae	346 Alismatales	in Najadaceae
300 Juncaginaceae	376 Najadales	in Scheuchzeriaceae	297 Helobiae	349 Juncaginales	in Najadaceae
a in Juncaginaceae	in Juncaginaceae	(in Scheuchzeriaceae)	in Juncaginaceae	350 Lilaeaceae (Juncaginales)	in Najadaceae
301 Najadaceae	384 Najadales	277 Najadales	300 Helobiae	357 Najadales	197 Apocarpae
302 Potamogetonaceae	380 Najadales	273 Zosterales	298 Helobiae	354 Potamogetonales	in Najadaceae
303 Ruppiaceae	381 Najadales	(in Potamogetonaceae)	in Potamogetonaceae	355 Potamogetonales	in Najadaceae
304 Zannichelliaceae	382 Najadales	275 Zosterales	299 Helobiae	356 Najadales	in Najadaceae
305 Zosteraceae	378 Najadales	276 Zosterales	in Potamogetonaceae	353 Aponogetonales	in Najadaceae
a in Zosteraceae	379 Posidoniaceae (Najadales)	274 Zosterales	in Potamogetonaceae	351 Juncaginales	in Najadaceae
b in Zosteraceae	383 Cymodoceaceae (Najadales)	(in Zannicheliceae)	in Zannichelliaceae	in Zannichelliaceae	in Najadaceae
Order 60: TRIURIDALES					
306 Petrosaviaceae	in Liliaceae	in Liliaceae	in Liliaceae	347 Alismatales	in Liliaceae?
307 Triuridaceae	385 Triuridales	278 Triuridales	301 Triuridales	348 Triuridales	195 Apocarpae
Subclass VIII: COMMELINIDAE					
Order 61: COMMELINALES					
308 Rapateaceae	426 Commelinales	295 Commelinales	325 Commelinales	363 Xyridales	186 Coronarieae
309 Xyridaceae	425 Commelinales	296 Commelinales	324 Commelinales	362 Xyridales	183 Coronarieae
[a Abolbodaceae (AS+): in Xyridaceae (? Takh., Cronq.)]					
310 Mayacaceae	424 Commelinales	302 Commelinales	323 Commelinales	361 Commelinales	184 Coronarieae
311 Commelinaceae	423 Commelinales	301 Commelinales	322 Commelinales	358 Commelinales	185 Coronarieae
a in Commelinaceae	in Commelinaceae	in Commelinaceae	in Commelinaceae	359 Cartonemataceae (Commelinales)	in Commelinaceae
Order 62: ERIOCAULALES					
312 Eriocaulaceae	427 Eriocaulales	303 Commelinales	326 Commelinales	364 Eriocaulales	198 Glumaceae
Order 63: RESTIONALES					
313 Flagellariaceae	430 Restionales	304 Commelinales	329 Commelinales	360 Commelinales	187 Calycinae
[a Joinvilleaceae: in Flagellariaceae]		(in Flagellariaceae)	in Flagellariaceae?	--	in Flagellariaceae
b in Flagellariceae	431 Hanguanaceae (?Restionales)				
314 Anarthriaceae	in Restionaceae?	in Restionaceae	in Restionaceae	in Restionaceae	in Restionaceae
315 Ecdeiocoleaceae	in Restionaceae?	in Restionaceae	in Restionaceae	in Restionaceae	in Restionaceae
316 Restionaceae	428 Restionales	305 Commelinales	327 Commelinales	409 Juncales	200 Glumaceae
317 Centrolepidaceae	429 Restionales	306 Commelinales	328 Commelinales	408 Juncales	199 Glumaceae

CRONQUIST	TAKHTAJAN	THORNE	ENGLER	HUTCHINSON	BENTHAM & HOOKER
Order 64: JUNCALES					
318 Juncaceae	419 Juncales	299 Commelinales	319 Juncales	406 Juncales	188 Calycinae
319 Thurniaceae	420 Juncales	in Juncaceae	320 Juncales	407 Juncales	in Juncaceae[B]
Order 65: CYPERALES					
320 Cyperaceae	421 Cyperales	300 Commelinales	338 Cyperales	410 Cyperales	201 Glumaceae
[a Kobresiaceae (AS+): in Cyperaceae]					
321 Gramineae (or: Poaceae)	432 Poales	307 Commelinales	330 Graminales	411 Graminales	202 Glumaceae
[a Anomochloaceae (AS+), b Bambusaceae (AS+), c Streptochaetaceae (AS+): in Gramineae]					
Order 66: TYPHALES					
322 Sparganiaceae	438 Typhales	289 Arales	336 Pandanales	383 Typhales	in Typhaceae
323 Typhaceae	439 Typhales	290 Arales	337 Pandanales	384 Typhales	192 Nudiflorae
Order 67: BROMELIALES					
324 Bromeliaceae	422 Bromeliales	294 Commelinales	321 Bromeliales	365 Bromeliales	173 Epigynae
Order 68: ZINGIBERALES					
325 Strelitziaceae	410 Zingiberales	in Musaceae	in Musaceae	367 Zingiberales	
326 Lowiaceae	413 Zingiberales	309 Zingiberales	343 Scitamineae	368 Zingiberales	
327 Heliconiaceae	412 Zingiberales	in Musaceae	in Musaceae	in Strelitziaceae	
328 Musaceae	411 Zingiberales	308 Zingiberales	339 Scitamineae	366 Zingiberales	172 Epigynae
329 Zingiberaceae	415 Zingiberales	310 Zingiberales	340 Scitamineae	369 Zingiberales	(as Scitamineae)
330 Costaceae	414 Zingiberales	in Zingiberaceae	in Zingiberaceae	in Zingiberaceae	
331 Cannaceae	416 Zingiberales	311 Zingiberales	341 Scitamineae	370 Zingiberales	
332 Marantaceae	417 Zingiberales	312 Zingiberales	342 Scitamineae	371 Zingiberales	
Subclass IX: ARECIDAE					
Order 69: ARECALES					
333 Palmae (or: Arecaceae)	433 Arecales	291 Arecales	331 Principes	393 Palmales	189 Calycinae
[a Nypaceae (AS+): in Palmae]					
Order 70: CYCLANTHALES					
334 Cyclanthaceae	434 Cyclanthales	292 Cyclanthales	332 Synanthae	395 Cyclanthales	191 Nudiflorae
Order 71: PANDANALES					
335 Pandanaceae	437 Pandanales	293 Pandanales	335 Pandanales	394 Pandanales	190 Nudiflorae
Order 72: ARALES					
336 Araceae	435 Arales[nn]	287 Arales	333 Spathiflorae	381 Arales	193 Nudiflorae
337 Lemnaceae	436 Arales	288 Arales	334 Spathiflorae	382 Arales	194 Nudiflorae

(On the printed page, the Bentham & Hooker entry "172 Epigynae (as Scitamineae)" stands beside a brace that covers rows 325–332.)

CRONQUIST	TAKHTAJAN	THORNE	ENGLER	HUTCHINSON	BENTHAM & HOOKER
Subclass X: LILIIDAE					
Order 73: LILIALES					
338 Philydraceae	405 Liliales	298 Commelinales	318 Liliiflorae	401 Haemodorales	182 Coronarieae
339 Pontederiaceae	404 Liliales	297 Commelinales	313 Liliiflorae	375 Liliales	181 Coronarieae
340 Liliaceae	386 Liliales	279 Liliales	302 Liliiflorae	372 Liliales	180 Coronarieae
[a Convallariaceae: in Liliaceae]					
b in Liliaceae	391 Amaryllidaceae (Liliales)	in Liliaceae	308 Liliiflorae	385 Amaryllidales	176 Epigynae
c in Liliaceae	388 Aphyllanthaceae (Liliales)	in Liliaceae	in Liliaceae	in Liliaceae	in Liliaceae
d in Liliaceae	389 Alliaceae (Liliales)	in Liliaceae	in Liliaceae	in Amaryllidaceae	in Liliaceae
e in Liliaceae	392 Alstroemeriaceae (Liliales)	in Liliaceae	in Liliaceae	378 Alstroemeriales	in Amaryllidaceae
f in Liliaceae	394 Hypoxidaceae (Liliales)	in Liliaceae	309 Liliiflorae	397 Haemodorales	in Amaryllidaceae
g in Liliaceae	396 Philesiaceae (Liliales)	in Liliaceae	in Liliaceae	380 Alstroemeriales	in Liliaceae
h in Liliaceae	in Philesiaceae?	in Liliaceae	in Liliaceae	379 Petermanniaceae (Alstroemeriales)	in Dioscoreaceae
i in Liliaceae	399 Asparagaceae (Liliales)	in Liliaceae	in Liliaceae	in Liliaceae	in Liliaceae
j in Liliaceae	in Asparagaceae	in Liliaceae	in Liliaceae	377 Ruscaceae (Liliales)	in Liliaceae
k in Liliaceae	in Liliaceae	in Liliaceae	in Liliaceae	374 Trilliaceae (Liliales)	in Liliaceae
[l Colchicaceae: in Liliaceae]					
[m Melanthiaceae: in Liliaceae]					
341 Iridaceae	406 Iridales	284 Liliales	314 Liliiflorae	386 Iridales	175 Epigynae
342 Agavaceae	390 Liliales	in Liliaceae	305 Liliiflorae	392 Agavales	in Amaryllidaceae & Liliaceae
[a Dracaenaceae: in Agavaceae]					
343 Xanthorrhoeaceae	387 Liliales	in Liliaceae	303 Liliiflorae	391 Agavales	in Juncaceae
344 Velloziaceae	395 Liliales	283 Liliales	310 Liliiflorae	398 Haemodorales	in Amaryllidaceae
345 Haemodoraceae	393 Liliales	in Liliaceae	306 Liliiflorae	396 Haemodorales	174 Epigynae
a in Haemodoraceae	397 Tecophilaeaceae (Liliales)	in Liliaceae	in Haemodoraceae	373 Liliales	in Haemodoraceae
346 Taccaceae	403 Liliales	282 Liliales	311 Liliiflorae	400 Haemodorales	177 Epigynae
347 Cyanastraceae	398 Liliales	in Liliaceae	307 Liliiflorae	in Tecophilaeaceae	--
348 Stemonaceae (or: Roxburghiaceae)	401 Liliales	280 Liliales	304 Liliiflorae	389 Dioscoreales	179 Coronarieae
[a Croomiaceae (AS+): in Stemonaceae]					
349 Smilacaceae	400 Liliales	in Liliaceae	in Liliaceae	376 Liliales	in Liliaceae
350 Dioscoreaceae	402 Liliales	281 Liliales	312 Liliiflorae	390 Dioscoreales	178 Epigynae
a in Dioscoreaceae	in Dioscoreaceae[t]	in Dioscoreaceae	in Dioscoreaceae	387 Stenomeridaceae (Dioscoreales)	in Dioscoreaceae

CRONQUIST	TAKHTAJAN	THORNE	ENGLER	HUTCHINSON	BENTHAM & HOOKER
b in Dioscoreaceae[s]	in Dioscoreaceae[t]	in Dioscoreaceae[ii]	in Dioscoreaceae	388 Trichopodaceae (Dioscoreales)	in Dioscoreaceae
Order 74: ORCHIDALES					
351 Geosiridaceae	407 Iridales	in Iridaceae	315 Liliiflorae	--	--
352 Burmanniaceae	408 Iridales	285 Liliales	316 Liliiflorae	402 Burmanniales	170 Microspermae
a in Burmanniaceae	in Burmanniaceae	in Burmanniaceae	in Burmanniaceae	403 Thismiaceae (Burmanniales)	in Burmanniaceae
353 Corsiaceae	409 Iridales	in Burmanniaceae	317 Liliiflorae	404 Burmanniales	in Burmanniaceae
354 Orchidaceae	418 Orchidales	286 Liliales	344 Microspermae	405 Orchidales	171 Microspermae
a in Orchidaceae	in Orchidaceae	in Orchidaceae	in Orchidaceae	399 Apostasiaceae (Haemodorales)	in Orchidaceae

INDEX TO FAMILY NAMES

INDEX TO FAMILY NAMES, continued

INDEX TO FAMILY NAMES, continued

INDEX TO FAMILY NAMES, continued

AUTHOR INDEX TO CLASSIFICATION SYSTEMS

CLASSIFICATION LITERATURE

I. General References

Airy Shaw, H. K. 1966. Willis' A Dictionary of the Flowering Plants and Ferns. 7th ed., revised. Cambridge University Press. Cambridge.

Arber, A. 1938. Herbals, Their Origin and Evolution. Cambridge University Press. Cambridge.

Benson, L. 1957. Plant Classification. D. C. Heath and Company. Boston.

Core, E. L. 1955. Plant Taxonomy. Prentice-Hall, Inc. Englewood Cliffs, New Jersey.

Cronquist, A. J. 1968. The Evolution and Classification of Flowering Plants. Houghton Mifflin. Boston.

Dalla Torre, C. G. de, and H. Harms. 1900-1907. Genera Siphonogamarum ad Systema Englerianum Conscripta.

Davis, P. H., and V. Heywood. 1963. Principles of Angiosperm Taxonomy. D. Van Nostrand Co., Inc. Princeton, New Jersey.

Gundersen, A. 1950. Families of Dicotyledons. Chronica Botanica. Waltham, Mass.

Hutchinson, J. 1959. The Families of Flowering Plants. 2 vols. Oxford.

Lawrence, G. H. M. 1951. Taxonomy of Vascular Plants. Macmillan. New York.

________. 1955. An Introduction to Plant Taxonomy. Macmillan. New York.

Porter, C. L. 1959. Taxonomy of Flowering Plants. W. H. Freeman and Company. San Francisco.

Rendle, A. B. 1967 (reprinting of 1930 book). The Classification of Flowering Plants. 2nd ed. Cambridge.

Swingle, D. B. 1934. Textbook of Systematic Botany. McGraw-Hill. New York.

Takhtajan, A. 1969. Flowering Plants--Origin and Dispersal. (English translation by C. Jeffrey). Oliver and Boyd. Edinburgh.

II. Ancients

Albertus Magnus. ca. 1250. De Vegetabilis.

Dioscorides. 64 AD. Materia Medica.

"Gart der Gesundheit." 1485. Mainz. } many editions and printings

Hortus Sanitatis. 1491. Mainz. }

Pliny. 77 AD. Natural History.

Theophrastus. ca. 300 BC. Enquiry into Plants and the Causes of Plants (Historia Plantarum).

III. Herbalists

Bankes, R. 1525. Herball.

Bock, J. 1539. Neu Kreuter Buch.

Brunfels, O. 1530-1536. Herbarum vivae eicones.

Cordus, V. 1561. Historia Stirpium.

Dodoens, R. 1583. Stirpium Historiae Pemptades; Cruydeboeck.

Fuchs, L. 1542. De Historia Stirpium; Neu Kreuter Buch.

Gerard, J. 1597. Herball or General Historie of Plants.

l'Ecluse, C. de. 1601. Rariorum plantarum historia.

l'Obel, M. de. 1581. Stirpium adversaria nova.

Mattioli, P. 1544. Commentarii in sex libros Pedacii Dioscoridis (plus many editions).

Turner, W. 1551, 1562, 1568. A New Herball.

IV. Mechanical Systems

Bauhin, G. 1623. Pinax theatri botanici.
Bauhin, J. 1650. Historia plantarum universalis (posthumous).
Caesalpino, A. 1583. De plantis libri.
Camerarius, R. 1694. De Sexu Plantarum Epistola.
Linnaeus, C. 1735. Systema Naturae.
________. 1737. Genera Plantarum (and subsequent editions).
________. 1751. Philosophia Botanica.
________. 1753. Species Plantarum.
Magnol, P. 1689. Prodromus historiae generalis.
Morison, R. 1680. Plantarum Historia Universalis.
Ray, J. 1686-1704. Historia plantarum.
________. 1703. Methodus plantarum.
Rivinus, A. Q. 1690. Introductio generalis in rem herbariam.
Tournefort, J. P. de. 1694. Elements de Botanique.
________. 1716. Institutiones Rei Herbariae.

V. Natural Systems

Adanson, M. 1763. Familles des Plantes.
Bentham, G., and J. D. Hooker. 1862-1883. Genera Plantarum. 3 vols. London.
Braun, A. 1851. Betrachtungen über die Erscheinung der Verjüngung in der Natur. Leipzig.
________. 1853. Das Individuum der Pflanze in seinem Verhältniss zur Species. Berlin.
Brongniart, A. 1843. Enumération des Genres des Plantes, &c. Paris.
Brown, R. 1866-1867. (The miscellaneous botanical works of Robert Brown. Edited by J. J. Bennett. Published for the Ray Society by R. Hardwicke. London.).
Candolle, A. P. de. 1813. Théorie Élémentaire de la Botanique. Paris.
Candolle, A. P., A., and C. de. 1824-1873. Prodromus systematis naturalis regni vegetabilis. 17 vols. Paris.
Endlicher, S. L. 1836-1840. Genera Plantarum. Vienna.
Hofmeister, W. 1851. Vergleichende Untersuchungen... F. Hofmeister. Leipzig.
Jussieu, A. L. de. 1789. Genera plantarum secundum ordines naturales deposita.
Lamarck, J. 1778. Flore Françoise; and other works.
Lindley, J. 1830. Introduction to the Natural System of Botany. London.
________. 1846. Vegetable Kingdom.

VI. Phylogenetic Period

Arber, E. A. N., and J. Parkin. 1907. On the origin of Angiosperms. Journal of the Linnean Society (Botany) 38: 28-80.
Baillon, H. 1867-1895. Histoire des Plantes. 13 vols. Paris.
________. 1871-1888. Natural History of Plants. 8 vols. London.
Benson, L. 1957. Plant Classification. D. C. Heath and Company. Boston.
Bessey, C. E. 1897. Phylogeny and taxonomy of the Angiosperms. Botanical Gazette 24: 145-178.
________. 1915. The phylogenetic taxonomy of flowering plants. Annals of the Missouri Botanical Garden 2: 109-164.

Boivin, B. 1956. Les Familles de Tracheophytes. Bulletin de la Societe Botanique de France 103: 490-505.

Burtt-Davy, J. 1937. On the primary groups of dicotyledons. Annals of Botany, New Series 1: 429-437.

Clements, F. E., and H. M. Hall. 1923. The Phylogenetic Method in Taxonomy. Carnegie Institute of Washington Publ. No. 326.

Copeland, H. F. 1957. Forecast of a system of the dicotyledons. Madrono 14: 1-9.

Cronquist, A. 1957. Outline of a new system of families and orders of dicotyledons. Bulletin du Jardin Botanique de l'état, Brussels 27: 13-40.

________. 1965. The status of the general system of classification of flowering plants. Annals of the Missouri Botanical Garden 52: 281-303.

________. 1968. The Evolution and Classification of Flowering Plants. Houghton Mifflin. Boston.

Darwin, C. 1859. Origin of Species.

Deyl, M. 1955. The evolution of plants and the taxonomy of the Monocotyledons. Acta Musei Nat. Pragae, V. 11, B, no. 6: 1-143.

Eichler, A. B. 1876. Syllabus der Vorlesungen über Phanerogamenkunde. Kiel.; 1890. Syllabus, 5 Aufl. Berlin.

________. 1875-1878. Blüthendiagramme. 2 vols. Leipzig.

Emberger, L. 1960. Les Végétaux Vasculaires (Vol. 2 of M. Chadefaud and L. Emberger, Traité de Botanique Systematique). Maisson et Cie. Paris.

Engler, A., and K. Prantl. 1887-1915. Die natürlichen Pflanzenfamilien. 23 vols. Leipzig.

________. Syllabus der Pflanzenfamilien ed. 2: 1898; ed. 3: 1903; ed. 4: 1904; ed. 5: 1907; ed. 6: 1909; ed. 7: 1912; ed. 8: 1919).

________, and E. Gilg. 1924. Syllabus der Pflanzenfamilien. (eds. 9, 10). Berlin.

________, and L. Diels. 1936. Syllabus der Pflanzenfamilien. (ed. 11). Berlin.

Gundersen, A. 1950. Families of Dicotyledons. Chronica Botanica. Waltham, Mass.

Hallier, H. 1905. Provisional scheme of the natural (phylogenetic) system of flowering plants. The New Phytologist 4: 151-162.

________. 1912. L'orgine et le système phylétique des Angiosperms. Archives néerlandaises des sciences exactes et naturelles, sér. 2, I: 146-234.

Hayata, B. 1921. The natural classification of plants according to the dynamic system. Icones Plantarum Formosanarum 10: 97-234.

Hutchinson, J. 1926, 1934. The Families of Flowering Plants. 2 vols. Macmillan. London.

________. 1959. ibid. ed. 2. 2 vols. Clarendon Press. Oxford.

________. 1969. Evolution and Phylogeny of Flowering Plants. Academic Press. London and New York.

Kimura, Y. 1956. Système et Phylogénie des Monocotylédones. Notulae Systematicae 15: 137-159. Paris.

Kuntze, O. 1891. Revisio Generum Plantarum. H. Stürtz. Würzburg.

Melchior, H. (ed.). 1964. A. Engler's Syllabus der Pflanzenfamilien. (ed. 12). Vol. II. (Vol. I, 1954. Bacteria to Gymnosperms, E. Werdermann, ed.) Gebrüder Borntraeger. Berlin.

Mez, C. 1926. Die Bedeutung der Serodiagnostik für die Stammesgeschictliche Forschung. Botanisches Archiv 16: 1-23.

Novák, F. A. 1954. Systém Angiosperm. Preslia 26: 337-364.

Pool, R. J. 1941. Flowers and Flowering Plants. McGraw-Hill. New York.

Pulle, A. A. 1938. Compendium van de Terminologie, Nomenclatuur en Systematiek der Zaadplanten.

Pulle, A. A. 1950. ibid., ed. 2. Utrecht.
Rendle, A. B. 1930, 1967. The Classification of Flowering Plants. ed. 2. Cambridge.
Schaffner, J. H. 1932. Revised catalog of Ohio Vascular Plants. Ohio State University. Columbus.
Skottsberg, C. 1940. Växternas Liv, vol. 5. Stockholm.
Soó, R. 1953. Die modernen Grundsätze der Phylogenie im neunen System der Blütenpflanzen. Acta biologica academiae scientiarum Hungaricae 4: 257-306.
Takhtajan, A. 1943. Sootnosheniia ontogeneza i philogeneza u vysshikh rastenii. (English summary) V. M. Molotov University Press. Erevan.
________. 1953. Phylogenetic principles of the system of higher plants. Botanical Review 19: 1-45.
________. 1959. Die Evolution der Angiospermen. Jena.
________. 1966. Systema et Phylogenia Magnoliophytorum. Soviet Sciences Press. Moscow and Leningrad.
________. 1969. Flowering Plants--Origin and Dispersal. (English translation by C. Jeffrey) Oliver and Boyd. Edinburgh.
Thorne, R. 1963. Some problems and guiding principles of angiosperm phylogeny. American Naturalist 97(896): 287-305.
________. 1968. Synopsis of a putatively phylogenetic classification of the flowering plants. Aliso 6: 57-66.
van Tieghem, Ph. 1898. Elements de Botanique. ed. 3.
Warming, E. 1895. A Handbook of Systematic Botany (English edition by M. C. Potter). Macmillan. New York.
Wernham, T. J. 1911. Floral evolution. New Phytologist 10: 109-120.
Wettstein, R. von. 1911. Handbuch der Systematischen Botanik. Gebrüder Borntraeger. Berlin.
________. 1935. ibid. ed. 4. Leipzig and Vienna.

VII. Recent Systems for Vascular Plants

Banks, H. P. 1968. The early history of land plants. Symposium on Evolution and Environment. E. T. Drake (ed.). Yale University Press. New Haven, Conn.
Bierhorst, D. W. 1971. Morphology of Vascular Plants. Macmillan. New York.
Lam, H. J. 1948. A new system of the Cormophyta. Blumea 6: 282-289.
Tippo, O. 1942. A modern classification of the plant kingdom. Chronica Botanica 7: 203-206. Waltham, Mass.

Chapter 29. FAMILIES OF VASCULAR PLANTS

Traditionally the family has been the basic study unit in systematics teaching and training. Expedience in identification comes with the recognition of family characteristics. Comprehension of present classification systems is dependent upon a fundamental understanding of the family diagnostic features. Three approaches are used in this chapter for understanding the family conceptually and identifiably: (1) diagnostic characters are given in summary form as aids in field identification, (2) synoptic keys to subclasses, orders and families are presented as conceptual aids for the fundamental understanding of families, and (2) plates illustrating features of the structurally more difficult families have been designed for visual help. For convenient reference the families, orders, and subclasses have been indexed. Detailed family descriptions will have to be procured from other sources (see references at the end of the chapter). See Family Plate Illustrations, pages 852-877.

Section A. AIDS FOR FIELD IDENTIFICATION OF COMMON FAMILIES OF FLOWERING PLANTS IN NORTH AMERICA NORTH OF MEXICO (Arranged according to the Engler System)

Monocotyledoneae

1. Typhaceae. Flowers densely crowded in a terminal spike; perianth of bristles.

2. Zosteraceae. Aquatics; flowers in spikes; no distinct perianth.

3. Najadaceae. Aquatics; flower solitary in axils of leaves; no perianth.

4. Alismataceae. Perianth present; parts in 3's; carpels form disk or ring on receptacle.

5. Poaceae. Culm hollow, terete; leaves flat, 2-ranked; sheath usually open; fruit a grain.

6. Cyperaceae. Culm solid, triangular; leaves channeled, 3-ranked; fruit a nutlet.

7. Araceae. Inflorescence a spadix; perianth wanting or inconspicuous.

8. Xyridaceae. Plant scapose; flowers in dense globose to cylindrical heads.

9. Commelinaceae. Herbs; leaves with a tubular sheath; floral parts in 3's; calyx green; corolla colored; ovary superior; capsules several-seeded. (3; 3; 2-6; 2-3)*

*Floral formula for flower; with the first number referring to the # of sepals; second, to the # of petals; third, to the # of stamens; and fourth, to the # of carpels. Formula is given only when it is considered of some value in family identification.

10. Juncaceae. Grass-like; floral parts in 3's; fruit a capsule. (3; 3; 3-6; 3)

11. Liliaceae. Perianth conspicuous; floral parts in 3's - stamens 6; ovary superior. (3; 3; 6; 3)

12. Amaryllidaceae. Perianth conspicuous; plants scapose; floral parts in 3's - stamens 6; ovary inferior. (3; 3; 6; 3)

13. Iridaceae. Perianth conspicuous; floral parts in 3's; ovary inferior. (3; 3; 3; 3)

14. Orchidaceae. Perianth conspicuous; floral parts in 3's - stamens 1 or 2; gynandrium and pollinia present; corolla zygomorphic; ovary inferior. (3; 3; 1-2; 3)

Dicotyledoneae

Apetalae (without petals)

15. Salicaceae. Trees or shrubs; dioecious; flowers in catkins; bracts of female catkin deciduous.

16. Juglandaceae. Trees; monoecious; fruit with a thick husk; leaves pinnately compound.

17. Betulaceae. Trees or shrubs; monoecious; female flowers usually in catkins; 2-carpellate.

18. Fagaceae. Trees or shrubs; monoecious; female flowers usually not in catkins; 3-carpellate.

19. Urticaceae. Herbs; ovary superior; stinging hairs usually present.

20. Santalaceae. Semiparasitic herbs, shrubs, or trees; leaves alternate or opposite; ovary inferior; fruit a drupe.

21. Aristolochiaceae. Herbs or vines; calyx petaloid; fruit a 6-locular capsule.

22. Polygonaceae. Herbs or vines; nutlet trigonous; stipules sheathing.

23. Chenopodiaceae. Herbs; perianth and bracts not scarious; filaments separate; fruit a utricle.

24. Amaranthaceae. Herbs; perianth and bracts scarious; filaments usually connate below; fruit a utricle.

Apopetalae (with separate petals)

25. Caryophyllaceae. Herbs; leaves opposite; inflorescence cymose; placentation free-central. (5; 5; 5-10; 1-5)

26. Nymphaeaceae. Aquatic; leaves long-petioled, peltate; floral parts many except sepals, 3; vessels absent; placentation laminar.

27. Ranunculaceae. Mostly herbs; floral parts spirally arranged; stamens numerous; pistil unicarpellate; endosperm present. (5; 5; ∞; ∞)

28. Magnoliaceae. Trees; floral parts numerous - spirally arranged; gynoecium apocarpous.

29. Papaveraceae. Herbs; sepals 2, caducous or fugacious; juice milky or colored; perianth conspicuous; gynoecium syncarpous.

30. Brassicaceae. Herbs; floral parts in 4's; fruit a silique or silicle; stamens tetradynamous. (4; 4; 6; 2-4)

31. Sarraceniaceae. Insectivorous perennials; leaves hollow, basal; flowers 5-merous; style umbraculate, ovary superior; fruit a capsule. (5; 5; ∞; 5)

32. Saxifragaceae. Herbs, shrubs, or vines; floral parts in 5's except carpels mostly 2; endosperm present. (5; 5; 5-10; 2)

33. Hamamelidaceae. Monoecious trees or shrubs; leaves alternate, petiolate; perianth present or absent; stamens numerous; carpels 2, ovary partly inferior; fruit a capsule. (0-7; 0-4; 8-∞; 2)

34. Rosaceae. Herbs, shrubs, or trees; floral parts in 5's; ovary superior to inferior; leaves stipulate; endosperm absent. (5; 5; 5-∞; 1-∞)

35. Fabaceae. Herbs, shrubs, trees, vines; floral parts in 5's - carpels 1; corollas mostly papilionaceous; stamens monadelphous or diadelphous; fruit a legume. (5; 5; 5-10; 1)

36. Geraniaceae. Herbs; floral parts in 5's; carpels coiled at maturity; leaves lobed or dissected. (5; 5; 10; 5)

37. Polygalaceae. Herbs; flowers showy, irregular; petals 3; stamens 6-8. (3; 3; 6-8; 2)

38. Euphorbiaceae. Herbs, shrubs, small trees; apetalous; monoecious 3-carpellate; inflorescence a cyathium in *Euphorbia*.

39. Anacardiaceae. Shrubs or small trees; leaves pinnately compound; stigmas 3; fruit a drupe.

40. Aquifoliaceae. Shrubs or trees; flowers dioecious or polygamo-dioecious, small; inflorescence axillary.

41. Aceraceae. Trees; flowers dioecious or polygamous, small; leaves palmately lobed; fruit a samara.

42. Rhamnaceae. Shrubs or small trees; stamens alternate with sepals; leaves pinnately veined; fruit a drupe.

43. Vitaceae. Woody vines; leaves palmately veined; fruit a berry.

44. Malvaceae. Perianth conspicuous; stamens monadelphous; sepals persistent; anthers 1-locular; carpels 5 or more; pubescence frequently stellate. (5; 5; ∞; 2-∞)

45. Hypericaceae. Herbs or shrubs; leaves opposite; stamens usually in fascicles; carpels 3 or 5; petals mostly yellow; conspicuous secretory canals or vescicles in all organs.

46. Violaceae. Herbs; corolla spurred; placentation parietal; fruit a capsule. (5; 5; 5; 3)

47. Cactaceae. Fleshy, prickly plants; floral parts many; ovary inferior.

48. Elaeagnaceae. Shrubs; leaves alternate, simple, usually lepidote below; flowers perfect, 4-merous; carpel 1, ovary superior; fruit drupaceous with a fleshy hypanthium. (4; 4; 4; 1)

49. Lythraceae. Herbs or shrubs; leaves opposite or whorled; ovary enclosed in a persistent calyx.

50. Melastomataceae. Herbs; petals conspicuous; floral parts in 4's; anthers poricidal; ovary inferior. (4; 4; 8; 4)

51. Onagraceae. Herbs; floral parts in 4's; ovary inferior; anthers longitudinally dehiscent. (4; 4; 8; 4)

52. Araliaceae. Herbs or shrubs; flowers in umbels; leaves compound; ovary inferior; fruit a several seeded berry-like drupe; carpophore usually absent.

53. Apiaceae. Herbs; flowers in umbels; stylopodium present; fruit a schizocarp; ovary inferior; carpophore usually present. (5; 5; 5; 2)

54. Cornaceae. Shrubs or small trees; leaves usually opposite; flowers 4-merous; ovary inferior; fruit a drupe. (4; 4; 4; 2)

<u>Sympetalae</u> (with fused petals)

55. Ericaceae. Mostly shrubs or trees; floral parts in 5's; petals usually conspicuous. (5; 5; 5-10; 3-5)

56. Primulaceae. Herbs; placentation free-central.

57. Styracaceae. Leaves simple; hairs stellate.

58. Oleaceae. Trees or shrubs; leaves opposite, frequently pinnately compound; fruit a samara or drupe; stamens 2. (4; 4; 2; 2)

59. Gentianaceae. Herbs; leaves opposite, exstipulate; ovary 1-locular; flowers showy; placentation parietal.

60. Apocynaceae. Herbs or shrubs; leaves opposite; milky latex present; fruit a follicle; pollen not in pollinia. (5; 5; 5; 2)

61. Asclepiadaceae. Herbs or vines; crown, gynostegium, pollinia present; milky latex present; fruit a follicle. (5; 5; 5; 2)

62. Convolvulaceae. Twining vines; corolla plicate; flowers axillary. (5; 5; 5; 3)

63. Polemoniaceae. Herbs; corolla regular; ovary 3-locular; carpels 3; corolla lobes convolute in bud. (5; 5; 5; 3)

64. Hydrophyllaceae. Herbs; corolla regular; ovary one-locular; carpels 2; corolla lobes usually imbricate in bud. (5; 5; 5; 2)

65. Verbenaceae. Herbs, shrubs, or small trees; leaves opposite; ovary 2- or 4-locular separating into 2 or 4 nutlets; style terminal. (5; 5; 4-5; 2)

66. Boraginaceae. Herbs; ovary 2-locular developing into a 4-nutlet fruit; leaves alternate. (5; 5; 5; 2)

67. Lamiaceae. Herbs or shrubs; corolla zygomorphic; style gynobasic; leaves opposite, aromatic. (5; 5; 2-4; 2)

68. Solanaceae. Herbs or shrubs; anthers frequently poricidal; ovary 2-locular or falsely 3-5 locular. (5; 5; 5; 2-5)

69. Scrophulariaceae. Herbs; corolla zygomorphic; stamens frequently 2; fruit a capsule. (4-5; 4-5; 2-5; 2)

70. Plantaginaceae. Herbs; flowers inconspicuous; scapose; perianth scarious; fruit a circumscissile capsule. (4; 4; 4; 2)

71. Rubiaceae. Herbs or shrubs; floral parts mostly in 4's, carpels 2; ovary inferior; leaves opposite or whorled and stipulate. (4; 4; 4; 2)

72. Caprifoliaceae. Mostly shrubs; floral parts in 5's, carpels 2 or 3; ovary inferior; leaves exstipulate. (5; 5; 5; 2-3)

73. Cucurbitaceae. Tendril-bearing succulent vines; monoecious; ovary 3-locular; fruit a berry.

74. Campanulaceae. Herbs; corolla usually zygomorphic; anthers occasionally united into a tube; ovary inferior. (5; 5; 5; 2-3)

75. Asteraceae. Herbs or shrubs; inflorescence a capitulum; pappus, bracts, syngenesium present; carpels 2; ovary inferior; ovule 1; fruit a nutlet or cypsela. (pappus; 5; 5; 2)

Section B. *SYNOPTICAL KEY TO FAMILIES, ORDERS, AND SUBCLASSES OF PINOPHYTA (GYMNOSPERMS) NATIVE AND CULTIVATED IN NORTH AMERICA NORTH OF MEXICO

1. Leaves pinnately compound, terminating a simple or seldom-branched stem; seeds borne on sporophylls. (Cycadopsida, Cycadales)
   2. Leaflets with single midrib and no lateral veins; megasporophylls crowded on an indeterminate axis; cultivated only....1. <u>Cycadaceae</u>
   2. Leaflets with numerous parallel longitudinal veins dichotomously branched near the base; megasporophylls forming a dense determinate cone; cultivated and native....................2. <u>Zamiaceae</u>

*Contributed by James W. Hardin, North Carolina State University, Raleigh.

1. Leaves simple, throughout branched stems; seeds terminal or borne on ovuliferous scales in cones.
   3. Leaves fan-shaped, bilobed at apex, dichotomously veined, deciduous, borne on ends of long shoots or on indeterminate short shoots; seeds 1-2 terminating an elongated stalk and with fleshy outer coat; tree, cultivated only (Gnetopsida, Ginkgoales).............3. Ginkgoaceae
   3. Leaves acicular, linear, awl-shaped, or scale-like, evergreen or deciduous; seeds terminal on short lateral branches or in cones.
      4. Stems jointed and photosynthetic; secondary wood with vessels; leaves reduced to scales, opposite or in whorls of 3; pollen cones compound; seeds dry with winged bracts or fleshy with wingless bracts (Gnetopsida, Ephedrales).........4. Ephedraceae
      4. Stems not jointed and without vessels; leaves photosynthetic and variable; pollen cones simple; seeds terminal without bracts or borne on scales in cones. (Pinopsida)
         5. Seeds one or rarely 2, drupe-like or berry-like; leaves linear and persistent.
            6. Pollen-sacs 2 per sporophyll; leaves mostly 1-4 in. long; seed on a lateral branch, drupe-like with fleshy coat, also bearing an enlarged fleshy aril at base of seed; cultivated only (Pinales)..............5. Podocarpaceae
            6. Pollen-sacs 3-9 per sporophyll; leaves 1/2 to 3 in. long.
               7. Leaves 1-3 in. long with 2 broad glaucous lines beneath; seed drupe-like with fleshy outer coat; basal aril lacking or rudimentary; cultivated only (Pinales)......................6. Cephalotaxaceae
               7. Leaves 1/2-1 1/4 in. long, either with pale green to tawny bands beneath or with 2 very narrow glaucous lines; seeds partially or completely enclosed by a fleshy aril (Taxales).................7. Taxaceae
         5. Seeds one to many in a woody or berry-like cone; leaves variable, persistent or deciduous (Pinales).
            8. Leaves ovate-lanceolate, 1/3 in. or more wide at base; one ovule per ovulate scale; scale and bract fused; cultivated only...........................8. Araucariaceae
            8. Leaves usually narrower at base; ovuliferous scale with 2 or more ovules; scales and bracts separate or partially to completely fused.
               9. Ovuliferous scales spirally arranged, imbricate and flat to somewhat peltate; leaves usually alternate (opposite or apparently whorled in cultivated *Metasequoia* and *Sciadopitys*).
                  10. Female cones with flat scales and 2 rounded seeds; bract subtending ovuliferous scale distinct; pollen with 2 air-bladders; leaves acicular or linear................9. Pinaceae
                  10. Female cones with somewhat peltate scales and 2-9 triangular seeds; bract fused to scale; pollen without air-bladders; leaves and twigs deciduous or persistent, leaves linear, awl- or scale-like.........................10. Taxodiaceae
               9. Ovuliferous scales opposite or whorled and peltate or flat; bract fused with scale; pollen without air-bladders; leaves opposite or whorled, awl- or scale-like, persistent.................11. Cupressaceae

Section C.

SYNOPTICAL KEYS TO FAMILIES, ORDERS, AND SUBCLASSES OF ANGIOSPERMS

(Adapted from Cronquist's "Evolution and Classification of Flowering Plants [1968]; with permission.)

Class: Magnoliopsida (Dicots)	Class: Liliopsida (Monocots)
Cotyledons 2 (seldom 1, 3, or 4)	Cotyledon 1 (or the embryo sometimes undifferentiated)
Leaves mostly net-veined	Leaves mostly parallel-veined
Intrafascicular cambium usually present	Intrafascicular cambium usually absent; usually no cambium of any sort
Vascular bundles usually borne in a ring which encloses a pith	Vascular bundles generally scattered, or in 2 or more rings
Floral parts, when of definite number, typically borne in sets of 5, less often 4, seldom 3 (carpels often fewer)	Floral parts, when of definite number typically borne in sets of 3, seldom 4, never 5 (carpels often fewer)
Pollen typically triaperturate, or of triaperturate-derived type, except in a few of the more primitive families	Pollen of uniaperturate or uniaperturate-derived type
Mature root system either primary or adventitious, or both	Mature root system wholly adventitious

SYNOPTICAL KEY TO THE SUBCLASSES OF MAGNOLIOPSIDA (DICOTS)

1. Plants relatively primitive, the flowers typically apocarpous, always polypetalous or apetalous and nearly always with an evident perianth, usually with numerous, centripetal stamens, the pollen always binucleate and often uniaperturate; ovules always bitegmic and crassinucellate..I. Magnoliidae
1. Plants more advanced in one or more respects than the Magnoliidae, and never with uniaperturate pollen.
   2. Flowers more or less strongly reduced and often unisexual, the perianth, poorly developed or wanting, often borne in catkins, but never forming bisexual pseudanthia, and never with numerous seeds on parietal placentae................................II. Hamamelidae
   2. Flowers usually more or less well developed and with an evident perianth, but if not so, then usually either grouped into bisexual pseudanthia or with numerous seeds on parietal placentae, only rarely with all of the characters of the Hamamelidae as given above.
      3. Flowers polypetalous or less often apetalous or sympetalous, if sympetalous then usually either with more stamens than corolla-lobes, or with the stamens opposite the corolla lobes, or with bitegmic or crassinucellate ovules; carpels 1 to many, free or more often united.
         4. Stamens, when numerous, developing in centrifugal sequence; placentation various, often parietal or free-central or basal, but often also axile; species with few stamens and axile placentation usually either bearing several or many ovules per locule, or with a sympetalous corolla, or both.

5. Pollen trinucleate; ovules bitegmic, crassinucellate, most often campylotropous or amphitropous; seeds very often with perisperm; plants usually either with betalains instead of anthocyanins, or with free-central to basal placentation, or both; only a few species are ordinary trees.............................III. Caryophyllidae
5. Pollen usually binucleate (notable exception: Cruciferae); ovules various, but seldom campylotropous or amphitropous; seeds seldom with perisperm; no betalains; placentation rarely free-central or basal, except in the Primulales; many species are ordinary trees........... .... IV. Dilleniidae
4. Stamens, when numerous, developing in centripetal sequence; flowers seldom with parietal placentation (notable exception: many Saxifragaceae) and also seldom (except in parasitic species) with free-central or basal placentation in a unilocular, compound ovary, but very often (especially in species with few stamens) with 2-several locules that have only 1 or 2 ovules each; flowers polypetalous or less often apetalous, only rarely sympetalous.........V. Rosidae
3. Flowers sympetalous; stamens generally isomerous with the corolla lobes or fewer, never opposite the lobes; ovules unitegmic and tenuinucellate; carpels most commonly 2, occasionally 3-5 or more..VI. Asteridae

I. MAGNOLIIDAE

1. Plants with ethereal oil cells.
   2. Plants woody; flowers normally developed, with an evident perianth of separate tepals which may or may not be differentiated into sepals and petals...1. Magnoliales
   2. Plants differing in one or more respects from the above, most species herbaceous.
      3. Flowers with much reduced or no perianth, tending to be crowded into a spadix; ovules orthotropous; seeds often with perisperm. ....2. Piperales
      3. Flowers with a well developed perianth typically consisting of a gamosepalous, more or less corolloid calyx, not crowded into a spadix; ovules anatropous; seeds never with perisperm......... ....3. Aristolochiales
1. Plants without ethereal oil cells.
   4. Plants aquatic, lacking vessels; placentation laminar; most species with uniaperturate or uniaperturate-derived pollen...4. Nymphaeales
   4. Plants terrestrial or occasionally aquatic, with vessels; placentation marginal; pollen triaperturate or triaperturate-derived.
      5. Gynoecium mostly apocarpous or seemingly or actually unicarpellate, seldom evidently syncarpous; sepals usually more than 2....... ....5. Ranunculales
      5. Gynoecium syncarpous; placentation parietal; sepals 2 or occasionally 3.....................................6. Papaverales

1. MAGNOLIALES

1. Flowers hypogynous; endosperm well developed; most families with several or numerous ovules per carpel; pollen mostly uniaperturate or triaperturate.

2. Pollen uniaperturate; stamens often more or less laminar.
   3. Nodes unilacunar, two-trace; stamens 6-9.
      4. Stamens laminar; tepals numerous; carpels 6-14; lianas with opposite, exstipulate leaves............1. Austrobaileyaceae
      4. Stamens with normal filament and anther; tepals 3; carpels 3; erect shrubs with alternate, stipulate leaves...2. Lactoridaceae
   3. Nodes trilacunar or multilacunar; stamens few to more often numerous.
      5. Leaves stipulate...............................3. Magnoliaceae
      5. Leaves exstipulate.
         6. Vessels absent; stamens usually laminar; endosperm not ruminate.................................4. Winteraceae
         6. Vessels present; stamens and endosperm various.
            7. Stamens laminar; perianth not trimerous.
               8. Carpel solitary, pluriovulate, unsealed; endosperm ruminate.........................5. Degeneriaceae
               8. Carpels several, uniovulate, largely or wholly closed; endosperm not ruminate.6.Himantandraceae
            7. Stamens not laminar; perianth most commonly trimerous.
               9. Stamens distinct, with short, thick filament and with the connective surpassing the anther and enlarged above it; endosperm ruminate; ovules 1-many...........................7. Annonaceae
               9. Stamens with the filaments united; endosperm not ruminate.
                  10. Carpels separate; ovule solitary; flowers unisexual; filaments united into a solid solumn ....8. Myristicaceae
                  10. Carpels united into a compound pistil with parietal placentation; ovules 2-many; flowers perfect; filaments united into a tube.........................9. Canellaceae
2. Pollen triaperturate; stamens with normal filament and anther; nodes unilacunar; leaves alternate, exstipulate.
   11. Flowers perfect; trees; carpels in one whorl; fruit follicular ....10. Illiciaceae
   11. Flowers unisexual; lianas; carpels spirally arranged, becoming fleshy in fruit..............................11. Schisandraceae

1. Flowers perigynous to epigynous (except in Amborellaceae and Trimeniaceae); ovules only 1 or 2 per carpel (except in Eupomatiaceae); pollen various but seldom either uniaperturate or triaperturate; stipules none.
   12. Nodes multilacunar; ovules numerous; endosperm ruminate; leaves alternate; pollen biaperturate....................12. Eupomatiaceae
   12. Nodes unilacunar; ovule 1 (or 2 but only 1 maturing); endosperm, when present, not ruminate.
      13. Vessels absent; stamens numerous; leaves alternate...13. Amborellaceae
      13. Vessels present; stamens numerous or often few.
         14. Endosperm present; leaves opposite.
            15. Ovary superior; flowers mostly unisexual.
               16. Flowers hypogynous; carpel 1 (2)......14. Trimeniaceae
               16. Flowers perigynous; carpels (1) several to many..15. Monimiaceae

15. Ovary inferior; flowers perfect; carpels 2-3..16. Gomortegaceae

14. Endosperm wanting.

17. Carpels numerous; pollen biaperturate or rarely triaperturate; leaves opposite..............17. Calycanthaceae

17. Carpel 1 (or seemingly so); pollen nonaperturate or rarely uniaperturate; leaves mostly alternate.

18. Ovary superior.........................18. Lauraceae

18. Ovary inferior......................19. Hernandiaceae

2. PIPERALES

1. Nodes unilacunar; leaves opposite; ovary more or less inferior, unicarpellate, with a single ovule; seed with copious endosperm and no perisperm...1. Chloranthaceae

1. Nodes trilacunar or multilacunar; leaves mostly alternate; ovary superior, except in some Saururaceae; seeds with copious perisperm and scanty endosperm.

2. Ovules (1) 2-10 per carpel; carpels distinct or united into a compound ovary with parietal or modified laminar placentation...2. Saururaceae

2. Ovule solitary in the single locule of the compound or pseudomonomerous pistil.......................................3. Piperaceae

3. ARISTOLOCHIALES

Represented by...1. Aristolochiaceae

4. NYMPHAEALES

1. Plants with roots; flowers perfect; carpels 1-many; leaves alternate, entire to dissected; integuments 2.

2. Pollen of uniaperturate or uniaperturate-derived type; seeds with small embryo, some endosperm, and abundant perisperm...1. Nymphaeaceae

2. Pollen triaperturate; seeds with large embryo and no endosperm or perisperm....................................2. Nelumbonaceae

1. Plants rootless, free-floating; flowers unisexual; carpel 1; leaves whorled, cleft with slender segments; integument 1; pollen uniaperturate...3. Ceratophyllaceae

5. RANUNCULALES

1. Gynoecium mostly of separate carpels, or seemingly or actually of a single carpel; endosperm copious to scanty or none; herbs and woody vines, seldom shrubs.

2. Leaves mostly alternate; petals not adhering to the carpels; flowers variously organized, often trimerous in part or with numerous stamens and carpels, never pentamerous throughout.

3. Flowers perfect; herbs (including herbaceous vines) or shrubs; endosperm well developed.

4. Leaves with dichotomous, free venation; nodes unilacunar; ovules orthotropous; flowers reduced, without petals or petaloid nectaries.........................2. Circaeasteraceae

4. Leaves with net venation; nodes trilacunar to multilacunar; ovules anatropous; flowers usually well developed, often with nectariferous petals.
    5. Carpels usually 2 or more and distinct, seldom solitary or weakly united; stamens usually numerous, the anthers opening by longitudinal slits; herbs, rarely shrubs....................................1. Ranunculaceae
    5. Carpel seemingly solitary (the pistil probably actually 2-3 carpellate with all but one carpel obsolete); stamens typically 1-several sets of 3, usually opening by valves; herbs and shrubs............3. Berberidaceae

3. Flowers unisexual; nearly all woody plants, mostly vines.
    6. Leaves mostly compound; seeds with small embryo and abundant endosperm; petals when present nectariferous; mostly monoecious...............................4. Lardizabalaceae
    6. Leaves mostly simple; seeds mostly with large embryo and scanty endosperm; petals not nectariferous; dioecious...5. Menispermaceae

2. Leaves opposite or whorled; petals persistent, enlarging in fruit and adhering to the individual carpels to produce a set of pseudodrupes; flowers wholly pentamerous; nodes unilacunar.........6. Coriariaceae

1. Gynoecium obviously of united carpels; endosperm scanty or none; trees and shrubs.
    7. Flowers diplostemonous, the inner set of stamens modified into staminodes; carpels 2 but only one fertile, this with one ovule...7. Corynocarpaceae
    7. Flowers haplostemonous; carpels 2-3, each with 1 or 2 ovules...8. Sabiaceae

6. PAPAVERALES

1. Stamens numerous; flowers regular; petals neither spurred nor saccate; latex system more or less well developed................1. Papaveraceae
1. Stamens 4 or 6; flowers usually irregular; usually some of the petals spurred or saccate; latex system wanting or occasionally represented by scattered elongate secretory cells.......................2. Fumariaceae

II. HAMAMELIDAE

1. Flowers either with distinct carpels (a single carpel in Didymeles, of the Hamamelidales), or with dehiscent fruits, or both, variously perfect or unisexual; ovules 1-many per carpel; seeds endospermous, with small to large embryo.
    2. Vessels wanting; some or all of the flowers perfect; carpels 4 or more, in a single whorl, laterally coherent but scarcely forming a compound pistil, the group of carpels ripening into a follicetum; ovules several to many in each carpel; stipules wanting or represented only by a pair of flanges on the petiole; leaves with unique, elongate (often branched) idioblasts.............1. Trochodendrales
    2. Vessels present, flowers perfect or more often unisexual; carpels either 1-several and separate, or 2-4 and united (at least below) to form a compound pistil; ovules 1-many; stipules nearly always present except in Euptelea and Didymeles; no conspicuous idioblasts...................................2. Hamamelidales

1. Flowers either with united carpels, or with a pseudomonomerous, seemingly unicarpellate pistil; fruits indehiscent; ovules not more than 2 per carpel; flowers in most families unisexual; seeds with or without endosperm, but the embryo always fairly large.
   3. Leaves pinnately compound or trifoliolate; ovary 1-2 locular, with 1-2 ovules per locule, or one locule empty..........6. Juglandales
   3. Leaves simple (though sometimes deeply lobed or cleft); gynoecium various.
      4. Flowers, or at least the male flowers, not in catkins; ovary unilocular or rarely bilocular; calyx often present; plants woody or herbaceous.
         5. Ovules 2, unitegmic, intermediate between crassinucellate and tenuinucellate; stipules none; nodes unilacunar; flowers solitary, without a perianth................3. Eucommiales
         5. Ovule solitary, bitegmic, crassinucellate; stipules usually present; nodes trilacunar to multilacunar; flowers clustered, usually with a vestigial perianth.......4. Urticales
      4. Flowers, or at least the male flowers, usually in catkins (except some Fagaceae); ovary with 1-several locules; calyx mostly wanting or very much reduced; plants woody.
         6. Pistil pseudomonomerous, seemingly composed of a single carpel, with a single style and locule; ovary superior...5. Leitneriales
         6. Pistil composed of more than one carpel, as evidenced by the presence of more than one style and often also more than one locule; ovary inferior or nude.
            7. Leaves opposite or alternate, seldom whorled, always more or less well developed; ovules each with a single embryo sac.
               8. Ovule solitary, orthotropous (except in Canacomyrica), unitegmic; plants aromatic, the leaves resinous-dotted..............................7. Myricales
               8. Ovules more than one, anatropous, bitegmic or unitegmic; plants not notably aromatic......8. Fagales
            7. Leaves whorled, reduced to scales; ovules with multiple embryo-sacs.............................9. Casuarinales

1. TROCHODENDRALES

1. Perianth present; stipular flanges present; stamens 4, opposite the sepals; leaves palmately veined; idioblasts simple or branched, secretory...1. Tetracentraceae
1. Perianth none; stipules none; stamens numerous; leaves pinnately veined; idioblasts branched, not secretory.................2. Trochodendraceae

2. HAMAMELIDALES

1. Carpels separate (or solitary).
   2. Ovules numerous; fruit follicular; plants dioecious; leaves stipulate, palmately veined, entire................1. Cercidiphyllaceae
   2. Ovules 1-3; fruit indehiscent; flowers variously perfect or unisexual; leaves various.
      3. Carpels several; stamens 3-many; plants monoecious or with perfect flowers.

4. Stamens numerous; ovules more or less anatropous; flowers perfect; connective of ordinary type, not apically enlarged and peltate; leaves exstipulate, pinnately veined..2. Eupteleaceae
4. Stamens 3-4 (7); ovules more or less orthotropous; plants monoecious; connective apically enlarged and peltate; leaves stipulate, palmately veined...........3. Platanaceae
3. Carpel solitary; stamens 2, united by their filaments; plants dioecious......................................4. Didymelaceae
1. Carpels united (at least below) into a compound pistil.
5. Leaves alternate (rarely opposite); carpels 2 (rarely 3); flowers perfect or unisexual, with at least a vestigial calyx and often with a corolla as well; filaments free; ovules 1-many; trees or shrubs.......................................5. Hamamelidaceae
5. Leaves opposite; carpels 3-4; plants dioecious; perianth none; filaments united; ovules many; undershrubs............6. Myrothamnaceae

3. EUCOMMIALES

Represented by...1. Eucommiaceae

4. URTICALES

1. Trees, without milky juice; seeds with straight embryo and no endosperm; ovule pendulous, anatropous; anthers erect in bud.
2. Leaves alternate, stipulate; styles 2....................1. Ulmaceae
2. Leaves opposite, exstipulate; style 1..................2. Barbeyaceae
1. Plants either not woody, or with milky juice, or with endospermous seeds, often with 2 or even all 3 of these features; anthers inflexed in bud, or less often erect.
3. Styles or style branches 2, sometimes one of them reduced; ovules mostly anatropous and pendulous; embryo straight or more often curved.
4. Woody plants, seldom herbs, mostly with milky juice; anthers usually inflexed in bud...........................3. Moraceae
4. Herbs, without milky juice; anthers erect in bud....4. Cannabaceae
3. Style 1, unbranched; ovule orthotropous, basal; embryo straight; herbs, seldom shrubs or soft-wooded trees, without milky juice...5. Urticaceae

5. LEITNERIALES

Represented by...1. Leitneriaceae

6. JUGLANDALES

1. Flowers in triplets, the middle one perfect and fertile, with 4 sepals, 6 stamens, and a superior, bilocular ovary, one locule empty, the other with a single ovule attached to the partition; lateral flowers female but sterile; leaves pinnately compound, stipulate; fruit samaroid...1. Rhoipteleaceae
1. Flowers unisexual, the staminate ones with 3-many stamens, borne in catkins, the calyx wanting or adnate to the bract; female flowers solitary or spicate, with 1-4 sepals or calyx teeth; fruit drupaceous or nutlike.

2. Ovules 4 (2 in each of 2 locules), apical; ovary superior; leaves trifoliate, minutely stipulate..................2. Picrodendraceae
2. Ovule solitary, basal, in a unilocular, inferior ovary; leaves pinnately compound, exstipulate.......................3. Juglandaceae

7. MYRICALES

Represented by...1. Myricaceae

8. FAGALES

1. Seeds with endosperm; leaves exstipulate; ovules erect, basal or nearly so, unitegmic; plants dioecious; pistillate flower solitary in a multibracteate involucre which ripens into a hull subtending the fruit...... ....1. Balanopaceae
1. Seeds without endosperm; leaves stipulate; ovules pendulous, apical or nearly so; plants monoecious or seldom dioecious, or the flowers rarely perfect.
   2. Styles and locules 3 (seldom 6); integuments 2; female flowers mostly not in catkins, subtended individually or in small groups by an involucre of many bracts which are concrescent at maturity into an indurated hull...2. Fagaceae
   2. Styles and locules 2 (seldom 3); integument 1; female flowers mostly borne in catkins; fruits without a hull of bracts, or sometimes with a foliaceous hull derived from 2 or a few bracts.3. Betulaceae

9. CASUARINALES

Represented by..1. Casuarinaceae

III. CARYOPHYLLIDAE

1. Seeds with very scanty or no true endosperm, but very often with an evident perisperm; betalains often present; plants often with anomalous secondary thickening.
   2. Ovules campylotropous or amphitropous, seldom anatropous; seeds with a curved embryo that is typically peripheral to a well defined perisperm; pollen trinucleate; flowers seldom very much reduced or borne in catkins, but if so (some Chenopodiaceae) then with unilocular, uniovulate ovary.....................1. Caryophyllales
   2. Ovules anatropous; seeds with nearly or quite straight embryo and neither endosperm nor perisperm; pollen binucleate; flowers very much reduced, borne in unisexual catkins; ovary 4-locular, with one ovule per locule; plants lacking both anthocyanins and betalains... ....2. Batales
1. Seeds with more or less copious endosperm, but without perisperm; embryo straight or curved, peripheral or often embedded in the endosperm; ovary unilocular, with a single basal ovule; no betalains; no anomalous secondary thickening.
   3. Perianth undifferentiated, or with 2 sets of more or less similar tepals, not dichlamydeous and sympetalous; carpels (2) 3 (4); stamens mostly 6-9, in 2 or 3 cycles; plants often with sheathing stipules...3. Polygonales
   3. Perianth differentiated into calyx and corolla, the corolla sympetalous; carpels 5; stamens 5, opposite the petals; leaves mostly exstipulate...................................4. Plumbaginales

1. CARYOPHYLLALES

1. Gynoecium apocarpous, monocarpous, or sometimes syncarpous, if syncarpous then usually with as many locules as carpels; ovules 1 per locule, or some locules empty.
   2. Locules all ovuliferous (though not always fertile); perisperm well developed; flowers perfect or less often unisexual; plants not cactuslike.
      3. Sepals mostly free or nearly so, seldom petaloid; carpels mostly 2-several, occasionally only one; inflorescence usually racemose or spicate; leaves alternate; stem seldom with anomalous secondary thickening..........................1. Phytolaccaceae
      3. Sepals joined into a tube which resembles a sympetalous corolla, and often subtended by sepaloid bracts; carpel 1; inflorescence mostly cymose; leaves opposite or sometimes alternate; stem very often with anomalous secondary thickening..2. Nyctaginaceae
   2. Two locules empty, the third with a single ovule; perisperm apparently wanting; flowers unisexual; inflorescence cymose; spiny, cactuslike trees..3. Didiereaceae
1. Gynoecium mostly syncarpous, either unilocular or with the ovules more numerous than the locules, or both.
   4. Tepals and stamens usually more or less numerous; plants succulent; ovary very often inferior.
      5. Ovary unilocular, nearly always inferior, with parietal (rarely basal) placentation; mostly spiny stem-succulents with much reduced leaves; New World..........................4. Cactaceae
      5. Ovary plurilocular, or occasionally unilocular with a free-central placenta; ovary superior to inferior; leaf-succulents, not spiny; chiefly Old World..........................5. Aizoaceae
   4. Tepals and stamens mostly few and cyclic; plants sometimes with succulent leaves, but not spiny or cactuslike; ovary superior.
      6. Perianth usually dichlamydeous or seemingly so (except notably the Molluginaceae, with axile placentation).
         7. Sepals 4 or 5; plants without betalains, but usually with anthocyanins; stamens not at once of the same number as and opposite the petals; plants sometimes with anomalous secondary thickening.
            8. Ovary 2-5-locular, with axile placentation; petals much reduced or commonly wanting; leaves opposite, alternate or whorled.......................6. Molluginaceae
            8. Ovary unilocular (sometimes partitioned toward the base, with free-central or basal placentation; petals usually more or less well developed; leaves opposite...........
            ....7. Caryophyllaceae
         7. Sepals (or sepaloid bracts) mostly 2, seldom more; plants with betalains, lacking anthocyanins; stamens usually as many as and opposite the petals (or petaloid sepals), or sometimes more numerous; no anomalous secondary thickening.
            9. Ovules (1) 2-many; plants not climbing; internal phloem usually wanting; fruit usually capsular.8. Portulacaceae
            9. Ovule solitary; climbers; internal phloem present; fruit indehiscent............................9. Basellaceae

6. Perianth monochlamydeous, seldom petaloid, often small or sometimes even obsolete; ovules mostly solitary (rarely several) on a basal placenta; stem nearly always with anomalous secondary thickening.
   10. Perianth not scarious, often green; filaments mostly free; flowers perfect or not infrequently unisexual..............
   ....10. Chenopodiaceae
   10. Perianth more or less scarious; filaments mostly connate below (the androecial tube sometimes even simulating a sympetalous corolla); flowers mostly perfect, seldom unisexual
   ....11. Amaranthaceae

2. BATALES

Represented by..1. Bataceae

3. POLYGONALES

Represented by..1. Polygonaceae

4. PLUMBAGINALES

Represented by..1. Plumbaginaceae

IV. DILLENIIDAE

1. Carpels separate; stamens numerous; ovules bitegmic, crassinucellate; seeds arillate and with a well developed endosperm.......1. Dilleniales
1. Carpels united to form a compound pistil; other characters various, the stamens numerous to few, the ovules bitegmic or unitegmic, crassinucellate to tenuinucellate, the seeds with or without an aril and with or without endosperm.
   2. Flowers mostly polypetalous or apetalous, seldom (mainly some Violales) sympetalous; stamens numerous to few; ovules chiefly bitegmic and crassinucellate, but sometimes unitegmic, or tenuinucellate, or both.
   3. Plants insectivorous herbs or shrubs............5. Sarraceniales
   3. Plants not insectivorous; habit various.
      4. Placentation mostly axile, seldom parietal.
         5. Sepals imbricate; filaments free or connate in groups; mucilage cells and sacs mostly wanting.......2. Theales
         5. Sepals valvate; filaments often connate into a tube or column; mucilage cells or sacs very often present.
            6. Ovary superior; pubescence very often stellate or lepidote..............................3. Malvales
            6. Ovary inferior; no stellate or lepidote hairs........
            ....4. Lecythidales
      4. Placentation mostly parietal, seldom axile.
         7. Flowers normally developed, with an evident perianth, perfect or less often unisexual, not borne in catkins.
            8. Plants usually without myrosin cells, very often woody; flowers hypogynous to often perigynous; or epigynous; leaves mostly simple; carpels mostly 3 or more, seldom only 2; perianth seldom tetramerous..............................6. Violales

8. Plants commonly with myrosin cells, herbaceous or less often woody; flowers hypogynous; leaves simple to more often compound or variously dissected; carpels most commonly 2; perianth often tetramerous..8. Capparales

7. Flowers much reduced, unisexual, without an evident perianth, borne in catkins; woody plants.......7. Salicales

2. Flowers mostly sympetalous (except mainly some Ericales, these with unitegmic, tenuinucellate ovules); stamens seldom more than 2 or 3 times as many as the corolla lobes.

9. Placentation axile or less often parietal; stamens various, often more numerous than or alternate with the corolla lobes.

10. Plants with the pollen in tetrads (sometimes only one grain of the tetrad maturing), or with poricidal anthers, or lacking chlorophyll, often with more than one of these characters; ovules unitegmic, tenuinucellate.................9. Ericales

10. Plants otherwise, the pollen in monads, the anthers opening by longitudinal slits, the shoot always chlorophyllous; ovules bitegmic or unitegmic, crassinucellate or tenuinucellate.

11. Plants dwarf (commonly prostrate) shrubs; functional stamens as many as and alternate with the corolla lobes; ovules unitegmic.......................10. Diapensiales

11. Plants shrubs or trees; stamens usually either more numerous than or opposite the corolla lobes; ovules bitegmic or unitegmic............................11. Ebenales

9. Placentation free-central or basal; functional stamens as many as and opposite to the corolla lobes; ovules chiefly bitegmic...12. Primulales

1. DILLENIALES

1. Flowers hypogynous, with neither a hypanthium nor an intrastaminal disk; raphides and crystal sand present, often in long sacs or tubes; leaves mostly entire; trees, woody vines, and shrubs, seldom herbs, of tropical and subtropical regions, especially Australia........1. Dilleniaceae

1. Flowers perigynous, or with an intrastaminal disk; raphides and crystal sand absent; North Temperate regions.

2. Seed with copious endosperm and small embryo; flowers with an intrastaminal disk, but without an evident hypanthium; mesophytic shrubs or herbs with large, usually cleft leaves............2. Paeoniaceae

2. Seed with thin endosperm and large embryo; flowers with an evident hypanthium, but without an intrastaminal disk; xerophytic shrubs with small, entire leaves.......................3. Crossosomataceae

2. THEALES

1. Leaves mostly alternate; endosperm present or absent.

2. Leaves mostly stipulate.

3. Leaves simple, or seldom pinnately compound; embryo of normal proportions.

4. Connective not exserted; calyx not winged; endosperm present in most species.

5. Style mostly gynobasic; anthers mostly opening by terminal pores; plants lacking mucilage cells; mostly not of Madagascar.............................1. Ochnaceae

5. Style not gynobasic (except in one sp. of Rhopalocarpaceae); anthers opening by longitudinal slits; plants with mucilage cells; confined to Madagascar.
   6. Flowers with a prominent intrastaminal disk; filaments irregularly connate toward the base; pollen in monads......................2. Rhopalocarpaceae
   6. Flowers without an intrastaminal disk, the stamens inserted within a ring of staminodes; filaments free; pollen in tetrads...........3. Sarcolaenaceae

4. Connective prominently exserted; calyx mostly winged in fruit; endosperm mostly wanting.........4. Dipterocarpaceae

3. Leaves palmately compound; embryo with enlarged, thickened hypocotyl and reduced cotyledons..................5. Caryocaraceae

2. Leaves exstipulate, except in some Theaceae.
   7. Erect trees or shrubs.
      8. Petals usually 5; stamens mostly numerous, seldom only one or 2 cycles; stigmas (often also the styles) distinct...6. Theaceae
      8. Petals 4; stamens 8; style with a capitate, merely lobed stigma.................................7. Stachyuraceae
   7. Trailing or climbing shrubs or vines.
      9. Endosperm copious; integument 1; bracts not highly modified...8. Actinidiaceae
      9. Endosperm scanty or none; integuments 2; bracts highly modified, becoming pitcherlike, saccate, or spurred.......................................9. Marcgraviaceae

1. Leaves mostly opposite or whorled; endosperm wanting.
   10. Leaves stipulate.
      11. Trees; stamens 15 or more; fruit a berry...........10. Quiinaceae
      11. Herbs or low shrubs; stamens 10 or fewer; fruit capsular....................................11. Elatinaceae
   10. Leaves exstipulate.
      12. Ovary 17-25-celled; ovules 2 per cell; no secretory canals...12. Medusagynaceae
      12. Ovary (1) 3-5 (15)-celled; ovules 1-many per cell; conspicuous secretory canals or cavities in all organs.......13. Guttiferae

3. MALVALES

1. Petals valvate; plants mostly without mucilage cells and always without mucilage sacs or canals; pubescence not stellate or lepidote.
   2. Leaves stipulate; sepals separate to the base or nearly so; petals mostly separate; integuments 2...................1. Elaeocarpaceae
   2. Leaves exstipulate; sepals connate into a cup or tube; petals often connate below; integument 1....................2. Scytopetalaceae
1. Petals imbricate or contorted; plants mostly with mucilage cells or often with mucilage sacs or canals; pubescence mostly stellate or lepidote.
   3. Anthers bilocular; filaments free or connate; mostly trees and shrubs, a few herbs.
      4. Stamens numerous, mostly free, arranged in a single whorl...3. Tiliaceae
      4. Stamens (including staminodes) 10-numerous, in two or more whorls, mostly connate by their filaments, seldom only 5 and then always connate...............................4. Sterculiaceae
   3. Anthers mostly unilocular; filaments connate.

5. Pollen smooth, triporate; fruit loculicidally dehiscent, or indehiscent; mostly trees and shrubs, a few herbs......5. Bombacaceae
5. Pollen minutely spiny, mostly multiporate; fruit typically septicidally dehiscent or schizocarpic, seldom loculicidally dehiscent; plants herbaceous, or occasionally softly woody..6. Malvaceae

4. LECYTHIDALES

Represented by...1. Lecythidaceae

5. SARRACENIALES

1. Leaves, or some of them, modified to form pitchers; placentation axile; style solitary, or sometimes very short or none.
 2. Flowers perfect; filaments distinct; pollen grains in monads; ovules unitegmic, tenuinucellate; herbs, not climbing; New World..1. Sarraceniaceae
 2. Flowers unisexual; filaments united into a column; pollen grains in tetrads; ovules bitegmic, crassinucellate; herbs or shrubs, often climbing; Old World.................................2. Nepenthaceae
1. Leaves not forming pitchers; placentation parietal; styles several...3. Droseraceae

6. VIOLALES

1. Ovary usually superior; plants of various habit, most families with a well developed endosperm.
 2. Flowers polypetalous or sometimes apetalous; endosperm mostly well developed.
  3. Endosperm oily; many species cyanogenetic.
   4. Stamens mostly numerous, seldom only 5 or fewer; trees or shrubs, seldom climbing.
    5. Ovules mostly on parietal placentae; petals usually present (sometimes scarcely distinguishable from the sepals); carpels 2-10, with free or united styles..1. Flacourtiaceae
    5. Ovules hanging from an apical placenta; petals none; carpels 3-4; styles free..................2. Peridiscaceae
   4. Stamens mostly numerous, seldom only 5 or fewer; trees or sometimes climbing.
    6. Carpels 9-12; stamens 3; flowers unisexual; ovules numerous on a broad, basal placenta; woody plants..3. Scyphostegiaceae
    6. Carpels 2-5; stamens 5; flowers mostly perfect; ovules on parietal placentae.
     7. Flowers without a corona; trees, shrubs, or herbs, not climbing.
      8. Flowers hypogynous, with a single style, often irregular..........................4. Violaceae
      8. Flowers perigynous, with separate styles, regular...............................5. Turneraceae
     7. Flowers with a corona; herbs and shrubs, often climbing.

9. Seeds arillate; plants often climbing by tendrils; flowers hypogynous to somewhat perigynous; sepals imbricate; petals imbricate or seldom wanting.......................6. Passifloraceae
9. Seeds exarillate; undershrubs or herbs, not climbing; flowers strongly perigynous; sepals and petals valvate...........7. Malesherbiaceae
3. Endosperm starchy; few species cyanogenetic.
10. Plants variously woody or herbaceous, but not climbing; ovules several or many, included in the fruit.
11. Stamens mostly numerous; style 1; leaves small or more often of ordinary size.
12. Leaves alternate, stipulate, palmately lobed or veined; ovules anatropous; upright shrubs or small trees..................................8. Bixaceae
12. Leaves opposite, exstipulate, entire, not palmate; ovules orthotropous or seldom anatropous; herbs or low shrubs.........................9. Cistaceae
11. Stamens mostly 4-10, seldom numerous; styles 1-several; leaves small, often ericoid or reduced to scales; herbs or more often shrubs or trees.
13. Leaves alternate; sepals connate below; styles 3-4... ....10. Tamaricaceae
13. Leaves opposite; sepals free; style 1............... ....11. Frankeniaceae
10. Plants woody climbers with stout, hooked branch tips or leaf tips.
14. Ovules numerous, on 2-5 parietal placentae, conspicuously exserted from the capsule at maturity; styles several; plants climbing by hooked leaf-tips.................. ....12. Dioncophyllaceae
14. Ovule solitary on a basal placenta, included in the mature nut; style solitary; plants climbing by hooked branch-tips.......................13. Ancistrocladaceae
2. Flowers sympetalous; endosperm scanty or none except in the Achariaceae.
15. Flowers perfect; leaves simple, entire or nearly so; stamens 10 or more.
16. Thorny, xerophytic shrubs with small leaves; carpels 3; fruit capsular; stamens 10-17............14. Fouquieriaceae
16. Unarmed, large-leaved trees; carpels 2; fruit drupaceous; stamens numerous.....................15. Hoplestigmataceae
15. Flowers unisexual; leaves simple or often lobed or compound; stamens 10 or less.
17. Endosperm well developed; plants herbaceous or nearly so and without a latex system; stamens 3-5; style 1.............. ....16. Achariaceae
17. Endosperm scanty or none; plants woody and with a well developed latex system; stamens 10 (5); styles free........ ....17. Caricaceae
1. Ovary mostly inferior; plants mostly herbs or herbaceous vines, a few trees and shrubs; endosperm scanty or none, except in some Loasaceae.
18. Flowers perfect; stamens mostly numerous; style 1; integument 1...... ....18. Loasaceae
18. Flowers mostly unisexual; stamens few or less often numerous; styles 1-several; integuments 2.

19. Stamens 4-many, all with bilocular anthers; plants without tendrils; styles free except in some Begoniaceae; petals separate or none.
    20. Sepals 2 (-5); leaves stipulate; carpels 2-5...19. <u>Begoniaceae</u>
    20. Sepals 3-9; leaves exstipulate; carpels 3-8....20. <u>Datiscaceae</u>
19. Stamens 1-5, typically 3 with one unilocular and two bilocular anthers; mostly tendriliferous vines; style mostly single; corolla mostly sympetalous....................21. <u>Cucurbitaceae</u>

7. SALICALES

Represented by...1. <u>Salicaceae</u>

8. CAPPARALES

1. Placentation axile or seemingly so; sepals, petals, and stamens 8 each; ovary closed, with 5-8 carpels; endosperm present; coarse herbs or soft shrubs with trifoliate leaves......................1. <u>Tovariaceae</u>
1. Placentation nearly always parietal (sometimes with intruded placentae) or seemingly so; sepals, petals, and carpels usually fewer; seeds with little or no endosperm.
    2. Flowers hypogynous.
        3. Flowers mostly regular or only slightly irregular; sepals mostly 4; carpels mostly 2 (or seemingly so); style short or elongate, with capitate or slightly lobed stigma.
            4. Ovary mostly 1-celled; stamens 4-many, never tetradynamous as in the next family; flowers almost always with an evident gynophore or androgynophore; pollen binucleate; leaves simple to trifoliolate or palmately compound, not much dissected; herbs, shrubs, or trees, chiefly tropical and subtropical.................................2. <u>Capparaceae</u>
            4. Ovary mostly 2-celled, with the ovules attached to the margins of the partition; stamens typically 6 and tetradynamous (the outer ones shorter than the 4 inner ones), seldom fewer; flowers rarely with a gynophore or androgynophore; pollen trinucleate; leaves often more or less dissected in pinnate fashion, but without distinct, articulated leaflets; herbs, seldom shrubs, chiefly of temperate regions....3. <u>Cruciferae</u>
        3. Flowers usually evidently irregular; sepals (4) 5-6 (-8); carpels 2-6; stigmas sessile, separate, as many as the carpels; ovary often open at the top; herbs.....................4. <u>Resedaceae</u>
    2. Flowers perigynous; sepals, petals and functional stamens each 5; carpels 3; trees with pinnately decompound leaves........5. <u>Moringaceae</u>

9. ERICALES

1. Embryo normally developed, with 2 cotyledons; plants more or less woody, always chlorophyllous.
    2. Pollen grains borne singly; stamens mostly twice as many as the petals, commonly opening by terminal pores.
        3. Ovules 1 (-3) per locule; petals commonly joined at the base; seed without a seed coat; carpels 2-5.................1. <u>Cyrillaceae</u>
        3. Ovules numerous; petals free; seed with a seed coat; carpels 3.... ....2. <u>Clethraceae</u>

2. Pollen grains nearly always borne in tetrads (in some Epacridaceae only one grain of the tetrad matures).
    4. Sepals and petals each 4-7; corolla mostly sympetalous; habit various.
        5. Stamens mostly twice as many as the corolla lobes; leaves rarely palmately veined; anthers mostly opening by terminal pores, often appendiculate; widespread.....3. Ericaceae
        5. Stamens mostly as many as the corolla lobes; leaves mostly palmately veined; anthers opening by longitudinal slits, not appendaged; chiefly Australian..........4. Epacridaceae
    4. Stamens and petals each 1-3, the petals separate; stamens opening by longitudinal slits, not appendaged; dwarf, often prostate, evergreen shrubs..................................5. Empetraceae

1. Embryo very small and scarcely differentiated, without cotyledons; herbs or half-shrubs, often without chlorophyll.
    6. Plants usually with green leaves; anthers opening by pores; pollen usually in tetrads; petals separate; placentation axile...6. Pyrolaceae
    6. Plants without chlorophyll, the leaves reduced to mere scales; anthers opening by longitudinal slits; pollen grains borne singly; petals separate or united; placentation axile or parietal...7. Monotropaceae

10. DIAPENSIALES

Represented by...1. Diapensiaceae

11. EBENALES

1. Plants with a well developed latex system; pubescence of 2-armed hairs, one arm sometimes reduced or obsolete; nodes mostly trilacunar...1. Sapotaceae
1. Plants without a latex system; pubescence not of 2-armed hairs except in some Ebenaceae; nodes unilacunar.
    2. Flowers mostly unisexual; styles separate at least distally...2. Ebenaceae
    2. Flowers perfect; style solitary, with a capitate or slightly lobed stigma.
        3. Pubescence of stellate or peltate hairs; fruit mostly dry; ovary superior to less often inferior; anthers more or less linear...3. Styracaceae
        3. Pubescence of simple hairs (or none); fruit more or less fleshy; ovary inferior (sometimes only half inferior).
            4. Anthers linear; flowers with a corona; stamens 8 (twice as many as the corolla lobes)................4. Lissocarpaceae
            4. Anthers ovate or rotund; flowers without a corona; stamens usually more than 8........................5. Symplocaceae

12. PRIMULALES

1. Plants woody, mostly arborescent, largely tropical; fruits mostly fleshy, often 1-seeded (even though there are several or many ovules in the ovary.
    2. Flowers with staminodes alternate with the corolla-lobes; plants without an evident secretory system, the leaves not gland-dotted...1. Theophrastaceae

2. Flowers without staminodes; plants with a schizogenous secretory system and gland-dotted leaves........................2. Myrsinaceae

1. Plants herbaceous or occasionally half-shrubby, chiefly of temperate regions or altitudes; fruit capsular, with (2-) many seeds.............
....3. Primulaceae

V. ROSIDAE

1. Plants relatively primitive, the flowers commonly either apocarpous or with the carpels united only toward the base or by their styles, or with more or less numerous ovules per carpel, or with more or less numerous stamens, often with more than one of these features; vascular bundles never with internal phloem; plants never parasitic (though sometimes insectivorous).....................................1. Rosales

1. Plants more advanced in one or another respect, the flowers mostly syncarpous (or pseudomonomerous), seldom apocarpous, typically with not more than twice as many stamens as sepals or petals (stamens numerous in many Myrtales and in some others), very often with only 1 or 2 ovules per carpel; vascular bundles sometimes with internal phloem; plants autotrophic or sometimes parasitic.

   2. Vascular bundles mostly with internal (as well as external) phloem; flowers strongly perigynous to epigynous, often with numerous stamens and numerous ovules................................4. Myrtales

   2. Vascular bundles without internal phloem; flowers various.

      3. Flowers pseudomonomerous, strongly perigynous.........5. Proteales

      3. Flowers evidently with more than one carpel, or not strongly perigynous, or both.

         4. Plants mostly parasitic or hemiparasitic (wholly autotrophic in some Santalales).

            5. Ovules 1-8; plants with or without chlorophyll...........
            ....7. Santalales

            5. Ovules very numerous; plants without chlorophyll..........
            ....8. Rafflesiales

         4. Plants (with few exceptions) autotrophic.

            6. Flowers of ordinary type, not markedly reduced except in a few small families such as the Garryaceae (Cornales).

               7. Leaves mostly compound or conspicuously lobed or cleft, less often simple and unlobed.

                  8. Ovary superior; pollen variously binucleate or trinucleate.

                     9. Plants mostly woody.............12. Sapindales

                     9. Plants mostly herbaceous or nearly so.........
                     ....13. Geraniales

                  8. Ovary inferior; pollen trinucleate..16. Umbellales

               7. Leaves simple and entire or merely toothed, only seldom compound.

                  10. Ovary mostly superior.

                     11. Flowers regular.

                        12. Stamens basally connate, usually more than 5; disk usually wanting; pollen trinucleate...................14. Linales

                        12. Stamens free from each other, seldom more than 5; disk often present.

13. Stamens alternate with the petals, or the flowers rarely diplostemonous; pollen variously binucleate or trinucleate.............9. Celastrales
13. Stamens opposite the petals; pollen binucleate...........11. Rhamnales
11. Flowers irregular (except in Tremandraceae)..........................15. Polygalales
10. Ovary inferior.......................6. Cornales
6. Flowers more or less reduced and often unisexual, the perianth poorly developed or wanting.
14. Endosperm wanting; submerged aquatics.2. Podostemales
14. Endosperm present; habitat various.
15. Herbs with an inferior ovary and without milky juice; flowers not forming pseudanthia...................................3. Haloragales
15. Trees, shrubs, or herbs, with superior or naked ovary and often with milky juice; flowers often grouped into pseudanthia.......10. Euphorbiales

1. ROSALES

1. Seeds mostly with endosperm (except notably the Davidsoniaceae and a few Grossulariaceae); leaves with or without stipules.
2. Plants more or less distinctly woody, not succulent (herbaceous only in some Byblidaceae, marked by the conspicuous, glandular, insect-catching hairs); carpels nearly always united into a compound pistil (some exceptions among the Cunoniaceae) (segregate order Cunoniales).
3. Leaves mostly compound (simple in some Cunoniaceae and Eucryphiaceae), stipulate; flowers hypogynous, with axile placentation, or the carpels rarely free.
4. Leaves opposite or whorled; seeds with endosperm.
5. Carpels 5-12 (18), multiovulate; stamens numerous...1. Eucryphiaceae
5. Carpels 2-5, 1-many-ovulate; stamens as many or twice as many as the sepals, seldom numerous......2. Cunoniaceae
4. Leaves alternate; seeds without endosperm...3. Davidsoniaceae
3. Leaves simple, entire to deeply cleft, exstipulate, or seldom with vestigial stipules; flowers hypogynous to epigynous, with axile or parietal placentae.
6. Flowers hypogynous or slightly perigynous; ovules unitegmic and, so far as known, tenuinucellate.
7. Stamens 5; corolla polypetalous to more or less sympetalous.
8. Leaves without insect-catching hairs; corolla usually more or less sympetalous; half-shrubs or more often shrubs or trees; anthers opening by elongate slits or less often by pores...........4. Pittosporaceae
8. Leaves with insect-catching hairs; corolla strictly polypetalous; herbs or half-shrubs; anthers opening by apical pores or short slits.......5. Byblidaceae
7. Stamens 2; corolla sympetalous..........6. Columelliaceae
6. Flowers usually partly or wholly epigynous, seldom hypogynous; ovules diverse.

9. Leaves opposite; stamens 10-numerous, seldom only 5; ovules few to many, unitegmic, tenuinucellate...7. Hydrangeaceae

9. Leaves alternate; stamens 4-6, sometimes accompanied by as many staminodes; ovules various.

10. Petals separate, or sometimes wanting.

11. Ovules mostly numerous, on axile or parietal placentae, variously unitegmic or bitegmic, crassinucellate or tenuinucellate; leaves normally developed, the plant not heatherlike.......................................8. Grossulariaceae

11. Ovules 1-2 (8) in each cell of a 2-3-celled ovary, axile, unitegmic, crassinucellate; leaves mostly small, the plant often heatherlike.9. Bruniaceae

10. Petals connate, the corolla sympetalous.......................................10. Alseuosmiaceae

2. Plants either herbaceous, or succulent, or both, without insect-catching hairs; leaves exstipulate, alternate or seldom opposite; stamens as many or twice as many as the sepals; ovules mostly bitegmic and crassinucellate (segregate order Saxifragales).

12. Plants succulent; carpels as many as the petals, free or united only at the base.............................11. Crassulaceae

12. Plants scarcely or not at all succulent; carpels seldom the same number as the petals.

13. Carpels 6, free, uniovulate; sepals 6; petals none; stamens 12; some of the leaves modified into pitchers...12. Cephalotaceae

13. Carpels 2-5, usually more or less united, at least below, and usually with numerous ovules on parietal placentae; flowers mostly 4-5-merous, with 1 or 2 sets of stamens and usually with petals; leaves not modified into pitchers...13. Saxifragaceae

1. Seeds mostly without endosperm, except in a few Rosaceae; leaves mostly stipulate.

14. Carpels 1-many, when solitary the plants with simple leaves and only 1 or 2 ovules per carpel; stamens free, numerous or less often only 5 or 10 (Rosales, sens. strict.).

15. Gynoecium of 1-many free or less often united carpels, never with the carpels united only by a common style, nor with a solitary carpel that has a gynobasic style.

16. Gynoecium apocarpous, or syncarpous with 2-5 styles; stamens usually more than 10...........................14. Rosaceae

16. Gynoecium syncarpous, with 10 styles; stamens 10...15. Neuradaceae

15. Gynoecium of 2-3 carpels united only by a gynobasic style, or seemingly of a single carpel with a gynobasic style...16. Chrysobalanaceae

14. Carpel nearly always solitary, and usually with more than 2 ovules; leaves usually compound; stamens (5) 10-many, often connate by the filaments (segregate order Fabales)................17. Leguminosae

2. PODOSTEMALES

Represented by..1. Podostemaceae

3. HALORAGALES

1. Ovule apical; embryo straight; stamens 1-8.
    2. Styles 2-4; ovule bitegmic; stamens (1) 2-8.
        3. Ovary 2-4-locular; terrestrial or aquatic herbs with small, alternate or more often opposite or whorled, exstipulate leaves and small inflorescences; fruit not drupaceous; petals present or absent....................................1. Haloragaceae
        3. Ovary unilocular; terrestrial herbs with large, alternate, stipulate leaves and large inflorescences; fruit drupaceous; petals none...2. Gunneraceae
    2. Style 1; ovule unitegmic; stamen 1; emergent aquatics; leaves whorled, exstipulate..............................3. Hippuridaceae
1. Ovule basal, unitegmic; embryo strongly curved; stamens 6-28; terrestrial herbs with stipulate leaves that are opposite below and alternate above; ovary unilocular, with a single style...........4. Theligonaceae

4. MYRTALES

1. Ovary superior.
    2. Plants woody or less often herbaceous, mostly terrestrial, when aquatic not free-floating.
        3. Ovules 2-many per carpel; ovary 2-several-locular; petals present or absent; fruit capsular.
            4. Stamens numerous (seldom only 12); carpels 4-20; ovules numerous................................1. Sonneratiaceae
            4. Stamens 4-8 (12 in some Lythraceae).
                5. Flowers perfect; petals present or absent.
                    6. Filaments elongate; carpels 2-3, seldom more numerous, each with 2-many ovules; mostly diplostemonous...............................2. Lythraceae
                    6. Filaments very short; carpels 4, each with 2-4 ovules; haplostemonous................3. Penaeaceae
                5. Plants polygamo-dioecious; petals none; carpels 2, multiovulate............................4. Crypteroniaceae
        3. Ovule solitary in a pseudomonomerous pistil, or the ovules solitary in each of 2 or more locules; petals scale-like or none; fruit seldom capsular; mostly diplostemonous...5. Thymelaeaceae
    2. Plants free-floating annual aquatics; ovary bilocular, with a single pendulous ovule in each locule, only one locule maturing; haplostemonous...6. Trapaceae
1. Ovary in most species inferior.
    7. Placentation axile, seldom parietal; fruit seldom 1-seeded and indehiscent.
        8. Stamens numerous.
            9. Carpels 2-3, seldom more; leaves glandular-punctate (inconspicuously so in Dialypetalanthaceae).
                10. Fruit a many-seeded capsule; anthers opening by terminal pores; no internal phloem........7. Dialypetalanthaceae
                10. Fruit few-seeded (even when the ovary has many ovules), a berry, drupe, or capsule; anthers opening by slits or less often by terminal pores; mostly with internal phloem....................................8. Myrtaceae
            9. Carpels more or less numerous, commonly about 9; leaves not glandular-punctate; fruit a many-seeded berry.9. Punicaceae

8. Stamens up to twice as many as the petals (or sepals).
   11. Anthers opening by longitudinal slits; connective not much modified; leaves mostly without subparallel longitudinal ribs.
      12. Embryo sac 4-nucleate; endosperm diploid; most species herbaceous; ovules usually many; fruit usually capsular, sometimes a berry or drupe..........10. Onagraceae
      12. Embryo sac 8-nucleate; endosperm triploid; plants woody; ovules 2-3 per carpel; fruit drupaceous...11. Oliniaceae
   11. Anthers mostly opening by terminal pores; connective variously modified; leaves mostly with prominent subparallel longitudinal ribs...........................12. Melastomataceae
7. Placentation apical in a 1-celled ovary; ovules 2 (-6); fruit 1-seeded, indehiscent..13. Combretaceae

5. PROTEALES

1. Plants with a pubescence of peltate scales; flowers not aggregated into Compositae-like bracteate heads; fruit a pseudodrupe, the dry achene surrounded by the persistent fleshy hypanthium or calyx-tube; nodes unilacunar...1. Elaeagnaceae
1. Plants without peltate scales; flowers often aggregated into Compositae-like bracteate heads; fruit diverse, but not as in the Elaeagnaceae; nodes trilacunar...2. Proteaceae

6. CORNALES

1. Flowers perfect or unisexual, with more or less well developed sepals or petals or both, not borne in catkins.
   2. Ovules 2 (seldom more) per locule; petals folded or convolute in bud; leaves opposite, stipulate; stamens more numerous than the petals; fruit a berry.....................................1. Rhizophoraceae
   2. Ovules solitary in each locule (or some locules empty); petals valvate or imbricate; leaves opposite or alternate, exstipulate; fruit mostly drupaceous.
      3. Stamens more numerous than the petals (or sepals) except in some Alangiaceae; leaves strictly alternate; embryo large.
         4. Ovary 6-9-locular.............................2. Davidiaceae
         4. Ovary 1-2-locular.
            5. Petals imbricate; flowers perfect or often unisexual; ovary unilocular...........................3. Nyssaceae
            5. Petals valvate; flowers perfect; ovary 1-2-locular........ ....4. Alangiaceae
      3. Stamens as many as the petals; leaves opposite, seldom alternate; embryo small....................................5. Cornaceae
1. Flowers unisexual (plants dioecious), the pistillate ones with vestigial or no perianth, the staminate ones with a perianth of 4 sepals, both types borne in catkins.....................................6. Garryaceae

7. SANTALALES

1. Fruit capsular; ovules 6-8; leaves with small stipules; plants autotrophic.
   2. Fertile stamens opposite the petals, alternating with well developed staminodes; petals and sepals well differentiated; flowers in slender, catkin-like racemes.......................1. Medusandraceae

2. Fertile stamens alternate with the petals, alternating with a set of nectary glands; petals and sepals very much alike; flowers in globose umbels..2. Dipentodontaceae

1. Fruit indehiscent; ovules 1-5; leaves without stipules; plants autotrophic or more often partly or wholly parasitic.
   3. Plants with chlorophyll; flowers perfect or sometimes unisexual; fruit dry or fleshy.
      4. Ovules mostly well differentiated; leaves alternate except in the Grubbiaceae and most Santalaceae; flowers mostly small and not very showy; plants rooted in the ground.
         5. Petals present; stamens, when in a single cycle, opposite the petals; ovary superior to occasionally inferior.
            6. Ovary 2-5-celled below, with 1 ovule per locule; integuments 0-2..................................3. Olacaceae
            6. Ovary 1-celled, with 1 apical or basal ovule; integument none.......................................4. Opiliaceae
         5. Petals absent; stamens, when in a single cycle, opposite the sepals; ovary inferior.
            7. Plants fully autotrophic; stamens 8, in 2 cycles; ovules unitegmic..................................5. Grubbiaceae
            7. Plants partly or wholly parasitic; stamens 3-6, in a single cycle; ovules without integument.....6. Santalaceae
      4. Ovules scarcely differentiated from the large, central placenta of the 1-celled ovary, without integument; leaves opposite or whorled; perianth often large and showy; plants rarely rooted in the ground..................................7. Loranthaceae
   3. Plants without chlorophyll; flowers mostly unisexual; fruit dry, nutlike.
      8. Plants aerial, attached to the stem of the host; fruit 3-winged; ovules without integument.....................8. Misodendraceae
      8. Plants terrestrial, attached to the roots of the host; fruit not winged.
         9. Ovary superior; ovules 1-5, without integument; stamens usually more than one..................9. Balanophoraceae
         9. Ovary inferior; ovule solitary, unitegmic; stamen 1...10. Cynomoriaceae

8. RAFFLESIALES

1. Ovary superior; scale-leaves opposite; flowers perfect...1. Mitrastemonaceae
1. Ovary inferior or half-inferior; scale-leaves alternate or none.
   2. Flowers mostly perfect; leaves none...................2. Hydnoraceae
   2. Flowers unisexual; scale-leaves alternate............3. Rafflesiaceae

9. CELASTRALES

1. Flowers diplostemonous; petals none; disk none; habit ericoid...1. Geissolomataceae
1. Flowers haplostemonous (except a few Celastraceae); petals generally well developed; disk often present; habit not ericoid.
   2. Ovules erect and basal, or several and superposed in 2 rows; disk present except in Salvadoraceae; integuments 2 except in many Salvadoraceae; leaves opposite or less often alternate.

3. Flowers with a disk; carpels more than 2 except in some Celastraceae; flowers pentamerous except in some Celastraceae.
    4. Woody plants; flowers hypogynous; leaves opposite or alternate.
        5. Disk extrastaminal; stamens mostly 3 (2-5); leaves mostly opposite; seeds without endosperm; latex system well developed............................2. Hippocrateaceae
        5. Disk intrastaminal, or the stamens seated on the disk; stamens 4-5 (seldom 6-10); seeds mostly with endosperm; latex system seldom well developed.
            6. Disk of ordinary type, merely surrounding the ovary at the base; leaves opposite or sometimes alternate; locules 1-5...................3. Celastraceae
            6. Disk much enlarged, enclosing the ovary; leaves alternate; locules 5.................4. Siphonodontaceae
    4. Herbs; flowers distinctly perigynous, the disk lining the hypanthium; leaves alternate; seeds with endosperm.......... ....5. Stackhousiaceae
3. Flowers without a disk, but the stamens sometimes alternating with nectariferous glands; carpels 2; flowers tetramerous; seeds without endosperm..............................6. Salvadoraceae
2. Ovules pendulous, apical or nearly so, 1-2 per locule, unitegmic (except in Dichapetalaceae); leaves alternate.
    7. Stipules wanting or vestigial; seeds with endosperm except in some Icacinaceae; disk wanting.
        8. Corolla mostly polypetalous; fruit drupaceous; woody plants without milky juice.
            9. Locules 4-6 or more; petals imbricate; pollen binucleate.. ....7. Aquifoliaceae
            9. Locules 3, or more often 1 by abortion; petals valvate; pollen trinucleate......................8. Icacinaceae
        8. Corolla sympetalous; fruit samaroid; climbing herbs with milky juice...............................9. Cardiopteridaceae
    7. Stipules present; endosperm wanting; disk present and intrastaminal, or represented by nectariferous glands alternating with the stamens.................................10. Dichapetalaceae

10. EUPHORBIALES

1. Fruit mostly dehiscent and with more than 1 seed, seldom drupaceous; seeds usually with a caruncle.
    2. Ovules apotropous, the raphe dorsal; fruit a loculicidal capsule, or sometimes drupaceous; small trees and shrubs, with neither stipules nor a milky juice; disk wanting........................1. Buxaceae
    2. Ovules epitropous, the raphe ventral; fruit mostly schizocarpic, each carpel eventually dehiscing ventrally to release its solitary seed; plants of very diverse habit, often with a milky juice, and usually with stipules; disk often present..................2. Euphorbiaceae
1. Fruit indehiscent, drupaceous, often 1-seeded; seeds mostly without a caruncle.
    3. Ovules 2 in each locule, or 1 locule empty; leaves without stipules.
        4. Ovules epitropous, present in each of the 2 (-4) locules; petals none; stamens 6-12; disk wanting; embryo minute..3. Daphniphyllaceae

4. Ovules apotropous, borne in only 1 of the 2 locules; petals 5; stamens 5, alternating with the disk-glands; embryo not especially small....................................4. Aextoxicaceae
3. Ovules solitary in each of the 2-5 locules; leaves with stipules; petals 5; stamens 5, 10, or 15; disk none.............5. Pandaceae

11. RHAMNALES

1. Ovule solitary in each locule; carpels mostly 3-8, seldom only 2; shrubs or small trees, sometimes climbing, but seldom with tendrils.
2. Stamens free; leaves simple, entire or merely toothed, usually stipulate; embryo large; flowers hypogynous to perigynous or epigynous; fruits diverse.................................1. Rhamnaceae
2. Stamens with the filaments united into a tube; leaves simple to compound, exstipulate; embryo small; flowers hypogynous; fruit a berry ....2. Leeaceae
1. Ovules 2 in each locule; carpels 2; mostly tendril-bearing climbers; leaves stipulate, usually compound or lobed; fruit a berry..3. Vitaceae

12. SAPINDALES

1. Ovules usually more than 2 per locule or carpel; endosperm well developed; ovules apotropous; flowers usually perfect.
2. Leaves stipulate, mostly pinnately compound; flowers haplostemonous, with 2-4 carpels and axile placentation.
3. Leaves mostly opposite; disk annular, intrastaminal, or the stamens seated on the disk; flowers regular; carpels 2-3, seldom 4; pollen binucleate..........................1. Staphyleaceae
3. Leaves alternate; disk unilateral, extrastaminal; flowers irregular; carpels mostly 4; pollen trinucleate......2. Melianthaceae
2. Leaves exstipulate, simple, alternate; flowers diplostemonous, with mostly 5 carpels, the placentation parietal; disk annular, extrastaminal..3. Greyiaceae
1. Ovules seldom more than 2 per carpel or locule; endosperm in most families wanting or scanty; ovules variously apotropous or epitropous; flowers perfect or unisexual.
4. Flowers strictly apocarpous.
5. Style terminal; fruit dehiscent; seeds arillate; leaves mostly compound...4. Connaraceae
5. Style basal; fruit indehiscent; seeds exarillate; leaves simple.. ....5. Stylobasiaceae
4. Flowers nearly always more or less strongly syncarpous (some exceptions in Simaroubaceae).
6. Disk mostly extrastaminal (and often unilateral) or wanting (intrastaminal in some Aceraceae); ovules apotropous except in Akaniaceae; flowers very often irregular.
7. Leaves mostly alternate; stamens typically 8 (5-many).
8. Seeds with endosperm and without an aril; wood rays very wide; disk wanting; ovules epitropous, 2 in each locule ....6. Akaniaceae
8. Seeds without endosperm and commonly with an aril; wood rays narrow; disk present, often unilateral; ovules apotropous, 1 or less often 2 in each locule.......... ....7. Sapindaceae
7. Leaves opposite.

9. Leaves palmately compound; fruit a tricarpellate, usually 1-seeded capsule; flowers irregular; stamens 5-8........ ....8. Hippocastanaceae

9. Leaves simple and palmately lobed, or pinnately compound; fruit a bicarpellate, 2-seeded samara; flowers regular, stamens 4-10, most often 8.................9. Aceraceae

6. Disk intrastaminal, annular or often developed into a gynophore (wanting in the Julianaceae); ovules epitropous except in the Anacardiaceae and Julianaceae.

10. Bark and wood with prominent, specialized resin ducts.

11. Ovary 2-5 celled, with 2 (seldom only 1) epitropous ovules per locule.............................10. Burseraceae

11. Ovary 1-celled (by abortion), with a single apotropous ovule.

12. Flowers normally developed, perfect or sometimes unisexual, with an evident perianth; intrastaminal disk present..........................11. Anacardiaceae

12. Flowers reduced, unisexual, the pistillate ones without a perianth; disk none...........12. Julianaceae

10. Bark and wood without prominent resin ducts.

13. Leaves mostly exstipulate, the stipules, if any, small and soon deciduous; leaves mostly alternate.

14. Stamens mostly free.

15. Leaves not glandular-punctate.

16. Seeds without endosperm; leaves mostly compound, seldom simple; bark bitter; styles often separate; flowers often 5-merous; pollen binucleate..........13. Simaroubaceae

16. Seeds with endosperm; leaves simple; bark not bitter (?); styles united; flowers 3-merous or seldom 4-merous; pollen trinucleate...... ....14. Cneoraceae

15. Leaves mostly glandular-punctate......15. Rutaceae

14. Stamens mostly connate by their filaments............ ....16. Meliaceae

13. Leaves with well developed, persistent stipules, mostly opposite; seeds often with endosperm...17. Zygophyllaceae

13. GERANIALES

1. Flowers regular or slightly irregular, not spurred, or the spur inconspicuous and adnate to the pedicel; stamens 2 or 3 times as many as the sepals or petals, sometimes some of them staminodial; leaves mostly deeply cleft.

2. Style not gynobasic; integuments 2; ovules, when few, pendulous; filaments basally connate; annual or more often perennial herbs, or seldom woody plants.

3. Fruit a beakless, loculicidal capsule or seldom a berry, with (1) 2-many seeds per carpel; styles free; endosperm well developed; pollen binucleate; leaves with or more often without stipules... ....1. Oxalidaceae

3. Fruit a beaked, mostly septicidally dehiscent capsule or a schizocarp, the mericarps 1-seeded and mostly again dehiscent; styles usually united, at least below, into a prominent column; endosperm mostly scanty or none; pollen trinucleate; leaves with stipules.......................................2. Geraniaceae

2. Style gynobasic; integument 1; ovules 1 per carpel, erect; filaments free; annual herbs; seeds without endosperm........3. Limnanthaceae
1. Flowers strongly irregular, one of the sepals with a conspicuous free spur; stamens 5 or 8, less than twice as many as the basic (unmodified) number of sepals or petals; leaves simple, not deeply lobed.
4. Leaves peltate, palmately nerved; stamens 8, carpels 3, each uniovulate, the fruit a schizocarp; plants with myrosin and without raphides...4. Tropaeolaceae
4. Leaves not peltate, the venation pinnate; stamens 5; carpels 5, each with several or many ovules; fruit an elastically dehiscent capsule, or seldom a berry; plants with raphides and without myrosin.. ....5. Balsaminaceae

14. LINALES

1. Plants woody, often arborescent; fruit drupaceous.
2. Petals exappendiculate; carpels 4-7, commonly 5; flowers with an intrastaminal disk; leaves without stipules..........1. Humiriaceae
2. Petals internally appendiculate; carpels 3, 1 fertile; flowers without a disk; leaves with stipules.................2. Erythroxylaceae
1. Plants herbaceous or occasionally shrubby, seldom arborescent; fruit mostly capsular, seldom drupaceous or nutlike; petals exappendiculate; leaves mostly stipulate; floral disk usually wanting........3. Linaceae

15. POLYGALALES

1. Carpels mostly 3, seldom 2-5; stamens mostly dehiscent by longitudinal slits, seldom by terminal pores; leaves usually stipulate.
2. Fertile stamens several; filaments usually connate toward the base.
3. Petals 5, commonly fringed or toothed; fruit indehiscent; ovules 1 per carpel; stamens 10, some sterile.........1. Malpighiaceae
3. Petals 3, entire; fruit capsular; ovules 1-many per carpel; stamens 3-12.....................................2. Trigoniaceae
2. Fertile stamen 1, but staminodes also usually present; filaments free; fruit usually capsular; ovules 1-many per carpel............ ....3. Vochysiaceae
1. Carpels 2, or seemingly only 1; stamens dehiscent by terminal pores; leaves usually exstipulate.
4. Pistil more or less evidently bicarpellate, but not always bilocular; seeds with endosperm; stamens 8-10.
5. Stamens free.
6. Flowers regular; placentation axile in a bilocular ovary; heatherlike shrubs or subshrubs with small leaves.......... ....4. Tremandraceae
6. Flowers irregular; placentation apical or parietal in a unilocular ovary which may show traces of a partition; small trees with normal leaves................5. Xanthophyllaceae
5. Stamens mostly monadelphous; flowers irregular; ovary evidently bilocular; herbs, vines, or shrubs.............6. Polygalaceae
4. Pistil pseudomonomerous and with a pair of apical ovules; seeds without endosperm; stamens 4 or seldom 3; low shrubs or seldom herbs... ....7. Krameriaceae

16. UMBELLALES

1. Carpels 1-many, typically 5; fruit fleshy or dry, usually a berry, the carpels sometimes separating, but without a carpophore; trees and shrubs, seldom herbs or lianas..............................1. Araliaceae
1. Carpels consistently 2; fruit a dry schizocarp, the mericarps separated by a usually persistent carpophore; herbs, rarely shrubs or trees.......... ....2. Umbelliferae

VI. ASTERIDAE

1. Ovary chiefly superior.
  2. Plants usually with well developed internal phloem; leaves opposite; endosperm nearly always nuclear; no integumentary tapetum; corolla regular, its lobes usually contorted in bud...........1. Gentianales
  2. Plants usually either without internal phloem, or with alternate leaves, or both; endosperm nuclear or more often cellular; integumentary tapetum often present; corolla regular or irregular, with various sorts of aestivation.
    3. Fruit typically of 4 or more or less distinct (or separating) nutlets, derived from a bicarpellate ovary that often has a gynobasic style..3. Lamiales
    3. Fruit otherwise; style rarely gynobasic.
      4. Corolla scarious, persistent; flowers mostly anemophilous, tetramerous as to calyx, corolla, and androecium; leaves usually all basal..........................4. Plantaginales
      4. Corolla otherwise; flowers mostly entomophilous or ornithophilous, variously pentamerous or tetramerous, with isomerous or anisomerous stamens; leaves seldom all basal.
        5. Flowers mostly regular and with as many functional stamens as corolla lobes, typically 5 of each............. ....2. Polemoniales
        5. Flowers mostly irregular and with fewer functional stamens than corolla lobes, or sometimes with regular, tetramerous corolla and 2 or 4 stamens, or the corolla rarely wanting.................................5. Scrophulariales
1. Ovary chiefly inferior.
  6. Flowers borne in various sorts of inflorescences, but if borne in involucrate heads then the ovules either pendulous or more than one; endosperm present or absent.
    7. Leaves nearly always alternate; stamens free from the corolla, or attached at the base of the tube; anthers in most families connivent or connate around the style, which pushes out the pollen; plants nearly always herbaceous..................6. Campanulales
    7. Leaves usually opposite or whorled (notable exception: Calyceraceae); stamens attached to the corolla tube, generally well above the base; flowers mostly without a specialized pollen presentation mechanism (notable exceptions: Calyceraceae, some Rubiaceae); plants woody or herbaceous.
      8. Stipules usually present and interpetiolar (sometimes reduced to a mere interpetiolar line, or enlarged into leaves); corolla typically regular and with isomerous stamens; endosperm nuclear; plants chiefly tropical and woody, but some members herbaceous, or of temperate regions, or both........ ....7. Rubiales

8. Stipules typically none, when present usually small and adnate to the petiole; corolla regular or often irregular, the stamens often fewer than the corolla lobes; endosperm cellular; plants woody or often herbaceous, of temperate or less often tropical regions................8. Dipsacales

6. Flowers borne in involucrate, centripetally flowering heads; ovary 1-celled, with a solitary, erect ovule; anthers connate or connivent into a tube around the style, which pushes out the pollen; endosperm wanting from mature seeds...................9. Asterales

1. GENTIANALES

1. Plants without a latex system; style not especially thickened and modified distally; carpels fully united, only the stigmas sometimes separate.
   2. Plants mostly woody and with more or less well developed stipules, often containing alkaloids, but lacking gentiopicrin; placentation axile..1. Loganiaceae
   2. Plants mostly herbaceous, without stipules, containing the bitter glucoside gentiopicrin but lacking alkaloids; placentation mostly parietal.......................................2. Gentianaceae
1. Plants with a well developed latex system; style usually more or less thickened and modified at the tip; carpels typically united only by the style and/or stigma, but sometimes wholly united; stipules none; alkaloids often present, but gentiopicrin lacking.
   3. Androecium free from the stigma and without translators; pollen granular, not forming pollinia; carpels often united by part or all of the style as well as by the stigma, or even wholly united; corona mostly poorly developed or wanting; trees, shrubs, vines, or less often herbs......................................3. Apocynaceae
   3. Androecium concrescent to the stigma and provided with translators, the pollen grains borne in dense masses (pollinia); carpels united only by the stigma; corona more or less well developed; herbs, vines, or shrubs, rarely trees...................4. Asclepiadaceae

2. POLEMONIALES

1. Plants with internal phloem, and always autotrophic.
   2. Carpels 5; ovary with a terminal style and ripening into a schizocarp, or with a gynobasic style and ripening into 5 or more 1- to several-seeded nutlets; endosperm cellular; more or less succulent herbs and shrubs, confined to the Pacific slope of S. Am.......... ....1. Nolanaceae
   2. Carpels 2 (-5); fruit not schizocarpic and style never gynobasic; plants widespread, seldom succulent.
      3. Ovules and seeds mostly numerous, on axile placentae; cotyledons not plicate; style barely or scarcely lobed; endosperm mostly cellular...2. Solanaceae
      3. Ovules 2 (rarely many) per carpel, basal, erect, the partition present or absent; cotyledons plicate; style entire or often cleft, or the styles sometimes wholly distinct; endosperm nuclear.................................3. Convolvulaceae
1. Plants without internal phloem, and either autotrophic or parasitic.
   4. Plants parasitic, without chlorophyll; embryo scarcely differentiated.

5. Carpels 2; fruit capsular; twining stem-parasites, not rooted in the ground at maturity; endosperm nuclear; pollen trinucleate...4. Cuscutaceae

5. Carpels 6-14; fruit of 1-seeded nutlets; erect root-parasites; endosperm probably cellular; pollen binucleate.....8. Lennoaceae

4. Plants autotrophic; embryo normally developed.

6. Corolla lobes valvate or induplicate-valvate in bud; plants more or less aquatic; placentation parietal; endosperm cellular...5. Menyanthaceae

6. Corolla lobes convolute or imbricate in bud; plants terrestrial, though sometimes of wet places.

7. Carpels 3 (2-4); corolla lobes convolute in bud; placentation axile; endosperm nuclear...................6. Polemoniaceae

7. Carpels 2; corolla lobes imbricate or rarely convolute in bud; placentation parietal (often with intruded placentae) or occasionally axile; endosperm nuclear or cellular...7. Hydrophyllaceae

3. LAMIALES

1. Leaves mostly alternate, usually entire; flowers mostly regular and with 5 stamens; style typically but not always gynobasic, the fruit typically but not always of 4 distinct nutlets; stems not square..1. Boraginaceae

1. Leaves mostly opposite (or whorled), and often toothed or cleft; flowers mostly more or less irregular (or apetalous) and with 1-4 stamens, the exceptions (regular corolla and/or 5 stamens) being found chiefly in the Verbenaceae; stems commonly square.

2. Flowers apetalous, unisexual; styles 2, separate; stamen solitary; endosperm well developed; plants mostly aquatic, often free-floating..2. Callitrichaceae

2. Flowers petaliferous, usually perfect; style usually solitary; stamens 2 or more; endosperm usually scanty or wanting; mostly terrestrial plants, never free-floating aquatics.

3. Style terminal, the ovary only slightly or not at all lobed; plants seldom aromatic.

4. Ovary with 2 (seldom more) usually biovulate carpels, with a false partition between the ovules of each carpel, ordinarily ripening into 4 separating nutlets or into a 4-stoned drupe...............................3. Verbenaceae

4. Ovary pseudomonomerous, with a single ovule and locule...4. Phrymaceae

3. Style mostly gynobasic, the ovary more or less deeply 4-cleft and ripening into 4 nutlets; plants mostly aromatic......5. Labiatae

4. PLANTAGINALES

Represented by..1. Plantaginaceae

5. SCROPHULARIALES

1. Flowers perfect or less often unisexual, almost always with a calyx and usually also with a corolla; stamens mostly 2-4 (5); habit and habitat various.

2. Placentation axile to parietal or apical; plants not insectivorous.
  3. Corolla with mostly 4 (5) lobes and only slightly (or not at all) irregular, or sometimes wanting; woody plants (rarely herbs) with mostly opposite or whorled leaves.
    4. Stamens 4, rarely 2 or 5; ovules mostly numerous in each locule....................................1. Buddlejaceae
    4. Stamens 2, rarely 5; ovules mostly 2 (seldom 4-many) in each locule..2. Oleaceae
  3. Corolla with mostly 5 (seldom 4) lobes, or 2-lipped with scarcely lobed lips; habit and leaves various.
    5. Mature seeds usually with a well developed endosperm.
      6. Placentation mostly axile or apical.
        7. Ovules 2 or more in each locule; anthers (1) 2-locular.
          8. Fruit dry and mostly dehiscent, rarely fleshy; herbs, rarely shrubs; ovules mostly more or less numerous in each locule, seldom only 2; leaves opposite or alternate.................. ....3. Scrophulariaceae
          8. Fruit drupaceous, indehiscent; woody plants, sometimes arborescent; ovules 2 (4-8) in each locule; leaves mostly alternate..4. Myoporaceae
        7. Ovules mostly solitary in each locule, sometimes only one locule developed; anthers unilocular at least at maturity; herbs or shrubs, mostly with alternate leaves; fruit small, indehiscent, the carpels sometimes separating..................5. Globulariaceae
      6. Placentation mostly parietal.
        9. Plants chlorophyllous, not parasitic, with well developed, mostly opposite or whorled leaves; ovary often more or less inferior; embryo well developed. ....6. Gesneriaceae
        9. Plants parasitic, without chlorophyll, the leaves reduced and alternate; ovary superior; embryo minute, scarcely differentiated....7. Orobanchaceae
    5. Mature seeds mostly with little or no endosperm; leaves chiefly opposite or whorled.
      10. Trees or large shrubs, seldom herbs; leaves mostly compound, seeds usually winged.............8. Bignoniaceae
      10. Herbs, rarely shrubs or trees; leaves mostly simple; seeds not winged.
        11. Seeds with an enlarged and specialized funiculus; plants without specialized mucilage hairs; fruits fleshy to more often capsular, with axile placentation.............................9. Acanthaceae
        11. Seeds with normal, unspecialized funiculus; plants with a unique type of mucilage hair; fruits capsular or drupaceous to more often nutlike, often with conspicuous appendages.............10. Pedaliaceae
2. Placentation free-central; herbs, aquatic or of wet places, mostly insectivorous...................................11. Lentibulariaceae

1. Flowers unisexual, without perianth; stamen solitary; aquatic herbs...... ....12. Hydrostachyaceae

6. CAMPANULALES

1. Style without an indusium, but often with collecting hairs.
    2. Stamens as many as the corolla lobes, typically 5, free or connate only by their anthers; anthers introrse.
        3. Style glabrous, with a solitary stigma; plants apparently without a latex system.
            4. Petals valvate; fruit a berry; leaves asymmetric; pollen binucleate..............................1. Pentaphragmataceae
            4. Petals imbricate; fruit a circumscissle capsule; leaves not notably asymmetric; pollen trinucleate.....2. Sphenocleaceae
        3. Style with well developed collecting hairs below the usually separate stigmas; plants with a well developed latex system..3. Campanulaceae
    2. Stamens 2 or 3, the filaments adnate to the style or connate into a tube around it; anthers extrorse; no latex system....4. Stylidiaceae
1. Style with a more or less cupulate indusium just beneath the stigma(s), but without collecting hairs; no latex system.
    5. Flowers regular or nearly so, borne in involucrate heads; endosperm none; ovary superior, unilocular, with a single basal ovule..5. Brunoniaceae
    5. Flowers mostly irregular, not borne in involucrate heads; endosperm well developed; ovary generally inferior, 1-2 locular, with 2 or more ovules..6. Goodeniaceae

7. RUBIALES

Represented by..1. Rubiaceae

8. DIPSACALES

1. Filaments and anthers distinct; leaves nearly always opposite; pollen trinucleate; inflorescence various.
    2. Plants mostly woody, seldom herbaceous; stamens mostly as many as the corolla lobes, typically 5, seldom only 2; ovules often more than 1 per locule; endosperm well developed; fruits diverse, often with several or many ovules per locule.................1. Caprifoliaceae
    2. Plants herbaceous or nearly so; stamens seldom (except some Dipsacaceae) of the same number as the corolla lobes; ovules pendulous, not more than 1 per locule.
        3. Stamens 8-10, twice as many as the corolla lobes, paired in the sinuses of the corolla, each with only a single pollen sac; fruit a dry drupe with several stones; endosperm well developed; flowers in compact heads..........................2. Adoxaceae
        3. Stamens 1-4 (rarely 5), as many as or usually fewer than the corolla lobes, each with 2 pollen sacs; fruit dry, 1-seeded; endosperm scanty or none.
            4. Ovaries not individually enclosed or subtended by a gamophyllous involucel; flowers not in compact, involucrate heads; ovary mostly trilocular, with 1 fertile and 2 sterile locules; corolla mostly 5-lobed.....3. Valerianaceae
            4. Ovaries individually enclosed or subtended by a gamophyllous involucel; flowers mostly borne in compact, involucrate heads; ovary strictly unilocular; corolla mostly 4-5 lobed..4. Dipsacaceae

1. Filaments and anthers connate into a tube around the style; leaves alternate; pollen binucleate; flowers in involucrate, centripetally flowering heads, with regular corolla........................5. Calyceraceae

9. ASTERALES

Represented by..1. Compositae

LILIOPSIDA

1. Flowers mostly apocarpous, the syncarpous forms with a modified type of laminar placentation; plants mostly aquatic or subaquatic, always herbaceous; pollen consistently trinucleate; endosperm mostly wanting, not starchy when present; subsidiary cells mostly 2..........I. Alismatidae
1. Flowers mostly syncarpous, the apocarpous forms borne on terrestrial, woody plants; plants terrestrial or less often aquatic; pollen binucleate or trinucleate; endosperm present or absent, often starchy, subsidiary cells none to several.
    2. Seeds mostly with a starchy endosperm; flowers either with well differentiated sepals and petals (the sepals mostly not petaloid), or more or less reduced and with chaffy or bristly (or no) perianth, not aggregated into a spadix......................II. Commelinidae
    2. Seeds mostly with a nonstarchy endosperm (notable exceptions: Araceae, Pontederiaceae, Philydraceae) or without endosperm; flowers usually either with the petals and sepals much alike (commonly both petaloid), or reduced, collectively subtended by a spathe, and aggregated into a spadix.
        3. Flowers usually numerous, small, subtended by a prominent spathe, and often aggregated into a spadix; plants often arborescent; leaves in most species broad, petiolate, and not with typical parallel venation; supporting cells typically 4, less often 3 or 2; ovary superior (sometimes sunken in the axis of the spadix)....................................III. Arecidae
        3. Flowers few to numerous and small to large, but generally not subtended by a spathe and never aggregated into a spadix; plants seldom arborescent; leaves narrow and parallel-veined, or less often broader and net-veined; supporting cells typically none, less often 2, rarely more; ovary superior or more often inferior.................................IV. Liliidae

I. ALISMATIDAE

1. Seeds without endosperm at maturity; plants mostly aquatic or semiaquatic, autotrophic.
    2. Perianth differentiated into evident sepals and petals; flowers often bracteate.
        3. Flowers hypogynous; carpels free or nearly so; embryo sac, except in Butomaceae, bisporic.........................1. Alismatales
        3. Flowers epigynous, with compound ovary and modified laminar placentation; embryo sac monosporic.............2. Hydrocharitales
    2. Perianth, when present, not differentiated into sepals and petals; bracts usually wanting, except in Scheuchzeriaceae, but the perianth sometimes consisting of only a single, bractlike tepal; embryo sac monosporic......................................3. Najadales
1. Seeds with well developed endosperm; plants terrestrial, mycotrophic, without chlorophyll......................................4. Triuridales

1. ALISMATALES

1. Plants without secretory canals; pollen uniaperturate; embryo sac monosporic; embryo straight; placentation laminar; leaves linear, not differentiated into blade and petiole....................1. Butomaceae
1. Plants with schizogenous secretory canals; pollen multiaperturate; embryo sac bisporic; embryo horseshoe-shaped; placentation and leaves various.
   2. Ovules several or many, scattered over the inner surface of the carpel (i.e., the placentation laminar)............2. Limnocharitaceae
   2. Ovules 1 or 2, seldom more, basal or on a marginal placenta.......... ....3. Alismataceae

2. HYDROCHARITALES

Represented by..1. Hydrocharitaceae

3. NAJADALES

1. Ovules 2 or more, basal, anatropous; stamens 6 or more, free, often more numerous than the tepals; fruits dehiscent, follicular; endosperm helobial.
   2. Tepals 0-3, sometimes petaloid; inflorescence a simple or basally forking spike, bractless, although the perianth sometimes consists of a single bractlike tepal; pollen uniaperturate, the grains borne singly; aquatic plants with submerged or floating leaves........... ....1. Aponogetonaceae
   2. Tepals 6, never petaloid; inflorescence a raceme, each pedicel subtended by a bract; pollen nonaperturate, in dyads; emergent marsh-plants...................................2. Scheuchzeriaceae
1. Ovule solitary; stamens either fewer than 6, or as many as and adnate to the tepals, or both; fruits indehiscent or seldom schizocarpic or irregularly dehiscent; tepals 0-6, never petaloid; pollen mostly nonaperturate.
   3. Ovule basal, erect, anatropous; endosperm nuclear.
      4. More or less emergent marsh-plants, scapose or subscapose, with chiefly or wholly basal leaves and a terminal raceme or spike; carpels and stamens usually more than 1.........3. Juncaginaceae
      4. Submerged aquatics with branching stems and opposite or whorled leaves, the flowers solitary in the axils; carpel 1; stamen 1... ....4. Najadaceae
   3. Ovule apical or lateral, pendulous, more or less orthotropous.
      5. Carpels usually 2 or more, separate; pollen not threadlike; endosperm chiefly helobial; plants not truly marine.
         6. Flowers in spikes or racemes, perfect or less often unisexual; habit various.
            7. Tepals 4; stamens 4; spikes axillary; fruiting carpels sessile; pollen ellipsoid or spheroid; plants of fresh water..............................5. Potamogetonaceae
            7. Tepals 0; stamens 2; spikes terminal; fruiting carpels stipitate; pollen of a unique, bilateral type; plants of brackish water...........................6. Ruppiaceae
         6. Flowers in axillary cymes, or solitary in the axils, unisexual; submerged plants of fresh or brackish water........... ....7. Zannichelliaceae
      5. Carpel 1; pollen threadlike; endosperm often or usually nuclear; submersed marine plants..........................8. Zosteraceae

4. TRIURIDALES

1. Flowers perfect; carpels 3; seeds numerous..............1. Petrosaviaceae
1. Flowers unisexual; carpels several; seeds 1 per carpel....2. Triuridaceae

II. COMMELINIDAE

1. Flowers without nectaries or nectar; ovary superior; vessels well developed in all vegetative organs.
    2. Flowers with more or less showy petals that are well differentiated from the sepals, nearly always perfect; pollination mostly by insects..1. Commelinales
    2. Flowers perfect or often unisexual, without showy petals, the perianth sometimes in 2 series, but the 2 series much alike and not petaloid.
        3. Ovules various, but never at once solitary, pendulous from the apex of the ovary, and anatropous; pollen mostly trinucleate; embryo embedded or more often peripheral to the endosperm; plants terrestrial or less often aquatic; flowers perfect or unisexual, borne in various sorts of inflorescences.
            4. Ovary with 1-3 fertile locules and as many stigmas, each fertile locule with a single, apical, pendulous, orthotropous ovule; embryo peripheral to the endosperm; flowers in most species and families unisexual.
                5. Flowers aggregated into dense, pseudanthial, involucrate heads, often pollinated by insects; ovules tenuinucellate; anthers usually bilocular; leaves all basal...2. Eriocaulales
                5. Flowers seldom aggregated into dense heads, normally wind-pollinated or self-pollinated; ovules crassinucellate; anthers often unilocular; leaves cauline or (in some small families) all basal................3. Restionales
            4. Ovary otherwise, either with more stigmas than locules, or with more than one ovule per locule, the ovules of various form and position; embryo peripheral (Gramineae) or embedded (Juncales, Cyperaceae) in the endosperm; flowers perfect or unisexual.
                6. Ovary with 1-3 locules and 3-many ovules; fruit capsular; pollen in tetrads; flowers in open or compact inflorescences unlike those of the Cyperales......4. Juncales
                6. Ovary with a single locule and ovule; fruit indehiscent; pollen in monads; flowers in characteristic spikes or spikelets...............................5. Cyperales
        3. Ovules solitary (or solitary in each locule), pendulous from the apex of the ovary, anatropous; pollen binucleate; embryo embedded in the endosperm; flowers unisexual; borne in dense spikes or heads; plants aquatic.....................6. Typhales
1. Flowers with septal or septal-derived nectaries; ovary inferior except in some Bromeliaceae; vessels chiefly confined to the roots, except mainly in some Bromeliaceae.
    7. Stamens 6; flowers regular or sometimes somewhat irregular; xerophytes and epiphytes with firm, narrow, often spiny-margined leaves ....7. Bromeliales
    7. Stamens fewer than 6 (6 only in Ravenala, of the Strelitziaceae); flowers distinctly irregular; mesophytes with unarmed, pinnately veined, often broad leaves............................8. Zingiberales

1. COMMELINALES

1. Leaf sheath open, often not well differentiated from the blade.
   2. Plants mostly terrestrial, often of wet or marshy places, but seldom truly aquatic; leaves generally all clustered at the base, the flowers in a compact inflorescence terminating a long scape or peduncle.
      3. Stamens 6, opening by terminal pores; inflorescence a head of spikelets, each spikelet with several bracts subtending its single flower....................................1. Rapateaceae
      3. Stamens 3, often accompanied by 3 staminodes; anthers opening by longitudinal slits; inflorescence a simple racemose head........ ....2. Xyridaceae
   2. Plants aquatic; leaves numerous, linear or threadlike, distributed along the stem; stamens 3, the anthers opening by terminal pores or short slits..3. Mayacaceae
1. Leaves differentiated into a closed sheath and a well defined, commonly somewhat succulent blade...............................4. Commelinaceae

2. ERIOCAULALES

Represented by...1. Eriocaulaceae

3. RESTIONALES

1. Flowers perfect; stamens 6; leaves chiefly or wholly cauline, with closed sheath...1. Flagellariaceae
1. Flowers unisexual; stamens 1-3; leaves cauline or basal, the sheath mostly open or poorly developed.
   2. Stamens (1) 2-3; tepals usually present; mostly coarse perennials.
      3. Leaves chiefly or wholly basal; anthers bilocular, free.
         4. Inflorescence open, cymose; flowers bracteate; ovary trilocular; plants mostly dioecious...............2. Anarthriaceae
         4. Inflorescence a dense spike suggesting that of Xyris; flowers ebracteate; ovary bilocular; plants monoecious............. ....3. Ecdeiocoleaceae
      3. Leaves wholly cauline; anthers unilocular, or rarely bilocular but then laterally connate; inflorescence more or less open......... ....4. Restionaceae
   2. Stamen 1; tepals none; diminutive, annual herbs with chiefly basal leaves; anthers unilocular......................5. Centrolepidaceae

4. JUNCALES

1. Inflorescence of diverse sorts, but not as in the Thurniaceae; vascular bundles of ordinary nature, with abaxial phloem, not as in the Thurniaceae; cells without silica bodies.........................1. Juncaceae
1. Inflorescence of one or more dense heads subtended by spreading, leafy bracts; vascular bundles of the leaf in vertical pairs, the lower (and smaller) bundle of a pair with the phloem on top (adaxial), facing the phloem of the upper bundle; some cells of the leaf epidermis containing silica bodies...2. Thurniaceae

5. CYPERALES

1. Flowers spirally or less often distichously arranged on the axis of the spike or spikelet, usually each flower seemingly or actually subtended by only a single bract, without an evident bract between the flower and the axis; seed coat generally free from the pericarp; leaf sheath usually closed, stem usually solid, often triangular; flowers often with a perianth of evident bristles; carpels 3 or less often 2; embryo embedded in the endosperm................................1. Cyperaceae
1. Flowers distichously arranged on the axis of the spikelet (or only one per spikelet), each flower ordinarily subtended by a pair of bracts (lemma and palea), the palea inserted between the flower and the axis; seed coat generally adnate to the pericarp; leaf sheath usually open; stem usually hollow, never triangular; flowers without a perianth, unless the lodicules are so interpreted; carpels 2, rarely 3; embryo peripheral to the endosperm................................2. Poaceae

6. TYPHALES

1. Inflorescence of globose heads; vestigial perianth usually present; achenes sessile or nearly so, not wind-distributed.....1. Sparganiaceae
1. Inflorescence of dense, cylindrical spikes; perianth apparently none; achenes slenderly long-stipitate, with long hairs on the stipe, wind-distributed..2. Typhaceae

7. BROMELIALES

Represented by..1. Bromeliaceae

8. ZINGIBERALES

1. Functional stamens 5 or rarely 6, each with 2 pollen sacs; plants with raphide sacs; guard cells symmetrical except in Lowiaceae.
   2. Flowers perfect; leaves and bracts distichously arranged; plants without laticifers; fruit capsular or schizocarpic.
      3. Ovules numerous in each locule; fruit capsular; seeds arillate; stigma trifid.
         4. Flowers without an evident hypanthium; leaves obviously penniveined; guard cells symmetrical; most species caulescent or even arborescent ....1. Strelitziaceae
         4. Flowers with a long hypanthium much surpassing the ovary; leaves seemingly parallel-veined (actually penniveined with narrowly ascending veins); guard cells asymmetrical; acaulescent herbs................................2. Lowiaceae
      3. Ovules solitary in each locule; fruit schizocarpic; seeds not arillate; stigma capitate......................3. Heliconiaceae
   2. Flowers unisexual; leaves and bracts spirally arranged; plants with laticifers; fruit fleshy, indehiscent, the seeds not arillate...... ....4. Musaceae
1. Functional stamen 1; plants without raphide sacs; guard cells asymmetrical except in Cannaceae.
   5. Stamen with 2 pollen sacs, not petaloid; endosperm helobial.
      6. Leaves and bracts distichously arranged; sheaths open; plants aromatic, with abundant oil cells..............5. Zingiberaceae

6. Leaves and bracts spirally arranged; sheaths closed; plants without oil cells..6. Costaceae

5. Stamen with a single functional pollen sac, the other sac modified and petaloid; endosperm nuclear.

7. Ovules numerous in each of the 1-3 locules; stem with mucilage canals; embryo straight; seeds not arillate.........7. Cannaceae

7. Ovules solitary in each of the (1-2-) 3 locules; stem without mucilage canals; embryo folded; seeds mostly arillate...8. Marantaceae

III. ARECIDAE

1. Flowers well developed (though small), with an evident, biseriate perianth of 6 members, not crowded into a spadix; carpels free or united; plants very often arborescent and with a terminal crown of large leaves, the leaves in any case with sheath, petiole, and expanded, plicate blade; endosperm nuclear..1. Arecales

1. Flowers more or less reduced and very often crowded into a spadix; carpels united to form a compound ovary; plants of various habit, but never at once arborescent and with a terminal crown of broad leaves; leaves not plicate except in some Cyclanthaceae.

2. Stamens mostly 10-numerous; vessels generally present in leaves and stems as well as in roots; endosperm nuclear or helobial; flowers unisexual; leaves usually either with bifid blade or not divided into blade and petiole.

3. Leaves with blade, petiole, and basal sheath, the blade usually bifid; plants monoecious; endosperm helobial; neotropical herbs, seldom woody.................................2. Cyclanthales

3. Leaves elongate and narrow, somewhat sheathing at the base, but not divided into blade and petiole, not bifid; plants dioecious; endosperm nuclear; paleotropical woody plants...3. Pandanales

2. Stamens mostly 1-8 (flowers rarely produced in Lemnaceae); vessels confined to the roots, or wholly wanting; endosperm cellular; flowers perfect or unisexual; leaves (when present) often divided into blade and petiole; herbs, or seldom woody climbers....4. Arales

1. ARECALES

Represented by..1. Arecaceae

2. CYCLANTHALES

Represented by..1. Cyclanthaceae

3. PANDANALES

Represented by..1. Pandanaceae

4. ARALES

1. Plants with roots, stems, and leaves, terrestrial or sometimes more or less aquatic, but only rarely free-floating; inflorescence a spadix; plants usually with vessels in the roots and tracheids in the roots, stems, and leaves..1. Araceae

1. Plants thalloid, free-floating, with or without 1-several short, slender roots; flowers (rarely produced) not forming a definite spadix; plants lacking both vessels and tracheids........................2. Lemnaceae

IV. LILIIDAE

1. Plants not obviously mycotrophic; seeds of ordinary number and structure, usually with a well differentiated embryo and well developed endosperm; most families and genera with septal nectaries, sometimes with other kinds of nectaries in addition or instead; ovary superior or inferior.. ....1. Liliales
1. Plants strongly mycotrophic, sometimes without chlorophyll; seeds very numerous and tiny, with undifferentiated embryo and very little or no endosperm; nectaries diverse, but not (or at least not usually) septal; ovary inferior...2. Orchidales

1. LILIALES

1. Endosperm starchy (sometimes also with some fat); stomates with 2 subsidiary cells (4 or more in some Philydraceae); perianth more or less distinctly irregular; plants either aquatic, or with only a single stamen.
   2. Tepals 4 (one member of the outer set derived by connation of 2, one member of the inner set missing), not united into a tube; plants terrestrial; sepal nectaries wanting; stamen solitary............. ....1. Philydraceae
   2. Tepals (4) 6, usually connate below into a perianth tube; plants aquatic; septal nectaries present; stamens (including staminodia) 3 or 6...2. Pontederiaceae
1. Endosperm (when present) with reserve cellulose and/or fat or protein, sometimes also with some starch; stomates in most families usually without subsidiary cells; flowers with more than one stamen; plants seldom aquatic.
   3. Leaves typically narrow and parallel-veined, sometimes broader and more net-veined, but only seldom with both a broad, net-veined blade and a distinct petiole.
      4. Vessels chiefly or wholly confined to the roots; stomates typically without subsidiary cells; perianth and ovary seldom conspicuously hairy or glandular.
         5. Habit liliaceous (mostly soft-leaved herbs from a bulb or rhizome); stamens 6 or 3.
            6. Stamens (including staminodes) mostly 6, seldom more than 6 or only 4; ovary superior or less often inferior.................................3. Liliaceae
            6. Stamens 3; ovary inferior...................4. Iridaceae
         5. Habit agavaceous (mostly coarse, often shrubby or arborescent plants with firm, perennial leaves); stamens 6.
            7. Perianth usually more or less corolloid, not chaffy; X = 16 or more, usually with a few large chromosomes and more numerous small ones..................5. Agavaceae
            7. Perianth dry and chaffy; X = 11......6. Xanthorrhoeaceae
      4. Vessels well distributed in all vegetative organs; stomates mostly with 2 subsidiary cells; perianth, or ovary, or both, commonly densely hairy or glandular.

8. Shrubs or subshrubs with the leaves crowded at the top of a usually branching stem, the old leaf-bases persistent and clothing the stem; plants without branched hairs...7. Velloziaceae
8. Scapose or subscapose herbs; inflorescence often with branched hairs..8. Haemodoraceae
3. Leaves with a definite petiole and an expanded (often large), more or less net-veined blade.
9. Plants acaulescent; vessels confined to the roots.
10. Ovules numerous on parietal placentae; endosperm well developed, rich in protein; chalazosperm wanting; leaves often much cleft....................................9. Taccaceae
10. Ovules 2 in each of the 3 locules (but usually only one locule fertile), on axile placentae; endosperm none; chalazosperm abundant, rich in starch and fat; leaves undivided..10. Cyanastraceae
9. Plants leafy-stemmed, usually climbing.
11. Flowers dimerous; vessels confined to the roots...11. Stemonaceae
11. Flowers trimerous; vessels well distributed in all vegetative organs.
12. Ovary superior; nectaries staminal or staminodial; plants not twining, mostly climbing by tendrils and without prominent tubers........................12. Smilacaceae
12. Ovary inferior; nectaries septal; plants mostly twining, without tendrils, usually with large tubers...13. Dioscoreaceae

2. ORCHIDALES

1. Stamens 3 or more, free from the style; pollen grains not cohering in pollinia; flowers regular or less often irregular; terrestrial plants, most species with reduced leaves and without chlorophyll; nectaries, at least in the Burmanniaceae, ovarian.
2. Flowers regular or nearly so; stamens 3 or 6.
3. Stamens 3, opposite the sepals; sepals valvate....1. Geosiridaceae
3. Stamens 6, or more often 3 and opposite the petals; sepals imbricate...2. Burmanniaceae
2. Flowers highly irregular; stamens 6.....................3. Corsiaceae
1. Stamens 1 or 2, rarely 3, mostly adnate to the style; pollen grains in most genera cohering in pollinia; flowers more or less strongly irregular; terrestrial or often epiphytic plants, with or less often without chlorophyll; nectaries diverse, but seldom ovarian........4. Orchidaceae

Section D.
DIAGNOSTIC KEY TO THE FAMILIES OF LOWER VASCULAR PLANTS IN NORTH AMERICA (NORTH OF MEXICO)*

1. Plants consisting of a dichotomously branching system of stems; leaves and roots lacking, but possessing minute enations scattered on aerial stems; sporangia large (1-2 mm across), trilocular........1. Psilotaceae
1. Plant consisting of stem, roots (except Salviniaceae), and leaves; sporangia large or small, but never trilocular.

*Contributed by John T. Mickel, The New York Botanical Garden.

2. Leaves with a single unbranching vein, small (less than 1.5 cm long) or long grass-like aquatic; stem protostelic or eustelic, lacking leaf and branch gaps (except Equisetaceae); sporangia borne in cones, on stem in axils of leaves, or embedded in the bases of grass-like leaves.
   3. Leaves whorled, reduced to toothed, generally non-green sheaths at nodes; stems hollow, jointed, fluted; sporangia borne under peltate scale-like structures that comprise the cone...2. Equisetaceae
   3. Leaves green, spirally arranged or rarely whorled with four leaves per whorl, never forming sheaths; stems never hollow, jointed, or fluted; sporangia always borne in association with leaves.
      4. Plants terrestrial; leaves less than 1 cm long; sporangia in the axils of either vegetative leaves or specialized leaves in cones; homosporous or heterosporous; spores all tetrahedral.
         5. Plants homosporous; sporangia in axils of vegetative leaves or in cones that are round in cross section..3. Lycopodiaceae
         5. Plants heterosporous, megaspores four per megasporangium; sporangia borne in cones that are four-sided in cross section, rarely round.......................4. Selaginellaceae
      4. Plants aquatic or semi-aquatic; leaves grass-like (3-40 cm long, 0.5-2.0 mm wide), tufted, arising from a short, erect, corm-like stem; sporangia embedded (adaxially) in the leaf near its base; heterosporous; megaspores tetrahedral, hundreds per megasporangium; microspores bilateral................5. Isoetaceae
2. Leaves generally with branching veins; leaves small to large; stems usually solenostelic with leaf and branch gaps; sporangia never borne in cones or in axils of leaves but on the leaf, rarely in nut-like sporocarps.
   6. Plants aquatic or semiaquatic; sporangia enclosed in nut-like sporocarps; heterosporous.
      7. Plants rooting in mud; leaves alternate, petiolate, more than 2 cm long, with four leaflets at tip of petiole or with no leaflets; sporocarps 2-6 mm across, arising from base of petiole.....................................6. Marsileaceae
      7. Plants free floating, rarely on mud; leaves simple, sessile, round or bilobed, opposite or whorled, less than 1.5 cm long; sporocarps 1 mm or less across, borne on leaf or root-like skeletonized leaf.
         8. Leaves 1 mm or less across, opposite and bilobed; sporocarps borne on lower leaf lobes; hairs lacking from leaf surface; plants inhabited by blue-green alga *Anabaena*..7. Azollaceae
         8. Leaves 6-15 mm across, whorled with one member of whorl trailing beneath, appearing as a root and bearing the sporocarps; upper leaf surface covered with hairs that are fused in clusters; plants not inhabited by *Anabaena*...8. Salviniaceae
   6. Plants terrestrial, rarely aquatic; sporangia borne on under side of leaf or on upright spike or panicle arising from the petiole, never enclosed in sporocarps; homosporous.
      9. Sporangia borne on upright spikes or panicles that arise from the petiole.
         10. Fertile panicles two per leaf; sporangia ca. 0.3 mm across; rhizome creeping at or just below ground level; strengthening tissue present throughout plants, making plant parts tough...................................9. Anemiaceae

10. Fertile spike or panicles one per leaf; sporangia ca. 1 mm across; rhizome upright, fleshy, well beneath the surface of the ground; strengthening tissue generally lacking, plant parts soft...10. Ophioglossaceae

9. Sporangia borne on the leaf blade, either on normal vegetative leaves or strongly modified leaves; sori on the blade margin or the under surface.

11. Leaves extremely delicate and filmy, only one cell thick, 1-15 cm long; sporangia borne on elongate receptacle in cups or tubes on leaf margin; rhizome often fine and hair-like, 1 mm or less in thickness....................................11. Hymenophyllaceae

11. Leaves more substantial, more than 1 cell thick; sori dorsal or marginal, but if in marginal cups, leaves more than 30 cm long; rhizomes generally more than 1 mm thick.

12. Leaves regularly dividing in half with dormant bud at the point of forking; pinnae pectinate; sori with few sporangia, (4-7) all developing at one time.......................12. Gleicheniaceae

12. Leaves not regularly dividing dichotomously; sori with many sporangia or only 1, usually not developing simultaneously.

13. Leaves climbing; sporangia borne singly under epidermal flaps on constricted segment lobes................13. Lygodiaceae

13. Leaves not climbing; sporangia not borne singly under epidermal flaps.

14. Leaves filiform; sporangia borne on pinnate (2-4 mm long) or digitate (7-8 mm long) divisions at end of leaf...14. Schizaeaceae

14. Leaves not filiform; sporangia not borne on small projections from end of petiole.

15. Fertile leaves or fertile portions of leaves lacking any leaf tissue; brown fertile portion at tip or middle of vegetative fronds or on totally distinct fronds; sporangia sessile with annulus an irregular lateral patch of thick-walled cells..15. Osmundaceae

15. Fertile leaves or fertile portions of leaves possessing at least a small amount of leafy tissue which may be green or brown; sporangia stalked, with a vertical annulus.........(Polypodiaceae sensu lato) go to 16 couplet

16. Rhizome clothed with hairs; rhizome long-creeping; branches arise regularly either by dichotomous forking of rhizome and/or from buds on the base of the petiole; sori marginal, protected either by an outer and inner indusium or only by a recurved marginal flap....16. Dennstaedtiaceae

16. Rhizome clothed with scales; rhizome erect or creeping; branches arise only from the stem, never from the petiole; sori marginal or on the lower side of the leaf, protected by a true indusium, a recurved marginal flap, or not at all.

17. Sori marginal, or if dorsal, elongate along veins or completely covering lower surface and without true indusium; spores tetrahedral; petiole bundles 1, 2, or rarely many (20 or more).

18. Plants aquatic, either floating or rooted in mud; leaves less than 0.5 mm long; often proliferating by buds on the leaves ....17. Parkeriaceae
18. Plants terrestrial, or if rooting in mud then leaves more than 1 m long; not proliferating by laminar buds.......... ....18. Adiantaceae

17. Sori dorsal (marginal in *Paltonium*); spores bilateral (tetrahedral in *Grammitis*); petiole bundles 1 (x shape), 2, or 3-7.
19. Indusium lacking; sori round (elongate along margin in *Paltonium*); blades mostly simple or pectinate, less commonly pinnate; spores lacking perispore......................... ....Polypodiaceae *sensu strictu*
19. Indusium generally present; sori round or elongate along veins or midrib, rarely covering blade surface; blades mostly once pinnate or more dissected, rarely pectinate or simple; spores with perispore.............19. Aspleniaceae*

*The name "Aspidiaceae" has more often been applied to this group, but that name is illegitimate.

FAMILY REVIEW EXERCISES

Family Key Exercise: An excellent review for diagnostic family study is to take the family key in a manual, guide, or synopsis, appropriate to your area, copy or mark the key leads to the families you have studied and then summarize the critical characters used in delimiting each family studied.

Family Comparison Chart Exercise: Make a list of families studied according to a system or alphabetically in a left hand column and across the top list characters used in your family analysis and then fill in columns with appropriate character states from your family analyses. See example below.

Family	Inflorescence	Sex	Floral Formula	Corolla Type	Etc.
Alismataceae					
Poaceae					
Cyperaceae					
Etc.					

FAMILY LITERATURE

(See references at end of Chapter 28)

Bessey, C. E. 1915. The phylogenetic taxonomy of flowering plants. Annals of the Missouri Botanical Garden 2: 109-164.

Cronquist, A. 1968. Evolution and Classification of Flowering Plants. Houghton-Mifflin Company. Boston.

Lawrence, G. H. M. 1951. Taxonomy of Vascular Plants. The Macmillan Company. New York.

Melchior, H. 1964. A. Engler's Syllabus Der Pflanzen-familien II. 12th ed. Gebrüder Borntraeger. Berlin.

FAMILY LABORATORY EXERCISES

During the course of a student's training in systematics he should become acquainted with the following families due to their phylogenetic or economic significance: Magnoliidae-(1) Magnoliaceae, (2) Annonaceae, (3) Lauraceae, (4) Saururaceae, (5) Aristolochiaceae, (6) Nymphaeaceae, (7) Ranunculaceae, (8) Berberidaceae, (9) Papaveraceae, (10) Fumariaceae; Hamamelidae-(11) Hamamelidaceae, (12) Ulmaceae, (13) Urticaceae, (14) Juglandaceae, (15) Myricaceae, (16) Fagaceae, (17) Betulaceae, (18) Casuarinaceae; Caryophyllidae-(19) Caryophyllaceae, (20) Portulacaceae, (21) Amaranthaceae, (22) Bataceae, (23) Polygonaceae, (24) Plumbaginaceae; Dilleniidae-(25) Dilleniaceae, (26) Theaceae, (27) Clusiaceae, (28) Malvaceae, (29) Droseraceae, (30) Violaceae, (31) Cistaceae, (32) Cucurbitaceae, (33) Salicaceae, (34) Brassicaceae, (35) Ericaceae, (36) Diapensiaceae, (37) Symplocaceae, (38) Primulaceae; Rosidae-(39) Saxifragaceae, (40) Rosaceae, (41) Fabaceae, (42) Podostemaceae, (43) Lythraceae, (44) Onagraceae, (45) Elaeagnaceae, (46) Cornaceae, (47) Santalaceae, (48) Celastraceae, (49) Aquifoliaceae, (50) Buxaceae, (51) Rhamnaceae, (52) Vitaceae, (53) Aceraceae, (54) Rutaceae, (55) Geraniaceae, (56) Polygalaceae, (57) Araliaceae, (58) Apiaceae; Asteridae-(59) Gentianaceae, (60) Asclepiadaceae, (61) Convolvulaceae, (62) Polemoniaceae, (63) Boraginaceae, (64) Verbenaceae, (65) Lamiaceae, (66) Plantaginaceae, (67) Oleaceae, (68) Scrophulariaceae, (69) Lentibulariaceae, (70) Campanulaceae, (71) Rubiaceae, (72) Caprifoliaceae, (73) Asteraceae; Alismatidae-(74) Alismataceae, (75) Hydrocharitaceae, (76) Najadaceae, (77) Potamogetonaceae; Commelinidae-(78) Xyridaceae, (79) Commelinaceae, (80) Eriocaulaceae, (81) Juncaceae, (82) Cyperaceae, (83) Poaceae, (84) Typhaceae, (85) Bromeliaceae, (86) Marantaceae; Arecidae-(87) Arecaceae, (88) Araceae, (89) Lemnaceae; Liliidae-(90) Pontederiaceae, (91) Liliaceae, (92) Iridaceae, (93) Dioscoreaceae, (94) Burmanniaceae, (95) Orchidaceae. At least 35-40 of these families should be studied in the basic course in systematics.

Each of the families listed above should be analyzed for diagnostic and phylogenetically significant characters. A sample family work sheet is provided which can be reproduced as it is or appropriately modified. Each family analysis should pertain to the species studied in that family with supplementary remarks to be added from the literature or other observations when the species does not adequately represent the diagnostic or phylogenetic characteristics of the family. A median longitudinal section of the flower and a cross section of the ovary should be drawn for each family. Unique features for the family; e.g., syngenesia, pollinia, arillate seeds, should be illustrated. Floral diagrams should be optional.

A desirable aid for family study is to have a class herbarium available with the families arranged alphabetically with mounted specimens of at least one species of each genus within each family from the area. To show intrafamilial relationships, demonstration material could be arranged according to sub-family and/or tribe.

SAMPLE FAMILY WORK SHEET

Species ______ Family ______

1. Plant Type ______
2. Plant Duration ______
3. Plant Sex ______
4. Root Type ______
5. Stem Type ______
6. Leaf Type ______
7. Inflorescence Type ______
8. Flower Type ______
9. Androecial Type ______
10. Stamen Type ______
11. Gynoecial Type ______
12. Carpel Type ______
13. Stigma Type ______
14. Style Type ______
15. Ovule Type ______
16. Fruit Type ______
17. Seed Type ______
18. Embryo Type ______
19. Location/Plant Organ ______
20. Position/Plant Organ ______
21. Arrangement/Plant Organ ______
22. Floral Merosity ______
23. Venation ______
24. Sex ______

Flower L.S.

Special Features

General Family Comments:

FAMILY AND HIGHER TAXA INDEX

Chapter 30. TAXONOMIC LITERATURE AND BIBLIOGRAPHIC AIDS

The descriptive nature of taxonomy, the type method of documentation, and the principle of priority have contributed greatly to the necessity that taxonomists be bibliographers as well as plant scientists. The degree to which one becomes acquainted with and is able to handle bibliography will affect, in large measure, one's success as a taxonomist. Most problems to be solved by the taxonomist will involve some reference to the voluminous and poorly indexed literature.

A classification scheme of taxonomic literature is presented in Section A, and the highly selected references in Section B are arranged according to this system. The aim of the presentation of references in Section B, to borrow a phrase from B. D. Jackson's introduction to A Guide to the Literature of Botany, is "suggestive, not exhaustive." The emphasis is on basic references and literature guides and types of references to be consulted. Study of the classification system in Section A,II and analysis of a question to be answered should enable one to choose from the selected bibliography (Section B) either specific references or types of references to be consulted in the library. Many references are annotated as a guide to their contents, purpose and use. Only taxonomic literature and basic literature guides are included here. References specifically treating morphology, anatomy, cytology, ecology, etc. may be found in the appropriate chapters in this book. In Section C some of the references useful in organizing and preparing taxonomic papers and suggestions for the collection and preparation of a reference file are given.

The following outline with page numbers of appropriate sections of this chapter is provided to facilitate the location of references.

Section A. CLASSIFICATION

I. CLASSIFICATION USED IN LIBRARIES

Two systems of classification are currently used in libraries--the Dewey Decimal and the Library of Congress. In fact, many libraries are in the process of changing from the older Dewey System to the LC System, and familiarity with both systems is advantageous. The following lists include examples of some of the major categories of each system.

DEWEY DECIMAL		LIBRARY OF CONGRESS	
Natural Science	500	Natural History	QH
Biology	570	Botany	QK
Botany	580	Zoology	QL
Zoology	590	Agriculture	S
Agriculture	630	Forestry	SD
		Horticulture	SB
		Bibliography	Z

For a comparison of the two systems and other pertinent information on library classification systems and theory and botanical bibliography see Swift, L. H., 1970, Botanical Bibliographies, a guide to the bibliographic materials applicable to Botany, Burgess Publishing Company, Minneapolis, Minnesota.

Since the LC System is so widely used, the following examples are given to acquaint the student with the disposition of taxonomic literature and the meaning of call numbers in this system.

A. Call Number for A Specific Reference

Conard, Henry S. 1905. The Waterlilies. A monograph of the genus *Nymphaea*. Carnegie Institution of Washington.

QK 495.N97 C7

Science	Botany	Spermatophyta-Systematics Division	Systematic Group	Author
Q	K	495	.N97	C7

B. General Groups and Examples

QK 1 - PERIODICALS, SOCIETIES
- QK 1.C2 - California University Publications in Botany
- QK 1.C55 - Chronica Botanica
- QK 1.F4 - Field Museum of Natural History - Botanical Series (Chicago Natural History Museum - Fieldiana: Botany)

QK 11 - INDICES, NOMENCLATORS, CHECKLISTS
- QK 11.L45 - Lemée's Dictionnaire des Genres des Plantes Phanérogames

QK 31 - BIOGRAPHY (Individual)
- QK 31.G8 D8 - Dupree, A. H. 1959. Asa Gray.

QK 91 - BOTANICAL WORKS OF LINNAEUS
- QK 91.S69 - Stafleu, F. A. Linnaeus and the Linnaeans.

QK 95 - CLASSIFICATION (DISCUSSION AND COMPARISON OF SYSTEMS)
- QK 95.L3 - Lawrence, G. H. M. 1951. An introduction to Plant Taxonomy.

QK 96 - NOMENCLATURE
- QK 96.I58 - International Code of Botanical Nomenclature.

QK 97 - COMPREHENSIVE SYSTEMATIC WORKS
- QK 97.C2 - de Candolle's Monographiaë Phanerogamarum.
- QK 97.E62 - Engler's Das Pflanzenreich.

QK 101-474 - GEOGRAPHIC DIVISIONS (FLORAS AND DISTRIBUTION)
- QK 110 - North America
  - QK 110.A1 - North American Flora.
- QK 125 - Southeastern U. S.
  - QK 125.S64 - Small, J. K. Manual of the Southeastern Flora.
- QK 146 - Alaska
  - QK 146.H84 - Hultén's Flora of Alaska and Yukon.
- QK 211 - Mexico
  - QK 211.M3 - Martinez, M. Catalogo de Plantas Mexicana.
- QK 281 - Europe
  - QK 281.H4 - Hegi, Flora Von Mittel-Europa.
  - QK 281.T8 - Flora Europaea.
- QK 321 - Russia
  - QK 321.A36 - Flora URSS.

QK 495 - SYSTEMATIC DIVISIONS A - Z
- QK 495.A17 - Acer
- QK 495.Q4 - Quercus
- QK 495.N97 - Nymphaeaceae
  - QK 495.N97 C7 - Conard, H. S. The Waterlilies.

QK 520-532 - PTERIDOPHYTA
- QK 524 - Systematic Divisions
- QK 525 - U. S.
- QK 527 - Great Britain
- QK 529 - Asia

Z - BIBLIOGRAPHY
- Z5356.T8R42 - Rehder, A. - Bibliography of Cultivated Trees and Shrubs

II. CLASSIFICATION OF TAXONOMIC LITERATURE

A. Nomenclature. Nomenclature is used here to refer to references dealing not only with rules of nomenclature (i.e., the Code) and works of value in solving nomenclatural problems but also those indices which include the names of species, genera and families. References cited in this class should assist primarily in locating original descriptions, authors, place of publication and synonomy and, secondarily, for finding information on monographs, revisions, distribution, etc.

B. Terminology and Description of Taxa. References helpful in the translation and/or writing of descriptions (e.g., glossaries and dictionaries) are placed here. It is often important to use the older treatments for changes in usage of terms. For an extensive classification and definitions of phytographic terminology and the meanings of specific epithets see Chapters 4 and 6.

C. Identification. Identification is one of the basic aspects of taxonomy. The references cited include a variety of kinds of taxonomic literature useful in identification as well as guides to references for identification. Selected references also include those treating native and cultivated plants; local, state, regional and international floras; and both technical and popular works. Such references are also good sources of other information such as distribution, variation, phenology, ecology, etc.

D. Distribution. The range or geographic distribution of taxa has long been of concern to the taxonomists (see Chapter 16), but such information has become of increasing importance in a variety of disciplines. The references given not only include major works dealing specifically with distribution but include examples of types of works which should be examined for distribution data.

E. Illustrations. Illustrations are particularly useful in the absence of a specimen of a particular plant in question, in interpreting type or other descriptions and may also be selected as lectotypes (see Chapter 3). Several sources of information on published illustrations are cited.

F. Classification Systems (Phylogeny). Classification systems are treated in Chapter 28 and an extensive bibliography is included there.

G. Botanical Institutions, Depositories and Collectors. It is often desirable and necessary to locate specific collections (e.g., types), borrow representative material in preparation of monographs and revisions, locate publications and field notebooks of collectors, compare samples of handwriting, and secure living specimens and other data housed in botanical institutions (herbaria, botanic gardens and libraries). Selected indices to such institutions and types of information are included here.

H. Current Literature (Periodicals and Guides). Current publications such as books are rather easily located in subject catalogs in libraries, booksellers' and publishers' catalogs, and book reviews, but periodical literature remains the most elusive. No single current, comprehensive and readily usable index or abstracting service specifically for taxonomic literature exists. The references treated here include the common abstracting and indexing services, selected guides, and journals of interest to taxonomists.

I. Reference Lists, Bibliographic Guides, Catalogs and Indices. Included here are references to the older literature, special approaches, library holdings, etc. with annotations on possible uses of specific references or types of references.

J. General Texts. The number of general texts in Taxonomy is rather limited, but these have a variety of uses. They represent one of the best ways of getting an overview of a specific subject and include valuable bibliographies to original sources of information. In addition to general taxonomic texts several texts treating specific areas (chemotaxonomy, taximetrics, etc.) are also included.

K. History and Exploration. Biographies, itineraries, collections (herbaria), personal libraries, diaries, fieldbooks, correspondence, handwriting samples, etc. of botanists and explorers are often important to present-day taxonomists not only in locating specimens, deciphering labels, and determining type localities, but also in understanding the philosophy and concepts of taxa of earlier workers (see discussion and bibliography in Chapter 2).

L. Monographs, Revisions, and Synopses. These comprehensive works of specific groups, often representing years of work by the most competent authorities, also are the most difficult to locate. No single index to such works exists but some of the basic sources, many of which are discussed elsewhere, are included here.

Section B. REFERENCES

I. NOMENCLATURE

A. Rules (Code)

Stafleu, F. A. (chairman of editorial committee). 1972. International Code of Botanical Nomenclature, adopted by the Eleventh International Botanical Congress, Seattle, August 1969. Regnum Vegetabile 82. Utrecht. [The Code is published in English, French, and German; in case of inconsistency the English wording has been arbitrarily decided as correct. See Chapter 3 for discussion of Code and its use. See pp. 395-397 in the Code for list of editions 1-10 and the appendices of the Code for conserved names.]

McVaugh, R., R. Ross, & F. A. Stafleu. 1968. An annotated glossary of botanical nomenclature. Regnum Vegetabile 56: 1-31. [An unofficial glossary of the more difficult terms used in the Code (ICBN)--a valuable aid in interpreting the Code.]

International Code of Nomenclature for Cultivated Plants. 1958. Regnum Vegetabile 10. Utrecht. [The rules and recommendations governing the naming of cultivated plants are presented. The aim of this work, according to Article 3, p. 11, is "to promote uniformity, accuracy and fixity in the naming of agricultural, silvicultural and horticultural cultivars (varieties)... which are normally given fancy names... ." The ICBN (cited above), however, still governs the use of scientific or Latin names of plants whether wild or cultivated.]

Synopsis of proposals on botanical nomenclature. [Prior to an International Botanical Congress a synopsis of proposals to amend or change the Code is published in Regnum Vegetabile. A list of Nomina Conservanda Proposita is also included (e.g., see Stafleu, F. A., & E. G. Voss, 1969, Synopsis of proposals on botanical nomenclature--Seattle, 1969, Regnum Vegetabile 60, Utrecht.).]

B. Selected References, Guides and Indices to Plant Names

Bentham, G., & J. D. Hooker. 1862-1883. Genera Plantarum ad exemplaria imprimis in herbariis kewensibus servata difinita. 3 vols. London. [A valuable reference to names of tribes and genera and generic descriptions. Many new genera proposed. Addenda et Corrigenda in each volume. Keys and descriptions in Latin; for dates of publication of parts and references to commentaries see Stafleu, F. A., 1967, Taxonomic Literature.]

Burbidge, N. T. 1963. Dictionary of Australian Plant Genera, Gymnosperms and Angiosperms. Angus and Robertson. Sydney, Australia. [A dictionary of the genera of Australia region including Tasmania. The information treated includes generic names (arranged alphabetically); citation (author and place of publication); synonomy; family or families (as treated in systems of Bentham & Hooker, Engler & Prantl, Hutchinson); phytogeographic data (general distribution--classed as endemic, native or naturalized); and bibliographies. An alphabetical list of families with their inclusive genera is also included.]

Candolle, A. P., A., and C. de. 1824-1873. Prodromus systematis naturalis regni vegetabilis. 17 vols. Paris. [See Vol. 17 (1873): 303-314 for history and analysis and pp. 323-493 for general index. Many new species described and valuable monographs included. See Stearn, W. T. (1939) Candollea 8: 1-4; and also Bueck, H. W. (1842-1874) Genera, species et synonyma Candolleana alphabetico ordine deposita, seu index generalis et specialis ad A. P. de Candolle Prodromum systematis naturalis regni vegetabilis (index to the Prodromus); and Stafleu, F. (1967) Taxonomic Literature, Regnum Vegetabile 52. Utrecht.]

Christensen, C. F. 1905-1965. Index Filicum sive enumeratio omnium generum specierumque filicum et hydropteridum ab anno 1753 ad finem anni 1905 descriptorium, adjectis synonymis principalibus, area geographica... . Hagerup, Hafniae. [Issued in 12 fascicles 1905-1906 with 4 supplements. This reference is to pteridology what Index Kewensis is to the taxonomy of flowering plants. A taxonomic conspectus of all genera and species of ferns with bibliographic reference, synonomy, and geographic distribution. Supplement 1, issued 1913 (treating names published 1906-1912); Supplement 2, issued 1917 (treating the years 1913-1916); Supplement 3, issued 1934 (including the years 1917-1933).]

The 4th supplement was prepared by R. E. G. Pichi-Sermolli and collaborators (1965) and differs from previous works in that it treats only generic, infrageneric and specific names validly published between 1934 and 1960 with geographic distribution of new taxa with cross-references to basionyms. A catalog of literature and indices to abbreviations of names of authors and periodicals are also present.]

Copeland, E. B. 1947. Genera Filicum. The genera of ferns. Annales Cryptogamici et Phytopathologici. Vol. 5. Chronica Botanica. Waltham, Massachusetts. [A treatment of the families and genera of the Filicineae arranged in a phylogenetic sequence. In addition to descriptions, keys to genera, many references to numbers of species, synonyms, types and references to the literature are included.]

Dalla Torre, C. G., & H. Harms (eds.). 1900-1907. Genera Siphonogamarum ad Systema Englerianum Conscripta. Engelmann. Leipzig. [An enumeration of taxa above the rank of species arranged according to the Engler System.
Families and genera are numbered consecutively throughout and these numbers are sometimes used in herbaria for filing and locating materials arranged according to this system. All names are followed by author, literature citation, and synonyms. The inclusion of subfamilies, tribes, etc., as well as the number of species in each genus, make this a valuable reference. A supplement, appendix and generic index are included. Users should always check the supplement and appendix. The index nominum has been updated and entries in the supplement and appendix added and issued as Register zu De Dalla Torre et Harms Genera Siphonogamarum (1958). This includes genus, author, Dalla Torre-Harms numbers, family, number of genera within family. Synonyms are also included with an indication of the genus in which they are placed. This is one of the most useful references, particularly when consulting an herbarium with specimens filed according to the Englerian system.]

Dandy, J. E. 1967. Index of Generic Names of Vascular Plants, 1753-1774. Regnum Vegetabile 51. Utrecht. [A list of all generic names first published from 1753 to 1774 whether in binomial or non-binomial works. The 2,998 entries of the index were checked with the original publication or a photocopy of the original work. Arrangement of entries (generic names) is alphabetical followed by author, date and place of publication, annotation, and name of family to which the genus belongs. In addition to a list of works admissible as sources of generic names, a list of sources which are not valid sources of names is also given. A systematic arrangement of genera according to Dalla Torre & Harms, Genera Siphonogamarum (with modifications), is presented. Both nomenclatural and taxonomic synonyms are indicated.]

Engler, A. 1954, 1964. Syllabus der Pflanzenfamilien. Ed. 12. Vol. 1. Bakterien bis Gymnospermen (H. Melchior & E. Werdermann, eds.). Vol. 2. Angiospermen (H. Melchior, ed.). Gebrüder Borntraeger. Berlin. [This is a valuable source of many kinds of information. In addition to names, the systematic arrangement, description of orders, families, subfamilies, etc. and notes on the number of genera and the number of species they include, there is an extensive bibliography including information on anatomy and morphology, phylogeny, floras, monographs and revisions as well as many valuable illustrations. This is a reference which should be widely used by students of systematics. Text in German.]

Engler, H. G. A., & K. A. E. Prantl. 1887-1914. Die natürlichen Pflanzenfamilien. Leipzig. Ed. 1. 23 vols; 2nd ed. 1924-(incomplete). [A survey of all plant groups (except bacteria), with keys to genera, descriptions, selected bibliography and many notes on taxonomy and nomenclature. A useful summary of information on morphology, anatomy, paleobotany, embryology of groups included.]

Gray Herbarium Card Index (Gray Card Index). [A card index (also available in volumes) issued quarterly to subscribers, to all new names and new combinations of vascular plants, except fossils, native to the new world (see area described below). The starting date for inclusion is 1 January 1886 (see exception below) and the first issue was in 1894. The complete set includes approximately 280,000 cards. Information includes (1) new taxa--the name, author, reference to place of publication, statement of range; (2) new combinations--new name, bibliographic reference and basionym; (3) base cards--basionym, bibliographic references (with some exceptions), new name based on it and other nomenclatural synonyms based on same type; (4) new issues--include type, collector and herbarium.

1. Geographic area
    a. Continental Americas including the immediately adjacent islands (see e below).
    b. Greater and Lesser Antilles and Bermuda.
    c. Canadian Arctic Archipelago, Greenland, and Iceland (coverage not complete).
    d. Aleutian Islands, Pribilof Islands and St. Lawrence Island in the Bering Strait (follows political boundary between United States and the Soviet Union).
    e. In southern hemisphere island possessions of American nations, such as Easter Island, Juan Fernandez, Falkland Islands and South Georgia. [Hawaiian Islands are not included.]
2. Taxa included

    From its start the index has included names of genera, species and taxa of infraspecific rank, with the inclusion at some time in the past of subgenera. In 1945 (with issue 186) it was decided to include names of infraspecific taxa published between 1754 and 1886. Beginning with issue 261 the Index will include in addition to subgenera, new names of sections, series, and of subdivisions of these (i.e., starting with issue 261 new names of taxa of all ranks that are present affected by principles of priority will be included). In addition, a taxon above the rank of species is indexed if the taxon includes genera with species native to the Index area. Up to 1970 hybrid groups were indexed if given a name. Again, with issue 261, names of hybrid groups validly published according to article 40 and conforming to the Articles of Appendix I of the ICBN (1966) will be included. Nothomorphs designated by an epithet preceded by a binary name will also be included. (For additional details see Shaw, in Taxon 20: 333-336, 1971.).]

Heller, A. A. 1900. Catalogue of North American Plants North of Mexico. Exclusive of the Lower Cryptogams.

Hooker, J. D., & B. D. Jackson et al. 1893-1895. Index Kewensis plantarum phanerogamarum. 2 vols. & 14 supplements. Oxford. [An index to the place of publication of generic names or binomials of seed plants published since 1753. Coverage is worldwide. Volumes 1-2 include names from 1753-1885. Supplements are issued at 5- or 10-year intervals (more frequent recently). Arrangement of information: (1) genus, author, bibliographic reference, (2) specific names arranged alphabetically followed by author, place of publication, and native country of plant. Supplements include date of publication and double author entry, but indication of synonomy has been discontinued. Beginning with Supplement 10 this index has taken over the function of Index Londinensis, and references to illustrations are designated by an asterisk. This is a reference to be consulted before publishing a new name or combination and also a source to locate publication of a name and thus the original description. The greatest disadvantage is that names of infraspecific taxa are not included.]

See also Rouleau, E. (compiler). 1970. Guide to Index Kewensis and its supplements i-xiv. Montréal. [This index consists of 3 sections--(1) an alphabetical list of genera of Magnoliophyta and references to volumes of Index Kewensis and its supplements in which the genera appear; (2) same as section 1 but treats Equisetophyta, Lycopodiophyta, Polypodiophyta and Psilophyta; (3) list of families and their inclusive genera and an alphabetical list of families. (See also the errata and the incerta sedis at the end of the volume.) This is a welcome time-saving device in that one no longer needs to search all volumes and supplements for a name.]

Index Nominum Genericorum. [A project proposed in 1954 to produce a file of names of all genera in all plant groups, with each entry the result of painstaking bibliographic research and application of the ICBN. The first issue (1000 cards) contributed by 14 botanists was in 1955. Information on each of the cards consists of the validly published generic name, its author, citation of place of publication, as precise a date as possible, type species and its basionym (when appropriate). Numbers on each card are a serial number and a number designation of the person who prepared the card. Class and family in which a genus is placed are also indicated. As of November, 1972 (Stafleu, F. A., 1972, Taxon 21: 732) all card production was stopped! A computer printout reproduced by offset is to be avail able in 1974. This consolidated printout will contain 50-60,000 generic names.

The principles observed in the preparation of this valuable index according to R. Cowan (Taxon 19: 52-54, 1970) are as follows: "(1) To avoid repetition of existing errors in citations, the original literature for every entry is reviewed with great care. (2) Citation of type species is undertaken only when evidence for doing so is incontestable as, for example, when a genus is published with only a single species or...when a subsequent monographer has designated a type species. Where evidence is not so clear, the type species is not listed and the card is marked as 'provisional.' Such cards are completed and will be re-issued as taxonomic studies are published that provide the data required. (3) Each card lists the essential nomenclatural synonyms but validly published taxonomic synonyms are treated as separate entries. (4) Except for the citation of family names for each entry, the Index scrupulously avoids taxonomic decisions."]

Jackson, B. D. 1881. Guide to the literature of Botany. London. [See description on contents in B,VIII B.]

Kelsey, H. P., & W. A. Dayton (eds.). 1942. Standardized Plant Names. A revised and enlarged listing of approved scientific and common names of plants and plant products in American commerce or use. 2nd ed. J. H. McFarland Co. Harrisburg, Pa. [This edition includes approximately 90,000 entries of names of economic plants and plant products. The arrangement is alphabetical by plant and/or product name; e.g., drug plant names are together and include drug name, scientific (botanical) name and S.P.N. common name. Cross-referencing is good. Although useful in interpreting names in agricultural and horticultural works, these names are not necessarily those in common usage.]

Linnaeus, C. 1753. Species Plantarum. Ed. 1, Holmiae. [The starting point for priority of the names of vascular plants begins with this publication. Typification, however, remains a problem since no types were designated and not all specimens extant in the Linnaean Herbarium today can be taken as types. Some references and guides to the use of Species Plantarum (various editions) are: Heller, J. L, & W. T. Stearn, 1959, An Appendix to the Species Plantarum of Carl Linnaeus, appended to the facsimile edition of

Linnaeus's Species Plantarum, 2, London; Savage, S., 1945, A Catalogue of the Linnaean Herbarium, London; Stearn, W. T., 1959, An Introduction to the "Species Plantarum" and cognate botanical works of Carl Linnaeus Prefixed to the Ray Society Facsimile of Linnaeus' Species Plantarum; Stearn, W. T., 1961, A new photographic record of the Linnaean Herbarium, Taxon 10: 16-19; Stearn, W. T., 1966, The use of bibliography in natural history, In: Bibliography and Natural History, T. R. Buckman (ed.), University of Kansas Publication Library Series 27: 1-26.]

Merrill, E. D. 1949. Index Rafinesquianus--the plant names published by C. S. Rafinesque with reductions, and a consideration of his methods, objectives and attainments. Arnold Arboretum of Harvard University. Jamaica Plain, Massachusetts.

Pfeiffer, L. 1873-1874. Nomenclator botanicus. Nominum ad finem anni 1858 publici juris factorum, classes, ordines, tribus, familias, divisions, genera, subgenera vel sectiones designantium enumartio alphabetica. Adjectis auctoribus, temporibus, locis systematicis apud varios, notis literariis atque etymologicis et synonymis. 2 vols. in 4 parts. Kassel. [A valuable guide to names and places of publication of taxa, particularly above the rank of genus.

Reed, C. F. The following selected references prepared by Dr. Reed are valuable indices to names, synonyms, classification, bibliographies, monographs and distribution of ferns and fern allies:

Index to Equisetophyta. Part 1, Fossils; Part 2, Extantes, Index Equisetorum. Contribution of the Reed Herbarium No. 19. Baltimore, Maryland. 1971.

Index Isoetales. Boletim da Sociedade Broteriana. Vol. 27 (2.ª série). 1953.

Index Marsileata et Salviniata. Boletim da Sociedade Broteriana. Vol. 28 (2.ª série). 1954. Supplement, 1965. Vol. 39 (2.ª série).

Index Selaginellarum. Memorias da Sociedade Broteriana. Vol. 18 5-287. 1966.

Index Thelypteridis. Phytologia 17: 249-328. 1968. Supplement 1, Phytologia 17: 465-466.

Stafleu, F. A. 1967. Taxonomic Literature. A selective guide to botanical publications with dates, commentaries and types. Regnum Vegetabile 52. Utrecht. [The entries in this work are arranged alphabetically by authors and include location of herbarium and types, publications, bibliography and references. An index to authors and to publication data is included in this volume. This publication brings together much information (particularly location of types and publication dates) scattered through botanical literature, much of which is not readily available to workers who do not have access to a superior botanical library. This reference is very useful when used in conjunction with other references cited in this class and is a desideratum for every botanical library.]

Steudel, E. G. 1821-1841. Nomenclator botanicus synonymia plantarum universalis enumerans ordine alphabetico nomina atque synonyma tum generica tum specifica et a Linnaeo et recentioribus de re botanica scriptoribus plantis phanerogamis imposita. Ed. 1, Vol. 1 (1821), Vol. 2 (1824); Ed. 2 (1840-1841), in 13 parts. Stuttgart. [This is one of the most important post-Linnaean indices to flowering plant names (1753-1840) and was the forerunner of Index Kewensis. Entries are arranged alphabetically and include genus, author, family synonyms, habit, and references to other publications (e.g., Roemer & Schultes (RS), Sprengel (S), deCandolle (D). (See Stafleu, F., 1967, Taxonomic Literature, p. 458 for dates of publication, biographical data and references.)]

Willis, J. C. 1973. A Dictionary of the Flowering Plants and Ferns. 8th ed. rev. by H. K. Airy Shaw. Cambridge University Press. Cambridge. [The 7th edition departed somewhat from previous editions (see preface of 7th edition) in that entries deal strictly with taxonomic matters. The aim of the 8th edition, like the 7th, is to include every published generic name from 1753 to the present and every family name published from the appearance of Jussieu's Genera Plantarum in 1789. Some supra- and infra-familial names are also included. Conserved generic names are indicated by an asterisk; reductions to synonomy and many variant spellings are also present; brief descriptions of all accepted families are given. The arrangement of names is strictly alphabetical without regard to rank. A genus entry includes author(s), generic synonym (if so regarded), family, number of species, general distribution, and in some cases annotations treating special morphological features, economic value, etc. The included key to the families of flowering plants is based on Engler's Classification as given in Die natürlichen Pflanzenfamilien and revised in the 7th edition of his Syllabus. This is one of the most useful references for checking the family to which a genus belongs, variant spellings, synonyms, etc., and is used daily in most herbaria. In addition to the number of corrections, this edition also includes names of inter-generic hybrids of orchids with an indication of their status and a list of family equivalents between this dictionary, the 12th edition of Engler's Syllabus and Bentham & Hooker's Genera Plantarum.

C. Journals

Taxon. This journal is published for the International Association for Plant Taxonomy by the International Bureau for Plant Taxonomy and Nomenclature, Tweede Transitorium, Uithof, Utrecht, Netherlands, and contains many important papers on nomenclature (for indices see Rouleau, E. [compiler], 1972, Index to Volumes 1-20, Part I, Scientific Names; and Lowden, R., 1973, Index to Taxon Volumes 1-20, Part II, Authors and Subjects).

II. TERMINOLOGY AND DESCRIPTION OF TAXA (See also Chapters 4 and 6)

Candolle, A.L.P. de. 1880. La Phytographie, ou l'art de décrire les végétaux considérés sous différents points de vue. Paris.

Featherly, H. I. 1954. Taxonomic Terminology of the Higher Plants. Iowa State College Press. Ames, Iowa.

Font Quer, P. 1963. Diccionario de Botánica. Editorial Labor, S. A. Barcelona.

Gray, A. 1879. Structural Botany. Macmillan & Company. London.

Jackson, B. D. 1928. A Glossary of Botanic Terms, 4th ed. J. B. Lippincott Co. Philadelphia. [A widely used basic glossary, reissued a number of times.]

Lawrence, G. H. M. 1951. Taxonomy of Vascular Plants. The Macmillan Company. New York. [See the illustrated glossary of taxonomic terms, pp. 737-775.]

Lindley, J. 1848. Illustrated Dictionary of Botanical Terms. London. [Appears as Book III Glossology in An Introduction to Botany and later issued separately. Excerpt of 40 pages with Introduction by Alice Eastwood was issued in 1938 under the auspices of the California Academy of Sciences and the Garden Club of San Francisco; this excerpt was reprinted in 1964 with a foreword by J. J. Graham, School of Earth Sciences, Stanford University.]

Stearn, W. T. 1966. Botanical Latin. Thomas Nelson & Sons. London. [A valuable aid to translation or writing of descriptions.]

Steinmetz, E. F. 1953. Vocabularium Botanicum. 2nd ed. Amsterdam. [Scientific terms are in Latin, Greek, French, Dutch, German and English. A very useful source of equivalents in these six languages.]

Systematics Association Committee for Descriptive Terminology. 1960. I. Preliminary list of works relevant to Descriptive Biological Terminology. Taxon 9: 245-257.

III. IDENTIFICATION

A. Floras

1. Guides to Floras and Floristic Works

Blake, S. F., & A. C. Atwood. 1942-1961. Geographical guide to floras of the world. Parts 1-2. U. S. Department of Agriculture Miscellaneous Publications 401 & 797. Washington, D. C. (Part 1 reprinted by Hafner Publishing Co., 1963). ["An annotated selected list of floras and floristic works relating to vascular plants, including bibliographies and publications dealing with useful plants and vernacular names." Introduction includes many helpful notes on bibliographies, guides, dictionaries, etc. used in compiling this work. Part I--Africa, Australia, Insular floras, North America (including Canada, Mexico and Central America, Greenland, W. Indies) and South America. Part II--Western Europe, European USSR and all of Asia including Turkey (not completed). An index to abbreviations and to authors is included in each part.]

Frodin, D. G. 1964. Guide to the standard floras of the world. Department of Botany. University of Tennessee, Knoxville. [Mimeographed. This guide, to be published later (personal communication), is for the most part limited to publications dealing with complete floras of seed plants for a given area. It includes much information from Blake and Atwood (1942-1961), Jackson (1881) and Goodale (1879). References are grouped into nine geographic regions--Arctic and Antarctic; North and Central America; the West Indies; South America; Africa; Europe; Asia; Malaysia; Australasia and Oceania; Oceanic Islands (including eastern Pacific, Atlantic and Indian Oceans). An index to authors is also included.]

Goodale, G. L. 1879. The floras of different countries. Bibliographical Contributions, Library, Harvard University 9: 1-12.

Shetler, E. R. 1966. Selected Guides to the Wildflowers of North America. Department of Botany, Smithsonian Institution. Washington, D. C. (Mimeographed). [A guide to selected popular, illustrated books in print as of 1966.]

Shetler, S. G. (ed.). 1973. Appendix D. Provisional List of Basic Floras. In: A Guide for Contributors to Flora North America (FNA). Flora North America Report 65: D1.

Tutin, T. G., V. H. Heywood, N. A. Burgess, D. H. Valentine, S. M. Walters, & D. A. Webb (eds.). 1964+. Flora Europaea. Cambridge University Press. Cambridge. [Volumes include list of basic and standard floras for Europe.]

2. Selected Regional, State and Local Floras of North America, North of Mexico

a. Arctic and Subarctic North America

Hultén, E. 1968. Flora of Alaska and Neighboring Territories; A Manual of the Vascular Plants. Stanford University Press. Stanford, California.
Polunin, N. 1959. Circumpolar Arctic Flora. Clarendon Press. Oxford.
Wiggins, I. L, & J. H. Thomas. 1962. A Flora of the Alaskan Arctic Slope. Arctic Institute of North America Special Publication No. 4. University of Toronto Press. Toronto.

b. Eastern North America

Batson, W. T. 1972. A Guide to the Genera of Native and Commonly Introduced Ferns and Seed Plants of the Southeastern United States excluding Peninsula Florida. Published by the author. University of South Carolina. Columbia, South Carolina.
Fernald, M. L. 1950. Gray's Manual of Botany. 8th ed. American Book Company. New York. [Corrected printing 1970, Van Nostrand Reinhold Company.]
Gleason, H. A. 1963. The New Britton and Brown Illustrated Flora of the Northeastern United States and Adjacent Canada. 3 vols., 3rd printing, slightly revised. Published for the New York Botanical Garden by Hafner Publishing Company. New York.
________, & A. Cronquist. 1963. Manual of Vascular Plants of Northeastern United States and Adjacent Canada. D. Van Nostrand Company, Inc. Princeton, New Jersey.
Harvill, A. M., Jr. 1970. Spring Flora of Virginia. Privately published by the author. Farmville, Virginia.
Marie-Victorin, Frère. 1935. Flore Laurentienne (Québec). Imprimerie de La Salle. Montréal. [In French.]
Radford, A. E., H. E. Ahles, & C. R. Bell. 1968. Manual of the Vascular Flora of the Carolinas. University of North Carolina Press. Chapel Hill, North Carolina.
Small, J. K. 1933. Manual of the Southeastern Flora. Published by the author, New York. [Reissued in 1953 by the University of North Carolina Press, Chapel Hill, North Carolina; facsimile reprint issued in 2 vols. in 1972 by Hafner Publishing Company.]
Strausbaugh, P. D., & E. L. Core. 1952-1964. Flora of West Virginia. Introduction and Pts. I-IV. West Virginia University. Morgantown, West Virginia. [Parts I, II revised 1970, 1971, respectively.]
Wood, C. E., Jr., & Collaborators. 1958+. Generic Flora of the Southeastern United States. Journal of the Arnold Arboretum. [Keys to genera, illustrations, and superb bibliographies.]

c. Central North America

Correll, D. S., & M. C. Johnston. 1970. Manual of the Vascular Plants of Texas. Texas Research Foundation. Renner, Texas.
Deam, C. C. 1940. Flora of Indiana. Department of Conservation. Indianapolis, Indiana.
Goodman, G. J. 1958. Spring Flora of Central Oklahoma. University of Oklahoma Duplicating Service. Norman, Oklahoma.
Jones, G. N. 1963. Flora of Illinois. 3rd ed. University of Notre Dame Press. South Bend, Indiana.
Mohlenbrock, R. H. (gen. ed.). 1967+. The Illustrated Flora of Illinois. Southern Illinois University Press. Carbondale, Illinois. [A multivolume, illustrated flora of Illinois to cover all plant groups--algae and fungi through flowering plants--in progress; volumes appear as completed.]

Rydberg, P. A. 1932. Flora of the Prairies and Plains of Central North America. The New York Botanical Garden. New York. [Reprinted in 1965 by Hafner Publishing Company, New York.]
Scoggan, H. J. 1957. Flora of Manitoba. National Museum of Canada Bulletin 140, Biological Series 47.
Shinners, L. H. 1958. Spring Flora of the Dallas-Fort Worth Area, Texas. Published by the author. Southern Methodist University, Box 473, Dallas, Texas. [2nd edition edited by W. E. Mahler published in 1972 by Prestige Press, Fort Worth, Texas.]
Steyermark, J. A. 1963. Flora of Missouri. Iowa State University Press. Ames, Iowa.
Voss, E. G. 1972. Michigan Flora. Part 1. Gymnosperms and Monocots. Cranbrook Institute of Science. Bloomfield Hills, Michigan.
Waterfall, U. T. 1969. Keys to the Flora of Oklahoma. 4th ed. Published by the author and distributed by Oklahoma State University Bookstore. Stillwater, Oklahoma.

d. Rocky Mountains, Great Basin, Southwestern States

Cronquist, A., A. H. Holmgren, N. H. Holmgren, & J. L. Reveal. 1972. Intermountain Flora. Vol. 1. Published for the New York Botanical Garden by Hafner Publishing Company. New York. [Other volumes in progress.]
Davis, R. J. 1952. Flora of Idaho. W. C. Brown Company. Dubuque, Iowa.
Harrington, H. D. 1954. Manual of the Plants of Colorado. Sage Books. Denver, Colorado.
Kearney, T. H., & R. H. Peebles. 1960. Arizona Flora. 2nd ed., with supplement by John Thomas Howell and Elizabeth McClintock. University of California Press. Berkeley, California.
Moss, E. H. 1959. Flora of Alberta. University of Toronto Press. Toronto.
Rydberg, P. A. 1922. Flora of the Rocky Mountains and Adjacent Plains. 2nd ed. Published by the author, New York. [Reprinted in 1954, Hafner Publishing Company, New York.]
Weber, W. A. 1972. Rocky Mountain Flora. A field guide for the identification of the Ferns, Conifers, and Flowering Plants of the Southern Rocky Mountains from Pikes Peak to Rocky Mountain National Park and from the plains to the Continental Divide. Colorado Associated University Press. Boulder, Colorado.
Wooten, E. O., & P. C. Standley. 1915. Flora of New Mexico. Contributions from the United States National Herbarium 19.

e. Pacific States

Abrams, L. 1923-1960. An Illustrated Flora of the Pacific States, Washington, Oregon, and California. 4 vols. Stanford University Press. Stanford, California.
Hitchcock, C. L., A. Cronquist, M. Ownbey, & J. W. Thompson. 1955-1969. Vascular Plants of the Pacific Northwest. Parts 1-5. University of Washington Publications in Biology Vol. 17. Seattle, Washington.
Mason, H. L. 1957. A Flora of the Marshes of California. University of California Press. Berkeley, California.
Munz, P. A., in collaboration with D. D. Keck. 1959. A California Flora. Supplement, 1968. Published for the Rancho Santa Ana Botanic Garden by the University of California Press. Berkeley, California. [A combined edition of manual and supplement issued in 1973.]

Peck, M. E. 1961. A Manual of the Higher Plants of Oregon. 2nd ed. Binfords and Mort, Publishers. Portland, Oregon.

Shreve, F., & I. L. Wiggins. 1964. Vegetation and Flora of the Sonoran Desert. Stanford University Press. Stanford, California.

f. Hawaiian Islands.

Degener, O. 1932+. Flora Hawaiiensis or the New Illustrated Flora of the Hawaiian Islands. Honolulu. [Privately published by the author, in progress.]

Hillebrand, W. 1888. Flora of the Hawaiian Islands; a description of their phanerogams and vascular cryptogams. B. Westermann and Company. New York.

3. Other Selected Floras (not cited or indexed in Frodin's Guide to the Floras of the World)

a. Africa

Brenan, J. P. M., & A. Exell (eds.). 1961. Flora Zambesiaca. Crown Agents for Overseas Governments and Administration. London. [Issued in parts.]

Fernandes, A. (ed.). 1969+. Flora de Moçambique. Junta Investigações do Ultramar, Centro de botânica. Lisboa. [In progress, review Taxon 19: 430-431.]

Hepper, F. N. (ed.). 1954-1972. Flora of West Tropical Africa. 3 vols. Crown Agents for Overseas Governments and Administration. London. [A revised edition of Hutchinson & Dalzier (1927-1936) Flora of West Tropical Africa. Treats plants of all territories of West Africa south of latitude 18°N to the west of Lake Chad and Fernando Po.]

Merxmüller, H. 1970. Prodromus einer Flora von Sudwestafrika. J. Cramer. 3301 Lehre, Germany. [Treats 166 families of flowering plants and ferns of Southwest Africa. This is issued in parts and is almost complete.]

b. Arctic and Antarctic Regions

Löve, Á. 1970. Íslenzk Ferdaflóra... Almenna Bókafélagid, Reykjavík. [In Icelandic; review Taxon 20: 181-182, 1971.]

c. Asia

Davis, P. H. (ed.). 1972. Flora of Turkey and the East Aegean Islands. Vol. 4. University of Edinburgh. Edinburgh at University Press. Edinburgh. [Issued in parts, 4 vols. issued to date.]

Iconographia Cormophytorum Sinicorum. Vol. 1. 1972. Botanical Institute of the Academy of Sciences. Peking. [1157 pages--Bryophytes, Pteridophytes, Gymnosperms and some Angiosperms; illustrated; see Galston, A. W., 1973, Flora North America, Science 180(4081): 9.]

Komarov, V. L., & B. K. Schischkin. 1934-1964. Flora U.R.S.S. 30 vols. Academy of Sciences of the U.S.S.R., Moscow and Leningrad. [Flora of the USSR--English translation in progress; see review in Taxon 18: 685-708, 1969.]

Nasir, E., & S. I. Ali (eds.). 1970?. Flora of West Pakistan. Steward Herbarium. Gordon College, Rawalpindi and University of Karachi, Karachi. [An illustrated flora issued in parts each treating a separate family. To date 26 parts have been issued.]

Ohwi, J. 1965. Flora of Japan. Smithsonian Institution. Washington, D. C. [Translation and revision of 1953 Flora of Japan and 1957 Flora of Japan--Pteridophyta edited by F. G. Meyer & E. H. Walker (English translation). In addition to keys, description includes phytographic resumé and historical sketch of floristic work in Japan, maps, reference list of authors' names, index to Japanese plant names, index to scientific names and index to English names (mostly names of families).]

Smitinand, T., & K. Larsen (eds.). 1970. Flora of Thailand. Applied Scientific Research Corporation of Thailand. Bangkok. [Issued in parts, incomplete; review in Taxon 20: 171, 1971.]

d. Europe

Clapham, A. R., T. G. Tutin, & E. F. Warburg. 1968. Excursion Flora of the British Isles. 2nd ed. Cambridge University Press. Cambridge.

Kuzmanov, B. A., & S. I. Kozuharov. 1963+. Flora Reipublicae Popularis Bulgaricae. Bulgarian Academy of Sciences. Sofia. [See review in Taxon 17: 205-206, 1968.]

Malagarriga, T. 1965. Flora analítica de Barcelona. I. Fanerogamas. La Salle Bonanova. Barcelona.

Tutin, T. G., V. H. Heywood, N. A. Burgess, D. H. Valentine, S. M. Walters, & D. A. Webb (eds.). 1964+. Flora Europaea. Cambridge University Press. Cambridge. Vol. 1 (1964) Lycopodiaceae to Platanaceae; Vol. 2 (1968) Rosaceae to Umbelliferae; Vol. 3 (1972) Diapensiaceae to Myoporaceae; Vols. 4-5 in preparation. [Volumes also include notes on Flora Europaea organization, list of contributors, list of basic and standard floras, appendices of keys to author abbreviations, book and periodical title abbreviations, glossary and list of English-Latin equivalents, and maps.]

e. Malaysia, Australasia and Oceania

Burbidge, N. T., & M. Gray. 1970. Flora of the Australian Capital Territory. Australian National University Press. Canberra. [Eastern Australia; see review in Taxon 20: 182, 1971.]

Moore, L. B., & E. Edgar. 1970. Flora of New Zealand. Vol. II. Government Printing, Wellington, New Zealand. [Vol. I by Allan, H. H. et al.; review in Taxon 20: 630, 1971.]

f. North and Central America

Flora Neotropica. 1968+. Published for the Organization for Flora Neotropica by Hafner Press. Riverside, New Jersey. [A series of "monographs" of <u>all</u> tropical plants growing spontaneously in the Western Hemisphere. Monographs issued separately without regard to sequence; e.g., Monograph 1--Cowan, R. S., 1968, Swartzia (Leguminosae, Caesalpinoideae, Swartzieae); monograph 2--Cuatrecasas, J., 1970, Brunelliaceae. Thirteen parts issued to date.

Sanchez, O. S. 1969. La flora del Valle de México. Editorial Herrero S. A., México.

g. Islands

Adams, C. D., & contributors. 1972. Flowering Plants of Jamaica. University of the West Indies. Mona, Jamaica.

Wiggins, I. L., & D. M. Porter & contributors. 1971. Flora of the Galápagos Islands. Stanford University Press. Stanford, California. [Introduction to physiography, soils, climate, vegetation of the Islands. Systematic treatment has keys, descriptions, distribution maps, illustrations, a glossary of botanical terms, literature cited and index to names.]

h. South America

Correa, M. N. (ed.). 1972. Flora Patagonica. Instituto Nacional de Tecnologia Agropecuaria. Buenos Aires, Argentina. [Only Compositae, monocotyledons except Gramineae published to date.]

Lasser, T. 1964. Flora de Venezuela. Instituto Botanico. Caracas. [Incomplete.]

B. General References for Identification of Families

Bailey, L. H. 1917. The Standard Cyclopedia of Horticulture. Vol. 1. The Macmillan Company. New York.

Benson, L. D. 1957. Plant Classification. D. C. Heath and Company. Boston. [Includes keys to orders and families of North American plants, north of Mexico, but with descriptions and discussion of other plant families.]

Bentham, G., & J. D. Hooker. 1862-1883. Genera Plantarum. 3 vols. London. [In Latin.]

Davis, P. H., & J. Cullen. 1965. The Identification of Flowering Plant Families, Including a Key to Those Native and Cultivated in North Temperate Regions. Oliver and Boyd. London.

Engler, A. (H. Melchior, ed.). 1964. Syllabus der Pflanzenfamilien. Ed. 12. Gebrüder Borntraeger. Berlin. [Text in German.]

________, & K. Prantl. 1924. Die natürlichen Pflanzenfamilien. Ed. 2. Leipzig.

Hutchinson, J. 1959. The Families of Flowering Plants. Ed. 2. Clarendon Press. Oxford.

________. 1967. Key to the Families of Flowering Plants of the World. Clarendon Press. Oxford.

Willis, J. C. 1973. A Dictionary of the Flowering Plants and Ferns. 8th ed. revised by H. K. Airy Shaw. Cambridge University Press. Cambridge. [Synoptic key at end of the dictionary is based on Engler's Classification as given in the <u>Pflanzenfamilien</u> and revised in the 7th ed. of his <u>Syllabus</u>.]

C. General References for Identification of Genera

Bentham, G., & J. D. Hooker. 1862-1883. Genera Plantarum. 3 vols. London. [In Latin.]

Engler, A., & K. Prantl. 1924. Die natürlichen Pflanzenfamilien. 2nd ed. Leipzig. [In German.]

Hutchinson, J. 1964, 1967. The Genera of Flowering Plants. Vols. I, II. Clarendon Press. Oxford. [See also Hutchinson's Families of Flowering Plants cited above.]

Lemée, A. 1929-1959. Dictionnaire Descriptif et Synonimique des Genres des Plantes Phanérogames. 10 vols. (9 and 10 are supplements). [See Vol. 8A for keys to genera; in French; also see appendix--Vocabulaire Français-Anglais--for a French-English dictionary to aid in translation.]

D. Monographs, Revisions, Synopses

(See Section B XB for guides to monographs and revisions. See also Bibliographies (B VIII).

E. Selected Handbooks and Specialized Works

1. Cultivated Plants

Bailey, L. H. 1900-1902. Cyclopedia of American Horticulture. 4 vols. The Macmillan Company. New York. [Includes keys, descriptions, illustrations, bibliographic references.]

________. 1914-1917. The Standard Cyclopedia of Horticulture. 6 vols. (ed. 2 in 3 vols.); new edition also issued in 1933. The Macmillan Company. New York. [A revision of the reference above. Synopsis of plant kingdom and revised keys to families and genera of cultivated plants.]

________. 1949. Manual of Cultivated Plants. 2nd rev. ed. The Macmillan Company. New York. [A convenient manual for identification of cultivated plants of the United States.]

________, & E. Z. Bailey. 1941. Hortus second. The Macmillan Company. New York. [An annotated dictionary of North American gardening and horticulture; reprinted in 1947.]

Graf, A. B. 1963. Exotica 3. Pictorial Cyclopedia of Exotic Plants. Roehrs Company. Rutherford, New Jersey. [This work has 12,025 illustrations (some in color) arranged by family. Brief descriptions, cultural notes, glossary of terms, a section on plant geography, an index to common names, and an index to scientific names make this a useful reference; no keys.]

________. 1970. Exotic Plant Manual. Roehrs Company. Rutherford, New Jersey. [Pictorial guide (3600 photographs) of exotic indoor plants commonly found in commercial cultivation. Cultural notes and plant geography, brief descriptions, index to names and a glossary are included.]

McDonald, E. 1963. The World Book of House Plants. The World Publishing Company. New York. [Cultural notes and illustrated encyclopedia of indoor plants; no keys.]

Neal, M. C. 1948. In Gardens of Hawaii. Bernice P. Bishop Museum Special Publication 40. Honolulu. [Keys to genera (1,150) or species and color key to flowers; common names and notes on uses, folklore, descriptions of plants in Hawaiian gardens.]

Rehder, A. 1940. Manual of Cultivated Trees and Shrubs Hardy in North America. 2nd ed. The Macmillan Company. New York. [This work includes a synopsis of orders and families; keys to families, genera, species; descriptions; a glossary, index, and references to illustrations. Only Latin named taxa are treated. Plants of subtropical and warmer temperate regions are excluded. (See also Rehder, A., 1949, Bibliography of Cultivated Trees and Shrubs, Jamaica Plain, Massachusetts).]

Wyman, D. 1971. Wyman's Gardening Encyclopedia. The Macmillan Company. New York. [A general reference treating cultivated plants.]

2. Aquatics

Arber, A. 1920. Water Plants; A Study of Aquatic Angiosperms. University Press. Cambridge, Massachusetts. (Reprinted as Historiae Naturalis Classica, 1963, with introduction by W. T. Stearn). [Not a taxonomic treatment, but excellent for interpretation of morphology and anatomy.]

Biswas, K., & C. C. Calder. 1955. Handbook of Common Water and Marsh Plants of India and Burma. 2nd ed. Health Bulletin No. 24. Calcutta.

Correll, D. S., & H. B. Correll. 1972. Aquatic and Wetland Plants of Southwestern United States. Environment Protection Agency. U. S. Government Printing Office. Washington, D. C. [Stock No. 5501-0177.]

Eyles, D. E., & J. L. Robertson, Jr. 1963. A Guide and Key to the Aquatic Plants of the Southeastern United States. U. S. Department of the Interior, Fish and Wildlife Service, Bureau of Sport Fisheries and Wildlife Circular 158. Washington, D. C. [Reprint of Public Health Bulletin 286 (1944).]

Fassett, N. C. 1957. A Manual of Aquatic Plants. University of Wisconsin Press. Madison, Wisconsin. [Revision of 1940 edition with appendix by E. C. Ogden, 1957.]

Hotchkiss, N. 1964. Pondweeds and Pondweedlike Plants of Eastern North America. Circular 187. U. S. Department of the Interior. Fish and Wildlife Service. Washington, D. C.

________. 1965. Bulrushes and Bulrushlike Plants of Eastern North America. Circular 221. U. S. Department of the Interior. Fish and Wildlife Service. Washington, D. C.

________. 1967. Underwater and Floating-leaved Plants of the United States and Canada. Resource Publication 44. U. S. Department of the Interior. Bureau of Sport Fisheries and Wildlife. Washington, D. C.

Mason, H. L. 1957. A Flora of the Marshes of California. University of California Press. Berkeley, California.

Matsumara, Y., & H. D. Harrington. 1955. The True Aquatic Vascular Plants of Colorado. Technical Bulletin of Colorado Agricultural Experiment Station 57.

Monson, P. H. 1954. A Preliminary Report on Marsh and Aquatic Dicotyledonous Flora of Iowa. Proceedings of the Iowa Academy of Science 60.

Muenscher, W. C. 1944. Aquatic Plants of the United States. Comstock Publishing Company, Inc. Ithaca, New York. [Keys, illustrations, distribution maps.]

Sculthorpe, C. D. 1967. The Biology of Aquatic Vascular Plants. St. Martin's Press. New York. [Not a taxonomic treatment but a valuable source of information for the taxonomist.]

Stewart, A. N., L. J. Dennis, & H. M. Gilkey. 1963. Aquatic Plants of the Pacific Northwest with Vegetative Key. 2nd ed. Oregon State University Press. Corvallis, Oregon.

Subramanyam, K. 1962. Aquatic Angiosperms, A Systematic Account of Common Indian Aquatic Angiosperms. Botanical Monograph 103, Council of Scientific and Industrial Research. New Delhi.

Winterringer, G. S., & A. C. Lopinot. 1966. Aquatic Plants of Illinois. Illinois State Museum Popular Science Series Vol. VI. Department of Registration and Education. Illinois State Museum Division and Department of Conservation, Division of Fisheries. Springfield.

3. Poisonous Plants

Arnold, H. L. 1968. Poisonous Plants of Hawaii. Charles E. Tuttle, Rutland, Vermont.

Beath, O. A. et al. 1953. Poisonous Plants and Livestock Poisoning. Wyoming Agricultural Experiment Station Bulletin 324.

Brown, R. G. 1955. Plants Poisonous to Livestock. Maryland Agricultural Extension Service Fact Sheet 91.

Choguill, H. S. 1958. Some Poisonous Plants of Kansas. Kansas Academy of Science Transactions 61(1).

Core, E. L. et al. 1961. The Poisonous Plants of West Virginia. West Virginia Department of Agriculture. Charleston, West Virginia.

Duncan, W. H. 1958. Poisonous Plants in the Southeastern United States. Published by the author. Athens, Georgia.

________, & T. J. Jones. 1949. Poisonous Plants of Georgia. University of Georgia School of Veterinary Medicine Bulletin XLIX, No. 13.

Featherly, H. I. 1945. Some Plants Poisonous to Livestock in Oklahoma. Oklahoma Agricultural Experiment Station Circular C-118.

Fyles, F. 1920. Principal Poisonous Plants of Canada. Canada Department of Agriculture, Dominion Experimental Farms Bulletin 39 (2nd ser.).

Gilkey, H. M. 1958. Livestock-Poisoning Weeds of Oregon. Oregon Agricultural Experiment Station Bulletin 564.

Hardin, J. W. 1973. Poisonous Plants of North Carolina. Agricultural Experiment Station Bulletin 414 (rev. ed.). Raleigh, North Carolina.

________, & J. M. Arena. 1974. Human Poisoning from Native and Cultivated Plants (ed. 2 rev.). Duke University Press. Durham, North Carolina.

Heller, C. A. 1953. Wild Edible and Poisonous Plants of Alaska. University of Alaska and USDA Cooperative Extension Service Bulletin F-40.

Hyatt, M. T., R. G. Brown, & J. W. Herron. 1953. Some Plants of Kentucky Poisonous to Livestock. Kentucky Agricultural Extension Service Circular 502.

Kingsbury, J. M. 1964. Poisonous Plants of the United States and Canada. Prentice-Hall. Englewood Cliffs, New Jersey.

________. 1965. Deadly Harvest: A Guide to Common Poisonous Plants. Holt, Rinehart and Winston. New York.

Massey, A. B. 1954. Poisonous Plants. Virginia Polytechnic Institute Extension Service Bulletin 222.

Muenscher, W. C. 1939. Poisonous Plants of the United States. The Macmillan Company. New York.

________, & W. T. Winne. 1955. Common Poisonous Plants. Cornell Agricultural Extension Service Bulletin 538.

Pammel, L. H. 1911. A Manual of Poisonous Plants. The Torch Press. Cedar Rapids, Iowa.

Schaffner, J. H. 1903. Poisonous and other injurious plants of Ohio. Ohio Naturalist IV: 32.

Sperry, O. E. et al. 1955. Texas Range Plants Poisonous to Livestock. Texas Agricultural Experiment Station Bulletin 796.

Stevens, O. A. 1933. Poisonous Plants and Plant Products. North Dakota Agricultural Experiment Station Bulletin 265.

West, E. 1957. Poisonous Plants around the Home. Florida Agricultural Experiment Station Circular S-100.

4. Trees and Shrubs

Bean, W. J. 1970. Trees and Shrubs Hardy in British Isles. 8th ed. John Murray, Ltd. London. [Descriptions but no keys.]

Benson, L., & R. A. Darrow. 1954. The Trees and Shrubs of the Southwestern Deserts. 2nd ed. University of Arizona Press, Tucson, and University of New Mexico Press, Albuquerque.

Braun, E. L. 1969. The Woody Plants of Ohio. Hafner Publishing Company. New York. [A facsimile of 1961 edition.]

Brown, H. P. 1937. Trees of the Northeastern United States. Boston.

Brown, R. G., & M. L. Brown. 1972. Woody Plants of Maryland. [Distributed by Student Supply Store, University of Maryland, College Park.]

Clark, R. C. 1971. The woody plants of Alabama. Annals of the Missouri Botanical Garden 58: 99-242.

Coker, W. C., & H. R. Totten. 1934. Trees of the Southeastern States. University of North Carolina Press. Chapel Hill, North Carolina.

Core, E. L., & N. P. Ammons. 1958. Woody Plants in Winter: A Manual of Common Trees and Shrubs in Winter in the Northeastern United States and Southeastern Canada. The Boxwood Press. Pittsburgh, Pennsylvania.

Dayton, W. A. 1952. United States Tree Books: A bibliography of tree identification. U. S. Department of Agriculture Bibliography Bulletin 20. [Guide to references.]

Grimm, W. C. 1957. The Book of Shrubs. Stackpole Company. Harrisburg, Pennsylvania.

Harlow, W. M. 1957. Trees of Eastern and Central States and Canada. Dover Publications. New York. [Corrected edition of 1942 work.]

________. 1959. Fruit Key and Twig Key to Trees and Shrubs. [Reprinted by Dover Publications, New York (trees of eastern North America).]

Harrar, E. S., & J. G. Harrar. 1962. Guide to Southern Trees. 2nd ed. Dover Publications. New York.

Hosie, R. C. 1969. Native Trees of Canada. Canadian Forest Service, Department of Fisheries and Forestry. Available from Queen's Printer. Ottawa. [Keys, descriptions, excellent photographs and illustrations and discussion of forest regions of Canada.]

Kurz, H., & R. K. Godfrey. 1962. Trees of Northern Florida. University of Florida Press. Gainesville, Florida.

Muenscher, W. C. 1946. Keys to Woody Plants. 5th ed. Comstock Publishing Company. Ithaca, New York.

Petrides, G. A. 1958. A Field Guide to Trees and Shrubs. Houghton Mifflin Company. Boston.

Preston, R. J., Jr. 1961. North American Trees exclusive of Mexico and Tropical United States. 2nd ed. Iowa State University Press. Ames, Iowa.

Rehder, A. 1940. Manual of Cultivated Trees and Shrubs Hardy in North America. 2nd ed. The Macmillan Company. New York. [This work excludes those plants of subtropical and warmer temperate regions. Only Latin named taxa are treated. (See also Rehder, A., 1949, Bibliography of Cultivated Trees and Shrubs, Jamaica Plain, Massachusetts).]

Rosendahl, C. O. 1955. Trees and Shrubs of the Upper Midwest. University of Minnesota Press. Minneapolis, Minnesota.

Sargent, C. S. 1891-1902. The Silva of North America. A description of the trees which grow naturally in North America exclusive of Mexico. 14 vols. Houghton Mifflin Company. New York.

________. 1905-1949. Manual of the Trees of North America. 2 vols. Dover Publications. New York. [Reprinted 1961 and 1965, the 1965 reprint with a table of changes in nomenclature prepared by E. S. Harrar.]

Standley, P. C. 1920-1926. Trees and Shrubs of Mexico. Contributions from the United States National Herbarium 23.

Stephens, H. A. 1969. Trees, Shrubs and Woody Vines in Kansas. University of Kansas Press. Lawrence, Kansas.

Trelease, W. 1931. Winter Botany. 3rd ed. reprinted in 1967 by Dover Publications. New York.

Treshow, M. T., S. L. Welsh, & G. Moore. 1970. Guide to the Woody Plants of the Mountain States. Brigham Young University Press. Provo, Utah. [Includes summer and winter keys to genera and key to species.]

Vines, R. A. 1960. Trees, Shrubs and Woody Vines of the Southwest. University of Texas Press. Austin, Texas.

In the United States cooperation between the U. S. Forest Service and the state forestry departments has resulted in a number of illustrated handbooks of forest trees (e.g., Forest Trees of Florida, 9th ed., Florida Forest Service). Most contain the same illustrations and have a similar format. No keys are included. These are available from the forest services of individual states.

5. Weeds

Chancellor, R. J. 1966. The Identification of Weed Seedlings of Farm and Garden. Blackwell's Scientific Publications. Oxford. [A guide to the identification of important weeds of the British Isles and includes keys, brief descriptions and illustrations.]

Delorit, R. J. 1970. An Illustrated Taxonomy Manual of Weed Seeds. Agronomy Publications. River Falls, Wisconsin. [Photographs of seeds.]

Frankton, C. 1963. Weeds of Canada. Botany and Plant Pathology Division, Science Service, Canada Department of Agriculture. Ottawa. Publication 948. [Descriptions, distribution, illustrations; no keys.]

Gilkey, H. M. 1957. Weeds of the Pacific Northwest. Oregon State College. Corvallis. [A treatment of weedy members of 46 plant families found in Oregon, Washington, Idaho and the western portion of Montana plus the agricultural areas of British Columbia and Alaska. Keys (to families and to species), descriptions and illustrations are presented.]

Henderson, M., & J. G. Anderson. 1966. Botanical Survey Memoir 37. Department of Agricultural Technical Services. Republic of South Africa. [Descriptions, distribution, illustrations; no keys.]

Isely, D. 1960. Weed Identification and Control in the North Central States. 2nd ed. Iowa State University Press. Ames, Iowa. [Keys, descriptions, illustrations, notes on control, etc.]

Kummer, A. P. 1951. Weed Seedlings. University of Chicago Press. Chicago. [Keys to 300 weed seedlings with illustrations, descriptions, and notes on distribution.]

Meadly, G. R. W. 1965. Weeds of Western Australia. Department of Agriculture, Western Australia. Perth. [A general consideration of W. Australian weeds and their control. Color plates and drawings and descriptions of common weedy species are included.]

Montgomery, F. H. 1964. Weeds of Canada and the Northern United States. Ryerson Press. Toronto. [An illustrated field guide.]

Muenscher, W. C. 1943. Weeds. The Macmillan Company. New York. [A general reference with keys, descriptions, bibliography and a discussion of dissemination, importance, habitat and control of weeds.]

Reed, C. F. 1970. Selected Weeds of the United States. Agriculture Handbook No 366. Agricultural Research Service, USDA. Washington, D. C. [A collection of drawings, descriptions, and distribution maps of 224 species of weeds. The work is indexed by both scientific and common names and a glossary of phytographic terms is included. No keys are included.]

6. Selected References to Specific Taxonomic Groups Commonly Treated in Separate Handbooks and Manuals

a. Cactaceae

Backeberg, C. 1958-1964. Die Cactaceae. 6 vols. G. Fisher. Jena.

Benson, L. 1969. The Cacti of Arizona. 3rd ed. University of Arizona Press. Tucson, Arizona.

Bravo, H. H. 1937. Las Cactaceas de México. Imprenta Universitaria. Universidad Naçional de México.
Britton, N. L., & J. N. Rose. 1919-1923. The Cactaceae. Descriptions and Illustrations of Plants of the Cactus Family. 4 vols. Publication 248. Carnegie Institution of Washington. (Reprinted 1963 in 2 vols. by Dover Publications, New York). [Keys to tribes, genera, species; well illustrated.]
Craig, R. T. 1945. The Mammillaria Handbook with Descriptions, Illustrations and Keys to the Species of the Genus Mammillaria of the Cactaceae. Abbey Garden Press. Pasadena, California.
Earle, W. H. 1963. Cacti of the Southwest: Arizona, Western New Mexico, Southern Nevada, and Eastern California. Arizona Cactus and Native Flora Society Bulletin 4. Phoenix, Arizona.
Marshall, W. T., & T. M. Bock. 1941. Cactaceae with Illustrated Keys of All Tribes, Subtribes and Genera. Abbey Garden Press. Pasadena, California.
Weniger, D. 1969. Cacti of the Southwest. University of Texas Press. Austin, Texas. [Includes native cacti of Texas, New Mexico, Oklahoma, Arkansas and Louisiana; color plates, keys to genera and species, descriptions, range, notes on taxonomy, literature, field observations, etc.]

b. Fabaceae (Leguminosae)

Fassett, N. C. 1939. The Leguminous Plants of Wisconsin. University of Wisconsin Press. Madison, Wisconsin.
Gambill, W. G., Jr. 1953. The Leguminosae of Illinois. Illinois Biological Monographs 22(4): 1-117.
Harborne, J. B., D. Boulter, & B. L. Turner (eds.). 1971. Chemotaxonomy of the Leguminosae. Academic Press. New York. [A compendium of symposium papers with considerable information and extensive bibliographies on the taxonomy of legumes.]
Mahler, W. F. 1970. Manual of the Legumes of Tennessee. Journal of Tennessee Academy of Science 45: 65-96.
Turner, B. L. 1959. The Legumes of Texas. University of Texas Press. Austin, Texas.
Wilbur, R. L. 1963. The Leguminous Plants of North Carolina. Technical Bulletin No. 151. North Carolina Agricultural Experiment Station. Raleigh, North Carolina.

c. Ferns

Brown, M. (compiler & editor). 1938. Index to North American Ferns. Published by the compiler. Orleans, Massachusetts. [A catalog of ferns and fern allies of North America, north of Mexico, with notes on habitats and general distribution.]
Copeland, E. B. 1947. Genera Filicum. Chronica Botanica Company. Waltham, Massachusetts.
Correll, D. S. 1956. Ferns and Fern Allies of Texas. Contributions from the Texas Research Foundation. Renner, Texas.
Holttum, R. E. 1968. Ferns of Malaya. In: A Revised Flora of Malaya. Vol. II. 2nd ed. Government Printing Office. Singapore.
Knobloch, I. W., & D. S. Correll. 1962. Ferns and Fern Allies of Chihuahua, Mexico. Texas Research Foundation. Renner, Texas.
Maxon, W. R. 1909. North American Flora 16(1): 31-88. (Schizaeaceae, Gleicheniaceae, Cyatheaceae (pars). [See also Vol. 16(1) for Ophioglossaceae, Marattiaceae, Osmundaceae and Ceratopteridaceae by other authors.]

Ogden, E. C. 1948. The Ferns of Maine. The Maine Bulletin 51. (University of Maine Studies, II. 62). University of Maine Press. Orono, Maine.

Small, J. K. 1938. Ferns of the Southeastern States. Science Press. Lancaster, Pennsylvania.

Wagner, W. H., Jr. 1965. Pteridophytes. In: Bibliography. Bioscience 15: 809 810. [A good general bibliography of Pteridophytes.]

Wherry, E. T. 1964. The Southern Fern Guide. Doubleday. Garden City, New York.

d. Orchidaceae

Case, F. W. 1964. Orchids of the Western Great Lakes Region. Cranbrook Institute of Science Bulletin 48. Bloomfield Hills, Michigan.

Correll, D. S. 1950. Native Orchids of North America (North of Mexico). Chronica Botanica Company. Waltham, Massachusetts.

Dodson, C. H., & R. J. Gillespie. 1967. The Biology of the Orchids. The Mid-America Orchid Congress, Inc. [Includes keys to tribes and subtribes of cultivated orchids, a comparison of orchid classification systems, a bibliography of "orchid floras" and a systematic bibliography of the family.]

Henry, L. K., & W. E. Baker. 1955. Orchids of Western Pennsylvania. Annals of the Carnegie Museum 330: 299-346.

Jones, D. L., & T. B. Muir (eds.). 1969. Orchids of Australia. (The complete edition drawn in natural color by W. H. Nicholls with descriptive text). Thomas Nelson (Australia) Ltd. Sydney. [Keys to genera, generic and specific descriptions, synonomy, distribution and flowering periods plus color illustrations.]

van der Pijl, L., & C. H. Dodson. 1966. Orchid Flowers, their Pollination and Evolution. Fairchild Tropical Garden and the University of Miami Press. Coral Gables, Florida. [Although not a taxonomic treatment, this work includes a synopsis of taxonomic relations in orchids, a glossary, and a bibliography.]

Schultes, R. E., & A. S. Pease. 1963. Generic Names of Orchids, Their Origin and Meaning. Academic Press. New York. [Illustrated and with bibliography.]

Wallace, J. E. 1951. The Orchids of Maine. University of Maine Bulletin Vol. 53. (University of Maine Studies II. 65). University of Maine Press. Orono, Maine.

Winterringer, G. S. 1967. Wild Orchids of Illinois. Illinois State Museum Popular Science Series. Vol. VI. Springfield, Illinois.

Withner, C. L. 1959. The Orchids; A Scientific Survey. The Ronald Press. New York.

e. Poaceae

Booth, W. E. 1964. Agrostology. The Endowment and Research Foundation at Montana State College. Bozeman, Montana. [A text with considerable information on morphology, anatomy, nomenclature, etc. and many useful bibliographies.]

Bor, N. L. 1960. The Grasses of Burma, Ceylon, India and Pakistan (excluding Bambuseae). (A manual issued as Vol. 1 of International Series of Monographs on Pure and Applied Biology, Division of Botany, R. C. Rollins & G. Taylor, general editors). Pergamon Press. New York.

Burbidge, N. T. 1966. Australian Grasses. 2 vols. Angus & Robertson, Ltd. Sydney. [A layman's guide, profusely illustrated.]

Chase, A. 1964. First Book of Grasses. 3rd ed. Smithsonian Institution Press. Washington, D. C. [Good introduction to morphology of grasses and to keys.]

Gould, F. W. 1951. Grasses of the Southwestern United States. University of Arizona Bulletin.] University of Arizona Press. Tucson, Arizona.

________. 1968. Grass Systematics. McGraw-Hill, Inc. New York. [A textbook of agrostology, including criteria for classification of grasses, tribal arrangement, key to genera of U. S. grasses, extensive bibliography, specimen preparation, nomenclature, etc.]

Hackel, E. 1882. Monographia Festucarum Europaearum. Berlin. [Reprinted 1964 by MicroMethods Ltd., Yorkshire, England, and Johnson Reprint Corporation, New York.]

Hitchcock, A. S., & A. Chase. 1951. Manual of Grasses of the United States. (2nd ed. revised by A. Chase). U. S. Department of Agriculture Miscellaneous Publication. Washington, D. C.

Kucera, C. L. 1961. The Grasses of Missouri. University of Missouri Studies 35. University of Missouri Press. Columbia, Missouri.

Lazarides, M. 1970. The Grasses of Central Australia. Australia National University Press. Canberra.

Pohl, R. W. 1966. The grasses of Iowa. Iowa State Journal of Science 40: 341-566.

Rotar, P. P. 1968. Grasses of Hawaii. University of Hawaii Press. Honolulu.

7. Popular Treatments

Works of this type which include books, pamphlets and service leaflets are so numerous and regional that readers are encouraged to consult local book stores, literature guides, state extension services, state museums, university presses, and the Government Printing Office. The references below are examples of the various types of popular works available.

Greene, W. F., & H. L. Blomquist. 1953. Flowers of the South. University of North Carolina Press. Chapel Hill, North Carolina.

Hardin, J. W., & D. M. Dumond. 1971. Guide to the Literature on Plants of North Carolina. Miscellaneous Extension Publication 66, North Carolina Agricultural Extension Service. Raleigh, North Carolina.

Justice, W. S., & C. R. Bell. 1968. Wild Flowers of North Carolina. The University of North Carolina Press. Chapel Hill, North Carolina.

Lemmon, R. S., & C. C. Johnson. 1961. Wildflowers of North America in Full Color. Hanover House. Garden City, New York.

Moldenke, H. N. 1949. American Wild Flowers. D. Van Nostrand Company, Inc. New York.

Rickett, H. W. 1966-. Wild Flowers of the United States. 5 vols. McGraw-Hill. Vol. 1, The Northeastern States (2 parts); Vol. 2, The Southeastern States (2 parts); Vol. 3, Texas (2 parts); Vol. 4, The Southwestern States (3 parts); Vol. 5, The Northwestern States; Vol. 6, The Rocky Mountain Region (to be published). [A series of volumes of illustrations and descriptions of selected wild flowers.]

Shetler, S. G., & F. Montgomery. 1965. Insectivorous Plants. Smithsonian Institution, Museum of Natural History Information Leaflet 447. Washington, D. C.

IV. DISTRIBUTION

A. General (Generic Distribution)

Dalla Torre, C. G. de, & H. Harms. 1900-1907. Genera Siphonogamarum ad Systema Englerianum Conscripta. W. Engelmann. Leipzig. [Information similar to that of Willis given below.]

Hutchinson, J. 1959. The Families of Flowering Plants. 2nd ed. Clarendon Press. Oxford. [e.g., Strelitziaceae....Tropical America, S. Africa and Madagascar. Strelitzia (S. Africa).]

_______. 1964, 1967. Genera of Flowering Plants. Vols. I, II. Clarendon Press. Oxford. [e.g., Monopetalanthus...8 spp. trop. Africa.]

Willis, J. C. 1973. A Dictionary of the Flowering Plants and Ferns. 8th ed. Revised by H. K. Airy Shaw. Cambridge University Press. Cambridge. [Lists the number of species in each genus, name of family, and general distribution; e.g., Harfordia Greene and Parry. Polygonaceae. 2 California.]

B. Specific

1. Monographs and Revisions (See Section B X)

Most modern treatments include either a distribution map of most taxa treated or a statement of range. In some cases a map may be prepared from the cited specimens examined in a systematic treatment.

2. Floras and Checklists (See Section B III)

A common feature in checklists and floras is a statement of species distribution or the inclusion of distribution maps. In some cases the general generic and/or species distribution is indicated but the introduction or format of each flora should be read carefully for criteria for inclusion and exclusion of distribution data.

3. Atlases and Indices of Plant Distribution

a. Index (World)

Tralau, H. (ed.). 1969, 1972. Index Holmensis, a world phytogeographic index. Vol. 1--Equisetales-Gymnospermae; Vol. 2--Monocotyledonae A-I. Scientific Publishers, Ltd. Zurich. [An index to distribution maps arranged alphabetically by species but also includes family, tribe, genus and subgenus, where applicable. Information includes author, title, bibliographic reference, date, page and/or map numbers, and geographic area (area covered by map). This is the first attempt at a comprehensive bibliography with worldwide coverage. The editor, in the preface to Vol. 1, cautions users that "the aim of the index is mere literature guidance."]

b. Specific Areas and Plant Groups

Critchfield, W. B., & E. L. Little, Jr. 1966. Geographic distribution of Pines of the World. U. S. Department of Agriculture Miscellaneous Publication 991. Washington, D. C.

Hannig, E., & H. Winkler (eds.). 1926-1940. Die Pflanzenareale. 5 vols. Fischer. Jena.

Hara, H. 1958-1959. Distribution maps of flowering plants in Japan. Inoue Book Company. Tokyo.

Hultén, E. 1958. The Amphi-Atlantic Plants and their Phytogeographic Connections. Almquist and Wiksell. Stockholm. [Reprinted by O. Koeltz Antiquariat, W. Germany.]

________. 1964, 1971. The Circumpolar Plants. Vol. 1--Vascular Cryptogams, Conifers, Monocotyledons (2nd ed.); Vol. II--Dicotyledons. Almquist and Wiksell. Stockholm. [These volumes include maps and descriptions of, or comments on, variability, taxonomy, nomenclature and distribution. Each volume has an extensive bibliography, i.e. basis for the distribution shown.]

________. 1971. Atlas of the Distribution of Vascular Plants in Northwestern Europe. 2nd revised and enlarged edition. Almquist and Wiksell. Stockholm.

Jalas, J., & J. Suominen. 1972. Atlas Florae Europaeae. Vol. 1--Pteridophyta (1972); Vols. 2-5 (1973-1976). Academic Bookstore. Helsinki. [This is to be an up-to-date work on the distribution in Europe of all species and subspecies of vascular plants with comments on taxonomy, nomenclature and total range. Volume 1 consists of 150 detailed maps showing the distribution of all species and subspecies of European Pteridophytes.]

Little, E. L., Jr. 1971. Atlas of the United States Trees. Vol. 1--Conifers and Important Hardwoods. Miscellaneous Publications No. 1146. U. S. Forest Service, U. S. Printing Office. Washington, D. C. [This volume includes distribution maps of 200 native trees of continental U. S. (including Alaska). For wide ranging species, their distribution in Mexico and Canada is also given.]

Mirov, N. T. 1967. The Genus Pinus. The Ronald Press Company. New York.

Perring, F. H., & S. M. Walters (eds.). 1962. Atlas of the British Flora. Thomas Nelson and Sons, Ltd. London. [Critical supplement to the Atlas of the British Flora, F. H. Perring (ed.), 1968.]

Radford, A. E., H. E. Ahles, & C. R. Bell. 1965. Atlas of the Vascular Flora of the Carolinas. North Carolina Agricultural Experiment Station Technical Bulletin 165. [This atlas "shows by means of dot maps, on a county basis, documented distribution of each species of vascular plant known to occur without cultivation in the Carolinas."]

4. Computer-Mapped Floras

Cadbury, D. A., J. G. Hawkes, & R. C. Readett. 1971. A Computer-Mapped Flora. A Study of the County of Warwickshire. Academic Press. New York.

5. Regional Bibliographies.

Regional bibliographies often have references treating species distribution and maps (see Section B VIIIB for additional bibliographies).

Adams, J. 1928. A Bibliography of Canadian Plant Geography I-VI. Transactions of Royal Canadian Institute 16: 293-355 (1928); 17: 103-145 (1929); 17: 227-265 (1930); 17: 267-295 (1930); 18: 343-373 (1932); 21: 19-135 (1936).

Hamann, U., & G. Wagenitz. 1970. Bibliographie zur Flora von Mitteleuropa. Carl Hanser Verlag. München. [References arranged geographically.]

Senn, H. A. 1951. A Bibliography of Canadian Plant Geography. Transactions of Royal Canadian Institute 1946-1947. Vol. 26, Publ. 863. Department of Agriculture. Ottawa.

V. ILLUSTRATIONS

A. Indices

Pritzel, G. A. 1854-1855. Iconum Botanicarum Index Locupletissimus. Berlin. (Supplement issued in 1866). [Most of the references are included and emended in Index Londinensis.]

Stapf, O. (ed.). 1929-1931. Index Londinensis to Illustrations of Flowering Plants, Ferns and Fern Allies. 6 vols. with supplements by W. C. Worsdell, 1941. [This work with supplements is an index to illustrations of flowering plants and ferns from 1753 to 1935, and includes most of the references in Pritzel's Iconum Botanicarum Locupletissimus (1855, 1866). Arrangement of this index is alphabetical by genus and species with some cross-referencing of generic synonyms. Care should be exercised in accepting illustrations as being authentic as to name. Symbols and abbreviations include: (1) no symbols = black and white illustration of entire plant; (2) asterisk = colored or partially colored plate; (3) Veg. = vegetative parts (stems, leaves, roots); (4) Fl. = flower or its parts; (5) Fr. = fruit or part of a fruit; (6) Hab. = habit sketch; (7) x = hybrid; (8) ! = teratological form. Since 1935 references to illustrations have been incorporated in the references in Index Kewensis (beginning with Supplement 10) and are so indicated by an asterisk at the end of an entry.]

B. Journals

Addisonia. 1916-1964. [A publication of the New York Botanical Garden made possible through a bequest from Addison Brown to establish a journal including colored plates of plants of the United States, its territorial possessions and other plants in the New York Botanical Garden and its conservatories with descriptions in popular language, notes and synonymy and a brief statement of known properties and uses of the plants illustrated. Publication suspended in 1964.]

Curtis's Botanical Magazine, containing coloured figures with descriptions and observations on the botany, history and culture of choice plants. 1787+. Royal Horticulture Society. London (as of January 1971 copyright was assigned to Bentham-Moxon Trust, Royal Botanic Gardens, Kew). [In addition to the excellent illustrations, descriptions, synonomy, distribution, origin of material for the illustrations, place of cultivation, etc. are also included. (See Chittenden, F. J., 1956, Curtis's Botanical Magazine Index, 1787-1947).]

Huntia. 1964+. Publication of Hunt Botanical Library (now Hunt Institute for Botanical Documentation, Carnegie-Mellon University, Pittsburgh). [Treats works on illustrations and botanical portraiture.]

Taxon. 1951+. Journal of the International Association for Plant Taxonomy. Utrecht, Netherlands. [Often includes papers treating illustrations and beginning with Vol. 21(1) 1972 now includes a series of portraits of botanists with biographical data and selected references pertinent to biographical literature.

C. Manuals

Many manuals and floras have excellent illustrations, and both A. Rehder's Manual of Cultivated Plants and L. H. Bailey's Cyclopedia of American Horticulture include references to illustrations of cultivated plants.

VI. BOTANICAL INSTITUTIONS, DEPOSITORIES AND COLLECTORS
(See also Chapters 31-34)

A. Indices and Guides to Herbaria

Clokie, H. N. 1964. An Account of the Herbaria of the Department of Botany in the University of Oxford. Oxford University Press. London. [An aid in locating collections and herbaria (e.g., Dillineus, Sherard); includes an excellent discussion on some of the outstanding "makers of herbaria," an index to collectors with biographical notes, and references to bibliographies. Areas where collectors collected are indicated, as are samples of their handwriting from labels and manuscripts.]

Dandy, J. E. (ed.). 1958. The Sloane Herbarium. British Museum (Natural History). London. [This guide to the contents and historical information about contributors to the herbarium of Sir Hans Sloane preserved at the Department of Botany at the British Museum (Natural History) is divided into two parts. Part 1 consists of an annotated list of the Horti Sicci, and Part 2 the biographical sketches of the contributors. The herbarium, contained in 265 volumes, ranges from important types to specimens of minor interest. The indices to geographical localities and botanical names and the collection of facsimiles of handwriting of many contributors to the herbarium make this an important and useful guide for botanists throughout the world.]

Kent, D. H. (compiler). 1957. British Herbaria: An Index to the Location of Herbaria of British Vascular Plants with Biographical References to Their Collectors. Botanical Society of the British Isles. London. [The division of this index into the following five sections makes it a readily usable guide to locating collections: (1) Abbreviations of names of institutions; (2) Universities, museums and other institutions (arranged geographically with full address of institution and notes on their herbaria and literature relating to collections); (3) Collectors (dates, reference to biographical details, location of institutions where collections are preserved, details of areas where plants collected); (4) Classified index to the location of collections with strong representation of local flora; (5) Classified index to location of collections of critical groups and genera. A bibliography is also included.]

Lanjouw, J., & F. A. Stafleu (compilers). 1964. Index Herbariorum. Part 1. The Herbaria of the World. 5th ed. Regnum Vegetabile 31. Utrecht. [This valuable guide is divided into four sections: (1) List of herbaria arranged alphabetically by city in which they are located and entries include such information as name of institution, date of founding, number of specimens, type of collection, important collections, activities (e.g., teaching, research), staff specialties, and availability of loans and exchange; (2) Herbarium abbreviations (acronyms); (3) Geographical arrangement of herbaria--arranged by country and also city; (4) General index to personal names (curators and staff).

The acronyms for herbaria have become standard and are used in most taxonomic papers when specimens are cited. This is a valuable reference for locating institutions, specific herbaria and specialists. Although the index is somewhat out of date with respect to the size of collections and staff, it is an important reference. The sixth edition is expected soon!]

Microfiche editions of herbaria. The Inter Documentation Company, Zug, Switzerland, has prepared microfiches of a number of herbaria (e.g., The Willdenow Herbarium and a Systematic Index to the Willdenow Herbarium) as well as books, catalogs and lists. New editions are announced in Taxon and a catalog may be requested from IDC, Postrasse 14, 6300 Zug, Switzerland. This method of dissemination of information is gaining favor and makes it possible for the contents of many herbaria and books to be consulted by those working outside the large botanical centers.

Savage, S. 1945. A Catalogue of the Linnaean Herbarium. Taylor and Francis. London. [List of specimens and inscriptions to 1828. Sequence of catalog is the same as that of the herbarium which is essentially the way it was left by the younger Linnaeus in 1783. The generic numbers are those added by Keppst in 1836 and sheet numbers indicate the sequence of specimens in the herbarium. This catalog also has 70+ samples of handwriting and a discussion of symbols and enumerations.]

Stafleu, F. A. 1967. Taxonomic Literature. Regnum Vegetabile 52. Utrecht. [Although basically a guide to dates of publicaton of a number of important older botanical works and other information, the location of herbaria and types are also indicated. The work is arranged by author, making it a readily usable guide.]

B. Guides to Collectors and Type Specimens

(See also references cited above)

Candolle, A. de. 1880. La Phytographie. Chapter XXX, pp. 391-462. Paris. [An alphabetical list of authors and collectors and location of their collections as of 1880. Important source of information on location of types. Much of this information is incorporated into Index Herbariorum.]

Chaudhri, M. N., I. H. Vegter, & C. del Wal. 1972. Index Herbariorum. Part 2(3) Collectors I-L. Regnum Vegetabile 86. Utrecht.

Hedge, I. C., & J. M. Lamond (eds). 1970. Index of Collectors in the Edinburgh Herbarium. Edinburgh. [A valuable source for locating specimens prepared by a specific collector or from a specific area. Also includes many valuable references.]

Hepper, F., & F. Neate. 1971. Plant Collectors in West Africa, A Biographical Index. Regnum Vegetabile 74. Utrecht. [Information included for collectors is name, nationality, date and place of birth, date and place of death, references to obituaries and to itineraries, countries in which collections were made, and herbaria in which materials are deposited. Selected handwriting facsimiles are also included.]

Hitchcock, A. S. 1934, 1935. Location of Type Specimens. List 1, 19 pp. (1934); List 2, 30 pp. (1935). Washington, D. C. (Mimeographed.) [Alphabetical list of 500-600 authors and location of their type specimens of phanerogams and cryptogams. Although this is a list of authors, not collectors, in many cases they are the same. The concern of the compiler was location of holotypes, not isotypes, etc. Many large herbaria and botanical libraries have copies of these works.]

Jackson, B. D. 1901. A list of the collectors whose plants are in the herbarium of the Royal Botanic Gardens, Kew. Kew Bulletin 190: 1-80.

Lanjouw, J., & F. A. Stafleu. 1954. Index Herbariorum. Part II(1). Index to Collectors A-D. Regnum Vegetabile 2. Utrecht. [Based on questionnaire sent to herbaria all over the world and on catalogs and other published works, as well as actual searches through some herbaria. Arrangement of information consists of a list of herbarium abbreviations, general abbreviations, and an alphabetical list of collectors with notes on dates, country where collections were made and herbarium in which specimens are deposited.]

Lanjouw, J., & F. A. Stafleu. 1957. Op.cit.(2): E-H, l.c. 9. [For collectors I-L see Chaudhri et al., 1972.]

Leeuwenberg, A. J. M. 1965. Isotypes of which holotypes were destroyed in Berlin. Webbia 19: 861-863.

Shetler, S. G. 1973. An Introduction to the Botanical Type Specimen Register. Smithsonian Contributions to Botany Number 12. Washington, D. C. [Includes concept, procedures and standards, and examples of files which are designed to become a union registry.]

C. Guides to Botanical Gardens

(See Chapter 32 for discussion and references on botanical gardens)

Fletcher, H. R., D. M. Henderson, & H. T. Prentice. 1969. International Directory of Botanic Gardens. 2nd ed. Regnum Vegetabile 63. Utrecht.

D. Collection Centers (other than for specimens)

Ewan (1970) in his Opportunities in botanical history (Taxon 19: 825-964) points out the need for searching out and saving historically valuable materials as well as compiling informational retrieval aids and writing interpretive accounts. He lists the following repositories for the preservation of the papers of botanists and other botanical records:

"1. Biohistorical Institute, Utrecht: Records of all aspects of the history of botany and horticulture.
2. Smithsonian Institute, Washington, D. C.: Apart from its Division of Archives, the Department of Botany (which includes the Hitchcock-Chase Agrostological Collection) separately maintains records of a wide spectrum of botanical topics.
3. Hunt Botanical Library, Pittsburgh: Records of all aspects of the history of botany and horticulture.
4. University of Wyoming Archives, Laramie: Records of botanists active in the Rocky Mountain Region.
5. University of Texas Archives, Austin: Records of the Botanical Society of America and of botany texts early and modern."

In addition to these sources most herbaria, many libraries and historical societies have many valuable botanical documents such as field notes, letters, manuscripts, etc. which may be of consequence in taxonomic studies.

VII. CURRENT LITERATURE

A. Selected Journals of Interest to Taxonomists

(See also Lawrence, 1951, Taxonomy of Vascular Plants, The Macmillan Company, New York, and guides mentioned below)

Aliso. 1948+. Journal of the Rancho Santa Ana Botanic Garden. Claremont, California.

American Fern Journal. 1910+. Published by the American Fern Society. Port Richmond, New York.

American Journal of Botany. 1914+. Published by the Botanical Society of America. Lancaster, Pennsylvania.

American Midland Naturalist. 1909+. Notre Dame, Indiana.

Annals of the Missouri Botanical Garden. 1914+. St. Louis, Missouri.

Baileya. 1953+. Journal of horticultural taxonomy. Ithaca, New York.

Bartonia. 1908+. A botanical annual. Philadelphia, Pennsylvania.

Bauhinia. 1955+. Zeitschrift der Basler Botanischen Gesellschaft. Basel.
Biota. 1954+. Magdalena del Mar, Peru.
Blumea. 1934+. A journal of plant taxonomy and plant geography. Leiden.
Boletim da Sociedade Broteriana II. 1922+. Coimbra, Portugal.
Botanical Review (Interpreting botanical progress). 1935+. Lancaster, Pennsylvania.
Botanische Jahrbücher für Systematik, Pflanzengeschichte und Pflanzengeographie. 1881+. Leipzig.
Botanisk Tidsskrift. 1880+. Copenhagen. Bruken series.
Bothalia. 1921+. A record of contributions from the national herbarium, Union of South Africa. Pretoria.
Brittonia. 1931+. Published by the New York Botanical Garden for the American Society of Plant Taxonomists. New York.
Bulletin of the Botanical Survey of India. 1959+. Calcutta.
Bulletin of the British Museum, Botany. 1951+. London.
Bulletin of the Torrey Botanical Club. 1870+. Lancaster, Pennsylvania.
Canadian Journal of Botany. 1951+. Ottawa.
Candollea; Organe du conservatoire et du jardin botaniques de la ville de Genève. 1922+. Geneva.
Castanea. 1936+. Published by the Southern Appalachian Botanical Club. Morgantown, West Virginia.
Contributions of the Gray Herbarium, Harvard University. 1891+. Cambridge, Massachusetts.
Folia Geobotanica and Phytotaxonomica. 1966+. Czechoslovak Academy of Science. Praha.
Journal of the Arnold Arboretum. 1920+. Harvard University. Cambridge, Massachusetts.
Kew Bulletin. 1946+. Royal Botanic Gardens. Kew, England.
Madroño. 1916+. Published by the California Botanical Society. Berkeley, California.
Memoirs of the Gray Herbarium, Harvard University. 1917+. Cambridge, Massachusetts.
Memoirs of the New York Botanical Garden. 1900+. New York.
Memoirs of the Torrey Botanical Club. 1889+. New York.
Phytologia. 1933+. (Designed to expedite botanical publication). New York.
Reinwardtia. 1950+. Journal on taxonomic botany, plant sociology and ecology. Published by Herbarium Bogoriense. Borgor, Indonesia.
Rhodora. 1899+. Published by the New England Botanical Club. Lancaster, Pennsylvania.
Sida. 1962+. (A taxonomic journal started by Lloyd H. Shinners and published privately at Southern Methodist University, W. F. Mahler and John Thieret, eds.). Dallas, Texas.
Svensk Botanisk Tidskrift utgifven Asvenska Botaniska Föreningen. 1907+. Stockholm.
Taxon. 1951+. Journal of the International Association for Plant Taxonomy. Utrecht, Netherlands.
Watsonia. 1949+. Journal of the Botanical Society of the British Isles. Oxford.

(See also proceedings of state academies of science.)

B. Guides to Titles of Periodicals

Lawrence, G. H. M., A. F. G. Buchheim, G. S. Daniels, & H. Dolezal (eds). 1968. Botanico-Periodicum-Huntianum. Hunt Botanical Library. Pittsburgh,

Pennsylvania. ["Botanico-Periodicum-Huntianum (B-P-H) is a compendium of information on all periodical publications that regularly contain (or in some period of their history, included) articles dealing with plant sciences and botanical literature and with the persons who have contributed to botany and its literature." The work includes more than 12,000 titles in 45 languages. Information includes abbreviation, full title as given on title page of Volume 1, city of publication of Volume 1, volumation (number of volumes published), dates of publication, report of change in title and location of entry in Union List of Serials. The two appendices include a list of abbreviations of words used in title abbreviations (Appendix I) and a list of language equivalents for geographic names found in the titles or imprint of periodicals treated in B-P-H (Appendix II). This is an excellent source for locating correct and complete titles, number of volumes, and standard abbreviations and is extremely helpful in deciphering abbreviations of titles used in other publications. The cross-referencing with the Union List of Serials is an aid in locating the name of the library where a particular serial is on deposit.]

Schwarten, L., & H. W. Rickett. 1958. Abbreviations of titles of serials cited by botanists. Bulletin of the Torrey Botanical Club 85: 277-300. [See also Supplement I, 88: 1-10, 1961.]

C. Abstracts, Guides and Indices to Current Literature

A.E.T.F.A.T.-Index. 1953+. (Association pour l'Etude Taxonomique de la Flore d'Afrique Tropicale). Laboratoire de botanique systématique de l'Université. 28 Avenue P. Héger. Bruxelles, Belgique. [This is an index to worldwide publications on the taxonomic botany (phanerogams) of the regions in Africa south of the Sahara, including South Africa and Madagascar. Arrangement is alphabetical by family, genus, and species. Papers of general interest concerning topics other than new names are listed at the beginning of each genus or family.

The index includes: "(1) all new taxa with geographic distribution: families, genera, species and infraspecific divisions, (2) new combinations and new names, (3) the basionyms of the new combinations and all other combinations in which the epithet has occurred, with reference to the combination, in order to avoid overlooking the new combination, (4) revisions of families, genera and critical species, (5) any references to papers of taxonomic interest." This is a thorough and valuable index of considerable use to botanists throughout the world.]

Ascher's Guide to Botanical Periodicals. 1973. A. Asher and Company, B.V. Keizersgracht 526 Amsterdam 1002. The Netherlands. [A new index to the contents of selected botanical periodicals. This guide is to appear in 15 three-weekly issues and will include: (1) a list of titles of periodicals with page references, (2) a list of the tables of contents of each of these periodicals with every entry provided with a reference number, (3) an index of authors' names with references to entry numbers, (4) a cumulative authors' names index, as well as an index of all plant names derived for articles listed during a year. The guide is to list the contents of about 5000 periodicals relating to botanical studies and treat about 100,000 articles. The subject index is to contain over 100,000 references to scientific names.

Biological Abstracts - BioResearch Index*

[Biological Abstracts (BA) is a semi-monthly publication consisting of abstracts and indexes published by BioSciences Information Service (BIOSIS), Philadelphia, Pennsylvania. Abstracts of more than 140,000 papers grouped under 623 subject categories, derived from more than 7600 journals originating in 100 countries are produced annually.

BioResearch Index (BioI) is a monthly publication which provides access to more than 100,000 research reports annually, in addition to those covered in Biological Abstracts. Types of literature reported in BioResearch Index include: symposia, meetings and congresses, reviews, letters, notes, bibliographies, etc.

Each issue of both publications contains four indexes: Author, Biosystematic, Cross and Subject--color coded and keyed to the abstracts and bibliographic references.

Each issue of Biological Abstracts contains the following in order of appearance:

(1) Subject Guide--alphabetical listing of subject categories (with synonyms).

(2) List of new books received--includes authors' names, title of publication, date published and price.

(3) Approximately 6,000 abstracts arranged according to alphabetical subject categories and accompanied by full bibliographic citations. Subject areas of particular interest to plant taxonomists are Botany--General and Systematic; General Biology-Taxonomy, Nomenclature and Terminology; Museums, Botanical Gardens, New Journals; Evolution; Plant Genetics, Plant Ecology; Plant Physiology--Chemical Constituents (Chemotaxonomy).

Each issue of BioResearch Index includes in order of appearance:

(1) Table of Contents

(2) List of the Publications Indexed with CODEN (an alphabetical code that represents that title in the computer files) and Abbreviated Title.

(3) Bibliography--each citation consists of the journal source, listed once and set off by dotted lines, author(s) title, added keywords, volume number and pagination.

Format and use of indexes--Indexes to both BA and BioI MUST BE SEARCHED for comprehensive coverage of the systematic literature.

(1) Biosystematic Index: This index is included in each issue of Biological Abstracts and BioResearch Index and enables retrieval of references according to gross taxonomic categories (e.g., for the vascular plants by Division, Class, and Family). An entry appears in the index for each organism mentioned in a paper and articles including descriptions of new taxa have an asterisk after each reference number. Reference numbers are to the abstracts in Biological Abstracts and to the bibliographic citations in BioResearch Index. Common, generic and specific names are listed in the Subject Indexes (see discussion below). The format of the Biosystematic Index consists of a three column entry--the Gross Taxonomic Category,

*Adapted from BioSciences Information Service, Guide to the Indexes, 1972 ed., Biological Abstracts, Inc., Philadelphia, with permission. The help of Ann Farren, Professional Relations Officer of BioScience Information Service and her staff in adapting, updating and checking this section is gratefully acknowledged.

a Major Subject Area (in Biological Abstracts the subject heading under which the abstract will appear) and an Abstract or Bibliographic Citation Number.

Use: (a) Where interest is in identifying all studies relating to a single taxonomic category, check all reference numbers next to that category in the Biosystematic Index.

(b) Where interest is in locating all reports describing new taxa, check the appropriate taxonomic category and note all reports with asterisks to the right of the number(s).

(c) Where interest relates to two groups of organisms (e.g., fungi parasitic to a vascular plant) match all numbers appearing next to both taxonomic categories and retain duplicates.

(d) where interest relates to a gross taxonomic category and a major subject category (e.g., Gramineae--Growth Studies), a match between each Gramineae reference in the Biosystematic Index with those appearing under the Cross Index Heading--Plant Physiology, Growth, Differentiation, offers a faster and more comprehensive technique--numbers duplicating under both headings will provide the desired references.

(e) Where interest is in locating all taxonomic studies, search using the taxonomic category and the major subject headings (e.g., Monocot Gramineae Monocot Syst) including the abbreviated name of the taxonomic category and the word Syst (Systematic) to identify 99% of these reports.

(2) Cross Index: This index is arranged alphabetically by major categories with reference numbers being repeated under every Cross heading that indicates the content of a specific article. Numbers are arranged under each Cross Heading in ten columns according to the last digit of the abstract and bibliographic citation number. The Cross Index can serve both as a bibliography for all reports on a particular topic area or can be used for match searches between the Subject or Biosystematic and the Cross Indexes.

In earlier volumes of BA and BioI, taxonomic papers were identified primarily in the Biosystematic Index. In 1969 the subheading--Floristics and Distribution, under General and Systematic Botany also appeared as a Cross Index heading in BA and BioI. Since 1970 the Cross Index has also included all reports under the subheadings for Botany, General and Systematic. Example: A search for Systematic studies on the Dicot family Labiatae could be obtained by checking the Biosystematic Index under Dicot Labiatae (Synonym Lamiaceae) Dicot Syst--take all numbers. If, on the other hand, you were interested in all Systematic Studies on Dicotyledones, it would be faster to check the Cross Index heading Botany, General and Systematic-Dicotyledones. All numbers will be listed together in each issue and cumulated by volume.

(3) Subject Index: This index, appearing in both Biological Abstracts (B.A.S.I.C.) and BioResearch Index, is a natural language index comprised of significant terms from the author's title and supplementary terms such as names of organisms, geographic location, compounds and other terms added by the editor to provide indexing in depth. The terms are arranged alphabetically in the approximate center of the 59 character portion of the supplemented title terms which precede and follow each key word.

*Genus, species and common names for organisms are found in this index. For those who are interested in locating systematic studies and/or new taxa for specific organisms, match searching of the numbers next to genus-species names in the Subject Index with the higher taxonomic category-Systematic assignments in the Biosystematic Index would identify taxonomy studies; (e.g., peas--taxonomy studies Subject Index--match numbers next to peas and Pisum-Sativum against the numbers in the Biosystematic Index next to Dicot Leguminosae Dicot Syst: retain duplicates).

(4) Author Index: This index to authors, both personal and corporate, is included in each issue of Biological Abstracts and BioResearch Index. Multi-authored articles are indexed under each author's name. For articles covered in Biological Abstracts, an author address is also provided--for requesting reprints or other information.

(5) Combination of Indexes: It is often desirable to use several of the above-mentioned indexes in combination. This can save time as well as assure more comprehensive searches.

(a) Subject Index with Cross Index--effective in locating specific information in a broad subject field.
Example: Effects of air pollution on a specific organism.

SUBJECT INDEX		CROSS INDEX
Reference numbers under specific term(s) *(genus-species and common names)	Match for duplicate numbers	Reference numbers under broad category--Public Health--Environmental Health-Air, Water and Soil Pollution

(b) Biosystematic Index with Cross Index
Example: Cellular studies on Labiatae

BIOSYSTEMATIC INDEX	CROSS INDEX
Specific Group (above genus-species or common name) e.g. Labiatae	Broad Subject Area e.g. Cytology and Cytochemistry-Plant

Match all numbers next to Labiatae from Biosystematic Index with those appearing under the Plant Cytology Cross heading. Retain duplicates.

For more details on other searches of one or several of the indexes, refer to the Guide to the Indexes 1972 Biological Abstracts and BioResearch Index; copies available in libraries; complimentary copies may be obtained from BioSciences Information Service - 2100 Arch Street, Philadelphia, Pennsylvania 19103. U.S.A.

Selected references used at BIOSIS for Vascular Plant Taxonomy Nomenclature are Engler's Syllabus (1964), Willis's Dictionary of Flowering Plants and Ferns (1966), Cronquist's The Evolution and Classification of Flowering Plants (1968), and Index Kewensis (and Supplements).

*A fifth index--the generic index--was issued in January, 1974. All generic names (up to 20) mentioned in an abstract are listed here rather than the Subject Index. The index lists taxa alphabetically and each entry includes an abbreviated subject heading to which a paper relates and a reference number. New taxa or records are indicated with an asterisk.

Biological and Agricultural Index. 1964+. H. W. Wilson and Company. New York. (formerly Agricultural Index, 1916-1964). [A subject index to selected lists of agricultural periodicals and bulletins.]

Biosystematic Literature; Contributions to a biosystematic literature index (1945-1964). Compiled by Solbrig, O. T., & T. W. J. Gadella. 1970. Regnum Vegetabile 69. Utrecht. [A source of references in biosystematics for 1945-1964 but not a comprehensive index; world coverage although coverage for some countries is uneven. A list of references to chromosome numbers is presented. References are cited under family and/or genus names with families arranged alphabetically. Genera are listed alphabetically under each family. Some cross-referencing of families (e.g., Agavaceae see Liliaceae, Amaryllidaceae) makes this a readily usable reference. Only flowering plants are included! Hopefully these authors will issue supplements to bring this work up to date and keep it current. This is an exceedingly useful reference and should be available to all students of systematics.]

Excerpta Botanica. A reference journal to literature of taxonomy and chorology and of phytosociology started in 1959 as a result of cooperation between Gustav Fisher Verlag of Stuttgart and the International Association for Plant Taxonomy, Utrecht.

[The two major sections--A. Taxonomy and Chorology, and B. Sociology are issued separately (only Section A is treated here). The four major classes and subclasses are given below:

Taxonomia, Phylogenia: (1) Taxonomia et Phylogenia generalis, (2) Nomenclature et Terminologia, (3) Cryptogamae, Scripta miscellanea, (4) Algae, (5) Fungi, (6) Lichenes, (7) Bryophyta, (8) Pteridophyta, (9) Spermatophyta, Scripta miscellanea, (10) Gymnospermae, (11) Dicotyledonae, (12) Monocotyledonae, (13) Palynologia, (14) Cytotaxonomia.

Chorologia: (1) Scripta miscellanea incl. Itineraria, (2) Europa, (3) Asia, (4) Africa, (5) America, (6) Australia et Oceania, (7) Regiones arcticae.

Palaeobotanica: (1) Scripta miscellanea, (2) Thallophyta, (3) Bryophyta, (4) Pteridophyta, (5) Spermatophyta.

Varia: (1) Herbaria, Horti, Musea, Concilia, (2) Biographiae, Bibliographia, (3) Praehistoria et Ethnobotanica, (4) Personalia.

The materials are compiled by geographic contributors and reviewers and are edited under the direction of an editorial board. The references are arranged under each category by author and consist of author, title, bibliographic reference and short review in English, German or French. Although quite useful it is incomplete and should be used with other sources.]

Horticultural Abstracts. 1931+. Compiled from World Literature on Temperate and Tropical Fruits, Vegetables, Ornamentals, Plantation Crops. Commonwealth Agricultural Bureaux. Farnham Royal. England. [Issued quarterly with annual author and subject indices. Each issue has an author index and subject classification. Particularly a good source for references to taxonomy, cultivated plants and to references which are pertinent to handling of plants for biosystematic research.]

Index to American Botanical Literature. Bulletin of the Torrey Botanical Club. 1886+. [This index was started in the early volumes of the Bulletin and appeared as notes on publications or under Botanical Literature and was presented as a review, but beginning with Volume 13 it became more inclusive and was presented in the format given below:

(1) Taxonomy, Phylogeny and Floristics
   (a) Algae
   (b) Bryophytes
   (c) Fungi
   (d) Pteridophytes
   (e) Spermatophytes
(2) Ecology and Plant Geography
(3) Paleobotany
(4) Morphology
(5) Genetics (including cytogenetics)
(6) General Botany (including biography)

References are arranged under the above classes alphabetically by author. Botanical literature written or published by Americans or based on American material is included. In 1947 the combined efforts of the editors of Taxonomic Index and the Index to American Literature were acknowledged and the Taxonomic Index as it appeared in Brittonia, Vol. 20, was the same as that in the IABL. The index appears in each issue of the Bulletin of the Torrey Botanical Club and is also available on cards and in bound volumes. For abbreviations of periodicals see Schwarten, L., & H. W. Rickett, 1947, Bulletin Torrey Botanical Club 74: 348-356.

Index to Australasian Taxonomic Literature for 1968. Regnum Vegetabile 66. 1970. Compiled by I. K. Ferguson. [This index to current literature, including names of new taxa, which is thought to be relevant to the study of vascular plant systematics in Australasia and Polynesia, is the first in a proposed series to be issued annually. The index is modeled on the Index to European Taxonomic Literature (cited below) to which Dr. Ferguson also contributes.

The area covered by the index includes: "Australia, New Zealand, New Guinea, the Pacific Islands west of 135°W to 105°E and northward to the Tropic of Cancer, i.e., including Micronesia, the Hawaiian Islands and Easter Island but excluding the Galapagos, San Fernandes and the Islands of the east coast of Asia. New Guinea was at first excluded but the close affinities between the flora of this area and those of the Solomon Islands and northern Australia seemed to warrant its inclusion." The New Guinea coverage is therefore probably not as complete as is desirable.

Disposition of references and notes is essentially the same as that in Index to European Taxonomic Literature for 1966 (Regnum Vegetabile 53, 1968 (cited below). The appendix includes a list of the periodicals cited and their abbreviations.

Continuation of this valuable index should be encouraged by workers throughout the world.]

Index to European Taxonomic Literature for 1965, 1966, 1967, 1968, 1969, 1970. Regnum Vegetabile 45, 53, 61, 70, 80, 83. Compiled by R. K. Brummitt et al. and published under the auspices of the Flora Europaea Organization by the International Bureau for Plant Taxonomy and Nomenclature. Utrecht.

[The goal is a convenient annual "index to current literature relevant to the study of systematics of vascular plants of Europe and adjacent areas" including references to papers, books and all proposed new names.

This index is based on books, journals and reprints received in the library at Kew, journals received at the British Museum (Natural History) and not taken at Kew, with some references being compiled from abstracting journals and other indices. Total number of periodicals cited in the Index for 1965 and 1966 is 712 and an additional 196 were added to the 1967 issue. Each subsequent issue lists some additional ones.

The language of the text of all papers is that of the title unless indicated to the contrary. Translation of titles is indicated and Cyrillic scripts are transliterated.

References in each group are arranged alphabetically by author and consist of title, bibliographic reference, date and brief notes; language of text and summary; presence of maps or photographs, figures; names of new taxa, etc.

The geographic areas and subject classes treated in this work are as follows:

(1) Area: "All of Europe as defined by Flora Europaea (Vol. 1, p. xvi) together with Madeira and the Canary Islands, Africa north of the Sahara, the countries at the east end of the Mediterranean (Israel, Syria, Lebanon, and Jordan), Cyperus, Turkey and the Caucasus."

(2) Subjects and Arrangement:

(a) General.

(b) Biography (Personalia). This section is arranged subject by subject and "includes references to any systematic botanists and collectors, native of, or working in, the area covered by the index."

(c) Bibliography (Vol. 45 only).

(d) Herbaria, Botanical Institutes, Botanic Gardens. This section is arranged alphabetically by city and does not include references to personal herbaria which are indexed in the biography section.

(e) Phytogeography. This section includes papers on vegetation and plant geography arranged alphabetically by author.

(f) Mapping (Not in Vol. 45). This section of references on mapping was put in the General section in the Index for 1965 but has been used as a separate class in subsequent issues. References are arranged alphabetically by authors.

(g) Floras and Floristic Studies, etc. Arranged by country, with Europe divided as in Flora Europaea. Includes notes, illustrations and lists relevant to area covered.

(h) Chromosome Surveys. This group includes those references which are of such scope that they cannot be conveniently listed under systematic groups.

(i) Pteridophyta. Divided into major groups--Psilopsida, Sphenopsida, Lycopsida and Filicopsida with genera arranged alphabetically in each group. General references appear at the beginning of the section.

(j) Gymnospermae. Arranged alphabetically by genus.

(k) Angiospermae. Major divisions are based on family with genera arranged alphabetically under appropriate family. This section includes some cross-referencing of family and generic names.

(l) Appendix. A key to the abbreviation of the periodicals cited in the index in recent volumes only.

Subjects excluded are Phytosociological papers [but does include autecological references] and Genetics (pure) [but does include papers treating cytotaxonomy and cytogenetics.]

Taxonomic Index. 1938-1967. [This subject index was first compiled and distributed (in mimeograph form) by W. H. Camp as an outgrowth of a card system started at the New York Botanical Garden. It was issued only to members of the American Society of Plant Taxonomists. Subjects covered include (1) taxonomy, (2) floristics, (3) phytogeography and (4) phylogeny.

It originally included only lichens, bryophytes and vascular plants and was based on literature pertaining to North American plant taxonomy which was received by N.Y.B.G. The inclusive geographic area is Panama northward; including the islands of the Caribbean (expanded later to W. Hemisphere). References in each subject class are arranged alphabetically by author with a few notes by the editor. The first nine volumes were mimeographed. Later an arrangement between the Torrey Botanical Club and the American Society of Plant Taxonomists resulted in a joint staff compiling and editing this bibliographic digest of all current botanical literature either originating in the Western Hemisphere or pertinent to its plants. Both the Taxonomic Index and Index to American Botanical Literature (published in the Bulletin of the Torrey Botanical Club) were derived from this compilation, and the Index appeared as a regular feature in Brittonia beginning with Vol. 20 (1951). In 1967, the ASPT voted to discontinue the Taxonomic Index but the Index to American Botanical Literature is still published in each issue of the Bulletin of the Torrey Botanical Club.]

VIII. REFERENCE LISTS, BIBLIOGRAPHIC GUIDES, CATALOGS AND INDICES

A. Catalogs (Library Holdings)

These sources are useful in locating specific books and other references (particularly old literature) which need to be consulted, but they may also be used as bibliographic indices.

Bibliographic Contributions from the Lloyd Library. Vol. 1 (1911-1914), Bibliography relating to the Floras; Vol. 2 (1914-1917), Bibliography relating to Botany, exclusive of Floras. Periodicals and books and pamphlets of authors A-M; Vol. 3, continuation of Vol. 2, Authors N-Z. Cincinnati, Ohio. [See also Bulletin of the Lloyd Library and Museum of Botany, Pharmacy and Materia Medica, Vol. 34 (1936), Catalogue of Periodical Literature of the Lloyd Library by Aiken, W. H., & S. Waldbott. The catalog of periodicals treats those received between 1914 and 1935.]

Catalogue of Botanical Books in the Collection of Rachel McMasters Miller Hunt. Vol. 1, Printed Books 1477-1700, compiled by Jane Quinby (1958); Vol. 2, Part 1, Introduction to Printed Books 1701-1800, Part 2, Printed Books 1701-1800, compiled by Allan Stevenson (1961).

Catalogue of the Library of the Arnold Arboretum of Harvard University. Compiled by Ethelyn M. Tucker. Vol. 1 (1914) Serial Publications and Author Catalogue; Vol. 2 (1917) Supplement to Volume 1 and Subject Catalogue [very valuable guide to older literature, particularly of woody plants and includes all the materials in Vol. 1 plus the additions during 1914-1917]; Vol. 3 (1933) Serial Publications and Author-Title Supplement Covering Accessions of 1917-1933.

Catalogue of the Library of the Royal Botanic Gardens, Kew. Bulletin Miscellaneous Information (Kew Bulletin). Additional Series 3 (1899 and supplement 1898-1915).

Catalogue of the Printed Books and Pamphlets in the Library of the Linnaean Society of London, 1925. [No manuscripts catalogued; listed by author or journal title, supplement at end. Works from Linnaeus's own library are so indicated.]

Hall, E. C. 1970. Printed Books 1481-1900 in the Horticultural Society of New York. The Horticultural Society of New York. New York. [Although the 3000 works in this collection of references are in the horticultural field, many rare and valuable books of great importance to taxonomic research are present. This "short-title" catalog is arranged alphabetically by author and should be consulted when trying to locate books published between 1481-1900. The library also has holdings of many reference sources such as bibliographies, library catalogs, etc. which are listed on pp. 243-277. This list includes many of the basic bibliographic references to be consulted in library research, many of which are mentioned in this chapter.]

Woodward, B. B., & A. C. Towsend. 1903-1940. Catalogue of Books, Manuscripts, Maps and Drawings in the British Museum (Natural History). 8 vols. [Vol. 1-6 facsimile reprint in 1964 by Stechert Hafner Service Agency, Inc. New York.]

B. Bibliographies (Selected)

1. General Bibliographies, Guides and Subject Indices

Bibliographia Huntiana (in preparation). [This is to be a multivolume (16?) annotated guide to the world's botanical literature published during 1730 to 1840. It will include comprehensive treatments of all pertinent books, pamphlets, reprints, and effectively published exsiccatae and will also cite the periodical literature of the authors concerned and published during this period. An attempt will be made to treat all books and articles dealing with plants from a botanical point of view. Base list of references has been compiled from the following sources: Pritzel, Thesaurus literaturae botanicae, Eds. 1 & 2 (1851, 1872); Haller, Bibliotheca botanica (1771-1772); Nissen, Die botanische Buchillustration (1951); Krok, Bibliotheca botanica suecana (1923); Soulsby, A catalogue of works of Linnaeus, ed. 2 (1933); Jackson, Guide to literature of Botany (1881); and the Catalogue of botanical books in the library of Rachel McMasters Miller Hunt.

In addition to title, biographical sketch of authors, references to published biographical materials, facsimiles of title pages of certain works, etc., much information on artists, printers, publishers, engravers and the like is also to be included. [See McVaugh, R., 1964, Announcement: Bibliographia Huntiana, Huntia 1: 17-24.]

Botany Subject Index. Compiled by USDA library, on index cards, reproduced in 1958 in 15 volumes by Microphotography Company, Boston, Massachusetts, currently available from G. K. Hall and Company, Boston. [Information is worldwide in scope and references are arranged by subject, names of countries, scientific names and some vernacular names. Genera are listed under the family, but are also cross-referenced. Each volume has a table of contents.]

Jackson, B. D. 1881. Guide to the Literature of Botany; being a classified selection of botanical works including nearly 6,000 titles not given in Pritzel's 'Thesaurus'. London. (facsimile reprint by Hafner Publishing Company, New York, 1964). [This classified index covers much of the literature between 1871 and 1881 plus the addition of 6,000 titles to the period covered by Pritzel's Thesaurus. The detailed classification of literature and the scope of literature make the index a valuable source for information in a number of botanical disciplines. The morphology, descriptive botany, and flora sections are particularly useful. This should be used with Pritzel for best coverage. Jackson characterized this work as "suggestive, not exhaustive."

Lawrence, G. H. M. 1951. Taxonomy of Vascular Plants. Chapter 14, pp. 284-331. The Macmillan Company. New York.

Pritzel, G. A. 1851. Thesaurus Literaturae Botanicae omnium gentium inde a rerum botanicarum initiis ad nostra usque tempora quindecim millia operum recensens. Ed. II (partly by C. W. J. Jessen, 1871-1877). [An outstanding guide to chief botanical literature up to 1851. Part I--authors and titles and reference numbers (note addenda); Part II--systematic (subject) index; Part III--author index. Ed. 1 (1847-1882); ed. 2 (cited above); see also Taxon 22: 119-126, 1973, for biographical notes on Pritzel and his Thesaurus by Frans Stafleu. The 'Thesaurus' should be used with Jackson (1881).]

Stafleu, F. A. 1967. Taxonomic Literature. A selective guide to botanical publications with dates, commentaries and types. Regnum Vegetabile 52. Utrecht. [A guide to selected older literature of significance to present-day taxonomy. It includes much valuable information on dates of publication, commentaries on works and their authors, as well as data on the location of types and herbaria. The bibliographies and indices to titles and authors make this a valuable and usable reference.]

Swift, L. H. 1970. Botanical Bibliographies; A guide to bibliographic materials applicable to botany. Burgess Publishing Company. Minneapolis, Minnesota. [An outstanding guide to bibliographic techniques and sources for botany. Parts I and II are particularly useful to taxonomists.]

2. Regional Bibliographies

Many of these sources are not of as limited usage as might be construed from their titles, e.g., many include important monographs, revisions, distribution maps, etc. for taxa of interest which do not even occur in the area surveyed (see also guides to current literature in the preceding section).

Boivin, B. 1967(?). Énumération des Plantes du Canada. Provancheria No. 6. Mémoires de l'Herbier Louis-Marie. Université Laval. Québec. (extracted from Naturaliste Canadien, Vols. 93(3,4,5,6), 24(1,4,5). ["An enumeration of the species, varieties, forms and hybrids known as spontaneous in Canada. Nearly all genera are preceded by an essential bibliography of recent monographs and other main papers relevant to each genus in Canada. Each taxon is followed by an abbreviated summary of its distribution north of the U.S.A. A distinction has been made between verified and unverified distributions and, in the latter case, whether or not justifying collections have been located yet."

In addition to being a valuable catalog of Canadian plants, this reference includes an excellent bibliography and is of considerable aid in locating monographs, revisions and other valuable taxonomic papers. References precede each genus.]

Core, E. L, W. H. Gillespie, & B. J. Gillespie. 1962. Bibliography of West Virginia Plant Life. Scholar's Library, Box 60, Gracie Station, New York. [A bibliography of more than 1,000 references on plant life of West Virginia. References listed alphabetically by author in the following classes: algae, fungi, lichens; bryophytes, vascular plants, fossil plants. Although selected for relevance to West Virginia, complete references are given for monographs and revisions which were serially published, even if only one part contains a reference to the West Virginia plants. A valuable reference for much of the eastern United States, particularly the Southeast.

Ewan, J. 1967. A Bibliography of Louisiana Botany. In: Flora of Louisiana by J. W. Thieret and collaborators. Southwestern Louisiana Journal 7: 2-81. [Annotated references are arranged chronologically. An index to authors, botanical names and subjects is included.]

Greene, H. C., & J. T. Curtis. 1955. A Bibliography of Wisconsin Vegetation. Milwaukee Public Museum Publications in Botany No. 1. Milwaukee Public Museum. Milwaukee. [This bibliography covers the period 1667-1953 with emphasis on post 1900 papers. Subjects treated include reports dealing "with the native plants and plant communities of Wisconsin, their distribution, taxonomy, ecology, sociology, economic uses and conservation." The bibliography is divided into 15 subject areas with references listed in chronological order. An author index is also provided.

Langman, I. K. 1964. A Selected Guide to the Literature on the Flowering Plants of Mexico. University of Pennsylvania Press. Philadelphia, Pennsylvania. [The aim of this guide is to bring existing bibliographic material on flowering plants of Mexico up to date and to organize the materials so desired references can be located with minimal effort. The references are arranged by author with their publications in chronological order with many entries annotated. Anonymous articles are arranged at the end by date. The elaborate subject index (systematic group, geography, etc.) makes this a very useful guide.

Lebedev, D. V. 1956. Guide to Botanical Literature of the USSR. 382 pp. Academy of Science USSR. Leningrad.

Meisel, M. 1924-1929. A Bibliography of American Natural History, the Pioneer Century 1769-1865. 3 vols. Premier Publishing Company. Brooklyn, New York. [Vol. 1 (1924) An Annotated Bibliography of the Publications Relating to the History, Biography and Bibliography of American Natural History and Its Institutions during Colonial Times and the Pioneer Century, which have been published to 1924, with a classified subject and geographic index and a bibliography of bibliographies; Vol. 2 (1926) The Institutions which have contributed to the rise and progress of American Natural History which were founded or organized between 1769 and 1844 (includes contents of their journals and proceedings relating to natural history); Vol. 3 (1929) The Institutions founded or organized between 1845 and 1865. Bibliography of Books, Chronological Tables, Index of Authors and Institutions, Addenda to Volume 1.]

Merrill, E. D., & E. H. Walker. 1938. A Bibliography of Eastern Asiatic Botany. The Arnold Arboretum of Harvard University. Jamaica Plain, Massachusetts. [This work includes references published to December, 1936. The area includes China, Japan, Formosa, Korea, Manchuria, Mongolia, Tiber and eastern and southern Siberia in addition to major papers of adjacent areas such as the Philippines, Indo-China, Siam, Burma, India and Central and Northern Asia. Although the point of view of selection was taxonomic, papers "on ecology, geography, exploration, economic botany, history, bibliography, phylogeny, pathology, agriculture, horticulture, forestry, materia medica, pharmacy, biography" and others are included. All plant groups are treated. Major categories are bibliography (authors and titles), a reference list of oriental serials (with characters), a list of oriental authors (with their oriental characters), and a subject index divided into 3 indices (genera, regional, and systematic).

Supplement 1, by E. H. Walker, 1960, American Institute of Biological Sciences, Washington, D. C. This supplement extends coverage to 1958.]

Merrill, E. D. 1947. A Botanical Bibliography of the Islands of the Pacific. Contributions from the United States National Herbarium 30:1-322. [A revision of the following papers: Merrill, E. D., 1924, Bibliography of Polynesian Botany, Bishop Museum Bulletin 13: 1-68; and Polynesian Botanical Bibliography 1773-1935, Bishop Museum Bulletin 144: 1-194.

The work therefore consists of about 3,850 author entries covering the period from 1773-1946. The region covered in this bibliography is "the islands of the Pacific between latitude 30°N and 30°S (excluding the Bonin Islands) extending from Juan Fernandez and Hawaii in the east to the extreme western limits of the Marianas, Caroline, and Palau Islands. It includes all of Polynesia and Micronesia and the eastern parts of Melanesia--as Fiji, New Caledonia, the New Hebrides, Lord Howe Island, Norfolk Island, the Loyalty Islands, and Santa Cruz--but not the larger archipelagoes contiguous to New Guinea..."

Indexed information includes (1) papers in which new genera or species are described from the region covered; (2) those papers involving transfers of names of Polynesian species; (3) species lists, all monographic works in which species of the Pacific Islands are mentioned; (4) general works (e.g., Bentham & Hooker's Genera Plantarum, and Engler & Prantl's Die natürlichen Pflanzenfamilien); (5) appropriate papers on ecology and phytogeography; (6) pathology, forestry and some agriculture; and (7) books on travel. Plant groups treated include vascular plants and cellular cryptogams. References are arranged by author chronologically. (See Walker, E. H., in this same volume for a subject index which is arranged by general subject--e.g., systematic groups.).]

Reed, C. F. 1969. Bibliography to Floras of Southeast Asia. Baltimore, Maryland. ["A bibliography to the flora of Southeast Asia ranging from Burma and Tonkin, south through Annam, Laos, Thailand and Cambodia to Cochinchina, the Malay Peninsula and Singapore." "Publications dealing with floristic works and taxonomic and monographic treatments of both vascular and nonvascular plants, as well as those dealing with the vernacular names are included." Other subjects include cultivated, agricultural or medicinal plants, forestry, forest floras and properties of forest timbers, climatic and edaphic phenomena and fossil flora. The entries are arranged alphabetically by author. No subject index is present. Note supplement on pp. 189-191. One valuable feature of this index is the inclusion of author, volumation, date of publication of specific treatments in major floras.

Sachet, M. H., & F. R. Fosberg (compilers). 1955. Island Bibliographies: Micronesian Botany, Land Environment and Ecology of Coral Atolls, Vegetation of Tropical Pacific Islands. Compiled under the auspices of the Pacific Science Board. National Academy of Sciences, National Research Council Publication 335. Washington, D. C. [A series of annotated bibliographies arranged alphabetically by author. Each bibliography is followed by a subject index including geographic and systematic classes. The bibliographies are as follows:

(1) Annotated Bibliography of Micronesian Botany. Area covered includes the Marianas, Caroline, Marshall and Gilbert Archipelagoes, besides the isolated islands of Marcus, Wake, Mapia, Nauru and Ocean.

(2) Bibliography of the Land Ecology and Environment of Coral Atolls. For the coral atolls the area covered is that part of the tropical seas where these islands are found, principally the central and western Pacific, the Indian Ocean, and the Caribbean and Gulf of Mexico. Rocas, in the Atlantic near Brazil and Clipperton Island in the eastern Pacific are also included.

(3) Selected Bibliography of Vegetation of the Tropical Pacific Islands. The coverage of this bibliography is the area extending from Guadalupe, the Revillagigedos, Clipperton, Cocos, the Galapagos and Desventuradas westward to the Bonins, Palau, Mapia, the Admiralties, Bismark Archipelago, Louisiades and islands of the Coral Sea. Excluded are Japan, the Riukius, Formosa, Philippines, East Indies, New Guinea, Australia, Kermadecs, Norfolk, Lord Howe, New Zealand, and Juan Fernandez.

A list of serial abbreviations for the titles of serials used in the references for the three bibliographies, with the full titles that they represent are included. Library of Congress numbers and libraries where these volumes are known are given in many cases. Note also addenda to each bibliography and to the list of serials. Taxonomic treatments are readily located by searching the list of references under the appropriate family name in the subject index. Other subjects include bibliography, expeditions and voyages, geology and phytogeography to mention a few. (See also Sachet, M., & F. R. Fosberg, 1971, Island Bibliographies Supplement 1971, National Academy of Sciences, Washington, D. C.).]

Santapau, H. 1958. History of Botanical Researches in India, Burma and Ceylon. Part II. Systematic Botany of Angiosperms. Bangalore Press. Bangalore City. [This work includes a brief history of botany of the area, notes on herbaria and botanic gardens, and a bibliography of floras, botanical explorations, and monographs. It includes treatments in journals which are not widely circulated.]

Schischkin, B. K., & A. A. Fedorov. 1963. On the taxonomic and floristic works published in the USSR during the last fifteen years. Webbia 18: 501-562.

Simpson, N. D. (compiler). 1960. A Bibliographic Index of the British Flora. Privately printed, in limited edition of 750 copies. Bournemouth, England.

["This work is devoted to phanerogams, vascular cryptogams and charophytes and its aim is to provide references to sources of information relevant to the identification, history and geographical range of taxa which occur or have occurred in Britain. Information is also provided on plant-lore, local names, poisonous plants and weeds."

The book is divided into the following classes: Part 1--Catalogs of Plants (mostly by author), Distribution, Economic Plants, Families, Genera and Species (nomenclature follows Clapham, A. R., 1940, in J. Ecol. 33, unless otherwise indicated); Floras (handbooks, keys, etc.), Herbals; Periodicals; Plant Lore; References (including bibliographies, biographies, collecting techniques, dictionaries, glossaries, herbaria), History; Teratology; Topography; Trees. Part II--Plant Records (geographically arranged).]

Steenis, C. G. G. J. van. 1955. Annotated Selected Bibliography in Flora Malesiana. ser. 1, 5, i-cxliv.

Zeybek, N. (compiler). 1972. A Bibliography of the Papers on the Taxonomy and Ecology of Turkish Flora. 1841-1971. Available from Head of Systematic Botany, University of Ege, Bornova-Izmir. (Mimeographed) [Arranged alphabetically by author, language of papers indicated; also includes names and addresses of taxonomists and ecologists of Turkey.]

3. Bibliographies (Specific Groups and Approaches)

A practice which should be encouraged is the publication of bibliographies on specific subjects. Selected examples of such compilations are given below. It is indeed unfortunate that the extensive card files on various subjects often involving many years of compilation by specialists remain unavailable to many taxonomists. See Chapters 22 & 23 for discussion on the use of electronic data processing as one solution to this problem.

Adams, J. 1928-1936. A Bibliography of Canadian Plant Geography. I. (1928) Transactions Royal Canadian Institute 16: 293-355; II. (1929) 17: 103-145; III. (1930) 17: 227-265; IV. (1930) 17: 267-295; Adams, J., & M. H. Norend, V. (1932) 18: 343-373; VI. (1936) 21: 19-135.

Bailey, L. H. 1933. The Standard Cyclopedia of Horticulture. 3 vols. The Macmillan Company. New York. [Volume 2 includes a bibliography of horticulture treating books, journals and societies (pp. 1520-1562) and biographical sketches of North American horticulturists (pp. 1563-1603).]

Jones, G. N. 1966. An Annotated Bibliography of Mexican Ferns. University of Illinois Press. Urbana, Illinois. [This work treats both ferns and fern allies of Mexico and, in addition to author and title, includes "finding" indexes. This is a valuable source of basic references on ferns in addition to the specific references to Mexican ferns.

Rehder, A. 1911-1918. The Bradley Bibliography. A guide to the literature of the woody plants of the world, published before the beginning of the twentieth century. Arnold Arboretum. Jamaica Plain, Massachusetts. [A comprehensive subject and author index to books, pamphlets and journal articles published to 1900 which treat woody plants.]

________. 1949. Bibliography of Cultivated Trees and Shrubs Hardy in the Cooler Temperate Region of the Northern Hemisphere. Arnold Arboretum. Jamaica Plain, Massachusetts. [The "purpose of this bibliography is to give references to the sources of the botanical names, valid names and synonyms, of the woody plants cultivated in the cooler regions of the temperate zone of the northern hemisphere." All names found in the 2nd edition of Rehder's manual (1940) are given in this treatment "with full synonymy and with additional genera, species and varieties introduced or described, up to the end of 1947." Lectotypes or standard-species are cited. In addition to the "systematic enumeration and synonymy of cultivated trees and shrubs," lists of abbreviations of serials and titles, abbreviations and signs, additions and emendations, and an index to scientific names are given.]

Senn, H. A. 1951. A Bibliography of Canadian Plant Geography. Transactions Royal Canadian Institute 1946-1947. Vol. 26. Publication 863, Department of Agriculture. Ottawa.

Solbrig, O. T., & T. W. J. Gadella. Biosystematic Literature; Contributions to a biosystematic literature index 1945-1964. Regnum Vegetabile 69. Utrecht. [See discussion under B VIIC, Current Literature.]

Swift, L. H. 1970. Botanical Bibliographies. Burgess Publishing Company. Minneapolis, Minnesota. [See Part III, pp. 417-425 for "Bibliographic Keys for literature on individual taxa, special groups."]

Wood, C. E., Jr., and contributors. 1958+. Generic Flora of Southeastern United States. [Has excellent bibliography for each group treated. See Journal of Arnold Arboretum, 1958+.]

Yuncker, T. G. 1964. A bibliography of the family Piperaceae. Candollea 19: 97-144.

4. Cumulative Indices to Journals

When conducting a literature search these indices can save considerable time. In botanical libraries where a series of specific journals is incomplete and funds are short, the purchase of cumulative index volumes might be the most expedient solution.

Index to Bulletin of Torrey Botanical Club, Vols. 1-74 (1870-1948). Compiled by H. W. Rickett, 1955, Science Press. Lancaster, Pa. [Includes index to authors with title of article and a systematic index to every botanical name mentioned in the 75 volumes.]

Index to Taxon, Vols. 1-20. Part 1, Scientific names, Rouleau, E. (compiler), 1972; Part 2, Authors and subjects, Lowden, R., 1973.

Nolan, E. J. (ed.). 1913. An index to scientific contents of the Journal of the Proceedings of the Academy of Natural Sciences of Philadelphia 1817-1910. Academy of Natural Sciences. Philadelphia. [Index to genera, species, etc. described or referred to in the proceedings and journal.]

Rouleau, E. 1953. Rhodora, Index to Vols. 1-50. New England Botanical Club.

5. Index to Dissertations

Dissertation Abstracts International, Section B. Science. University Microfilms, Inc. Ann Arbor, Michigan. (See also Index to American dissertations). [For dissertations prior to 1956 see Doctoral dissertations accepted by American Universities, Wilson, New York. For discussion and more bibliographies see Swift, L. H., 1970, Botanical Bibliographies, Burgess Publishing Company, Minneapolis, Minnesota.]

6. Research Indices and Directories of Plant Systematists

BioResearch Index (BioI). Biosciences Information Service of Biological Abstracts. [See discussion under Biological Abstracts in Section B VIIC.]

Indices to Chromosome Numbers. [See references in Chapter 10.]

Jackson, R. C. (compiler). 1966. ASPT, IOPB Index of Current Taxonomic Research. Regnum Vegetabile 43. Utrecht. [This index is now under the auspices of the Botany Department of the Smithsonian Institution and a current edition is long overdue. The research projects, including name and address of researcher, are arranged by systematic groups; state floras of the U. S. (also includes smaller areas such as counties, vegetation types within confines of a single state, etc.); regional floras of North and South America; floristic studies in Europe, Asia, Africa; paleobotany; phytogeography; indices; anatomy (arranged by systematic group); and bibliography, and miscellaneous studies.]

Keuken, W. (compiler). 1971. Dictionary of Plant Systematists. 1970. Regnum Vegetabile 72. Utrecht. [IAPT membership list for 1970 with indication of membership in affiliate societies (IABG, IOPB, IOP, IOS) arranged by country; members are listed alphabetically and addresses given. Institutional members (libraries, herbaria, botanic gardens, etc.) are also listed.]

Roon, A. C. de (compiler). 1958. International Directory of Specialists in Plant Taxonomy with a Census of their Current Interests. Regnum Vegetabile 13. Utrecht. [Arranged alphabetically by specialist and by systematic group and geography. Rather out of date but still useful.]

IX. GENERAL TEXTS

Alston, R. E., & B. L. Turner. 1963. Biochemical Systematics. Prentice Hall, Inc. Englewood Cliffs, New Jersey.

Benson, L. D. 1962. Plant Taxonomy: Methods and Principles. The Ronald Press Company. New York.

Blackwelder, R. E. 1967. Taxonomy: A Text and Reference Book. John Wiley and Sons, Inc. New York. [A zoology text but good discussion of general aspects of taxonomy.]

Briggs, D., & S. M. Walters. 1969. Plant Variation and Evolution. World University Library, McGraw-Hill Book Company. New York.

Core, E. L. 1955. Plant Taxonomy. Prentice-Hall, Inc. Englewood Cliffs, New Jersey.

Cronquist, A. 1968. The Evolution and Classification of Flowering Plants. Houghton Mifflin Company. Boston.
Datta, S. C. 1970. A Handbook of Systematic Botany. 2nd ed. Asia Publishing House. New York.
Davis, P. H., & V. H. Heywood. 1963. Principles of Angiosperm Taxonomy. D. Van Nostrand, Inc. Princeton, New Jersey.
Heslop-Harrison, J. 1964. New Concepts in Flowering Plant Taxonomy. Harvard University Press. Cambridge.
Heywood, V. H. (ed.). 1968. Modern Methods in Plant Taxonomy. Botanical Society of British Isles Conference Report No. 10. Academic Press. New York. [A collection of papers given at a conference on Modern Methods in Plant Taxonomy and the reviews include a major survey of the major fields of taxonomy with emphasis on new techniques.]
Hitchcock, A. S. 1925. Descriptive Systematic Botany. John Wiley and Sons, Inc. New York.
Jeffrey, C. 1968. An Introduction to Plant Taxonomy. J. & A. Churchill, Ltd. London.
Johnson, A. M. 1931. Taxonomy of the Flowering Plants. The Century Company. New York.
Lawrence, G. H. M. 1951. Taxonomy of Vascular Plants. The Macmillan Company. New York.
Pool, R. J. 1941. Flowers and Flowering Plants. 2nd ed. McGraw-Hill Book Company, Inc. New York.
Porter, C. L. 1967. Taxonomy of Flowering Plants. 2nd ed. W. H. Freeman and Company. San Francisco.
Sneath, P. H. A., & R. R. Sokal. 1973. Numerical Taxonomy: The Principles and Practice of Classification. W. H. Freeman and Company. San Francisco.
Solbrig, O. T. 1970. Principles and Methods of Plant Biosystematics. The Macmillan Company. New York.
Swingle, D. B. 1946. A Textbook of Systematic Botany. 3rd ed. McGraw-Hill Book Company, Inc. New York.
Takhtajan, A. 1969. Flowering Plants--Origin and Dispersal. Smithsonian Institution Press. Washington, D. C. [Authorized and translated from the Russian by C. Jeffrey.]

X. MONOGRAPHS AND REVISIONS

A. Serial Monographs and Revisions
(including examples of regional "monographs" or revisions in "floras")

Boissiera, Mémoires des Conservatoire et Jardin botaniques de la villa de Genéve. [A series of monographs which arose as a supplement to Candollea. See note in Taxon 22: 309, 1973.]
Candolle, A. P. de. 1823-1873. Prodromus systematis naturalis regni vegetabilis... Paris. [See discussion in Section B IB.]
Candolle, A. de, & C. de Candolle (eds.). 1878-1896. Monographiaë Phanerogamarum. 9 vols. Paris. Monographic series to succeed the Prodromus. Vol. 1 (1878) Smilacées, Restiaceae, Meliaceae; Vol. 2 (1879) Araceae; Vol. 3 (1881) Philydraceae, Alismaceae, Butomaceae, Juncaginaceae, Commelinaceae, Cucurbitaceae; Vol. 4 (1883) Burseraceae, Anacardiaceae, Pontederiaceae; Vol. 5 (1883-1887) Cyrtandreae, Ampelidées; Vol. 6 (1889) Andropogoneae; Vol. 7 (1891) Melastomacées; Vol. 8 (1893) Guttiferae; Vol. 9 (not seen). [Specimens cited.]
Engler, A. 1900-1968. Das Pflanzenreich. Nos. 1-108. Berlin. [A series of monographs of the plant kingdom by various authors. Although incomplete,

the parts issued are significant works. The works were published as completed and the arrangement is often confusing. Basically the plan was to divide the plant kingdom into four major groups (I-IV). Under each group families were also numbered (e.g., Butomaceae IV 16). Each issue (Heft) was given a number (i.e., Hefte 1-108) based on order of appearance. For a detailed index see Davis, M. T., 1957, A Guide and Analysis of Engler's "Das Pflanzenreich," Taxon 6: 161-182, and also Stafleu, F., 1967, Taxonomic Literature. The various ways these monographs have been cited in the literature are also discussed by both Davis and Stafleu.

Engler, H. G. A., & K. A. E. Prantl. Die natürlichen Pflanzenfamilien. Ed. 1 (1887-1915); ed. 2 (1924+). [See discussion in Section B IB.]

Flora Neotropica. 1968+. Organization for Flora Neotropica. Hafner Press. Riverside, New Jersey. [A series of monographs of tropical plants of the western hemisphere. See discussion in section B IIIA.]

North American Flora. 1905+. Britton, N. L., and contributors. New York Botanical Garden.

Phanerogamarum monographiae. 1969+. Verlag von J. Cramer. Lehre, Germany. [A new series of lengthy monographs, revisions and indepth treatments of limited numbers of taxa of phanerogams. Vol. 1, Bocquet, G. (1969) Revisio Physolychnidum (Silene section Physolychnis) (world-wide coverage); Vol. 2, van Royen, P., The genus Rubus in New Guinea (regional) (see review in Taxon 18: 456.]

Steenis, C. G. G. J. van (general ed.). 1948+. Flora Malesiana. Wolters-Noordhoff Publishing. Groningen, The Netherlands. Series 1--Spermatophyta: Vol. 1 Cyclopaedia of Collectors and Collections (1950); Vol. 2 Malesian Vegetation (in preparation); Vol. 3 Malesian Plant Geography (in preparation); Vol. 4 General Chapters and Revisions (1955-1958); Vol. 5 Bibliography, Specific Delimitation, & Revisions (1955-1958); Vol. 6 Systematic Revisions, 6 pts. (1960-1972); Vol. 7 Systematic Revisions, pts. 1, 2 (1971-1972). [See inside front cover of Vol. 7, pt. 2 (1972) for Index to Revised Families in Series 1; volumation and page numbers are also given.]

B. Guides

Many excellent guides have been mentioned in other sections of this chapter (particularly the sections dealing with bibliographies and current literature) which should be consulted. The titles of some of these and others which are commonly used are cited here.

A.E.T.F.A.T. Index (See Section B VIIC).

Biological Abstracts (See Section B VIIC).

Boivin, B. 1967(?). Énumération des plantes du Canada. Provancheria No. 6. Mémoires de l'Herbier Louis-Marie. Université Laval. Québec.

Botany Subject Index. Compiled by USDA (See Section B VIIIB).

Davis, M. T. 1957. A guide and analysis of Engler's Das Pflanzenreich. Taxon 6: 161-182.

Excerpta Botanica (See Section B VIIC).

Gray Card Index. Although this index has a rather specific use (See Section B IB) it can be useful in locating monographs and revisions since such works usually include new combinations and descriptions of new taxa.

Index to American Botanical Literature (See Section B VIIC).

Index to Australasian Taxonomic Literature (See Section B VIIC).

Index to European Taxonomic Literature. 1965-1969. (See Section B VIIC).

Kent, D. H. 1967. Index to Botanical Monographs. Published for the Botanical Society of the British Isles. Academic Press. London and New York.

["A guide to monographs and taxonomic papers relating to phanerogams and vascular cryptogams growing wild in the British Isles." Mostly includes works published since 1800.

Lawrence, G. H. M. 1951. Taxonomy of Vascular Plants. The Macmillan Company. New York. [See literature cited following description of each family.]

Merrill, E. D., & E. H. Walker. 1938. A Bibliography of Eastern Asiatic Botany. Supplement by E. H. Walker, 1960. (See discussion in Section B VIIIB).

Reed, C. F. 1969. Bibliography to Floras of Southeast Asia. (See Section B VIIIB).

Sachet, M. H., & F. R. Fosberg. 1955. Island Bibliographies and Supplement. 1971. (See Section B VIIIB).

Section C. SUGGESTIONS FOR THE PREPARATION OF TAXONOMIC TREATMENTS AND REFERENCE FILES

The preparation of any taxonomic treatment involves considerable reference to the literature. Sound taxonomic work requires not only an acquaintance with literature but also careful documentation. In the preparation of a monograph, revision, flora or manual, one must account for all the names published for plants in a specific group or in a particular area. Original descriptions and type specimens must be located, pertinent literature surveyed, representative herbarium specimens examined, taxonomic decisions made, and results published.

Some means of compilation of references, data and other records is necessary. Several "how to" references of value in the organization and preparation of such treatments are Leenhouts, P. W., 1968, A Guide to the practice of herbarium taxonomy, Regnum Vegetabile 58, Utrecht; Benson, L., 1962, Plant Taxonomy, The Ronald Press Company, New York; and Jacobs, M., 1969, Large families--not alone!, Taxon 18: 253-262. Ewan (Taxon 18: 194 ff., 1969) points out some of the historical problems for taxonomists, and Stafleu (Taxon 12: 43 ff., 1963) discusses the dates of publication of early works and some of the means of determining the date of effective publication and associated problems.

Students are encouraged to begin a reference (card) file early in their careers. The references to be collected and the system of cataloging are matters of individual interest and preference. The system should be one that best complements an individual's thinking (e.g., does one tend to remember authors, systematic divisions or special subjects). Whatever the system, it should be flexible, should not require that numerous reference cards be made for a single reference, and should be readily useable. Most students will keep such a file for specific research projects (class projects, theses, etc.). This file may be expanded as other interests develop, or separate research and general reference files may be kept. The usual 3 x 5 or 5 x 8 card files are convenient and provide space for the references and for annotations. Several time-saving steps in the compilation of a bibliography or a reference file include accurate citation (author, date, title, journal or publisher), record of call number, notes if reference is studied (comments on contents, concepts, etc.), and notation on sources of references recorded but not consulted (will expedite interlibrary loans).

Systematic perusal of current issues of literature guides and indices such as Index to American Botanical Literature, Excerpta Botanica, etc., as well as bibliographies in journal articles, is very helpful not only in compiling references of interest but in directing one's attention to current papers which might otherwise be overlooked.

Hopefully, widespread availability and use of electronic data processing for retrieval of such information will soon reduce the time and effort spent on making and using such a primitive form of information retrieval as card files, or better yet, make them obsolete.

TAXONOMIC LITERATURE EXERCISE

Select any species of vascular plant described 25 or more years ago and give the following information:

1. Citation of the original publication.
2. List of synonyms (to 10) and references.
3. Copy and translate (if necessary) the original description.
4. List of references (to 10) to illustrations.
5. Location of type specimen.
6. Collector of type and type locality.
7. Correct generic name, authority and bibliographic reference.
8. Type species of the genus.
9. Number of species in genus.
10. Infraspecific taxa described since 1885.
11. Distribution of this species in US and Canada and general world distribution for the genus.
12. Most recent monograph of the group.
13. List the key characters used in at least 3 manuals to separate this species (or genus) from closely related ones.
14. An economically or horticulturally important member of the group.
15. Chromosome number.
16. List current papers or research projects published within the last 3 years treating this species (or genus).
17. Current authority or worker on the group.

SAMPLE FORM FOR RECORDING INFORMATION

No.	Indexes & Guides Consulted (e.g. Index Kewensis)	LC or Dewey No.	Answer	Documentation (citation of specific references)
1.				
2.				
3.				
4.				
5.				
6.				
7.				
8.				
9.				
10.				
11.				
12.				

AUTHOR INDEX

INDEX TO TAXONOMIC INDEXES AND GUIDES

Chapter 31. THE HERBARIUM

An herbarium is both a collection of dried plants and an educational and research institution. As a basic resource for the study of systematic botany and related fields, the herbarium serves as a reference center, a documentation facility and a data storehouse. The organization, use, operation and maintenance of herbaria are discussed below.

Section A. HERBARIUM IN GENERAL

I. HERBARIUM: CONCEPT AND DEFINITION

The term *herbarium*, used in the strictest sense today, is simply a collection of dried specimens. Lawrence (1951) and others include in their definitions the arrangement of specimens in the sequence of an accepted classification, and that the specimens are available for reference or other scientific study. "Herbarium" used in its original sense, however, referred not to a collection of plants, but to a book about medicinal plants. Tournefort, in about 1700, used the term as an equivalent to *hortus siccus* (Stearn, 1957), and this use was taken up by Linnaeus who also adopted it as a substitute for *hortus siccus*, *hortus mortus*, and others. It was largely through his influence that the word herbarium superceded the former terms.

The procedure of pressing and drying specimens for storage has been an amazingly successful one in terms of preservation of detail and specimen longevity, and the plants so preserved provide a concrete basis for past, present, and future studies. In its more than four-hundred-year history the herbarium has become an institution. Today one associates the term herbarium not only with preserved plant specimens but also with certain botanical activities. The herbarium is the basic reference source of the taxonomist and has become a center for research as well as teaching and public information.

An herbarium, a special kind of museum, can also be regarded as a data bank with vast quantities of raw data. Each specimen has information content and therefore value which will, of course, vary depending on the completeness of the specimen and data and the source of the material. Each specimen has information about the vegetation of an area, a population, and the taxon to which it belongs (Rollins, 1965). The collection, therefore, represents a source of primary information about man's explorations and observations of the earth's vegetation and documents the results of much of the past inquiry into the nature and relationships of plants. Herbarium specimens are now used for studies in disciplines which were probably never even dreamed of at the time early collections were made and herbaria organized. These studies include such fields as cytogeography, biochemical systematics, palynology and genecology.

II. HISTORICAL DEVELOPMENT

The beginning of the herbarium as a collection of dried specimens affixed to paper for a lasting record is attributed to Luca Ghini (1490?-1556). According to Arber (1938) Ghini seems to have been the sole initiator of the art of herbarium-making and this art was disseminated over Europe by his students. Gherards Cibo, a pupil of Ghini, began collecting and preserving specimens as early as 1532 and his herbarium is extant today. John Falconer, an Englishman, is mentioned in the writings of Lusitanus in 1553 and William Turner in 1569 as possessing an herbarium, and it is believed that he also learned of herbarium making either directly or indirectly from Ghini (Arber, 1938). Although the herbarium technique was a well-known botanical practice at the time of Linnaeus, he departed from the convention of the day (mounting specimens and binding them into volumes) by mounting his specimens on single sheets and storing them horizontally much as is the practice today (Stearn, 1957; Dewolf, 1968). Although this method became general during the second half of the 18th century, perhaps, as Stearn (1957) believes, due to Linnaeus's example and teaching, it was by no means universal. In fact as late as 1833 Asa Gray was offering bound volumes of grasses and sedges for sale (Dewolf, 1968).

In the United States several herbaria are known to have existed in the mid-1700's. Many of these and later ones found their way to Europe, where they are preserved. According to Lanjouw and Stafleu (1964) and Shetler (1969) the oldest institutional herbarium in the United States is that of Salem College started in 1772. Other early institutions include the Academy of Natural Science of Philadelphia, Amherst College, Boston Society of Natural History, Charleston Museum, and others, all founded before 1860 (Jones and Meadows, 1948).

In addition to changes in mounting, early workers also began depositing specimens in established collections, as well as exchanging specimens. It is indeed fortunate for botanists of today that such practices developed so early. Many herbaria have been destroyed by fire, insects, war and ignorance, and all that remains of some collections are the duplicates sent to other institutions on exchange. An unfortunate policy was the scanty field notes made by early collectors and the lack of concern for original labels as early herbarium specimens were exchanged or sold. For an interesting account of the sale, transfer, and handling of several early private North American herbaria, see Stuckey (1971).

III. PRESENT SIZE AND STATE OF HERBARIA

Herbaria range from small personal collections (mostly of a few hundred specimens with a few notable exceptions) to large collections of colleges, universities, private foundations and governmental agencies involving millions of specimens. In 1964 ten herbaria reported collections in excess of three million specimens (Lanjouw and Stafleu, 1964; Shetler, 1969). Table 31-1 lists the five largest herbaria in the world and the eight herbaria in the United States with one million or more specimens.

Shetler (1969) estimates that the herbarium resources of the world may include as many as 250 million specimens, although the total number actually reported by the various institutions was 148 million. Of this estimate about 78 million are in European collections and 36 million are in North America. The difficulty is, of course, the number of private and small public herbaria which often do not report their holdings or whose holdings are not generally accessible.

Today collections are more accessible than ever before due to the efforts of many individuals and national and international organizations. Two

Table 31-1. Largest Herbaria in the World (After Shetler, 1969).

Location	Abbreviation	No. of Specimens	Rank by Size
Kew, England	K	6,500,000	1
Leningrad, USSR	LE	6,000,000	2
Paris, France	P	5,000,000	3
Geneva, Switzerland	G	4,000,000	4
Lyon, France	LY	3,800,000	5
Cambridge, Mass. (Combined herbaria of Harvard Univ.)	A, AMES, ECON, FH, GH, NEB	3,540,000	
New York (Bot. Gard.), New York	NY	3,000,000	8
Washington, D.C. (U.S. Nat. Herb.)	US	3,000,000	8
Chicago, Illinois	F	2,350,000	11
St. Louis, Mo. (Mo. Bot. Gard.)	MO	1,700,000	14
Berkeley, California	UC	1,225,000	21
Ann Arbor, Mich.	MICH	1,000,000	24
Philadelphia, Pa.	PH	1,000,000	24

extremely useful indices for locating collections are Index Herbariorum (Lanjouw and Stafleu, 1965) which includes location and curators of herbaria of the world, size and type of collections, type of institution, publications (serials), and important historical collections and Index Xylariorum (Stern, 1967), an index to wood collections of the world. The principal wood collections of the United States include those of the Arnold Arboretum (Aw), Cambridge, Massachusetts; Rancho Santa Ana Botanic Garden (RSAw), Claremont, California; U. S. Forest Products Laboratory (MADw,SJRw), Madison, Wisconsin; and Smithsonian Institution (USw), Washington, D. C. A current project underway which should also be of tremendous aid to systematists is the "Type Registry" being compiled by the Smithsonian Institution. Various herbaria, particularly those with an abundance of type material and other historic collections, often issue accounts of their collections; e.g., The Sloane Herbarium in the British Museum (Dandy, 1958), Herbaria of the Department of Botany in the University of Oxford (Clokie, 1964), and Linnean Herbaria (Stearn, 1957).

IV. THE CHANGING NATURE OF THE HERBARIUM AND ITS FUTURE

The nature of the herbarium as an institution and collection remained essentially unchanged until about fifty years ago. Developments in genetics, biochemistry, cytology, ecology, and other fields leading to less descriptive and more experimental work have been largely responsible for the changes that have occurred, but unfortunately there has been some resistance to this change. The herbarium, while it serves its former goals as a center for documentation, a source of information for monographs and floristic works, and as a resource in education and other fields, has the opportunity and the responsibility to expand to meet the needs of the future. Most systematists now use numerous approaches to problem solving, and the herbarium should remain as the central facility, as indeed it is in many institutions.

The status, operational aspects, and future of herbaria have been reviewed by Beaman (1965) and Shetler (1969). Rollins (1965) has also pointed out some of the roles of the herbarium in research and teaching. Shetler (1969) has cast some doubt on the future of the herbarium as a scientific institution and resource. It seems desirable that herbarium goals and practices be reviewed, particularly in the light of the recent developments in taxonomy and botany in general. Each herbarium administrator, staff member, and all of those who use the herbarium should consider the role of the herbarium not only at their institution but also on the international scene, and guidelines should be set up to insure maximum availability of information in herbaria both at present and in the future. Turrill (1964) has stated that there are two sets of problems facing herbaria--one is how herbaria can be maintained within reasonable limits and, two, how they can be improved. Some of his suggestions include division into special and general types. Special herbaria would include local ones (those devoted to a given area and serving local interests), special research herbaria (e.g., cytological and cytogenetic vouchers) and ecologic herbaria (including not only specimens but photographs, field notes, schemes of ecological analysis). Admittedly the general herbarium presents additional and more difficult problems, and Turrill (1964) proposed the limiting of such herbaria in number with each having some specialization; e.g., systematic or regional. He further summarized the purposes of a general herbarium as follows: "(1) to provide facilities for determination on any material, including new taxa; (2) to enable new monographs and floras to be prepared; (3) to preserve specimens of historic importance, such as those dating discoveries, introductions, and increases or restrictions of ranges; (4) to assemble data for working out ranges and ecological distributions; (5) to bring together in a relatively permanent form specimens for comparative morphological or phylogenetic studies and to provide material for special researches, as in plant anatomy and palynology." For an interesting account of the relevance of national, regional and local herbaria see Cronquist (1968), Brenan (1968), and McNeil (1968).

The basic problems of general herbaria still remain. How large should they become? What kind of materials are to be included? How can information and materials be efficiently retrieved? Shetler (1969) has pointed out some of the idiosyncrasies of herbaria as well as their shortcomings. He proposes that herbaria employ business-like methods and that computers be used for much of the laborious retrieval now needed and demanded by scientists and laymen. Such reorganization and innovation requires broad cooperation and considerable planning. Curators should not blindly become slaves to the "machine" as many have to "their collections." The first step in solving the basic problems of herbaria is to accept the challenge of the future with an open mind and a resolve to deal with these problems. This may mean a complete reorganization, dissolution, expansion, or combination of existing collections. As mentioned above, the establishment of special and regional herbaria is an attractive possibility. Regional centers with adequate support from all levels--local, regional, national, and perhaps international sources--and proper commitment could, through the consolidation of collections and the consequent reduction in useless and expensive competition and duplication, serve as clearing houses for exchanges and loans, provide safe depositories for documentation and assist local and special herbaria by routing special materials and duplicates to the proper institution. Certainly such centers should be able to supply materials and assist in building superior teaching and reference collections. These institutions should use efficient means of specimen handling (loans, location of types, and special materials) and information retrieval and could well provide research space and equipment for use of visiting botanists. In addition to the collection, a superior library, gardens, and technical staff could become available to more people than is currently possible.

In July, 1972, the Association of Systematics Collections (ASC) was established and began to address itself to many of the problems facing all systematics collections. The goal of the ASC, according to the constitution and by-laws, is "to foster care, management, preservation and improvement of systematics collections and to facilitate their utilization in science and society by:

1. Providing representation for institutions housing systematics collections;
2. Encouraging direct interaction among those concerned with systematics collections and their use;
3. Providing a forum for consideration of mutual problems;
4. Promoting the role of systematics collections in research, education, and public service through:
    a. the coordination of information concerning needs of users,
    b. the planning of advisory services,
    c. the development and implementation of national goals and priorities;
5. Other means and devices which shall be determined from time to time by the membership" (Association of Systematics Collections, 1972: 1).

At the time this association was formed at a Systematics Biology Symposium held in July, 1972, by the Smithsonian Institution in collaboration with the National Academy of Sciences, six reports were presented by a Committee of the Conference of Directors of Systematics Collections. The six reports and related materials, including considerable input from the symposium participants, were considered by a writing committee of twenty-five systematists and administrators who met to plan for the coordinated management of national systematics resources.

This committee in their prepublication draft of America's Systematics Collections: A National Plan, compiled and edited by H. S. Irwin, W. W. Payne, D. M. Bates and P. S. Humphrey (1973) presents a "statement of the primary goals of the systematics collections community," and a "description of the specific goals of the systematics collections community with respect to improving the condition of the collections and the services they provide;" a "discussion of systematics in science and society today" and "of the problems affecting the systematics collections community;" a list of "specific recommendations (The National Plan) that will move to resolve the problems and enable the systematics collections community to realize its goals" and a "statement of priorities and estimated costs of implementing them."

The primary goals of the systematics collections community are "to improve the condition of systematics collections as a National resource" and "to improve the efficiency of services associated with systematics collections resources. Specific goals and specific service-related goals include:

"1. Management of the specimen inventory and associated documentation to insure: (a) permanent conservation of the specimens themselves; (b) ready access to them and their documentation; and (c) space, facilities, and library resources enabling systematists to improve the information content of the collections through identification, classification, and elaboration of the intrinsic information carried by each specimen.

2. Addition of new specimens and associated information that: (a) reflect the goals and priorities of science and applied science; and (b) improve the quality and quantity of specimen and taxon related data, so that the information load of each specimen is enhanced.
3. To make available nonproprietary, specimen- or taxon-related information in a variety of useful configurations.
4. To enable incorporation of specimens and associated data in the information management system.
5. To enable ready access to the specimens themselves, and to associated documentation and library materials."

The National Plan includes the following recommendations which, when implemented, should solve many of the problems of systematics collections.

"1. Support and strengthen an organization dedicated to improving the condition of systematics collections and their services.
2. Establish a series of Councils to study specific problems relating to systematics collections.
3. Establish Advisory Committees representing those elements of the scientific community concerned with systematics collections.
4. Identify systematics collections of importance as national resources and designate National Resource Centers.
5. Develop standards for systematics collections: physical facilities, collection storage, preservation, specimen and data acquisition and documentation, collection growth, interinstitutional loans.
6. Implement electronic data processing in collection management procedures.
7. Develop technical training programs for professional service personnel and establish mechanisms for personnel placement.
8. Develop mechanisms whereby the resources of systematics collections can be used more effectively in studying and resolving problems affecting the quality of the environment and develop programs for improving public awareness of the importance of correct identification of species to indicate environmental changes and their effects on human welfare.
9. Improve the services and other contributions of systematics collections to graduate education in systematic biology.
10. Address the problems of libraries and publications associated with systematics collections.
11. Develop and implement programs for improving public awareness and use of systematics collections as a national resource important to science and to the solution of problems affecting human welfare.
12. Study financial resources available and potentially available to the systematics collections community and seek means to implement priorities..."

Section B. PROCESSING AND ACCESSIONING OF SPECIMENS

I. THE SPECIMEN

Traditionally the typical herbarium specimen has been a vascular plant affixed to a 11 1/2 inch x 16 1/2 inch sheet, but recent developments necessitate a change in the concept of herbarium specimens. Wood samples, fossils, pollen and spores, microslides, liquid-preserved materials, photographs, and drawings should be considered as specimens and therefore part of the herbarium. These changes require new housing facilities and record keeping, as well as cross referencing and retrieval systems. Today most herbarium curators are urging collectors to make a concerted effort to make their collections as complete as possible by including material for chromosome counts, wood samples, pollen slides, pollinators (or at least observational records), associated species and detailed habitat data. Weeds and cultivated species are at last being considered worthy of inclusion in the collections. If the herbarium is to be the source of information and documentation it should be, the materials used in biosystematic studies (e.g., natural and artificial hybrids, mass collections, etc.) as well as other types of studies should be included in the herbarium. Biosystematists and others should confer with directors and curators concerning the handling of materials.

II. SOURCES OF MATERIALS

The influx of material into an herbarium is usually from the following sources:

A. Staff and Student Collections. These vary according to staff research interests, specialities, and herbarium goals. Collections may be for floristic, monographic, biosystematic, or taximetric studies. Staff and students at institutions which have no herbarium, or only teaching or reference collections, may choose to document their studies by sending specimens to other institutions. These specimens may be sent either as gifts or loans and published works should always include the name of the herbarium where vouchers are deposited.

B. Exchange. Duplicates or special collections are exchanged among various institutions, usually on a one-for-one basis. This is one of the major means of adding to a collection and an efficient means of getting needed materials at minimal cost to all concerned.

C. Gifts. These are variously interpreted in different herbaria. Some treat receipt of all specimens other than exchange or loans as gifts while others include only those for which no staff services are required. These may range from an entire herbarium to a few specimens sent to a staff specialist.

D. Loans. Loans are generally either temporary (short term; e.g., a study of specific taxa for preparation of a monograph) or indefinate (permanent; e.g., the loan of an entire herbarium from one institution to another). The latter is a particularly desirable procedure for historic and other collections which may be for the most part inaccessible to most botanists or where proper housing facilities are lacking. These can often be loaned to another institution and still have the original name maintained.

E. Purchases. The purchase of an herbarium is an uncommon event today. Usually private or institutional herbaria are donated or are deposited on permanent loan at an institution. Most purchases today involve collections from specific areas (particularly the tropics) with prices ranging from ten cents to over one dollar per sheet. In some cases an institution may help to sponsor a collecting trip in exchange for a set of specimens.

F. Identification Service. In most herbaria specialists are willing to determine specimens for other workers and interested amateurs. Some herbaria have staff members whose specific job is service identifications. In general the specimens sent for identification are kept by receiving institutions unless other arrangements are made in advance. These specimens should be of such quality that they are desirable as specimens (see Chapter 18 on collecting), and should not be sent without some prior agreement. Requests for identification services should be reasonable in terms of the time required and the number of specimens to be identified.

III. HANDLING OF INCOMING SPECIMENS

Regardless of the source, most materials are first unpacked and variously treated to rid them of insect pests which are a serious threat to specimen longevity. Such pests include the larvae of cigarette and carpet beetles, moths, booklice and silverfish. Procedures vary, but several methods are commonplace:

A. Fumigation. This involves the use of such compounds as methyl bromide, carbon bisulphide, carbon tetrachloride, ethylene dichloride, hydrocyanic gas, lindane, dichlorvos strips, or paradichlorobenzene. While it is necessary to treat all new additions and loans upon arrival, the entire collection should be treated on a regular basis (see Section C--Herbarium Use, Operation and Maintenance). Several drawbacks to fumigation given by Fosberg and Sachet (1965) are:

1. Commonly used fumigants are also harmful to humans.
2. Many fumigants are inflammable.
3. Insect eggs and pupae are often not killed unless fumigation is done in a vacuum chamber.

B. Heating. The use of specially constructed chambers with heating elements which can keep specimens at 44° C for some hours is an effective method. While this is a good method for new acquisitions, it is usually not practical in most cases for treating entire collections (for details and references see Fosberg and Sachet, 1965).

C. Poisoning. There are several methods of treating specimens making them either permanently poisonous or unpalatable to herbarium pests. Specimens may be poisoned by dipping or painting them with an alcoholic solution of mercuric chloride. Specimens so treated should always be clearly labelled since this compound is extremely poisonous. It is also said to make some paper brittle and crumbly, blacken labels, and attack aluminum.

Another method employs lauryl pentachlorphenate (a 3.75% solution of LPCP in pure white kerosene free from high boiling paraffins). This solution is used at the British Museum Herbarium where specimens are dipped in the solution in a galvanized iron tank. The excess liquid is allowed to drain off and the specimens are placed in a drying cabinet where the kerosene evaporates leaving the plants impregnated with LPCP. For details see Fosberg and Sachet (1965) and Whitmore (1965). This technique has also been used on mounted specimens with success. Mounted specimens may be sprayed with a 5% lauryl pentachlorphenate (Mystox) in varsol. An adjustable spray gun is used to saturate the plant materials while a cardboard shield is held over the specimen label to prevent fading of inks. Drying time is 24 hours (for details of spraying apparatus and handling of specimens see Lundell and Kirkham, 1966).

IV. ACCESSIONING OF COLLECTIONS

Accessioning is defined by Fosberg and Sachet (1965) as the recording of the receipt and origin of lots of specimens coming into the herbarium. Each lot is assigned an accession number. The minimum data that is recorded is name of sender, date of receipt, type of transaction (gift, exchange, purchase), number of sheets, place (country, province, etc.), and kinds of plants (algae, fungi, vascular plants, etc.). Space may be left to record the inclusive sheet numbers after the specimens are mounted (Fosberg and Sachet, 1965). Other systems used include the registration of each specimen (mostly abandoned but for an example see Millspaugh, 1925) or simply mounting specimens and assigning sheet numbers. A practical and efficient way is to stamp sheets with the official herbarium seal and number them before specimens are mounted. Since exchange, loan, and gift records are kept, as well as all correspondence, the separate accession record may be considered unnecessary. Sheet numbers, in addition to being used to assess growth and efficiency in herbarium operations, have importance directly related to taxonomic research. They are an ideal means of referring to a particular specimen in a collection. The exclusive use of collectors' numbers may not be sufficient, particularly in designating types.

V. SPECIMEN PREPARATION

Mounting procedures vary but at present only a few methods are popular. It cannot be emphasized enough that the mounting procedure is a critical step in specimen preparation. A carelessly mounted specimen can be essentially useless. Mounting, like pressing, is indeed an art.

A. Strapping. A specimen is placed on mounting paper and the label pasted squarely in the lower right-hand corner. Strips of linen tape (gummed cloth or Holland cloth) are moistened and placed around stems, across leaves, etc. until the specimen is firmly attached to paper. (Cellulose or plastic tapes which deteriorate or soften with age or which decompose when specimens are fumigated should not be used.) Care should be taken so that important details are not covered by the strips. Length and width of strips depend on the part to be secured.

Another method of strapping or stripping (as it is often called) involves the use of liquid plastic. This method, the Archer Method, uses a mixture of ethyl-cellulose and Dow resin in toluene and methanol, prepared according to the schedule of Rollins (1955) given below.

Toluene - 880 cc.
Methanol - 220 cc.
Ethocel - 10 cps. standard
Dow resin - 75 gms.

Toluene and methanol are mixed first and the resin added and dissolved completely. The ethocel is added slowly and stirred until partly dissolved. The solution should be allowed to stand 24 hours before using. Store in airtight containers and use a 4:1 mixture of toluene and methanol as a solvent. The plastic is usually dispensed from a pistol-grip pressure oil can or a polyethylene (mustard or catsup) squeeze bottle.

B. Pasting or Gluing. Specimens are often attached to their sheets by gluing. An efficient and effective way is to coat the surface of a glass or copper plate with glue (use a paint brush), place the specimen on the plate, and tap gently with forceps or cards. The specimen is then lifted, using forceps, inspected to check distribution of glue, and placed on the sheet. If the glue is not evenly distributed, more may be applied using a soft brush. Tap specimens down gently using relatively lint-free tissue. Labels generally are applied using the same glue and put on before the specimen. If any glue is on the exposed surface, the sheet may be covered with waxed paper. This is a particularly good procedure for grasses and sedges. All fragments, fruits, and seeds should be placed in packets or envelopes and these affixed to the herbarium sheet at this time. In cases where materials are too bulky, the packet should be numbered with an herbarium accession number and/or collector's number and set aside to be mounted later. The specimen is then covered with a sheet of newspaper (usually the one containing the specimen, if not too dirty) and placed in a stacking box or simply placed on the table. Specimens are so stacked until 20 to 25 specimens are in a stack. A board and weight are then placed on the stack. If mounting is slow, the weight should be applied sooner. Stacks should be allowed to dry for 24 hours. The papers are then removed and specimens are ready for the next step. Several glues or pastes are available. The "Special A Tin Paste" is used in some herbaria, but library pastes such as "IPI" or glues such as "Elmers," "Nicobond," or "Wilhold" are also used. Some herbaria use the Archer formula (plastic) for gluing specimens to the sheet. The plastic, dispensed from a "squeeze" bottle or pressure gun, is applied to the back of a specimen and material is placed on the sheet, weighted with washers or other suitable weights and the plastic allowed to dry. All of these glues, but not the plastic, are water soluble.

C. Sewing. Except where strapping is the only process, most specimens, after gluing, are strapped in some manner--linen tape, liquid plastic, white glue, and/or sewing with a heavy linen thread. A very efficient process combines sewing and strapping (with glue or Archers). Sewing is restricted to heavy stems, overlapping leaves, rhizomes, matted bases of grasses, large fruits, cones or heads, or other places where the use of plastic or glue is impossible or impractical. Threads on the back of sheets should be covered with linen or paper tape. Specimens are placed on a sheet of cardboard and strapped. Each specimen is then stacked one upon the other using 4-5 wooden blacks to separate each specimen. This requires little work space and allows proper ventilation.

D. Special Handling. One case of special handling is the use of serial sheets for a single specimen, long a necessity in tropical areas.

Although several techniques are used and much depends on the notes of the collector, the series of sheets should bear a single accession number and an annotation noting on each sheet the number of sheets in the series; e.g., sheet 1 of 5. Such a series is much more desirable than the practice of insisting that regardless of size, an entire plant should be on a sheet even though critical examination of the material is impossible.

Inflorescences of Typha, Cirsium, etc. often shatter and may be held intact by enclosing them in bags (cheese cloth or netting). Bulky fruits, cones, etc. may be removed and stored in special boxes or cases. Specimens should clearly indicate that both sheets and accessory materials are present in the collection. Each sheet should also be stamped with the distinctive seal and/or name of the herbarium. The use of letters (abbreviations) only should be avoided (e.g., specimens marked OU Herbarium may mean University of Oxford, Oregon, Oklahoma). Some herbaria use a stamp with the name of the institution and its official abbreviation (see Lanjouw and Stafleu, 1964; Fosberg and Sachet, 1965).

Proper mounting depends greatly on careful collecting. Suggestions for collecting and pressing of herbarium specimens are given in Chapter 18.

VI. SORTING AND FILING OF SPECIMENS

Once the specimens have been glued, sewn and/or strapped and allowed to dry (in the cases where glue and/or plastic are used), they should be sorted for filing. Techniques and physical arrangements vary, but ultimately specimens are sorted to family, genus, species, and infraspecific taxa. If geographic provinces are recognized in the collection, additional sorting is required. Regardless of the filing system used, specimens should be arranged in the order they will occur in the herbarium. Considerable time and energy can be saved if this procedure is followed. In some collections, once the specimens are sorted, they are routed to curators of special groups for checking and/or filing.

Specimens are filed on shelves in herbarium cases in 24 x 16 1/2 inch folders called genus covers. These heavy manilla or bristol folders are often color coded. Specimens are either filed directly in these folders or in a thinner species folder. Today there seems to be a breakdown in terminology for folders. The genus cover often in reality serves as a species folder if the representation of a species warrants an entire folder. In small collections the commercial "genus cover" may serve as a family folder.

Specimens are placed in folders face up with labels in the lower righthand corner. The genus name usually appears in the lower lefthand corner on the face of the cover. Specific names or alphabetical sequence appear in the lower righthand corner. Names may be printed or stamped on the folder or on a label which is then glued to the cover.

VII. ARRANGEMENT OF SPECIMENS IN THE HERBARIUM

A. The General Collection. Although there are numerous possible systems of arrangement of specimens, the schemes most often encountered are given below:

1. Bentham and Hooker. This system is rarely used in the United States. See Franks (1965) for a list of larger European herbaria using this system.
2. Dalla Torre and Harms numerical arrangement of the Englerian System. A common system in the United States and other countries although it is often modified so that only family sequence is followed with genera and species being alphabetically arranged. (See Dalla Torre, C. G. de, and H. Harms, 1900-1907).
3. Besseyan System. This system became popular in the 1920's and is still used in some herbaria (particularly those organized after the above date).
4. Alphabetical. This system varies from strictly alphabetical (by family, then genus, species, etc.) to some modified system (e.g., segregation of ferns, gymnosperms, monocotyledons, dicotyledons, with alphabetical arrangement of families in each of these groups).

In addition to these basic systems, many herbaria also employ a geographic arrangement. Local areas are in one color folder, with U. S., other New World countries, Old World, etc. in other colors. Such schemes range from few geographic categories to quite elaborate ones. These divisions are exceedingly useful for floristic studies. Even though colored folders are more expensive and fading colors occasionally are confusing, the reduction of handling of all specimens of a species to examine only those specimens from a particular area (e.g., county, state, or physiographic province) may justify the added cost.

Undetermined materials (of which every herbarium has its fair share) are generally filed at the end of genus or family and are usually in specially marked folders (e.g., "undet."). In addition to specimens and special materials (discussed below) non-plant materials are often of great value and are included in general collections. These include photographs (e.g., a type specimen, a habitat), drawings, correspondence concerning a specimen, keys, notes on synonymy and cross-reference markers ("dummy" sheet or cover).

B. Special Collections. In addition to the general collection, many herbaria maintain various special collections.

1. Type Collection. Type specimens such as holotypes, isotypes, and co-types are often housed separately from the general collection, where unnecessary handling and risk of damage can be greatly reduced. These are commonly filed in the same sequence as the general collection, or alphabetically. Another solution is to place types in the general collection, but in specially and conspicuously marked genus covers.
2. Synoptic Collection (Teaching or identification). Special collections for routine identifications and student use are sometimes separated from general collections. The synoptic collection can greatly reduce time spent consulting specimens in a herbarium as well as specimen wear. These collections are usually small and may be arranged and rearranged (particularly for teaching purposes) according to recent schemes

of classification, e.g., Cronquist, Thorne, Hutchinson.

3. Special Research Collection. Mass collections and artificial hybrids, as well as pollen and spores, anatomical preparations and bud materials may often be housed separately or in special folders or cabinets. If this is the case, some notation in the form of a dummy sheet or annotation should be made in the general collection so that the existence of these materials can be readily known by those consulting the herbarium. The arrangement of such systems varies according to the type of materials and inclination of the curators. Alphabetical, community, numerical (according to accession number) or phylogenetic systems are commonly used. Where the system differs from that of the general collection it should be clearly explained in the "Herbarium Guide" (see section C).
4. Historical Collection. Many earlier collections, as mentioned above, were bound in volumes and these are often not amenable to inclusion in general collections. Other collections include those loans or gifts with specification that they are to be maintained as separate entities. The only other reason for separate housing is for protection. Some prefer to include these materials in the general collection in special folders, as mentioned earlier concerning types. In fact many group types and other historical materials (including letters, notes, etc.) into one special collection. Regardless of physical arrangement, a catalog of materials and their arrangement should be included in the "Herbarium Guide" and/or appropriate "dummies" inserted into the general collection.

Section C. HERBARIUM USE, OPERATION AND MAINTENANCE

I. HERBARIUM USE

Herbaria exist only for use regardless of the type of collection--research, identification (synoptic), teaching, or other. Every effort for maximum efficiency in locating material in a collection (information retrieval) should be made without endangering the collection.

A. Herbarium Guide. Every herbarium should have a printed guide to the collection. This guide should explain the systematic and geographic arrangement; color code of folders; list of families with numbers and case number (including segregate and combined families; e.g., Ericaceae vs. Pyrolaceae, Monotropaceae); map of case arrangement; list, location and arrangement of special collections; and a generic catalog. The guide should also contain herbarium policy statements concerning reshelving of materials, handling of specimens, use of equipment and other facilities, loan procedures, and other pertinent information. Such guides can save considerable time on the part of visitors as well as the herbarium staff.

B. Specimen Use. Specimens are preserved for use but should be handled with extreme care since they are of scientific value and generally irreplaceable. The following suggestions are to aid users in handling specimens properly.

1. Keep sheets flat, do not shuffle or leaf through a folder like a book. Specimens are brittle and easily damaged.
2. Store properly. Preferably, specimens should be stored in herbarium cases or at least on shelves. Do not crowd too many

specimens into a box or "pigeon hole." Keep materials in covers when not in use.

3. Do not lay books or other heavy objects on specimens.
4. Place loose fragments in packets or envelopes if the specimen to which the fragment belongs can be ascertained.
5. Use materials sparingly. It is often necessary to dissect flowers or fruits. Use no more than is required and place all fragments in a fragment envelope on the sheets. For dissection, flowers and fruits may be softened by boiling or treating with a softening or wetting agent. Pohl's softening agent (Pohl, 1965) is an excellent wetting material and may be prepared according to the following formula:

1% solid "Aerosol OT"	1.6 ml 75% aqueous "Aerosol OT"
74% distilled water	73.4 ml distilled water
25% methyl alcohol	25 ml methyl alcohol

This solution will not stain or discolor sheets but is best used in a watch glass. Place materials to be softened in a watch glass and add several drops of the solution. Many materials are sufficiently softened in a few minutes. Fragments may be blotted dry and returned to the fragment envelope.

6. Use a long-armed dissecting microscope. An entire specimen can be studied with this instrument without bending the sheet.
7. Support the specimens with a ventilator or other stiff board when carrying specimens, even for short distances.
8. Call to the attention of curators those specimens in need of repair.
9. Do not write on the herbarium sheet unless permission to do so has been given by someone in charge.
10. Reshelve specimens only with permission and always with extreme care.
11. For additional information on annotations, borrowing specimens, etc., see Section C ID (Loans).

C. Visitors. Visiting an herbarium is, of course, one of the best possible ways to study material. An investigator can examine the exact materials desired as well as search "undetermined" folders, consult types and other special collections (some not available on loan), and use the library and botanic garden as well as exchange ideas with fellow botanists. Visitors would do well to follow common courtesy if they wish to have access to collections, particularly at unusual hours, such as evenings, weekends, or holidays. A note in advance can mean the difference in having or not having a comfortable place to work and access to a microscope, as well as gaining admittance to special collections and libraries.

D. Loans. Most herbaria lend materials to other institutions (rarely to individuals). Reciprocal agreements between many institutions often result in transit costs only one way. One should, however, be sure that such agreements exist before requesting a loan. Otherwise, the borrower pays all transit costs. For the most part loans should be requested after some preliminary work has been done so that unduly long-term loans are avoided (loan periods are generally for six months). Requests should be for reasonable amounts of material and for specific taxa rather than communities or floristic provinces. Loans should be requested by the curator or director of the herbarium for study by a specific person, and should be addressed to the director or curator of the lending institution. Special permission should be obtained if fragments, pollen, or other materials are to be removed. Even with permission the utmost discretion should be exercised. Many herbaria ask that duplicate slides, chromatograms, and other preparations made from their material be included when the specimens are returned. When in doubt, the borrower should consult the curator of the lending institution for specific instructions. A commonly accepted procedure is that all specimens be annotated whether they are used in ecological, floristic, monographic, anatomical, biosystematic, or other studies. This should include even those specimens determined correctly. A convenient way to agree is by using an exclamation point (!) followed by the name of the investigator and date (at least the year). Printed annotation labels affixed above the original label are the most desirable. The original label should not be altered. Some institutions allow the use of a rubber stamp with small, yet readable print (preferably using black ink); e.g., "Examined in a survey of parasitic angiosperms of Southwestern U. S., John Doe, 1971." The name of the institution where work is being conducted is also quite desirable. Annotations should reflect the history of use of an individual specimen. Materials should be returned promptly on the date due or an extension should be requested. Most institutions honor requests as soon as staff resources permit and at considerable time and expense. Loan forms to be returned are generally sent out before or at the time the loans are shipped. Upon receipt of a loan, the borrower should examine, count and acknowledge the shipment. Any damage occurring in transit due to mishandling or insects should be noted on the loan acknowledgment form. Each sheet should be checked for the stamp of the lending institution. Since sheets may have the seal of more than one institution, careful notes should be made, particularly if the borrower has material from several institutions. Specimens were packed by the lender in a certain order and an effort should be made to return them the same way.

Borrowed materials should be housed in "fireproof," dust- and pest-free cases. Materials may be fumigated and repellents used provided they do not damage specimens, labels, and photographs. Great care should be used in all cases. Plastic strapping is softened by strong fumigants and repellents such as paradichlorobenzene. This and other information is often included on loan forms. Most specimens are sent "library rate" by parcel post at considerable savings to all; however, the lending institution determines the mode of transit and this should be followed unless permission to do otherwise is obtained.

In addition to the annotations, the borrower can perform many other services:

(1) Determining duplication.
(2) Supplying additional data (obtained from duplicates from other institutions.
(3) Correcting locality data (by annotation).
(4) Pointing out mixed labels.
(5) Determining specimens as to type--isotype, paratype, etc.
(6) Indicating a mixed collection.

In most cases both parties should profit from a loan.

II. OPERATION AND MAINTENANCE

An herbarium, contrary to the opinion of some, requires constant attention if materials, information and facilities sought by botanists and others are to be available and in proper condition.

A. Specimen Care.

1. Backlog of materials. Keeping up with the influx of materials is probably one of the greatest problems in most herbaria. This is often due to lack of support, staff and space and to inefficiency. Where possible, backlogs, which in turn create storage and fumigation problems and consequently endanger specimens, are to be avoided. Inaccessible materials are of no value.
2. Specimen repair. This is an endless problem. Specimens in need of repair and remounting should receive prompt attention as soon as they are noted. Some institutions repair all material before it is loaned or before reshelving it after it has been used. If staff resources permit, repairs should be made as they are noted by persons filing new acquisitions, inserting new genus covers, etc.
3. Fumigation. As mentioned earlier (Section B) fumigation (or other means of protection) of the entire collection is a necessity in most if not all herbaria. Annual fumigation may not be sufficient to insure a collection against damage, particularly if all new accessions are not first fumigated. Constant alertness is mandatory. Those cases housing members of the Asteraceae, Brassicaceae, Ranunculaceae and Apiaceae should be scrutinized often since members of these families are commonly attacked by herbarium pests. Visitors would be of great service to curators by calling attention to signs of insect damage.

    Fumigation of a collection may be done by a commercial firm or by the herbarium staff. The entire collection may be fumigated at one time or only part of it. The method used will, of course, vary depending upon physical arrangement and size of the collection, financing, and staff preferences. A 3:1 mixture of ethylene dichloride and carbon tetrachloride is a commonly used fumigant which is placed in each case and the case then sealed. This solution is dangerous and should be used with extreme care. Stern *et al*. (1969) and Lellinger (1972) have discussed the advantages and disadvantages of using lindane and dichlorvos as fumigants and repellants. Lellinger (1972) indicated that dichlorvos ("No Pest" or "Vapona") is probably the safer material when used as a fumigant rather than a repellent. This method involves placing one 0.25 strip/22.3 ft.^{3} of case space for 7-10 days with the case remaining closed. This treatment should be used twice each year assuming all incoming materials are fumigated before insertion and that there is no infestation. Curators should consult both the Stern *et al*. and Lellinger papers before using this method. Initial fumigation of small lots of specimens, e.g., new accessions or loans, may be carried out in specially constructed cabinets or chambers. Methyl bromide is the fumigant often used in this operation.

B. Insertion of Materials (other than new accessions).

1. Genus covers. As collections grow new species and/or genus covers must be added. Division of material into smaller groups (systematic or geographic) results in easier access and a reduction in needless handling.
2. Dummy folders. As a result of changes in nomenclature and taxonomic revision, materials may in time be lost unless some precautions are taken. The problem can easily be solved by inserting a dummy folder in the appropriate place in the herbarium; e.g., *Hymenoxys* (see *Tetraneuris*); *Calylophus serrulatus* (see *Oenothera serrulata*). These may also be used to refer to additions to special collections--wood samples, cones, types. Temporary dummies are useful for indicating materials on loan or those being repaired.
3. Photographs, Drawings, Maps, etc. These may be mounted on sheets or placed in envelopes and are usually treated as specimens. Care should be taken to make sure such materials will not be damaged in the course of regular fumigation.

C. Supplies and Equipment. All equipment, including herbarium cases, should be checked and cleaned carefully. In addition to having a clean and attractive place conducive to work as well as functional equipment, many of the sources of insect infestation, dust catchers and fire traps are eliminated. Supplies for mounting, repairs, labels, genus covers, etc. must be inventoried and ordered. Fortunately, mounting paper (100% rag) is used in sufficient quantities that it is usually readily available from paper companies and biological supply houses. In fact, some paper companies issue special catalogs and supply lists for herbarium supplies--mounting paper, genus covers, paste, glue, etc. Most herbaria maintain a file on sources of supplies.

D. Correspondence and Record Keeping. (See also Fosberg and Sachet, 1965). All records and correspondence concerning exchanges, loans, gifts, and accessions should be kept and filed. Methods vary, but correspondence is often filed alphabetically according to person or institution. Exchange records are commonly entered in a ledger or journal or in a special card file. Information often recorded includes date, number of specimens received or sent, origin of material, and balance. Duplicate or triplicate exchange forms are usually sent with an exchange lot and the receiving institution acknowledges receipt of shipment, verifies specimen count and condition, and account balance. One copy is returned to the sender as a record. Loan forms generally are similar to exchange forms. In fact, many herbaria use the same form and simply check the appropriate type of transaction. Loan agreements should be read carefully for they may specify conditions under which specimens should be housed, type of annotation required, and other particular requirements.

Information of considerable importance concerning specimens, types, etc. may be in herbarium correspondence which should be preserved. Fosberg and Sachet (1965) suggest binding letters from one person or institution in chronologically arranged volumes.

E. Annual Report. The preparation of an annual report can be exceedingly profitable for herbaria of numerous kinds and sizes. It not only provides an opportunity to reflect on progress and shortcomings of the past year but is useful in planning for future growth and development. Such reports usually include summaries of various accounts and records; e.g., loans and exchanges, special requests, and gifts. These reports also often include information not available from account ledgers and journals and which consequently may be of historical and other importance; e.g., research in progress, field trip reports, library and herbarium requisitions, new equipment purchases, staff changes, and development of new techniques and procedures. This information may be useful in preparing grant proposals and annual budgets and supplying information to various administrators and botanical institutions. For examples of detailed reports see director's annual report in current volumes of Aliso and Arnoldia.

Section D. HERBARIUM GLOSSARY
(Largely adapted from Fosberg and Sachet, 1965)

Accession Number. The number applied to each lot of specimens received by the herbarium or the sheet number assigned to a specific specimen.

Accessioning. Recording the receipt and origin of lots of specimens coming into the herbarium; also includes assigning of sheet numbers.

Annotation. A note written on or attached to an herbarium sheet indicating a correction or change in identification or a point or points of interest about the specimen; any note attached to a specimen.

Annotation Labels (annotation slip). A small slip of paper on which an annotation may be written and then glued to an herbarium sheet (see also determinavit and approbavit slips).

Approbavit Slip or Label. A special annotation label, indicating that the name on the label is correct.

Arrangement (of specimens in the herbarium). The system of classification or scheme followed in the placement of specimens in the herbarium; e.g., Bentham and Hooker, Engler, Cronquist, alphabetical.

Archer Method. Affixing specimens to mounting sheets by means of small strips of liquid plastic extruded from a container with a narrow nozzle.

Board (of plant press). One of the two stiff sheets of wallboard, cardboard or wood between which the blotters or ventilators, and the folders of plants are laid and which are tied together to form a press.

Carpological Collection. Separate collection of fruits and seeds.

Case. The cabinet in which herbarium specimens are stored.

Collecting. The gathering of specimens in the field.

Collection. The accumulation of specimens in an herbarium; the specimens collected on a single expedition; a single gathering of a particular species at a given time and place; a specimen plus its replicates (many would prefer the term "duplicate" rather than "replicate.")

Collection Number. The number assigned to a collection in the field, when it is collected or when the notes are written up, the same for all of its replicates; it is associated with the specimen(s) and identifies them from then on, and refers to the data recorded in the collector's notebook. (This is often also called the collector's number and should not be confused with the sheet number which is assigned to a particular specimen in the herbarium.)

Corrugate. A sheet of pasteboard or thin metal with fluted ducts extending across the sheet (not lengthwise) used in presses when drying plants by means of artificial heat; these are often called ventilators.

Determination. Ascertaining the correct name for a specimen; identification.

Determinavit Slip. A type of annotation bearing the name of the plant and the name of the person who identified the plant; date of identification should also be included.

Distribution. A confusing term used to refer to the filing of specimens as well as the sending out of duplicates on exchange, loan, etc.

Documentation. The deposition of, or reference to, voucher specimens in an herbarium.

Dryer or Plant Dryer. An apparatus for drying plant specimens by artificial heat; term used for a sheet of blotting paper used in drying plants.

Dummy Sheet. A blank herbarium sheet or manilla paper of the same size as a herbarium sheet or genus cover inserted in the herbarium for cross reference purposes.

Duplicate. One of two or more specimens collected at the same time and place, under the same collection number, to represent a particular species; more appropriately called a replicate if the collection contained more than two sheets; an extra sheet of a collection.

Envelope. Another name for packet, or pocket; a piece of paper folded and affixed to an herbarium sheet to contain fragments.

Exchange. The process of distributing duplicates (replicates) or other materials to other institutions in return for their duplicates.

Felt Driers. Sheets of heavy blotting paper or builder's "deadening felt," cut 12 x 17 inches, used to absorb moisture in a press, with or without corrugated metal ventilators; also for tying piles of plants in folders to make light packages.

Fide. Ablative singular of fides, "according to or by the assurance of"--a term used to refer to another author; e.g., Neptunia pubescens Bentham, fide Windler, 1966.

Field Book. A notebook used to record data in the field at the time of collection or a book containing data collected in the field. Each collection usually is assigned a number which refers to notes in the field book.

Field Label. A special label for recording data in the field.

Field Notes. Information recorded about a collection or specimen in the field and usually including locality, habitat description, as well as those features of the plant not discernible from the dry specimen; e.g., height, if entire plant is not collected; bark and branching characteristics, if a tree or shrub; flower color; odor; etc.

Field Press or Portfolio. A light plant press carried in the field when collecting, into which the specimens may be placed as they are gathered.

Filing. The insertion of mounted specimens, dummy sheets, and other materials into the herbarium cases.

Flimsies. Folds of thin absorbent paper into which plant specimens are collected and in which they may be dried and stored.

Fragment. A part of a plant; any detached portion of a specimen; also used to refer to an incomplete or poor specimen; e.g., a sterile twig with a few leaves; extra materials collected for dissection and placed in the packet or envelope on a mounted specimen.

Fumigant. A volatile substance used to kill insect pests in the herbarium.

Fumigation. The process of killing or getting rid of insect pests by subjecting them to a lethal concentration of a volatile poisonous chemical.

General Herbarium. Synonymous with general collection; all materials excluding special collections.

Genus Cover. A heavy manilla folder slightly larger than an herbarium sheet in which specimens belonging to one genus are filed; often in various colors indicating geographic regions.

Glue. An adhesive used to affix specimens to the herbarium paper; see also white glue.
Gluing. Fastening specimens to mounting sheets by means of glue or other adhesive.
Herbarium. A collection of dried plants; an institution built around a collection of dried plants.
Herbarium Abbreviation. The abbreviation assigned to an herbarium in Index Herbariorum.
Herbarium Number. The sheet number of a specimen.
Index Herbariorum. A series of indices to herbaria and collectors published in Regnum Vegetabile by the International Association for Plant Taxonomy, Utrecht.
Manilla or manila. A coarse, unbleached paper of which folders are made.
Mass Collection. A population sample, composed of a number large enough to be statistically significant, of critically selected corresponding plant fragments collected at a particular time and place, to show the range of variation in selected characteristics of the population sampled.
Merrill Case. A cardboard container 48 cm. long, 34.5 cm. wide and 24 cm. high (outside dimensions) with a door on one end. Used as temporary storage for filing of specimens; developed by E. D. Merrill; not dust or insect proof.
Microfiche. Greatly reduced transparent positive photographs of printed material or herbarium specimens designed for ready filing and for reading with a special magnifying projector (reader) or with a binocular microscope.
Mounting. The process of affixing dried and pressed specimens of plants to herbarium sheets of heavy paper.
Newsprint. The paper on which newspapers are printed and often used for pressing plants; also called specimen paper.
Numbering. The process of stamping or printing sheet numbers (accession numbers) on mounting paper or herbarium specimens.
Pack Frame. A light wooden frame with a shoulder strap for back-packing.
Mounting Paper. The heavy herbarium paper to which specimens are affixed.
Packet. An envelope folded and mounted on a herbarium sheet to contain extra materials and fragments.
Paper Folder. A smooth strip of bone (wood), rounded at all corners, used in creasing paper, pressing down glued paper, etc.
Para-dichlorobenzene (PDB or moth crystals). An insecticide or repellent commonly used in herbaria.
Pigeon-hole. Compartment in a herbarium case in which the folders of specimens are inserted.
Plant Dryer or Dryer. An apparatus for drying plant specimens by artificial heat (commonly consists of a frame to support the presses above an electric heater, series of light bulbs, camp stove, or lantern).
Plant Press. An apparatus for flattening and drying plant specimens, usually consisting of two lightweight boards or frames and a pair of straps or sash cords.
Pocket. An envelope that is pasted to an herbarium sheet with the specimen to contain extra pieces or detached fragments of the specimen.
Pohl's Softening Agent. An excellent detergent solution for softening flowers and fruits for dissection.
Portfolio or Field Press. A light plant press for carrying in the field while collecting, into which the specimens may be placed immediately as they are gathered.
Preparation (specimen). The process of getting specimens mounted and ready for insertion in the herbarium.

Press. An apparatus for flattening and drying plant specimens.
Protectant. See repellent.
Quire. A number of sheets of paper folded once, resulting in a number of folds one inside another.
Registration (of specimens). The recording of each specimen to be inserted in the herbarium, a procedure now rarely followed.
Repellent. A chemical used continuously in herbarium cases to prevent insect infestations (also called protectant).
Replicate. One of the specimens of a series or suite collected at the same time and place, under a single collection number, to represent a particular species, often called a duplicate; an extra sheet of a collection.
Reprint. A separate copy of a journal article, specially printed for private distribution.
Scheda, Schedula. A label; *in scheda*--on an herbarium label; *ex confusione schedae* - from confusion of a label.
Secateurs (clippers). Small hand clippers for cutting tough or woody stems.
Set (of specimens). When a collection is made with more than one sheet (replicate) of each number, these are sorted into lots containing one sheet of each number, called sets.
Sewing. Reinforcing the attachment of herbarium specimens to mounting sheets by loops of thread around the stem and through the paper.
Sheet. A standard sheet of heavy paper 11 1/2 x 16 1/2 inches (29 x 41 cm.) or other standard size on which a dried plant specimen is mounted; often used colloquially to refer to the specimen that is or will be attached to a sheet.
s.n. (*sine numero*). Literally without a number; a designation for citing a collection lacking a collection or collector's number; e.g., John J. Doe, *s.n.*
Sorting (of specimens). Arranging of specimens for insertion in the herbarium.
Space Treatment. See fumigant.
Special Collection. Any materials kept apart from the general collection; e.g., wood specimens, carpological collection, type collection, microslide collection.
Species Cover. A light manilla folder in which specimens belonging to one species are filed within a genus cover.
Specimen. A plant or fragment of a plant, taken and preserved to represent a species or other taxon or a population, or to serve as a voucher to document observations made on a plant or population of plants.
Spirit Collection (liquid). Separate collection of plants or plant parts preserved in liquid preservatives.
Staff. Collective term to apply to all the employees of an herbarium.
Sterile Specimen. A specimen lacking reproductive parts (flowers, fruits, sporangia).
Strapping. Affixing specimens to herbarium sheets by means of strips of gummed cloth, plastic, or glue.
Stripping. Colloquial term applied to strapping, particularly when plastic is used.
Stripping or Strapping Plastic. A plastic used to affix specimens to herbarium sheets; see Archer Method.
Type. See type specimen.
Type Specimen. The specimen to which a name is permanently attached.
Type Collection. A special or separate collection of type specimens (holotypes, isotypes); the type specimen (holotype) and its duplicates.
Unicate. A specimen of which there are no duplicates.

Undetermined (undet.). Refers to unidentified materials in the herbarium; special genus covers may be so marked.
Vasculum. A collecting container or can, which is usually made of sheet metal; standard vascula are 18-24 inches long, 8 inches wide and 10 inches high and have a hinged door provided with a secure fastener on one side.
Ventilated Press. A plant press so made that warm air may pass through it to hasten the drying process.
Ventilator. A 12 x 17 inch sheet of corrugated metal, preferably aluminum, or of corrugated cardboard, or of lattice bamboo splints, which are laid between blotters and folders of plants to facilitate the passage of air through the leaves.
Voucher. A specimen preserved to substantiate recorded observations, and to which reference may be made in the future to verify the identity of the plant on which studies were made.
V-shaped Rest. A frame for holding genus covers and specimens when in use in the herbarium. Large ones on wheels are excellent for holding and moving specimens about the herbarium when filing.
Webbing Strap. A strap made of heavy cotton or linen fabric used on plant presses.
Wetting Agent. A variety of solutions used to soften flowers, fruits, etc. for dissection.
White Glue. Gelatin base or synthetic liquid used to paste specimens to herbarium paper; e.g., Elmer's Glue, Wilhold, Fish Glue.

HERBARIUM LITERATURE

Arber, A. 1938. Herbals, Their Origin and Evolution. 2nd ed. University Press, Cambridge.
Archer, W. A. 1950. New plastic aid in mounting herbarium specimens. Rhodora 52: 298-299.
Beaman, J. H. 1965. The present status and operational aspects of university herbaria. Taxon 14: 127-133.
Brenan, J. P. M. 1968. The relevance of the national herbaria to modern taxonomic research [in Great Britain]. In: Modern Methods in Plant Taxonomy. V. H. Heywood (ed.). Botanical Society of British Isles and Linnean Society. Academic Press. London.
Clokie, H. N. 1964. An Account of the Herbaria of the Department of Botany in University of Oxford. Oxford University Press. London.
Cronquist, A. 1968. The relevance of the national herbaria to modern taxonomic research in the United States. In: Modern Methods in Plant Taxonomy. V. H. Heywood (ed.). Botanical Society of British Isles and Linnean Society. Academic Press. London.
Dalla Torre, C. G., and H. Harms. 1900-1907. Genera Siphonogamarum ad systema Englerianum conscripta. Leipzig.
Dandy, J. E., editor & compiler. 1958. The Sloane Herbarium. British Museum (Natural History). London.
Davis, P. H., and V. H. Heywood. 1963. Principles of Angiosperm Taxonomy. D. van Nostrand Company, Inc. New York.
DeWolf, G. P., Jr. 1968. Notes on making an herbarium. Arnoldia 28: 69-111.
Fosberg, F. R., and M. Sachet. 1965. Manual for tropical herbaria. Regnum Vegetabile 39: 5-132.
Franks, J. W. 1965. A Guide to Herbarium Practice. Handbook for Museum Curators, Part E, Section 3. The Museums Association. London.

Irwin, H. O., W. Payne, D. Bates, & P. Humphrey (compilers & editors). 1973. America's Systematics Collections: A National Plan (Prepublication Draft).

Jones, G. N., and E. Meadows. 1948. Principal institutional herbaria of the United States. American Midland Naturalist 40: 724-740.

Lanjouw, J., and F. A. Stafleu. 1964. Index Herbariorum. Part 1. The herbaria of the world. 5th ed. Regnum Vegetabile 31: 1-251.

Lawrence, G. H. M. 1951. Taxonomy of Vascular Plants. The Macmillan Company. New York.

Lellinger, D. B. 1972. Dichlorvos and lindane as herbarium insecticides. Taxon 21: 91-95.

Lundell, C. L., and R. Kirkham. 1966. A method of applying Mystox (lauryl pentachlorphenate) to protect mounted herbarium specimens. Wrightia 3: 1-7-180.

Millspaugh, C. F. 1925. Herbarium organization. Field Museum Technical Services 1(3): 1-18.

McNeill, J. 1968. Regional and local herbaria. In: Modern Methods in Plant Taxonomy. V. H. Heywood (ed.). Botanical Society of British Isles and Linnean Society. Academic Press. London.

Pohl, R. W. 1965. Dissecting equipment and materials for the study of minute plant structures. Rhodora 67: 96-96.

Porter, C. L. 1959. Taxonomy of Flowering Plants. W. H. Freeman and Company. San Francisco.

Rollins, R. C. 1955. The Archer method for mounting herbarium specimens. Rhodora 57: 294-299.

________. 1965. The role of the university herbarium in research and teaching. Taxon 14: 115-120.

Shetler, S. G. 1969. The herbarium: past, present and future. Proceedings of the Biological Society of Washington 86: 687-758.

________, et al. 1973. An introduction to the botanical type specimen register. Smithsonian Contributions to Botany 12: 1-184.

Smith, D. E., Jr. 1971. Preparing herbarium specimens of vascular plants. Agriculture Information Bulletin No. 348. Superintendent of Documents. Washington, D. C.

Stearn, W. T. 1957. An introduction to the "Species Plantarum" and cognate botanical works of Carl Linnaeus. Prefixed to the Ray Society facsimile of Linnaeus's Species Plantarum, 1. London.

________. 1971. Sources of information about botanic gardens and herbaria. Biological Journal of the Linnean Society 3: 225-233.

Stern, W. L. 1967. Index Xylariorum. Regnum Vegetabile 49. Utrecht.

________, H. B. Gillenwater, G. Eason, A. Garcia-Quintana, and R. S. Cail. 1968. Lindane and dichlorvos for protection of herbarium specimens against insects. Taxon 17: 629-632.

Stuckey, R. L. 1971. The first public auction of an American herbarium including an account of the fate of the Baldwin, Collins, and Rafinesque herbaria. Taxon 20: 443-459.

Turrill, W. B. 1964. Plant taxonomy, phytogeography and plant ecology. In: Vistas in Botany. W. B. Turrill (ed.). IV: 187-224.

Herbarium of the University of North Carolina

TEXAS
Bee Co.

Pyrrhopappus multicaulis DC.

Fine hard sandy loam, grassy roadside, eight miles northeast of Beeville at Medio Creek.

Flowers pale sulfur yellow, tinged rose-violet outside; anthers black (11 AM); common.

C. A. Lawson 928 28 March 1967

HERBARIUM OF THE UNIVERSITY OF NORTH CAROLINA

Det. Date

HERBARIUM OF THE UNIVERSITY OF CALIFORNIA

See pollen slide collection for permanent mount prepared from this specimen.

Prepared by: Date

PLANTAE EXSICCATAE GRAYANAE

1427. Sibara virginica (L.) Rollins

In Rhodora, xliii. 481 (1941) et in Contrib. Gray Herb. clxv. 135 (1947); Fernald, Gray's Man. (ed. 8) 723 (1950).

Syn. *Cardamine virginica* L. Sp. Pl. 656 (1753). *Arabis virginica* (L.) Poir. Encycl. Suppl. i. 413 (1810); Hopkins in Rhodora, xxxix. 80 (1937). *Cardamine ludoviciana* Hook. in Journ. Bot. i. 191 (1834). *Arabis ludoviciana* (Hook.) C. A. Mey. in Fisch. & Mey. Ind. Sem. Hort. Petrop. ix. 60 (1843). *Planodes virginicum* (L.) Greene, Leafl. Bot. Obs. ii. 221 (1912).

Waste places; Tobler Road, off Lowe's Ferry Pike, Knox County, Tennessee.

Coll. A. J. Sharp (no. 2444) May 6, 1947

ANNOTATION LABEL

Det. Ronald L. McGregor

Pressed by P. C. Hutchison;

field number ; cultivation number

Grown from imported plants of the above collection at University of California Botanic Garden, Berkeley.

UNIVERSITY OF ARIZONA HERBARIUM

BAJA CALIFORNIA, MEXICO

FLORA of QUEBEC, CANADA
The Gaspe' Peninsula
Herbarium of Keene State College

County

No.
July 1972
Collector, D. E. Boufford

PLANTS OF THE ROCKY MOUNTAIN REGION
Colorado

Alt. ft.

Reed C. Rollins, No.

Figure 31-1. Sample herbarium and annotation labels. (Actual size of the UNC label is 2 3/4 x 4 1/2 inches.)

Chapter 32. BOTANIC GARDENS AND ARBORETA*

The purpose of the general survey of botanic gardens as presented here is to give the student of natural sciences an appreciation of the heritage of botanic gardens, and an understanding of the important scientific and public services they provide today, as well as an idea of their potential for future service. Some of the material on the organization and development of botanic gardens may be of little value to the average student; but it is included here to give the professional botanist a feeling for the practical and philosophical problems basic to the operation of botanic gardens. Botanic gardens may be utilized by many different individuals and groups with general or specific interests, such as:

Botanists	Other Botanic Gardens	Artists and Photographers
Zoologists	Industry	Horticultural Societies
Ornithologists	Landscape Architects	Garden Clubs
Arborists	Teachers	Home Owners
Conservationists	Students	General Public
Nurserymen		

Each will be specifically concerned with perhaps only one aspect within the botanic garden program. The extent to which a botanic garden is able to fulfill each special interest will depend on many factors as will be pointed out in the following pages.

Strict definitions of arboreta and botanic gardens are difficult because of their varying degrees of development and overlapping purposes and functions. Arboreta and botanic garden definitions in this chapter are extracted from the National Recreation and Park Association Bulletin No. 90, Arboreta, Botanical Gardens, Special Gardens (Bunce, 1971) which was prepared by professionals in the field of botanic garden structure and development.

1. Arboretum. "An example area set aside for the growing and effective display of all the different kinds of woody ornamental trees and shrubs, vines, and other plants which may be grown in a given area, their maintenance, proper labeling and study" (Wyman, 1959).
2. Botanic Garden. "An institution organized to maintain plant collections, usually representing a large number of genera and species...to serve educational, aesthetic, scientific or economic purposes. Recently, botanic gardens offer a source of recreation" (from Fred Widmoyer, Secretary-Treasurer, American Association of Botanical Gardens and Arboreta contributing to American Encyclopedia, 1967).

Maintained plant areas include many other types of developments such as private estate gardens open to the public, recreational parks, historical restorations, public and commercial display gardens, and even landscaped highway right-of-way borders and median strips. These various developments are generally designed to provide aesthetically pleasing surroundings. Arboreta and botanic gardens assume the additional responsibilities of education and research. Since they are frequently quite similar in design and function, they will be treated as one and the same in this present discussion (with the exception of comments concerning the historical development of the botanic garden). Botanic gardens and arboreta share the following basic objectives:

*Contributed by J. Kenneth Moore, University of North Carolina, Chapel Hill.

1. Display of plant material for the general public.
2. Introduction of new plants into cultivation.
3. Plant exploration of remote natural regions.
4. Scientific investigations in such areas as plant breeding, hybridization, and taxonomy.
5. Development and maintenance of an herbarium.
6. Development and maintenance of a library.
7. Development and maintenance of various research laboratories.
8. Maintenance of representatives of the entire plant kingdom from the poles to the tropics grown outside or under glass.
9. General advancement of botanical studies and horticulture.
10. Conservation of natural areas for ecological studies.
11. Establishment and maintenance of gene pools.
12. Dissemination of information gained through plant sciences.

I. HISTORY

W. T. Stearn (1971) has summarized the significant sources of information relating to plant collectors, herbaria, and botanic gardens. The general historical surveys of the development of botanic gardens by A. W. Hill (1915) and Stafleu (1969) are the foundation for the following summary account.

A. The Earliest Botanic Gardens. Though gardens existed in ancient Egypt and Mesopotamia for the growing of herbs, food plants, ornamentals, or for pleasure and as status symbols, they cannot in the proper sense be designated as botanic gardens, that is, a garden in which plants are collected and maintained for scientific purposes. Bearing in mind the primary function of science and education, the first botanic garden may be considered to have been that of the "father of botany," Theophrastus, whose garden, attached to his school, the "Lyceum," near Athens, was bequeathed to him by his former teacher, Aristotle.

The Romans utilized small gardens as sources of medicine and aids to medical studies. The first such gardens in Europe are attributed to the necessity for medicine and drugs. The medieval monastic gardens originated in the late 8th century during the reign of Charlemagne, who assigned the specific task of medical training to the monasteries. The typical monastic garden included the "hortus," for vegetables and fruit, and the "herbularis," for various herbs, the latter being the precursor of the physic gardens which were often established in connection with the medical faculties of the universities during the 16th and 17th centuries.

The credit for the establishment of the first modern botanic garden belongs to the Italian, Luca Ghini (ca. 1490-1556), a professor of botany called from Bologna to Pisa in 1543 or 1544, at which time he immediately began planting a botanic garden connected with the university. Other university botanic gardens immediately followed at Padua and Florence in 1545; all three of these gardens were aided by the patronage of the Medici family. The establishment of other important gardens followed in succession: Bologna, Italy (1567); Leyden, Netherlands (1587); Montpellier, France (1593); Heidelburg, Germany (1593); Strasbourg, France (1619); Oxford, England (1621); Paris, France (1653); Groningen, Netherlands (1642); Berlin, Germany (1646); Uppsala, Sweden (1655); Edinburgh, Scotland (1670); Chelsea, England (1673); Amsterdam, Netherlands (1682); Vienna, Austria (1754); Kew, England (1760); Cambridge, England (1762); and Coimbra, Portugal (1773). Luca Ghini is also credited with having established the first herbarium; he taught scientific courses in plant taxonomy not limited simply to medicinal herbs.

Though the Pisa Botanic Garden does not exist today, records of its design demonstrate the geometric arrangements of plantings originating with the monastic gardens and characteristic of many continental gardens up into the present century.

The history of botanic gardens logically falls into distinct periods coincident with world exploration and subsequent introduction into Europe of plants from distant lands (Stearn, 1971, Stafleu, 1969).

B. European Period (to 1560). Plant collections of this time were principally of indigenous European species and those collected from southern and southeastern Europe and the adjacent Mediterranean lands, including Egypt. Some species had to be maintained in tubs as pot-plants. The pot-plants were kept in "cubicula tepida" during the winter; these glass-houses were simple rooms with windows facing south, another important innovation attributed to Luca Ghini.

C. The Near East Period (1560-1620). This oriental period of plant introductions from southeastern Europe and adjacent Asia may be characterized by a few groups spectacular for the fragrance and the brilliant colors of their flowers such as hyacinths, tulips, and lilies. The Flemish-Austrian botanist and father of the bulb industry, Carolus Clusius, an inquisitive natural scientist, botanical traveler, scientific horticulturist, and plant taxonomist, was very interested in this group of introduced plants. His study and travels carried him through the Netherlands, Vienna, Frankfurt, Leiden, Hungary, Italy, and Spain; all the while he was engaged in obtaining and describing new plants. Clusius became a dominant figure in botany during the last half of the 16th century, and as director of the botanic gardens at Vienna and Leiden, he greatly influenced the development of botanic gardens in Europe. By the end of the Near East period the center of plant activity had shifted from Italy to Austria, Germany, and the Netherlands.

D. The Period of Canadian and Virginian Herbaceous Plants (1620-1687). During this period the center of botanical activity turned to France, coincident with French exploration following the decline of Spanish sea power. While the Dutch were involved in the tropics and the African Cape, and the English with their Virginia territories, the French exploited the Canadian wilderness. Introductions into the botanic garden in Paris included such familiar plants as arborvitae, sumac and poison ivy, black locust, and perennials such as Black-eyed Susan, Dutchman's Breeches, and Goldenrod. The botanic garden at Paris, known as the "Jardin du Roi" or "Jardin des plantes," was in no way a palace garden; it was an independent educational and scientific institution with a royal endowment, originally founded as an establishment to promote the teaching of pharmaceutical botany. Presently its collections, particularly its herbarium and paleobotanical and paleontological departments, are among the world's best; it is the oldest and most important non-university botanic garden still in existence. Some of the private gardens of England received their first North American plants by way of French introductions. A few decades later private gardens of England, such as that of the Tradescants at Lambeth, had early Virginian introductions which included red maple and tulip poplar, thus marking a second phase of this botanical period. The introduction of North American species to the Netherlands from Dutch settlements began a third phase.

E. Cape Period (1687-1772). Dutch sea traffic around the Cape of Good Hope during the golden age of Holland is responsible for the most characteristic plant introductions of this period. By now glass houses and conservatories had become much more sophisticated so that the introductions of geraniums and numerous succulents could be better accommodated. After travel in South Africa and India, Paul Hermann, a German immigrant, was appointed professor of botany at Leiden, where he was instrumental in the great increase in the plant collection at the Leiden botanic garden. In Amsterdam at about the same time, 1682, Jan Commelin was appointed to create a botanic garden. The efforts of these two botanists stimulated a strong interest for plants in the Dutch sea merchants which resulted in vigorous and enthusiastic plant collecting. Linnaeus completed his botanical training in Holland, utilizing the latest collections which often contained plants representative of families previously unknown to him. From 1735 to 1737 he was employed by George Clifford, a wealthy Dutch banker, to describe the collections in his own private botanic garden, De Hartecamp, which included a permanent herbarium now at the British Museum. With access to Holland's excellent botanical libraries and printers, Linnaeus was able to complete and publish "Systema naturae" (1735), "Bibliotheca" and "Fundamenta botanica" (1736), "Genera plantarum" (1737), and "Hortus Cliffortianus" (1737). The latter described the collections of Clifford's garden, many of which were introductions from the Cape. In 1741 Linnaeus became director of the botanic garden at Uppsala, Sweden. Prior to this time the garden had been allowed to deteriorate to such an extent that plant collections were down to fewer than 300 species. Through Linnaeus' recommendation, the garden obtained the services of the chief gardener of Clifford's estate, with the result that by the time Linnaeus was stationed at Uppsala, the garden was beginning to improve. Under Linnaeus' supervision plant collections had grown to over 3,000 species within seven years time (Blunt, 1971).

The Cape plants reached England almost exclusively through Holland, but during the course of the 18th century, direct introductions into England rapidly increased. By the end of this period the English gardens began to rival those of Holland; the gardens of both nations had surpassed the important German, French, and Italian gardens of earlier periods.

F. Period of North American Trees and Shrubs (1687-1772). Through the efforts of Lancelot Brown, English gardening experienced a revolution in the development of an open landscape style, free from the rigid geometrical formality which had persisted in the continental gardens from the monastic influence. Introductions of new woody species from North America gained prominence. Many of the introductions reached England through the efforts of Peter Collinson, whose private gardens at Peckam and Mill Hill were significant botanic collections. Collinson corresponded extensively with the American naturalist, John Bartram, who is credited with establishing the first botanic garden in America in 1731 on his estate on the Schuylkill River near Philadelphia. Bartram studied and propagated the native plants prior to sending them to his friend Collinson. Through Collinson many of the American introductions reached other gardens such as Philip Miller's at Chelsea, and the new gardens of Princess Augusta at Kew which were supervised by William Aiton.

The great botanic gardens of England were preceded by the establishment of several private gardens devoted to the cultivation of medicinal herbs and other plants of botanical interest. The establishment of the botanic garden at Oxford in 1621 was specifically for the purpose of cultivating "physic" plants and for the demonstration of principles in the study of botany. The Chelsea Physic Garden was founded in 1673 in London as the garden of the Society of Apothecaries. Sir Hans Sloane, who deeded the land for the Chelsea garden, served botanical study well when he stipulated that fifty preserved specimens of distinct plants from the garden be presented to the Royal Society of London each year until two thousand distinctly different types of plants were attained. The appointment in 1723 of Philip Miller as Head Gardner was an important event for the garden. Under his practical skill and botanical knowledge, many of the various plants newly acquired from all over the world were being grown and flowered for the first time in cultivation. Miller was also the teacher of William Aiton, the first curator of the Royal Botanic Gardens at Kew. Kew owes its establishment, in 1760, to the interest of Princess Augusta, Princess Dowager of Wales, who set aside a portion of the Royal Garden at Kew House for a physic garden. William Aiton was in charge of this garden for thirty-four years. After the death of the Princess in 1772, George III inherited the Kew property and united the gardens with those lying contiguously which had formed the gardens of the Palace of Richmond, thus producing the extensive area known as the Royal Botanic Gardens, Kew. Kew differs from other botanic gardens in that it has no connection with any university or educational organization as such, thus its usefulness has largely been with the economic aspect of botany, assisting and encouraging botanists, travelers, merchants, and manufacturers in various botanical investigations.

G. The Australian Period (1772-1820). Kew Garden's Sir Joseph Banks accompanied Captain Cook on his first voyage (1768-1771) marking the beginning of the Australian period of development of botanical gardens. Species of numerous unfamiliar representatives of southern hemisphere families of plants were discovered and introduced to England and from there to the continent. The tropical botanic gardens were developed during the Australian period. Great Britain is credited with establishing the first of these economic gardens (for the cultivation of spices and other commercial plants) on the island of St. Vincent, British West Indies, in 1764. The Calcutta Botanic Garden, founded in 1786 for cultivation of spices, is noted for potato cultivation and the introduction of tea, mahogany, jute, sugar cane, and the quinine-yielding cinchonas. The Botanic Garden at Buitenzorg in Java, Indonesia, founded in 1817, is noted for the cultivation of rubber and coffee. Other tropical gardens were subsequently established in Malaya and Ceylon during this period.

Botanic gardens lost the restrictive character of physic gardens and became more extensive in size and function, with the result that every well known university without a botanic garden began making plans for one. It was near the end of this period that true botanic gardens began to appear in North America. The Elgin Botanic Garden, established in 1801, accumulated a valuable plant collection and became the Botanic Garden of the State of New York in 1810. It was subsequently granted to Columbia College. Although it does not exist today, several of this garden's associates, Thomas Eaton, John Torrey, and Asa Gray, will be remembered as some of America's leading botanists.

H. Subsequent Periods.

1. Period of tropical glasshouse plants and hardy plants from Japan and North America (1820-1900).

2. Period of West Chinese plants (1900-1930).
3. Period of hybrids (1930 to present).

It was during the last three periods, and principally the last two, that American botanic gardens were established. Before 1900 the Missouri Botanical Garden or Shaw's Garden, St. Louis (1859), the Arnold Arboretum, Harvard University (1872), and the New York Botanical Garden, New York City (1891), were the major botanic gardens established. American botanic gardens and arboreta did not appear in significant numbers until after 1920. Presently there are more than 155 such gardens throughout America. In spite of the distinction between botanic gardens and arboreta, it is necessary to use the two designations interchangeably when referring to American operations, particularly because the arboreta, like botanic gardens, often boast greenhouses and displays of herbaceous plants.

II. ORGANIZATION AND DEVELOPMENT

This section pertaining to organization and development is reduced to outline form because it does not bear directly upon the significance of botanic gardens in taxonomy and related botanical research. The basic information presented here is included as an effort to demonstrate in general the total machinery necessary for botanic garden activities. Students of botany as well as plant researchers who utilize garden facilities and materials often fail to see beyond their own specialized interests. It is important that they be aware that their individual interests are only a small part of the total program essential to a botanic garden.

A. Personnel. Depending upon resources available, botanic gardens, whether public or private, are similarly structured as follows (Richmond, 1971):

Botanic Garden Board - - - - - - - - - and - - - - - - - - - - - - - Support Group
Governing Body

Director

Professional Staff	Clerical Staff	Labor Force
Assistant Director	Secretary	Nurseryman-Foreman
Education and Program Specialists	Clerks	Gardeners
Horticulturists		
Librarian		

1. The Board.
   a. Holds full authority for, and responsibility in, all matters pertaining to the operation.
   b. Establishes garden policies.
   c. Approves garden budgets.
   d. Receives and expends all garden funds.

2. Support Group.
   The support group may be parent to a trust fund established to receive and expend money for operational purposes or may be active in raising funds for garden use.
   a. Fiscal support for the garden.
   b. Establishment of a trust fund for garden use.
   c. Publication of newsletters and periodicals.

d. Volunteer work in the garden.
e. Public relations.

An active and vigorous support group is not only one of the most valuable assets a botanic garden or arboretum can have, in many cases it is essential to the funding necessary for survival of the garden.

3. The Director.
   a. Recommends policy to the Board.
   b. Implements the policies of and is responsible to the Board.
   c. Responsible for the staff's day-to-day operation.
   d. Plans further development for consideration by the Board.
   e. Prepares and submits budgets to the Board for its consideration.
   f. Maintains professional liaison with other botanic gardens, arboreta, and similar institutions.
   g. Promotes public relations through speaking engagements, the press, and other media.
   h. Procures funds through active solicitation of grants, contributions, and other monies from public and private sources.

4. The Staff.
   Carry out day-to-day operations and maintenance under the general supervision of the Director. Professional staff members also advise the Director in matters of their specialization.

B. Financing the Botanic Garden. With increasing costs and responsibilities, even some of the most liberally endowed organizations must seek additional operational funds from both public and private sources.

1. Government Sources.
   A number of Federal agencies such as The National Science Foundation, The Agricultural Research Service, Soil Conservation Service, and the Forest Service have funds set aside for special projects for which arboreta and botanic gardens may qualify. State funds aid through university budgets, special boards, special legislation, and in cases where instruction is a significant part of the program, through the State Department of Education.

2. Private Sources.
   a. Contributions. Varying amounts may be acquired from individual membership or actively solicited from large industrial concerns, particularly those based upon the extraction of nonrenewable resources.
   b. Philanthropic Organizations. Societies and Foundations active in advancing education, science, conservation, or cultural activities are logical sources.
   c. Bequests. Infrequently, substantial sources of funds are bequests from individuals who have had long and pleasant associations with the botanic garden.

C. Planning. Significant botanic gardens are based upon thoughtful prior planning. Individuals qualified in this area have published various master plans emphasizing planning before development (Wyman, 1970, and Longenecker, 1964). The Bulletin of the American Association of Botanical Gardens and Arboreta is the primary vehicle of distribution of information pertaining to developing and maintaining garden facilities. In addition to formulating the general purposes of a botanic garden, planning includes basic matters such as complete contour and vegetational mapping of available land areas, soil studies, initial trail, path, and road systems, nursery areas, and designation of existing plant groups to be set aside, undisturbed.

D. Maintenance. Maintenance, the single most important factor (aside from a good preliminary plan) is often overlooked and under-budgeted during considerations of botanic garden functions and activities. In addition to the more obvious mechanical efforts involving utilization of tractors, chainsaws, and various gardening tools, are such indispensible activities as:

1. Watering (including installation and repair of watering system).
2. Protective fencing.
3. Plant propagation.
4. Replacement of old or dead plants.
5. Mapping and construction of new trails and maintenance of old ones.
6. Maintaining records pertaining to all collections of plants.
7. Labeling of plants.
8. Soil maintenance; fertilizing, fumigations, sterilization, etc.
9. Police or other guard force.

E. Design. General designs of botanic gardens have been molded largely by fashion. The gardens of the continent developed before the 19th century mainly conformed to the geometrical arrangements dictated by the medieval monastic gardens. Later, garden design, though still confined within a geometrical framework, began to reflect the growing understanding of plant relationships. Emphasis was placed on groupings of similar types of plants or plant families such as found at Kew, or of various phylogenetic systems as exemplified by A. L. de Jussieu in the gardens of Trianon, France, at the close of the 18th century. This system of display design was imitated by others including de Candolle at Geneva. As world exploration opened up distant lands, display designs were modified to exhibit plants in geographical groups. These general design frameworks including the open English designs, represent a rich botanic garden heritage upon which American gardens are designed. Though small geometric designs are still present in tribute to our heritage, the general trend in the American design is an open landscape, which is logical, since until relatively recently space in America was not scarce. However, within the open landscape design are to be found the more traditional arrangements of plants grouped (1) as to type or genus as at the Arnold Arboretum or the National Arboretum, Washington, D. C., (2) according to geographical origin as demonstrated at the Strybing Arboretum and Botanical Garden, San Francisco, Calif., (3) illustrating plant phylogeny from primitive to advanced families as designed by the Arthur Hoyt Scott Horticultural Foundation on the campus proper of Swarthmore College, Swarthmore, Pa., and (4) more recently with emphasis on plant community or habitat arrangements such as developed at Rancho Santa Ana Botanic Garden, Claremont, California.

The following outline (Teuscher, 1933) demonstrates not only the opportunities for esthetic landscape design, but also the possible arrangements which facilitate various taxonomic or other botanical research based on living material.

1. Geographical or Ecological Groups. The emphasis is on ecology; an effort is made to reproduce in a natural manner and on an extensive enough scale to achieve examples of various characteristic types of vegetation.

2. The Flower Garden.
   a. The perennial garden, demonstrating the best selected varieties of perennials for continuously blooming borders from spring through fall.
   b. The annual garden, demonstrating the best varieties in annual plants.
   c. The water and bog garden, demonstrating water-loving plants such as water lilies grown in basins or in natural or artificial pools.
   d. The rock garden, demonstrating not only collections typical of the local area, but possibly plantings to illustrate the flora of distant areas, as for example, Asiatic or European alpines.

3. Woody Plants for Special Effect.
   a. The rose garden, for beauty, hybridization, or as a testing site for new varieties.
   b. The spring flower garden, grouping the most ornamental spring-flowering shrubs and trees perhaps in combination with spring-flowering bulbs.
   c. The autumn garden, grouping shrubs and trees outstanding for their fall leaf color and colorful fruit in combination with fall-flowering bulbs.
   d. The color-garden, grouping the best of trees and shrubs for display of fall foliage best exhibited against a background of evergreens.
   e. The winter color-garden, a collection of trees and shrubs which display conspicuously colored branches during the winter.
   f. The form or freak-garden, being a grotesque or humorous display of weeping, fastigiate, globe-headed, and dwarf forms of trees and shrubs.
   g. Collection of vines, perhaps demonstrating potential as natural fences or utilization with other types of plants.

4. Plants of Special Botanical Interest.
   a. The medicinal garden, featuring an ancient herb-garden or a demonstration of spice and poisonous plants.
   b. The economic garden, featuring a vegetable garden and demonstrating various grain crops, fodder plants, berry shrubs, fruit trees, fiber plants, and oil plants.
   c. The morphological or biological garden, demonstrating any designated plant function or design, such as vegetative propagation, protective mechanisms, pollinating designs, and seed dispersal.
   d. The taxonomic garden, concentrating on the demonstration of particular genera or families, or groups of families in a particular system of relationships. This and the preceding type display are perhaps best closed to the public except to students and interested persons with permission.

5. Special Displays.
   a. A desert garden.
   b. An evergreen garden.
   c. A hedge garden.
   d. A rhododendron dell.
   e. A flowering Crabapple collection.
   f. A fragrance garden for the blind.
   g. Others.

III. FUNCTIONS OF THE BOTANIC GARDEN

Activities of botanic gardens are many and variable, limited only by the policies of the organization, the talents of the personnel, and the resources of the facilities. These functions vary from general civic contributions to disciplined botanical research.

A. Public Service.

1. Social. Facilities of the botanic garden may be utilized for meetings of various plant-associated groups and service organizations. Many such organizations make their headquarters at facilities furnished by botanic gardens.
2. Cultural. If facilities are adequate, various lectures, art exhibits, or musical programs may be presented in the pleasing surroundings of a botanic garden. As a consequence of proximity to botanic gardens, community businesses and private homes are often more carefully landscaped and maintained.
3. Economic. The presence of a botanic garden enhances the community through its needs for personnel, goods, and services, and depending upon visitor levels, it benefits local businesses. Through cooperation with the nursery industry new ornamental plants of possible commercial value are introduced and through display of new plant material available in nurseries, gardens create a market for desirable ornamentals.
4. Recreational. Depending upon total structure of the garden program and particularly on the space available, limited areas may be open to the public for picnic and play areas. However, such activities are best restricted to public park areas and not included as a program of botanic gardens.
5. Special Events. Other activities may include flower shows, garden club projects, special or seasonal displays, and use of flowers for hospitals or local functions.

B. Education. Many botanic gardens are adjunct facilities to, or an actual part of, an educational institution. As such, the particular garden organization must provide courses or programs in plant related subjects, varying from formal graduate level courses for college credit to popular one-day, non-credit plant workshops. Most significant independent botanic gardens are organized to support their own plant related courses serving much the same group of people. General types of courses are listed below.

1. Graduate and undergraduate level botany and horticulture.
2. Short courses and workshops on many phases of plant study for the general public and professionals from other disciplines.
3. Nature study programs for school groups.
4. Environmental awareness programs for pre-school age groups.
5. Special workshops for teachers of natural sciences.
6. Classes and demonstrations for professional gardeners.

C. Conservation. With the increased pressure of population and development upsetting or destroying natural areas, botanic gardens are in a position of public and political influence not yet fully realized.

1. Acquisition of natural areas (habitat tracts).
2. Planting of representative native plants, cultivation of rare and endangered species, and preserving various portions of the botanic garden grounds in their natural state.

3. Various courses offered to the general public which stress the importance of natural areas.
4. Preservation of various ornamental forms and varieties which may be neglected by nurseries because of current public unpopularity.
5. Cooperative efforts with other conservation groups to lobby for particular legislation or governmental protection.

D. Research. A recent survey of botanic gardens and arboreta indicates an extensive range of botanic research being conducted at garden facilities and by garden personnel and associates (Chan, 1968). Only the most extensively researched disciplines are listed under each category; many others exist.

1. Biochemical Programs
   - Biochemical mechanisms
   - Amino acids, peptides, proteins
   - Plant Growth Regulators
2. Biophysical Programs
   - Climatology
   - Crop-Weather relationship
   - Micrometeorology
3. Ecological Programs
   - General Ecology
   - Biogeography - distribution
   - Ecosystems
   - Control (chemical and others)
   - Community or population dynamics
4. Genetical - Breeding Programs
   - Breeding, hybridization, testing
   - Cytogenetics
5. Morphological Programs
   - General anatomy - Morphology
   - Comparative
   - Cell tissue control
   - Embryology - development
6. Physiological Programs
   - Development - growth
   - General plant physiology
   - Frost Hardiness
   - Photosynthesis
   - Water relations
7. Taxonomical/Classification Programs
   - General taxonomy - classification (the largest single discipline)
   - Floristic Surveys
   - Phylogeny, evolution, adaptation
   - Experimental taxonomy
   - Chemotaxonomy - paleobiochemistry
   - Palynology
   - Numerical taxonomy
8. Forestry Programs
   - General
9. Pathological Programs
   - Disease control
   - Host resistance
   - General Pathology
10. Horticultural Programs
    - Cultural
    - Testing and evaluation
    - Propagation - techniques
    - Winter Hardiness
11. Field Crop Programs
    - Cereals
    - Forages

The following is a list of those programs which are most likely to be emphasized during the next decade:

1. Expansion of collections
2. Conservation
   a. Preservation of natural areas
   b. Conservation of biotypes or gene pools
3. Taxonomic studies of cultivated plants
4. Studies on deterioration of the quality of the environment
   a. Air
   b. Water
   c. Soil
   d. Relation of man to his environment
5. Analysis of vegetation dynamics by computer techniques
6. Species biology studies of native plants
7. Ecology and ecosystems
8. Native plant research (economically useful)
9. Breeding and evaluation of ornamental shrubs
10. Breeding and evaluation of shade trees
11. Plant introduction
12. Propagation techniques

13. Plant physiology
    a. Physiology of adventitious root formation
    b. Physiology of fire-retardant plants
14. Virology - resistance to virus infections
15. Pathology of ornamental plants
16. Plant geography
17. Ethnobotany
18. Phenology
19. Cytogenetics

As applied to urban vegetation, the results of studies within many of the disciplines listed above will significantly benefit future efforts in establishing pollutant resistant plant material on our public landscapes and highways.

E. <u>Herbaria and Publications</u>. Botanic gardens throughout the world which are active in research usually maintain herbarium collections or are closely associated with an active herbarium. The major botanic gardens distribute the results of plant investigations at their own facilities and elsewhere through scientific publication. Representative research botanic gardens which maintain herbaria are listed below with publications where applicable.

Garden and Important Collections	Publication
Arnold Arboretum, Harvard University, Mass.; Woody ornamentals	Journal of the Arnold Arboretum
Botanic Gardens of Indonesia, Bogor, Indonesia; Malaysian flora	Annales Bogoriensis
Botanic Gardens, Peradeniya, Ceylon; Flora of Ceylon	Ceylon Journal of Science, Section A - Botany
Indian Botanic Garden, Calcutta, India; Flora of India and neighboring countries	Annales of the Indian Botanic Garden
Jardim Botanico de Rio de Janeiro, Brazil; Brazilian flora	Rodriguesia
Main Botanical Garden, Moscow, USSR; Native flora of USSR	Proceedings of the Main Botanical Garden
Missouri Botanical Garden, St. Louis; Temperate and tropical American flora	Annals of the Missouri Botanical Garden
Montreal Botanical Garden, Canada; Economic, Native, and Alpine plants	Memoires du Jardin Botanique de Montreal
National Botanic Gardens of South Africa, Capetown; Flora of South Africa	Journal of South African Botany
Rancho Santa Ana Botanic Garden, Claremont, Calif.; Native flora of California	Aliso (Journal of the Rancho Santa Ana Botanic Garden)
Royal Botanic Gardens, Kew, England; Herbaceous, alpine, insectivorous, economic and succulent plants	Kew Bulletin
Singapore Botanic Garden, Singapore; Malaysian flora	Gardens Bulletin, Singapore

In addition to scholarly publications of restricted circulation, many botanic gardens publish popular journals or bulletins such as The Morton Arboretum Quarterly (Morton Arboretum), Garden Journal (The New York Botanical Garden), and Arnoldia (Arnold Arboretum) which bridge the gap between the public and professional botanists and plantsmen. Complete lists of botanic gardens and herbaria and associated research and publications are included in Index Herbariorum and the International Directory of Botanical Gardens.

IV. THE FUTURE OF BOTANIC GARDENS

Addressing the Geneva Symposium in 1968, Richard A. Howard, Director, Arnold Arboretum, discussed the botanic garden as an unexploited source of information. He singularly honored the achievements of the Jardin de Botanique at Geneva; the work of the three generations of de Candolles (Augustin-Pyramus, Alphonse, and Casimir), all intimately associated with the botanic garden as perhaps unexcelled in the history of botanic achievements. The founder and first director, Augustin-Pyramus de Candolle, was much concerned that the public should be participant in the development of the new garden. He energetically traveled and corresponded with botanists and travelers, constantly introducing new plants to his garden for his research and for the enjoyment of the people, giving publicity to new and rare plants thriving at the garden. While he apologized for the fact that the director was required to pay more attention to the general foundation and establishment of the organization than to the component plants, his own studies added considerable insight into the morphology of plants and the homology of their parts. His son Alphonse concentrated on economic plants and was particularly interested in observations on their anatomical structure. The grandson Casimir concentrated on monographic works stressing vascular bundle patterns in the stems and leaves. All of these studies utilized, in many instances, the very same living and preserved material maintained at the Geneva garden. The botanical heritage shown in this single institution, The Botanic Garden at Geneva, is exemplary of the botanic garden of the future.

Botanical research has been greatly aided by the establishment of the Plant Records Center (see Appendix for historical account and explanation). For example, plants from all locations of the world which have been cultivated in various botanic gardens and arboreta throughout the United States for many years have been neglected in several monographic treatments or anatomical-morphological studies because, through no fault of the researcher, the locations of particular plants or groups of plants are not generally known. The Plant Records Center can function in locating all the material available in this country for future studies. Other functions of botanic gardens for future research include the maintenance in pure form of primitive strains of known provenance (gene pools) to be utilized in the elucidation of the origins of cultivated plants. Special gene-banks and plantations of important agricultural and horticultural strains can be maintained. The garden operation can continue to serve the distribution of suitable strains of economic plants, and perpetuate and distribute rare and novelty type plants. Conservation should be a major responsibility of botanic gardens, whether it be the preservation of single species or a large scale operation of administering the protection of an extensive natural area containing an entire ecosystem as a "biological garden."

In addition to the purely scientific and conservation efforts supported by botanic gardens and arboreta, there is another yet not fully realized responsibility, as expressed by George S. Avery, Jr., former director of the Brooklyn Botanic Garden: public education. Courses designed to teach a wide range of plant related subjects to the general public may aid in broadening the base of American culture, particularly as the public is increasingly faced with more leisure time. Through public courses, botanic gardens and arboreta, in cooperation with the professional botanists and horticulturists associated with them, are in a vital position to educate and influence the thinking of the general population in areas of conservation and environmental concepts. Contemporary environmental necessity forces a major duty upon botanic gardens. Utilizing their staff and living plant collections, they must serve as interpretive centers for the mass of public citizens, demonstrating the significance of our natural heritage and of ecological principles which have been unheeded far too long. Two recent publications present good statements of the changing role of botanic gardens and arboreta in our modern world: The Urban Arboretum in a Time of Crisis--A summary of University of Washington Arboretum Symposium in 1972 published by the Arboretum Foundation, Seattle, Washington, and The Prospective Role of an Arboretum prepared for Holden Arboretum, Ohio, by the Institute of the Study of Science in Human Affairs, Columbia University.

The basic programs of botanic gardens will differ according to the sources of financial support, and personal and professional emphasis of the personnel instrumental in the operation. The result of all the variables will be that, although all botanic gardens and arboreta may not completely fulfill all of their responsibilities or functions, each should make significant contributions through specialization in one or more of its areas of potential.

APPENDIX

PLANT RECORDS CENTER (MacDonald, 1971)

In 1964, coincident with the establishment of the University of Tennessee Arboretum, a need was realized for the efficient retrieval of plant records information. In 1966 a formal research project, Electronic Data Processing of Plant Records, was approved by the University of Tennessee Agricultural Experiment Stations. The results of this research were reported in the April, 1966, issue of the Association of American Botanical Gardens and Arboreta (AABGA) Newsletter by Robert D. MacDonald, who had been instrumental in the research project and the development of the University of Tennessee Arboretum. At the request of Dr. Harold R. Fletcher, President of the International Association of Botanical Gardens (IABG) the research was reported by MacDonald before the meeting of the IABG during the XVII International Horticultural Congress held in Maryland in late summer of 1966. Enthusiasm for the potential of the study spread and by October, 1967, the American Horticultural Society, which had been considering a data processing center of its own, and the AABGA united their efforts and through the generosity of the Longwood Foundation a two-year pilot program was initiated.

Success followed and the American Horticultural Society Plant Records Center is now permanently located with the American Horticultural Society at Mt. Vernon, Virginia, 22121. Plant records of the following institutions were in data processing form by the end of 1971:

Denver Botanic Garden, Denver, Colorado
Fairchild Tropical Garden, Coral Gables, Florida
Longwood Gardens, Kennett Square, Pennsylvania
Plant Research Institute, Ottawa, Ontario
Scott Horticultural Foundation, Swarthmore, Pennsylvania
U. S. National Arboretum, Washington, D. C.

Records of other institutions already completed or in various stages of preparation include:

Los Angeles State and County Arboretum, California
Rancho Santa Ana Botanic Garden, Claremont, California
Santa Barbara Botanic Garden, Santa Barbara, California
University of California Botanic Garden, Los Angeles
University of California Botanic Garden, Berkeley
University of Minnesota Landscape Arboretum, Chaska
Planting Fields Arboretum, Oyster Bay, New York
Royal Botanic Gardens, Hamilton, Ontario
South Coast Botanic Garden, Palos Verdes, California
Strybing Arboretum, San Francisco, California
University of California Arboretum, Davis
University of Washington Arboretum, Seattle
Mt. Cuba Botanical Park, Wilmington, Delaware
John J. Tyler Arboretum, Lima, Pennsylvania

The objective of the PRC and its services centers on three areas.

1. The PRC is developing a central data bank to contain information about the living plant collections in arboreta, botanic gardens, and similar institutions. As such, PRC serves as a botanic garden records-management consulting organization, aiding in the development of new record systems and improving existing systems.
2. The PRC is developing a system to provide a means to manipulate, analyze, and evaluate statistical data about these collections.
3. The PRC will disseminate information to the scientific, professional, and amateur communities which draw on the botanical disciplines for information and materials. Special compilations will be available such as checklists and indices to important horticultural plant groups and a directory to plant sources. Presently, the PRC has 21 computer programs which generate reports, inventories, or data listings. As a research facility for scientific study and a practical aid for locating plant source material for agencies such as nurseries and horticultural research groups, the PRC will become more and more irreplaceable as its storehouse of information increases.

BOTANIC GARDEN AND ARBORETA LITERATURE

Avery, G. S., Jr. 1957. Botanic gardens--what role today? American Journal of Botany 44(3): 268-271.

Blunt, W. 1971. The Compleat Naturalist--A Life of Linnaeus. Viking Press. New York.

Brown, R. A., and R. D. MacDonald. 1971. The A. H. S. plant records center. Arboretum and Botanical Garden Bulletin 5(4): 108-117.

Bunce, F. H. 1971. Arboreta, Botanical Gardens, Special Gardens. Bulletin No. 90. National Recreation and Park Association. Washington, D. C.

Chan, A. (ed.). 1968. Arboretum and Botanical Garden Bulletin 2(4): 86-97. (A special issue on research programs).

Fletcher, H. R., D. M. Henderson, and H. T. Prentice. 1969. International Directory of Botanical Gardens. International Bureau for Plant Taxonomy and Nomenclature. Utrecht, Netherlands.

Fogg, J. M., Jr. 1970. The nature and functions of an arboretum. Arboretum and Botanical Garden Bulletin 4(2): 47-54.

Hill, A. W. 1915. The history and functions of botanic gardens. Annals of the Missouri Botanic Garden 2: 185-223.
Howard, R. A. 1969. The botanical garden--an unexploited source of information. Boissiera 14: 109-117.
Lanjouw, J., and F. A. Stafleu. 1959. Index Herbariorum. Utrecht, Netherlands.
Longenecker, G. W. 1964. Arboretum master plan. American Association of Boranical Gardens and Arboretums Quarterly Newsletter 61: 6-9.
MacDonald, R. D. 1971. A history of the Plant Records Center. Arboretum and Botanical Garden Bulletin 5(4): 107-108.
Pyle, R. 1937. Report on the Committee on Botanical Gardens and Arboreta. American Association of Nurserymen. Chicago.
Richmond, G. B. 1971. General comments on arboretum organization, financing, and administration. Arboretum and Botanical Garden Bulletin 5(3): 75-80.
Stafleu, F. A. 1969. Botanical gardens before 1818. Boissiera 14: 31-46.
Stearn, W. T. 1971. Sources of information about botanic gardens and herbaria. Biological Journal of the Linnean Society 3(3): 225-233.
Taylor, G. 1969. Opening address. Symposium International de Geneve. Boissiera 14: 23-29.
Teuscher, H. 1933. The botanical garden of the future. Journal of the New York Botanic Garden 34: 49-62.
Thompson, P. A. 1972. The role of the botanic garden. Taxon 21(1): 115-119.
Tsitsin, N. V. 1969. The system of botanical gardens of the USSR and its role in investigating vegetable resources of the country. Boissiera 14: 135-140.
Wyman, D. 1947. The Arboretums and Botanical Gardens of North America. Chronica Botanica, Vol. 10, No. 5/6.
________. 1959. The Arboretums and Botanical Gardens of North America. Arnold Arboretum of Harvard University. Jamaica Plain, Massachusetts.
________. 1970. How to establish an arboretum or botanic garden. Arboretum and Botanical Garden Bulletin 4(52): 52-60.

Chapter 33. INFORMATION SYSTEMS AND DATA BANKING*

Biology can answer many complex questions about nature, but the simple questions often confound the scientific machine. How many species? Where do they grow, and what are their habitat preferences? When do they flower and fruit? Which ones are rare and endangered? Such questions usually are unanswerable, not because the basic data never have been collected, but because the data are locked up in libraries and museums and cannot be compiled and evaluated in a timely, economical manner. The means of storing and retrieving new findings have not kept pace with discovery. Furthermore, man by nature has always seemed to derive more pleasure and challenge from the process of discovery itself than from the process of synthesizing the results of previous discoveries.

Today's biologist, faced with the staggering overburden of historical documentation in the world's libraries and museums and with the annual avalanche of new publications, specimens, and other documents, is tending more and more to ignore the existing information store and instead to go back to nature for direct answers to his questions. The risks and penalties of rediscovery and duplication are proving more tolerable than the crushing burdens of dealing with the information retrieval problem. Yet mankind cannot afford to keep duplicating scholarship and documentation generation after generation. Our libraries and museums already are filled to overflowing, and the world that we seek to discover is being consumed or destroyed in the very process.

Nowhere in science is the problem of redundant scholarship and documentation more keenly realized or the need for better means of retrieving existing information more evident than in systematic biology, in this case plant systematics. The plant systematist must deal with an enormous library of Floras, monographs, bibliographies, indexes, journal articles, and other documents which often overlap to the point of repeating large amounts of the same information word for word. This is especially true of descriptive works like Floras. He must also contend with the millions of specimens in the world's herbaria, many of which carry basically redundant information.

The computer offers a breakthrough in information technology to rival the invention of the printing press in the 15th century. From commercial banking to law enforcement, the "machine" already has pervaded our daily lives and begun to make an impact on scientific research and communication. Yet the computer revolution has only begun.

Computer processing, from the applications point of view, is of two types: (1) <u>numeric processing</u> (<u>computing</u>) and (2) <u>word</u> (<u>information</u>) <u>processing</u>. Broadly speaking, the term (<u>electronic</u>) <u>data processing</u> (<u>EDP</u>) covers both types. (The term <u>automatic data processing</u> (<u>ADP</u>) may refer to mechanical and/or electronic data processing, but often is used synonymously with the term EDP.) The <u>computer</u>, as the name implies, achieved its widest usage initially as a computing machine. As machine memory capacity expanded, word-processing capacity expanded, and the science and technology of information storage and retrieval developed. Today's big machines are used more and more for information processing.

*Contributed by Stanwyn G. Shetler, Department of Botany, Smithsonian Institution.

In the late 1950s a few taxonomists began to use the computer for numeric processing at least in a limited way, and now it is used widely in systematic biology as a tool for statistical and classificatory studies. Its usefulness as a numeric processor was at once obvious to biostatistically oriented taxonomists. The new subdiscipline of numerical taxonomy quickly developed as a logical outgrowth of the biostatistical application. Formally heralded in 1963 by the appearance of the Sokal and Sneath book, Principles of Numerical Taxonomy, this new approach offered the philosophical basis and practical methodology for using the computer to correlate numerical data on a large scale to analyze group structure and relationships. The goal was a more objective means of classifying organisms based on actual numerical values rather than subjective terminology. Numerous professional papers and several books have been published. Today numerical taxonomic methods, which initially encountered fierce opposition at both the philosophical and methodological levels, are generally accepted and are being adopted as part of the new orthodoxy of systematics; the methods now are being used routinely in many standard systematic investigations.

Information processing has developed more slowly than numeric processing in taxonomy despite the headstart given the former by the pioneering taxonomic applications of Sydney W. Gould in the late 1940s and early 1950s when the computer was brand new. The situation in taxonomy, however, merely reflected the situation in computer processing generally. Users of the computer had to resolve major problems in designing compact file structures and efficient processing strategies. Computer technology, especially with respect to memory capacity, probably was less limiting than the slow-to-develop expertise of the users. To overcome the limitations, the pioneers in word processing resorted to extensive numerical coding of information. Perhaps for this reason more than any other Gould's laudable efforts were premature and unconvincing to the worldwide taxonomic fraternity as a whole. Taxonomists, no less than others, exhibited the typical human aversion to the substitution of numbers for words and generally refused to accept the computer as a tool as long as this appeared necessary.

During the mid-1960s, David J. Rogers and his associates made the first attempt to develop a generalized computer program for handling taxonomic data. His program package, TAXIR (TAXonomic Information Retrieval), is being used for an ever-increasing number of applications, some completely outside of biology.

The most ambitious effort thus far to process taxonomic information by computer has been undertaken by the Flora North America (FNA) Program, organized in late 1966 by the American Society of Plant Taxonomists under the auspices of the American Institute of Biological Sciences and headquartered at the Smithsonian Institution. Just as numerical taxonomy was the logical, inevitable application to systematics of the numeric processing capability of the computer, the current data-banking developments epitomized by FNA are the logical, inevitable application of the information processing capability.

The purpose of this chapter is to consider the principles and elements of information systems and data banking, especially computer-based systems, as they apply to plant systematics. The reader who is interested in further information should consult the general references cited at the end of the chapter and particularly their bibliographies which list numerous papers and books.

I. DEFINITIONS

The term information system (i.e., information storage and retrieval system) has various meanings, and its usage must be defined. Although it has gained special meaning in the context of EDP and nowadays tends to be used exclusively in this context, the term does not mean implicitly that a computer is involved. In fact, it may be applied quite aptly to conventional schemes for storing and retrieving information.

An information system may be defined in general terms as a total strategy and methodology for gathering, storing, retrieving, and exchanging (interchanging, transferring) information to serve a specified purpose or set of purposes. Included is the concrete organization of people, procedures, facilities, equipment, media, and data required to perform the functions of the system routinely. At the core of the system is the data bank, the collection or library of files of data created and maintained to support the information functions of the system. Usually the data bank is such an integral part of the system, fundamentally shaping and governing the system, that it is impractical if not meaningless from an operational point of view to speak of the system or the data bank apart from one another. In fact, the term data bank as used in common parlance today not only denotes the data but also presumes a computerized system to process the data; in many contexts, therefore, the term has come to mean information system, as here defined. Conceptually, nevertheless, the data bank should not be confused with the other elements of the system.

An information system may be as simple as a desk-top card file or office filing cabinet, or as complex as a national library system or weather data network. It may be specialized to serve a single user or category of users, or generalized to serve many purposes and users. The more generalized the system, the greater is the flexibility to exploit it for new uses or users as they appear, but also the less responsive it can be to the specific needs of the individual user. Customizing a system at the outset before the uses are fully defined and the design is fixed through actual experience with the system may prove unnecessarily costly and inefficient in the long run. While a system is being evolved, the builder usually cannot afford to keep tailoring it to the needs of the moment and should be satisfied to adopt reasonably generalized procedures until the system's specific purpose and function stabilize. Even in the long run he may be well advised to stay with a general purpose system that will keep his options open and optimize the efficiency and costs of storage and retrieval over several to many applications. As a general rule, an information system should be designed to answer whole classes of questions, not particular questions. On the other hand, a general purpose system can become so generalized that it serves the purpose of no one in particular, and there are many circumstances in which the use of a system tailored to answer specific questions constitutes the only practical alternative. The developer of a system always is confronted by a spectrum of opportunities and limitations, and charting a course to follow is a subtle and complex process. In actual practice, he strikes the best compromise possible between the idealized goals and the real resources and circumstances.

The process of operating the information system to create or enhance the data bank, exercising all the input functions of the system including review and editing, may be defined as data banking. Those who store and/or retrieve data may be defined collectively as the user community. Either storing or retrieving exercises some of the functions of the system and therefore constitutes "using" the system and its data bank. If it be necessary or desirable to distinguish between persons storing and persons retrieving, those who store data, i.e., add to the data bank, may be called contributors or, collectively, the contributor community, and the term user community can be reserved for the users in the strict sense, i.e., those who retrieve and use the data.

The terms "data" and "information" have similar connotations in many contexts and often are used interchangeably, but they are not synonymous. Let us define the terms for our purposes as they frequently are used in computing circles. Data are the individual measurements, observations, etc., expressible either as words or numbers, that constitute the factual elements of knowledge. Information is organized and evaluated data--data in context that convey a message. Information is a function of context, therefore, and the same data organized different ways may convey different information. Thus, by this distinction, data are compiled, whereas information must be authored, although authorship may be as limited as organizing a set of questions or question-defining parameters to query the computer and then editing the output to constitute evaluated, coherent answers to the questions. From the computer processing point of view, precise usage of the terms data and information depends on whether one is focusing on the data as elementary facts or as organized and evaluated blocks of facts, but often making a distinction is unnecessary and unduly pedantic, either word sufficing to convey the practical sense.

II. CONVENTIONAL SYSTEMS

Modern "computerized" (computer-based) systems largely are logical extensions of conventional systems man has developed and perfected through long use.

The human mind with its phenomenal memory and powers of correlation and evaluation is nature's own, most fantastic computer and information system. Since man first learned to write, however, he has turned his ingenuity to devising schemes for extending his own mental capacity to record and recall information of use to him. Modern man is surrounded in everyday life, privately and publicly, by countless systems for storing and retrieving information. Birth records, health records, job records, property records, school records, military records, banking records, shopping records, credit records, police records, library records--these are only a few of the most important files of vital statistics for which he has devised information systems.

Through the long course of human history, civilized man has evolved and institutionalized two great, complementary systems for storing and retrieving his accumulated wisdom--his record of history and scientific and cultural achievement. These two colossal systems or institutions, which epitomize civilization and subsume all other lesser systems, are the library and the museum. He has evolved the first to cope with his written and, more recently, spoken record and the second to cope with his material record, including his artifacts and his voucher specimens of nature. The pictorial record--illustrations, photographs, paintings--is divided between them. Each system documents and supports the other, and each is characterized by a vast data bank.

To the library, in this broad sense, belongs all of man's written archives, published and unpublished, including the innumerable files of vital statistics mentioned above, as well as his tape and disk recordings and many of his illustrations, photographs, and paintings. The library, as the repository for the world's literature and the system for exchanging it, symbolizes communication through publication as it developed in the 500 years since the invention of movable type. It comprises many systems within the total system. Every book or journal in itself can be viewed as a kind of information system, containing selected, organized, and evaluated data displayed and indexed in a particular format for fast, convenient retrieval. Customarily, the printed page is regarded in static terms as merely a source of information, but any volume also constitutes a means of communicating or transferring information and as such satisfies the criteria of a dynamic system.

Like the library, the museum as an institution is a hierarchy of systems. The functions are storing, curating, displaying, loaning, and exchanging collections of objects and specimens. Apart from mere curiosities, these collections not only vouch for the published record but also provide a vast reservoir of untapped data for future generations to study. There are many classes of museums, e.g., art museums, historical museums, technological museums, and there are museums within museums, such as herbaria (sing., herbarium) within natural history museums. The system also includes countless informal and private collections beyond the walls of the great formal institutions which constitute the visible museum establishment.

The library and the museum are as generalized as any systems man has been able to conceive. They exist for the benefit of society as a whole, and no individual could ever need or be capable of taking advantage of either system in its entirety. The particular needs of individuals are met through specialization within the systems. In the course of history, numerous special-purpose systems have evolved within the larger systems to accommodate the specific needs of the individual disciplines and branches of human thought.

Systematic botany, among the scientific disciplines, has developed its own share of special-purpose information systems through the years. New ones appear constantly as new needs are perceived, but few persist for long as measured by the centuries. The special prominence of historical scholarship in taxonomy gives a special prominence to the taxonomic libraries and museums (herbaria) where the essential objects of this scholarship--the literature and the specimens--are preserved, and to the systems devised for retrieving information from these repositories. A few of the more durable systems or categories of systems deserve mention here.

Of the many types of taxonomic publications that botanists have always produced, certain ones stand out in their significance as information retrieval devices: (1) Floras and taxonomic monographs, (2) literature indexes, and (3) name indexes and nomenclators.

The Flora brings together selected information about the plants of a given geographical region, while the monograph brings together the essence of what is known on a worldwide basis about the systematics of a given taxonomic group of plants. In both cases, the information is organized and formatted to answer a time-tested set of predefined questions, and effective use is made of a limited vocabulary, as standardized as is ever achieved in written communication. The binomial names are used as the ultimate addresses for the information, and these are grouped hierarchically according to a systematic classification. Diagnostic traits are presented in the form of dichotomous keys to facilitate identification of unknown plants. In short, as a class of publications Floras and taxonomic monographs constitute one of the most effective specialized information systems devised in science. There has been a perennial need for them.

Various systems for indexing and abstracting the taxonomic literature by subject have been introduced through the years, and, although old systems keep giving way to newer ones, out of sheer necessity some form of literature retrieval continues to survive. Notable among present systems, from the viewpoint of North American vascular plant systematists, are the Bibliography of Agriculture and Biological Abstracts, which cover the broad areas of agriculture (including botany) and biology, respectively, and have been running for many years; Excerpta Botanica, which abstracts the literature of plant taxonomy, distribution, and sociology on a worldwide basis; and the various bibliographic indexes being published by the International Association for Plant Taxonomy.

One of the perennial problems that the systematic biologist faces is to keep track of all the new scientific names that are published. For many years flowering plant taxonomists have been served by two key name-indexing systems which cite the original publications: Index Kewensis, which is produced in book form at the Royal Botanic Garden, Kew, England, and covers the world; and the Gray Herbarium Index, which is produced in card form at Harvard University, Cambridge, Massachusetts, and covers only the New World. In the realm of nomenclators, special mention should be made of J. C. Willis' A Dictionary of the Flowering Plants and Ferns, now in its eighth edition as revised by H. K. Airy Shaw. A nomenclator seeks to establish the acceptable names, which this Dictionary does for family and generic names. The Dictionary also includes synoptical descriptions and geographical distributions for all accepted names. For a single volume, Willis is perhaps the most valuable information retrieval aid available to the vascular plant systematist.

The herbarium, which traces its formal history as a specialized branch of the museum back more than four centuries, has always been an indispensable information system of the plant systematist. Upwards of 200 million specimens, including at least 35 million in the United States alone, have accumulated in the world's herbaria during this long period. Today, this immense storehouse of vouchers of the plant world is managed by a worldwide network of more than a thousand public herbaria, among them nearly two dozen institutions with 2 million or more specimens each and a dozen with 3 million specimens each.

The thousands of transactions that take place every day in the world's herbarium system sustain a multimillion-dollar enterprise of scientific research and public education. Every herbarium is in itself a filing system designed for rapid retrieval of information by kind of plant. Specimens are filed by their taxonomic names under the appropriate species, and the species are grouped by genera and families. The families are then arranged alphabetically or, in most cases, according to some well known systematic classification. Sometimes there is a secondary classification by geographic region within genera or species.

Finally, it should be noted that phylogenetic and other taxonomic systems of classification are themselves a type of metaphysical information system, because they enable the systematist to organize his knowledge for efficient retrieval, whether in his mind, on paper, or in the herbarium, by grouping like organisms and all information pertaining to them according to their intrinsic relationships, i.e., their shared characteristics. In another sense, of course, the taxonomic classification is not the system but a part of the system, namely, the scheme whereby the data bank is arranged or structured. Regardless of the emphasis, a taxonomic system of classification is an intellectually thrifty device for storing and retrieving information about plants.

III. COMPUTERIZED SYSTEMS

A. Philosophical Basis. For 500 years the printed page has ruled as the medium of information exchange. Now the high-speed electronic computer introduces new media and new methods to information handling, making a whole new mode of operation possible. The computer can free the written word from the physical limitations of the printed page. In the data bank of the future, all data can be equally accessible, and virtually any comparison, correlation, or combination can be made at will. Promising new ways of using data are possible, as, for example, in the use of computer programs to "converse" directly with the machine to identify plants by matching the characteristics of unknown specimens with the stored values of known species. By taking advantage of the machine's power to permute data and to afford multidimensional querying, the user can define a completely new functional relationship to his data.

Just as scholarly man first took to writing and then to printing, today's scientist is communicating with the aid of machines. As for the systematic botanist, surely concern for the future if not the present welfare of his science will impel him to exploit the computer fully as a tool for storing and retrieving literature and original data. Such exploitation must be done in concert with development in other scientific disciplines, of course, because the plant systematist's retrieval problems are not unique.

Today's taxonomist can hardly cope with the new literature of his own specialty as it appears; yet he continues to publish with abandon, casting his contributions into the swelling sea of literature in the hope that he has truly added to the store of knowledge. With every passing year, however, the chances diminish that his contributions will ever be found by those who need them. Likewise, no taxonomist can claim realistically that he is able to take full advantage of the mass of information locked up among the millions of specimens in the world's herbaria; yet every year taxonomists go into the field in ever greater numbers to collect more specimens for the stockpile.

If they are to continue to discharge their responsibilities to science and society, therefore, plant systematists urgently need better ways and means of keeping permanent documentation of what has already been discovered about organisms in a form that permits ready comparison with new data so that truly more and more is learned about them in an efficient manner as discovery continues. This calls for better ways and means of dealing with the world's 200 million herbarium specimens and deciding realistically about the needs for more specimens, and vastly improved means of retrieving and analyzing the literature. Using computer processing systems now in existence, systematic botanists can hope to develop a botanical data bank and information system to cope with these staggering problems. There also is hope of breaking out of the Linnaean taxonomic mold.

The present taxonomic system is an outgrowth of the Linnaean taxonomic system, which was an outgrowth of earlier folk taxonomies, and it now is quite inadquate for dealing with the hundreds of thousands of organisms already described, not to mention the millions still to be discovered. Linnaeus knew a much simpler world than we know today, and his taxonomic system was designed to handle a memorizable number of genera. The human mind can cope easily with a thousand different kinds (taxa) of organisms at any one time and tends to deal in generics (oak, maple, hickory), as seen in various folk taxonomies existing in the world today and in historical times. (Some would say that trained taxonomists can cope with upwards of ten thousand different kinds of organisms at any one time.) When knowledge about organisms is forced into a system that cannot cope with large numbers, it tends to get lost or dealt with artificially and superficially, with the result that the system of names itself receives undue attention. The computer can extend the taxonomist's capacity to remember the names and traits of organisms and can allow him, therefore, to deal with virtually any number of taxonomic units on an equal and objective basis and to store the information so that new constructs built from it will not destroy, distort, or obscure the information itself. Accordingly, the modern systematist can begin to examine taxonomic discontinuities at any level he chooses and bring them into sharper focus using the other modern methods of investigation. This should advance our very perception of the living world.

The Flora or other taxonomic treatise of the future will be a dynamic data bank written on the "pages" of a computer memory which are available via telecommunications to the entire world and from which can be printed in hard copy any number of subsets and permutations of data for individual or mass consumption. The very process of research and communication will be altered. Today research culminates in publication. Information retrieval, if any, comes after publication. Tomorrow the results of research will go directly to the data bank, and conventional publication, if any, will come as output from the data bank. Today the would-be user of scientific information must wait until the contributor has collected enough data to justify publication. Tomorrow the user will be able to ask the system at any time to compile the latest information on a given topic, and the contributor will be able to submit his data a datum at a time if he so chooses.

B. EDP Terminology. Any computer system has two major components: (1) hardware, and (2) software. The hardware consists of the computer itself and all associated (peripheral) equipment. Included in this equipment are data conversion devices and computer terminals.

A computer terminal is a typewriter or other device for transmitting data and/or operating instructions directly to the computer. A local terminal is wired to the computer and must be relatively near the computer, while a remote terminal is connected to the computer via telephone or other data transmission lines and can be located wherever the necessary telecommunications are available. Through the use of satellite transmission continents and oceans can be spanned. The common terminal is some type of teletypewriter. With increasing frequency typewriter terminals are being linked to display systems using a television screen (CRT: cathode ray tube) and even to mini-computers.

Data conversion (automation, reduction) is the process of converting (reducing) data to machine-readable form by mechanically punching holes in cards or paper tape or by electronically encoding on magnetic tape or a direct storage device (e.g., disk) associated with the computer. The most widely used data conversion device is the card punch (keypunch), but various paper tape and magnetic tape machines also are now widely used. Recently, optical character reading (OCR) and other optical scanning devices have come into use. Some of the newest data conversion devices are built to double as computer terminals. They can be used off-line, i.e., as self-contained units, to convert data, or on-line, i.e., connected to the computer, to convert or transmit data. On-line data conversion systems, which are more costly to operate than off-line systems, transmit data directly from the keyboard to a computer where the data are encoded on the magnetic storage deivces.

The software of a computer system is the total set of encoded instructions--apart from those "hard-wired" into the circuitry of the machine--required to run the computer and process data. Software which accomplishes functions such as equipment (hardware) allocation, job scheduling, and program compilation makes up the operating system, and individual programs of this software are known as system programs. Software designed to process user data is made up of application programs. Some application programs are generalized to make them usable on a wide variety of data and are controlled entirely or in part by options specified at run time in lieu of submitting specially coded instructions. These programs are attractive because they eliminate or greatly minimize program development for whole classes of applications. Frequently used application programs may be held, as the operating system, on storage devices that are connected to the computer's central processor at all times. Application programs tell the computer what and how to process in order to execute jobs. They are written in one of the standard programming languages, e.g., COBOL, FORTRAN, ALGOL, that the particular machine can interpret. Often the programs are encoded on cards or tape in the same manner as data and submitted with the data at run time.

The capabilities of the hardware are limited by the capabilities of the software, because the software controls the hardware and constitutes the means whereby its potential is realized. Software even of a routine application type rarely if ever is totally machine-independent and capable, therefore, of running without any modification on different computers of the same or of different manufacturers; however, programs running on a particular manufacturer's computer usually are more or less compatible upward through his line, i.e., from his smaller to his larger machines.

Modern third-generation computers, which can link to the latest telecommunications to support teleprocessing (processing by remote control), offer the user several processing options. In the multiprogramming mode, the computer can execute more than one job and handle more than one processing option simultaneously. The traditional option is batch processing, in which the user delivers his jobs on cards or tape to the computer center and then collects his results after the jobs have been run in their turn. This is the most economical way of processing, especially when large jobs are involved, and much computing is still being done this way.

A variation of the batch-processing option (or mode) is called remote batch processing or remote (job) entry (RJE), which allows the user to submit his jobs over a telephone line or other data-transmission system from a remote terminal, removed perhaps by hundreds or even thousands of miles from the processing center. The jobs enter an electronic queue at the center and are processed in turn. Turn-around on these jobs may be very rapid, and processing often is accomplished without anyone being directly involved at the computer center. Some computer systems permit the user to govern the speed of turn-around in part by assigning a priority to the job from a relative scale established for the system. To get faster turn-around he sets a higher priority but then pays a higher unit charge. The output can be directed back to the originating remote terminal, to a printer or other output device in the computer center itself, or to some other remote terminal or computing facility, as specified by the user. Both in batch and remote batch processing there is a discrete, though perhaps extremely short, time lag between submitting the job and receiving back the processed results. The computer is not executing jobs instantaneously, in so-called "real-time," under the control of the user, but sequentially or in order of priority as controlled by the computer operator or automatically by the system itself.

On-line or real-time processing is the ultimate option because it puts the computer under the direct control of the user himself, if only for a split second, and enables him to execute his job immediately. He can communicate or "converse" with the computer continuously as his job is being executed to guide the processing from step to step, adding or modifying instructions as necessary. This processing option often is called the conversational mode, for obvious reasons. Output can be obtained in any of the ways available for RJE.

Sometimes an entire computer system is dedicated to continuous on-line use by a single user, but most users cannot justify or afford real-time access to a computer system except on a shared-time basis. In the time-sharing mode, the computer shares its central processing unit (CPU) with several to many users simultaneously by allocating each user short bursts or "slices" of processing time in turn until his job is done. The process might be likened to juggling balls and keeping all in the air but one at any one time. This time juggling usually is so rapid and precise that each user gets virtually instantaneous response to his commands and has the illusion of exclusive and continuous control of the entire computer system. Speed of response degrades only when the system becomes overloaded with users, which may happen momentarily or for extended periods of time during the busiest part of the operating day or week. Time-sharing access to the computer ordinarily is via remote terminals.

The practical effect of the newer options that enable multiprogramming, real-time processing, and teleprocessing is to put distance, both figuratively and literally, between the user and the computer and to reduce his dependence on understanding the computer's operation, thereby making it possible for him to use the computer simply as a utility like the telephone.

C. Literature vs. Data Retrieval. Literature retrieval is the process of identifying and locating, by manual or automated means, books, journals, individual articles, or other documents within the total mass of published literature that are of interest to a particular user. The means may be analytical reference works, bibliographic indexes, card catalogs, or other retrieval systems. As it is used, especially in computing circles, the concept of literature retrieval does not necessarily include the act of physically retrieving the items, although this obviously is implied by the term. In the broadest sense, however, a literature retrieval system deals with all functions and processes required to transfer documents between author and user. Document retrieval is perhaps a more apt term because the word "document" covers unpublished as well as published material, including manuscripts, research notes, illustrations, films, tapes, and even biological specimens. Some would generalize even further by using the term document transfer because the word "retrieval" tends to emphasize only one aspect of the transfer process. The term information retrieval as used in most contexts refers to document (literature) retrieval. Although literature retrieval might be said to be the original and ordinary meaning of information retrieval, the latter term has come to have a general connotation since the advent of computers. Furthermore, this term, if taken literally, implies that the information itself is being retrieved when in fact the documents that contain the information or references to them are what is being retrieved.

Computerized document retrieval systems are now a commonplace, and they are multiplying rapidly. Obviously, this has been one of the most urgent areas in which to apply EDP. Anyone who has faced the task of searching the past literature or of keeping up with current literature understands the fundamental inadequacies of the present retrieval methods and appreciates the rush to computerization. The *Bibliography of Agriculture* is produced with a computer-based system. *Biological Abstracts* has been computerizing gradually, with the result that a whole new range of bibliographic services are being offered under the umbrella of BioSciences Information Service. Only the most tentative steps have been taken in the specific area of plant systematics. The Flora North American Program developed the "FNA Automated Bibliography" on a pilot basis, but for lack of funds this system was never implemented. Within the framework of systematic botany, it was designed to be a general purpose system, serving all of the taxonomist's needs from retrieval of current titles by several parameters, including author and subject, to retrieval by Latin name of up-to-date lists of new names, new distribution records and maps, new chromosome counts, etc. Many individual scientists have developed personalized systems, using local computing centers.

Data retrieval is the process of identifying, locating, and retrieving, by manual or automated means, specific facts in stored documents and specimen collections which are of interest to a particular user. Computerized data systems are still in their infancy, but they hold far greater potential than document systems as a tool for solving our modern information problem. They first demonstrated their utility in the intelligence area, and today no self-respecting intelligence or police agency is without its data system. In biology, data banking has just begun. Where document systems present the user only with references to the information desired, data systems deliver the actual data or elements of information. Obviously, the scientist who is harassed by too much literature to read will opt for a system that presents him with the precise data he needs in preference to one that gives him only references to these data. In an FNA survey of nearly 500 biologists, conducted in 1971-72, data retrieval was favored over literature retrieval by almost 2:1 as being the more urgently needed kind of information service today.

The principles of machine retrieval are basically the same for data systems and document systems because the elements of a document record are the same as data to the computer, but in practice the two types of systems present quite different operational problems, from data preparation and input to report generation. Consequently, different file designs and processing strategies often are used, and information retrieval specialists tend to be specialized in one or the other type of system but not both. The distinction is far from black and white, however, and many hybrid systems can be found. Indeed, the pilot FNA Automated Bibliography was a hybrid system, primarily for document retrieval but also including actual data, e.g., new chromosome counts, extracted from the contents of the documents recorded.

The remaining sections of this chapter are concerned only with data systems. The subject of document systems is being left to the numerous publications on literature retrieval both in and out of biology.

D. Components of Generalized System. A generalized representation of an information system is illustrated in Figure 33-1. This could be either a data system or a document system, depending on the contents of the data bank. It is only one of the many schematic representations that could be drawn, depending on the points of emphasis, and it does not purport to show all components, functions, and relationships of an operating system in their every detail. The schematic is simplified deliberately, and in particular it does not indicate all of the potential feedback points nor the basically circular nature of information flow through such a system. The flow of data through the system is from the sources at the left to the data bank at the core and from there to the users at the right who can pose direct questions to the data bank and receive direct answers. If contributors or other sources interact with the system, they in effect become users. Likewise, users who submit corrections become contributors.

According to this schematic, the contributors are the source of new research that cannot be compiled more or less mechanically from existing literature, specimen collections, or other information systems. The information specialists are the scientists and technical experts who provide the critical human interface between the machine and its processing system on the one hand and the contributors and users on the other hand. In this category there may be several to many types of specialists, depending on the size and complexity of the system--e.g., systems analysts, programmers, computer operators, data base specialists, data conversion operators, technical editors, scientific editors, managers. Management information specialists, who are trained and experienced in the whole process and know the characteristics and parameters of the data base thoroughly, manage the system and interpret its operations and its products and services to the "public," the contributors and users. Scientific editors are responsible for upholding the scientific standards of the data bank and must review and edit all data before they enter or leave the hands of the technical specialists. All technical and scientific functions may be performed by one or two individuals in the case of small systems. A more detailed representation of the data collection process is shown in Figure 33-3, discussed in the next section.

The "information processing system" in Figure 33-1 symbolizes all aspects of machine processing, including data conversion as well as what happens in the computer "black box." Thus in this broad sense the box in the chart represents both the hardware and the software. In the strict sense, however, the information processing system is only the software for performing the standard functions of file creation, file maintenance and updating, retrieval, and report generation. This is the technical meaning of the term "information system" in computing circles, but in contexts where there is likely to be confusion with the more general meanings of the term it is best to use the full expression "information processing system" to denote the software.

To qualify as generalized, an information processing system must be able to perform the standard functions enumerated above and in addition must meet these two criteria:

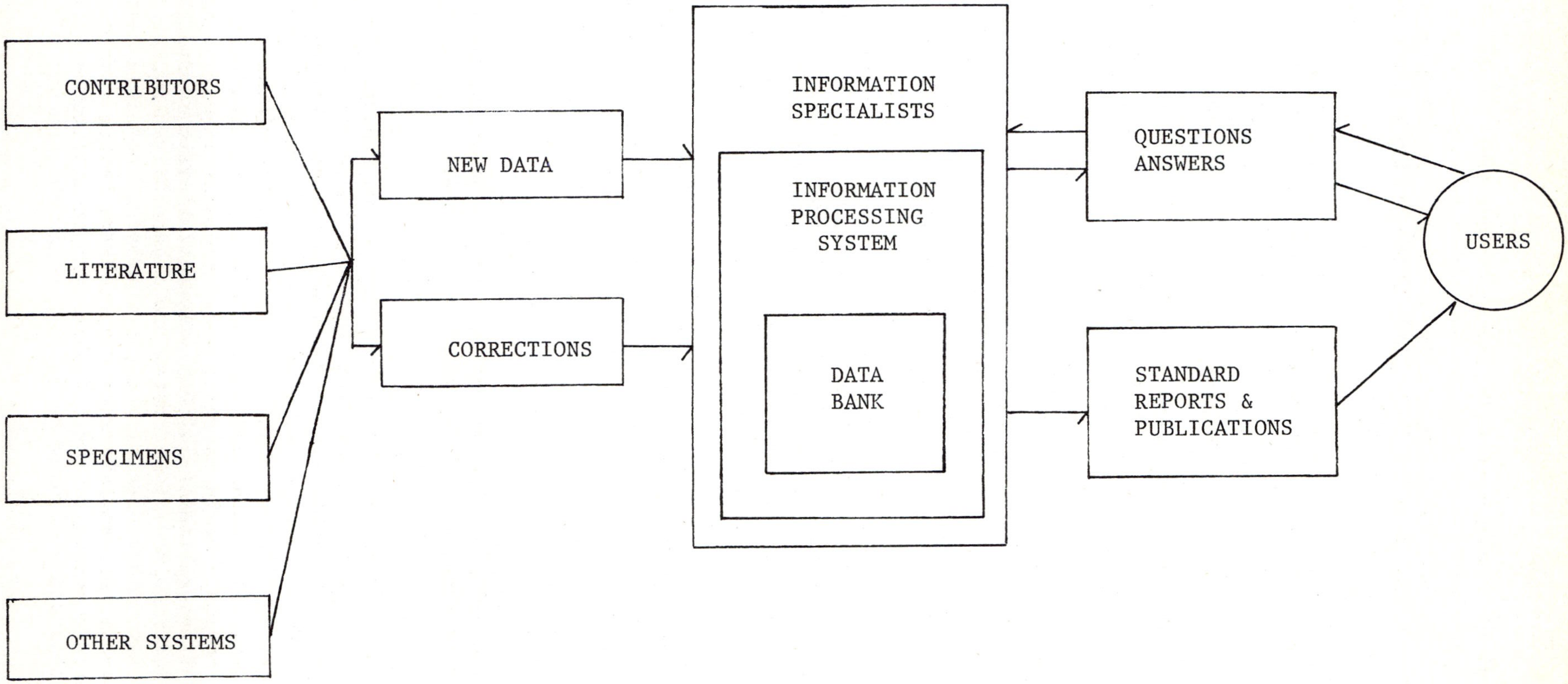

Figure 33-1. Highly generalized schematic representation of an information system.

1. The system must be characterized by "data independence," i.e., it must be designed without regard to any particular type or file of data.
2. The system should be controlled by a high-level language that is easy to learn and use (i.e., user-oriented) for defining processing requirements and thus for invoking and directing the functions of the system.

Other highly desirable features of a generalized software package for information processing are the following:

3. The system should stand alone as self-contained software, performing all functions without dependence on a host language such as COBOL.
4. The system should be capable of processing hierarchical data structure, e.g., multiple collections within a species or multiple specimens within a collection.
5. The system should be capable of multi-file search and retrieval, allowing for retrieval of related information from two or more files automatically in response to a single query.
6. The system should operate in both the batch and remote-entry modes but not necessarily on-line, although the potential for this is desirable.

The "data bank" at the heart of the system may contain not only the primary data files but also authority files and specialized application programs. Examples of data files pertinent to plant systematics are an index of chromosome numbers, a register of type specimens, a directory of specialists in the field, an herbarium catalog of specimens, and a species-by-species file of morphological, ecological, and geographical descriptions.

An authority file is an authoritative dictionary of names or terms for standardizing usage in the data bank. Examples are a checklist of species names, a list of author names and abbreviations, and a vocabulary of morphological or ecological terms. These files, in addition to providing guidelines for contributors, editors, and users, also become tools of the software which enable the system to edit automatically for conformity to standards during the processing of input or output. Any deviation from the standards in the dictionaries will be flagged for the attention of the editor or user. The authority file certainly is not a new invention, but the computer is a powerful tool for applying such a file consistently. Without properly maintained authority files, data banks soon get out of hand.

A good data bank will include a library of application programs to perform specialized operations on retrieved data. Examples are programs for computing statistics, performing numerical taxonomic analyses, mapping geographic distributions, constructing dichotomous keys, and identifying plants on-line. These programs should not be confused with the information processing system per se.

The aim of the retrieval strategy of any information system should be maximum user-accessibility to the stored data, and a generalized system can offer a wide range of products and services. The number of possible permutations of even a relatively few data elements can be astronomical. For every data bank, it is necessary to define the standard reports and formal publications that are to be issued according to some schedule for the benefit of the entire user community, but beyond this formal output the system can provide informal answers to an unlimited variety of personalized queries.

E. Systems Compatibility. Every effort should be made in developing a system to achieve a high degree of compatibility with other systems having similar objectives, but total compatibility is unattainable. To say that two systems are "compatible" is to say nothing unless the nature of the compatibility is defined, because there are so many elements to make compatible in any total system. Pushed to its extreme, compatibility would mean identical systems with identical data banks. As surely as the purposes of systems are different, the systems themselves must be different at some point. What is more pertinent to stress than compatibility is translatability. As long as the procedures followed allow for easy data translation from one system to another, the conceptual ideal of compatibility is achieved.

IV. STEPS IN DEVELOPING A DATA SYSTEM

A. Defining Purpose. The purpose(s) of a data retrieval system should be defined without regard for whether the system will be computerized. The computer tends to invest a system with an aura of significance and authority regardless of the system's intrinsic worth. It should always be kept in mind, therefore, that computerization in itself can never give worth to worthless data or purpose to purposeless functions. On the other hand, a computer may spell the difference between feasibility and unfeasibility and make a system worth building, not because it bestows purpose, but because it allows the purpose to be realized within the limits of the time and resources available.

The chief reason for computerizing any system should be to increase the efficiency and flexibility of access to the data, hence to enhance the effectiveness of the data bank, not to reduce the absolute cost or increase the absolute efficiency. Computerizing will not necessarily reduce the cost of operating the system or increase the efficiency of operation, as measured by the conventional products and services or the functions of the conventional operation. In fact, the out-of-pocket cost in all probability will increase, perhaps many fold, because new types of personnel (e.g., systems analysts, programmers, data conversion operators) are needed, and machine costs even if modest represent a new cash outlay. Overall system efficiency is likely to decrease, even dramatically, at first, because new procedures must be instituted. Many would-be builders of computerized systems have been led down the rosy path by promises of greatly reduced costs and increased efficiency which did not materialize. In the final analysis, the cost and efficiency of a computerized system must be measured in terms of products and services, by comparing the new array with the conventional array. This is a critical point especially because it means that cost-effectiveness cannot be determined before the system has been put into operation and the new use pattern established, which usually happens long after the case must be made for financial support to build the system.

Finally, it is important to keep in mind that the main purpose for taking a generalized approach is to have a system that will be capable of answering whole classes of questions of concern to many users. Systems designed to answer specific questions or a single set of predefined questions seldom justify the costs of computerization. The system should be measured by the types of questions it can answer. A well designed system will be capable of responding to new types of questions that were not anticipated when the system was first built. The layman who watches retrieval from a data bank for the first time often jumps to the conclusion that the system was built to answer the specific query he has posed, and not surprisingly he then may conclude that the result is not worth the cost. This is tantamount to concluding that the library system was developed for the express purpose of answering the specific question that one may be asking on a particular day and then judging the worth of the investment thereby.

B. Identifying User Community. It rarely if ever is possible to know the exact limits of the user community, because the primary users pass along information to secondary users, secondary users to tertiary users, etc. For this reason there is great danger in drawing conclusions about the value of a system from an analysis of the circle of primary users who, though more or less visible and identifiable, may form only a small fraction of the real user community. Thus, while a floristic or other taxonomic data bank is designed to serve primarily systematic botanists, the findings of the systematist sooner or later make their way, sometimes quite indirectly, into the handbooks and reference works of scientists and the public at large. Anticipated direct or primary use, nevertheless, is the only concrete basis on which to build a system. If primary uses and users cannot be identified, then presumably there is no need for the system. Despite the difficulty, therefore, of defining the primary user community, especially in advance of having had experience with an operational system and real users, some effort must be made to do this as a basis for designing the system and justifying its development. In most cases, a previously existing system is being computerized, and its pattern of use can serve as a preliminary guide to the potential new pattern of use.

As a general rule, the size of the user community will be directly proportional to the size and scope of the data bank and to the degree of generalization of the information service, in the same manner as the user community of a library or museum is governed.

C. Defining Data Bank. The natural tendency in developing a computerized system is to want everything in the data bank immediately. The old adage, "Walk before trying to run," certainly applies here. The first task in defining a machine-readable data bank is to cut the universe of possibilities down to a practical number. If, as is often the case, an existing manual system is to be converted to a computer-based system, then it is advisable to stick as closely as possible initially to the parameters of the current data base, resisting the urge to enlarge the number and variety of data elements. This approach of using the old data bank as the model for the new one takes maximum advantage of past experience in defining useful categories of data. If a totally new data bank is to be created, then it must be defined in careful detail at the outset, and subsequent changes must be made in an orderly way supported by thorough documentation.

A taxonomic data bank dealing with flora (or fauna) can be represented as a taxon X data-category matrix of data, as depicted in Figure 33-2. Every datum is keyed on one axis to a taxon, usually a species, and on the other axis to a type or category of data. The data category may be a general class of features, e.g., morphological features, or a specific characteristic, e.g., flower color. The data bank can grow in either direction indefinitely and thus is open-ended. The monographic approach to building the data bank is to

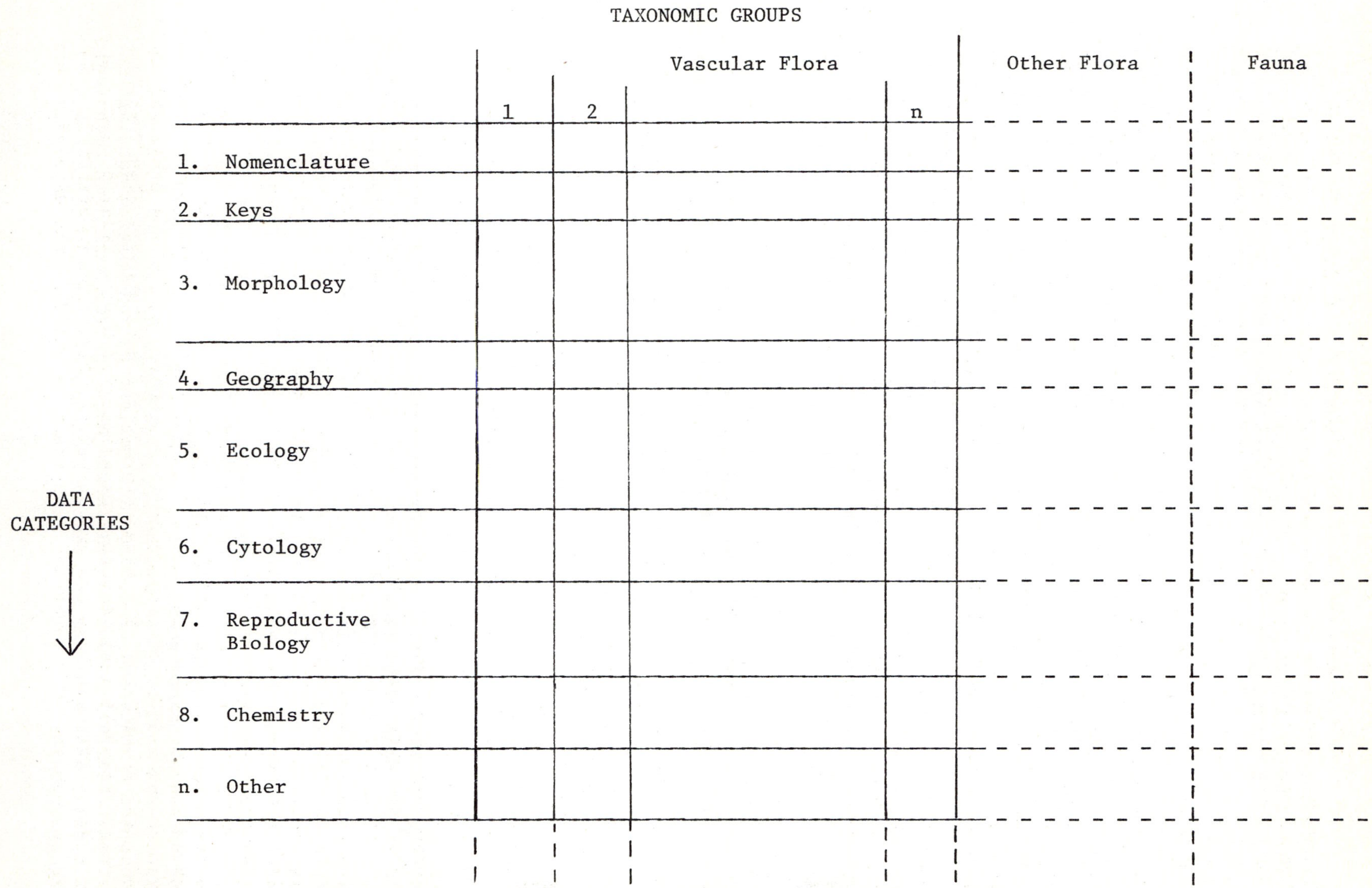

Figure 33-2. Matrix representation of a taxonomic data bank

collect data in many categories for a relatively few taxa, while the floristic (faunistic) approach is to collect data in few categories but for many taxa. It is important to decide at the outset whether the initial effort will be directed toward taxonomic or data comprehensiveness.

D. Collecting Data. The major functions of the data collection process, from the sources of the raw data to the publication of output ready for use, may be represented schematically as in Figure 33-3. Some data collection systems will be more, others less complex than the system depicted here. The major sources of data should be identified concretely before any data banking is begun. The collecting and editing procedures may differ significantly between small one-man efforts, such as in the computerizing of a single university herbarium, and large network efforts involving many people and institutions, such as in the FNA Program. In the first instance, one person is in total control and may be able to maintain consistency without establishing a closed system of standards at the outset, but in the second instance it is critical, because of the different parties involved, that all standards be fixed at the outset and then be held to firmly. In either case, no significant data collection should be started until a precise format, down to the punctuation, has been established. An example of the data collection form being used for the Type Specimen Register, a network system centered at the Smithsonian Institution, is shown in Figure 33-4.

Two types of editing are always required--content editing and technical editing. As in conventional publishing, editing for scientific content is necessary to control the reliability and authority of the data. The facts may be reliably collected but not authoritatively synthesized as scientific information by an appropriate specialist; contrariwise, an authoritative presentation may be riddled with unreliable data. The botanical editor, who is responsible for overall quality control, must keep these two aspects of the data distinct in his mind when setting up data collection and editing procedures. He will realize that the person who is best qualified to prepare an authoritative synthesis of data is not necessarily the one best qualified to collect reliable raw data for synthesis. Thus, although only a specialist can produce an authoritative taxonomic treatment, an experienced professional technician or research assistant may do a more thorough and accurate job of compiling routine specimen data and nomenclatural data from the herbarium and library, respectively. The botanical editor also will make sure that an appropriate scientific review by other specialists forms part of the editorial process in his data collection system.

The technical editor in the data collection system enforces the established formatting standards of the conversion and processing systems. After data have been processed initially, the machine records must be recycled to the technical editor and perhaps also to the botanical editor for verification, to ensure that the original data have been transcribed into the system faithfully, without typographical errors and in conformity with all standards of the system. Several cycles of editing and updating may be required before a "clean" data base is produced.

Once the formats and procedures for collecting and editing data are stabilized, then full documentation should be prepared, including technical specifications of the system and instruction manuals for the editors and operating personnel.

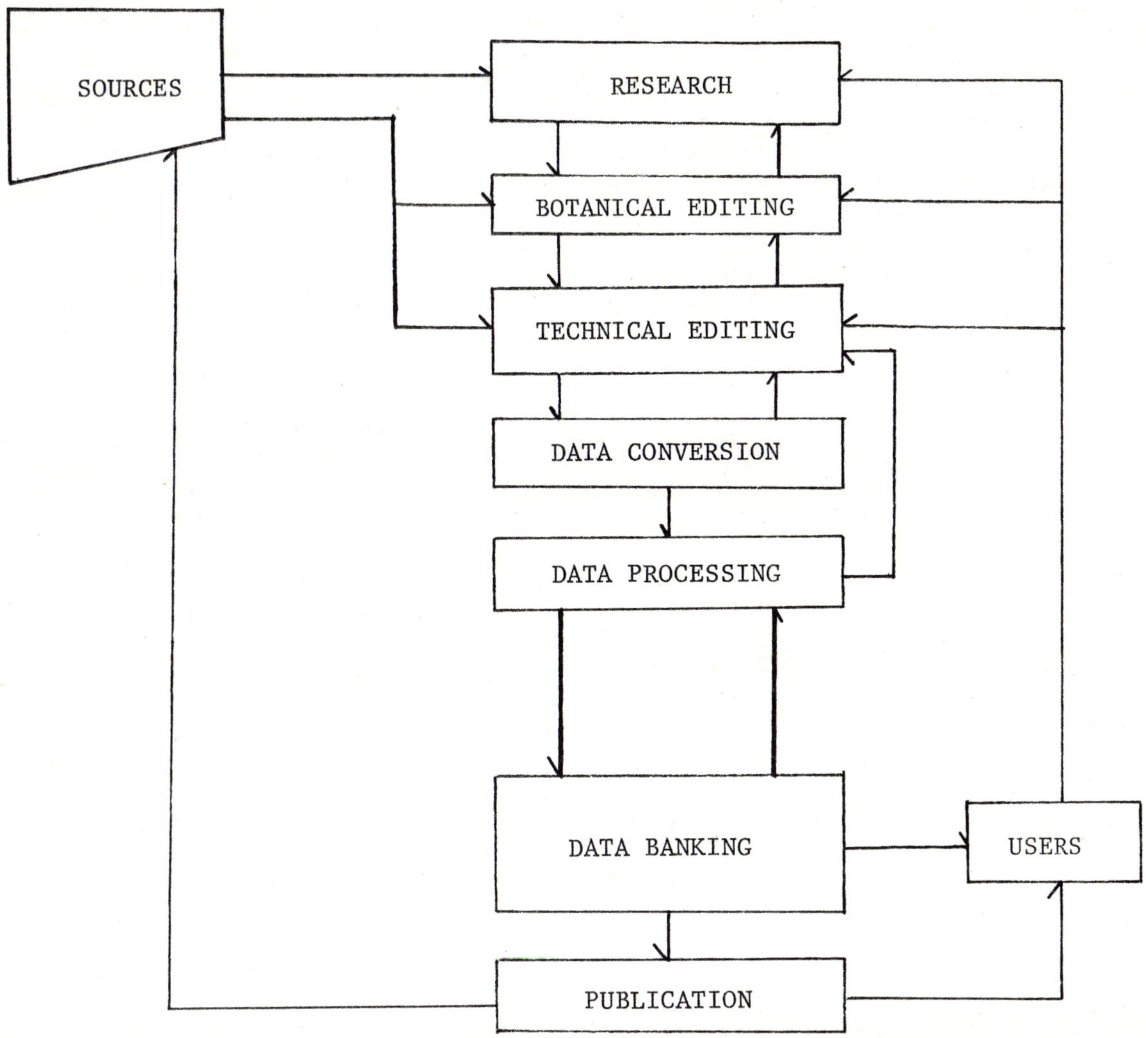

Figure 33-3. Schematic representation of a data collection system

Please type. Enter new names only. *Essential fields--information required.

1. FAMILY*____

2. GENUS* ____

3. SPECIES*____

4. INFRASPECIFIC TAXON____________________________________
   (Indicate rank: ssp, var, svr, for, sfm)

5. AUTHOR(S)*____

6. CITATION*____
   (Cite periodicals and serials according to standards of B-P-H.)

7. COLLECTOR(S)*____

8. COLLECTION NO.__________________ 9. COLLECTION DATE__________
   (Indicate whose series if not collector's series.)

LOCALITY:

10. COUNTRY*____
    (Use modern name and cite original as follows: Ethiopia ("Abyssinia").)

11. STATE, PROVINCE, DEPARTMENT, OR
    EQUIVALENT____

12. COUNTY OR EQUIVALENT__________________________________

13. TOWN OR LOCAL REFERENCE_______________________________
    (Place important words first and omit unnecessary words.)

SHEETS:	14. HERB. ACRONYM(S)*	15. SHEET NO(S).	16. KIND(S) OF TYPE(S)
1st	____________	____________	____________
2nd	____________	____________	____________
3rd	____________	____________	____________

17. REMARKS____
    (If more than 2 sheets, indicate to which sheet remarks apply.)

Note: For additional sheets, continue in "Remarks"; for additional collections (e.g., syntypes), continue on back.

Source of information_____________________________________

References checked:

B-P-H____ Gray Card Index____ Index Kewensis____ Other?__________

Information provided by:___________________________Date__________

Figure 33-4. Data collection form of Botanical Type Specimen Register.

E. <u>Converting Data</u>. Collected and edited, the data must be converted into computer-readable form. In selecting a data conversion medium and procedure, the botanist should seek the advice of a systems specialist where the data are to be processed.

Today there are many devices on the market for converting data, and there are several basic types of data conversion systems:

1. Card System: The tried-and-true, 80-column, keypunch card, commonly known as the "IBM card," is still the best medium for many applications where a limited number of data fields are involved and a physical record in the form of a card file also is desired. The card system works best when all data for a unit record can be punched on a single card. This permits easy filing on the one hand and inexpensive sorting by machine on the other hand. Valuable listings can be obtained without any computer programming by using a card sorter and then printing so-called 80/80 listings, i.e., printing column for column as given on the card. Before any computer processing can be done, however, even if only to reformat data or decode data on the printout, the cards must be read into the computer. Often cards are used only as an interim storage medium until the data can be copied onto magnetic tape or other magnetic storage.

    The chief limitations are the fixed number of 80 columns per card and the bulkiness of the card decks when they become large. Nevertheless, the punched card should be used more often by biologists. Like other scientists, biologists are accustomed to maintaining various types of card files, and such files can be made much more useful by punching them on machine-readable cards. It is easy to learn to punch and sort cards and to design simple computer programs for permuting the information more elaborately. Once the data are punched on cards the door is open for any type of information processing. Moreover, there are few data processing centers in the world that cannot read punch cards.
2. Paper Tape System: For lengthy unit records that would require several to many cards per record, as in most data-banking applications, paper tape is a relatively cheap and effective medium. Paper tape is analogous to cards attached in an endless string and thus without column limitation, and in fact this is the way the computer "perceives" a deck of cards. A specially designed typewriter with tape-reading and tape-punching attachments is used to encode the data on a strip of whatever length is required to include the record. Many individual records can be punched on a single reel of tape, as long as they are separated from each other by machine-readable codes, or individual records can be kept on individual pieces of tape folded or rolled for filing. In the latter case, usually the pieces are spliced together into large spools or the records are copied by machine onto continuous reels before processing, because generally it is more efficient to feed one long tape to the computer than to feed many short tapes.

    As with cards, the mechanical coding on paper tape is visible and can be read by an experienced user, but unlike with cards the codes cannot be "interpreted," i.e., typed in black and white on the top of the tape itself. Some paper tape devices can be programmed to punch automatically invariable data, such as the field names of a data collection form or the fixed information of a form letter. The operator needs only to enter the variable data from the keyboard.

Paper tape systems are useful not only for inputting data to the computer but also as office machines for automatically replicating specimen labels, manuscripts, letters, or standard forms. Each copy is in effect an original. Once a manuscript is on tape, for example, revisions can be incorporated by producing a corrected tape and then playing it back to get a corrected copy automatically, without manual typing.

A paper tape system is not without drawbacks. Some tape-punching typewriters are subject to frequent mechanical or electronic failures. In a large data operation, the tape, which in quantity is bulky, may be used in such volume as to create a serious handling and storage problem. Splicing or duplicating smaller pieces to produce larger reels, whenever this is necessary, can be a time-consuming nuisance in a data-banking operation. Paper tape tears or mutilates more easily than cards, and such tearing can be a frustrating problem, particularly if it occurs while the paper tapes are being read at high speed by the computer.

Despite its drawbacks, paper tape offers important advantages over cards in many applications, and it is widely used. Numerous examples could be cited of effective applications of a paper tape system of data conversion. Under many circumstances, in fact, a paper tape system is the only practical and, at least in terms of cash outlay, economical alternative to a card system. Because paper tape has been in use for quite a few years, industry has been able to develop sophisticated, programmable paper tape machines, which offer many features, e.g., tape-editing options, that can enhance a data conversion system. Programmable paper machines are especially adaptable for applications that require the production of some type of finished copy, e.g., specimen labels, in parallel with the conversion of data to computer-readable form. For such applications, there probably is no more flexible data conversion device on the market short of a terminal supported directly by a computer of some type.

Paper tape remains, however, a relatively cumbersome conversion medium, and the trend in the information processing industry is away from paper tape systems to systems of the types described below. Furthermore, when all elements of the system, including personnel, are considered, a paper tape system is not necessarily the cheapest. Whenever extensive editing is required during input, a computer-linked conversion system that offers text-editing features may prove cheaper in the long run if it allows for more effective personnel utilization.

3. Magnetic Tape System: Functionally, a magnetic tape system is analogous to a paper tape system except for the medium of conversion. Instead of being encoded as holes in paper tape, the data are electronically recorded on magnetic tape. Magnetic tape permits enormous data compression and thus reduces the handling and storage problem several orders of magnitude by comparison with paper tape. For further ease of handling, the tapes for magnetic tape typewriters are packaged in compact cartridges or cassettes, which can be reused many times. Thus, while cards and paper tape are cheaper storage media, they can be used only once, whereas magnetic

tapes, which are costly to purchase, not only allow for vastly greater data content on a given section of tape but also are reusable. In the end, therefore, the cost of the storage medium per unit of record converted for processing probably is about the same whether card, paper tape, or magnetic tape is used.

It is not advisable or practical, however, to archive data on the input cartridges or cassettes, because of their high unit cost and their uncertain life expectancy as a reliable magnetic storage medium. If the data conversion medium itself must be preserved for some reason, then it is better to use cards or paper tape, which, though susceptible to fire and mutilation, nonetheless preserve the data record in visible physical code rather than in invisible electromagnetic code. If records are to be stored indefinitely on magnetic tape, then they should be copied promptly onto standard computer tape, which is far more reliable as a storage medium than the tapes of the various input devices. In fact, a data conversion system, regardless of the medium used, should be designed to transfer the data from the input medium to standard computer storage devices in as short a time interval as possible.

One of the chief advantages of the newer magnetic tape machines is that they can be used off-line as stand-alone conversion devices or on-line as computer terminals. Although paper tape machines also can be built to function this way (e.g., the Teletype), magnetic tape terminals offer a faster, more flexible interface with the computer. Increasingly, in fact, magnetic tape devices are being marketed with built-in minicomputers, which provide the data conversion system with dedicated, programmable memory.

Magnetic tape systems are not without their drawbacks. Chief among these is the relatively high cost of purchasing or renting the equipment and the tapes. Operations may be quite expensive if the system is supported by a computer link. Some magnetic tape machines are just as susceptible, if not more so, to breakdown as paper tape machines. At least some magnetic tape devices on the market lose synchrony too easily, and they may not read tapes they have produced previously, or they may not read tapes produced on another machine of the same make. This problem adds to the hazard of storing data on the cartridges or cassettes for any length of time, because a perfectly good tape may read like gibberish on a machine that has gone out of synchronization. The tapes do wear out and cannot be reused indefinitely. Finally, in some applications it is a decided disadvantage to use magnetic tape rather than cards or paper tape because the coding cannot be read by eye.

All considered, magnetic tape systems almost invariably are preferred over paper tape systems by those who have used both. The best magnetic tape systems offer most of the advantages of paper tape systems and few of the serious disadvantages, and in addition they can furnish options that are difficult if not impossible to provide with paper tape systems. For many applications, the logistics of producing and processing tapes can be simplified greatly by switching to magnetic tape.

4. Scanning and Sensing Systems: Various machines are on the market today that can scan typescript or graphic materials optically or can sense special markings or coordinate (X,Y) positions and then translate what they "see" or "sense" into codes on a tape or other magnetic storage device, ready for processing. Most optical character readers (OCRs) are capable of accurately reading only one to several type faces. Usually, therefore, it is necessary to prepare the data by typing

them on precisely registered data forms with a special type font, which, however, can be done with an ordinary Selectric typewriter provided with the right head. Although OCR machines are being improved rapidly and some now can read printed pages directly, such as from a telephone book, in general many problems remain to be solved. Given their present weaknesses, scanning systems tend to cause an extra editorial burden in proofreading and correcting the initial input.

Mark-sensing has been used for years to grade examinations in colleges and universities and in some kinds of data capture.

Machines for digitizing the points in an X-Y grid are being used to create maps and other graphics in machine-processable form.

Scanning and sensing systems clearly are gaining in importance as ways of converting data to machine-readable form automatically, thereby eliminating the manual keystroking element of the data conversion process; however, although they hold great promise for the future, such systems are useful at present only for specialized applications particularly suited to the scanning or sensing procedures and involving very large data sets.

5. On-Line System: As explained above, some data conversion devices are terminals linked directly or by telephone to mini-computers or large computer systems to which the data can be transmitted directly from the keyboard. In such systems the data can be put into a processing cycle at once if desired. Often special software packages that enable the terminal operator, at his own command, to direct the computer to perform various editing functions have been written for data conversion systems of this type. Sophisticated text-editing systems are available which, though relatively expensive to use, constitute powerful tools for getting data into machine-readable form accurately. Output and input may be printed on paper or, in some systems, displayed on a screen.

   Persons accustomed to using a good on-line system supported by text-editing software become addicted to this method of data conversion and are reluctant to return to any card or tape system unless it is an economic necessity. Because on-line data capture invariably is more costly than off-line data capture, at least in terms of actual cash outlay for equipment and machine time, it is important to develop convincing cost-effective justification before opting for an on-line system. Cost-effectiveness in this instance is likely to hinge on the efficiency of manpower utilization.

A general caution about multi-purpose data conversion systems is in order at this point. If the main purpose is to capture data for computer processing, then the system should be designed to do this in the most straightforward, efficient manner possible, and it should not be ladened with other peripheral functions. It may seem logical at first to build a system that can produce various byproducts such as file cards or specimen labels in the process of encoding input cards or tapes for the computer, or produce encoded cards or tapes that can be used both to feed the computer and then to replicate file cards or labels mechanically. Inevitably, however, compromises are required when a system is designed for several purposes. Each additional function places unnecessary constraints on the others, to some extent reducing the system's efficiency in performing the other functions. Even if the same equipment is to be used, therefore, in most applications it is likely to be more efficient in the long run to design independent systems for the separate functions, although in practice this is a matter that the individual user will have to determine for himself. In practical terms, this means that labels should be produced as one operation and input cards or tapes for the computer as another operation. Another solution is to produce all necessary by-products, e.g., file cards, labels, as direct or indirect output from the computer. In conclusion, the multi-purpose system often proves to be more desirable in theory than in practice.

F. Processing Data. Once the data are converted into machine-readable form they are ready for processing and incorporation into the data bank. The processing strategy and methods should be worked out before any significant data collecting begins. The processing step is the final and most critical one in creating a data bank and it must not be left to chance. The advice of an information systems expert should be sought and followed carefully, particularly with respect to choosing an information processing system and to deciding what if any new computer programs need to be written. It is not the time to accept the advice of an enthusiastic first-year programming student, or of someone who has made his mark in computing as opposed to information processing.

The state of the art in the development of generalized information processing software is highly advanced today. Before seriously considering the costly course of writing his own software, anyone who contemplates building a data bank should look carefully at software packages presently on the market, and if at all possible he should rent or buy one of the many now available "off the shelf." The software manufacturer or vendor will maintain his package to keep it compatible with the current generation of hardware, and from time to time he will upgrade the software by adding new features. An individual or an institutional computing center hardly can compete with industry in providing such service. The day is past when one would think of building one's own processing hardware, and the day is virtually past when one should consider building one's own processing software.

Many computer service bureaus exist which can rent hardware and software support on an as-needed basis, just as, for example, large telephone systems exist to rent telephone service. There are, of course, weaknesses to this approach and certain pitfalls to avoid if possible. The chief weakness is that the fate of the customer's data processing is in the hands of the service bureau, which will be looking after its own interests on a commercial basis, and these interests will not necessarily coincide with those of the customer. Decisions will be taken by the firm with respect to hardware and software without consulting the customer, whose project may be affected profoundly, perhaps disastrously, by the consequences. The service bureau may abruptly quit offering the software being used by the customer, for example, thereby forcing the customer to make a costly mid-stream conversion to another software package or to make a disruptive transfer to another service bureau offering the original software. Computer centers, whether public or private, are notorious for promising more than they can deliver at a given cost, and the would-be customer has to select his service bureau wisely. In particular, he must try to avoid getting involved with a fly-by-night or financially unstable firm, which has a strong probability of going out of business or drastically curtailing service at a crucial moment in the customer's development effort. This is the major pitfall to avoid because the software vending and renting industry is still in a dynamically growing state, and the rate of business failure in this industry is high. It should be kept in mind, however, the same weaknesses and pitfalls are likely to be encountered, though perhaps in different forms, in dealing with in-house computing installations. In short, the manager of a data bank, wherever he elects to process his data, must guard against the basic capriciousness of the information processing industry at this stage in its history.

A number of generalized information processing systems currently are on the market and are being used for developing biological information systems. These include notably TAXIR (TAXonomic Information Retrieval), the system developed by David J. Rogers and his associates; GIPSY (General Information Processing SYstem), developed by the University of Oklahoma Computing Center under the guidance of James W. Sweeney; SELGEM (SELf GEnerating Master), the Smithsonian Institution's generalized processing system developed under the direction of Reginald Creighton; and GIS (Generalized Information System), developed by IBM. The Flora North America Program adopted GIS as the basis for its data banking.

V. CURRENT DATA SYSTEMS

To give more than a mention of the many data systems now operating or under development would require another chapter. The interested student is encouraged to consult the references cited at the end of the chapter for pertinent details. Most of these systems are in such a state of flux that to describe their present characteristics in detail here would be pointless in any event.

Flora North America aims to develop a truly generalized information system to do data banking on a broad network basis. Few other systems of equally generalized character appear to be underway.

Various botanists have developed systems for mapping plant distributions by computer. James H. Soper of Canada's National Museum of Natural Sciences has pioneered in this area. Recently, in a project initiated at Colorado State University, William M. Klein and Robert P. Adams have developed some exciting new approaches to distribution-mapping by computer in relation to critical environmental factors. Mapping by machine was done for the entire British flora some years ago by Franklyn H. Perring and S. M. Walters, but this was done with a mechanical tabulator rather than a computer. Several other British botanists have just completed and published a study entitled *A Computer-Mapped Flora: A Study of the County of Warwickshire* in which the computer was used for some of the mapping and a mechanical tabulator for the rest of it.

In the realm of the herbarium, a number of plant taxonomists are now using computer-based systems to store and retrieve specimen data. Notable herbarium projects are the Smithsonian Institution's Type Specimen Register, an ambitious effort to create a computerized registry of botanical type specimens on a worldwide basis, indicating where the names were published and where the specimens are deposited; Notre Dame University's project, under the direction of Theodore J. Crovello, to computerize the specimen data in the E. L. Greene Herbarium; and Soper's effort at the National Museum of Natural Sciences in Ottawa to create a specimen data bank for the Province of Ontario.

The Plant Records Center of the American Horticultural Society is well along in the development of a data bank of pertinent botanical and horticultural data concerning living plant collections in North American botanical gardens and arboretums.

The Hunt Institute of Botanical Documentation, located at the Carnegie-Mellon University in Pittsburgh, Pennsylvania, has concentrated on the development of an information system to handle botanical literature, and it has undertaken several pacesetting retrieval projects in this area. Using its system, the Hunt Institute processes nomenclatural data for the Index Nominum Genericorum, a project of the International Bureau for Plant Taxonomy and Nomenclature in Utrecht. The project, which is far advanced in the creation of a computerized index of validly published generic names for all plant groups of the world, is headquartered at the Smithsonian Institution. A similar world-wide index of species names is now contemplated.

Finally, in the area of specialized application programs, mention should be made of the promising program packages that have been developed for using the computer to assist in plant identification. The imaginative taxonomic matrix system of Larry E. Morse is particularly noteworthy because it enables the botanist to sit at a remote terminal and use the computer on-line in the conversational mode to identify unknown plants or to construct dichotomous identification keys on the basis of stored data matrices linked to stored lists of taxonomic names and characters. This system can be used to great advantage in the classroom as a teaching aid or in the laboratory as a research tool. It is only one example of the numerous application programs that are available to the initiated scientist today.

VI. CONCLUSION

Of necessity botanists will turn increasingly to the computer in the years to come for help in solving their information retrieval problems, and scientific data banks of all sizes and descriptions will result. It is incumbent upon plant systematists to make full use of the computer as a tool and to guide the development of botanical data banks in an orderly and coordinated way. Surely no present-day student of plant systematics can afford to omit from his training the study of information processing technology as it pertains to scientific data banking. We have only begun to exploit this technology, and yet today's technology is a mere prelude to the state of the art as it will be in the future.

INFORMATION SYSTEMS AND DATA-BANKING REFERENCES

I. General Applications References

Bossert, W. 1969. Computer techniques in systematics. In: Systematic Biology, p. 595-614, publication 1692, National Academy of Sciences. Washington, D. C.

Crovello, T. J. 1967. Problems in the use of electronic data processing in biological collections. Taxon 16: 481-494.

Cutbill, J. L. (ed.). 1971. Data Processing in Biology and Geology [symposium]. Systematics Association Special Vol. 3. Academic Press. New York and London. 346 p. (See especially Shetler et al., p. 93-128, "Pilot Data Processing Systems for Floristic Information.")

EDUCOM. 1973. Networks and disciplines. Proceedings of the EDUCOM Fall Conference, 11-13 October 1972. EDUCOM (Interuniversity Communications Council). Princeton, New Jersey. 209 p.

Gates, D. M. 1971. Flora North America: a data bank for systematic biology [editorial]. BioScience 21: 507.

Raven, P. H. 1968. Electronic data processing and the taxonomic system. Flora North America Report 19: 1-12.

________, B. Berlin, and D. E. Breedlove. 1971. The origins of taxonomy. Science 174: 1210-1213.

Shetler, S. G. 1974. Demythologizing biological data banking. Taxon 23: 75-104.

Sneath, P. H. A., and R. R. Sokal. 1973. Numerical Taxonomy. The Principles and Practice of Numerical Classification. W. H. Freeman. San Francisco. xvi + 574 p.

Sokal, R. R., and P. H. A. Sneath. 1963. Principles of Numerical Taxonomy. W. H. Freeman. San Francisco. 359 p.

________. 1966. Efficiency in taxonomy. Taxon 15: 1-21.

II. General Systems References

American National Standard Vocabulary for Information Processing. 1970. American National Standards Institute. New York. ANSI X3.12-1970: 1-128.

Annual Review of Information Science and Technology (Carlos A. Cuadra, ed.). 1966+. Series of review volumes published by the American Society for Information Science. Chicago.

Federal Information Processing Standards Publications. A series issued at irregular intervals of publications setting technical standards for data banking. National Bureau of Standards, U. S. Department of Commerce. Washington, D. C.

Hunt Institute for Botanical Documentation Programming Notes (G. S. Daniels, ed.). 1970+. An irregularly issued series of technical notes on the Hunt Institute's processing system. Hunt Institute for Botanical Documentation, Carnegie-Mellon University. Pittsburgh, Pennsylvania.

King, D. W., and E. C. Bryant. 1971. The Evaluation of Information Services and Products. Information Resources Press. Washington, D. C. 306 p.

Meadow, C. T. 1967. The Analysis of Information Systems, a Programmer's Introduction to Information Retrieval. John Wiley and Sons. New York. 301 p.

________. 1970. Man-Machine Communication. John Wiley and Sons. New York. 422 p.

Smithsonian Institution Information Systems Innovations. 1969+. An irregularly issued series of technical bulletins describing computer advances in handling museum information. Information Systems Division, Smithsonian Institution. Washington, D. C.

III. Data Conversion and Processing Systems

Addison, C. H., P. W. Blackwell, W. E. Smith, Jr., R. W. Shields, and J. W. Sweeney. 1969. GIPSY--General Information Processing System. Remote terminal users guide. University of Oklahoma Information Science Series, Monograph No. 3. 26 p.

Creighton, R. A., and J. J. Crockett. 1971. SELGEM: a system for collection management. Smithsonian Institution Information Systems Innovations 2(3): 1-24 + appendix.

Krauss, H. M. 1973. The use of generalized information processing systems in the biological sciences. Taxon 22: 3-18. (FNA Report 67). [Layman's explanation of IBM's Generalized Information System (GIS).]

________. 1973. The information system design for the Flora North America Program. Brittonia 25: 119-134. (FNA Report 69).

Pankhurst, R. J. 1972. A method of data capture. Taxon 21: 549-558.

IV. Nomenclatural Applications

Cowan, R. S. 1970. The Index Nominum Genericorum Project--past, present, and future. Taxon 19: 52-54.

Flora North America Program. 1972. A provisional checklist of species for Flora North America. Flora North America Report 64: viii + 1-648.

Gould, S. W. 1972. Family Names of the Plant Kingdom. International Plant Index, Vol. 1. New York Botanical Garden and Connecticut Agricultural Experiment Station. New York and New Haven. 111 p.

________, and D. C. Noyce. 1965. Authors of Plant Genera. International Plant Index, Vol. 2. New York Botanical Garden and Connecticut Agricultural Experiment Station. New York and New Haven. 336 p.

Shetler, S. G. et al. 1973. An introduction to the Botanical Type Specimen Register. Smithsonian Contributions to Botany 12: 1-186 (FNA Report 70).

V. Museum, Herbarium, and Botanical Garden Applications

Argus, G. W., and J. W. Sheard. 1972. Two simple labeling and data retrieval systems for herbaria. Canadian Journal of Botany 50: 2197-2209.

Beschel, R. E., and J. H. Soper. 1970. The automation and standardization of certain herbarium procedures. Canadian Journal of Botany 48: 547-554.

Brown, R. A., and R. D. MacDonald. 1971. The A.H.S. [American Horticultural Society] Plant Records Center. Arboretum and Botanical Garden Bulletin 5(4): 4-13.

Chenhall, R. G. In press. Computerized Museum Catalogs. American Association for State and Local History. Nashville, Tennessee. (Sponsored by Museum Data Bank Coordinating Committee).

Cohen, D. M., and R. F. Cressey (eds). 1969. Natural history collections: past, present, future [symposium]. Proceedings of the Biological Society of Washington 82: 559-762.

Crovello, T. J. 1972. Computerization of specimen data from the Edward Lee Greene Herbarium (ND-G) at Notre Dame. Brittonia 24: 131-141.

Cutbill, J. L. 1973. Computer Filing Systems for Museums and Research. Published by the author, Sedgwick Museum, Cambridge University. Cambridge. 23 p.

Hale, M. E., and R. A. Creighton. 1970. An automated system for recording exchanges. Flora North America Report 32: 1-9.

Hall, A. V. 1972. Computer-based data banking for taxonomic collections. Taxon 21: 13-25.

Scheinvar, L., and A. Gómez-Pompa. 1969. Some automatic methods of preparing herbarium labels. Boletín de la Sociedad Botánica de México 30: 73-93.

Soper, J. H., and F. H. Perring. 1967. Data processing in the herbarium and museum. Taxon 16: 13-19.

Taylor, R. L., S. Sziklai, and N. J. Turner. 1970. An automated information filing system for plants desired for botanical garden components. Botanical Garden of the University of British Columbia Technical Bulletin 1: 1-37 + appendices.

VI. Machine Mapping

Cadbury, D. A., J. G. Hawkes, and R. C. Readett. 1971. A Computer-Mapped Flora: a study of the county of Warwickshire using computer techniques. Academic Press. New York and London. 788 p. + 960 maps.

Klein, W. M., and R. P. Adams. 1971. A computerized system for inventorying floristic resources [abstract]. American Journal of Botany 58: 464-465.

Perring, F. H. 1963. Data processing for the atlas of the British flora. Taxon 12: 183-190.

________, and S. M. Walters (eds.). 1962. Atlas of the British Flora. Thomas Nelson and Sons. London and Edinburgh. 432 p. (See also Perring, F. H., assisted by P. D. Sell, Critical Supplement to the Atlas of the British Flora. Thomas Nelson and Sons. 159 p., 1968).

Soper, J. H. 1964. Mapping the distribution of plants by machine. Canadian Journal of Botany 42: 1087-1100.

________. 1966. Machine-plotting of phytogeographical data. Canadian Geographical Journal 10: 15-26.

VII. Computer-Assisted Identification

Furlow, J. J., and J. H. Beaman (compilers). 1971. Sample taxonomic data matrices for vascular plants prepared by students at Michigan State University. Flora North America Report 56: 1-118.

Morse, L. E. 1971. Specimen identification and key construction with time-sharing computers. Taxon 20: 269-282. (FNA Report 52).

________, J. A. Peters, and P. B. Hamel. 1971. A general data format for summarizing taxonomic information. BioScience 21: 174-181. (FNA Report 54).

Pankhurst, R. J. 1971. Botanical keys generated by computer. Watsonia 8: 257-268.

VIII. Teaching Applications

Beaman, J. H. (ed.). 1971. Some applications for the taxonomic data matrix: six term papers by students at Michigan State University. Flora North America Report 63: 1-101.

Crovello, T. J. 1974. Computers in biological teaching. BioScience 24: 20-23.

Furlow, J. J., L. E. Morse, and J. H. Beaman. 1971. Computers in biological systematics, a new university course. Taxon 20: 283-290. (FNA Report 53).

Morse, L. E., J. J. Furlow, and J. H. Beaman. 1971. Computers in systematic biology: a course syllabus. Flora North America Report 62: 1-58.

IX. Current-Awareness Aids

Shetler, S. G., and R. W. Read (eds.). 1973. International index of current research projects in plant systematics, no. 7. Flora North America Report 71: xii + 118 p.

X. Current Data Banking Projects

Crovello, T. J., and R. D. MacDonald. 1970. Index of EDP-IR projects in systematics. Taxon 19: 63-76.

Flora North America Report Series (S. G. Shetler, ed.). 1966+. Reports on the development of the FNA Program, issued at irregular intervals. Smithsonian Institution. Washington, D. C.

Gómez-Pompa, A., and L. I. Nevling, Jr. 1973. The use of electronic data processing methods in the Flora of Veracruz Program. Contributions from the Gray Herbarium 203: 49-64.

Porter, D. M., R. W. Kiger, and J. E. Monahan. 1973. A guide for contributors to Flora North America, part II: An outline and glossary of terms for morphological and habitat description (provisional edition). Flora North America Report 66: 1-210 + G1-32.

Shetler, S. G. (Program Director), and H. R. Meadow (Systems Development Manager). 1971. Flora North America: a comprehensive program of biological research, information systems development, and data banking concerned with the vascular plants of North America north of Mexico. Flora North America Report 61: 1-126 + Appendices A-I.

________ et al. 1973. A guide for contributors to Flora North America (FNA) (provisional edition). Flora North America Report 65: 1-28 + Appendices A-E.

XI. Conventional Systems

Edwards, P. I. (convenor). 1971. Storage and retrieval of biological information [series of lectures for a course]. Biological Journal of the Linnean Society 3: 165-299. See especially Meikle, p. 295-299, "The History of Index Kewensis." (Some of systems discussed are computerized at least to a degree.)

International Bureau for Plant Taxonomy and Nomenclature. Publishes and distributes, sometimes with the aid of a computer, indexes and other valuable information retrieval aids for the International Association for Plant Taxonomy, in the series Regnum Vegetabile (F. A. Stafleu, ed.).

Shaw, E. A. 1971. The Gray Herbarium Card Index. Taxon 20: 333-336.

Chapter 34. SOCIETIES, BIOLOGICAL STATIONS, AND RESEARCH SUPPORT*

Herbaria, botanic gardens, arboreta, data banks, libraries (limited sense), record centers, taxonomic literature, bibliographic aids, and information systems have been treated in the previous four chapters of this division on facilities and general information. In this chapter a miscellaneous assemblage of material is presented on societies, systematic centers, biological stations and supportive organizations in an effort to make information on taxonomic facilities and general information comprehensive and complete. The treatment of biological stations and funding sources is made with a little trepidation since some of these facilities and agencies may be relatively short-lived. The facts in this chapter on biological stations and research support indicate the diversity of types of stations and financial support at this time. An effort should always be made to procure the most recent information on field stations and grant sources.

I. SOCIETIES

It is a real problem for any professional to keep abreast of the current literature and rapid development in his specialty. The problem is compounded in taxonomy since it is such a synthetic field involving so many peripheral disciplines--each expanding rapidly in techniques and data. The mass of current publications has confronted the taxonomist with an almost insurmountable task of deciding what journals to get for his own library, what to read regularly in addition, and how to learn about important papers elsewhere.

One should form a habit of surveying the current literature by regularly and systematically checking certain abstracting journals or indices (see Chapter 30). The "Index to American Botanical Literature," published in the Bulletin of the Torrey Botanical Club, is especially helpful in this regard. In addition, make a list of current journals in your school library which regularly have articles of interest to you, then survey all of these at least once a year. Such a list could easily include fifty or more journals.

Personal membership in professional and honorary societies is one way of making certain that important journals are seen regularly and also permanently available in your own library. However, getting the journal is not the only reason to belong to professional societies. After all, these organizations represent us and our field of endeavor on a regional, national, or worldwide basis, and can sometimes be effective spokesmen in matters of concern to all. We all have an obligation to support the societies or associations which represent our area of specialty and areas of interest. Attendance at the annual meetings of these societies gives one an opportunity to have an input into the business affairs, programs, and activities of the organization, as well as to know the other members.

Each individual must make his own choice of memberships, but consideration should be given to at least one or more in each of the following broad categories.

1. General science or biology in the U. S.
   (American Association for the Advancement of Science, American

*Contributed by James W. Hardin, North Carolina State University, Raleigh.

Institute of Biological Sciences--effective "voices," hopefully, of science and biology in this country)

2. Botany in the U. S.
   (Botanical Society of America--the "mother society" for all botanists in this country)
3. Taxonomy in the U. S. and internationally
   (American Society of Plant Taxonomists, International Association for Plant Taxonomy--the chief societies in this discipline)
4. Regional or state organizations
   (This category includes numerous societies that publish such journals as Rhodora, Bulletin of the Torrey Botanical Club, Castanea, Journal of the Elisha Mitchell Scientific Society, Madroño, Southwestern Naturalist, Bulletin of the California Academy of Science; most states have an Academy of Science.)
5. Special areas of interest
   (Society for the Study of Evolution, American Fern Society, American Bryological and Lichenological Society, American Forestry Association, or those representing peripheral disciplines)
6. Honor societies (membership is by invitation only)
   (e.g., Sigma Xi, Phi Kappa Phi, Gamma Sigma Delta, Phi Sigma)

An important organization for taxonomy, with institutional memberships only, is the Association of Systematics Collections. This organization was begun in 1972 to "foster the care, management, preservation, and improvement of systematics collections and to facilitate their utilization in science." One significant result of this organization is the writing and presentation of a document entitled "America's Systematics Collections: A National Plan." This report is addressed principally to systematists concerned with the nature and quality of America's systematics collections as a national resource, and describes and recommends ways whereby the collections can be more responsive to the needs of the basic and applied sciences as well as society in general.

This organization will be an effective voice for all systematists concerned with the preservation and utilization of biological collections, and it is therefore most important for all museums and herbaria to be represented in the ASC.

II. BIOLOGICAL STATIONS

Most important for the study of taxonomy or taxonomic research is the ability to get out in the field with the plants. Taxonomists must not lose sight of the fact that the living plant in the field is the ultimate source of data. Many colleges and universities are in growing urban areas and field experience for most students is rapidly becoming limited. It is also increasingly difficult to obtain travel money for research. Biological stations are often the answer to these problems.

Summer courses and/or research at biological stations have obvious advantages over most campuses. The locations are generally in environments with unique or representative biotas, or at least within easy access to natural areas. It is a good way to get into different areas and learn a different flora or be in an ideal location for research on a particular group. Courses are generally field oriented and often different from those available on most campuses. Money is sometimes available from the station to help defray expenses. Students and faculty are enthusiastic biologists and a delightful group with which to be; and, it is generally a most pleasant way to spend a summer of learning and research.

Biological stations are easily divisible into either "marine stations" specializing in oceanography and marine species, or "inland field stations" specializing in terrestrial or fresh-water biotas. Vascular plant taxonomists should become familiar with the inland stations and the facilities available.

Facilities, programs, and opportunities vary tremendously among stations. Some have research facilities only, such as the Highlands Biological Station, while others have teaching programs in addition, such as Mountain Lake Biological Station. All have something unique to offer, either in program or the biota or ecology of the area. In 1945 there were 53 inland field stations, in 1966 there were 42 (Arvey and Riemer, 1966), and in 1970 there were about 100 in the United States (Hunsaker and Dalgleish, 1970). This growth probably refelects the recent emphasis on environmental biology as well as the decrease in available natural areas close to university campuses.

To learn about the facilities available and programs offered at a particular station, write to the director of a station in the area of interest (Figure 34-1). Addresses can be located in the publications cited below or through a university in the area. Many of the stations also belong to the OIBFS (Organization of Inland Biological Field Stations) which can be contacted c/o The E. N. Huyck Preserve, Inc., P. O. Box 87, Rensselaerville, N. Y. 12147. Most of the larger stations have brochures which are available upon request.

References on Biological Stations

Arvey, M. D., and W. J. Riemer. 1966. Inland Biological Field Stations of the United States. BioScience 16: 249-254.

Hunsaker, D., and R. Dalgleish. 1970. Biological Field Stations of North America. MSS Educational Publishing Company, Inc. New York.

Palmer, A. M. (ed.). 1972. Research Centers Directory, 4th ed. Gale Research Company. Detroit, Michigan.

III. RESEARCH SUPPORT, AND PRE- AND POST-DOCTORAL FELLOWSHIPS

Whether the desire or need is for a stipend, a major piece of equipment, travel for field work or to examine type specimens, or page cost for a publication, the one vexing question is "where to get the money." It sometimes takes real ingenuity and aggressiveness to find funds. Don't be afraid or embarrassed to ask! But, have a good plan and justification in mind first.

For extended stays in various parts of the country or world, you might consider teaching in secondary schools or colleges in the area. Often there are American schools abroad that need teachers. Some of the addresses below will put you in contact with these. Also the Peace Corps programs may be a good way to stay in an area for awhile.

Another possibility for some limited travel funds is the sale of specimens. This is something to consider particularly if you are going to some remote area which has not been well collected. First, get an agreement with

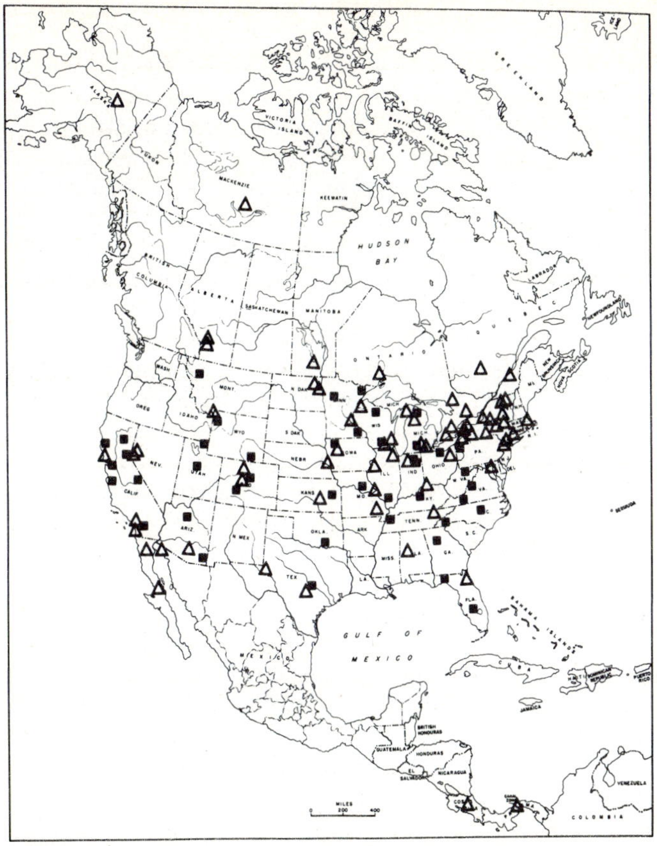

Figure 34-1. Inland Biological Field Stations of North America. Solid blocks: stations reported in 1966; open triangles: additional stations reported in 1970.

the herbarium curator, museum director, or researcher regarding type of collections, number, identifications and labels, methods of preservation, and cost per specimen. Think beyond regular herbarium specimens to such things as pollen samples, wood sections, roots or leaves for chemical extracts, photographs, pathogenic fungi, pollinating insects, or damaging insects. The research division of some petroleum companies may be interested in pollen samples; certain chemical or pharmaceutical companies or the USDA may want plant material for extracting alkaloids, glycosides, or fats. Identifiable voucher specimens must accompany any such collections.

Fellowships or grants may be obtained from private, state or federal sources. Start by asking your colleagues who may have already been successful in this venture, then the administration in your own school (department head, deans of your school, graduate school, or of research). Someone in the administration usually keeps up with programs available through the various granting agencies (National Science Foundation [NSF], National Institute of Health [NIH], Department of Health, Education and Welfare [HEW], National Aeronautics and Space Administration [NASA], National Defense Education Act [NDEA], Atomic Energy Commission [AEC]), and the various fellowship programs available. It is always possible that you might find the money right in your own institution.

Beyond this, there are numerous possibilities as indicated below.

A. <u>Non-state or non-federal agencies</u>

1. Biological field stations (see discussion in this chapter).
2. Organization of Tropical Studies (OTS) North American Office, P. O. Box 1499, South Miami, FL 33143).

3. Sigma Xi Grants-in-Aid of Research (Sigma Xi, 155 Whitney Ave., New Haven, CT 06510).
4. Private foundations. See: "The Foundation Directory" available in the reserve room of most libraries; look under Life Sciences in the Index of Fields of Interest. Some do give grants or fellowships to individuals. Some examples are:
   a. The Danforth Foundation, 222 South Central Ave., St. Louis, MO 63105 (graduate fellowships, graduate fellowships for women, Kent fellowships).
   b. Ford Foundation Doctoral Fellowships, 320 East 43rd St., New York, NY 10017 (graduate study leading to Ph.D. for Black Americans, American Indians, Mexican Americans, and Puerto Ricans).
   c. John Ray Whitney Foundation Opportunity Fellowships. Write: Opportunity Fellowships, John Ray Whitney Foundation, 111 W. 50th St., New York, NY 10020 (for Black Americans, Spanish Americans, American Indians, residents of the southern Appalachian and Ozark Mt. areas, Guam, Puerto Rico, Samoa, Pacific Trust Territory, and Virgin Islands).
   d. Woodrow Wilson National Fellowship Foundation, Box 642, Princeton, NJ 08540 (first-year graduate study; dissertation fellowship).
   e. John Simon Guggenheim Memorial Foundation, 90 Park Ave., New York, NY 10016 (fields unrestricted).
5. National Geographic Society, Committee for Research and Exploration, 17th & M St. NW, Washington, DC 20036 (post-doctoral travel grants).
6. The Inter-University Council for Higher Education Overseas, 90/91 Tottenham Ct. Rd., London, England (assists in recruitment of academic staff of smaller colleges in British Commonwealth).
7. Near East College Association, 305 E. 45th St., New York, NY 10017 (teaching positions in private American colleges in Greece, Turkey, and Lebanon).
8. Smithsonian Institution Visiting Research Associate Program, or Visiting Postdoctoral Research Associate Program. Write: Director, Division of Graduate Studies, Office of Academic Programs, Smithsonian Institution, Washington, DC 20560 (fellowships for study at the Smithsonian).
9. Environmental consulting firms (possible job or subcontract research with various groups involved with ecological studies or environmental impact studies).
10. Conference Board of Associated Research Councils, 2101 Constitution Ave. NW, Washington, DC 20418 (university lecturing and research; postdoctoral fellowships).
11. Institute of International Education, 809 United Nations Plaza, New York, NY 10017 (graduate study abroad; predoctoral fellowships).
12. Council of International Educational Exchange, 777 United Nations Plaza, New York, NY 10017 (aid and information on study programs abroad).
13. American Association of University Women Educational Foundation Fellowships. Write: Director, AAUW Fellowship Program, 2401 Virginia Ave., NW, Washington, DC 20036 (pre- and post-doctoral fellowships).
14. East-West Center, 1777 East-West Road, Honolulu, HI 96822 (senior specialist awards; scholarships for graduate study at the University of Hawaii).

15. Foreign Area Fellowship Program, 110 E. 59th St., New York, NY 10022 (research in Africa, Near and Middle East, South Asia, Southeast Asia, East Asia, Latin America and the Caribbean, Soviet Union, Eastern and Western Europe).
16. Latin American Teaching Fellowships, The Fletcher School of Law and Diplomacy, Tufts University, Medford, MA 02155 (predoctoral fellowships in social and natural sciences).
17. National Wildlife Federation Fellowships, 1412 16th St. NW, Washington, DC 20036 (predoctoral fellowships in conservation and related areas).
18. Oak Ridge Graduate Fellowships. Write: University Programs Office, Oak Ridge Associate Universities, Oak Ridge, TN 37830 (for work at Oak Ridge or other AEC sites).
19. Organization of American States, 17th St. & Constitution Ave. NW, Washington, DC 20006 (OAS Fellowship for advanced specialized study, training, or research in Latin America).
20. Southern Fellowship Fund, 745 Peachtree St. NE, Atlanta, GA 30308 (predoctoral fellowships for black students).
21. NATO fellowships. Write: Fellowship Office, National Research Council, 2101 Constitution Ave. NW, Washington, DC 20418 (postdoctoral fellowships for study primarily, but not restricted to, countries of the NATO alliance).
22. The American Philosophical Society, Independence Square, Philadelphia, PA 19106 (postdoctoral grants-in-aid for research costs; not for stipends, publication costs, or permanent equipment).
23. The Phi Kappa Phi Fellowships. Write: Phi Kappa Phi, 3001 Plymouth Rd., Ann Arbor, MI 48105 (first and second year graduate fellowships in all areas).
24. State academy of science (some have modest grants-in-aid of research).

B. State agencies

1. State agencies such as museums, parks department, wildlife agencies, science and technology boards, etc. (for summer jobs or modest support of research.)
2. University facilities such as phytotrons, computing centers, agricultural experiment stations. (Often these have state funds or federal grants for support and can allow use by faculty and students.)

C. Federal agencies (for a comprehensive listing see "Catalog of Federal Domestic Assistance," U. S. Printing Office 1969-0-328-264).

1. Grants for Improving Doctoral Disseration Research in the Field Sciences. National Science Foundation (funds for travel to libraries, museums, or field research locations; special equipment; computer time; no stipends given).
2. Institute of International Studies, Office of Education, Department of HEW, Washington, DC 20202 (research or teaching in foreign schools under the Fulbright-Hays Act; also, request information on foreign travel, work, and study).
3. Division of Student Financial Aid, Bureau of Higher Education, Office of Education, Department of HEW, Washington, DC 20202 (loans for study abroad).
4. Fellowship Office, Office of Scientific Personnel, National Research Council, 2101 Constitution Ave. NW, Washington, DC 20418 (Publication: "A Selected List of Major Fellowship Opportunities and Aids to Advanced Education for United States Citizens").

Chapter 35. QUESTIONS AND PROBLEMS

A series of questions are presented for thought and discussion which will provide the student with an introduction to some of the basic problems and issues in systematics. The species biology, ecosystematics, biosystematics, and phylosystematics problems require a summary application of the various facets of taxonomy. These problems can be used as class exercises, individual student projects, or parts of each can be used in various combinations for particular systematics training emphases by the teacher. Any one of the problems could provide the basis for a student research project. Sample reviews are presented to show how character or topical summaries of the features presented in the text can be used as bases for class or individual reviews or quizzes in the different aspects of systematics. No effort has been made to show test methods since systematics test methods should be similar to those found in all fields of education. A basic course midterm and final examination have been included to show what can be expected from students at that taxonomic level rather than for quizzing methodology.

Section A. QUESTIONS FOR THOUGHT AND DISCUSSION

I. GENERAL SYSTEMATICS QUESTIONS

1. What are the principles of taxonomy?
2. Why is the holistics approach in a basic course in systematics desirable?
3. How can taxonomists justify their type of work or study in this time of environmental crisis and socio-political turmoil?
4. How does a systematics reference text or a general taxonomic text help and hinder the development of the field of systematics?
5. Why should a basic systematics course be a core course in a liberal education curriculum? A fundamental course in botanical training?
6. Why should plant taxonomy be considered the acme and pedestal of plant sciences?

II. HISTORY AND CLASSIFICATION QUESTIONS

1. How does a study of history of taxonomy give the student perspective in systematics?
2. How is the organization of history into "Periods of Development of Classification" misleading?
3. Why is the development of systematics dependent upon the development of other fields of science?
4. What is the practical use of a classification system which divides groups of organisms into hierarchies; e.g., species, genera, families, and higher taxa?
5. Why should plant taxonomists work vigorously for the production of an effective phenetic classification of plants rather than a phylogenetic one?
6. What are the present trends of systematics research and what is the future of taxonomic work?

III. NOMENCLATURE AND EPITHETS QUESTIONS

1. What are the advantages and disadvantages of a stable or fixed nomenclature?

2. Why did synonymy arise and why can it not be eliminated in the future?
3. What are the difficulties in applying the type concept to all kinds of plants and plant remains (fossils)?
4. Even though there are only several hundred epithets with their meanings listed in the text, how can the meanings of thousands be determined from the information in the text?
5. Why should species names be conserved? Generic names? Any names?
6. What should the fundamental components of a species definition be?
7. What precautions should be taken in selecting a good specific epithet? A generic name? A higher taxon name?
8. Why is binomial nomenclature considered basic to all of taxonomy? To all of botany?
9. What are the reasons for the rejection of names?

IV. EVIDENCE QUESTIONS

1. What is meant by character or evidence type, character state, ordering of characters and what are some of the problems in determining type, state, and order?
2. Under what conditions can a meaningful character be reliable, or diagnostic, or homologous, or analytic, or synthetic, or two-state, etc.; or reliable and analytic, or analytic and synthetic, etc.?
3. Why should morphological evidence or characters be the basic type used in identification and classification?
4. What are the advantages and disadvantages of a refined and definitive descriptive morphological terminology based on measurements; e.g., angles and ratios?
5. Should ontogeny-phylogeny be the basis for definitions of morphological and anatomical terms and characters? Explain.
6. What are the dangers in using one type of evidence almost exclusively; e.g., anatomical, morphological, or chemical in classification?
7. At what level(s) in the taxonomic hierarchy or system are the following types of data taxonomically most useful: anatomical, embryological, palynological, cytological, chemical? Explain why for each type of data.
8. What are some of the precautions to take in using each of the following types of data for taxonomic purposes: anatomical, embryological, palynological, cytological, chemical?
9. What are some of the fundamental problems involved in procuring data of the following types for taxonomic purposes: anatomical, embryological, palynological, cytological, chemical?
10. Are all types of data of equal weight and value for classification? Identification? Explain.
11. How are the following features relevant or useful in taxonomic research: polyploidy, karyological features, breeding patterns, meiotic pairing, habitats, types of reproduction?
12. Why are ecological characteristics not used more frequently in taxonomic descriptions?
13. What are the fundamental problems involved in working out a satisfactory classification of trichomes, venation patterns, inflorescence types, fruit types?
14. Physiological characteristics are seldom ever used in any phase or aspect of systematics. Explain why not.
15. Should basic courses in each of the fields of evidence be required training for advanced taxonomy students?

V. EVOLUTION AND PHYLOGENY QUESTIONS

1. What is meant by evolutionary evidence and how is it used in systematics? Phylogenetic evidence?
2. In the study of anatomy and morphology, how can phylogenetic or evolutionary advancement be distinguished from ecological specialization?
3. If a plant possesses a single or a very few primitive characters, can it be assumed to be totally primitive? Explain.
4. What is the basis for determination of "primitive or advanced"? Cite a few examples.
5. What are some of the evolutionary "pitfalls" in morphological and anatomical phylogenetic study?
6. What is adaptation? Preadaptation?
7. Why do all parts of an organism not evolve at the same rate?
8. What is the evolutionary role of each of the following: phenotypic variation, fitness, competition, gene flow barriers?
9. Is mutation or recombination the more important source of individual variation? Explain.
10. Why is evolution unidirectional under strong abiotic selections?
11. In phylogeny should some characters be given more weight than others? Explain.
12. What are the relationships between mutation rate and rate of evolution?
13. What is meant by chemical phylogeny and what is its significance to systematics?
14. Why does the search for evolutionary relationships inspire the majority of taxonomists today in their studies?
15. Why are angiosperm classification systems based on phylogeny unattainable?
16. Why is vessel element evolution considered a major phylogenetic tool?

VI. GEOGRAPHY, ECOLOGY, FIELD QUESTIONS

1. Of what significance is plant distribution in elucidating evolutionary histories?
2. What is meant by center of distribution? Why can it not be determined readily for most taxa?
3. What are some of the chief factors that make some plants widely dispersed and some very limited in their distribution?
4. What is meant by endemism? Why is there so much confusion in applying the concept?
5. What are the fundamental problems in defining or classifying the following: dispersal types, community types, habitat types, biogeographic regions, ecological characteristics?

VII. DATA AND METHODOLOGY QUESTIONS

1. What is meant by documentation of taxonomic research?
2. What are the components of good experimental design in biosystematics research?
3. Why are different analytical procedures necessary for a comprehensive study of local populations?
4. What can be done to make breeding systems and transplant studies practical for graduate student research?
5. What are the general problems associated with selecting the best method of data presentation?

6. Why are experimental studies and populational analyses considered fundamental to modern systematics research?
7. What should the role of the botanical garden and herbarium be in taxonomic research, conservation, and teaching?
8. What are some of the problems in classifying evidence for deposit in a data bank?
9. What is the relevance of data banking and data retrieval to the development or demise of herbaria and botanical libraries?
10. How will the computer reduce taxonomic labor and the need for taxonomists? How will the computer and data bank be of service to the taxonomist; of taxonomic use to the public?

Section B. PROBLEMS

I. SPECIES BIOLOGY PROBLEM

Select one species and work out the character state(s) for each of the following:

PHYTOGRAPHIC CHARACTERS
(See Chapter 6)

1. Plant Type
2. Plant Duration
3. Root Types
4. Root Structural Type
5. Stem Type(s)
6. Stem Structural Type
7. Stem Location
8. Branch Position
9. Branch Arrangement
10. Branch Orientation
11. Branch Transverse Posture
12. Branch Longitudinal Posture
13. Bud Type
14. Bud Structural Type
15. Bud Scale Arrangement
16. Bud Covering
17. Leaf Type
18. Leaf Structural Type
19. Petiole Structural Type
20. Petiolule Structural Type
21. Stipule Type
22. Stipel Type
23. Leaf Location
24. Leaf Position
25. Leaf Arrangement
26. Leaf Orientation
27. Leaf Transverse Posture
28. Leaf Longitudinal Posture
29. Leaf Ptyxis
30. Leaf Number
31. Leaf Division
32. Leaf Size
33. Leaf Shape
34. Leaf Margin
35. Leaf Apex
36. Leaf Base
37. Leaf Surface Configuration
38. Leaf Surface
39. Leaf Vestiture
40. Leaf Venation
41. Leaf Texture
42. Inflorescence Type
43. Inflorescence Position
44. Flower Type
45. Corolla Type
46. Calyx Type
47. Hypanthium Type
48. Perianth Arrangement
49. Perianth Position
50. Hypanthium Adnation
51. Androecial Type
52. Androecial Position
53. Stamen Structural Type
54. Stamen Arrangement
55. Anther Type
56. Anther Shape
57. Anther Attachment
58. Gynoecial Type
59. Carpel Type
60. Carpel Arrangement
61. Stigma Type
62. Style Type
63. Ovary Position
64. Ovule Type

65. Ovule Position
66. Placentation
67. Flower Parts Number
68. Flower Cycly
69. Flower Merosity
70. Flower Parts Fusion
71. Flower Parts Division
72. Petal Shape
73. Petal Margin
74. Petal Apex
75. Petal Base
76. Sepal Shape
77. Sepal Margin
78. Sepal Apex
79. Sepal Base
80. Petal Surface Configuration
81. Petal Surface
82. Petal Vestiture
83. Petal Venation
84. Petal Texture
85. Sepal Surface Configuration
86. Sepal Surface
87. Sepal Vestiture
88. Sepal Venation
89. Sepal Texture
90. Fruit Type
91. Fruit Position
92. Fruit Shape
93. Fruit Surface
94. Seed Type
95. Seed Shape
96. Seed Surface Feature
97. Embryo Type
98. Embryo Shape
99. Cotyledon Position
100. Seedling Type
101. Flower Sex
102. Inflorescence Sex
103. Plant Sex
104. Pollination Type
105. Fertilization Type
106. Flower-Pollinator Type
107. Periodicity (By Parts)
108. Maturation (By Parts)
109. Duration (By Parts)
110. Growth Region (By Parts)
111. Growth Periodicity (By Parts)
112. General Development (By Parts)
113. Developmental Shoot Types
114. Symmetry (By Parts)
115. Arrangement System (By Parts)
116. Branching Pattern
117. Secretory and Excretory Type(s)
118. Secretory and Excretory Structure Position
119. Variegated Color (By Parts)
120. Solid Color (By Parts)

ANATOMICAL CHARACTERS

(See Chapter 7)

Secondary Xylem

1. Imperforate Tracheary Element Type
2. Vessel Element Shape and Size
3. Vessel Pore Distribution Type
4. Vessel Distribution Type
5. Vessel Perforation Plate Type
6. Vessel Pitting Type
7. Ray Type
8. Specialized Ray Cell Type(s)
9. Axial Parenchyma Distribution Type
10. Specialized Axial Parenchyma Cell Type(s)
11. Crystal Type(s)
12. Other Secondary Xylem Features

Leaf Blade

13. Cuticular Character Distribution
14. Cuticular Character Orientation
15. Epidermal Cell Shape
16. Epidermal Cell Orientation
17. Stomatal Type
18. Stomatal Ontogeny Type
19. Sclereid Type
20. Fiber Type
21. Mesophyll Type
22. Major Venation Pattern Type
23. High Order Venation Pattern Type
24. Vein Sheathing Type
25. Trichome Type(s)

Petiole

26. Sclerenchyma Distribution
27. Vasculation Pattern

Young Stem

28. Nodal Anatomy Type
29. Vascular Bundle Type
30. Ground Tissue Cell Type
31. Sieve Element Type
32. Sclerenchyma Distribution
33. Anomalous Structure(s)

Stelar Anatomy

34. Primary Xylem Maturation Pattern
35. Stelar Pattern

PALYNOLOGICAL CHARACTERS
(See Chapters 8 & 20)

1. Pollen Aggregation Type
2. Pollen Grain Shape, Polar View
3. Pollen Grain Shape, Equatorial View
4. Pollen Grain Aperture Condition
5. Pollen Grain Wall Sculpture
6. Pollen Grain Wall Structure
7. Pollen Grain Nuclear State

EMBRYOLOGICAL CHARACTERS
(See Chapter 9)

Anther

1. Number of Locules
2. Endothecial Structure
3. Tapetal Type
4. Microspore Mother Cell Division Type

Ovule

5. Integument Number
6. Vasculation Pattern
7. Aril Nature
8. Nucellar Type
9. Archesporial Type
10. Megaspore Arrangement
11. Embryo Sac Type
12. Endosperm Formation Type
13. Embryo Development Type

CYTOLOGICAL CHARACTERS
(See Chapter 10)

1. Chromosome Meiotic Behavior Pattern
2. Chromosome Morphology Type
3. Chromosome Number

REPRODUCTIVE BIOLOGY
(See Chapters 11, 12, & 26)

1. Pollination Agent
2. Compatibility
3. Isolating Mechanism
4. Breeding System
5. Dormancy Type
6. Germination Type
7. Seed Potential
8. Variability Source(s)
9. Vegetative Patterns of Variation
10. Reproductive Patterns of Variation

ECOLOGICAL CHARACTERS
(See Chapter 15)

1. Habitat
2. Nutritional Relations
3. Sociability
4. Symbiotic Relations
5. Temperature Regime
6. Air Moisture Regime
7. Light Regime
8. Soil Texture Regime
9. Soil Aeration Regime
10. Soil Nutrients Regime
11. Soil pH Regime
12. Stratificational Relations
13. Phenological Relations
14. Successional Relations
15. Life Form
16. Leaf Size
17. Diaspore Type
18. Fidelity
19. Vitality

GEOGRAPHIC CHARACTERS
(See Chapters 16 & 17)

1. Biogeographic Region
2. Biome
3. Flora Province
4. Community Type
5. Habitat Type
6. Genus Range Pattern

7. Genus Range Size
8. Genus Discontinuity Distribution Pattern
9. Genus Center of Distribution
10. Origin or Establishment in Local Area
11. Plant Dispersal Type

II. ECOSYSTEMATICS PROBLEM

After consultation with your instructor, select a small area for local community or flora study. Identify and list the species found in the area in a flora summary chart similar to the one below according to some system of your choosing, preferably an alphabetical one. Determine and list the characteristics of each species in the chart and then make a spectral analysis of the flora. Consult and review Chapters 15, 16, and 17 for basic information and meanings of symbols. (The analysis form for this problem is on the next page.)

FLORA CHART

							Phe					
Family	Genus	Species	Lf	Dia	Soc	Dur	lf	Fl	Fr	Tax	Com	Dis
1.												
2.												
3.												
4.												
5.												
6.												
7.												
8.												
9.												
10.												
11.												
12.												
13.												
14.												
15.												
16.												
17.												
etc.												

FLORISTIC OR COMMUNITY SPECTRAL ANALYSIS
(A sample form)

1. Life Form (Lf)

	P-t	P-s	P-v	Ch	G	Th	E	S	Pa	Sa	H
# of Species											
% Total Flora											

2. Diaspore (Dia)

	Spo	Pte	Pog	Cyc	Epz	Syz	Enz	Bal	Bar	Hyd	Ant	Ate	Pol
# of Species													
% Total Flora													

3. Sociability (Soc)

	Soc 1	Soc 2	Soc 3	Soc 4	Soc 5
# of Species					
% Total Flora					

4. Duration (Dur)

	Ann	Bien	Per	Decid	Evergreen Broad	Evergreen Needle
# of Species						
% Total Flora						

5. Phenology (Phe) Leaf (L) - Flower (Fl) - Fruit (Fr)

	V	Ae	Au	H	All	Ds	Ws
# of Species							
% Total Flora							

6. Taxon (Tax)

Div./Class	Psi	Lyco	Sphe	Pter	Pine	Magn	Magt	Lilt
# of Species								
% Total Flora								

Sub-Class	Mag	Ham	Car	Dill	Ros	Ast	Alis	Arec	Com	Lil
# of Species										
% Total Flora										

7. Community (Examples)

	DG	SF	DS	SA	SDF	BHF	TG	LPF	etc.
# of Species									
% Total Flora									

8. Distribution (Di)

	Bor	Cp	App	Tro	Pra	All	Aut	Dis	Rel	Spo	Nat	Dom
# of Species												
% Total Flora												

Note: Many other spectral analyses could be made; e.g., those based on fidelity, vitality, leaf size, leaf shape, corolla types, fruit types, pollinating vectors, etc.

III. BIOSYSTEMATICS PROBLEM

Read over carefully all of the information given; study the figures and tables in detail; read over the following questions; think of probable answers; check your answers with the data given here (and, if you wish, any literature source pertinent); and then answer the questions posed briefly and to the point.

1. Work out the evolutionary relationships within the hypothetical genus Thothera and summarize your decisions in a single, neat, one-page diagram indicating the most likely origins of the species where possible, the chromosome number and terminology, and the genome or genomes involved (DD for T. deserticola, CC for T. chinensis, perhaps DDCC, etc.). Brief comments (10 words or less) that you think helpful in explaining your conclusions may be included as marginal notes adjacent to the appropriate entry on your chart. In a box in one corner indicate the names and taxonomic rank you would assign each of the entities considered.
2. Answer any 5 of the following questions. Wherever you include an evolutionary process in your answer, give, at the end of your answer by actual complete citation, one or two pertinent references to experimental work or documented observations that support your answer.

---

1. When (in evolutionary time relative to species formation) and in what species is it most likely that the Thothenine producing mutant first occurred? How do you account for the 100% occurrence of this character throughout T. deserticola?
2. What is the likely origin of the two pairs of SAT chromosomes in T. sibirica, T. arctica and T. yukonensis? Would any other evidence help validate your theory?
3. Is the keel length more uniform in T. yukonensis or in T. deserticola? Why do you say so? Also, how can you account for the rather large variation in keel length in T. arctica?
4. How do you account for the occurrence of one member of this essentially alpine or arctic genus in the deserts of the southwestern U. S. and Mexico?
5. Compare briefly the evolutionary potential of T. yukonensis and T. Holmgrenii.
6. What series of conditions account for the rarity of T. yukonensis?
7. From your experience with measurements, what question might be raised in regard to the legume measurements given for the genus as a whole? Also, how do you account for the variation in seed number through the genus as compared with the uniformity in legume size?

Plants of the circumboreal genus Thothera closely resemble those of the genus Lupinus in all vegetative and floral characters, but the two genera are easily separable on the basis of a seed character. Thothera seeds invariably have only one normal cotyledon, the second cotyledon being reduced to a mere scale on one side of the embryo in T. chinensis and completely lacking in all other species. Within the genus the taxa are all distinct as regards "key" characters and there are never any problems of identification, however, some of the taxa have been given different ranks by various workers as indicated: T. arctica L. and T. yukonensis Barnaby have both been considered subspecies of T. sibirica L.; T. Holmgrenii Cronquist has been considered a variety of

T. montana Rydberg; both T. deserticola Jepson and T. chinensis L. have been considered only on the species level. Since none of the changes in taxonomic rank were based on experimental studies of any sort and since some preliminary chromosome counts showed considerable variation in number, it appeared that a thorough biosystematic study of the genus would be both interesting and rewarding as regards our understanding of the evolutionary relationships and possible origins of these plants.

A review of the literature turned up the following items of interest concerning the plants of this genus:

1. In 1930 an article in the Journal of Anthropology noted that the distinctive seeds of T. chinensis were found with the skeletal pieces and artifacts of Peking Man and apparently were used as food.
2. In some experimental studies on the photoperiods of desert plants Went (1941) found that T. deserticola from the Mojave Desert bloomed normally under a photoperiod of 24 hours of light whereas in its desert environment blooming was dependent on rainfall and independent of photoperiod.
3. An article on "The alkaloids in leaves and stems of members of the Fabaceae" (published in 1965 in the American Chemical Ingredient Digest) showed Thothenine, a bitter lethal alkaloid, to be present in 100% of the plants of T. deserticola tested, in 12% of the T. arctica plants, 5% of the T. yukonensis plants, 22% of the T. montana plants and 9% of the T. Holmgrenii plants. Thothenine was not found in any plants of T. chinensis or T. sibirica.

General Morphology

All plants in the genus Thothera are perennial, or weakly perennial, from deep, fibrous-fleshy roots. The mature legumes are extremely uniform in shape and size throughout the genus with a mean length of 2.76 cm ± 0.58 (sample size: 5 pods per plant, 10 plants per species, total 350 measurements). The leaves are palmately compound but there is considerable variability in leaflet shape and texture as shown in the summary of diagnostic characters of the species in Table 35-2.

Thothera L.

Stems terete.
- Stems puberulent; mts. of central Asia..................1. T. chinensis.
- Stems glandular-pubescent; arctic from Norway to Vladivostock............................2. T. sibirica.

Stems angled or flattened.
- Leaves orbicular and membranous, or absent; keel 26-30 mm long.............................3. T. deserticola.
- Leaves linear to lanceolate, coriaceous; keel less than 24 mm long.
  - Stems glandular pubescent.
    - Pollen 50% or more abortive......................4. T. yukonensis.
    - Pollen 75% or more viable..........................5. T. arctica.
  - Stems glabrous or puberulent, never glandular.
    - Seed coat brown; petals white.......................6. T. montana.
    - Seed coat black; petals yellow....................7. T. Holmgrenii.

Chromosome Pairing and Morphology

Pairing: See Table 35-1.

Inversions: Through controlled crosses it was shown that *T. deserticola* was homozygous for an inversion in each of chromosomes I, III and V; *T. montana* was also found to be homozygous for the inversions I, III and V and in addition was about 20% heterozygous for an inversion in chromosome X; *T. Holmgrenii* is homozygous for the inversion in chromosome X and has no other inversions. Inversions are not known in any other taxa of the genus.

SAT Chromosomes: The chromosomes of all plants examined were essentially similar except that examination of the somatic chromosomes (root tip feulgen squashes) showed four chromosomes with satellites in *T. sibirica*, *T. arctica* and *T. yukonensis*. In crosses involving only these three taxa it can be shown that the SAT chromosomes always pair at meiosis. However, in crosses between plants of these taxa and plants of any other taxa in the genus SAT chromosomes show up as two, single, unpaired chromosomes at meiosis.

Table 35-1. Pairing at meiosis in the F_1 plants from experimental *Thothera* crosses.

C X C	12_{II}	D X D	13_{II}	H X M	$11_{II} + 1_{I}$
C X D	$8_{II} + 9_{I}$	D X H	$11_{II} + 2_{I}$	H X S	$6_{II} + 23_{I}$
C X H	$6_{II} + 11_{I}$	D X M	$11_{II} + 1_{III}$	H X A	$11_{II} + 13_{I}$
C X M*	$9_{II} + 6_{I}$	D X S	-----	H X Y***	$11_{II} + 36_{I}$
C X S	$12_{II} + 12_{I}$	D X A	$11_{II} + 1_{III} + 12_{I}$	M X M	12_{II}
C X A	36_{I}	D X Y	-----	M X S	$9_{II} + 18_{I}$
C X Y	$23_{II} + 14_{I}$	H X H	11_{II}	M X A	$12_{II} + 12_{I}$
S X S**	24_{II}	A X A	24_{II}	M X Y***	$12_{II} + 35_{I}$
S X A	$12_{II} + 24_{I}$	A X Y	$24_{II} + 23_{I}$	Y X Y	$24_{II} + 11_{IV} + 2_{I}$
S X Y	$24_{II} + 23_{I}$				

*Crosses between some populations give up to 24_{I}

**Occasionally some quadrivalents.

***Cross possible only when Y used as pollen parent.

Table 35-2. Summary of diagnostic characters of the species of <u>Thothera</u>.

	T. deserticola	T. chinensis	T. sibirica	T. arctica	T. yukonensis	T. montana	T. Holmgrenii
Habitat and Distribution (See Fig. 1)	Deserts; Calif., Ariz. Mexico	High mts; Central Asia	Margin of tundra; Norway to Vladivostock	Margin of tundra; Alaska to Labrador	Sand bars below river gorges;* Yukon Terr.	High open slopes; Rocky Mts.	Steep slopes above old Bonneville shoreline, Utah
Leaflets	orbicular membranous (rainy season only)	linear coriaceous	lanceolate coriaceous	lanceolate coriaceous	narrowly lanceolate subcoriaceous	linear coriaceous	linear coriaceous
Stems	green flattened villous	green terete puberulent	green terete pubescent	green angled pubescent	green weakly angled glandular-pubescent	green angled puberulent	green angled glabrous
Flowers	white sweet scent at dusk	blue odorless	pale blue, rarely white odorless	dark blue to white, rarely yellow odorless	pale blue odorless	white sweet scent	pale yellow faint scent or odorless
$\overline{X}$ Keel length (mm)	28.3 $\pm$ 1.9	12.7 $\pm$ 3.6	15.1 $\pm$ 2.2	17.1 $\pm$ 4.6	16.8 $\pm$ 1.9	17.4 $\pm$ 2.7	16.8 $\pm$ 3.5
Pollinators	Long tongued moth	Small bee	self	self	self incompatible, pollinators unknown	moth	bees
Legume Seed #	3	4-5	4-6	4-5	4-5	10-12	9-12
Seed Coat	Gray to tan; soft, smooth 0.7 mm thick	Black soft, smooth 0.1mm thick	Gray to tan; hard, pitted 0.3mm thick	Gray to tan hard, pitted 0.5mm thick	Light gray hard, pitted 0.5mm thick	Brown soft, smooth 0.2mm thick	Black soft, smooth 1mm thick
% Pollen Viability	97	99	94	96	18	87	98

*These plants are quite rare and are always found as single plants widely scattered within their very limited range.

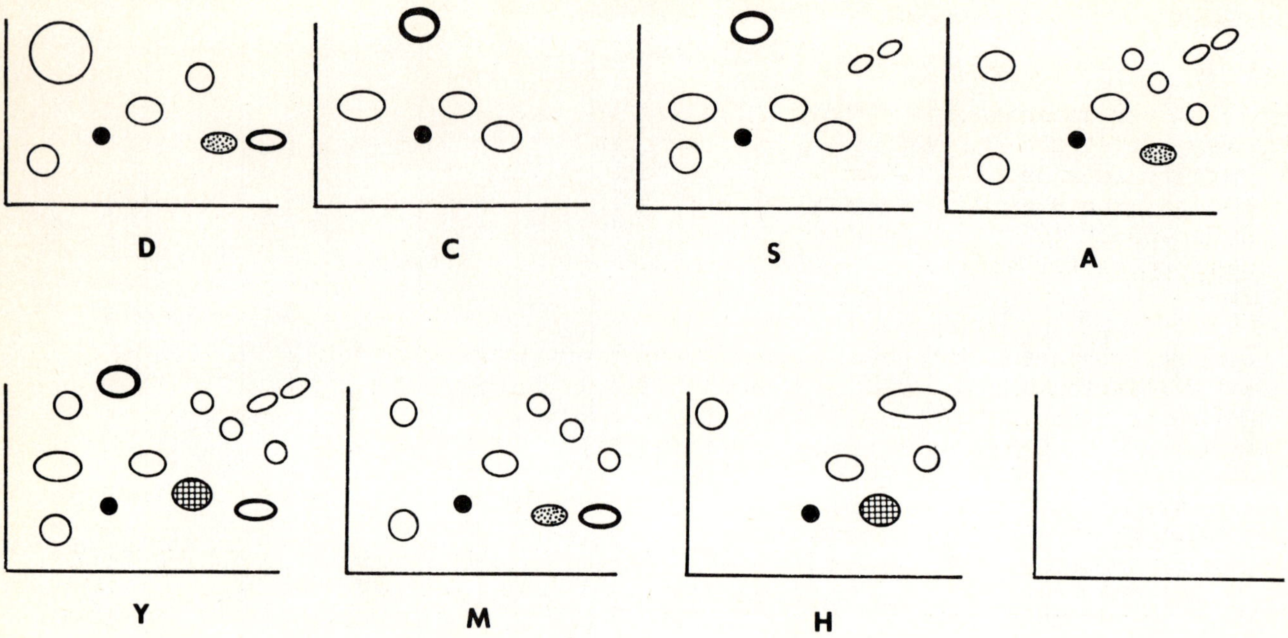

Figure 35-1. Chromatographs of the seven taxa of Thothera.

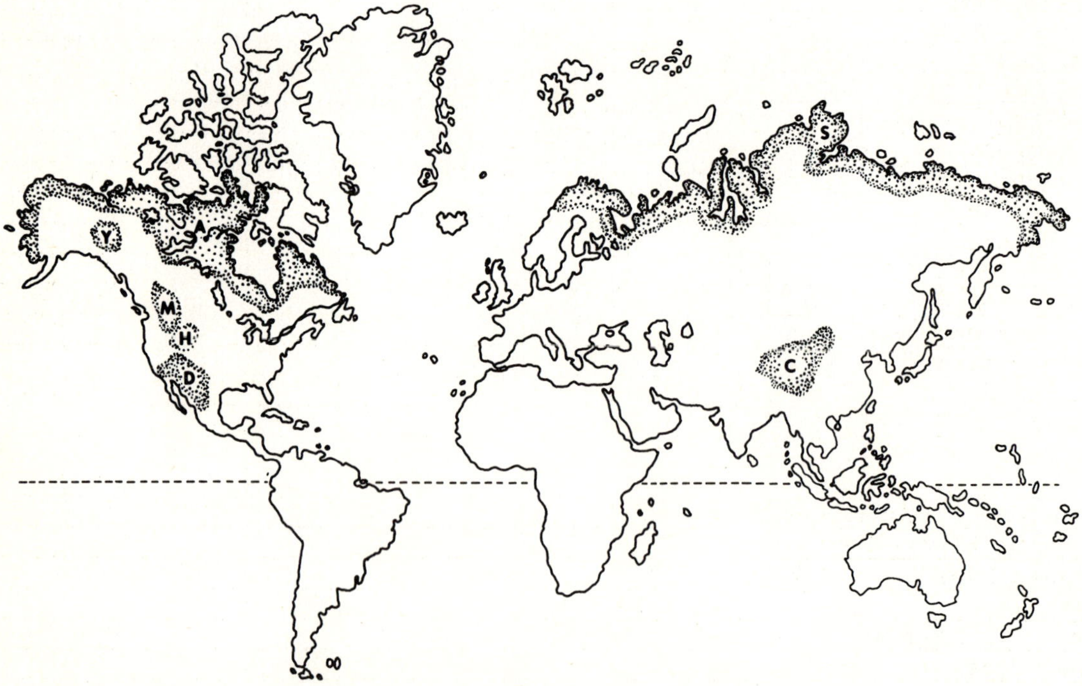

Figure 35-2. Thothera distribution.

IV. PHYLOSYSTEMATICS PROBLEM

After consultation with your instructor, select one to three taxa for study. List the characteristics of each taxon and decide whether the character is advanced (+), advanced and primitive in Taxon (±) or primitive (-). The features that do not apply, cannot be determined, or are unknown should be indicated (0). Upon completion of the chart determine the Index of Advancement for each taxon and the Coefficient of Relationship between the taxa.

For character information consult Chapters 6-18, your family work sheets, and Darlington, C. D., and A. P. Wylie (1955), Erdtman (1962), Cave (1958), Cronquist (1968), Lawrence (1951), Metcalfe, C. R. (1960, 1971), Metcalfe, C. R., and L. Chalk (1950), and Hegnaur (1961).

TAXON / FEATURE	I.______		II.______		III.______	
	CHARACTER STATE	SYMBOL	CHARACTER STATE	SYMBOL	CHARACTER STATE	SYMBOL
1. Plant Habit						
2. Plant Duration						
3. Plant Sex						
4. Root Type						
5. Stem Type or Habit						
6. Leaf Type						
7. Leaf Arrangement						
8. Leaf Position						
9. Leaf Duration						
10. Leaf Venation						
11. Leaf Shape						
12. Leaf Apex						
13. Leaf Base						
14. Leaf Margin						
15. Leaf Attachment						
16. Leaf Surface						
17. Leaf Venation						
18. Inflorescence Type						
19. Inflorescence Position						
20. Inflorescence Develop.						
21. Flower Symmetry						
22. # Floral Parts						
23. # Whorls of Floral Pts.						
24. # of Members within Whorl						
25. Insertion of Floral Pts.						
26. Fusion of Floral Parts						
27. Perianth Type						
28. Stamen Type						
29. Stamen Arrangement						
30. Anther Type						
31. Filament Feature						
32. Pollination Type						
33. Pollen Aggregation Type						

TAXON FEATURE	I.		II.		III.	
	CHARACTER STATE	SYMBOL	CHARACTER STATE	SYMBOL	CHARACTER STATE	SYMBOL
34. Pollen Aperture Type						
35. Pollen Nuclear Condition						
36. Pollen Wall Sculpture						
37. Carpel Arrangement						
38. Stigma Type						
39. Style Type						
40. Gynoecial Type						
41. Ovary Position						
42. Ovule Type						
43. # Ovules per Ovary						
44. # Integuments						
45. Nucellus Size Type						
46. Embryo Sac Type						
47. Embryogeny Type						
48. Endosperm Develop. Type						
49. Fruit Type						
50. Fruit Dehiscence Type						
51. Placentation Type						
52. Seed Surface Feature						
53. Seed Type						
54. Embryo Type						
55. # Cotyledons						
56. Cotyledon Position						
57. Germination Type						
58. Eophyll Type						
59. Epidermis Thickness						
60. Mature Stomatal Type						
61. Epidermal Cell Type						
62. Stomatal Ontogeny Type						
63. Mesophyll Type						
64. Vascular Bundle Type						
65. Nodal Type						
66. Stelar Type						
67. Protoxylem Pattern of Diff.						
68. Phloem Type						
69. Sieve Element Type						
70. Phloem Fiber Type						
71. Vessel Element Type						
72. Tracheid Type						
73. Fiber Type						
74. Axial Parenchyma Type						
75. Ray Type						
76. Pith Type						
77. Floral Trace Feature						
78. # Root Xylem Archs						
79. Nectary Type						
80. Breeding Mechanism						
81. Isolating Mechanism						
82. # Chromosomes						
83. Chromosome Size						
84. Chromosome Type						

TAXON FEATURE	I. CHARACTER STATE	SYM-BOL	II. CHARACTER STATE	SYM-BOL	III. CHARACTER STATE	SYM-BOL
85. Chromosome Structural Feature						
86. Bud Type						
87. Bud Cover Type						
88. Nutritional Relations						
89. Symbiotic Relations						
90. Temperature Regime						
91. Moisture Regime						
92. Light Regime						
93. pH Regime						
94. Soil Texture Regime						
95. Mineral Nutritional Reg.						
96. Periodicity Type						
97. Successional Type						
98. Phenological Type						
99. Growth Pattern Type						
100. Food Reserve Structure						
Etc.						

$$\text{Index of Advancement} = \frac{\text{Total \# of advanced characters (+)}}{\text{Total \# of advanced (+) \& primitive characters (-)}} \times 100$$

Coefficient of Relationship
between two taxa (See Coefficient of Similarity in Chapter 23)

Section C. REVIEWS
(Samples only)

I. Morphology. Review or determine the character states of the morphological character for each of the following parts or organs:

ORGANISM PARTS OR ORGANS

A. Plant
B. Root
C. Stem
D. Bud
E. Leaf
F. Inflorescence
G. Flower
H. Perianth
I. Androecium
J. Gynoecium
K. Fruit
L. Seed
M. Embryo

CHARACTER

1. Parts
2. Habit
3. Types and Specialized Types
4. Duration
5. Position
6. Arrangement
7. Texture
8. Surface
9. Vestiture
10. Shape-Body
11. Shape-Apex
12. Shape-Margin
13. Shape-Base

II. Sexual Reproduction and Pollination. Review or determine the states or types of the following in flowering plants:
A. Sexual Conditions B. Pollination Types C. Fertilization Types

III. Dispersal and Dehiscence. Review or determine the types of the following found in seed plants:
A. Diaspore Types B. Dehiscence Types in Fruits and Anthers

IV. Growth and Development. Review or determine the types or conditions of each of the following in vascular plants:
A. Periodicity States B. Maturation Times C. Symmetry Patterns
D. Development Patterns E. Growth Regions

V. Anatomy (Selected not comprehensive). Review or determine the character states for each of the following characters:
A. Tracheid and Fiber Types B. Ray Types C. Axial Parenchyma Cell Types D. Angiosperm Stomatal Types E. Major Venation Pattern Types F. Sclereid Types G. Nodal Types H. Vessel Perforation Plate Types--and many others.

Section D. EXAMINATION SAMPLES
(For Basic Systematics Course Only)

Sample Midterm Examination

Basic Systematics Course Name____________________

I. Diagram or sketch: 1. Septate Pith 2. Ring Porous Wood 3. Anomocytic Stomata 4. Stellate Trichome 5. Primitive Vessel 6. Vasicentric Parenchyma 7. Trilacunar and three-trace node 8. Diacytic Stomata 9. Pore Multiple Vessels 10. Bilateral or Isolateral Mesophyll.

II. Give an example of a Family or Genus with each of the following: 1. Silique 2. Zygomorphic Corolla 3. Valvular Dehiscent Anthers 4. Apetaly 5. Epigyny 6. Polygamy or Imperfect Flowers 7. Syngenesium 8. Parietal Placentation 9. Tetradynamy 10. Apocarpy 11. Gynandrium 12. Ligulate Corolla 13. 3;3;1-2;3 inferior 14. 5;5;3-10; 3 cyme 15. 5;5;5;5-3 sympetaly, inf. 16. 3;3;6;3 inferior

III. Give four ways in which cross-pollination is insured in angiosperms.

IV. List four types of author citations and explain each.
Give an example of a specific epithet for each of the following: 1. Habit 2. Habitat 3. Commemoration 4. Sexual Condition
Give the rank of each of the following: 1. Magnoliales 2. Rosoideae 3. Bryophyta 4. Acer

V. Diagram the development (megasporogenesis and megagametogenesis) of a primitive type and an advanced type embryo sac. Label all parts of one embryo sac with surrounding ovular parts.

VI. A. Write a dichotomous key or define and give examples of ten inflorescence types or ten fruit types.
B. Answer one of the following:
1. Distinguish between: a. stolon, rhizome, tuber. b. Paripinnately, imparipinnately, palmately compound leaves. c. Scale, bract, bracteole types of leaves. d. Rotate, salverform, campanulate corolla types. e. Hastate, sagittate, auriculate leaf shapes or bases.
2. Describe ten sexual conditions in plants.
3. Sketch and name ten types of leaf bases or attachments.

VII. Give four phases in the development of systematics.
List three basic axioms or principles of taximetrics.
Describe the ways in which one character can be diagnostic, reliable, and meaningful, respectively.

VIII. Describe each of the following: 1. Tricolporate aperture 2. Zonoaperturate condition 3. Linear tetrad
Define each of the following: 1. Allotetraploidy 2. Aneuploidy 3. Trisomic 4. Polyploidy

Name the most primitive aperture condition in pollen.
Give the number of chromosomes of a type in a diploid cell.
Give the nucellus or female sporangium type in a crassinucellate ovule.

IX. Give the author of each of the following:
1. "Materia Medica" 2. "Origin of Species"
3. "Species Plantarum" 4. "Pinax theatri botanici"
Name: 1. a German herbalist noted for his illustrations; 2. the "Father of Botany"; 3. the author of the first detailed plant synonymy; 4. the author who first consistently used binomial nomenclature.

Sample Final Examination

Ia. List ten fruit types and give an example of a family in which each is found.

b. List ten corolla, calyx, or perianth features and give an example of a family in which each is found.

c. List five staminal features or five placentation types and give an example of a family in which each is found.

IIa. List five different communities and give two indicator or character species found in each and the habitat of each.

b. List five different communities found in the piedmont or mountains and name the next most xeric community and the next higher community found in the same area (in the order noted below).

Community	Xeric	Higher	Community	Xeric	Higher

III. a. Write a dichotomous key to ten types of sexual conditions found in plants or a key to ten inflorescence types.

IV. Name one type of each of the following and either define or sketch.

1. Nodal Type
2. Leaf Margin
3. Embryo Sac Type
4. Pollen Aperture Type
5. Leaf Vernation
6. Stem Habit
7. Embryogeny Type
8. Axial Parenchyma Type
9. Inflorescence Development
10. Stigma Feature
11. Ovule Type
12. Fruit Dehiscence Type
13. Stomatal Type
14. Flower Symmetry
15. Ray Type

V. a. Give three basic differences between the Linnaean, Englerian, and Besseyan Systems.
b. Describe five types of isolating mechanisms involved in evolution.
c. Give the most primitive conditions or antithesis of each of the following: 1. Epigyny 2. Campylotropy 3. Berry 4. Helobial Endosperm 5. <u>Adoxa</u> type Embryo Sac

VI. a. List, give an example, and define five types of diaspores.
b. List and define or explain five distribution principles or plant origins.

VII. a. Define or explain five of the following: 1. Basionym 2. Nomenclatural Priority 3. Commemorative Epithet 4. Valid Publication 5. Lectotype 6. Author Citation in ( ) 7. Tautonym

VIII. Explain how each of the following is used or give significance of each: (Answer ten only)
1. Index Kewensis, 2. Species Plantarum, 3. Rhodora, 4. Microfiche, 5. Davis & Heywood, 1963, 6. Origin of Species, 7. Die natürlichen Pflanzenfamilien, 8. Phytopinax, 9. Metcalfe & Chalk, 10. M. L. Fernald, 8th ed., 11. Materia medica, 12. Index Nominum Genericorum

Sample Midterm Lecture Examination

I. Principles (Answer 5 only)
1. What are the basic reason(s) for naming any object, thing, or organism?
2. Which identification principle(s) are involved in keys, polyclaves, color charts, advertisements?
3. Which principle(s) of classification are involved in the grading of apples, ranking of basketball teams, and the categorizing of plants?
4. What is the basic premise in botanical nomenclature? The basic premise in the naming of anything, i.e., brand names, dogs, nuts, bolts?
5. Describe the principles of description which are basic to the principles of nomenclature, identification, classification and relationships.
6. How are the principles of relationship and classification related?

II. Distinguish between the following: (Answer 10)
1. Evolution and phylogeny, 2. Naming and nomenclature, 3. Polyclave and key, 4. Natural and phylogenetic systems of classification, 5. Character state and character class, 6. Position and arrangement, 7. Abiotic and biotic relationships, 8. Epithet and name, 9. Lectotype and syntype, 10. Autonym and tautonym, 11. Validly published and effectively published name, 12. Monothetic and polythetic taxa.

III. Explain each of the following: (Answer three in each group)

A. Citations
1. Chamaecrista aspera (Muhl.) Green 2. Phyllanthus emend. Mull. Arg.
3. Cassia fasciculata var. robusta (Poll.) MacBride 4. Lupinus [Tourn.] L.

B. Meanings of Epithets
1. Magnolia grandiflora L. 3. Rhynchospora microcephala Britton ex Small
2. Veratrum parviflorum Michx. 4. Vaccinium macrocarpon Aiton

C. Types of Epithets
1. Erythronium americanum Ker 3. Oxydendrum arboreum (L.) DC.
2. Veratrum viride Aiton 4. Amorpha schwerini Schneider

D. Accenting and syllabalizing
1. ter-rés-tris 3. À-cer 5. prì-ce-ae
2. ca-na-dén-se 4. a-mòe-nus 6. glà-brus

IV. Answer three of the following:
1. Under what condition(s) can an analytical character become a synthetic one; a descriptive character, a diagnostic one; an analogous character a homologous one; a reliable character, an unreliable one?
2. What is meant by the development of systems of classification? Cite example(s) of development within specific systems.
3. List five suggestions for users of keys or five precautions to take in using keys.
4. Select two classification systems and give the bases for each.

V. 1. Select and name four indexes to taxonomic literature and four major non-index references; indicate how or for what each is used.

Chapter 36. PRINCIPLES AND CONCLUSIONS

Section A. PRINCIPLES

These principles represent an attempt to provide a broad conceptual summary of taxonomy. The general systematics principles are essentially idealistic, the taxonomic principles are relatively realistic, and the notes on supporting field principles are simply references to principles for the fields of evidence and aspects of data handling used in systematics. The ideas and basic tenets included in this section are from many, mostly unknown, sources over many years with the only possible originality being in some of the recombinations of concepts within the statements of principles.

I. GENERAL SYSTEMATICS PRINCIPLES

1. Systematics is based on precise description and exact definition of character states.
2. Systematics uses many characters from all types of evidence and all fields of botany for the description, identification, and classification of plants.
3. Systematics involves characterization and delimitation of taxa which make identification and classification schemes possible.
4. Systematics involves hierarchical arranging or ordering of data and taxa which make classification systems feasible.
5. Systematics is involved in the determination of relationships of taxa which are basic to a phylogenetic classification.
6. Systematics provides the system for naming, and names for all taxonomic groups.

7. Systematics uses the species as the base and a hierarchical arrangement of named taxa as the skeletal framework to which all taxonomic information is attached.

8. Systematics involves all phases and levels of learning: recall, recognition, application, interpretation, generalization, and evaluation for the development, communication, and use of taxonomic information.
9. Systematics is a holistic science involving field, laboratory, computer, garden, herbarium, and library studies in the accumulation and use of quantitative and qualitative data for taxonomic purposes.
10. Systematics uses the intellectual approaches of observation and description, experiment and analysis, synthesis and theory for the procurement and interpretation of taxonomic data.
11. Systematics is an intellectual pursuit that makes meaningful explanations for the diversity of nature.
12. Systematics is a fundamental discipline to botanical endeavor and is an integrative field, using all aspects of botany, in its own development.
13. Systematics is a science which contributes to the advancement of related disciplines.
14. Systematics is a historically-based but dynamic and ever-changing field. A steady stream of data incorporated into the taxonomy of plants requires reordering of descriptions, revision of identification schemes, reevaluation of classification systems, and perception of new relationships.

II. BASIC TAXONOMIC PRINCIPLES
(See Figure 1-1, page 12)

A. Principles of Nomenclature (Principles 1-6 from the "International Code")

1. Correct Name. Each taxonomic group with a particular circumscription, position, and rank can bear only one correct name, the earliest that is in accordance with the Rules, except in specified cases. (Principle IV)
2. Latin. Scientific names of taxonomic groups are treated as Latin regardless of their derivation. (Principle V)
3. Priority. The nomenclature of a taxonomic group is based upon priority of publication. (Principle III)
4. Type. The application of names of taxonomic groups is determined by means of nomenclatural types. (Principle II)
5. Retroactive. The rules of nomenclature are retroactive unless expressly limited. (Principle VI)
6. Independent. Botanical nomenclature is independent of zoological nomenclature. The Code applies equally to names of taxonomic groups treated as plants whether or not these groups were originally assigned to the plant kingdom. (Principle I)
7. Application of the code of nomenclature results in names for taxa. Names are the communication symbols and reference bases for information storage, retrieval, and use.

B. Principles of Description (See Chapters 5 and 24)

1. Description is the orderly recording of characters, with appropriate character states, for plant parts, plants, and plant taxa.
2. The basic unit in description is the attribute or characteristic which is the character state; character states are distinct phases of characters; characters are grouped as character classes which are fundamental components of types of evidence.
3. Character states have to be precisely defined for accuracy and ease of application as descriptors.
4. Characters have to be hierarchically arranged and classified for comparative accuracy in relevant and diagnostic description.
5. Character classes have to represent all types of evidence for the total diversity of characters required for a comprehensive descriptive terminology.
6. Characterization provides the descriptors and descriptive system for information storage and retrieval for all taxa.

C. Principles of Identification (See Chapter 25)

1. Taxa have to be accurately characterized, diagnostically delimited, hierarchically arranged, and named to make identification schemes possible.
2. Determination of a group to which a specimen belongs is done by comparing the nature and qualities of the unknown specimen with the nature and qualities of known groups.
3. Determination of the name of a taxon is done by comparing the nature and qualities of the unknown with those of named taxa.
4. Identification, process and system, provides the key to information retrieval.

----

5. "No more groups can be distinguished than the product of the number of character states." Sokal and Sneath, 1963 (see page 745 for ref.).
6. Monothetic taxa are basically easier to identify than polythetic taxa; taxa with limited diagnostic characteristics are easier to identify than taxa with numerous combinations of similarities.
7. Expert determination precedes recognition which precedes comparison in identification reliability.
8. A system of identification in which a choice of many characteristics can be used, as in the polyclave method, is more desirable than one in which a rigidly limited number of characteristics, as in a key, is used.

D. Principles of Classification (See Chapter 28)

1. Classification is based on character correlations and discontinuities of variation which are necessary for characterization and delimitation of taxa.
2. Classification is the ordering of plants into a hierarchy of taxa with the species being the fundamental basic unit, rank or category.
3. Classification results in an orderly arrangement or system designed to express interrelationships in a practical or natural way.
4. Classification provides a system for efficient and effective information storage.

5. No character, per se, is more important than another in classification, but one character may be more significant within a group.
6. The limits of a taxon cannot usually be defined quantitatively, or qualitatively by a single characteristic.
7. Assignments of natural populations or of population systems to taxa and the hierarchical arrangement of those taxa involve judgment.

E. Principles of Relationship (See Chapters 15, 26 and 27)

1. Relationships of individuals and populations are biotic, abiotic, spatial, and temporal.
2. Systematic relationships are biotic.
3. Biotic relationships are genetic, phyletic, and phylogenetic.
4. Genetic, phyletic and phylogenetic relationships in organisms are reflected in the analyzable and determinable differences and resemblances in many types of characters.
5. The basic evolutionary mechanisms of mutation, recombination and selection in different environments result in subsequent interrelationships and change in relationships.
6. A study of relationships results in an understanding of the basic continuities for practical or "natural" classification systems and the discontinuities necessary for identification systems.

7. All parts of organisms at all stages of their development may produce valuable evidence in the establishment of relationships.
8. No single character from any type of evidence is more important than another in the establishment of relationships, but one character, or group of characters, is usually more significant within a limited group.

III. NOTES ON PRINCIPLES OF SUPPORTING FIELDS

1. Each major field of evidence has its own set of principles. (See the basic information in Chapters 6-19 or a major reference at the end of each chapter for principles pertinent to that field.)
2. Data principles have been devised for data accumulation, analysis, use, documentation and presentation. (See Chapters 20-24, particularly Chapter 22.)
3. Phyletic and phylogenetic studies are based on the principles of evolution and phylogeny. (See Chapters 26 and 27.)

Section B. CONCLUSIONS

New approaches, better definition of objectives, more realistic evaluation of curricula and course presentations, and better organization of materials and methods are greatly needed for improved systematics education. Taxonomic training is still basically a local flora study, one advanced course, and a research apprenticeship. More comprehensive and more problem and principles oriented courses should be developed. New techniques, new procedures, and application-designed modules and minicourses need to be incorporated with the best of the traditional training for more effective instruction. The course(s) in taxonomy need to be made more attractive, while being made sounder pedagogically, to attract the brain-power necessary for critical research and the general development of the field.

Taxonomy should make its training socially relevant. Students should get basic training in systematics that will prepare them for impact studies, for natural area inventories, as nature counselors, as consultants on the most effective use of local plant resources, as workers in systematics centers, and as research associates and technicians. All of our training does not have to be directly related to the Ph.D. program. Applied courses could be added to undergraduate botanical curricula with little effort. Terminal masters degree programs could be designed within the framework of most taxonomic offerings that would prepare students better for impact and inventory work. Since systematics courses frequently are a requirement for field-oriented environmental programs, they should be geared more toward the applied aspects of ecology. Professional systematists are obligated to do a better job of making their training and/or research germane to the botanical problems of our culture. Knowing the plants, being able to determine the biotic, abiotic, spatial, and temporal relationships of the individuals and populations, and having a knowledge of the classification of plants are fundamentals of taxonomic training that can be made applicable, with a little imagination and innovation, to some of the basic problems of our times.

The present nomenclature is relatively stable, with that pertaining to higher taxa increasingly more so. Only minor tinkering is anticipated in the improvement of the "International Code," although some individuals advocate eliminating the present system of nomenclature for a more rationally based code.

Our descriptive terminology is extensive but the precision and clarity of definition needed for accurate application in this computer age is far from totally acceptable. Observations of plant parts, plants, and populations have to be made more extensive as well as more consistent with the resulting descriptors being fed into a storage and retrieval system that is more usable and more efficient than the library.

The dichotomous key, long the trade-mark of plant identification, needs to be made more effective by increased precision in diagnostic terminology and more consistent comparisons of characters in key couplets. Keys or systems of identification need to be devised which are based on micro structural characters (palynological, embryological, cytological, anatomical), ultrastructural characters, chemical characters, and physiological characters for a larger user community. Great improvements in identification lie ahead with the development of mechanical and computer-assisted means and methods.

In the development of systems of classification in the last two decades many more characters from many types of evidence have been used in their synthesis. With the resultant increase in relationships shown, more natural classifications have evolved. The processing of taxonomic information by computer and the development of data banking with a concomitant increase in botanical data and information should provide the classificationists with untold opportunities in the future for making the systems more "natural" or more practical.

Determination of plant relationships by experimental and analytical means has barely begun even though some of the methodologies are relatively sophisticated. Systematic research needs a drastic improvement in problem definition, experimental design, and selection of analytical procedures for the procurement and interpretation of taxonomic data.

Much taxonomic research remains to be done. New species await discovery, description, and naming. Monographs of many genera have to be made. Revisions of many genera treated more than a century ago are sorely needed. Relatively unknown floristic provinces and vegetation types, particularly in the tropics, still exist which need exploration and study. Floristics based upon edaphic conditions is hardly underway in most of the world. Detailed floristic spectral and diversity analyses as aids in the determination of evolutionary relationships and as bases for the understanding of species and higher taxa diversity and distribution have just begun. The basic fields of evidence used in taxonomic studies need redefinition, new conceptual bases, and the use of modern techniques for increased precision, greater accuracy, and easier application. Many types of evidence, other than morphological, remain to be applied to taxonomic problems of all sorts. The principles of taxonomy need much thought and reevaluation. From a world standpoint, biological (Species Biology) studies have not been made for the vast majority of the species of the earth. Mathematically based schemes for determination of evolutionary relationships have been well developed but not widely used. Classification shows some maturity but new relationships are, and will continue to be, discovered with the incorporation of new data and ideas into systems synthesis and theory. Many anomalous genera remain to be accurately placed in family classification schemes.

Taxonomy will continue to develop as it assimilates the pertinent and useful from the many expanding fields of evidence on which it is built. The science must incorporate more mathematical logic, improved methods, and more refined intuition into its problem solving. If systematics is to provide a sound foundation for biology and if it is to be an integrating discipline in the future progress of the life sciences, it must be <u>creative</u> and <u>dynamic</u>.

ILLUSTRATED ANGIOSPERM FAMILIES

The 25 plates that follow are part of a series that Carroll E. Wood, Jr., has selected for use in teaching from among more than 350 illustrations prepared for A Generic Flora of the Southeastern United States, which is under his direction. One-hundred twenty of these specially labeled illustrations, including representatives of 70 families of dicotyledons in eastern North America, are being produced as a separate publication by Harper and Row, the publishers of this book.

All of the illustrations prepared for the Generic Flora have been intended to illustrate as many biological details as possible, and not to be merely "recognition" drawings. Most of the illustrations have been made either from living plants or from material preserved in alcohol, and all have been checked for accuracy by one or more botanists who have worked on the Flora. The original illustrations are ten inches wide; the length depends on the size and number of details that are figured; and all have been drawn for reduction by one-half. For teaching purposes these plates have been modified by placing the labels directly on the illustration, rather than in a conventional legend below it, and they are published here with reductions of only 12 to 35 per cent. The combination of larger format and labeled illustration makes it possible to pinpoint various structures that are difficult to explain otherwise (e.g., the flower of Asclepias). It has not been feasible, however, to give the magnification of each component figure.

The labels have been prepared by Professor Wood with the collaboration of Dr. Elizabeth A. Shaw and the help of Dr. Kenneth R. Robertson. Of the 25 plates on pages 853-877, eight were originally published (with a greater reduction and with conventional legends) in the Journal of the Arnold Arboretum, in which the extensive series of papers treating the genera of flowering plants in the southeastern United States has been appearing since 1958 (volume 39). The six artists whose work is represented on the following pages are Karen Stoutsenberger Velmure (initials KS or KSV on illustrations), Dorothy H. Marsh (DHM), Arnold D. Clapman (ADC), Virginia Savage (VS), Rachel A. Wheeler (RAW), and Laverne Trautz (LT). The Generic Flora is a project of the Arnold Arboretum and the Gray Herbarium of Harvard University that has been made possible by the generosity of Dr. George R. Cooley and by grants from the National Science Foundation. Many individuals have played a part in the production of the drawings and in various aspects of the research for the Flora. Even to name them would take up more than the allotted space, but their help has contributed in uncountable ways to these illustrations. It is the hope of all who have worked on the Generic Flora that these detailed illustrations will be useful to both the beginning student and the experienced botanist.

Index to Illustrations

(Family arrangement is that of Diels, A. Engler's Syllabus der Pflanzenfamilien, ed. 11. Berlin. 1936)

ARACEAE: Arisaema. a-j, A. atrorubens; k-m, A. stewardsonii; n, A. dracontium

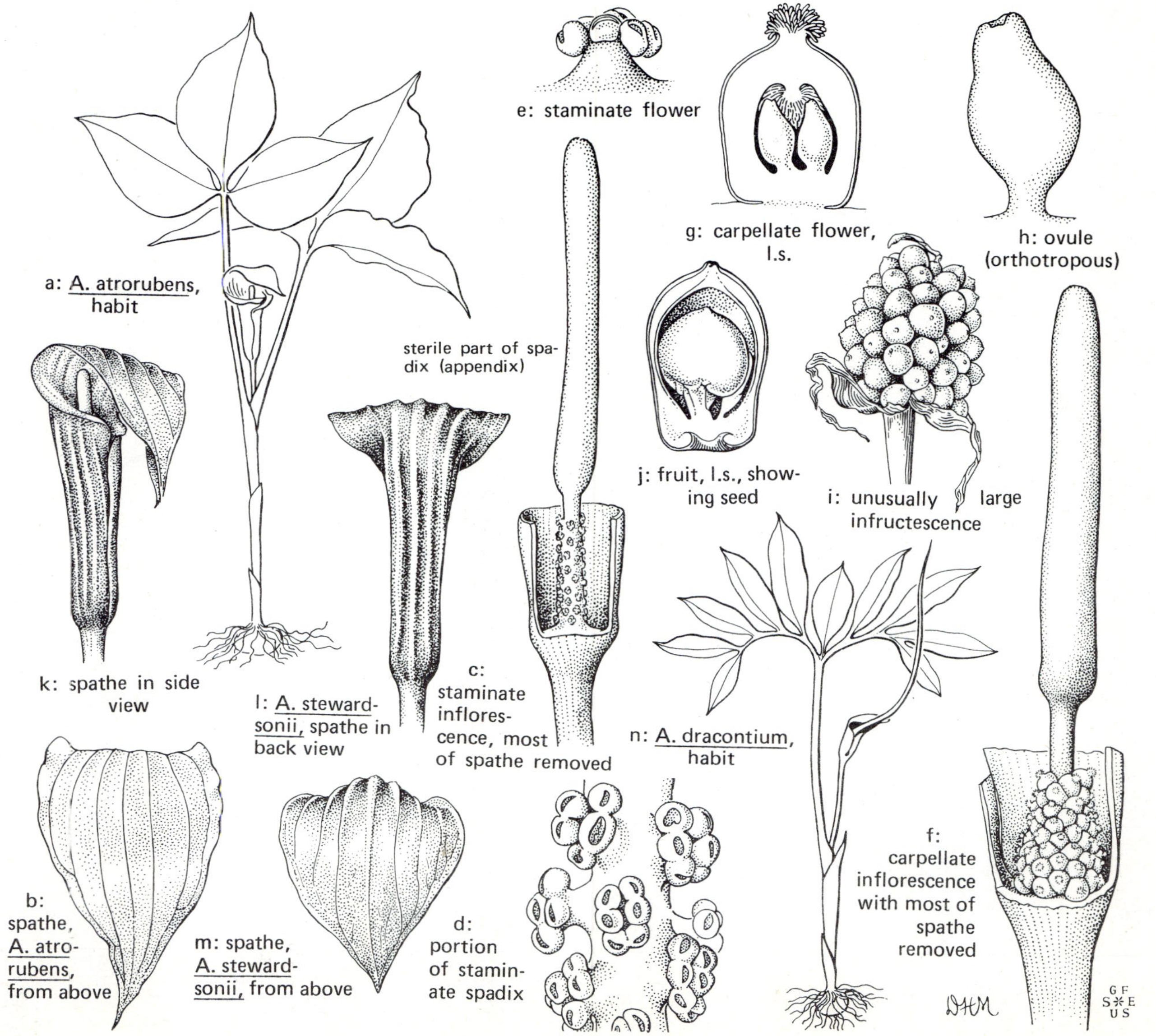

LILIACEAE: Smilacina and Maianthemum. a-g, S. racemosa; h-l, M. canadense

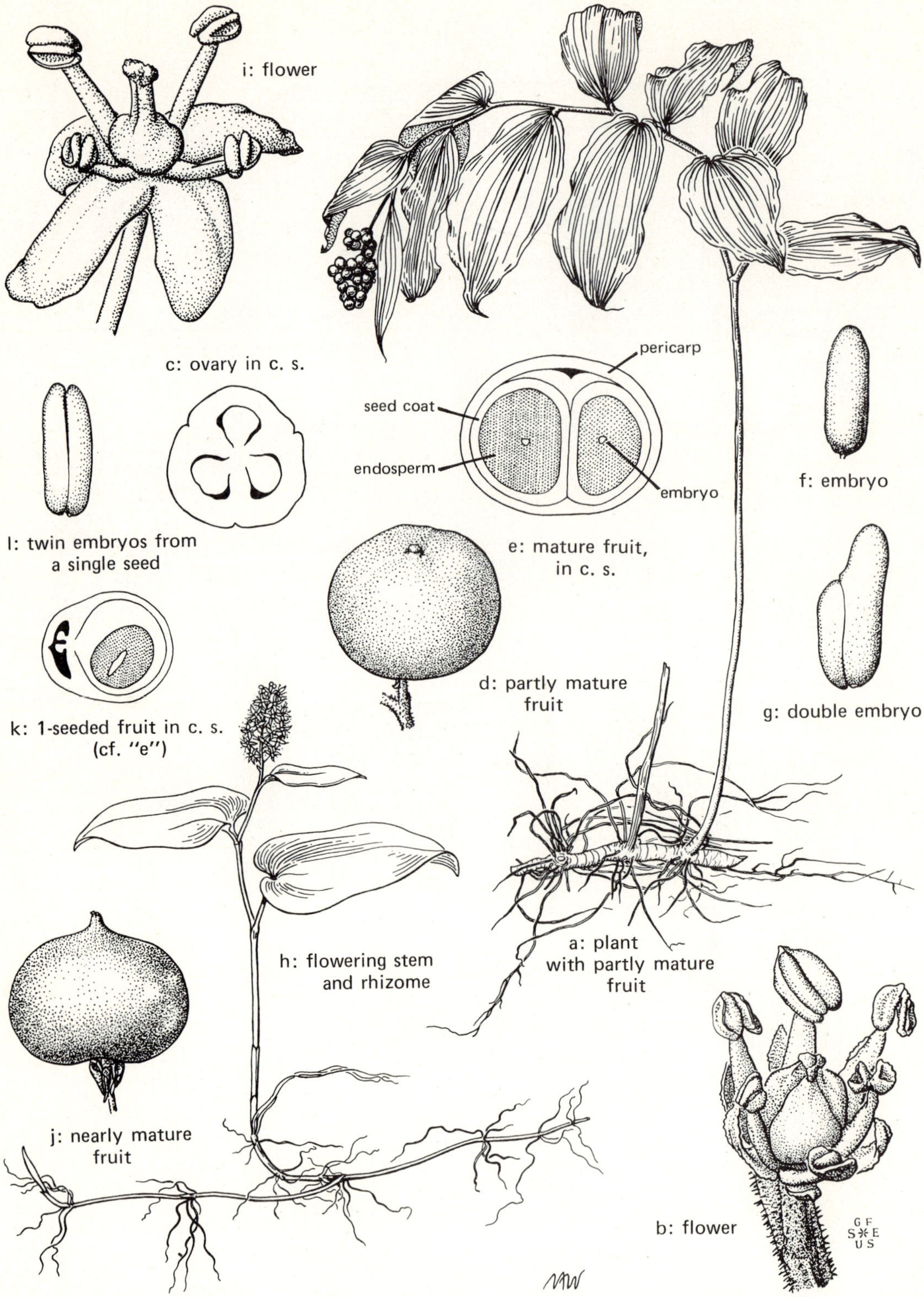

ORCHIDACEAE: Epidendrum section Encyclia. a-m, E. tampense

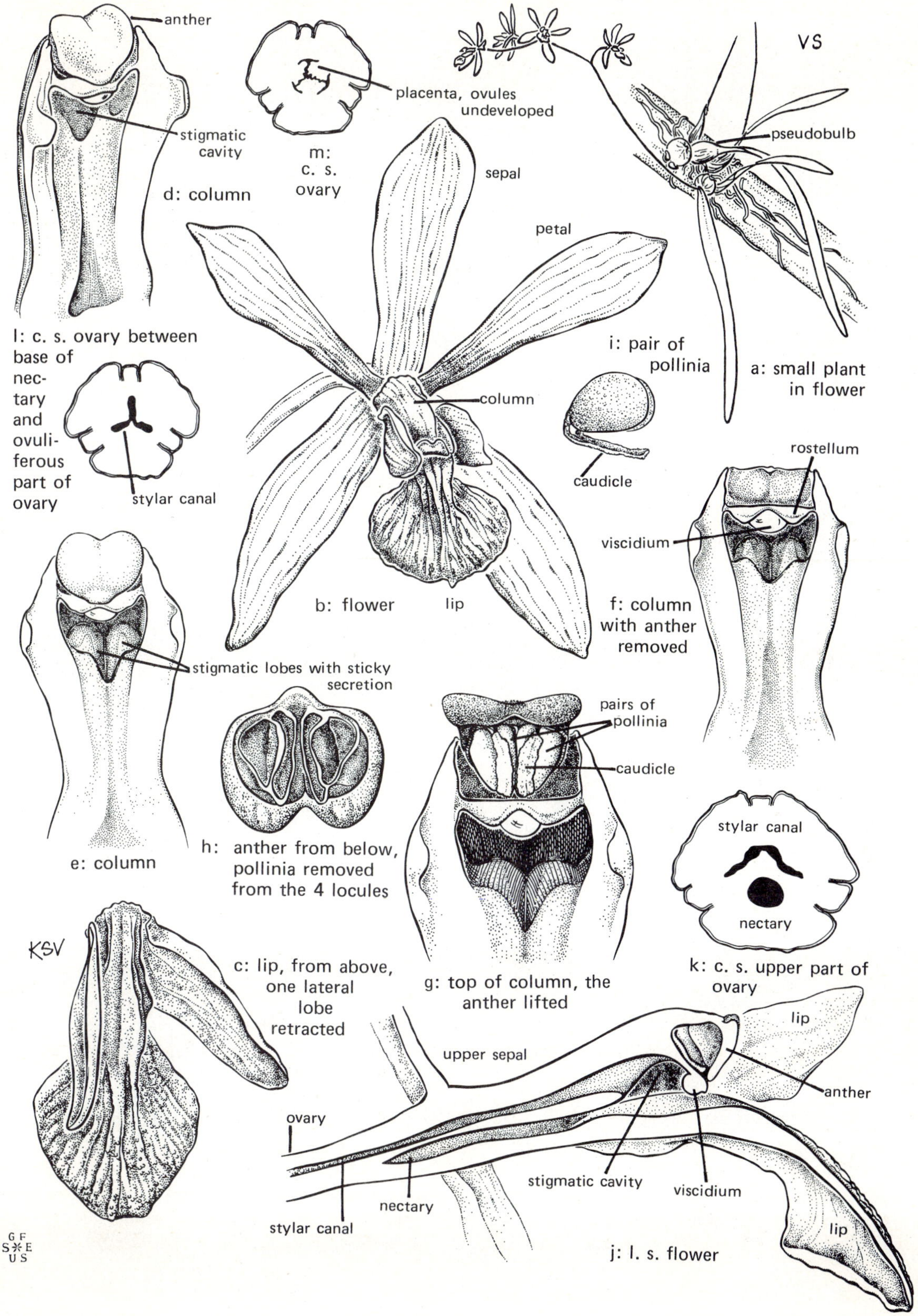

JUGLANDACEAE: Carya. a-m, C. ovata var. australis; n-s, C. laciniosa

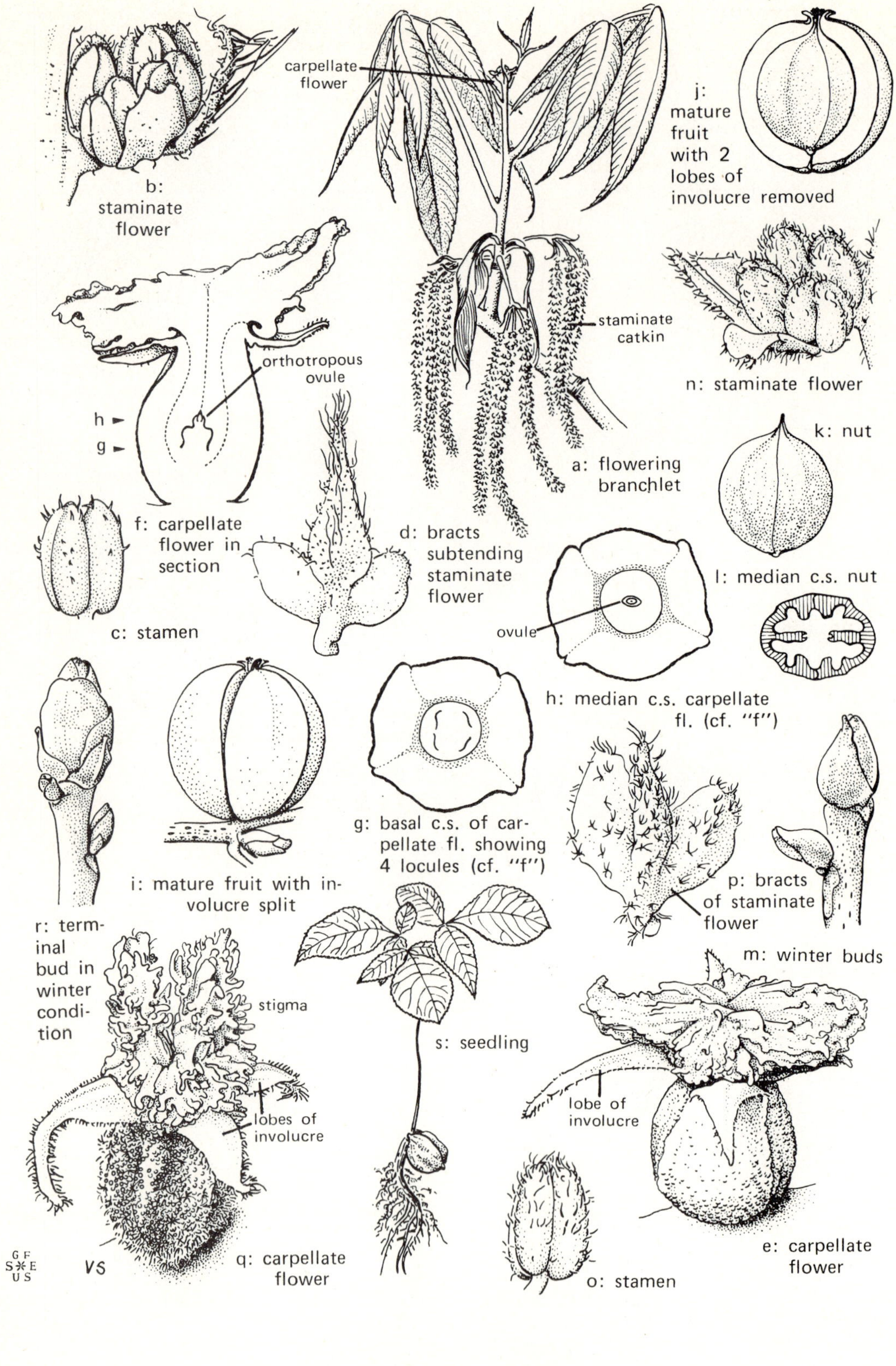

FAGACEAE: Fagus. a-l, F. grandifolia

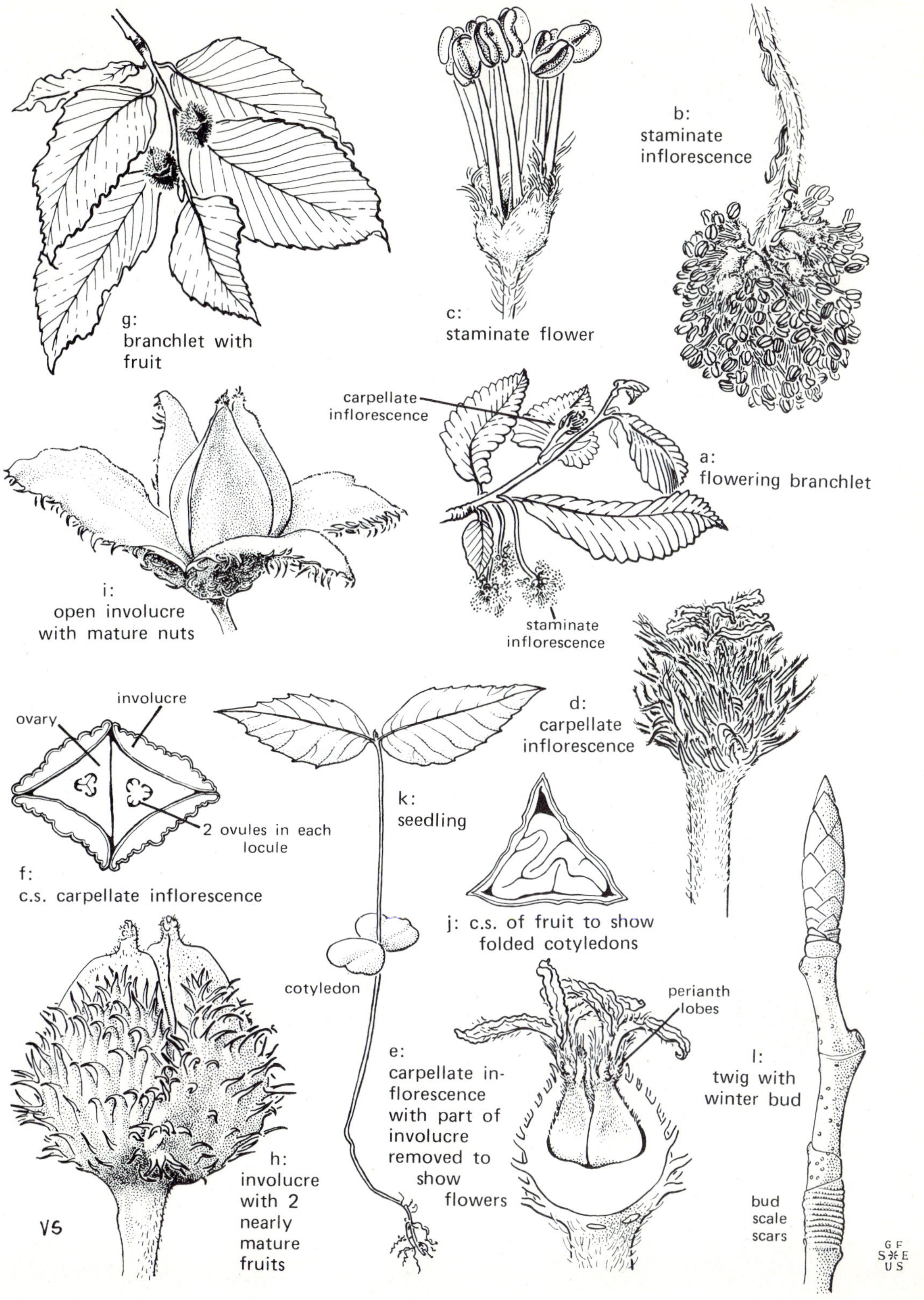

AMARANTHACEAE: Amaranthus. a-h, A. spinosus; i-j, A. retroflexus

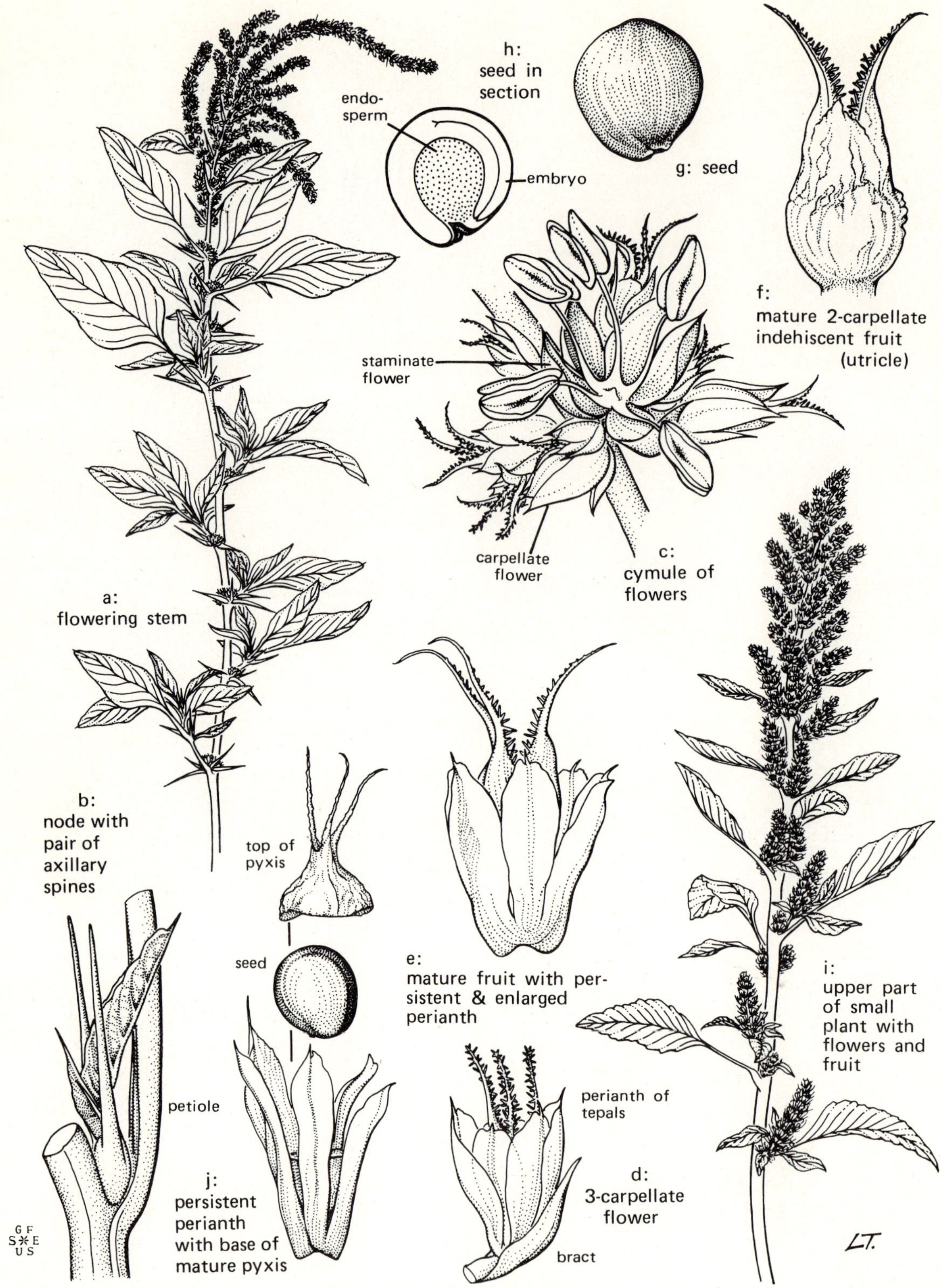

CARYOPHYLLACEAE: Silene: a-j, S. virginica; k, S. caroliniana; l, S. ovata; m, n, S. antirrhina

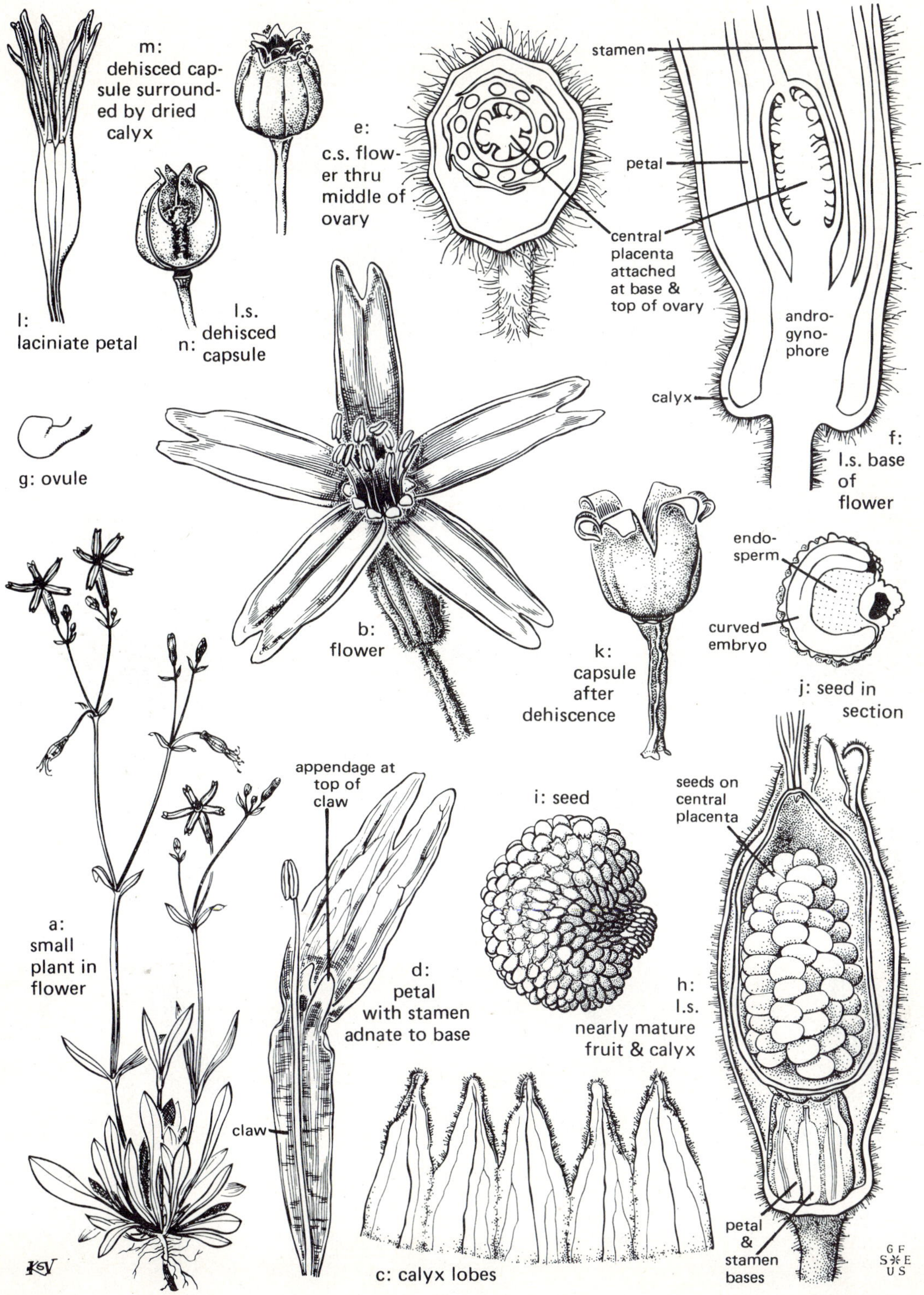

RANUNCULACEAE: Caltha. a-i, C. palustris

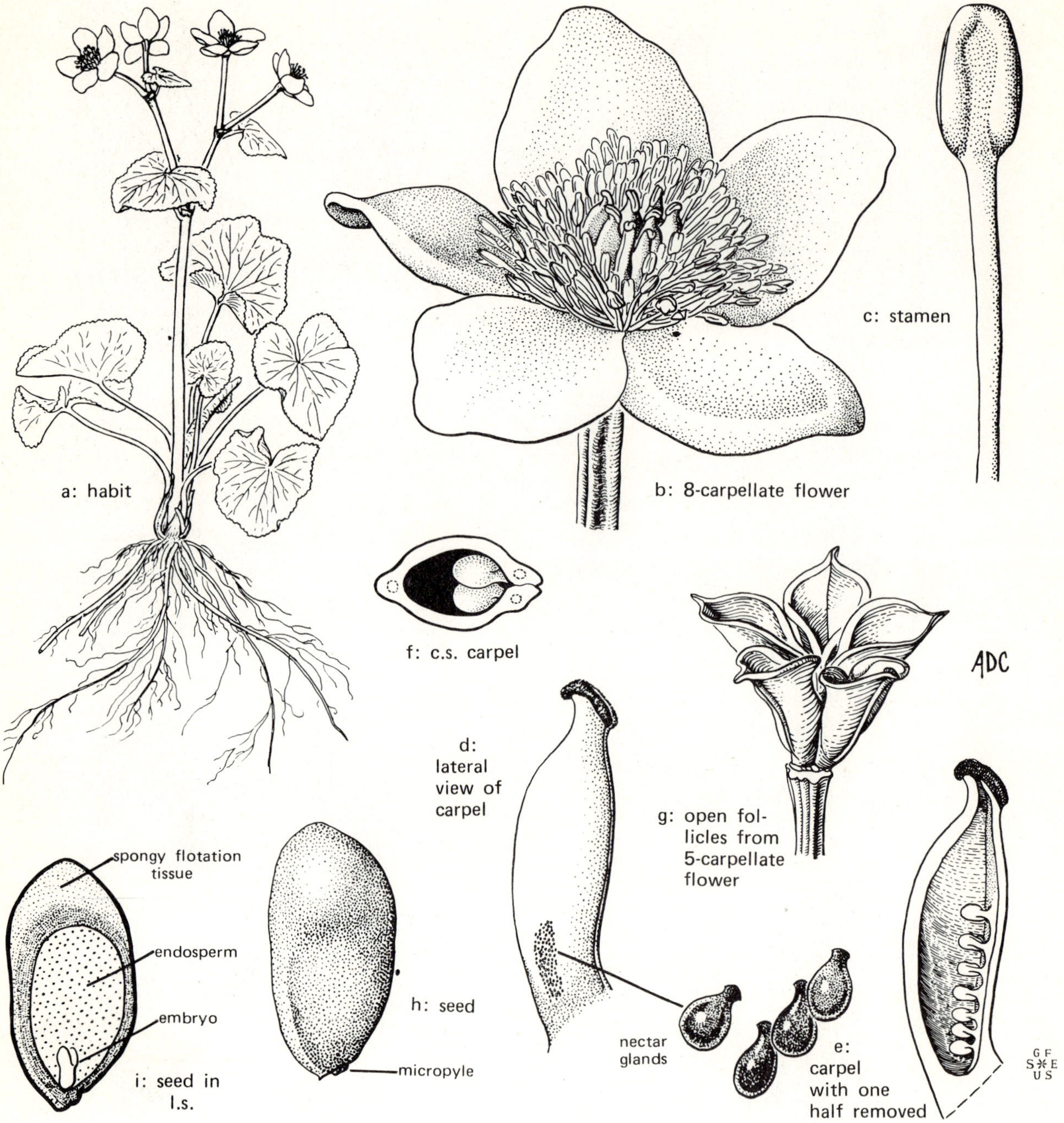

MAGNOLIACEAE: Magnolia. a-i, M. virginiana; j-l, M. grandiflora; m, M. tripetala; n-q, M. acuminata

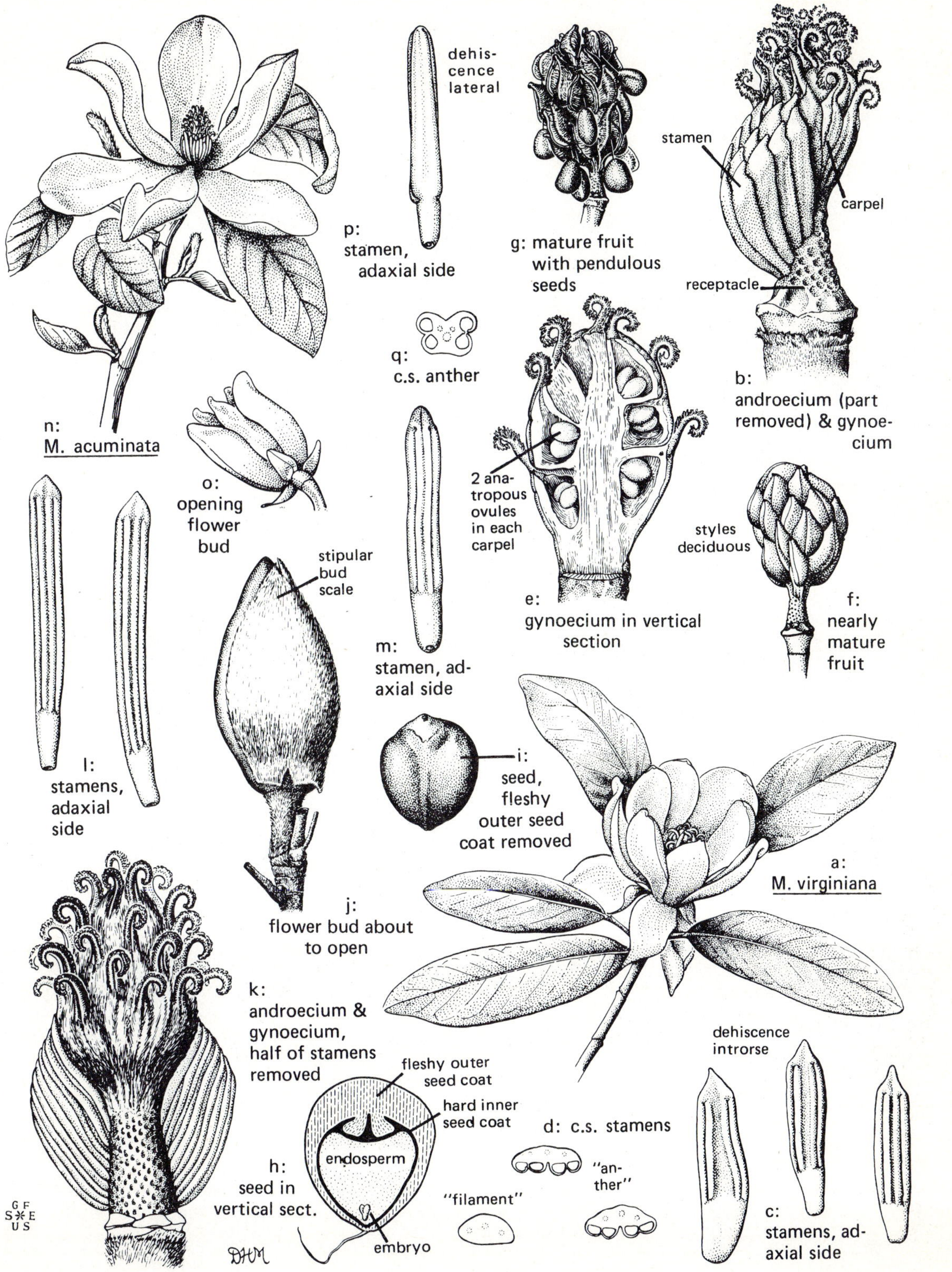

BRASSICACEAE (CRUCIFERAE): a-h, Capsella bursa-pastoris; i-t, fruits, seeds, & embryos of various species

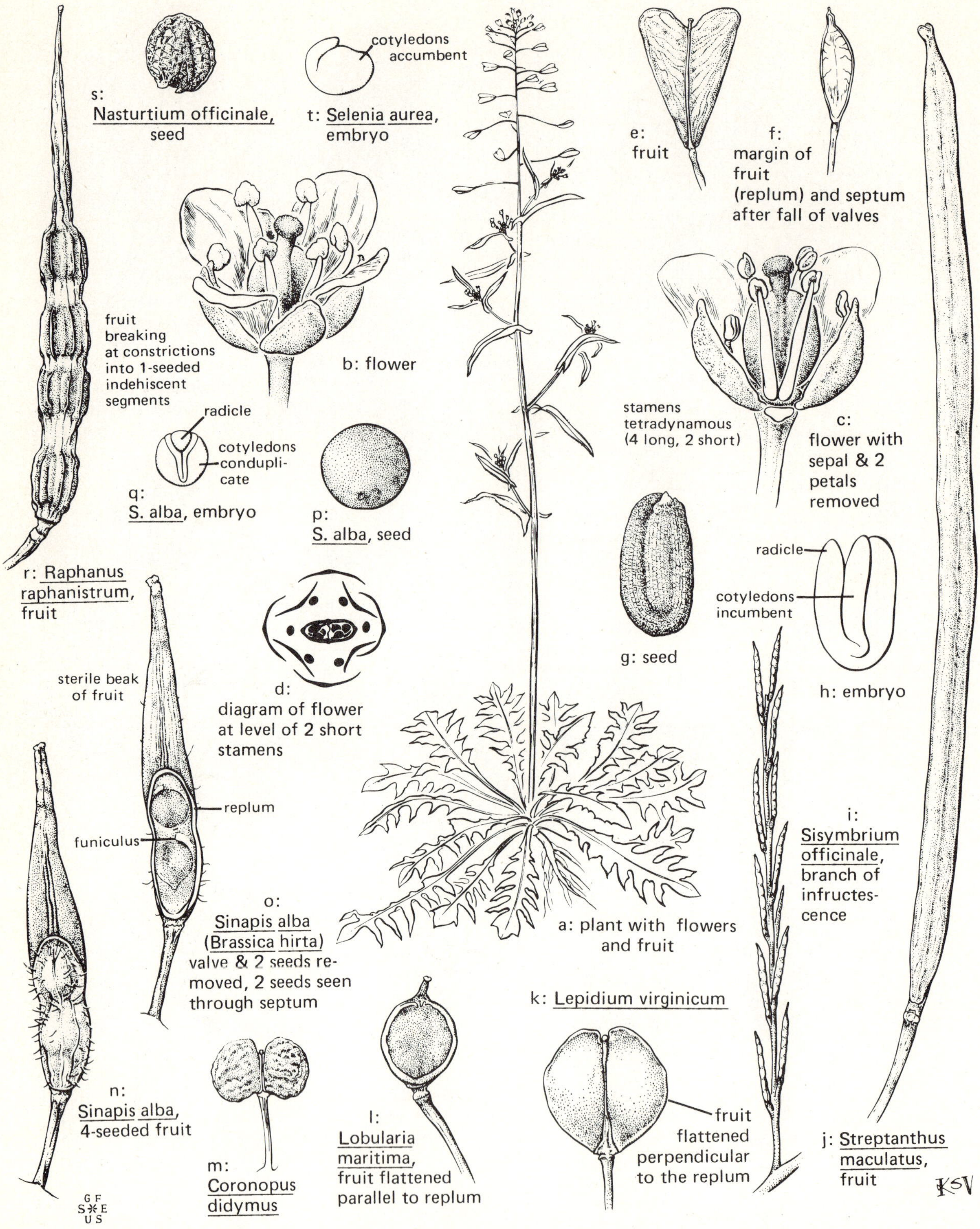

SAXIFRAGACEAE: Saxifraga. a-i, S. michauxii; j-n, S. virginiensis

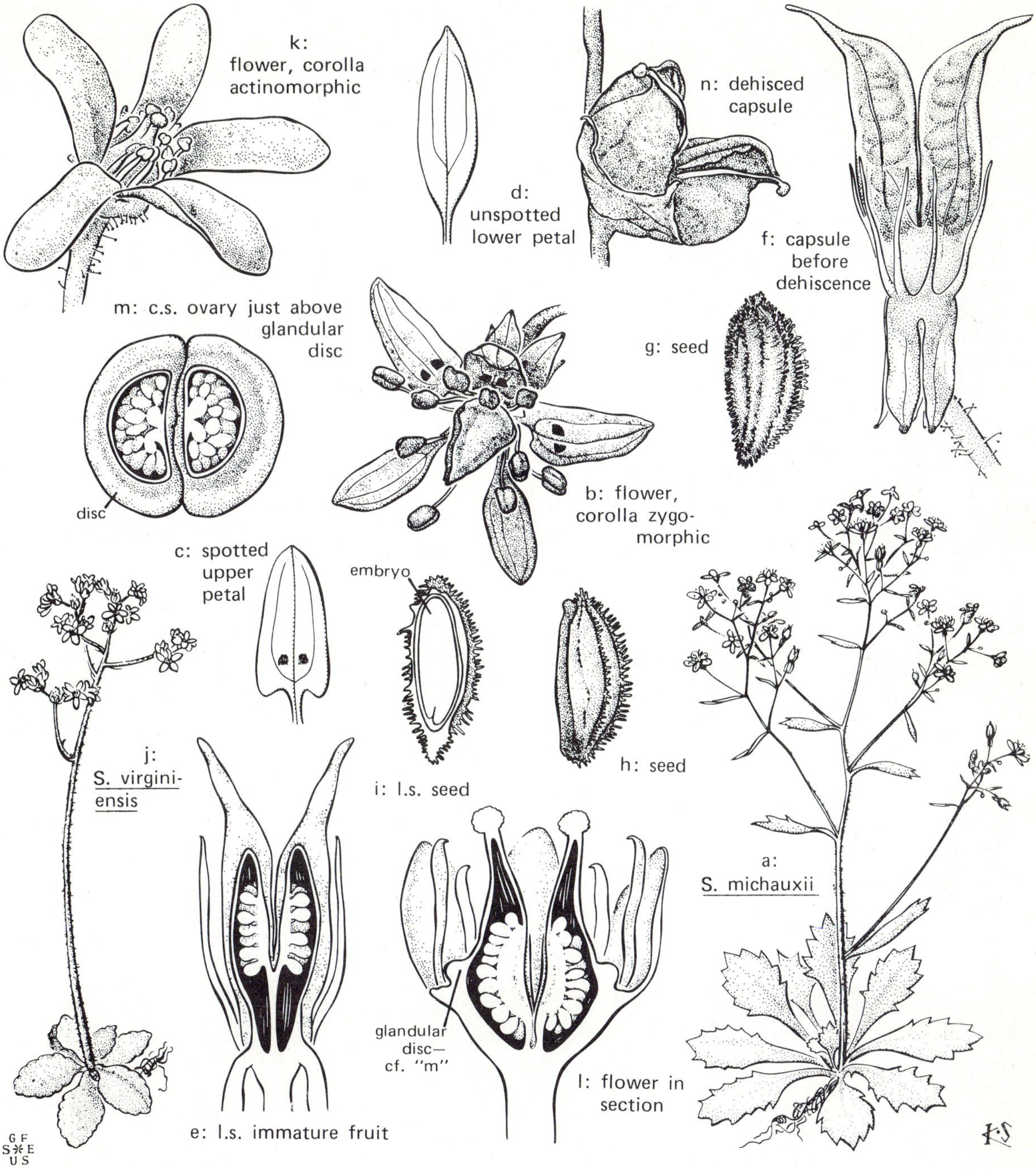

HAMAMELIDACEAE: Hamamelis. a-k, H. vernalis; l, H. virginiana

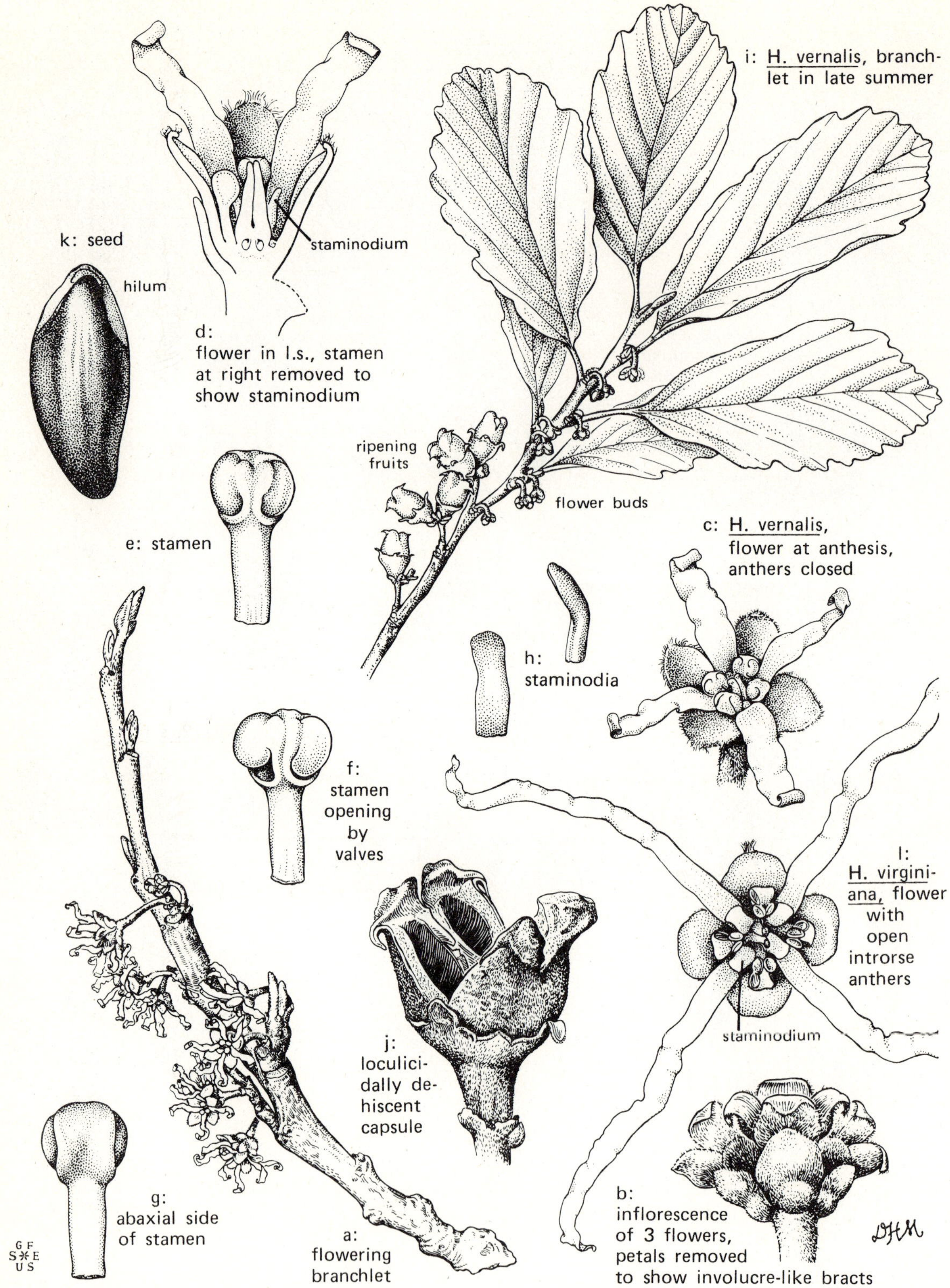

ROSACEAE subfamily ROSOIDEAE: Potentilla. a-h, P. canadensis; i, P. simplex; j, P. recta; k-m, P. tridentata

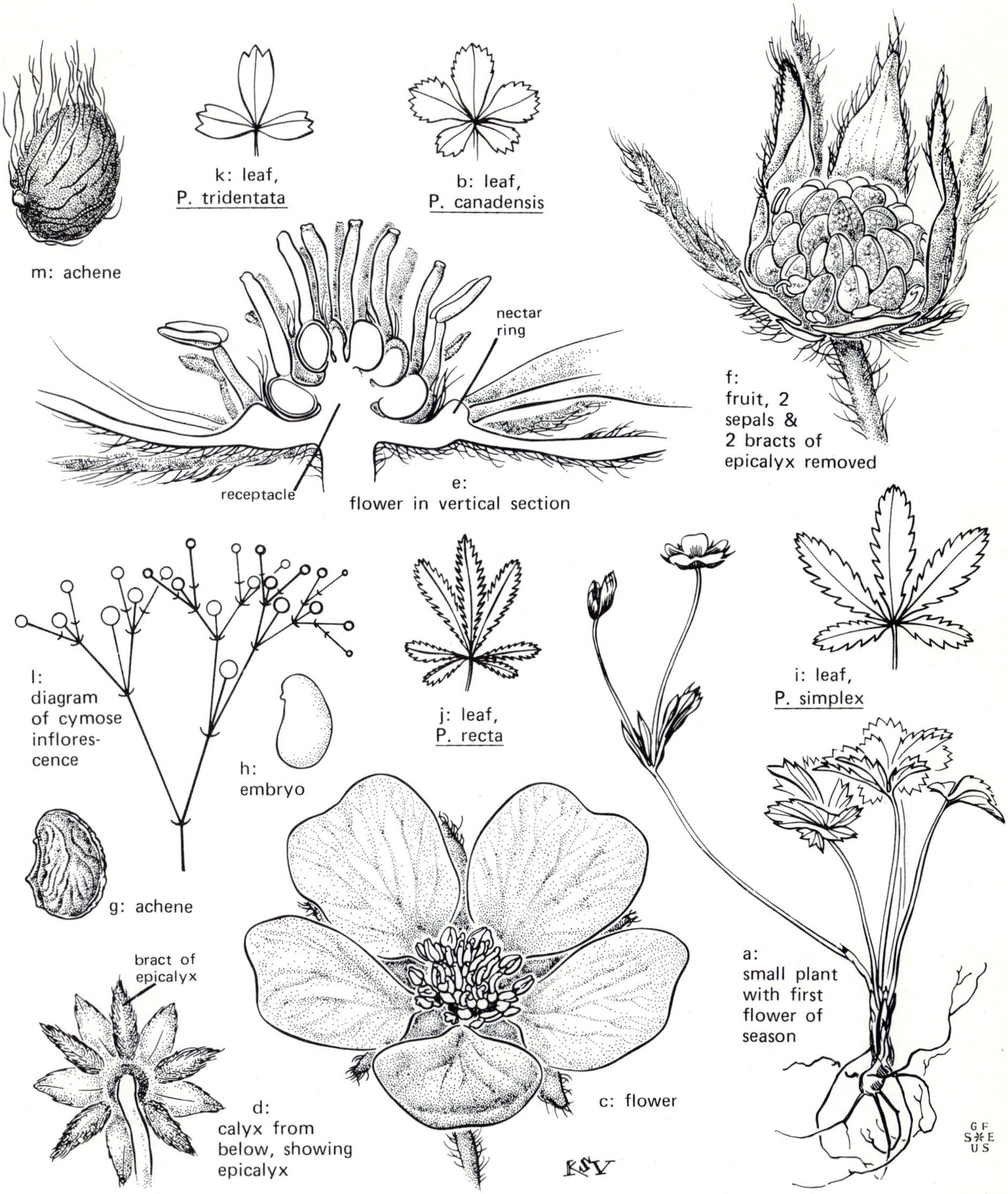

FABACEAE (LEGUMINOSAE) subfamily FABOIDEAE: Baptisia. a-m, B. australis; n, B. tinctoria; o, B. arachnifera

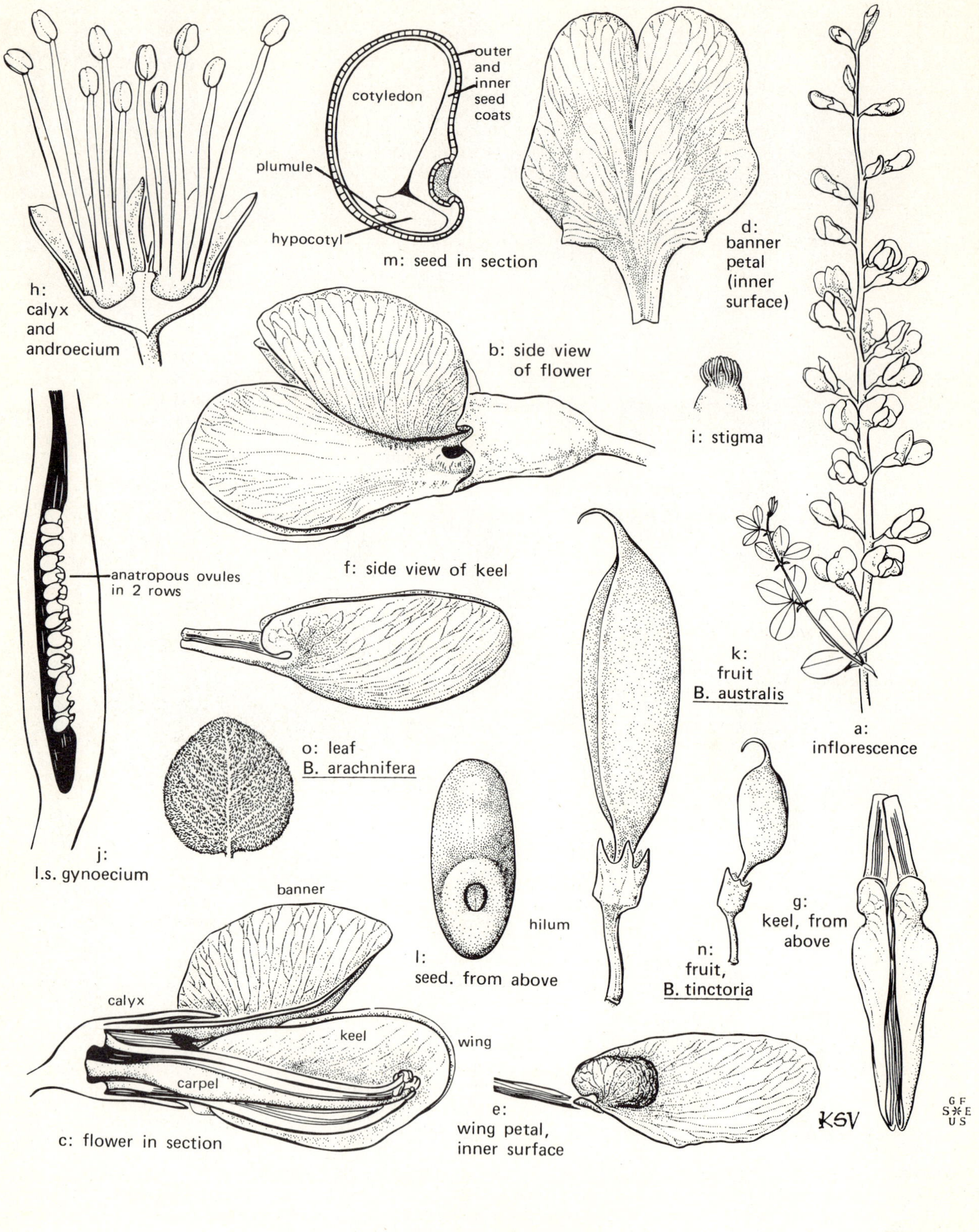

GERANIACEAE: Geranium. a-j, G. maculatum; k-m, G. carolinianum

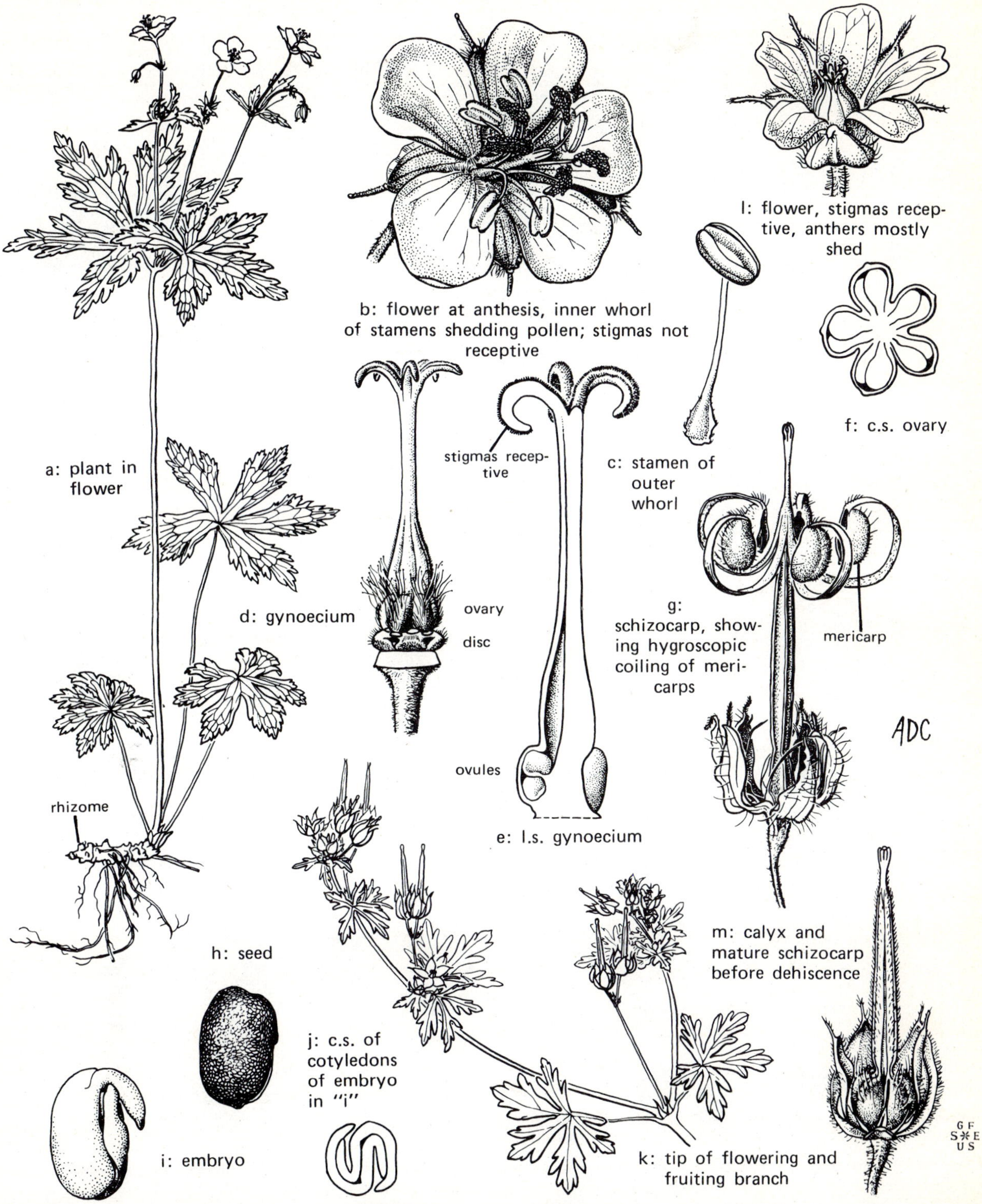

EUPHORBIACEAE: a-i, Euphorbia corollata; j, E. inundata; k, l, E. commutata; m, E. dentata; n-p, Chamaesyce maculata

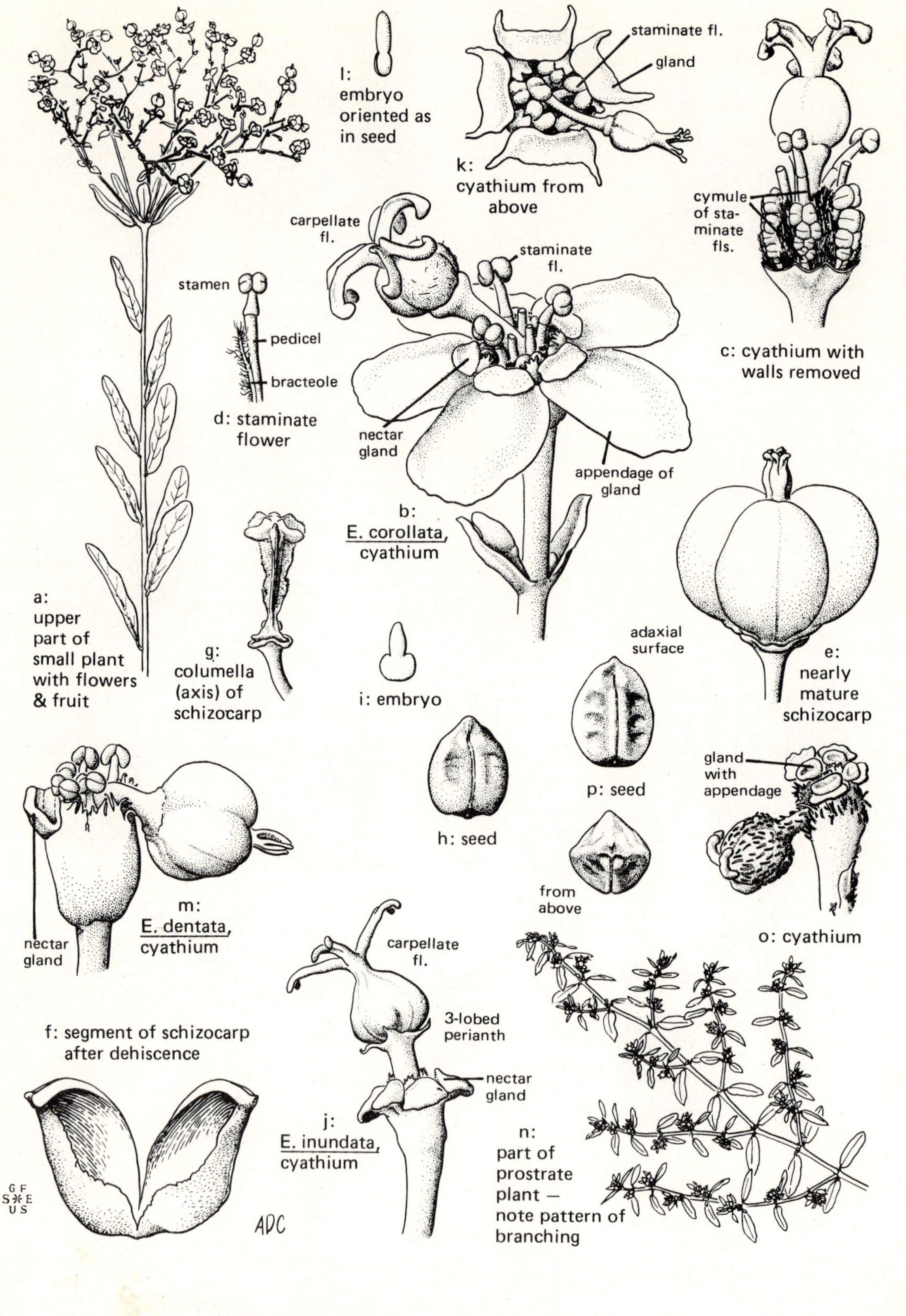

MALVACEAE: Kosteletzkya. a-f, K. virginica var. altheifolia

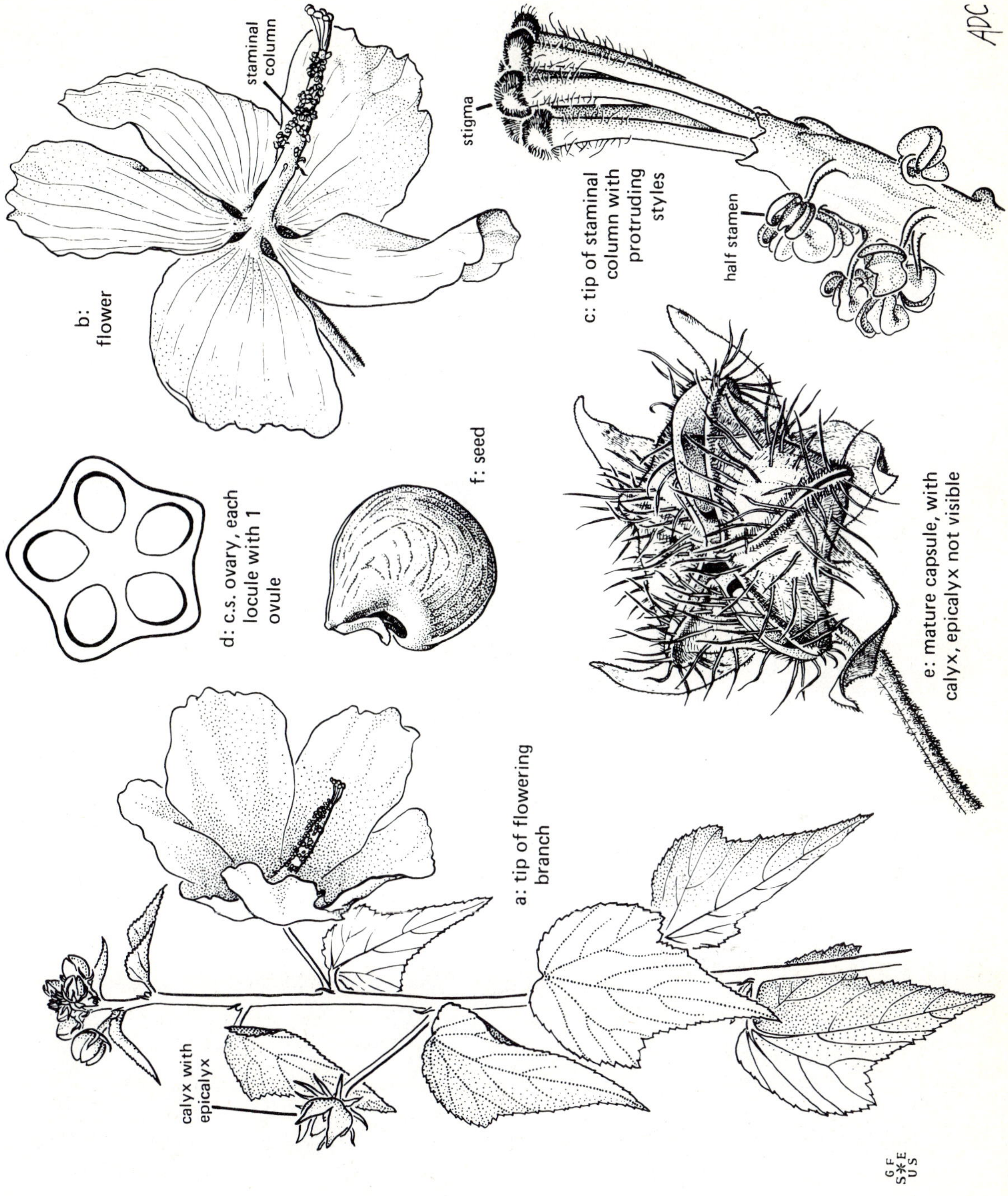

ONAGRACEAE: Oenothera. a-k, O. missouriensis

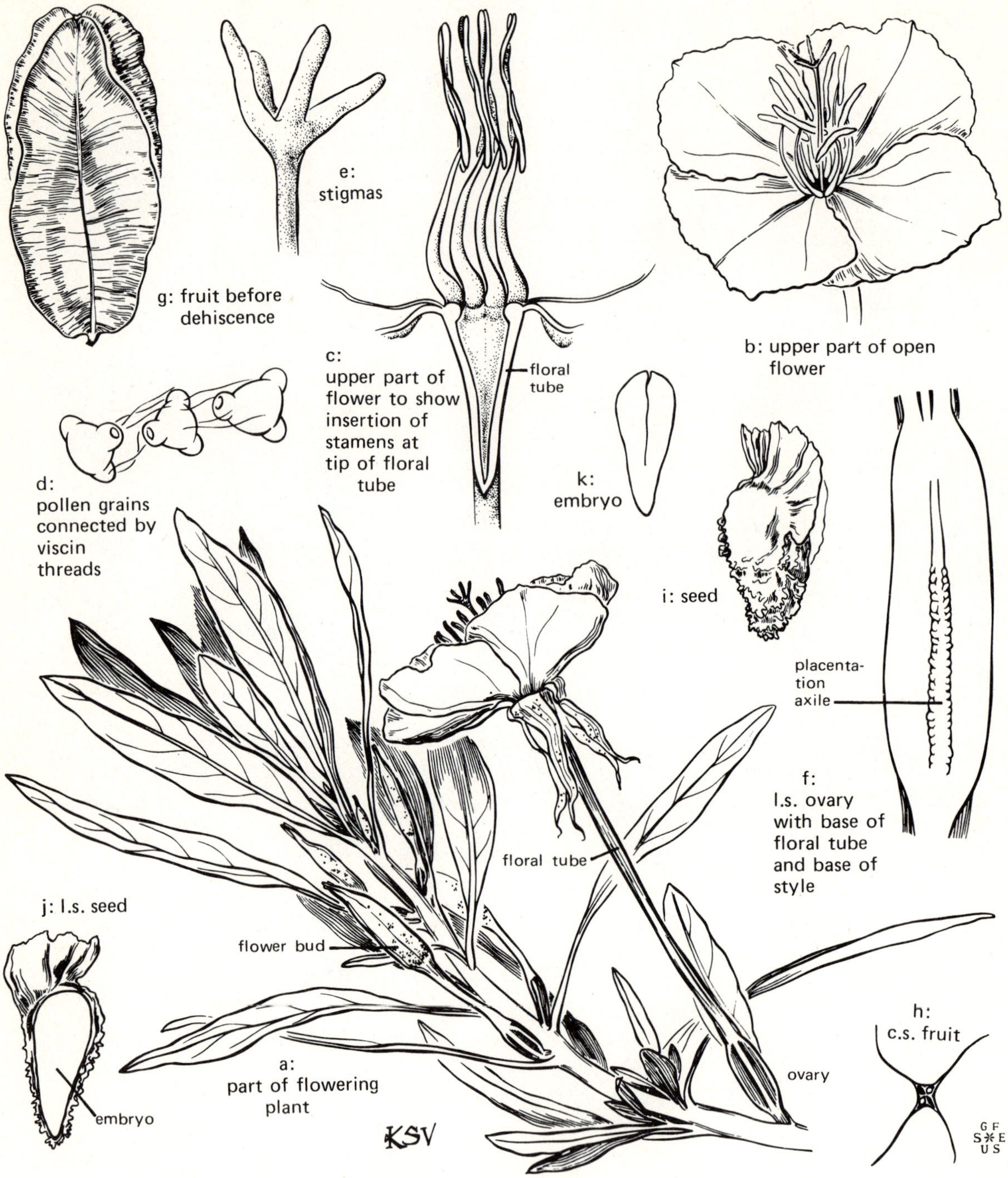

APIACEAE (UMBELLIFERAE): Daucus. a-m, D. carota; n, D. pusillus

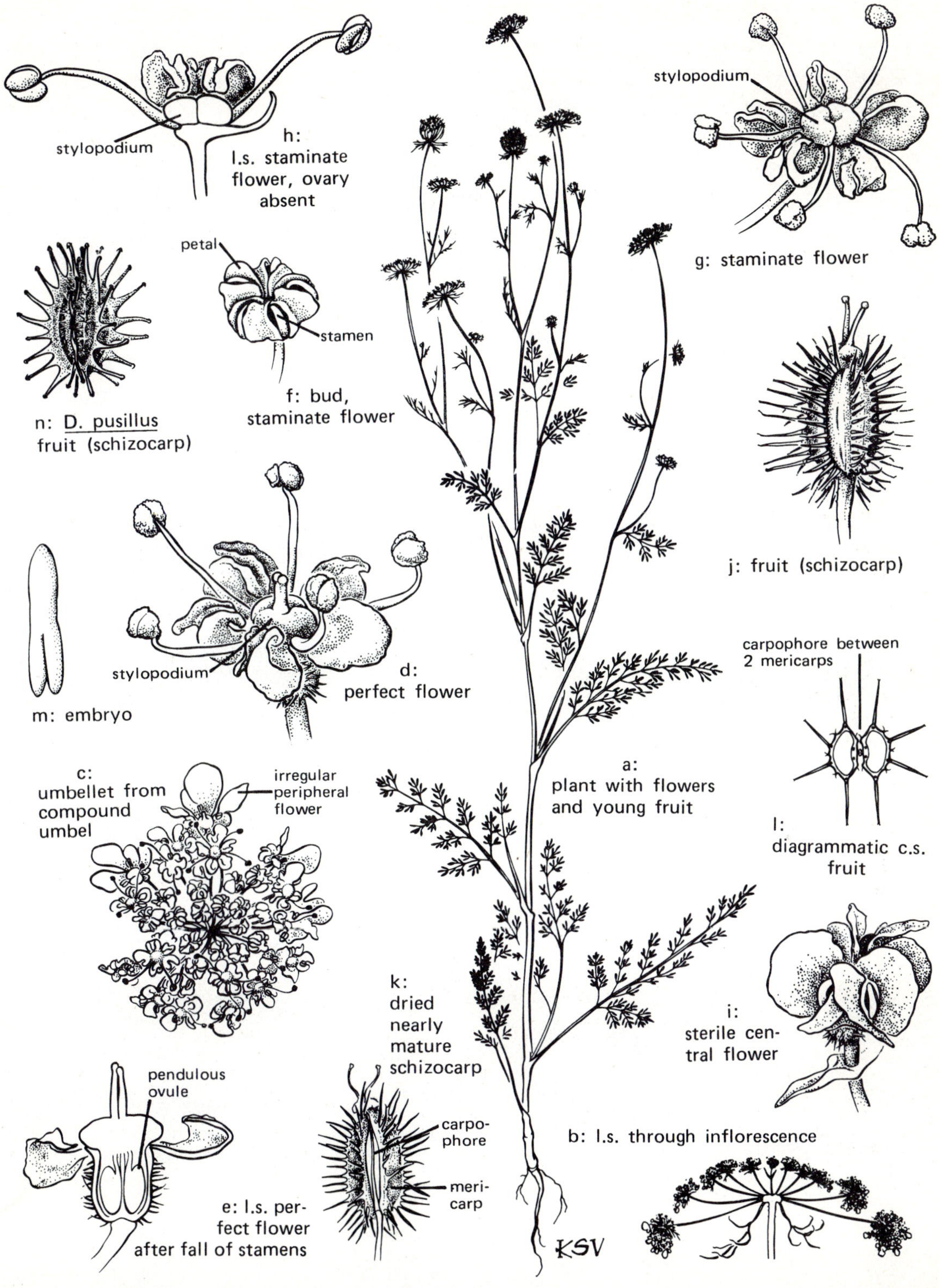

ERICACEAE: Oxydendrum. a-g, O. arboreum

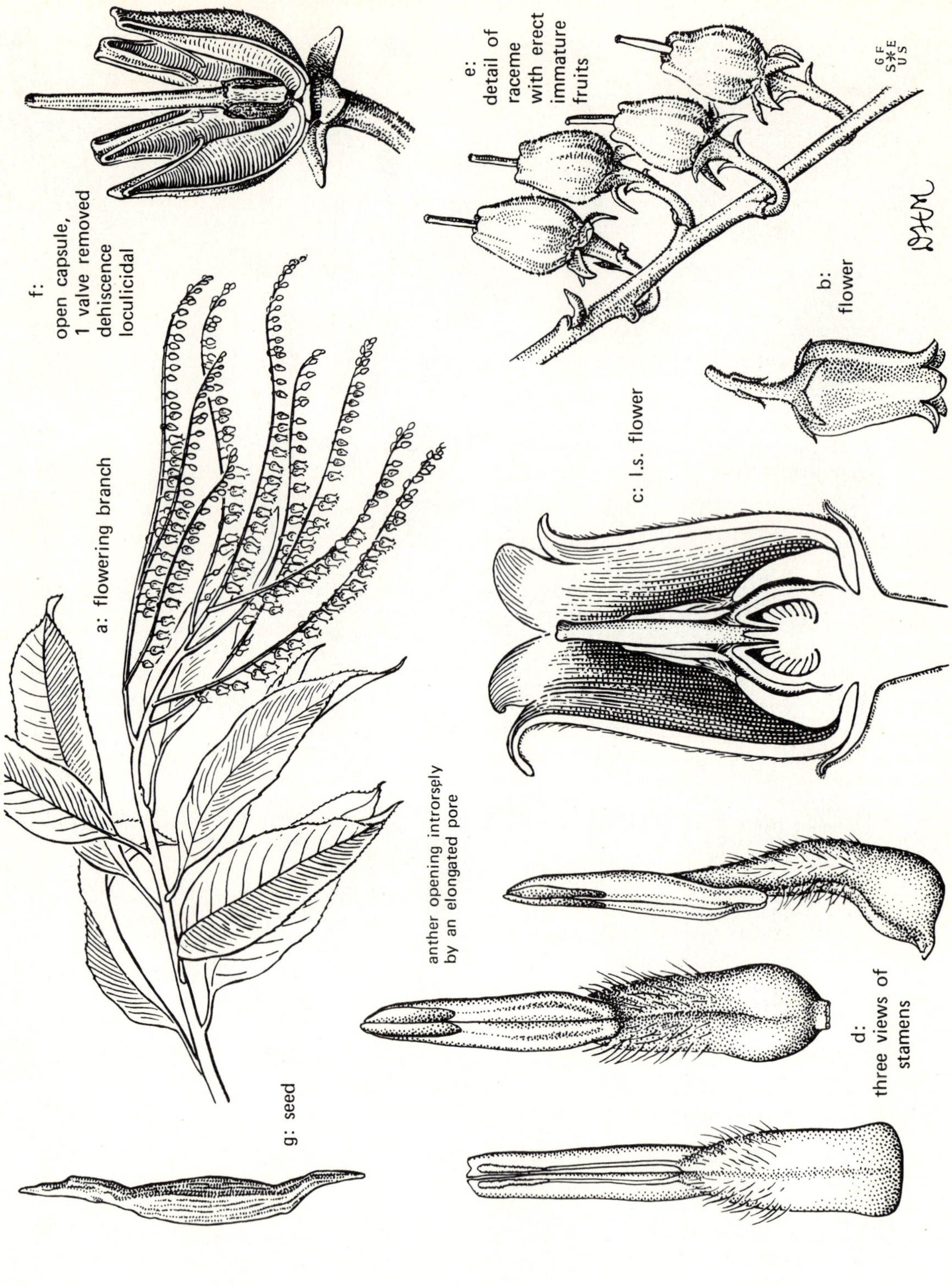

ASCLEPIADACEAE: Asclepias. a-f, A. syriaca; g, A. connivens; h, A. pedicellata

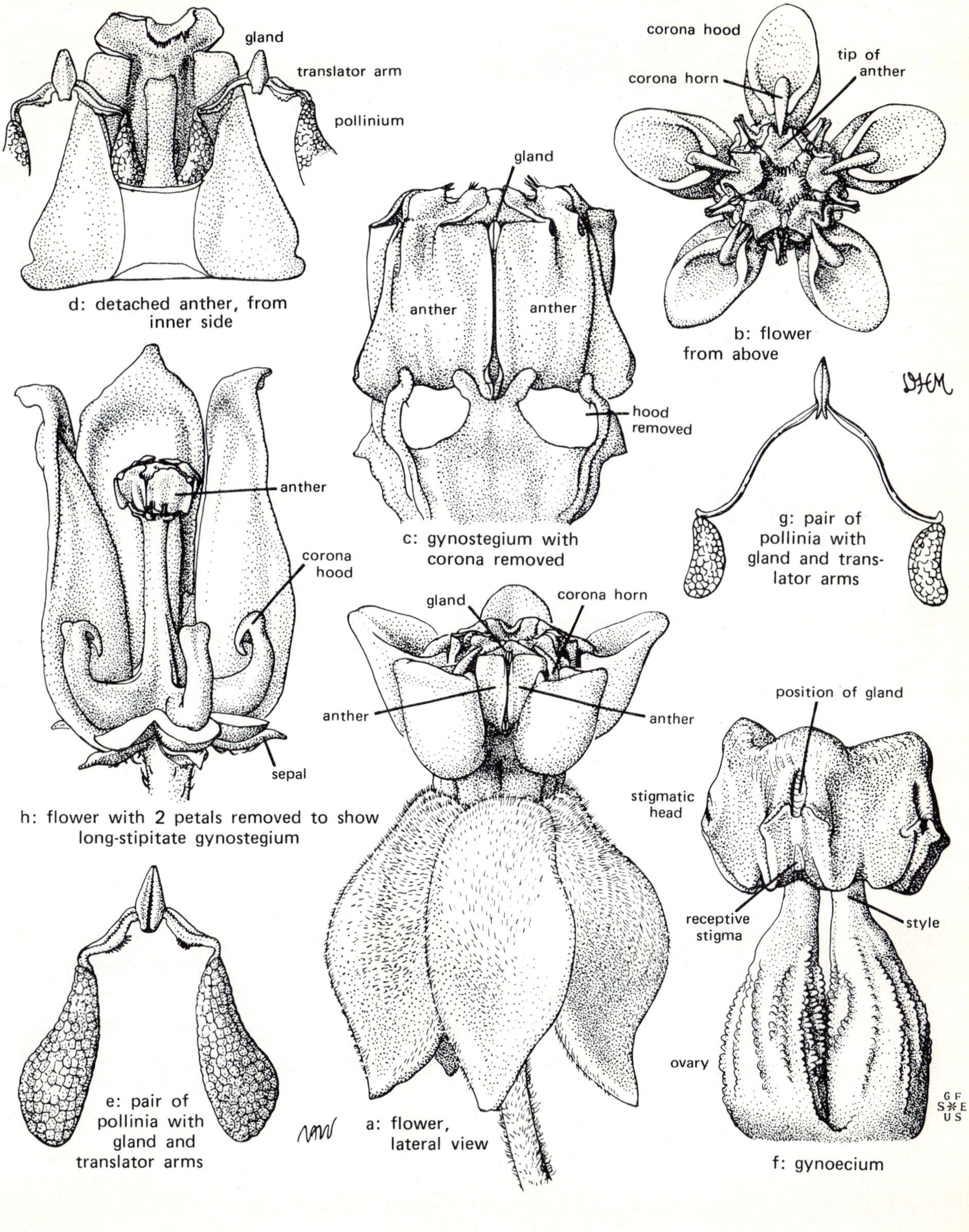

d: detached anther, from inner side

c: gynostegium with corona removed

b: flower from above

g: pair of pollinia with gland and translator arms

h: flower with 2 petals removed to show long-stipitate gynostegium

e: pair of pollinia with gland and translator arms

a: flower, lateral view

f: gynoecium

ASCLEPIADACEAE: Asclepias. a, b, A. syriaca; c-f, A. incarnata

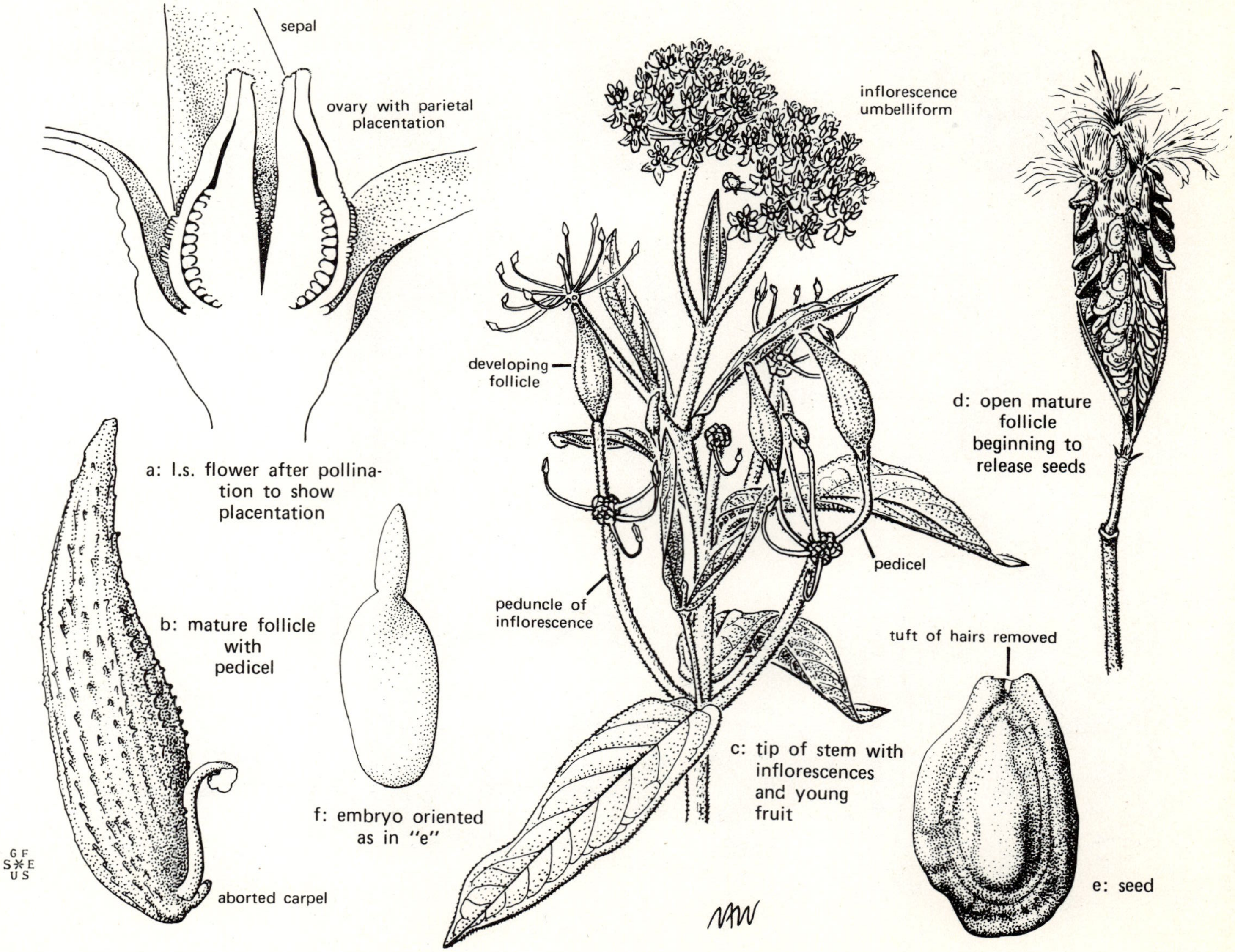

LAMIACEAE (LABIATAE): Salvia. a-j, S. urticifolia; k-o, S. lyrata

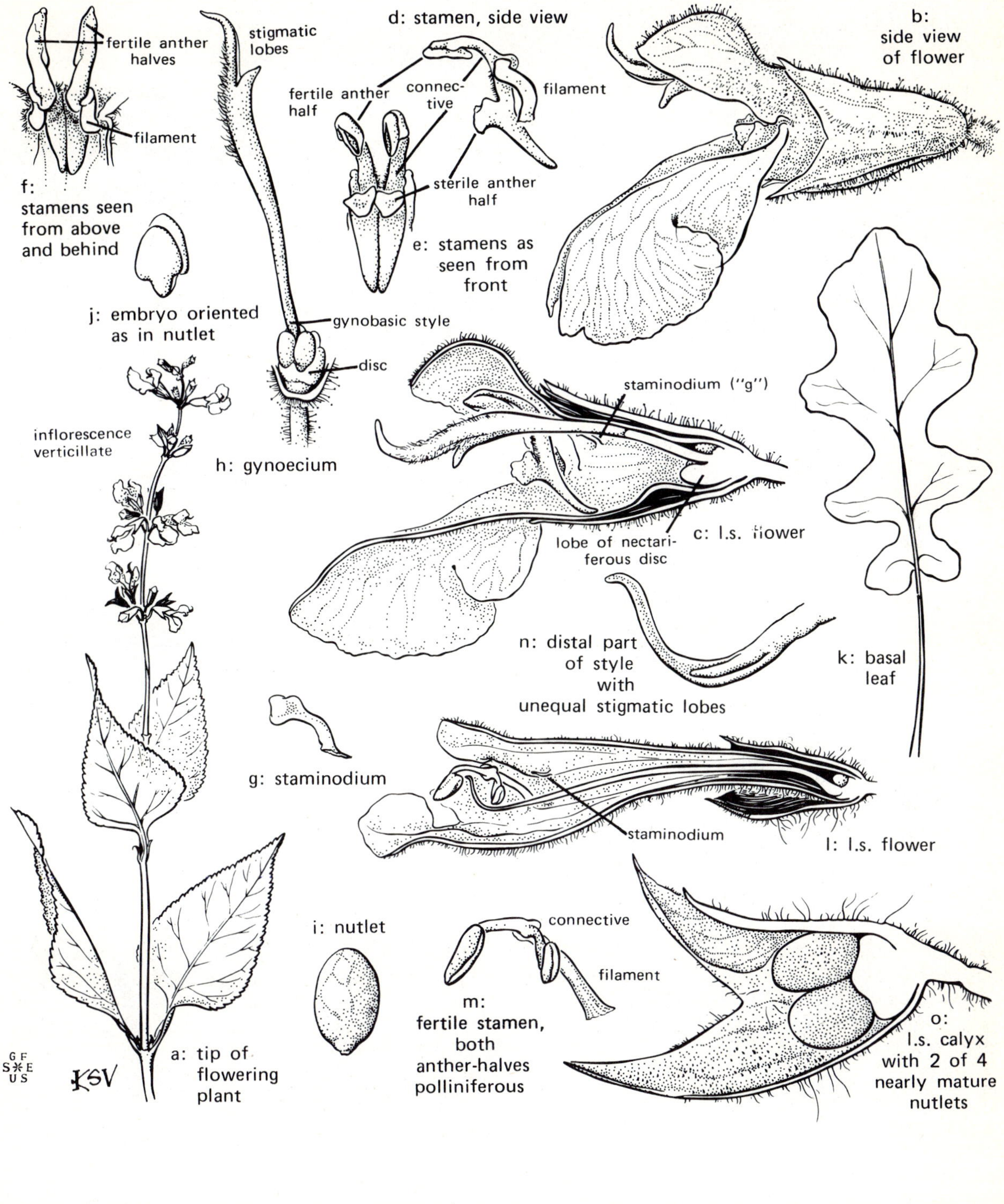

SCROPHULARIACEAE: Penstemon. a-j, P. canescens

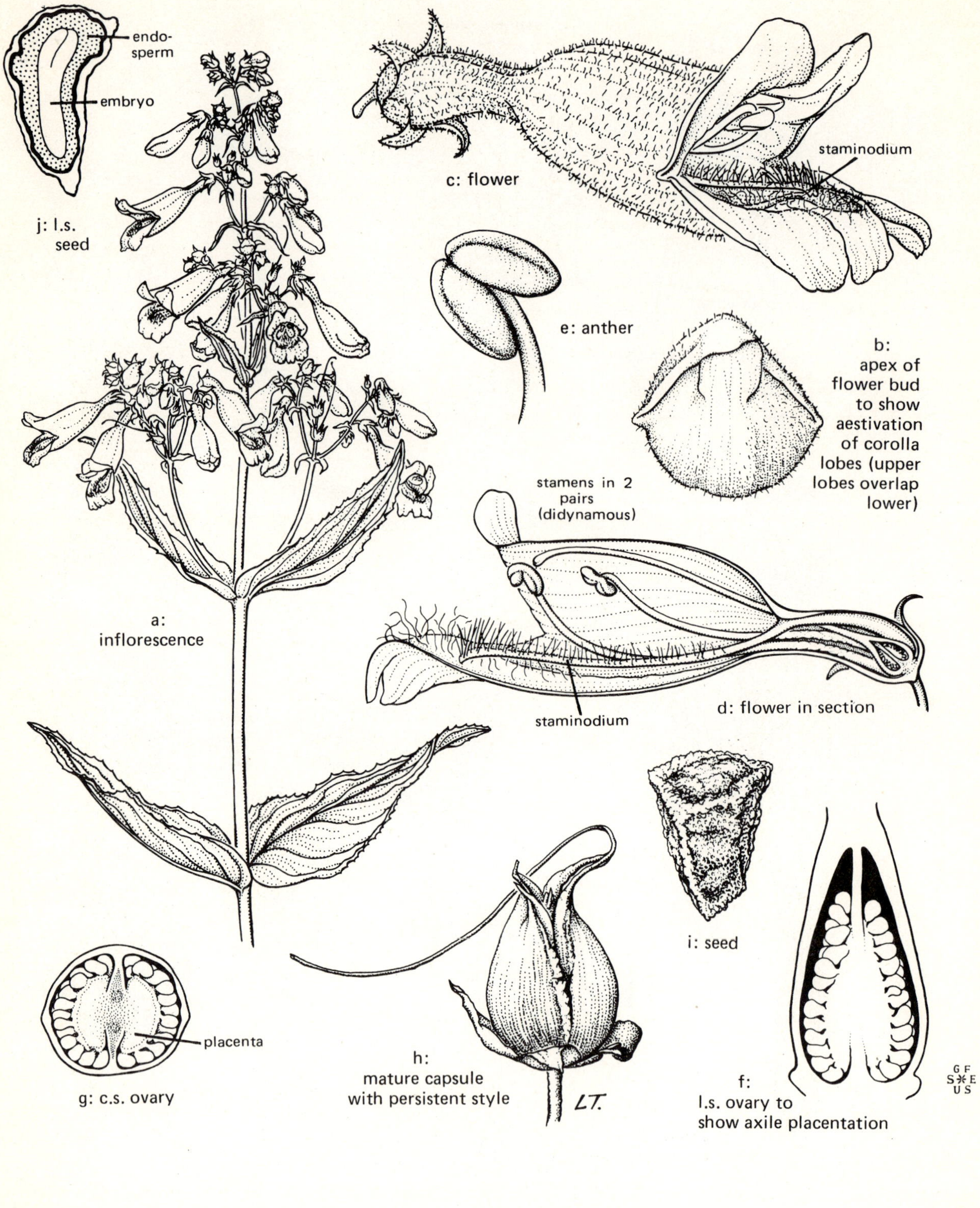

ASTERACEAE (COMPOSITAE): Inflorescence, florets, and stamens

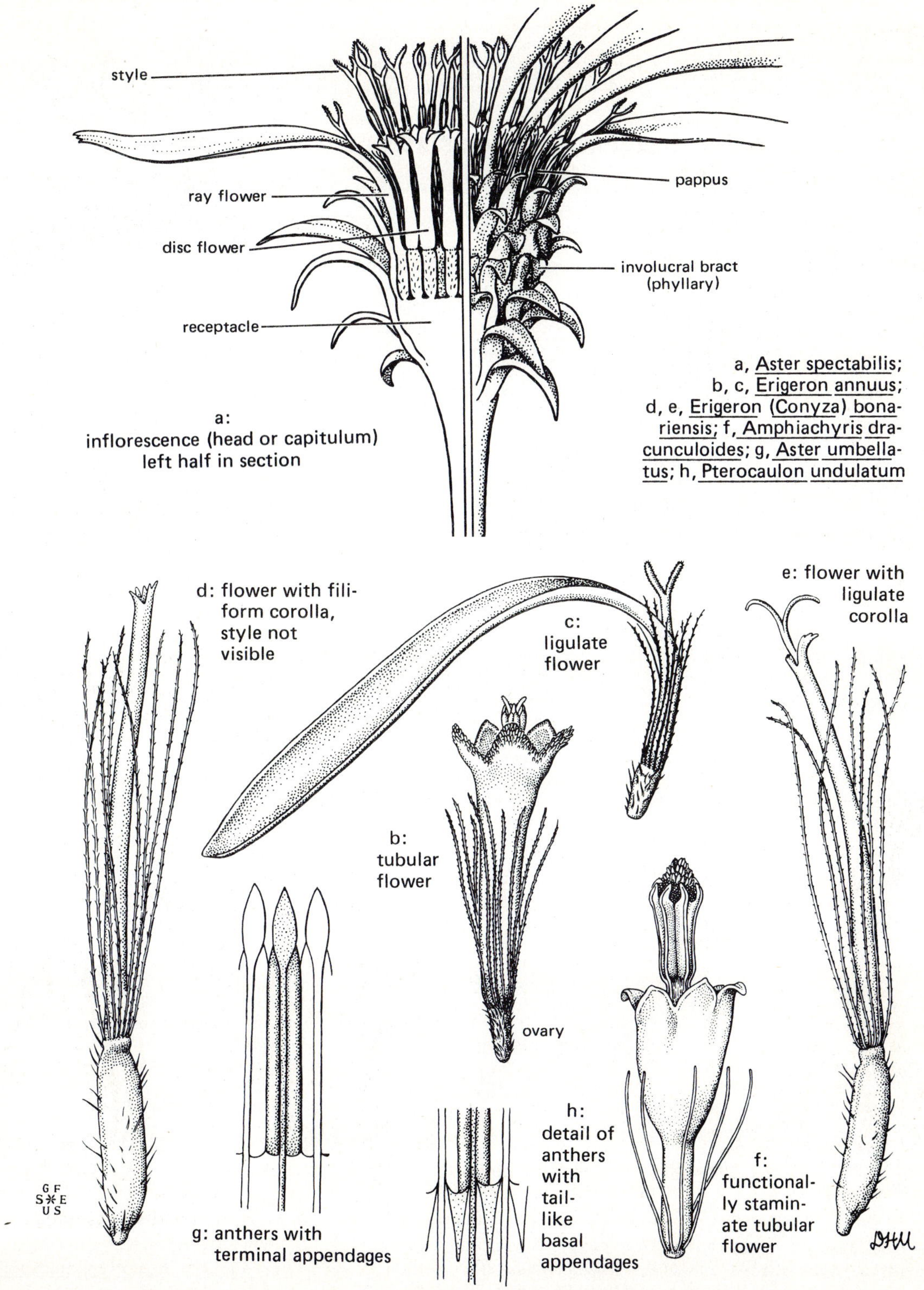

INDEX TO INDEXES

INDEX TO EXERCISES, PROBLEMS AND TESTS (SAMPLES)

INDEX TO GLOSSARIES

INDEX TO LITERATURE

INDEX TO TABLES AND COMPARATIVE SUMMARIES

INDEX TO TECHNIQUES, STUDY METHODS, RESEARCH ORGANIZATION

GENERAL SUBJECT AND TERMS INDEX

Subjects underlined; terms not underlined
(Also see phytographic terms index, page 159, and index to indexes, page 878)

-T-

INDEX TO FIGURES

74 75 76 77 9 8 7 6 5 4 3 2 1